Mechanical Engineering Series

Frederick A. Leckie and Dominic J. Dal Bello, **Strength and Stiffness of Engineering Systems**

G. Genta, **Vibration Dynamics and Control**

R. Firoozian, **Servomotors and Industrial Control Theory**

G. Genta and L. Morello, **The Automotive Chassis, Volumes 1 & 2**

F. A. Leckie and D. J. Dal Bello, **Strength and Stiffness of Engineering Systems**

Wodek Gawronski, **Modeling and Control of Antennas and Telescopes**

Makoto Ohsaki and Kiyohiro Ikeda, **Stability and Optimization of Structures: Generalized Sensitivity Analysis**

A.C. Fischer-Cripps, **Introduction to Contact Mechanics, 2nd ed.**

W. Cheng and I. Finnie, **Residual Stress Measurement and the Slitting Method**

J. Angeles, **Fundamentals of Robotic Mechanical Systems: Theory Methods and Algorithms, 3rd ed.**

J. Angeles, **Fundamentals of Robotic Mechanical Systems: Theory, Methods, and Algorithms, 2nd ed.**

P. Basu, C. Kefa, and L. Jestin, **Boilers and Burners: Design and Theory**

J.M. Berthelot, **Composite Materials: Mechanical Behavior and Structural Analysis**

I.J. Busch-Vishniac, **Electromechanical Sensors and Actuators**

J. Chakrabarty, **Applied Plasticity**

K.K. Choi and N.H. Kim, **Structural Sensitivity Analysis and Optimization 1: Linear Systems**

K.K. Choi and N.H. Kim, **Structural Sensitivity Analysis and Optimization 2: Nonlinear Systems and Applications**

G. Chryssolouris, **Laser Machining: Theory and Practice**

V.N. Constantinescu, **Laminar Viscous Flow**

G.A. Costello, **Theory of Wire Rope, 2nd ed.**

K. Czolczynski, **Rotordynamics of Gas-Lubricated Journal Bearing Systems**

M.S. Darlow, **Balancing of High-Speed Machinery**

W. R. DeVries, **Analysis of Material Removal Processes**

J.F. Doyle, **Nonlinear Analysis of Thin-Walled Structures: Statics, Dynamics, and Stability**

J.F. Doyle, **Wave Propagation in Structures: Spectral Analysis Using Fast Discrete Fourier Transforms, 2nd ed.**

P.A. Engel, **Structural Analysis of Printed Circuit Board Systems**

A.C. Fischer-Cripps, **Introduction to Contact Mechanics**

A.C. Fischer-Cripps, **Nanoindentation, 2nd ed.**

J. García de Jalón and E. Bayo, **Kinematic and Dynamic Simulation of Multibody Systems: The Real-Time Challenge**

W.K. Gawronski, **Advanced Structural Dynamics and Active Control of Structures**

W.K. Gawronski, **Dynamics and Control of Structures: A Modal Approach**

G. Genta, **Dynamics of Rotating Systems**

D. Gross and T. Seelig, **Fracture Mechanics with Introduction to Micro-mechanics**

K.C. Gupta, **Mechanics and Control of Robots**

R. A. Howland, **Intermediate Dynamics: A Linear Algebraic Approach**

D. G. Hull, **Optimal Control Theory for Applications**

J. Ida and J.P.A. Bastos, **Electromagnetics and Calculations of Fields**

M. Kaviany, **Principles of Convective Heat Transfer, 2nd ed.**

M. Kaviany, **Principles of Heat Transfer in Porous Media, 2nd ed.**

E.N. Kuznetsov, **Underconstrained Structural Systems**

P. Ladevèze, **Nonlinear Computational Structural Mechanics: New Approaches and Non-Incremental Methods of Calculation**

(continued after index)

Frederick A. Leckie • Dominic J. Dal Bello

Strength and Stiffness of Engineering Systems

Springer

Frederick A. Leckie
University of California Santa Barbara
Department of Mechanical and
  Environmental Engineering
Santa Barbara, CA
USA

Dominic J. Dal Bello
Allan Hancock College
Department of Engineering
Santa Maria, CA
USA

ISBN: 978-0-387-49473-9   e-ISBN: 978-0-387-49474-6
DOI 10.1007/978-0-387-49474-6

Library of Congress Control Number: 2008921723

Printed on acid-free paper

springer.com

Mechanical Engineering Series

Frederick F. Ling
Editor-in-Chief

The Mechanical Engineering Series features graduate texts and research monographs to address the need for information in contemporary mechanical engineering, including areas of concentration of applied mechanics, biomechanics, computational mechanics, dynamical systems and control, energetics, mechanics of materials, processing, production systems, thermal science, and tribology.

Series Preface

Mechanical engineering, an engineering discipline born of the needs of the industrial revolution, is once again asked to do its substantial share in the call for industrial renewal. The general call is urgent as we face profound issues of productivity and competitiveness that require engineering solutions, among others. The Mechanical Engineering Series is a series featuring graduate texts and research monographs intended to address the need for information in contemporary areas of mechanical engineering.

The series is conceived as a comprehensive one that covers a broad range of concentrations important to mechanical engineering graduate education and research. We are fortunate to have a distinguished roster of series editors, each an expert in one of the areas of concentration. The names of the series editors are listed on page vi of this volume. The areas of concentration are applied mechanics, biomechanics, computational mechanics, dynamic systems and control, energetics, mechanics of materials, processing, thermal science, and tribology.

Note to Instructors

The full solutions manual for the book can be found on the book's webpage at www.springer.com. Numerical only solutions can be found on both the book's webpage at www.springer.com as well as the author's webpage www.ah-engr.com/strengthandstiffness. Supplementary information can also be found on the author's webpage.

Preface

Overview and Goals

This text is concerned with how engineering components and systems support loads without suffering failure or excessive deformation. Such topics are traditionally addressed in texts entitled *Strength of Materials* or *Mechanics of Materials*. The aim of this text is to develop a simple and comprehensive approach which recognizes today's wide range of technical applications and materials. Such an approach is well-suited to meet the demands of interdisciplinary and rapidly changing technologies.

The scale of engineering systems considered varies from 10^{-6} to 10^3 meters. The basic concepts are applicable to micro-electromechanical systems (MEMS, $\sim 10^{-6}$ m), piezoelectric devices, electronic and computer hardware, tools, machines, vehicles, buildings, space structures and bridges ($\sim 10^3$ m). The materials considered include metals, ceramics, polymers, composites, piezoelectric materials and shape memory alloys.

The work is intended as a basic text and learning tool for undergraduate and graduate students, and as a comprehensive resource – for review and introduction – for researchers, scientists and engineers of all fields.

With these goals in mind, an attempt is made to keep the approach simple, direct and concise. The initial chapters – *Chapters 1* through *10* – are developed by means of basic theory and illustrated with examples designed to solidify understanding of the fundamental principles. Mastering these basic principles is vital in solving practical problems and in creating innovative and safe designs.

Chapters 11 through *16* cover topics such as energy methods, plasticity, fracture, composites and smart systems, each usually the subject of specialized texts and upper division or graduate courses. As in the early chapters, these topics are approached in a succinct and straightforward manner so that the text is comprehensive to an advanced level.

Supplemental materials are included in an accompanying web site **www.ah-engr.com/ strengthandstiffness**, referred to as *Online Notes* in the text. The *Online Notes* include: (1) basic interactive problems used to test basic knowledge and skills; (2) additional material and information, which while important in their own right, are not necessary for the development of the written text; and (3) basic computing tools, resources and data in support of the text.

Organization

The first ten chapters cover topics generally taught in a sophomore- or junior-level Strength of Materials course. These chapters can be broken into roughly four sections:

- ***Chapters 1-3***. *Chapter 1* is both an overview of the main topics of the text, and a review of topics studied in previous courses (e.g., units, significant figures). *Chapter 2* is a brief review of Statics; applying the conditions of equilibrium is usually the first step in solving any Strength of Materials problem. *Chapter 3* introduces the fundamental concepts of stress and strain, and the material laws that relate them.
- ***Chapters 4-6***. *Chapters 4* through *6* develop approaches to determine the stresses, strains and deformations in basic engineering components: axial members, pressure vessels, torsion members and beams. The first three sections of *Chapter 14* – basic bolted joints – may be studied directly after *Chapter 4*.
- ***Chapters 7-9***. *Chapter 7* considers situations when a structural member is loaded by a combination of loads: axial, torsion, and/or bending about one or two axes, primarily to determine the general state of stress at a material point. *Chapter 8* develops the stress and strain transformation equations. *Chapter 9* illustrates how to determine if a general state of stress causes material failure.
- ***Chapter 10*** covers the buckling of columns, including such topics as the effect of transverse forces, column shortening, and buckling on an elastic foundation.

The last six chapters cover advanced topics that may be taught at any class-level. Select chapters or sections may be included in a first Strength of Materials course, or the latter chapters may compose the curriculum of an advanced upper-division or graduate course. These chapters are succinct introductions to entire fields of study in engineering.

- ***Chapter 11*** covers energy methods, expanding on approaches briefly introduced in *Chapter 4*.
- ***Chapters 12*** considers the effect of plasticity (yielding) in ductile materials and its impact on design.
- ***Chapter 13*** introduces fracture mechanics, and the statistical approach used to evaluate the strength of ceramic components.
- ***Chapter 14*** covers stresses in bolted connections and adhesive joints.
- ***Chapter 15*** introduces unidirectionally-reinforced composite materials, including estimating the elastic properties of a single ply and of two-ply laminates, determining system integrity, and basic design considerations.
- ***Chapter 16*** introduces smart systems, including micro-electromechanical systems, piezoelectric devices and shape memory alloys.

Problems for the student to work out are all grouped together in a single "chapter" entitled ***Problems***, placed directly before the ***Appendix***.

Acknowledgements

The authors would like to express their thanks to Dr. David Hayhurst and Dr. Darrell Socie for their thoughtful feedback on early drafts of this text. In addition, we wish to thank the many students at UCSB and Allan Hancock College who studied through the various typos of early versions.

We have tried to eliminate errors, but as in any human endeavor, this text is not perfect. Comments and suggestions may be directed to domdalbello@yahoo.com.

For Liz,	For my father and mother, J.J. and Irene,
F.A. Leckie	D.J. Dal Bello

The Authors

Frederick A. Leckie, Ph.D., is Professor Emeritus of Mechanical Engineering at the University of California Santa Barbara, and Professor Emeritus of Theoretical and Applied Mechanics and Mechanical Engineering at the University of Illinois at Urbana-Champaign. He is a Fellow in the American Society of Mechanical Engineers, and the recipient of the A.S.M.E. Nadia Medal in 2000. Professor Leckie has also taught at the University of Leicester and the University of Cambridge in Great Britain.

Professor Leckie has taught a full range of Mechanical Engineering subjects (Statics, Dynamics, Strength of Materials, Materials, Thermodynamics, and Design), from large undergraduate classes to specialized graduate courses on advanced technology including MEMS. His research interests have included material performance and design at high temperature, material modeling and computation, and composite materials.

Dominic J. Dal Bello is Assistant Professor of Engineering at Allan Hancock College. He has also taught engineering courses at UC Santa Barbara, Cal Poly San Luis Obispo and Santa Barbara City College. He is a member of the American Society of Engineering Education and the American Society of Mechanical Engineers. He was the first recipient of the Allan Hancock Foundation Outstanding Faculty Award in Spring 2006. He was a National Science Foundation Fellow as a graduate student at UC Santa Barbara.

Professor Dal Bello teaches a full range of fundamental engineering subjects: Statics, Dynamics, Strength of Materials, Materials Science, Circuit Analysis, Circuits and Devices, MATLAB, Excel, Engineering Drafting, and Introduction to Engineering.

Image Credits

All illustrations, graphs and photographs were created or taken by Dominic J. Dal Bello, except as noted in the text.

Table of Contents

Chapter 1   Opening Remarks

1.0 Introduction

Stepping on a car's accelerator increases the power produced by the engine. The generated torque must be transmitted from the engine to the wheels without any part of the system failing. When the car is loaded with passengers, the chassis distorts. The frame deflections must be small so that the doors can continue to open and close.

A bridge must be sufficiently strong to support loads due to traffic and the environment without any component breaking. The bridge should also be stiff so that its deflections are relatively small, otherwise commuters would feel apprehensive about using the bridge.

Micro-electromechanical systems (MEMS) must be designed with the proper stiffness (force–deflection response) to perform their functions as sensors and actuators. Biomedical implants such as hip and heart valve replacements must be biocompatible and strong enough to function for long periods of time under cyclic loads.

Figure 1.1. Galileo Galilei (1564–1642) is recognized as the first to attempt to understand mathematically how structures and materials respond to loads. This is his drawing of a tip-loaded cantilever beam (*Dialogues Concerning Two New Sciences*, 1638. Translated by H. Crew and A. de Salvio, 1914. Reissued by Northwestern University Press, 1968).

How do engineers design a physical system that is strong enough and that has the appropriate stiffness? The purpose of this text is to introduce the methods used to ensure the reliability of engineering systems subjected to various loading conditions.

1.1 Strength and Stiffness

The Golden Gate Bridge in San Francisco (*Figure 1.2*) was designed to be strong enough to support its daily traffic load, as well as the environmental loads placed on it

F.A. Leckie, D.J. Dal Bello, *Strength and Stiffness of Engineering Systems*, Mechanical Engineering Series,   **1**
DOI 10.1007/978-0-387-49474-6_1, © Springer Science+Business Media, LLC 2009

(e.g., wind). The bridge's deflection was also considered in the design. Under a full six-lane load of traffic and full sidewalks, the bridge was designed to deflect downward 10 ft.

In your *Statics* course, the primary goal was to determine the reaction forces that acted on a system, as well as the loads internal to the system. In engineering design and analysis, the next questions that need to be asked are as follows:

Is the system STRONG enough? No component of the bridge can fail; e.g., the cables must not break. In a building, the floor joists (the beams that support floors) must be strong enough to support the loads of people, furniture, etc. Artificial heart valves must sustain many millions of cycles.

Figure 1.2. The Golden Gate Bridge. Photo by David A. Emery, ©2002.

Is the system STIFF enough? The components must not deflect excessively under load. The distance between the Golden Gate Bridge's towers – the *span* – is 4200 ft. Since it was designed to deflect 10 ft under its maximum service load, the deflection-to-span ratio is 1/420. In buildings, the ratio specified for roof beams that support plastered ceilings is 1/360 in order to avoid cracks. Relative to the length of the structure, such deflections are difficult to observe. When deflections become readily visible, structures that are sufficiently strong seem unsafe.

Strength is the maximum load a system can support without failure. **Stiffness** is the ability of a system to resist deformation when subjected to load. By determining the strength and stiffness of a system, and comparing it to the design requirements, the engineer can determine if the system is adequate or if it needs to be redesigned.

1.2 Size and Shape

A structural component's **size** and **shape** contribute to its *strength* and *stiffness*. Component geometry may be specified or constrained by the requirements of the customer, or may result from the design process.

Size

The range of sizes of engineering structures is great. Three systems of significantly different size are shown in *Figures 1.2* and *1.3*.

- The distance between towers of the Golden Gate Bridge is 1280 m;
- the distance between axles of a bicycle is about 1 m;

- the distance between supports of a V-beam atomic force microscope (AFM) is about 80×10^{-6} m.

The bridge is about 1280 times, or three orders of magnitude (10^3), larger than the bicycle. The bicycle is about 12,500, or four orders of magnitude (10^4) larger than the AFM. The bridge is therefore seven orders of magnitude larger than the AFM.

The analysis methods described in the text apply over these ranges of scale. However, as objects get smaller than the AFM, atomic interactions become significant. Analysis using nano-mechanics (beyond the scope of this text) must then be considered.

Shape

Wooden structural members have traditionally had rectangular cross-sections. In steel construction, I-beam cross-sections are generally used (*Figure 1.4*). Although I-beams are more costly to manufacture than rectangular cross-sections, the I-beam shape is more efficient (per unit weight) in supporting bending loads, the primary type of load in a beam. The cost savings in material compensates for the cost of forming the I-beam. It is now common to see wooden joists with I-beam cross-sections.

Circular cross-sections are generally used to transmit torsional (twisting) loads in rotating shafts. Circular cross-sections are more efficient than square cross-sections in supporting such loads. Hollow tubes are used in bicycles due to their strength and stiffness efficiency, weight savings, as well as being inexpensive to manufacture.

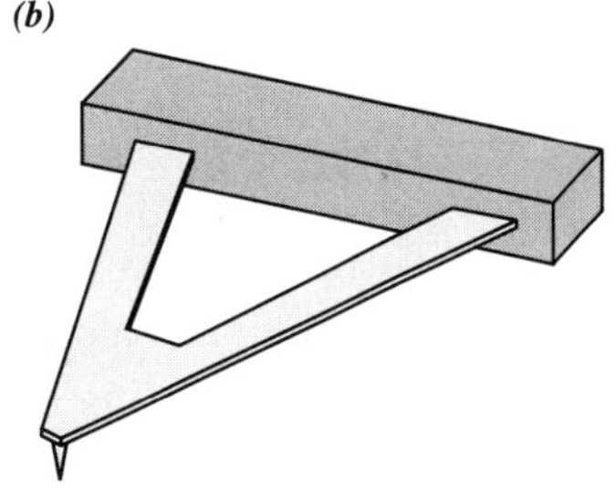

Figure 1.3. (a) A bicycle and **(b)** a V-beam atomic force microscope. Copyright ©2008 Dominic J. Dal Bello and licensors. All rights reserved.

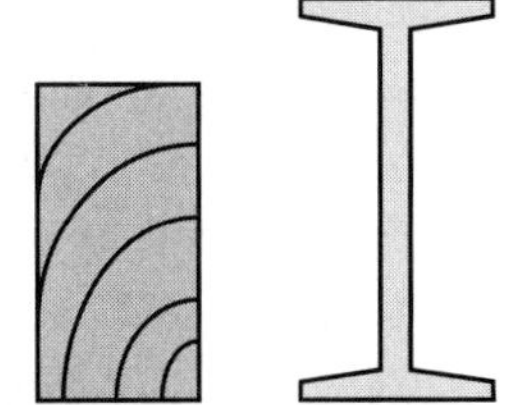

Figure 1.4. Common cross-sections for beams. A wooden rectangular beam and a steel I-beam.

1.3 Loads

Building loads are generally well-defined in building codes. They include:

- Dead loads: the self-weight of the structure and any permanent equipment such as fire suppression and air conditioning equipment.

- Live loads: people, office equipment, the weight of stored materials, vehicle loads and other moveable loads. A few representative values of live loads are given in *Table 1.1*.

- Wind loads: wind causes pressure loading on a building's surfaces. The equivalent *static pressure p* in pounds per square foot (lb/ft^2, commonly called psf) of wind traveling with velocity V in miles per hour (mph) is:

$$p = 0.00256V^2 \quad \text{[Eq. 1.1]}$$

This pressure is modified (multiplied by a correction factor) depending on the building surface being considered. Wind speeds vary throughout the United States, and wind maps are available in the various building codes. A typical design value for the steady-state wind speed is 90 mph.

- Snow loads: snow loads also vary throughout the United States. In snowy regions, the design snow load on roofs is typically 30 psf or more.

- Seismic loads: dynamic seismic forces are especially important in earthquake-prone regions such as California and the Pacific Rim.

Loads in engines. In an engine, the combustion of fuel releases energy as an expanding gas that pushes against the piston. The linear motion of the piston is converted into rotational motion by the crankshaft.

The maximum torque output of the gasoline engine of a typical pickup truck is 266 ft-lb at 4000 rpm (2008 Toyota Tacoma, 4.0 L, V-6). The maximum torque output of the diesel engine of a larger truck is 650 ft-lb at 2000 rpm (2008 Ford F-150 SuperDuty 6.4L, V-8). The power of both engines operating at their maximum torques are comparable (203 and 248 hp, respectively). However, since diesel systems transfer higher torques, they must be made stronger and therefore heavier than gasoline systems.

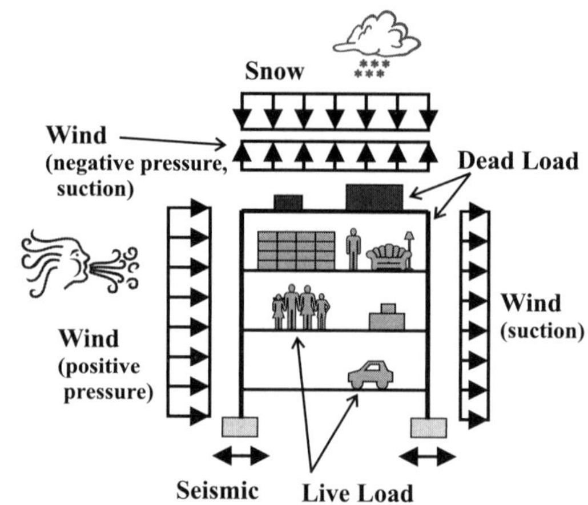

Figure 1.5. Buildings must support various types of loads.

Table 1.1. Minimum Uniformly Distributed Live Loads on Floors, *2000 International Building Code* (SI units converted, rounded up).

Building type	Distributed load	
	lb/ft^2	kN/m^2
Residential, basic floor	40	2.0
Classrooms	40	2.0
Offices	50	2.4
Lobbies; first floor corridors	100	4.8
Corridors above first floor	80	3.9
Library reading rooms	60	2.9
Library stack rooms	150	7.2
Retail stores, first floor	100	4.8
Garages (passenger cars)	50	2.4
Light manufacturing; storage	125	6.0

The operating conditions in gasoline and diesel engines are also different because of the form of combustion. In gasoline engines combustion is caused by an electric spark; in diesel engines combustion is caused by high pressure. The high pressure in diesel cylinders requires them to be made of sufficient strength and size, a second reason that diesel engines are more substantial than gasoline engines.

Cyclic loads. In a four-stroke engine, pressure is applied to each piston and cylinder once every two cycles (two strokes per cycle). The high-pressure stroke is followed by three low-pressure strokes. Such repetitive loading is known as *cyclic loading.* Cyclic loading can cause failure by a phenomenon known as *fatigue.* Fatigue must be considered in the design of engines, rotating parts, and other systems subjected to repeated loads.

Loads in micro-electromechanical devices. Micro-structures of dimensions on the order of 100 μm (= 100×10^{-6} m) long and 10 μm thick are used as micro-sensors and actuators. An example of a micro-sensor is an accelerometer used to detect large negative accelerations that trigger air bag deployment in automobiles. The behavior of micro-cantilever beams is the key to the operation of the atomic force microscope. The forces in these devices can be 10 nN (1 nanonewton = 10^{-9} N); in biological systems, the loads can be 10 pN (1 piconewton = 10^{-12} N). Micro-mechanisms are useful tools in medicine, environmental engineering, and micro-scale manufacturing.

1.4 Failure Modes

When creating a new product, the designer must anticipate the many ways – *modes* – in which it might fail. *Failure* is when a component or system can no longer safely perform its intended function within the specified requirements. Time spent thinking through possible failure modes is time well spent. A lack of good information does not stifle the creative process, but encourages it. With experience, information is gradually filled out and is often incorporated into design practices and building codes. Experience gained from structural failures of the Loma Prieta (1989) and Northridge (1997) earthquakes in California, and other earthquakes, along with experimental and theoretical studies facilitated by computer modeling, is used to continually improve the earthquake requirements of building codes.

The following is a partial list of failure modes.

Excessive Deflection of a Component

While a structure may not break into two parts, it may become unusable because it deflects too much. The requirements on deflection are rather severe and deflection can be the limiting failure condition in many structural applications.

Deflection of bridges. Bridges are often designed so that the maximum deflection is no more than 1/240 of the span. This translates into a 12 in. deflection for a 240 ft bridge. If the deflection is too large, drivers would become aware of the sag and believe the bridge to be unsafe even though it does not break into two parts or permanently deform. The criterion is therefore based partly on psychological factors. The Golden Gate Bridge was designed to have a maximum deflection-to-span ratio of 1/420.

Deflection of buildings. In buildings, the deflection of roof beams that support non-plaster ceilings must be less than 1/240 of the span, while the deflection of floor beams due to live loads must be less than 1/360 of the span (*2000 International Building Code*, Table 1604.3). Doors must still open when the wind load changes direction and magnitude, or as the building's occupancy weight increases. Pre-fabricated units such as shelves and ceramic plumbing fixtures must continue to fit in place. Brittle plaster ceilings must not crack; their supports are limited in deflection to 1/360 of the span.

Deflection of vehicles. The distortion of vehicles such as automobiles, trains and subway cars must be small. When passengers enter or exit a vehicle, it is essential that the doors continue to operate. A requirement for the stiffness of an automobile chassis is that the relative displacement of any corner point must be less than about 0.5 in. Since a chassis is typically 10 ft long, the ratio of displacement to the span is 1/240.

The above examples of the performance of load-bearing systems demonstrate that the requirements across different industries are often similar.

Deflection of aircraft wings. By contrast with the previous examples, deflection is rarely a limiting factor in aircraft. The maximum tip deflection/wing span ratio is on the order of 1/12. The deflections in large aircraft are large enough to be observed visually, which can be a little disconcerting to nervous passengers. Although there are large deflections, airplanes still function properly and safely.

Overloading a Component

Fracture. In any manufacturing or construction program, it is impossible to avoid flaws and cracks. One source of flaws occurs during welding. If the size of a crack introduced during manufacture of a component is greater than a critical crack length, then when a sufficient load is applied, the component will *fracture* or break into two parts. The larger the crack or flaw, the smaller the applied load required to cause fracture.

Yielding. Although a component may not break into two parts, the loads may cause the material to permanently deform or *yield*. The component will not return to its initial dimensions and/or shape when unloaded. *Yielding* or *plastic deformation* is easily demonstrated by bending a paper clip out of shape; the permanently deformed paper clip no longer serves its intended purpose. Materials that plastically deform can absorb a large amount of energy. Automobile designs include a crumple zone. During a collision, the kinetic energy of a moving vehicle is partially dissipated by the plastic deformation of the crumple zone structure.

Fatigue. Fatigue is one of the most common modes of failure and occurs when the applied loads are cyclic over time. Such failure is common in engines, and rotating shafts and axles. An unfortunate feature of fatigue is that there is little indication that failure is close at hand. One moment all seems well, and the next moment, the component has broken in two.

Fatigue failures tend to occur at locations where there is a change in component geometry that causes the stress to be locally higher than the average stress.

Corrosion. When steel is exposed to water and oxygen, the iron in the steel combines with hydrogen and oxygen to form a weaker surface layer of ferric oxide – rust. The layer becomes thicker with time and eventually spalls off. Consequently, less material is available and the load-carrying capacity of the component is diminished. Steel surfaces are often painted or coated to provide protection in a corrosive environment (the Golden Gate Bridge is painted continuously).

1.5 Materials

Various systems are made from different ***materials***. The choice of material depends on the requirements and constraints of the application. A bridge is primarily made of steel and concrete; a bicycle is made of aluminum or a composite material; an atomic force microscope (AFM) is made of silicon nitride; hip replacements are made of titanium. Each material was chosen for each application due to its particular properties or characteristics. In this text, the most important properties include:

- ***density***: the mass per unit volume of a material;
    - *weight density*: the weight per unit volume of a material;
- ***strength***: the maximum force per unit area that a material can support without failure;
- ***stiffness***: the ability of a material to resist deformation.

If two bicycle frames with the same dimensions are built, one of steel and the other of aluminum, the qualitative responses are similar, but the quantitative responses differ:

- steel is three times *stiffer* than aluminum; under the same load (e.g., the weight of the biker), the aluminum frame deflects three times more than the steel frame; however,
- steel is three times *denser* than aluminum; the steel frame is three times heavier than aluminum frame.

Density is a ***physical property***, a characteristic of a material's physical nature. Strength and stiffness are ***mechanical properties*** that characterize the material's response when the material system is subjected to applied forces and moments.

Other factors, such as ductility (material elongation at failure), corrosion resistance, thermal expansion, heat and electrical conductivity, ease of manufacture, cost of raw material, overall weight of the structure, etc., are also issues to consider when choosing a material for a particular application.

Representative material properties are given throughout the text, in *Appendix B*, and in the *Online Notes*. These approximate properties should not be used for actual design. Properties may also be found in printed references and on various web sites. When designing and analyzing actual systems, the properties used should be those provided by manufacturers or found through mechanical tests on the actual material.

1.6 Factor of Safety, Proof Testing

Factor of Safety

The load at which a system fails is the ***failure load*** or ***ultimate load*** P_f, which is determined by calculations or by actual tests on the component. The ***design load*** or ***working load*** P_D is the maximum load that the system is expected to support during service. This load is also referred to as the ***allowable load***. The design load is determined from the performance requirements, e.g., building codes.

The *factor of safety* FS is the ratio of the failure load to the design load:

$$\text{FS} = \frac{P_f}{P_D} \qquad\qquad \text{[Eq. 1.2]}$$

Both loads in the equation must have the same units. A factor of safety is used in design to account for uncertainties in loads, construction methods, and materials.

Factor of safety values depend on the industry. In construction, a typical factor of safety is 2.0; the structure should support twice the maximum load that it is expected to carry. Where the manufacturing environment is very controlled and the service loads are well known, such as in aircraft engines, the FS can be made closer to 1.0.

Example 1.1 Factor of Safety

Given: A number of specimens are tested from a production run of Steel A36 bars. Each bar has a cross-sectional area $A = 1.0$ in.^{2} (*Figure 1.6*), and on average, each supports a tensile force of $P = 36{,}000$ lb before *yielding* (before starting to permanently deform). The steel designation A36 indicates a material with a yield strength of 36,000 lb/in.^{2} (psi).

Required: If the required *factor of safety* is 2.0 against yielding, determine the allowable load (the design load, or maximum working load), P_D.

Solution: $P_D = \dfrac{P_f}{\text{FS}} = \dfrac{36{,}000 \text{ lb}}{2.0} = 18{,}000 \text{ lb}$

$\qquad\qquad$ *Answer:* $P_D = 18$ kips

Note: 1 kip = 1000 lb.

In service, the bar should not be loaded to more than $P_D = 18{,}000$ lb.

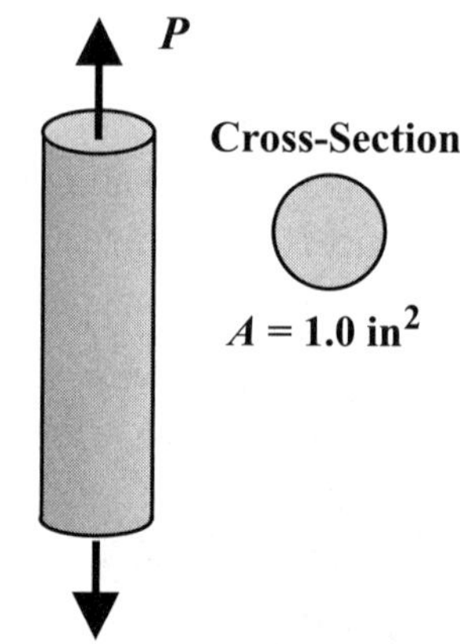

Figure 1.6. A steel bar loaded by tensile force P.

Margin of Safety

In the aerospace industry, a common measure is the *margin of safety*. The margin of safety m is the fraction of the design load that the failure load P_f is above the design load P_D :

$$m = \frac{P_f - P_D}{P_D} = \frac{P_f}{P_D} - 1 = \text{FS} - 1 \qquad \text{[Eq. 1.3]}$$

For an acceptable design, the values of m must be positive since FS > 1.0 to avoid failure. In the above example problem, FS = 2.0, so m = 1.0. If FS = 1.5, which is common in aircraft applications, then m = 0.5.

Proof Testing

Before purchasing an automobile, it is common practice to give it a test drive to determine if it meets your requirements. Likewise, an engineering system is typically tested before it is placed into service to prove that it works. A *proof test* is especially necessary when the system or construction methods are new, the structure is difficult to analyze, or when the failure of the system is dangerous to human life or can cause significant damage to other systems.

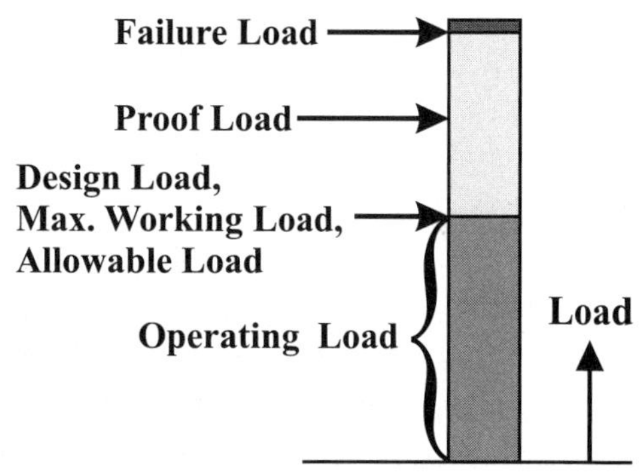

Figure 1.7. Graph showing the relationship between the various loads.

The load at which a component or system is tested is called the ***proof load*** P_P (*Figure 1.7*). The proof load is greater than the design load P_D (to prove that the system will work at its maximum expected load), but less than the failure load P_f (the purpose of the test is to prove the component works, not to break it).

Pressure vessels, such as propane tanks or steam pipes, often contain gasses at high pressure. Fracture of these vessels releases a large amount of energy that can tear a vessel apart and result in serious injuries to people nearby. Even if the vessel does not catastrophically fail, a leak in a toxic-gas vessel is dangerous.

The American Society of Mechanical Engineers (ASME) Design Code for pressure vessels specifies that pressure vessels are, in general, to be tested at a proof load of 1.5 times the maximum working pressure. Proof testing of gas pressure vessels is often done with water because the energy stored in a pressurized liquid is much less than that in a compressed gas under the same pressure. If the vessel breaks during testing, the water simply gushes out of the vessel as opposed to there being a violent gas-explosion.

Experience and modern manufacturing and analysis techniques reduce, if not eliminate, the need for proof testing. However, whenever there is any uncertainty, a proof test should be performed.

1.7 Unit Systems

There are two systems of measurement used by American engineers: (1) the *US Customary System, English System* or *British Imperial System* and (2) the *International System of Units* or simply, the *SI System* (from the French *Système International d'Unités*). The most important dimensions in this text are force and length.

The US Customary System

The unit of force in the US Customary System (USCS) is the *pound* (lb). One thousand pounds is known as a *kilopound*, abbreviated as kip (1 kip = 1000 lb). The unit of length is the *foot* (ft). The *inch* (1 ft = 12 in.) is used for dimensioning parts and specifying cross-sectional areas. The unit of time is the *second* (sec).

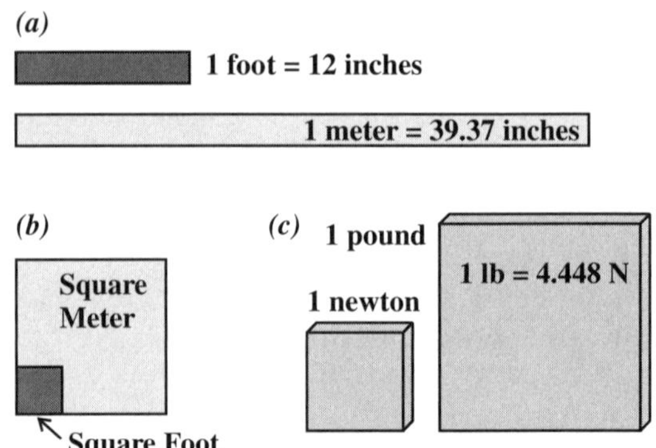

Figure 1.8. Comparison of length, area, and force between US and SI systems.
(a) Length. **(b)** Area. **(c)** Two plates of the same material and same thickness, one weighing 1 newton, one weighing 1 pound.
Note: *Figures are to scale within each comparison* (length, area, weight) only.

The SI System

The unit of force in the SI system is the *newton* (N). One thousand newtons is known as a *kilonewton*, abbreviated as kN. The unit of length is the *meter* (m). The unit of time is the *second* (s).

Order of magnitude in the SI system is efficiently indicated by adding a preface to the base unit. For example, 0.001 meter (m) is 1 *milli*meter (mm); 1000 newtons (N) is 1 *kilo*newton (kN); 1,000,000 watts (W) is 1 *mega*watt (MW). It is much easier to write and read 1.0 MW than 1,000,000 W. In this way, the numerical values are kept at a reasonable size – generally between 0.1 and 1000 – while the prefix (k, M, etc.) gives the order of magnitude. The prefixes for the standard orders of magnitude in the SI system are given in *Table 1.2*. Results are typically given using the standard SI prefix form.

Conversion

Four basic conversion factors for length and force are listed in *Table 1.3*. For example, 1.000 m equals 39.37 in., and 1.000 lb equals 4.448 N. For back-of-the-envelope approximations, 1 m is about 40 in. or 3.25 ft (~1.5% error) or 3.0 ft (~8.6% error), and 1 lb is about 4.5 N (~1.2% error).

Table 1.2. Standard prefixes of the SI system between 10^{-12} and 10^{12}.

Letter preface	Preface	Value	Value	Example
T	tera-	10^{12}	trillion	Saturn is on average, about 1.43 Tm from the Sun*
G	giga-	10^{9}	billion	3.0 GHz (processor speed)
M	mega-	10^{6}	million	50 MW (electrical power)
k	kilo-	10^{3}	thousand	8.288 kN = 2,000 lb = 2 kips = 2 kilopounds (a force), 1.28 km (length of central span of Golden Gate Bridge)
		1	one	m (meter), N (newton)
m	milli-	10^{-3}	one-thousandth	1.00 in. equals 25.4 mm
μ	micro-	10^{-6}	one-millionth	100 μm (diameter of human hair, or composite fiber)
n	nano-	10^{-9}	one-billionth	the size of an atom is between 0.1 and 0.5 nm
p	pico-	10^{-12}	one-trillionth	200 pN (approximate strength of a chemical bond)**

* http://www.nasa.gov/worldbook/saturn_worldbook.html   Accessed May 2008.
** http://physicsweb.org/articles/world/12/9/9   Accessed May 2008.

Table 1.3. Basic conversion factors for length and force.

To convert from	multiply by	to get. To convert from	multiply by	to get
inches, in.	0.0254	*meters*, m	39.37	*inches*, in.
pounds, lb	4.448	*newtons*, N	0.2248	*pounds*, lb

A table of *conversion factors* for length, force, and temperature, and for other quantities, is given in the *Online Notes*. Conversion calculators are also available at various Internet sites and in computer analysis tools.

Example 1.2 Converting Units 1

Required: How fast is 60.0 miles per hour (mph) in meters per second (m/s)?

Solution: The velocity $V = 60.0$ mph is multiplied by various conversion factors as follows:

$$V = 60 \, \frac{\text{miles}}{\text{h}} \left(\frac{5280 \text{ ft}}{\text{mile}} \right) \left(\frac{12 \text{ in.}}{\text{ft}} \right) \left(\frac{2.54 \text{ cm}}{\text{in.}} \right) \left(\frac{1 \text{ m}}{100 \text{ cm}} \right) \left(\frac{1 \text{ h}}{60 \text{ min}} \right) \left(\frac{1 \text{ min}}{60 \text{ s}} \right)$$

Answer: $V = 26.8$ m/s

In this example, a step-by-step conversion has been performed, using unit equalities that the reader should be familiar with. The term in each set of parentheses is simply equal to unity (1), as shown here:

$$5280 \text{ ft} = 1 \text{ mile} \rightarrow \frac{5280 \text{ ft}}{1 \text{ mile}} = 1 \; ; \quad 1 \text{ h} = 60 \text{ min} \rightarrow \frac{1 \text{ h}}{60 \text{ min}} = 1$$

The unit names cancel just as if they were variables. For example, *miles* in the numerator of the first term cancels with *miles* in the denominator of the second term:

$$V = 60 \, \frac{\text{mile}}{\text{h}} \left(\frac{5280 \text{ ft}}{\text{mile}} \right) = (60 \times 5280) \, \frac{\text{ft}}{\text{h}}$$

Likewise, the other units cancel except for *meters* in the numerator and *seconds* in the denominator.

The *conversion factor* to convert from mph to m/s can be given directly as:

$$\frac{V_{\text{m/s}}}{V_{\text{mph}}} = \frac{26.8 \text{ m/s}}{60.0 \text{ mph}} = 0.447 \, \frac{\text{m/s}}{\text{mph}}$$

To convert *any* velocity from mph to m/s, multiply the numerical value of the velocity in mph by the conversion factor, 0.447; e.g., 75 mph is:

$$(75 \text{ mph}) \left(0.447 \, \frac{\text{m/s}}{\text{mph}} \right) = 33.5 \text{ m/s}$$

Example 1.3 Converting Units 2

Given: *Stress* is an important quantity in this text. Stress is the intensity of force per unit area, and so has units of lb/in.^2 or N/m^2.

Required: Convert a stress of $\sigma = 10.0$ ksi (kilopounds per square inch) into its SI equivalent (MN/m^2).

Solution: The stress of $\sigma = 10.0$ ksi is multiplied by conversion factors from *Table 1.3*:

$$\sigma = 10.0 \times 10^3 \, \frac{\text{lb}}{\text{in.}^2} \left(\frac{39.37 \text{ in.}}{1.00 \text{ m}} \right)^2 \left(\frac{4.448 \text{ N}}{1.00 \text{ lb}} \right) = 6.894 \times 10^7 \, \frac{\text{N}}{\text{m}^2}$$

$$\textit{Answer:} \quad \sigma = 68.9 \text{ MN/m}^2$$

A good rule of thumb is illustrated here: to convert from lb/in.^2 to N/m^2 (a N/m^2 is a *pascal*, Pa), the conversion factor is approximately 7,000; e.g., 10 ksi ~ 70,000 kN/m^2 = 70 MN/m^2. Alternatively, multiply lb/in.^2 by 7 and increase the prefix by three orders of magnitude (e.g., 1 psi ~ 7 kN/m^2, 1 ksi ~ 7 MN/m^2).

Most of the world currently uses the SI system (meters, kilograms, and seconds). In the United States, most scientific work is done in SI units. However, most US industrial work and day-to-day measurements are done using the US system (feet, pounds, and seconds). While there has been a long-term effort to convert the United States to the SI system, and the Federal government requires the use of SI units in technical reports, an

American engineer will need to work in both systems – USCS and SI – for years to come. Engineers must be aware of the differences between the two systems, and always communicate which system they are working with.

Mass and Weight

The difference between mass and weight (a force) can be a source of confusion, so a word may be appropriate here.

Mass

Mass is the amount of *matter* that an object is composed of. The mass of an object is constant regardless of where it is in the universe.

Newton's Second Law states that the vector acceleration **a** of an object is proportional to the resultant force **F** acting on it, and that **a** is in the direction of that force. For an object of mass m, the relationship is:

$$\mathbf{F} = m\mathbf{a} \qquad\qquad \text{[Eq. 1.4]}$$

Mass is therefore a measure of an object's *inertia* – the resistance the object has to changes in velocity. The more mass an object has, the more force is required to give it a certain acceleration (change in velocity).

In the SI system, the unit of mass is the *kilogram* (kg). The kilogram is one of the *base* or *fundamental units* in the SI system (the meter and second are two others). The SI unit of force, the newton (N), is a *derived unit*. One newton is required to accelerate 1 kg by 1 m/s^2:

$$1 \text{ N} = (1 \text{ kg})(1 \text{ m/s}^2) \qquad\qquad \text{[Eq. 1.5]}$$

In the USCS, the unit of mass is the *slug*. The slug is a *derived unit*, defined in terms of the US *base units* of pound, foot and second. One pound is required to accelerate 1 slug by 1 ft/s^2:

$$1 \text{ lb} = (1 \text{ slug})(1 \text{ ft/s}^2) \qquad\qquad \text{[Eq. 1.6]}$$

Weight

Weight is the *force* of gravitational attraction exerted by the earth on an object:

$$W = mg \qquad\qquad \text{[Eq. 1.7]}$$

where g is the acceleration of gravity, taken as constant at the surface of the earth. In the SI system, the standard value of g is 9.81 m/s^2; in the US system, the standard value is 32.2 ft/s^2. Since g has units of acceleration, weight has units of force. In addition, weight is proportional to mass.

In the SI system, the unit of weight (gravitational force) is the newton. A mass of 1 kg weighs 9.81 N. In the US system, a mass of 1 slug weighs 32.2 lb.

$$9.81 \text{ N} = (1 \text{ kg})(9.81 \text{ m/s}^2) \qquad\qquad \text{[Eq. 1.8]}$$

$$32.2 \text{ lb} = (1 \text{ slug})(32.2 \text{ ft/s}^2)$$ [Eq. 1.9]

Away from the earth – in space, on the moon, or on another planet – the weight of an object is still proportional to its mass, but the effective value of g varies per Newton's Universal Law of Gravitation. Therefore, weight depends on where an object is in the universe; the mass of an object is the same everywhere.

When to Multiply by g

The weight of an object in the metric system is often given in kilograms (or *kilogram force*, kgf), which in a scientific context is its *mass*. If the weight (mass) of an object is given in kilograms, it is necessary to multiply by the acceleration of gravity $g = 9.81$ m/s^2 to obtain the weight (gravitational force) in newtons.

When weight is given in newtons, it is not necessary to multiply by g.

In the US system, weight is given in pounds, the unit of force; it is not necessary to multiply by g.

1.8 Coordinate Systems

Most problems in this text use the standard 3D cartesian $(x–y–z)$ coordinate system, and many problems can be reduced to a planar or 2D $(x–y)$ coordinate system (*Figure 1.9*). In 2D, the x-direction is commonly drawn horizontal, the y-direction vertical, and the z-direction is out of the plane of the paper (*Figure 1.9b*).

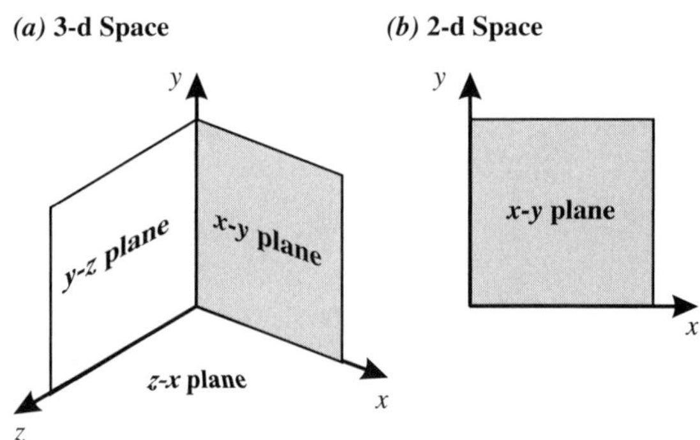

Figure 1.9. (a) 3D and (b) 2D coordinate systems.

There are times when aligning the x- and y-directions with the horizontal and vertical is not convenient. It is often useful to have the x-axis correspond to the axis of a rod or a beam, and this may not coincide with a horizontal line on the paper (*Figure 1.10b*). To avoid confusion, the coordinate system *triad*, or at least the two in-plane axes, should be included on each drawing.

Sometimes more than one coordinate system is necessary to describe a system and its components. A set of axes associated with the entire structure is called the **global coordinate system** (*X–Y, Figure 1.10a*), while a set of axes specific to a particular member is a **local coordinate system** (*x–y, Figure 1.10b*).

The Right-Hand Rule

The cartesian coordinate system consists of three orthogonal (mutually perpendicular) axes. The positive directions of the *x*- and *y*-axes are 90° to each other, with the +*y*-axis being found by rotating counterclockwise from the +*x*-axis. The positive direction of the *z*-axis (third axis) is defined by the *Right-Hand Rule*.

To determine the direction of the +*z*-axis, the Right-Hand Rule is performed as shown in *Figure 1.11*. Referring to *Figure 1.11a*, the fingers of the right-hand are initially pointed in the +*x*-direction (palm upward). The fingers are then closed upon themselves (counterclockwise) toward the

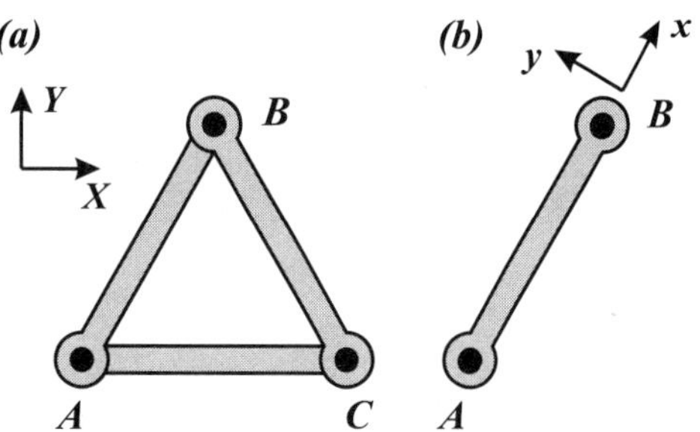

Figure 1.10. (a) Truss *ABC* in the traditional *X–Y* set of axes. **(b)** A set of axes based on the orientation of a particular member is a *local coordinate system*, while a set of axes describing the entire system is a *global coordinate system*, e.g., *Figure (a)*.

+*y*-axis, wrapping around the *z*-axis. The direction that the right thumb points is along the +*z*-axis.

The Right-Hand Rule is necessary to define the positive and negative sense of torques and moments (loads that cause twisting or bending about an axis). Mathematically, the Right-Hand Rule gives the direction of the *cross-product* result of two vectors. The cross-product of the *x*- and *y*-direction unit vectors, **i** and **j**, is the unit vector in the *z*-direction: **k** = **i** × **j**.

Position, Displacements, and Angles

The *position* of a point is indicated by its *x–y–z* coordinates. The *displacement* – the change in position – of a point in the *x*-, *y*-, and *z*-directions are *u*, *v*, and *w*, respectively. Displacements are *positive* if they are in the same direction as the positive axis of interest; otherwise they are *negative*.

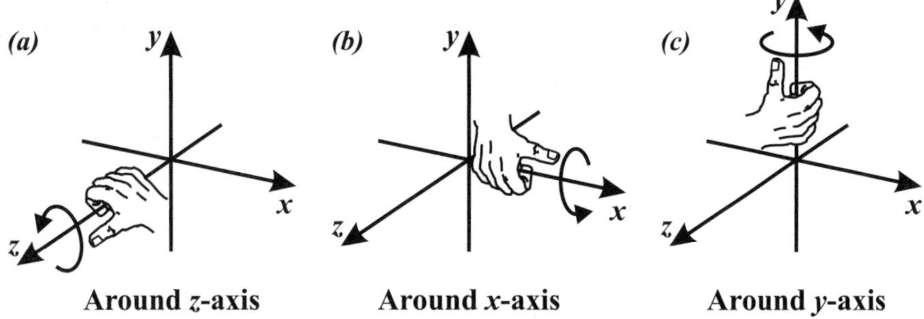

Figure 1.11. The Right-Hand Rule. Direction of the positive moment about each axis, found by taking the cross-product of the unit vectors: **(a)** **i** × **j** = **k**, **(b)** **j** × **k** = **i**, and **(c)** **k** × **i** = **j**.

Angles are *positive* counterclockwise about each positive axis – in the same sense as the Right-Hand Rule (*Figure 1.11*). In the 2D x–y coordinate system, the angle is generally measured from the x-axis.

Forces

A *force* **F** is a vector:

$$\mathbf{F} = F_x\mathbf{i} + F_y\mathbf{j} + F_z\mathbf{k} \qquad \text{[Eq. 1.10]}$$

with scalar components in the x-, y-, and z-directions, F_x, F_y, and F_z, respectively. If the x-component of the force acts in the positive x-direction, then scalar component F_x is *positive*. If the x-component of the force acts in the negative direction, then scalar component F_x is negative. The same rules hold for the y- and z-components.

Moments and Torques

A *moment* **M** is also a vector:

$$\mathbf{M} = M_x\mathbf{i} + M_y\mathbf{j} + M_z\mathbf{k} \qquad \text{[Eq. 1.11]}$$

with scalar components M_x, M_y, and M_z (M_x acts about the x-axis). When the x-axis corresponds to the axis of a shaft or beam, the x-component M_x is known as the **torque**.

A scalar component of the moment (e.g., M_z) is positive if it causes counterclockwise rotation about its corresponding positive axis (e.g., the positive z-axis). The positive sense about each axis is shown in *Figure 1.11*.

A moment or torque is represented in this text in one of the two ways: by a curved arrow about an axis, or by a straight vector with a double arrow. Both methods are shown in *Figure 1.12*.

In *Figure 1.13*, a planar slab rotates about the z–z axis, which is perpendicular to the slab and passes through point O. Force **F** acts in the plane of the slab at point A, vector position **r** from point O. The moment arm – the perpendicular distance from the *moment center* (point O) to the *line of action* of the force – is $d = r\,\sin\theta$. The moment about point O due to force **F** is:

$$\mathbf{M} = \mathbf{r} \times \mathbf{F} \qquad \text{[Eq. 1.12]}$$

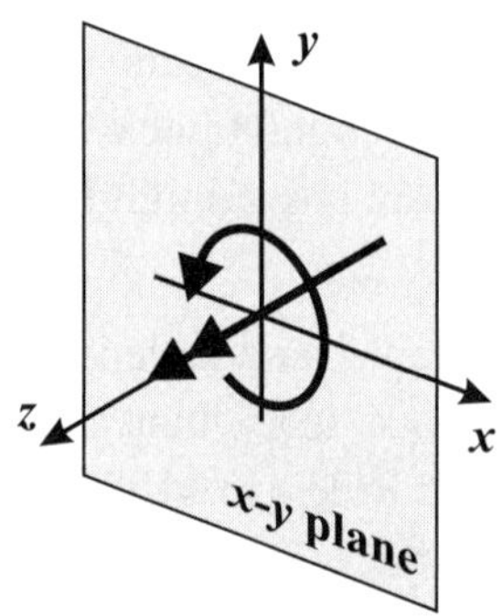

Figure 1.12. The circular arrow in the x–y plane indicates twisting about the positive z-axis. The double arrowhead vector along the $+z$-axis represents the same action.

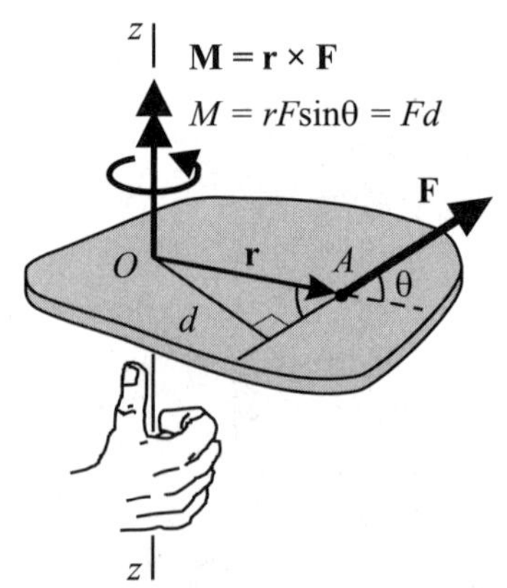

Figure 1.13. Using the Right-Hand Rule to determine the direction of moment **M** caused by force **F**.

The moment has magnitude $M = Fd$, and its direction is represented by either the circular arrow or the double-headed vector. The curved arrow is the path that the fingers of the right-hand follow when performing the Right-Hand Rule on vectors $\mathbf{r}$ and $\mathbf{F}$. The double-headed vector defines the axis and direction of rotation, the arrow coinciding with the direction of the thumb during the Right-Hand Rule.

1.9 Numbers

Significant Digits (Significant Figures)

When using a calculator, it can be tempting to write down all 10 digits that the calculator might display. However, engineering solutions should conform to the rules of **significant digits** or **significant figures**. These rules may be reviewed in any basic chemistry or physics textbook. The number of significant digits that a quantity is written with indicates the precision to which that quantity is known. For example, reporting a force of 34.32 N implies that the force is known to within 0.01 N; writing it as 34 N implies that it is known to 1 N.

A calculated answer should contain no more significant digits than the number of significant digits in the given data. For example, if a rectangle is measured to be 5.72 by 4.39 in. (three significant digits each), and the area is to be determined, a calculator will display six digits: 25.1108 in.^{2}. Since there are at most three significant digits in the given data, the area should be given to three digits: 25.1 in^2. Any more digits in the result would imply that the calculated area is known to a better degree of precision than the actual measurements. The six-digit answer implies a precision of 0.0001 per 25.1108 (0.0004%), while the best measurement only has a precision of 0.01 per 4.39 (0.2%).

Care should be taken not to round intermediate answers too soon during a solution. This will cause a loss of accuracy in the final answer. If the final answer is to have three significant digits, intermediate steps should use at least four, if not more, significant digits.

Engineering Data, Estimates and Error Calculations

Most engineering data are only reliable to two or three significant digits, so practical answers generally have no more than two or three significant digits. Typical textbook problems are usually assumed to be good to three digits.

Estimates, error calculations, and other rough calculations using approximate data, are given to one, or at most two, significant digit(s).

Angles

Since angles are a measure of direction rather than of magnitude, they do not follow the standard rules of significant digits. The final answers for angles should consistently be given to the same number of digits beyond the decimal point; e.g., to one digit beyond the decimal point, 21.3° and 221.3°, not to the same number of digits, 21.3° and 221°.

To avoid round-off error when using trigonometric functions, angles in intermediate calculations should generally be taken to at least two places beyond the decimal point.

Engineering Notation

When writing very large or very small numbers, it is convenient to use ***engineering notation***. Engineering notation is scientific notation except that the exponent must be divisible by three. In scientific notation 23,500 is 2.35×10^4; in engineering notation it is 23.5×10^3. In engineering notation, the numerical value is generally written so that it is between 0.1 and 1000.

Engineering notation coincides with the standard SI prefixes, which change every three orders of magnitude (e.g., kilo = 10^3; mega = 10^6, etc.). Engineering notation is thus preferable to scientific notation. Final results should be written using either engineering notation or SI prefixes.

Finally, when writing a number less than 1.0, a zero should be placed before the decimal point, e.g., one-fourth is "0.25," not ".25." Not including a zero may lead to the decimal point being overlooked, or may leave the reader wondering if there should be a non-zero number in front of the decimal point. Engineering communication should be clear and without sources of possible confusion.

1.10 Analysis and Problem Solving

The first step in analyzing any problem is to define the problem as accurately as possible. What is the intended function of the system? What is its overall geometry and shape? What materials are to be used? What are the loads? How are they applied to the system? How is the system supported? What is known? What is unknown? What are the constraints?

It is vital to make a sketch, no matter how simple, of the basic system geometry and expected loads. A sketch or drawing helps in visualizing the problem, and in getting ideas across to other engineers reviewing the problem, as well as to managers, customers, etc. ***Free Body Diagrams*** showing the forces acting on part of the system or on the entire system, are invaluable to the solution process. An error in the FBD will result in the entire analysis being incorrect.

It is wise to stop and think about a problem before using an equation that happens to have similar variables as the quantities listed in the problem statement. There are many equations that solve for the quantity known as *stress*, but only one will fit a particular situation. First understand what is physically going on in the problem, and then select the appropriate solution method and equation(s) that model the physics.

While the correct content (the solution process) and the correct result are the goals, presentation is important. Homework grades can sometimes be based on the neatness and clarity of student work. There is little value in a solution that cannot be clearly presented to others. The care taken as a student to do presentable work that logically leads the reader through the solution process will help develop those skills necessary to do acceptable and presentable work as an engineering professional.

In this text, the examples follow a consistent format, already illustrated. The **Given** information (i.e., the system and the knowns) and the **Required** result (i.e., what needs to be solved) are first stated (**Given, Required**). A figure of the problem at hand is given, followed by a free body diagram and other sketches. The problem *solution* is then presented (**Solution**). The final answer, including appropriate units, is underlined. Your instructor may ask for a similar format to be followed; your solution will then be a stand-alone document, without need for the reader or yourself to refer back to the original text.

1.11 Modeling

Modeling

The process involved in deciding which college to attend, which career to follow or which house to buy, is complicated. To make a good decision requires a good strategy supported by information. The process of developing the strategy combined with information is referred to as *modeling*. A model is used to predict what will happen in the real situation.

Business models and economic models are used to help make financial decisions. In the same way, an engineering model is needed to design, develop, and analyze a new product. The modeling process is rarely clear cut, with advantages and disadvantages associated with different types of models and levels of complexity. However, even simple (but correct) models help to clarify understanding and guide a rational decision-making process. A good model is one that can be used repeatedly in different circumstances.

A good drawing or sketch helps to formulate a good model. Many projects involve people from different disciplines working together. Communication is most effective when the objective is clear and the participants can explain their work to others, both in pictures and in words.

In the early stages of the design of a new product, it is most productive to use the simplest model possible. Basic design decisions can then be made with little cost and within a reasonable time. As the design evolves, more details can be introduced. In this way, the progress of the design remains in the hands of the designer. It is easy to make models and problems complicated. It is more difficult to keep them simple.

Effective models are detailed enough to realistically describe the physical phenomenon, but simple enough to keep the calculations cost-effective.

An Example of a Simple Model

A simple model that leads to simple calculations, and that is useful in practice, is the ***two-bar model*** shown in *Figure 1.14*. The model consists of two bars, Bars 1 and 2, each supported at one end by a solid foundation which holds them in place, and at the other end by a rigid boss that is constrained to move in the same direction as the axes of the bars. Force F is applied to the boss, and the system elongates by distance δ. The bars share in supporting the load, and both elongate by the same amount.

Two examples for which the two-bar model is useful are in modeling composite materials and in modeling surface coatings on metal components (*Figure 1.15*).

A *composite material* is one in which two discrete materials are combined to form a new material with superior properties. One of the most common is reinforced concrete, where steel bars are strong in tension, and the concrete matrix is less expensive, formable, and strong in compression. Modern aircraft and space vehicles are made of composite materials in which strong small-diameter fibers are set in a polymer matrix. This combination makes for a strong and light-weight structure. In 1986, Burt Rutan's *Voyager*, a plane with a structure made entirely of composites, made the first-ever non-stop, un-refueled flight around the world.

Turbine blades in aircraft engines consist of a metal structure with a ceramic coating. The metal provides the strength, while the ceramic thermal barrier coating protects the metal from exposure to high temperatures.

The performance of both composite and coated systems can be studied with the two-

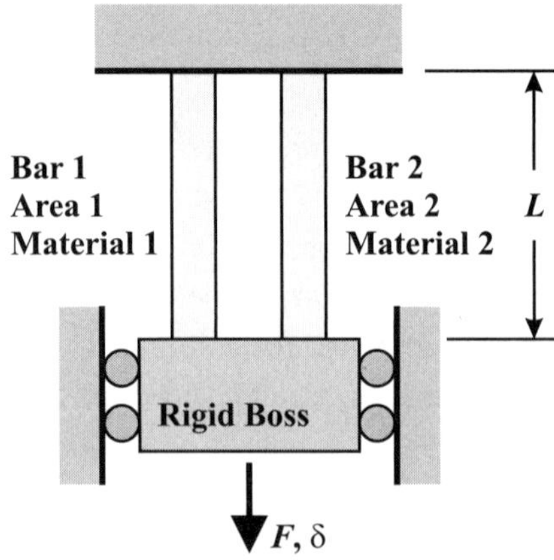

Figure 1.14. A two-bar model. The bars may have different cross-sectional areas and be made of different materials.

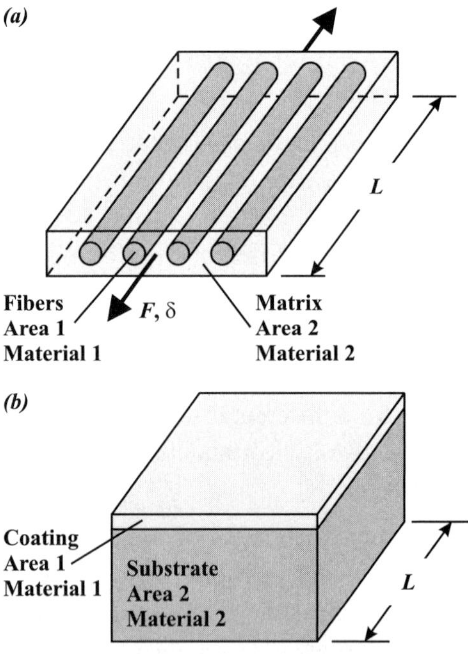

Figure 1.15. (a) Unidirectional composite. Using the two-bar model (shown in *Figure 1.14*), the fibers can be modeled as Bar 1 and the matrix as Bar 2. **(b)** Coated substrate. In a 1D analysis, the coating can be modeled as Bar 1 and the substrate as Bar 2.

bar model. For the composite, one bar represents the fibers and the other the matrix; for the coated substrate, one bar represents the coating and the other the substrate. Once the model has been set up, application of loads to these two different technical problems is simple and computer design studies are relatively easy. Basic design decisions based on these calculations can then be made before more expensive and time-consuming analyses are performed.

1.12 Computer Tools

Provided that a model is sufficiently simple, it is possible to carry out the design and analysis of an engineering system with a pencil, paper, and a hand-calculator. A design notebook containing calculations, graphs, and hand-sketches tracks the history and progress of the design and analysis.

However, engineering systems can be very complex, requiring many design iterations, each iteration possibly requiring many difficult and/or lengthy calculations. Computers are very efficient in performing mathematical operations, and thus provide an invaluable tool for the modern engineer in determining the reliability of a system.

Software packages such as MATLAB and Mathcad have been developed to model the pencil–paper–calculator approach to analyze a system. Spreadsheet programs such as Microsoft Excel can be configured to automatically perform numerous calculations and compare data. Many pages of the traditional design notebook have been replaced by electronic files. However, there is still great value in being able to draw an appropriate sketch, perform hand-calculations, and have a feeling for how the system will respond.

There are tremendous advantages in using computer software. Many basic and complex calculations are easily carried out, and results and graphs are quickly generated. MATLAB and Mathcad can be used to symbolically manipulate algebraic equations and to perform calculus operations. The grunt work of traditional pencil-and-paper methods and the time to analyze a system can both be greatly reduced. However, it takes time and commitment to learn how to correctly and effectively use any software package, and to understand its limitations.

Also available are finite element analysis (FEA) packages that are capable of analyzing very complex systems in 2D and 3D, containing thousands of structural members or elements. The use of these powerful tools should be delayed until the step-by-step traditional practice of carefully solving and understanding problems is complete and the results of a complex computer analysis can be properly understood and appreciated.

The engineer should have a rough idea from paper calculations what output to expect; there should be no major surprises. An unexpected result may mean an error in formulating the model ("garbage-in, garbage-out"), an analysis that is beyond the limits of the software, or a misunderstanding of the physical principles that govern the problem. It is very embarrassing to present the results of complex work using an expensive software package only for someone to notice a very simple error. Computer tools rarely come with a guarantee of reliability. If a design fails because of error in the software, the responsibility for failure rests with the engineer.

Software is also available on the Internet for specialized calculations. Calculations required to rotate the 2D stresses that develop in a material occur commonly in practice. Software to complete these repetitive *Stress Transformation* calculations is available in the *Online Notes*.

Another example of an online tool covering a specialized topic is at www.fatiguecalculator.com. Software is available to determine the life of systems subjected to cyclic loading. This site is a reliable source for material fatigue properties, fatigue life predictions, and stress concentration factors.

In addition to printed references, each engineer should build a library of software packages and Internet sites that fit his or her particular needs. These are the electronic tools and resources that help in the design and analysis of new systems.

2.0 Introduction

A *Statics* analysis is generally the first step in determining how an engineering system deforms under load, and how the system supports the load internally. For ***equilibrium***, the vector sum of all the forces $\mathbf{F}_i$ that act on a system must equal zero. Likewise, the vector sum of all the moments $\mathbf{M}_i$ that act on a system must equal zero. However complex the system, it must support the applied loads without accelerating.

In mathematical terms, *equilibrium of forces* and *equilibrium of moments* are represented by the vector equations:

$$\sum_i \mathbf{F}_i = 0 \qquad\qquad\text{[Eq. 2.1]}$$

$$\sum_i \mathbf{M}_i = 0 \qquad\qquad\text{[Eq. 2.2]}$$

Free body diagrams (FBDs) of the entire system, of individual components, and of parts of individual components, are vital to the solution of any problem. In a FBD, the body of interest is first isolated from its surroundings. All of the forces and moments that act on the body from its surroundings are then represented with force and moment vectors. The coordinate system is also indicated. The examples in this chapter, and throughout the text, illustrate the importance of FBDs.

An example of a FBD is given in *Figure 2.1*. A highway sign is acted on by wind load F_W; the weight of the sign and support mast are presently ignored. A FBD of the support mast is shown in *Figure 2.1b*. The mast has

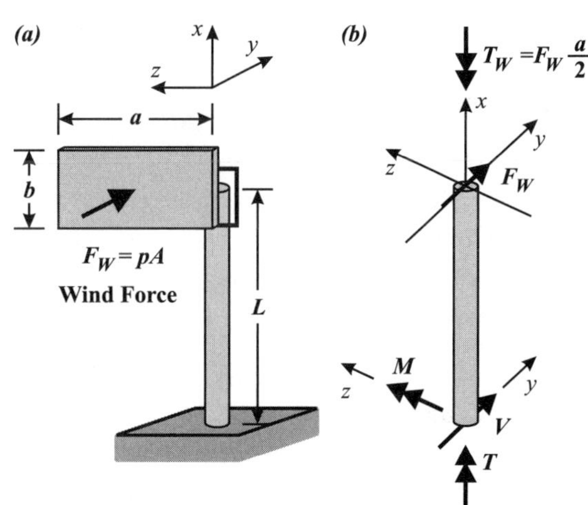

Figure 2.1. A free body diagram isolates a system or part of a system, and shows the forces acting on it. **(a)** A highway sign and support mast under a wind load. **(b)** FBD of the support mast.

F.A. Leckie, D.J. Dal Bello, *Strength and Stiffness of Engineering Systems*, Mechanical Engineering Series,   **23**
DOI 10.1007/978-0-387-49474-6_2, © Springer Science+Business Media, LLC 2009

been removed from its surroundings, and all the forces and moments that act on it are represented. Here, the wind load causes a force F_W at the top of the mast, requiring a reaction shear force at the base V. The wind causes torque T_W, requiring reaction torque T. In engineering systems, a torque is a moment that causes twisting of an axial member or shaft. Finally, reaction moment M at the base keeps the mast from falling over due to the wind load. All the reactions have been drawn in the positive directions of the axes, not necessarily in the directions that they physically act. This problem is studied in more detail in *Example 2.9*.

Carefully drawn FBDs make the loads acting on a structure easier to visualize, and help to communicate to others how the system operates. The FBDs are the basis for the equations of equilibrium. Every problem and solution should include an FBD.

2.1 Axial Members

An ***axial member*** is a straight component that only supports a force P parallel to its axis (*Figure 2.2*). This ***axial force*** must pass through the ***centroid*** (center of area) of the component's cross-section so that the response at any cross-section is uniform (the same at every point). Loads that stretch the component are ***tensile loads***, and those that shorten the component are ***compressive loads***. *Figure 2.2a* shows a tensile force P and *Figure 2.2b* shows a compressive force.

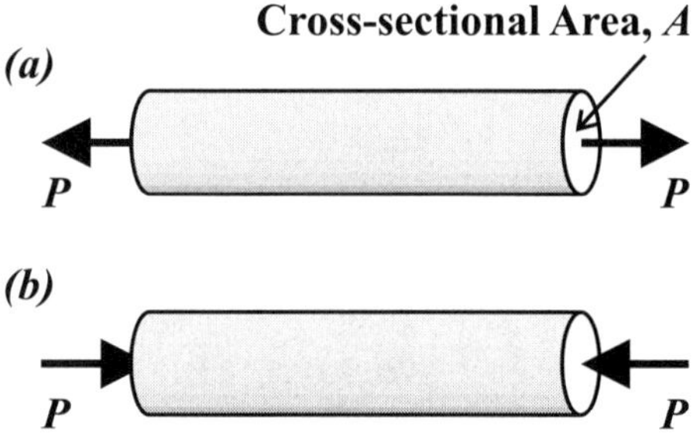

Figure 2.2. An axial member under **(a)** a tensile load and **(b)** a compressive load.

If axial force P is consistently drawn assuming that it is a tensile force, as in *Figure 2.2a*, then a calculated positive value for P ($P > 0$) means that it is tensile, while a negative value for P ($P < 0$) means that it physically acts opposite drawn, i.e., the force is compressive.

An axial member is typically a ***two-force member***. The forces that act at each end of a two-force member are *equal*, *opposite*, and *co-linear*. A straight two-force member supports a constant force P normal to any cross-section along its entire length. An assembly of straight two-force members appropriately pinned together is a *truss*.

The internal force that an axial member supports may vary along its length. These changes may occur at discrete locations due to point loads, or continuously due to distributed loads. The cross-sectional area may also change abruptly or continuously.

A few statics examples with axial members follow.

Example 2.1 Flower Pots on Hanging Shelves

Given: Three shelves are hung from the ceiling by means of four rope cords. Plants are evenly distributed on each shelf as shown in *Figure 2.3a*. The total weight of the plants, pots, and earth on each shelf is $W = 100$ lb, which is assumed to be uniformly distributed over the shelf area (*Figure 2.3b*).

Required: Determine the force in each segment of each cord.

Solution: Assuming that weight W is evenly distributed on each shelf, that the cords are symmetrically placed, and that the shelves are level, then the four cords share equally in supporting each shelf.

Step 1. A FBD of the entire system (*Figure 2.3b*) is made by taking a cut at A, replacing the physical ceiling supports with reaction forces T; the plants are replaced with total weight W on each shelf. Equilibrium in the vertical (y-) direction gives the reaction at each ceiling support:

$$\sum F_y = 0: \ 4T - 3W = 0$$

$$Answer: \ T = \frac{3}{4}W = 75 \text{ lb}$$

Step 2. By taking a cut between levels A and B, the same equation is used to show that each cord segment AB carries tensile load $T_{AB} = 75$ lb. The FBD in 2D is shown in *Figure 2.3c*.

Step 3. By taking a cut through the cords below the top shelf (*Figure 2.3d*), equilibrium gives the tension in each cord between B and C:

$$\sum F_y = 0: \ 4T_{BC} - 2W = 0$$

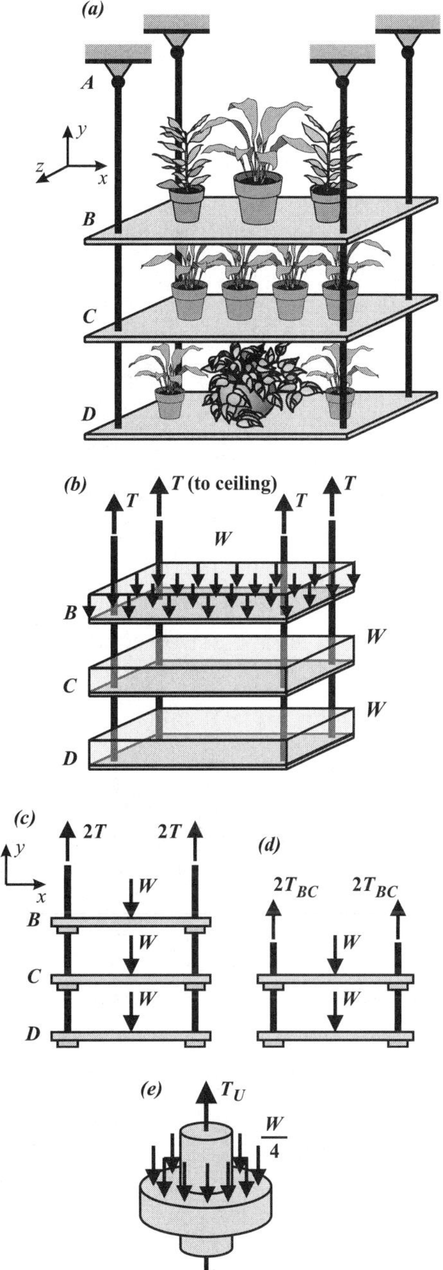

Figure 2.3. (a) Three hanging shelves. (b) 3D FBD of entire system. (c) 2D FBD of entire system. (d) 2D FBD cut at middle cords. (e) FBD of connector. Copyright ©2008 Dominic J. Dal Bello and licensors. All rights reserved.

$$Answer:\ T_{BC} = \frac{2}{4}W = 50\ \text{lb}$$

Step 4. The reader should verify by cutting the cords between C and D that the tension in each of the lowest cords is $T_{CD} = 25$ lb.

A connector is necessary to transfer the load from each shelf to the cord (*Figure 2.3e*). Each connector supports a load $W/4$ applied by the shelf, and the connector system must be strong enough to transfer that load to the cord. The tension in the cord above the connector T_U equals its share of the weight of the shelf $W/4$ and the tension in the cord below the connector T_D.

Example 2.2 Tower Crane – Method of Joints

Given: The tower crane shown in *Figure 2.4* consists of tower *DCE* fixed at the ground, and two jibs *AC* and *CB*. The jibs are supported by tie bars *AD* and *DB*, and are assumed to be attached to the tower by pinned connections. The counterweight W_C weighs 390 kips and the crane has a lifting capacity of $W_{max} = 250$ kips. Neglect the weight of the crane itself.

Required: Determine (a) the reactions at the base of the tower when the crane is lifting its capacity, (b) the axial forces in tie bars *AD* and *DB*, and jibs *AC* and *CB*, and (c) the internal forces and moment in the tower at point F, 40 ft below joint D.

Solution: *Step 1. Reactions.* The FBD of the entire crane lifting load W is shown in *Figure 2.4b*. Equilibrium requires:

$$\sum F_y = 0:\ R - W_C - W = 0$$

$$\sum M_z = 0:$$

$$M + (W_C \times a) - (W \times b) = 0$$

In the moment calculation, moments that cause counterclockwise rotation are taken as positive.

Substituting the values of a, b, W_C, and $W = W_{max}$ into the above equations gives:

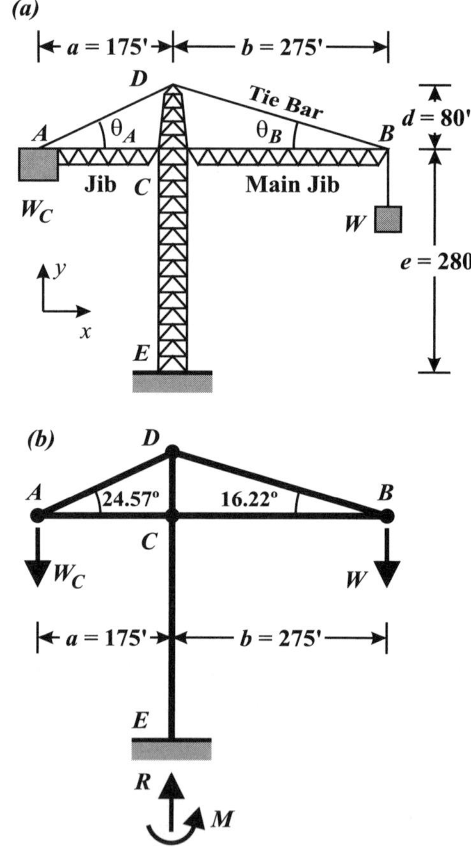

Figure 2.4. (a) Tower crane lifting load *W* with the counterweight W_C at its maximum distance, *a*. **(b)** FBD of entire crane.

Answer: R = 640 kips

Answer: M = 500 kip-ft

Note that the calculated moment is relatively small. If the base of the crane is 25 ft wide, then the forces of the equivalent couple are 20.0 kips (25×20=500), which is small when compared with R = 640 kips. For any general load W, the counterweight W_C is moved along AC to balance the moment caused by the load. This action minimizes the moment at the base, reducing the tendency for the tower to overturn. Ideally, the moment at the base is M = 0.

*Step 2. Forces in members AC, AD, BD and BC are solved using the **Method of Joints** by isolating joints A and B, and considering their FBDs (Figures 2.4c and d). The forces* in the tie bars and jibs are assumed to be in *tension,* so are drawn *acting away* from the joint. If a calculation results in a negative force, then the force actually acts opposite that drawn; i.e., the force is *compressive.*

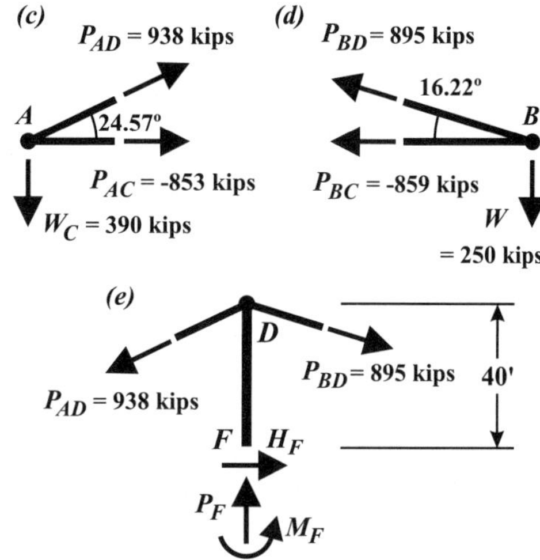

Figure 2.4. (c) FBD of joint *A*. **(d)** FBD of joint *B*. **(e)** FBD of *DF*; point *F* is 40 ft below point *D*.

From the geometry, angles θ_A and θ_B are found:

$$\tan\theta_A = \frac{80 \text{ ft}}{175 \text{ ft}} = 0.4571 \Rightarrow \theta_A = \tan^{-1}(0.4571) = 24.57°$$

$$\tan\theta_B = \frac{80 \text{ ft}}{275 \text{ ft}} = 0.2909 \Rightarrow \theta_B = \tan^{-1}(0.2909) = 16.22°$$

Applying force equilibrium at joint A:

$$\sum F_y = 0: \ -W_C + P_{AD} \sin\theta_A = 0$$

Answer: $P_{AD} = \dfrac{390 \text{ kips}}{\sin(24.57°)} = 937{,}940 \text{ lb} \Rightarrow P_{AD} = 938 \text{ kips}$

$$\sum F_x = 0: \ P_{AC} + P_{AD}\cos\theta_A = 0$$

Answer: $P_{AC} = -(937.9 \text{ kips})\cos(24.57°) \Rightarrow P_{AC} = -853 \text{ kips}$

The value of the force in jib AC, P_{AC}, is *negative,* meaning that the force is *compressive* (the jib pushes against joint A; if the jib was not there, the counterweight would swing downward). Force P_{AD} is *positive,* meaning the force acts in the direction drawn; the tie bar force is *tensile.*

Step 3. Applying equilibrium at joint B:

$$\sum F_y = 0: \quad -W + P_{BD}\sin\theta_B = 0$$

Answer: $P_{BD} = \dfrac{250 \text{ kips}}{\sin(16.22°)} = 895.0 \text{ kips} \Rightarrow \underline{P_{BD} = 895 \text{ kips}}$

$$\sum F_x = 0: \quad -P_{BC} - P_{BD}\cos\theta_B = 0$$

Answer: $P_{BC} = -(895.0 \text{ kips})\cos(16.22°) \text{ kips} \Rightarrow \underline{P_{BC} = -859 \text{ kips}}$

At point *F*, 40 ft below joint *D* on member *DC* (*Figure 2.4e*), equilibrium requires that:

$$\sum F_y = 0: \quad -P_{AD}\sin\theta_A - P_{BD}\sin\theta_B + P_F = 0$$

Answer: $P_F = (938)\sin(24.57°) + (895)\sin(16.22°) \Rightarrow \underline{P_F = 640 \text{ kips}}$

$$\sum F_x = 0: \quad -P_{AD}\cos\theta_A + P_{BD}\cos\theta_B + H_F = 0$$

Answer: $H_F = (938)\cos(24.57°) - (895)\cos(16.22°) \Rightarrow \underline{H_F = -6.3 \text{ kips}}$

$$\sum M_{z,F} = 0: \quad (P_{AD}\cos\theta_A)(DF) - P_{BD}(\cos\theta_B)(DF) + M_F = 0$$

Answer: $\underline{M_F = 252 \text{ kip-ft}}$

Note that $P_F = 640$ kips is the vertical reaction force at the base. The downward forces of weights W and W_C are carried up through the tie bars to joint *D*, and the tower carries the vertical load to ground. The negative result for H_F indicates that it acts in the opposite direction drawn.

Example based on the K10000 tower crane by Kroll Giant Towercranes, as cited at: http://www.towercrane.com/   Accessed May 2008. Values are approximate.

Example 2.3 Truss System – Method of Joints and Method of Sections

Background: *Trusses* are used in such applications as cranes, railway bridges, supermarket roofs, ships, aircraft, and space structures. Axial members are pinned together to form a beam-like structure. Trusses are a very effective means of spanning large distances.

Given: A simple truss bridge is shown in *Figure 2.5a*. Two plane trusses are constructed of 15 ft long axial members assembled with pins into equilateral triangles. Crossbeams connected to the trusses at the pin (node) locations maintain the spacing between the trusses, while diagonal bracing keeps them from moving laterally with respect to each other. The lower crossbeams support a roadway (deck) 20 ft wide. For the design, the roadway is to carry a uniformly distributed load of 80 lb per square foot (psf). Neglect the weight of the bridge itself.

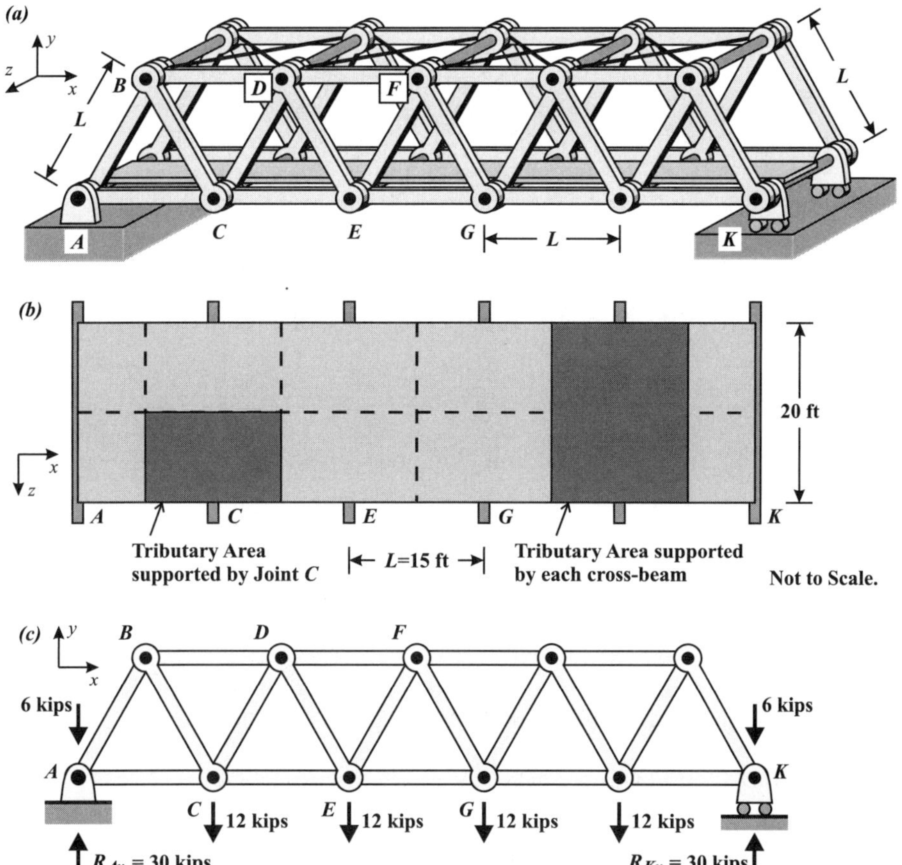

Figure 2.5. (a) A truss-bridge. The roadway rests on crossbeams that are connected to the trusses at the lower joints (e.g., joints *A, C, E, G, K*, etc.). **(b)** Top view of roadway supported by the lower crossbeams, showing *tributary area* of each crossbeam and of each pin. **(c)** 2D view of front truss.

Required: Considering only the roadway load, determine the forces (a) in members *AB* and *AC* using the **Method of Joints** and (b) in *DF, EF* and *EG* using the **Method of Sections**.

Solution: *Step 1. Load path.* Each lower crossbeam supports a deck area of 15×20 ft – a *tributary area* of 300 ft^2 (*Figure 2.5b*). The force on each crossbeam is

$$(300 \text{ ft}^2)(80 \text{ psf}) = 24,000 \text{ lb} = 24.0 \text{ kips}$$

Each end of the crossbeam is supported by a truss. Since the load is uniformly distributed on the roadway, and the geometry is symmetric about the center of the roadway, both trusses carry the same load; only one truss needs to be analyzed.

The force supported at any lower pin is half the value of the load on each crossbeam, or 12,000 lb = 12.0 kips. At the supports, the tributary area on the crossbeams is half the standard tributary area, so the downward force at each support is 6.0 kips (*Figure 2.5c*).

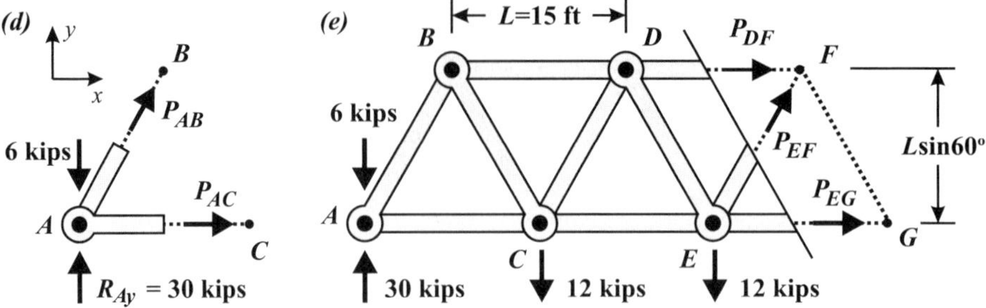

Figure 2.5. (d) FBD of joint A. **(e)** FBD of truss cut through members DF, EF, and EG.

Step 2. Reactions. From equilibrium and symmetry considerations, the vertical reaction loads are:

$$R_{Ay} = R_{Ky} = 30{,}000 \text{ lb} = 30.0 \text{ kips}$$

Because one end of the truss is supported by a roller and no horizontal loads are applied to the truss, the horizontal reactions are $R_{Ax} = R_{Kx} = 0$.

Step 3. Forces in members AB and AC (Figure 2.5d). Applying the *method of joints* at joint A, the forces in members AB and AC are determined:

$$\sum F_y = 0: \quad 30 \text{ kips} + P_{AB}\sin(60°) - 6 \text{ kips} = 0$$

$$\textit{Answer: } P_{AB} = \frac{(6 - 30 \text{ kips})}{\sin(60°)} \Rightarrow P_{AB} = -27.7 \text{ kips}$$

$$\sum F_x = 0: \quad P_{AC} + P_{AB}\cos(60°) = 0$$

$$\textit{Answer: } P_{AC} = -\frac{1}{2}P_{AB} \Rightarrow P_{AC} = 13.86 \text{ kips}$$

Since P_{AB} is negative, member AB is in compression.

Step 4. Forces in DF, EF, and EG, (Figure 2.5e). The *method of joints* can be used to solve for the forces in all of the members, one joint at a time. However, using the *method of sections*, the force in any inner member can be determined directly. For example, take a cut through members DF, EF and EG, as shown in *Figure 2.5e*.

Considering moment equilibrium about joint E to eliminate forces P_{EF} and P_{EG}:

$$\sum M_{z,E} = 0: \quad (6 - 30 \text{ kips})(30 \text{ ft}) + (12 \text{ kips})(15 \text{ ft}) - P_{DF}(15 \text{ ft})\sin 60° = 0$$

$$\textit{Answer: } P_{DF} = \frac{-540 \text{ kips-ft}}{(15 \text{ ft})\sin 60°} \Rightarrow P_{DF} = -41.6 \text{ kips (compression)}$$

To solve for P_{EG}, take moments about joint F, eliminating P_{EF} and P_{DF} from the calculation. Note that the horizontal distance from point A to point F is $2.5L = 37.5$ ft.

$$\sum M_{z,F} = 0: (6-30)(37.5) + (12)(22.5) + (12)(7.5) + P_{EG}(15)\sin 60° = 0$$

$$\textit{Answer: } P_{EG} = -\frac{-540 \text{ kip-ft}}{(15 \text{ ft})\sin 60°} \Rightarrow P_{EG} = 41.6 \text{ kips (tension)}$$

Vertical equilibrium requires:

$$\sum F_y = 0: (30-6-12-12) + (P_{EF}\sin 60°) = 0$$

$$\textit{Answer: } P_{EF} = 0 \text{ kips}$$

In general, the force in diagonal member *EF* is not zero; it is zero here because the shear force goes to zero at the center of a symmetrically loaded simply-supported truss (beam).

As a check, consider horizontal equilibrium in *Figure 2.5e*:

$$\sum F_x = 0: P_{EG} + P_{DF} + P_{EF}\cos 60° = -41.6 + 41.6 + 0 = 0 \quad \textbf{OK}$$

2.2 Torsion Members

A ***torsion member*** is a component that transmits torque *T* (*Figure 2.6*). The torque twists the member about its axis, which passes through the *centroid* (center of area) of its cross-section.

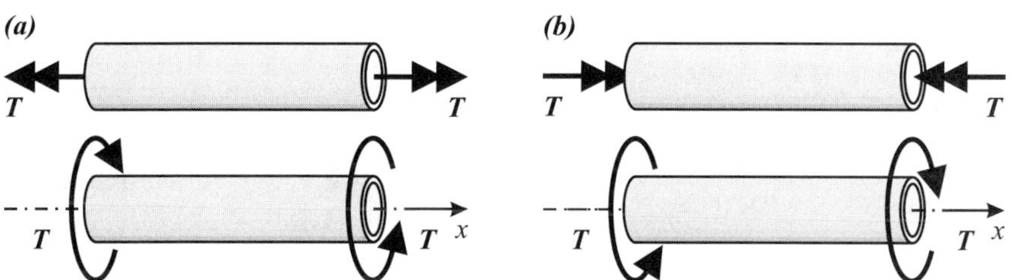

Figure 2.6. Torsion members. **(a)** A positive torque represented alternatively by a double-headed vector or a curved arrow. A torque within a torsion member is termed *positive* if it points in the same direction as the outward-pointing normal vector of the cross-section on which it acts (e.g., on a positive face in a positive direction or on a negative face in a negative direction). If torque is consistently drawn positive, a negative value indicates that it physically acts as shown in **(b)**.

Example 2.4 Drive Shaft in a Machine Shop

Given: The individual machines of classical machine shops were powered by belts driven by drive shafts. An example is shown in *Figure 2.7*, in which three machines, *B*, *C*, and *D*, draw torque from the main shaft according to *Table 2.1*. Bearing *E* is assumed to be frictionless, and therefore draws no torque.

Required: (a) Determine the torque anywhere along the drive shaft and (b) draw the torque diagram, $T(x)$ vs. x.

Table 2.1. Torque drawn by each machine.

Machine	Torque (lb-ft)
B	15
C	30
D	20

Solution: *Step 1.* The FBD of the entire drive shaft is shown in *Figure 2.7b*. Torque T_A is the input torque. From equilibrium, the sum of the torques about the x-axis must be zero:

$$\sum T_x = 0: \quad -T_A + T_B + T_C + T_D = 0$$

$$\Rightarrow T_A = (15 + 30 + 20)\text{lb-ft} = 65 \text{ lb-ft}$$

Step 2. The internal torque supported at any cross-section is found by taking a cut at that section, and a FBD of the remaining structure is considered. The torque carried inside the shaft between A and B, T_{AB}, is found by taking a cut between A and B (*Figure 2.7c*) and applying equilibrium to the external and internal torques.

$$\sum T_x = 0: \quad -T_A + T_{AB} = 0$$

$$\textit{Answer: } \underline{T_{AB} = T_A = 65 \text{ lb-ft}}$$

Step 3. Likewise, the internal torque between B and C (*Figure 2.7d*) is:

$$\sum T_x = 0: \quad -T_A + T_B + T_{BC} = 0$$

$$\Rightarrow T_{BC} = (65 - 15) \text{ lb-ft}$$

$$\textit{Answer: } \underline{T_{BC} = 50 \text{ lb-ft}}$$

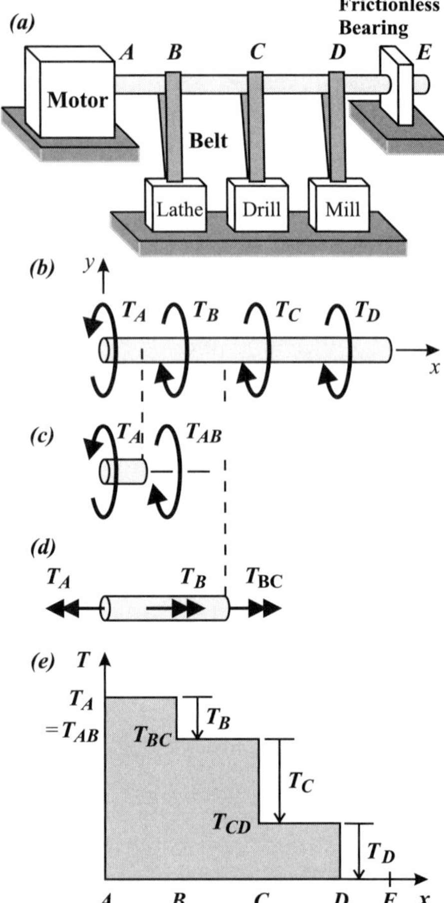

Figure 2.7. (a) Rotating shaft powering three machines. **(b)** FBD of the entire shaft. **(c)** FBD of shaft cut between points *A* and *B*. Internal torque T_{AB} is drawn in its positive sense – counterclockwise about the positive *x*-axis. **(d)** FBD of shaft cut between points *B* and *C*; torques represented with double-headed vectors. **(e)** Torque diagram.

Step 4. Verify for yourself that the torque in segment *CD* is $T_{CD} = 20$ lb-ft.

Note that cuts to determine the torque in a torsion member should never be taken at the point of application of a point torque; always take cuts between the point loads.

Step 5. The torque diagram in *Figure 2.7e* is used to display the internal torque carried by the shaft. The torque diagram is analogous to the shear force and moment diagrams for beams.

Example 2.5 Classic Lug Wrench

Given: A lug nut is tightened by applying a downward force F on the lug wrench's right arm and an upward force F on the wrench's left arm (*Figure 2.8a*). Linear motion is converted into angular motion; force is converted into torque. The forces are assumed to be of equal magnitude.

Required: Determine the magnitude of the torque applied to the lug nut.

Solution: The torque, or *couple*, applied to the wrench stem at point A is:

$$T_A = 2(F)\left(\frac{d}{2}\right) = Fd$$

and is clockwise with respect to the $+x$-axis. The reaction torque T_B applied by the lug nut against the wrench's stem is shown in *Figure 2.8b*. The torque applied to the lug nut by the stem is equal and opposite to the reaction torque T_B.

The torque applied by the wrench to the nut is the same as that applied by the user to the wrench. The magnitude of this torque is:

$$\text{Answer: } |T_{nut}| = Fd$$

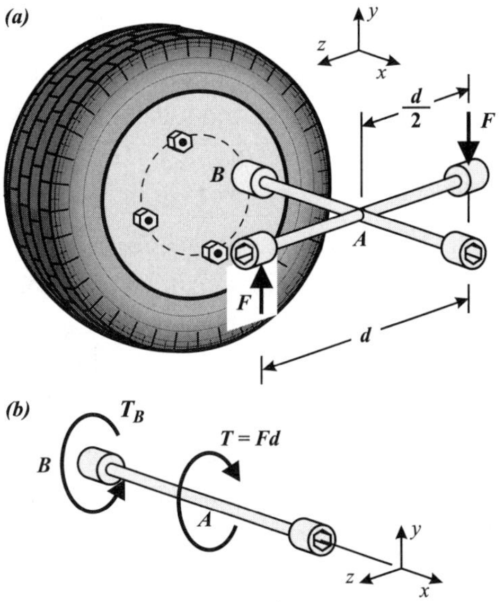

Figure 2.8. (a) Tightening a lug nut with a classic lug wrench. **(b)** Torque $T = Fd$ is applied at point A. The curved arrow at point A represents the direction that the applied torque physically acts, so its value is written as positive. Copyright ©2008 Dominic J. Dal Bello and licensors. All rights reserved.

Note that since the applied forces are equal and opposite, there is no shear force acting in the lug wrench stem; the two forces form a *couple*.

2.3 Beams

Beams are components that support loads transverse to their main structural axis. Examples include aircraft wings, floor, and ceiling joists in buildings, bridges, atomic force microscopes, robotic arms in space structures, tree branches, etc. (*Figure 2.9*). The internal loads in beams are **bending moments** and **shear forces** (*Figure 2.10*).

In general, the internal bending moment M and shear force V vary with distance x along the beam. In *Figure 2.10b*, they are drawn in their **positive senses** as defined by the convention of this text, and described in the following paragraphs.

The internal bending moment is *positive* if it causes *compression* on the top of the beam; the moment is *negative* if it causes compression at the bottom of the beam.

The shear force is *positive* if it acts on a *positive face* in a *positive direction*, or on a *negative face* in a *negative direction*. Otherwise, the shear force is negative.

Figure 2.10b is a FBD of a length of the beam exposing a cross-section that faces in the $+x$-direction – a *positive face*. At the cut, a *positive moment M* is drawn acting about the $+z$-axis, out of the paper (check this with the right-hand rule), and a *positive shear force V* is drawn acting in the $+y$-direction. Drawn in their *positive senses*, moment and shear force both act on a *positive face* in a *positive direction*.

Figure 2.10c is the complementary FBD of *Figure 2.10b*. The FBD of L–x exposes a cross-section that faces in the *negative* x-direction. Drawn in their *positive senses*, moment and shear force both act on a *negative face* in a *negative direction*: the moment about the $-z$-axis and the shear force in the $-y$-direction.

Calculations that determine the sign of the internal moment and shear force thus determine the directions in which they act. Internal axial forces and torques follow the same convention.

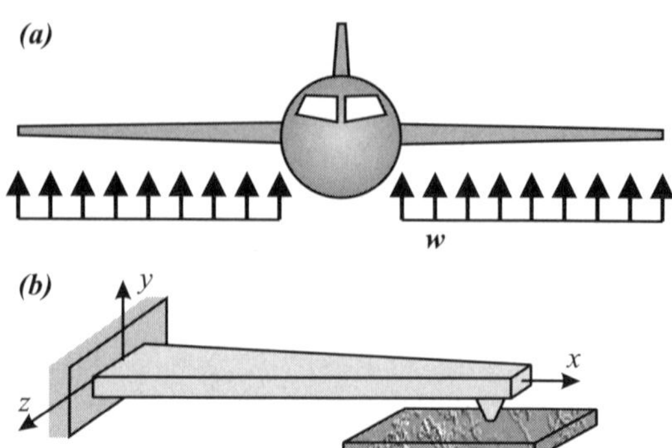

Figure 2.9. (a) Airplane wings act as beams loaded by air pressure to keep the plane aloft. **(b)** An atomic force microscope is a cantilever beam (built-in at one end, free at the other) loaded at its tip.

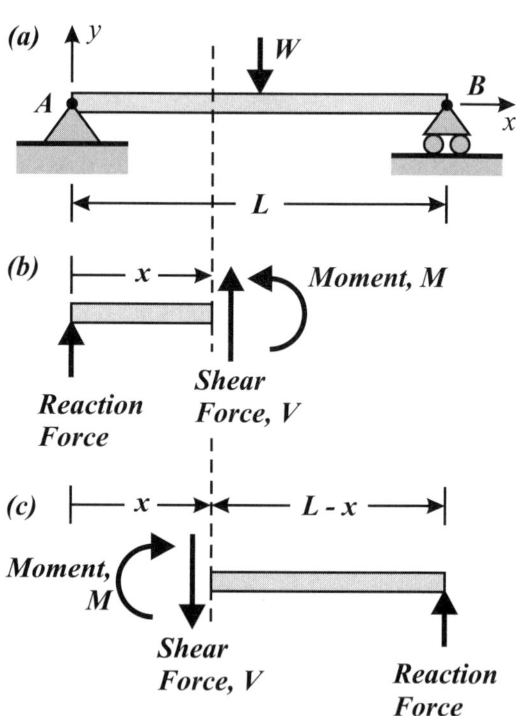

Figure 2.10. (a) A simply-supported beam (supported by a pin and a roller) loaded by central force W. **(b)** A FBD of length x of the beam, and **(c)** a FBD of complementary length L–x. Internal moment M and shear force V are drawn in their *positive senses* per the convention of this text: **(b)** positive face–positive direction and **(c)** negative face–negative direction.

Example 2.6 **Park Bench: Modeled as a Simply-Supported Beam under a Point Load**

Given: A person sits in the middle of a park bench (*Figure 2.11*). The slats of the bench are supported at each end by a set of legs. The bench is modeled as a beam supported by a *pin* at the left end and a *roller* at the right end, with a point load applied at the center (*Figure 2.11b*). Pinned supports allow rotation and cannot support or resist a moment. A beam with pinned supports (e.g., a pin and a roller) is called a *simply-supported beam.*

Required: (a) Determine the *shear force* and the *bending moment* as functions of x along the length of the beam. (b) Draw the *shear force* and *bending moment diagrams*.

Solution: *Step 1.* The FBD of the entire beam is shown in *Figure 2.11c*. Since the loading and geometry are both *symmetric*, then the vertical reactions are equal:

$$R_{Ay} = R_{By} = R = \frac{W}{2}$$

Since there is no load applied horizontally, the horizontal reaction at point A is zero.

Step 2. Shear force and *bending moment* for $0 < x < L/2$.

The shear force and bending moment at any section D to the left of the load are found from the FBD in *Figure 2.11d*. From equilibrium of segment AD (taking moments about point D):

$$\sum F_y = 0: \ \frac{W}{2} + V_D = 0$$

$$\Rightarrow V_D = -\frac{W}{2}$$

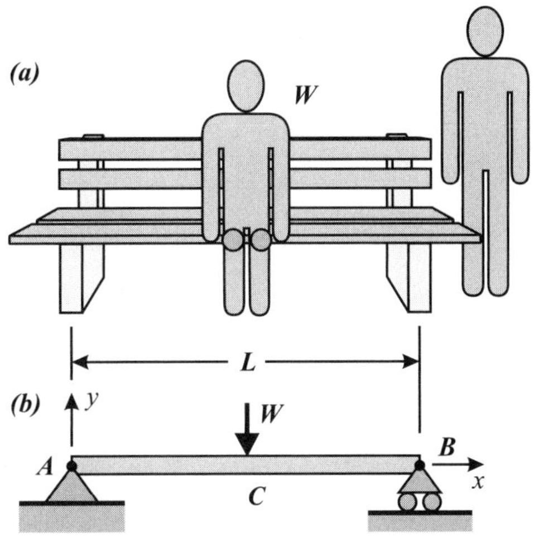

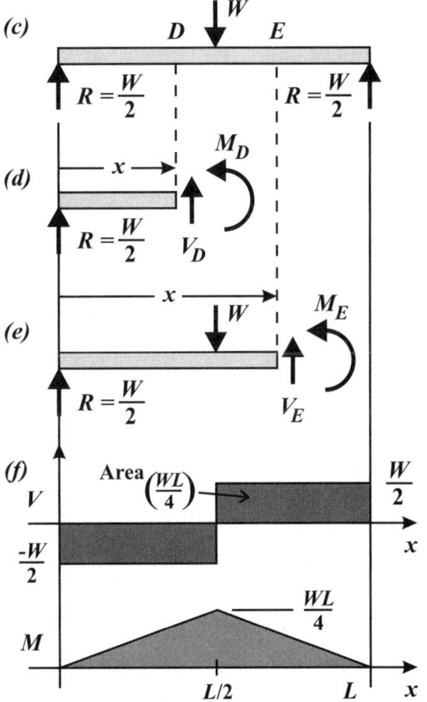

Figure 2.11. (a) A person sitting on a park bench. **(b)** The system modeled as a simply-supported beam under a central point load W. **(c)** FBD of entire beam. **(d)** FBD for $x < L/2$. **(e)** FBD for $x > L/2$. **(f)** Shear force diagram $V(x)$ and bending moment diagram $M(x)$.

$$\sum M_{z,D} = 0: \quad -\frac{W}{2}x + M_D = 0$$

$$\Rightarrow M_D = \frac{W}{2}x$$

If equilibrium of the right-hand FBD (a FBD from point D to point B) is considered, then the same results are obtained. Check this statement.

Step 3. Shear force and *bending moment* for $L/2 < x < L$.

The *shear force* and *bending moment* on any cross-section E to the right of the load are found from the FBD in *Figure 2.11e*. From equilibrium of segment AE:

$$\sum F_y = 0: \quad \frac{W}{2} - W + V_E = 0$$

$$\Rightarrow V_E = \frac{W}{2}$$

$$\sum M_{z,E} = 0: \quad -\frac{W}{2}x + W\left(x - \frac{L}{2}\right) + M_E = 0$$

$$\Rightarrow M_E = \frac{W}{2}(L - x)$$

In summary:

Answer:

$$\bullet \text{ For } 0 < x < \frac{L}{2}: \quad V(x) = -\frac{W}{2}; \quad M(x) = \frac{W}{2}x$$

$$\bullet \text{ For } \frac{L}{2} < x < L: \quad V(x) = \frac{W}{2}; \quad M(x) = \frac{W}{2}(L - x)$$

Note that cuts to determine the shear and moment in a beam should never be taken at the point of application of a point load (force or moment); always take cuts between point loads.

Step 4. The variations of *shear force* and *bending moment* along the beam are shown in *Figure 2.11f*. These plots are the *shear force* and *bending moment diagrams*.

A simply-supported beam under a central point load is referred to as *three-point bending*. This form of loading is often used in experiments to determine the strength of a material.

Example 2.7 Park Bench: Modeled as a Simply-Supported Beam; Uniformly Distributed Load

Given: The park bench in the previous example is now completely full (*Figure 2.12*). The beam is assumed to have the same geometric boundary conditions (simple supports). Since the beam is full, the load is modeled as a *uniformly distributed load* (force per unit length). The distributed load is:

$$w = \frac{nW}{L}$$

where n is the number of people on the bench, W is the weight of each person (assumed to be the same), and L is the distance between supports. The beam model is shown in *Figure 2.12b*.

Required: (a) Determine the *shear force* and the *bending moment* along the length of the beam. (b) Draw the shear force and bending moment diagrams.

Solution: *Step 1.* The FBD of the entire beam is shown in *Figure 2.12c*. Since the loading and geometry are both symmetric, $R_{Ay} = R_{By} = R$. From vertical equilibrium:

$$\sum F_y = 0: \quad -wL + R_{Ay} + R_{By} = 0$$

$$\Rightarrow R_{Ay} = R_{By} = R = \frac{wL}{2}$$

Step 2. The internal shear force and bending moment at any section D distance x from the origin may be found from the FBD shown in *Figure 2.12d*. From vertical equilibrium:

$$\sum F_y = 0: \quad \frac{wL}{2} - wx + V_D = 0$$

Answer: $V_D = V(x) = w\left(x - \frac{L}{2}\right)$

Moment equilibrium about point D gives:

$$\sum M_{z,D} = 0$$

$$\left(-\frac{wL}{2}\right)x + (wx)\left(\frac{x}{2}\right) + M_D = 0$$

Answer: $M_D = M(x) = \frac{w}{2}(Lx - x^2)$

Note that the second term in the moment equilibrium equation is the product of the equivalent force wx due to the distributed

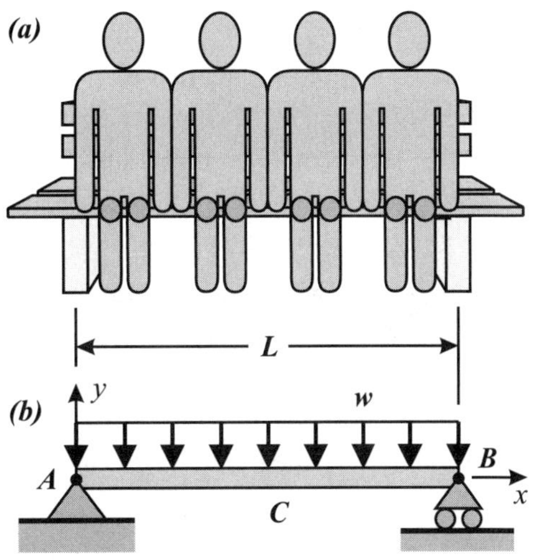

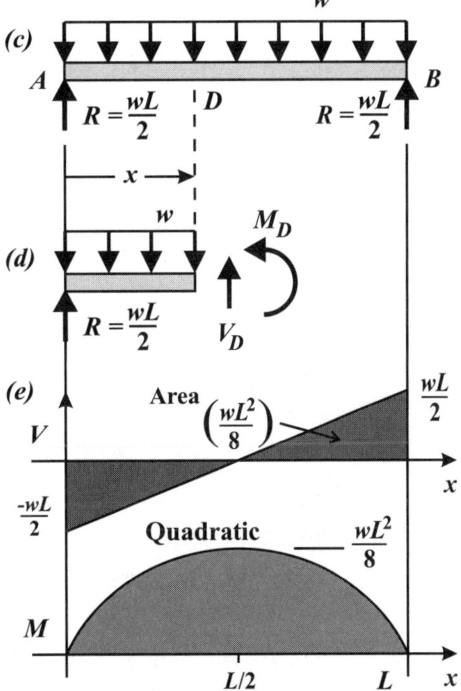

Figure 2.12. (a) A fully-loaded park bench. **(b)** The bench modeled as a simply-supported beam under uniformly distributed load w (force per unit length). **(c)** FBD of entire beam. **(d)** FBD at any distance x from the left end. **(e)** Shear force diagram $V(x)$ and bending moment diagram $M(x)$.

load and its lever arm ($x/2$) with respect to the cut at point D.

Step 3. The *shear force* and *bending moment diagrams* are shown in *Figure 2.12e*.

For this problem, the shear force is *linear*, with a maximum magnitude of $wL/2$ that occurs at each support. The bending moment is *parabolic*, with a maximum value of $wL^2/8$ at the center of the beam. The maximum bending moment occurs when the shear force is equal to zero. Because of the symmetry of the geometry and applied load, the response (shear force and moment) are symmetric.

Note that only one equation each for the shear and moment was required. In the previous example (*Example 2.6*), two equations for each load type were required. The difference is that the loading in this example is constant over the entire length of the beam. Whenever there is a sudden change in the beam's loading (e.g., at the point load in *Example 2.6*), an additional set of shear and moment equations must be considered.

Example 2.8 Atomic Force Microscope: A Cantilever with a Point Load

Background: The principal component of the atomic force microscope (AFM), used to measure the micro-geometry of surfaces and the forces in biological systems, consists of a *cantilever beam*. A cantilever beam is *built-in* (*fixed* against displacement and rotation) at one end and free at the other end (*Figure 2.13*).

Given: Force P is applied at the free end of the AFM cantilever. A representative load at this scale is $P = 20$ nN (20×10^{-9} N) and the beam length is $L = 60$ μm (60×10^{-6} m).

Required: **(a)** Determine the *shear force* and the *bending moment* along the length of the beam. **(b)** Draw the shear force and moment diagrams.

Solution: *Step 1*. The FBD of the entire beam is shown in *Figure 2.13c*.

From force equilibrium in the y-direction and moment equilibrium about the z-axis, the reaction force and reaction moment are:

$$\sum F_y = 0: \quad R + P = 0$$

$$\Rightarrow R = -P$$

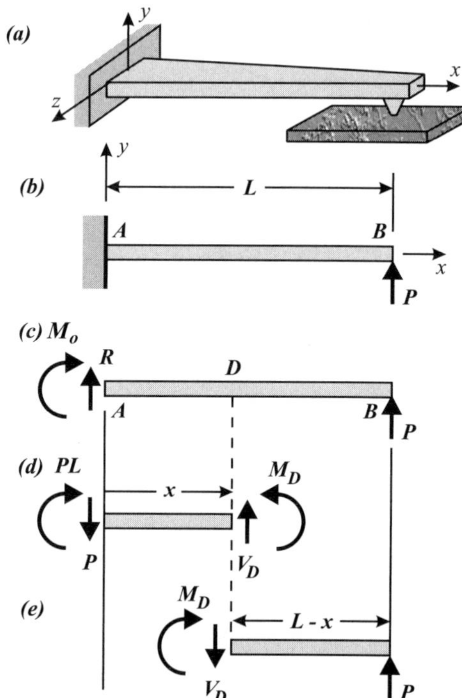

Figure 2.13. (a) An atomic force microscope scans a material surface. **(b)** The AFM modeled as a cantilever beam under tip load P. **(c)** FBD of entire beam. **(d)** Left-hand FBD at any distance x from the left end. **(e)** Right-hand FBD at any distance x from the left end.

$$\sum M_{z,A} = 0: \quad -M_o + PL = 0$$

$$\Rightarrow M_o = PL$$

Step 2. To investigate how the shear force and moment vary with distance x along the beam, a cut is taken at an arbitrary cross-section D. Since the load on the beam does not change over its length, only one cut needs to be taken.

Taking equilibrium of the *right-hand* FBD, segment DB (*Figure 2.13e*):

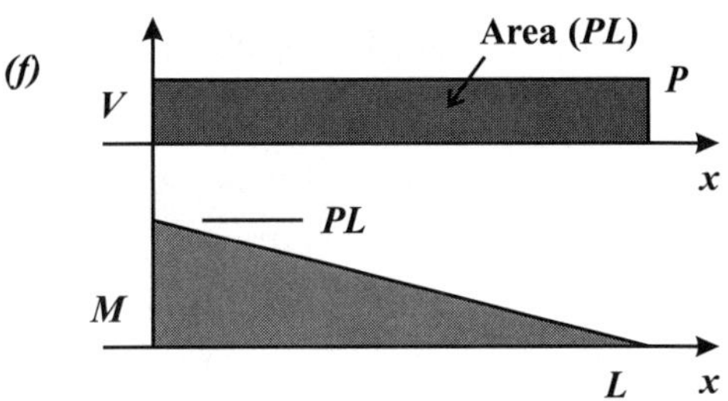

Figure 2.13. (f) shear force diagram $V(x)$, and bending moment diagram $M(x)$.

$$\sum F_y = -V_D + P = 0$$

$$\text{Answer: } \underline{V_D = V(x) = P}$$

The shear force is constant throughout the beam.

The moment along the beam is given by:

$$\sum M_{z,D} = -M_D + P(L-x) = 0$$

$$\text{Answer: } \underline{M_D = M(x) = P(L-x)}$$

Note that the moment equation checks with the expected values at each end of the beam: at the clamped end $M(x=0) = PL$ and at the free end $M(x=L) = 0$. The general FBD of *Figure 2.13e* reduces to *Figure 2.13c* for $x = 0$.

If equilibrium of the left-hand side of the beam AD was considered (*Figure 2.13d*), then the same results for the shear force and moment would be obtained.

Step 3. The shear force and moment diagrams are shown in *Figure 2.13f*. Using the given representative values, the maximum bending moment is $M_{max} = PL = (20 \text{ nN})(60 \text{ μm}) = 1.2 \times 10^{-12}$ N·m. Shear force V_D is plotted as positive since it acts upward on a $+x$-face (or downward on a $-x$-face). Moment M_D is plotted positive as it causes compression on the top of the beam (it is a $+z$-moment on the $+x$-face or a $-z$-moment on a $-x$-face).

2.4 Combined Loading

Components are frequently subjected to several types of loading at the same time. Two examples of combined loading follow.

Example 2.9 Highway Sign with Wind Load

Given: Signs overhanging highways are often supported by steel masts as shown in *Figure 2.14*. The sign is $b = 4.0$ ft high and $a = 12$ ft wide, and its center is $L = 16$ ft above the road. Wind blows against the sign at $V = 100$ mph. The sign weighs $W_S = 400$ lb and the mast weighs $W_M = 1000$ lb.

Required: Determine the reactions at the base of the mast.

Solution: *Step 1. Loading.* For wind, the equivalent static pressure is (*Equation 1.1*):

$$p = 0.00256V^2$$

where p is in pounds per square foot (psf) and wind velocity V is in miles per hour (mph). For a wind speed of 100 mph, the static pressure is:

$$p = 0.00256(100)^2 = 25.6 \text{ psf}$$

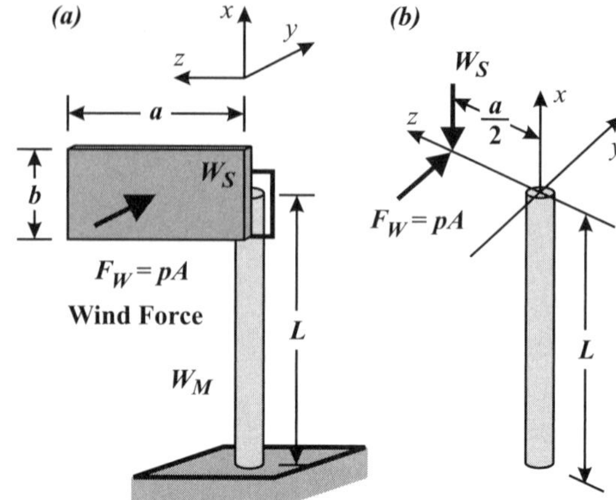

Figure 2.14. (a) Highway sign under wind load F_W. **(b)** Sketch of the wind load and weight of the sign acting at distance $a/2$ from the axis of the mast. Note that this is not a FBD.

Assuming the entire static pressure acts against the sign, the wind load is:

$$F_W = pA = (25.6 \text{ psf})(48 \text{ ft}^2) = 1229 \text{ lb} = 1.23 \text{ kips}$$

and acts at the centroid of the sign. The x-axis is taken to coincide with the axis of the mast and the wind force is taken to act in the positive y-direction.

The wind force F_W and weight of the sign W_S, both act distance $a/2$ from the x-axis of the mast (*Figure 2.14b*). The wind force causes a shear force F_W and a torque $T_W = F_W[a/2]$ (clockwise about the $+x$-axis) on the mast. The weight of the sign causes an axial force W_S in the mast and a moment $M_S = W_S[a/2]$ (clockwise about the $+y$-axis) applied at the top of the mast.

Step 2. Reactions. The FBD of the entire mast is shown in *Figure 2.14c*. The reactions at the ground due to the wind load are: shear force V, bending moment M_2, and torque T. The reactions due to the weight of the sign are: part of the axial reaction force R, and bending moment M_1. The reaction due to the weight of the mast also makes up part of the axial reaction R.

Applying equilibrium to the FBD of the entire structure (*Figure 2.14c*):

$$\sum F_x = 0: \quad -W_S - W_M + R = 0$$

Answer: $R = 400 + 1000 \Rightarrow \underline{R = 1.40 \text{ kips}}$

$$\sum F_y = 0: \quad F_W + V = 0$$

Answer: $V = -F_W \Rightarrow \underline{V = -1.23 \text{ kips}}$

A negative sign indicates that force V acts opposite drawn.

Taking moments about the x-axis:

$$\sum M_x = 0: \quad -F_W\left(\frac{a}{2}\right) + T = 0$$

$$\Rightarrow T = (1.23 \text{ kips})(6 \text{ ft})$$

Answer: $\underline{T = 7.38 \text{ kip-ft}}$

Taking moments about the base, first about the y-axis, and then about the z-axis:

$$\sum M_y = 0: \quad -W_S\left(\frac{a}{2}\right) + M_1 = 0$$

$$\Rightarrow M_1 = (400 \text{ lb})(6 \text{ ft})$$

Answer: $\underline{M_1 = 2.40 \text{ kip-ft}}$

$$\sum M_z = 0: \quad F_W L + M_2 = 0$$

$$\Rightarrow M_2 = -(1.23 \text{ kips})(16 \text{ ft})$$

Answer: $\underline{M_2 = -19.7 \text{ kip-ft}}$

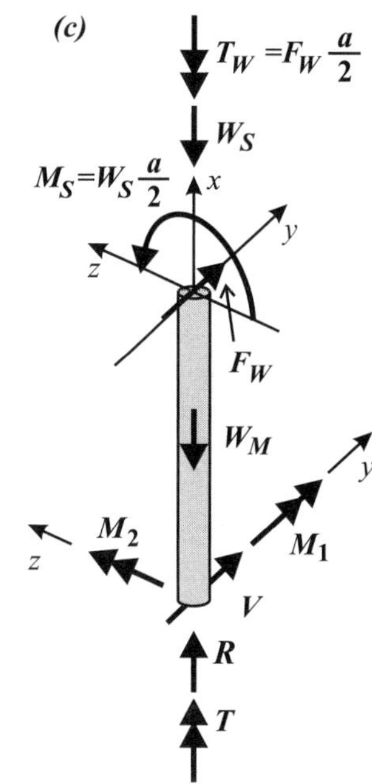

Figure 2.14. (c) FBD of the mast. Although the weight of the mast W_M is distributed along its length, its equivalent force acts at the center of gravity. At the ground, all the reactions are drawn acting in the positive direction of the appropriate axis, not necessarily in the directions that they physically act.

Moment M_2 acts opposite drawn. Note that the shear force, torque, and bending moment about the y-axis are constant along the length of the mast. The bending moment about the z-axis increases from zero at the top of the mast to its maximum at ground level.

Example 2.10 Single-Arm Lug Wrench

Given: The loading on the compact lug wrenches that come in modern automobiles is similar to the wind loading on the sign of the previous example (*Figure 2.15*). In these lug wrenches, the lug nut is tightened by applying a downward force F at point C when the wrench arm is on the right side of the stem AB.

Required: Determine the reactions at the lug nut.

Solution: To support force F, the nut–wrench interface must have the following shear force, torque, and moment reactions (*Figure 2.15c*):

Answer: $R_B = F$

Answer: $T_B = -\dfrac{Fd}{2}$;

Answer: $M_B = -Fl$

A negative sign indicates that the reaction load acts opposite drawn.

In classic lug wrenches (*Example 2.5*), essentially a pure torque is applied to the lug nut; the bending moment and shear force at the nut are zero. The new lug wrenches cause extra loads on the nut–wrench interface. The bending moment tends to cause the single-arm lug wrench to slip off the nut. Additionally, the force applied to the single-arm lug wrench must be about twice that applied to the classic lug wrench to obtain the same torque. This leads too often to the unsafe practice of standing on the lug wrench to get a large enough torque to loosen or tighten the nut.

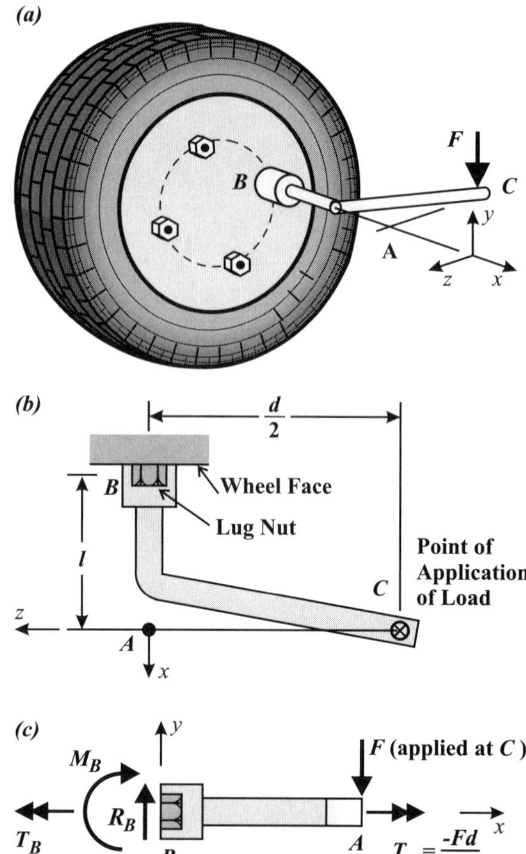

Figure 2.15. (a) The single-arm lug wrench as supplied in modern automobiles. **(b)** Top view of lug wrench. **(c)** Side view of single-arm wrench with reaction loads at lug nut. Unlike the classic lug wrench of *Example 2.5*, the nut–wrench interface must now transfer a shear force and a moment. Copyright ©2008 Dominic J. Dal Bello and licensors. All rights reserved.

3.0 Introduction

The concepts of **strain** and **stress** are introduced in this chapter. **Strain** and **stress** allow the engineer to judge if a component is stiff enough or strong enough for its intended application. The component must not deflect too much, permanently deform, or break into two parts.

Two basic tests help to define *strain* and *stress*, and to determine the material properties that relate them:

Figure 3.1. (a) Bar under tension. **(b)** Thin-walled shaft under torsion.

- the *tension test*: a straight uniform bar is subjected to a force along its axis (*Figure 3.1a*), and
- the *torsion test*: a straight thin-walled circular shaft is subjected to a torque about its axis (*Figure 3.1b*).

3.1 The Tension Test – Axial Properties

In the tension test, a straight uniform bar of length L and constant cross-sectional area A is subjected to an axial force P applied through the *centroid* of its cross-section (*Figure 3.2*). The load causes the axial *member* to elongate by distance Δ. Such a response is readily seen when stretching a rubber band. Elongation is not so easy to observe when stretching a steel bar, but it does occur.

When force P is proportional to *elongation* or *extension* Δ (the relative *displacement* of the ends of the bar), then the response is **linear**. If the bar returns to its

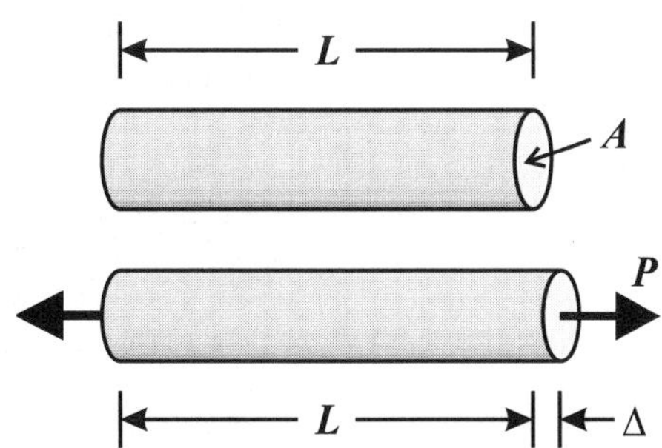

Figure 3.2. Bar of length L and cross-sectional area A. Under axial load P, the bar elongates (extends) by Δ.

F.A. Leckie, D.J. Dal Bello, *Strength and Stiffness of Engineering Systems*, Mechanical Engineering Series,   **43**
DOI 10.1007/978-0-387-49474-6_3, © Springer Science+Business Media, LLC 2009

original length when the load is completely removed, the response is *elastic*. A *linear-elastic* response is observed in most materials for small loads and elongations.

The results of a tensile test are used to determine the mechanical properties of a material. The force–elongation (P–Δ) response of the test is generalized by defining the terms *strain* and *stress*.

Strain

The ***axial strain*** ε (Greek "epsilon") – the strain parallel to the axial load – is the ratio of the bar's extension Δ to its original length L. Hence, the *strain* of the axial bar is:

$$\varepsilon = \frac{\Delta}{L} \qquad \text{[Eq. 3.1]}$$

This strain is also called the ***normal strain*** since it describes movement perpendicular (normal) to the bar's cross-section.

This definition of strain – based on the original length of the bar – is called *engineering strain*. Engineering strain is used when the system remains *elastic*, or when strain is otherwise small, as it is in most practical applications. When strains become large, a different definition – based on the bar's length at any instant – must be used for detailed calculations (see *Chapter 12*). For now, engineering strain will be used as the measure of *strain*.

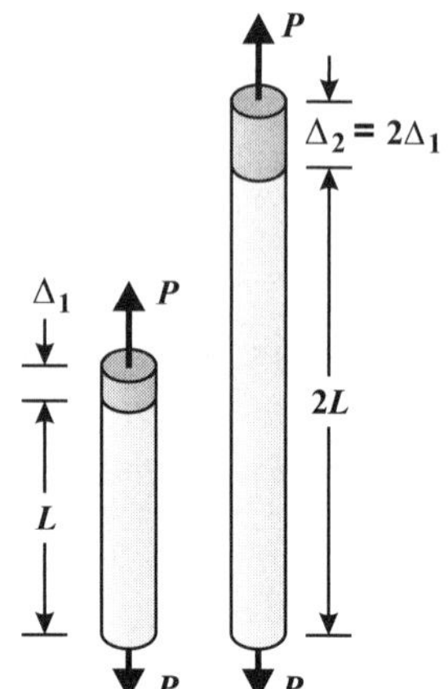

Figure 3.3. Two otherwise identical bars of different lengths having the same strain.

Strain is *dimensionless* since it is the ratio of two quantities whose dimensions are both length. A typical value in practice is on the order of $\varepsilon = 0.0005$. Strain is generally expressed as a percentage, e.g., 0.05%. Another form is $\varepsilon = 500 \times 10^{-6}$ or 500 micro-strain.

If the length of the bar is doubled, the same load P will elongate the bar by 2Δ (*Figure 3.3*); the $2L$-long bar is essentially two L-long bars in series. Applying the definition of strain gives:

$$\varepsilon = \frac{2\Delta}{2L} = \frac{\Delta}{L} \qquad \text{[Eq. 3.1b]}$$

For the same load P, cross-sectional area A and material properties, strain is independent of length.

If the bar elongates ($\Delta > 0$), the strain is *positive* ($\varepsilon > 0$). The bar is said to be in *tension* and the force is a *tensile force* ($P > 0$). When the force is in the opposite direction, pushing against the bar, the bar is in *compression*, and the elongation Δ is *negative*. The bar shortens ($\Delta < 0$), so the strain is also *negative* ($\varepsilon < 0$). The force is a *compressive force* ($P < 0$ with the force drawn in tension, as in *Figure 3.3*).

Example 3.1 **Strain in a Pipe**

Given: The Trans-Alaska Pipeline, diameter $D = 1.22$ m, transports oil under pressure. The pressure causes an increase in pipe diameter (*Figure 3.4*). Field measurements at a certain location show that the circumferential strain – the strain around the pipe circumference – is $\varepsilon = 0.050\%$.

Required: Determine the change in pipe diameter ΔD.

Solution: The strain ε is change in length divided by original length. Take L as the original pipe circumference: $L = \pi D$. Due to pressure, the new circumference is $(L+\Delta L) = \pi(D+\Delta D)$. The circumferential strain is then:

$$\varepsilon = \frac{\Delta L}{L} = \frac{\pi(D + \Delta D) - \pi D}{\pi D} = \frac{\Delta D}{D}$$

The measured strain is $\varepsilon = 0.0005$, so the increase in diameter is:

$$\Delta D = \varepsilon D = (0.0005)(1.22 \text{ m})$$

Answer: $\underline{\Delta D = 0.61 \text{ mm}}$

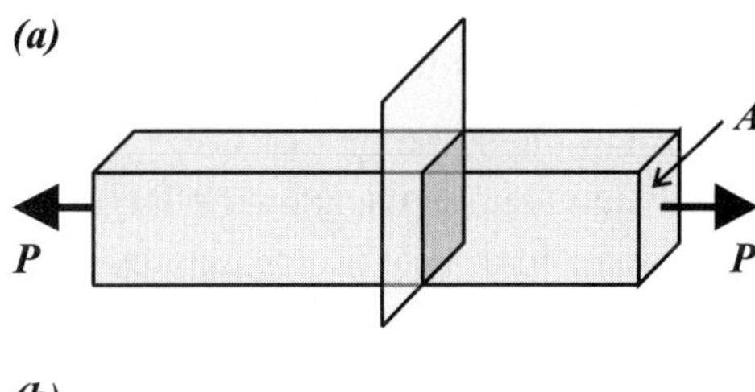

Figure 3.4. Under internal pressure, the diameter of a pipe increases (not to scale).

Stress

The bar in *Figure 3.5* has constant cross-sectional area A, and is subjected to axial force P. The cross-section need not be square, but the load must act through the centroid of the cross-section so that the response at any cross-section is *uniform* (the same over the entire cross-sectional area).

The **axial stress** or **normal stress** σ ("sigma") in a bar is the axial force divided by the cross-sectional area over which the force acts. Thus, the *stress* normal to the cross-section is:

$$\sigma = \frac{P}{A} \qquad \text{[Eq. 3.2]}$$

The *normal stress* acts perpendicular (normal) to an interior cross-section of the bar (*Figure 3.5b*). The units of stress are force per area.

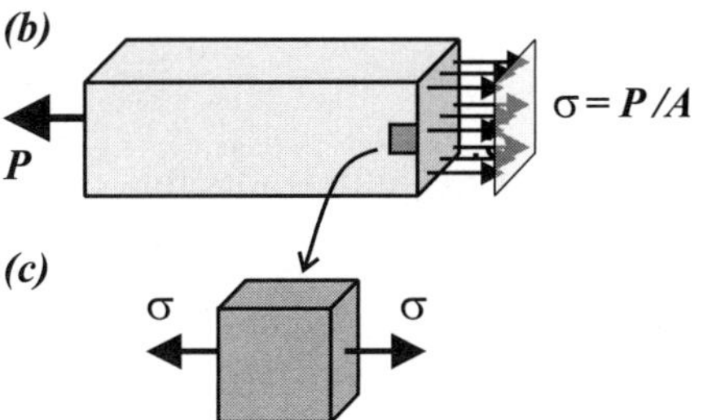

Figure 3.5. (a) Axial bar in tension with plane cutting the bar normal to its axis. **(b)** The axial stress on an interior cross-section is uniform if the axial load is applied through its centroid. **(c)** A *stress element*: a material point represented by an infinitessimal cube under stress.

This definition of stress – based on the original cross-sectional area of the bar – is called *engineering stress*, and is used when the system remains *elastic*, or when strain is otherwise small. Here, engineering stress is simply called *stress*. When strain is large, a different definition for stress – based on the area at any instant – must be used for detailed calculations (see *Chapter 12*).

If the cross-sectional area is doubled from A to $2A$, then the stress due to load P is:

$$\sigma = \frac{P}{2A} \qquad\qquad \text{[Eq. 3.2b]}$$

The intensity of the force per area has been reduced by a factor of 2.

Tension describes the condition when the bar is stretched. In *tension*, force and stress are both *positive* ($P > 0$, $\sigma > 0$), and act as drawn in *Figure 3.5*. In *compression*, force P physically presses against the bar, shortening it. In *compression*, the force and stress are both *negative* ($P < 0$, $\sigma < 0$), and act opposite to those drawn in *Figure 3.5*. If the force and stress are consistently drawn assuming that the bar is in tension, then a positive stress is a *tensile stress* and a negative stress is a *compressive stress*.

The dimensions of stress are *force per unit area*. In SI units, the unit of force is the *newton*, N, and the unit of length is the *meter*, m. Thus, the unit of stress is *newtons per square meter*, called a *pascal* (Pa):

$$\text{Stress} = \frac{\text{Force}}{\text{Area}} = \frac{\text{N}}{\text{m}^2} = \text{Pa}$$

In practical situations, stresses are typically on the order of 100×10^6 Pa. For convenience, stress is generally given in *megapascals*, MPa, where $1.0\,\text{MPa} = 10^6$ Pa. A common convention is to keep the numerical value between 0.1 and 1000 and use SI prefixes to indicate the magnitude.

In US units, the unit of force is the *pound*, lb, and the unit of length is the *foot*, ft. A distributed load that acts over a large area (e.g., a snow load on a roof) is given in pounds per square foot, psf. Stress acts over an area having dimensions of inches (e.g., a bar's cross-section), so stress is given in *pounds per square inch*:

$$\text{Stress} = \frac{\text{Force}}{\text{Area}} = \frac{\text{lb}}{\text{in.}^2} = \text{psi}$$

In practice, the magnitude of stress is large and is given in *kilopounds per square inch*, ksi, where $1.0\,\text{ksi} = 1000\,\text{psi}$ (1.0 kilopound $= 1.0$ kip $= 1000$ lb).

To convert between the US and SI systems, 1.0 ksi is equal to 6.895 MPa. For a rough conversion from ksi to MPa (or psi to kPa, etc.), multiply by 7 and increase the metric prefix by a factor of 3; e.g., $10\,\text{ksi} \sim 70\,\text{MPa}$. The error in simply using 7 is less than 2%, sufficient for an approximation.

Typical design values of stress in engineering applications are on the order of $\sigma \sim 10\,\text{ksi} \sim 100\,\text{MPa}$.

Example 3.2 Hanging Lamp

Given: A lamp weighing $W = 10.0$ lb hangs from the ceiling by a steel wire of diameter $D = 0.100$ in. (*Figure 3.6*).

Required: Determine the stress in the wire.

Solution: The normal stress σ in an axial member is the applied force divided by the cross-sectional area over which it acts. Using statics, the force everywhere in the wire is $P = W$. Thus:

$$\sigma = \frac{P}{A} = \frac{10.0\ \text{lb}}{\pi(0.100\ \text{in.})^2/4} = 1273\ \frac{\text{lb}}{\text{in.}^2}$$

Answer: $\sigma = 1.27$ ksi

The answer has been rounded to three significant digits, since the data were given to three digits.

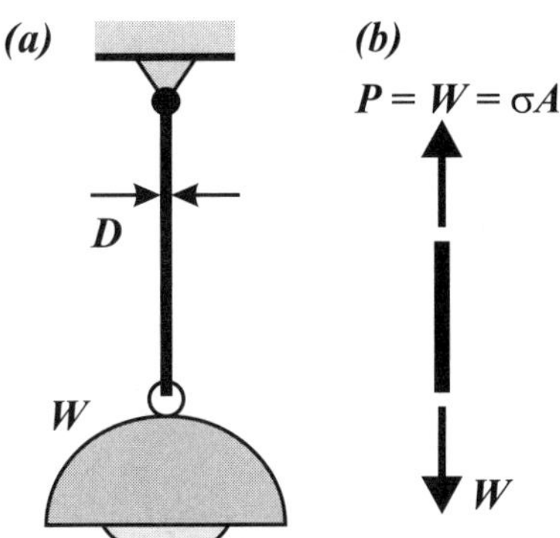

Figure 3.6. (a) Lamp of weight W supported by a single wire of cross-sectional area A. **(b)** FBD of the wire.

Example 3.3 Lamp Hanging by Two Wires

Given: A lamp of weight W hangs by two wires, each of length L (*Figure 3.7*). Each wire has a cross-sectional area A and makes an angle of θ with the horizontal.

Required: Determine the stress σ in each wire.

Solution: *Step 1.* The FBD of point B is given in *Figure 3.7b*. Since the geometry and loading are both symmetric:

$$T_{AB} = T_{BC} = T$$

In *Figure 3.7c*, T_{BC} is decomposed into its x- and y-components. Vertical equilibrium of point B requires that:

$$W = T_{ABy} + T_{BCy} = 2T\sin\theta$$

so:

$$T = \frac{W}{2\sin\theta}$$

Step 2. The stress in each wire is then:

$$\sigma_{AB} = \sigma_{BC} = \frac{T}{A}$$

Answer: $\sigma = \dfrac{W}{2A\,\sin\theta}$

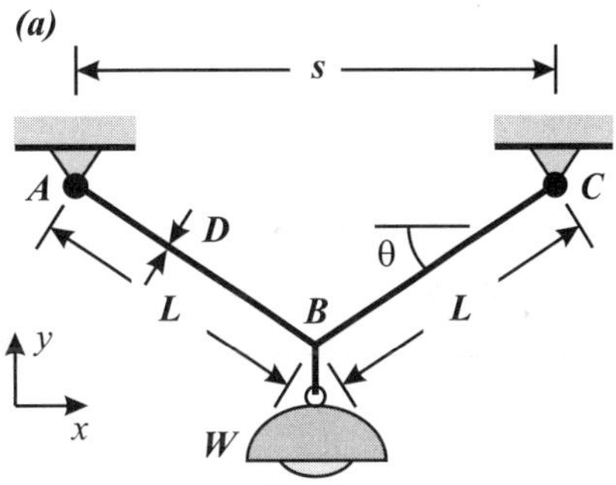

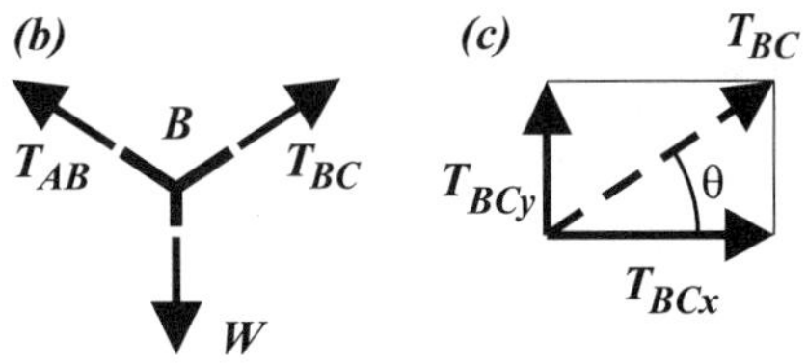

Figure 3.7. (a) Lamp supported by two wires. **(b)** FBD of point B. **(c)** Decomposition of tensile force T_{BC} into its x- and y-components.

Young's Modulus

Stiffness is a measure of a system's resistance to deformation. A rubber band is easily deformed by hand; but a steel wire or rod of the same cross-sectional area is not. The stiffness of steel is greater than that of rubber.

Tensile force P applied to a linear spring causes it to elongate by Δ (*Figure 3.8a*). When the measured values of force and elongation are plotted on a graph, the result is a straight line (*Figure 3.8b*). The line is defined by *Hooke's Law for Springs*:

$$P = K\Delta \qquad \text{[Eq. 3.3]}$$

where K is the *spring constant* or *spring stiffness*. If the elongation (displacement) is too large, this linear relationship is no longer valid, which can be demonstrated by stretching a spring so much that it does not return to its original shape.

Tensile force P applied to a bar of initial cross–sectional area A and initial length L causes it to elongate by Δ (*Figure 3.8c*). For small displacements, the bar behaves like a linear spring, i.e., $P = K\Delta$. Dividing force by area and elongation by length gives the stress and strain: $\sigma = P/A$ and $\varepsilon = \Delta/L$.

A plot of *stress* versus *strain* is known as a ***stress–strain curve*** (*Figure 3.8d*). Since A and L are constants, the stress–strain curve is also linear for small displacements.

By normalizing (dividing) force by area, and elongation by length, the *force–displacement* (force–elongation, P–Δ) curve becomes a *stress–strain curve* (σ–ε). Since the bar's volume is AL, the stress–strain response is independent of bar size and depends only on the material.

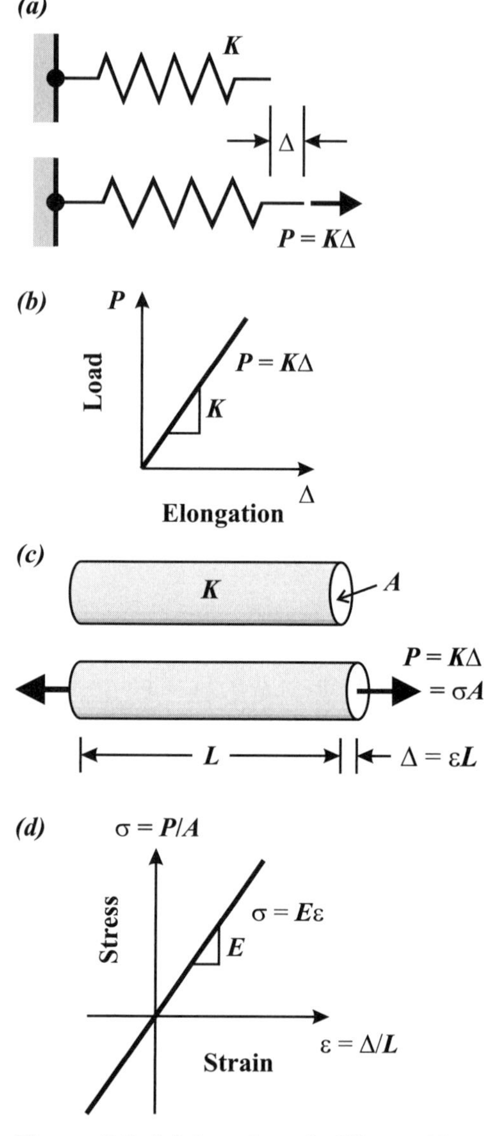

Figure 3.8. (a) A spring of stiffness K, under force P, deflects $\Delta = P/K$.
(b) Force–displacement curve for a linear spring. **(c)** A bar in tension acts like a spring. **(d)** Initial stress–strain response of a bar in tension or compression.

The *slope E* of the stress–strain curve is the *stiffness of the material*. The material stiffness is called **Young's modulus**, the **elastic modulus**, or the **modulus of elasticity**.

Hooke's Law for Stress–Strain is:

$$\sigma = E\varepsilon \qquad \text{[Eq. 3.4]}$$

Since strain ε is dimensionless, Young's modulus has the same units as stress (Pa, lb/in.^{2}).

Representative moduli are given in *Table 3.1*. The modulus is generally large and expressed in SI units as GPa (gigapascals, 1 GPa = 10^9 Pa), and in US units as ksi or Msi (megapounds per square inch, 1 Msi = 10^3 ksi = 10^6 psi).

In *compression* ($\sigma < 0$), the initial material response is generally linear with a slope of E (*Figure 3.8d*).

Young's modulus depends on the nature of the atomic bonds of the material. Although there are many types of steels, all have a modulus of about $E = 190$–215 GPa (28–31 Msi);

Table 3.1. Representative values of Young's Modulus (SI and US units).

Material	E (SI) GPa $= 10^9$ Pa	E (US) Msi $= 10^6$ psi
Steels	207	30*
Titanium alloys	115	17
Aluminum alloys	70	10
Nickel alloys	215	31
Cast irons	180	26
Douglas fir (parallel to grain)	12.4	1.8
Glass	70	10
Rubbers	0.01–0.1	0.0015–0.015
Polymers	0.1–5	0.015–0.75
Engineering ceramics	300–450	44–65
Carbon fiber/polymer matrix composite	70–200	10–30

* The American Institute of Steel Construction (AISC) recommends 29 Msi = 29,000 ksi (200 GPa) for structural steel.

steel is primarily iron with a small amount of carbon and a few additional elements. The amorphous nature of rubbers and polymers, and their various degrees of atomic bonding, results in the low and relatively widespread values of their moduli. Engineering ceramics have covalent and ionic bonds – the strongest atomic bonds – so their moduli are high.

Because steel has a large modulus, it is used in buildings and other structures where deflections must be small. The modulus of aluminum (70 GPa, 10 Msi = 10,000 ksi) is one-third that of steel. If aluminum is used in place of steel with the same geometry, the resulting strains and deflections would be three times greater for the same loads. The designer must be aware of this response.

The values of E for rubbers and polymers are very small; these materials are seldom used to support large loads as they would have large deflections. Rubber is often used as a cushion. Polymers are used to cover automobile panels and to provide protective enclosures for electronic components.

In high-tech aerospace applications, polymers are used to keep high-strength/high-stiffness fibers in alignment and to protect them. Such a combination of materials is called a *composite*. One class of composites is carbon fiber reinforced polymers (CFRP), which have high elastic moduli but low density. This combination of properties is very attractive in aerospace applications, but composite materials are generally expensive due to the cost of manufacture.

Example 3.4 **Displacement of Hanging Lamp Due to Weight**

Given: The wire supporting the lamp in *Example 3.2* is $L = 5.00$ ft long (*Figure 3.9*). The wire is steel with elastic modulus $E = 30{,}000$ ksi $= 30 \times 10^6$ lb/in^2 .

Required: Determine the elongation Δ of the wire due to the lamp's weight.

Solution: *Step 1.* From *Example 3.2*, the stress is $\sigma = 1.273$ ksi. The strain ε is calculated using Hooke's Law:

$$\varepsilon = \frac{\sigma}{E} = \frac{1273 \text{ psi}}{30 \times 10^6 \text{ psi}} = 42.43 \times 10^{-6}$$

Step 2. The elongation is:

$$\Delta = \varepsilon L = \left(42.43 \times 10^{-6}\right)[(5.00 \text{ ft})(12 \text{ in./ft})]$$

Answer: $\underline{\Delta = 2.55 \times 10^{-3} \text{ in.}}$

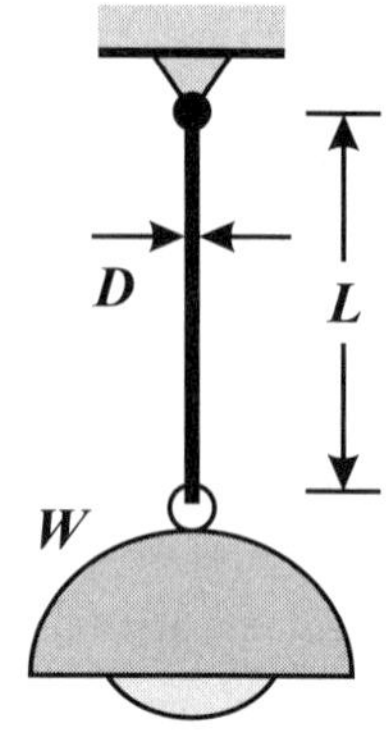

Figure 3.9. Lamp of weight *W* supported by a single wire.

As expected, the displacement is small. If, instead of steel, nylon string of the same diameter and having modulus $E \sim 10^6$ psi is used, then the displacement would be 30 times greater than that of the steel wire, or $\Delta = 76.4 \times 10^{-3}$ in.

Stiffness and Flexibility of an Axial Member

Since Young's modulus E for many materials is tabulated, it may be used to calculate the *stiffness* and *flexibility* of any axial bar. Consider a bar of constant cross-sectional area A, length L, and modulus E, subjected to load P applied at each end through the centroid of the cross-section. The bar's *stiffness K* is found using the definitions of stress and strain:

$$P = \sigma A = (E\varepsilon)A = E\frac{\Delta}{L}A = \frac{EA}{L}\Delta = K\Delta \qquad \text{[Eq. 3.5]}$$

The stiffness of an axial bar is therefore:

$$K = \frac{EA}{L} \qquad \text{[Eq. 3.6]}$$

Equation 3.5 can be rearranged to solve for displacement in terms of the applied load. The inverse of stiffness K is the *flexibility F*, defined by:

$$\Delta = \frac{1}{K}P = FP \qquad \text{[Eq. 3.7]}$$

$$F = \frac{L}{AE} \qquad \text{[Eq. 3.8]}$$

Thus, the elongation of a bar due to axial load P is given by this very useful equation:

$$\Delta = \frac{PL}{AE}$$ [Eq. 3.9]

The terms *stiffness* and *flexibility* are commonly used in practice. In racing car suspensions, stiffness is high to avoid excessive sway in tight turns at high speed. By contrast, highway riding requires comfort and a more flexible suspension, meaning a lower stiffness or higher flexibility.

Example 3.5 Stiffness and Flexibility of an Axial Member

Given: The lamp from *Examples 3.2* and *3.4* (*Figure 3.10*).

Required: Determine the stiffness K and flexibility F of the 5.0 ft long steel wire.

Solution: The cross-sectional area of the wire is $A = 7.854 \times 10^{-3}$ in.^{2}, the length is $L = 60.0$ in., and the modulus is $E = 30{,}000$ ksi. Thus, the stiffness and flexibility are:

$$K = \frac{EA}{L} = \frac{(30 \times 10^{6}\ \text{psi})(7.854 \times 10^{-3}\text{in.}^{2})}{(60.0\ \text{in.})}$$

Answer: $K = 3927$ lb/in.

$$F = K^{-1}$$

Answer: $F = 0.255 \times 10^{-3}$ in./lb

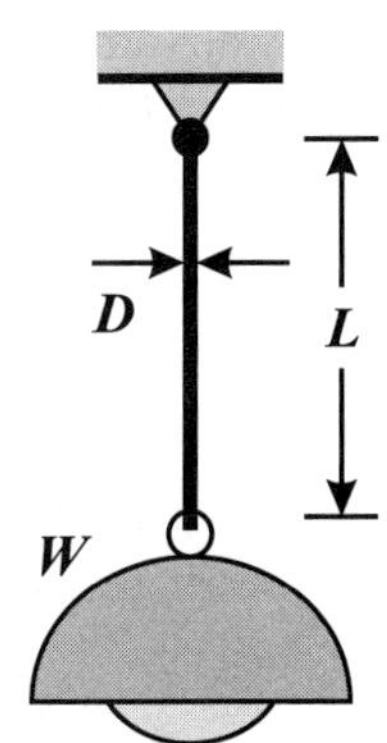

Figure 3.10. Lamp.

As an example, to achieve a displacement of $\Delta = 50 \times 10^{-3}$ in., the required force is:

$$P = K\Delta = (3927\ \text{lb/in.})(50 \times 10^{-3}\ \text{in.}) = 196.4\ \text{lb}$$

Due to a force of 10.0 lb (the weight of the lamp), the displacement is:

$$\Delta = FP = (0.255 \times 10^{-3}\ \text{in./lb})(10\ \text{lb}) = 2.55 \times 10^{-3}\ \text{in.}$$

in agreement with *Example 3.4*.

Stress–Strain Curves for Ductile Materials

Stress–strain curves are determined from experiments on axial bars (*Figure 3.11*). A bar with cross-sectional area A is clamped into a set of grips of a testing machine. One grip is fixed and the other moves by known displacement δ; the force P required to cause the displacement is measured. Such an experiment is known as a *displacement-* or *strain-controlled* experiment.

Ideally, the bar is in a pure state of tension (or compression). Since the grips of a testing machine cause a complex state of stress in the nearby material, a length of the bar of constant cross-section is identified over which the load is considered to be purely axial.

This length is the *gage length L*. The change in length Δ of the *gage section* is often measured with a device called an *extensometer*, which has a typical gage length of 1.0 in.

Measurements are taken of the change in length Δ and the associated force P. By dividing the applied force P by area A and the elongation Δ by gage length L, the stress required to cause a certain strain is found:

$$\sigma = \frac{P}{A} \quad ; \quad \varepsilon = \frac{\Delta}{L} \qquad \text{[Eq. 3.10]}$$

Strain can also be measured with *strain gages*, which are typically about 10 mm in length (see *Chapter 8, Page 269*). A strain gage is epoxied to the specimen, and is part of an electronic circuit. As the specimen – and thus the strain gage – changes length, the resistance of the strain gage changes, providing an electric signal that indicates the strain of the specimen.

Plotting stress against strain gives the *stress–strain curve*. A representative stress–strain curve for a **ductile material** (most metals) is given in *Figure 3.12*.

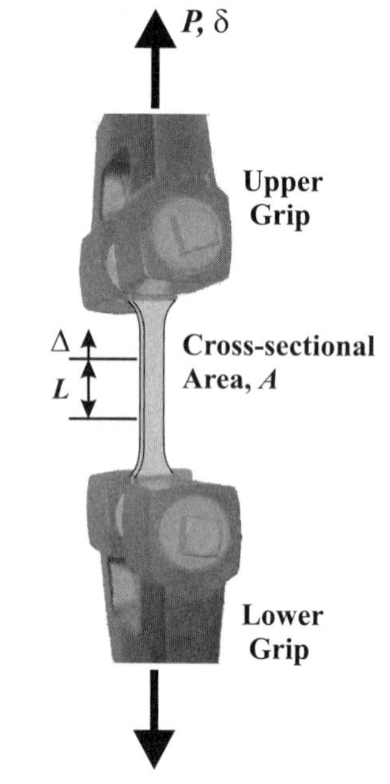

Figure 3.11. A tensile specimen placed in the grips of a testing machine.

Linear–Elastic Loading and Unloading

As the bar is initially loaded (line OAB, *Figure 3.12*), experimental observations indicate that *stress is linearly proportional to strain* – Hooke's Law – with slope equal to the material's Young's Modulus E:

$$\sigma = E\varepsilon \qquad \text{[Eq. 3.11]}$$

When the load on the bar is removed, the bar returns to its original length; i.e., the strain returns to zero. The ability for a material to recover its original shape when it is *unloaded* is known as **elasticity**. Most engineering components are designed to remain elastic in service. When an elastic material obeys Hooke's Law (*Equation 3.11*), the behavior is **linear–elastic**.

Yielding

Linear–elastic behavior is observed up to a certain value of stress called the **proportional limit** (point B). Above this point, the stress–strain curve becomes *non-linear* and the slope reduces (line CD).

The value of stress where the curve transitions from linear to non-linear behavior is the *proportional limit* S_p. It is difficult to determine the exact value of S_p from the experimental data, so an engineering approximation – which gives the *yield strength* – is used.

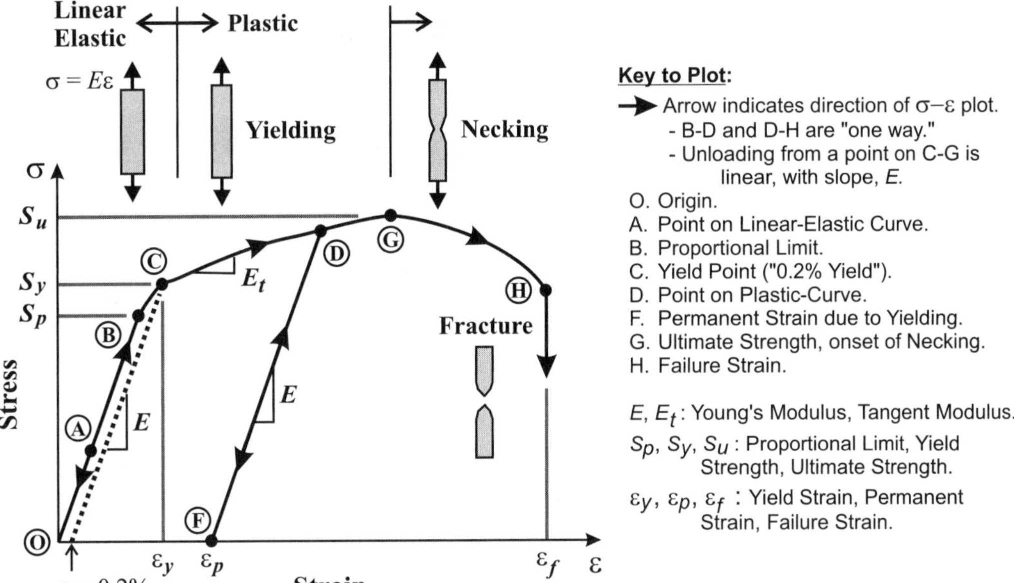

Figure 3.12. Typical stress–strain curve of a ductile material (not to scale). A material is classified as ductile if the strain-to-failure ε_f is much greater than the yield strain ε_y (e.g., $\varepsilon_y \sim 0.4\%$, $\varepsilon_f \sim 15\%+$).

By convention, the ***yield strength*** is determined by constructing a line parallel to the linear portion of the curve, but displaced by 0.2% on the strain axis – the dotted line in *Figure 3.12*. The intersection of this line with the *experimental stress–strain curve* is defined as the *yield strength* S_y (sometimes σ_y) of the material (point C). The yield strength is used as the proportional limit; Hooke's Law is considered valid for stresses up to S_y. Representative values of S_y are given in *Table 3.2* on *Page 55*.

The *strain at yielding* is $\varepsilon_y = S_y/E$, and is taken to be the maximum elastic strain. Typical values of *yield strain* ε_y are also given in *Table 3.2*. The values of ε_y can have design implications as noted below.

Plastic Deformation, Necking, and Failure

As strain is further increased (from point C towards point G, *Figure 3.12*), the slope of the stress–strain curve – the ***tangent modulus*** E_t – decreases. For each additional increment of strain $d\varepsilon$, a smaller increment of stress $d\sigma$ is required ($E_t = d\sigma/d\varepsilon$). The slope eventually becomes zero when the stress reaches a maximum. This stress is the ***ultimate strength*** S_u (or *ultimate tensile strength*), the maximum stress that the material can support.

After reaching the *ultimate strength*, stress decreases with increasing strain (from point G to point H), meaning that the applied force ($P = \sigma A$) needed to cause further extension decreases. Force P decreases because somewhere along the bar, its cross-sectional area begins to decrease significantly. This localized reduction in area is called *necking* (*Figure 3.13*). At the *neck*, the stress is higher than the nominal stress, so the strain and elongation are concentrated there. Failure (fracture into two pieces) finally occurs in the neck at the *failure strain* ε_f.

A material is generally classified as *ductile* if the **strain to failure** ε_f is much greater than its yield strain ε_y (more than an order of magnitude). Ductile materials typically have failure strains on the order of 15% or more. The ductility of metals allows them to be bent into various shapes without breaking.

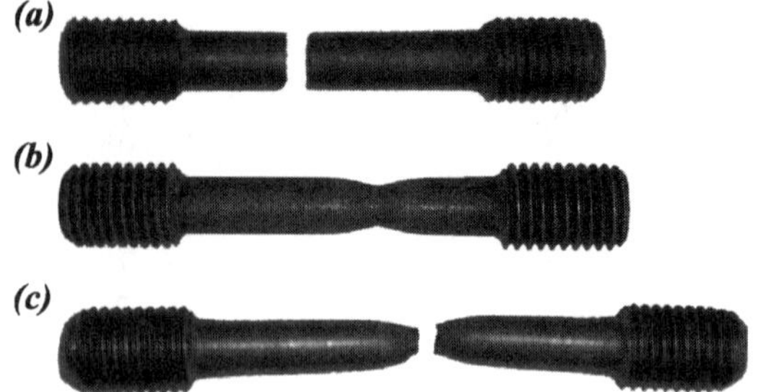

Figure 3.13. Failed tensile specimens (0.5 in. diameter), all originally the same length. **(a)** Cast Iron is brittle and shows little plastic deformation. The fracture surface is flat. **(b)** C1045 Hot-Rolled Steel; broken specimen pieced together to illustrate *necking* of a ductile material. **(c)** A36 Steel typically fails at $\varepsilon_f \sim 24\%$.

Unloading After Plastic Deformation and Reloading

If the load is removed after the specimen has yielded, but before it necks (i.e., between points C and G in *Figure 3.12*), the stress–strain response follows an elastic *unloading line* (line DF). The unloading response has the same slope as the linear–elastic loading line (line OAB). When the stress is completely removed, the bar does not return to its original length, but suffers a **permanent strain** or **plastic strain** ε_p.

When the load is reapplied, the stress–strain response begins at point F and is linear up to point D, where it rejoins the overall curve and moves towards point G. The material remains linear up to a greater stress than the original yield strength S_y. This phenomenon is known as **strain-hardening** or **work-hardening**. By mechanically processing a ductile material, its yield strength can be increased. However, there is always a price to pay. A bar that has been strain-hardened is less ductile – the strain to failure is reduced from its original value of ε_f to $\varepsilon_f - \varepsilon_p$ (although in most cases, there is still sufficient strain to failure). Once necking occurs, the bar can no longer be strain-hardened.

General Comments

In stress–strain experiments, it is standard practice to apply a displacement (strain) and measure the required force (stress) – the *displacement-controlled* (*strain-controlled*) test just discussed. The alternative is to apply a force (stress) and measure the resulting strain – a *force-controlled* test. Data collection in the force-controlled test is difficult because beyond the proportional limit, the slope of the stress–strain curve decreases; small increments of stress cause large changes in strain. Better results during yielding are achieved using the strain-controlled test; small increments of strain require very small changes in stress. Additionally, since force continuously increases in the force-controlled test, the decrease in stress at necking is not captured.

Unlike the modulus, the yield strength of a metal can be significantly increased by the addition of atoms of another element (*alloying*), by mechanical processing, or by heat treatment. Metals can, therefore, have a wide range of yield strengths, as shown in *Table 3.2*. By understanding processing techniques (beyond the scope of this text), a metal alloy can be engineered to have a specific yield strength S_y. Increasing the yield strength does not generally influence the value of Young's modulus E.

Of course, the more complex the processing route, the more expensive the material. High strength metals are expensive and are generally only used in specialized applications where the cost is justified. The yield strength of steel used in buildings and bridges is at the low end of the range with a value of 250 MPa (36 ksi). The selection of low strength steel is based on cost. By contrast, modern pressure vessel technology is possible using high-strength steels with yield strengths nearing 1900 MPa (275 ksi).

High-strength, high-temperature nickel-based alloys are used in the manufacture of jet engines. The forces acting on the rotating engine discs are large and a design with compact dimensions is only possible using strong nickel alloys. If only yield strength were considered in the design, the system could be made strong enough. However, larger allowable stress levels mean larger elastic deflections. An elastic extension is not necessarily negligible. For example, a strain of $\varepsilon = 0.6\%$ is

Table 3.2. Representative values of Modulus, Yield Strength, Yield Strain and Failure Strain.

Material	E GPa (Msi)	$S_y{}^*$ MPa (ksi)	ε_y (%)	ε_f (%)
Steels	207 (30)	250–1900 (36–270)	0.12–0.95	25+
Titanium and alloys	115 (17)	200–1300 (29–185)	0.17–1.2	20
Aluminum and alloys	70 (10)	100–600 (15–90)	0.14–0.86	15
Nickel alloys	215 (31)	200–1600 (30–40)	0.1–0.74	30
Cast irons	180 (26)	220–1000 (31–145)	0.12–0.55	0 (gray) 15 (ductile)
Douglas fir	12.4 (1.8)	100 (S_u) (15)	N/A	–
Glass**	70 (10)	N/A	Very small	~0
Rubbers	0.01–0.1 (0.0015–0.015)	30 (S_u) (4)	>10	500+
Polymers	0.1–5 (0.015–0.75)	20–30 (3–4)	0.5–2	–
Engineering ceramics**	300–450 (44–65)	N/A	Very small	~0
Carbon fiber polymer matrix composite	70–200 (10–30)	1800 (260) (S_u, in fiber direction)	N/A	N/A

* Values of yield strength in metals depend on chemical composition, mechanical processing, thermal processing, etc.

** Ceramics and glasses are brittle and exhibit little, if any, plastic strain.

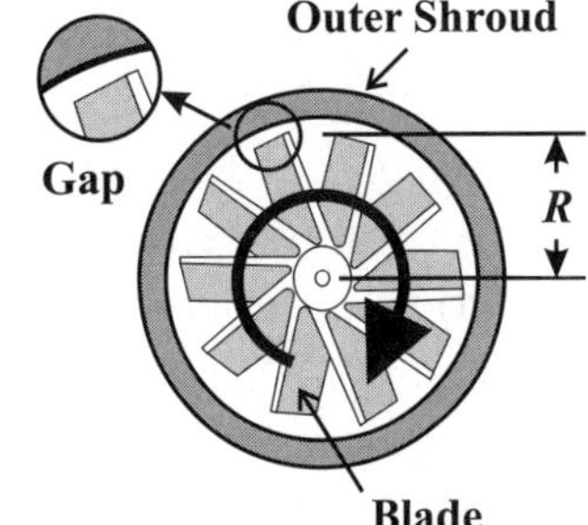

Figure 3.14. The turbine blades in a jet engine must not expand and hit the housing. Copyright ©2008 Dominic J. Dal Bello and licensors. All rights reserved.

within the elastic region for high-strength nickel alloys ($\varepsilon_y = 0.75\%$ for $S_y = 1600$ MPa, $E = 215$ GPa). If the radius of the engine's compressor disc is $R = 1.0$ m and the dynamic loading causes a strain of $\varepsilon = \Delta R / R = 0.006$, then the increase in radius is $\Delta R = 6$ mm. The gap between the blades and the outer shroud (which itself may deform) must be large enough to accommodate this expansion (*Figure 3.14*). Elastic deformation must be considered.

The majority of materials used in practice are ductile. Design methods are often based on the assumption that the stress in a material is limited by its yield strength S_y so that the material remains elastic. The ability of a metal to yield before it breaks is a useful property, since plastic deformation provides a visible warning of impending failure.

Stress–Strain Curves for Brittle Materials

When a ceramic or glass component is subjected to tensile stress, it breaks by snapping suddenly (with a *bang*! so be careful when testing these materials; they can shatter into dangerous flying pieces). Materials that show little, if any, plastic deformation are classified as **brittle**, such as the cast iron specimen in *Figure 3.13a*. The stress–strain curve of a brittle material is essentially limited to the elastic region, as shown in *Figure 3.15*. The strength is usually defined by the *ultimate strength S_u*, i.e., the stress at failure (fracture into two pieces).

It is generally difficult to specify a single value of S_u for a brittle material because there is so much scatter in tests.

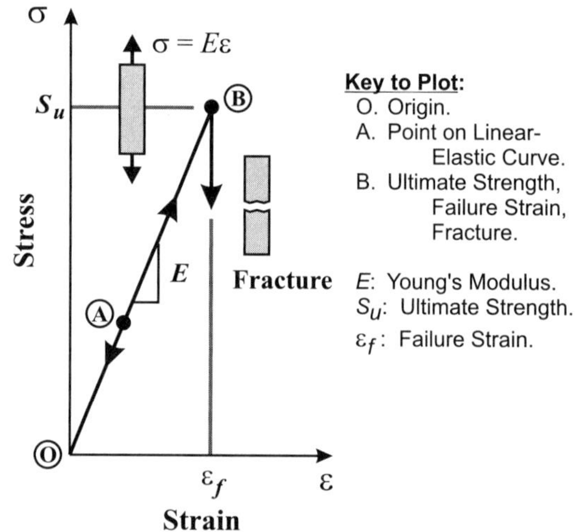

Figure 3.15. Representative stress–strain curve of a brittle material. There is little, if any, plastic deformation.

The measured strength depends on the size of the specimen being tested, and the probability distribution of pre-existing flaws or cracks in the material. Consequently, values of S_u for brittle materials are often not tabulated, are given with conservative values, or are given with a broad range. Special methods are necessary to design components made of brittle materials, requiring an understanding of Fracture Mechanics and Statistics (introduced in *Chapter 13*).

Gray cast iron is used extensively in castings of engine blocks for automobiles and diesel engines. When tested in tension, gray cast iron breaks into two with little warning. Gray cast iron does, however, exhibit some of the characteristics of ductile materials with a failure strain ε_f typically between $[2S_y/E]$ and $[5S_y/E]$. Nevertheless, because the failure strain is small by comparison with ductile materials, gray cast iron is sometimes described as *semi-brittle,* and this word sends a signal to the designer to proceed with caution.

Example 3.6 Factor of Safety of Hanging Lamp

Given: The wire holding up the lamp of *Example 3.2* has a yield strength of $S_y = 60$ ksi. The factor of safety against yielding is to be 2.5 (just in case someone pulls down on it, etc. The diameter of the wire is $D = 0.10$ in. (*Figure 3.16*).

Required: Determine the allowable (design) load P_D.

Solution: *Step 1.* The factor of safety against yielding is:

$$\text{FS} = \frac{\text{Load at yield}}{\text{Allowable load}} = \frac{P_y}{P_D} = 2.5$$

where P_y is the load at yielding:

$$P_y = AS_y = \frac{\pi(0.10 \text{ in.})^2}{4}(60 \text{ ksi}) = 471.2 \text{ lb}$$

Step 2. The allowable load on the lamp is then:

$$P_D = \frac{P_y}{\text{FS}} = \frac{471.2 \text{ lb}}{2.5} = 188.5$$

Answer: $P_D = 188$ lb

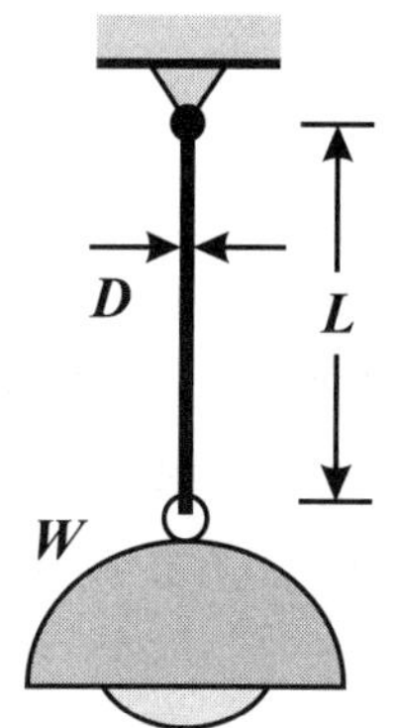

Figure 3.16. Lamp.

P_D must include the weight of the lamp and any additional loads expected.

Note: Allowable loads are not rounded up; 189 lb would technically exceed the calculated allowable load of 188.5 lb.

Example 3.7 Elongation of Tower Crane Tie Bar

Given: Tie bar *DB* supports the main jib (*CB*) of the tower crane discussed in *Example 2.2*. The crane is shown in *Figure 3.17*. When the crane is operating at its maximum working load, the force in the tie bar is 895 kips. The tie bar is made of a high-strength steel with yield strength $S_y = 60$ ksi and modulus $E = 30$ Msi.

Required: (a) If the factor of safety against yielding is 2.0, determine the minimum cross-sectional area A of tie bar *DB*. (b) Using the area calculated in Part (a), determine the change in length Δ of *DB*.

Solution: *Step 1.* From the definition of stress:

$$\sigma = \frac{P}{A} \rightarrow A = \frac{P}{\sigma}$$

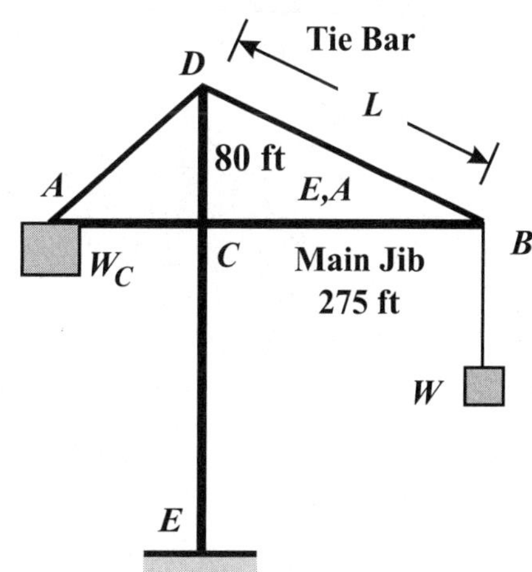

Figure 3.17. Schematic of crane.

The maximum service (working) load in *DB* is 895 kips. Applying the factor of safety, the required strength of the tie bar – in this case, the force at yielding – is:

$$P_y = P_D(\text{FS}) = (895 \text{ kips})(2.0) = 1790 \text{ kips}$$

Hence, the minimum required area is:

$$A = \frac{P_y}{S_y} = \frac{1790 \times 10^3 \text{ lb}}{60 \times 10^3 \text{ psi}}$$

Answer: $\underline{A = 29.8 \text{ in.}^2}$

Design the tie bar cross-section to be a full 30.0 in^2.

An alternative method that is commonly used is allowable stress design (ASD). The factor of safety is applied to the yield strength. The allowable or design stress σ_D is then:

$$\sigma_D = \frac{S_y}{\text{FS}} = \frac{60 \text{ ksi}}{2.0} = 30 \text{ ksi}$$

so:

$$A = \frac{P_D}{\sigma_D} = \frac{895 \text{ kips}}{30 \text{ ksi}} = 29.8 \text{ in.}^2$$

Step 2. The elongation of the tie bar Δ at the maximum operating load P_D is:

$$\Delta = \frac{P_D L_{DB}}{AE}$$

From the crane geometry, the length of the tie bar L_{DB} is:

$$L_{DB} = \sqrt{(L_{CD})^2 + (L_{CB})^2} = \sqrt{80^2 + 275^2} = 286 \text{ ft}$$

Then:

$$\Delta = \frac{(895 \text{ kips})(286 \text{ ft} \times 12 \text{ in./ft})}{(30 \text{ in.}^2)(30{,}000 \text{ ksi})}$$

Answer: $\underline{\Delta = 3.41 \text{ in.}}$

The elongation is small in relation to the length of the tie bar (3.41 in. to 286 ft); the strain is 0.00099, or approximately 0.1%.

Poisson's Ratio

When a rubber band is stretched, it not only elongates in the direction of the load but also becomes thinner. Such behavior occurs in most materials, although it is difficult to see with the naked eye. This behavior is referred to as the ***Poisson effect***.

In *Figure 3.18*, the tensile bar of length L and square cross-section of side b elongates by Δ. The cross-section need not be square, but is used here to ease the calculations. The ***longitudinal strain*** ε (or *direct strain*) is the strain in the direction of the load:

$$\varepsilon = \frac{\Delta}{L} \qquad\qquad \text{[Eq. 3.12]}$$

The **_transverse strain_** ε_T is normal to the load direction. The transverse strain is the change in bar width Δb divided by the original width b (or the change in length divided by the original length of any line perpendicular to the load direction):

$$\varepsilon_T = \frac{\Delta b}{b} \qquad \text{[Eq. 3.13]}$$

Poisson's ratio ν ("nu") is defined as the negative ratio of the transverse strain to the longitudinal strain:

$$\nu = -\frac{\varepsilon_T}{\varepsilon} \qquad \text{[Eq. 3.14]}$$

Poisson's ratio is an *elastic property* of a material; selected values are given in *Table 3.3*. In general, when a material is stretched $(\varepsilon > 0)$, it contracts in the transverse direction $(\varepsilon_T < 0)$, hence Poisson's ratio is typically positive.

For **_homogeneous, isotropic materials_** such as metals, the range of Poisson's ratio is $0.0 < \nu < 0.5$. *Homogeneous* means the material is the same at every point; *isotropic* means the material responds the same in every direction. For some materials such as composites and other very non-isotropic materials, where the response is different in different directions, the value of ν may be less than zero, or greater than 0.5. Metals undergoing plastic deformation (yielding) have a Poisson's ratio of 0.5.

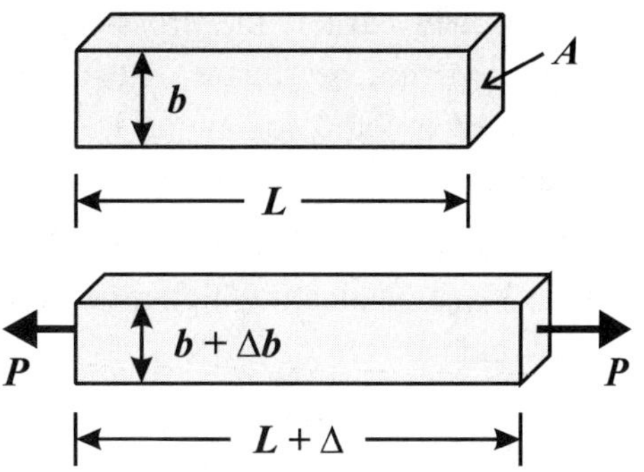

Figure 3.18. As a bar is stretched, its width is reduced – the *Poisson effect.*

Table 3.3. Poisson's Ratio, Elastic Deformation.

Materials	ν
Cork	~ 0
Concrete	0.1
Ceramics	0.2
Steel	0.3
Aluminum	0.33
Rubber	~ 0.5

Area and Volume Change Due to the Poisson Effect

A homogeneous, isotropic elastic bar loaded in tension increases in length and reduces in cross-sectional area (*Figure 3.19*). The original area and volume are $A = b^2$ and $V = AL = b^2 L$. Under load, the longitudinal strain is ε and the transverse strain is $\varepsilon_T = -\nu\varepsilon$. The new width is $b(1+\varepsilon_T)$, so the new cross-sectional area A_{new} is:

$$A_{new} = [b(1 + \varepsilon_T)][b(1 + \varepsilon_T)] = b^2(1 - \nu\varepsilon)^2$$
$$= A(1 - 2\nu\varepsilon + \nu^2\varepsilon^2) \qquad \text{[Eq. 3.15]}$$

Since ε is small compared with 1.0 (e.g., $\varepsilon \sim 0.005$ and $\varepsilon^2 \sim 0.000025$), the higher order term ε^2 can be neglected. The new area is then:

$$A_{new} = A(1 - 2\nu\varepsilon) \qquad \text{[Eq. 3.16]}$$

As mentioned earlier, the cross-section of the bar need not be square, and the result just derived is valid for any axial member with any cross-sectional shape.

A similar calculation gives the new volume V_{new}. Neglecting higher order terms (i.e., ε^2 and ε^3):

$$V_{new} = V[1 + \varepsilon(1 - 2\nu)] \qquad \text{[Eq. 3.17]}$$

The *dilation*, or change in volume, is:

$$\Delta V = \varepsilon(1 - 2\nu)V \qquad \text{[Eq. 3.18]}$$

The **volumetric strain**, or *dilational strain*, is defined as:

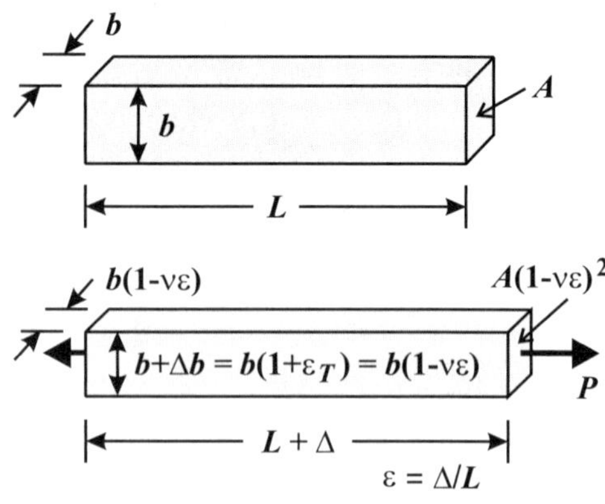

Figure 3.19. The Poisson effect.

$$\varepsilon_V = \frac{\Delta V}{V} = \varepsilon(1 - 2\nu) \qquad \text{[Eq. 3.19]}$$

The range of ν for *homogeneous, isotopic* materials is governed by *Equations 3.16* and *3.18*. From *Equation 3.16*, if $\nu < 0$, then the bar would fatten as it is stretched ($\varepsilon > 0$), or become thinner as it is compressed, which makes little physical sense; ν must be non-negative. The upper bound, $\nu = 0.5$, is deduced from *Equation 3.18* by noting that if the applied strain is tensile ($\varepsilon > 0$), then the volume should increase since the atomic bonds are being stretched.

Example 3.8 Poisson's Ratio

Given: A 1.2 m long aluminum bar has a circular cross-section of diameter 20.0 mm ($A = 314.2 \times 10^{-6}$ m^2), and is subjected to an axial tensile load $P = 30.0$ kN. For aluminum, the modulus is 70 GPa and Poisson's ratio is 0.33.

Required: Determine (a) the longitudinal and transverse strains and (b) the new cross-sectional area.

Solution: *Step 1.* Due to the load, the longitudinal strain is:

$$\varepsilon = \frac{\sigma}{E} = \frac{P}{AE} = \frac{30 \times 10^3 \text{ N}}{(314.2 \times 10^{-6} \text{ m}^2)(70 \times 10^9 \text{ Pa})} = 1.364 \times 10^{-3}$$

$$\underline{\textit{Answer: } \varepsilon = 1364 \times 10^{-6} = 0.136\%}$$

Step 2. The transverse strain is:

$$\varepsilon_T = -\nu\varepsilon = -(0.33)(1364 \times 10^{-6}) = -450 \times 10^{-6}$$

Answer: $\varepsilon_T = -450 \times 10^{-6} = -0.045\%$

Step 3. The new area is:

$$A_{new} = A(1 - 2\nu\varepsilon) = (314.2 \times 10^{-6}\ m^2)[1 - 2(0.33)(1364 \times 10^{-6})]$$

Answer: $A_{new} = 313.9 \times 10^{-6}\ m^2$

The reduction of area is small, approximately 0.03%.

Elastic Strain Energy of a Tensile Bar

The relationship between axial force P and elongation Δ for an elastic bar is:

$$P = \frac{EA}{L}\Delta = K\Delta \qquad \text{[Eq. 3.20]}$$

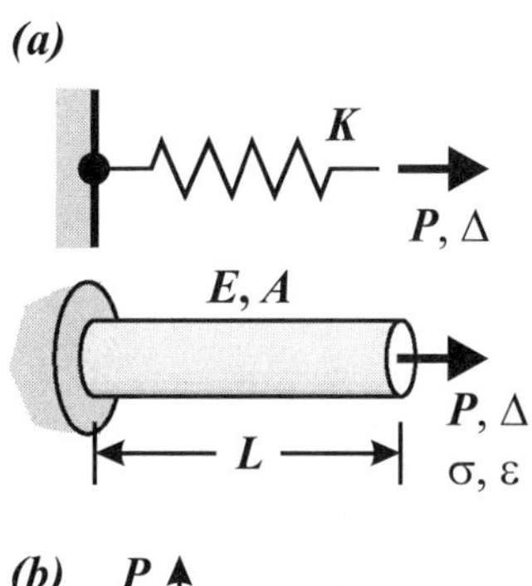

The bar behaves like a spring of stiffness K (*Figure 3.20*). The increment of work dW done by force P in deflecting the spring by an additional increment of displacement $d\Delta$ is:

$$dW = Pd\Delta = (K\Delta)d\Delta \qquad \text{[Eq. 3.21]}$$

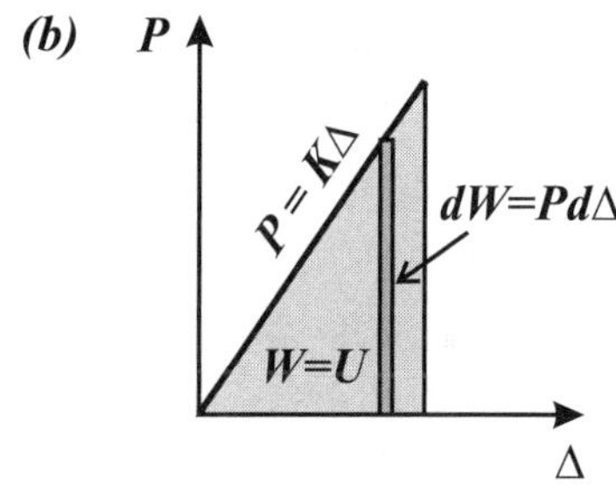

as shown in *Figure 3.20b*. The total *work* done to elastically deform the bar is determined by performing the integral:

$$W = \int_0^{\Delta}(K\Delta)d\Delta = \frac{1}{2}K\Delta^2 = \frac{1}{2}\frac{EA}{L}\Delta^2$$

$$\text{[Eq. 3.22]}$$

Using *Equation 3.20*, the total work done can also be expressed as:

$$W = \frac{1}{2}\frac{P^2 L}{EA} = \frac{1}{2}\frac{P^2}{K} \qquad \text{[Eq. 3.23]}$$

The work done W is stored internally as **elastic strain energy** U:

$$W = U \qquad \text{[Eq. 3.24]}$$

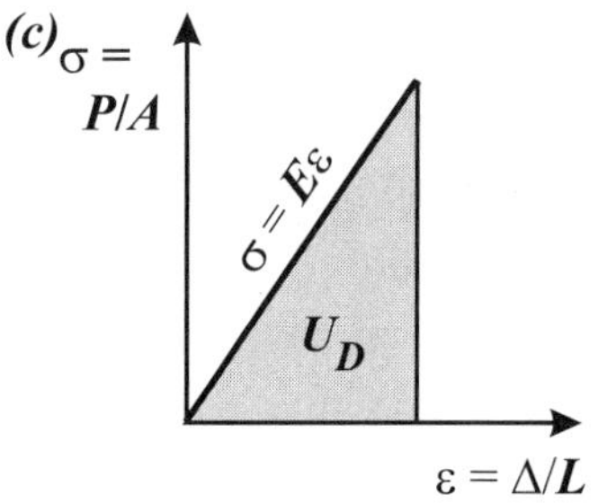

Using the relationships for stress and strain:

$$\sigma = \frac{P}{A}\ ;\ \ \varepsilon = \frac{\Delta}{L} \qquad \text{[Eq. 3.25]}$$

Figure 3.20. (a) Spring and axial member, both subjected to axial force P. **(b)** Energy stored in spring or axial member. **(c)** Elastic strain energy density in an axial member.

the expression for the *internal energy* is:

$$U = W = \frac{1}{2}E\varepsilon^2(AL) = \frac{1}{2}\frac{\sigma^2}{E}(AL) \qquad \text{[Eq. 3.26]}$$

Since the bar volume is $V = AL$, the **elastic strain energy density** – the *elastic strain energy per unit volume* – U_D, is:

$$U_D = \frac{1}{2}E\varepsilon^2 = \frac{1}{2}\frac{\sigma^2}{E} = \frac{1}{2}\sigma\varepsilon \qquad \text{[Eq. 3.27]}$$

The elastic strain energy density is the area under the linear–elastic region of the stress–strain curve (*Figure 3.20c*).

The maximum value of the elastic strain energy density is when the stress reaches the yield strength. The maximum elastic strain energy density is known as the **modulus of resilience** U_R:

$$U_R = \frac{1}{2}\frac{S_y^2}{E} \qquad \text{[Eq. 3.28]}$$

The *resilience* is the maximum energy per unit volume that can be applied to the material without plastic deformation occurring.

Example 3.9 Car Bumper Design

Given: Car bumpers are often protected with a strip or pad of rubber approximately 2.0 m long, 5 mm thick, and 100 mm high (*Figure 3.21*). The practical purpose of the strip is to absorb energy in low-speed accidental crashes such as in parking lots or when parallel parking. Assume the strip supports the entire load uniformly.

Required: (a) From an elastic energy standpoint, why might rubber be a good choice of material compared to steel and aluminum? (b) If a 1100 kg (2430 lb) car traveling at 2.2 m/s (5 mph) in a parking lot hits a wall, and comes to a complete stop, can the rubber pad (2.0 m long, 5 mm thick, and 100 mm wide) absorb the energy without exceeding the elastic limit?

Solution: *Step 1.* While remaining linear–elastic, the maximum energy that a material can store per unit volume is the resilience:

$$U_R = \frac{1}{2}\frac{S_y^2}{E}$$

Thus, for a given volume, to maximize the elastic energy stored, it is desirable to choose a

(a)

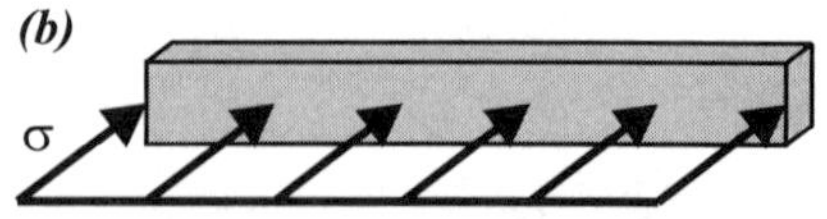
(b)

Figure 3.21. (a) Car with rubber pad on bumper. **(b)** Uniformly applied stress on pad. Copyright ©2008 Dominic J. Dal Bello and licensors. All rights reserved.

material that maximizes the resilience. Consider representative values for steel, aluminum, and rubber, and assume each material remains linear–elastic up to the listed yield point:

Material	S_y (MPa)	E (GPa)	U_R (MN·m/m³)
Steel	250	200	0.16
Aluminum	240	70	0.41
Rubber	20	0.05	4.0

For the same volume, *rubber* absorbs more elastic energy than steel or aluminum (without yielding).

Step 2. The kinetic energy of the car is:

$$KE = \frac{1}{2}mv^2 = \frac{1}{2}(1100 \text{ kg})(2.2 \text{ m/s})^2 = 2662 \text{ N·m}$$

The total elastic energy that the rubber pad can absorb is:

$$U = U_R[\text{Volume}] = \left(4.0\frac{\text{MN·m}}{\text{m}^3}\right)[2.0 \times 0.005 \times 0.1 \text{ m}^3] = 4000 \text{ N·m}$$

The maximum elastic strain energy is greater than the energy of the car moving at 2.2 m/s. The rubber strip can absorb the energy elastically; the bumper should theoretically suffer no damage. *Please do not try this on your own!*

In this example, the ratio of the maximum energy density for steel to that of rubber is 25 (= 4.0/0.16). A volume of steel that is 25 times as great as rubber is required to absorb the same energy without plastically deforming. A large piece of steel is reminiscent of the heavy bumpers on classic automobiles.

Cyclic Loading

Engineering systems are often subjected to *cyclic loading*. This is especially true in automobiles and other machines that have continuous rotary motion. Automobile parts are commonly subjected to millions of load cycles. In electronic components, cyclic stresses are caused by fluctuating temperatures.

Cyclic loading can cause a material to degrade with time, a phenomenon known as *fatigue*. Fatigue can cause materials to fail – break into two – at stress levels well below their uniaxial strength. In many steels, the *fatigue strength* is less than half of the ultimate strength of the material.

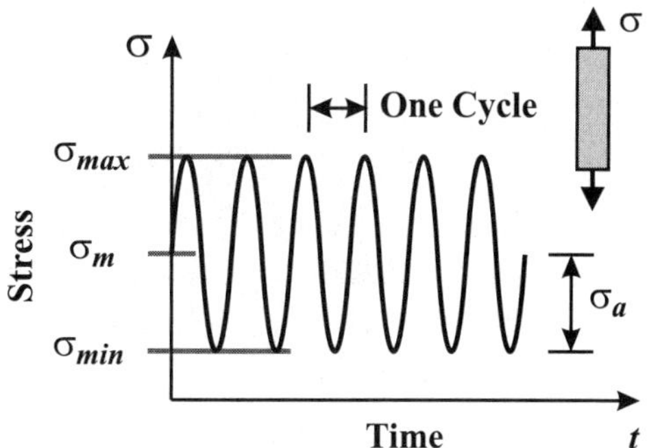

Figure 3.22. Typical stress–time graph for cyclic loading.

To understand fatigue, a few definitions must be given to describe the nature of cyclic loading. A sample cyclic loading history is shown in *Figure 3.22*. The *maximum stress* and *minimum stress* are σ_{max} and σ_{min}, respectively. The *mean stress* is:

$$\sigma_m = \frac{\sigma_{max} + \sigma_{min}}{2} \qquad \text{[Eq. 3.29]}$$

and the *stress amplitude* is:

$$\sigma_a = \frac{\sigma_{max} - \sigma_{min}}{2} \qquad \text{[Eq. 3.30]}$$

The ratio of the minimum to maximum stresses is the *R-ratio*:

$$R = \frac{\sigma_{min}}{\sigma_{max}} \qquad \text{[Eq. 3.31]}$$

S–N Curves

The standard baseline *fatigue test* is shown in *Figure 3.23*. The mean stress is zero, $\sigma_m = 0$, the amplitude is $\sigma_a = \sigma_{max}$, and the *R*-ratio is $R = -1.0$. A specimen is fatigued at a specific stress amplitude σ_a until it breaks into two parts. The number of **cycles to failure** is N_f, also known as the *fatigue life*. The test is repeated for various stress amplitudes σ_a.

When the stress amplitude is plotted against the number of cycles to failure on a log–log set of axes (or on a semi-log set of axes), the graph is known as the **S–N curve** (*Figure 3.24*).

The **fatigue strength** is the stress amplitude $\sigma_a = \sigma_{max}$ corresponding to a specified number of cycles to failure.

When the data are plotted, the relationship between stress amplitude σ_a and cycles to failure N_f can be expressed in the following form:

$$\sigma_a = S_f'(N_f)^b \qquad \text{[Eq. 3.32]}$$

where S_f' and b depend on the material, and are determined from a best-fit line of the material's *S–N curve* since:

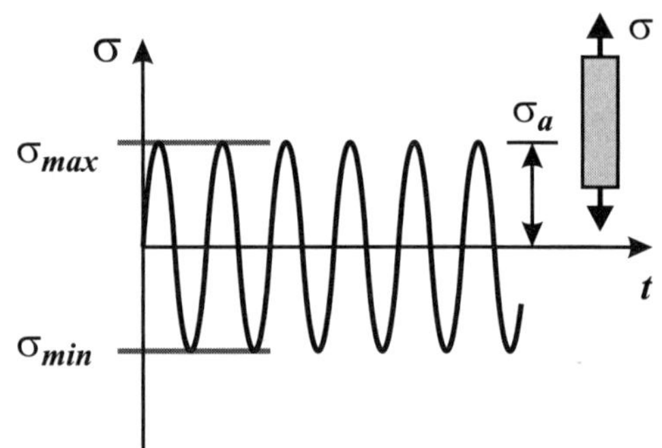

Figure 3.23. Stress–time graph for standard fatigue test: $\sigma_a = \sigma_{max} = |\sigma_{min}|$, $\sigma_m = 0$, $R = -1.0$.

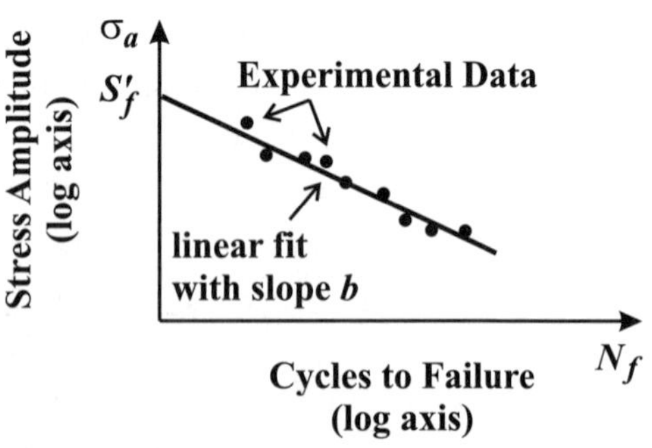

Figure 3.24. *S–N curve*. The curve is plotted on a set of log–log axes.

$$\log(\sigma_a) = \log(S_f') + b \log(N_f)$$

[Eq. 3.33]

Quantity S_f' corresponds to the curve fit's intercept at $N_f = 1$ cycle, and b is the slope of the curve in the log–log axes. Values for structural aluminum and steel are given in *Table 3.4*.

A very useful reference and set of tools for fatigue analysis is Dr. Darrell F. Socie's web site, www.fatiguecalculator.com (accessed May 2008).

Table 3.4. Fatigue properties for structural aluminum and steel.

Property	Al 6061-T6	Steel A36
S_y	240 MPa	250 MPa
S_u	314 MPa	540 MPa
S_f'	505 MPa	1035 MPa
b	–0.082	–0.11
E	70 GPa	200 GPa

Source: www.fatiguecalculator.com/ Accessed May 2008.

Example 3.10 Fatigue Strength of Structural Aluminum and Steel

Given: A tensile bar is subjected to cyclic loading, with zero mean stress (*Figure 3.23*).

Required: Determine the fatigue strength, $\sigma_a = \sigma_{max}$, (a) for an aluminum bar (Al 6061-T6) at 10^7 cycles and (b) for a steel bar (A36) at 10^6 cycles.

Solution: *Step 1.* For aluminum, the fatigue strength corresponds to 10^7 cycles. The fatigue variables are given in *Table 3.4*: $S_f' = 505$ MPa and $b = -0.082$. Therefore:

$$\sigma_{a,\,al} = S_f'(N_f)^b = (505 \text{ MPa})(10^7)^{-0.082}$$

Answer: $\underline{\sigma_{a,\,al} = 134 \text{ MPa}}$

For aluminum 6061-T6, $S_y = 240$ MPa and $S_u = 314$ MPa. Here, the fatigue strength σ_a is 56% of the yield strength S_y and 43% of the ultimate strength S_u.

Step 2. For steel, the fatigue strength corresponds to 10^6 cycles, therefore:

$$\sigma_{a,\,st} = (1035 \text{ MPa})(10^6)^{-0.11}$$

Answer: $\underline{\sigma_{a,\,st} = 226 \text{ MPa}}$

For A36 steel, $S_y = 250$ MPa and $S_u = 540$ MPa. The fatigue strength σ_a is 90% of S_y and 42% of S_u.

The values of *fatigue strength* just calculated for the aluminum (with $N_f = 10^7$) and steel (with $N_f = 10^6$) are also known as their *fatigue limit* S_{FL}, as discussed below. The fatigue limit for many steels is typically on the order of 35 - 50% of the ultimate strength.

Fatigue Limit

The S–N curve for most steels does not decrease to zero (*Figure 3.25*). Below a certain value of σ_a, steel has essentially an infinite fatigue life. The stress amplitude below which fatigue failure does not occur is the *fatigue limit* S_{FL}, also known as the *endurance limit*.

For steels, the *fatigue limit* typically corresponds to a fatigue life of about 10^6 cycles. The S–N curve exhibits no further reduction and becomes horizontal (*Figure 3.25*).

Aluminum does not exhibit such limiting behavior. No matter how small the stress amplitude σ_a, aluminum will eventually fail by fatigue. Accordingly, when designing with aluminum, the fatigue limit S_{FL} is generally taken to be the stress amplitude for $N_f = 10^7$ cycles.

The fatigue limits S_{FL} for structural steel ($N_f = 10^6$) and aluminum ($N_f = 10^7$) were (conveniently) calculated in *Example 3.10*.

Effect of Mean Stress on Fatigue Strength

In general, cyclic stresses are applied with a non-zero mean stress σ_m (*Figure 3.26*). When this is the case, the fatigue strength for a given number of *cycles to failure* N_f is determined using the **Goodman Diagram** (*Figure 3.27*). For a specified number of cycles to failure N_f, the fatigue strength is given by:

$$\frac{\sigma_a}{S_{FL}} + \frac{\sigma_m}{S_u} = 1 \qquad \text{[Eq. 3.34]}$$

which defines a line on the Goodman diagram. All points on a line have the same fatigue life. The *ultimate strength* of the material S_u is plotted on the abscissa, and the *fatigue limit* S_{FL} (for a specified N_f cycles to

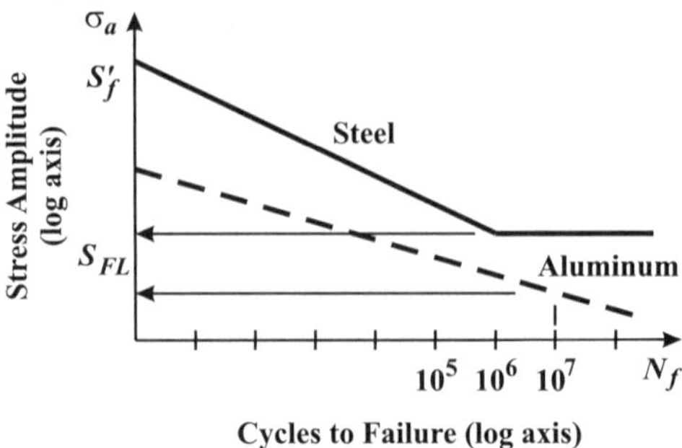

Figure 3.25. *S–N curves* for steel (solid) and aluminum (dashed). Steel has a certain stress level below which no fatigue failure occurs. This is the *fatigue limit S_{FL}*. Aluminum eventually fails by fatigue no matter how small the stress amplitude.

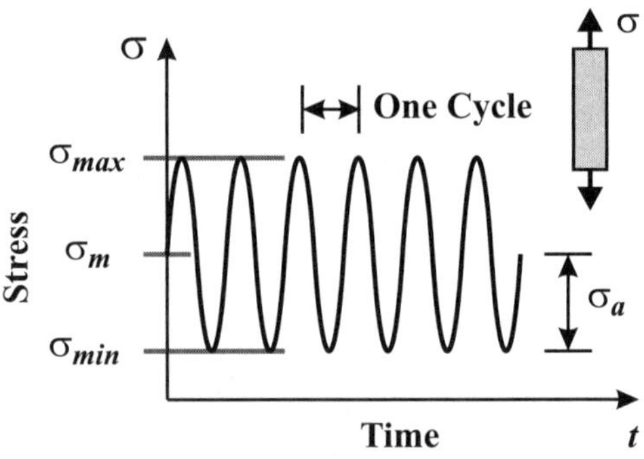

Figure 3.26. Typical stress–time graph.

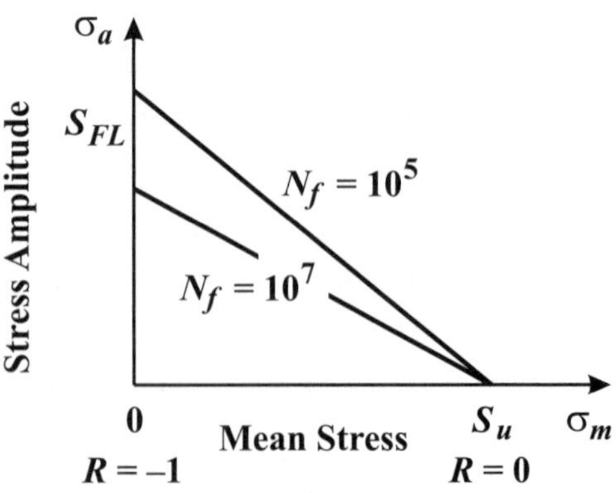

Figure 3.27. Goodman Diagram.

failure with no mean stress) is plotted on the ordinate, and a line drawn between them. The greater the specified cycles to failure N_f, the lower the value of S_{FL} (*Equation 3.32*).

The new fatigue strength σ_a – the amplitude of the cyclic loading with mean stress σ_m for the specified cycles to failure N_f – is then:

$$\sigma_a = S_{FL}\left(1 - \frac{\sigma_m}{S_u}\right) \qquad \text{[Eq. 3.35]}$$

When $\sigma_m = S_u$, the materials breaks into two upon first loading; there can be no alternative stress σ_a ($\sigma_a = 0$). The effect of the non-zero mean stress is to reduce the amplitude of the cyclic stress, as shown in the following example. If the mean stress is zero, then the stress amplitude is the *fatigue limit* S_{FL} for the specified number of cycles to failure.

Example 3.11 Fatigue Strength With Non-Zero Mean Stress

Given: A tensile bar is subjected to cyclic loading with a mean stress of $\sigma_m = 100$ MPa.

Required: With the mean stress applied, determine the fatigue strength (a) for an Al 6061-T6 bar at 10^7 cycles and (b) for an A36 steel bar at 10^6 cycles.

Solution: For the aluminum bar, the fatigue limit (fatigue strength) corresponding to 10^7 cycles with $\sigma_m = 0$ is $S_{FL} = 134$ MPa (*Example 3.10*). The ultimate strength is $S_u = 314$ MPa (*Table 3.4*). So, the fatigue strength with $\sigma_m = 100$ MPa is:

$$\sigma_{a,\,al} = S_{FL}\left(1 - \frac{\sigma_m}{S_u}\right) = (134 \text{ MPa})\left(1 - \frac{100}{314}\right)$$

$$\textit{Answer: } \underline{\sigma_{a,\,Al} = 91 \text{ MPa}}$$

Due to the mean stress of 100 MPa, the fatigue strength of aluminum – the amplitude of the cyclic stress for failure at 10^7 cycles – is reduced by 32%, from 134 to 91 MPa.

For the steel bar, the fatigue limit (fatigue strength) corresponding to 10^6 cycles with $\sigma_m = 0$ is $S_{FL} = 226$ MPa (*Example 3.10*). The ultimate strength is $S_u = 540$ MPa, so:

$$\sigma_{a,\,st} = S_{FL}\left(1 - \frac{\sigma_m}{S_u}\right) = (226 \text{ MPa})\left(1 - \frac{100}{540}\right)$$

$$\textit{Answer: } \underline{\sigma_{a,\,st} = 184 \text{ MPa}}$$

Due to the mean stress of 100 MPa, the fatigue strength of steel – the amplitude of the cyclic stress for failure at 10^6 cycles – is reduced by 19%, from 226 to 184 MPa.

The results of *Examples 3.10* and *3.11* illustrate that aluminum is more susceptible to fatigue than is steel.

3.2 The Torsion Test – Shear Properties

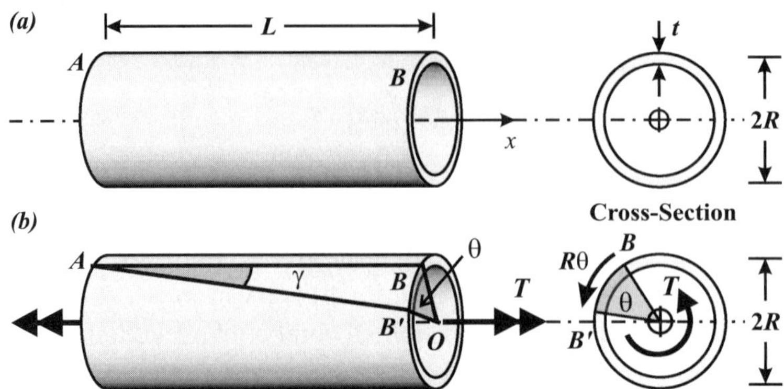

Figure 3.28. (a) Thin-walled circular shaft of radius R, thickness t ($t << R$), length L. **(b)** Shaft under torque T. With the left-end fixed, line AB rotates by angle γ to AB'; cross-section B displaces through angle θ, the angle of twist.

Thin-walled circular shaft AB, of average radius R, thickness t and length L, is shown in *Figure 3.28*. A shaft is considered ***thin-walled*** if $t << R$, e.g., $t \leq 0.1R$. For the thin-walled shaft, the inner, outer, and average radii are taken to be equal and the material response is constant across the thickness.

Consider the left end (section A) of the shaft to be fixed. When torque T is applied to the shaft, the right end (section B) rotates with respect to the left-end by angle θ. point B moves to point B'. The relative rotation between sections A and B is the ***angle of twist*** θ of the shaft.

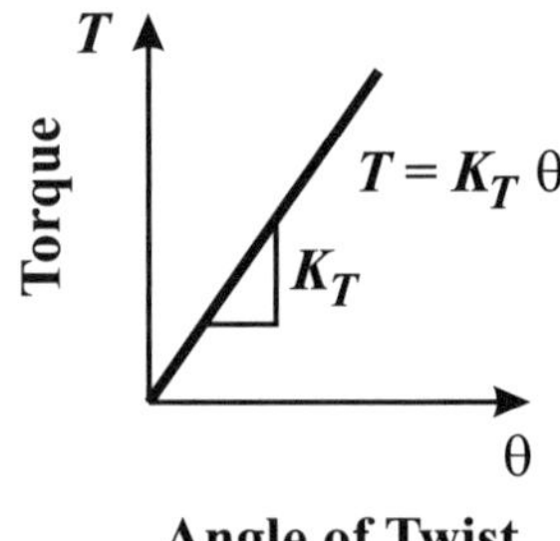

Figure 3.29. Torque-Angle of Twist response for small angles.

Experiments show that torque increases linearly with θ for small (elastic) rotations. The shaft behaves as a torsional spring:

$$T = K_T\theta \qquad \text{[Eq. 3.36]}$$

where K_T is the torsional spring stiffness (*Figure 3.29*).

The results of the torsion test are used to determine the shear properties of a material. With the shear properties known, the torsional strength and stiffness of a thin-walled circular shaft of any R, t, and L can be determined. The torque-angle of twist response is investigated using *shear strain* and *shear stress*.

Shear Strain

The ***shear strain*** γ ("gamma", *Figure 3.28b*) is the ratio of the displacement (movement) of point B to the shaft length L, with section A taken as fixed. The distance point B moves is:

$$BB' = R\theta = \gamma L \qquad \text{[Eq. 3.37]}$$

Hence the relationship between the shear strain γ and the angle of twist θ is

$$\gamma = \frac{BB'}{L} = \frac{R\theta}{L} \qquad \text{[Eq. 3.38]}$$

Displacement BB' is *perpendicular* to the axis of the shaft, length L. Since section A is fixed, BB' is also the *relative displacement between the two ends*. Recall that the analogous definition for the normal (axial) strain is the relative displacement of the two ends, *parallel* to the axis of the component, over length L: $\varepsilon = \Delta / L$.

Both γ and θ are measured in radians.

Now consider a square material element on the surface of the shaft originally bound by two dashed axial lines and two solid circumferential lines (*Figure 3.30*). The torque causes the dashed lines to deform to the position of the solid angled lines. The square is transformed into a rhombus.

The *shear strain* γ is the change in right angle of the square element as it deforms into a rhombus, measured in *radians*. In *Figure 3.31*, the rhombus has been rotated so that one side is aligned with the horizontal x-axis; the top of the rhombus is shifted to the right by distance w (exaggerated). Like normal strain ε, the shear strain γ is generally small (~0.001). Thus, $\sin \gamma \sim \tan \gamma \sim \gamma$, and the change in right angle – the shear strain – is:

$$\gamma = \frac{w}{h} \qquad \text{[Eq. 3.39]}$$

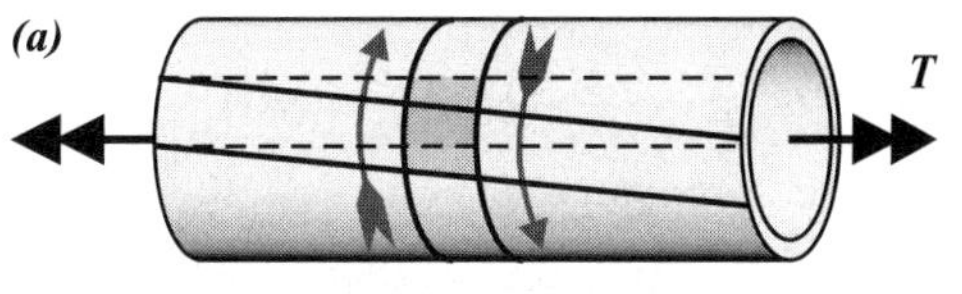
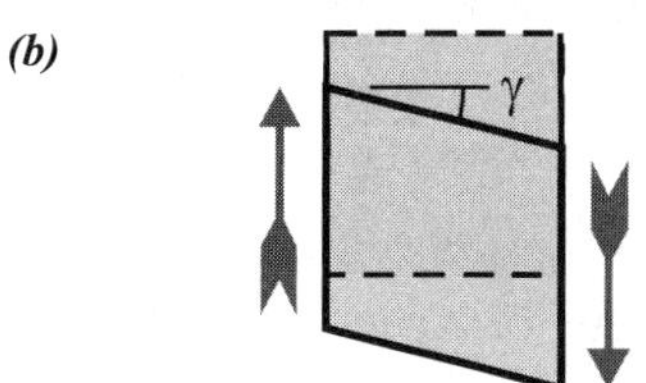

Figure 3.30. A square element on the surface of a shaft in torsion is deformed into a rhombus.

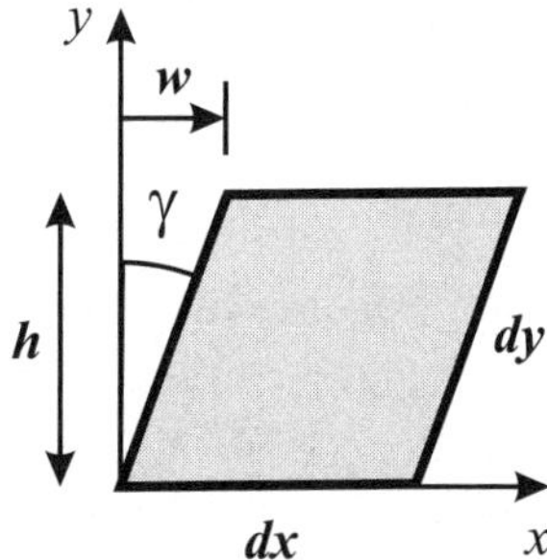

Figure 3.31. The square has been deformed into a rhombus.

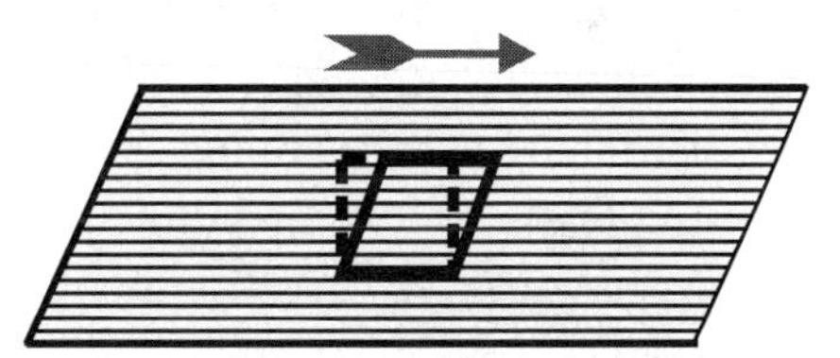

Figure 3.32. The square on the side of a telephone book becomes a rhombus when the top cover is displaced.

In axial members, the Poisson effect causes a change in volume. *There is no change in volume due to shear.* To model and illustrate this, consider a square drawn on the side of a phone book (*Figure 3.32*). Displacing the cover parallel to the pages (applying a shear force) turns the square into a rhombus. Like planes of atoms, each page is slightly shifted to the right with respect to the one below it, but its height and width do not change. Hence, material volume is unchanged by shear.

Example 3.12 **Shear Strain**

Given: A rectangular piece of Jello^{TM}, $h = 100$ mm tall by $b = 200$ mm wide, sits on a plate with a piece of aluminum foil on top (*Figure 3.33*). The foil is moved to the right by $w = 4$ mm, pulling the top of the Jello^{TM} along with it.

Required: Assume the small angle approximation holds. Determine the shear strain in the Jello^{TM}.

Solution: The shear strain is the reduction in right angle (in radians):

$$\gamma = \frac{w}{h} = \frac{4 \text{ mm}}{100 \text{ mm}}$$

Answer: $\gamma = 0.04 = 4\%$

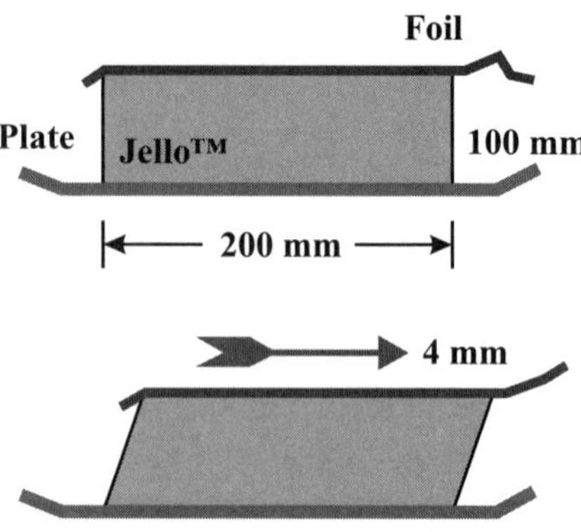

Figure 3.33. The top of the Jello is shifted to the right.

Shear Stress

Consider again the thin-walled shaft of thickness t and average radius R $(t \ll R)$, subjected to torque T (*Figure 3.34*). Take a cut perpendicular to its axis, exposing an interior cross-section. The torque carried by the cross-section is T. The torque is not carried at a single point, but is supported uniformly over cross-sectional area A at distance R from the axis. The torque is thus $T = RF$, where F is an equivalent force distributed evenly around the cross-section, and continually changing direction. A good approximation for the cross-sectional area of a thin-walled shaft is $A = 2\pi Rt$.

The ***shear stress*** τ ("tau") in the thin-walled shaft is then:

$$\tau = \frac{F}{A} = \frac{T/R}{A} = \frac{T}{R(2\pi Rt)} = \frac{T}{2\pi R^2 t}$$

[Eq. 3.40]

Shear stress has units of force per unit area (ksi, MPa, etc. Unlike the normal stress, the shear stress acts across, or parallel to, the interior material surface.

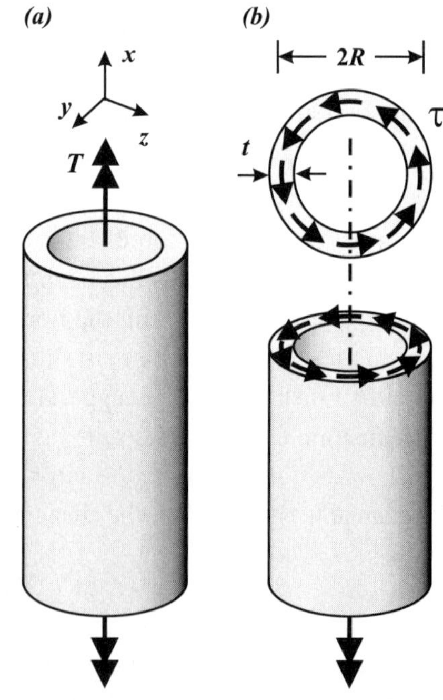

Figure 3.34. (a) Thin-walled circular shaft of average radius R and thickness t, subjected to torque T. **(b)** Shear stress τ is carried on the cross-section of the shaft.

Note that the shear stress due to a torque constantly changes direction as it moves around the cross-sectional area (*Figure 3.34b*). Also, the *thin-wall* assumption means that the thickness is so small that the shear stress can be considered constant through the wall thickness.

Example 3.13 Thin-walled Pipe Tightened with a Wrench

Given: A pipe of diameter $D = 1.0$ in. and thickness $t = 0.10$ in. is twisted by a wrench with torque $T = Pd = 200$ lb-in., as shown in *Figure 3.35a*.

Required: Use the thin-wall formula to estimate the shear stress in the thin-walled pipe due only to torque T.

Solution: The shear stress for a thin-walled pipe is:

$$\tau = \frac{T}{2\pi R^2 t} = \frac{2T}{\pi D^2 t}$$

$$= \frac{2(200 \text{ lb-in.})}{\pi[(1 \text{ in.})^2 (0.1 \text{ in.})]}$$

Answer: $\tau = 1270$ psi $= 1.27$ ksi

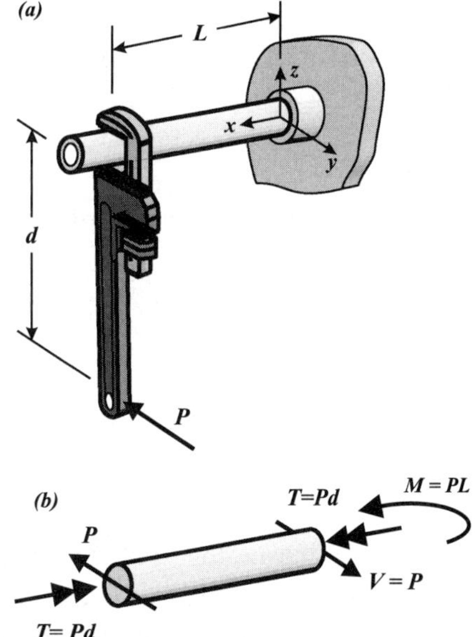

Figure 3.35. (a) Wrench tightening a thin-walled pipe. **(b)** FBD of pipe.

Complementary Shear Stress

The shear stress on a cross-section of the thin-walled circular shaft in torsion is:

$$\tau = \frac{T}{2\pi R^2 t} \qquad \text{[Eq. 3.41]}$$

However, the shear stress on the cross-section is not the only shear stress that is active. Consider a 3D element with sides dx, dy, and t removed from the shaft, showing only the shear stress on the exposed cross-section (*Figure 3.36a*).

The element is drawn in 2D in *Figure 3.36b*, showing only the shear stresses acting on cross-sectional planes *AB* and *CD*. Equilibrium of forces requires that these stresses be equal but opposite.

The horizontal stresses cause equal but opposite forces $\tau(t\,dx)$ on surfaces *AB* and *CD*. These forces produce a clockwise couple:

$$M = [\tau(t dx)](dy) \qquad \text{[Eq. 3.42]}$$

The only means of maintaining rotational equilibrium is to apply a counter-acting couple in the form of shear forces on sides AD and BC, F_{AD} and F_{BC}, which are dx apart (*Figure 3.36c*). Forces F_{AD} and F_{BC} must be equal and opposite to maintain vertical equilibrium. To balance the couple M due to the horizontal stresses:

$$F_{AD} = F_{BC} = \frac{M}{dx} = \tau[t(dy)]$$

[Eq. 3.43]

The shear stresses acting on AD and BC, both of area $t(dy)$, are then (*Figure 3.36d*):

$$\tau_{AD} = \tau_{BC} = \frac{F_{AD}}{[t(dy)]} = \tau$$

[Eq. 3.44]

The shear stresses on sides AD and BC are therefore equal and perpendicular to those on AB and CD.

At a point in the material, the shear stresses at right angles to the applied shear stress are referred to as ***complementary shear stresses***. Whenever a shear stress acting on a plane is identified at a point, there is automatically a complementary shear stress on a plane at right angles to the first plane. At any point, shear stresses on perpendicular planes are equal.

Note the directions of the direct shear stresses acting on the cross-section (sides AB and CD in *Figure 3.36*) and the directions of the balancing complementary shear stresses (on sides AD and BC). The shear stresses on a material element always act to cause counteracting couples. The arrowheads – and tails – of the shear stresses must meet at the corners of the element.

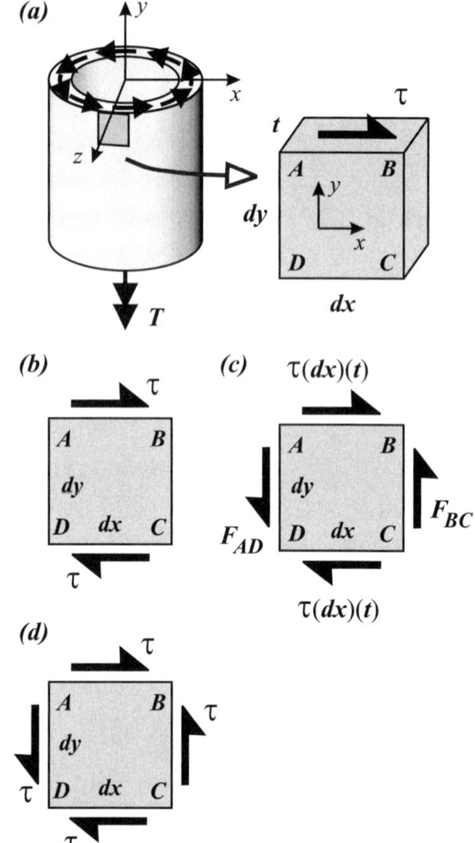

Figure 3.36. (a) The 3D element taken out of a thin-walled shaft. **(b)** The 2D element, *t* thick into paper, showing stresses acting on surfaces *AB* and *CD*. **(c)** The 2D element with forces on each side to enforce equilibrium. **(d)** The element showing equal complementary shear stresses required to keep element from rotating.

Example 3.14 Complementary Shear Stress – A Welded Pipe

Given: A thin-walled pipe (radius R, thickness $t << R$) is formed by rolling a plate into a cylinder and welding the edges together to form a continuous shaft. The shaft is subjected to a torque T (*Figure 3.37*).

Required: Determine the shear stress on the weld, $\tau_W = \tau_{CD}$ (along line *CD*).

Solution: *Step 1.* The shear stress *on the cross-section* is:

$$\tau_{AD} = \frac{T}{2\pi R^2 t}$$

This is the shear stress on *AD* and *BC* of element *ABCD*, due directly to the applied torque *T*.

Step 2. Complementary shear-stress means that the weld is also subjected to shear-stress τ_{AD}, so:

$$Answer:\ \ \tau_W = \tau_{CD} = \tau_{AD} = \frac{T}{2\pi R^2 t}$$

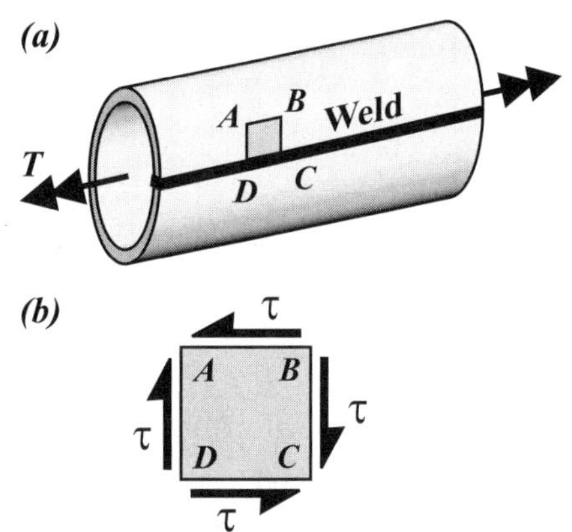

Figure 3.37. (a) A thin-walled circular shaft manufactured from a plate rolled into a cylinder and welded. **(b)** Shear stress acting on element *ABCD*. Side *CD* is on the weld.

Elastic Shear Modulus

The relationship between normal stress and strain is given by Hooke's Law, $\sigma = E\varepsilon$. A similar relationship exists between *shear stress* τ and *shear strain* γ (*Figure 3.38*):

$$\tau = G\gamma \qquad \text{[Eq. 3.45]}$$

where G is the **shear modulus**. Typical values for the shear modulus are found in *Table 3.5*.

The values of G are not always readily accessible. A good approximation for the value of G for *homogeneous* (same at every point) and *isotropic* (same in every direction) materials is:

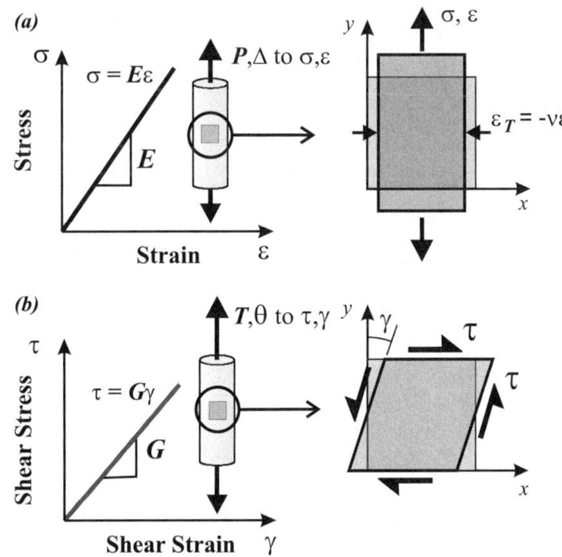

Figure 3.38. (a) Elastic response of axial member and deformation of material element. **(b)** Elastic response of torsional member and deformation of material element.

$$G = \frac{E}{2(1 + \nu)} \qquad \text{[Eq. 3.46]}$$

E, G, and ν are the ***elastic properties*** of a material.

Since $\nu \sim 1/3$ for many metals, then G can be approximated:

$$G \approx \frac{3}{8}E$$

[Eq. 3.47]

For aluminum, $E = 70$ GPa and $\nu = 0.33$; the "3/8 rule" predicts $G = 26$ GPa, as given in *Table 3.5*. For steel, $E = 207$ GPa, the rule predicts $G = 78$ GPa.

Shear Stress–Strain Curves

The shear stress–strain response observed in strain-controlled tests on metal shafts is similar to the behavior observed in the tension tests described in *Section 3.1*. A typical shear stress–strain curve is shown in *Figure 3.39*.

Initially, the behavior is *linear–elastic* with $\tau = G\gamma$. However, when the shear stress exceeds a critical value, the linear relationship no longer applies. Additional small increments in plastic strain result in smaller additional increments of stress.

To define the **shear yield strength**, the procedure described in *Section 3.1* is again used. A line is constructed parallel to the elastic line, displaced on the strain-axis by 0.2%. Where the line intersects the experimental stress–strain curve is the *shear yield strength* τ_y.

Table 3.5. Representative values of Young's Modulus E, and Shear Modulus G.

Materials	E, GPa (Msi)	G, GPa (Msi)
Steels	207 (30)	80 (12)
Titanium alloys	115 (17)	43 (6.3)
Aluminum alloys	70 (10)	26 (3.8)
Nickel alloys	215 (31)	83 (12)
Cast iron	180 (26)	70 (10)
Douglas fir	12.4 (1.8) parallel to grain	Depends on grain direction
Glass	70 (10)	29 (4)
Rubbers	0.01–0.1 (0.0015–0.015)	0.003–0.03 (0.0005–0.005)
Polymers	0.1–5 (0.015–0.75)	0.03–1.7 (0.056–0.2)
Engineering ceramics	300–450 (44–65)	125–190 (18–28)
Carbon fiber/ polymer matrix composite	10–150 requires composite calculation	7–40 requires composite calculation

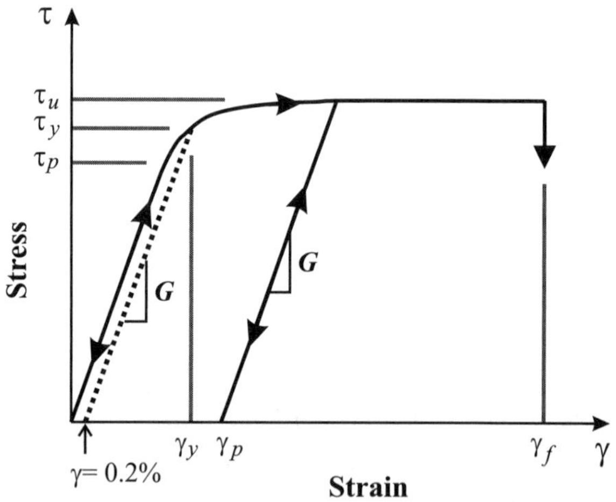

Figure 3.39. Shear stress–strain curve of a ductile material. The failure strain γ_f can be very large; solid ductile shafts tested in torsion can usually be completely twisted around several times before failure.

From experiments on ductile metals, it is observed that τ_y is often related to the axial yield strength S_y as follows:

$$\tau_y = \frac{S_y}{\sqrt{3}} = \frac{S_y}{1.73} \qquad \text{[Eq. 3.48]}$$

This means that τ_y for many metals can be deduced from S_y, a very valuable result. For example, the tabulated values of yield strength for aluminum 6061-T6 are: $S_y = 35$ ksi and $\tau_y = 20$ ksi (Aluminum Association *Aluminum Design Manual*, 2005). Using *Equation 3.48* gives a value in agreement with the tabulated value:

$$\tau_y = \frac{S_y}{\sqrt{3}} = \frac{35 \text{ ksi}}{\sqrt{3}} = 20 \text{ ksi}$$

For advanced composites, non-isotropic materials and brittle materials, this relation is not generally valid.

Torsional Stiffness of a Thin-Walled Circular Shaft

With G known, the torsional stiffness of any straight, thin-walled circular shaft of length L, radius R, and wall thickness t, can be calculated. The thin-walled shaft of *Figure 3.40* is subjected to torque T. Substituting the expressions for shear stress τ (*Equation 3.40*) and shear strain γ (*Equation 3.38*) into *Equation 3.45* results in:

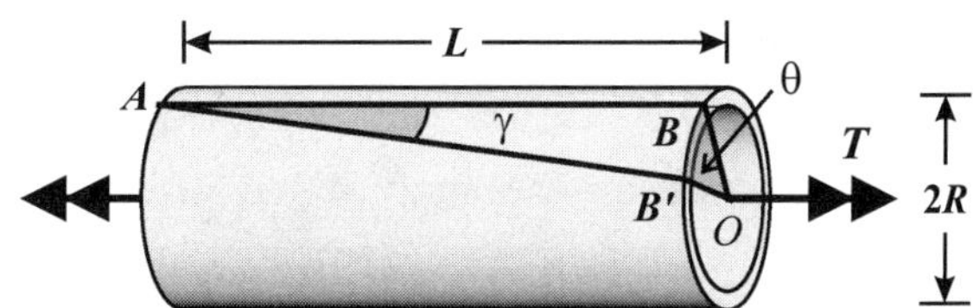

Figure 3.40. The thin-walled circular shaft.

$$\tau = G\gamma \Rightarrow \frac{T}{2\pi R^2 t} = G\frac{R\theta}{L} \qquad \text{[Eq. 3.49]}$$

Solving for torque T:

$$T = \left(\frac{2\pi R^3 t G}{L}\right)\theta \qquad \text{[Eq. 3.50]}$$

The ***torsional stiffness*** K_T (*Figure 3.41a, b*) of the thin-walled shaft is:

$$K_T = \frac{T}{\theta} = \frac{2\pi R^3 t G}{L} \qquad \text{[Eq. 3.51]}$$

Note the similarity of the form of the torsional stiffness to the form of the axial stiffness of a bar, $K = EA/L$ (*Equation 3.6*). Both are proportional to a function of cross-sectional geometry and to material stiffness, and inversely proportional to length.

If the applied torque is T, the *angle of twist* θ over length L can be calculated:

$$\theta = \left(\frac{L}{2\pi R^3 tG}\right) T \qquad \text{[Eq. 3.52]}$$

The *torsional flexibility* F_T of the shaft is:

$$F_T = \frac{\theta}{T} = \frac{L}{2\pi R^3 tG} \qquad \text{[Eq. 3.53]}$$

Elastic Strain Energy of a Thin-Walled Circular Shaft

The thin-walled circular shaft is a torsional spring of stiffness K_T (*Figure 3.41*). The work done by torque T rotating through differential angle $d\theta$ is $dW = Td\theta$. The total work to twist the shaft by angle θ during elastic deformation is the area under the linear–elastic curve of *Figure 3.41b*:

$$W_T = \int_0^\theta T\,d\theta = \int_0^\theta (K_T\theta)d\theta = \frac{1}{2}K_T\theta^2 \qquad \text{[Eq. 3.54]}$$

The work done on the shaft W_T is stored as *elastic shear strain energy U*. Hence:

$$U = W_T = \frac{1}{2}K_T\theta^2 = \frac{1}{2}\left(\frac{2\pi R^3 tG}{L}\right)\left(\frac{\gamma L}{R}\right)^2$$

$$= \frac{1}{2}(2\pi RtL)\gamma^2 G = \frac{\gamma^2 G}{2}[\text{Volume}] \qquad \text{[Eq. 3.55]}$$

The elastic energy per unit volume – the **elastic shear strain energy density** $U_{D,\tau}$ – is:

$$U_{D,\tau} = \frac{1}{2}\gamma^2 G = \frac{1}{2}\frac{\tau^2}{G} \qquad \text{[Eq. 3.56]}$$

The maximum shear stress to avoid yielding is τ_y. The maximum elastic shear strain energy density is then:

$$U_{R,\tau} = \frac{1}{2}\frac{\tau_y^2}{G} \approx \frac{4}{9}\frac{S_y^2}{E} \qquad \text{[Eq. 3.57]}$$

The approximate equality in *Equation 3.57* is valid when $\tau_y = S_y/\sqrt{3}$ and $G = 3E/8$; i.e., for homogeneous, isotropic ductile metals.

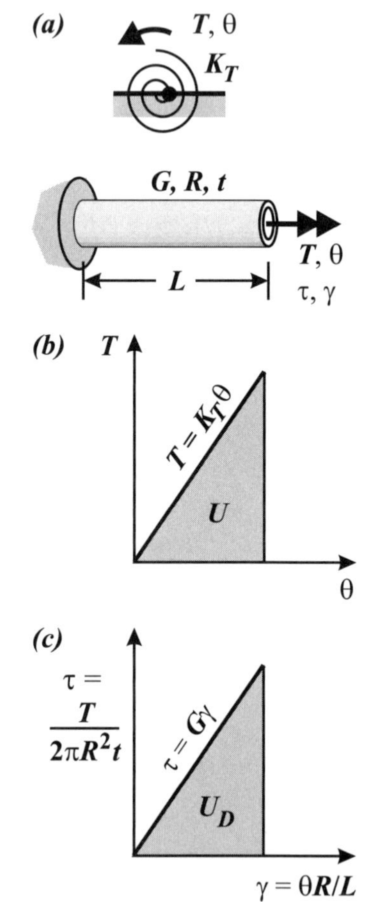

Figure 3.41. (a) Torsional spring and thin-walled circular shaft, both under torque *T*. **(b)** Energy stored in a torsional spring or a thin-walled shaft. **(c)** Elastic shear strain energy density stored in a thin-walled shaft.

Example 3.15 **Thin-Walled Circular Shaft in Torsion**

Given: A thin-walled circular shaft has average diameter $D = 2R = 150$ mm, thickness $t = 10.0$ mm, and length $L = 2.0$ m (*Figure 3.42*). The material is a high-strength steel with shear modulus $G = 82$ GPa and tensile yield strength $S_y = 600$ MPa. The applied torque is $T = 42.0$ kN·m.

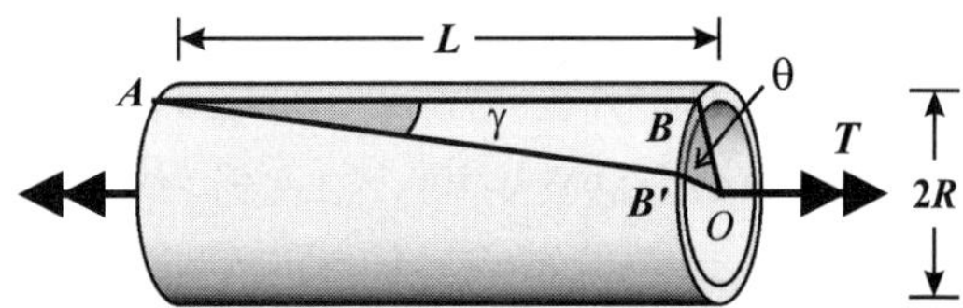

Figure 3.42. Thin-walled circular shaft.

Required: Determine (a) the shear stress in the shaft, (b) the angle of twist between the ends of the shaft, (c) the torsional stiffness, (d) the factor of safety against yielding, and (e) the elastic shear strain energy density.

Solution: *Step 1.* The shear stress in a thin-walled shaft is:

$$\tau = \frac{T}{2\pi R^2 t} = \frac{42.0 \times 10^3 \text{ N·m}}{2\pi (0.075 \text{ m})^2 (0.010 \text{ m})} = 118.8 \times 10^6 \text{ N/m}^2$$

Answer: $\tau = 119$ MPa

Step 2. The angle of twist θ is given by *Equation 3.52*:

$$\theta = \left(\frac{L}{2\pi R^3 t G}\right) T = \frac{\tau L}{GR} = \frac{(118.8 \times 10^6 \text{ Pa})(2.0 \text{ m})}{(82 \times 10^9 \text{ Pa})(0.075 \text{ m})} = 0.0386 \text{ rad}$$

Answer: $\theta = 0.0386$ radians $= 2.21°$

Step 3. The torsional stiffness K_T is given by *Equation 3.51*:

$$K_T = \frac{2\pi R^3 t G}{L} = \frac{2\pi (0.075 \text{ m})^3 (0.010 \text{ m})(82 \times 10^9 \text{ psi})}{2.0 \text{ m}}$$

Answer: $K_T = 1.09 \dfrac{\text{MN·m}}{\text{rad}} = 19.0 \dfrac{\text{kN·m}}{\text{degree}}$

The stiffness can also be calculated from the torque and the angle of twist:

$$K_T = \frac{T}{\theta} = \frac{42.0 \text{ kN·m}}{0.0386 \text{ rad}} = 1.09 \frac{\text{MN·m}}{\text{rad}}$$

Step 4. The *factor of safety* is the torque to cause yielding divided by the maximum working load. The magnitude of the torque when the material begins to yield, assuming $\tau_y = S_y/\sqrt{3}$, is:

$$T_y = (2\pi R^2 t)\tau_y = (2\pi R^2 t)(S_y/\sqrt{3})$$

$$= [2\pi (0.075 \text{ m})^2 (0.010 \text{ m})](600 \text{ MPa}/\sqrt{3}) = 122 \text{ kN·m}$$

The factor of safety is:

$$FS = \frac{T_y}{T} = \frac{122}{42.0}$$

Answer: $\underline{FS = 2.9}$

Alternatively, applying the factor of safety to the stresses gives:

$$FS = \frac{\tau_y}{\tau} = \frac{600/\sqrt{3}}{119} = 2.9$$

Step 5. The shear strain energy density is:

$$U_{D,\tau} = \frac{1}{2}\frac{(118.8 \times 10^6\,\text{Pa})^2}{(82 \times 10^9\,\text{Pa})} = 86.06 \times 10^3\,\frac{\text{N·m}}{\text{m}^3}$$

Answer: $\underline{U_{D,\tau} = 86.1\,\dfrac{\text{kN·m}}{\text{m}^3}}$

3.3 General Stress and Strain

An axial bar has uniform axial stress and strain; the thin-walled circular shaft also has uniform shear stress and shear strain. *Uniform* stress (strain) means that the stress (strain) is *the same at every point* in a body. For the axial and torsion members studied in this chapter, the stress state at any point can be visualized on 2D (square) elements since the stresses act in a single plane.

In general, however, *stresses and strains vary from point-to-point* in a body. Referring to *Figure 3.43*, it is evident that every point in the arbitrary body is affected differently; each point experiences a different stress, and thus a different strain. Stresses are, in general, 3D, but in many cases can be reduced to 2D.

A material *point* may be visualized as a very small cube or element. The stresses at that point are the average stresses acting on that cube, as shown in *Figure 3.44*. A cube is a natural stress element as it provides a built-in coordinate system (cartesian). Any object can be thought of as being made of many infinitesimally small cubes.

Each face of the **stress element** has three stresses acting on it: *a normal stress* σ, which is tensile (positive) or compressive (negative) and acts perpendicular to the face; and *two shear stresses* τ, which act parallel

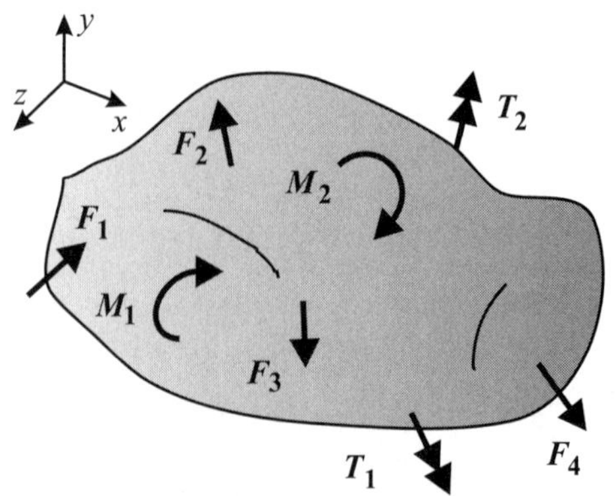

Figure 3.43. General body under various loads.

to the face in the other two directions. All of the stresses acting at a point are shown on the cube in *Figure 3.44*; they are drawn in their *positive senses* as defined below.

The values of the stresses specify a **state of stress**. Equilibrium of forces and moments on the cube can be used to show that there are six unique stresses:

$$\sigma_x, \sigma_y, \sigma_z, \tau_{xy}, \tau_{yz}, \tau_{zx}$$

Note that $\tau_{xy} = \tau_{yx}$, $\tau_{yz} = \tau_{zy}$, and $\tau_{zx} = \tau_{xz}$, since at any point, shear stresses on perpendicular planes are equal (complementary shear stress).

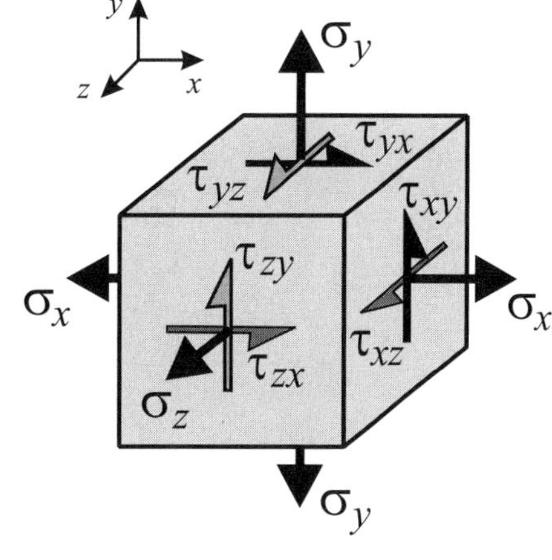

Figure 3.44. Cubic element showing all the stresses in their positive directions.

Stress Subscripts

The subscripts on the stress symbols represent:

1. the face on which the stress acts, and
2. the direction in which the stress acts.

Thus, τ_{xy} is a shear stress (τ) on the *x*-face acting in the *y*-direction. Likewise, τ_{yz} is a shear stress on the *y*-face in the *z*-direction.

While two subscripts may be used for normal stresses (e.g., σ_{xx}), this is somewhat redundant and the normal stresses are simply given here with one subscript, e.g., σ_x is the normal stress on the *x*-face in the *x*-direction.

Positive Sense/Sign Convention

Figure 3.44 shows all of the 3D stresses, each drawn in its **positive sense**.

A *positive* stress physically acts *on a positive face in a positive direction*, or *on a negative face in a negative direction*. Two such stresses are in equilibrium with each other. For example, the left-side of the element in *Figure 3.44* is a negative *x*-face, since an outward pointing normal vector to that face points in the negative *x*-direction. The normal stress on that face is drawn in the negative *x*-direction, in agreement with the definition of *positive sense*. If the value of σ_x is positive, it acts on both *x*-faces in the directions drawn; σ_x is tensile. If τ_{zy} is positive, it acts in the direction drawn (upward on the +*z*-face).

A *negative* stress physically acts *on a positive face in a negative direction*, or *on a negative face in a positive direction*. If the value of any stress is negative, it physically acts opposite to the direction drawn in *Figure 3.44*. A negative normal stress is a compressive stress. A negative τ_{zy} physically acts opposite drawn (downward on the +*z*-face).

Internal forces, torques, and moments are defined positive/negative in the same manner as the stresses. For example, the force in an axial member is positive (tensile), if it acts on a positive face in a positive direction, or on a negative face in a negative direction.

General Stress-Strain Relationship

For a *homogeneous* (the same at every point) and *isotropic* (the same in every direction) material, as are most materials considered in this text (e.g., metals, ceramics, etc.), the strains caused by the stresses are defined by Hooke's Law.

The normal strains are given by the general *3D Hooke's Law*:

$$\varepsilon_x = \frac{1}{E}[\sigma_x - v(\sigma_y + \sigma_z)]$$

$$\varepsilon_y = \frac{1}{E}[\sigma_y - v(\sigma_z + \sigma_x)] \qquad \text{[Eq. 3.58]}$$

$$\varepsilon_z = \frac{1}{E}[\sigma_z - v(\sigma_x + \sigma_y)]$$

The normal strain in any direction is caused by the stress in that direction – the *direct stress* – and by the normal stresses in the two perpendicular directions due to the Poisson effect.

The shear strains are:

$$\gamma_{yz} = \frac{\tau_{yz}}{G} \quad ; \quad \gamma_{zx} = \frac{\tau_{zx}}{G} \quad ; \quad \gamma_{xy} = \frac{\tau_{xy}}{G} \qquad \text{[Eq. 3.59]}$$

The stress–strain relationship can be written in matrix form as follows:

$$\begin{bmatrix} \varepsilon_x \\ \varepsilon_y \\ \varepsilon_z \\ \gamma_{yz} \\ \gamma_{zx} \\ \gamma_{xy} \end{bmatrix} = \begin{bmatrix} 1/E & -v/E & -v/E & 0 & 0 & 0 \\ -v/E & 1/E & -v/E & 0 & 0 & 0 \\ -v/E & -v/E & 1/E & 0 & 0 & 0 \\ 0 & 0 & 0 & 1/G & 0 & 0 \\ 0 & 0 & 0 & 0 & 1/G & 0 \\ 0 & 0 & 0 & 0 & 0 & 1/G \end{bmatrix} \begin{bmatrix} \sigma_x \\ \sigma_y \\ \sigma_z \\ \tau_{yz} \\ \tau_{zx} \\ \tau_{xy} \end{bmatrix} \qquad \text{[Eq. 3.60]}$$

The square matrix is the *flexibility matrix* **F** of the material. The inverse of the *flexibility matrix* is the *stiffness matrix* **K**:

$$\mathbf{K} = \mathbf{F}^{-1} \qquad \text{[Eq. 3.61]}$$

Matrix methods are useful in efficiently dealing with general states of stress and strain. Advanced topics in engineering mechanics, such as Structural Analysis, Continuum Mechanics, the Theory of Elasticity and Fluid Mechanics, all rely on matrix methods. Matrix methods are also useful in coordinate-system transformations. Stiffness and flexibility matrices (*Chapter 11*), as well as coordinate transformations (introduced briefly in *Chapter 8*), are used to analyze the response of composite materials in *Chapter 15*.

Special Stress and Strain States

Plane Stress

In many cases, loads applied to a system act in a single plane. When the stresses on an element act only in one plane (e.g., the *x–y* plane in *Figure 3.45a*), the state of stress is called *plane stress*. Plane stress generally exists when the loaded element is relatively thin in the out-of-plane direction. The plane stress element is usually drawn in 2D (e.g., in the *x–y* plane).

Taking the *z*-axis as the *out-of-plane* direction, then the non-zero stresses are: σ_x, σ_y, and τ_{xy}. The *out-of-plane* stresses (with *z* in their subscripts) are all zero: $\sigma_z = \tau_{yz} = \tau_{zx} = 0$. The non-zero strains are:

$$\varepsilon_x = \frac{1}{E}(\sigma_x - \nu\sigma_y)$$

$$\varepsilon_y = \frac{1}{E}(\sigma_y - \nu\sigma_x)$$

$$\varepsilon_z = \frac{-\nu}{E}(\sigma_x + \sigma_y)$$

[Eq. 3.62]

$$\gamma_{xy} = \frac{\tau_{xy}}{G}$$

Note that there is a non-zero normal strain in the *z*-direction ε_z due to the Poisson effect. However, the shear strains with components in the *z*-direction are zero ($\gamma_{yz} = \gamma_{zx} = 0$) since there are no out-of-plane shear stresses.

Plane Strain

For *plane strain* problems, no deformation is allowed in the out-of-plane direction. Plane strain generally occurs when the out-of-plane thickness is comparable to, or larger than, the in-plane dimensions (*Figure 3.47*), or where a part is constrained between rigid objects (objects made of a much stiffer material).

With the *z*-axis as the out-of-plane direction: $\varepsilon_z = \gamma_{yz} = \gamma_{zx} = 0$. In order for there to be no out-of-plane normal strain:

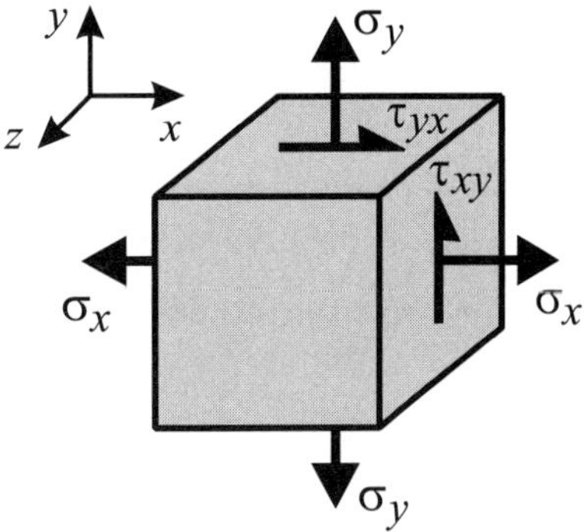

Figure 3.45. Plane stress element.

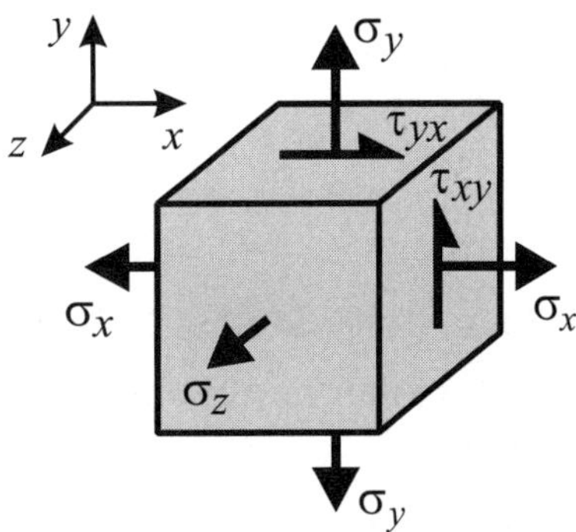

Figure 3.46. Plane strain element. $\sigma_z = \nu(\sigma_x + \sigma_y)$.

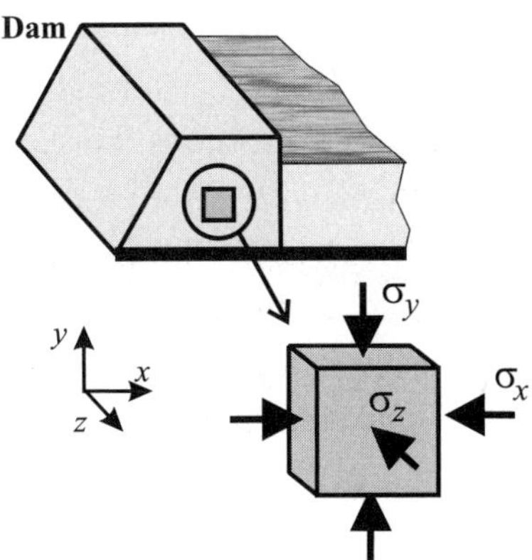

Figure 3.47. The *z*-direction of a dam has a large dimension relative to the in-plane dimensions. It is considered a plane strain case.

$$\varepsilon_z = \frac{1}{E}[\sigma_z - \nu(\sigma_x + \sigma_y)] = 0 \qquad \text{[Eq. 3.63]}$$

an out-of-plane normal stress must be applied (*Figure 3.46*):

$$\sigma_z = \nu(\sigma_x + \sigma_y) \qquad \text{[Eq. 3.64]}$$

When the sum of σ_x and σ_y is positive, the Poisson effect would cause the z-thickness to decrease. A positive stress σ_z is therefore required so that the z-thickness does not change.

The non-zero strains are:

$$\varepsilon_x = \frac{1}{E}[\sigma_x - \nu(\sigma_y + \sigma_z)]$$

$$= \frac{(1+\nu)}{E}[\sigma_x(1-\nu) - \nu\sigma_y]$$

$$\varepsilon_y = \frac{1}{E}[\sigma_y - \nu(\sigma_z + \sigma_x)] \qquad \text{[Eq. 3.65]}$$

$$= \frac{(1+\nu)}{E}[\sigma_y(1-\nu) - \nu\sigma_x]$$

$$\gamma_{xy} = \frac{\tau_{xy}}{G}$$

Other Stress States

The following are special stress states of interest (*Figure 3.48*):

- *Uniaxial stress*: Normal stress on an element that acts in only one direction; no shear stress acts.

- *Biaxial stress*: Normal stresses in two directions; no shear stress.

- *Triaxial stress*: Normal stresses in all three directions; no shear stress.

- *Hydrostatic stress*: Equal normal stresses in all three directions; no shear stress.

- *Pure shear*: Only shear stresses act on an element (usually 2D).

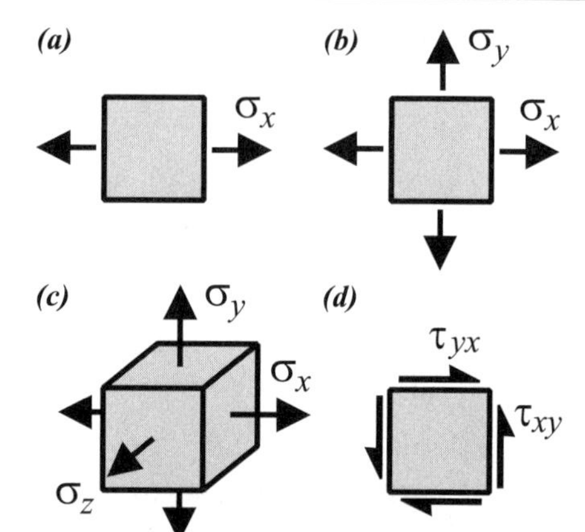

Figure 3.48. (a) Uniaxial stress. **(b)** Biaxial stress. **(c)** Triaxial stress. **(d)** Pure shear.

Chapter 4   Axial Members and Pressure Vessels

4.0 Introduction

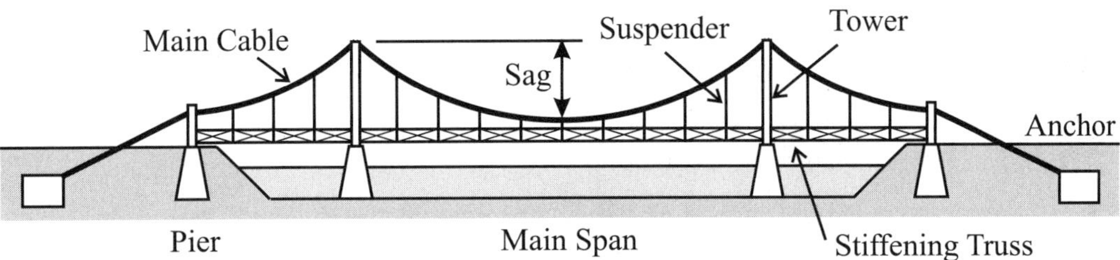

Figure 4.1. Main components of a suspension bridge. The main cables and suspenders are axial members in *tension*; the towers are primarily axial members in *compression*.

The strength of suspension bridges is provided by two *main cables* that extend from the anchor at one end, over the two central support towers, to the anchor at the opposite end (*Figure 4.1*). The cables are subjected to tensile loads. The roadway is supported by vertical cables called *suspenders* attached at one end to the main cable and at the other to the roadway's support system. The suspenders are loaded in tension. The towers, which hold up the main cable, are primarily loaded in compression.

Trusses, such as those used in bridges, space structures, and supermarket or warehouse roof systems, are made of straight members that are subjected to axial loads (*Figure 4.2*).

In automobiles, the transmission of power from engine to wheels is achieved with rotating shafts coupled by gears.

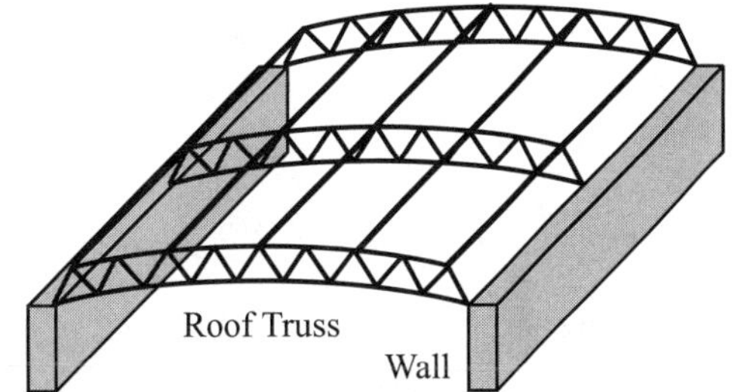

Figure 4.2. Roof trusses spanning the distance between two walls. The trusses are separated by purlins.

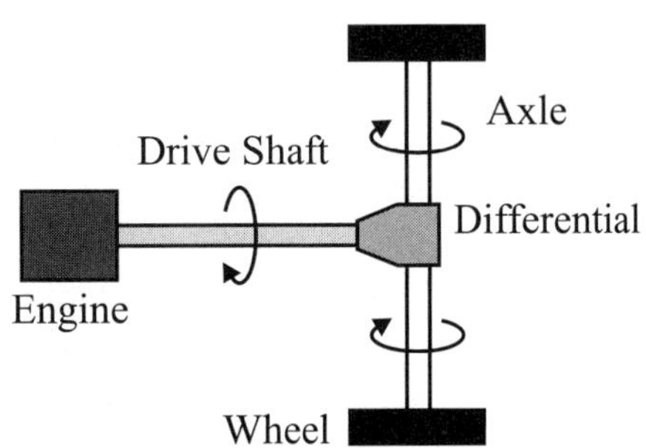

Figure 4.3. Schematic of the assembly of torsion members to propel automobiles (top view).

F.A. Leckie, D.J. Dal Bello, *Strength and Stiffness of Engineering Systems*, Mechanical Engineering Series, **83**
DOI 10.1007/978-0-387-49474-6_4, © Springer Science+Business Media, LLC 2009

Referring to *Figure 4.3*, at least three shafts are necessary to drive the rear wheels. The main drive shaft transmits torque from the engine to the differential gears. From the differential, two half-shafts transmit torque to the wheels to drive the vehicle forward.

The above examples illustrate that complex technical challenges can be solved by joining together components of simple shapes. A particularly vivid example of this philosophy is the Space Shuttle, which is a complex system constructed from many simple components (*Figure 4.4*).

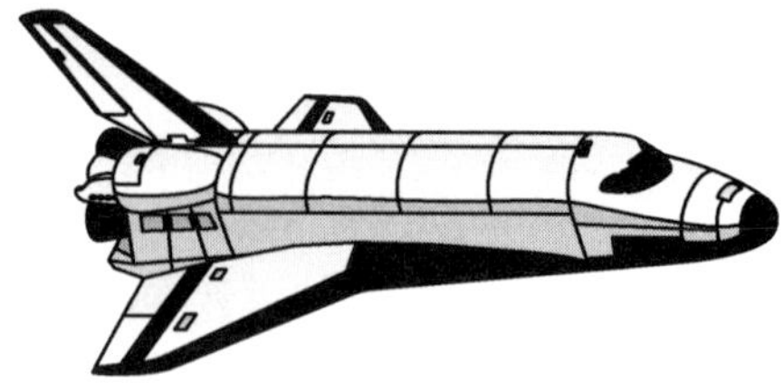

Figure 4.4. The Space Shuttle is a complex assembly of many basic components. Copyright ©2008 Dominic J. Dal Bello and licensors. All rights reserved.

The aims of this chapter are to determine:

1. the *stresses*, *strains*, and *elongations* of individual components,
2. the *stiffness* (force–deflection relationship) of an assembly of several axial components using compatibility and energy methods, and
3. the stresses in axial components due to thermal loading and basic stress concentrations.

The basic components studied in this chapter are **axial members** and **pressure vessels**. Each of these components supports a single load (force or pressure) that results in only *normal stresses* (axial or biaxial). Bolts and rivets, which join members together, are discussed in *Chapter 14*.

Methods of Analysis

There are two general approaches to determine the response of an assembly of components – the *force method* and the *displacement method*.

Force Method

The steps of the *force method* are as follows:

1. Apply the conditions of *equilibrium* to the applied loads, reactions, and internal forces to determine the stresses in each component; e.g., for a bar in tension: $\sigma = P/A$.
2. Calculate the strains from the stresses using *Hooke's Law*; e.g., $\varepsilon = \sigma/E$.
3. Integrate the strain in each component to find its elongation; e.g., $\Delta = \varepsilon L$.
4. Relate the elongation of each component to the overall displacement or deflection of the assembly; the elongation of each member must be **compatible** with the elongations of the other members.

Displacement Method

The steps of the **displacement method** follow the reverse sequence:

1. Determine the elongation of each member in terms of the overall displacement of the assembly.
2. Determine the strain in each component in terms of its elongation; e.g., $\varepsilon = \Delta/L$.
3. Calculate the stresses from the elastic law; e.g., $\sigma = E\varepsilon$.
4. Apply the conditions of equilibrium to determine the internal and external forces; e.g., $P = \sigma A$.

For any given problem, either method can be used; which method to chose is a matter of convenience.

Types of Systems

There are two types of systems: *statically determinate* and *statically indeterminate systems*.

Statically Determinate Systems

In *statically determinate* systems, all of the reactions and internal forces can be calculated using only the equations of *equilibrium* (statics). The *force method* is then the primary method to calculate the stresses, strains, and displacements.

The lamp supported by the two wires in *Figure 4.5* is a *statically determinate* system (recall *Example 3.3*). Equilibrium in the *x*- and *y*-directions on the FBD of point *B* solves for the forces in both wires.

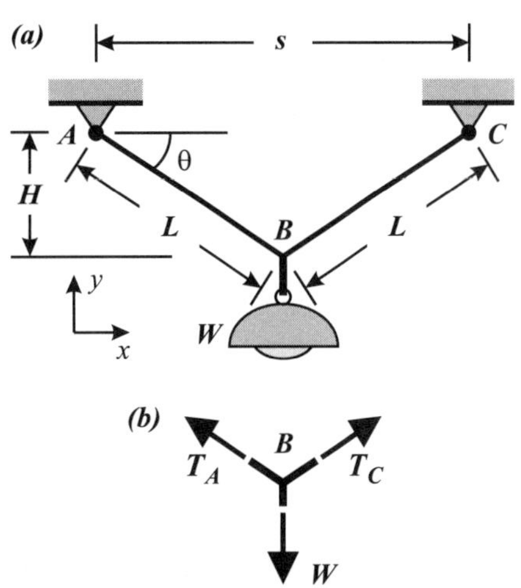

Figure 4.5. The forces in the wires supporting the lamp are found by applying equilibrium at point *B*.

Statically Indeterminate (Redundant) Systems

In *statically indeterminate systems* (*redundant systems*), the reactions and internal forces cannot be calculated from statics alone. The rigid bar system in *Figure 4.6* is an example of such a system.

Bar *AB* is rigid, while identical bars *C* and *D* are elastic. Force *P* is applied at point *B* which displaces upward distance δ. The FBD of bar *AB* is given in *Figure 4.6b*. There are two useful equilibrium equations:

$$\sum F_y = 0: \quad R_A - R_C - R_D + P = 0$$

$$\sum M_A = 0: \quad -R_C\left(\frac{L}{3}\right) - R_D\left(\frac{2L}{3}\right) + PL = 0$$

[Eq. 4.1]

but three unknowns: reaction R_A and member forces R_C and R_D. There are more unknowns than equilibrium equations; the system cannot be solved by statics alone. The forces beyond those that are required for equilibrium are called *redundant forces*. This problem is solved in detail in *Example 4.9*.

A *statically indeterminate* system may be solved using the *force method* by considering the redundant forces as temporarily known, and solving the statics problem in terms of the known and redundant forces. The necessary relationships between the elongations (deflections) of various members are then used to relate the redundant forces to the known forces. This solution method for statically indeterminate problems is demonstrated in *Examples 4.6–4.8.*

Statically indeterminate systems are usually best solved using the *displacement method,* where the relationships between the elongations of the members are found first. In *Figure 4.6,* since *AB* is rigid, bar *C* must elongate by $\Delta_C = \delta/3$ and bar *D* by $\Delta_D = 2\delta/3$. The displacements of the individual components must be *compatible* with the overall system. **Compatibility** requires the system to continue to fit together under the applied load. Since $\Delta_D = 2\Delta_C$, and bars *C* and *D* are identical, then the force in bar *D* must be twice that of bar *C*:

$$R_D = 2R_C \qquad \text{[Eq. 4.2]}$$

This relationship provides the third equation required to solve the three unknowns in *Equation 4.1.*

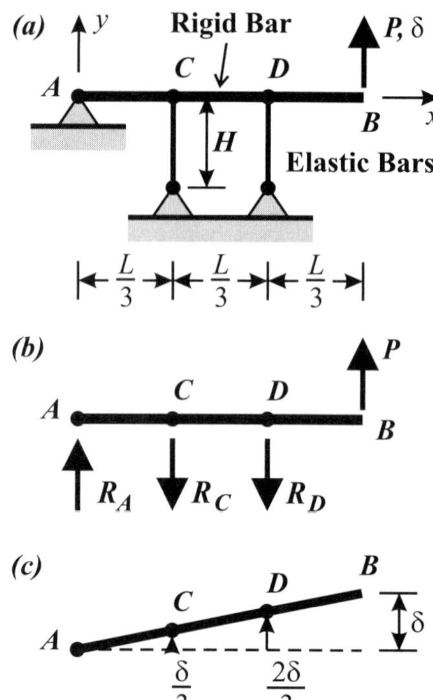

Figure 4.6. **(a)** Rigid bar *AB* under load *P.* **(b)** FBD of *AB.* **(c)** *AB* pivots about point *A.* The displacements (exaggerated) of points *C* and *D* are proportional to tip displacement δ.

Modern computational methods, such as the *finite element method,* are developed on the basis of the displacement method and energy methods. While the force method is often the primary solution method for statically determinate systems, the displacement method can also be used.

The following two sections (*Sections 4.1* and *4.2*) illustrate the *force method* and the *displacement method* with a series of examples analyzing axial members.

4.1 Axial Members – Force Method

Statically Determinate Systems

Examples 4.1–4.5 are *statically determinate* systems that are solved using the *force method.*

Example 4.1 Uniform Bar in Tension

Given: Load P is applied at the ends of a bar of constant cross-section A and length L (*Figure 4.7*). The bar is made of an elastic material with modulus E.

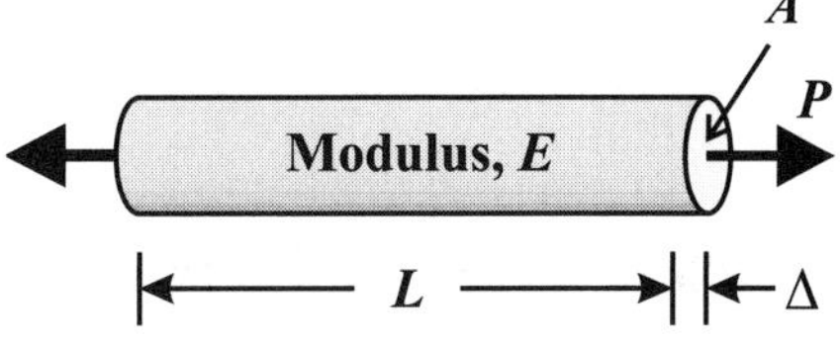

Figure 4.7. An elastic bar under load P.

Required: Determine the *stress* σ and *strain* ε in the bar, and its elongation Δ in terms of $P, L, A,$ and E.

Solution: Following the steps of the *force method*:

Step 1. Equilibrium. The relationship between the applied force and the internal stress is:

$$P = \sigma A$$

Answer: $\sigma = \dfrac{P}{A}$

Step 2. Elasticity (*Stress–Strain relationship*). From Hooke's Law, the strain is:

$$\varepsilon = \sigma/E$$

Answer: $\varepsilon = \dfrac{P}{AE}$

Step 3. Strain–Elongation. The elongation as a function of strain is:

$$\Delta = \varepsilon L = \frac{\sigma L}{E}$$

Answer: $\Delta = \dfrac{PL}{AE}$

The elongation Δ of a bar of constant cross-sectional area A and constant modulus E, subjected to constant axial force P (positive in tension, negative in compression) along its entire length L is:

$$\Delta = \frac{PL}{AE} \qquad \text{[Eq. 4.3]}$$

The force method directly gives the ***flexibility*** F of the bar, where $\Delta = FP$. Thus:

$$F = \frac{L}{AE} \qquad \text{[Eq. 4.4]}$$

The ***stiffness*** K of the bar is defined: $P = K\Delta$. Thus:

$$K = \frac{EA}{L} \qquad \text{[Eq. 4.5]}$$

Equations 4.3–4.5 are valid when P, A, and E are all constant over the entire length of the bar L, resulting in a uniform strain.

To determine the elongation when P, A, and/or E change over the length of the bar, the bar must be broken up into shorter segments L_i, where P, A, and E are all constant. The total elongation is the sum of the elongations of each segment Δ_i:

$$\Delta = \sum \Delta_i = \sum \frac{P_i L_i}{A_i E_i} \qquad \text{[Eq. 4.6]}$$

where P_i, A_i, E_i, and L_i are the force, area, modulus, and length of the ith segment.

Example 4.2 Rods Supporting Elevated Walkways

Given: A set of six stepped rods supports two elevated walkways overlooking a hotel atrium (*Figure 4.8*). The upper segment of each rod, AB, has area A_1, length L_1, and modulus E_1. The lower segment, BC, has properties A_2, L_2, and E_2. The load of the upper walkway is supported at joint B, the junction of the two bars. The upper walkway applies a load to each rod of W_U at point B, and the lower walkway applies a load to each rod of W_L at point C (*Figure 4.8c*). Each rod is assumed to carry the same load.

Required: Determine (a) the internal stresses in each segment of the rod AB and BC and (b) the total elongation Δ of stepped rod AC.

Solution: *Step 1. Equilibrium* of rod ABC requires the reaction to equal the applied loads (*Figure 4.8d*):

$$T = W_U + W_L$$

The internal loads in the rods are found by taking cuts between A and B, and between B and C (*Figures 4.8e, f*). The internal loads in the upper and lower segments of the stepped rod are:

$$P_{AB} = T = W_U + W_L$$

$$P_{BC} = W_L$$

The stresses in the segments are then:

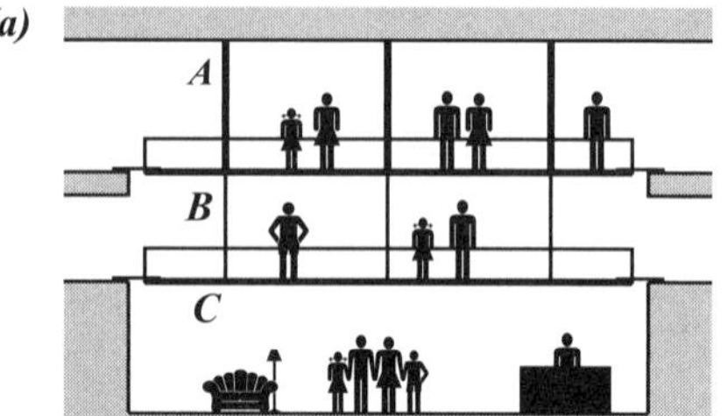

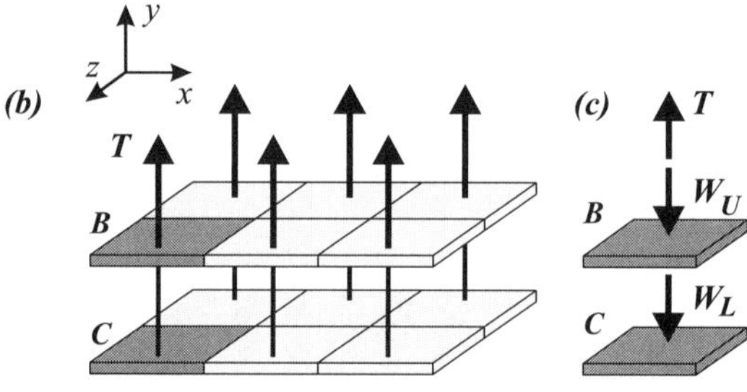

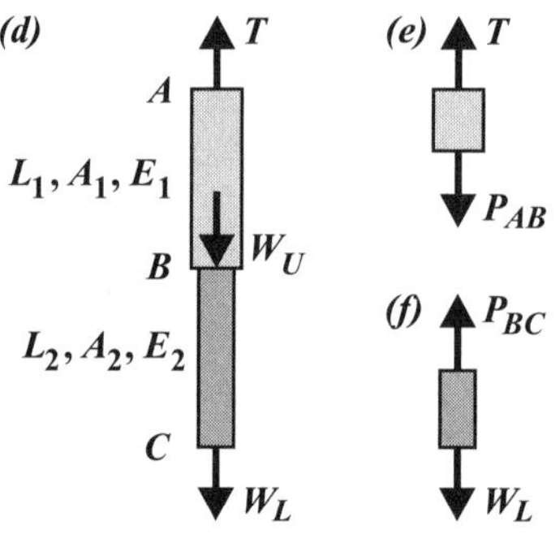

Figure 4.8. **(a)** Two elevated walkways overlooking a hotel atrium. **(b)** A sketch of the walkways to determine the reaction load T at the ceiling. The tributary area contributing to the load on the left-front rod is shaded on each walkway. **(c)** Forces and tributary areas of each rod. **(d)** FBD of each rod. The load of the upper walkway W_U is applied at point B, and the load of the lower walkway W_L is applied at point C. **(e)** FBDs to determine the internal forces in the upper and **(f)** lower segments of the rods.

$$\textit{Answer: } \sigma_{AB} = \frac{P_{AB}}{A_1} = \frac{W_U + W_L}{A_1} \quad \text{and} \quad \sigma_{BC} = \frac{P_{BC}}{A_2} = \frac{W_L}{A_2}$$

Step 2. Elasticity relates the stresses to the strains in each segment of the rod:

$$\varepsilon_{AB} = \frac{\sigma_{AB}}{E_1} = \frac{W_U + W_L}{A_1 E_1} \quad \text{and} \quad \varepsilon_{BC} = \frac{\sigma_{BC}}{E_2} = \frac{W_L}{A_2 E_2}$$

Step 3. Strain–elongation gives the elongation of each segment:

$$\Delta_{AB} = \varepsilon_{AB} L_{AB} = \frac{(W_U + W_L)L_1}{A_1 E_1} \quad \text{and} \quad \Delta_{BC} = \frac{W_L L_2}{A_2 E_2}$$

Step 4. Elongation–displacement. The total elongation Δ is the sum of the elongations of each segment:

$$\textit{Answer: } \delta = \Delta_{AB} + \Delta_{BC} = \frac{(W_U + W_L)L_1}{A_1 E_1} + \frac{W_L L_2}{A_2 E_2}$$

Note that point B displaces (moves) downward by $\delta_B = \Delta_{AB}$, and point C displaces downward by $\delta_C = \Delta_{AB} + \Delta_{BC}$.

Continuously Varying Stress in Axial Members

The tapered column in *Figure 4.9* supports a compressive load F. While the force through the column is constant, the cross-sectional area is not. Hence, stress and strain vary continuously over the length of the member.

Even though the cross-section may be constant over the length of a rod or column, stress can still vary continuously. This is generally due to external friction-like forces acting on the surface of the member, such as a foundation pile in the ground, fibers in the matrix of a composite, or a nail being extracted from a piece of wood. Friction forces transfer the internal axial force from the rod to the surrounding material (e.g., dirt, composite matrix, wood). The internal force of a column also varies continuously when its self-weight is taken into account.

Examples with varying stress follow.

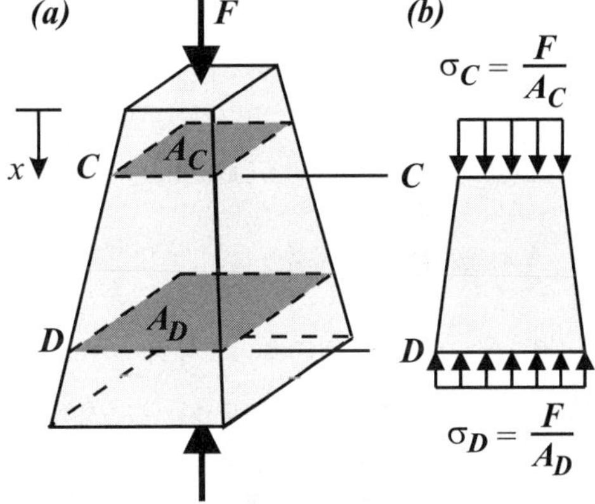

Figure 4.9. (a) 3D view of tapered column. **(b)** Side-view of column showing stresses at levels *C* and *D*.

Example 4.3 Tapered Column under Compressive Load

Given: A tapered concrete column has a square cross-section that varies from side $a = 125$ mm at the top to side $2a = 250$ mm at the bottom (*Figure 4.10a*). The total length of the column is $L = 1.20$ m, and it carries a compressive load of $F = 200$ kN. The modulus of concrete is $E = 30$ GPa. Neglect the weight of the concrete.

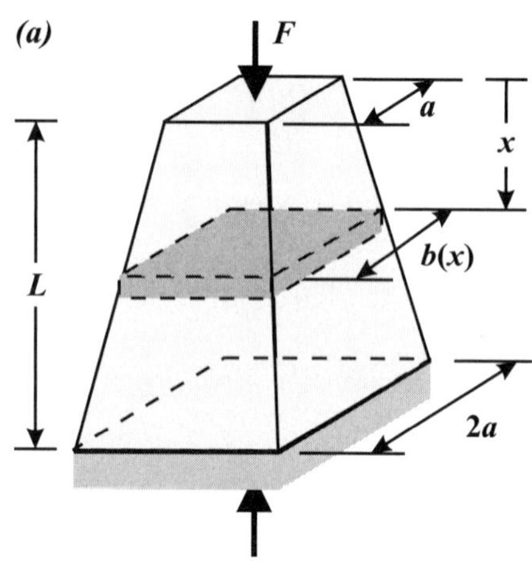

Since the force is constant, and area increases from top $(x = 0)$ to bottom $(x = L)$, the axial compressive stress varies; it is maximum at the top and minimum at the bottom.

Required: Determine (a) the variation of stress in the column $\sigma(x)$, (b) the variation of strain $\varepsilon(x)$, and (c) the change in length of the column Δ.

Figure 4.10. (a) The tapered column under constant load.

Solution: Since the internal axial force is known throughout, $P(x) = -F$, the *force method* is utilized.

Step 1. Equilibrium requires the force–stress relationship at any cross-section to be:

$$P(x) = \sigma(x)A(x) = -F$$

To determine how the stress varies, $A(x)$ must be determined. Since area is always positive, $\sigma(x)$ will always be negative (compressive).

At any section distance x from the top of the column, the length of the side of the cross-section is:

$$b(x) = a + a\frac{x}{L} = a\left(1 + \frac{x}{L}\right) = 0.125\left(1 + \frac{x}{1.2}\right) \text{ m}$$

Side $b(x)$ increases linearly from a (at $x = 0$) to $2a$ (at $x = L = 1.2$ m).

Hence, the stress at any cross-section is

$$\sigma(x) = \frac{P(x)}{A(x)} = \frac{-F}{[b(x)]^2} = \frac{-200 \times 10^3}{[0.125(1 + x/1.2)]^2} \frac{\text{N}}{\text{m}^2}$$

$$\textit{Answer: } \sigma(x) = \frac{-12.80}{(1 + x/1.2)^2} \text{ MPa}$$

Step 2. Stress–Strain. Young's modulus for concrete is $E = 30$ GPa, so the strain at any position x is:

$$\varepsilon(x) = \frac{\sigma(x)}{E}$$

$$\textit{Answer: } \varepsilon(x) = \frac{-426.7 \times 10^{-6}}{(1 + x/1.2)^2}$$

Step 3. Strain–Elongation. To determine the displacement $u(x)$ of a cross-section at location x, it is necessary to do some calculus. A slice of the column, dx thick, is considered, where dx is so small that $A(x)$ is essentially *constant* (*Figure 4.10b*). Thus, over length dx, the force, area, and modulus are all constant. Slice dx is a very short axial member under constant force $P(x) = -F$.

Under compression, the top face of slice dx displaces downward $u(x)$ and the bottom face displaces downward $u(x)+du$ (*Figure 4.10c*). The change in length of the slice is:

$$du = \varepsilon(x)dx = \frac{P(x)}{A(x)E(x)}dx$$

Step 4. Elongation–Displacement. Performing the indefinite integral of du gives the downward displacement of the cross-section at any x:

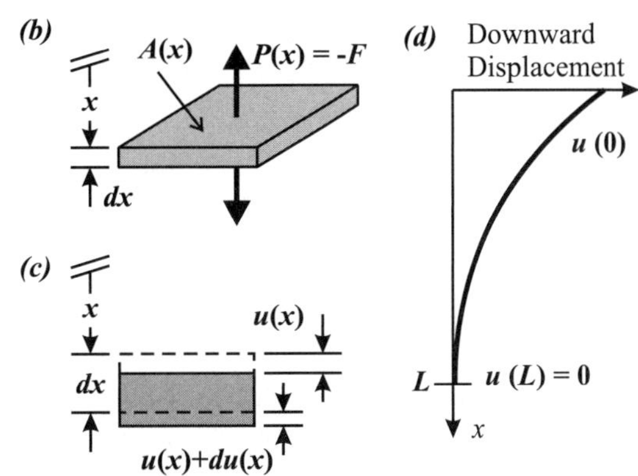

Figure 4.10. **(b)** A slice of the column dx. Over length dx, the area is considered to be constant. **(c)** The top of the slice moves down by $u(x)$, the bottom by $u(x)+du$; the strain of the slice is then $\varepsilon(x) = du/dx$. **(d)** Qualitative plot of the downward displacement $u(x)$; the column is fixed at the bottom ($x = L$), so $u(L) = 0$. The displacement of the top is the sum of the elongations of all infinitessimal slices dx.

$$\int du = \int \varepsilon(x)dx = u(x) = -426.7 \times 10^{-6} \int \left(1 + \frac{x}{1.2}\right)^{-2} dx = \frac{0.512 \times 10^{-3}}{(1 + x/1.2)} + C$$

The constant of integration C is found by applying the condition that the displacement of the base of the column – the cross-section at $x = 1.2$ m – is zero:

$$u(x = 1.2 \text{ m}) = 0 = \frac{0.512 \times 10^{-6}}{2} + C \rightarrow C = -0.256 \times 10^{-3}$$

Hence, the displacement $u(x)$ of any cross-section distance x from the top of the column is:

$$u(x) = 0.256 \times 10^{-3}\left[\frac{2}{(1 + x/1.2)} - 1\right] \text{ m} = 0.256\left[\frac{2}{(1 + x/1.2)} - 1\right] \text{ mm}$$

The change in length of the entire column corresponds to the downward displacement of the top cross-section $x = 0$:

$$u(0) = 0.256 \text{ mm}$$

A positive value of u means the cross-section moves downward (in the $+x$-direction), which is the case for all sections. The column shortens, so the change in length is negative:

Answer: $\Delta = -0.256$ mm

Example 4.4 Extraction of a Nail

Given: A nail is pulled out of a piece of wood (*Figure 4.11a*). The nail diameter is $D = 2R$, and its embedded length is L. Force T applied to extract the nail is resisted by an *interfacial* shear stress τ acting between the nail surface and the wood; τ is assumed to be constant (*Figure 4.11b*). The tip of the nail provides no resistance to pull-out. The nail begins to slide (pull-out) when the sliding stress τ_s is reached (τ_s is the stress to overcome the nail–wood friction).

Required: Determine the change in length of the embedded part of the nail just before it starts to slide, in terms of τ_s, L, D, and E.

Solution:

Step 1. Equilibrium. Applying equilibrium to *Figure 4.11b* gives the relationship between force T and interfacial shear stress τ:

$$T = \tau A_S = \tau(\pi D L)$$

A_S is the surface area of the nail supporting the shear stress.

The variation in axial force $P(x)$ is found by taking a FBD of the nail from 0 to x (*Figure 4.11c*). The internal force at any distance x from the bottom is therefore:

$$P(x) = \tau(\pi D x)$$

At $x = L$, $P = T = \tau[\pi D L]$, so $P(x)$ can be rewritten:

$$P(x) = T\frac{x}{L}$$

The corresponding stress variation is:

$$\sigma(x) = \frac{P(x)}{A} = \frac{4}{\pi D^2}\left(T\frac{x}{L}\right)$$

Step 2. Stress–Strain gives:

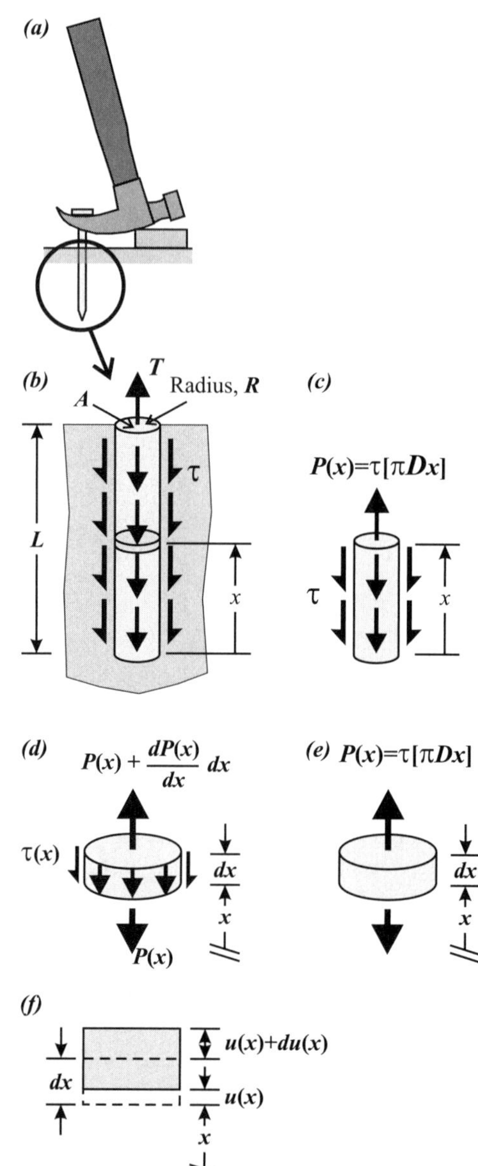

Figure 4.11. **(a)** Hammer extracting nail. **(b)** The shear stress resisting pull-out. **(c)** A slice of the nail with internal force $P(x)$ and shear stress τ. **(d)** A slice of the nail dx. **(e)** Length dx is taken so small that the axial force is considered to be constant. **(f)** The bottom of the slice moves up by $u(x)$, the top by $u(x)+du$; its strain is then $\varepsilon(x) = du/dx$.

$$\varepsilon(x) \;=\; \frac{\sigma(x)}{E} \;=\; \frac{4Tx}{E\pi D^2 L}$$

Step 3. Strain–Elongation. Consider a thin slice dx (*Figure 4.11d*), so thin that the change of $P(x)$ over dx can be neglected (*Figure 4.11e*). The change in length of slice dx is:

$$du \;=\; \varepsilon(x)dx \;=\; \frac{P(x)}{A(x)E(x)}dx$$

as shown in *Figure 4.11f*.

Step 4. Elongation–Displacement. Taking the indefinite integral of du gives the displacement (movement) of the cross-section at x:

$$u(x) \;=\; \int \varepsilon(x)dx \;=\; \int \frac{4Tx}{E\pi D^2 L}dx \;=\; \frac{2Tx^2}{E\pi D^2 L} + C$$

Constant C is found by noting that the displacement at the bottom ($x = 0$) is zero, since the nail has yet to move. Hence $C = 0$. The expression for the displacement at any point x, before the nail slides, is:

$$u(x) \;=\; \frac{2Tx^2}{E\pi D^2 L}$$

The change in length of the embedded part of the nail is equal to the displacement of the cross-section at $x = L$:

$$\Delta \;=\; u(L) \;=\; \frac{2TL}{E\pi D^2}$$

When the nail begins to slide, $\tau = \tau_s$, so $T_s = \tau_s(\pi DL)$, and the change of the embedded length is:

$$\Delta = u(L) \;=\; \frac{2T_s L}{E\pi D^2}$$

$$Answer\!: \; \Delta = \frac{2\tau_s L^2}{ED}$$

Compatibility

Under load, the components of a system must deform and deflect in such a way that the system remains intact. The axial members of a loaded truss must each elongate in such a way that they continue to be pinned together. In other words, the members of an assembly are geometrically constrained to deform together. This condition is called **compatibility**. Applying the concept of compatibility is a key step when solving systems where several components are joined together, whether the systems are *statically determinate* or *statically indeterminate*.

Example 4.5 Hanging Lamp

Given: A lamp weighing $W = 14.0$ lb is supported by two wires, both of length $L = 5.0$ ft and diameter $D = 0.10$ in. (*Figure 4.12*). The distance between the two cable mounts is $s = 8.0$ ft so that point B is $H = 3.0$ ft below horizontal line AC. The wires are made of steel with modulus $E = 30 \times 10^6$ psi and yield strength $S_y = 50$ ksi. Assume the wire below point B is rigid and of sufficient strength.

Required: Determine (a) the downward displacement of the lamp δ (i.e., of point B) due to its own weight, (b) the stiffness of the wire assembly in the vertical direction, $K = W/\delta$, (c) the load $W = W_y$ when yielding occurs in the wires, and (d) the factor of safety against yielding for the lamp load $W = 14$ lb.

Figure 4.12. (a) Lamp supported by wires. **(b)** FBD of point B.

Solution: *Step 1.* Vertical *equilibrium* of point B and symmetry (*Figure 4.12b*) require that:

$$T = T_A = T_C = \frac{W}{2 \sin \theta} = \frac{W}{2(H/L)} = \frac{14.0}{2(3/5)} = 11.67 \text{ lb}$$

The stress in each wire is thus:

$$\sigma = \frac{T}{A} = \frac{4T}{\pi D^2} = \frac{4(11.67 \text{ lb})}{\pi(0.100 \text{ in.})^2} = 1.486 \text{ ksi}$$

Step 2,3 (combined). The *stress–strain* and *strain–displacement* steps are combined into a single *force–elongation* step. The elongation Δ of each wire is:

$$\Delta = \frac{TL}{AE} = \frac{\sigma L}{E} = \frac{(1486 \text{ psi})(60 \text{ in.})}{(30 \times 10^6 \text{ psi})} = 0.002972 \text{ in.}$$

Step 4. Elongation–displacement. Compatibility requires that wires AB and BC elongate such that their lower endpoints – joint B on each member – stay connected. Wire elongations Δ must be *compatible* with the overall system displacement δ (*Figure 4.12c*). By symmetry, point B moves downward δ and the wires move from their dotted position ABC to the solid position $AB'C$ (greatly exaggerated).

Figure 4.12d magnifies the displacement. If AB were to elongate by itself, point B would move to D. If BC were to elongate by itself, point B would move to D'. Since AB and BC must remain connected, endpoints D and D' adjust to position B'. This is done by AD rotating clockwise and CD' rotating counterclockwise, each by $\delta\theta$. Since Δ is very small compared to length L, $\delta\theta$ is very small, and the changes in wire slopes are negligible; i.e., the slopes of *AB* and *AB'* are the same, which means the forces in the wires do not change.

The dashed circular arc *DB'* in *Figure 4.12d* is the path of point *D* as *AD* rotates. The arc is normal to radius *AD*. Solid line *DB'* is also normal to *AD*. Since radius *AD* is very large compared to both Δ and distance *DB'*, then the difference between dashed arc *DB'* and solid line *DB'* over the distance of interest is negligible. A similar argument can be made for dashed arc and solid line *D'B'*.

Thus, point *B'* can be located by constructing perpendiculars to *AD* and *CD'*. Point *B'* is the intersection of the perpendiculars.

Due to symmetry, either triangle *BDB'* or *BD'B'* can be used to determine δ, the new position of the joint below point *B*.

The relationship between displacement δ and wire elongation Δ is (*Figure 4.12d*):

$$\delta = \frac{\Delta}{\sin\theta} = \frac{0.002972 \text{ in.}}{3/5}$$

Answer: δ = 0.00495 in.

Step 5. The *stiffness* is:

$$K = \frac{W}{\delta} = \frac{14 \text{ lb}}{0.00495 \text{ in.}}$$

Answer: *K* = 2830 lb/in.

Step 6. The *load to yield* each wire is:

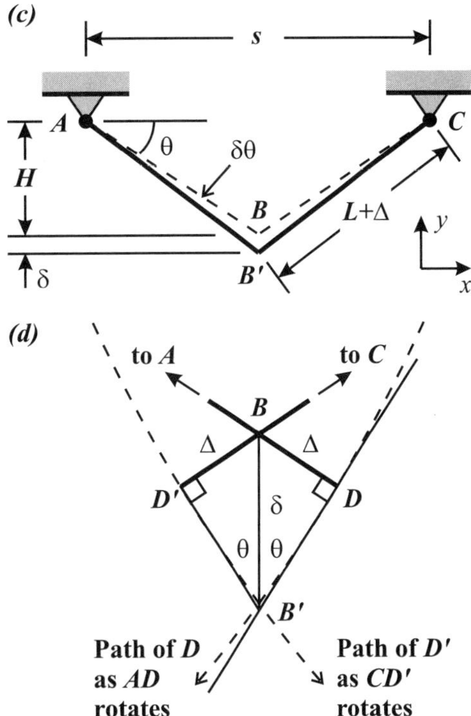

Figure 4.12. **(c)** Each wire elongates by Δ as point *B* deflects downward distance δ to point *B'*. **(d)** Geometry of the deflection: Δ = *BD* = δsinθ.

$$T_y = S_y A = (50 \text{ ksi})\frac{\pi(0.10 \text{ in.})^2}{4} = 393 \text{ lb}$$

By symmetry, the wires yield at the same time, so the load that causes yielding W_y is:

$$W_y = 2T_y \sin\theta = 2(393 \text{ lb})(3/5)$$

Answer: W_y = 471 lb

Step 7. The *factor of safety* against yielding is the ratio of the load at yield (the failure load) to the working load (here the lamp's weight):

$$FS = \frac{W_y}{W} = \frac{471 \text{ lb}}{14.0 \text{ lb}}$$

Answer: FS = 33

The following example illustrates the compatibly principle on a more complex system.

Example 4.6 Truss Deflection

Given: Aluminum truss *ABC* is loaded at joint *B* by a point load of $F = 10.0$ kips (*Figure 4.13a*). The cross-sectional areas of the bars are: $A_{AB} = 0.5$ in.^{2} and $A_{BC} = 0.6$ in.^{2}. The modulus of aluminum is $E = 10,000$ ksi.

Required: Determine the horizontal and vertical displacements of joint *B*, *u*, and *v*.

Solution: *Step 1. Equilibrium* of point *B* (*Figure 4.13b*) requires that:

$$\sum F_x = -\frac{1}{\sqrt{5}}P_{AB} + \frac{3}{5}P_{BC} = 0$$

$$\sum F_y = \frac{2}{\sqrt{5}}P_{AB} + \frac{4}{5}P_{BC} - F = 0$$

Solving for P_{AB} and P_{BC} gives:

$$P_{AB} = 6708 \text{ lb}$$

$$P_{BC} = 5000 \text{ lb}$$

Steps 2,3. Elongation–strain–force. From geometry, the length of each bar is:

$$L_{AB} = \sqrt{5}(1.5 \text{ ft}) = 40.25 \text{ in.}$$

$$L_{BC} = (5/3)(3.0 \text{ ft}) = 60.0 \text{ in.}$$

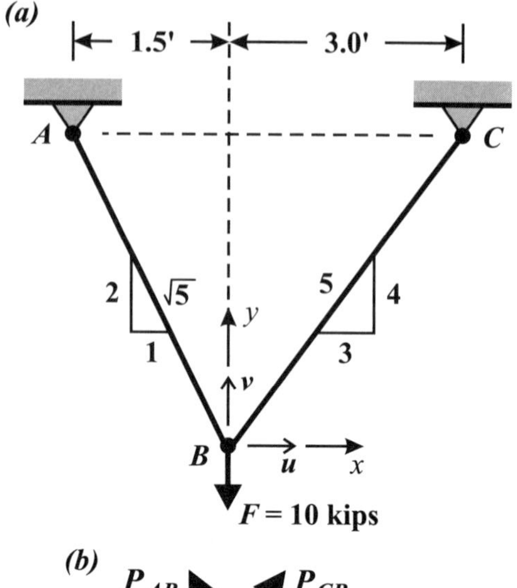

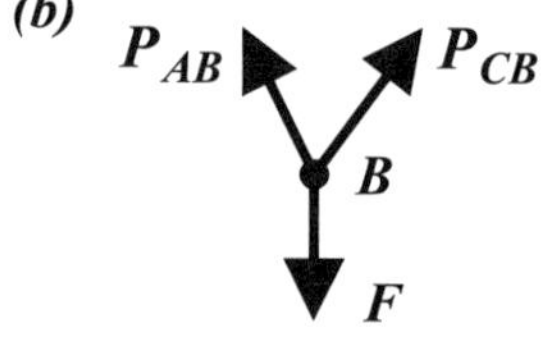

Figure 4.13. (a) Truss *ABC* loaded at joint *B*. **(b)** FBD of joint *B*.

The elongation of each bar is then (*Figure 4.13c*):

$$\Delta_{AB} = \frac{P_{AB}L_{AB}}{A_{AB}E} = \frac{(6708 \text{ lb})(40.25 \text{ in.})}{(0.5 \text{ in.}^2)(10 \times 10^6 \text{ psi})} = 0.0540 \text{ in.}$$

$$\Delta_{BC} = \frac{P_{BC}L_{BC}}{A_{BC}E} = \frac{(5000 \text{ lb})(60.0 \text{ in.})}{(0.6 \text{ in.}^2)(10 \times 10^6 \text{ psi})} = 0.0500 \text{ in}$$

Step. 4. Compatibility. The change in length of each bar is small (~0.14%), so the change in slope of each bar is negligible. Due to the forces, bar *AB* extends by $\Delta_{AB} = BD$ and bar *BC* extends by $\Delta_{BC} = BD'$ (*Figure 4.13c*).

Compatibility requires that the ends of each bar, points *D* and *D'*, continue to fit together. In order for this to happen, *AD* and *CD'* must rotate towards each other, point *D* moving on a circular arc of radius *AD* and point *D'* moving on a circular arc of radius *CD'*. The circular arcs are perpendicular to *AD* and *CD'*. Because the elongations are small, the required rotations are small. Over the distance of interest, the circular path of point D, and a line perpendicular to AD at D, are the same; the difference is negligible. Thus, D moves along a line perpendicular to AD. Likewise, point D' moves on a line perpendicular to CD' (*Figure 4.13c*).

Point B' can be located by constructing perpendiculars to AD and CD'. Point B' is the intersection of the perpendiculars. To determine this point requires a bit of geometry using *Figures 4.13c* and *d*.

A vertical dashed line is dropped from point B, and elongations $\Delta_{AB} = BD$ and $\Delta_{BC} = BD'$ are constructed from point B. Perpendiculars are drawn from D and D'; the intersection of these lines is the new point B'. From the scale drawing in *Figure 4.13c*, joint B moves down and to the left; how far it moves must be determined.

Line DB' crosses the vertical dashed line at point G. Right triangle BDG is used to determine the vertical distance of point G below point B:

$$BG = \frac{2}{\sqrt{5}}BD + \frac{1}{2}\left(\frac{1}{\sqrt{5}}BD\right)$$

$$= \frac{\sqrt{5}}{2}BD = 0.060374 \text{ in.}$$

Line $D'B'$ crosses the vertical dashed line at point H. From the right triangle $BD'H$:

$$BH = \frac{4}{5}BD' + \frac{3}{4}\left(\frac{3}{5}BD'\right)$$

$$= \frac{5}{4}BD' = 0.0625 \text{ in.}$$

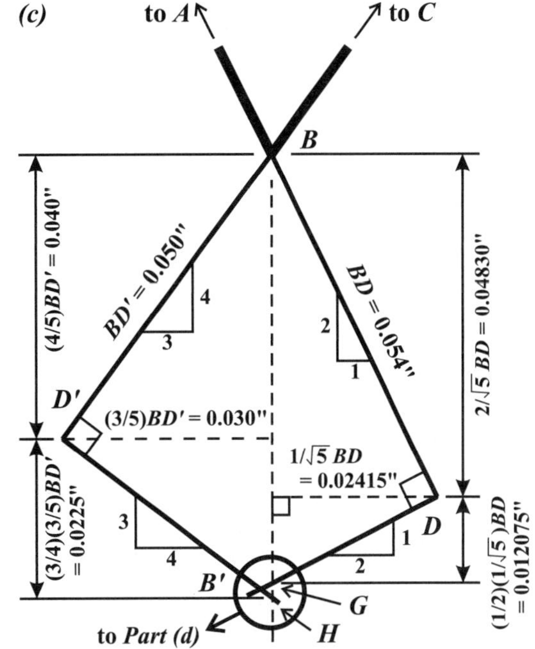

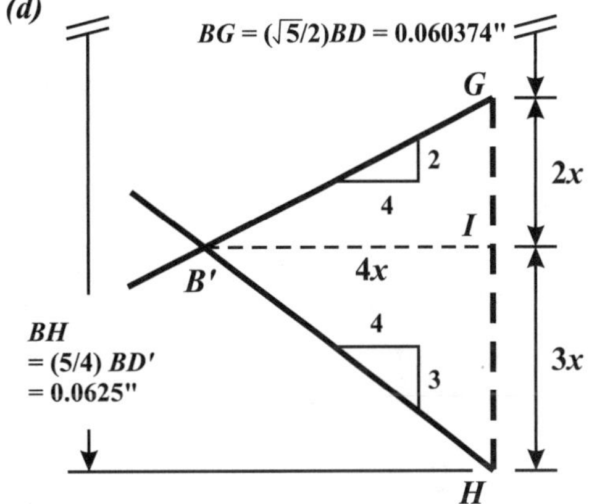

Figure 4.13. **(c)** Elongation of *AB* and *BC*. Point *B* on *AB* moves to *D*; point *B* on *BC* moves to *D'*. Compatibility requires that points *D* and *D'* coincide, so the bars must rotate towards each other. **(d)** Magnification near new joint *B'* to determine its location.

From the geometry of the magnified view in *Figure 4.13d*, *GH* can be divided into five equal parts, x, with B' being $2x$ below and $4x$ to the left of point G. Solving for x:

$$x = \frac{GH}{5} = \frac{1}{5}\left(\frac{5}{4}BD' - \frac{\sqrt{5}}{2}BD\right) = 0.000425 \text{ in.}$$

The displacement of joint B to its new location joint B' is to the left and down:

$$u = -4x \Rightarrow v = -BG - 2x$$

Answer: $u = -0.0017$ in.

Answer: $v = -0.0612$ in.

Other mathematical methods may be used to solve for the location of B'. For example, once point D is located (in the x–y plane), an equation for line DB' can be developed (the slope and a point are known). Likewise, once D' is located, line $D'B'$ can be described. The resulting linear equations intercept at B'. Or, magnified drawings constructed to scale can actually be measured to determine the displacement.

Statically Indeterminate (Redundant) Systems

In the chapter introduction, it was noted that some systems cannot be solved by *Statics* alone; this is the case when there are more unknowns than equilibrium equations. These systems are *statically indeterminate* or *redundant*. When applying the force method, the idea of a **redundant force** must be introduced to complete such a problem. *Examples 4.7– 4.9* are applications of the force method to redundant systems.

Example 4.7 Two Parallel Bars

Given: Bars 1 and 2 are each attached to a rigid base and a rigid boss (*Figure 4.14*). The boss is constrained to move vertically only. The bars have lengths, cross-sectional areas, and moduli as shown in the diagram. Downward load F is applied to the boss, which displaces (deflects) downward distance δ. Assume the system remains elastic.

Required: For the particular case $L_2 = 2L_1$, $A_2 = 4A_1$, and $E_1 = E_2 = E$, determine expressions for (a) the stresses in each bar, σ_1 and σ_2 and (b) the downward deflection δ of the rigid boss.

Solution: *Step 1. Equilibrium.* If the forces in the bars are P_1 and P_2, then from equilibrium of the FBD in *Figure 4.14b*:

$$F = P_1 + P_2$$

There is one equilibrium equation, but two unknowns, P_1 and P_2. The system is *redundant*.

To proceed, assume that $P_1 = R$ (the *redundant*). Force R is unknown, but is temporarily treated as known. The internal forces are then:

$$P_1 = R \quad \text{and} \quad P_2 = F - R$$

The stress in each bar is written:

$$\sigma_1 = \frac{R}{A_1} \quad \text{and} \quad \sigma_2 = \frac{F-R}{A_2}$$

Step 2,3. Force–elongation. The elongation of each bar is:

$$\Delta_1 = \frac{RL_1}{A_1 E_1} \quad \text{and} \quad \Delta_2 = \frac{(F-R)L_2}{A_2 E_2}$$

Step 4. Compatibility requires that the elongations of the individual parts be *compatible* with the system displacements. Here, the bars both elongate the same amount δ; this provides the second needed equation:

$$\delta = \Delta_1 = \Delta_2 \Rightarrow \frac{RL_1}{A_1 E_1} = \frac{(F-R)L_2}{A_2 E_2}$$

Solving for the redundancy R gives:

$$\frac{R}{F} = \left[1 + \frac{L_1 A_2 E_2}{L_2 A_1 E_1} \right]^{-1}$$

For the particular case $L_2 = 2L_1$, $A_2 = 4A_1$ and $E_1 = E_2 = E$ (*Figure 4.14c*):

$$P_1 = R = \frac{F}{3}$$

$$P_2 = F - R = \frac{2F}{3}$$

The force in each member is now known. Expressing the stresses in the bars in terms of the total cross-sectional area, $A = A_1 + A_2$, so that $A_1 = A/5$ and $A_2 = (4/5)A$, gives:

$$\textit{Answer: } \sigma_1 = \frac{R}{A_1} = \frac{5F}{3A} \quad \text{and} \quad \sigma_2 = \frac{5F}{6A}$$

The displacement of the boss is equal to the elongation of each bar:

$$\delta = \Delta_1 = \Delta_2 = \varepsilon_1 L_1 = \frac{\sigma_1}{E}L_1 = \frac{\sigma_2}{E}L_2$$

$$\textit{Answer: } \delta = \frac{5FL_1}{3AE} = \frac{5FL_2}{6AE}$$

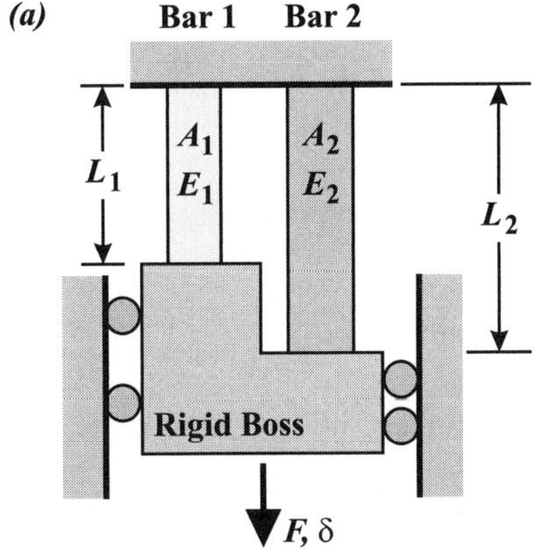

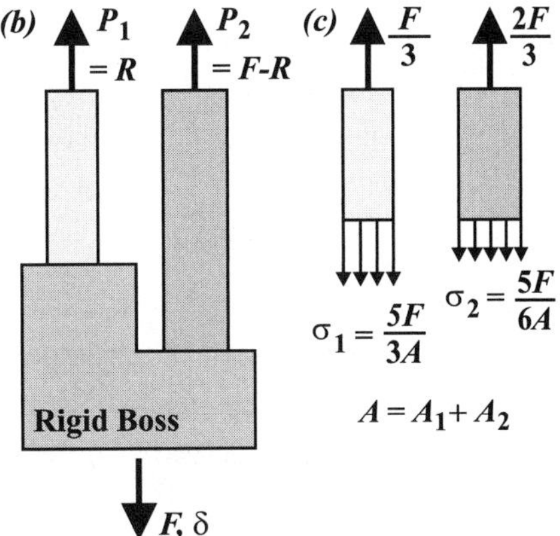

Figure 4.14. **(a)** Two parallel bars constrained to move in the vertical direction. **(b)** FBD of system considering only vertical loads. **(c)** After solving for the internal forces, the stresses are determined.

For this particular case, the short bar carries one-third of the load F, and the stress in the short bar is twice that of the long bar. This *concentration of stress* into the shorter bar is a well-known phenomenon observed in many situations where there are changes in geometry. The **stress concentration factor** is defined as the maximum stress in the system divided by an average stress in the system. Letting $\sigma_{ave} = F/A$, then:

$$SCF = \frac{\sigma_{max}}{\sigma_{ave}} = \left(\frac{5F}{3A}\right)\left(\frac{A}{F}\right) = 1.67$$

Stress concentration is discussed further in *Section 4.5*.

Example 4.8 Two In-Line Bars

Given: Another redundant system consists of two in-line bars, Bars 1 and 2, which are joined together as shown, and fixed at the top and bottom (*Figure 4.15*). Load W is applied at the junction of the bars.

Required: For the particular case $L_2 = 2L_1$, $A_2 = 4A_1$, and $E_1 = E_2 = E$, determine (a) the internal force and stress in each bar and (b) the deflection δ of the junction (where the load is applied).

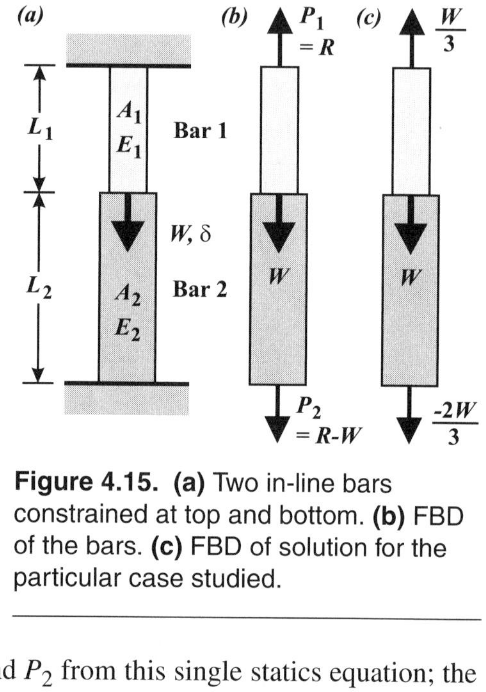

Solution: *Step 1. Equilibrium.* Considering the FBD in *Figure 4.15b*, the reaction forces at the top and bottom equal the internal forces of Bars 1 and 2, P_1 and P_2, respectively. Both internal forces are drawn in tension, so:

Figure 4.15. **(a)** Two in-line bars constrained at top and bottom. **(b)** FBD of the bars. **(c)** FBD of solution for the particular case studied.

$$P_1 - W - P_2 = 0 \Rightarrow W = P_1 - P_2$$

It is not possible to calculate the values of P_1 and P_2 from this single statics equation; the system is *redundant*.

Let force $P_1 = R$ be the redundant force, so that: $P_2 = R - W$.

Steps 2 and 3. The *force–elongation* relationship gives the elongation of each bar:

$$\Delta_1 = \frac{P_1 L_1}{A_1 E_1} = \frac{R L_1}{A_1 E_1} \quad \text{and} \quad \Delta_2 = \frac{P_2 L_2}{A_2 E_2} = \frac{(R-W)L_2}{A_2 E_2}$$

Step 4. The *compatibility* condition is used to find the redundant force R. The downward displacement δ of load W is equal to the extension of Bar 1 Δ_1, and the shortening (negative elongation) of Bar 2 Δ_2. The elongations of the bars, Δ_1 and Δ_2, are therefore equal but opposite. Or, since the top and bottom of the assembly are fixed, the total elongation of the stepped-bar-assembly is zero:

$$\Delta_1 + \Delta_2 = 0 = \frac{RL_1}{A_1E_1} + \frac{(R-W)L_2}{A_2E_2}$$

This expression is rearranged to solve for R:

$$R = W\left[1 + \frac{L_1A_2E_2}{L_2A_1E_1}\right]^{-1}$$

The internal forces are then:

$$P_1 = R = W\left[1 + \frac{L_1A_2E_2}{L_2A_1E_1}\right]^{-1} \quad \text{and} \quad P_2 = R - W = -W\left[1 + \frac{L_2A_1E_1}{L_1A_2E_2}\right]^{-1}$$

With the forces known, for the particular case $L_2 = 2L_1$, $A_2 = 4A_1$, and $E_1 = E_2 = E$ (*Figure 4.15c*):

$$\text{Answer: } P_1 = \frac{W}{3} \quad \text{and} \quad P_2 = -\frac{2W}{3}$$

The stresses in the bars are:

$$\sigma_1 = \frac{W}{3A_1} \quad \text{and} \quad \sigma_2 = -\frac{2W}{3A_2} = -\frac{W}{6A_1}$$

The downward displacement of the load is $\delta = \Delta_1 = -\Delta_2$:

$$\delta = \Delta_1 = \frac{P_1L_1}{A_1E_1} = -\Delta_2 = -\frac{P_2L_2}{A_2E_2}$$

$$\text{Answer: } \delta = \frac{WL_1}{3A_1E} = \frac{2WL_2}{3A_2E}$$

Example 4.9 **Rigid Bar**

Given: Bar AB is rigid with length L, while bars C and D are elastic, both with modulus E, cross-sectional area A, and length H (*Figure 4.16*). Force P is applied at point B which displaces upward by distance δ.

Required: Determine the force in each bar, R_C and R_D, in terms of P.

Solution: *Step 1. Equilibrium.* The FBD of Bar AB is given in *Figure 4.16b*. There are two equilibrium equations:

$$\sum F_y = 0: \ R_A - R_C - R_D + P = 0$$

$$\sum M_A = 0: \ -R_C\left(\frac{L}{3}\right) - R_D\left(\frac{2L}{3}\right) + PL = 0$$

and three unknowns: reaction R_A and tension forces R_C and R_D. There are more unknowns than equations; the system is redundant.

In the *force method*, one force (or as many as necessary) is selected as the *redundant force*, and is assumed for the time to be known. Selecting R_D as the redundant, and solving for R_A and R_C in terms of P and R_D gives:

$$R_A = -R_D + 2P$$

$$R_C = -2R_D + 3P$$

Steps 2 and 3. Force–elongation. The elongation of each elastic bar is:

$$\Delta_C = \frac{(-2R_D + 3P)H}{AE} \quad \text{and} \quad \Delta_D = \frac{R_D H}{AE}$$

Step 4. Compatibility. Since Bar AB is rigid, the elongations of Bars C and D are related to the displacement of point B:

$$\Delta_C = \frac{1}{3}\delta \quad \text{and} \quad \Delta_D = \frac{2}{3}\delta$$

Eliminating δ between the last two sets of equations gives the redundant R_D in terms of force P:

$$2(-2R_D + 3P) = R_D$$

$$\textit{Answer: } R_D = \frac{6}{5}P$$

Solving for R_A and R_C:

$$R_A = \frac{4}{5}P$$

$$\textit{Answer: } R_C = \frac{3}{5}P$$

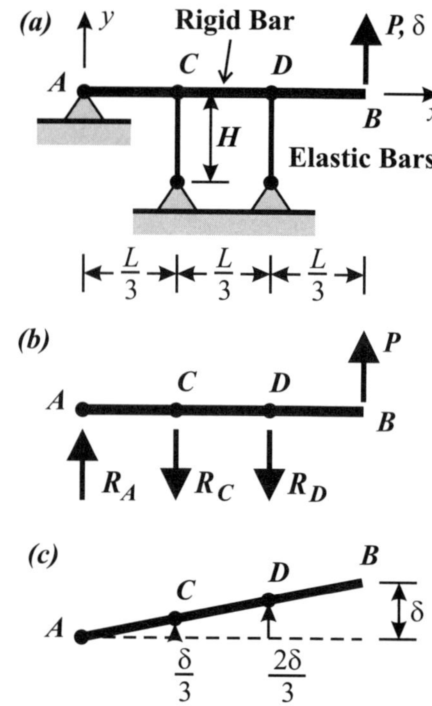

Figure 4.16. **(a)** Rigid Bar *AB* under load *P.* **(b)** FBD of *AB.* **(c)** *AB* pivots about Point *A.* The displacements of Points *C* and *D* are proportional to tip displacement δ.

R_C and R_D are both positive, corresponding to tension forces and positive elongations. Since the properties of Bars C and D are the same, and $\Delta_B = 2\Delta_C$, then $R_D = 2R_C$.

The displacement of point B, δ, as a function of P is: $\delta = \dfrac{9PH}{5AE}$

Note that the use of the *force method* gives the *flexibility* of the system directly, since $\delta = FP$.

4.2 Axial Members – Displacement Method

The ***displacement method*** is used in this section to solve four of the previous examples in *Section 4.1*, two statically determinate and two statically indeterminate (redundant). A new example, with many redundant members, is also solved. From kinematic (displacement) relationships, the internal strains are first determined, which lead to the stresses via Hooke's Law, from which internal and external forces are determined using equilibrium. The steps are as follows:

1. Determine the elongation of each member in terms of the overall displacement (deflection) of the assembly; the elongation of each member must be *compatible* with the elongations of the other members.
2. Determine the strain in each member in terms of its elongation; e.g., $\varepsilon = \Delta/L$.
3. Calculate the stresses from the elastic law; e.g., $\sigma = E\varepsilon$.
4. Apply the conditions of equilibrium to determine the internal and applied forces; e.g., $P = \sigma A$.

Example 4.10 Uniform Bar in Tension

Given: The bar in *Example 4.1* elongates by Δ when axial force P is applied at its ends. The bar has constant cross-sectional area A and length L (*Figure 4.17*). The material of the bar is elastic with modulus E.

Required: Determine the force P needed to elongate the bar by Δ.

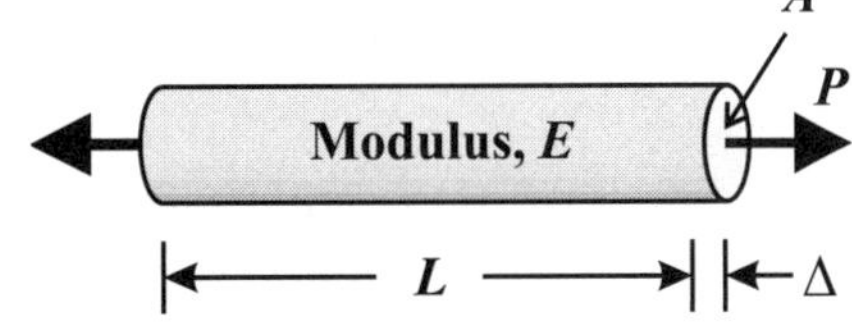

Figure 4.17. Bar under tension.

Solution: Following the steps of the *displacement method*:

Steps 1 and 2 (combined). *Strain–elongation–displacement.* The elongation Δ causes axial strain:

$$\varepsilon = \frac{\Delta}{L}$$

Step 3. Elasticity (the *stress–strain relationship*) requires the stress to be:

$$\sigma = E\varepsilon = E\frac{\Delta}{L}$$

Step 4. Equilibrium. From equilibrium, the internal force P throughout the bar is:

$$\text{Answer: } P = \sigma A = \frac{EA}{L}\Delta$$

The stiffness of the axial bar under constant tension is, as noted previously:

$$K = \frac{P}{\Delta} = \frac{EA}{L}$$

The displacement method immediately gives the *stiffness* of the system; the result directly gives force as a function of displacement.

Example 4.11 Parallel Bars With Applied Displacement

Given: The two-bar structure of *Example 4.7* is shown in *Figure 4.18*. The system is statically redundant. Due to load F, the rigid boss displaces downward by δ.

Required: Using the *displacement method*, for the particular case $L_2 = 2L_1$, $A_2 = 4A_1$, and $E_1 = E_2 = E$, determine (a) the relationship between force F and displacement δ and (b) the stress in each bar, σ_1 and σ_2.

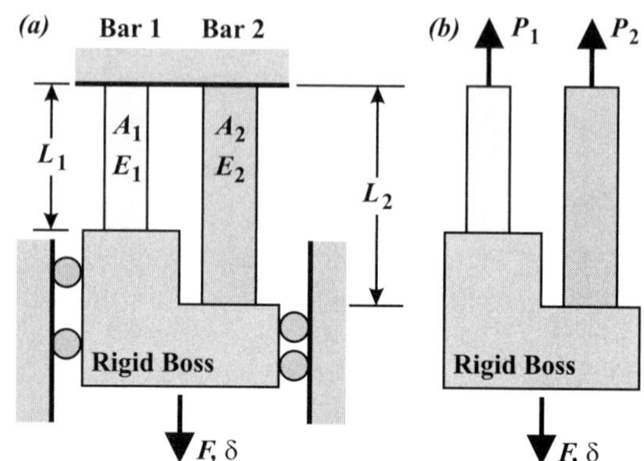

Figure 4.18. **(a)** Two parallel bars constrained to move vertically. **(b)** FBD of system considering only vertical loads.

Solution: *Step 1. Elongation– displacement. Compatibility* requires that the elongation of each bar, Δ_1 and Δ_2, equal the downward displacement of the rigid boss δ:

$$\Delta_1 = \delta \quad \text{and} \quad \Delta_2 = \delta$$

Steps 2 and 3. The *force–elongation* relationship for axial members gives:

$$P_1 = \frac{E_1 A_1}{L_1}\Delta_1 = \frac{E_1 A_1}{L_1}\delta \quad \text{and} \quad P_2 = \frac{E_2 A_2}{L_2}\Delta_2 = \frac{E_2 A_2}{L_2}\delta$$

Step 4. Equilibrium. Applying equilibrium to *Figure 4.18b*, the sum of the internal forces must equal to the applied force F:

$$F = P_1 + P_2 = \left(\frac{E_1 A_1}{L_1} + \frac{E_2 A_2}{L_2}\right)\delta$$

The stiffness of the system is within the set of parentheses.

For the particular case $L_2 = 2L_1$, $A_2 = 4A_1$, and $E_1 = E_2 = E$, the applied force is:

$$F = \left[\frac{EA_1}{L_1} + \frac{E(4A_1)}{(2L_1)}\right]\delta = \frac{3EA_1}{L_1}\delta$$

With the total cross-sectional area $A = A_1 + A_2$, the load–displacement relationship reduces to:

$$\text{Answer: } F = \frac{3EA}{5L_1}\delta \quad \rightarrow \quad \delta = \frac{5FL_1}{3EA} = \frac{5FL_2}{6EA}$$

The load–displacement relationship is the same as found using the *force method* in *Example 4.7*. However, it is apparent that the *displacement method* applied to this redundant system is easier to use than the *force method*.

The stresses are the same as in *Example 4.7*:

$$\sigma_1 = \frac{P_1}{A_1} = \frac{1}{A_1}\frac{EA_1}{L_1}\delta = \frac{E_1}{L_1}\frac{5FL_1}{3EA}$$

$$\text{Answer: } \sigma_2 = \frac{5}{3}\sigma_{ave} \quad \text{and} \quad \sigma_2 = \frac{5}{6}\sigma_{ave}$$

where $\sigma_{ave} = F/A$. The value of $\sigma_1/\sigma_{ave} = 5/3 = 1.67$ is the stress concentration factor on the short bar.

Example 4.12 Two In-Line Bars

Given: The redundant system of *Example 4.8* consists of two bars in series: Bars 1 and 2, fixed at their ends (*Figure 4.19*). The applied load W at the junction displaces it by distance δ.

Required: For the particular case $L_2 = 2L_1$, $A_2 = 4A_1$, and $E_1 = E_2 = E$, determine (a) the load W to cause displacement δ, and (b) how the load is distributed to the individual bars.

Step 1. Displacement–elongation. Compatibility requires that Bar 1 elongate by the downward displacement of the junction δ, and that Bar 2 shorten by the same amount:

$$\Delta_1 = \delta \quad \text{and} \quad \Delta_2 = -\delta$$

Step 2 and 3. Force–elongation:

$$P_1 = \frac{E_1 A_1}{L_1}\Delta_1 = \frac{E_1 A_1}{L_1}\delta$$

$$P_2 = \frac{E_2 A_2}{L_2}\Delta_2 = \frac{E_2 A_2}{L_2}(-\delta)$$

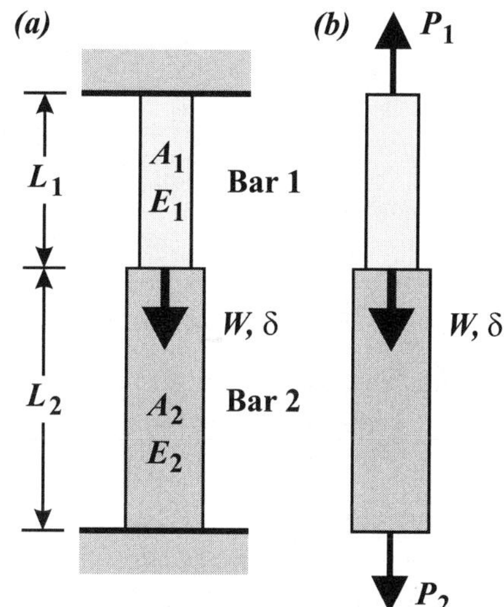

Figure 4.19. (a) Two in-line bars constrained at top and bottom. **(b)** FBD of the bars.

Step 4. Equilibrium of the joint alone, or of the overall system, requires that:

$$W = P_1 - P_2 = \frac{E_1 A_1}{L_1}\delta - \frac{E_2 A_2}{L_2}(-\delta)$$

$$\textit{Answer: } W = \left(\frac{E_1 A_1}{L_1} + \frac{E_2 A_2}{L_2}\right)\delta$$

The *stiffness* of the system is: $K = \dfrac{W}{\delta} = \dfrac{E_1 A_1}{L_1} + \dfrac{E_2 A_2}{L_2}$

For the particular case $L_2 = 2L_1$, $A_2 = 4A_1$, and $E_1 = E_2 = E$:

$$K = \frac{E A_1}{L_1} + \frac{4 E A_1}{2 L_1} = \frac{3 E A_1}{L_1}$$

Step 5. The force W to cause displacement δ is:

$$\textit{Answer: } W = K\delta = \frac{3 E A_1}{L_1}\delta$$

Step 6. The internal forces are:

$$P_1 = \frac{E_1 A_1}{L_1}\delta = \frac{E_1 A_1}{L_1}\frac{W}{K}$$

$$\textit{Answer: } P_1 = \frac{W}{3}$$

$$P_2 = -\frac{2W}{3}$$

Example 4.13 Hanging Lamp

Given: The hanging lamp problem (*Example 4.5*) is reexamined using the *displacement method*. Only the displacement figures are repeated here (*Figure 4.20*). A $W = 14$ lb lamp is supported by two wires both of length $L = 5.0$ ft. The distance between the two cable mounts is $s = 8.0$ ft. The wires have a diameter of $D = 0.10$ in., and are made of steel ($E = 30\times10^6$ psi, $S_y = 50$ ksi).

Required: Using the *displacement method*, determine (a) the downward displacement δ of the lamp and (b) the tension in each wire, $T_{AC} = T_{BC} = T$.

Solution: Apply the steps of the displacement method.

Step 1. Displacement–elongation. Because the load and geometry are symmetric, point B moves directly downward. The wires move from the dashed position ABC to the solid position $AB'C$

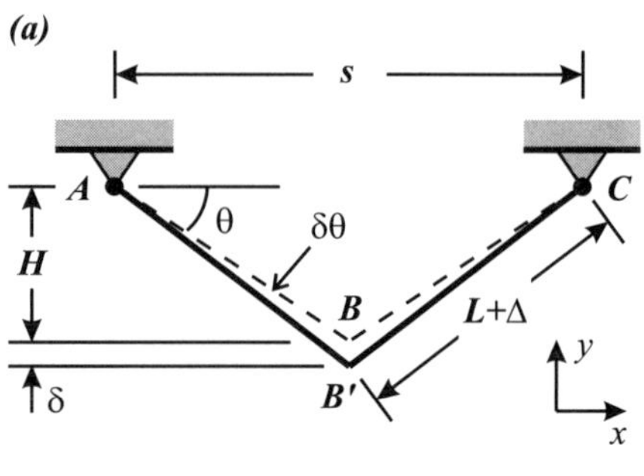

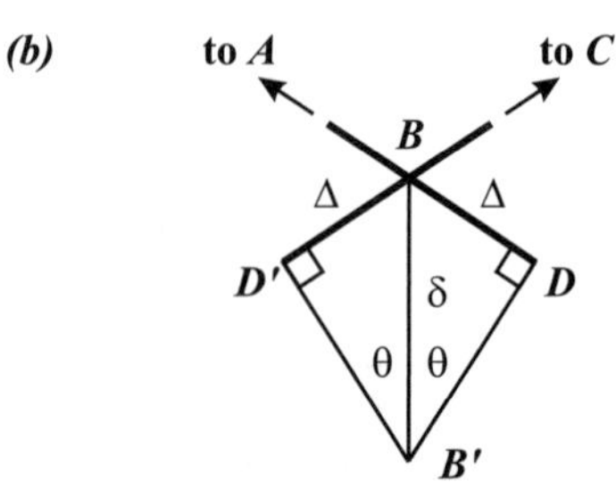

Figure 4.20. **(a)** Geometry of original (*dashed*) and displaced (*solid*) wires. **(b)** The triangle relating the downward displacement of point B, δ, the extension of each wire, $\Delta = BD = BD'$, and the angle θ.

in *Figure 4.20a* (the displacements are exaggerated). The displacement of wire *AB* can be described in two steps: it elongates by Δ becoming *AD*, and rotates clockwise to *AB'*. Wire *BC* elongates by Δ becoming *CD'*, and rotates counterclockwise to *CB'* (*Figure 4.20b*).

The relationship between the downward displacement δ of point *B* and the elongation Δ of either wire is determined from the triangle in *Figure 4.20b*:

$$\Delta = \delta \sin \theta = \delta \frac{H}{L}$$

From the geometry, $H = 3.0$ ft.

Steps 2 and 3. Force–elongation. The relationship between the force in each wire *T* and its elongation Δ is:

$$T = \frac{EA}{L}\Delta = \frac{EA}{L}\left(\delta \frac{H}{L}\right) = \frac{EAH}{L^2}\delta$$

Step 4. Equilibrium. From equilibrium, the force *W* needed to cause displacement δ is:

$$W = 2T \sin \theta = 2T\frac{H}{L} = 2\left(\frac{EA\delta H}{L^2}\right)\frac{H}{L} = \frac{2EAH^2}{L^3}\delta$$

The stiffness of the system is then:

$$K = \frac{W}{\delta} = \frac{2EAH^2}{L^3} = \frac{E\pi D^2 H^2}{2L^3} = \frac{(30 \times 10^6 \text{ psi})\pi(0.10 \text{ in.})^2(36 \text{ in.})^2}{2(60 \text{ in.})^3} = 2828 \text{ lb/in.}$$

Step 5. The displacement of the system is thus:

$$\textit{Answer: } \delta = \frac{W}{K} = \frac{14 \text{ lb}}{2828 \text{ lb/in.}} = \underline{\delta = 0.00495 \text{ in.}}$$

Step 6. The force in each member is:

$$T = T_{AB} = T_{BC} = \frac{EAH}{L^2}\delta = \frac{(30 \times 10^6 \text{ psi})\pi(0.10 \text{ in.})^2(36 \text{ in.})}{4(60 \text{ in.})^2}(0.00495 \text{ in.})$$

$$\textit{Answer: } \underline{T = 11.7 \text{ lb}}$$

The answers for the deflection δ and the wire tension *T* agree with the results of *Example 4.5*.

Notice the stiffness increases with $(H/L)^2 = \sin^2\theta$. When $\theta = 0^\circ$, the wires are both horizontal and the stiffness is zero. For example, the transverse stiffness of a guitar string is small; it is easy to deform a guitar string perpendicular to its axis. As soon as there is sideways displacement of the string, the stiffness increases. When the two wires are parallel $L = H$, the stiffness equation simplifies to that of two axial members *in parallel*:

$$K_{parallel} = 2\left(\frac{EA}{L}\right) = \frac{E\pi D^2}{2L}$$

Example 4.14 Stiffness of a Wheel with Many Spokes

Background: The displacement method is especially well suited for systems with many redundant members. For such systems, the force method is generally impractical.

Given: A bicycle wheel of radius R has N spokes, each of cross-sectional area A. The modulus of the spokes is E. The rim and hub are taken to be rigid. The weight and dynamic forces of the rider cause downward force F at the rigid wheel hub, displacing it downward by distance δ (*Figures 4.21a–c*).

Required: Determine the relationship between force F and displacement δ; i.e., the stiffness of the assembly of spokes for a downward load applied at the hub.

Solution: Consider a triangular-shaped element $d\theta$ at angle θ to the horizontal (*Figure 4.21d, dashed triangle*). The number of spokes represented by element $d\theta$ is:

$$dN = N\frac{d\theta}{2\pi}$$

The cross-sectional area of the spokes in $d\theta$ is:

$$dA = AN\frac{d\theta}{2\pi}$$

Step 1. Displacement–Elongation. From the geometry of the displacement, the relationship between the elongation of a spoke element $\Delta(\theta)$ and the downward displacement of the hub δ is (*Figure 4.21e*):

$$\Delta(\theta) = \delta \sin \theta$$

Steps 2 and 3. Force–elongation. The strain ε of a spoke element at angle θ is:

$$\varepsilon(\theta) = \frac{\Delta(\theta)}{R} = \frac{\delta \sin \theta}{R}$$

The force dP in each element is:

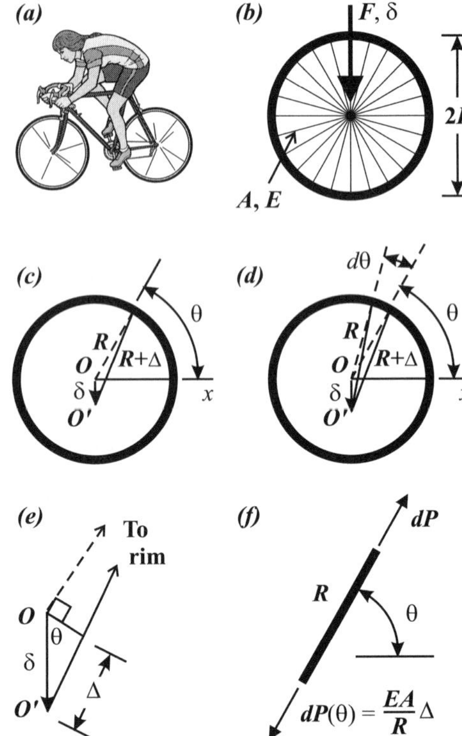

Figure 4.21. **(a)** The bicycle rider applies load F to the rear wheel. **(b)** The wheel is made up of N spokes, R long, each spoke of cross-sectional area A, and modulus E. **(c)** The hub displaces downward, so a single spoke at angle θ moves from its original (*dashed*) position to a new (*solid*) one, elongating by Δ. **(d)** An element $d\theta$ moves from its unloaded dotted position to the solid one. **(e)** Movement of point O to point O', giving the relationship between δ and Δ. **(f)** FBD of a spoke. Copyright ©2008 Dominic J. Dal Bello and licensors. All rights reserved.

$$dP(\theta) = \sigma(\theta)dA = E\varepsilon(\theta)dA = E\left(\frac{\delta \sin \theta}{R}\right)\left(\frac{AN\,d\theta}{2\pi}\right) = \frac{EAN}{2\pi R}\delta(\sin\theta\,d\theta)$$

Step 4. Equilibrium. The vertical force component of each spoke element is: $dV = dP \sin \theta$. Applying equilibrium in the vertical direction requires an integral:

$$F = \int_0^{2\pi} dV = \int_0^{2\pi} dP \sin \theta = \int_0^{2\pi} \frac{EAN}{2\pi R}\delta(\sin\theta\,d\theta)\sin\theta = \frac{EAN\delta}{2R\pi}\int_0^{2\pi}\sin^2\theta\,d\theta$$

Solving for the integral alone: $\displaystyle \int_0^{2\pi} \sin^2\theta\,d\theta = \pi$

Thus, the force–displacement relationship is:

$$\textit{Answer: } F = \frac{EAN\delta}{2R\pi}\pi = \frac{EAN}{2R}\delta$$

The stiffness of the spoke system is:

$$\textit{Answer: } K = \frac{F}{\delta} = \frac{NEA}{2}\frac{1}{R}$$

which is equal to one-half of the total stiffness of the system if all of the spokes were aligned vertically.

The stress in any spoke at angle θ is: $\sigma(\theta) = E\varepsilon(\theta) = \dfrac{E\Delta(\theta)}{R} = \dfrac{E\delta \sin \theta}{R}$

For $0 < \theta < \pi$ (the upper half of the wheel), the stress is tensile ($\sin\theta > 0$). For $\pi < \theta < 2\pi$ (the bottom half), the stress is compressive ($\sin\theta < 0$). These results make sense physically; the upper spokes are being pulled downward, and the lower spokes are being compressed. In order to avoid *buckling* of the spokes due to compression, they are generally pretensioned.

Summary of Equations for Stress, Strain and Elongation

When the axial force, the cross-sectional area, and the modulus are *all* constant throughout the length of an axial member, the strain is constant. Only for a *constant–strain* component may its elongation – the relative displacement between its endpoints – be expressed as:

$$\Delta = \frac{PL}{AE} \qquad\qquad \text{[Eq. 4.7]}$$

If any of the values of force, area, and modulus vary over the length of an axial system, the system must be broken up into smaller elements where all the values are constant in each element. The elements may be shorter lengths, L_i, or even differential lengths dx. The total elongation is the summation (integral) of the displacement of each small constant–strain element. A summary of the relevant equations is given in *Table 4.1*.

Table 4.1. General equations for in-line axial members.

Axial Force, Cross-sectional Area, Modulus	Stress on any Cross-section	Strain at any Cross-section	Elongation of System	Remark; Constant Strain Element
All Constant	$\sigma = \dfrac{P}{A}$	$\varepsilon = \dfrac{\Delta}{L} = \dfrac{\sigma}{E}$	$\Delta = \dfrac{PL}{AE}$	L
Discretely Varying	$\sigma_i = \dfrac{P_i}{A_i}$	$\varepsilon_i = \dfrac{\Delta_i}{L_i} = \dfrac{\sigma_i}{E_i}$	$\Delta = \sum \dfrac{P_i L_i}{A_i E_i}$	Break component into lengths L_i, where P, A, and E are all constant.
Continuously Varying	$\sigma(x) = \dfrac{P(x)}{A(x)}$	$\varepsilon(x) = \dfrac{\sigma(x)}{E(x)}$	$\displaystyle\int_L \dfrac{P(x)}{A(x)E(x)} dx$	Consider length dx, so small that over its length, P, A, and E are all constant.

4.3 Thermal Loading

Changes in temperature cause materials to expand or contract. Consider a bar of length L that is free to expand. When the temperature is increased by an amount ΔT, the length of the bar increases by:

$$\Delta = L\alpha\,\Delta T \qquad \text{[Eq. 4.8]}$$

where α is the *coefficient of thermal expansion*. The *thermal strain* ε_t of the bar is then:

$$\varepsilon_t = \frac{\Delta}{L} = \alpha\Delta T \qquad \text{[Eq. 4.9]}$$

Table 4.2. Coefficient of Thermal Expansion for common structural materials.

Material	α (1/°C)	E (GPa)
Steel	14×10^{-6}	200
Aluminum	23×10^{-6}	70
Concrete	7×10^{-6}	30

The coefficient of thermal expansion is a material property. The units of α are the inverse of temperature (e.g., 1/°C, 1/°F). Representative values of α can be found in *Appendix B* and in the *Online Notes*.

For convenience in the following examples, *Table 4.2* provides representative material properties for steel, aluminum, and concrete. Although they actually vary with temperature, the coefficient of thermal expansion α and modulus E are taken as constant with temperature in this treatment.

When two materials with different expansion coefficients must deform together, internal *thermal stresses* will develop within the system.

Example 4.15 Unconstrained Expansion of a Steel Bar

Given: An unconstrained (free to expand) steel bar of length $L = 1.0$ m is heated from room temperature (25°C) to 100°C.

Required: Determine (a) the thermal strain ε_t and (b) the elongation Δ of the bar.

Solution: *Step 1.* The thermal strain is:

$$\varepsilon_t = \alpha\ \Delta T = (14 \times 10^{-6}\ °C^{-1})[(100 - 25)°C]$$

$$\textit{Answer: } \varepsilon_t = 1.05 \times 10^{-3}$$

Step 2. The elongation is:

$$\Delta = \varepsilon_t L = (1.05 \times 10^{-3})(1.0\ \text{m}) = 1.05 \times 10^{-3}\ \text{m}$$

$$\textit{Answer: } \Delta = 1.05\ \text{mm}$$

Thermal and Mechanical Loading (Temperature and Applied Stress)

The unconstrained bar of length L (*Figure 4.22*) is now subjected to a constant axial stress σ. The temperature is then increased by ΔT. The total strain in the bar is the sum of the *mechanical* and *thermal* strains, ε_m and ε_t :

$$\varepsilon = \varepsilon_m + \varepsilon_t = \frac{\sigma}{E} + \alpha\ \Delta T \qquad \text{[Eq. 4.10]}$$

The change in length Δ is:

$$\Delta = \varepsilon L = \left(\frac{\sigma}{E} + \alpha\ \Delta T\right) L \qquad \text{[Eq. 4.11]}$$

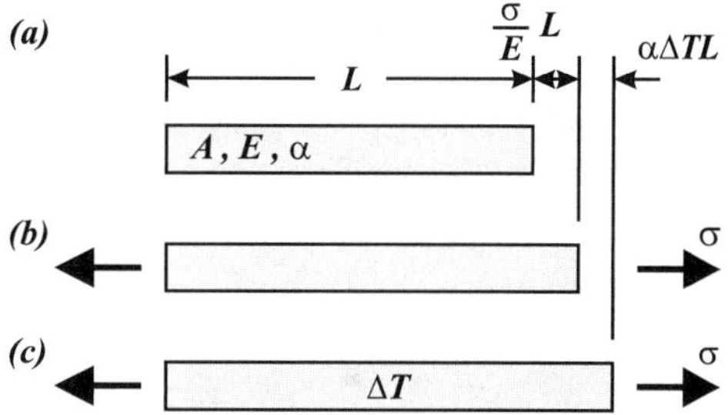

Figure 4.22. **(a)** Original length of bar. **(b)** Elongation due to mechanical load P. **(c)** Additional elongation due to thermal load ΔT.

Example 4.16 Steel Bar under Applied Stress and Temperature

Given: An unconstrained steel bar (*Figure 4.23*) of length $L = 1.0$ m and square cross-section of side $b = 20$ mm is subjected to a compressive axial load $P = 20$ kN ($\sigma = 50$ MPa). The modulus is $E = 200$ GPa.

Required: Determine the temperature increase ΔT that must be applied to the loaded bar to return it to its original length.

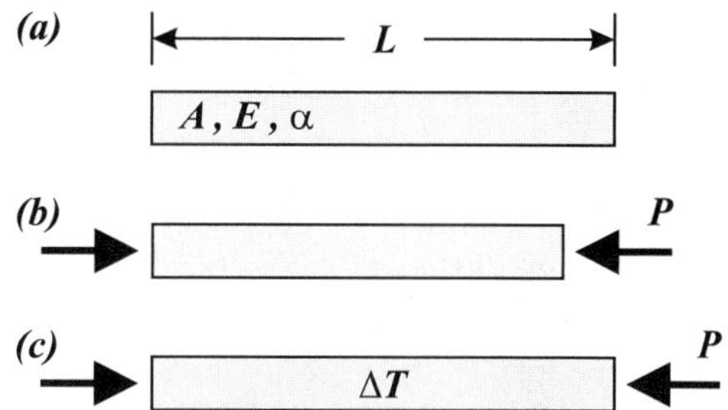

Figure 4.23. **(a)** Original length of bar. **(b)** Elongation (here negative) due to mechanical load P. **(c)** Additional elongation due to thermal load ΔT.

Solution: By applying temperature, the compressed bar is to expand to its original length, so the total elongation of the bar due to the mechanical and thermal loads is zero:

$$\Delta = \varepsilon L = \left(\frac{\sigma}{E} + \alpha\,\Delta T\right)L = 0$$

Thus:

$$\Delta T = -\frac{\sigma}{E\alpha} = -\frac{P}{b^2 E\alpha} = -\frac{(-20 \times 10^3\ \text{N})}{(20 \times 10^{-3}\ \text{m})^2(200 \times 10^9\ \text{Pa})(14 \times 10^{-6}\ {}^\circ\text{C}^{-1})}$$

Answer: $\underline{\Delta T = 17.8\,{}^\circ\text{C}}$

Example 4.17 Aluminum Rod with Fixed Ends

Given: An aluminum rod fixed between two rigid supports (*Figure 4.24*).

Required: Determine the stress in the rod when it is heated by 35°C.

Solution: The total change in length is zero:

$$\Delta = \varepsilon L = \left(\frac{\sigma}{E} + \alpha\,\Delta T\right)L = 0$$

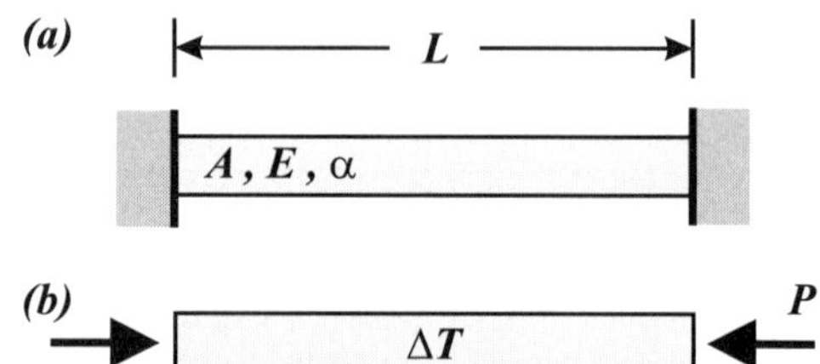

Figure 4.24. **(a)** Bar between rigid supports. **(b)** FBD of bar at temperature ΔT.

Thus:

$$\sigma = -E\alpha\,\Delta T = -(70 \times 10^9\ \text{Pa})(23 \times 10^{-6}\ {}^\circ\text{C}^{-1})(35\,{}^\circ\text{C})$$

Answer: $\underline{\sigma = -56.4\ \text{MPa}}$

The bar is in compression.

Example 4.18 Loss of Prestress in Reinforced Concrete under Thermal Load

Given: A representative element of reinforced concrete (square cross-section 40×40 mm) surrounds a single high strength steel rebar (diameter $D = 10$ mm), as shown in *Figure 4.25*. The system is *prestressed* by tightening the rebar endcaps, which places the steel rebar in tension and the concrete in compression. No external load is applied to the system. The purpose of prestressing is to prevent tensile stresses in the concrete – and thus avoid fracture or cracking – by preloading the concrete in compression. Here, the rebar is under a tensile stress of $\sigma_{s,p} = 200$ MPa. Assume that the rebar and concrete remain the same length.

Required: Determine (a) the stress in the concrete after the prestressing process and (b) the loss of prestress in the concrete when the temperature increases from 20 to 40°C.

Solution: *Step 1. Stress due to prestressing.* The area of the steel is:

$$A_s = \frac{\pi}{4}(0.01 \text{ m})^2 = 78.54 \times 10^{-6} \text{ m}^2$$

The area of the concrete is thus:

$$A_c = [0.04 \times 0.04 \text{ m}^2] - \left[\frac{\pi}{4}(0.01 \text{ m})^2\right]$$

$$= 1521 \times 10^{-6} \text{ m}^2$$

The stress in the concrete due to the mechanical prestress is found from equilibrium of *Figure 4.25c*:

$$\sigma_{s,p} A_s + \sigma_{c,p} A_c = 0$$

where $\sigma_{s,p}$ and $\sigma_{c,p}$ are the stresses in the steel and concrete due to prestressing; there is no external load. Thus:

$$\sigma_{c,p} = -\sigma_{s,p}\frac{A_s}{A_c}$$

$$= -(250 \text{ MPa})\left(\frac{78.54 \times 10^{-6} \text{ m}^2}{1521 \times 10^{-6} \text{ m}^2}\right)$$

Answer: $\underline{\sigma_{c,p} = -12.90 \text{ MPa}}$

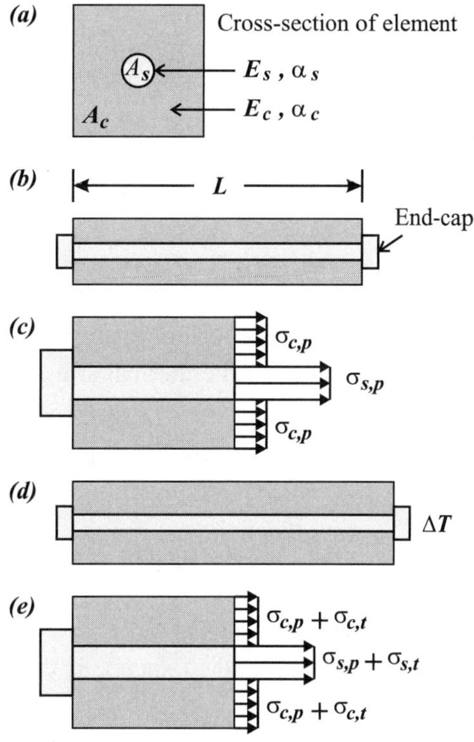

Figure 4.25. (a) Cross-section of element. **(b)** Side view. **(c)** Internal stresses due to prestress only. **(d)** Expansion under thermal load ΔT. **(e)** Internal stresses due to prestress and temperature.

The concrete is initially in compression as intended.

Step 2. Stress due to thermal loading. When temperature is applied, both materials expand (*Figure 4.25d*). Unconstrained, the steel would expand more than the concrete. Here, since the materials are constrained to remain the same length, *thermal stresses* – stresses due to the thermal load – are *induced* in each material $\sigma_{s,t}$ and $\sigma_{c,t}$.

During thermal loading, no external mechanical load is applied to the system. Thus, the thermal stresses are in equilibrium:

$$\sigma_{s,t} A_s + \sigma_{c,t} A_c = 0$$

or:

$$\sigma_{c,t} = -\sigma_{s,t}\frac{A_s}{A_c}$$

Consider the prestressed length as the reference for the thermally induced strain. Since the concrete and steel must expand or contract together, their strains due to thermal loading are the same. These strains are a combination of the free thermal expansion of each material, $\alpha \, \Delta T$, and the elastic strain due to the induced thermal stresses:

$$\varepsilon_{s,t} = \alpha_s \Delta T + \frac{\sigma_{s,t}}{E_s} = \varepsilon_{c,t} = \alpha_c \Delta T + \frac{\sigma_{c,t}}{E_c}$$

Rearranging and solving for the thermal stress in the steel gives:

$$\sigma_{s,t} = E_s(\alpha_c - \alpha_s)\Delta T + \frac{E_s}{E_c}\sigma_{c,t}$$

Substituting the equilibrium expression for the thermal stress in the concrete $\sigma_{c,t}$ into the last expression gives the thermal stress induced in the steel $\sigma_{s,t}$:

$$\sigma_{s,t} = E_s(\alpha_c - \alpha_s)\Delta T \left[1 + \frac{E_s A_s}{E_c A_c}\right]^{-1}$$

$$= [200 \times 10^9 \text{ Pa}][(7 - 14) \times 10^{-6} \text{ °C}^{-1}][(40 - 20) \text{ °C}]\left[1 + \frac{(200)(78.54)}{(30)(1521)}\right]^{-1}$$

$$= -20.83 \text{ MPa}$$

The thermal stress in the concrete is then:

$$\sigma_{c,t} = -\sigma_{s,t}\frac{A_s}{A_c} = -(-20.83 \text{ MPa})\left(\frac{78.54}{1521}\right)$$

Answer: $\underline{\sigma_{c,t} = 1.07 \text{ MPa}}$

Since the coefficient of thermal expansion of steel is higher than that of concrete, the steel would expand more than the concrete if both were free to expand. Since both must remain the same length, the thermal stress in the steel is compressive, and that in the concrete is tensile.

Step 3. Total stress due to prestressing and thermal loading. The total stress in the steel and in the concrete due to both prestressing and thermal loading is (*Figure 4.25e*):

Answer: $\underline{\sigma_s = \sigma_{s,p} + \sigma_{s,t} = 229 \text{ MPa}}$

Answer: $\underline{\sigma_c = \sigma_{c,p} + \sigma_{c,t} = -11.8 \text{ MPa}}$

Here, the applied temperature decreases the prestress by 8%. Although not significant in this example, loss of prestress can become a real concern in practice as the concrete becomes more susceptible to fracture under applied tensile loads. With time, the prestress can also decrease as the system relaxes.

Example 4.19 **Two-Bar Structure under Mechanical and Thermal Loads**

Given: A two-bar structure, Bars 1 and 2. Each bar has length L, cross-sectional area A, modulus E, yield strength S_y, and thermal expansion coefficient α (*Figure 4.26*). The system is subjected to a tensile load P. Bar 2 is subjected to a thermal load ΔT greater than Bar 1. The bars are constrained to remain the same length. The material properties are assumed to be constant with temperature.

Required: (a) Determine the stress in each bar due to mechanical and thermal loading. (b) Determine the conditions to avoid yielding in terms of force P and temperature increase ΔT. Present the result on a plot.

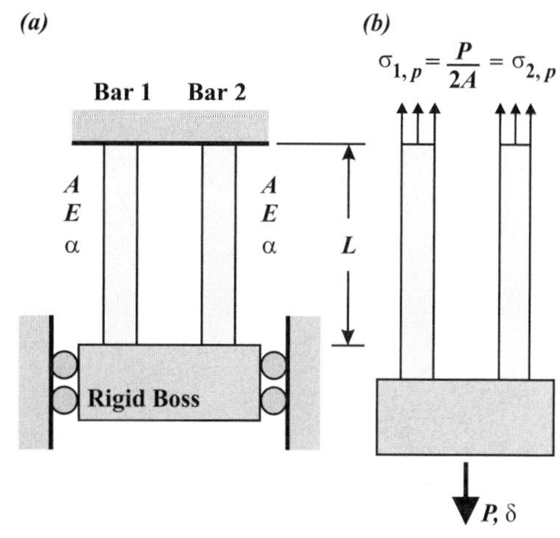

Figure 4.26. **(a)** Two-bar structure. **(b)** Stress in bars due to mechanical load P only.

Solution: *Step 1. Mechanical loading.* Since the bars are identical, due to applied load P, they support the same mechanical stress and have the same strain (*Figure 4.26b*):

$$\sigma_{1,p} = \sigma_{2,p} = \frac{P}{2A} \quad \text{and} \quad \varepsilon_{1,p} = \varepsilon_{2,p} = \frac{P}{2AE}$$

Step 2. Thermal loading. Apply temperature ΔT to Bar 2. Compatibility requires that the additional strain in each bar be the same:

$$\varepsilon_{1,t} = \frac{\sigma_{1,t}}{E} = \varepsilon_{2,t} = \frac{\sigma_{2,t}}{E} + \alpha\,\Delta T$$

where $\sigma_{1,t}$ and $\sigma_{2,t}$ are the additional stresses induced in the bars by the increase in temperature of Bar 2. Equilibrium relates the *thermal stresses*:

$$\sigma_{1,t}A + \sigma_{2,t}A = 0 \rightarrow \sigma_{2,t} = -\sigma_{1,t}$$

From compatibility of the thermal strains, and equilibrium, the induced thermal stresses are:

$$\sigma_{1,t} = \frac{E\alpha\,\Delta T}{2} \quad \text{and} \quad \sigma_{2,t} = -\frac{E\alpha\,\Delta T}{2}$$

Step 3. The total stress in each bar as a function of P and ΔT is found by superimposing the stresses from the mechanical and thermal cases (*Figure 4.26c*):

$$\textit{Answer: } \sigma_1 = \sigma_{1,p} + \sigma_{1,t} = \frac{1}{2}\left[\frac{P}{A} + E\alpha\,\Delta T\right]$$

$$Answer: \sigma_2 = \sigma_{2,p} + \sigma_{2,t} = \frac{1}{2}\left[\frac{P}{A} - E\alpha\,\Delta T\right]$$

Step 4. Yielding. Since P and ΔT are positive in this case, then $\sigma_1 > \sigma_2$. Tensile yielding occurs when σ_1 reaches the yield strength S_y. To avoid yielding:

$$Answer: \sigma_1 = \frac{1}{2}\left[\frac{P}{A} + E\alpha\,\Delta T\right] < S_y$$

When $\Delta T = 0$, yielding occurs when:

$$P = P_y = 2AS_y$$

When $P = 0$, yielding occurs when:

$$\Delta T = \Delta T_y = \frac{2S_y}{E\alpha}$$

Normalizing σ_1 by the yield strength S_y, the equation to avoid yielding in Bar 1 reduces to:

$$Answer: \frac{P}{2AS_y} + \frac{E\alpha\,\Delta T}{2S_y} < 1.0$$

or

$$Answer: \frac{P}{P_y} + \frac{\Delta T}{\Delta T_y} < 1.0$$

A Temperature-Force Failure Map for the system can be plotted as shown in *Figure 4.26d*. The solid line is the boundary at which yielding occurs. Provided that the operating condition – temperature change ΔT and load P – lies within this boundary, yielding does not occur.

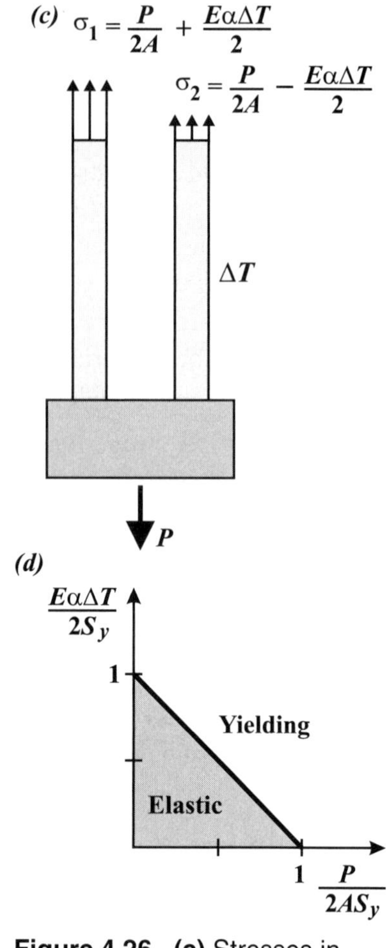

Figure 4.26. **(c)** Stresses in bars under mechanical and thermal load ΔT. **(d)** Failure Map for yielding of present two-bar structure.

4.4 Thin-Walled Pressure Vessels

Pressure vessels are used to contain gasses and pressurized liquids. The 48-in. diameter Trans-Alaska Pipeline transports oil from northern Alaska to Valdez, Alaska, over a distance of 800 miles; the maximum oil pressure is 1180 psi. The brake cylinder in an automobile contains the pressure of brake oil leading to the brake drums. In agricultural areas, domestic heating systems and gas stoves are fueled by propane gas stored in cylindrical vessels capped with hemispherical ends; their working pressures are on the order of 150 psi. The huge grain silos in America's heartland are pressure vessels (exerting pressure in the radial direction only). Even a shaken can of soda is a pressure vessel.

Two common pressure vessel geometries are ***cylindrical*** and ***spherical*** (*Figure 4.27*). The thickness of a vessel wall is often small compared to its diameter. For example, the typical wall thickness of the Alaska Pipeline is $t = 0.462$ in. and the diameter is $D = 48$ in. The ratio of thickness t to diameter D is $t/D = 1/104 = 0.0096$. In general, if t is less than $D/20 = R/10$, the vessel is described as being ***thin-walled***. This condition is satisfied in most practical situations.

The outward pressure of the contained gas or liquid is resisted by tensile stresses in the walls of the pressure vessel. Considering the vessel to be *thin-walled* ($t \ll R$) implies that these stresses are constant through the wall thickness. Any differences from the inner radius to the outer radius are taken as negligible.

On occasion, pressure vessels fail and release their contents with explosive force. It is therefore crucial that they are designed to maintain their structural integrity.

Cylindrical Pressure Vessels

A thin-walled cylindrical vessel has outer radius R, wall thickness t, and contains pressure p. The walls of the pressure vessel are subjected to a biaxial state of stress (*Figure 4.28a*). These stresses can be calculated from equilibrium of FBDs associated with cuts taken through the cylindrical part of the vessel (*Figure 4.28b*).

Hoop (Circumferential) Stress

The ***hoop stress*** σ_H is caused by the pressure acting to expand the circumference of the vessel. The *hoop stress* is calculated by taking a horizontal cut through the diametrical plane of *Figure 4.28b*, resulting

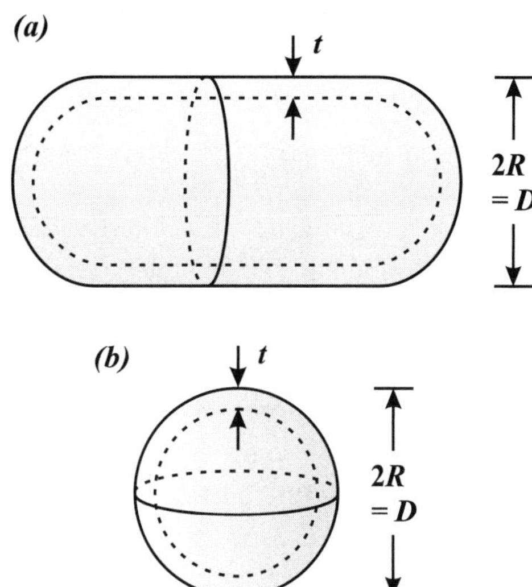

Figure 4.27. **(a)** A cylindrical pressure vessel with hemispherical end caps. **(b)** A spherical pressure vessel.

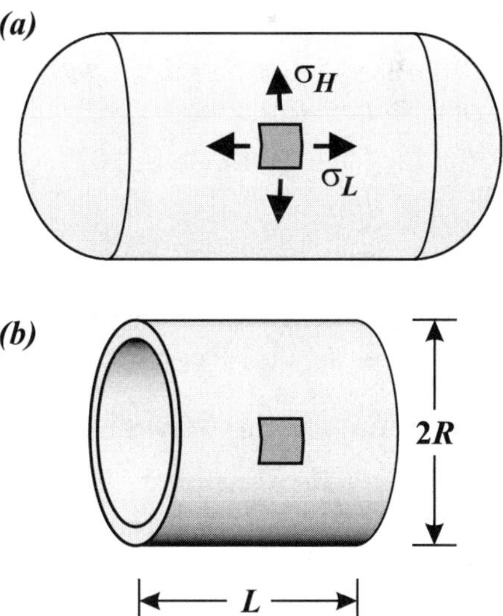

Figure 4.28. **(a)** An element in a pressure vessel is in a state of biaxial stress. **(b)** Main body of a cylindrical vessel.

in the half-cylinder in *Figure 4.29*. The downward force F_p acting on the cut plane is due to the pressure p and is:

$$F_p = p[2(R - t)L] \qquad \text{[Eq. 4.12]}$$

The pressure may be considered to act on a membrane across the cylinder diameter since a FBD can be made using any boundary.

The pressure force is counteracted by the *hoop stress* in the pressure vessel wall. The corresponding force F_w in the wall is:

$$F_w = \sigma_H[2tL] \qquad \text{[Eq. 4.13]}$$

Equating the two forces to satisfy equilibrium:

$$p[2(R - t)L] = \sigma_H[2tL] \qquad \text{[Eq. 4.14]}$$

and solving for the hoop stress:

$$\sigma_H = \frac{pR}{t}\left[1 - \frac{t}{R}\right] \qquad \text{[Eq. 4.15]}$$

Since t is small compared to R, the term t/R is much smaller than unity (1.0) and is dropped to give the **hoop** or **circumferential stress**:

$$\sigma_H = \frac{pR}{t} \qquad \text{[Eq. 4.16]}$$

Since R/t is large, the hoop stress in the vessel is generally 10 or more times larger than the enclosed pressure.

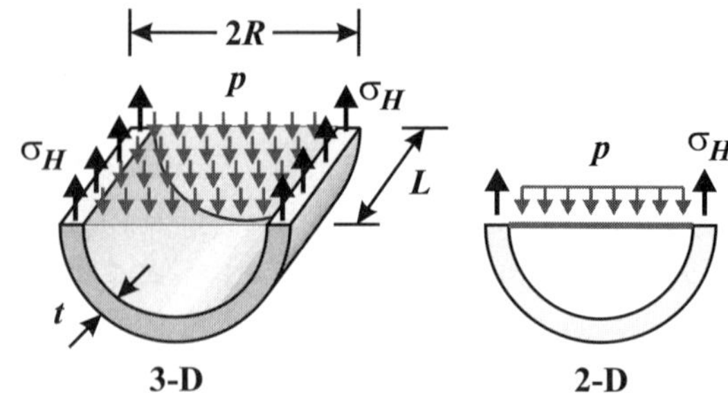

Figure 4.29. Determination of the **hoop** or **circumferential stress**. At the diametrical cut, the downward force is caused by the pressure, and the upward force is caused by the *hoop stress*.

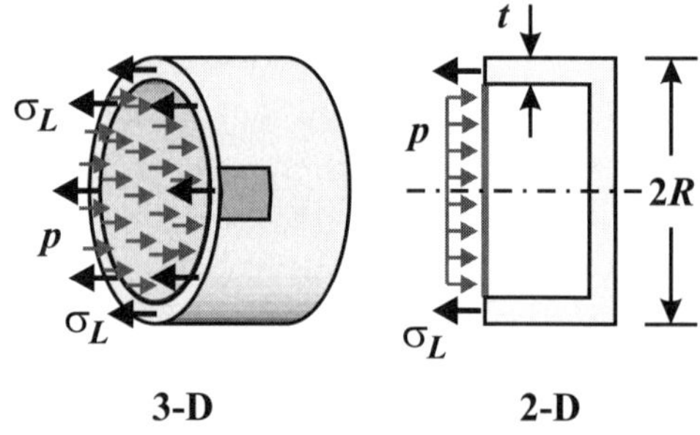

Figure 4.30. Determination of the **longitudinal** or **axial Stress**. At the vertical cut, the force acting to the right is caused by the pressure, and the force acting to the left is caused by the *longitudinal stress*.

Longitudinal (Axial) Stress

The **longitudinal stress** σ_L is caused by the pressure acting against the cylinder end caps. The *longitudinal stress* is calculated by considering the forces on the cross-section of the cylinder by taking a cut perpendicular to the cylinder axis (*Figure 4.30*). The pressure force F_p acts to the right over area $A = \pi(R{-}t)^2$:

$$F_p = p\pi(R - t)^2 = p\pi R^2\left[1 - \frac{t}{R}\right]^2 \qquad \text{[Eq. 4.17]}$$

Since $t/R \ll 1$, the pressure force is approximately:

$$F_p = p[\pi R^2]$$ [Eq. 4.18]

This force is counteracted by the longitudinal stress σ_L in the pressure vessel wall. The projected cross-sectional area of the wall is:

$$A = \pi[R^2 - (R-t)^2] = \pi[2Rt - t^2] = 2\pi Rt\left[1 - \frac{t}{2R}\right]$$ [Eq. 4.19]

Since $t/R \ll 1$, the expression for the area reduces to $A = 2\pi Rt$ (the product of the circumference and thickness). The longitudinal force F_w in the vessel wall is therefore:

$$F_w = \sigma_L[2\pi Rt]$$ [Eq. 4.20]

Equating the two forces to satisfy equilibrium:

$$p[\pi R^2] = \sigma_L[2\pi Rt]$$ [Eq. 4.21]

and solving for the **longitudinal** or **axial stress**:

$$\sigma_L = \frac{pR}{2t}$$ [Eq. 4.22]

By symmetry, no shear stresses act on a material element oriented with the hoop- and axial-directions. The pressure vessel material is in a **biaxial state of stress**, with stress components:

$$\sigma_H = \frac{pR}{t} \quad \text{and} \quad \sigma_L = \frac{pR}{2t}$$ [Eq. 4.23]

Note that the hoop stress is *twice* as large as the axial stress; it is the hoop stress that governs the strength of a cylindrical pressure vessel (*Figure 4.31*).

On the outer surface of the vessel, the normal stress is zero. On the inner surface, pressure p acts against the wall, but because R is much larger than t, the *in-plane* stresses σ_H and σ_L are much larger than pressure p (*Equation 4.23*). The stress element in a thin-walled pressure vessel is thus considered to be in a state of *plane stress*.

The stresses of *Equation 4.23* are for material elements in the main (cylindrical) part of the vessel, away from any openings or geometric irregularities which cause *stress concentrations*. Also, the internal pressure must be greater than the external pressure, i.e., a gas held in compression, stretching the walls in tension. The equations are not valid for an outside pressure (e.g., submarines) since outside pressure causes a different failure condition

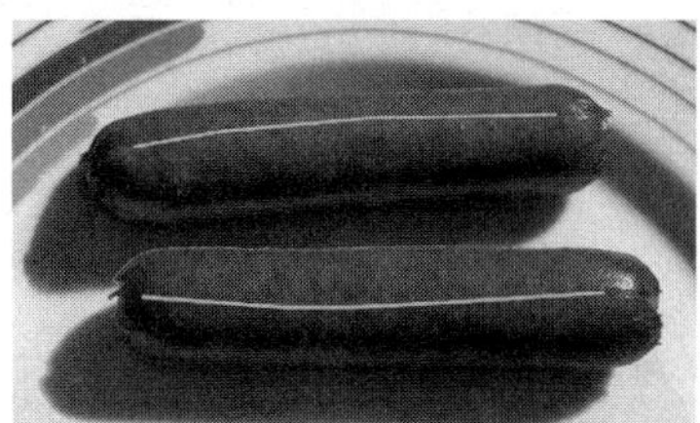

Figure 4.31. Hot dogs are cylindrical vessels pressurized with steam. Since the hoop stress is twice the axial stress, the dog's skin splits along its axis.

(buckling, as when the walls of an empty aluminum can are crushed).

The stresses developed in *thick-walled pressure vessels* (e.g., cannon barrels, nuclear pressure vessels, etc.) are beyond the scope of this text. They are found using the Theory of Elasticity.

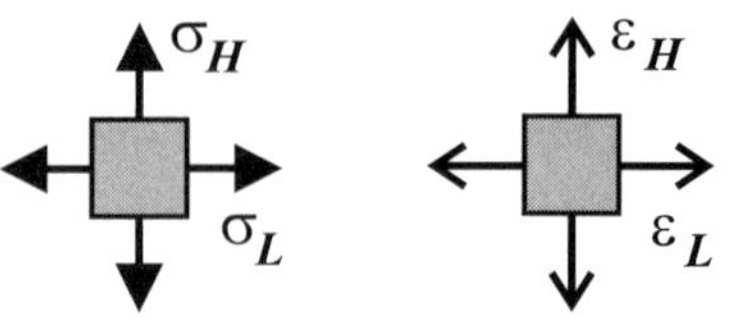

Figure 4.32. Stress and strain states for a cylindrical pressure vessel.

Strains in Cylindrical Vessels

Once the stresses are known, the *hoop* and *longitudinal strains* can be determined using Hooke's Law. Since there are stress components in two directions, the strains must include the *Poisson effect* (*Figure 4.32*). The elastic strains caused by:

- the *hoop stress* σ_H acting alone are:
$$\varepsilon_{H,H} = \frac{\sigma_H}{E} \quad ; \quad \varepsilon_{L,H} = -v\frac{\sigma_H}{E}$$

- the *axial stress* σ_L acting alone are:
$$\varepsilon_{H,L} = -v\frac{\sigma_L}{E} \quad ; \quad \varepsilon_{L,L} = \frac{\sigma_L}{E}$$

Strain $\varepsilon_{i,j}$ represents the strain in the *i*-direction due to stress in the *j*-direction.

The total strains at a particular point due to stresses σ_H and σ_L acting at the same time are found by *superimposing* the strains due to each stress acting independently. Thus:

$$\varepsilon_H = \frac{\sigma_H}{E} - v\frac{\sigma_L}{E} \quad \text{and} \quad \varepsilon_L = \frac{\sigma_L}{E} - v\frac{\sigma_H}{E} \qquad \text{[Eq. 4.24]}$$

Substituting the expressions for σ_H and σ_L gives:

$$\varepsilon_H = \frac{pR}{tE}\left(1 - \frac{v}{2}\right) \quad \text{and} \quad \varepsilon_L = \frac{pR}{tE}\left(\frac{1}{2} - v\right) \qquad \text{[Eq. 4.25]}$$

Strain–Displacement

A cylindrical pressure vessel expands radially by ΔR and axially by ΔL. The change of radius ΔR is calculated from the hoop strain:

$$\varepsilon_H = \frac{\text{Change in circumference}}{\text{Orginal circumference}} = \frac{2\pi(R + \Delta R) - 2\pi R}{2\pi R} = \frac{\Delta R}{R} \qquad \text{[Eq. 4.26]}$$

The increase in radius is:
$$\Delta R = R\varepsilon_H = \frac{pR^2}{tE}\left(1 - \frac{v}{2}\right) \qquad \text{[Eq. 4.27]}$$

The change in length of the cylinder is:
$$\Delta L = L\varepsilon_L = \frac{pRL}{tE}\left(\frac{1}{2} - v\right) \qquad \text{[Eq. 4.28]}$$

Example 4.20 Alaska Pipeline

Given: The Trans-Alaska Pipeline is constructed of steel pipe ($E = 29$ Msi, $S_y = 65$ ksi, $v = 0.3$), with an outside diameter of $D = 48$ in. (*Figure 4.33*). The typical operating pressure is $p_D = 840$ psi.

Required: (a) Estimate the thickness t of the Alaska Pipeline using a factor of safety of 1.5. (b) Estimate the increase in pipe diameter ΔD due to the operating pressure. Neglect any Poisson effect due to the axial stress.

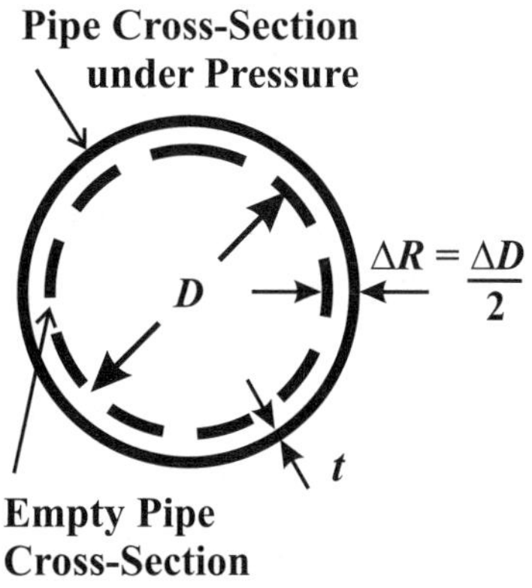

Solution: *Step 1.* Thickness t. With a FS = 1.5 (actual value), the allowable stress in the pipe is:

$$\sigma_A = \frac{S_y}{FS} = \frac{65 \text{ ksi}}{1.5} = 43.33 \text{ ksi}$$

The maximum stress in the wall is the hoop stress:

$$\sigma_H = \frac{pR}{t}$$

Pipe Cross-Section under Pressure

Empty Pipe Cross-Section

Figure 4.33. Cross-section of pipeline. Not to scale.

Setting the hoop stress equal to the allowable stress and the pressure equal to the operating pressure, then solving for the required thickness gives:

$$t = \frac{p_D R}{\sigma_A} = \frac{(840 \text{ psi})(24 \text{ in.})}{43.33 \text{ ksi}}$$

Answer: $t = 0.465$ in.

The typical wall thickness in the Alaska Pipeline is actually 0.462 in.
Source: http://www.alyeska-pipe.com/Pipelinefacts/Pipe.html   Accessed May 2008.

Step 2. The increase in diameter under operating conditions is found from the hoop strain. Neglecting the Poisson effect due to axial stress:

$$\varepsilon_H \approx \frac{pR}{tE} = \frac{(840 \text{ psi})(24 \text{ in.})}{(0.465 \text{ in.})(29 \times 10^6 \text{ psi})} = 0.001495$$

The increase in diameter is: $\Delta D = D\varepsilon_H = (48 \text{ in.})(0.001495)$

Answer: $\Delta D = 0.072$ in.

Spherical Pressure Vessels

Spherical Stress

In spherical pressure vessels (*Figure 4.34*), the *spherical stress* σ_S is calculated by considering a diametrical cut through the vessel (*Figure 4.35*). This cut results in the same configuration as the *longitudinal stress* in the cylindrical pressure vessel.

The pressure force F_p acts over area $\pi(R{-}t)^2$, and recalling that $t << R$, then:

$$F_p = p\pi(R - t)^2 = p\pi R^2\left(1 - \frac{t}{R}\right)^2$$

$$= p\pi R^2$$

[Eq. 4.29]

The force in the vessel wall that resists the pressure force is:

$$F_w = (2\pi R t)\sigma_S \qquad \text{[Eq. 4.30]}$$

Equating the forces and solving for the *spherical stress*:

$$\sigma_S = \frac{pR}{2t} \qquad \text{[Eq. 4.31]}$$

No matter the direction of the diametrical cut, the sphere looks the same. Thus, the normal stress on a surface element oriented in any direction is the same. Due to symmetry, no shear stress acts, so a material point is under a *biaxial state of stress*, with both stresses being equal (*Figure 4.36*).

On the outer surface of the vessel, the normal stress is zero. On the inner surface, pressure p acts against the wall, but because R is much larger than t, the *in-plane* stresses σ_S are much larger than pressure p (*Equation 4.31*). The stress element in a thin-walled pressure vessel is considered to be in a state of *plane stress*.

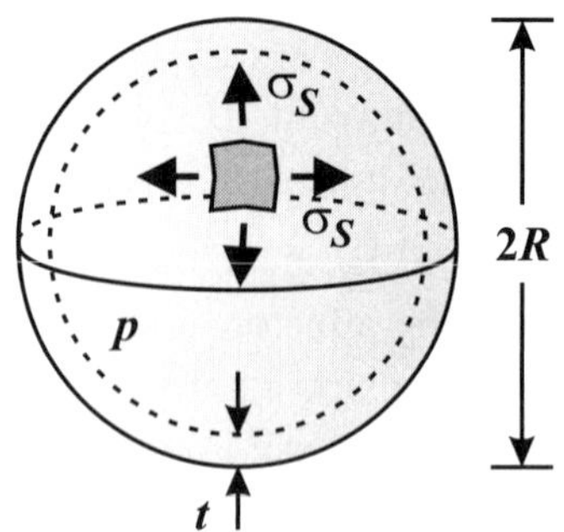

Figure 4.34. A spherical pressure vessel.

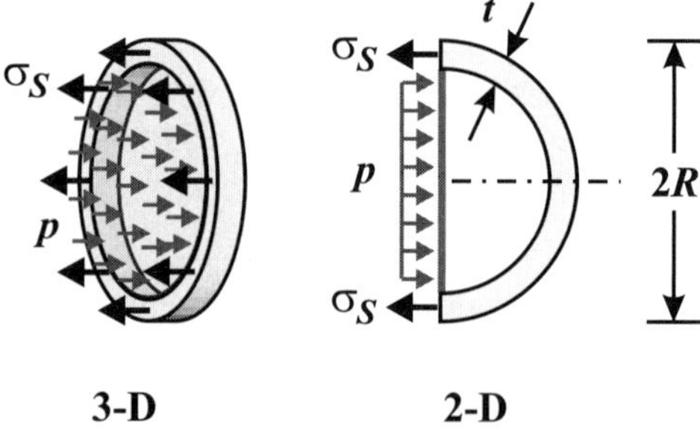

Figure 4.35. Force across a diametrical cut of a spherical pressure vessel.

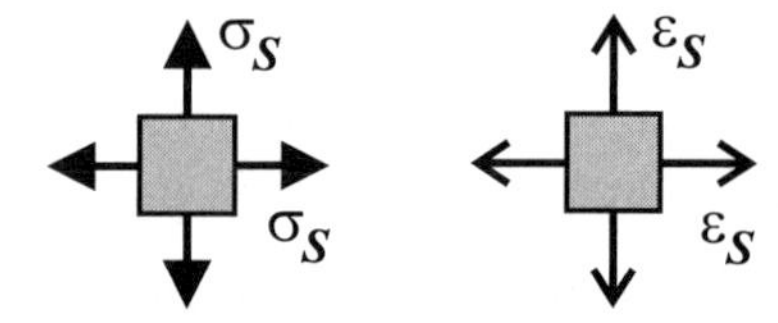

Figure 4.36. Stress and strain states.

Strains and Displacements in Spherical Vessels

The spherical strain ε_S is calculated from Hooke's Law:

$$\varepsilon_S = \frac{\sigma_S}{E} - v\frac{\sigma_S}{E} = \frac{pR}{2tE}(1 - v) \qquad \text{[Eq. 4.32]}$$

By symmetry, this is the strain in any direction in the wall (*Figure 4.36*).

The change in radius is:

$$\Delta R = \varepsilon_S R = \frac{pR^2}{2tE}(1 - v) \qquad \text{[Eq. 4.33]}$$

Example 4.21 Propane Tank

Given: A propane tank for a barbecue (BBQ) is assumed to be spherical. The tank is made of steel ($S_y = 40$ ksi) and has a diameter of $D = 16$ in. The propane is stored at $p_D = 30$ psi.

Required: Determine the required thickness t if the design factor of safety is 10.0.

Solution: The allowable stress is:

$$\sigma_A = \frac{S_y}{\text{FS}} = \frac{40 \text{ ksi}}{10} = 4.0 \text{ ksi}$$

The spherical stress is: $\quad \sigma_S = \frac{pR}{2t}$

Setting the spherical stress equal to the allowable stress and the pressure equal to the design pressure, the required thickness is:

$$t = \frac{p_D R}{2\sigma_A} = \frac{(30 \text{ psi})(8.0 \text{ in.})}{2(4000 \text{ psi})}$$

Answer: $t = 0.030$ in.

This is the minimum thickness required to contain the pressure at the design factor of safety. Other types of loadings, such as just moving the vessel, hole geometry, etc., must also be considered; they will require a thicker wall.

Proof Testing of Pressure Vessels

It is a matter of principle that major components are proof tested before going into service. For safety, proof testing of pressure vessels is especially important.

The design or working pressure is p_D, while the ultimate or failure pressure is p_f. The *proof load* p_P lies between the working and failure pressures. The American Society of Mechanical Engineers (ASME) Pressure Vessel Code requires that pressure vessels be tested to a proof load of at least $1.5p_D$. The vessel is loaded to 50% more than the maximum pressure expected in service.

Even though a vessel is designed to contain pressurized gas, proof testing is usually performed with compressed water. For the same pressure, the energy of pressurized gas is over 200 times greater than the energy of pressurized water. Failure of a pressure vessel filled with water is not as violent as a test conducted with gas; the water will simply leak out, while the gas may send metal shrapnel flying.

4.5 Stress Concentration Factors

It is often necessary to drill holes in a component, or to design a component with a change in cross-section. When component geometry changes, such as at a hole in a plate (*Figure 4.37*), the stress at that location is elevated. The ratio of the maximum stress σ_{max} to an average stress σ_{ave} in the system is the **stress concentration factor** *SCF*:

$$SCF = \frac{\sigma_{max}}{\sigma_{ave}} \qquad \text{[Eq. 4.34]}$$

Equivalently, the maximum stress in the system is:

$$\sigma_{max} = (SCF)\sigma_{ave} \qquad \text{[Eq. 4.35]}$$

Understanding how stress is elevated due to local changes in component geometry is important in creating successful designs. The following introductory discussion is valid for materials that remain linear–elastic.

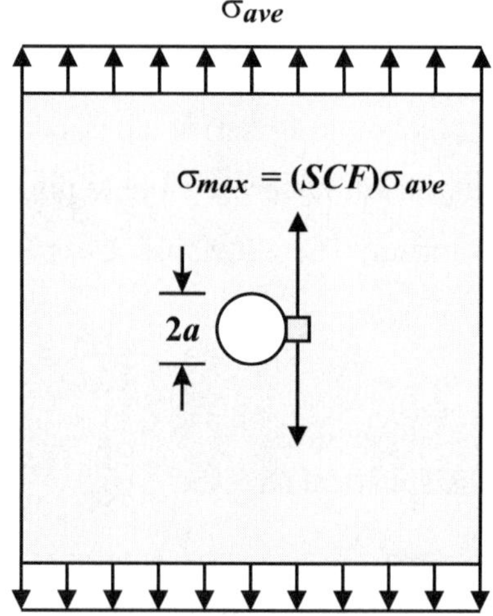

Figure 4.37. Hole in a plate subjected to stress σ_{ave}.

Elliptical Hole in an Infinite Plate

Figure 4.38a represents an **infinite plate** of finite thickness *t*, with an elliptical hole of major diameter *2a* and minor diameter *2b*. An *infinite plate* has width and height dimensions much greater than the plate thickness and hole dimensions. Thus, the effect of plate size on the material response near the hole is negligible. The plate is loaded by stress σ_{ave}, normal to diameter *2a*.

For an ellipse, the radius of curvature ρ of the hole at $x = \pm a$ is:

$$\rho = \frac{b^2}{a} \qquad \text{[Eq. 4.36]}$$

For large values of *b/a* (*b/a* > 1, *Figure 4.38b*), ρ is large, and the change in geometry at the sides of the hole in the stress direction is gradual. For small values of *b/a* (*b/a* < 1,

Figure 4.38a), ρ is small, and the hole is relatively sharp. Sharp corners give rise to large stress concentration factors.

If the plate material remains linear elastic, then the stress concentration factor for a stress element at the side of the ellipse is:

$$SCF = \frac{\sigma_{max}}{\sigma_{ave}} = 1 + 2\frac{a}{b} \qquad \text{[Eq. 4.37]}$$

or:

$$SCF = 1 + 2\sqrt{\frac{a}{\rho}} \qquad \text{[Eq. 4.38]}$$

The more rapid the change in geometry – the larger the ratio a/b or a/ρ – the greater the stress concentration factor.

Circular Hole in an Infinite Plate

A circle is an ellipse with radius $a = b = \rho$ (*Figure 4.37*). The stress concentration factor at the surface of a circular hole in an infinite plate loaded by stress σ_{ave} is:

$$SCF = \frac{\sigma_{max}}{\sigma_{ave}} = 3 \qquad \text{[Eq. 4.39]}$$

A complete elastic stress analysis (beyond the scope of this text) determines the stress on the cross-section through the center of the circle (*Figure 4.39*). The stress on this cross-section as a function of distance r ($r \geq a$) from the center of the circle is:

$$\sigma(r) = \sigma_{ave}\left[1 + \frac{1}{2}\left[\left(\frac{a}{r}\right)^2 + 3\left(\frac{a}{r}\right)^4\right]\right]$$

$$\text{[Eq. 4.40]}$$

Equation 4.40 reduces to $3\sigma_{ave}$ at the hole surface $r = a$. In an infinite plate, the stress far from the hole ($r \gg a$) is σ_{ave}.

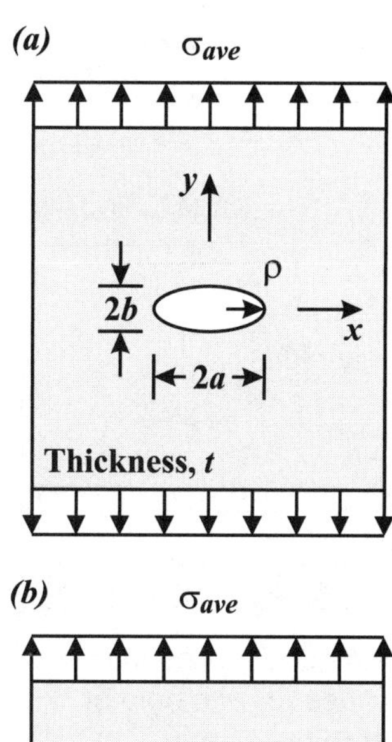

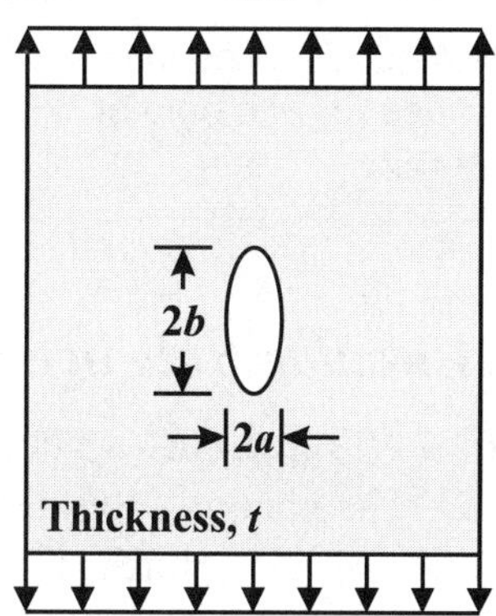

Figure 4.38. **(a)** Plate with an elliptical hole, $a > b$. **(b)** Plate with a hole, $a < b$.

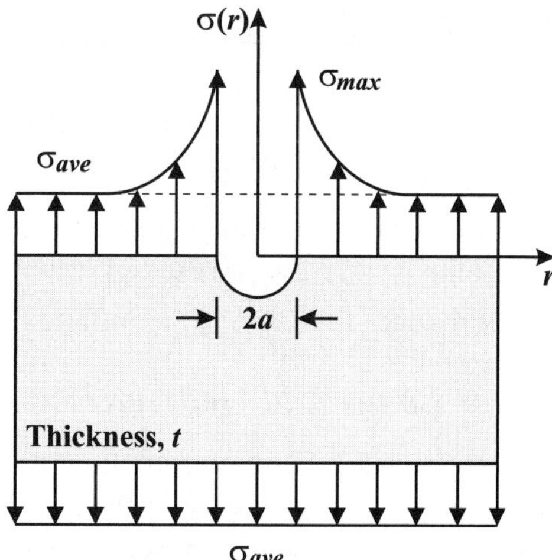

Figure 4.39. Stress distribution in an infinite plate with a circular hole.

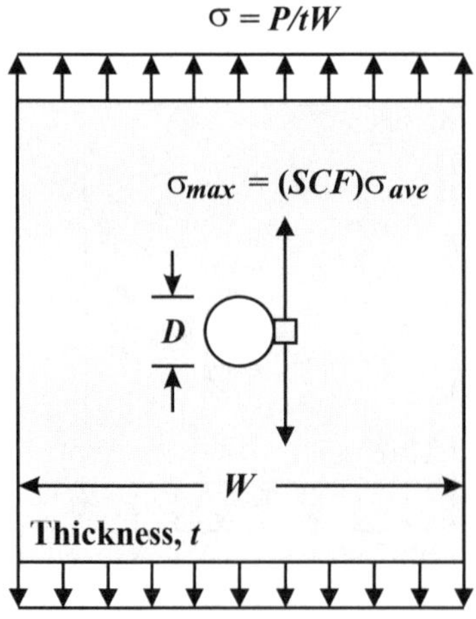

Figure 4.40. Plate of finite width W, thickness t, with hole of diameter D.

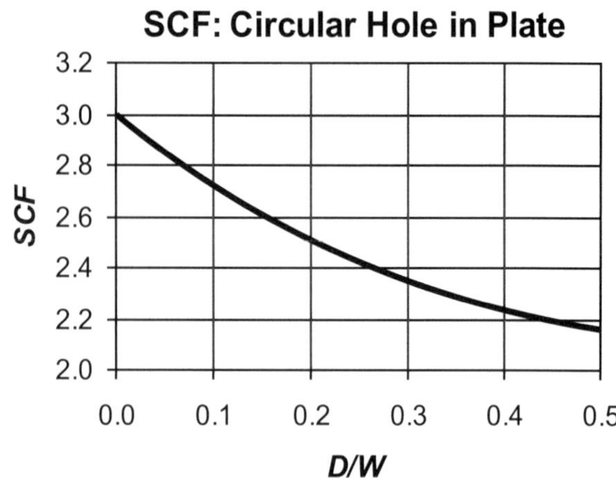

Figure 4.41. *SCF* for tensile stress at hole of diameter D at center of a plate of width W.

Center Circular Hole in a Plate of Finite Width

Figure 4.40 represents a plate of finite width W and thickness t, having a hole of diameter D. The plate is subjected to axial load P, causing *applied stress* $\sigma = P/tW$. The *average stress* σ_{ave} is defined at the **net section** (the cross-section with the least area):

$$\sigma_{ave} = \frac{P}{t(W-D)} \qquad \text{[Eq. 4.41]}$$

The stress concentration factor for this case is (from R. Roark and W. Young, *Formulas for Stress and Strain*, 1975):

$$SCF = \frac{\sigma_{max}}{\sigma_{ave}} = 3.00 - 3.13\left(\frac{D}{W}\right) + 3.66\left(\frac{D}{W}\right)^2 - 1.53\left(\frac{D}{W}\right)^3 \qquad \text{[Eq. 4.42]}$$

The stress concentration factor *SCF* is plotted against *D/W* in *Figure 4.41*. For large values of W ($D \ll W$, $D/W \to 0$), σ_{ave} reduces to P/tW, and the stress concentration factor reduces to *SCF* = 3, the infinite plate solution (*Equation 4.39*).

Other Loads and Geometries

Stress concentration factors for other load types (bending, torsion, and shear), and for other changes in geometry (reduction in area, notches, etc.) are available in reference books and online. One such website is www.fatiguecalculator.com.

Example 4.22 **Stress Concentration Factor: Elliptical Hole in an Infinite Plate**

Given: An elliptical hole $2a = 20$ mm wide by $2b = 5.0$ mm high, in an infinite plate (*Figure 4.38a*). The plate is subjected to a stress of $\sigma = 20$ MPa.

Required: Determine (a) the stress concentration factor and (b) the maximum stress in the plate.

Solution: *Step 1.* The stress concentration factor is:

$$SCF = 1 + 2\frac{a}{b} = 1 + 2\frac{(20/2)}{(5/2)}$$

Answer: $\underline{SCF = 9}$

Step 2. The maximum stress is therefore:

$$\text{Answer: } \sigma_{max} = 9(20 \text{ MPa}) = 180 \text{ MPa}$$

Example 4.23 **Stress Concentration Factor: Circular Hole in an Infinite Plate**

Given: A circular hole of diameter $D = 0.5$ in. in an infinite plate (*Figures 4.37* and *4.39*). The plate is subjected to a stress of $\sigma = 5.0$ ksi.

Required: Determine (a) the stress at the hole surface and (b) the stress on the net section at $r = 0.5$ in. (one diameter from the center of the circle).

Solution: *Step 1.* The stress concentration factor at the hole surface is $SCF = 3$. Hence:

$$\text{Answer: } \sigma_{max} = 3\sigma = 15.0 \text{ ksi}$$

Step 2. At distance $r = D = 2a$ from the hole center, the stress on the net section $\sigma\,(r)$ is:

$$\sigma(r = 2a) = \sigma\left[1 + \frac{1}{2}\left[\left(\frac{a}{r}\right)^2 + 3\left(\frac{a}{r}\right)^4\right]\right] = \sigma\left[1 + \frac{1}{2}\left[\left(\frac{a}{2a}\right)^2 + 3\left(\frac{a}{2a}\right)^4\right]\right] = 1.22\sigma$$

Answer: $\underline{\sigma(2a) = 6.1 \text{ ksi}}$

Example 4.24 **Stress Concentration Factor: Circular Hole in a Finite Plate**

Given: A hole of diameter $D = 0.5$ in. is drilled in a steel plate of finite width $W = 6.0$ in. and thickness $t = 0.25$ in. (*Figure 4.40*). The plate is subjected to force P. The yield strength is $S_y = 45$ ksi.

Required: Determine the load $P = P_y$ that will cause yielding.

Solution: *Step 1.* The stress concentration factor at the hole surface, for $D/W = 1/12$, is:

$$SCF = \frac{\sigma_{max}}{\sigma_{ave}} = 3.00 - 3.13\left(\frac{D}{W}\right) + 3.66\left(\frac{D}{W}\right)^2 - 1.53\left(\frac{D}{W}\right)^3 = 2.76$$

Step 2. For yielding to occur, $\sigma_{max} = S_y$, so:

$$\sigma_{ave} = \frac{\sigma_{max}}{SCF} = \frac{45 \text{ ksi}}{2.76} = 16.3 \text{ ksi}$$

$$P_y = t(W - D)\sigma_{ave} = (0.25)(6.0 - 0.5)(16.3)$$

Answer: $\underline{P_y = 22.4 \text{ kips}}$

4.6 Energy Methods

It was noted earlier that axial members are essentially linear springs. One of the properties of a spring is that as it supports load P and elongates Δ, it stores energy (*Figure 4.42*). The elastic energy stored in a spring is:

$$U = \frac{1}{2}P\Delta = \frac{1}{2}\frac{P^2}{K} = \frac{1}{2}K\Delta^2 \qquad \text{[Eq. 4.43]}$$

Whether a spring is in tension ($P > 0$, $\Delta > 0$) or compression ($P < 0$, $\Delta < 0$), the elastic energy is always positive.

Recall that the spring stiffness K of an axial member of constant area A and modulus E subjected to constant force P over its length L is:

$$K = \frac{EA}{L} \qquad \text{[Eq. 4.44]}$$

so the elastic energy stored in an axial bar is:

$$U = \frac{1}{2}P\Delta = \frac{1}{2}\frac{P^2L}{AE} = \frac{1}{2}\frac{EA}{L}\Delta^2 \qquad \text{[Eq. 4.45]}$$

Now, consider force P applied to an assembly of elastic springs (*Figure 4.43*). The ith spring in the assembly has stiffness K_i, elongates Δ_i, supports load P_i, and stores energy U_i:

$$U_i = \frac{1}{2}P_i\Delta_i = \frac{1}{2}\frac{P_i^2}{K_i} = \frac{1}{2}K_i\Delta_i^2 \qquad \text{[Eq. 4.46]}$$

The total elastic energy stored in the assembly of springs is:

$$U = \sum \frac{1}{2}P_i\Delta_i = \sum \frac{1}{2}\frac{P_i^2}{K_i} = \sum \frac{1}{2}K_i\Delta_i^2 \qquad \text{[Eq. 4.47]}$$

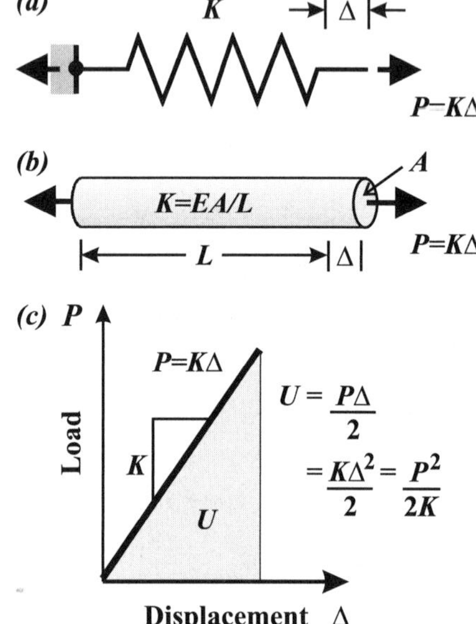

Figure 4.42. **(a)** Spring under force *P.* **(b)** Axial member under force *P.* **(c)** The area under the *P–Δ* curve equals the energy stored in the system.

Substituting the axial stiffness expression:

$$U = \sum \frac{1}{2} \frac{P_i^2 L_i}{A_i E_i} = \sum \frac{1}{2} \frac{E_i A_i}{L_i} \Delta_i^2$$

[Eq. 4.48]

Thus, knowing either all of the internal forces P_i (from the force method), or all of the elongations Δ_i (from the displacement method), the total internal energy can be determined.

The total energy stored is equal to the work done by the applied load:

$$W = \frac{1}{2} P \delta$$

[Eq. 4.49]

where P and δ *must be in the same direction.*

Equating the applied work to the total elastic energy stored:

$$\frac{1}{2} P \delta = U$$

[Eq. 4.50]

Knowing either P or δ, the other can be solved from the stored energy U.

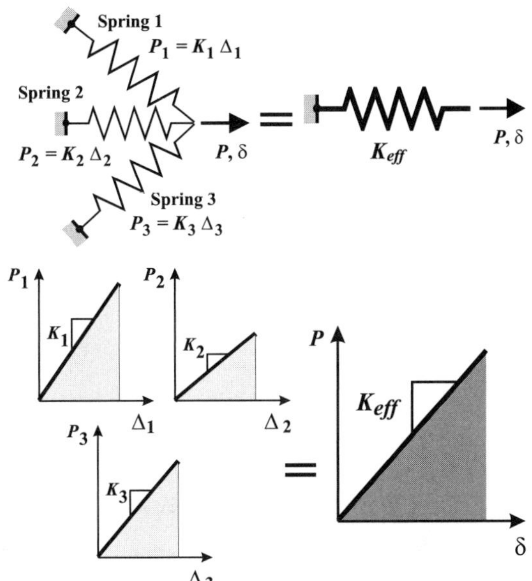

Figure 4.43. As far as the applied load is concerned, an assembly of springs is equivalent to a single spring. The sum of the energies stored in the individual springs of an assembly is equal to the work done on the assembly.

The *effective stiffness* of the system *in the direction of the applied load* is the ratio of the load to the displacement in the direction of the load:

$$K_{eff} = \frac{P}{\delta}$$

[Eq. 4.51]

From the point of view of the load P, how the springs are assembled does not matter. Load P is simply acting on a structure that has an effective stiffness of K_{eff}.

Example 4.25 The Lamp Problem, Using the Energy Method
(Example 4.5, Example 4.13)

Given: A lamp weighing $W = 14.0$ lb is supported by two wires, both of length $L = 5.0$ ft (*Figure 4.44*). The distance between the two cable mounts is $s = 8.0$ ft, and $H = 3.0$ ft. The wires have a diameter of $D = 0.10$ in. ($A = 0.007854$ in.^{2}), and are made of steel ($E = 30 \times 10^6$ psi, $S_y = 50$ ksi). Assume the wires remain elastic.

Required: Determine (a) the energy stored in the wires, (b) the downward displacement of the lamp due to its weight, and (c) the stiffness of the wire assembly in the vertical direction $K = W/\delta$.

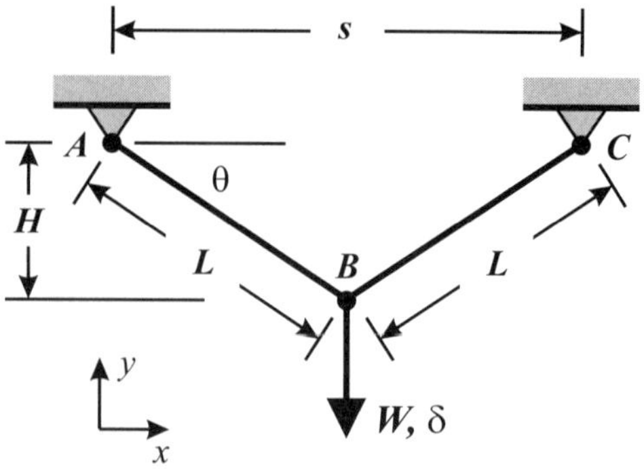

Solution: Since this problem is *statically determinate*, the *force method* is used. From equilibrium equations in *Example 4.5*, the tensile force in each wire was found to be:

$$T = T_{AB} = T_{BC}$$

$$= \frac{W}{2 \sin \theta} = \frac{W}{2(3/5)} = 11.67 \text{ lb}$$

where θ is the angle of the wires measured from the horizontal.

Figure 4.44. The lamp supported by two wires.

The total energy stored in the two wires is:

$$U = \sum \frac{1}{2}\frac{P_i^2 L_i}{AE}$$

$$= 2 \times \frac{1}{2}\left[\frac{(11.67 \text{ lb})^2(60 \text{ in.})}{(0.007854 \text{ in.}^2)(30 \times 10^6 \text{psi})}\right]$$

Answer: $U = 0.0347$ in-lb

The downward displacement is therefore:

$$\delta = \frac{2U}{W} = \frac{2(0.0359 \text{ in-lb})}{14.0 \text{ lb}}$$

Answer: $\delta = 0.00495$ in.

This is the same result obtained for the displacement as in *Examples 4.5* and *4.13*.

The stiffness of the wire assembly is:

$$K = \frac{W}{\delta} = \frac{14.0 \text{ lb}}{0.00495 \text{ in.}} \qquad \text{[Eq. 4.52]}$$

Answer: $K = 2830$ lb/in.

Example 4.26   The Bike Wheel Problem, Using the Energy Method
(*Example 4.14*)

Given: The bicycle wheel of radius R has N spokes, each of cross-sectional area A (*Figure 4.45*). The elastic modulus of the spokes is E. The rim and hub are assumed to be rigid. The weight and dynamic forces of the rider cause downward force F at the rigid wheel hub, displacing it downward distance δ. Assume the system remains elastic.

Required: Determine the relationship between force F and displacement δ; i.e., the stiffness of the assembly of spokes for a downward load applied at the hub. Use the *energy method.*

Solution: Since this problem is *statically indeterminate*, the *displacement method* is used to set up the *energy method*. Consider a triangular-shaped element $d\theta$ at angle θ to the horizontal (*Figure 4.21d, dashed triangle*). The number of spokes represented by element $d\theta$ is:

$$dN = N\frac{d\theta}{2\pi}$$

The cross-sectional area of the spokes in $d\theta$ is:

$$dA = AN\frac{d\theta}{2\pi}$$

Step 1. Displacement–elongation. From the geometry of the displacement, the elongation of a spoke element $\Delta(\theta)$ with respect to hub displacement δ is (*Figure 4.45e*):

$$\Delta(\theta) = \delta \sin \theta$$

Step 2. The internal energy stored in the *i*th spoke element at angle θ is:

$$dU = \frac{1}{2}\left[\frac{E_iA_i}{L_i}\right]\Delta_i^2$$

$$= \frac{1}{2}\left[\frac{E}{R}\left(AN\frac{d\theta}{2\pi}\right)\right](\delta \sin\theta)^2$$

Integrating from 0 to 2π to find the total internal elastic energy in the spokes:

$$U = \sum\frac{1}{2}\frac{E_iA_i}{L_i}\Delta_i^2$$

$$= \int_0^{2\pi}\frac{1}{2}\left[\frac{E}{R}\left(\frac{AN}{2\pi}\right)\right](\delta \sin\theta)^2 d\theta$$

$$= \frac{EAN\delta^2}{4\pi R}(\pi) = \frac{EAN\delta^2}{4R}$$

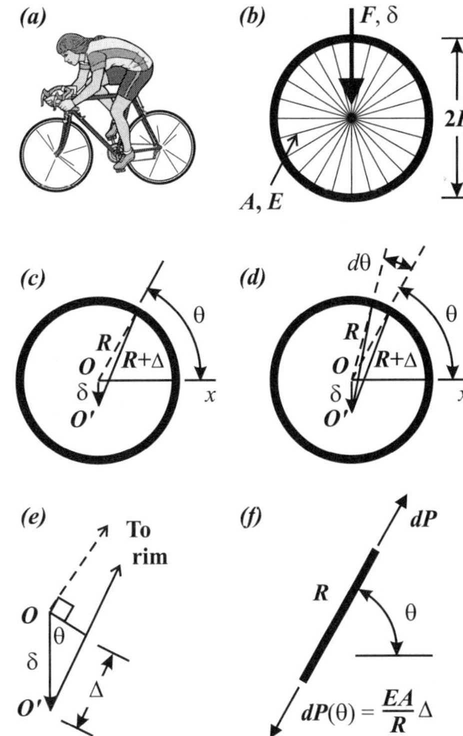

Figure 4.45. **(a)** The bicycle rider applies load *F* to the rear wheel. **(b)** The wheel is made up of *N* spokes, *R* long, each spoke of cross-sectional area *A*, and modulus *E*. **(c)** The hub displaces downward, so a single spoke at angle θ moves from its original (dashed) position to a new (solid) one, elongating by Δ. **(d)** An element $d\theta$ moves from its unloaded dotted position to the solid one. **(e)** Movement of point *O* to point *O'*, giving the relationship between δ and Δ. **(f)** FBD of a spoke. Copyright ©2008 Dominic J. Dal Bello and licensors. All rights reserved.

Step 3. The downward displacement is therefore:

$$F = \frac{2U}{\delta} = \frac{2}{\delta}\left(\frac{EAN\delta^2}{4R}\right) = \frac{EAN}{2R}\delta$$

The stiffness of the spoke system is:

$$Answer:\ K = \frac{F}{\delta} = \frac{EAN}{2R}$$

which agrees with the result of *Example 4.14* using the *displacement method* alone. Note that the *energy method* was much simpler to apply. In *Example 4.14*, both the force and the elongation of each spoke element needed to be determined. Using the energy method, only the elongation of each member was required.

The *energy method* is a very useful and powerful tool. Mathematical operations dealing with energy (a scalar) are easier to apply than those dealing with forces (vectors). The energy method is the basis for most computer structural analysis tools. *Chapter 11* covers the energy method in more detail.

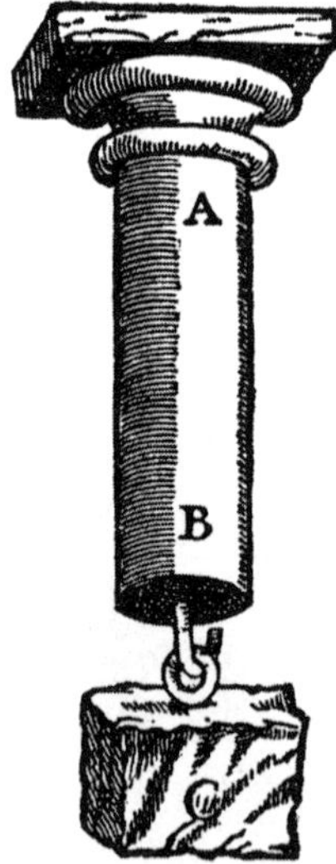

Figure 4.46. Galileo's sketch of an axial member under a tensile load (*Dialogues Concerning Two New Sciences*, 1638. Translated by H. Crew and A. de Salvio, 1914. Reissued by Northwestern University Press, 1968).

5.0 Introduction

A common method of transmitting power is by means of *torque* in a rotating shaft. Shafts in turbines, motors, and automobile drive systems are of circular cross-section. Aircraft wings are also subject to torsion, and the torque is transmitted by a torsion box that runs along the interior of the wing. The operation of micro-mechanical devices often depends on the torsional properties of a micro-shaft.

These types of components are considered here under the general heading of *torsion members*.

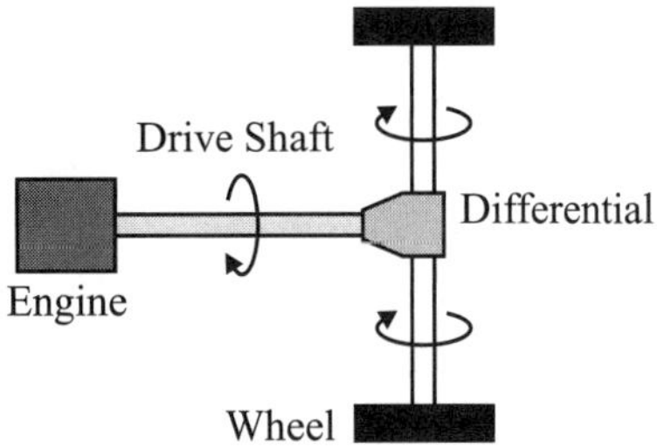

Figure 5.1. Schematic of the assembly of torsion members to propel automobiles.

5.1 Shafts of Circular Cross-Section

Thin-Walled Shaft in Torsion

Recall from *Chapter 3*, the thin-walled circular shaft of thickness t and average radius R ($t \ll R$), subjected to torque T (*Figure 5.2*). The **shear stress** τ, **shear strain** γ, and **angle of twist** θ between two cross-sections distance L apart, are:

$$\tau = \frac{T}{2\pi R^2 t} \qquad\qquad \text{[Eq. 5.1]}$$

$$\gamma = \frac{\tau}{G} = \frac{T}{2\pi R^2 t G} \qquad\qquad \text{[Eq. 5.2]}$$

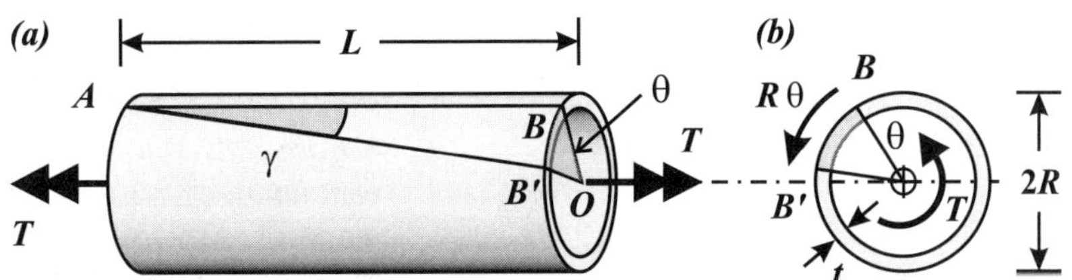

Figure 5.2. (a) Thin-walled shaft of radius R and thickness t, subjected to torque T. **(b)** Cross-section of thin-walled shaft.

F.A. Leckie, D.J. Dal Bello, *Strength and Stiffness of Engineering Systems*, Mechanical Engineering Series, **133**
DOI 10.1007/978-0-387-49474-6_5, © Springer Science+Business Media, LLC 2009

$$\theta = \frac{\gamma L}{R} = \frac{TL}{2\pi R^3 t G} \qquad \text{[Eq. 5.3]}$$

Since the shaft is thin-walled, the shear stress is uniform (constant) across the wall thickness (*Figure 5.3*). Also, the outer, inner, and average radii are approximately all equal.

In practice, most shafts are solid or have a thick wall. The *displacement method* is used to investigate these cases.

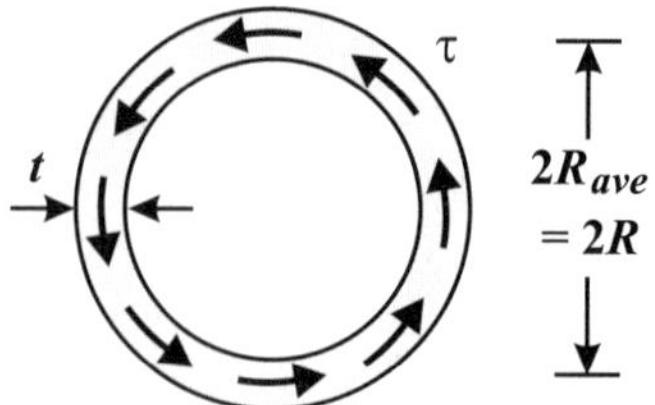

Figure 5.3. Shear stress τ acting on cross-section of thin-walled circular shaft due to torque $T = (2\pi R^2 t)\tau$.

Solid Shaft in Torsion

A solid shaft of radius R and length L (*Figure 5.4*) can be considered as an assembly of infinitely many thin-walled circular shafts or elements (*Figure 5.5*). Each thin-walled shaft has radius r, thickness dr, and length L, and supports shear stress $\tau(r)$ on its cross-section (*Figure 5.6*).

Due to applied torque T, the *angle of twist* of the solid shaft of length L is θ. Due to symmetry, any radial line on the cross-section must remain straight. Since radii remain straight, the thin-walled elements must all rotate together by angle θ. In other words, the rotation of each thin-walled element must be *compatible* with the rotation of the solid shaft as a whole. Hence, the displacement method is used.

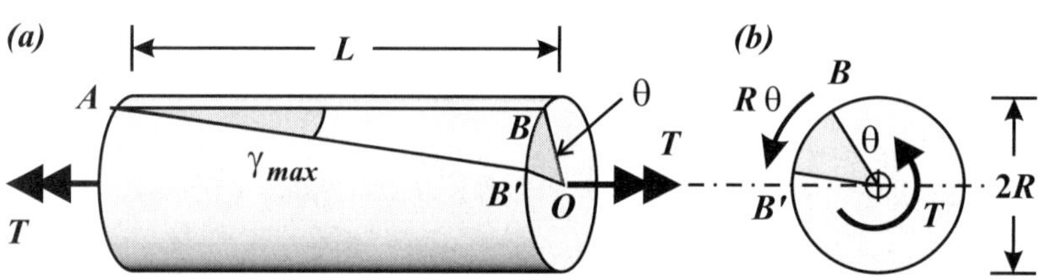

Figure 5.4. **(a)** Solid circular shaft in torsion. **(b)** Cross-section of solid shaft.

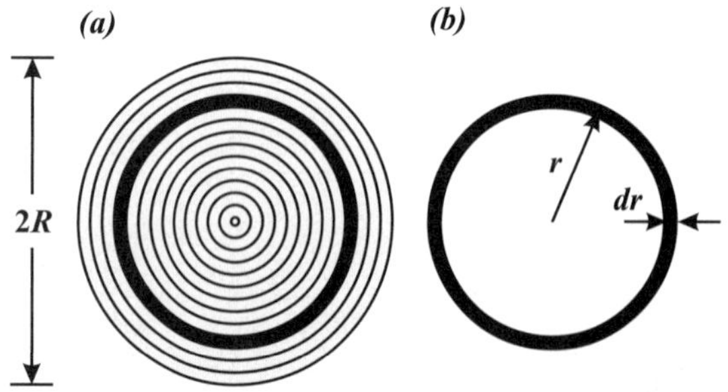

Figure 5.5. **(a)** A solid circular shaft can be considered as an assembly of many thin-walled shafts. **(b)** A thin-walled shaft of radius r, thickness dr.

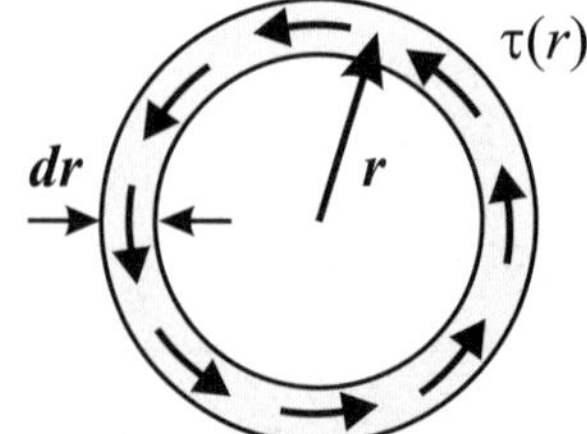

Figure 5.6. Shear stress varies with distance r. Each cylindrical element of radius r and thickness dr contributes to the total torque on the shaft: $dT = rF = r\,[\tau(r)\,(2\pi r\,dr)]$.

Each thin-walled element makes a contribution dT to support the applied torque T. The angle of twist of any thin-walled element is expressed using *Equation 5.3*, replacing T with dT, R with r, and t with dr:

$$\theta = \frac{(dT)L}{2\pi(r)^3(dr)G} \qquad \text{[Eq. 5.4]}$$

Solving for dT:

$$dT = \frac{2\pi G\theta}{L} r^3 \, dr \qquad \text{[Eq. 5.5]}$$

To solve for the total torque T supported at any cross-section of the solid shaft, integrate dT from $r = 0$ to R. Noting that rotation θ is not a function of r, then:

$$T = \int_0^R dT = \frac{2\pi G\theta}{L}\int_0^R r^3 \, dr = \frac{\pi G\theta}{2L}R^4 \qquad \text{[Eq. 5.6]}$$

The angle of twist of the solid shaft due to torque T is then:

$$\theta = \frac{2TL}{\pi R^4 G} \qquad \text{[Eq. 5.7]}$$

From the geometry of a thin-walled shaft (*Figure 5.2*), the shear strain of a thin-walled element at radius r within the solid shaft is:

$$\gamma(r) = \frac{r\theta}{L} \qquad \text{[Eq. 5.8]}$$

The shear stress, $\tau = G\gamma$, thus varies linearly with r:

$$\tau(r) = G\gamma(r) = G\frac{r\theta}{L} \qquad \text{[Eq. 5.9]}$$

Substituting θ from *Equation 5.7* into *Equation 5.8* gives the shear strain with r:

$$\gamma(r) = \frac{2Tr}{\pi R^4 G} \qquad \text{[Eq. 5.10]}$$

The shear stress as a function of r is therefore:

$$\tau(r) = \frac{2Tr}{\pi R^4} \qquad \text{[Eq. 5.11]}$$

The shear strain and shear stress increase linearly with distance r from the center of the shaft. The shear stress is zero at the center – the axis of rotation – and reaches its maximum at $r = R$. The ***maximum shear stress*** for the solid shaft is then:

$$\tau_{max} = \tau(R) = \frac{2T}{\pi R^3} \qquad \text{[Eq. 5.12]}$$

Combining *Equations 5.11* and *5.12*, the shear stress at any distance r is:

$$\tau(r) = \frac{r}{R}\tau_{max} \qquad \text{[Eq. 5.13]}$$

as shown in *Figure 5.7*.

The *torsional stiffness* of a solid shaft in torsion – the ratio of applied torque to angle of twist – is:

$$K_T = \frac{T}{\theta} = \frac{\pi G R^4}{2L} \qquad \text{[Eq. 5.14]}$$

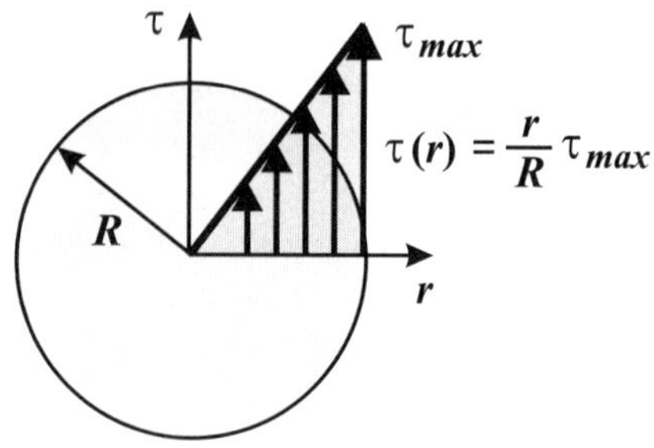

Figure 5.7. Shear stress distribution in a solid shaft in torsion.

Hollow (Thick-Walled) Shaft in Torsion

A hollow shaft of outer radius R, inner radius R_i, and length L (*Figure 5.8*) can also be considered as an assembly of infinitely many thin-walled shafts or elements. Each thin-walled shaft has radius r, thickness dr and length L, and supports stress $\tau(r)$ (*Figure 5.9*).

Due to torque T, the angle of twist of the hollow shaft is θ. Due to symmetry, any radial line remains straight. The rotation of each thin-walled element must be *compatible* with the rotation of the shaft as a whole, so each element has an angle of twist of θ.

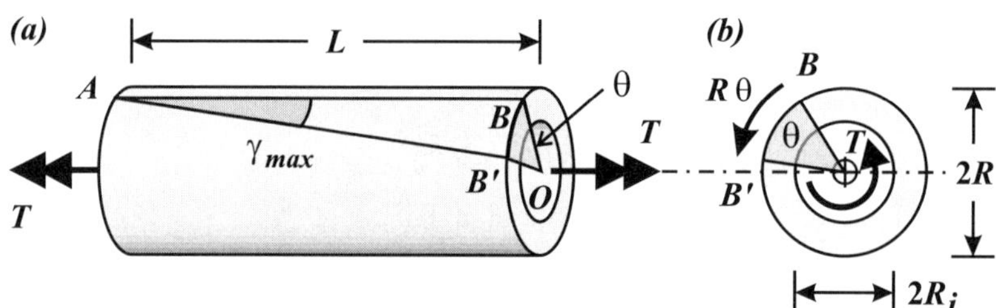

Figure 5.8. **(a)** Thick-walled shaft under torsion. **(b)** Cross-section of thick-walled shaft.

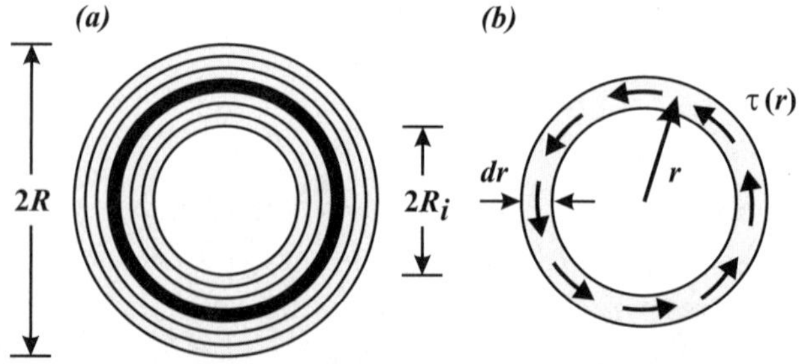

Figure 5.9. **(a)** A thick-walled shaft acts as an assembly of many thin-walled shafts. **(b)** Each cylindrical element of radius r and thickness dr, contributes to supporting the total applied torque.

Each thin-walled element makes a contribution dT to support the applied torque T (*Figure 5.9*), and repeating *Equation 5.5*:

$$dT = \frac{2\pi G\theta}{L}r^3 dr \qquad \text{[Eq. 5.15]}$$

Integrating over the solid area of the hollow shaft, i.e., from R_i to R, gives the torque:

$$T = \int_{R_i}^{R} dT = \frac{\pi G\theta}{2L}\left(R^4 - R_i^4\right) \qquad \text{[Eq. 5.16]}$$

The angle of twist θ is then:

$$\theta = \frac{2TL}{\pi(R^4 - R_i^4)G} \qquad \text{[Eq. 5.17]}$$

Substituting θ from *Equation 5.17* into *Equation 5.8*, gives the shear strain with r:

$$\gamma(r) = \frac{r\theta}{L} = \frac{2Tr}{\pi(R^4 - R_i^4)G} \qquad \text{[Eq. 5.18]}$$

The shear stress is linear with r:

$$\tau(r) = \frac{2Tr}{\pi(R^4 - R_i^4)} \qquad \text{[Eq. 5.19]}$$

As is the case for the solid shaft, the shear stress increases linearly with radius r; it is minimum at $r = R_i$, and reaches its maximum value at $r = R$ (*Figure 5.10*). For $r < R_i$, there is no material so there is no shear stress. On the inner surface of the cylinder, $r = R_i$, so the shear stress is:

$$\tau(R_i) = \frac{2TR_i}{\pi(R^4 - R_i^4)} \qquad \text{[Eq. 5.20]}$$

The *maximum shear stress* occurs at the outer radius, $r = R$:

$$\tau_{max} = \tau(R) = \frac{2TR}{\pi(R^4 - R_i^4)} \qquad \text{[Eq. 5.21]}$$

In general, the shear stress can be written as:

$$\tau(r) = \frac{r}{R}\tau_{max} \qquad \text{[Eq. 5.22]}$$

with $r > R_i$.

The torsional stiffness of the hollow shaft is:

$$K_T = \frac{T}{\theta} = \frac{\pi G}{2L}(R^4 - R_i^4) \qquad \text{[Eq. 5.23]}$$

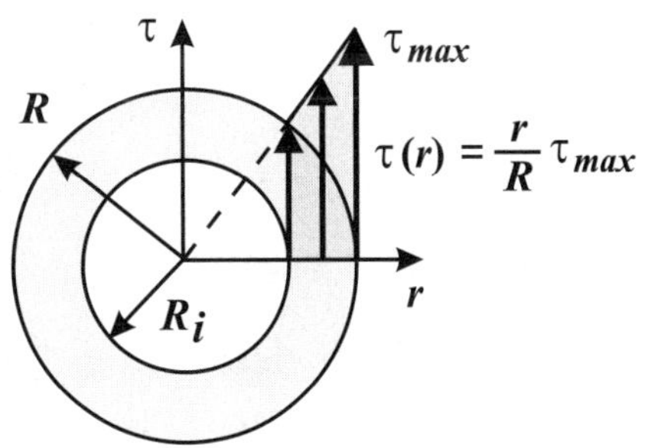

Figure 5.10. Shear stress distribution in a hollow circular shaft in torsion.

Polar Moment of Inertia

For any circular cross-section in torsion, the shear stress and strain distributions, the angle of twist, and the torsional stiffness can each be written in a general form:

$$\text{Shear Stress: } \tau(r) = \frac{Tr}{J}\;;\; \tau_{max} = \frac{TR}{J} \qquad\qquad \text{Shear Strain: } \gamma(r) = \frac{Tr}{JG} = \frac{\tau(r)}{G}$$

$$\text{Angle of Twist: } \theta = \frac{TL}{JG} \qquad\qquad\qquad \text{Torsional Stiffness: } K_T = \frac{T}{\theta} = \frac{JG}{L}$$

$$\text{[Eq. 5.24]}$$

Variable J is the **polar moment of inertia** and is a function of the cross-section geometry. The general expression for the polar moment of inertia is:

$$J = \int_A r^2 \, dA \qquad\qquad \text{[Eq. 5.25]}$$

The polar moment of inertia of a cross-section is a measure of its resistance to twisting. For a given torque, the larger the value of J, the larger the torsional stiffness, and the smaller the angle of twist. Shear stress is proportional to R/J, and since J increases faster than R, stress decreases with increasing J. The units of J are $mm^4 = 10^{-12}$ m^4 or $in.^4$.

Expressions for the polar moment of inertia for various circular cross-sections are given in *Table 5.1*.

Table 5.1. Polar Moment of Inertia for Circular Cross-sections; R is the outer radius.

Cross-section	Cross-section	Polar Moment of Inertia, J	
	$2R$	$\dfrac{\pi R^4}{2} = \dfrac{\pi D^4}{32}$	[Eq. 5.26]
Thick-walled (Hollow) Shaft $R = D/2,\ R_i = D_i/2$	$R_i \quad R$	$\dfrac{\pi(R^4-R_i^4)}{2} = \dfrac{\pi(D^4-D_i^4)}{32}$	[Eq. 5.27]
Thin-walled Shaft * $t \ll R$ $t = R-R_i = (D-D_i)/2$ $R_{ave} = (R+R_i)/2;\ D_{ave} = (D+D_i)/2$	$R_i \quad R$ t	$2\pi R_{ave}^3 t = \dfrac{\pi D_{ave}^3 t}{4}$	[Eq. 5.28]

* Use R_{ave} for the best results for J in the *thin-walled formula*. The *thin-walled formula* can be derived from the *thick-walled formula* by letting $t = R-R_i$ approach zero; i.e., $t \ll R, R \sim R_i$:

$$J = \frac{\pi}{2}(R^4 - R_i^4) = \frac{\pi}{2}(R^2 + R_i^2)(R^2 - R_i^2) = \frac{\pi}{2}(R^2 + R_i^2)(R + R_i)(R - R_i) = \frac{\pi}{2}(2R^2)(2R)(t) = 2\pi R^3 t$$

Example 5.1 **Effect of Geometry**

Given: A circular steel shaft, $L = 2.0$ m long, has an outside diameter $D = 80.0$ mm and transmits a torque of $T = 4000$ N·m. The shear modulus of steel is $G = 77$ GPa.

Required: Using the same outer diameter for each case, determine the maximum shear stress and angle of twist for (a) a solid shaft, (b) a hollow shaft with $D_i = 30.0$ mm, and (c) a thin-walled shaft with $t = 5.00$ mm. (d) Compare the results of the different geometries. Use the actual outer radius in the stress formula: $\tau = TR/J$.

Solution: The various cross-sectional geometries are: (a) $D = 80.0$ mm, (b) $D = 80.0$ mm and $D_i = 30.0$ mm, and (c) $D_{ave} = [D + (D - 2t)]/2 = 75.0$ mm and $t = 5.0$ mm.

Using the equations for the *polar moment of inertia* from *Table 5.1*, and the general formulas for the maximum shear stress and angle of twist (*Equation 5.24*), gives the following results:

Shaft	Formula for J	J ($\times 10^{-6}$ m^4)	Max. Stress $\tau_{max} = TR/J$ (MPa)	Angle of Twist $\theta = TL/JG$	Cross-sectional Area ($\times 10^{-3}$ m^2)
Solid Shaft	$\dfrac{\pi D^4}{32}$	4.021	39.8	25.8×10^{-3} rad $= 1.48°$	5.026
Hollow Shaft	$\dfrac{\pi(D^4 - D_i^4)}{32}$	3.942	40.6	26.4×10^{-3} rad $= 1.51°$	4.320
Thin-walled Shaft	$\dfrac{\pi D_{ave}^3 t}{4}$	1.657	96.6	62.7×10^{-3} rad $= 3.59°$	1.178

The area of the hollow (thick-walled) shaft is 14% less than that of the solid shaft, but the performance is essentially unchanged (the stress and angle of twist have only increased by about 2% each). The central core of the circular shaft adds little to the strength and stiffness of a torsion member.

In the thin-walled shaft, all of the material is acting near the maximum value of the stress. The area of the thin-walled shaft is 23% (~1/4) that of the solid shaft, while the stress is only 2.4 times that of the solid shaft. For high-performance engines, where weight is important, thin-walled shafts offer an attractive alternative design since they are very efficient in supporting torsional loads.

5.2 Torsion Members – Force Method

The following examples demonstrate the *force method* applied to torsion members. The torque everywhere in the system is first determined, followed by the shear stress and the angle of twist.

Example 5.2 Solid Shaft under Torsion

Given: A solid shaft, $L = 3.0$ ft long, with a diameter $D = 2R = 2.0$ in., is subjected to a torque $T = 4000$ lb-in. (*Figure 5.11*). The shaft is made of steel with shear modulus $G = 12 \times 10^6$ psi and yield strength $S_y = 50$ ksi.

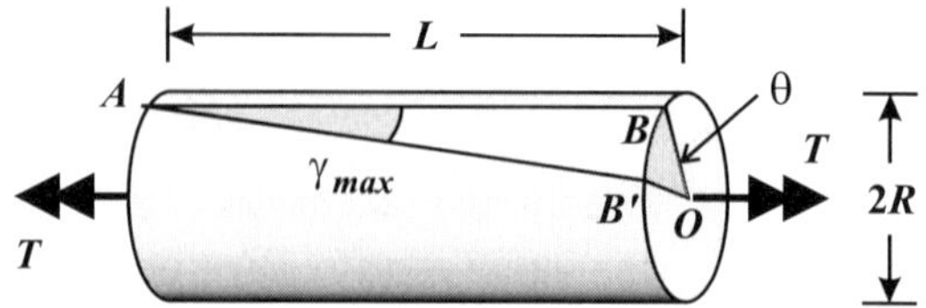

Figure 5.11. Solid shaft in torsion.

Required: Determine (a) the maximum shear stress, (b) the angle of twist of the shaft, and (c) the factor of safety FS against yielding.

Solution: *Step 1.* The polar moment of inertia is:

$$J = \frac{\pi D^4}{32} = \frac{\pi (2.0 \text{ in.})^4}{32} = 1.571 \text{ in.}^4$$

Step 2. The maximum shear stress is:

$$\tau_{max} = \frac{TR}{J} = \frac{(4000 \text{ lb-in.})(1.0 \text{ in.})}{1.571 \text{ in.}^4}$$

Answer: $\tau_{max} = 2.55$ ksi

Step 3. The angle of twist is:

$$\theta = \frac{TL}{JG} = \frac{(4000 \text{ lb-in.})(36.0 \text{ in.})}{(1.571 \text{ in.}^4)(12 \times 10^6 \text{ psi})}$$

Answer: $\theta = 0.00764$ rad $= 0.44°$

Step 4. The shear yield strength is not given, but for *ductile materials* it can be approximated from the yield strength S_y:

$$\tau_y = \frac{S_y}{\sqrt{3}} = \frac{50 \text{ ksi}}{\sqrt{3}} = 28.9 \text{ ksi}$$

The factor of safety is therefore:

$$FS = \frac{\tau_y}{\tau} = \frac{28.9 \text{ ksi}}{2.55 \text{ ksi}}$$

Answer: FS $= 11.3$

The design of shafts is often limited by the angle of twist.

Example 5.3 **Hollow Shaft under Torsion**

Given: A hollow shaft, $L = 3.0$ ft long, with outer diameter $D = 2.0$ in., and inner diameter $D_i = 1.0$ in., is subjected to a torque $T = 4000$ lb-in. (*Figure 5.12*). The shaft is made of steel with shear modulus $G = 12 \times 10^6$ psi and yield strength $S_y = 50$ ksi.

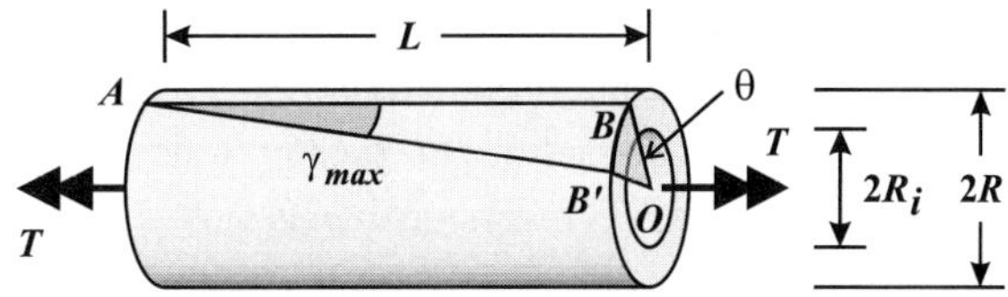

Figure 5.12. Hollow shaft in torsion.

Required: Determine (a) the shear stresses at the outer and inner surfaces, (b) the angle of twist of the shaft, and (c) the factor of safety FS against yielding.

Solution: *Step 1.* The polar moment of inertia is:

$$J = \frac{\pi(D^4 - D_i^4)}{32} = \frac{\pi[(2.0 \text{ in.})^4 - (1.0 \text{ in.})^4]}{32} = 1.473 \text{ in.}^4$$

Step 2. The maximum shear stress is at the outer surface of the shaft, and is:

$$\tau(R) = \frac{TR}{J} = \frac{(4000 \text{ lb-in.})(1.0 \text{ in.})}{1.473 \text{ in.}^4}$$

Answer: $\underline{\tau_{max} = 2.72 \text{ ksi}}$

The shear stress at the inner surface is:

$$\tau(R_i) = \frac{TR_i}{J} = \frac{(4000 \text{ lb-in.})(0.5 \text{ in.})}{1.473 \text{ in.}^4}$$

Answer: $\underline{\tau(R_i) = 1.36 \text{ ksi}}$

Step 3. The angle of twist is:

$$\theta = \frac{TL}{JG} = \frac{(4000 \text{ lb-in.})(36.0 \text{ in.})}{(1.473 \text{ in.}^4)(12 \times 10^6 \text{ psi})}$$

Answer: $\underline{\theta = 0.00815 \text{ rad} = 0.47°}$

Step 4. The shear yield strength is:

$$\tau_y = \frac{S_y}{\sqrt{3}} = \frac{50 \text{ ksi}}{\sqrt{3}} = 28.9 \text{ ksi}$$

so the factor of safety is:

$$FS = \frac{\tau_y}{\tau} = \frac{28.9 \text{ ksi}}{2.72 \text{ ksi}}$$

Answer: $\underline{FS = 10.6}$

Note that there is not much difference in the responses of the current hollow shaft and the solid shaft of *Example 5.2*. Both shafts have the same outer diameter, but the hollow shaft provides a weight savings of 25% with a modest increase in stress of 6.7%.

Example 5.4 Pre-torqued Suspension Torsion Bar

Given: In high-performance cars, the steering and handling characteristics are improved by introducing a prestressed circular torsion bar. The torsion bar is pre-torqued by rotating one end of the bar with respect to the other end. One end of the torqued bar is then attached to the frame, and the other end to the wheel suspension.

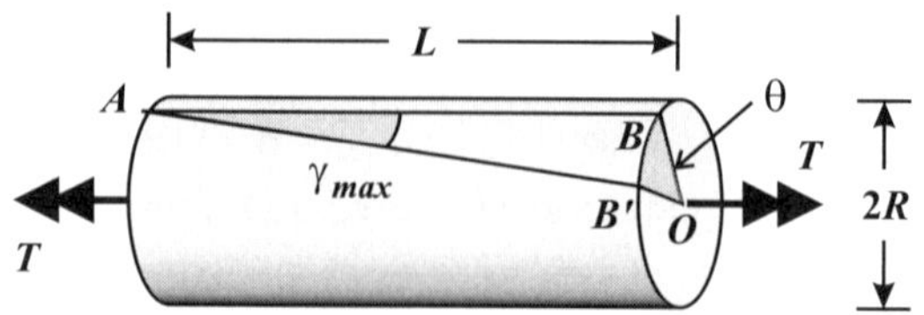

Figure 5.13. Solid shaft in torsion.

Such a solid shaft, with diameter $D = 30$ mm and length $L = 1.3$ m, is made of steel ($G = 77$ GPa). The pre-torque causes a relative angular displacement between the ends of the bar of $\theta = 3.0°$ (*Figure 5.13*).

Required: Determine (a) the pre-torque required, and (b) the maximum shear stress.

Solution: *Step 1.* The polar moment of inertia is:

$$J = \frac{\pi D^4}{32} = \frac{\pi (0.030 \text{ m})^4}{32} = 79.52 \times 10^{-9} \text{ m}^4$$

Step 2. Solving for the torque, and noting that $3.0° = 0.05236$ rad:

$$T = \frac{JG\theta}{L} = \frac{(79.52 \times 10^{-9} \text{ m}^4)(77 \times 10^{9} \text{ Pa})(0.05236 \text{ rad})}{1.3 \text{ m}}$$

Answer: $T = 247$ N·m

Step 3. The maximum shear stress is:

$$\tau_{max} = \frac{TR}{J} = \frac{(247 \text{ N·m})(0.015 \text{ m})}{(79.5 \times 10^{-9} \text{ m}^4)}$$

Answer: $\tau_{max} = 46.5$ MPa

Example 5.5 Stepped Shaft in Torsion

Given: The solid stepped shaft, ABC, is fixed at A, and is subjected to torques $T_B = 880$ N·m and $T_C = 275$ N·m (*Figure 5.14*). Segment AB has length $L_{AB} = 1.5$ m and diameter $D_{AB} = 50$ mm. Segment BC has length $L_{BC} = 1.0$ m and diameter $D_{BC} = 30$ mm. The material is steel with a shear modulus of $G = 77$ GPa.

Required: Determine (a) the maximum shear stress in AB, (b) the maximum shear stress in BC, and (c) the total angle of twist of shaft ABC, θ_{AC}.

Solution: *Step 1.* The polar moment of inertia of each cross-section is:

$$J_{AB} = \frac{\pi (D_{AB})^4}{32} = \frac{\pi (0.05 \text{ m})^4}{32} = 613.6 \times 10^{-9} \text{ m}^4$$

$$J_{BC} = \frac{\pi(D_{BC})^4}{32} = \frac{\pi(0.03 \text{ m})^4}{32}$$

$$= 79.52 \times 10^{-9} \text{ m}^4$$

Step 2. From a FBD of the shaft (*Figure 5.14b*), the reaction at the wall is:

$$T_A = T_B + T_C = 1155 \text{ N·m}$$

The torques carried in *AB* and *BC* are (*Figure 5.14c*):

$$T_{AB} = T_A = 1155 \text{ N·m}$$

$$T_{BC} = T_C = 275 \text{ N·m}$$

Step 3. The maximum shear stress and angle of twist of segment *AB* are:

$$\tau_{AB, \, max} = \frac{T_{AB}R_{AB}}{J_{AB}}$$

$$= \frac{(1155 \text{ N·m})(0.025 \text{ m})}{613.6 \times 10^{-9} \text{ m}^4}$$

Answer: $\underline{\tau_{AB, \, max} = 47.0 \text{ MPa}}$

$$\theta_{AB} = \frac{T_{AB}L_{AB}}{J_{AB}G}$$

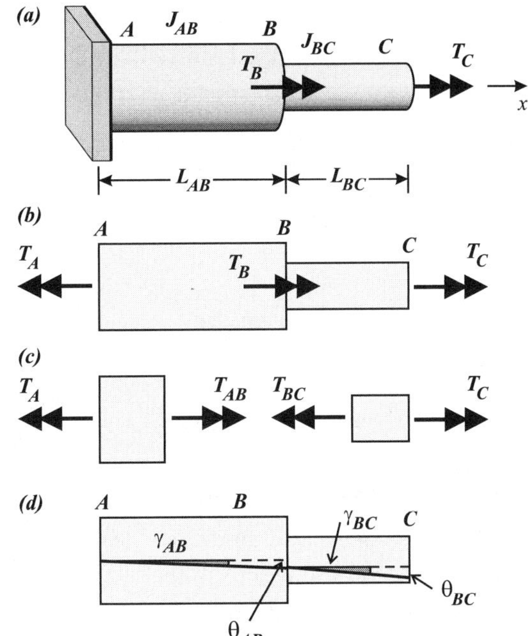

Figure 5.14. **(a)** A stepped-shaft subjected to torques T_B and T_C. **(b)** FBD of entire system. **(c)** FBDs to find internal torques T_{AB} and T_{BC}. **(d)** Strain γ, and angle of twist $\theta = \gamma R$, of *AB* and *BC*. Total angular displacement is $\theta_{AC} = \theta_{AB} + \theta_{BC}$.

$$= \frac{(1155 \text{ N·m})(1.5 \text{ m})}{(613.6 \times 10^{-9} \text{ m}^4)(75 \times 10^9 \text{ Pa})} = 0.0376 \text{ rad} = 2.16°$$

Step 4. The maximum shear stress and angle of twist of segment *BC* are:

$$\tau_{BC, \, max} = \frac{TR}{J} = \frac{(275 \text{ N·m})(0.015 \text{ m})}{79.52 \times 10^{-6} \text{ m}^4}$$

Answer: $\underline{\tau_{BC, \, max} = 51.9 \text{ MPa}}$

$$\theta_{BC} = \frac{TL}{JG} = \frac{(275 \text{ N·m})(1.0 \text{ m})}{(79.52 \times 10^{-9} \text{ m}^4)(75 \times 10^9 \text{ Pa})} = 0.0461 \text{ rad} = 2.64°$$

Step 5. Since the internal torques twist both segments in the same direction, the total angle of twist is (*Figure 5.14d*):

$$\theta_{AC} = \theta_{AB} + \theta_{BC} = 2.16° + 2.64°$$

Answer: $\underline{\theta_{AC} = 4.8°}$

Example 5.6 Torsional Vibration Damper

Given: A damper used to reduce the torsional vibration in a machine consists of a solid shaft AB, set inside hollow shaft BC (*Figure 5.15*). For AB:

D_{AB} = 60 mm,
L_{AB} = 750 mm.

For BC:

D_o = 110 mm, D_i = 80 mm,
L_{BC} = 600 mm.

The shafts are joined at cross-section B. The vibrating machine is at point A and the assembly is anchored at fixed base point C. The damper is made of steel with G = 77 GPa.

The damper is designed to occupy minimal space by doubling back upon itself.

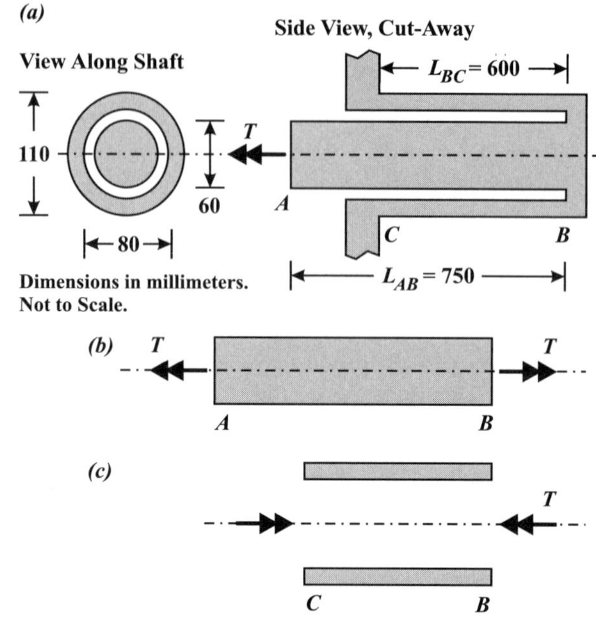

Figure 5.15. **(a)** Vibrational damper. Dimensions in mm. **(b)** FBD of *AB*. **(c)** FBD of *BC*.

Required: Determine (a) the maximum shear stress in each segment for an applied torque T = 8.60 kN·m, (b) the rotations of cross-sections A and B with respect to fixed cross-section C, and (c) the torsional stiffness $K_T = T/\theta$ of the damper system.

Solution: *Step 1.* The polar moments of inertia are:

$$J_{AB} = \frac{\pi D^4}{32} = \frac{\pi (60 \times 10^{-3}\ \text{m})^4}{32} = 1.27 \times 10^{-6}\ \text{m}^4$$

$$J_{BC} = \frac{\pi (D^4 - D_i^4)}{32} = \frac{\pi [(110 \times 10^{-3}\ \text{m})^4 - (80 \times 10^{-3}\ \text{m})^4]}{32} = 10.4 \times 10^{-6}\ \text{m}^4$$

Step 2. Equilibrium requires that the torque be the same throughout. Although the damper is turned back upon itself, it is the same as two shafts joined together in series. The torque in each segment is:

$$T_{AB} = T_{BC} = T = 8.6\ \text{kN·m}$$

The maximum shear stress in each segment is then:

$$\tau_{AB,\,max} = \frac{TR_{AB}}{J_{AB}} = \frac{(8.6\ \text{kN·m})(0.03\ \text{m})}{1.27 \times 10^{-6}\ \text{m}^4}$$

$$\textit{Answer: } \tau_{AB,\,max} = 203\ \text{MPa}$$

$$\tau_{BC,\,max} = \frac{TR_{BC}}{J_{BC}} = \frac{(8.6 \text{ kN·m})(0.055 \text{ m})}{10.4 \times 10^{-6} \text{ m}^4}$$

Answer: $\underline{\tau_{BC,\,max} = 45.5 \text{ MPa}}$

Step 3. Torque–twist (as in *force–displacement*) gives the angle of twist of each segment to be:

$$\theta_{AB} = \frac{TL_{AB}}{J_{AB}G} = \frac{(8.6 \text{ kN·m})(0.750 \text{ m})}{(1.27 \times 10^{-6} \text{ m}^4)(77 \text{ GPa})} = 0.0660 \text{ rad} = 3.78°$$

$$\theta_{BC} = \frac{TL_{BC}}{J_{BC}G} = \frac{(8.6 \text{ kN·m})(0.600 \text{ m})}{(10.4 \times 10^{-6} \text{ m}^4)(77 \text{ GPa})} = 0.0064 \text{ rad} = 0.37°$$

The angle of twist is the relative rotation of one end of a shaft with respect to the other end. θ_{AB} is the angle of twist between cross-sections A and B; θ_{BC} is the angle of twist between cross-sections B and C.

Step 4. Compatibility (rotation–twist). Section C is fixed, so the absolute rotation of section B, θ_B, is equal to the angle of twist of BC, θ_{BC} (from point C to point B):

Answer: $\underline{\theta_B = \theta_{BC} = 0.0064 \text{ rad} = 0.37°}$

The total rotation of section A is the absolute rotation of section B and the angle of twist of A with respect to B:

$$\theta_A = \theta_{BC} + \theta_{AB}$$

Answer: $\underline{\theta_A = 0.0724 \text{ rad} = 4.1°}$

Step 5. The torsional stiffness K_T of the assembly is:

$$K_T = \frac{T}{\theta_A} = \frac{8.6 \times 10^3 \text{ N·m}}{0.0724 \text{ rad}}$$

Answer: $\underline{K_T = 119 \text{ kN·m/rad} = 2.1 \text{ kN/deg}}$

Example 5.7 **Drill Bit**

Given: A drill bit is stuck in a piece of wood, but the user attempts to continue to operate it (*Figure 5.16*). Assume that the bit is a simple cylinder of radius R and embedded length L. The drill applies torque T. The surrounding wood applies a uniform interfacial shear stress τ_i on the surface of the cylinder, reducing the internal torque $T(x)$ carried in the bit linearly from the drill-end ($x = L$) to the tip of the bit ($x = 0$). No torque is transferred across the free-end (tip) of the bit.

Required: Determine the angle of twist of the embedded length of the drill bit. Measure x from the tip (free end) of the bit.

Solution: *Step 1. Equilibrium* of the embedded bit (*Figure 5.16b*). Assuming that the bit attempts to rotate clockwise as it cuts, the value of the torque T at the drill end is:

$$T = \tau_i(2\pi RL)R = \tau_i 2\pi R^2 L$$

and acts in the direction drawn.

Equilibrium on a FBD from the tip of the bit to the cross-section at x (*Figure 5.16c*) gives the internal torque $T(x)$:

$$T(x) = -\tau_i\left(2\pi R^2\right)x = -T\frac{x}{L}$$

$T(x)$ is drawn in its positive sense, but physically acts in the opposite direction, so it has a negative value.

Step 2. Torque–twist. Consider a length dx so small that the change in $T(x)$ over dx is negligible. Over dx (*Figure 5.16d*), the angle of twist is:

$$d\theta(x) = \frac{T(x)dx}{JG} = \frac{-\tau_i(2\pi R^2)x\ dx}{JG}$$

Step 3. Compatibility (rotation–twist). The angle of twist from the tip ($x = 0$) to any cross-section at x is found by integrating $d\theta\ (x)$ from 0 to x:

$$\theta(x) = \int_0^x \frac{T(x)dx}{JG} = \int_0^x \frac{-\tau_i(2\pi R^2)x\ dx}{JG}$$

$$= \frac{1}{JG}\int_0^x\left(-T\frac{x}{L}\right)dx = \frac{-Tx^2}{2JGL}$$

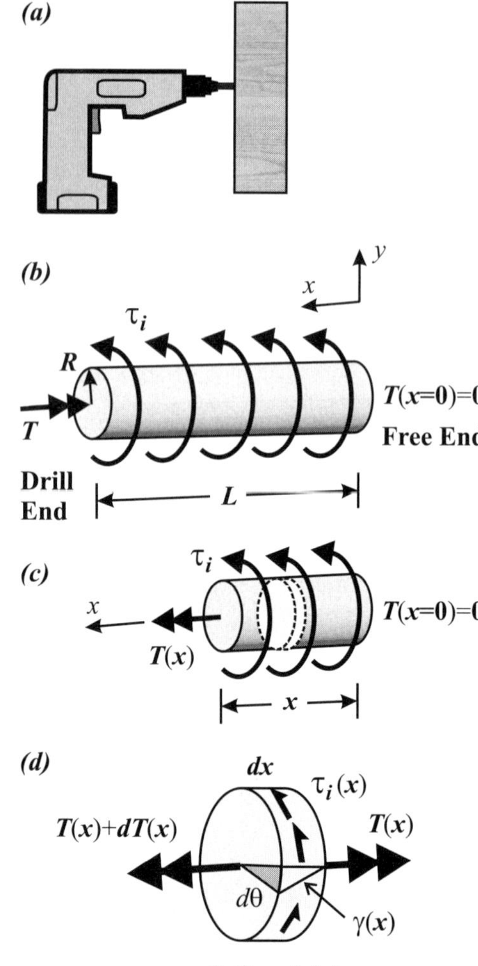

Figure 5.16. **(a)** Drill. **(b)** FBD of embedded drill bit, with interfacial shear stress τ_i. **(c)** FBD of length of drill from 0 to x. **(d)** FBD of cylindrical slice dx thick.

The angle of twist between the free end ($x = 0$) and the drill end ($x = L$) is:

$$\theta = \theta(L) = \frac{-TL^2}{2JGL}$$

Answer: $\theta = \dfrac{-TL}{\pi R^4 G} = \dfrac{-16TL}{\pi D^4 G}$

The angle of twist is negative because with respect to the x-axis, the rotation of the left (drill) end of the bit is clockwise.

The drill bit is the twisting analogy to the nail being extracted from a piece of wood (*Example 4.4*). Recall the nail's elongation under tensile force T was:

$$\Delta = u(L) = \frac{2TL}{\pi D^2 E}$$

The equations are the same except for the numerical constant, the definition of the load T (torque versus tension), the type of modulus (shear versus axial), and the exponent on the diameter (due to how the cross-section supports the applied load).

Example 5.8 Stepped-Shaft with Fixed Ends – Statically Indeterminate

Given: The stepped shaft ABC shown in *Figure 5.17* is fixed at ends A and C. Torque T is applied at the juncture point B. The shear modulus of both segments of the shaft is the same, equal to G.

Required: Using the force method, determine (a) the angular displacement θ_B of cross-section B as a function of T, and (b) expressions for the torques carried in each segment of the shaft, T_{AB} and T_{BC}. Take the system to remain elastic.

Solution:

Step 1. Equilibrium of the entire system provides only one equation – the torque about the x-axis (*Figure 5.17b*):

$$\sum T_x = 0:\quad T - T_A + T_C = 0$$

Since there are two unknowns but only one useful equilibrium equation, the system is *statically indeterminate*.

Let the redundant torque be $T_C = R$. In terms of the applied and redundant torques, the reactions are:

$$T_C = R \quad \text{and} \quad T_A = R + T$$

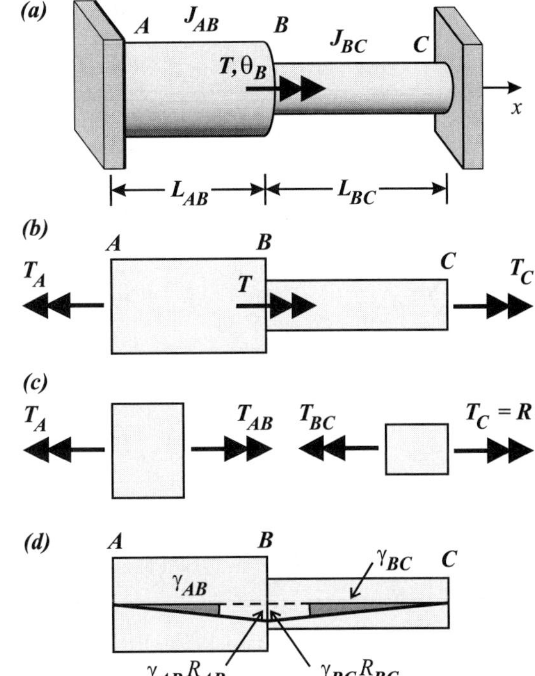

Figure 5.17. (a) A stepped-shaft with fixed ends. (b) FBD of entire system. (c) FBDs to find internal torques T_{AB} and T_{BC}. (d) Strain γ, and angular displacement $\theta = \gamma R$, of AB and BC.

From equilibrium, the internal torques equal the reaction torques (*Figure 5.17c*):

$$T_{AB} = T_A = R + T$$

$$T_{BC} = T_C = R$$

Step 2. The *twist–torque* elastic relationship gives the angles of twist for AB and BC:

$$\theta_{AB} = \frac{T_{AB}L_{AB}}{J_{AB}G} = \frac{(R+T)L_{AB}}{J_{AB}G} \quad \text{and} \quad \theta_{BC} = \frac{RL_{BC}}{J_{BC}G}$$

Step 3. Compatibility (rotation–twist relationship) relates the angle of twist of each member (θ_{AB} and θ_{BC}) to the displacement of the system θ_B (the rotation of section B):

$$\theta_{AB} = \theta_B \quad \text{and} \quad \theta_{BC} = -\theta_B$$

With reference to the $+x$-axis, section B rotates counterclockwise (positive) with respect to section A; section C rotates clockwise (negative) with respect to section B.

The total angle of rotation from A to C is zero:

$$0 = \theta_{AB} + \theta_{BC} = \frac{(R+T)L_{AB}}{J_{AB}G} + \frac{RL_{BC}}{J_{BC}G}$$

Therefore: $$R = -T\left(1 + \frac{J_{AB}L_{BC}}{J_{BC}L_{AB}}\right)^{-1}$$

Step 4. Solving for torques T_{AB} and T_{BC} :

$$T_{AB} = T + R = T - T\left(1 + \frac{J_{AB}L_{BC}}{J_{BC}L_{AB}}\right)^{-1}$$

Answer: $$T_{AB} = T\left(1 + \frac{J_{BC}L_{AB}}{J_{AB}L_{BC}}\right)^{-1}$$

Answer: $$T_{BC} = R = -T\left(1 + \frac{J_{AB}L_{BC}}{J_{BC}L_{AB}}\right)^{-1}$$

Step 5. The angle of rotation of section B, θ_B, is:

$$\theta_B = \frac{T_{AB}L_{AB}}{J_{AB}G} = T\left(1 + \frac{J_{BC}L_{AB}}{J_{AB}L_{BC}}\right)^{-1}\frac{L_{AB}}{J_{AB}G} = \frac{T}{G}\left[\frac{J_{AB}/L_{AB}}{(J_{AB}/L_{AB}) + (J_{BC}/L_{BC})}\right]\frac{L_{AB}}{J_{AB}}$$

Answer: $$\theta_B = \frac{T}{G}\left(\frac{J_{AB}}{L_{AB}} + \frac{J_{BC}}{L_{BC}}\right)^{-1}$$

The torsional flexibility and stiffness are:

$$F_T = \frac{\theta}{T} = \frac{1}{G}\left(\frac{J_{AB}}{L_{AB}} + \frac{J_{BC}}{L_{BC}}\right)^{-1} \quad \text{and} \quad K_T = \frac{T}{\theta} = G\left(\frac{J_{AB}}{L_{AB}} + \frac{J_{BC}}{L_{BC}}\right)$$

Note that the stiffnesses of the torsion elements are additive since the rotation is resisted by both AB and BC acting together.

5.3 Torsion Members – Displacement Method

The following examples illustrate the *displacement method*. In the *displacement method* for torsion members, a rotation θ is imposed on the shaft at a key point. Through the angle of twist–torque relationships, the torques and shear stresses are determined.

Example 5.9 Stepped-Shaft with Fixed Ends – Statically Indeterminate

Given: The stepped-shaft problem of *Example 5.8* and *Figure 5.17* is repeated.

Required: Use the displacement method to determine (a) the angular displacement θ_B of cross-section B as a function of T and (b) the expressions for the torques carried in each segment of the shaft, T_{AB} and T_{BC}, in terms of T, J, and L.

Solution: *Step 1.* Rotation θ_B is imposed on the shaft at cross-section B. The *displacement–twist relationship* requires that the angles of twist of segments AB and BC be:

$$\theta_{AB} = \theta_B \quad \text{and} \quad \theta_{BC} = -\theta_B$$

Step 2. The *torque–twist relationship* requires:

$$T_{AB} = \frac{J_{AB}G}{L_{AB}}(\theta_{AB}) = \frac{J_{AB}G}{L_{AB}}\theta_B \quad \text{and} \quad T_{BC} = \frac{J_{BC}G}{L_{BC}}(\theta_{BC}) = \frac{J_{BC}G}{L_{BC}}(-\theta_B)$$

Step 3. Equilibrium requires:

$$T = T_A - T_C = \frac{J_{AB}G}{L_{AB}}\theta_B - \frac{J_{BC}G}{L_{BC}}(-\theta_B) = \left(\frac{J_{AB}}{L_{AB}} + \frac{J_{BC}}{L_{BC}}\right)G\theta_B$$

Step 4. The stiffness of the system is:

$$K_T = \frac{T}{\theta_B} = G\left(\frac{J_{AB}}{L_{AB}} + \frac{J_{BC}}{L_{BC}}\right)$$

The torsional stiffness K_T and the torsional flexibility $F_T = K_T^{-1}$ agree with the results of *Example 5.8*.

The angular displacement of cross-section B is:

$$\textit{Answer: } \theta_B = \frac{T}{G}\left(\frac{J_{AB}}{L_{AB}} + \frac{J_{BC}}{L_{BC}}\right)^{-1}$$

Step 5. The torque in each segment of the bar is:

$$T_{AB} = \frac{J_{AB}G}{L_{AB}}\left[\frac{T}{G}\left(\frac{J_{AB}}{L_{AB}} + \frac{J_{BC}}{L_{BC}}\right)^{-1}\right]$$

$$\textit{Answer: } T_{AB} = T\left(1 + \frac{J_{BC}L_{AB}}{J_{AB}L_{BC}}\right)^{-1}$$

$$T_{BC} = \frac{J_{BC}G}{L_{BC}}(-\theta_B)$$

$$\text{Answer:}\quad T_{BC} = -T\left(1 + \frac{J_{AB}L_{BC}}{J_{BC}L_{AB}}\right)^{-1}$$

Note that the *displacement method* solves for T in terms of θ_B, immediately giving the *stiffness* of the system, while the *force method* (*Example 5.8*) solves for θ_B in terms of T, immediately giving the *flexibility*.

Example 5.10 Micro-Shaft with Fixed Ends – Statically Indeterminate

Given: A 3000 μm long silicon carbide shaft has a diameter of $D = 10$ μm. The ends of this MEMS (micro-electromechanical system) structure are fixed (*Figure 5.18*). The shear modulus of silicon carbide is $G = 195$ GPa.

Required: Determine the torsional stiffness of the shaft when a torque is applied at point B, one-third of the way along the shaft from A.

Solution: *Step 1.* The polar moment of inertia is:

$$J = \frac{\pi D^4}{32} = \frac{\pi(10 \times 10^{-6}\ \text{m})^4}{32}$$

$$= 982 \times 10^{-24}\ \text{m}^4$$

Step 2. The stiffness of a shaft ABC, fixed at both ends, with a torque applied at point B is, from *Example 5.8*:

$$K_T = \frac{T}{\theta} = G\left(\frac{J_{AB}}{L_{AB}} + \frac{J_{BC}}{L_{BC}}\right)$$

Here, $J_{AB} = J_{BC}$, $L_{AB} = 1000$ μm and $L_{BC} = 2000$ μm, so:

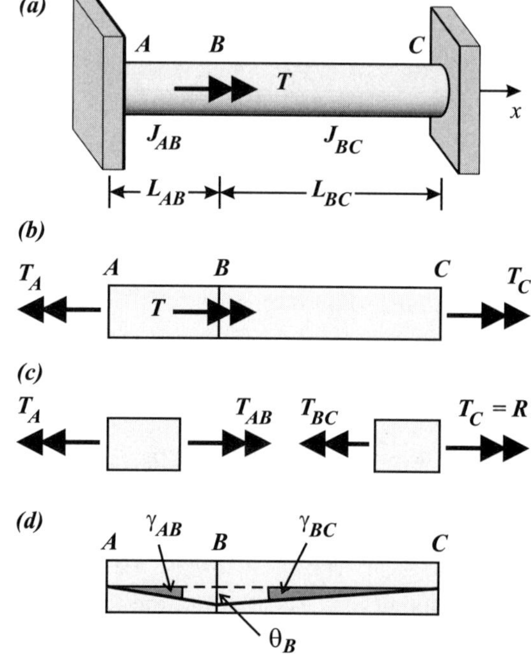

Figure 5.18. **(a)** A solid shaft with fixed ends. **(b)** FBD of entire system. **(c)** FBDs to find internal torques T_{AB} and T_{BC}. **(d)** Strain γ and angular displacement $\theta_B = \gamma\,R$.

$$K_T = GJ\left(\frac{1}{L_{AB}} + \frac{1}{L_{BC}}\right) = (195 \times 10^9\ \text{Pa})(982 \times 10^{-24}\,\text{m}^4)\left(\frac{1}{1000 \times 10^{-6}\,\text{m}} + \frac{1}{2000 \times 10^{-6}\,\text{m}}\right)$$

$$\text{Answer:}\quad K_T = 287 \times 10^{-9}\ \text{N·m/rad}$$

Summary of Equations for Shear Stress, Strain, and Angle of Twist

Only when torque, polar moment of inertia, and shear modulus are all constant over the entire length of a circular shaft can its angle of twist (angular displacement) be expressed as:

$$\theta = \frac{TL}{JG}$$ [Eq. 5.29]

If any of the values of torque, polar moment of inertia and shear modulus vary over the length of a shaft, the system must be broken up into smaller lengths or elements, where all three values are constant in each element. The elements may be shorter lengths L_i, or even differential lengths dx. The total angle of twist is the summation (integral) of the angle of twists of each small element. A summary of the relevant equations is given in *Table 5.2*.

Note the similarity of the torsion member equations in *Table 5.2* to the axial member equations in *Table 4.1* on *Page 110*. For example, the *angular displacement* equation (*Equation 5.29*) is analogous to the elongation, or *axial displacement*, of an axial bar:

$$\Delta = \frac{PL}{AE}$$ [Eq. 5.30]

Table 5.2. General equations for in-line torsion members.

Torque, Polar Moment of Inertia, Shear Modulus	Maximum Shear Stress on Cross-section $(r = R)$	Maximum Shear Strain at Cross-section	Angle of Twist of System	Remark/ Constant Strain Element
All Constant	$\tau = \dfrac{TR}{J}$	$\gamma = \dfrac{R\theta}{L} = \dfrac{\tau}{G}$	$\theta = \dfrac{TL}{JG}$	L
Discretely Varying	$\tau_i = \dfrac{T_i R_i}{J_i}$	$\gamma_i = \dfrac{R_i \theta_i}{L_i} = \dfrac{\tau_i}{G_i}$	$\theta = \sum \dfrac{T_i L_i}{J_i G_i}$	Break component into lengths L_i, where T, J, and G are all constant.
Continuously Varying	$\tau(x) = \dfrac{T(x)R}{J(x)}$	$\gamma(x) = \dfrac{\tau(x)}{G(x)}$	$\displaystyle\int_L \dfrac{T(x)}{J(x)G(x)}dx$	Consider length dx, so small that over its length, T, J, and G are all constant.

Error When Using the Thin-Walled Formula for J

In *Chapter 3*, the shear stress in a *thin-walled* shaft due to torque T was given as:

$$\tau = \frac{T}{2\pi R^2 t}$$

where R is the average radius and t $(<< R)$ is the wall thickness. The *thin-walled approximation* means that the variation in shear stress from *inner* to *outer radius* is negligible, i.e., the stress over the wall thickness is constant. Geometrically, the *inner, average*, and *outer radii* are nearly equal: $R_i \sim R_{ave} \sim R$ (R is now the outer radius).

In this chapter, the polar moment of inertia J was introduced. The maximum shear stress in a shaft with outer radius R is:

$$\tau_{max} = TR/J$$

For a hollow shaft of inner radius R_i and outer radius R, the exact expression for J is:

$$J_{exact} = \frac{\pi}{2}\left(R^4 - R_i^4\right)$$

while the thin-walled approximation for J is:

$$J_{thin} = 2\pi R_{ave}^3 t$$

J_{thin} is easier to manipulate than J_{exact}, a useful attribute in initial design calculations. If the outer radius is nearly equal to the average, $R \sim R_{ave}$, then R can be used in J_{thin}. Outer radius R sizes the shaft with respect to other components.

The preferred radius to use in the thin-walled approximation J_{thin} is the *average radius* R_{ave}. R_{ave} is the best value is shown in the graphs below, which show the error in J and τ_{max}:

$$J_{thin} = 2\pi r^3 t \quad \text{and} \quad \tau_{max} = TR_{max}/J_{thin}$$

when using the inner $(r = R_i)$, average $(r = R_{ave})$, and outer radius $(r = R)$, for the value of r in J_{thin}. For τ_{max}, R_{max} is always taken as the outer radius R.

Using the *outer radius* $r = R$ in J_{thin} overestimates J_{exact}, and *underestimates* the actual τ_{max}. Using the *inner radius* $r = R_i$ in J_{thin} underestimates J_{exact}, and *overestimates* the actual τ_{max}.

Using the *average radius* $r = R_{ave}$ in J_{thin} *slightly underestimates* J_{exact}, and *slightly overestimates* the actual τ_{max}. Even at $t/R = 0.4$, the error in the maximum shear stress is less than 7%; for $t/R < 0.2$, the error is negligible. The slight overestimation of $\tau_{max} = TR/J_{thin}$ is a good result for a design calculation.

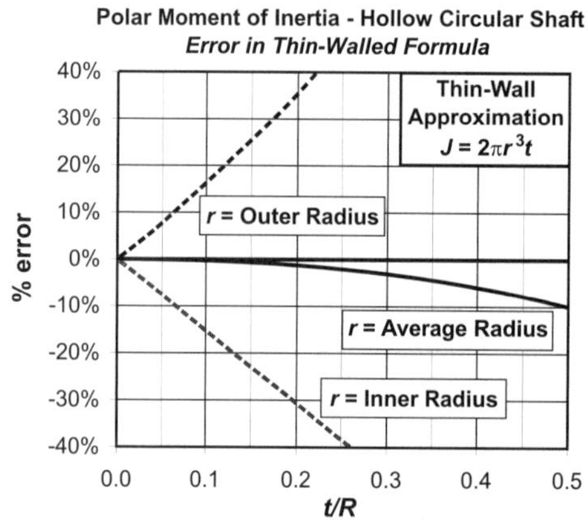

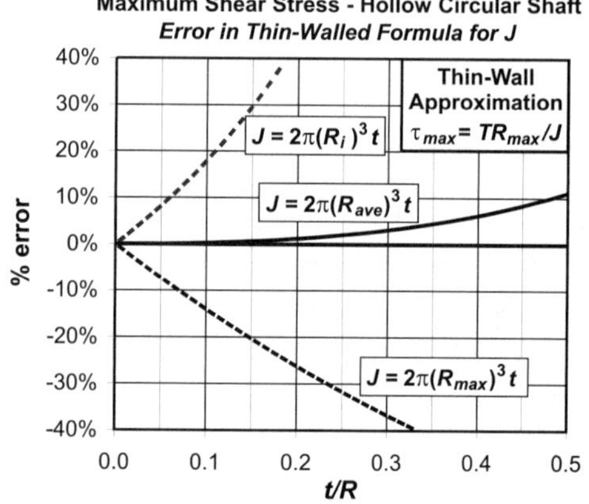

Error in $J_{thin} = 2\pi r^3 t$ and $\tau_{max} = TR/J_{thin}$ compared to exact values when the value for r used in J_{thin} is $r = R_i$, R_{ave} and R (inner, average and outer radii, respectively).

5.4 Closed Thin-Walled Members in Torsion

Closed thin-walled members of arbitrary cross-sectional shape are often used to carry torsion (*Figure 5.19*). A thin-walled section has a thickness t much smaller than the overall dimensions of the cross-section. The thickness may vary around the cross-section perimeter.

Non-circular thin-walled members are used in aircraft wings. Part of the structural system in the wings is the so-called torsion box, which supports the torque applied to the wings.

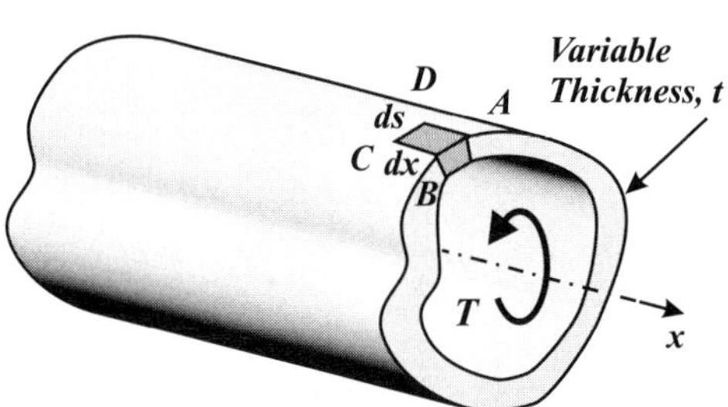

Figure 5.19. A closed thin-walled member in torsion.

Shear Stresses and Shear Flow

Consider the cross-section of a closed thin-walled member subjected to constant torque T and varying thickness t (*Figure 5.19*). The cross-section is constant over length L of the component. The aim is to determine the distribution of shear stress τ around the cross-section.

Consider element $ABCD$, dx long by ds wide by t thick, removed from the thin-walled member (*Figures 5.19* and *5.20a*). Side AB is on the cross-section and its thickness varies from t_A to t_B. Because the thickness varies, the shear stress changes as well, from τ_A to τ_B. By complementary shear stress arguments (*Figure 5.20b*), the stresses on sides AD and BC are:

$$\tau_{AD} = \tau_A \quad \text{and} \quad \tau_{CB} = \tau_B \quad \text{[Eq. 5.31]}$$

Applying equilibrium in the x-direction requires that:

$$\tau_{AD}(t_A\,dx) = \tau_{CB}(t_B\,dx) \quad \text{[Eq. 5.32]}$$

Hence:

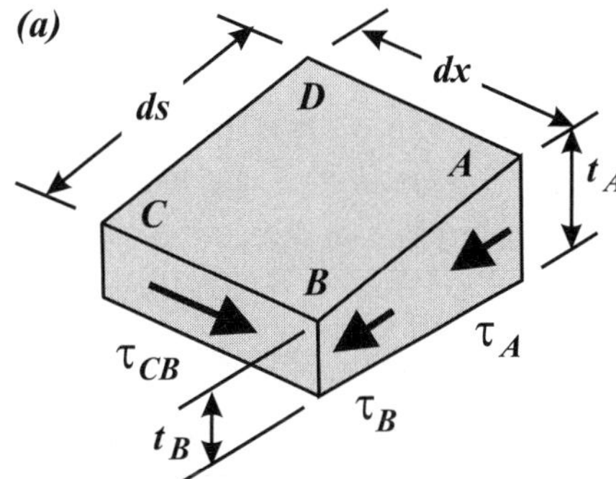

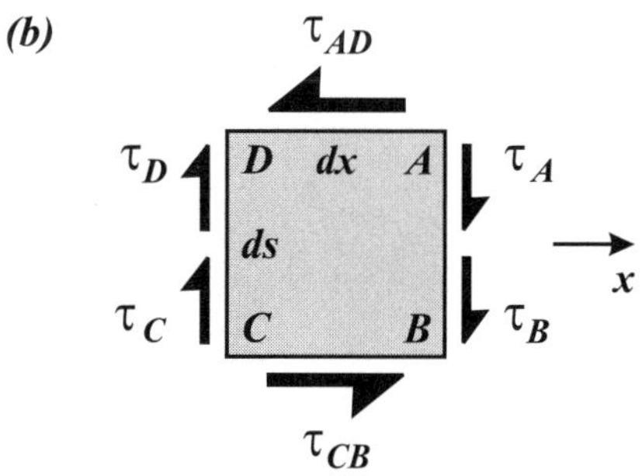

Figure 5.20. (a) Element $ABCD$, dx wide, ds long, and t thick. **(b)** Top view of $ABCD$.

$$\tau_A t_A \, dx = \tau_B t_B \, dx \qquad \text{[Eq. 5.33]}$$

The **shear flow** q is defined as:

$$q = \tau t \qquad \text{[Eq. 5.34]}$$

Since A and B are arbitrary points on the cross-section, *Equations 5.33* and *5.34* imply that the shear flow is constant at every point around the cross-section:

$$q = \tau_A t_A = \tau_B t_B = \tau t \qquad \text{[Eq. 5.35]}$$

The element, width ds, is at distance s on the perimeter of the cross-section from an arbitrary origin (*Figure 5.20c*). The shear force on the element face of area $dA = t \, ds$ is:

$$dF = \tau t \, ds = q \, ds \qquad \text{[Eq. 5.36]}$$

The torque caused by dF about any interior point O is therefore:

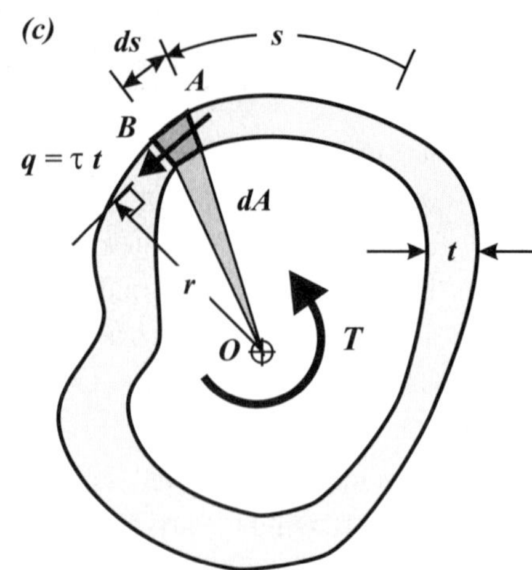

Figure 5.20. (c) Thin-walled cross-section under torque *T*. Shear flow *q* acting on element face *AB*.

$$dT = rq \, ds \qquad \text{[Eq. 5.37]}$$

where distance r is the lever arm of $q \, ds$ about point O, measured from O to the centerline of the wall (*Figure 5.20c*). From geometry, the triangular area bound by ds and the line segments from point O to either side of ds is:

$$dA = \frac{1}{2} r \, ds \qquad \text{[Eq. 5.38]}$$

Integrating dT about the entire perimeter gives the applied torque:

$$T = \int_s dT = \int_s rq \, ds = \int_s rq\!\left(\frac{2dA}{r}\right) = \int_s 2q \, dA = 2qA_o \qquad \text{[Eq. 5.39]}$$

where A_o is the *total area enclosed by the perimeter of the section*. The perimeter is measured at the centerline of the wall. For a given torque T, the shear flow q varies inversely with enclosed area:

$$q = \frac{T}{2A_o} \qquad \text{[Eq. 5.40]}$$

Since $q = \tau t$, the shear stress anywhere is given by:

$$\tau = \frac{T}{2A_o t} \qquad \text{[Eq. 5.41]}$$

Angle of Twist

The *angle of twist* θ is found using the *energy method*. The internal energy in element $ABCD$ is the product of the elastic strain energy density and the element volume $(t\,ds)dx$:

$$dU = \frac{1}{2}\frac{\tau^2}{G}[(t\,ds)dx] \qquad \text{[Eq. 5.42]}$$

where $t\,ds$ is the differential area on the cross-section. Since $\tau = q/t$, then:

$$dU = \frac{1}{2G}q^2\frac{ds}{t}dx = \frac{1}{2G}\left(\frac{T^2}{4A_o^2}\right)\frac{ds}{t}dx \qquad \text{[Eq. 5.43]}$$

In this case, torque T, modulus G, and the cross-sectional geometry are all constant over length L, so integrating with respect to x results in replacing dx with L. Quantities T, G, and A_o do not depend on perimeter distance s, but t does. Integrating $1/t$ around the perimeter gives the total elastic energy stored:

$$U = \frac{LT^2}{8GA_o^2}\int_s \frac{ds}{t} \qquad \text{[Eq. 5.44]}$$

The work done by torque T twisting an elastic shaft by angle θ is:

$$W = \frac{1}{2}T\theta \qquad \text{[Eq. 5.45]}$$

From energy balance, the energy stored equals the work done:

$$\frac{LT^2}{8GA_o^2}\int_s \frac{ds}{t} = \frac{1}{2}T\theta \qquad \text{[Eq. 5.46]}$$

Therefore, the angle of twist θ of the member is:

$$\theta = \frac{TL}{4GA_o^2}\int_s \frac{ds}{t} \qquad \text{[Eq. 5.47]}$$

and the angle of twist per unit length is:

$$\frac{\theta}{L} = \frac{T}{4GA_o^2}\int_s \frac{ds}{t} \qquad \text{[Eq. 5.48]}$$

The torsional stiffness K_T is:

$$K_T = \frac{T}{\theta} = \frac{4GA_o^2}{L}\left[\int_s \frac{ds}{t}\right]^{-1} \qquad \text{[Eq. 5.49]}$$

To solve for the angle of twist or the stiffness, it is necessary to find the thickness t as a function of distance s around the thin-walled cross-section.

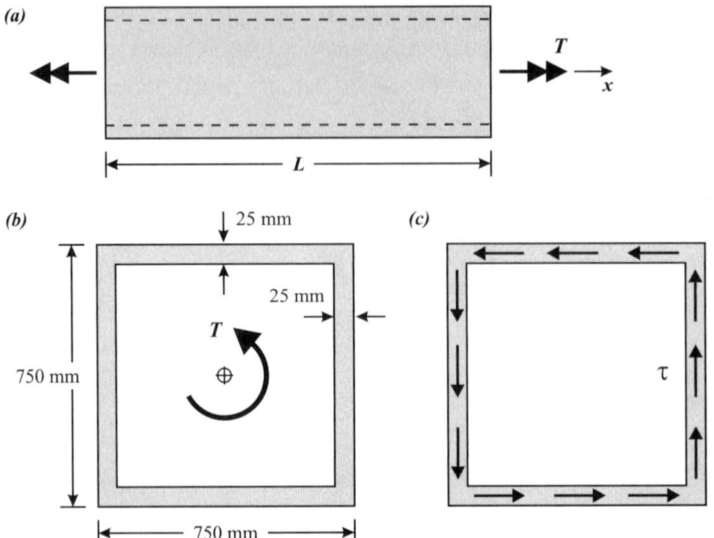

Figure 5.21. **(a)** Thin-walled member under torque T. **(b)** Square cross-section with constant thickness t. **(c)** Shear stress on cross-section τ.

Example 5.11 Box Beam in Torsion 1

Given: Steel box beams are often used in bridge construction. The length of a box beam is 10.0 m, and the cross-section is a hollow square with outside dimensions $d = 750$ mm and constant thickness $t = 25$ mm (*Figure 5.21*). The applied torque is 2000 kN·m. The shear modulus of the steel is 75 GPa.

Required: Estimate (a) the shear stress in the beam due to the torque and (b) the angle of twist.

Solution: *Step 1.* A_o is the area bound by the centerline of the walls:

$$A_o = (0.725 \text{ m})^2$$

Since t is constant, the shear stress in the walls is:

$$\tau = \frac{T}{2A_o t} = \frac{2.0 \times 10^6 \text{ N·m}}{2(0.725 \text{ m})^2 (0.025 \text{ m})}$$

Answer: $\tau = 76.1$ MPa

Step 2. The angle of twist is:

$$\theta = \frac{TL}{4GA_o^2} \int_s \frac{ds}{t} = \frac{TL}{4G(d-t)^4}\left[\frac{4(d-t)}{t}\right] = \frac{(2.0\times10^6 \text{ N·m})(10 \text{ m})}{(75\times10^9 \text{ Pa})(0.725 \text{ m})^3 (0.025 \text{ m})}$$

Answer: $\theta = 28.0 \times 10^{-3}$ rad $= 1.60°$

Since thickness t is constant, the integral term is simply the perimeter at the centerline divided by the thickness.

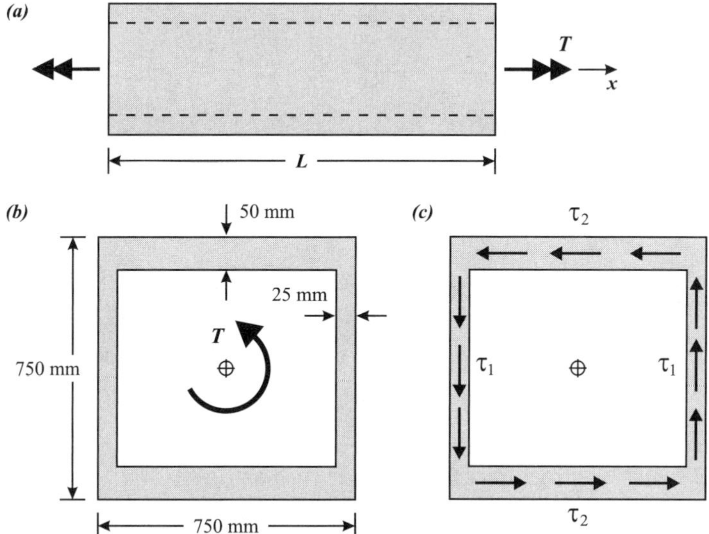

Figure 5.22. **(a)** Thin-walled member under torque T. **(b)** Square cross-section with variable thickness t. **(c)** Shear stress τ_1 and τ_2.

Example 5.12 Box Beam in Torsion 2

Given: A similar beam as in the previous example is considered. The length is again 10.0 m, and the cross-section is a hollow square with outside dimensions $d = 750$ m. The thickness of one set of opposite sides is $t_1 = 25$ mm, while the thickness of the other set of sides is $t_2 = 2t = 50$ mm (*Figure 5.22*). The applied torque is 2000 kN·m. The shear modulus of the steel is 75 GPa.

Required: Estimate (a) the shear stress in the beam in the thin walls, (b) the shear stress in the thick walls, and (c) the angle of twist.

Solution: *Step 1.* A_o is the area bound by the centerline of the walls:

$$A_o = (d - t_1)(d - t_2) = (0.725 \text{ m})(0.700 \text{ m}) = 0.5075 \text{ m}^2$$

The shear stress in the 25 mm wall is:

$$\tau_1 = \frac{T}{2A_o t_1} = \frac{2.0 \times 10^6 \text{ N·m}}{2(0.5075 \text{ m}^2)(0.025 \text{ m})}$$

Answer: $\tau_1 = 78.8$ MPa

The shear stress in the 50 mm walls is:

$$\tau_2 = \frac{T}{2A_o t_2} = \frac{2.0 \times 10^6 \text{ N·m}}{2(0.5075 \text{ m}^2)(0.05 \text{ m})}$$

Answer: $\tau_2 = 39.4$ MPa

Step 2. The angle of twist is:

$$\theta = \frac{TL}{4GA_o^2}\int_s \frac{ds}{t} = \frac{TL}{4GA_o^2}\left[\frac{2(d-t_2)}{t_1} + \frac{2(d-t_1)}{t_2}\right]$$

$$= \frac{(2.0 \times 10^6 \text{ N·m})(10 \text{ m})}{4(75 \times 10^9 \text{ Pa})(0.5075 \text{ m}^2)^2}\left[\frac{2(0.700 \text{ m})}{0.025 \text{ m}} + \frac{2(0.725 \text{ m})}{0.050 \text{ m}}\right]$$

Answer: $\theta = 22.3 \times 10^{-3}$ rad $= 1.26°$

Note that although more material was added to the cross-section, the maximum shear stress in the 25 mm wall is the same as in *Example 5.11* (actually the calculated value is slightly larger since the centerline perimeter is smaller). To reduce the maximum stress found in *Example 5.11*, the thickness of each wall must be increased.

Example 5.13 Thin-walled Circular Shaft in Torsion

Given: A thin-walled shaft of circular cross-section, *average* radius R, constant thickness t $(t \ll R)$, and length L, is subjected to a torque T (*Figure 5.23*).

Required: Using the general closed thin-walled torsion equations, determine (a) the shear stress on the cross-section and (b) the angle of twist. (c) Compare the results with the thin-walled circular shaft equations (*Equations 5.1 and 5.3*).

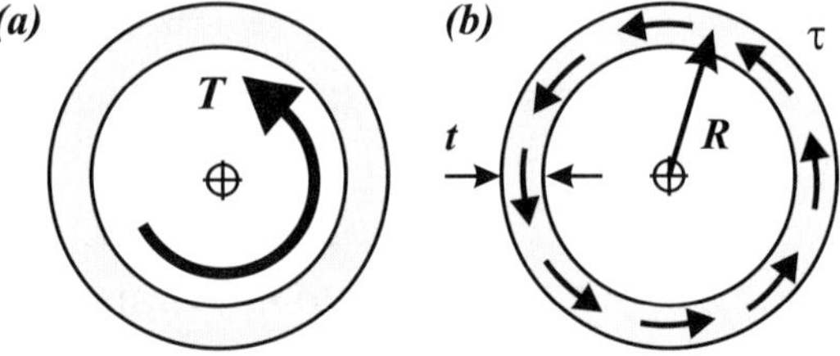

Figure 5.23. (a) Thin-walled circular shaft in torsion. **(b)** Shear stress distribution around a thin-walled shaft.

Solution: *Step 1*. The area bound by the average radius R is:

$$A_o = \pi R^2$$

The shear stress in the wall is therefore:

$$\tau = \frac{T}{2A_o t} = \frac{T}{2(\pi R^2)t}$$

Answer: $\tau = \dfrac{T}{2\pi R^2 t}$

Step 2. The angle of twist is:

$$\theta = \frac{TL}{4GA_o^2}\int_s \frac{ds}{t} = \frac{TL}{4G(\pi R^2)^2}\left[\frac{2\pi R}{t}\right]$$

Answer: $\theta = \dfrac{TL}{2\pi R^3 t G}$

These results are the same as *Equations 5.1 and 5.3*, the stress and angle of twist for a thin-walled circular shaft of average radius R and thickness t.

5.5 Power Transmission

Power is often transmitted through rotating shafts. For example, the drive shaft in an automobile transfers the engine's power to the wheels; the drive shafts in a ship transfer power to the propellers (*Figure 5.24*).

Power is the amount of work performed per unit time. Work has units of force–length, e.g., N·m = J (joules), or ft-lb. Power has units of work per time, e.g., N·m/s = J/s = W (watts), or ft-lb/s.

In the SI system, the power P (in W) transmitted by a shaft is the torque T (N·m) carried by the shaft multiplied by its angular velocity ω (rad/s):

$$P = T\omega \qquad \text{[Eq. 5.50]}$$

Rearranging *Equation 5.50*, the torque (N·m) supported by a rotating shaft that transmits power (W) at a given angular speed (rad/s) is:

$$T = \frac{P}{\omega} \qquad \text{[Eq. 5.51]}$$

When US units are used, torque is usually given in *pound–feet* (lb-ft), angular speed in *revolutions per minute* (rpm), and power in *horsepower* (hp). Horsepower is defined as:

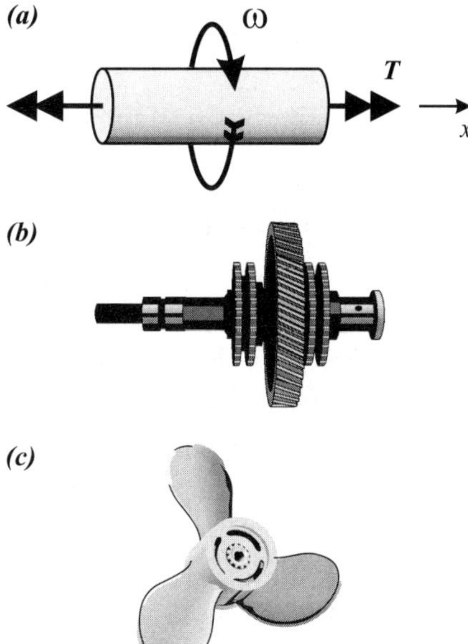

Figure 5.24. **(a)** Shaft rotating with constant angular velocity ω.
(b) Shafts transmit power through gears. **(c)** A ship's drive shaft transmits power to its propeller.
Copyright ©2008 Dominic J. Dal Bello and licensors. All rights reserved.

$$1 \text{ hp} = 550 \text{ ft-lb/sec} = 6600 \text{ in.-lb/sec} \qquad \text{[Eq. 5.52]}$$

The US unit equation for relating torque T in *pound–feet* to power HP is:

$$T(\text{lb-ft}) = \frac{33{,}000 \times HP}{2\pi N} \qquad \text{[Eq. 5.53]}$$

where HP is in *horsepower* and N is the angular velocity in *revolutions per minute*. The torque can also be given in *pound–inches*:

$$T(\text{lb-in.}) = \frac{63{,}000 \times HP}{N} \qquad \text{[Eq. 5.54]}$$

Take care when using *Equations 5.53* and *5.54*; power is in horsepower and angular velocity N is in revolutions per minute. Note the different units of torque in each equation.

To convert from rad/s to rpm:

$$N(\text{rpm}) = \omega(\text{rad/s})\left(\frac{1 \text{ rev}}{2\pi \text{ rad}}\right)\left(\frac{60 \text{ s}}{1 \text{ min}}\right) = 9.549 \times \omega \text{ (rad/s)} \qquad \text{[Eq. 5.55]}$$

and from rpm to rad/s:

$$\omega(\text{rad/s}) = 0.1047 \times N \text{ (rpm)} \qquad \text{[Eq. 5.56]}$$

Example 5.14 **Propeller Shaft**

Given: The solid-shaft $(D = 20.0 \text{ mm})$ of a scale-model boat transmits a torque of $T = 100$ N·m at an angular speed of 4.00 rad/s (*Figure 5.25*).

Required: Determine (a) the maximum shear stress in the shaft due to the torque and (b) the power transmitted by the shaft to the propeller.

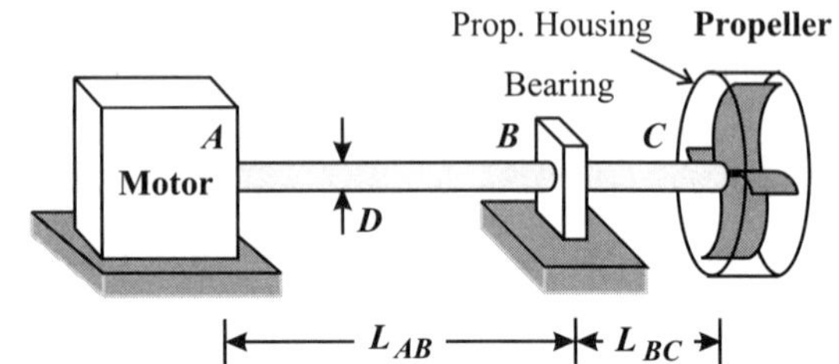

Figure 5.25. Shaft powering a propeller.

Solution: *Step 1.* The polar moment of inertia is:

$$J = \frac{\pi R^4}{2} = \frac{\pi(0.01 \text{ m})^4}{2} = 15.71 \times 10^{-9} \text{ m}^4$$

Step 2. The maximum shear stress is:

$$\tau_{max} = \frac{TR}{J} = \frac{(100 \text{ N·m})(0.01 \text{ m})}{15.71 \times 10^{-9} \text{ m}^4}$$

Answer: $\tau_{max} = 63.7$ MPa

Step 3. The power transmitted by the shaft is:

$$\text{Power} = T\omega = (100 \text{ N·m})(4 \text{ rad/s})$$

Answer: Power = 400 N·m/s = 400 W

Example 5.15 Design of a Hollow Shaft

Given: A hollow circular shaft is to transmit 240 hp at 2000 rpm. The allowable shear stress is $\tau_A = 12.0$ ksi.

Required: If the ratio of the outer- to inner-diameter of the shaft is to be approximately 1.5, design the shaft; i.e., select the outer and inner diameters D and D_i. Round the diameters to the nearest 0.10 in.

Solution: *Step 1.* The required torque is:

$$T\text{(lb-ft)} = \frac{33{,}000 \times HP}{2\pi N} = \frac{33{,}000(240 \text{ hp})}{2\pi(2000 \text{ rpm})} = 630 \text{ lb-ft} = 7560 \text{ lb-in.}$$

Step 2. The maximum shear stress for a hollow shaft (outer radius R, inner radius R_i) is:

$$\tau_{max} = \frac{TR}{J} = \frac{2TR}{\pi(R^4 - R_i^4)}$$

Since $D/D_i = R/R_i = 1.5$, then $R_i = \dfrac{2R}{3}$

Thus:

$$\tau_{max} = \frac{2TR}{\pi\left[R^4 - \left(\frac{2}{3}R\right)^4\right]} = \frac{2TR}{\pi R^4\left[1 - \frac{16}{81}\right]} = \frac{2T}{\pi R^3\left[\frac{65}{81}\right]} = 0.7933\frac{T}{R^3}$$

Step 3. Setting $\tau_{max} = \tau_A$, and solving for R:

$$R = \left[0.7933\frac{T}{\tau_A}\right]^{1/3} = \left[0.7933\left(\frac{7560 \text{ lb-in.}}{12{,}000 \text{ psi}}\right)\right]^{1/3} = 0.794 \text{ in.}$$

Taking the outer radius to be $R = 0.80$ in., then the inner radius is $R_i = 0.53$ in. Rounding the outer diameter up and the inner diameter down to the nearest 0.10 in. gives:

Answer: $\underline{D = 1.60 \text{ in.}}$

Answer: $\underline{D_i = 1.00 \text{ in.}}$

Check:

$$\tau_{max} = \frac{TR}{J} = \frac{2(7560 \text{ lb-in.})(0.80 \text{ in.})}{\pi[(0.80 \text{ in.})^4 - (0.50 \text{ in.})^4]} = 11.1 \text{ ksi} < 12 \text{ ksi} \quad \textbf{OK}$$

Chapter 6   Bending Members: Beams

6.0 Introduction

A ***beam*** is a structural component that supports loads applied transverse to its main dimensional axis (*Figure 6.1*). Floor joists support the vertical loads due to the occupants and furniture of a building. Sign-posts and trees resist the horizontal forces caused by wind. Diving boards support divers, tree limbs support children on tire-swings, and park benches support resting seniors. Beams must support the loads applied to them without breaking or deflecting excessively.

Since beams are subjected to transverse loads, they must support both internal ***bending moments*** and ***shear forces***. The *four-point bend test* (*Figure 6.2a*) is commonly used to determine the properties of high-tech materials. In this test, a *simply-supported beam* of constant cross-section is loaded *symmetrically* by two identical point loads P, one load applied distance a inside each support.

Due to symmetry of both the geometry and the load, the reaction at each support is $R = P$ (*Figure 6.2b*). The ***shear force diagram*** $V(x)$ vs. x, and the ***bending moment diagram*** $M(x)$ vs. x, are shown in *Figure 6.2c*, along with the *positive convention* of this text (an internal load is *positive* if it acts on a positive face in a positive direction or on a negative face in a negative direction). The shear and moment equations are:

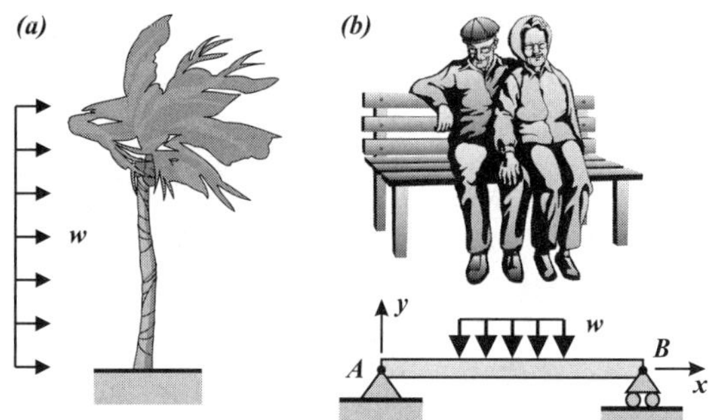

Figure 6.1. (a) A tree acts as a cantilever beam under the wind load *w* (force per length). (b) Seniors resting on a park bench. Copyright ©2008 Dominic J. Dal Bello and licensors. All rights reserved.

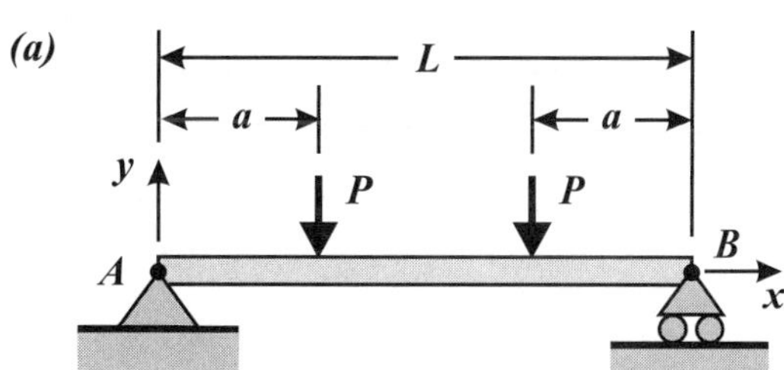

Figure 6.2. (a) A four-point bend test: a simply supported beam under two equal and symmetrically applied point loads.

F.A. Leckie, D.J. Dal Bello, *Strength and Stiffness of Engineering Systems*, Mechanical Engineering Series, **163**
DOI 10.1007/978-0-387-49474-6_6, © Springer Science+Business Media, LLC 2009

- $x < a$:

  $$V(x) = -P \qquad M(x) = Px$$

- $a < x < L - a$:

  $$V(x) = 0 \qquad M(x) = Pa$$

- $x > L - a$:

  $$V(x) = P \qquad M(x) = P(L - x)$$

Between the applied loads, the shear force is zero and the bending moment is constant.

The condition of *zero shear force* and *constant bending moment* is known as *pure bending*. Although most beams also support a shear force, the study of the *pure bending* case provides the basis of a straightforward theory to determine the *normal stresses* and *deflections* that arise in beams.

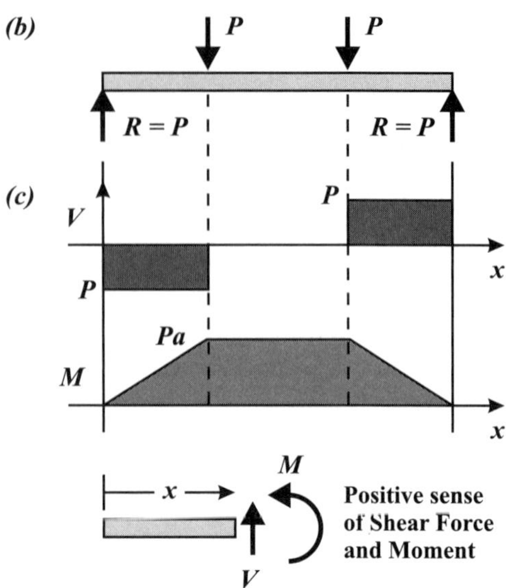

Figure 6.2. (b) Free body diagram of beam. **(c)** Shear force and bending moment diagrams for four-point bending.

6.1 Bending Strain and Stress

The behavior of a beam in bending is established using the *displacement method*. A constant radius of curvature R is impressed on a length L of the beam (*Figure 6.3*). The current aims are to determine:

1. the strains caused by the deformation,
2. the moment required to cause the deformation, and
3. the resulting stresses in the beam due to the moment.

In the following formulation, the cross-section of the beam is constant along the x-axis and *symmetric about the y-axis* (*Figure 6.3a*). The moment acts about the z-axis.

1. Radius of Curvature – Strain Relationship

Consider a beam of uniform cross-section and material properties subjected to bending moment M, constant along the x-axis (*Figure 6.3b*). This *pure bending* condition is the same loading as on the center segment of the four-point bend test. The moment is drawn in the positive sense, with the top of the beam in compression.

Since the bending moment and the cross-section are both constant with x, the beam at every point along its length must bend, or curve, in the same manner; e.g., cross-sections *AB* and *CD* are under the same loading and bending conditions (*Figure 6.3b*). The beam must therefore deform into the arc of a circle, since the circle is the only shape of constant curvature.

In pure bending, each cross-section must deform the same way, so each aligns with a radius of the circle (*Figure 6.3b*). The graphical extensions of the cross-sectional planes meet at a single point, O, called the **center of curvature**. A *cross-sectional plane must remain plane* after the moment is applied.

That cross-sectional *planes remain plane* can be demonstrated by considering center plane AB in *Figure 6.4*. The loading and geometry of the beam is the same viewed from either the front or the back (*Figures 6.4a* and *b*). Therefore, center plane AB must deflect the same way, viewed from either front or back. The only way that this is possible is for plane AB to remain plane (straight). This is true for all cross-sectional planes since the beam in *Figure 6.4* can be continually halved and the same argument used.

Under a positive moment, the top surface of the beam shortens, while the bottom surface elongates (*Figure 6.3b*). Somewhere between those surfaces, there is an interior surface that does not change length; this is called the **neutral plane** (at $y = 0$). Viewing the beam from the side, this plane is a line, called the **neutral axis**, indicated by the broken x-axis in *Figure 6.3*. Where the neutral axis (NA) lies – the location of $y = 0$ on the cross-section – depends on the shape of the cross-section.

The radius of the circular arc that the beam bends into, measured from the center of curvature O to the neutral axis, is the **radius of curvature** R (*Figure 6.3*). The inverse of R is the **curvature** κ (Greek "kappa"):

$$\kappa = \frac{1}{R} \qquad \text{[Eq. 6.1]}$$

Now consider the beam segment of length L in *Figure 6.5*. The beam is

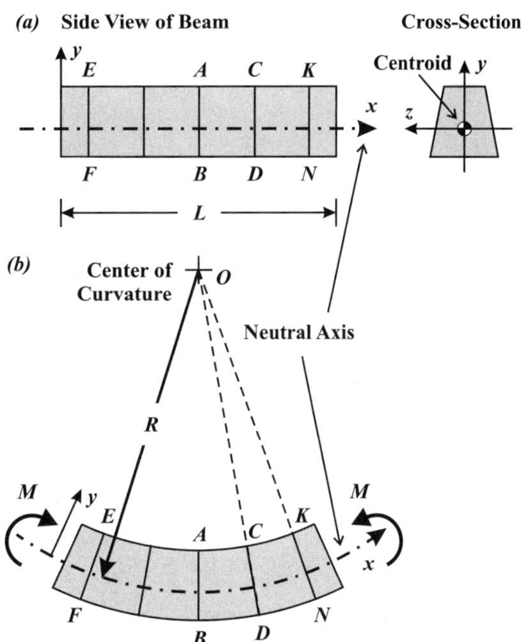

Figure 6.3. (a) Side view and cross-section of beam. Its constant cross-section is symmetric about the y-axis. **(b)** Side view of beam bent under a pure (and positive) bending moment about the z-axis. The length of the beam at the neutral axis does not change.

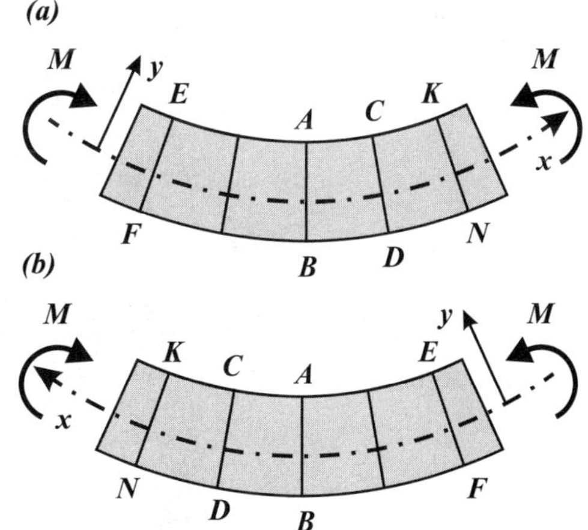

Figure 6.4. (a) Side view of beam in pure bending, viewed from the front. **(b)** Beam viewed from the back. Due to symmetry, cross-sectional planes remain plane; e.g., AB must look the same from front and back.

subjected to the impressed curvature caused by a pure moment. Cross-sectional planes *AB* and *CD* remain plane, moving to *A'B'* and *C'D'* (exaggerated in *Figure 6.5b*). At the neutral axis, distance *R* from the center of curvature, the beam remains length *L*.

Line *GH* (representing a plane parallel to the neutral plane) is distance *y* above the neutral axis. The original length of *GH* is *L*. The deformed length *G'H'* lies on a circular arc of radius $R - y$ from the center of curvature. From proportionality:

$$\frac{G'H'}{L} = \frac{R-y}{R} \qquad \text{[Eq. 6.2]}$$

The strain of *GH* is linear with *y*:

$$\varepsilon(y) = \frac{G'H' - L}{L} = \frac{(R-y) - R}{R} = -\frac{y}{R}$$

$$\text{[Eq. 6.3]}$$

For a positive moment, the strain is negative (compressive) above the neutral axis ($y > 0$), and positive below it ($y < 0$). At $y = 0$, the strain is zero – the neutral axis does not change length. Note that the neutral axis still has yet to be located.

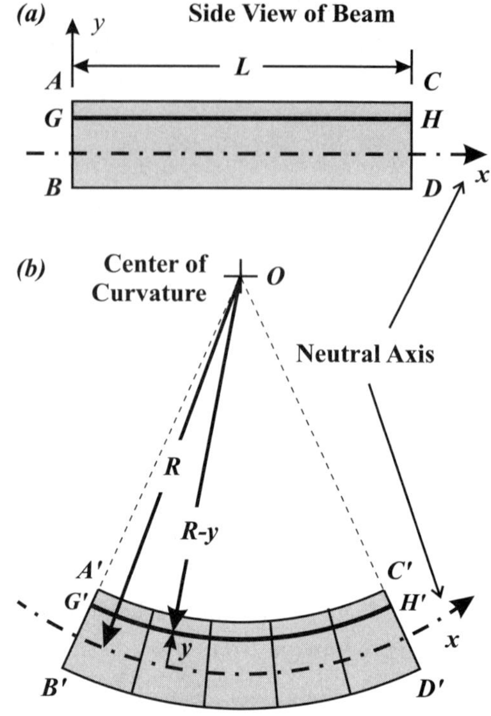

Figure 6.5. (a) Side view of length *L* of beam. Line *GH* is distance *y* above the Neutral Axis. **(b)** Beam under impressed curvature due to pure bending. *GH* deforms into *G'H'*.

2. Stress–Strain Relationship

From Hooke's Law, the normal stress due to bending varies linearly with *y*:

$$\sigma(y) = E\varepsilon(y) = -\frac{Ey}{R} \qquad \text{[Eq. 6.4]}$$

The stress is zero at the neutral axis ($y = 0$) and is greatest in magnitude where *y* is maximum, furthest away from the **bending axis** (i.e., the *z*-axis).

3. Equilibrium

The beam of cross-sectional area *A* can be considered as a stack of very thin axial members or elements, each supporting part of the applied load (*Figure 6.6*). Each element has height *dy* and width *t(y)*, where *t* is the beam width at *y*. The cross-sectional area of each thin element is thus $dA = t(y)dy$.

The force on each element *dF(y)* at height *y* is its stress multiplied by its area:

$$dF(y) = \sigma(y)dA \qquad \text{[Eq. 6.5]}$$

The total force F in the x-direction (along the beam's axis), is calculated by integrating dF over the cross-sectional area A:

$$F = \int_A \sigma(y)dA = \int_A -\frac{Ey}{R}dA \quad \text{[Eq. 6.6]}$$

Neither E nor R depends on area and can be removed from the integral.

The beam is subjected to pure bending only, so the resultant force F that acts on any cross-section is zero:

$$F = -\frac{E}{R}\int_A y\,dA = 0 \quad \text{[Eq. 6.7]}$$

Since neither E nor $1/R$ is zero, it follows that:

$$\int_A y\,dA = 0 \quad \text{[Eq. 6.8]}$$

The integral is the ***first moment of area*** of the cross-section, used to locate the cross-section's ***centroid*** or *center of area*. The first moment of area is zero when y is measured from a horizontal axis through the area's centroid. The intersection of the cross-section and the neutral plane – the z-axis in *Figure 6.6* – therefore passes through the cross-section's centroid. Since the cross-section is symmetric about the y-axis, the neutral axis also passes through the centroid. *Table 6.1* gives expressions for the locations of the centroids of a few common cross-sections.

The moment of $dF(y)$ about the bending axis, i.e., about the z-axis in *Figure 6.6*, is:

$$dM = -y\,dF(y) = -y\sigma(y)dA \quad \text{[Eq. 6.9]}$$

The negative sign is included in *Equation 6.9* since, as drawn, dF for positive values of y causes a clockwise (negative) moment about the z-axis. The total moment M about the z-axis is then:

$$M = -\int_A y\sigma(y)dA = \int_A y\frac{Ey}{R}dA = \frac{E}{R}\int_A y^2 dA \quad \text{[Eq. 6.10]}$$

The integral term is a geometric property of the cross-section called the ***second moment of area***:

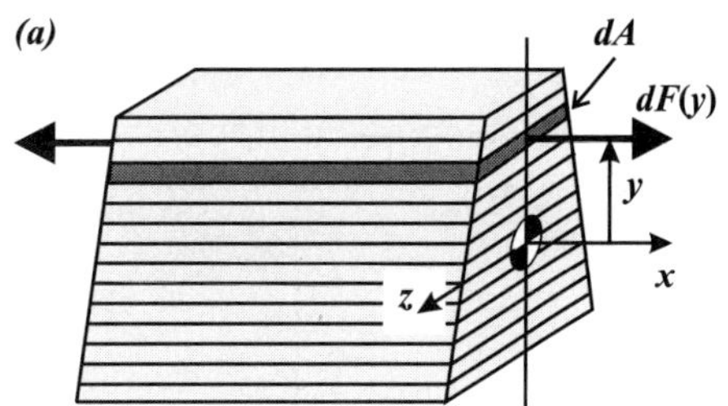

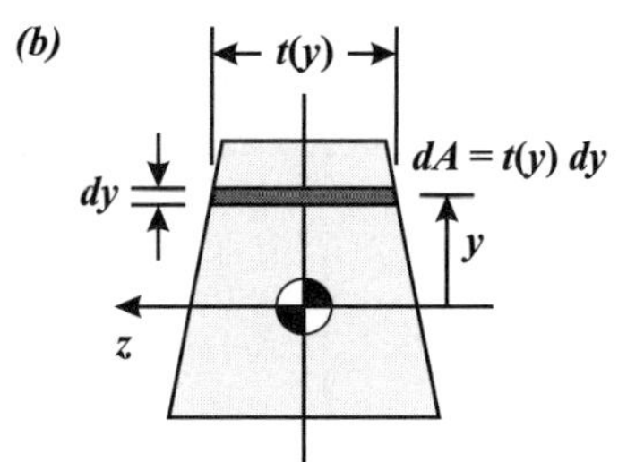

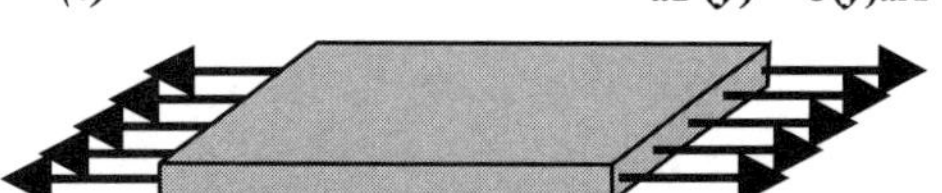

Figure 6.6. **(a)** Isometric view of beam visualized as a stack of axial members, each under load $dF(y) = \sigma(y)dA$. **(b)** Cross-section of beam. **(c)** A thin slice, cross-sectional area $dA = t(y)dy$ under stress $\sigma(y)$.

Table 6.1. Geometric properties: *Area*, *Moment of Inertia* and *Neutral Axis*.

Shape		Area	Moment of Inertia about horizontal axis through Centroid	Location of Neutral Axis (from bottom)
Rectangle, Square	d ▭ b	bd	$\dfrac{bd^3}{12}$	$\dfrac{d}{2}$
Hollow Rectangle	w d b }f	$bd-(b-2w)(d-2f)$ (*d* is total depth)	$\dfrac{bd^3-(b-2w)(d-2f)^3}{12}$	$\dfrac{d}{2}$
Circle, solid	D $=2R$	$\pi R^2 = \dfrac{\pi D^2}{4}$	$\dfrac{\pi R^4}{4} = \dfrac{\pi D^4}{64}$	$R = \dfrac{D}{2}$
Circle, thick wall	D $=2R$ $D_i=2R_i$	$\pi(R^2-R_i^2)$	$\dfrac{\pi(R^4-R_i^4)}{4} = \dfrac{\pi(D^4-D_i^4)}{64}$	$R = \dfrac{D}{2}$
Built up I-beam	w d b }f	$w(d-2f)+2bf$ (*d* is total depth)	$\dfrac{w(d-2f)^3}{12}+2\left[\dfrac{bf^3}{12}+bf\left(\dfrac{d}{2}-\dfrac{f}{2}\right)^2\right]$ (parallel axis theorem)	$\dfrac{d}{2}$
Chords of Truss	$\dfrac{A}{2}$ ○ $\dfrac{A}{2}$ ○ }d	A (total cross-sectional area of upper and lower chord)	$\dfrac{Ad^2}{4}$ (*d* is distance between chord centroids)	$\sim\dfrac{d}{2}$
T-beam	b }f }d w	$wd+bf$ (*d* is height of vertical rectangle only)	$\left[\dfrac{wd^3}{12}+wd\left(d_1^2\right)\right]+\left[\dfrac{bf^3}{12}+bf\left(d_2^2\right)\right]$	See below
T-Beam, inverted	w }d }f b	$wd+bf$ (*d* is height of vertical rectangle only)	$\left[\dfrac{wd^3}{12}+wd\left(d_1^2\right)\right]+\left[\dfrac{bf^3}{12}+bf\left(d_2^2\right)\right]$	See below

Notes: For circular beams: $I = J/2$.

For T-beams: d_1 = from neutral axis to centroid of vertical rectangular piece ($w{\times}d$).

d_2 = from neutral axis to centroid of horizontal rectangular piece ($b{\times}f$).

For T-beams: distance of centroid above bottom:

$$\text{Upright T: } y_c = \frac{\frac{d}{2}(wd)+\left(d+\frac{f}{2}\right)(bf)}{wd+bf} \; ; \quad \text{Inverted T: } y_c = \frac{\left(f+\frac{d}{2}\right)(wd)+\frac{f}{2}(bf)}{wd+bf} .$$

$$I = \int_A y^2 \, dA \qquad \text{[Eq. 6.11]}$$

The *second moment of area* is known as the **moment of inertia** of the cross-section. *Table 6.1* gives expressions for a few common cross-sections, where I is taken about the horizontal axis through the centroid. Hence, *Equation 6.10* reduces to the *moment-curvature relationship*:

$$M = \frac{EI}{R} = EI\kappa \qquad \text{[Eq. 6.12]}$$

Recalling the stress–strain relationship of *Equation 6.4*, and substituting the expression for R from *Equation 6.12*, gives the distribution of stress on the cross-section of the beam:

$$\sigma(y) = \varepsilon(y)E = -\frac{y}{R}E = -y\frac{M}{I}$$

$$\text{[Eq. 6.13]}$$

This normal stress due to the bending moment is the **bending stress**:

$$\sigma(y) = -\frac{My}{I} \qquad \text{[Eq. 6.14]}$$

The *bending stress* is zero at the centroid, $y = 0$ (at the neutral plane or neutral axis), increases linearly with y, and reaches maximum magnitudes at the top and bottom of the beam (*Figure 6.7*).

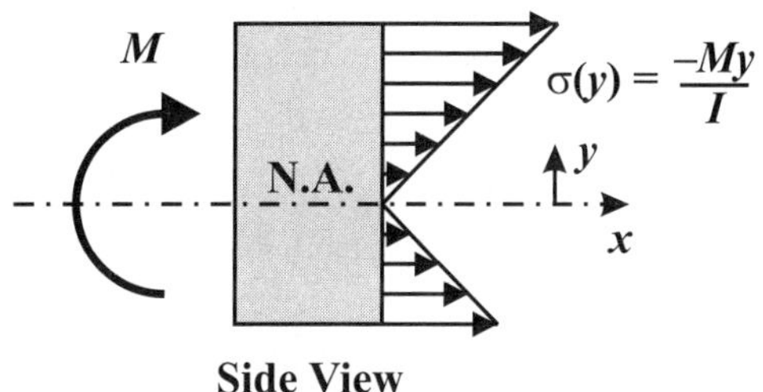

Side View

Figure 6.7. *Bending stress* varies linearly with distance y from the centroid. The centroid is not necessarily equidistant from the top and bottom of the beam, so the magnitudes of the bending stresses at those locations are, in general, not equal.

Key results for bending

- The neutral axis passes through the centroid of the beam's cross-section.
- The moment-radius of curvature relation is:

$$M = \frac{EI}{R}$$

- The strain is: $\quad \varepsilon(y) = \dfrac{-y}{R} = \dfrac{-My}{EI}$

- The stress is: $\quad \sigma(y) = -\dfrac{My}{I}$

- The positive and negative sense of the moment can be remembered by:

Bending stress is *positive* (tensile) on one side of the neutral plane and *negative* (compressive) on the other side. A positive moment causes compression ($\sigma < 0$) in the upper part of the beam ($y > 0$); a negative moment causes compression in the lower part of the beam ($y < 0$). Since M is generally a function of x, bending stress varies with both x and y.

The *magnitude* of the **maximum stress** at a given cross-section is:

$$\sigma_{max} = \frac{Mc}{I} \qquad \text{[Eq. 6.15]}$$

Variable c represents the distance to the material point furthest from the neutral axis, i.e., $c = y_{max}$. When the cross-section is symmetric about the z-axis (e.g., rectangles, circles, I-beams), the centroid is vertically centered on the cross-section, so the distances from the centroid to the top and bottom of the beam are the same. In general, the centroid is *not* vertically centered on the cross-section, e.g., the trapezoid (*Figure 6.6*). For such cross-sections, there will be two values for $c = y_{max}$, one at the top of the beam and the other at the bottom; the maximum positive and negative bending stresses will have different magnitudes (*Figure 6.7*).

Example 6.1 Moment of Inertia of a Rectangle

Given: A rectangular cross-section has depth d and width b (*Figure 6.8*).

Required: Determine the moment of inertia I about the z-axis.

Solution: By symmetry, the centroid of the cross-section, and thus the neutral axis, is centered vertically and horizontally. Coordinate y varies from $-d/2$ to $+d/2$, and differential area on the cross-section is $dA = b(dy)$. Thus:

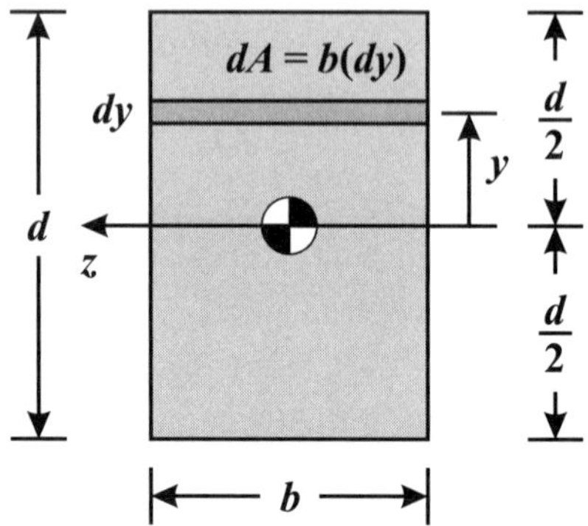

Figure 6.8. Rectangular cross-section.

$$I = \int_A y^2\, dA = \int_{-d/2}^{d/2} y^2(b\, dy) = b\frac{y^3}{3}\bigg|_{-d/2}^{d/2}$$

$$Answer:\quad I = \frac{bd^3}{12}$$

Expressions for the moment of inertia for common shapes have already been calculated. The expressions of I about the horizontal axis for a few shapes are given in *Table 6.1*.

Example 6.2 Rectangular Beam, Simply-Supported, Central Point Load

Given: A simply-supported beam of length L supports a central point load P (*Figure 6.9*). In experiments, this set-up is called a *three-point bend test*. The constant cross-section is rectangular with width b and depth d.

Required: Determine (a) the maximum bending stress and (b) the minimum radius of curvature for the case: $P = 250$ kN, $L = 3.0$ m, $b = 100$ mm, $d = 300$ mm, $E = 200$ GPa.

Solution: *Step 1.* The variation of shear force and bending moment may be found by constructing *shear force* and *bending moment diagrams* (SFD and BMD), as in *Figure 6.9c*. The bending moment increases linearly with x, and reaches a maximum value $M = PL/4$ at $x = L/2$. The BMD is symmetric about $x = L/2$.

Step 2. The moment of inertia of the cross-section is:

$$I = \frac{bd^3}{12} = \frac{(100\times10^{-3}\ \text{m})(300\times10^{-3}\ \text{m})^3}{12} = 225\times10^{-6}\ \text{m}^4 = 225\times10^6\ \text{mm}^4$$

Note that 10^{-6} m^4 = 10^6 mm^4.

Step 3. The *bending stress* is:

$$\sigma(y) = -\frac{My}{I}$$

Since I is constant, the bending stress is maximum where the moment is maximum (at $x = L/2$) and where y is maximum ($y_{max} = c = \pm d/2$):

$$M_{max} = \frac{PL}{4}$$

$$= \frac{(250\times10^3 \text{ N})(3 \text{ m})}{4}$$

$$= 187.5 \text{ kN·m}$$

Thus, the maximum stress is (*Figure 6.9d*):

$$\sigma_{max} = -\frac{M_{max}c}{I}$$

$$= -\frac{(187.5 \text{ kN·m})(\pm150\times10^{-3} \text{ m})}{(225\times10^{-6} \text{ m}^4)}$$

Answer: $\sigma_{max} = \mp 125$ MPa

The stress is compressive at the top ($\sigma = -125$ MPa at $y = +150$ mm) and tensile at the bottom, increasing linearly from the neutral axis.

Step 4. Radius of curvature. Moment M, and in turn radius of curvature R, varies with x. The radius of curvature is minimum – the beam has the tightest bend – when the moment is maximum:

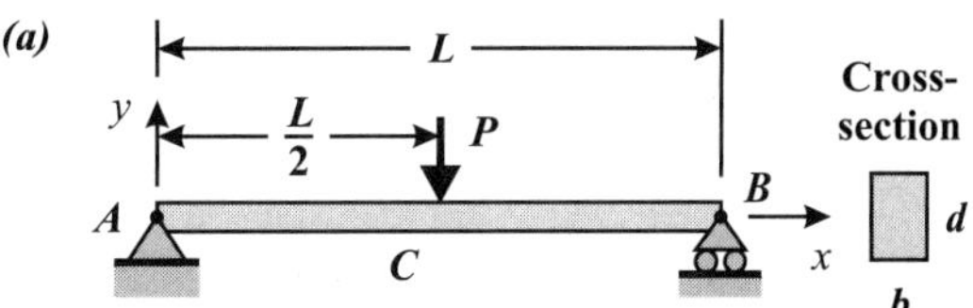

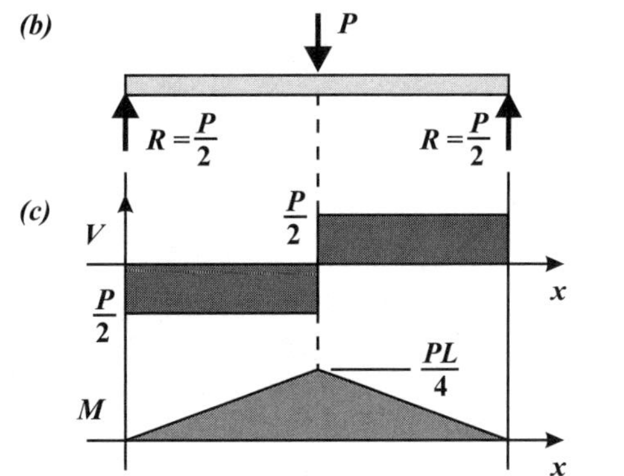

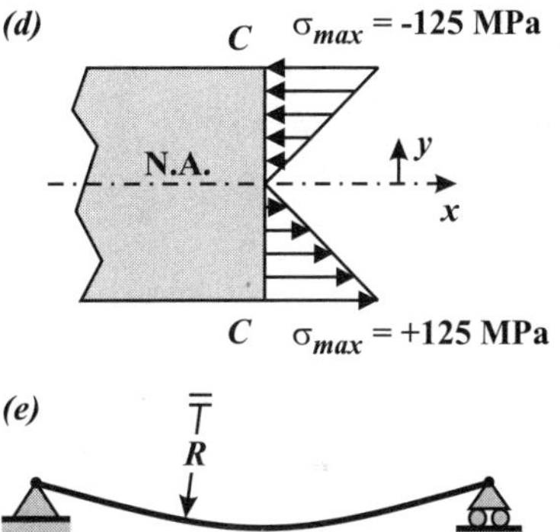

Figure 6.9. (a) Simply-supported beam of rectangular cross-section under central point load. (b) FBD of beam. (c) Shear and moment diagrams. (d) Bending stress distribution at $x = L/2$; stresses act as drawn. (e) Qualitative deflection of the neutral axis.

$$R_{min} = \frac{EI}{M_{max}} = \frac{(200\times10^9 \text{ Pa})(225\times10^{-6} \text{ m}^4)}{(187.5\times10^3 \text{ N·m})}$$

Answer: $R_{min} = 240$ m

The radius of curvature $R = 240$ m is much greater than the beam span $L = 3$ m and depth $d = 300$ mm. Physically, this means that the beam deflection is small compared to its length.

Example 6.3   Built-up T-beam, Simply-Supported, Uniformly Distributed Load

Given: A simply-supported wooden beam is made by nailing two planks together as shown in *Figure 6.10*. The beam supports a uniformly distributed load of $w = 3.0$ kN/m and the span is $L = 4.0$ m. The dimensions of the cross-section are $b = d = 200$ mm and $w = f = 50$ mm (*Figure 6.10e*). Take the modulus of wood to be $E = 1.2$ GPa.

Required: Determine (a) the maximum bending stress, and (b) the minimum radius of curvature.

Solution: *Step 1.* Determine the SFD and BMD. It is convenient (but not necessary) to measure x from the *center* of the span.

From equilibrium and symmetry, the reaction forces are $R = wL/2$. Due to the simple (pinned) supports, the moment at each end of the beam is zero.

The shear force from the FBD of *Figure 6.10c* is:

$$V(x) = wx$$

The maximum magnitude of the shear force occurs at the supports, $x = \mp L/2$:

$$V_{max} = \mp \frac{wL}{2}$$

The bending moment equation, with x measured from the center of the beam, is:

$$M(x) = \frac{w}{2}\left[\left(\frac{L}{2}\right)^2 - x^2\right]$$

The moment is zero at the supports and maximum at the center ($x = 0$):

$$M_{max} = \frac{wL^2}{8} = \frac{(3\times10^3 \text{ N/m})(4 \text{ m})^2}{8}$$

$$= 6.0\times10^3 \text{ N·m}$$

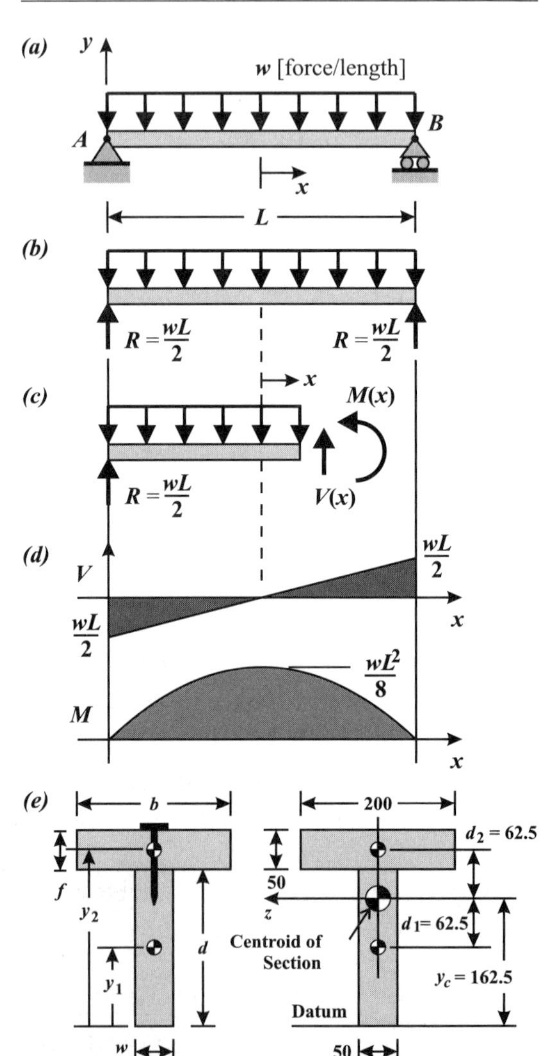

Figure 6.10. **(a)** A simply-supported T-beam under uniformly distributed load *w*. **(b)** FBD of beam. **(c)** FBD of segment of beam cut at $x > 0$. **(d)** Shear and moment diagrams. **(e)** Cross-section of T-beam and determination of location of centroid.

The general shear and moment diagrams for a simply-supported beam under uniformly distributed load are shown in *Figure 6.10d*.

Step 2. The *neutral axis* (NA) passes through the centroid of the cross-section. Taking the bottom of the beam as the datum (*Figure 6.10e*), the distance y_c from the datum to the centroid of an upright T-beam is given in *Table 6.1*:

$$y_c = \frac{\sum y_i A_i}{\sum A_i} = \frac{\frac{d}{2}(wd) + \left(d + \frac{f}{2}\right)(bf)}{wd + bf} = \frac{100(50 \times 200) + (200 + 25)(200 \times 50)}{(50 \times 200) + (50 \times 200)}$$

$$= 162.5 \text{ mm}$$

The centroid is 162.5 mm above the bottom surface.

Step 3. The *moment of inertia* about the horizontal (z-) axis through the centroid is found using the parallel axis theorem. Again, referring to *Table 6.1* for the T-beam:

$$d_1 = |y_c - y_1| = |162.5 - 100| = 62.5 \text{ mm}$$

$$d_2 = |y_c - y_2| = |225 - 162.5| = 62.5 \text{ mm}$$

$$I = \left[\frac{wd^3}{12} + wd\left(d_1^2\right)\right] + \left[\frac{bf^3}{12} + bf\left(d_2^2\right)\right] = 113.5 \times 10^6 \text{ mm}^4$$

By coincidence, distances d_1 and d_2 are equal; this is generally not the case.

Step 4. Maximum bending stresses. The bending stress is: $\sigma = -\dfrac{My}{I}$

Referring to *Figure 6.10f*, the top of the beam is at $y_t = +87.5$ mm, and the bottom is at $y_b = -162.5$ mm. The bending stresses at the top and bottom are:

$$\sigma_{max,t} = -\frac{M_{max}y_t}{I} = \frac{(6.0 \times 10^3 \text{ N·m})(87.5 \times 10^{-3} \text{ m})}{113.5 \times 10^{-6} \text{ m}^4}$$

$$\text{Answer: } \underline{\sigma_{max,t} = -4.6 \text{ MPa (compression)}}$$

and

$$\sigma_{max,b} = -\frac{M_{max}y_b}{I}$$

$$\text{Answer: } \underline{\sigma_{max,b} = +8.6 \text{ MPa (tension)}}$$

The bottom surface is further from the NA than the top surface, so the magnitude of the stress at the bottom is greater than at the top (*Figure 6.10f*).

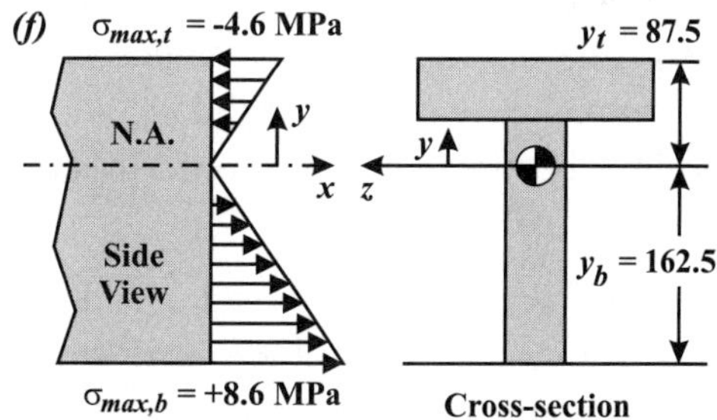

Figure 6.10. (f) Bending stress distribution at the center of the beam.

Step 5. The minimum *radius of curvature* occurs where the bending moment is maximum:

$$R_{min} = \frac{EI}{M_{max}}$$

$$= \frac{(1.2\times10^9 \text{ Pa})(113.5\times10^{-6} \text{ m}^4)}{6.0\times10^3 \text{ N·m}}$$

Answer: $R_{min} = 22.7$ m

The *curvature* κ is the inverse of the radius of curvature $\kappa = 1/R = 44.1\times10^{-3} \text{ m}^{-1}$.

The radius of curvature is much larger than the span, which is generally the case in practice.

Notes on T-beams

For T-beams, and other sections that are not symmetric about the bending-axis (e.g., the z-axis), the magnitude of the maximum bending stresses (at top and bottom) are different. This knowledge can be used to efficiently design a cross-section when a material has different strengths in tension and compression.

The tensile strength of cast iron is less than its compressive strength. Thus, cast iron beams, common during the 19th century, had **T**-shaped cross-sections, with the cap of the **T** designed to be in tension, where the magnitude of the bending stress is smaller since it is closer to the neutral axis than the foot of the **T**. A modern context is ceramic materials, which are strong in compression and weak in tension.

Woods tend to be stronger in tension than in compression, so wooden T-beams are designed so that the cap of the **T** is in compression, as in *Example 6.3*.

6.2 Beam Deflection

When sitting on a park bench, the wooden slats deflect downward. Such a movement is unsettling if the slats deflect by more than say, 1.0 inch. Likewise, bridges deflect when cars, trucks, and trains pass over them; an atomic force microscope (AFM) deflects as it scans a surface (*Figure 6.11*). In engineering systems, it is often necessary to know the magnitude of beam deflections under various loading conditions. Often, the maximum allowable beam deflection δ_{allow} is defined in terms of the span L of the beam.

The **deflection index** is defined as $f = \delta/L$. The allowable value of f is approximately $\delta/L \sim 1/240$ for a wide range of engineering applications. This small *displacement-to-span* restriction means that

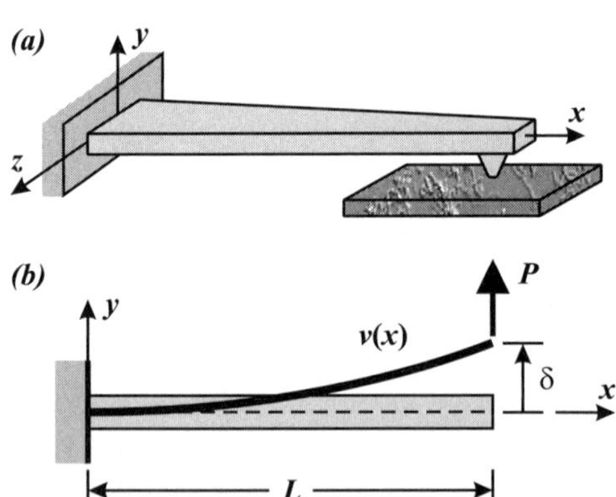

Figure 6.11. (a) Atomic force microscope scanning a surface. **(b)** Force and displacement at AFM tip.

the slope of the deflected shape is also small. The ratio 1/240 is equivalent to a deflection of 0.5 in. over 10 ft, or about 0.4%.

The Golden Gate Bridge, under full vehicle and pedestrian traffic load, was designed to deflect no more than 10 ft. The distance between towers – the central span – is 4200 ft, so the deflection index is $f = 1/420$, and rush-hour travelers should barely notice the deflection.

Deflection requirements are defined because excessive deflections may cause difficulties in service. Deflection may cause such problems as:

Misfit or interference of parts. Automobile, aircraft, and subway chassis and door frames must be stiff enough so that their doors can open and close whenever passengers congregate in the vehicle. Plastered ceilings must not deflect so much that the plaster begins to crack. Building codes require the deflection index for walls to be 1/240 for brittle coverings (e.g., plaster), and 1/120 for more pliable coverings. For beams that support ceilings, the index is 1/360 for brittle coverings and 1/240 for general coverings (*2000 International Building Code*).

Interference of moving parts. Rotating parts such as blades in engine motors have a large amount of kinetic energy. Excessive deformation of the housing of a turbine motor, and/or elongation of the blades, will cause a violent failure.

Psychological reaction. Although a system may be strong enough, large deflections can cause fear, if not physical discomfort. A pedestrian bridge built over the River Thames in England in 2000 was closed for modification because its excessive sway caused anxiety. Although the bridge was strong enough, its large deflection did not provide for a safe and comfortable passage; it was not stiff enough to assure the public of its strength.

Increased loading due to deflection. If a low-sloped roof deflects too much, say under a newly installed air-conditioning unit, rain water may begin to pond on the roof (*Figure 6.12*). This ponding increases the load on the roof, further increasing deflection, allowing more water to pond, and so on until failure.

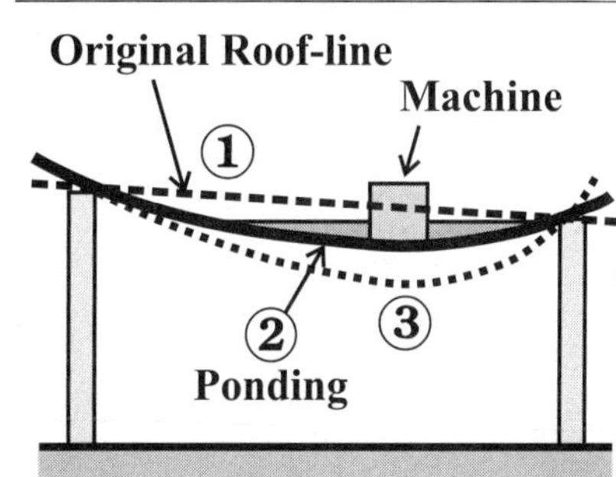

Figure 6.12. A low-sloped roof (1) is loaded with a heavy cooling unit, causing the roof to deflect to position (2). The deflection causes rain water to pond on the roof, further deflecting the roof to (3), allowing more water to pond, and so on.

Curvature–Displacement Relation

In the x–y–z coordinate system, the **displacement**, or **deflection**, of a beam in the x–y plane (with cross-section in the y–z plane) is $v(x)$, defined as positive in the positive y-direction (*Figure 6.13*). The **deflected shape** $v(x)$ describes how the neutral axis deflects with position x. Function $v(x)$ is also referred to as the *elastic curve*; it is the elastic deflection of the beam (yielding does not occur).

From analytical geometry, the expression for the radius of curvature R varies with deflection $v(x)$:

$$\frac{1}{R(x)} = \left[\frac{d^2v}{dx^2} \right]\left[1 + \left(\frac{dv}{dx}\right)^2 \right]^{-3/2} \qquad \text{[Eq. 6.16]}$$

In practical problems, the slope dv/dx is much smaller than unity (1.0) (the allowable displacement is typically less than the span divided by 240). Then, the second bracket of *Equation 6.16* reduces to 1.0, so that:

$$\frac{1}{R(x)} = \frac{d^2v}{dx^2} \qquad \text{[Eq. 6.17]}$$

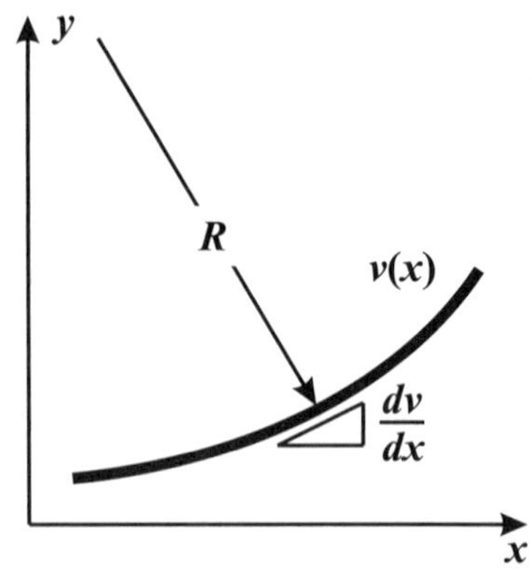

Figure 6.13. Geometry of beam deflection $v(x)$, slope dv/dx, and curvature $1/R(x)$.

The reader may recall from calculus that the second derivative of deflection with position is called the *curvature*. *Curvature* has already been defined in this text as $\kappa = 1/R$ (*Equation 6.1*).

Repeating *Equation 6.12*, the moment–radius of curvature, *M–R*, relationship is:

$$M(x) = \frac{EI}{R(x)} \qquad \text{[Eq. 6.18]}$$

Eliminating $1/R$ from the previous two equations gives the second derivative of displacement – the curvature – in terms of the moment:

$$\frac{d^2v}{dx^2} = \frac{M(x)}{EI} \qquad \text{[Eq. 6.19]}$$

The *curvature* of the beam at any point – the tightness of its bend – is proportional to the *moment* at that point $M(x)$, and inversely proportional to the beam's **bending stiffness EI**. *Equation 6.19* can be used to determine the deflected shape by integration.

The **slope** and **deflection** of the beam are found by taking the indefinite integral twice:

$$v'(x) = \frac{dv}{dx} = \theta(x) = \int \frac{M(x)}{EI}dx + C_1 \qquad \text{[Eq. 6.20]}$$

$$v(x) = \int \left[\int \frac{M(x)}{EI}dx \right]dx + C_1x + C_2 \qquad \text{[Eq. 6.21]}$$

Constants of integration C_1 and C_2 are determined by applying the **geometric boundary conditions** of the beam. A boundary condition must be at a specific location x along the beam where the beam slope or deflection is known. For example, at a wall, the slope and deflection are both zero; at a pin the deflection is zero (but the slope is not known).

Variable θ, usually used for an angle, is used as an alternative variable for the *slope*. The tangent of an angle is the slope, $\tan\theta = \Delta y/\Delta x$. For small angles, $\tan\theta = \theta$. Because the deflection and thus slope of a beam are small, the slope (tangent) is equal to the angle of the elastic curve: $v'(x) = \theta(x)$.

Example 6.4 Deflection of a Cantilever Beam under Tip Load

Given: A cantilever beam under a tip load P (*Figure 6.14*). The moment of inertia I is constant over beam length L. The elastic modulus is E.

Required: Determine (a) the slope and deflection of the beam as a function of x, (b) the slope and deflection at the point of application of the load, and (c) the stiffness of the cantilever beam loaded at its tip.

Solution: *Step 1.* To find the equation of the elastic curve, first determine the moment $M(x)$. Consider the FBD of *Figure 6.14c*. Taking moments about point D (at the cut), the moment as a function of x is:

$$M(x) = P(L - x)$$

so that:

$$\frac{d^2v}{dx^2} = \frac{M(x)}{EI} = \frac{P(L-x)}{EI}$$

Integrating the curvature gives the slope:

$$v' = \frac{P}{EI}\left(Lx - \frac{x^2}{2}\right) + C_1$$

Integrating again gives the deflection:

$$v(x) = \frac{P}{EI}\left(\frac{Lx^2}{2} - \frac{x^3}{6}\right) + C_1 x + C_2$$

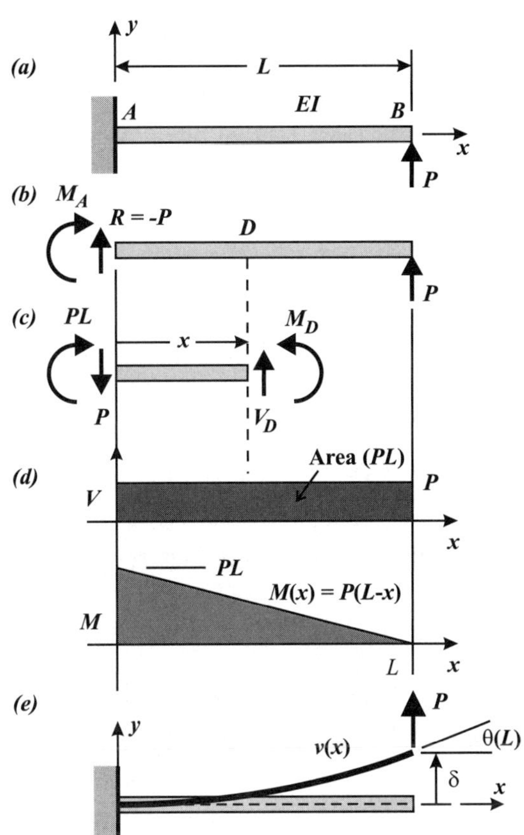

Figure 6.14. (a) Cantilever beam under tip load. (b) FBD of entire beam. (c) FBD of length x from left end. (d) Shear and moment diagrams. (e) Elastic (deflection) curve.

Constants C_1 and C_2 are determined by applying the *geometric boundary conditions*. At $x = 0$, the beam is built-in (fixed) to the wall, so the slope and deflection are both zero:

$$v'(x = 0) = C_1 = 0$$
$$v(x = 0) = C_2 = 0$$

CAUTION: C_1 and C_2 are not always zero.

The expressions for the slope and deflection are then:

$$Answer:\ v'(x) = \frac{P}{EI}\left(Lx - \frac{x^2}{2}\right)$$

$$Answer:\ v(x) = \frac{P}{EI}\left(\frac{Lx^2}{2} - \frac{x^3}{6}\right)$$

At the point of application of the load P, where $x = L$, the slope and deflection are:

$$Answer:\ v'(L) = \theta(L) = \frac{PL^2}{2EI}$$

$$Answer:\ v(L) = \delta = \frac{PL^3}{3EI}$$

The stiffness K is the ratio of the load to the deflection. The *stiffness of a cantilever beam under tip load* is:

$$Answer:\ K = \frac{P}{\delta} = \frac{3EI}{L^3}$$

Notice that the stiffness is inversely proportional to L^3; the beam's stiffness decreases rapidly with length.

Example 6.5 Atomic Force Microscope

Given: A cantilever beam is the key component in an atomic force microscope (AFM). A representative beam is $L = 140\ \mu m$ long and, for simplicity, is taken to have a rectangular cross-section $12\ \mu m \times 3\ \mu m$ ($b{\times}d$, *Figure 6.15*). The cantilever is manufactured from silicon carbide (SiC) for which $E = 450\ GPa$. The applied upward force at the tip is $P = 100\ \mu N$, and causes a positive moment everywhere along the beam. Note that AFM forces and dimensions are very small since they are used to measure small displacements.

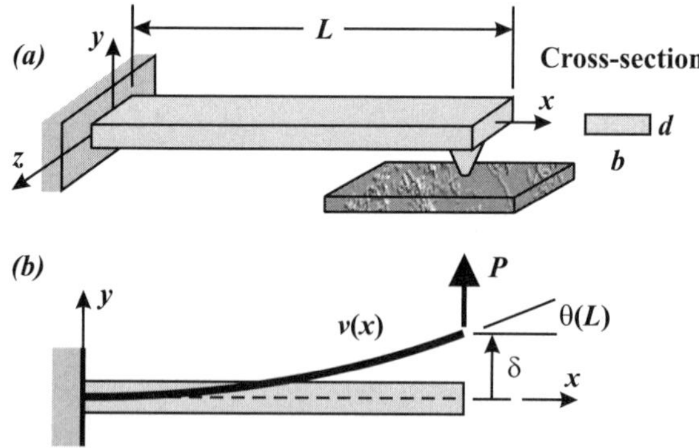

Figure 6.15. **(a)** AFM with rectangular cross-section. **(b)** Deflection of beam.

Required: Determine (a) the tip slope (rotation), (b) the tip displacement, and (c) the beam stiffness for tip loading.

Solution: *Step 1.* The moment of inertia is:

$$I = \frac{bd^3}{12} = \frac{(12 \times 10^{-6} \text{ m})(3 \times 10^{-6} \text{ m})^3}{12} = 27.0 \times 10^{-24} \text{ m}^4$$

Step 2. The slope at the tip of the cantilever is:

$$v'(x) = \frac{PL^2}{2EI} = \frac{(100 \times 10^{-6} \text{ N})(140 \times 10^{-6} \text{ m})^2}{2(450 \times 10^9 \text{ Pa})(27 \times 10^{-24} \text{ m}^4)}$$

Answer: $v'(x) = 0.0806 \text{ rad} = 4.6°$

Step 3. The tip deflection is:

$$\delta = \frac{PL^3}{3EI} = \frac{(100 \times 10^{-6} \text{ N})(140 \times 10^{-6} \text{ m})^3}{3(450 \times 10^9 \text{ Pa})(27 \times 10^{-24} \text{ m}^4)}$$

Answer: $\delta = 7.53 \times 10^{-6} \text{ m} = 7.53 \text{ μm}$

The stiffness is:

$$K = \frac{P}{\delta} = \frac{100 \times 10^{-6} \text{ N}}{7.53 \times 10^{-6} \text{ m}}$$

Answer: $K = 13.3 \text{ N/m}$

which is typical for such devices.

The maximum bending stress occurs where the moment is maximum, which here occurs at $x = 0$ ($M_{max} = PL$):

$$\sigma_{max} = \frac{M_{max}(d/2)}{I} = \frac{[(100 \times 10^{-6} \text{ N})(140 \times 10^{-6} \text{ m})](1.5 \times 10^{-6} \text{ m})}{27 \times 10^{-24} \text{ m}^4} = 778 \text{ MPa}$$

The stresses in micro-devices are often high.

Example 6.6 Simply-Supported Beam, Uniformly Distributed Load

Beams in bridges and buildings are designed to carry uniformly distributed loads.

Given: A simply-supported beam of length L is subjected to a uniformly distributed load w (force per unit length) in the negative y-direction (*Figure 6.16a*). The moment of inertia is I and the modulus is E. Point $x = 0$ corresponds to the left support of the beam.

Required: Develop expressions for (a) the slope $v'(x)$, (b) the deflection $v(x)$, and (c) the maximum deflection δ_{max}.

Solution: *Step 1.* Due to the symmetry of the loading and of the geometry, the vertical reaction at each support is $R = wL/2$ (*Figure 6.16b*). Considering the FBD of *Figure 6.16c*, equilibrium requires that:

$$\sum M_z = 0: \quad M(x) - Rx + wx\left(\frac{x}{2}\right) = 0$$

The bending moment as a function of x is then:

$$M(x) = \left(\frac{wL}{2}\right)x + \left(-\frac{wx^2}{2}\right) = \frac{w}{2}(Lx - x^2)$$

Step 2. Applying *Equation 6.19*:

$$\frac{d^2v}{dx^2} = \frac{M}{EI} = \frac{w}{2EI}(Lx - x^2)$$

and integrating twice gives:

$$v'(x) = \frac{w}{2EI}\left(\frac{Lx^2}{2} - \frac{x^3}{3}\right) + C_1$$

and

$$v(x) = \frac{w}{2EI}\left(\frac{Lx^3}{6} - \frac{x^4}{12}\right) + C_1 x + C_2$$

The values for C_1 and C_2 are found by applying the geometric boundary conditions that the displacements are zero at both ends of the beam, $x = 0$ and L:

$$v(0) = C_2 = 0$$

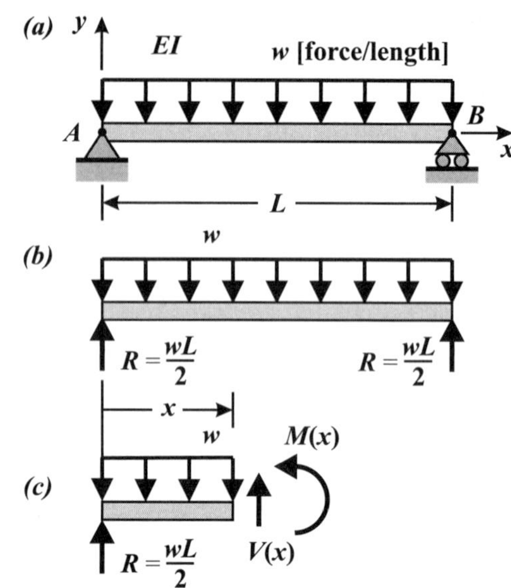

Figure 6.16. (a) Simply supported beam under uniformly distributed load. **(b)** FBD of entire system. **(c)** FBD of length x from left.

$$v(L) = \frac{w}{2EI}\left[\frac{L(L)^3}{6} - \frac{(L)^4}{12}\right] + C_1(L) = 0 \rightarrow C_1 = -\frac{wL^3}{24EI}$$

The slope and displacement of the beam are then:

Answer: $v'(x) = \dfrac{-w}{24EI}(4x^3 - 6Lx^2 + L^3)$

Answer: $v(x) = -\dfrac{w}{24EI}(x^4 - 2Lx^3 + L^3 x)$

By symmetry of the load and geometry, the maximum displacement is at the center:

Answer: $\delta_{max} = v(L/2) = -\dfrac{5wL^4}{384EI}$

The calculated displacement is negative, which means that the beam deflects downwards as expected.

Due to the symmetry of the applied loading and of the geometry, the slope must be zero at the center, i.e., $v'(L/2) = 0$; this condition could have been used to determine C_1.

Example 6.7 Bridge on Simple-Supported I-beams

Given: The deck of a bridge (*Figure 6.17*) is $L = 10$ m long and $B = 4.8$ m wide, and carries a uniformly distributed *area* load $p = 25$ kN/m^2. The deck is supported by four metric S510-141 beams that are $L = 10$ m long, and spaced so that each beam supports 1/4 of the roadway (*Figure 6.17b*). Beams with such cross-sections are known as *I-beams* since their cross-sections are shaped like the letter "I" (*Figure 6.17c*). The properties of I-beams are listed in tables (see *Appendix C* for some US I-beams).

For the US metric S510–128 I-beam (S 20×86 in US units):

$$b = 179 \text{ mm}; \ d = 516 \text{ mm};$$
$$I = 658 \times 10^6 \text{ mm}^4$$

Required: Determine (a) the maximum deflection δ_{max}, and (b) the maximum bending stress σ_{max}.

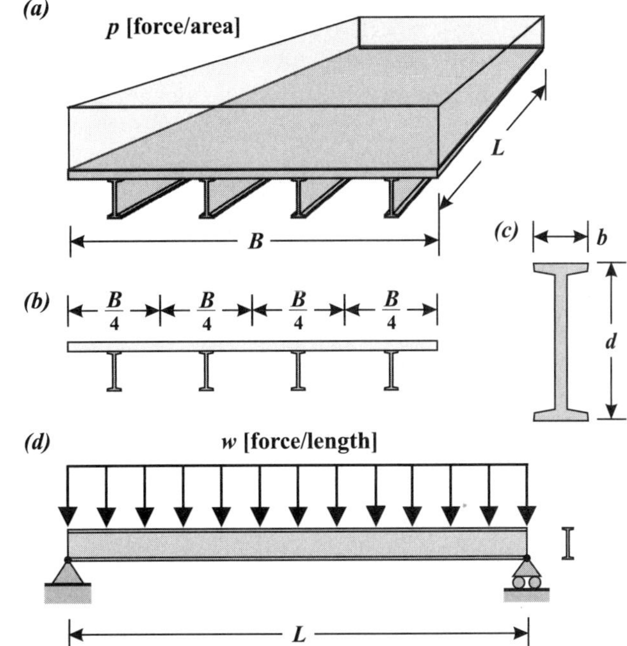

Figure 6.17. **(a)** A bridge roadway under a uniformly distributed area load is supported by 4 I-beams. **(b)** End view of roadway. The I-beams are spaced so that each supports 1/4 of the roadway. **(c)** I-beam cross-section. **(d)** Side view of a single beam under uniformly distributed load $w = p(B/4)$.

Solution: *Step 1.* Determine the load on each beam. Each beam supports one-fourth of the total load on the deck. The total load is:

$$W = pLB = (25 \text{ kN/m}^2) \, (10 \times 4.8 \text{m}^2) = 1200 \text{ kN}$$

Since the beams each support one-fourth of the load, the load on each beam is 300 kN. Alternatively, the *tributary area* A_t of each beam (the area of the road that each beam supports) is:

$$A_t = (L)(B/4) = (10 \text{ m})(1.2 \text{ m}) = 12 \text{ m}^2$$

so the load on each beam is:

$$F = pA_t = (25 \text{ kN/m}^2)(12 \text{ m}^2) = 300 \text{ kN}$$

This load is not applied at a point, but is uniformly distributed over the length of the beam. The distributed load on each beam is the load it supports divided by its length:

$$w = (W/4)/L = (300 \text{ kN})/(10 \text{ m}) = 30 \text{ kN/m}$$

Or, since each beam has a tributary area of $B/4$ wide, then:

$$w = p\frac{B}{4} = (25 \text{ kN/m}^2)\frac{(4.8 \text{ m})}{4} = 30 \text{ kN/m}$$

Step 2. The maximum deflection δ_{max} occurs at the center of the beam:

$$\delta_{max} = -\frac{5wL^4}{384EI} = -\frac{5(30\times10^3 \text{ N/m})(10 \text{ m})^4}{384(200\times10^9 \text{ Pa})(658\times10^{-6} \text{ m}^4)}$$

Answer: $\delta_{max} = -0.0297 \text{ m} = -29.7 \text{ mm}$

As an aside, the ratio of the maximum deflection δ_{max} to the span L is:

$$\frac{\delta_{max}}{L} = \frac{0.0297}{10.0} = \frac{1}{337}$$

Assuming the allowable deflection index is $f = \delta/L = 1/240$, the deflection is acceptable.

Step 3. Maximum bending stress. The maximum bending moment occurs at the center of the beam, and is:

$$M = \frac{wL^2}{8} = \frac{(30\times10^3 \text{ N/m})(10 \text{ m})^2}{8} = 375 \text{ kN·m}$$

The maximum stresses occur at the top and bottom of the cross-section: $c = \pm d/2 = \pm254 \text{ mm}$. The maximum bending stress at the top and bottom are, respectively:

$$\sigma_{max} = -\frac{Mc}{I} = -\frac{(375\times10^3 \text{ N·m})(\pm0.254 \text{ m})}{670\times10^{-6} \text{ m}^4}$$

Answer: $\sigma_{max} = \mp142 \text{ MPa}$

Due to the positive moment, the top of the beam is in compression ($\sigma = -142$ MPa at $y = +254$ mm), while the bottom is in tension ($\sigma = +142$ MPa at $y = -254$ mm).

Variation of Moment Equation Over the Length of the Beam

When the bending moment equation changes over the length of the beam, the same methods to find the slope and deflection may be used. Consider the beam in *Figure 6.18*; it has a different moment equation over each segment, *AB*, *BC* and *CD*. Different moment, slope, and displacement equations are valid over each segment, as shown in *Table 6.2*.

The *elastic curve* – the deflected shape of the neutral axis – must be **smooth** and **continuous** (*Figure 6.19*). *Smooth* requires that there be no kinks in the beam; at the *x*-position where two slope equations meet, they must have the same value. *Continuous* requires that the beam stays together; at the *x*-position where two deflection equations meet, they must have the same value.

The values of slope and displacement where two segments meet are generally not known beforehand. However, applying the *smooth* and *continuous* conditions between two segments gives two *geometric boundary conditions* at each point.

For the beam in *Figure 6.18*, integrating the three moment equations results in six constants (*Table 6.2*). The built-in end gives two boundary conditions: the slope and displacement are both zero. That the beam be *smooth* and *continuous* gives two boundary conditions at point B: $\theta_{AB} = \theta_{BC}$ and $v_{AB} = v_{BC}$, and two more boundary conditions at point C: $\theta_{BC} = \theta_{CD}$ and $v_{BC} = v_{CD}$. These six total boundary conditions are used to determine constants C_1 through C_6.

Table 6-3 gives a few common zero-value *boundary conditions*.

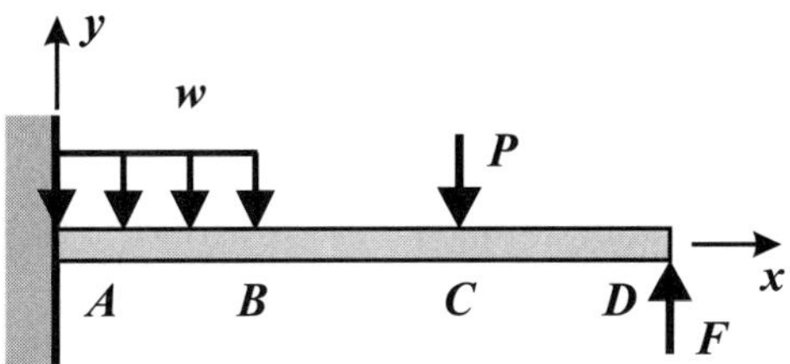

Figure 6.18. A beam that requires three moment equations, $M_{AB}(x)$, $M_{BC}(x)$, and $M_{CD}(x)$, to describe the moment over its entire length.

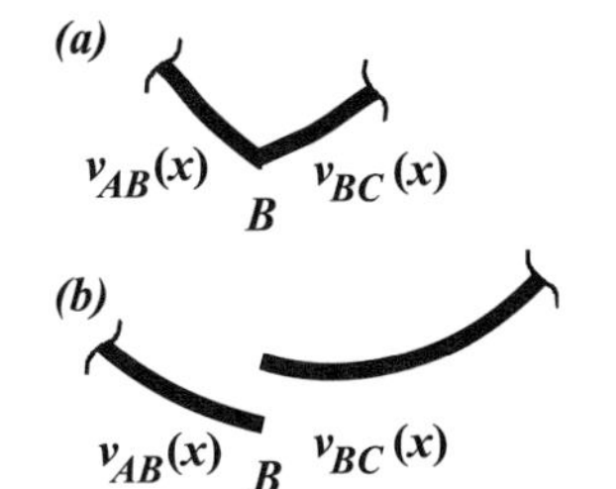

Figure 6.19. (a) Continuous but not smooth. **(b)** Smooth but not continuous.

Table 6.2. General equations for beam in *Figure 6.18*. When the moment equation changes, integration is still used to determine the slope and displacement equations. The slopes and displacements of adjacent beam segments must equal each other at their common point. Constants C_1 through C_6 are solved using the kinematic boundary conditions (geometric constraints at the supports, and matching slopes and deflections at common points).

Eqn	Valid over *AB*	Valid over *BC*	Valid over *CD*
$M(x)$	$M_{AB}(x)$	$M_{BC}(x)$	$M_{CD}(x)$
$EI\theta(x)$	$\int M_{AB}\,dx + C_1$	$\int M_{BC}\,dx + C_3$	$\int M_{CD}\,dx + C_5$
$EIv(x)$	$\int\left[\int M_{AB}\,dx\right]dx + C_1 x + C_2$	$\int\left[\int M_{BC}\,dx\right]dx + C_3 x + C_4$	$\int\left[\int M_{CD}\,dx\right]dx + C_5 x + C_6$

Table 6-3. Common boundary conditions with zero value.

End Conditions		Zero Value Boundary Conditions
Unloaded pinned support	$M = 0,\ v = 0$	Reaction moment and displacement
Built-in support	$v = 0,\ \theta = 0$	Displacement and slope
Unloaded free end	$M = 0,\ V = 0$	Reaction moment and reaction shear

Example 6.8 Cantilever Beam with Intermediate Point Load

Given: Cantilever beam *ABC* of length *L* supports intermediate load *P* (*Figure 6.20a*). Load *P* is applied at point *B*, distance *a* from the wall. The beam extends beyond point *B* by distance *b*. The moment of inertia is *I* and the modulus is *E*.

Required: Determine the deflection and slope (a) at the point of application of the load, δ_B and θ_B, and (b) at the tip of the cantilever beam, δ_C and θ_C.

Solution: *Step 1.* Determine the moment equation over segment *AB* (length *a*). A FBD from 0 to *x* (*x* < *a*, *Figure 6.20c*) is loaded in the same manner as the tip-loaded cantilever of length *L* studied in *Example 6.4*. With *x* measured from the left of the beam, as in *Example 6.4*, the work has already been done. Substituting *a* for *L*, the slope and deflection over segment *AB* are:

$$\theta_{AB}(x) = \frac{P}{EI}\left(ax - \frac{x^2}{2}\right)$$

$$v_{AB}(x) = \frac{P}{EI}\left(\frac{ax^2}{2} - \frac{x^3}{6}\right)$$

Hence the slope and deflection at point *B* (at *x* = *a*) are:

$$\text{Answer: } \theta_B = \theta_{AB}(a) = \frac{Pa^2}{2EI}$$

$$\text{Answer: } \delta_B = v_{AB}(a) = \frac{Pa^3}{3EI}$$

Step 2. Determine the moment equation over segment *BC* (*Figure 6.20d*). Segment *BC* carries no bending moment, so its curvature is:

$$\frac{d^2v}{dx^2} = \frac{M_{BC}(x)}{EI} = 0$$

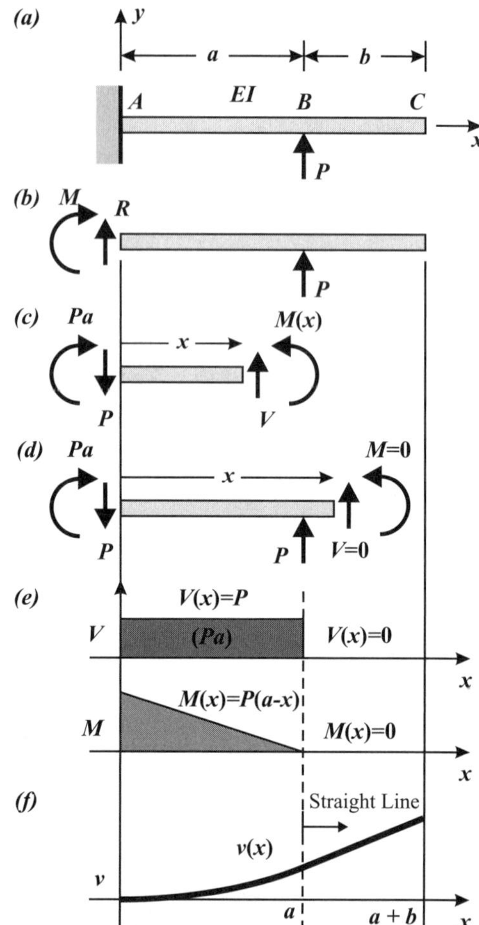

Figure 6.20. **(a)** Cantilever beam under intermediate point load. **(b)** FBD of entire beam. **(c)** FBD for 0 < *x* < *a*. **(d)** FBD for *a* < *x* < *a+b*. **(e)** Shear and moment diagrams. **(f)** Deflection curve.

Integrating gives the expressions for the slope and deflection over segment BC:

$$\theta_{BC}(x) = C_3$$

$$v_{BC}(x) = C_3 x + C_4$$

C_3 and C_4 are the constants of integration for the deflection curve for $a < x < a+b$. Since the curvature of BC is zero, its slope is constant; BC is a straight line. Constants C_3 and C_4 are found by matching the slopes and deflections of AB and BC at point B ($x = a$):

$$\theta_{AB}(a) = \theta_{BC}(a) \quad \rightarrow \quad C_3 = \frac{Pa^2}{2EI}$$

and $\quad v_{AB}(a) = v_{BC}(a)$

$$\frac{Pa^3}{3EI} = C_3 a + C_4 \quad \rightarrow \quad C_4 = \frac{-Pa^3}{6EI}$$

The slope of BC is therefore:

$$\theta_{BC}(x) = \frac{Pa^2}{2EI}$$

Because the slope of BC is constant, the displacement at any value of $x > a$ is related to the deflection and slope at B:

$$v_{BC}(x) = \delta_B + \theta_B(x - a) = \frac{Pa^3}{3EI} + \frac{Pa^2}{2EI}(x - a)$$

The deflection of point C is the deflection of point B, δ_B, plus the slope of BC, θ_C, multiplied by the length of BC, b.

$$Answer:\ \theta_C = \frac{Pa^2}{2EI}$$

$$Answer:\ \delta_C = \frac{Pa^3}{3EI} + \frac{Pa^2 b}{2EI} = \frac{Pa^2}{6EI}(2a + 3b)$$

To summarize the equations:

Eqn	Segment AB: $0 \le x \le a$	Segment BC: $a \le x \le a + b$
$M(x)$	$M_{AB}(x) = P(a - x)$	$M_{BC}(x) = 0$ (no curvature)
$\theta(x)$	$\theta_{AB}(x) = \dfrac{P}{EI}\left(ax - \dfrac{x^2}{2}\right)$	$\theta_{BC}(x) = \dfrac{Pa^2}{2EI}$ (constant slope)
$v(x)$	$v_{AB}(x) = \dfrac{Px^2}{6EI}(3a - x)$	$v_{BC}(x) = \dfrac{Pa^3}{3EI} + \dfrac{Pa^2}{2EI}(x - a) = \dfrac{Pa^2}{6EI}(3x - a)$

6.3 Statically Indeterminate (Redundant) Beams

The following two examples illustrate two methods of analyzing beams with redundant supports. These beams cannot be analyzed by the methods of *Statics* alone. Such systems are **statically indeterminate** or **redundant**. **Compatibility** – the necessary deflection of the beam – is used to complete the solution.

In *Example 6.9*, the problem is broken up into two statically determinate beams, and the deflection of each beam is calculated. Using the *method of superposition* (discussed further in *Chapter 7*), the deflections of the two individual cases are summed; the total deflection is the deflection of the original beam, assuming that the system remains linear–elastic. In *Example 6.10*, the deflection of a beam is studied to determine where its curvature – and thus moment – must be zero.

Example 6.9 Redundant Support

Given: Beam AB, of length L and moment of inertia I, is built-in at one end and simply supported at the other end (*Figure 6.21*). It is loaded by uniformly distributed load w. The modulus is E.

Required: Determine (a) the reactions at the wall and the roller support, (b) the expression for the beam deflection $v(x)$, and (c) the bending moment in the beam with the maximum magnitude and its location. Assume the system remains linear–elastic.

Solution: *Step 1.* From equilibrium of the entire beam (*Figure 6.21b*):

$$\sum F_y = 0: \quad R_A + R_B - wL = 0$$

$$\sum M_{z,A} = 0: \quad -M_A + R_B L - \frac{wL^2}{2} = 0$$

There are two equilibrium equations for three unknowns: R_A, R_B, and M_A. Hence, the system is redundant. Selecting $R_B = R$ as the redundant force, then:

$$R_A = wL - R$$

$$M_A = RL - \frac{wL^2}{2}$$

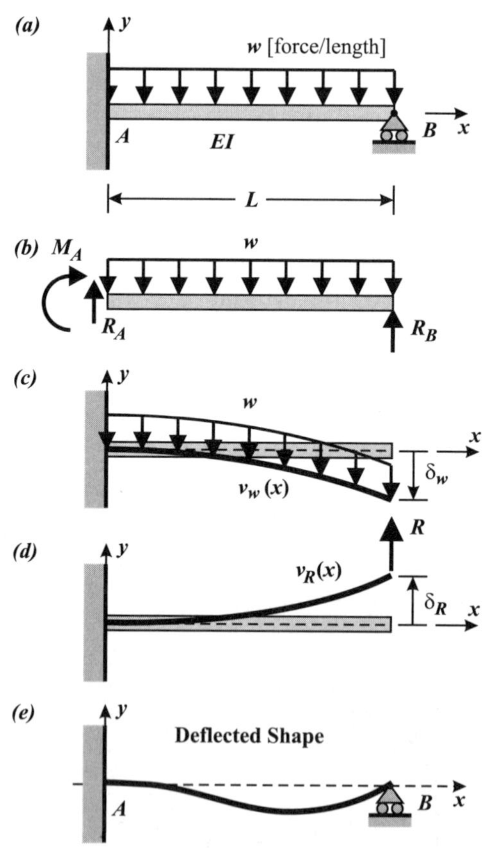

Figure 6.21. (a) Fixed-pinned beam under uniformly distributed load *w.* **(b)** FBD of beam. **(c)** Deflection of cantilever under UDL. **(d)** Deflection of cantilever under tip load *R.* **(e)** Deflected shape of fixed-pinned beam.

Step 2. Force–displacement. Consider the displacement at point B caused by each load acting separately.

Due to a downward uniformly distributed load w acting alone (*Figure 6.21c*), the tip displacement of a cantilever beam of length L can be found to be:

$$\delta_w = -\frac{wL^4}{8EI}$$

Due to an upward point load R (*Figure 6.21d*), the tip displacement of a cantilever is:

$$\delta_R = \frac{RL^3}{3EI}$$

Step 3. Compatibility. Since point B is pinned, its total displacement is zero. *Superimposing* (adding) the tip displacements due to each load gives:

$$\delta = \delta_q + \delta_R = 0 = -\frac{wL^4}{8EI} + \frac{RL^3}{3EI}$$

Thus:

$$Answer:\ R = R_B = \frac{3wL}{8}$$

Solving for the reactions at the wall:

$$Answer:\ R_A = \frac{5wL}{8} \quad \text{and} \quad M_A = -\frac{wL^2}{8}$$

Step 4. With the end forces known, the derivation of the deflection follows the procedure outlined above. The moment at any section x is:

$$M(x) = -M_A + R_A x - \frac{wx^2}{2} = -\frac{wL^2}{8} + \frac{5wLx}{8} - \frac{wx^2}{2}$$

Starting with the moment–curvature relationship (*Equation 6.19*) and integrating:

$$EIv''(x) = M(x) = -\frac{wL^2}{8} + \frac{5wLx}{8} - \frac{wx^2}{2}$$

$$EIv'(x) = -\frac{wL^2 x}{8} + \frac{5wLx^2}{16} - \frac{wx^3}{6} + C_1$$

The slope at $x = 0$ is zero, so $C_1 = 0$.

$$EIv(x) = -\frac{wL^2 x^2}{16} + \frac{5wLx^3}{48} - \frac{wx^4}{24} + C_2$$

The deflection at $x = 0$ is zero, so $C_2 = 0$. Thus:

$$Answer:\ v(x) = -\frac{wx^2}{48EI}(3L^2 - 5Lx + 2x^2)$$

Step 5. Check. At $x = L$, $v(L) = 0$.

Step 6. Maximum bending moment. The moment with the maximum magnitude occurs at the wall, M_A, or where the derivative of the moment is zero:

$$\frac{dM}{dx} = \frac{5wL}{8} - wx = 0 \Rightarrow x = \frac{5}{8}L$$

At $x = 5L/8$: $M(5/8L) = \frac{9}{128}wL^2$

which is smaller in magnitude than the moment at the wall ($x = 0$). Therefore:

$$\textit{Answer: } |M_{max}| = \frac{wL^2}{8} \text{ at } x = 0$$

Example 6.10   Fixed-Fixed Beam under Central Point Load

Given: Beam AB, of length L and moment of inertia I, is built-in at both ends (*Figure 6.22*). A central deflection δ is impressed at the center of the beam, point C, by force P. This arrangement is often used in MEMs devices.

Required: Determine (a) the force P necessary to cause displacement δ, and (b) the reactions at each wall.

Solution: *Step 1.* Since the load and geometry are both *symmetric* about the center, the reactions at each end are equal: $R_A = R_B = P/2$ and $M_A = M_B$ (*Figure 6.22b*). However, the moments can take on any value and still satisfy equilibrium. The system is redundant.

Step 2. Compatibility. Consider the deflection. Due to symmetry of the load and geometry, the displacement must be symmetric (*Figure 6.22c*). Furthermore, the deflection of each half-span is *anti-symmetric* about the quarter-points (point D). The curvature at the quarter-points is zero (they are inflection points), so the moment at those points is also zero. Since half the applied force is carried by each half of the beam, the problem is reduced to a

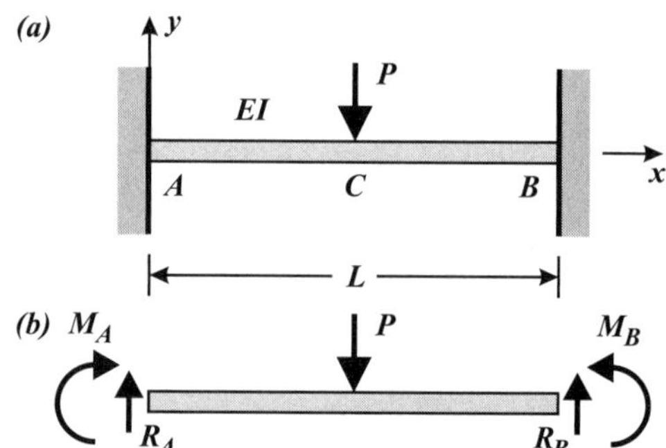

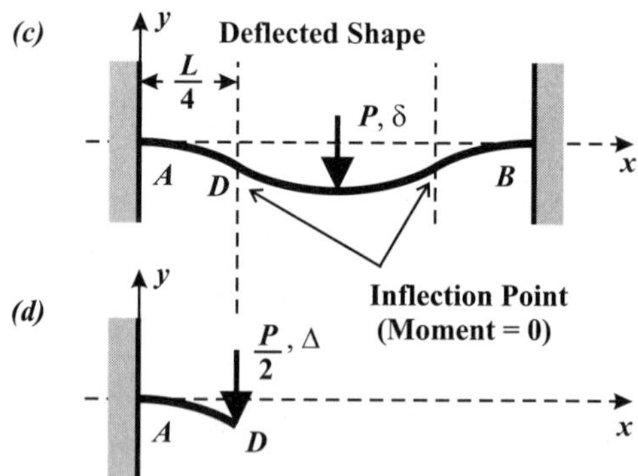

Figure 6.22. (a) Beam built-in at both ends under central load P. **(b)** FBD of beam. **(c)** The deflected shape of beam is *symmetric* about the center, and each half is *anti-symmetric* about the quarter-points. **(d)** The problem reduces to a cantilever beam of length $L/4$ under tip load $P/2$.

cantilever beam of length $L/4$ subjected to a tip load of $P/2$ (*Figure 6.22d*).

Step 3. The displacement of point D, Δ, is half of the impressed deflection δ (*Figures 6.22c, d*). The force–deflection relationship for the reduced cantilever beam AD (*Figure 6.22d*) is:

$$\Delta = \frac{\delta}{2} = \frac{(P/2)(L/4)^3}{3EI} = \frac{1}{384}\frac{PL^3}{EI}$$

The force P to cause central displacement δ is:

$$Answer:\ P = \frac{192EI}{L^3}\delta$$

The stiffness of the beam for a central point load is:

$$K = \frac{192EI}{L^3}$$

Since stiffness is inversely proportional to L^3, a great range of stiffnesses is possible.

The magnitude of the reactions at either wall is taken from the FBD of the quarter-length cantilever beam:

$$Answer:\ R_A = R_B = \frac{P}{2} \quad \text{and} \quad M_A = M_B = -\frac{PL}{8}$$

By anti-symmetry of the half-beam, the internal bending moment at the center, M_C, also has a magnitude of $PL/8$, but is positive.

6.4 Shear Stress

Average Shear Stress

In addition to bending moments, beams must support shear forces V (*Figure 6.23*). Shear forces cause **shear stresses**. The *average shear stress* acting on a cross-section of area A is:

$$\tau_{ave} = \frac{V}{A} \qquad \text{[Eq. 6.22]}$$

An assumption that might follow from this equation is that the shear stress is constant over the cross-section. However, the top and bottom surfaces of the beam are free of shear stress ($\tau = 0$). *Complementary shear stress* requires that at any point the

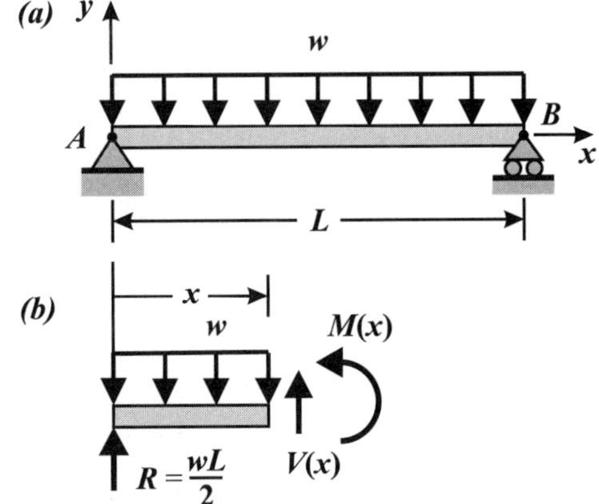

Figure 6.23. (a) Beam under uniformly distributed load. (b) In general, a beam carries shear force $V(x)$ and bending moment $M(x)$.

shear stresses on perpendicular planes are equal. Thus, the shear stress on the cross-section at the top and bottom of a beam must be zero. Shear stress cannot be constant over the cross-section.

A more detailed understanding of the shear stress distribution is required. How shear stress varies on a beam's cross-section is investigated below.

Distribution of Shear Stress

Consider a beam of constant cross-sectional area under a general distributed load $w(x)$ (*Figure 6.24a*). The cross-section has any shape that is symmetric about its vertical y-axis (the axis along which the shear force is applied).

A differential element dx of the beam, *CDFE*, with its loads, is shown in *Figure 6.24b*. Length dx is so small that $w(x)$ is constant over dx; the changes in shear force and bending moment, dV and dM, are also very small.

Moment equilibrium about any point on the left face of *CDFE* gives:

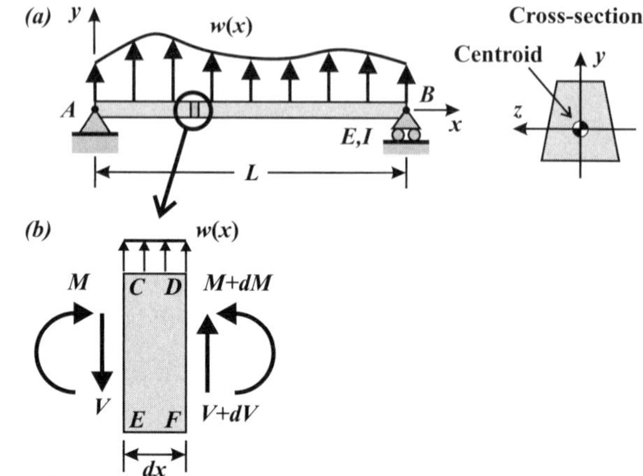

Figure 6.24. **(a)** Beam under a general distributed load. **(b)** Element of beam *CDFE*, length *dx*, at distance *x* from the left support. All loads on *dx* are drawn positive.

$$-M + (M + dM) + (V + dV)dx + (w\,dx)\left(\frac{dx}{2}\right) = 0 \qquad \text{[Eq. 6.23]}$$

Since the terms dx, dV, and dM are very small, any such terms multiplied together are negligible. *Equation 6.23* reduces to:

$$dM = -V\,dx \qquad \text{[Eq. 6.24]}$$

or

$$\frac{dM}{dx} = -V \qquad \text{[Eq. 6.25]}$$

Equation 6.24 (or *Equation 6.25*) states that whenever there is a non-zero shear force, there is a change in bending moment.

The bending stress for a given moment is:

$$\sigma(y) = -\frac{My}{I} \qquad \text{[Eq. 6.26]}$$

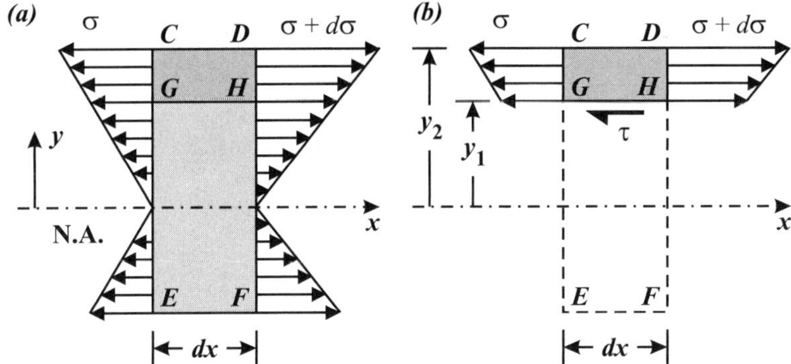

Figure 6.25. (a) Due to the shear force *V*, the bending moment and bending stresses change over length *dx*. **(b)** *CDHG* isolated. For equilibrium in the *x*-direction, shear stress τ must act on *GH* to counter the net force due to the change in bending stress. The stresses are drawn in the positive sense, e.g., positive face, positive direction; and negative face, negative direction.

The bending stresses on each cross-sectional face, *CE* and *DF*, are shown in *Figure 6.25a*, drawn in their positive senses.

Moment *M* acts on *CE* and *M+dM* acts on *DF*. Due to the change in moment with *x*, there is a change in bending stress *d*σ.

An isolated part of the beam element, *CDHG*, is shown in *Figure 6.25b*. Internal plane *GH* is distance y_1 above the neutral axis and $y_2 = y_{max}$. *Figure 6.25* shows only the stresses acting on *CDHG* in the *x*-direction.

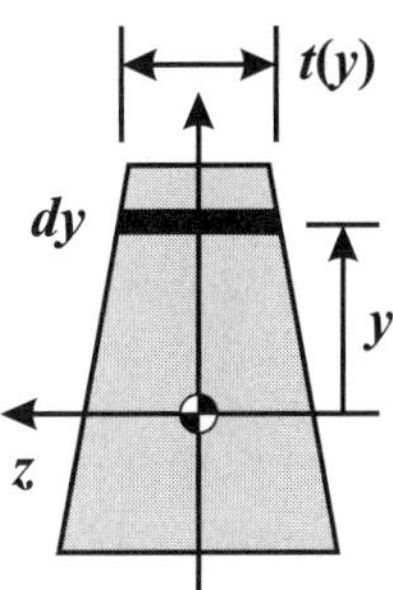

Figure 6.26. General cross-section symmetric about the *y*-axis. The dark differential area is $dA = t(y)dy$.

For equilibrium of *CDHG*, a shear stress τ must act on *GH*. The area over which τ acts is $[t(y_1)\,dx]$, where $t(y_1)$ is the width of the beam (into the paper) at height y_1. Shear stress τ is drawn to the left in its positive sense (negative *y*-face, negative *y*-direction).

Applying equilibrium in the *x*-direction on *CDHG*:

$$-\int_{y_1}^{y_2} \sigma(y)t(y)dy \; - \tau(y_1)[t(y_1)dx] \; + \; \int_{y_1}^{y_2}[\sigma(y)+d\sigma(y)]t(y)dy \; = \; 0$$

[Eq. 6.27]

where $t(y)dy = dA$ is the increment of area on the cross-section at *y* (*Figure 6.26*). Substituting the bending stress equation (*Equation 6.26*) into *Equation 6.27*:

$$\int_{y_1}^{y_2}\left[\left(\frac{M}{I}\right)y\right]t(y)dy \; - \; \tau(y_1)[t(y_1)dx] \; - \; \int_{y_1}^{y_2}\left[\left(\frac{M+dM}{I}\right)\right]yt(y)dy \; = \; 0$$

[Eq. 6.28]

Since the integrals are over the same limits, and $t(y)dy = dA$ is the same at each cross-section (the cross-section is constant), the two integrals can be combined:

$$\tau(y_1)[t(y_1)dx] = -\int_{y_1}^{y_2}\left(\frac{dM}{I}y\right)dA \qquad \text{[Eq. 6.29]}$$

Dividing by $[t(y_1)\,dx]$, and taking dM out of the y-integral as M is only a function of x:

$$\tau(y_1) = -\frac{dM}{dx}\frac{1}{It(y_1)}\int_{y_1}^{y_2}y\,dA \qquad \text{[Eq. 6.30]}$$

Since $dM/dx = -V$, the shear stress at $y = y_1$ is:

$$\tau(y_1) = \frac{V}{It(y_1)}\int_{y_1}^{y_2}y\,dA \qquad \text{[Eq. 6.31]}$$

This equation gives the shear stress τ_{yx} acting on GH, a y-face, x-direction stress. The initial intent was to solve for the shear stress on the cross-section τ_{xy}, an x-face, y-direction stress. Recall that at any material point, complementary shear stresses are equal. Hence, *Equation 6.31* also gives the shear stress on the cross-section τ_{xy} at $y = y_1$.

The integral in *Equation 6.31* is the *first moment of area* of the region on the cross-section bound by $y = y_1$ and $y = y_2$, with the moment taken about the centroidal z-axis of the entire cross-section, i.e., through $y = 0$. The first moment of area of a finite region is non-zero except when the moment of area is taken about an axis through the region's own centroid (e.g., the first moment of area of the cross-section about its own centroid is zero).

The first moment of area locates the **centroid** of an area. It is convenient to rewrite the integral:

$$\int_{y_1}^{y_2}ydA = y^*A^* = Q \qquad \text{[Eq. 6.32]}$$

where A^* is the area bound by $y = y_1$ and $y = y_2 = y_{max}$, and y^* is the vertical distance from the centroid (neutral axis) of the entire cross-section to the centroid of A^* (*Figure 6.27*). The first moment of area y^*A^* is often given the symbol Q.

Finally, replacing y_1 with y, the equation that gives the shear stress at distance y above the neutral axis is:

$$\tau(y) = \frac{VA^*y^*}{It} = \frac{VQ}{It} \qquad \text{[Eq. 6.33]}$$

where variables A^*, y^*, and t are defined in *Figure 6.27*. Force V is the shear force supported

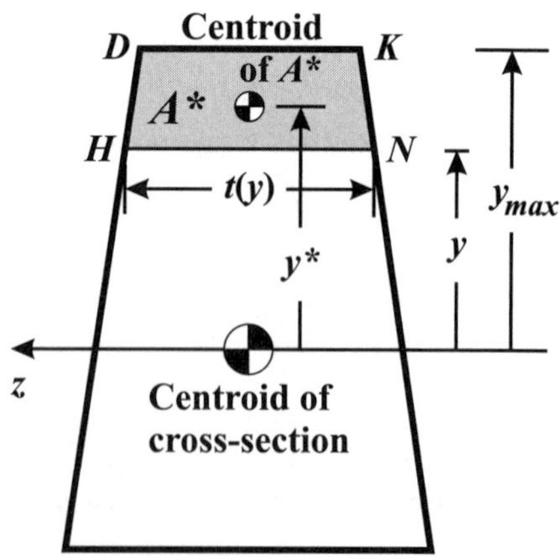

Figure 6.27. Trapezoidal cross-section showing pertinent geometry to find shear stress at y.

by the cross-section, and I is the moment of inertia of the entire cross-section. Both V and I are constant at a given cross-section along the beam.

It is vital to remember that y^* is not the distance to the y-value where the shear stress is to be determined. y^* is the distance from the centroid of the entire cross-section to the centroid of area A^*, which is bound by y and y_{max}.

To find the shear stress at a negative value of y, the same equation is used. Area A^* now crosses over the centroidal z-axis, bound by y and the positive y_{max}. Alternatively, A^* would be bound by the negative y and the negative y_{max}; both A^* and y^* would then be negative, since A^* is completely below the centroid. The sign of the shear stress is still that of the shear force.

> ***Review of Thought Process used in Deriving Equation 6.33***
>
> 1. A shear force V acting on a cross-section causes a change in bending moment dM over length dx.
> 2. dM causes the bending (normal) stress to change, which means a differential force dF is developed acting over area A^*. This force acts parallel to the neutral axis.
> 3. dF must be resisted by a shear force at y parallel to the beam axis. Thus, a shear stress $\tau_{yx}(y)$ is developed parallel to the neutral axis.
> 4. At any point, $\tau_{xy} = \tau_{yx}$, and the problem is solved. $\tau_{xy}(y)$ is the shear stress acting on the cross-section (on the x-face in the y-direction) at height y.

Example 6.11 Rectangular Cross-Section

Given: A rectangular cross-section b wide and d deep is subjected to an upward vertical shear force V on the positive x-face (*Figure 6.28*).

Required: Determine (a) the shear stress on the cross-section as a function of y and (b) the maximum shear stress.

Solution: *Step 1.* The geometry of a rectangular cross-section is:

$$A = bd; \quad I = \frac{bd^3}{12}$$

Equation 6.33 is the shear stress equation:

$$\tau(y) = \frac{VA^*y^*}{It}$$

The beam width $t = b$ is constant. A^* and y^* as functions of y are:

$$A^* = b\left(\frac{d}{2} - y\right); \quad y^* = \frac{1}{2}\left(\frac{d}{2} + y\right)$$

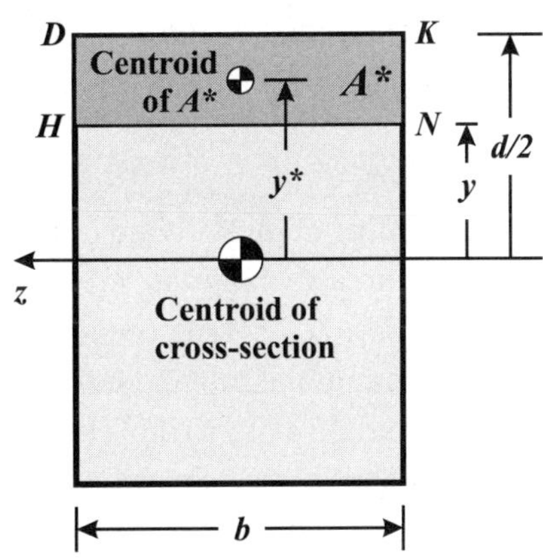

Figure 6.28. Rectangular cross-section.

The shear stress as a function of y is then:

$$\tau(y) = \frac{V\left[b\left(\frac{d}{2}-y\right)\right]\left[\frac{1}{2}\left(\frac{d}{2}+y\right)\right]}{\left(\frac{bd^3}{12}\right)(b)} = \frac{12}{2}\frac{V}{bd^3}\left[\frac{d^2}{4}\left(1-\frac{2y}{d}\right)\left(1+\frac{2y}{d}\right)\right]$$

$$\text{\textit{Answer:} } \tau(y) = \frac{3}{2}\frac{V}{A}\left[1-\left(\frac{2}{d}\right)^2 y^2\right]$$

The shear stress distribution is parabolic (*Figure 6.29a*).

Step 2. The shear stress on a rectangular cross-section has a maximum value at $y = 0$, where $A^* = bd/2$ and $y^* = d/4$:

$$\text{\textit{Answer:} } \tau_{max} = \tau(y=0) = \frac{3}{2}\frac{V}{A} = \frac{3}{2}\tau_{ave}$$

Note: At the top of the cross-section, $A^* = 0$, so:

$$\tau(y = +d/2) = 0$$

This result is consistent with the physics of the problem. The top of the beam is a free surface where no shear stress acts. By complementary shear stress, $\tau_{xy} = \tau_{yx}$, so the shear stress at the top of the cross-section is zero.

At the bottom, $A^* = bd$, but $y^* = 0$, so:

$$\tau(-d/2) = 0$$

The bending stress and shear stress distributions for a rectangular cross-section are shown in *Figure 6.29*. The bending stress is linear, maximum at the top and bottom, and zero at the centroidal axis. The shear stress is quadratic (parabolic), maximum at the centroid, and zero at the top and bottom.

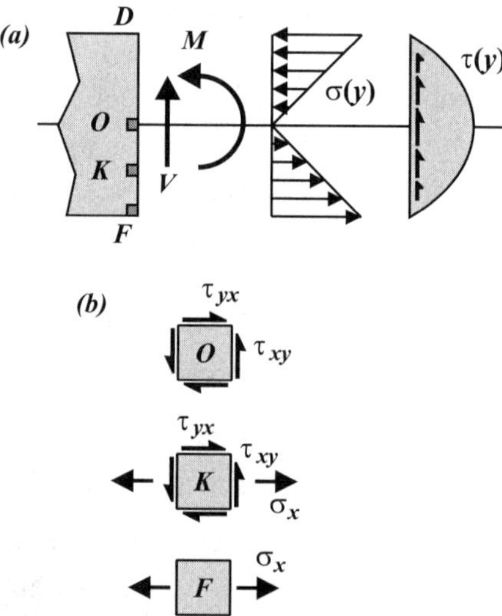

Figure 6.29. (a) Bending stress and shear stress distributions on a rectangular cross-section. $\tau_{max} = 1.5\,V/A$. (b) At point O (on the neutral axis), only shear stress acts. At point F (at the bottom or top) only bending (normal) stress acts. At general point K, both stresses act.

Example 6.12 **Laminated Beam**

In modern architecture, beams are often formed by laminating several wooden planks together. The glue holding the planks together must be sufficiently strong to prevent failure in shear.

Given: A simply supported beam is made from four pieces of lumber, each 2.0 by 4.0 in. in cross-section, and laminated together as shown in *Figure 6.30*. The beam is $L = 10$ ft long, and is to support a central point load of $P = 12.0$ kips.

Required: (a) Determine the required shear strength of the glue, using a factor of safety of FS = 3.0. (b) Determine the stress in the upper glue layer.

Solution: *Step 1*. The maximum shear force in a simply-supported beam loaded by a central point load is half of that load:

$$V_{max} = \frac{P}{2} = 6000 \text{ lb}$$

For a rectangular cross-section, the shear stress is maximum at $y = 0$:

$$\tau_{max} = \frac{3}{2}\frac{V_{max}}{A} = \frac{3}{2}\frac{(6000 \text{ lb})}{(4 \text{ in.})(8 \text{ in.})}$$

$$= 281 \text{ psi}$$

Applying the factor of safety, the glue must have a shear strength of:

$$\tau_f = \text{FS}(\tau_{max}) = 3(281 \text{ psi})$$

Answer: $\tau_f = 843$ psi

If the shear stress at the glue exceeds its shear strength, the laminate will no longer act as a single piece.

Step 2. At the upper glue level, $y = 2$ in., $A^* = (2{\times}4 \text{ in.}^2)$, $y^* = 3$ in. and $t = 4$ in. (*Figure 6.30c*). The shear stress is:

$$\tau(y = 2) = \frac{VA^*y^*}{It}$$

$$= \frac{(6000 \text{ lb})(2 \times 4 \text{ in.}^2)(3 \text{ in.})}{[(4 \times 8^3)/12 \text{ in.}^4](4 \text{ in.})}$$

Answer: $\tau = 211$ psi

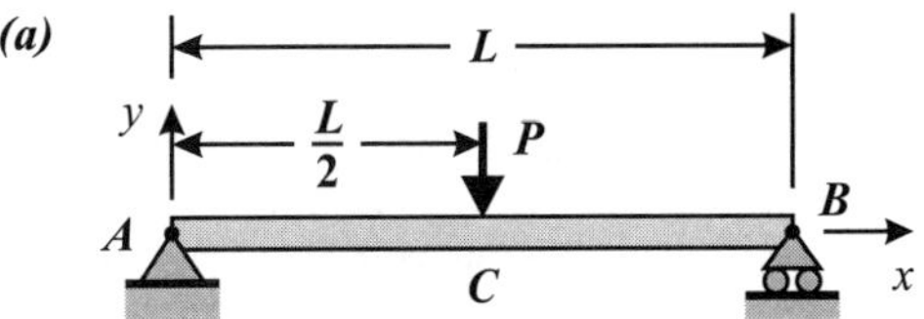

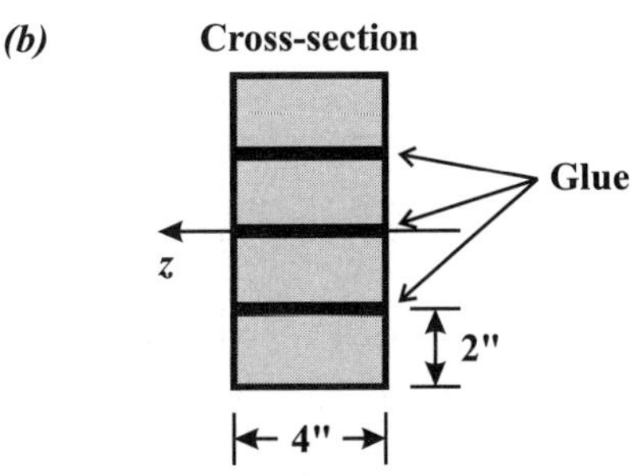

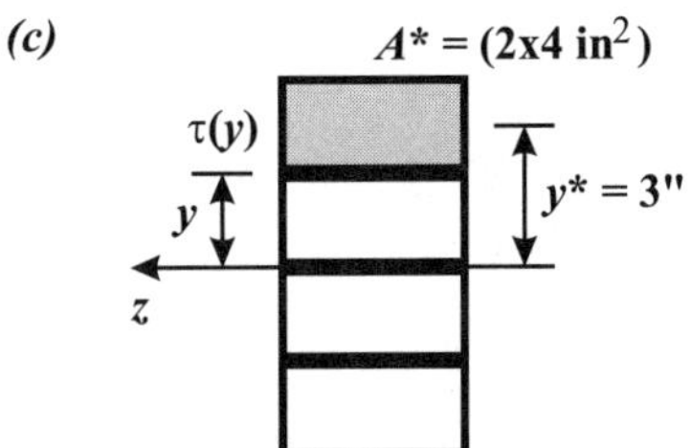

Figure 6.30. (a) Simply supported beam under central load *P*. **(b)** Cross-section of laminated beam built of four 2×4s (true), glued together. **(c)** *A** and *y** to find shear stress in upper glue layer.

Maximum Shear Stress Values for Common Sections

Table 6.4 gives the maximum shear stresses for basic cross-sections caused by shear force *V*. The maximum stress on these cross-sections occurs at the horizontal axis through the centroid, i.e., at the height of the neutral axis.

Table 6.4. Maximum Shear Stress due to shear force V for rectangular, solid circular and hollow circular cross-sections of area A.

Cross-sectional Shape	Geometry	Maximum Shear Stress	Equation No.
Rectangle	d b	$\dfrac{3}{2}\dfrac{V}{A}$	[Eq. 6.34]
Solid Circle	$D = 2R$	$\dfrac{4}{3}\dfrac{V}{A}$	[Eq. 6.35]
Hollow Circle R = Outer radius R_i = Inner radius	$D = 2R$ $D_i = 2R_i$	$\dfrac{4}{3}\dfrac{V}{A}\left[\dfrac{R^2 + RR_i + R_i^2}{R^2 + R_i^2}\right]$	[Eq. 6.36]

Shear Flow

A useful concept in shear stress analysis is ***shear flow*** q. Mathematically, *shear flow* is the product of the shear stress $\tau(y)$ and beam width $t(y)$ at height y above the neutral axis. From *Equation 6.33*:

$$q(y) = \tau(y)t(y) = \frac{VA^*y^*}{I} \qquad \text{[Eq. 6.37]}$$

Shear flow is independent of beam width t. Thus, while the cross-section may have abrupt changes in its width, causing discontinuities in the calculated shear stress, the shear flow is *continuous* over the cross-section. A comparison of shear stress and shear flow is given in *Example 6.13*.

Shear stress $\tau(y)$ also acts along the beam, parallel to the neutral axis (i.e., the complementary shear stress). Consider a length of beam Δs. At height y, the beam width is t and the shear stress $\tau(y)$ acts on a material plane that is parallel to the neutral plane. The force F_S that acts over area $t\Delta s$ is:

$$F_S = \tau(t\,\Delta s) = q\Delta s \qquad \text{[Eq. 6.38]}$$

Note that force F_S is not equal to the shear force V. Force F_S is parallel to the neutral axis, and depends on the length of the beam Δs being considered. Shear force V is perpendicular to the neutral axis and acts on the beam cross-section.

Shear flow q can then be interpreted as the shear force per unit beam length acting along the axis of the beam at height y:

$$q = \tau t = \frac{F_S}{\Delta s} \qquad \text{[Eq. 6.39]}$$

Example 6.14 illustrates shear flow as it pertains to force per length parallel to the neutral axis.

Example 6.13 Shear Stress versus Shear Flow

Given: The built-up I-beam is subjected to an upward shear force of $V = 30$ kips on its cross-section (*Figure 6.31a*).

Required: Calculate and plot (a) the vertical shear stress $\tau(y)$ and (b) the vertical shear flow $q(y)$ over the cross-section.

Solution: *Step 1.* The cross-section is doubly symmetric so the neutral axis is centered vertically and horizontally. The moment of inertia about the z-axis, using the *parallel axis theorem* for the two flanges, is:

$$I = \frac{(2)(4)^3}{12} + 2\left[\frac{6(2)^3}{12} + (6 \times 2)(3)^2\right] \text{ in.}^4$$

$$= 234.7 \text{ in.}^4$$

Step 2. The vertical *shear stress* is parabolic, and is given by:

$$\tau(y) = \frac{VA^*y^*}{It}$$

At the top and bottom of the beam, $\tau = 0$.

Just above and below the upper flange–web intersection, the calculated shear stresses $\tau(y)$ are:

$$\tau(2^+) = \frac{(30 \times 10^3 \text{ lb})(6 \times 2 \text{ in.}^2)(3 \text{ in.})}{(234.7 \text{ in.}^4)(6 \text{ in.})}$$

$$= 0.767 \text{ ksi}$$

$$\tau(2^-) = \frac{(30 \times 10^3 \text{ lb})(6 \times 2 \text{ in.}^2)(3 \text{ in.})}{(234.7 \text{ in.}^4)(2 \text{ in.})}$$

$$= 2.30 \text{ ksi}$$

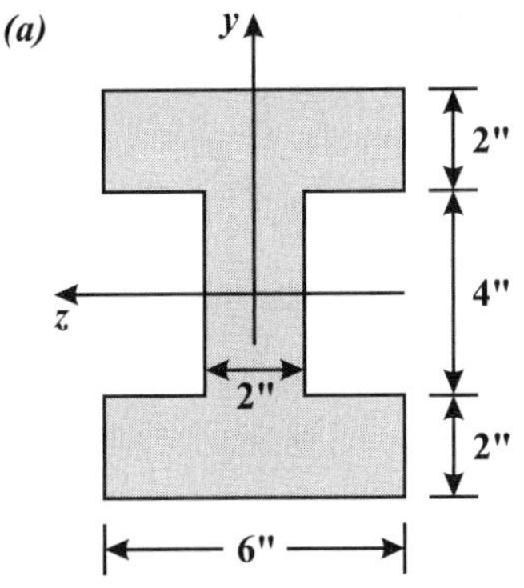

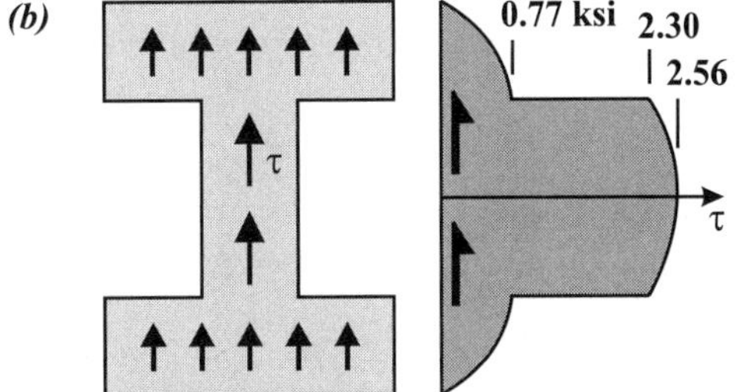

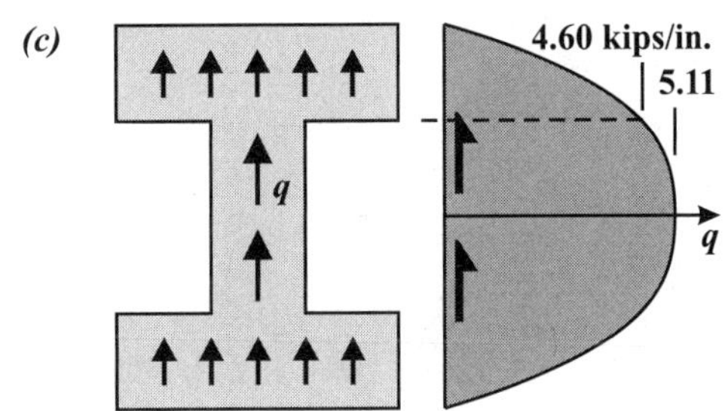

Figure 6.31. (a) Built-up I-beam.
(b) Vertical shear stress distribution.
(c) Vertical shear flow distribution.

See *Inset* concerning the shear stress just above the flange–web intersection, $\tau(2^+ \text{ in.})$.

At the centroid:

$$\tau(0) = \frac{(30 \times 10^3)[(6 \times 2)(3) + (2 \times 2)(1)]}{(234.7)(2)} = 2.56 \text{ ksi}$$

Step 3. The vertical shear stress is plotted in *Figure 6.31b*. The *calculated* shear stress is *discontinuous* due to the abrupt change in beam width. The shear stress in the web is nearly constant, varying by less than 6% from its middle value of 2.43 ksi.

Step 4. The vertical *shear flow* is also parabolic since $q(y) = \tau(y)t(y)$. Performing the calculations gives the shear flow at key y-values:

$$q(2^+) = 4.60 \text{ kips/in.}$$

$$q(2^-) = 4.60 \text{ kips/in.}$$

$$q(0) = 5.11 \text{ kips/in.}$$

Shear flow is *continuous*, e.g., $q(2^+) = q(2^-)$. Shear flow does not experience a jump when the beam width is discontinuous (*Figure 6.31c*).

Shear Stress at Flange–Web Intercept

In *Example 6.13*, two-thirds of the bottom of the upper flange ($y = 2$ in.) is actually stress-free (it is a free surface). The vertical shear stress at those locations must be zero due to complementary shear. The shear stress equation thus gives the average value of the vertical shear stress at y.

The top and bottom surfaces of a flange are stress free; any vertical shear stress that develops in the flange is therefore small. When an I-beam's flange is thin (not necessarily the case here), it does not contribute significantly to supporting the vertical shear force. I-beams are generally designed assuming that only the web supports shear.

Example 6.14   Built-Up T-beam under Shear Force

Given: The wooden beam of *Example 6.3* is again considered, with the two planks that make up the beam joined by two alternate methods. In Case 1, the planks are *glued* together (*Figure 6.32a*). In Case 2, the planks are *nailed* together with nail spacing Δs (*Figure 6.32b*).

The beam is to support a maximum shear force of $V = 2000$ N. The cross-sectional dimensions are:

$$b = d = 200 \text{ mm}$$

$$w = h = 50 \text{ mm}$$

as shown in *Figure 6.32c*.

Required: (a) For Case 1, estimate the shear stress τ that the glue must support. (b) For Case 2, assuming that the allowable shear force in each nail is $F_{allow} = 1200$ N, specify the maximum allowable nail spacing Δs_{max}.

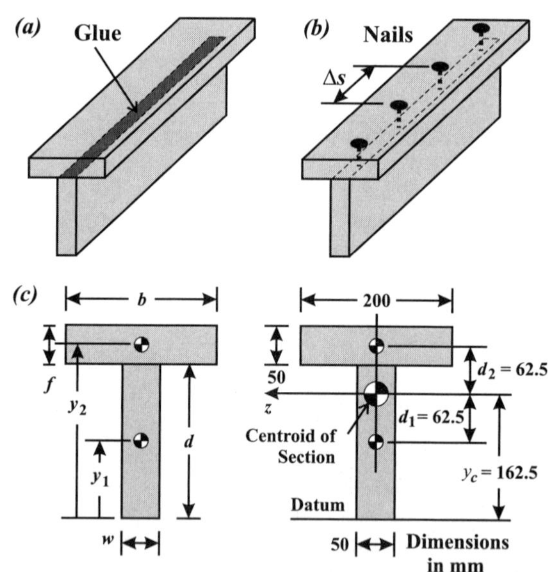

Figure 6.32. (a) Isometric view of planks glued together. **(b)** Isometric view of planks nailed together. **(c)** Cross-section of beam.

Solution: *Step 1.* Using the results of *Example 6.3*, the neutral axis is 162.5 mm above the foot of the "T," and the moment of inertia is:

$$I = 113.5 \times 10^{-6} \ m^4$$

Step 2. Force parallel to neutral axis. The two planks are joined together to act as a single beam. The glue or nails now must act in the same way that an interior material plane would act if the beam was made of one solid piece.

Whenever there is a shear force in a beam, there is a change in moment over length Δs. This change in moment causes a bending stress differential $d\sigma$ on the cross-section, and thus a force F_{dM} (*Figures 6.32e, f*). For equilibrium of the upper plank, the glue must provide a shear stress τ over area $t \, \Delta s$, or the nail must provide a shear force F_S over length Δs (*Figures 6.32e, f*). If the glue or nails fail, the planks will no longer act as a single unit.

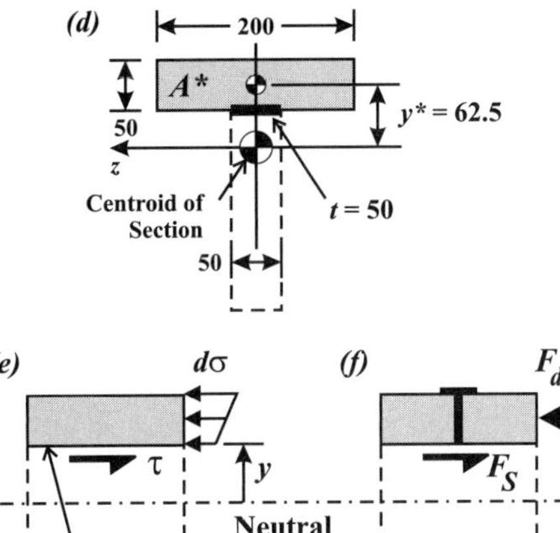

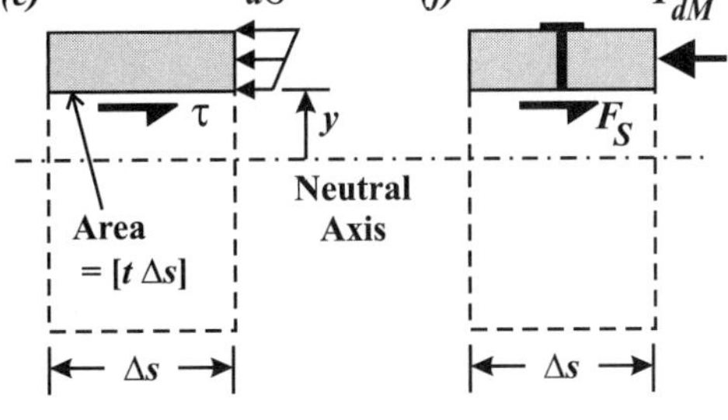

Figure 6.32. (d) Geometry of cross-section required to calculate τ or q at the plank interface. **(e)** Change in bending stress $d\sigma$ must be resisted by shear stress τ in the glue, or **(f)** by a shear force F_S in the nail.

Case 1: *Step 3. Shear stress in glue (Figure 6.32e).* The shear stress at any height y is:

$$\tau = \frac{VA^*y^*}{It}$$

The shear stress acts on the cross-section, as well as along the axis of the beam (complementary shear stress).

The glue is at $y = 37.5$ mm, where $t = 50$ mm (*Figure 6.32d*), and:

$$A^*y^* = (50 \times 200) \times (62.5) \ mm^3$$

$$= 625 \times 10^3 \ mm^3$$

For the maximum shear force, the shear stress in the glue is:

$$\tau = \frac{(2000 \ N)(625 \times 10^{-6} \ m^3)}{(113.5 \times 10^{-6} \ m^4)(0.050 \ m)}$$

Answer: $\tau = 220$ kPa

The glue must be able to support a shear stress of 220 kPa. If the glue is not strong enough, the two planks will no longer be joined together, and will slide with respect to each other. The beam will no longer act as a single unit.

Case 2: *Step 3. Shear force in nail (Figure 6.32f).* Each nail is Δs apart and the width of the plank interface is t. The shear force in each nail F_S is the product of the shear stress τ that would act at the plank interface and the area that each nail supports $t \, \Delta s$:

$$F_S = \tau[t \, \Delta s] = q \, \Delta s$$

For the maximum shear force, the shear flow at the plank interface is:

$$q = \frac{VA^*y^*}{I} = \frac{(2000 \text{ N})(625 \times 10^3 \text{ mm}^3)}{113.5 \times 10^6 \text{ mm}^4} = 11.0 \text{ N/mm}$$

A force of 11.0 N must be transferred in shear for every millimeter along the beam.

The equation that gives the nail spacing is:

$$\Delta s = \frac{F_S}{\tau t} = \frac{F_S}{q}$$

Since each nail is allowed to carry no more than 1200 N, the *maximum nail spacing* is:

$$\Delta s_{max} = \frac{F_S}{q} = \frac{1200 \text{ N}}{11.0 \text{ N/mm}}$$

$$\text{Answer: } \underline{\Delta s_{max} = 109 \text{ mm}}$$

The nails should be spaced no more than 109 mm apart.

In general, shear force V, and thus shear flow q, vary. In theory, if the shear force distribution is exactly known, nail-spacing may vary accordingly. However, attempting to actually implement this is problematic (and costly). In practice, nails are evenly spaced considering the worst case load. Building codes specify the maximum nail spacing for structural components under various conditions; nail spacing is especially important in high-wind regions.

Note that the effective shear stress at the joint is the shear flow divided by the thickness t of the joint:

$$\tau = \frac{q}{t} = \frac{11.0 \text{ N/mm}}{50 \text{ mm}} = 220 \text{ kPa}$$

Distribution of Shear Stress in Channels and I-Beams

Figure 6.33 shows a channel cross-section, breadth b, depth d, flange thickness f, and web thickness w. In general, the flange and web thicknesses are much less than the breadth and depth of the cross-section.

Although the channel section is not symmetric about the vertical (y-)axis, the bending stress is assumed to be calculated as before (*Equation 6.13*) and distributed as shown in *Figure 6.33b*. The shear stress can be calculated to a good approximation from the shear force using *Equation 6.33*, repeated here:

$$\tau(y) = \frac{VA^*y^*}{It} \qquad \text{[Eq. 6.40]}$$

Shear Stress in the Web

In channel sections and I-beams, the top and bottom surfaces of each flange are free surfaces; i.e., stress-free. Thus, the vertical shear stresses on the cross-section at the top and bottom of each flange are zero. The flange thickness is small, so any vertical shear stress that develops in the flange is also small and does not contribute significantly to supporting a vertical shear force V.

The shear force supported by the web is therefore taken to be equal to the applied shear force V. To find the shear stress in the web, a cut is taken at any point, e.g., point D in *Figure 6.33c*. Area A^* is the area above the cut:

$$A^* = bf + w\left(\frac{d}{2} - y\right) \qquad \text{[Eq. 6.41]}$$

Although y^* is the distance to the centroid of A^*, it need not be exactly determined since it is the product A^*y^* – the center of area equation – that is required:

$$A^*y^* = \sum A_i y_i$$

$$= bf\left(\frac{d}{2}\right) + \left[w\left(\frac{d}{2} - y\right)\right]\left[\frac{1}{2}\left(\frac{d}{2} + y\right)\right]$$

$$\text{[Eq. 6.42]}$$

Like the shear stress in a rectangular member, the shear stress in the web varies *parabolically* with y.

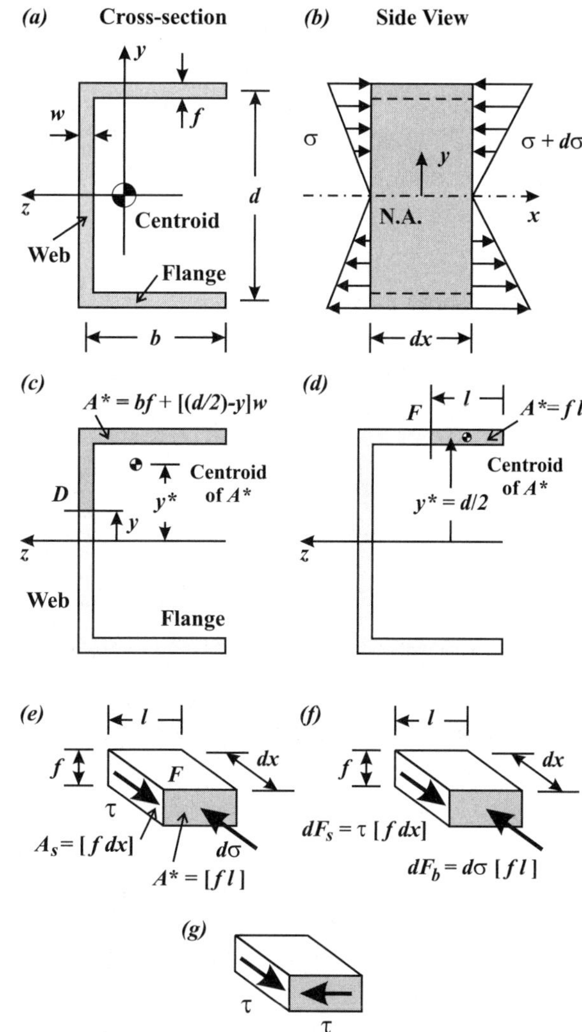

Figure 6.33. (a) Cross-section of channel. **(b)** Due to the shear force, a differential bending stress acts over beam length *dx*. **(c)** A^* and y^* to determine shear stress in the web. **(d)** A^* and y^* to determine shear stress in the flange. **(e)** Volume cut from flange, showing net stresses in *x*-direction – the unbalanced average bending stress *dσ*, and the shear stress τ. **(f)** FBD of flange volume. **(g)** Complementary shear requires that the shear stress on the cross-section of the flange equal that in the flange along the *x*-direction.

Shear Stress in the Flange

The shear stress in the flange is derived using the same arguments used to develop the original shear stress equation (*Equation 6.40*). *Vertical cuts* are taken in the flange, e.g., point F in *Figure 6.33d*. Such cuts must be perpendicular to the thin flange; the resulting shear stress on the cross-section is *horizontal*. Because the flange is thin, the horizontal shear stress is taken as constant across the flange thickness f.

Consider a volumetric element cut from the upper flange at point F (*Figures 6.33d,e*), where l is measured from the right (free) end of the flange and dx is measured along the axis of the beam. When shear force V acts at a cross-section, the moment changes with x, causing a differential bending stress $d\sigma(y)$ over distance dx (*Figures 6.33b,e*).

The differential bending stress causes a net force dF_b on $A^* = fl$ (*Figure 6.33f*). This net force must be resisted by the material on the left side of the flange element with shear force $dF_s = \tau\,[f\;dx]$. All of the other flange surfaces are free surfaces, so they cannot resist the differential bending stress.

Shear stress τ is developed in the flange along the axis of the beam (the x-axis), distance l from the free end (*Figure 6.33f*). Due to complementary shear stress, a horizontal shear stress is also developed on the cross-section in the flange (*Figure 6.33g*). The resulting shear stress in the flange is given by *Equation 6.40*. The shear stress is taken as constant across the thickness of the thin flange.

With l as the distance from the free-end of the flange to a cut at point F, then:

$$A^*y^* = (fl)(d/2) \qquad \text{[Eq. 6.43]}$$

The shear stress in the flange acts horizontally, and is equal to:

$$\tau(l) = \frac{VA^*y^*}{It} = \frac{V(fl)(d/2)}{If} = \frac{Vld}{2I} \qquad \text{[Eq. 6.44]}$$

The shear stress varies *linearly* with distance l; it is zero at the free end of the flange, and reaches a maximum at the intersection with the web (at $l = b$):

$$\tau_{max,\,f} = \frac{Vbd}{2I} \qquad \text{[Eq. 6.45]}$$

Since the shear stress in the flange is linear, the effective horizontal force in the flange is the product of the average shear stress in the flange (half the maximum value) and the flange area:

$$F_f = \tau_{ave}A_f = \frac{Vbd}{4I}(bf) \qquad \text{[Eq. 6.46]}$$

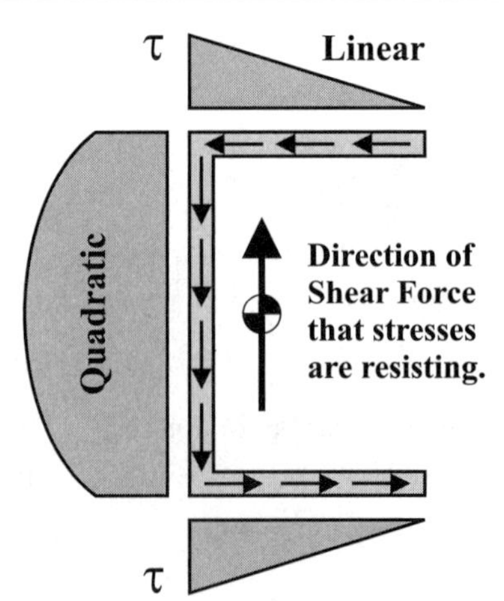

Figure 6.34. Qualitative response of shear stress on channel cross-section.

Shear Stress Distribution

The shear stress distribution on the channel cross-section is qualitatively shown in *Figure 6.34*. In each flange, the horizontal shear stress is *linear* since A^*y^* is linear ($A^* = fl$ is linear, y^* is constant). In the web, the vertical shear stress is *quadratic* since A^* and y^* are both linear. If the flange and web thicknesses are different, then at the flange–web junction, the flange and the web shear stresses are different.

I-Beams

I-beams are doubly symmetric, so their centroid is centered on the cross-section horizontally and vertically. I-beams are essentially two channel sections back-to-back. The shear stress distribution in an I-beam is determined using the same method used for a channel section. The shear stress is horizontal and linear in the flanges, and vertical and quadratic in the web.

Example 6.15 Shear Stress and Shear Flow in an I-beam

Given: The built-up I-beam cross-section in *Figure 6.35* is subjected to a 15-kip shear force applied upward on the cross-section.

Required: Due to the applied shear force, plot (a) the horizontal shear stress in the flange and the vertical shear stress in the web and (b) the corresponding shear flow.

Solution: *Step 1. Cross-section.* The moment of inertia is:

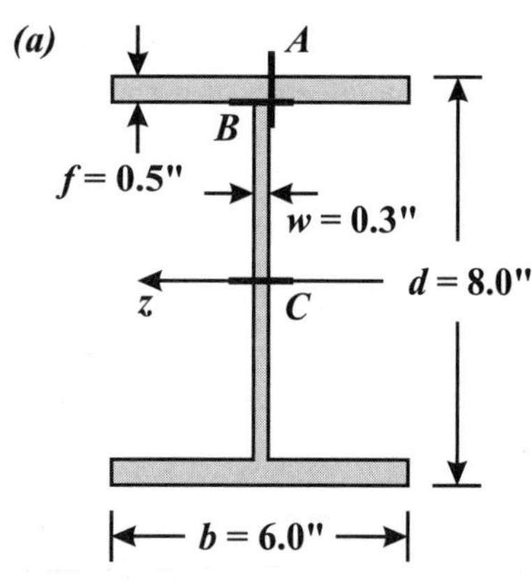

Figure 6.35. (a) I-beam.

$$I = \frac{(0.3)(7)^3}{12} + 2\left[\frac{6(0.5)^3}{12} + (6 \times 0.5)(3.75)^2\right]$$

$$= 93.08 \text{ in.}^4$$

Step 2. Shear Stress. At the end of the flange, the shear stress is zero. The shear stress is linear in the flange and quadratic in the web. Plotting the shear stress requires finding the stress at the web–flange intersection (cuts A and B) and at the centroid (C), and sketching the appropriate shape of the stress. The shear stress is:

$$\tau(y) = \frac{VA^*y^*}{It}$$

The shear stresses at cuts A and B are:

$$\tau_A = \frac{(15{,}000 \text{ lb})(3 \times 0.5 \text{ in.}^2)(3.75 \text{ in.})}{(93.08 \text{ in.}^4)(0.5 \text{ in.})} = 1.813 \text{ ksi}$$

$$\tau_B = \frac{(15{,}000 \text{ lb})(6 \times 0.5 \text{ in.}^2)(3.75 \text{ in.})}{(93.08 \text{ in.}^4)(0.3 \text{ in.})} = 6.04 \text{ ksi}$$

For cut C:

$$A^*y^* = (6 \times 0.5)(3.75) + (0.3 \times 3.5)(1.75)$$

$$= 13.09 \text{ in.}^3$$

so:

$$\tau_C = \frac{(15{,}000 \text{ lb})(13.09 \text{ in.}^3)}{(93.08 \text{ in.}^4)(0.3 \text{ in.})} = 7.03 \text{ ksi}$$

The shear stress is plotted in *Figure 6.35b*.

Step 3. The *shear flow* is $q = \tau t$, so:

$$q_A = 0.907 \text{ kips/in.}$$

$$q_B = 1.81 \text{ kips/in.}$$

$$q_C = 2.10 \text{ kips/in.}$$

The shear flow is plotted in *Figure 6.35c*.

At the flange–web junction, the shear flow in the web is the sum of the shear flows of each half of the flange:

$$q_B = q_A + q_A$$

Like water, the *flow* that enters a junction must leave that junction. This is generally not true for the shear stress since the flange and web thicknesses are usually different.

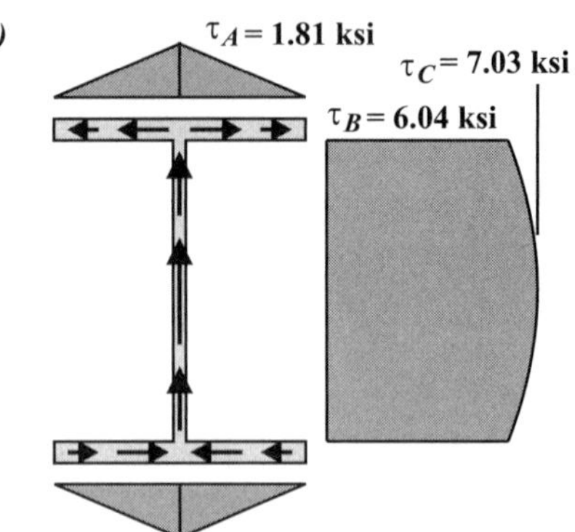

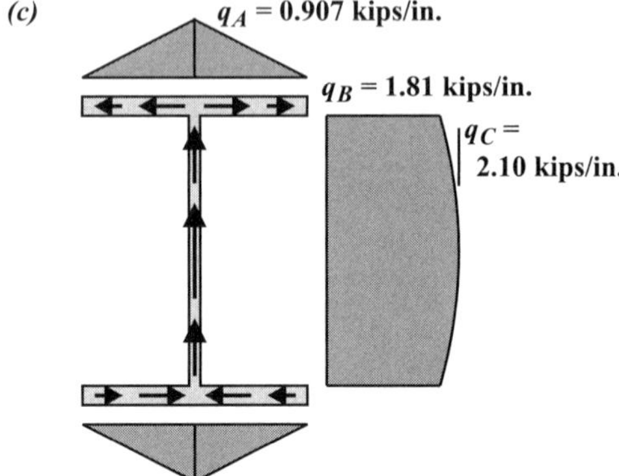

Figure 6.35. (b) Distribution of shear stress. **(c)** Distribution of shear flow.

Note: If the shear stress in the web is calculated by simply dividing the applied shear force by the web area (neglecting the flanges), then:

$$\tau_{web} = \frac{V}{A_{web}} = \frac{15{,}000 \text{ lb}}{(0.3 \text{ in.})[(8 - 1) \text{ in.}]} = 7.14 \text{ ksi}$$

This value is slightly larger than that calculated above at the centroid. I-beams are generally designed assuming only the web supports the shear force, and this straightforward calculation often provides a reasonable approximation for the maximum shear stress in an I-beam.

Shear Center

If the applied shear force V acts through the centroid of the channel section as shown in *Figure 6.36*, the beam rotates or twists about its axis. Rotation occurs because the moment caused by the applied force and the resisting shear stress in the web, and the moment caused by the shear stresses developed in the flanges, both act in the same direction (counterclockwise in *Figure 6.36*). To prevent twisting, the applied force must be moved a distance e away from the channel's centroid.

The ***shear center*** of a cross-section is the point that an applied shear force must pass through so that the beam does not twist. When the cross-section has an *axis of symmetry*, the shear center lies on that axis. A rectangle, a circle, an I-beam, etc., all have two axes of symmetry; they are *doubly symmetric*. The shear center is at the centroid of a doubly symmetric shape.

The channel section has only one axis of symmetry; the shear center does not coincide with the cross-section's centroid. An example of finding the shear center is given below.

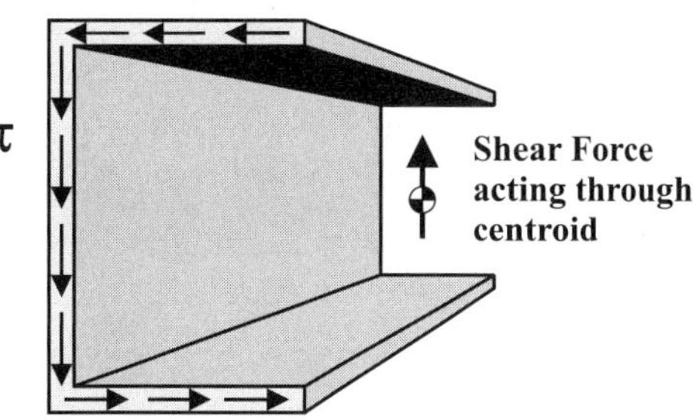

Figure 6.36. A length of a channel beam subjected to a shear force acting through the centroid of the rear cross-section. The resisting shear stress on the front cross-section has the qualitative response shown. The applied force and the resisting shear stresses cause the beam to twist.

Example 6.16 Shear Center of a Channel Section

Given: The channel section in *Figure 6.37a* has dimensions:

$$b = 6.0 \text{ in.}, \quad d = 12.0 \text{ in.},$$
$$f = w = 0.5 \text{ in.}$$

Distance b is the width of the flange from its free end to the centerline of the web, and d is the distance between the centerlines of the flanges. Distances f and w are the thicknesses of the flanges and web, respectively. The flange and web thicknesses are considered small compared to the overall dimensions. The shear force is V.

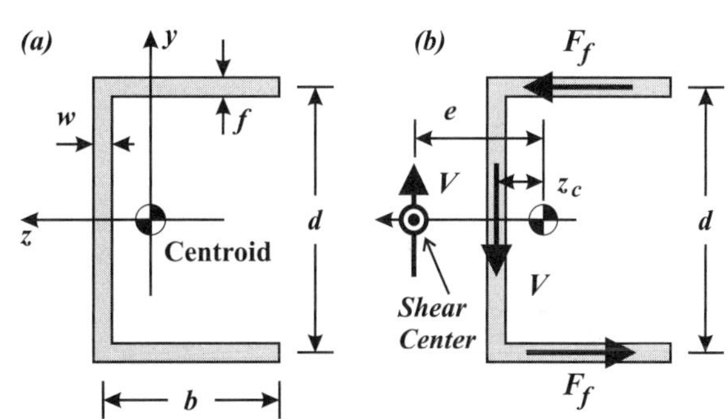

Figure 6.37. (a) Channel cross-section. **(b)** Locating the *shear center* of the cross-section.

Required: (a) Determine the expression for the location of the *shear center e* from the centroid as a function of b, d, w, and f. (b) For the values given above, determine e.

Solution: *Step 1.* Take the applied shear force to act at the *shear center,* distance e to the left of the centroid (*Figure 6.37b*). To balance the counterclockwise moment due to the shear forces in the flanges, the applied force must act to the left of the web. The total shear force in the web is equal to the applied force V.

Step 2. The shear force in each flange is (*Equation 6.46*):

$$F_f = \tau_{ave}(bf) = \frac{Vbd}{4I}(bf)$$

where I is the moment of inertia of the entire cross-section about the z-axis. Here:

$$I = \frac{wd^3}{12} + 2\left[\frac{bf^3}{12} + bf(d/2)^2\right] = 288.13 \text{ in.}^4$$

The counterclockwise moment caused by the shear in the flanges is:

$$M_f = F_f d = \frac{Vb^2d^2f}{4I}$$

Step 3. The applied force and the reacting force in the web cause a clockwise moment:

$$M_V = V[e - z_c]$$

where z_c is the distance from the web-centerline to the centroid of the cross-section. For the channel:

$$z_c = \frac{b^2f}{dw + 2bf} \quad \text{(prove this for yourself)}$$

For this channel, $z_c = 1.50$ in.

Step 4. Equating the moments and solving for e algebraically:

$$V[e - z_c] = \frac{Vb^2d^2f}{4I} \quad \rightarrow \quad e = z_c + \frac{b^2d^2f}{4I}$$

$$\text{Answer: } e = \frac{b^2f}{dw + 2bf} + \frac{b^2d^2f}{4}\left[\frac{wd^3}{12} + 2\left[\frac{bf^3}{12} + bf(d/2)^2\right]\right]^{-1}$$

Step 6. Substituting the values:

$$e = 1.5 + 2.249 \text{ in.}$$

$$\text{Answer: } \underline{e = 3.75 \text{ in.}}$$

The *shear center* is 3.75 in. to the left of the centroid, or 2.25 in. to the left of the web, outside of the channel (*Figure 6.37b*).

That the *shear center* is outside the channel has an interesting consequence. To lift a channel beam without causing it to twist, the shear force must be applied through the shear center. Since the shear center is outside the structure of the cross-section, external framework is needed to properly apply the force. Consider the difficulty in lifting a long and large channel-section beam (say $d = 10$ ft tall), that would make up the support structure of the walkway on either side of a large bridge.

Comparison of Bending and Shear: Deflection and Stress

Deflection

The deflection $v(x)$ due to bending moment $M(x)$ was determined in *Section 6.2*. The deflection due to shear force was not considered, the reason for which is discussed below.

Consider a cantilever beam, under tip load P. The cross-section is rectangular, with width b and depth d. The material moduli are E and G. For a homogeneous, isotropic material $G \sim 3E/8$.

The tip deflection of the beam due to bending (*Figure 6.38a*) is:

$$\delta_B = \frac{PL^3}{3EI} = \frac{PL^3}{3E}\left(\frac{12}{bd^3}\right) = \frac{4PL^3}{Ebd^3}$$

The tip deflection due to shear (*Figure 6.38b*) is:

$$\delta_V = \gamma L = \frac{\tau_{ave}}{G}L = \frac{P}{A}\frac{1}{G}L = \frac{P}{bd}\frac{8}{3E}L = \frac{8}{3}\frac{PL}{Ebd}$$

The ratio of the bending and shear deflections is:

$$\frac{\delta_B}{\delta_V} = \frac{4PL^3}{Ebd^3}\left(\frac{3Ebd}{8PL}\right) = 1.5\left(\frac{L}{d}\right)^2$$

The value of L/d for beams is generally $L/d \sim 10$. The ratio of the deflections is then:

$$\frac{\delta_B}{\delta_V} = 1.5\left(\frac{L}{d}\right)^2 = 1.5(10)^2 = 150$$

For most beams, the deflection due to shear is negligible. Very short beams (small L/d) are affected more by shear.

Stress

The maximum bending stress in the cantilever beam is:

$$\sigma_B = \frac{Mc}{I} = \frac{PL(d/2)}{[(bd^3)/12]} = \frac{6PL}{bd^2}$$

The maximum shear stress in a beam of rectangular cross-section is:

$$\tau_V = \frac{3}{2}\frac{V}{A} = \frac{3}{2}\frac{P}{bd}$$

The ratio of the normal and shear stresses is, taking $L/d \sim 10$:

$$\frac{\sigma_B}{\tau_V} = \frac{6PL}{bd^2}\left(\frac{2bd}{3P}\right) = 4\frac{L}{d} \sim 40$$

The ratio of the normal yield strength to shear yield strength for ductile metals is: $S_y/\tau_y = 1.73$. The ratio of the stresses is greater than the ratios of the strengths. Thus, the bending stress dominates and causes failure in most beams. The shorter the beam, the more susceptible it is to failure by shear.

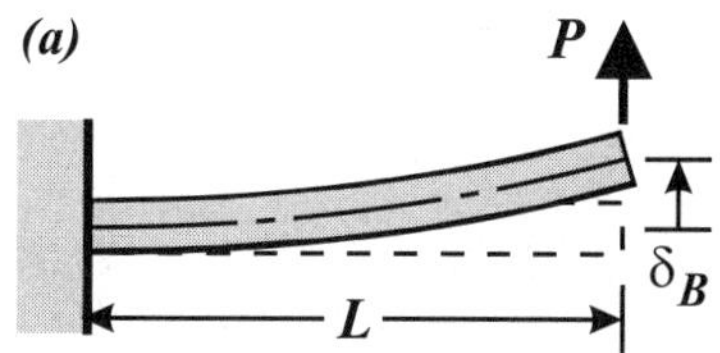

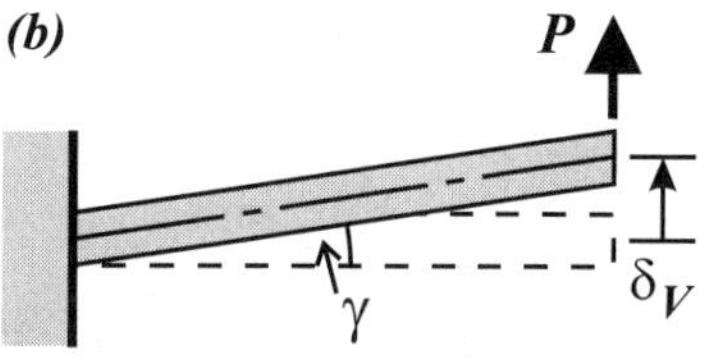

Figure 6.38. (a) Deflection of a cantilever beam due to bending. **(b)** Deflection of a cantilever beam due to shear force.

6.5 Shape: Section Modulus and Shape Factor

Throughout the discussion of beams, it becomes apparent that cross-sectional shape is an important factor in determining beam stresses and deflections. The response of a beam depends on how well its cross-sectional area is distributed about the axis of bending. For beams, the relevant property of the cross-section is the ***moment of inertia I***. The moment of inertia is a measure of how stiff a cross-section is with respect to bending about an axis in the plane of the cross-section through its centroid. For shafts in torsion, the analogous property is the *polar moment of inertia J*, a measure of how stiff a cross-section is with respect to twisting about the axis of the shaft (perpendicular to the cross-section).

Two additional geometric properties of a beam cross-section are introduced here: the *section modulus* and the *shape factor*.

Section Modulus

The bending stress in a beam is given by the formula:

$$\sigma = -\frac{My}{I} \qquad \text{[Eq. 6.47]}$$

where y is measured from the centroidal axis. The bending stress increases with y so that the magnitude of the maximum stress on the cross-section is:

$$\sigma_{max} = \frac{My_{max}}{I} = \frac{Mc}{I} \qquad \text{[Eq. 6.48]}$$

In general, the most important value of the bending stress is its maximum value.

It would be useful to have an equation similar to that for the normal stress in an axial member, where the stress on a cross-section is simply the load divided by a single cross-sectional property:

$$\sigma = \frac{P}{A} \qquad \text{[Eq. 6.49]}$$

For a given cross-section, I and y_{max} are both constant geometric properties; thus, their ratio is also a constant property. The ***section modulus*** Z of a cross-section is defined as:

$$Z = \frac{I}{y_{max}} = \frac{I}{c} \qquad \text{[Eq. 6.50]}$$

The magnitude of the maximum bending stress can then be written:

$$\sigma_{max} = \frac{M}{Z} \qquad \text{[Eq. 6.51]}$$

For a given moment M, as *section modulus* Z is increased, the maximum bending stress decreases. The units of section modulus are length cubed (in.^{3}, m^3, mm^3).

Formulas for the section modulus of common shapes are given in *Table 6.5* on *Page 212*. Section moduli for selected I-beams are given in *Appendix C*.

Example 6.17 **Solid Circular Cross-Section**

Required: Determine (a) the section modulus for a solid circular cross-section of radius R and (b) the maximum bending stress due to moment M.

Solution: For a solid circle: $I = \pi R^4/4$; $y_{max} = R$.

The section modulus is:

$$Answer:\ Z = \frac{I}{y_{max}} = \frac{\pi R^4/4}{R} \Rightarrow Z = \frac{\pi R^3}{4}$$

The maximum bending stress is:

$$Answer:\ \sigma_{max} = \frac{M}{Z} = \frac{4M}{\pi R^3}$$

Example 6.18 **Rectangular Cross-Section**

Required: Determine (a) the section modulus for a rectangular cross-section of depth d and width b and (b) the maximum bending stress due to moment M.

Solution: For a rectangle: $I = bd^3/12$; $y_{max} = d/2$.

The section modulus is:

$$Z = \frac{I}{y_{max}} = \frac{bd^3/12}{d/2}$$

$$Answer:\ Z = \frac{bd^2}{6}$$

The maximum bending stress is:

$$Answer:\ \sigma_{max} = \frac{M}{Z} = \frac{6M}{bd^2}$$

Example 6.19 **An S510-141 I-beam**

Given: A metric I-beam shape, S510-128, has a moment of inertia $I = 658 \times 10^6$ mm^4 and a total depth $d = 516$ mm (*Figure 6.39*).

Required: Determine (a) the section modulus of the I-beam and (b) the maximum bending stress due to moment $M = 200$ N·m.

Solution:

$$Z = \frac{I}{d/2} - \frac{658 \times 10^6\ \text{mm}^4}{258\ \text{mm}}$$

$$Answer:\ Z = 2.55 \times 10^6\ \text{mm}^3$$

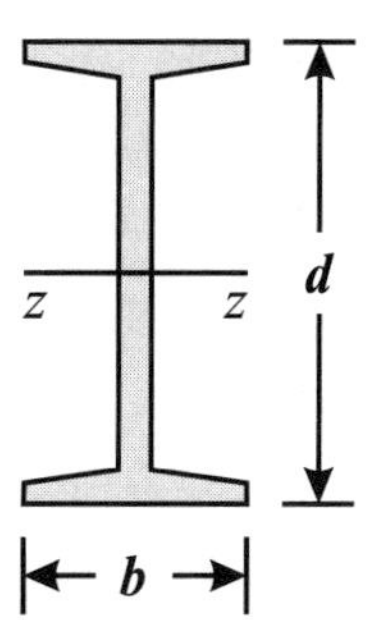

Figure 6.39. General shape of an I-beam.

$$\sigma_{max} = \frac{M}{Z} = \frac{200 \text{ N·m}}{2.55 \times 10^{-6} \text{ m}^3}$$

Answer: $\sigma_{max} = 78.4$ MPa

The section modulus of I-beams is generally included in tables of their geometric properties (see *Appendix C*).

Shape Factor for Elastic Bending

The following discussion is based on very innovative ideas developed by Prof. Michael F. Ashby at the University of Cambridge in the United Kingdom. Details may be found in his book, *Materials Selection in Mechanical Design* (1992, 1999).

Desire for improved fuel efficiency in automobiles and the power limitations of rocket engines used to launch space vehicles are examples where lightweight components are necessary. Ashby's **elastic shape factor** for bending is a useful concept in lightweight design.

The *shape factor* for bending stiffness – the *elastic shape factor* – is defined as:

$$\phi_B^e = \frac{4\pi I}{A^2} \qquad \text{[Eq. 6.52]}$$

where I is the moment of inertia of the beam cross-section and A is its area. The shape factor is dimensionless, and is a single variable that combines two important geometric properties of the cross-section, I and A.

The shape factor is a measure of the *stiffness* of the cross-section I per its *cross-sectional area squared* A^2 (weight squared when comparing cross-sections of the same material). The constant 4π is introduced so that the shape factor of a solid circle is unity (1.0).

The greater the shape factor ϕ_B^e, the more efficient the cross-section is in terms of its *bending stiffness to weight ratio*. For the same material and area A (weight), the greater the shape factor, the greater the value of I, the greater the bending stiffness EI, and the smaller the deflection.

Formulas for the elastic shape factor for bending stiffness of some common cross-sections are given in *Table 6.5*.

A shape factor for bending strength can also be developed, but will not be discussed in this treatment. The reader is referred to the work of Prof. Ashby.

Example 6.20 **Solid Circular Cross-Section**

Required: Determine the shape factor of a solid circle.

Solution:

$$\phi_B^e = \frac{4\pi I}{A^2} = \frac{4\pi(\pi R^4/4)}{(\pi R^2)^2}$$

Answer: $\phi_B^e = 1.0$

Example 6.21   Rectangular Cross-Section

Required: Determine the shape factor of a rectangular cross-section of depth d and width b.

Solution:

$$\phi_B^e = \frac{4\pi I}{A^2} = \frac{4\pi(bd^3/12)}{(bd)^2} = \frac{\pi d}{3b}$$

$$Answer:\ \phi_B^e = 1.05\frac{d}{b}$$

The ratio d/b defines the shape factor for a rectangular cross-section. Note that the absolute size of the cross-section does not matter; the factor is based only on the *shape* of the cross-section.

Example 6.22   American Standard Metric S510-128 I-beam

Given: A metric I-beam shape, S510-128, has a moment of inertia of $I = 658{\times}10^6$ mm^4 and a cross-sectional area of $16.32{\times}10^3$ mm^2.

Required: Determine the shape factor of the I-beam.

Solution:

$$\phi_B^e = \frac{4\pi I}{A^2} = \frac{4\pi(658{\times}10^6\ \text{mm}^4)}{(16.32{\times}10^3\ \text{mm}^2)^2}$$

$$Answer:\ \phi_B^e = 31$$

Typical values for the shape factor ϕ_B^e of I-beams range from 15 to 35. For rectangular cross-sections, the shape factor is $\phi_B^e = 1.05(d/b)$. A deep rectangular beam is a 2 by 12, which gives a shape factor of $\phi_B^e = 6.3$. Based on typical shape factor values, the efficiency of the I-beam over the rectangle in carrying bending loads is readily seen.

Table 6.5. Geometric properties of shapes subject to bending:
Area, *Moment of Inertia*, *Section Modulus* and *Elastic Shape Factor*.

Shape		Area, A (length^2)	Moment of Inertia about horizontal axis at Centroid, I (length^4)	Section Modulus, Z (length^3)	Elastic Shape Factor ϕ_B^e
Rectangle, Square		bd	$\dfrac{bd^3}{12}$	$\dfrac{bd^2}{6}$	$1.05\dfrac{d}{b}$
Hollow Rectangle		$bd - (b-2w)(d-2f)$ (d is total depth)	$\dfrac{bd^3 - (b-2w)(d-2f)^3}{12}$	$\dfrac{2I}{d}$	$\dfrac{4\pi I}{A^2}$
Circle, solid		πR^2	$\dfrac{\pi R^4}{4}$	$\dfrac{\pi R^3}{4}$	1.0
Circle, thick wall		$\pi(R^2 - R_i^2)$	$\dfrac{\pi(R^4 - R_i^4)}{4}$	$\dfrac{\pi(R^4 - R_i^4)}{4R}$	$\dfrac{(R^2 + R_i^2)}{(R^2 - R_i^2)}$
Circle, thin-wall		$2\pi R_{ave}t$ R = outer radius $R_{ave} = R - t/2$	$\pi R_{ave}^3 t$	$\dfrac{\pi R_{ave}^3 t}{R} \approx \pi R_{ave}^2 t$	$\dfrac{R_{ave}}{t}$
Triangle		$\dfrac{bh}{2}$	$\dfrac{bh^3}{36}$	$\dfrac{bh^2}{12}\ ;\ \dfrac{bh^2}{24}$	$\dfrac{4\pi h}{9 b}$
Built-up I-beam		$w(d-2f) + 2bf$ (d is total depth)	$\dfrac{w(d-2f)^3}{12} + 2\left[\dfrac{bf^3}{12} + bfx^2\right]$ $x = (d/2) - (f/2)$	$\dfrac{2I}{d}$	$\dfrac{4\pi I}{A^2}$
Space Frame; Truss		A (total cross-sectional area)	$\dfrac{Ad^2}{4}$ (d is distance between chord centroids)	$\dfrac{Ad}{2}$	$\dfrac{\pi d^2}{A}$

6.6 Design of Beams

Strength and Stiffness

Beams are generally designed to satisfy two requirements:

1. *Strength*: when subjected to load, the maximum stress should not exceed an allowable stress determined by the material strength and the factor of safety.

2. *Stiffness*: when subjected to load, the maximum deflection should not exceed an allowable amount. The allowable deflection is given in terms of the beam span or length L. The *deflection index f* is:

$$f = \frac{\text{deflection}}{\text{span of beam}} = \frac{\delta}{L} \qquad \text{[Eq. 6.53]}$$

The *allowable deflection index* is usually small. Building codes give the maximum allowable deflection index for structural members as $f_A = 1/240$ (about 0.4%). For plastered ceilings, the allowable deflection index is less: $f_A = 1/360$. For automobile chassis, the allowable deflection index is approximately $f_A = 1/240$.

The following examples illustrate basic design concepts, and selecting beam materials and cross-sectional shapes for light-weight design.

Example 6.23 Maximum Allowable Load

Given: A simply-supported beam of length L is subjected to a central point load P (*Figure 6.40*). The cross-section is doubly symmetric (e.g., a rectangle, I-beam, etc.) with depth D and moment of inertia I. The material properties are yield strength S_y and modulus E. the allowable bending stress is based on a factor of safety of 2.0 against yielding. The allowable deflection index is $f_A = 1/240$.

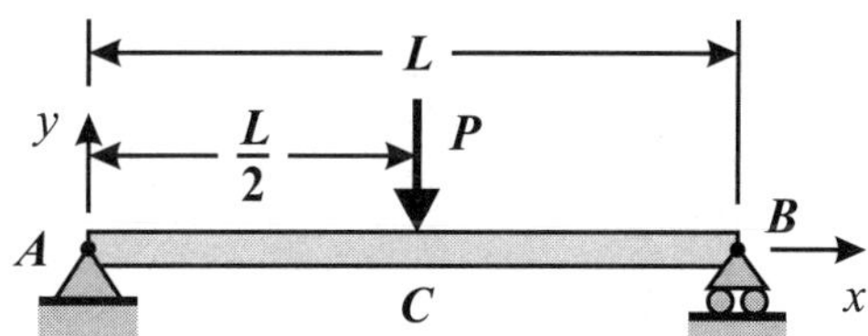

Figure 6.40. Simply-supported beam subjected to central load P.

Required: Based on bending strength and deflection requirements, determine how the allowable load P_A varies with span L.

Solution: *Step 1. Allowable load based on strength.* For a simply supported centrally loaded beam, the maximum moment is at the center:

$$M_{max} = \frac{PL}{4}$$

The maximum bending stress is:

$$\sigma_{max} = \frac{M_{max}y_{max}}{I} = \frac{PLD}{4I\,2} = \frac{PLD}{8I}$$

From the factor of safety requirement, the allowable stress is:

$$\sigma_A = \sigma_{max} = \frac{S_y}{FS} = \frac{S_y}{2}$$

The allowable load based on strength is:

$$\textit{Answer: } P_{A,\,\sigma} = \frac{4S_yI}{LD} = \frac{2S_yZ}{L}$$

where Z is the section modulus of the cross-section.

Step 2. Allowable load based on deflection. The central deflection for a simply supported centrally loaded beam is:

$$\delta = \frac{PL^3}{48EI}$$

The deflection index and allowable deflection are:

$$f = \frac{\delta}{L} = \frac{PL^2}{48EI} = f_A = \frac{1}{240}$$

The allowable load based on stiffness is:

$$\textit{Answer: } P_{A,\,\delta} = \frac{EI}{5L^2}$$

Step 3. Allowable load. Both strength and stiffness requirements must be satisfied, so:

$$\textit{Answer: } P_A = \min\left[\frac{2S_yZ}{L}, \frac{EI}{5L^2}\right]$$

Step 4. The shear stress, generally considered a secondary load, should also be checked against its allowable value to ensure a safe structure.

Example 6.24   Effect of Length on Allowable Loads for Structural Steel and Aluminum

Given: The simply-supported center-loaded beam of *Example 6.23* and *Figure 6.40*. The cross-section is the metric I-beam S510-128:

$$I = 658\times10^{-6}\text{ m}^4, \, D = 0.516\text{ m}, \, Z = 2.55\times10^{-3}\text{ m}^3$$

Required: For structural steel and structural aluminum, plot the allowable load P_A for both strength and deflection against beam length L. For steel: $E_{st} = 200$ GPa and $S_{y,st} = 250$ MPa; for aluminum: $E_{al} = 70$ GPa and $S_{y,al} = 240$ MPa (structural metals Steel A36 and Aluminum 6061-T6 have about the same yield strength). For strength, use a factor of safety of 2 against yielding. For the allowable deflection factor, take $f_A = 1/240$.

Solution: *Step 1*. For simplicity, the strength S_y of both metals is taken to be the same (250 MPa), the allowable load based on strength, including the factor of safety, is:

$$P_{A,\sigma} = \frac{2S_y Z}{L}$$

Step 2. The moduli of the two materials are different, so the allowable loads based on deflection are, from *Example 6.23*:

$$P_{A,\delta,st} = \frac{E_{st} I}{5L^2} \quad \text{and}$$

$$P_{A,\delta,al} = \frac{E_{al} I}{5L^2}$$

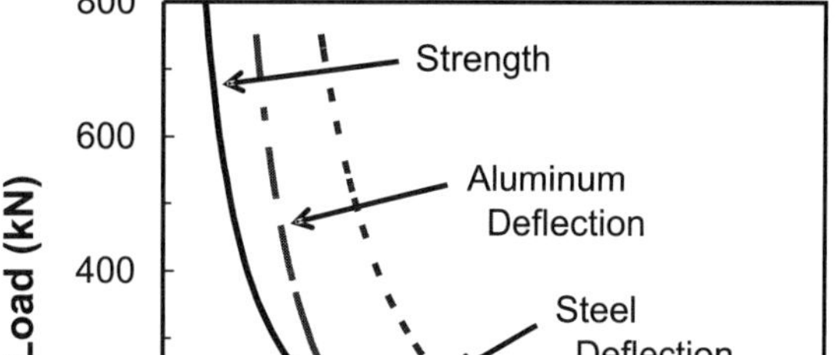

Figure 6.41. Allowable central point load plotted against beam length for allowable stress and allowable deflection. Beam cross-section: S510-128.

Step 3. The allowable loads for the S510-128 beam are plotted against length in *Figure 6.41*.

For the S510-128 steel beam, the allowable load is limited by its strength for $L < 20.6$ m, which is very long for a beam that is only 0.5 m deep. Steel has a high modulus so deflection is rarely a major concern in practical systems.

For the aluminum beam, the load is limited by strength for $L < 7.2$ m. Longer aluminum beams are limited by deflection. It is especially important to consider both strength and deflection criteria in aluminum structures. While replacing steel with aluminum of the same cross-section (here, S510-128) reduces the weight (the density of aluminum is about one-third that of steel), deflection can become the failure mode.

In aluminum design, the yield strength of Aluminum 6061-T6 is generally taken as the minimum expected from manufacture, which is $S_y = 240$ MPa. Thus, for a given cross-section, regardless of length and neglecting self-weight, structural steel beams can support more or equal load than structural aluminum beams of the same cross-sectional geometry.

Example 6.25　Selection of an I-beam

Given: A 10 ft long steel I-beam is to support a tip load of $P = 5.0$ kips (*Figure 6.42*). The allowable stress in bending is $\sigma_A = 15$ ksi, the modulus is $E = 30{,}000$ ksi, and the allowable deflection index is $f_A = \delta/L = 1/240$. The beam is to be made of an S-shape beam.

Required: Specify the required I-beam cross-section from the S-shapes listed in *Appendix C*. Select the shape that gives the least weight while satisfying the stress and deflection requirements.

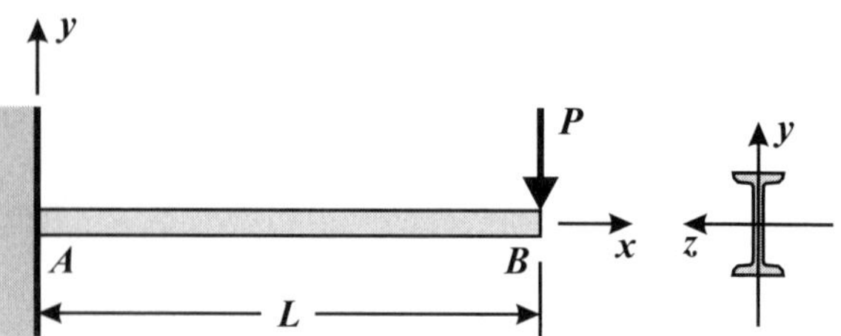

Figure 6.42. Cantilever beam of length *L* subjected to tip load *P*.

Solution: *Step 1*. For a cantilever beam under tip load, the maximum moment is:

$$M = PL = 50 \text{ kip-ft} = 600 \text{ kip-in.}$$

so the maximum bending stress is:

$$\sigma_{max} = \frac{M}{Z}$$

To satisfy the allowable stress:

$$Z \geq \frac{M}{\sigma_{max}} = \frac{600 \text{ kip-in.}}{15 \text{ ksi}} = 40 \text{ in.}^3$$

Step 2. The deflection due to a tip load is:

$$\delta = \frac{PL^3}{3EI}$$

The allowable deflection is:

$$\delta_A = \frac{L}{240} = 0.5 \text{ in.}$$

To satisfy the allowable deflection:

$$I \geq \frac{PL^3}{3E\delta_A} = \frac{(5 \text{ kips})(120 \text{ in.})^3}{3(30,000 \text{ ksi})(0.5 \text{ in.})} = 192 \text{ in.}^4$$

Step 3. Referring to *Appendix C*, the smallest S-shape I-beam that satisfies the strength requirement is S12×40.8. The smallest I-beam to satisfy the deflection requirement is S12×31.8. Since both requirements must be met:

Answer: Select: <u>S12×40.8</u>

Step 4. After the cross-section is selected, the system must be reanalyzed to include the weight of the beam itself. In this case, a uniformly distributed load $w = 40.8$ lb/ft must be applied in addition to the point load. If the applied load and self-weight causes the stress or deflection to exceed the allowables, a new cross-section must be selected, usually the next largest area of the same depth. In this case, S12×40.8 does support the applied load and its own weight without exceeding either the allowable stress or the allowable deflection.

Minimum Weight Design

After strength and stiffness, a third design consideration is weight. Weight reduction has always been of great concern for aeronautical and aerospace engineers. In addition, lighter automobiles increase fuel efficiency; lighter building materials reduce shipping costs and make construction easier; lighter golf shafts and tennis rackets give sports enthusiasts at least a perceived advantage. An efficient engineering structure uses less material. This makes the structure lighter and cheaper, and minimizes the requirements placed on natural resources.

In the next two examples, equations are developed that determine the minimum weight of a rectangular beam that must support a given load or have a certain stiffness. This development is also based on the ideas of Michael Ashby (*Materials Selection in Mechanical Design*, 1992, 1999).

Example 6.26 **For a Given Load, Design a Rectangular Beam of Minimum Weight**

Given: A cantilever beam of length L with rectangular cross-section of breadth b and depth $2b$ is to be subjected to tip load P (*Figure 6.43*). The material has weight density γ, yield strength S_y, and modulus E. Neglect the weight of the beam on the load, i.e., $W \ll P$.

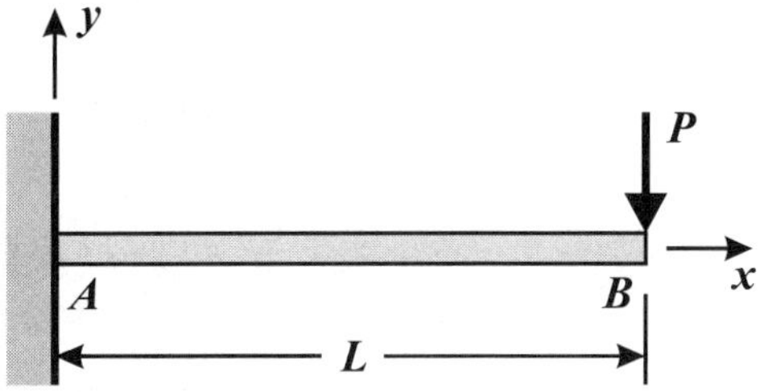

Figure 6.43. Cantilever beam of length L subjected to tip load P.

Required: For a given load P, develop an expression that can be used to design a beam (i.e., select the material) to minimize its weight. The length of the beam is specified to be L.

Solution: *Step 1.* The weight of the beam is:

$$W = \gamma A L = \gamma(b \times 2b)L = 2\gamma b^2 L$$

The maximum moment for a tip-loaded cantilever is: $M = PL$

The maximum stress for the $b{\times}2b$ rectangular beam is:

$$\sigma_{max} = \frac{My_{max}}{I} = \frac{Mb}{b(2b)^3/12} = \frac{3}{2}\frac{PL}{b^3}$$

from which: $b = \left(\dfrac{3}{2}\dfrac{PL}{\sigma_{max}}\right)^{1/3}$

Step 2. Substituting b into the weight equation, and setting the maximum stress to the yield strength, gives:

$$W = 2\gamma\left(\frac{3}{2}\frac{PL}{\sigma_{max}}\right)^{2/3}L = 2\left(\frac{3}{2}\right)^{2/3}P^{2/3}\left(\frac{\gamma}{S_y^{2/3}}\right)L^{5/3}$$

$$\textit{Answer: } W = 2\left(\frac{3}{2}\right)^{2/3}P^{2/3}\left(\frac{\gamma}{S_y^{2/3}}\right)L^{5/3}$$

The expression for weight can be described as follows:

$$W = \Big[\text{constant}\Big]\begin{bmatrix}\text{strength}\\\text{term}\end{bmatrix}\begin{bmatrix}\text{material properties}\\\text{term}\end{bmatrix}\begin{bmatrix}\text{length}\\\text{term}\end{bmatrix}$$

The required strength P is given and length L is usually constrained by overall geometric requirements. The only term that can be modified is the material properties term. Thus, to minimize the weight of the beam, a material should be selected that has a low value of $\gamma/S_y^{2/3}$ ($\gamma^{3/2}/S_y$), or a high value of $S_y/\gamma^{3/2}$.

The cross-sectional area A does not appear in the result as it was eliminated by the math. The actual size of the rectangular cross-section is not known, only that it has an aspect ratio (depth:height) of 2. After a material is chosen, the beam will have to be sized with strength and stiffness calculations to ensure it is of reasonable dimension.

Example 6.27   For a Given Stiffness, Design a Rectangular Beam of Minimum Weight

Given: A cantilever beam of length L with rectangular cross-section of breadth b and depth $2b$, is to be subjected to tip load P (*Figure 6.43*). The material has weight density γ, yield strength S_y, and modulus E. Neglect the weight of the beam on the load.

Required: For a given stiffness $K = P/\delta$, develop an expression that can be used to design a beam (i.e., select the material) to minimize its weight. The length of the beam is fixed.

Solution: *Step 1.* The weight of the beam is:

$$W = \gamma AL = 2\gamma b^2 L$$

The maximum deflection of a cantilever is:

$$\delta = \frac{PL^3}{3EI} = \frac{PL^3}{3E\left[\dfrac{b(2b)^3}{12}\right]} = \frac{PL^3}{2Eb^4}$$

Solving for b^2: $b^2 = \left(\dfrac{PL^3}{2E\delta}\right)^{1/2}$

Step 2. Substituting b^2 into the weight equation gives:

$$W = 2\gamma\left(\frac{PL^3}{2E\delta}\right)^{1/2}L$$

Answer: $W = 2^{1/2}\left(\dfrac{P}{\delta}\right)^{1/2}\left(\dfrac{\gamma^2}{E}\right)^{1/2}L^{5/2}$

The expression for weight is given by:

$$W = \Big[\,\text{constant}\,\Big]\begin{bmatrix}\text{stiffness}\\ \text{term}\end{bmatrix}\begin{bmatrix}\text{material properties}\\ \text{term}\end{bmatrix}\begin{bmatrix}\text{length}\\ \text{term}\end{bmatrix}$$

The required stiffness P/δ is given and the length L is usually constrained by overall geometric requirements. The only term that can be modified is the material properties term. To minimize the weight of the beam, a material should be selected that has a low value of γ^2/E, or a high value of E/γ^2. Once a material is chosen, the cross-sectional dimensions will need to be sized.

Performance Index

In *Examples 6.26* and *6.27*, the weight of a cantilever beam of cross-section $b\times2b$ was given by a general function:

$$W = \Big[\,\text{constant}\,\Big]\begin{bmatrix}\text{system response}\\ \text{or perfomance}\\ \text{term}\end{bmatrix}\begin{bmatrix}\text{material properties}\\ \text{term}\end{bmatrix}\begin{bmatrix}\text{length}\\ \text{term}\end{bmatrix} \qquad \text{[Eq. 6.54]}$$

The system response or performance is the required **strength** or **stiffness** of the beam. By analyzing other beams, it can be shown that the constant term depends on the type of load, the support system, and the chosen cross-sectional shape.

The material properties ratio is the **performance index**. Once a beam system (load, supports, and cross-section) is selected, to minimize beam weight, the appropriate material *performance index* must be minimized (or maximized) as shown in *Table 6.6*.

Table 6.7 compares the appropriate ratios for minimum weight design for various materials using representative values. The cross-sectional shape of each system must be the same, but not necessarily the size. For a given *stiffness*, the lightest beam has properties that maximize the *performance index* E/γ^2. For a given *strength*, the lightest beam maximizes $S_y/\gamma^{3/2}$.

Table 6.7 indicates that a wooden beam compares favorably to a structural steel beam of the same cross-sectional shape. For stiffness calculations, the

Table 6.6. Performance Indices for minimum weight beams; same cross-section and length.

For a given	Minimize	Maximize
Strength	$\dfrac{\gamma^{3/2}}{S_y},\ \dfrac{\rho^{3/2}}{S_y}$	$\dfrac{S_y}{\gamma^{3/2}},\ \dfrac{S_y}{\rho^{3/2}}$
Stiffness	$\dfrac{\gamma^2}{E},\ \dfrac{\rho^2}{E}$	$\dfrac{E}{\gamma^2},\ \dfrac{E}{\rho^2}$

γ: weight per unit volume, $\gamma = \rho g$
ρ: mass per unit volume

Table 6.7. Representative Performance Indices for Minimum Weight Design. The last two columns are calculated directly from the numbers in the first three; for simplicity, units are not included.

Material	γ (lb/in.^{3})	E (Msi)	S_y or S_u (ksi)	For Stiffness, maximize E/γ^2	For Strength, maximize $S_y/\gamma^{3/2}$
Steel A36 (structural)	0.284	30	36	0.37×10^3	0.24×10^3
Aluminum 6061-T6 (structural)	0.098	10	40	1.04×10^3	1.30×10^3
Wood, Douglas Fir (parallel to grain)	0.019	1.7	6	4.71×10^3	2.29×10^3
Ceramic, silicon carbide, SiC	0.116	65	200	4.83×10^3	5.06×10^3
Composite, CFRP (fiber direction)	0.058	20	300	5.94×10^3	21.5×10^3

performance index E/γ^2 of wood is 13 (=4.71/0.37) times as great a steel. Similarly, for the same load carrying capacity, the performance index $S_y/\gamma^{3/2}$ of wood is 9.6 (=2.29/0.24) times greater than steel. Aluminum is also superior to structural steel. A higher performance index means a lighter system. Aluminum was the primary material used in the development of the aircraft industry, which requires light structures. It should be noted that stronger steels can be made (A36 is a structural steel used in common construction applications).

The ceramic SiC would seem to be a good choice for a lightweight beam. However, this engineering ceramic is brittle. Large beams of ceramic materials are not feasible since there is a large probability of the system having a crack large enough to cause fracture. Small ceramic beams are used in micro-devices, and ceramic fibers (~100 μm in diameter) are used in composites; smaller volumes of a brittle material are less likely to fail under a given stress.

Lastly, composites such as carbon fiber reinforced polymers (CFRP) have great advantages. While the modulus and strength given in *Table 6.7* are for the uniaxial-only direction, which exaggerates the typical performance index of CFRP, the general trend is illustrated. The structural efficiency of composites – their high strength-to-weight and high stiffness-to-weight ratios – is why they are widely used in the aerospace industry.

Minimum Weight Design using Shape Factors

Applied loads and overall geometry (length) are often specified values in a design. In addition to material selection, the cross-sectional shape is usually a design choice. Ashby's *elastic shape factor* ϕ_B^e for beam bending (*Equation 6.52*, provides the most effective single-variable description of a cross-section for minimum weight stiffness problems. The shape factor relates the moment of inertia I (the bending stiffness due to the shape of the cross-section) to the cross-sectional area A (~weight):

$$\phi_B^e = \frac{4\pi I}{A^2} \qquad \text{[Eq. 6.55]}$$

A shape factor can also be developed for strength problems (as well as for torsion problems, etc.), but only beam deflection problems are considered in this presentation. The elastic shape factor will simply be written ϕ.

The shape factor method is demonstrated by repeating *Example 6.27* (a cantilever requiring a given stiffness). However, the cross-sectional shape is not specified.

Example 6.28 For a Given Stiffness, Design a Beam of Minimum Weight

Given: A cantilever beam of length L is to be subjected to tip load P (*Figure 6.43*). The material has weight density γ, yield strength S_y, and modulus E.

Required: For a given stiffness $K = P/\delta$, develop an expression that can be used to design a beam of minimum weight. The length of the beam is fixed. The beam material and cross-sectional shape can be selected.

Solution:

The weight of the beam is: $W = \gamma A L$

The maximum deflection of a cantilever is: $\delta = \dfrac{PL^3}{3EI}$

The shape factor is: $\phi = \dfrac{4\pi I}{A^2}$

Solving for the moment of inertia from the deflection and shape factor equations:

$$I = \frac{PL^3}{3E\delta} = \frac{\phi A^2}{4\pi}$$

The area can be rewritten: $A = \left(\dfrac{PL^3}{3E\delta}\dfrac{4\pi}{\phi}\right)^{1/2}$

Substituting the area into the expression for the weight, and rearranging, gives:

$$\textit{Answer: } W = \left(\frac{4\pi}{3}\right)^{1/2}\left(\frac{P}{\delta}\right)^{1/2}\left(\frac{\gamma^2}{E}\right)^{1/2}\left(\frac{1}{\phi}\right)^{1/2} L^{5/2}$$

Replacing weight density γ with mass density ρ gives the mass of the beam:

$$\textit{Answer: } m = \left(\frac{4\pi}{3}\right)^{1/2}\left(\frac{P}{\delta}\right)^{1/2}\left(\frac{\rho^2}{E}\right)^{1/2}\left(\frac{1}{\phi}\right)^{1/2} L^{5/2}$$

The weight equation just derived is more general than that found in *Example 6.27*. The general form of the equation now has five terms:

$$W = \left[\text{constant}\right]\left[\begin{array}{c}\text{stiffness}\\\text{term}\end{array}\right]\left[\begin{array}{c}\text{material properties}\\\text{term}\end{array}\right]\left[\begin{array}{c}\text{cross-section}\\\text{term}\end{array}\right]\left[\begin{array}{c}\text{length}\\\text{term}\end{array}\right] \qquad \text{[Eq. 6.56]}$$

The constant depends on the beam loading and its supports; the required stiffness (performance) and length are generally given. The materials and cross-sectional shape have yet to be determined. The area of the cross-section does not appear in the equation, but is incorporated in the shape factor ϕ.

To minimize beam weight for a given stiffness, the following expression is *minimized*:

$$\left(\frac{\gamma^2}{E}\right)\left(\frac{1}{\phi}\right) \quad \text{or} \quad \left(\frac{\rho^2}{E}\right)\left(\frac{1}{\phi}\right) \qquad \text{[Eq. 6.57]}$$

or, the following expression is *maximized*:

$$\left(\frac{E}{\gamma^2}\right)(\phi) \quad \text{or} \quad \left(\frac{E}{\rho^2}\right)(\phi) \qquad \text{[Eq. 6.58]}$$

Using *Examples 6.57* or *6.58*, a designer can choose *material* and *shape* to achieve the most efficient design. A low value of E/γ^2 can be compensated for with a high value of ϕ.

For example, based only on *Table 6.8*, steel is less desirable than either aluminum or wood. However, steel is often formed into a very efficient cross-section, the I-beam (a shape factor of $\phi \sim 15 - 30$), while wood has traditionally had a rectangular cross-section ($\phi = 1.05 \, d/b$). Appropriate choices of material and shape can result in a very efficient structural system, as illustrated in the following example.

Table 6.8. Material properties for minimum weight design based on beam stiffness (SI units).

Material	ρ (kg/m^3)	E (GPa)	$\dfrac{E}{\gamma^2}$
Steel, A36	7.9	200	3.2
Aluminum, 6061-T6	2.7	70	9.6
Wood, Douglas Fir	0.6	12	33.3

Example 6.29 Wood Rectangular Beam vs. Steel I-beam

Given: A wooden beam with a rectangular cross-section (aspect ratio $d/b = 2$) and a steel I-beam ($\phi = 25$) are being considered for an application requiring a given stiffness P/δ (*Figure 6.44*).

Required: Determine which combination of material and cross-sectional shape will give the lightest beam. For the material properties, use *Table 6.8*.

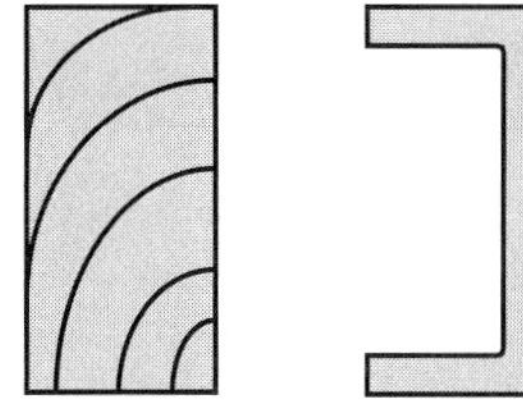

Figure 6.44. Wooden beam of rectangular cross-section and steel I-beam.

Solution: *Step 1.* The shape factor for a rectangle with an aspect ratio of 2 is:

$$\phi = 1.05\frac{d}{b} = 1.05(2) = 2.1$$

Step 2. To minimize weight, the following expression must be maximized:

$$\left(\frac{E}{\rho^2}\right)(\phi)$$

For the wooden beam:

$$\left(\frac{E}{\rho^2}\right)(\phi) = (33.3)(2.1) = 70$$

For the steel I-beam, $\phi = 25$:

$$\left(\frac{E}{\rho^2}\right)(\phi) = (3.2)(25) = 80$$

Since the expression for the steel I-beam is greater than that for the wooden rectangular beam:

Answer: the steel beam system is lighter

This is why steel is manufactured into I-beams with large values of ϕ.

In recent years, wood has also begun to be manufactured as I-beams, with wooden flanges and plywood webs. The I-beam shape makes a wooden beam a more efficient system, allowing less of this natural and increasingly expensive material to be consumed.

7.0 Introduction

In *Chapters 4–6*, the following components were analyzed:

- *Axial members* that support *axial forces* that cause *axial (normal) stresses*;
- *Torsional members* that support *torques* that cause *shear stresses*;
- *Pressure vessels* that contain fluids under *pressure* that cause *biaxial (normal) stresses*;
- *Beams* that support *bending moments* and *shear forces* that cause *bending (normal) stresses* and *shear stresses*, respectively.

For convenience, these structural components with their loadings, stresses, and strains are summarized in *Table 7.1*. The various loads cause only two types of stresses: **normal stresses** σ and **shear stresses** τ.

In practice, a system and its loadings can be quite complex. Many situations arise that cause a component to be loaded simultaneously by axial, torsional, shear, and/or bending loads. Such is the case for the support mast of the highway sign introduced in *Chapter 2*, and shown again in *Figure 7.1*. The mast acts as:

- an *axial member* supporting the weight of the sign and its own weight;
- a *torsional member* supporting a torque due to the wind force acting on the sign;
- a *beam* supporting a shear force due to the wind, and bending moments about two axes due to the weight of the sign (M_1) and the wind force (M_2).

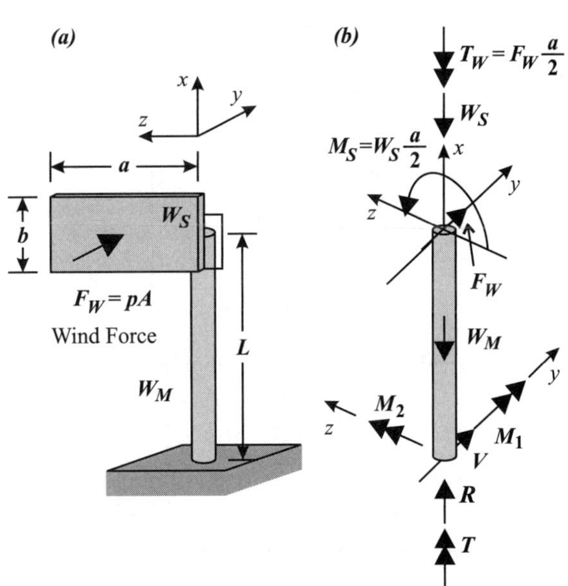

Figure 7.1. (a) A highway sign under wind load. **(b)** Considering the wind load (F_W) and the weight of the structure ($W_S + W_M$), the support mast acts as an axial member, a torsional member, and a beam in shear and in bending (about two axes).

Table 7.1. Components and their Loads, and resulting Stresses and Strains.

Member and Load	Resulting Stress	Maximum Stress and Strain
Bar, Rod, Column: axial force P	*Normal (axial) Stress σ_a*	$\sigma_a = \dfrac{P}{A}$; $\varepsilon = \dfrac{P}{AE}$ [Eq. 7.1]
Shaft in Torsion: torque T	*Shear Stress τ_T*	$\tau_T = \dfrac{TR}{J}$; $\gamma = \dfrac{TR}{JG}$ [Eq. 7.2]
Pressure Vessel: pressure p	*Normal (biaxial) Stress* cylindrical: σ_H, σ_L; spherical: σ_S	$\sigma_H = \dfrac{pR}{t}$; $\varepsilon_H = \dfrac{pR}{2Et}(2-dv)$ $\sigma_L = \dfrac{pR}{2t}$; $\varepsilon_L = \dfrac{pR}{2Et}(1-2v)$ $\sigma_S = \dfrac{pR}{2t}$; $\varepsilon_S = \dfrac{pR}{2Et}(1-v)$ [Eq. 7.3]
Beam: moment M	*Normal (bending) Stress σ_b*	$\sigma_b = -\dfrac{Mc}{I}$; $\varepsilon = -\dfrac{Mc}{EI}$ [Eq. 7.4]
Beam: shear force V	*Shear Stress τ_V*	$\tau_V = \dfrac{VA^*y^*}{It}$; $\gamma = \dfrac{VA^*y^*}{GIt}$ [Eq. 7.5]

At first glance, the analysis of a system subjected to several loads can be daunting. However, the solution method for linear systems is, in general, straightforward. The stresses and strains caused by each load acting alone are first determined. The total effect is found by combining the results due to each load.

Breaking a complex problem into a number of simpler problems, solving the simpler problems, and then adding the individual solutions together to get the total response, is called the ***method of superposition***. This method is a very powerful tool for solving linear systems. For stresses, strains, and displacements, the method of superposition only works when the material remains within the linear–elastic range everywhere, i.e., the material does not yield. If the stress anywhere exceeds the material yield strength (in tension or compression), plastic deformation occurs, and the stress–strain relationship is no longer linear. If yielding is predicted to occur, the component will likely need to be redesigned.

Stresses generally vary from point-to-point in a system. It is therefore important to find the stresses that act at critical points due to each load. A clear and orderly solution method is necessary to avoid confusion.

7.1 Superposition Examples

The *method of superposition* is illustrated with a series of examples.

Example 7.1 Axial Loads

Given: Two weights, W_1 and W_2, hang from a bar of cross-sectional area A and length L (*Figure 7.2*). The modulus of the bar is E. Assume the material remains elastic.

Required: Consider the weights separately (*Figure 7.2a, center, right*), and use *superposition* to determine the (a) stress and (b) strain states at point C, and (c) the elongation of the bar under both loads.

Solution: *Step 1.* The stress, strain, and elongation due only to W_1 are:

$$\sigma_1 = \frac{W_1}{A}; \quad \varepsilon_1 = \frac{W_1}{AE}; \quad \Delta_1 = \frac{W_1 L}{AE}$$

The stress, strain, and elongation due only to W_2 are:

$$\sigma_2 = \frac{W_2}{A}; \quad \varepsilon_2 = \frac{W_2}{AE}; \quad \Delta_2 = \frac{W_2 L}{AE}$$

Step 2. Superimposing (summing) the results for the bar loaded separately by W_1 and W_2 gives:

$$\sigma = \sigma_1 + \sigma_2 = \frac{W_1}{A} + \frac{W_2}{A} = \frac{W_1 + W_2}{A}$$

$$\varepsilon = \varepsilon_1 + \varepsilon_2 = \frac{W_1}{AE} + \frac{W_2}{AE} = \frac{W_1 + W_2}{AE}$$

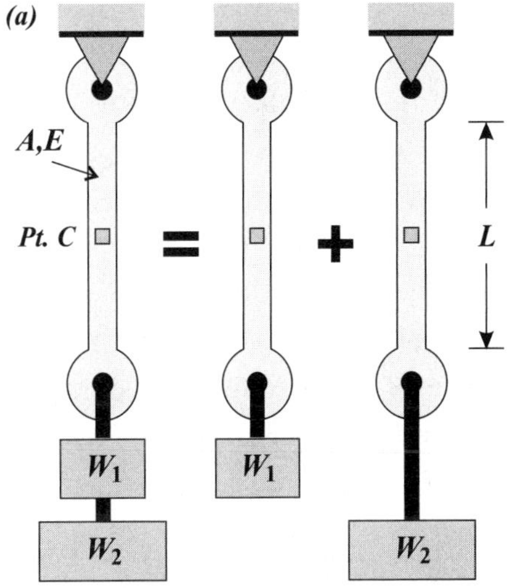

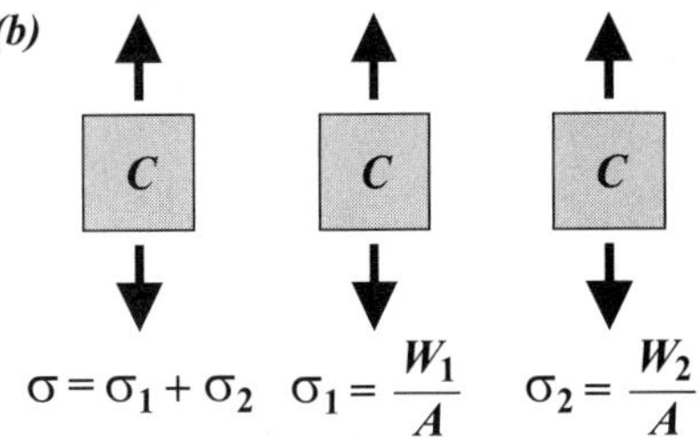

Figure 7.2. **(a)** A bar under load $W_1 + W_2$ is equivalent to the *superposition* of the bar loaded separately by W_1 and W_2. **(b)** The stress state at point C is the superposition of the stresses caused by the weights individually.

$$\Delta = \Delta_1 + \Delta_2 = \frac{W_1 L}{AE} + \frac{W_2 L}{AE} = \frac{(W_1 + W_2)L}{AE}$$

These results agree with those found by simply taking the total load, $W_1 + W_2$, and solving immediately for the stress, strain, and elongation:

$$Answer: \quad \sigma = \frac{(W_1 + W_2)}{A} ; \quad \varepsilon = \frac{(W_1 + W_2)}{AE} ; \quad \Delta = \frac{(W_1 + W_2)L}{AE}$$

Summary. This trivial example shows the process of *superimposing like-stresses* at a point due to separate loads. Here, the stresses are constant everywhere. In general, stresses vary with location, such as in beams that carry a non-constant bending moment.

Again, the method of superposition works if, and only if, the system remains linear; i.e., the material continues to obey Hooke's Law, $\sigma = E\varepsilon$. To find strain and elongation by superposition, the normal stress due to one type of load, as well as the combined values of stress, must remain less than the yield strength S_y:

$$\sigma_1 \le S_y; \quad \sigma_2 \le S_y; \quad \sigma \le S_y \qquad \text{[Eq. 7.6]}$$

Example 7.2 Superposition of Axial Force and Torque

Given: A steel shaft of solid circular cross-section ($R = 1.0$ in.) is subjected to an axial force $P = 5000$ lb, and transmits a torque $T = 4000$ lb-in. (*Figure 7.3*).

Required: Determine the stresses acting on element B on the surface of the shaft.

Solution: *Step 1.* The area and polar moment of inertia of the cross-section are:

$$A = \pi R^2 = \pi(1.0)^2 = 3.142 \text{ in.}^2$$

$$J = \frac{\pi R^4}{2} = \frac{\pi(1.0)^4}{2} = 1.571 \text{ in.}^4$$

Step 2. The stresses caused by force P and torque T are considered separately (*Figures 7.3b* and *c*).

The normal stress is:

$$\sigma = \frac{P}{A} = \frac{5.0 \text{ kips}}{3.142 \text{ in.}^2}$$

$$Answer: \quad \sigma = 1.59 \text{ ksi}$$

The shear stress is:

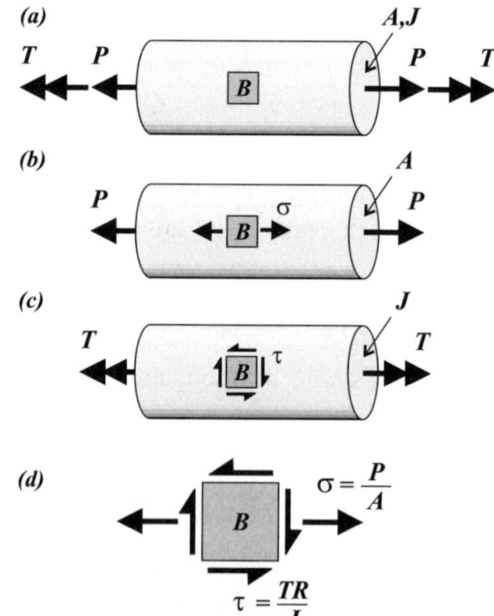

Figure 7.3. **(a)** A shaft under tensile load *P* and torque *T*. **(b)** The shaft as an axial member only. **(c)** The shaft as a torsion member only. **(d)** The stress state at point *B* due to the superposition of stresses.

$$\tau = \frac{TR}{J} = \frac{(4.0 \text{ kip-in.})(1.0 \text{ in.})}{1.571 \text{ in.}^4}$$

Answer: $\underline{\tau = 2.55 \text{ ksi}}$

The stresses acting on element B – the *state of stress* or *stress state* at point B – are shown in *Figure 7.3d*; the two stresses are *superimposed* on the element.

Note that these stresses are not added numerically because although they act at the same point, they are different kinds of stresses: a normal stress and a shear stress.

Examples 7.1 and *7.2* are trivial examples, showing how to superimpose:

1. like-stresses acting on the same face of an element, and
2. unlike stresses acting on an element.

The following examples are a survey of various types of combined loading.

> ***The Principle of Superposition is Valid when Systems Remain Linear***
>
> If a and b are valid inputs to a function $f(x)$, then for superposition to work, the function must be linear; i.e., if $f(x)$ is linear, then:
>
> $$f(a+b) = f(a) + f(b)$$
>
> This is demonstrated with the following two functions, one linear and one nonlinear:
>
> $$g(x) = 2x \quad \text{and} \quad h(x) = x^2$$
>
> Checking superposition for inputs a and b:
>
> $$g(x): \quad 2(a+b) = 2(a) + 2(b)$$
> $$h(x): \quad (a+b)^2 \neq (a)^2 + (b)^2$$
>
> Function $g(x)$ is linear and thus the Method of Superposition works. Function $h(x)$ is nonlinear, and superposition does not work.
>
> For Strength of Materials, the implication is that the stresses must remain in the linear-elastic range for superposition to be valid.

Example 7.3 Cantilever Beam Under Point Load and Its Own Weight

Given: A cantilever beam of length L is loaded by a uniformly distributed load w due to its own weight, and a tip load P. The moment of inertia is I and the modulus is E (*Figure 7.4*).

Required: Using superposition, determine (a) the moment everywhere along the beam $M(x)$, (b) the deflection everywhere along the beam $v(x)$, and (c) the tip deflection.

Solution: The beam is analyzed by considering each load separately.

Step 1. For the point load P, the moment and displacements along the cantilever beam are known from previous work (*Figure 7.4b*):

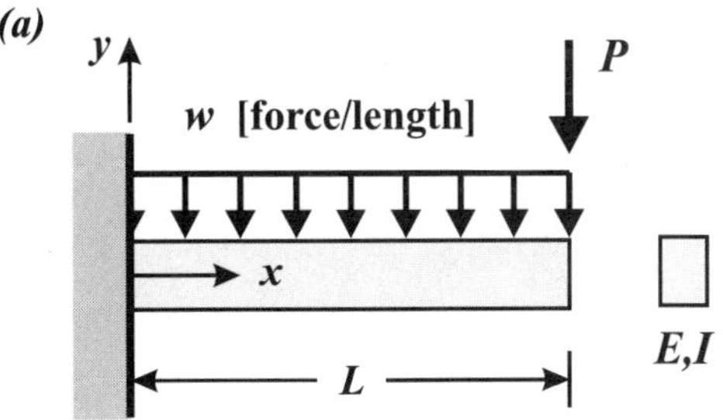

Figure 7.4. (a) Beam under uniformly distributed load w and tip load P.

$$M_P(x) = -P(L - x)$$

$$v_P(x) = -\frac{Px^2}{6EI}(3L - x)$$

$$v_{P,tip} = -\frac{PL^3}{3EI}$$

For a uniformly loaded cantilever (*Figure 7.4c*):

$$M_w(x) = -w\left(\frac{L^2}{2} - Lx + \frac{x^2}{2}\right)$$

$$v_w(x) = -\frac{wx^2}{24EI}(6L^2 - 4Lx + x^2)$$

$$v_{w,tip} = -\frac{wL^4}{8EI}$$

Step 2. If the system remains linear elastic, and the overall deflections are small, then the method of superposition is valid (*Figure 7.4d*). The moment along the beam is:

$$M(x) = M_P(x) + M_w(x)$$

Answer: $\underline{M(x)}$

$$= -P(L - x) - w\left(\frac{L^2}{2} - Lx + \frac{x^2}{2}\right)$$

The deflection is:

$$v(x) = v_P(x) + v_w(x)$$

Answer: $\underline{v(x)}$

$$= -\frac{x^2}{6EI}\left[P(3L - x) - \frac{w}{4}(6L^2 - 4Lx + x^2)\right]$$

Finally, the tip deflection is:

$$v_{tip} = v_{P,tip} + v_{w,tip}$$

Answer: $\underline{v_{max} = \frac{-L^3}{EI}\left(\frac{P}{3} + \frac{wL}{8}\right)}$

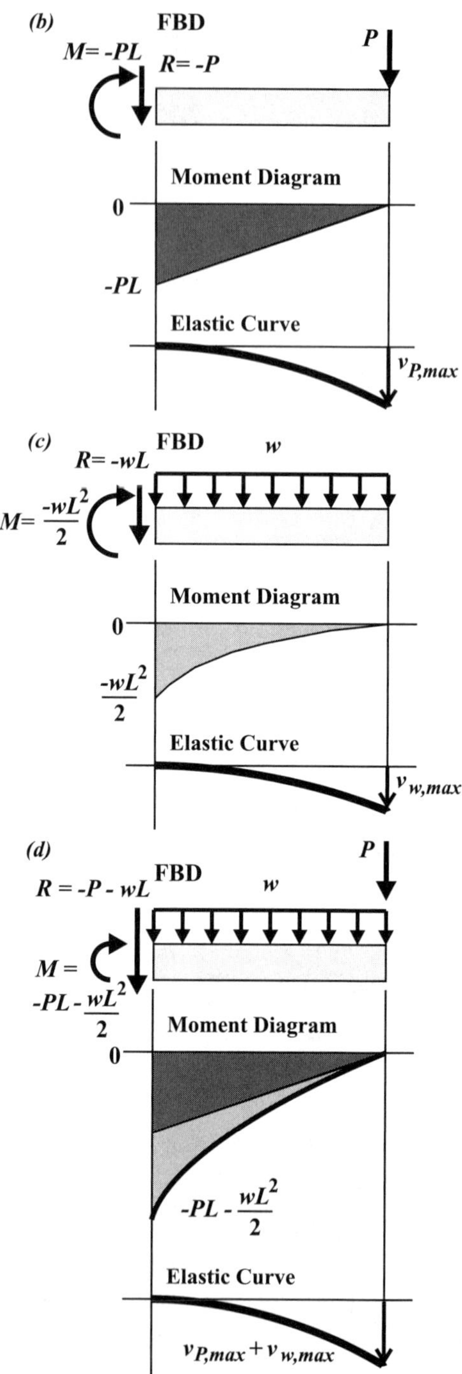

Figure 7.4. **(b)** Beam under tip load only. **(c)** Beam under UDL only. **(d)** Superposition of individual solutions.

Example 7.4 Eccentrically Loaded Axial Member

Background: Successful masonry (stone, concrete, etc.) structures are designed so that the masonry carries only compressive stresses. The tensile strength of such brittle materials is so small that any tensile stress is assumed to cause failure by cracking. **Columns** are axial members in compression. When the axial force is not applied through the centroid of the column's cross-section, both an axial force and a bending moment must be considered.

Given: A statue of weight P sits on a column of rectangular cross-section $b \times d$ (*Figure 7.5a, not to scale*). The statue is placed off-center, so the column is loaded by an off-axis load P applied on the $+y$-axis, distance e from the centroid. Distance e is the **eccentricity** of the load (*Figure 7.5b*). Neglect the weight of the column in the calculations.

Required: Determine (a) the stress on the column cross-section as a function of y and (b) the maximum magnitude of e to avoid tension in the column.

Solution: *Step 1*. Load P acts on the y-axis at $y = e$. This loading is statically equivalent to force P acting through the centroid and a moment acting about the centroidal z-axis (*Figure 7.5c*):

$$M_z = Pe$$

Step 2. The *axial stress* due to force P is constant throughout the system (*Figure 7.5d, left*):

$$\sigma_a = -\frac{P}{bd}$$

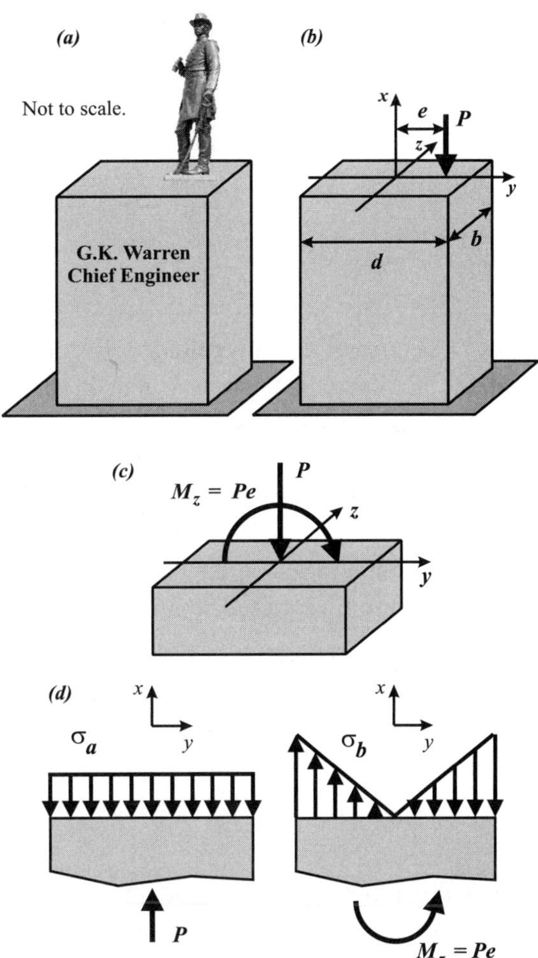

Figure 7.5. **(a)** A statue* on a masonry pedestal. **(b)** Statue weight P applied with eccentricity e. **(c)** The eccentric force is equivalent to force P applied through the centroid, and moment $M_z = Pe$ about the z-axis. **(d)** Stresses due to the axial force and bending moment, σ_a and σ_b, respectively.

* G.K. Warren, Chief Engineer of the Union Army of the Potomac, summer, 1863. His statue actually is on the rocks of Little Round Top at Gettysburg, Pennsylvania. Photo by Don Wiles, ©2006.

Step 3. The *bending stress* due to M_z varies with the *y*-coordinate (*Figure 7.5d, right*):

$$\sigma_b(y) = -\frac{M_z y}{I_z} = -\frac{(Pe)y}{bd^3/12} = -\frac{12Pey}{bd^3}$$

Step 4. Superimposing the stresses that act on the cross-section gives the *total normal stress* at any *y*-value:

$$\sigma(y) = \sigma_a + \sigma_b(y) = -\frac{P}{bd} - \frac{12Pey}{bd^3}$$

$$\text{Answer:} \quad \sigma(y) = -\frac{P}{bd}\left[1 + \frac{12e}{d^2}y\right]$$

Step 5. Determine the magnitude of *e* to avoid tension in the column. As drawn in *Figure 7.5d*, the bending stress is tensile for $y < 0$ and is maximum at $y = -d/2$. The total normal stress at $y = -d/2$ is:

$$\sigma(-d/2) = -\frac{P}{bd}\left[1 - \frac{6e}{d}\right]$$

For no tension to occur on the cross-section, the compressive stress due to the axial force must be greater in magnitude than the maximum tensile stress due to bending:

$$\frac{P}{bd} \geq \frac{12Pe(d/2)}{bd^3} \quad \text{or} \quad 1 - \frac{6e}{d} \geq 0$$

Thus, to avoid tension:

$$\text{Answer:} \quad e \leq \frac{d}{6}$$

Provided that $e \leq d/6$, regardless of the magnitude of compressive force *P*, the stress everywhere on the cross-section is compressive (*Figure 7.6e*).

For rectangles, provided that force *P* is applied on the *y*-axis in the middle third of the cross-section ($|y| \leq d/6$), then there can be no tensile stress on the cross-section. This is known as the **Middle Third Rule** (*Figure 7.6a*). Likewise, if the load is applied on the *z*-axis only, it must fall within the middle third ($|z| \leq b/6$), i.e., $e_z \leq b/6$.

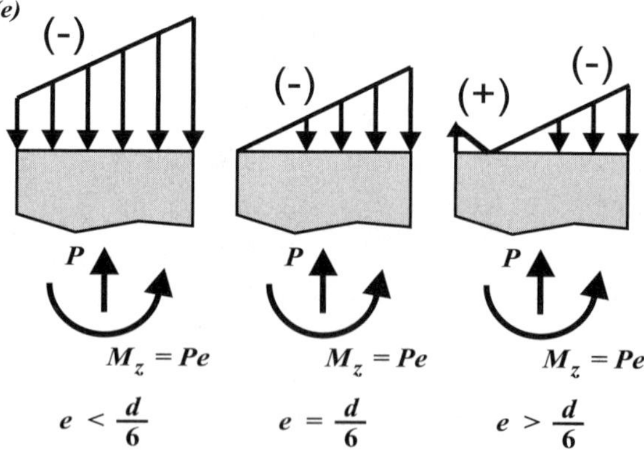

Figure 7.5. (e) Depending on the value of the eccentricity *e*, the cross-section may or may not be subjected to *tensile* stress. For *e* < *d*/6, no tensile stress is developed on the cross-section.

The Kern

Load P may be placed anywhere on a column's cross-section, causing moments about both the y- and z-axes. An axial stress and two bending stresses therefore contribute to the stress at any point.

To avoid developing tension in a a column of rectangular cross-section, the compressive axial force must be applied within the shaded diamond shown in *Figure 7.6b*. This area is known as the **kern**, which for a rectangle is a diamond with diagonals equal to one-third of each side of the rectangular cross-section.

Conversely, if a tensile axial force is applied through the kern, no compressive stresses will develop on the cross-section.

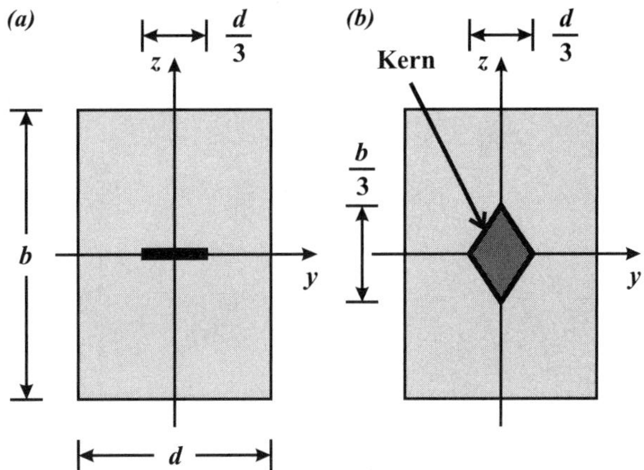

Figure 7.6. (a) If a compressive force is applied on the y-axis in the *middle third* of the cross-section, no tension will develop on the cross-section. **(b)** Provided a compressive force is applied within the *kern* – a diamond for a rectangular cross-section – no tensile stress will develop on the cross-section.

The kern for a solid circular cross-section is a circle of radius $0.25R$. Check this statement using the method demonstrated in *Example 7.4*.

Example 7.5 Pipe Tightened with a Wrench

Given: A pipe of length L is tightened with a pipe wrench (*Figure 7.7*). Force P is applied perpendicular to the wrench handle, distance d below the pipe axis, and the pipe has stopped turning. The pipe has inner radius R_i and outer radius R.

Required: Consider cross-section $ABCD$ at $x = 0$, where the pipe is fixed to the wall. (a) Determine the expression for the stresses on the surface of the pipe at the front (point A), top (B), back (C), and bottom (D) of the pipe, in terms of force P, and lengths d, L, R_i, and R. Neglect the weights of the pipe and wrench. (b) Draw the stress element at each point as viewed from the outside.

Solution: *Step 1.* Determine the forces acting at section $ABCD$. Equilibrium requires the following internal loads (reactions) at section $ABCD$ (the right side of the FBD in *Figure 7.7b*):

- a torque about the x-axis: $T = Pd$;
- a shear force along the y-axis: $V = P$; and
- a bending moment about the z-axis: $M = PL$.

The loads have been drawn in the directions in which they physically act on the pipe.

Step 2. The geometric properties of the cross-section are:

$$A = \pi(R^2 - R_i^2)$$

$$J = \frac{\pi}{2}(R^4 - R_i^4)$$

$$I = \frac{\pi}{4}(R^4 - R_i^4)$$

For simplicity, these geometric terms are represented by A, J, and I below.

Step 3. The maximum stresses caused by the loads, and the critical points at which they act, are given in *Table 7.2*.

Consider the effects of the various loads on a single point. At point A, the stresses are due to the *moment* and the *torque*, as drawn on the stress element in *Figure 7.7c*. Point A is viewed from the front of the pipe and the stress arrows are drawn in their positive senses with respect to the x–y–z coordinate system (i.e., positive-face, positive-direction). A

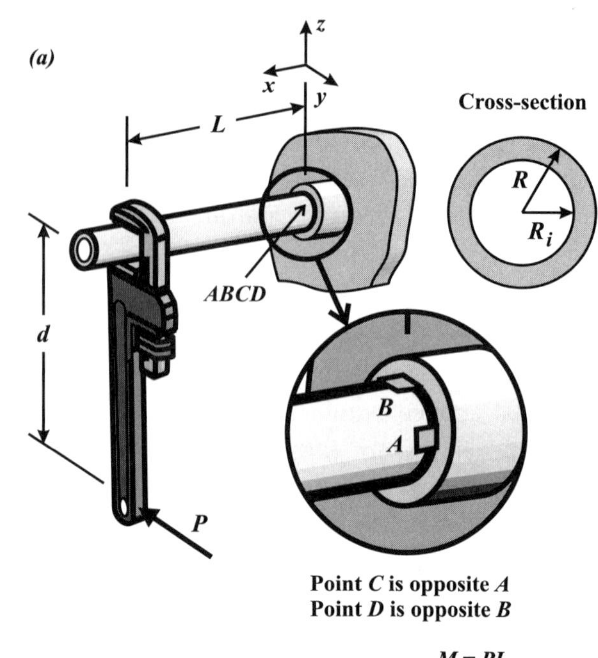

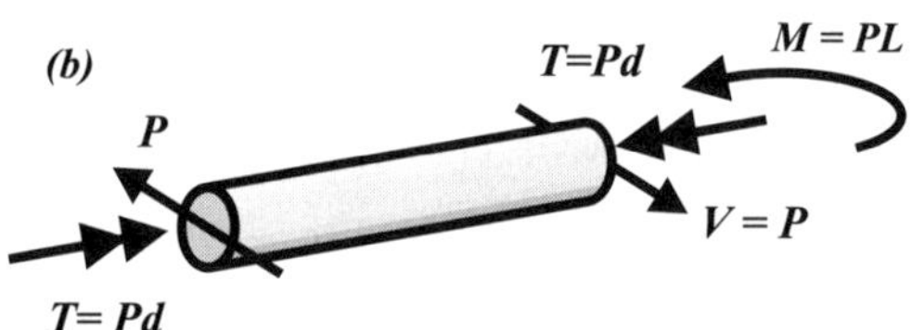

Figure 7.7. (a) Pipe wrench tightening a pipe. **(b)** FBD of pipe.

Table 7.2. Maximum Stresses acting at surface of pipe at section *ABCD*.

Load at *ABCD*	Magnitude of Maximum Stress at Surface of Pipe	Stress acts at Points
Shear force, $V = P$	$\tau_V = \dfrac{4P}{3A}\left[\dfrac{R^2 + RR_i + R_i^2}{R^2 + R_i^2}\right] = C_s\dfrac{P}{A}$	B and D (at axis of bending)
Torque, $T = Pd$	$\tau_T = \dfrac{(Pd)R}{J}$	A, B, C and D (at all surface points)
Moment about z-axis, $M = PL$	$\sigma_x = \dfrac{(PL)R}{I}$	A and C (at top/bottom of beam)

Note: The shear stress τ_V is the maximum shear stress that acts on a hollow circular cross-section due a shear force P.

negative sign indicates that the stress physically acts opposite drawn.

Points *A*, *B*, *C*, and *D*, as viewed from the outside, are shown in *Figure 7.7d*. Take care to note the direction of the *x*-axis in each figure. The value of C_s is, from *Table 7.2*:

$$C_s = \frac{4}{3}\left[\frac{R^2 + RR_i + R_i^2}{R^2 + R_i^2}\right]$$

Note also that the shear stresses at point *B* act opposite each other; those at point *D* act in the same direction.

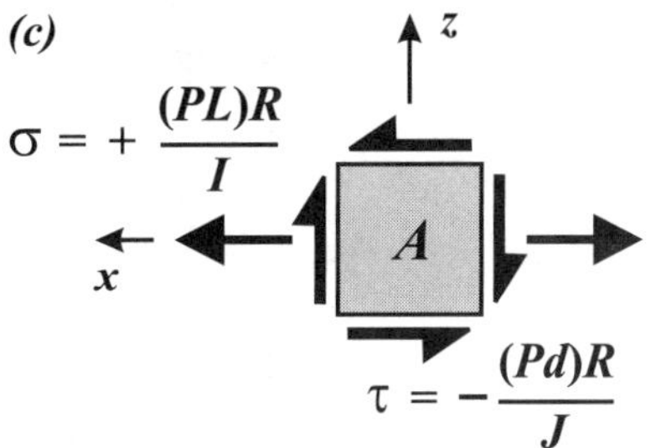

Figure 7.7. (c) Stresses acting on point *A*, viewed from the front. The bending stress is tensile and thus positive; the shear stress acts opposite drawn, and is thus negative.

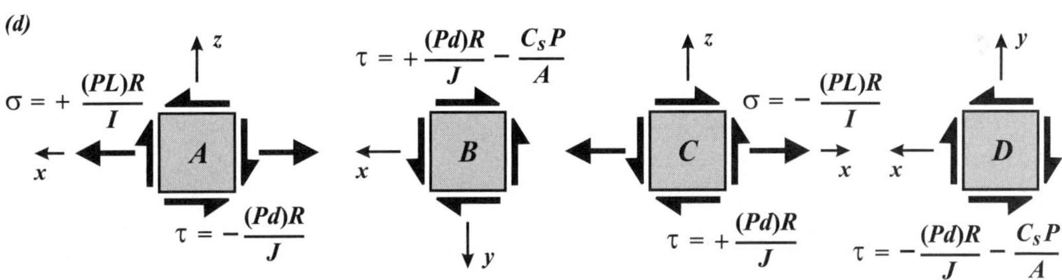

Figure 7.7. (d) Points *A* (viewed from *front*), *B* (from *top*), *C* (from *back*), and *D* (from *bottom*). Stress arrows are drawn in their positive senses (note the axes); a negative value indicates the stress acts opposite drawn.

Example 7.6 Highway Sign under Wind Load and its own Weight

Given: The highway sign introduced in *Chapter 2* is subjected to a wind load (*Figure 7.8*). The wind force is taken as the product of the wind pressure and the sign area $F_W = pA$. The weight of the sign is W_S and the weight of the mast is W_M.

Required: Consider the cross-section of the mast cut at plane *ABCD*, distance *h* below the top of the mast (assume the height of the center of the sign is at the top of the mast).

(a) Determine the *stress states* in the *x–y–z* coordinate system acting at plane *ABCD*, at points *A*, *B*, *C*, and *D* (*Figures 7.8b, c*). Point *A* is on the *windward* side (facing the wind); point *C* is on the *leeward* side (back side, away from the wind); and points *B* and *D* are on the *left* and *right* sides, respectively. (b) Draw

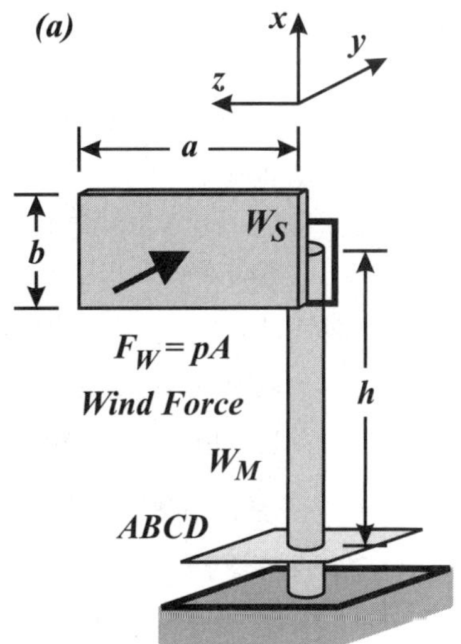

Figure 7.8. (a) A highway sign under wind load.

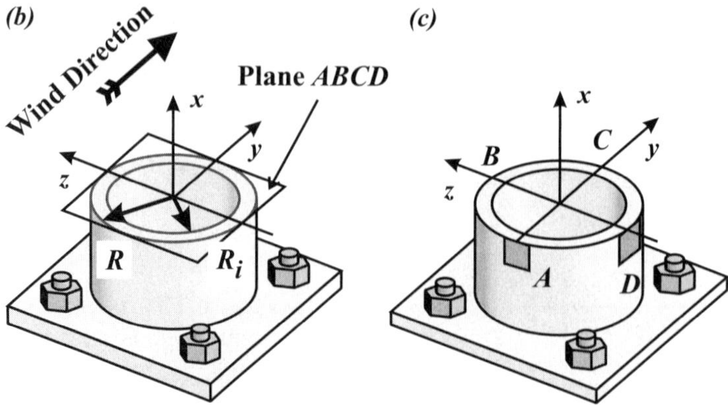

Figure 7.8. (d) Geometry of cross-section at *ABCD*.
(e) Elements *A*, *B*, *C*, and *D*.

the stress states on 2D stress elements that are viewed from the *outside* of the mast.

Solution: *Step 1.* Determine the equivalent loads being applied to the mast, and the internal loads that they cause at cross-section *ABCD*.

The wind force and weight of the sign act with respect to the mast as shown in *Figure 7.8d*. A FBD of the mast above *ABCD* is shown in *Figure 7.8e*.

The internal loads at *ABCD*, viewing the cross-section from above, are (*Figure 7.8f*):

- due to *wind force* F_W:

  - a clockwise torque about the +*x*-axis of the mast, T_x;

  - a shear force in the +*y*-direction, V_y; and

  - a bending moment about the +*z*-axis, M_z.

- due to the *weight of the sign* W_S:

  - a compressive force in the mast, W_S; and

  - a bending moment about the –*y*-axis, M_y.

- due to the *weight of the mast* itself W_M:

  - a varying axial force in the mast, $W_M(x) = W_M(-h)$ at *ABCD*. For simplicity, take the weight of the mast above $x = -h$ as W_M.

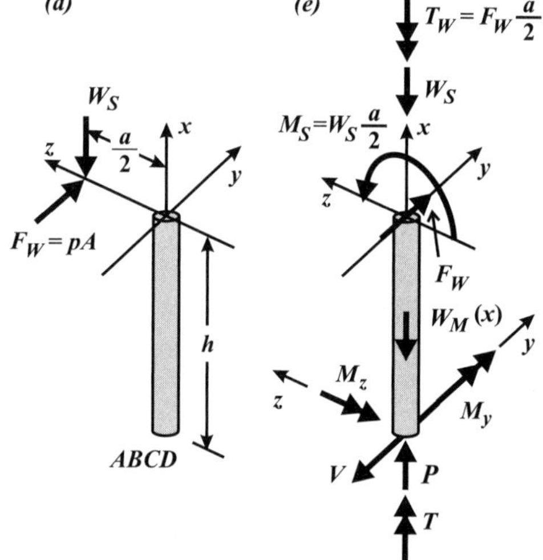

Figure 7.8. (d) The force of the wind pressure and the weight of the sign act at the centroid of the sign. **(e)** A FBD of the mast cut at section *ABCD*, including the weight of the mast above *ABCD*. Reactions at *ABCD* drawn in the direction in which they act.

The loads acting on cross-section *ABCD* are summarized in *Figure 7.8f* and *Table 7.3*.

Step 2. The geometric terms needed to find the stresses are (*Figure 7.8d*):

$$A = \pi(R^2 - R_i^2)$$

$$J = \frac{\pi}{2}(R^4 - R_i^4)$$

$$I_y = I_z = \frac{\pi}{4}(R^4 - R_i^4)$$

where R is the outer radius and R_i is the inner radius. The moments of inertia about the y- and z-axes are I_y and I_z, respectively.

Step 3. With the loads and geometry known, the stresses are summarized in *Table 7.3*.

Step 4. Consider how each internal load affects point A, at the *windward* side (front) of the mast. The y–z coordinates of point A are $y = -R$ and

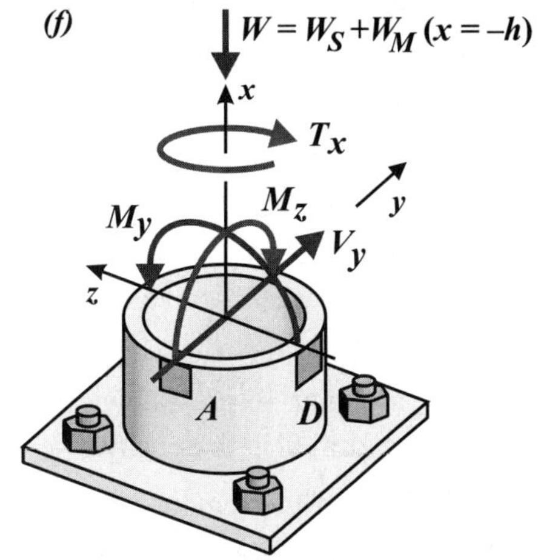

Figure 7.8. (f) Internal loads acting on the cross-section at *ABCD*.

Table 7.3. Loads at cross-section *ABCD* and the Stresses that they cause.

Load at *ABCD*	Magnitude of Load	Caused by	Action	Stress at Surface of Mast	Acts at Points
Torque about x-axis	$T_x = F_W(a/2)$	Wind	Twists mast clockwise about $+x$-axis	$\tau_T = \dfrac{T_x R}{J}$	A, B, C and D
Shear force along y-axis	$V_y = F_W$	Wind	Shear in direction of wind $(+y)$	$\tau_V = \dfrac{4V_y}{3A}\left[\dfrac{R^2 + RR_i + R_i^2}{R^2 + R_i^2}\right]$	B and D $(y = 0)$
Moment about z-axis	$M_z = F_W h$	Wind	Bends mast in the direction of wind, about $+z$-axis	$\sigma_x = -\dfrac{M_z y}{I_z}$	A and C $(y = \pm R)$
Moment about y-axis	$M_y = W_S(a/2)$	Weight of sign	Bends mast towards left, about $-y$-axis	$\sigma_x = -\dfrac{M_y z}{I_y}$	B and D $(z = \pm R)$
Axial force along x-axis	$W(x=-h)$ $= [W_S + W_M]$	Weight of sign + weight of mast above *ABCD*	Compresses mast along x-axis	$\sigma_x = -\dfrac{W}{A}$	A, B, C and D

Note: The shear stress τ_V is the maximum shear stress that acts on a hollow circular cross-section due to a shear force V.

$z = 0$. Stress element A is in an x–z plane and is subjected to three stresses:

1. a shear stress due to torque;
2. a normal (tensile) stress due to bending about the z-axis; and
3. a normal (compressive) stress due to the weight of the system above section $ABCD$.

The stress equations are given in *Table 7.3*, and the stresses at point A are shown in *Figure 7.8g*. The stresses are drawn in the positive sense.

Step 5. Each of the remaining elements, B, C, and D, are considered in turn. Point C has the same stresses as point A, but the bending stress is now compressive ($\sigma = -M_z R/I_z$).

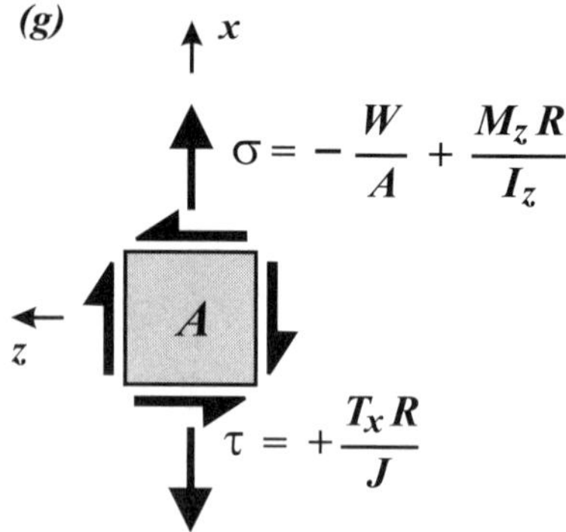

Figure 7.8. (g) The stress state at point A.

Points B and D also carry the compressive stress due to the weight, and the shear stress due to torque. The bending stress due to the wind force is zero at both B and D because B and D lie on the axis about which that moment acts. However, points B and D are subject to the bending stress due to the weight of the sign causing bending about the y-axis ($\sigma_{max} = \pm M_y R/I_y$), compressive at point B and tensile at point D. Also, a shear stress due to the shear force of the wind acts at both points B and D.

Figure 7.8h summarizes the stresses acting at the various points A, B, C, and D. All of the elements are viewed from the outside of the support mast. In *Figure 7.8h*, the value of constant C_s is:

$$C_s = \frac{4}{3}\left[\frac{R^2 + RR_i + R_i^2}{R^2 + R_i^2}\right]$$

C_s is the stress concentration factor with respect to the average shear stress caused by a shear force acting on a beam of hollow circular cross-section.

Shear stress notes. Shear stress τ_T due to torque T acts on every surface element, and constantly changes its direction with respect to the x–y–z coordinate system. Shear stress τ_V due to shear force V_y varies with y, is zero at points A and C, and maximum at B and D.

At point B, the two shear stresses act in the same direction, reinforcing each other. At point D, the shear stresses act in opposite directions.

Critical points. Points B and C are likely the most critical points in the mast. Point B is where two normal stresses act in the same direction, as well as the two shear stresses. Point C is where two normal stresses act in the same direction, along with one shear. To ensure that the mast is of sufficient strength, points B and C (and other critical points) should be checked using the methods discussed in the following chapters.

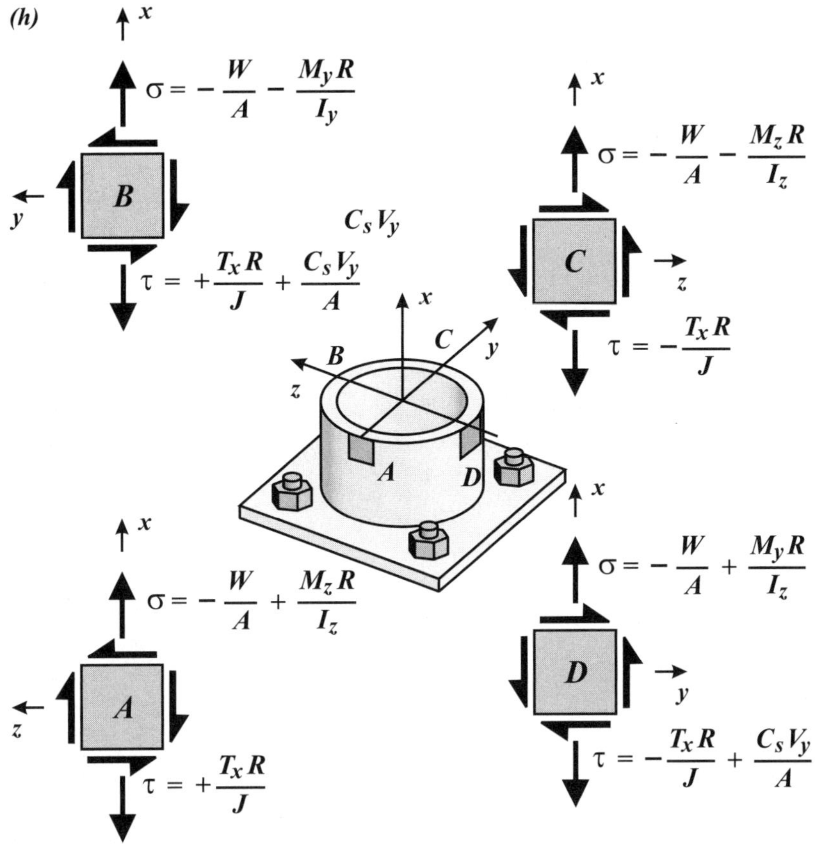

Figure 7.8. (h) Stresses at points *A, B, C,* and *D, as viewed from the outside of the mast.* The stresses are drawn in their positive senses. A negative value for a stress indicates that it acts opposite drawn.

Point *A* : Normal stress: *Compression* (*negative*) due to weight;
Tension (*positive*) due to bending caused by wind;
Shear stress: *Positive* due to torque caused by wind.

Point *B* : Normal stress: *Compression* due to weight;
Compression due to bending caused by sign;
Shear stress: *Positive* due to torque caused by wind;
Positive due to shear force caused by wind.
Shear stresses act in same direction.

Point *C* : Normal stress: *Compression* due to weight;
Compression due to bending caused by wind;
Shear stress: *Negative* due to torque caused by wind.

Point *D* : Normal stress: *Compression* due to weight;
Tension due to bending caused by sign;
Shear stress: *Negative* due to torque caused by wind;
Positive due to shear force caused by wind.
Shear stresses act in opposite directions.

Example 7.7 Bending about Two Axes

Given: A column has a rectangular cross-section $b = 8.0$ in. wide by $d = 12.0$ in. deep (*Figure 7.9a*). Compressive force $P = 96$ kips is applied through the point on the cross-section $(y, z) = (3.0, -2.0)$ in., so that the moments are positive about the z- and y-axes.

Required: (a) Determine the stress at each corner of the cross-section, A, B, C, and D and (b) draw the stress distribution on the cross-section.

Solution: *Step 1.* The equivalent load on the cross-section is axial force P acting through the centroid, and moments M_z and M_y acting about the z- and y-axis, respectively (*Figure 7.9b*):

$$P = 96 \text{ kips (compression)}$$

$$M_z = 96(3) = 288 \text{ kip-in.}$$

$$M_y = 96(2) = 192 \text{ kip-in.}$$

The relevant cross-sectional properties are the area A and the section moduli about the z- and y-axes, Z_z and Z_y ($Z = I/c$, where $c = d/2$ for a rectangle of height d):

$$A = 96 \text{ in.}^2$$

$$Z_z = \frac{I_z}{d/2} = \frac{bd^2}{6} = \frac{(8)(12)^2}{6} = 192 \text{ in.}^3$$

$$Z_y = \frac{I_y}{b/2} = \frac{db^2}{6} = \frac{(12)(8)^2}{6} = 128 \text{ in.}^3$$

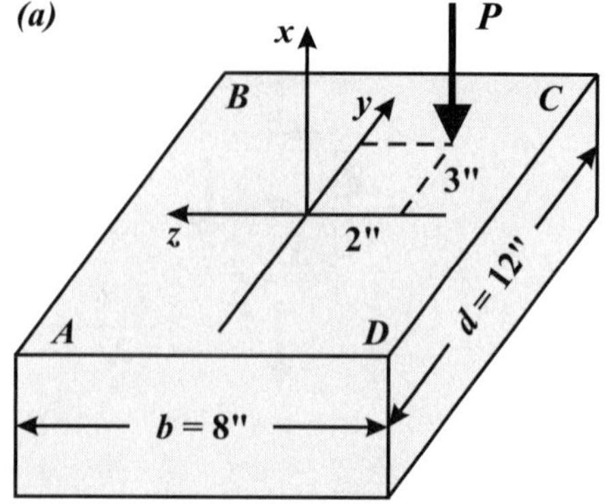

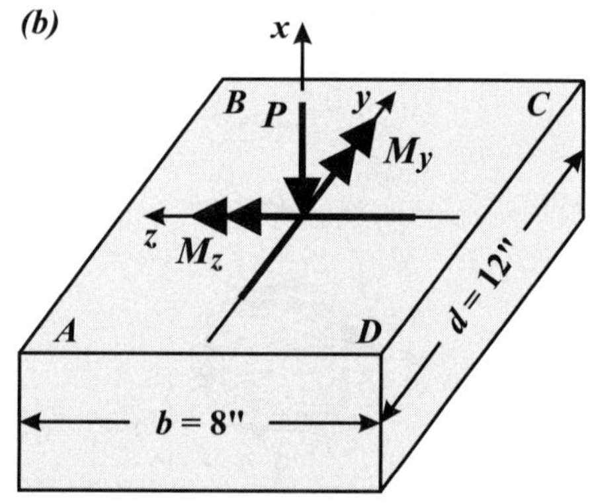

Figure 7.9. (a) Compressive force P applied at $(y, z) = (3.0, -2.0)$ in. **(b)** Equivalent load: force P applied through the centroid, and moments $M_z = 3P$ and $M_y = 2P$.

Step 2. Individual stresses. The axial compressive stress everywhere on the cross-section is:

$$\sigma_a = \frac{-96 \text{ kips}}{96 \text{ in.}^2} = -1.0 \text{ ksi}$$

The maximum bending stress about each axis is:

$$\sigma_{b,z} = \pm\frac{M_z}{Z_z} = \pm\frac{288}{192} = \pm 1.5 \text{ ksi} \quad \text{(tension at points } A, D; \text{ compression at } B, C)$$

$$\sigma_{b,y} = \pm\frac{M_y}{Z_y} = \pm\frac{192}{128} = \pm 1.5 \text{ ksi} \quad \text{(tension at } A, B; \text{ compression at } C, D)$$

Step 3. Superposition. At any corner point, the total normal stress is the sum of the axial and bending stresses at that point:

Point A: $\sigma_A = \sigma_a + \sigma_{b,\,z,\,A} + \sigma_{b,\,y,\,A} = (-1) + (1.5) + (1.5)$

Answer: $\sigma_A = +2.0$ ksi

Point B: $\sigma_B = \sigma_a + \sigma_{b,\,z,\,B} + \sigma_{b,\,y,\,B} = (-1) + (-1.5) + (1.5)$

Answer: $\sigma_B = -1.0$ ksi

Point C: $\sigma_C = \sigma_a + \sigma_{b,\,z,\,C} + \sigma_{b,\,y,\,C} = (-1) + (-1.5) + (-1.5)$

Answer: $\sigma_C = -4.0$ ksi

Point D: $\sigma_D = \sigma_a + \sigma_{b,\,z,\,D} + \sigma_{b,\,y,\,D} = (-1) + (1.5) + (-1.5)$

Answer: $\sigma_D = -1.0$ ksi

The corner stresses are plotted in *Figure 7.9c*.

Step 4. The stress at any point (y, z) on the cross-section is:

$$\sigma(y, z) = \frac{-P}{A} - \frac{M_z y}{I_z} + \frac{M_y z}{I_y}$$

where:

$$A = 96 \text{ in.}^2$$

$$I_z = \frac{bd^3}{12} = \frac{(8)(12)^3}{12} = 1152 \text{ in.}^4 \quad \text{and} \quad I_y = \frac{db^3}{12} = \frac{(12)(8)^3}{12} = 512 \text{ in.}^4$$

Note that the third term in the general stress equation is positive, since for $M_y > 0$, the resulting bending stress is tensile (positive) for $z > 0$.

Step 5. Plot (Figure 7.9c). The stress $\sigma(y, z)$ at any point is the superposition of a constant and two linear functions. Since the corner stresses are known, the plot of $\sigma(y, z)$ is made by constructing lines between the corner points on each side of the rectangular cross-section, taking care to note that the stress should go to zero between stresses of opposite sign (e.g., points F and H). Plotting along each side of the rectangle fixes the value of one of the linear functions.

The location of zero stress on an edge is found using similar triangles. The line that joins the edge zero-stress points, FH, is the locus of points on the cross-section where the stresses sum to zero.

Since tension exists on the cross-section, force P has been applied outside the kern.

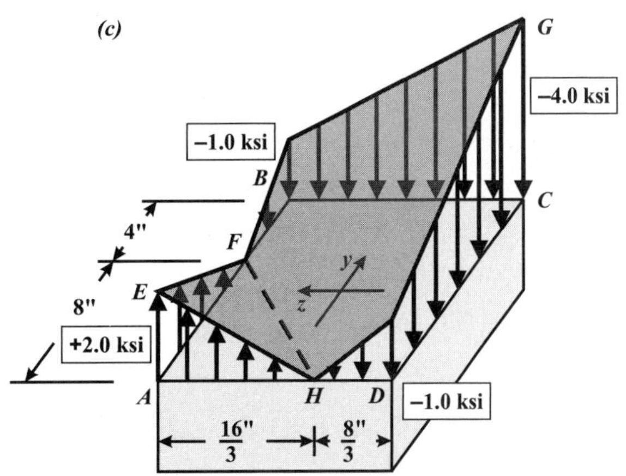

Figure 7.9. (c) Stress distribution on cross-section. Line *FH*, on the cross-section, is the locus of points having zero stress, and is the intersection of the planes labeled *FEH* and *FGH*.

Chapter 8　　　Transformation of Stress and Strain

8.0 Introduction

Consider a plate with a cross-sectional area A subjected to axial load P (*Figure 8.1a*). The plate is made of two pieces joined together by a diagonal weld. The direction normal to the weld, x', is at angle θ counterclockwise from the x-axis (*Figure 8.1b*). The stress in the applied load direction is $\sigma_x = P/A$ (*stress element B, Figure 8.1c*).

To ensure that the weld is strong enough, it is necessary to determine the stresses that act perpendicular and parallel to the weld line: normal stress $\sigma_{x'}$ and shear stress $\tau_{x'y'}$ (element C, *Figure 8.1d*). These stresses are caused by the normal and shear components, N and V, of applied force P (*Figure 8.1b*).

The normal and shear forces, N and V, and the area of the weld, A_w, are:

$$N = P\cos\theta; \quad V = -P\sin(\theta); \quad A_w = \frac{A}{\cos\theta}$$

$$[\text{Eq. 8.1}]$$

The negative sign for shear force V indicates that it acts in the opposite direction drawn. For this load condition, the normal and average shear stresses that act on the weld depend on angle θ as follows:

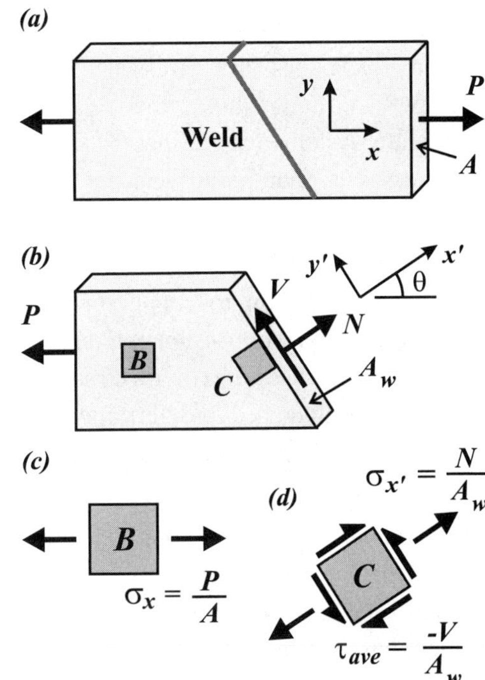

Figure 8.1. **(a)** A plate under axial load P; a weld holds the two parts of the plate together. **(b)** The left-side of the plate, with exposed weld area A_w; the forces on A_w are drawn in the positive directions. **(c)** Stress Element B oriented with the x–y axes. **(d)** Stress Element C oriented with the x'–y' axes, normal-parallel to the weld line.

$$\sigma_{x'} = \frac{N}{A_w} = \frac{P\cos^2\theta}{A} = \sigma_x\cos^2\theta$$

$$[\text{Eq. 8.2}]$$

$$\tau_{x'y'} = \frac{V}{A_w} = \frac{-P\sin\theta\cos\theta}{A} = -\sigma_x\sin\theta\cos\theta$$

F.A. Leckie, D.J. Dal Bello, *Strength and Stiffness of Engineering Systems*, Mechanical Engineering Series, **243**
DOI 10.1007/978-0-387-49474-6_8, © Springer Science+Business Media, LLC 2009

These new stresses – the *transformed stresses* – are compared against the allowable normal and shear stresses of the weld material (S_A, τ_A) to ensure that the system can support the applied loading.

The weld problem is one example of the need to calculate stresses in directions other than the directions of the applied loads. As the stress element – and thus the coordinate system that defines the stresses at that point – is rotated, the stresses are transformed (*Figure 8.1d*).

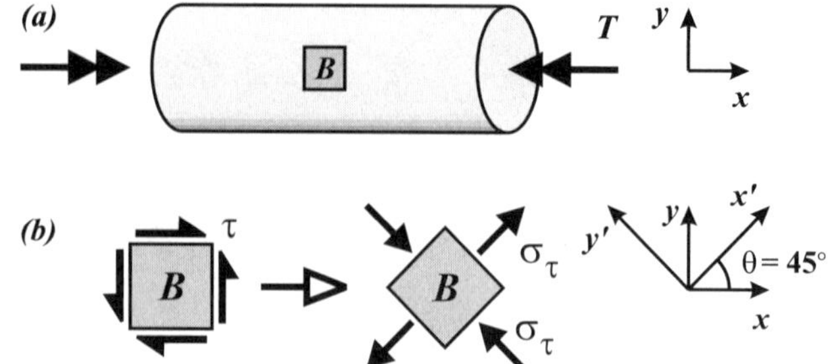

Figure 8.2. **(a)** A shaft under torque *T*. **(b)** The pure shear stress τ due to a torque-only load is equivalent to a biaxial state of stress ($\sigma_\tau = \tau$) on the element rotated by 45°.

A second reason to rotate the stress element (coordinate system) is to find the *maximum and minimum normal stresses* and the *maximum shear stress* that occur at a material point. These maximum stresses are generally oriented at angles different from the coordinate system used to determine the original stresses.

For example, consider a solid circular shaft under applied torque *T* (*Figure 8.2*). Due to the torque, shear stress τ acts on surface element *B*, causing it to elongate along the +45° diagonal and shorten along the –45° diagonal (*Figure 8.2b*). The *pure shear stress* state in the original orientation of the element is equivalent to a *biaxial* state of stress on the element rotated by 45°. For this particular case, both of the normal stresses on the rotated element have a magnitude of $\sigma_\tau = \tau$, one stress being tensile (here, oriented at +45°) and the other being compressive (oriented at –45°).

If the torsion member is made of a brittle material (e.g., a ceramic, chalk, etc.), the *maximum* (most tensile) *normal stress* σ_{max} is compared to the ultimate tensile strength S_u of the material to ensure that it does not fail. If the torsion member is made of a ductile material (e.g., steel, aluminum, etc.), one method of assessing if the material yields is to compare the maximum shear stress τ_{max} to the shear yield strength τ_y. Failure theories are introduced in *Chapter 9*.

The *stress transformation equations*, and the equations to determine the maximum and minimum normal stresses and the maximum shear stress, are developed below. While the transformed stresses have different values than the original stresses, the transformed stresses represent the same **state of stress** as the original stresses. The stresses that act on a material point are simply described in a rotated (transformed) coordinate system.

8.1 Stress Transformation (Plane Stress)

Consider a 3D stress element subjected to a general state of *plane stress*: σ_x, σ_y, and τ_{xy} (*Figure 8.3a*); there are no *out-of-plane* (*z*-direction) stresses. Recall the convention for the positive sense of stress: a positive stress physically acts on a positive face in a positive direction, or on a negative face in a negative direction; otherwise the stress is negative. Being consistent with the sense of

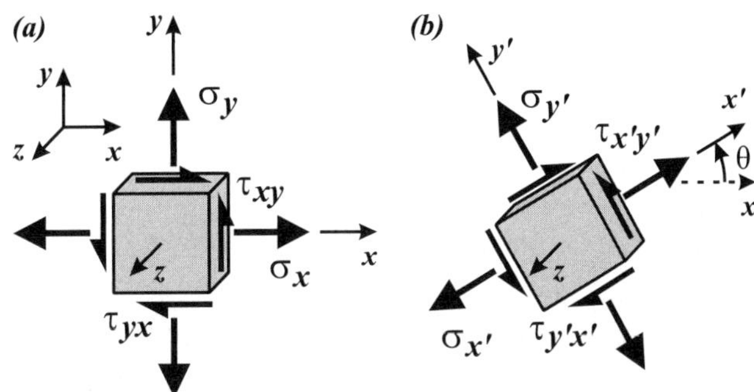

Figure 8.3. **(a)** A 3D element under a state of *plane stress*, oriented with the *x–y* axes. **(b)** The element rotated by angle θ and oriented with the *x′–y′* axes.

stress is key when applying the *stress transformation equations*. Also, recall that complementary shear stresses are equal, i.e., $\tau_{xy} = \tau_{yx}$.

The goal is to find the **in-plane transformed stresses** that act on the (2D) element when it is rotated in the *x–y* plane by angle θ (positive counterclockwise about the *z*-axis). The stresses on the rotated element are $\sigma_{x'}$, $\sigma_{y'}$ and $\tau_{x'y'}$ (*Figure 8.3b*).

The *stress transformation equations* are:

$$\sigma_{x'}(\theta) = \frac{\sigma_x + \sigma_y}{2} + \frac{\sigma_x - \sigma_y}{2}\cos 2\theta + \tau_{xy} \sin 2\theta \qquad \text{[Eq. 8.3]}$$

$$\sigma_{y'}(\theta) = \frac{\sigma_x + \sigma_y}{2} - \frac{\sigma_x - \sigma_y}{2}\cos 2\theta - \tau_{xy} \sin 2\theta \qquad \text{[Eq. 8.4]}$$

$$\tau_{x'y'}(\theta) = -\frac{\sigma_x - \sigma_y}{2}\sin 2\theta + \tau_{xy} \cos 2\theta \qquad \text{[Eq. 8.5]}$$

The transformation equations are derived using equilibrium, as presented below.

The following derivation of the *stress transformation equations* is only valid for *plane stress* conditions, and one additional case. An *out-of-plane* normal stress σ_z may exist if no shear stresses act *out-of-plane*, i.e., $\tau_{xz} = \tau_{yz} = 0$. When this is the case, the transformation equations in the *x–y* plane may still be used, with σ_z remaining constant with rotation θ about the *z*-axis, and the out-of-plane shear stresses remaining zero.

Stresses on New x′-Face

The new *x′*-axis is oriented at angle θ counterclockwise from the original *x*-axis. The *x′*-axis is perpendicular to the positive *x′*-face. The stresses acting on the *x′*-face are $\sigma_{x'}$ and $\tau_{x'y'}$. To derive the new stresses, each of the original stresses σ_x, σ_y, and τ_{xy}, is considered in turn. The results from each analysis are then *superimposed*.

Stresses on x'-Face Due to Stress σ_x

A stress element aligned with the x–y–z axes and subjected to only normal stress σ_x is shown in 2D in *Figure 8.4a* (in the x–y plane). A cut is taken through the element, creating a triangular wedge and exposing diagonal surface having area dA (*Figure 8.4b*).

The new x'-axis is normal to dA; the new y'-axis is parallel to dA. The new x'-axis is oriented at angle θ (positive counterclockwise) from the original x-axis. Normal and shear stresses $\sigma_{x'}$ and $\tau_{x'y'}$ act on dA. From trigonometry, the left (negative) x-face of the wedge has area $dA \cos \theta$, and the bottom (negative) y-face has area $dA \sin \theta$.

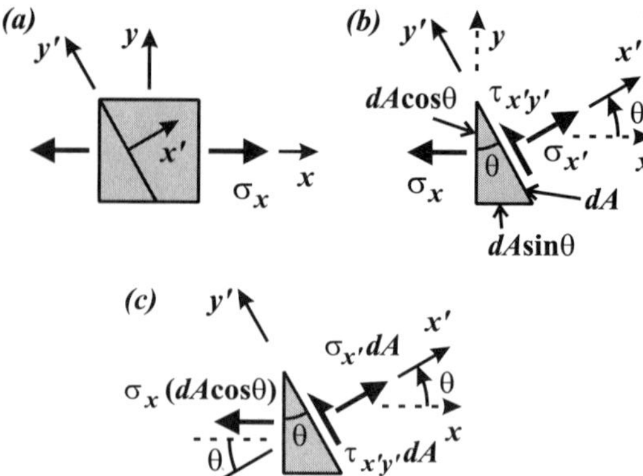

Figure 8.4. (a) A 2D stress element under normal stress σ_x only. (b) Wedge showing stresses acting on the x-, y-, and x'-faces. (c) Forces on each face.

Equilibrium of forces in the x'-direction, normal to dA (*Figure 8.4c*), gives:

$$\sum F_{x'} = 0 = \sigma_{x'}\, dA - [\sigma_x(dA \cos \theta)]\cos\theta \qquad \text{[Eq. 8.6]}$$

Equilibrium in the y'-direction, tangent to dA, gives:

$$\sum F_{y'} = 0 = \tau_{x'y'}\, dA + [\sigma_x(dA \cos \theta)]\sin\theta \qquad \text{[Eq. 8.7]}$$

Hence, due to σ_x only, the new stresses $\sigma_{x'}$ and $\tau_{x'y'}$ acting on the x'-face are:

$$\sigma_{x'} = \sigma_x\cos^2\theta$$
$$\tau_{x'y'} = -\sigma_x \cos \theta \sin \theta \qquad \text{[Eq. 8.8]}$$

As the element, loaded by σ_x only, is rotated by θ, the maximum magnitude of the shear stress occurs when $\theta = 45°$. The normal and shear stresses at $\theta = 45°$ are:

$$\sigma_{x'}(45°) = \frac{\sigma_x}{2}; \quad \tau_{max} = \tau_{x'y'}(45°) = \frac{-\sigma_x}{2} \qquad \text{[Eq. 8.9]}$$

When σ_x is positive, then the transformed shear stress on the x'-face for $\theta = 45°$ is negative; i.e., $\tau_{x'y'}$ physically acts on the positive x'-face in the negative y'-direction. At an angle of $\theta = 135°$ (with x'-face normal to the $\theta = 135°$ direction), the shear stress is positive; i.e., it acts on the positive x'-face in the positive y'-direction.

Stresses on x'-Face Due to Stress σ_y

Consider the stresses on the x'-face due only to normal stress σ_y (*Figure 8.5*). Equilibrium of forces acting on the wedge in the normal (x'-) and tangential (y'-) directions gives:

$$\sum F_{x'} = 0 = \sigma_{x'}\, dA - [\sigma_y(dA \sin\theta)]\sin\theta$$

$$\sum F_{y'} = 0 = \tau_{x'y'}\, dA - [\sigma_y(dA \sin\theta)]\cos\theta$$

$$[\text{Eq. 8.10}]$$

Due to σ_y, the stresses acting on the x'-face are:

$$\sigma_{x'} = \sigma_y \sin^2\theta$$
$$\tau_{x'y'} = \sigma_y \sin\theta \cos\theta$$

$$[\text{Eq. 8.11}]$$

As the element, loaded by σ_y only, is rotated by θ, the maximum magnitude of the shear stress occurs when $\theta = 45°$. The normal and shear stresses at $45°$ are:

$$\sigma_{x'}(45°) = \frac{\sigma_y}{2}$$

$$[\text{Eq. 8.12}]$$

$$\tau_{max} = \tau_{x'y'}(45°) = \frac{\sigma_y}{2}$$

Stresses on x'-Face Due to Shear Stress τ_{xy}

Consider the stresses on the x'-face due only to shear stress τ_{xy} (*Figure 8.6*). Equilibrium of forces acting on the wedge in the normal (x'-) and tangential (y'-) directions gives:

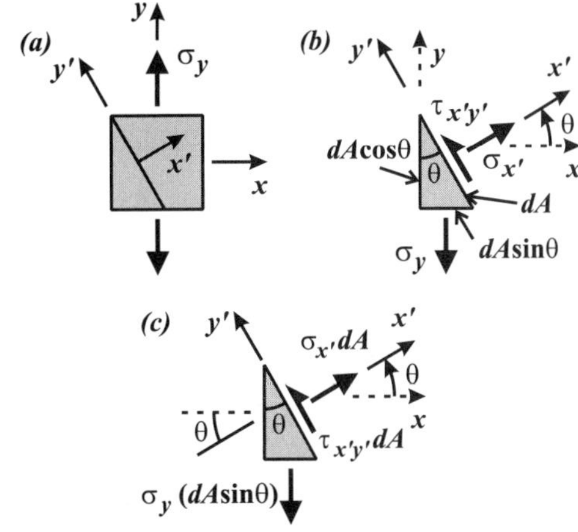

Figure 8.5. (a) A stress element under normal stress σ_y only. **(b)** Wedge with stresses on the x-, y-, and x'-faces. **(c)** Forces on each face.

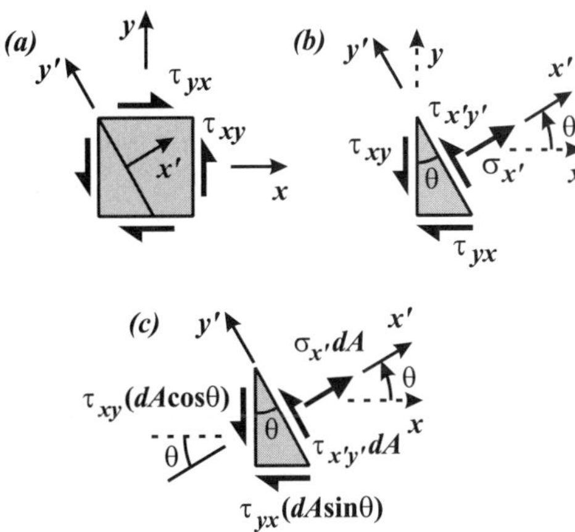

Figure 8.6. (a) A stress element under shear stress τ_{xy} only. **(b)** Wedge with stresses on the x-, y-, and x'-faces. **(c)** Forces on each face.

$$\sum F_{x'} = 0 = \sigma_{x'}\, dA - [\tau_{xy}(dA \cos\theta)]\sin\theta - [\tau_{xy}(dA \sin\theta)]\cos\theta$$

$$[\text{Eq. 8.13}]$$

$$\sum F_{y'} = 0 = \tau_{x'y'}\, dA - [\tau_{xy}(dA \cos\theta)]\cos\theta + [\tau_{xy}(dA \sin\theta)]\sin\theta$$

Due to τ_{xy}, the stresses acting on the x'-face are:

$$\sigma_{x'} = 2\tau_{xy}\sin\theta\,\cos\theta$$

$$\tau_{x'y'} = \tau_{xy}(\cos^2\theta - \sin^2\theta)$$

[Eq. 8.14]

As the element, loaded by τ_{xy} only, is rotated by θ, the maximum normal stress occurs when $\theta = 45°$. The normal and shear stresses at 45° are:

$$\sigma_{max} = \sigma_{x'}(45°) = \tau_{xy}; \quad \tau_{x'y'}(45°) = 0$$

[Eq. 8.15]

This is the first step of showing that the simple shear stress state (τ_{xy}, $\sigma_x = \sigma_y = 0$) is equivalent to a biaxial state of stress at 45° ($\sigma_{x'} = -\sigma_{y'} = \tau_{xy}$, $\tau_{x'y'} = 0$, *Figure 8.2*).

Superposition of Stresses on *x'*-Face

Superimposing the stress transformation results for the three individual stresses (σ_x, σ_y, τ_{xy}) gives the appropriate stress values on the new *x'*-face. For example, the normal stress in the new *x'*-direction is:

$$\sigma_{x'}(\theta) = \sigma_x\cos^2\theta + \sigma_y\sin^2\theta + 2\tau_{xy}\sin\theta\,\cos\theta$$

[Eq. 8.16]

Stresses on New y'-Face

Stresses on *y'*-Face Due to Stress σ_x

An element aligned with the *x*–*y* axes and subjected to only axial stress σ_x is shown in *Figure 8.7*. A cut is taken through the element, exposing diagonal surface *dA*. In this case, the normal to the surface is the new *y'*-axis, oriented θ from the original *y*-axis, $\theta +90°$ from the original *x*-axis, and 90° from the new *x'*-axis. The *x'*-axis is parallel to exposed area *dA*. The stresses on *dA* are normal and shear stresses, $\sigma_{y'}$ and $\tau_{y'x'}$, respectively. From trigonometry, the right *x*-face of the wedge has area *dA* sin θ, and the bottom *y*-face has area *dA* cos θ.

Force equilibrium in the tangential (*x'*-) and normal (*y'*-) directions gives:

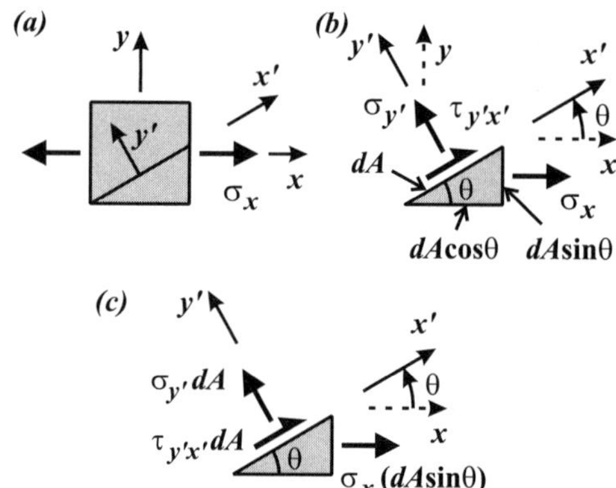

Figure 8.7. (a) A stress element under normal stress σ_x only. (b) Wedge showing stresses acting on the *x*-, *y*-, and *y'*-faces. The normal to the inclined face is the new *y'* axis, $\theta+90°$ from the original *x*-axis. (c) Forces on each face.

$$\sum F_{x'} = 0 = \tau_{y'x'}\,dA + [\sigma_x(dA\,\sin\theta)]\cos\theta$$

$$\sum F_{y'} = 0 = \sigma_{y'}\,dA - [\sigma_x(dA\,\sin\theta)]\sin\theta$$

[Eq. 8.17]

Due to σ_x, the stresses acting on the *y'*-face are:

$$\sigma_{y'} = \sigma_x \sin^2\theta$$

$$\tau_{y'x'} = \tau_{x'y'} = -\sigma_x \cos\theta \sin\theta \qquad \text{[Eq. 8.18]}$$

Recall that at a point shear stresses on normal planes are equal, i.e., $\tau_{x'y'} = \tau_{y'x'}$. The expression for $\tau_{y'x'}$ in *Equation 8.18* is the same as that for $\tau_{x'y'}$ in *Equation 8.8*.

Stresses on y'-Face Due to Stress σ_y and Shear Stress τ_{xy}

Similar equilibrium calculations may be made on the wedge exposing the y'-face (*Figure 8.7*) with only σ_y and then only τ_{xy} applied. With these two final contributions (not derived here), the equations for the new stresses with respect to the x'–y' axes are found using superposition.

Summary of Stress Transformation Equations

The stress components in the x'–y' coordinate system, and the original stress components in the x–y system that give rise to them, are given in *Table 8.1*. The transformed stresses are:

$$\sigma_{x'}(\theta) = \sigma_x \cos^2\theta + \sigma_y \sin^2\theta + 2\tau_{xy} \sin\theta\cos\theta \qquad \text{[Eq. 8.19]}$$

$$\sigma_{y'}(\theta) = \sigma_x \sin^2\theta + \sigma_y \cos^2\theta - 2\tau_{xy} \sin\theta\cos\theta \qquad \text{[Eq. 8.20]}$$

$$\tau_{x'y'}(\theta) = -\sigma_x \sin\theta\cos\theta + \sigma_y \sin\theta\cos\theta + \tau_{xy}[\cos^2\theta - \sin^2\theta] \qquad \text{[Eq. 8.21]}$$

In matrix form, the transformation equations may be written:

$$\begin{bmatrix} \sigma_{x'} \\ \sigma_{y'} \\ \tau_{x'y'} \end{bmatrix} = \begin{bmatrix} c^2 & s^2 & 2sc \\ s^2 & c^2 & -2sc \\ sc & sc & c^2 - s^2 \end{bmatrix} \begin{bmatrix} \sigma_x \\ \sigma_y \\ \tau_{xy} \end{bmatrix} \qquad \text{where } \begin{Bmatrix} c = \cos\theta \\ s = \sin\theta \end{Bmatrix} \qquad \text{[Eq. 8.22]}$$

The square matrix is the *transformation matrix* **T**, used to rotate the coordinate system by angle θ. In short form, the matrix equation is:

Table 8.1. Summary of *transformed stresses* due to *original stresses*. The new coordinate system is rotated counterclockwise by angle θ.

Transformed Stress	Due to Normal Stress σ_x	Due to Normal Stress σ_y	Due to Shear Stress τ_{xy}
$\sigma_{x'}$	$\sigma_x \cos^2\theta$	$\sigma_y \sin^2\theta$	$2\tau_{xy} \sin\theta\cos\theta$
$\sigma_{y'}$	$\sigma_x \sin^2\theta$	$\sigma_y \cos^2\theta$	$-2\tau_{xy} \sin\theta\cos\theta$
$\tau_{x'y'}$	$-\sigma_x \sin\theta\cos\theta$	$\sigma_y \sin\theta\cos\theta$	$\tau_{xy}[\cos^2\theta - \sin^2\theta]$

$$\underline{\sigma}' = \mathbf{T}\underline{\sigma} \qquad \text{[Eq. 8.23]}$$

where $\underline{\sigma}$ and $\underline{\sigma}'$ are the original and transformed stress vectors, respectively.

Products or squares of trigonometric terms can be difficult to work with. The expressions for the transformed stresses can be rewritten using the following trigonometric identities:

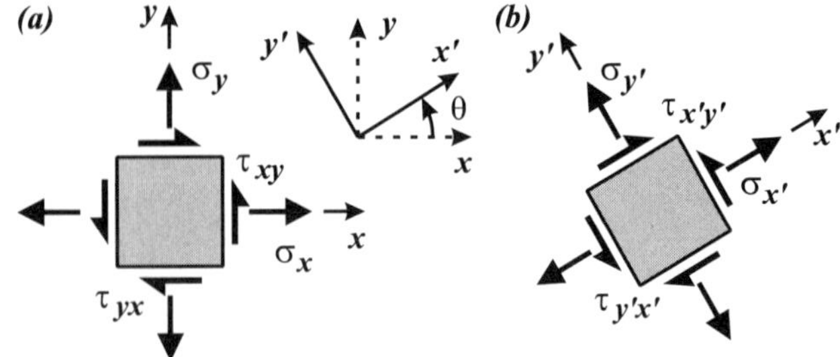

Figure 8.8. (a) Stresses in the *x–y* system. **(b)** Stresses in the *x'–y'* system, rotated counterclockwise θ from the *x–y* system.

$$\cos^2\theta = \frac{1 + \cos 2\theta}{2}; \quad \sin^2\theta = \frac{1 - \cos 2\theta}{2}; \quad \cos\theta\sin\theta = \frac{\sin 2\theta}{2} \qquad \text{[Eq. 8.24]}$$

These identities are substituted into *Equations 8.19 – 8.21* to give a more compact form.

The *plane stress transformation equations* for a stress element rotated by angle θ (positive counterclockwise, as shown in *Figure 8.8*) are:

$$\sigma_{x'}(\theta) = \frac{\sigma_x + \sigma_y}{2} + \frac{\sigma_x - \sigma_y}{2}\cos 2\theta + \tau_{xy}\sin 2\theta \qquad \text{[Eq. 8.25]}$$

$$\sigma_{y'}(\theta) = \frac{\sigma_x + \sigma_y}{2} - \frac{\sigma_x - \sigma_y}{2}\cos 2\theta - \tau_{xy}\sin 2\theta \qquad \text{[Eq. 8.26]}$$

$$\tau_{x'y'}(\theta) = -\frac{\sigma_x - \sigma_y}{2}\sin 2\theta + \tau_{xy}\cos 2\theta \qquad \text{[Eq. 8.27]}$$

Recall that the x'-axis is at angle θ from the original x-axis, and that the y'-axis is at angle θ + 90° from the original x-axis.

If $\tau_{x'y'} > 0$, then it physically acts on the positive x'-face in the positive y'-direction. If $\tau_{x'y'} < 0$, then it acts on the positive x'-face in the negative y'-direction. Because shear stresses on perpendicular planes at a point are equal, $\tau_{y'x'}(\theta) = \tau_{x'y'}(\theta)$. The direction of the shear stress on the new y'-face is in the $+x'$-direction if $\tau_{x'y'}$ is positive, and in the negative x'-direction if it is negative.

Invariance

The second and third terms of $\sigma_{x'}$ and $\sigma_{y'}$ (*Equations 8.25* and *8.26*) are equal but opposite. Adding the transformed normal stresses results in the following equation:

$$\sigma_{x'}(\theta) + \sigma_{y'}(\theta) = \sigma_x + \sigma_y \qquad \text{[Eq. 8.28]}$$

Since σ_x and σ_y are given constants determined by the system geometry and applied loads, the sum of the normal stresses acting on an element is the same regardless of the orientation θ of the axes. This is called *invariance*. If any of the three normal stresses are known, the fourth is also known. Invariance also means that the average value of the normal stresses is the same no matter what direction the element is oriented.

A software module in the *Online Notes* automates the stress transformation equations.

Example 8.1 Stress Transformation

Given: A stress element (*Figure 8.9a*) is subjected to the following stresses:

$$\sigma_x = 9.1 \text{ ksi}; \quad \sigma_y = -15.4 \text{ ksi}; \quad \tau_{xy} = 9.5 \text{ ksi}$$

The negative sign means σ_y acts opposite drawn; it is compressive.

Required: If the element is rotated $\theta = 33°$, determine the stresses on the element in the new orientation. Draw the stress element in its new orientation.

Solution: Apply the transformation equations (*Equations 8.25 – 8.27*). With $2\theta = 66°$:

$$\sigma_{x'}(33°) = \frac{9.1 + (-15.4)}{2} + \frac{9.1 - (-15.4)}{2}\cos(66°)$$

$$+ (9.5)\sin(66°)$$

$$= 10.5 \text{ ksi}$$

Applying the other equations:

$$\sigma_{y'}(33°) = \frac{9.1 + (-15.4)}{2}$$

$$-\left[\frac{9.1 - (-15.4)}{2}\right]\cos(66°) - (9.5)\sin(66°)$$

$$= -16.8 \text{ ksi}$$

$$\tau_{x'y'}(33°) = -\left[\frac{9.1 - (-15.4)}{2}\right]\sin(66°) + (9.5)\cos(66°)$$

$$= -7.3 \text{ ksi}$$

Thus:

$$\textit{Answer:} \quad \sigma_{x'} = 10.5 \text{ ksi}; \quad \sigma_{y'} = -16.8 \text{ ksi}; \quad \tau_{x'y'} = -7.3 \text{ ksi}$$

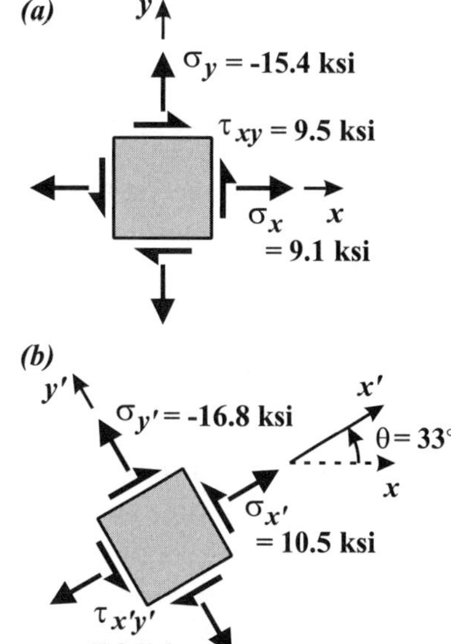

Figure 8.9. (a) The original stress state. The stress arrows are drawn in their positive senses. A negative stress physically acts in the opposite direction drawn, e.g., σ_y is compressive. **(b)** The rotated stress state.

The stresses are drawn in their positive sense on the rotated element (*Figure 8.9b*). The negative signs mean that $\sigma_{y'}$ and $\tau_{x'y'}$ act opposite drawn.

Note that since the y'-axis is $+90°$ from the x'-axis, the stress in the y'-direction could have been found by evaluating $\sigma_{x'}$ at $\theta = 33° + 90° = 123°$:

$$\sigma_{y'}(33°) = \sigma_{x'}(123°)$$

$$= \frac{9.1 + (-15.4)}{2} + \frac{9.1 - (-15.4)}{2} \cos(246°) + (9.5)\sin(246°) = -16.8 \text{ ksi}$$

Note also that the sum of the normal stresses is *invariant*:

$$\sigma_x + \sigma_y = (9.1) + (-15.4) = -6.3 \text{ ksi}$$

$$\sigma_{x'} + \sigma_{y'} = (10.5) + (-16.8) = -6.3 \text{ ksi}$$

Example 8.2 Plate with a Weld

Given: A welded plate is loaded by a force of $P = 150$ kN as shown in *Figure 8.10a*. The width of the plate is $w = 200$ mm and the thickness is $t = 10$ mm. The weld line is $55°$ from the horizontal, and is assumed to have the same thickness as the plate.

Required: (a) Determine the normal stress and average shear stress in the weld. (b) If the weld's yield strength is $S_y = 280$ MPa, and its shear yield strength is $\tau_y = 160$ MPa, determine the factor of safety of the current loading condition with respect to failure of the weld.

Solution: *Step 1.* Determine the stress state in the x–y plane. Since the load is purely axial, only σ_x is non-zero ($\sigma_y = \tau_{xy} = 0$, *Figure 8.10b*):

$$\sigma_x = \frac{P}{wt} = \frac{150 \times 10^3 \text{ N}}{(0.200 \text{ m})(0.01 \text{ m})} = 75 \text{ MPa}$$

Step 2. Using the stress transformation equations, determine the stresses normal and parallel to the weld. The weld is $55°$ from the horizontal. The angle that the weld's normal direction (the x'-axis) makes with the x-axis is $\theta = +35°$. The normal stress on the weld (on the x'-face) is $\sigma_{x'}$ and the shear stress is $\tau_{x'y'}$.

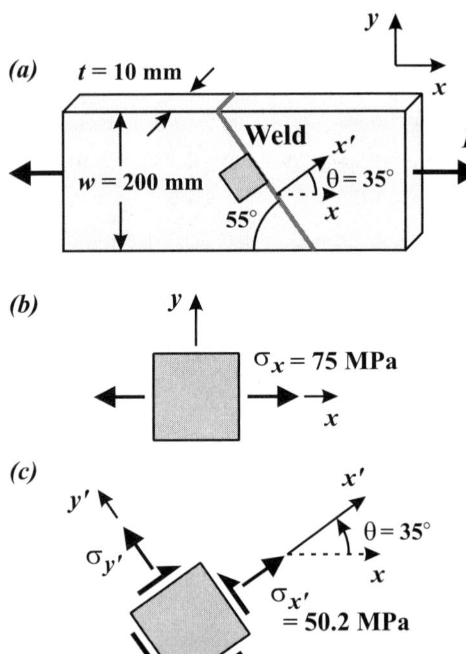

Figure 8.10. (a) Welded plate under axial load. **(b)** Stress state aligned with x–y axes. **(c)** Stress state aligned with axes normal and parallel to the weld.

The stress transformation equations give the stresses on the weld (on the x'-face):

$$\sigma_{x'}(35°) = \frac{\sigma_x}{2} + \frac{\sigma_x}{2}\cos 2\theta = \frac{75}{2}[1 + \cos(70°)]$$

Answer: $\sigma_{x'} = 50.3$ MPa

$$\tau_{x'y'}(35°) = -\frac{\sigma_x}{2}\sin 2\theta = -\frac{75}{2}[\sin(70°)]$$

Answer: $\tau_{x'y'}(35°) = -35.2$ MPa

The transformed stress state is shown in *Figure 8.10c*. Stress $\sigma_{y'}$ is not calculated; it does not act on the weld, but in the plate.

Step 3. The Factor of Safety for the weld is found by dividing the weld strengths by the actual stresses:

$$FS_{normal} = \frac{S_y}{\sigma_{x'}} = \frac{280 \text{ MPa}}{50.3 \text{ MPa}} \Rightarrow FS_{normal} = 5.6$$

$$FS_{shear} = \frac{\tau_y}{\tau_{x'y'}} = \frac{160 \text{ MPa}}{35.2 \text{ MPa}} \Rightarrow FS_{shear} = 4.5$$

The governing factor of safety is the smaller value: *Answer:* $\underline{FS = 4.5}$

8.2 Principal Stresses

It is often necessary to determine the maximum (most tensile) and minimum (most compressive) values of the in-plane normal stress $\sigma_{x'}$ (or $\sigma_{y'}$) as the stress element is rotated through angle θ. These extreme values are the **principal stresses**. The *principal stresses* occur on planes, and thus in directions, that are 90° from each other. The direction of the maximum or minimum normal stress is found by taking the derivative of $\sigma_{x'}(\theta)$ (*Equation 8.25*) with respect to angle θ:

$$\frac{d\sigma_{x'}}{d\theta} = -\frac{\sigma_x - \sigma_y}{2}(2\sin 2\theta) + 2\tau_{xy}\cos 2\theta = 0 \qquad \text{[Eq. 8.29]}$$

Solving for θ, and renaming it θ_p for **principal angle**, results in:

$$\tan 2\theta_p = \frac{2\tau_{xy}}{\sigma_x - \sigma_y} \qquad \text{[Eq. 8.30]}$$

The tangent function repeats every 180°. Therefore, within 360°, there are two solutions to *Equation 8.30*, which are $2\theta_p$ and $2\theta_p+180°$, or θ_p and $\theta_p+90°$. These are the *principal angles*.

The *principal angles* correspond to an x'–y' set of axes where the normal stresses are the *principal stresses*:

$$\sigma_{x'p} = \frac{\sigma_x + \sigma_y}{2} + \frac{\sigma_x - \sigma_y}{2}\cos 2\theta_p + \tau_{xy}\sin 2\theta_p \qquad \text{[Eq. 8.31]}$$

$$\sigma_{y'p} = \frac{\sigma_x + \sigma_y}{2} - \frac{\sigma_x + \sigma_y}{2}\cos 2\theta_p - \tau_{xy}\sin 2\theta_p \qquad \text{[Eq. 8.32]}$$

The principal stresses are 90° apart, along the principal angles (directions), θ_p.

From *Equation 8.30*, the sine and cosine of θ_p can be written:

$$\cos 2\theta_p = \frac{(\sigma_x - \sigma_y)/2}{\sqrt{\left(\dfrac{\sigma_x - \sigma_y}{2}\right)^2 + (\tau_{xy})^2}} \; ; \quad \sin 2\theta_p = \frac{\tau_{xy}}{\sqrt{\left(\dfrac{\sigma_x - \sigma_y}{2}\right)^2 + (\tau_{xy})^2}} \qquad \text{[Eq. 8.33]}$$

By substituting these expressions into *Equations 8.31* and *8.32*, the principal stresses can be compactly written:

$$\sigma_I = \frac{\sigma_x + \sigma_y}{2} + \sqrt{\left(\frac{\sigma_x - \sigma_y}{2}\right)^2 + (\tau_{xy})^2} \qquad \text{[Eq. 8.34]}$$

$$\sigma_{II} = \frac{\sigma_x + \sigma_y}{2} - \sqrt{\left(\frac{\sigma_x - \sigma_y}{2}\right)^2 + (\tau_{xy})^2} \qquad \text{[Eq. 8.35]}$$

As a matter of convention, principal stress σ_I is the *maximum* (most tensile or most positive) normal stress, and σ_{II} is the *minimum* (most compressive or most negative) normal stress, that an element is subjected to as it is rotated. The principal stresses and their respective angles are shown in *Figure 8.11*.

The question of which angle, θ_p or $\theta_p+90°$, corresponds to which stress, σ_I or σ_{II}, is not answered by *Equations 8.34* and *8.35*. With experience, which stress goes with which angle can often be readily seen. However, it is often best to substitute one of the *principal angles* into the general stress-transformation equation for $\sigma_{x'}(\theta)$ (*Equation 8.31*). The resulting stress is matched to the substituted angle. The principal angles are denominated θ_I and θ_{II} and are matched with their appropriate principal stresses, σ_I and σ_{II}, respectively (*Figure 8.11*).

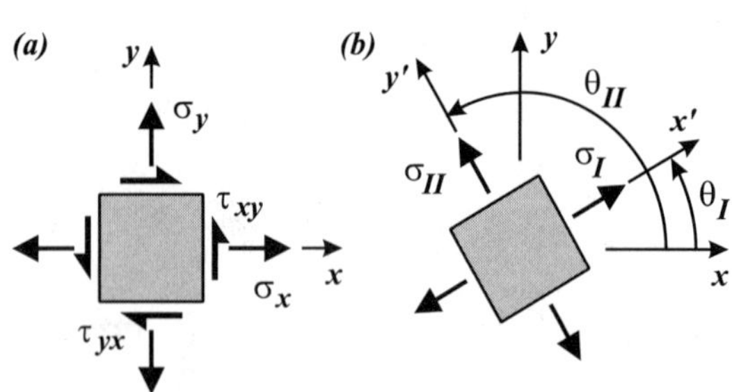

Figure 8.11. (a) The original stress state. (b) The principal stresses and their directions.

Equation 8.30 is also the condition for $\tau_{x'y'} = 0$. Substituting the sine and cosine of $2\theta_p$ (*Equation 8.33*) into the shear stress transformation equation (*Equation 8.27*) results in a shear stress of zero:

$$\tau_{x'y'}(\theta_p) = -\frac{\sigma_x - \sigma_y}{2}\sin 2\theta_p + \tau_{xy}\cos 2\theta_p = 0 \qquad \text{[Eq. 8.36]}$$

The shear stress corresponding to a principal angle is zero. In other words, when the normal stresses are *principal stresses*, the shear stress is zero.

The concept of principal stresses provides a compact way to represent any general state of plane stress σ_x, σ_y, and τ_{xy}, with only two values σ_I and σ_{II}. Is this much easier to visualize and deal with the stress state at a point by using *two principal stresses* rather than *three general stress* components.

Example 8.3 **Principal Stresses and Directions** (*Example 8.1*)

Given: An element (*Figure 8.12a*) is subjected to the following stresses:

$$\sigma_x = 9.1 \text{ ksi}; \quad \sigma_y = -15.4 \text{ ksi}; \quad \tau_{xy} = 9.5 \text{ ksi}$$

Required: Determine the principal stresses and their directions.

Solution: *Step 1.* Find the *principal angles*:

$$\tan 2\theta_p = \frac{2\tau_{xy}}{\sigma_x - \sigma_y} = \frac{2(9.5)}{9.1 - (-15.4)}$$

$$\Rightarrow 2\theta_p = 37.8°, \, 217.8°$$

$$\Rightarrow \theta_p = 18.9°, \, 108.9°$$

The principal angles are 18.9° and 108.9°.

Step 2. Take the smaller angle, 18.9°, and substitute it into the general stress transformation equations for $\sigma_{x'}$ and $\sigma_{y'}$ (*Equations 8.25 and 8.26*):

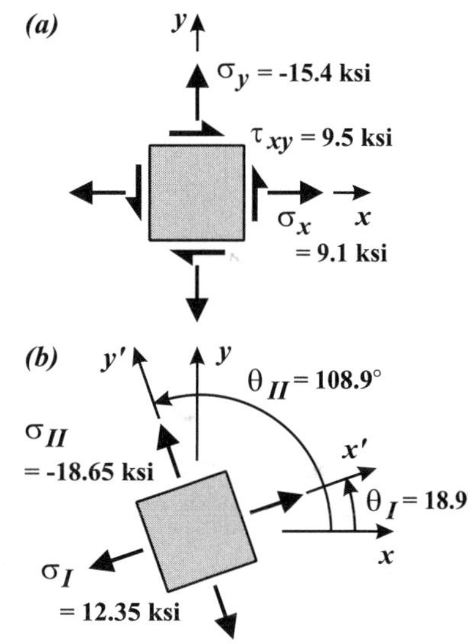

Figure 8.12. (a) The original stress state. **(b)** The Principal Stresses and their directions.

$$\sigma_{x'}(18.9°)$$

$$= \frac{(9.1) + (-15.4)}{2} + \frac{(9.1) - (-15.4)}{2}\cos(37.8°) + (9.5)\sin(37.8°) = 12.35 \text{ ksi}$$

$$\sigma_{y'}(18.9°) = \sigma_{x'}(108.9°)$$

$$= \frac{(9.1) + (-15.4)}{2} - \frac{(9.1) - (-15.4)}{2}\cos(37.8°) - (9.5)\sin(37.8°) = -18.65 \text{ ksi}$$

Step 3. Taking σ_I as the maximum stress, the principal stresses and their respective angles are (*Figure 8.12b*):

Answer: $\sigma_I = 12.35$ ksi at $\theta_I = 18.9°$

Answer: $\sigma_{II} = -18.65$ ksi at $\theta_{II} = 108.9°$

Again, the sum of the normal stresses is constant (invariant):

$$\sigma_I + \sigma_{II} = \sigma_x + \sigma_y = 12.35 + (-18.65) = 9.1 + (-15.4) = -6.3 \text{ ksi}$$

Note that the principal stresses could quickly have been found:

$$\sigma_I, \sigma_{II} = \frac{\sigma_x + \sigma_y}{2} \pm \sqrt{\left(\frac{\sigma_x - \sigma_y}{2}\right)^2 + (\tau_{xy})^2}$$

$$= \frac{(9.1) + (-15.4)}{2} \pm \sqrt{\left[\frac{(9.1) - (-15.4)}{2}\right]^2 + (9.5)^2} = 12.35, -18.65 \text{ ksi}$$

However, the directions of these stresses are not known until one of the angles (e.g., 18.9°) is substituted into $\sigma_{x'}(\theta)$, or if the angles can be matched to the stresses by observation.

Note also that the shear stress corresponding to either principal angle is zero:

$$\tau_{x'y'}(18.9°) = \tau_{x'y'}(108.9°) = -\left[\frac{9.1 - (-15.4)}{2}\right]\sin(37.8°) + (9.5)\cos(37.8°) = 0$$

Example 8.4 Principal Stresses and Directions

Given: An element (*Figure 8.13*) is subjected to the following stresses:

$$\sigma_x = 141 \text{ MPa}; \quad \sigma_y = 71 \text{ MPa}; \quad \tau_{xy} = -108 \text{ MPa}$$

Required: Determine the principal stresses and their directions.

Solution: *Step 1.* Find the *principal angles*:

$$\tan 2\theta_p = \frac{2\tau_{xy}}{\sigma_x - \sigma_y} = \frac{2(-108)}{141 - 71}$$

$$\theta_p = -36.0°, 54.0°$$

Step 2. Substituting the angles (−36° and 54°) into *Equations 8.25* and *8.26*:

$$\sigma_{x'}(-36°) = 219.5 \text{ MPa}$$

$$\sigma_{y'}(-36°) = \sigma_{x'}(54°) = -7.5 \text{ MPa}$$

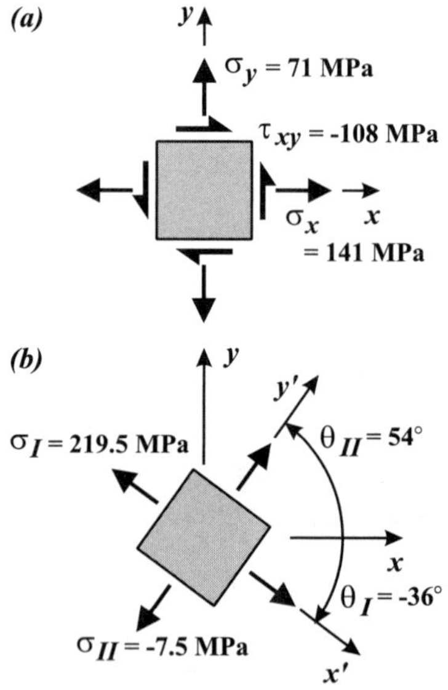

Step 3. The principal stresses and their angles are then:

Answer: $\sigma_I = 220$ MPa at −36.0° and $\sigma_{II} = -7.5$ MPa at 54.0°

Figure 8.13. (a) The original stress state. **(b)** The principal stresses and their directions.

8.3 Maximum (In-Plane) Shear Stress

It is also useful to know the ***maximum in-plane shear stress*** that an element experiences as it rotates through angle θ in the x-y plane. The shear stress transformation equation $\tau_{x'y'}(\theta)$ (*Equation 8.27*) is differentiated with respect to θ and the result set equal to zero to find the angle at which the *in-plane shear stress* is a maximum. The angle is given by:

$$\tan 2\theta_s = -\frac{\sigma_x - \sigma_y}{2\tau_{xy}} \qquad \text{[Eq. 8.37]}$$

Angle θ_s defines the direction (x'-axis) perpendicular to the plane (x'-face) in which the maximum shear stress acts, not the direction of the shear stress itself (along the y'-axis). As is the case for the principal angles, there are two angles within the complete circle that satisfy *Equation 8.37*, θ_{s1} and $\theta_{s2} = \theta_{s1} + 90°$.

Equation 8.37 is the negative reciprocal of *Equation 8.30*, the equation that defines $\tan 2\theta_p$. Since the tangent is the slope, and slopes that are negative reciprocals of each other define lines that are perpendicular, $2\theta_s$ defines a direction $\pm 90°$ from $2\theta_p$, meaning θ_s is $\pm 45°$ from θ_p. Thus the maximum shear stresses occur on planes that are $\pm 45°$ from the planes in which the principal stresses act.

The magnitude of the maximum shear stress can be shown to be:

$$\tau_{max} = \sqrt{\left(\frac{\sigma_x - \sigma_y}{2}\right)^2 + (\tau_{xy})^2} \qquad \text{[Eq. 8.38]}$$

Substituting τ_{max} into the equations for the principal stresses σ_I and σ_{II} (*Equations 8.34 and 8.35*), and taking their difference, it can also be shown that the magnitude of the maximum shear stress in the x–y plane is half the difference of the principal stresses:

$$\tau_{max} = \frac{\sigma_I - \sigma_{II}}{2} \qquad \text{[Eq. 8.39]}$$

The sense of the shear stress is found by substituting one of the values of θ_s into the shear stress transformation equation, $\tau_{x'y'}(\theta_s)$ (*Equation 8.27*). This gives the direction of the shear stress acting on the face normal to the new x'-axis (x_s-axis), oriented θ_s from the original x-axis. If the shear stress is positive, it acts on the new x'-face in the positive y'-direction.

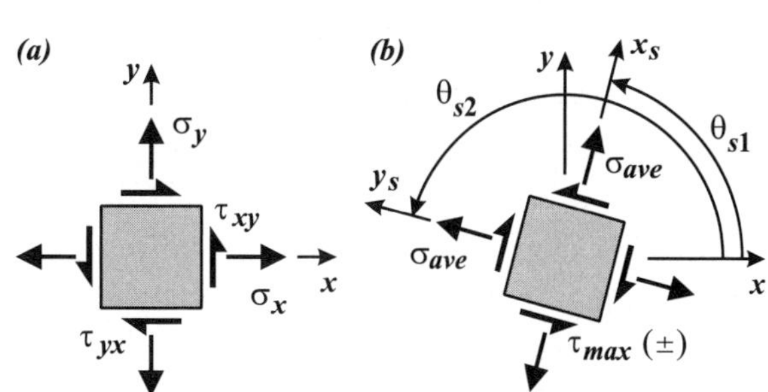

Figure 8.14. (a) The original stress state. (b) The maximum shear stresses and the angle of the planes in which they act. The angle θ_s defines the direction of the plane's normal vector.

Table 8.2. Principal Stresses, Maximum (in-plane) Shear Stress, and their Angles.

	Principal Stresses	Maximum In-plane Shear Stress
Maximum Stress(es)	σ_I, σ_{II} $= \dfrac{\sigma_x + \sigma_y}{2} \pm \sqrt{\left(\dfrac{\sigma_x - \sigma_y}{2}\right)^2 + (\tau_{xy})^2}$ $= \sigma_{ave} \pm \tau_{max}$	$\tau_{max} = \sqrt{\left(\dfrac{\sigma_x - \sigma_y}{2}\right)^2 + (\tau_{xy})^2}$ $= \dfrac{\sigma_I - \sigma_{II}}{2}$
Angle of normal vector (from x-axis)	$\theta_p = \dfrac{1}{2}\tan^{-1}\left(\dfrac{2\tau_{xy}}{\sigma_x - \sigma_y}\right)$	$\theta_s = \dfrac{1}{2}\tan^{-1}\left(-\dfrac{\sigma_x - \sigma_y}{2\tau_{xy}}\right)$
Value of other Stress(es)	$\tau = 0$	$\sigma_{x'} = \sigma_{y'} = \sigma_{ave} = \dfrac{\sigma_I + \sigma_{II}}{2}$

When the shear stress is maximum, the normal stresses, $\sigma_{x'}(\theta_s)$ and $\sigma_{y'}(\theta_s)$ are equal to each other, and are equal to the *average normal stress*:

$$\sigma_{x'}(\theta_s) = \sigma_{y'}(\theta_s) = \frac{\sigma_x + \sigma_y}{2} = \frac{\sigma_I + \sigma_{II}}{2} = \sigma_{ave} \qquad \text{[Eq. 8.40]}$$

This last result again demonstrates the *invariance* of the normal stresses.

The maximum shear stress and their angles are shown in *Figure 8.14*. The axes that define the planes of maximum shear stress have been labeled x_s–y_s, located at θ_{s1} and θ_{s2} from the original x-axis.

The principal and maximum shear stress equations are summarized in *Table 8.2*.

Example 8.5 Maximum In-Plane Shear Stresses and Their Directions
(Examples 8.1 and 8.3)

Given: An element (*Figure 8.15a*) is subjected to the following stresses:

$$\sigma_x = 9.1 \text{ ksi}; \quad \sigma_y = -15.4 \text{ ksi}; \quad \tau_{xy} = 9.5 \text{ ksi}$$

Required: Determine (a) the maximum in-plane shear stresses and the direction of the vectors normal to the planes in which they act and (b) the normal stresses when the shear stress is maximum.

Solution: *Step 1.* Find the angles that define the maximum shear stress (*Equation 8.37*):

$$\tan 2\theta_s = -\frac{\sigma_x - \sigma_y}{2\tau_{xy}} = -\frac{(9.1) - (-15.4)}{2(9.5)}$$

$$\Rightarrow 2\theta_s = -52.2°, \ 127.8°$$

Answer: $\underline{\theta_s = -26.1°, 63.9°}$

Note that these angles are $\pm 45°$ from the principal angles found in *Example 8.3*. Recall that the angles θ_s are not the directions of the shear stresses, but the directions of vectors normal to the planes on which the shear stresses act.

Step 2. Substitute the smaller angle into the shear stress transformation equation:

$$\tau_{x'y'}(-26.1°) = -\frac{(9.1)-(-15.4)}{2}\sin(-52.2°)$$
$$+ (9.5)\cos(-52.2°)$$

Answer: $\underline{\tau_{max} = +15.5 \text{ ksi}}$

$\theta_s = -26.1°$ is the direction of the x'-axis corresponding to a *positive* shear stress, i.e., the shear stress acts on the $+x'$-face in the $+y'$-direction. These axes are labeled x_s and y_s in *Figure 8.15b*.

When the shear stress is maximum, the normal stresses are equal, and equal to the average of the normal stresses σ_x and σ_y :

$$\sigma_{ave} = \frac{\sigma_x + \sigma_y}{2} = \frac{(9.1)+(-15.4)}{2}$$

Answer: $\underline{\sigma_{ave} = -3.15 \text{ ksi}}$

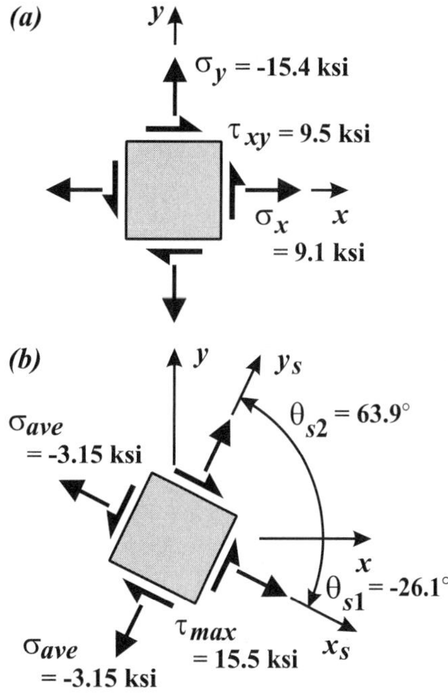

Figure 8.15. (a) The original stress state. **(b)** The maximum shear stresses.

Example 8.6 Shaft under Torque and Axial Force

Given: A propeller-driven boat moves forward at constant velocity against a drag force of 120 kN. The solid drive shaft transmits 2500 kW of power to the propeller at 1000 rpm (revolutions per minute). The radius of the shaft is 100 mm (*Figure 8.16a*).

Required: Determine (a) the stresses on a surface element oriented with the axis of the shaft, (b) the principal stresses, and (c) the maximum shear stress.

Solution: *Step 1.* Find the loads acting on the shaft. To move forward at constant velocity, the propeller pushes against the water with a force equal to the drag force. The water pushes against the propeller, so the shaft supports a *compressive* force having magnitude $P = 120$ kN.

The angular velocity of the shaft is:

$$\omega = \left(1000\ \frac{\text{rev}}{\text{min}}\right)\left(\frac{2\pi\ \text{rad}}{\text{rev}}\right)\left(\frac{1\ \text{min}}{60\ \text{s}}\right)$$

$$= 104.7\ \text{rad/s}$$

The torque is related to the power as:

$$T = \frac{\text{Power}}{\omega} = \frac{2500\times10^3\ \text{N·m/s}}{104.7\ \text{rad/s}}$$

$$= 23.9\ \text{kN·m}$$

Step 2. Since the loads are an axial force and a torque, the relevant cross-sectional properties are:

$$A = \pi R^2 = \pi(0.1)^2 = 0.03142\ \text{m}^2$$

$$J = \frac{\pi R^4}{2} = \frac{\pi(0.1)^4}{2} = 0.1571\times10^{-3}\ \text{m}^4$$

Step 3. On surface element B, oriented with the axis of the shaft (the x-axis), the stresses are (*Figure 8.16b*):

$$\sigma_x = \frac{F}{A} = \frac{-120{,}000\ \text{N}}{0.03142\ \text{m}^2} = -3.820\ \text{MPa}$$

Answer: $\underline{\sigma_x = -3.82\ \text{MPa}}$

and

$$\tau_{xy} = \frac{TR}{J} = \frac{(23.9\times10^3\ \text{N·m})(0.1\ \text{m})}{(0.1571\times10^{-3}\ \text{m}^4)} = 15.22\ \text{MPa}$$

Answer: $\underline{\tau_{xy} = 15.22\ \text{MPa}}$

Step 4. The principal stresses are:

$$\sigma_I, \sigma_{II} = \frac{\sigma_x + \sigma_y}{2} \pm \sqrt{\left(\frac{\sigma_x - \sigma_y}{2}\right)^2 + (\tau_{xy})^2} = -1.91 \pm 15.34\ \text{ksi}$$

Answer: $\underline{\sigma_I = 13.43\ \text{MPa}}$

Answer: $\underline{\sigma_{II} = -17.25\ \text{MPa}}$

The principal angles, without showing calculations, are $\theta_I = 48.6°$ and $\theta_{II} = 138.6°$ (or $\theta_{II} = -41.4°$) (*Figure 8.16c*).

Step 5. The maximum shear stress is:

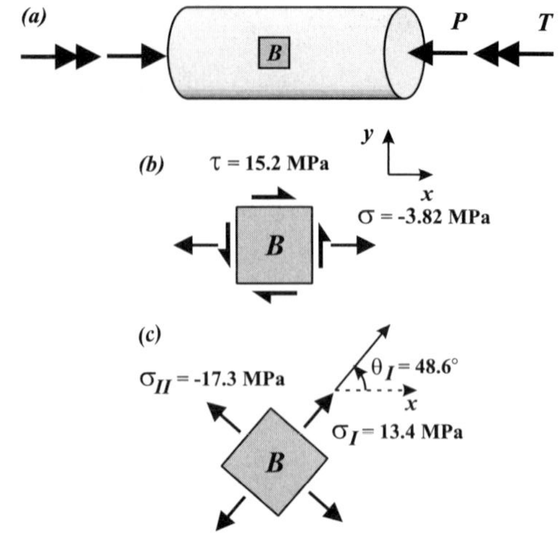

Figure 8.16. (a) Shaft under torque T and compressive force P. **(b)** Stresses in the x–y plane. **(c)** The principal stresses.

$$\tau_{max} = \sqrt{\left(\frac{\sigma_x - \sigma_y}{2}\right)^2 + (\tau_{xy})^2}$$

Answer: $\tau_{max} = 15.34$ MPa

The maximum shear stresses occur on planes that have normal vectors at $\theta_s = 3.6°$ and $93.6°$ to the original x-axis, which are $\pm 45°$ from the principal angles. The normal stresses on those planes are $\sigma_{xs} = \sigma_{ys} = \sigma_{ave} = -1.91$ MPa.

8.4 Mohr's Circle

The *plane stress* stress transformation equations define a circle. Otto Mohr, a German engineer in the late 1800s, recognized these relationships and developed a useful engineering tool now known as **Mohr's Circle**.

Consider the plane stress transformation equations for $\sigma_{x'}(\theta)$ and $\tau_{x'y'}(\theta)$, rearranged so that only trigonometric terms are on the right side:

$$\sigma_{x'}(\theta) - \frac{\sigma_x + \sigma_y}{2} = \frac{\sigma_x - \sigma_y}{2}\cos 2\theta + \tau_{xy}\sin 2\theta$$

$$\tau_{x'y'}(\theta) = -\frac{\sigma_x - \sigma_y}{2}\sin 2\theta + \tau_{xy}\cos 2\theta$$

[Eq. 8.41]

Squaring both sides of both equations, and adding them together, results in:

$$\left(\sigma_{x'} - \frac{\sigma_x + \sigma_y}{2}\right)^2 + (\tau_{x'y'})^2 = \left(\frac{\sigma_x - \sigma_y}{2}\right)^2 + (\tau_{xy})^2$$

[Eq. 8.42]

This is the equation of a circle plotted on a normal stress–shear stress set of axes (σ–τ). Stresses σ_x, σ_y, and τ_{xy} are given values, and the center of the circle is at (σ_{ave}, 0), where:

$$\sigma_{ave} = \frac{\sigma_x + \sigma_y}{2}$$

[Eq. 8.43]

The radius of the circle is equal to the maximum in-plane shear stress, τ_{max}, the square root of the right side of *Equation 8.42*:

$$\tau_{max} = \sqrt{\left(\frac{\sigma_x - \sigma_y}{2}\right)^2 + (\tau_{xy})^2}$$

[Eq. 8.44]

The points on the circle, ($\sigma_{x'}$, $-\tau_{x'y'}$) and ($\sigma_{y'}$, $\tau_{x'y'}$), represent the transformed stresses on the new x'- and y'-faces, as explained below.

Constructing Mohr's Circle

Mohr's circle is constructed as follows (*Figure 8.17*):

1. Construct (on graph paper) an x–y axis with the horizontal axis (the abscissa) labeled σ, and the vertical axis (the ordinate) labeled τ.

2. Plot two points: $X(\sigma_x, -\tau_{xy})$ and $Y(\sigma_y, +\tau_{xy})$. The points represent the stresses acting on the $+x$- and $+y$-faces of the element, respectively. The positive sense of stress follows the convention of this text (*Figure 8.17a*):

   - Stress is *positive* if it acts: (a) on a positive face in a positive direction, or (b) on a negative face in a negative direction (e.g., shear stress is positive if it causes an upward shear stress on the positive x-face).
   - Stress is *negative* if it acts: (a) on a positive face in a negative direction, or (b) on a negative face in a positive direction.

   Just as with the transformation equations, maintaining sign convention is important.

3. Construct a line connecting X and Y. Line X–Y is the diameter of the circle and has a length of $2R = 2\tau_{max}$. Where the line intercepts the σ-axis is the center of the circle, which is the average normal stress σ_{ave}.

4. Construct the circle, with center at $(\sigma_{ave}, 0)$. The circle intercepts the σ-axis at the *principal stresses* σ_I and σ_{II}. The radius of the circle is equal to the *maximum shear stress*, or half the difference of the principal stresses. τ_{max} is the magnitude of

---

Equation of a Circle

The general equation of a circle of radius R, with center at $(a,0)$, in the x–y coordinate system is:

$$(x - a)^2 + y^2 = R^2$$

The variables and constants used in Mohr's Circle are:

$$(x, y) = (\sigma_{x'}, -\tau_{x'y'}), (\sigma_{y'}, \tau_{x'y'})$$

$$a = \frac{\sigma_x + \sigma_y}{2} = \sigma_{ave}$$

$$R = \sqrt{\left(\frac{\sigma_x - \sigma_y}{2}\right)^2 + (\tau_{xy})^2} = \tau_{max}$$

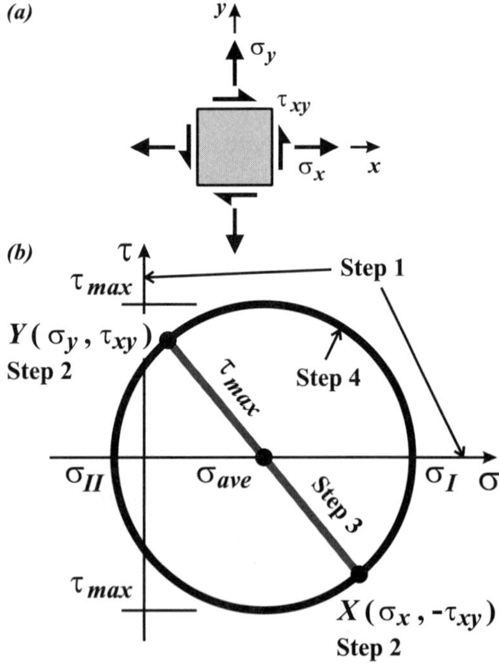

Figure 8.17. Constructing Mohr's circle.
1. Construct the σ–τ axes.
2. Plot $X(\sigma_x, -\tau_{xy})$ and $Y(\sigma_y, \tau_{xy})$.
3. Construct the diameter X–Y.
4. Construct Mohr's circle.

the shear stress at the top and bottom of the circle.

By constructing the circle to scale on graph paper with compass and ruler, these values can be read off the plot without any calculations required.

Using Mohr's Circle

To find the stresses acting on a stress element when it is rotated by any angle θ from the original x–y coordinate system, rotate the X–Y diameter of Mohr's circle by 2θ, positive counterclockwise (*Figure 8.18*).

A counterclockwise rotation θ of the *stress element* corresponds to a counterclockwise rotation of 2θ in *Mohr's circle*. With the rotation, Point X moves to Point X', corresponding to $(\sigma_{x'}, -\tau_{x'y'})$; Point Y moves to Point Y', corresponding to $(\sigma_{y'}, \tau_{x'y'})$.

Using a protractor to measure 2θ locates the new diameter and points X'

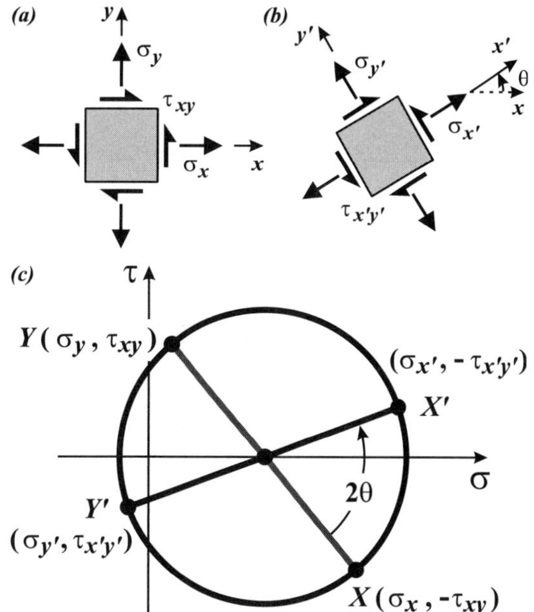

Figure 8.18. (a) The original stresses. **(b)** The transformed stresses on the element rotated by θ. **(c)** Mohr's circle; demonstrating the rotation of the diameter by 2θ to the transformed stresses.

and Y'. The stress values can be read directly off the plot. For more exact results, a few trigonometric calculations will give values for the new coordinates, or the transformation equations may be used.

The *principal stresses* correspond to the values where the Mohr's circle intercepts the σ-axis (the abscissa, *Figure 8.19*). The principal stresses are the average of the normal stresses (the center of the circle), plus or minus the circle's radius:

$$\sigma_{I, II} = \frac{\sigma_x + \sigma_y}{2} \pm \sqrt{\left(\frac{\sigma_x - \sigma_y}{2}\right)^2 + (\tau_{xy})^2} \qquad \text{[Eq. 8.45]}$$

$$= \sigma_{ave} \pm \tau_{max}$$

When the diameter is *horizontal*, its endpoints are at $(\sigma_I, 0)$ and $(\sigma_{II}, 0)$. The horizontal diameter, $X_{pI} - Y_{pI}$, corresponds to the principal stresses. The normal stresses are maximum and minimum, and the shear stress is zero (*Figure 8.19*).

The principal angles are found by determining the angles $2\theta_I$ and $2\theta_{II}$ that the original Mohr's circle X–Y diameter must be rotated by so that the diameter is horizontal. This may be done by measuring with a protractor, or by performing trigonometric calculations from the circle geometry, e.g.:

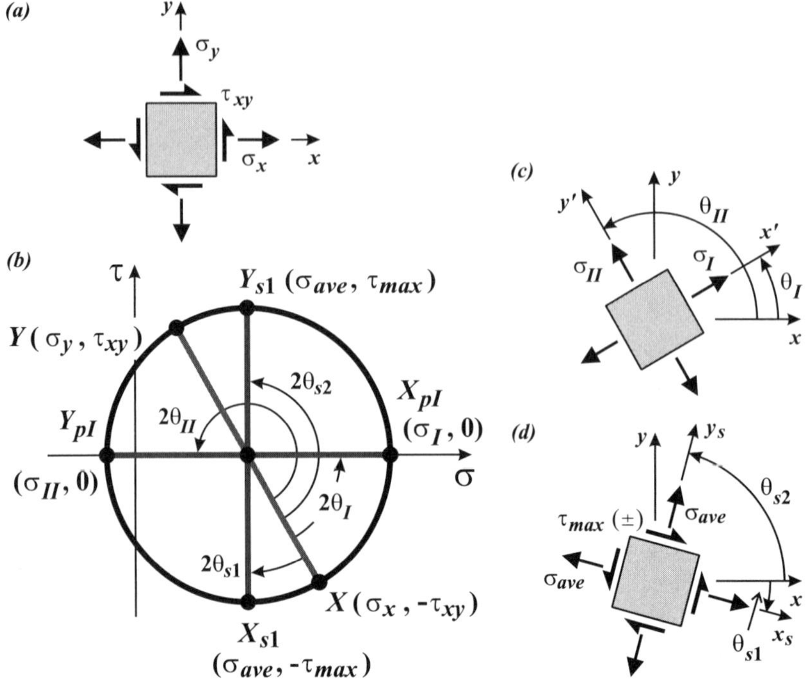

Figure 8.19. (a) The original stresses. **(b)** Mohr's circle with diameter X–Y rotated to the principal stress diameter X_{pI}–Y_{pI} and the maximum shear stress diameter X_{s1}–Y_{s1}. **(c)** The principal stresses and their directions. **(d)** The maximum shear stresses, their normal stresses, and directions.

$$\tan 2\theta_I = \frac{\tau_{xy}}{\sigma_x - \sigma_{ave}} = \frac{2\tau_{xy}}{\sigma_x - \sigma_y} \qquad \text{[Eq. 8.46]}$$

The *maximum in-plane shear stress* is the value of τ at the top and bottom of the circle, and is equal to the radius of the circle. When the Mohr's circle diameter is *vertical*, its endpoints are at $(\sigma_{ave}, \pm\tau_{max})$.

The maximum shear stress angles θ_{s1} and θ_{s2} are found in the same manner as the principal angles. In *Figure 8.19*, θ_{s1} is associated with the x'-face in which the maximum shear stress is positive (the value of $-\tau_{max}$ for X_{s1} is negative, so τ_{max} is positive).

***Example 8.7* Using Mohr's Circle** (*Examples 8.1, 8.3, and 8.5*)

Given: An element (*Figure 8.20a*) is subjected to the following stresses:

$$\sigma_x = 9.1 \text{ ksi}; \quad \sigma_y = -15.4 \text{ ksi}; \quad \tau_{xy} = 9.5 \text{ ksi}$$

Required: Using Mohr's circle, determine (a) the stresses when the element is rotated by $\theta = 33°$, (b) the principal stresses and their directions, and (c) the maximum shear stresses and their directions, and the normal stresses associated with the maximum shear stress.

Solution: *Step 1.* Plot Mohr's circle (*Figure 8.20c*). The endpoints of the original diameter are:

$$X = (\sigma_x, -\tau_{xy}) = (9.1, -9.5)\ \text{ksi}$$

$$Y = (\sigma_y, \tau_{xy}) = (-15.4, 9.5)\ \text{ksi}$$

Step 2. Rotate the Mohr's circle diameter, $X\text{–}Y$, by $2\theta = 2(33°) = +66°$ to $X'\text{–}Y'$, corresponding to a rotation of the stress element by $33°$. This angle may be measured with a protractor. The new stresses are the coordinates of X' and Y' (*Figures 8.20b, c*).

$$\sigma_{x'} = 10.5\ \text{ksi}$$

Answer: $\quad \sigma_{y'} = -16.8\ \text{ksi}$

$$\tau_{x'y'} = -7.3\ \text{ksi}$$

Step 3. Mohr's circle intercepts the σ-axis at the principal stresses and may be read off the graph (*Figure 8.20d*). The angles $2\theta_I$ and $2\theta_{II}$ may be found by measurement, or using *Equation 8.46*:

$$\tan 2\theta_p = \frac{\tau_{xy}}{\sigma_x - \sigma_{ave}} = \frac{2\tau_{xy}}{\sigma_x - \sigma_y}$$

Here, $2\theta_I = 37.8°$ and $2\theta_{II} = 217.8°$. The principal stresses and their angles are:

Answer:
$$\sigma_I = 12.35\ \text{ksi} \ \text{ at } \theta_I = 18.9°$$

Answer:
$$\sigma_{II} = -18.65\ \text{ksi} \ \text{ at } \theta_{II} = 108.9°$$

Recall that a rotation of 2θ in the circle is a rotation of θ in the stress element (*Figure 8.20e*).

Step 4. The maximum shear stress is the magnitude of τ at the top and bottom of the circle, equal to the radius:

Answer: $\tau_{max} = 15.5\ \text{ksi}$

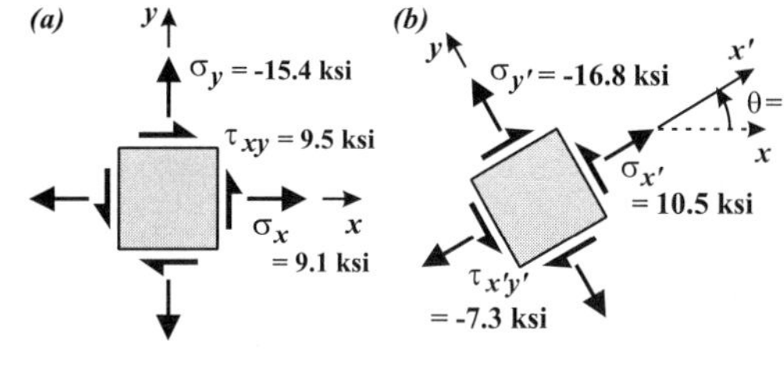

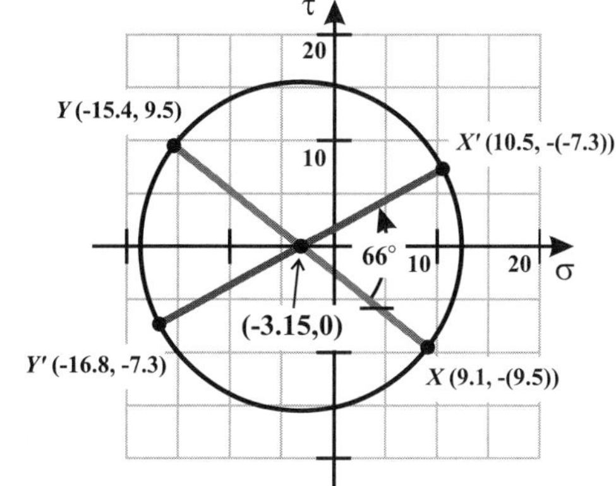

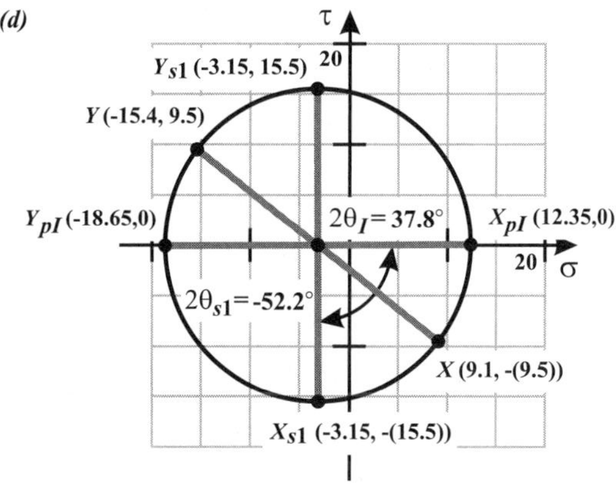

Figure 8.20. (a) The original stress state. **(b)** The stresses on element rotated by $\theta = 33°$. **(c)** Mohr's circle with diameter rotated by $2\theta = 66°$. **(d)** Mohr's circle to determine the principal stresses and maximum shear stress.

The direction of the vectors normal to the planes on which Maximum Shear Stresses act, θ_s, may be found by measurement, by trigonometry, or by using *Equation 8.37*:

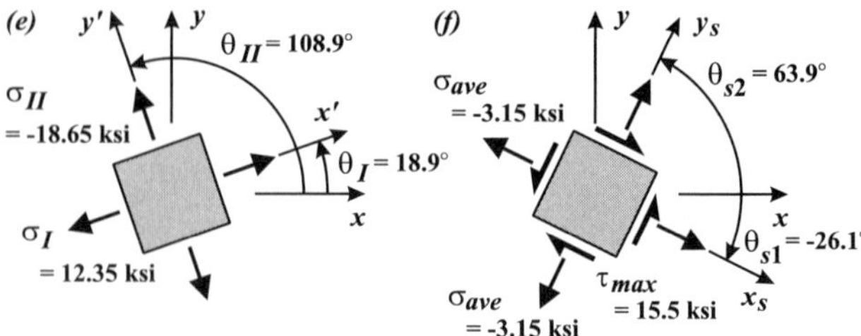

Figure 8.20. (e) Stress element aligned with principal axes. (f) Element aligned with maximum shear stress axes.

$$\tan 2\theta_s = -\frac{\sigma_x - \sigma_y}{2\tau_{xy}}$$

The angles that define the planes on which maximum shear stresses occur are:

Answer: $\theta_{s1} = -26.1°$ and $\theta_{s2} = 63.9°$

The maximum shear stress angles are ±45° from the principal angles (*Figure 8.20f*).

The normal stress that occurs on each face when the shear stress is maximum is:

Answer: $\sigma_{ave} = -3.15$ ksi

8.5 Strain Transformation

It has been demonstrated that a general state of plane stress, σ_x, σ_y, and τ_{xy}, can be reduced to its principal stresses, σ_I and σ_{II} (with zero shear stress). Using two principal stresses, instead of three general stress state terms, simplifies the description of that state of stress.

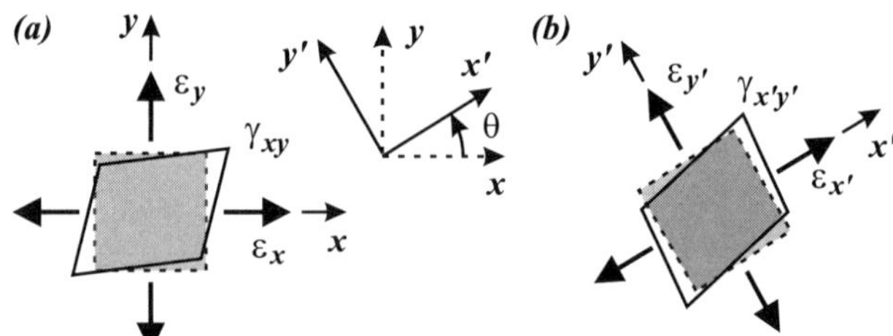

Figure 8.21. (a) State of strain with respect to *x–y* axes. (b) State of strain with respect to *x'–y'* axes.

With the condition that no *out-of-plane* shear strains exist ($\gamma_{zx} = \gamma_{zy} = 0$), strains may be transformed in a similar manner as the stresses (*Figure 8.21*). Simply substitute ε_x for σ_x and ε_y for σ_y. However, because of the way the mathematics works out, $\gamma_{xy}/2$ is used for the shear component. The **strain transformation equations** are:

$$\varepsilon_{x'}(\theta) = \frac{\varepsilon_x + \varepsilon_y}{2} + \frac{\varepsilon_x - \varepsilon_y}{2}\cos 2\theta + \frac{\gamma_{xy}}{2}\sin 2\theta \qquad \text{[Eq. 8.47]}$$

$$\varepsilon_{y'}(\theta) = \frac{\varepsilon_x + \varepsilon_y}{2} - \frac{\varepsilon_x - \varepsilon_y}{2}\cos 2\theta - \frac{\gamma_{xy}}{2}\sin 2\theta \qquad \text{[Eq. 8.48]}$$

Table 8.3. Principal Strains, Maximum Shear Strains, and Angles.

	Principal Strains (Maximum Normal Strains)	**Maximum In-Plane Shear Strain**
Maximum Strain(s)	$\varepsilon_I\,,\varepsilon_{II}$ $= \dfrac{\varepsilon_x + \varepsilon_y}{2} \pm \sqrt{\left(\dfrac{\varepsilon_x - \varepsilon_y}{2}\right)^2 + \left(\dfrac{\gamma_{xy}}{2}\right)^2}$	$\dfrac{\gamma_{max}}{2} = \sqrt{\left(\dfrac{\varepsilon_x - \varepsilon_y}{2}\right)^2 + \left(\dfrac{\gamma_{xy}}{2}\right)^2}$ $= \dfrac{\varepsilon_I - \varepsilon_{II}}{2}$
Angle of normal vector (from x-axis)	$\theta_{p\varepsilon} = \dfrac{1}{2}\tan^{-1}\left(\dfrac{\gamma_{xy}}{\varepsilon_x - \varepsilon_y}\right)$	$\theta_{s\varepsilon} = \dfrac{1}{2}\tan^{-1}\left(-\dfrac{\varepsilon_x - \varepsilon_y}{\gamma_{xy}}\right)$

$$\frac{\gamma_{x'y'}}{2}(\theta) = \quad -\frac{\varepsilon_x - \varepsilon_y}{2}\sin 2\theta + \frac{\gamma_{xy}}{2}\cos 2\theta \qquad\text{[Eq. 8.49]}$$

The **principal strains** – the maximum and minimum normal strains, ε_I and ε_{II} – and the **maximum shear strain**, γ_{max}, and their **angles**, $\theta_{sp\text{-}\varepsilon}$ and $\theta_{s\varepsilon}$, are found by differentiating the strain transformation equations. The maximum strains and their angles are summarized in *Table 8.3*. The transformation tool in the *Online Notes* may also be used.

Angles $\theta_{p\varepsilon}$ correspond to the *principal strain angles*, and $\theta_{s\varepsilon}$ to the directions of vectors normal to the maximum shear strain planes. Angles $2\theta_{p\varepsilon}$ and $2\theta_{s\varepsilon}$ are 90° apart; hence, $\theta_{p\varepsilon}$ and $\theta_{s\varepsilon}$ are 45° apart.

By adding *Equations 8.47* and *8.48*, the normal strains are shown to be *invariant*:

$$\varepsilon_{x'} + \varepsilon_{y'} = \varepsilon_x + \varepsilon_y \qquad\text{[Eq. 8.50]}$$

Example 8.8 Strain Transformation

Given: An element is subjected to the following strains:

$$\varepsilon_x = 91.0\times10^{-6}; \ \ \varepsilon_y = -154\times10^{-6}; \ \ \gamma_{xy} = -190\times10^{-6}$$

Required: If the element is rotated $\theta = 33°$ ($2\theta = 66°$), determine the strains in the new coordinate system.

Solution: Apply the strain transformation equations (*Equations 8.47 – 8.49*); e.g.:

$$\varepsilon_{x'}(33°) = \frac{91 + (-154)}{2} + \frac{91 - (-154)}{2}\cos(66°) + \left(\frac{-190}{2}\right)\sin(66°) = 105\times10^{-6}$$

Applying the other equations:

$$\textit{Answer: } \varepsilon_{x'} = 105 \times 10^{-6}; \quad \varepsilon_{y'} = -168 \times 10^{-6}; \quad \gamma_{x'y'} = -147 \times 10^{-6}$$

Example 8.9 Principal Strains and Directions

Given: An element is subjected to the following strains:

$$\varepsilon_x = 400 \times 10^{-6}; \quad \varepsilon_y = -600 \times 10^{-6}; \quad \gamma_{xy} = 800 \times 10^{-6}$$

Required: Determine (a) the principal strains and their directions and (b) the maximum shear strain.

Solution: *Step 1.* Find the *principal strain angles.*

$$\tan 2\theta_{p\varepsilon} = \frac{\gamma_{xy}}{\varepsilon_x - \varepsilon_y} = \frac{800}{400 - (-600)}$$

$$\rightarrow 2\theta_{p\varepsilon} = 38.66°, \, 218.66°$$

$$\rightarrow \theta_{p\varepsilon} = 19.3°, \, 109.3°$$

The principal angles are 19.3° and 109.3°.

Step 2. Take the smaller angle, 19.3°, and substitute it into the general strain transformation equations for $\varepsilon_{x'}$ and $\varepsilon_{y'}$. Equivalently, substitute both angles into the strain transformation equation for $\varepsilon_{x'}$.

$$\varepsilon_{x'}(19.3°) = \frac{(400) + (-600)}{2} + \frac{(400) - (-600)}{2}\cos(38.66°) + \frac{800}{2}\sin(38.66°) = 540$$

$$\varepsilon_{y'}(19.3°) = \varepsilon_{x'}(109.3°)$$

$$= \frac{(400) + (-600)}{2} - \frac{(400) - (-600)}{2}\cos(38.66°) - \frac{800}{2}\sin(38.66°) = -740$$

Step 3. Since $\varepsilon_{x'}(19.3°)$ is greater than $\varepsilon_{y'}(19.3°)$, setting $\varepsilon_{x'} = \varepsilon_I$ gives:

$$\textit{Answer: } \varepsilon_I = 540 \times 10^{-6} \text{ at } 19.3°$$

$$\textit{Answer: } \varepsilon_{II} = -740 \times 10^{-6} \text{ at } 109.3°$$

Note that the sum of the normal strains is *invariant*:

$$\varepsilon_I + \varepsilon_{II} = [540 + (-740)] \times 10^{-6} = -200 \times 10^{-6}$$

$$\varepsilon_x + \varepsilon_y = [400 + (-600)] \times 10^{-6} = -200 \times 10^{-6}$$

The principal strains could quickly have been found:

$$\varepsilon_I, \, \varepsilon_{II} = \frac{\varepsilon_x + \varepsilon_y}{2} \pm \sqrt{\left(\frac{\varepsilon_x - \varepsilon_y}{2}\right)^2 + \left(\frac{\gamma_{xy}}{2}\right)^2} = 540 \times 10^{-6}, \, -740 \times 10^{-6}$$

However, the directions of these strains are not known until one of the angles (e.g., 19.3°) is substituted into $\varepsilon_{x'}(\theta)$, unless the associations can be determined by observation.

The shear strain corresponding to either principal strain angle is zero:

$$\frac{\gamma_{x'y'}}{2}(19.3°) = -\frac{(400)-(-600)}{2}\sin(38.66°) + \left(\frac{800}{2}\right)\cos(38.66°) = 0$$

Step 4. The maximum shear strain is:

$$\frac{\gamma_{max}}{2} = \frac{\varepsilon_I - \varepsilon_{II}}{2} = \frac{540-(-740)}{2} = \frac{1280}{2}$$

$$Answer: \quad \gamma_{max} = 1280 \times 10^{-6}$$

8.6 Strain Gages

A *strain gage* is shown in *Figure 8.22a* (at four times its original size). Strain gages are typically about 10 mm in length. A gage consists of an electrical conductor (path) turned back upon itself several times and embedded in a flexible matrix material. In a mechanical test, the gage is bonded to a specimen, and then the gage is wired into an electronic circuit. When the specimen is strained, the gage is strained. As the conductor stretches (in the *x*-direction of *Figure 8.22a*), its length increases and its cross-sectional area decreases, which in turn increases its electrical resistance. By knowing how the resistance varies with strain, the circuit's output voltage can be related to the gage's strain.

Strain Gage Rosettes

A single strain gage can only measure strain in the direction of its axis. If only one gage is used on a specimen tested in tension, it must be perfectly aligned with the specimen, otherwise it measures an off-axis strain.

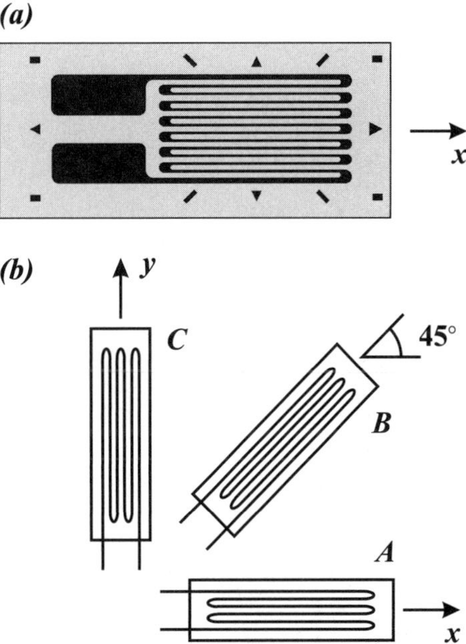

Figure 8.22. (a) Pictorial of a typical strain gage (~4× original size). Electrical conductor (shown in black) on flexible backing. Large tabs are to solder gage into a circuit. The arrows help with aligning the gage on a specimen. The gage measures strain along the *x*-axis shown. (b) Configuration of a 0/45/90 strain gage rosette.

A *strain gage rosette* is a set of gages manufactured as a single unit. A common configuration is the so-called 0/45/90 rosette, which consists of three separate gages, as shown in *Figure 8.22b*. Here, the strain gages are labeled A, B and C. Each gage measures the strain in its own axial direction, ε_A, ε_B and ε_C, respectively. For mathematical convenience, strain gage A is taken to be aligned with the x-axis (0°), gage C with the y-axis (90°), and gage B is at $\theta = 45°$. Strain $\varepsilon_A = \varepsilon_x$ is not necessarily in the direction of a Principal Stress.

Strains ε_x, ε_y and γ_{xy} can be determined from the normal strains measured in the three gage directions. From these strains, the principal strains and maximum shear strains can be calculated.

Here, strain gage A measures strain ε_x directly since it is aligned with the x-axis ($\theta = 0°$). This is also shown using the transformation equation:

$$\varepsilon_A = \varepsilon_{x'}(\theta = 0°) = \frac{\varepsilon_x + \varepsilon_y}{2} + \frac{\varepsilon_x - \varepsilon_y}{2}\cos(2\cdot 0°) + \frac{\gamma_{xy}}{2}\sin(2\cdot 0°) = \varepsilon_x \qquad \text{[Eq. 8.51]}$$

Strain gage B measures ε_B ($\theta = 45°$):

$$\varepsilon_B = \varepsilon_{x'}(\theta = 45°) = \frac{\varepsilon_x + \varepsilon_y}{2} + \frac{\varepsilon_x - \varepsilon_y}{2}\cos(2\cdot 45°) + \frac{\gamma_{xy}}{2}\sin(2\cdot 45°)$$

$$= \frac{\varepsilon_x + \varepsilon_y}{2} + \frac{\gamma_{xy}}{2} \qquad \text{[Eq. 8.52]}$$

Strain gage C measures ε_y directly since it is aligned with the y-axis. Hence, $\varepsilon_C = \varepsilon_y$.

Substituting $\varepsilon_x = \varepsilon_A$ and $\varepsilon_y = \varepsilon_C$ into *Equation 8.52* gives γ_{xy} in terms of ε_A, ε_B and ε_C. The strains ε_x, ε_y and γ_{xy} can then be written:

$$\varepsilon_x = \varepsilon_A \ ; \ \ \varepsilon_y = \varepsilon_C \ ; \ \ \gamma_{xy} = 2\varepsilon_B - (\varepsilon_A + \varepsilon_C) \qquad \text{[Eq. 8.53]}$$

Or, in matrix form:

$$\begin{bmatrix} \varepsilon_x \\ \varepsilon_y \\ \gamma_{xy} \end{bmatrix} = \begin{bmatrix} 1 & 0 & 0 \\ 0 & 0 & 1 \\ -1 & 2 & -1 \end{bmatrix} \begin{bmatrix} \varepsilon_A \\ \varepsilon_B \\ \varepsilon_C \end{bmatrix} \qquad \text{[Eq. 8.54]}$$

In practice, the readings ε_A, ε_B and ε_C can be directly fed into a computer and the strains in the x–y plane are then calculated.

Although using a strain gage rosette may seem to be extra work compared to a single gage, the rosette has several advantages. In a tensile test, if the axial member is properly aligned in the testing machine, then the maximum principal strain from the strain rosette should equal to the axial strain of the specimen, regardless of the rosette's orientation on the specimen. The exact alignment of the strain rosette on a tensile specimen is not as critical as when only one gage is used. The minimum principal strain in a tensile test is due to the Poisson effect.

Strain rosettes are also used in large-scale structures (e.g., bridges, buildings, airplanes, etc.) to measure the strains at key locations. The direction of the maximum strain can vary with time, but by using strain rosettes instead of a single gage, the principal strains and maximum shear strains at any point can be determined. If the material properties are known, the principal stresses and maximum shear stresses can be calculated to ensure that the structure remains within design limits. The structure's load and displacement can also be monitored with strain gages.

Example 8.10 Strain Rosette

Given: The strain readings from a 0/45/90 strain rosette (*Figure 8.22*) are:

$$\varepsilon_A = 1000 \times 10^{-6}; \quad \varepsilon_B = -600 \times 10^{-6}; \quad \varepsilon_C = 400 \times 10^{-6}$$

Strain gage A is aligned with the x-axis, gage C with the y-axis and gage B is at $\theta = 45°$.

Required: Determine (a) the strains in the x–y coordinate system and (b) the principal strains.

Solution: Apply the 0/45/90 transformation equations to the gage strains:

$$Answer: \ \varepsilon_x = 1000 \times 10^{-6}$$

$$Answer: \ \varepsilon_y = 400 \times 10^{-6}$$

$$\gamma_{xy} = 2\varepsilon_B - (\varepsilon_A + \varepsilon_C) = [2(-600) - (1000 + 400)]$$

$$Answer: \ \gamma_{xy} = -2600 \times 10^{-6}$$

The principal strains are:

$$\varepsilon_I, \varepsilon_{II} = \frac{\varepsilon_x + \varepsilon_y}{2} \pm \sqrt{\left(\frac{\varepsilon_x - \varepsilon_y}{2}\right)^2 + \left(\frac{\gamma_{xy}}{2}\right)^2}$$

$$= \frac{1000 + 400}{2} \pm \sqrt{\left(\frac{1000 - 400}{2}\right)^2 + \left(\frac{-2600}{2}\right)^2}$$

$$Answer: \ \varepsilon_I, \varepsilon_{II} = 2030 \times 10^{-6}, -634 \times 10^{-6}$$

8.7 Three-Dimensional Stress

Although the *stress transformation equations* were developed for *plane stress* states ($\sigma_z = \tau_{zx} = \tau_{zy} = 0$), they can still be used if σ_z is non-zero (e.g., *Figure 8.23a*). As long as the *out-of-plane* shear stresses are zero ($\tau_{zx} = \tau_{zy} = 0$), then σ_z is a *principal stress*. If the stress element is constrained to rotate in the x–y plane about the z-axis, then σ_z remains constant, and the out-of-plane shear stresses and strains remain zero.

We now consider the effect of the out-of-plane principal stress σ_z on the maximum shear stress. We begin the discussion by considering Mohr's circle for 3D.

First, the Mohr's circle for the *in-plane* (*x–y* plane) stresses is constructed as in *Section 8.4*. Then two additional Mohr's circles are constructed. Each new circle is bound on the σ-axis by one of the *in-plane* principal stresses and the *out-of-plane* principal stress σ_z (e.g., *Figure 8.23c*).

The circles can be quickly drawn by finding the *in-plane* principal stresses:

$$\sigma_I, \sigma_{II}$$

$$= \frac{\sigma_x + \sigma_y}{2} \pm \sqrt{\left(\frac{\sigma_x - \sigma_y}{2}\right)^2 + (\tau_{xy})^2}$$

[Eq. 8.55]

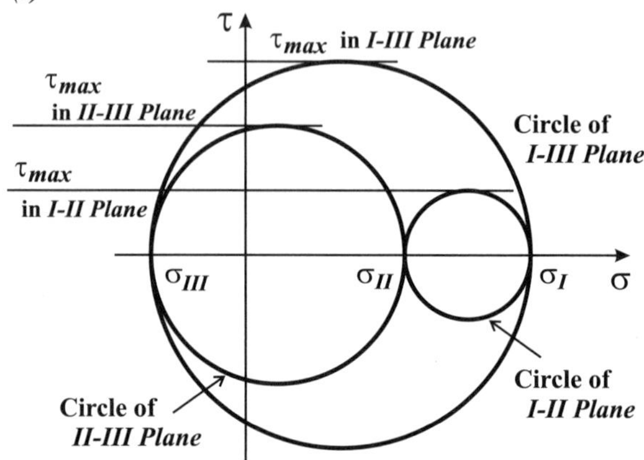

Figure 8.23. (a) The principal stresses in 3D. **(b)** The in-plane principal stresses only. **(c)** The 3D Mohr's circles.

and knowing the *out-of-plane* principal stress, $\sigma_z = \sigma_{III}$. If desired, the principal stresses may be relabeled so that σ_I is the most positive (tensile) stress and σ_{III} is the most negative (compressive). Regardless, with the three principal stresses plotted on the σ-axis, three Mohr's circles can be constructed as shown in *Figure 8.23c*.

The maximum shear stress of the stress element is the radius of the largest of the three Mohr's circles:

$$\tau_{max} = \max\left[\left|\frac{\sigma_I - \sigma_{II}}{2}\right|, \left|\frac{\sigma_{II} - \sigma_{III}}{2}\right|, \left|\frac{\sigma_I - \sigma_{III}}{2}\right|\right]$$

[Eq. 8.56]

If σ_I and σ_{III} are the maximum and minimum principal stresses, respectively, then half of their difference is the maximum shear stress. This shear stress may act in the *x–y* plane (*in-plane*) or may act *out-of-plane* (if either σ_I or σ_{III} is associated with σ_z).

If the stress state is *plane stress*, then only the two *in-plane* principal stresses are non-zero and the *out-of-plane* principal stress is zero (i.e., $\sigma_z = \sigma_{III} = 0$), resulting in:

$$\tau_{max} = \max\left[\left|\frac{\sigma_I - \sigma_{II}}{2}\right|, \left|\frac{\sigma_I}{2}\right|, \left|\frac{\sigma_{II}}{2}\right|\right]$$

[Eq. 8.57]

where σ_I and σ_{II} are in the x–y plane. For *plane stress*, if the *in-plane* principal stresses are of the *same sign*, the maximum shear stress is *out-of-plane*. If the *in-plane* principal stresses are of *opposite sign*, the maximum shear stress is *in-plane*.

Example 8.11 Cylindrical Pressure Vessel

Given: A cylindrical pressure vessel (*Figure 8.24*) contains a gas at 90 psi. The radius is $R = 24$ in. and the thickness is $t = 0.40$ in. The axes of the vessel are H-, L-, and r-, where H is around the cylinder (hoop-direction), L is along the axis of the cylinder (longitudinal), and r is normal to the cylinder (radial).

Required: Determine (a) the maximum *in-plane* shear stress in the cylinder wall (in the H–L plane), and the angle at which it acts and (b) the maximum shear stress in the system, whether *in-plane* or *out-of-plane*.

Solution: *Step 1.* The stresses acting on an element in the body of the vessel are found first. The wall is in a *biaxial state* of stress:

$$\sigma_H = \frac{pR}{t} = \frac{(90\ \text{psi})(24\ \text{in.})}{(0.40\ \text{in.})}$$

$$= 5.4\ \text{ksi} = \sigma_I$$

$$\sigma_L = \frac{pR}{2t} = 2.7\ \text{ksi} = \sigma_{II}$$

$$\sigma_r = 0\ \text{ksi} = \sigma_{III}$$

No shear stress acts on an element oriented in the H–L set of axis. The hoop and longitudinal stresses are the principal stresses in the cylinder's wall (*Figure 8.24c*). The surface is free of stress, so $\sigma_r = 0$.

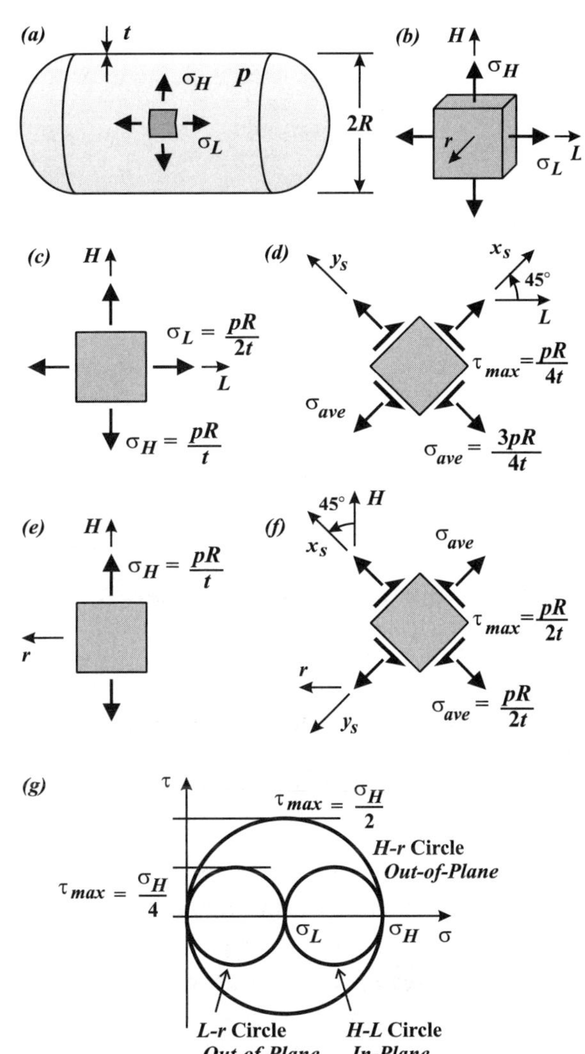

Figure 8.24. (a) Cylindrical pressure vessel. **(b)** Stress element on surface oriented with H–L–r axes. **(c)** A 2D element in H–L plane. **(d)** Maximum Shear Stress in H–L plane. **(e)** A 2D element in H–r plane. **(f)** Maximum Shear Stress in H–r plane. **(g)** A 3D Mohr's circles for surface element.

Step 2. The *maximum in-plane shear stress* is half the difference of the *in-plane* principal stresses, σ_I and σ_{II}. Thus:

$$\tau_{max,in} = \frac{\sigma_I - \sigma_{II}}{2} = \frac{1}{2}\left(\frac{pR}{t} - \frac{pR}{2t}\right) = \frac{pR}{4t}$$

Answer: $\underline{\tau_{max,in} = 1.35 \text{ ksi}}$

The normal stresses associated with the maximum in-plane shear stress are:

$$\sigma_{x's} = \sigma_{y's} = \frac{\sigma_I + \sigma_{II}}{2} = \frac{1}{2}\left(\frac{pR}{t} + \frac{pR}{2t}\right) = \frac{3PR}{4t} = 4.05 \text{ ksi}$$

Step 3. The angle of the plane in which the *in-plane* maximum shear stress occurs is defined by:

$$\tan 2\theta_s = -\frac{\sigma_x - \sigma_y}{2\tau_{xy}}$$

But $\tau_{xy} = 0$ in the original stress state, which means that $\tan(2\theta_s)$ is undefined. Thus, $2\theta_s = \pm 90°$ or:

Answer: $\underline{\theta_s = \pm 45°}$

The direction of the shear stress (positive or negative) on the plane defined by $\theta_s = 45°$ can be found by applying the general shear stress transformation equation. The result is shown in *Figure 8.24d*.

Step 4. In 3D, the maximum shear stress is given by:

$$\tau_{max} = \left|\frac{\sigma_I - \sigma_{III}}{2}\right|$$

where σ_I and σ_{III} are the most tensile (positive) and most compressive (negative) of the three principal stresses. Here, the maximum principal stress is the hoop stress, σ_H, and the minimum principal stress is the stress normal to the surface of the cylinder, $\sigma_r = 0$. One of these extreme stresses, σ_r, is *out-of-plane*, so the maximum shear stress is *out-of-plane* (*Figure 8.24e*).

The *maximum shear stress* is:

$$\tau_{max} = \left|\frac{\sigma_I - \sigma_{III}}{2}\right| = \left|\frac{\sigma_H - \sigma_r}{2}\right| = \frac{1}{2}\left[\frac{pR}{t} - 0\right] = \frac{pR}{2t}$$

Answer: $\underline{\tau_{max} = 2.7 \text{ ksi}}$

The *out-of-plane* maximum shear stress occurs on a plane oriented at 45° and passing through the hoop–radial (*H–r*) plane (*Figure 8.24f*). Not shown in this view is σ_L, which acts out of the plane of the paper. For a cylindrical pressure vessel, the maximum shear stress is *out-of-plane*, and is equal to half the hoop stress. The 3D set of Mohr's circles is shown in *Figure 8.24g*. It is made of three 2D Mohr's circles.

In general, when using the 3D Mohr's circle for stress transformation, only one circle may be traveled on at a time, the element rotating about one of the principal axes. With σ_r, σ_H, and σ_L as the principal stresses:

- σ_r is kept constant while rotating the element in the *H–L* plane about the *r*-axis, to get general stresses, $\sigma_{H'}$, $\sigma_{L'}$, and $\tau_{H'L'}$, in the *H–L* plane; or
- σ_H is kept constant while rotating the element in the *L–r* plane about the *H*-axis, to get general stresses, $\sigma_{L'}$, $\sigma_{r'}$, and $\tau_{L'r'}$, in the *L–r* plane; or
- σ_L is kept constant while rotating the element in the *H–r* plane about the *L*-axis, to get general stresses, $\sigma_{r'}$, $\sigma_{H'}$, and $\tau_{r'H'}$, in the *H–r* plane.

In general, 3D case, stresses are represented as *tensors*, and rotations in 3D space are developed using tensor mathematics. Such mathematics is beyond the scope of this text.

9.0 Introduction

In *Chapter 3*, two general classifications of materials were introduced: *brittle* and *ductile* (*Figure 9.1*). The stress criteria used to determine if a brittle or ductile material will fail are discussed in this chapter.

Brittle materials show little or no plastic deformation before they ***fracture*** (break into two parts, *Figure 9.2a*). Ceramics, chalk, cast iron, and concrete are brittle. The small amount of deformation beyond the proportional limit provides little warning of impending failure.

Ductile materials yield (permanently deform) long after the strain reaches the linear–elastic limit (*Figure 9.2b*). A paper clip can be bent by a large amount, resulting in a large permanent deformation, before it breaks in two. Aluminum, steel, and copper are ductile. The onset of *yielding* or *plastic deformation* is frequently used as a design limitation.

The question to be addressed in this chapter is how to determine if a material fails under a complex state of stress, such as a general state of plane stress, σ_x, σ_y, and τ_{xy}. In *Chapter 8*, it was learned that a plane stress state can be reduced to two principal stresses, σ_I and σ_{II}. It is much easier to develop failure criteria based on principal stresses and strains, and maximum shear stresses, and this is indeed the approach taken.

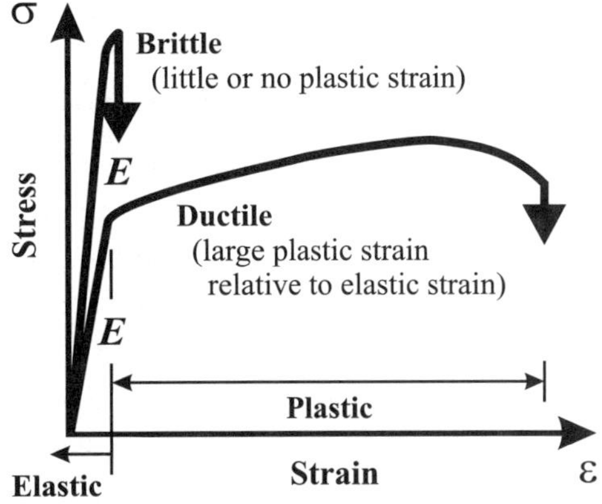

Figure 9.1. Representative stress–strain curves of brittle and ductile materials.

(a)

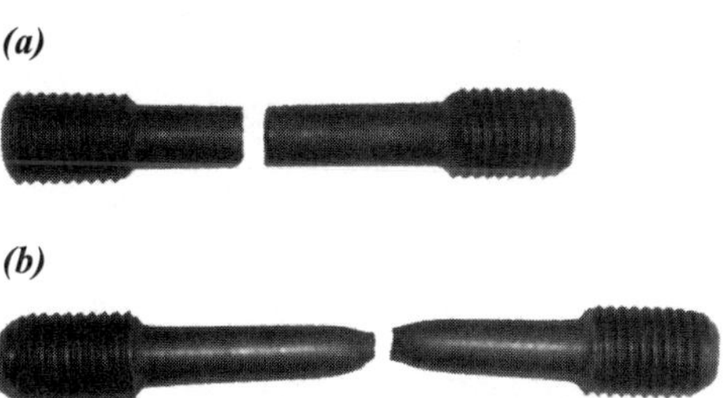

(b)

Figure 9.2. (a) Failure of a brittle (cast iron) tensile test specimen (0.5 in. diameter). Little plastic deformation occurs; the failure surface is flat. **(b)** Failure of a ductile (steel A36) tensile test specimen. Large plastic strains and necking occur before failure.

F.A. Leckie, D.J. Dal Bello, *Strength and Stiffness of Engineering Systems*, Mechanical Engineering Series, **277**
DOI 10.1007/978-0-387-49474-6_9, © Springer Science+Business Media, LLC 2009

9.1 Failure Condition for Brittle Materials

In a uniaxial tension test, a brittle material fractures when the applied stress reaches the *ultimate tensile strength* of the material, S_u. In a compression test, the material fails when the compressive stress reaches the *compressive strength* of the material, S_C. For materials such as concrete and ceramics, the magnitude of S_C can be approximately 10 times that of S_u.

Maximum Normal Stress Criterion

For many brittle materials, failure occurs when the maximum normal stress in any direction reaches either the tensile or compressive strength of the material. Thus, finding the principal stresses – the maximum (most tensile) and minimum (most compressive) normal stresses – at critical points is necessary to assess the integrity of a brittle system.

For a plane stress condition, a material element is subjected to stresses σ_x, σ_y and τ_{xy}. The principal stresses, σ_I and σ_{II}, can be found using the methods of *Chapter 8*. If either principal stress equals or exceeds the strength of the material, either S_u in tension or S_C in compression, the material fails:

$$\sigma_I \geq S_u \quad \text{or} \quad \sigma_{II} \geq S_u$$
$$|\sigma_I| \geq S_C \quad \text{or} \quad |\sigma_{II}| \geq S_C$$

[Eq. 9.1]

This is called the ***maximum normal stress failure criterion***. No restriction has been made on the relative values of σ_I and σ_{II} in *Equation 9.1*.

The square-shaded region shown in *Figure 9.3* represents coordinates (σ_I, σ_{II}) for which the normal stresses do not exceed the failure condition. If the principal stresses at a material point are plotted (σ_I, σ_{II}), and the plotted point lies on or outside the strength boundary, the material fails.

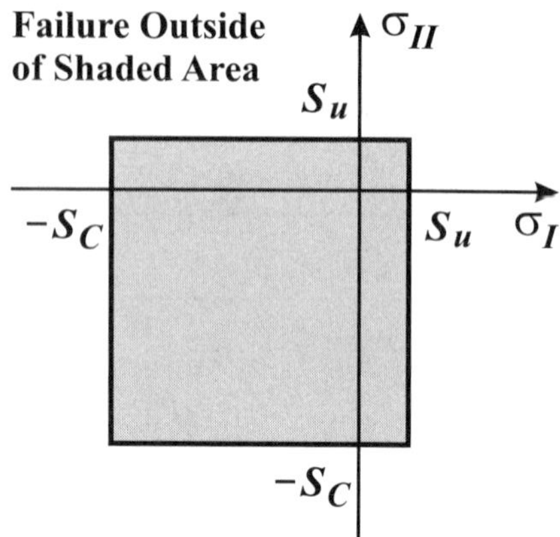

Figure 9.3. Failure Map for Maximum Normal Stress Failure Criterion. When the calculated principal stress point (σ_I, σ_{II}) is plotted on the $\sigma_I - \sigma_{II}$ axes, if it falls within the shaded region, there is no failure.

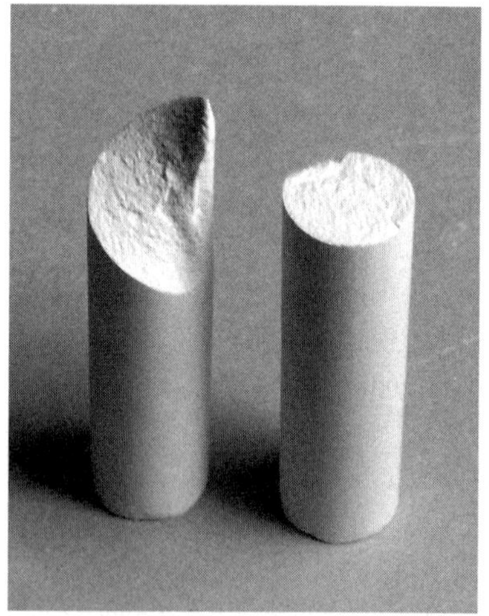

Figure 9.4. Pieces of chalk that have failed in torsion (*left*) and in bending (*right*). The analyses of these failures is given in *Examples 9.1* and *9.2*.

Examples of brittle materials that follow the maximum normal stress failure criterion are cast iron, cement, ceramics, and chalk (*Figure 9.4*).

Example 9.1 Twisting a Piece of Chalk

Given: Chalk is a brittle material that fails according to the maximum normal stress criterion. A piece of chalk is twisted with torque T until it fractures (*Figure 9.5*).

Required: Determine the angle of the fracture plane.

Solution: The torque causes a shear stress τ on a surface element of the piece of chalk in the x–y plane (parallel and normal to the chalk's axis). The state of stress is pure shear (*Figure 9.5b*).

The principal stresses are $\pm 45°$ from the x–y axes:

$$\sigma_I = +\tau \quad ; \quad \sigma_{II} = -\tau$$

Failure occurs when the maximum principal stress reaches the ultimate tensile strength S_u.

The maximum tensile stress occurs at $+45°$ from the x-axis, which defines the fracture plane. Failure occurs along the plane perpendicular to the maximum tensile stress; a crack will open along the $-45°/+135°$ diagonal, normal to σ_I (*Figures 9.5c, d*). Due to the twisting nature of the torsion load, the fracture surface spirals about the chalk's axis (*Figures 9.4, 9.5d*).

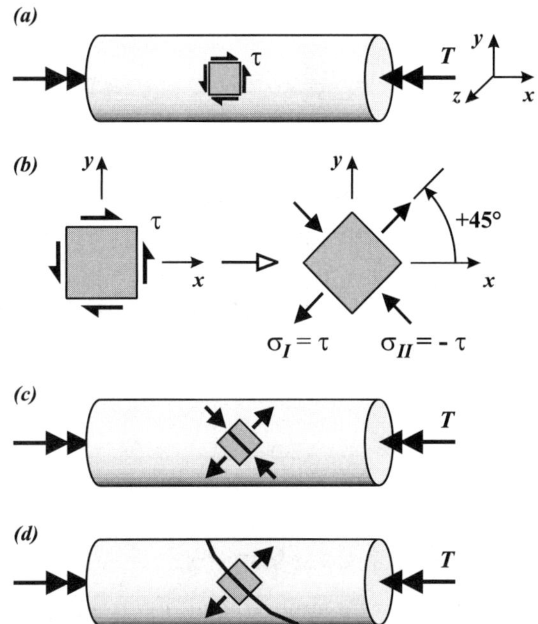

Figure 9.5. (a) A piece of chalk under torque *T*. A surface element is in a state of pure shear stress. **(b)** The element rotated 45° to the principal angles. **(c)** The tensile principal stress causes a crack to open along the −45°/+135° diagonal. **(d)** The fracture surface spirals about the chalk's axis.

Example 9.2 Bending a Piece of Chalk

Given: A piece of chalk is loaded as a beam by bending moment M causing compression at the bottom of the beam (a negative moment), as shown in *Figure 9.6*.

Required: Determine the angle of the fracture plane.

Solution: In pure bending, the only stresses acting in the piece of chalk are bending stresses. Since there is no shear stress, the maximum tensile bending stress is the maximum principal stress, which occurs at the top of the beam (*Figure 9.6b*):

$$\sigma_I = +\frac{MR}{I}$$

Failure occurs when σ_I reaches S_u.

The maximum tensile stress coincides with the x-axis. The crack opens and runs normal to the maximum tensile stress (*Figure 9.6c*). The fracture surface is flat and normal to the x-axis (*Figure 9.4*).

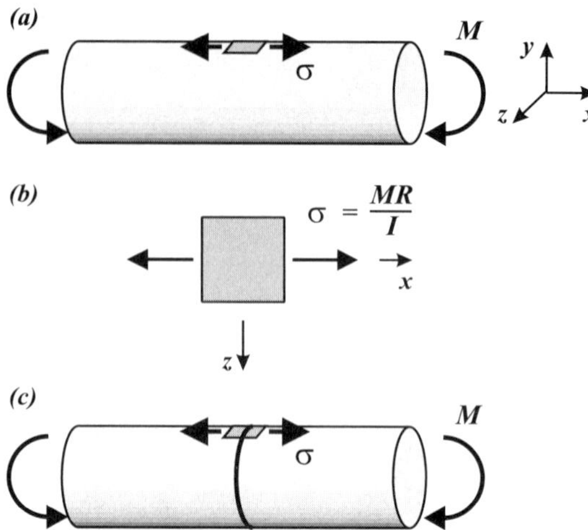

Figure 9.6. (a) Piece of chalk under moment *M*. A surface element at the top is under a state of uniaxial stress. **(b)** The surface element viewed from the top; the stresses are principal stresses. **(c)** The fracture surface runs normal to the chalk's axis.

Example 9.3 Cast Iron

Given: Under an applied load, the principal stresses at a critical point in a gray cast iron diesel engine are calculated to be $\sigma_I = 40$ ksi and $\sigma_{II} = -90$ ksi. Take the ultimate strength of cast iron to be $S_u = 30$ ksi and its compressive strength to be $S_C = 120$ ksi.

Required: (a) Plot the maximum normal stress criterion map. (b) Using the maximum normal stress criterion, does the cast iron fail under the applied state of stress?

Solution: The shaded area in *Figure 9.7* is the region in which a set of principal stresses, (σ_I, σ_{II}), does not cause failure.

The maximum tensile stress is:

$$\sigma_I = 40 \text{ ksi} > 30 \text{ ksi} = S_u$$

Since $\sigma_I > S_u$, the material fails in the *I*-direction due to tension.

If σ_I is reduced below 30 ksi, the system will not fail.

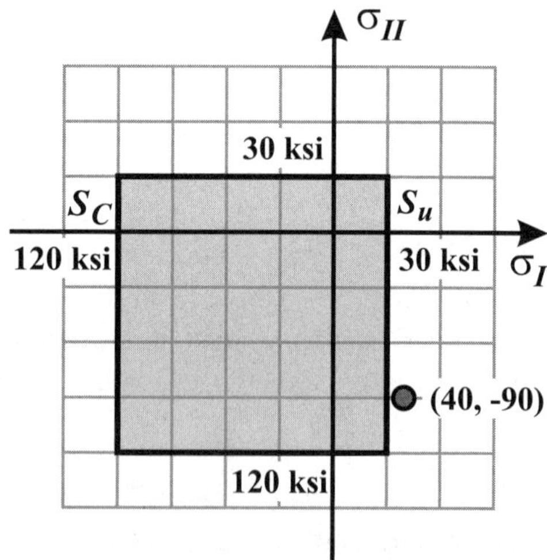

Figure 9.7. Failure Map for the cast iron of *Example 9.3*. Principal stress point (σ_I, σ_{II}) lies outside the shaded area, and the material is predicted to fail.

Example 9.4 Femur Bone

Given: The shaft of a femur (thigh bone) can be approximated as a hollow cylindrical shaft (*Figure 9.8*). The loads that tend to cause femur bones to fracture are torques and bending moments.

A particular femur has an outside diameter of $D = 24$ mm and an inside diameter of $D_i = 16$ mm. The ultimate tensile strength of bone is $S_u = 120$ MPa. During a strenuous activity, such as skiing, the femur is subjected to a torque of $T = 100$ N·m.

Required: Under the applied torque, determine the maximum bending moment M that the bone can support without failure. Consider only torsion and bending loads, and assume the bone to be a brittle material.

Solution: *Step 1.* Since the bone is brittle, it follows the maximum normal stress criterion. If the calculated maximum principal stress is greater than the material ultimate strength, $\sigma_I > S_u$, the bone fractures.

Step 2. The loading is a torque and a bending moment, so the pertinent geometric terms for the hollow cylinder (femur) are:

$$I = \frac{\pi}{64}\left(D^4 - D_i^4\right)$$

$$= \frac{\pi}{64}\left[(0.024 \text{ m})^4 - (0.016 \text{ m})^4\right]$$

$$= 13.1 \times 10^{-9} \text{ m}^4$$

$$J = \frac{\pi}{32}\left(D^4 - D_i^4\right)$$

$$= \frac{\pi}{32}\left[(0.024 \text{ m})^4 - (0.016 \text{ m})^4\right]$$

$$= 26.1 \times 10^{-9} \text{ m}^4$$

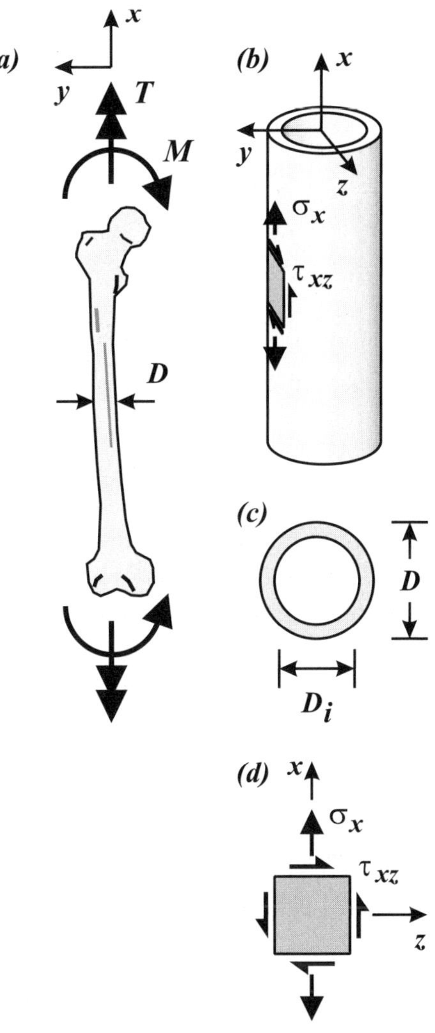

Figure 9.8. (a) Femur bone under torque *T* and moment *M*. **(b)** A length of femur modeled as a hollow cylindrical shaft. **(c)** Bone cross-section. **(d)** Stress element with the most tensile bending stress (on the *left side*).

Step 3. Consider a stress element on the left side of the femur (*Figure 9.8b*); the element is in the z–x plane. The maximum tensile bending stress occurs at this point, and combined with the shear stress, this is likely the location of the maximum principal stress (*Figure 9.8d*). The stresses on this element are:

$$\sigma_x = \frac{MR}{I} = \frac{M(0.012 \text{ m})}{(13.1 \times 10^{-9} \text{ m}^4)}$$

$$\sigma_z = 0$$

$$\tau_{xz} = \frac{TR}{J} = \frac{(100 \text{ N·m})(0.012 \text{ m})}{(26.1 \times 10^{-9} \text{ m}^4)} = 45.9 \text{ MPa}$$

The moment M that causes failure is to be determined.

The maximum principal stress in the x–z plane is:

$$\sigma_I = \frac{\sigma_x + \sigma_z}{2} + \sqrt{\left(\frac{\sigma_x - \sigma_z}{2}\right)^2 + (\tau_{xz})^2}$$

At fracture, three of the variables in the principal stress equation are known: $\sigma_I = S_u = 120$ MPa, $\sigma_z = 0$, and $\tau_{xz} = 45.9$ MPa. It remains to solve for σ_x, which is a function of bending moment M. At failure:

$$S_u = \frac{\sigma_x}{2} + \sqrt{\left(\frac{\sigma_x}{2}\right)^2 + \tau_{xz}^2}$$

Isolating the radical term and squaring both sides gives:

$$\left(S_u - \frac{\sigma_x}{2}\right)^2 = \left(\frac{\sigma_x}{2}\right)^2 + \tau_{xz}^2$$

Solving for σ_x:

$$\sigma_x = \frac{S_u^2 - \tau_{xz}^2}{S_u} = \frac{120^2 - 45.9^2}{120} = 102.4 \text{ MPa}$$

When the bending stress reaches 102.4 MPa, the femur will break. The moment to cause fracture is:

$$M = \frac{\sigma_x I}{R} = \frac{(102.4 \text{ MPa})(13.1 \times 10^{-9} \text{ m}^4)}{(0.012 \text{ m})}$$

Answer: $\underline{M = 112 \text{ N·m}}$

Modified Maximum Stress Criterion

Some brittle materials, such as concrete and rock, exhibit a more complex failure criterion in which shear stress also has an influence. This modifies the failure map as shown in *Figure 9.9*. The principal stress must now lie within the boundaries defined by the *normalized* equations summarized in *Table 9.1*.

When the *in-plane* principal stresses are of opposite sign, the in-plane shear stress becomes important in failure. Thus, the diagonal modifications to *Figure 9.3*, resulting in *Figure 9.9*. The source of the sloped lines – shear failure – is discussed further in *Section 9.2*.

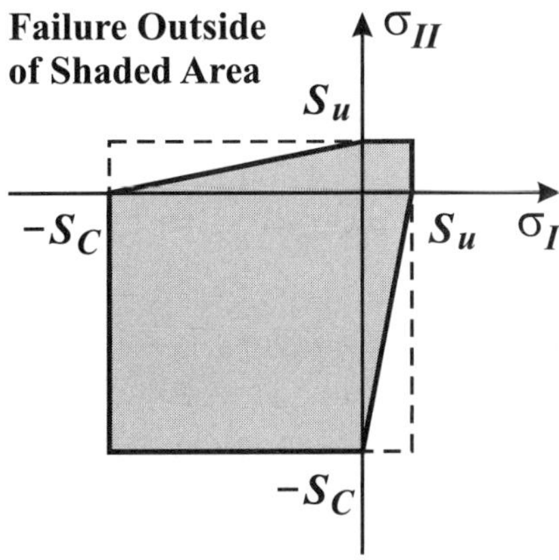

Figure 9.9. Modified maximum normal stress failure map for brittle materials that also fail by in-plane shear stress, i.e., when principal stresses are of opposite signs.

Table 9.1. Conditions for modified maximum stress criterion.

Principal Stresses, Quadrant	Failure Type	Equations that must be satisfied for Principal Stresses to lie within the Boundaries of *Figure 9.9*	
$\sigma_I > 0,\ \sigma_{II} > 0$ 1st Quadrant	Tensile	$\dfrac{\sigma_I}{S_u} < 1;\quad \dfrac{\sigma_{II}}{S_u} < 1$	[Eq. 9.2]
$\sigma_I < 0,\ \sigma_{II} < 0$ 3rd Quadrant	Compressive	$\dfrac{\left\|\sigma_I\right\|}{S_C} < 1;\quad \dfrac{\left\|\sigma_{II}\right\|}{S_C} < 1$	[Eq. 9.3]
$\sigma_I < 0,\ \sigma_{II} > 0$ 2nd Quadrant	Shear-dominated	$\dfrac{\left\|\sigma_I\right\|}{S_C} + \dfrac{\sigma_{II}}{S_u} < 1$	[Eq. 9.4]
$\sigma_I > 0,\ \sigma_{II} < 0$ 4th Quadrant	Shear-dominated	$\dfrac{\sigma_I}{S_u} + \dfrac{\left\|\sigma_{II}\right\|}{S_C} < 1$	[Eq. 9.5]

9.2 Failure Condition for Onset of Yielding of Ductile Materials

In a uniaxial tension test, a ductile material begins to yield when the applied axial stress reaches the yield strength S_y. In a compression test, a ductile material yields when $\sigma = -S_y$. In a torsion test on a thin-walled shaft, the material yields when $\tau_{max} = \tau_y$, the shear yield strength.

Stress states are usually more complicated than simple tension or pure shear. In *plane stress*, a material point is subjected to a general state of stress, σ_x, σ_y, and τ_{xy}. The goal here is to predict when a ductile material yields under a general state of stress. There are two basic theories to determine when a ductile material yields: the **Tresca** and the **von Mises** *failure criteria*.

Maximum Shear Stress Criterion (Tresca)

Based on experimental observations, Tresca proposed that yielding occurs when the maximum shear stress τ_{max} reaches the *shear yield strength* τ_y. In a *tension test*, yielding occurs when the applied stress σ_x reaches the uniaxial yield strength S_y. The maximum shear stress in a uniaxial test occurs on a plane at $45°$ to the axial direction, and $\tau_{max} = \sigma_x/2$ (*Figure 9.10*).

It, therefore, follows that the relationship between the shear yield strength and the axial yield strength is:

$$\tau_y = \frac{S_y}{2} \qquad \text{[Eq. 9.6]}$$

The shear stress is parallel to an interior surface. The direction of the

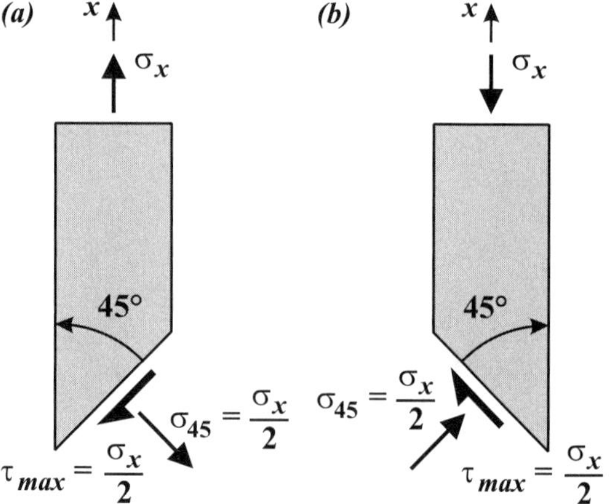

Figure 9.10. The magnitude of the maximum shear stress is $\sigma_x/2$ for both **(a)** the uniaxial tension test, and **(b)** the uniaxial compression test. Per Tresca, yielding occurs when $\tau_{max} = \tau_y$.

shear stress does not matter; when the magnitude of τ_{max} reaches τ_y, the material yields. Thus, for a *compression test*, the material yields at the same magnitude of stress as in tension, i.e., $|\sigma_x| = S_y$ or $\sigma_x = -S_y$ (*Figure 9.10b*).

For the plane stress condition ($\sigma_z = \tau_{yz} = \tau_{zx} = 0$), with principal stresses σ_I and σ_{II} in the x–y plane, the maximum shear stress is:

$$\tau_{max} = \max\left[\frac{|\sigma_I - \sigma_{II}|}{2}, \frac{|\sigma_I|}{2}, \frac{|\sigma_{II}|}{2} \right] \qquad \text{[Eq. 9.7]}$$

Whenever τ_{max} reaches τ_y, yielding occurs.

The failure map for the *plane stress Tresca yield criterion* is shown in *Figure 9.11*. The principal stresses (σ_I, σ_{II}) at a critical point are calculated and then plotted on the map. Yielding does not occur if the principal stress point (σ_I, σ_{II}) lies within the Tresca boundary.

The sloped boundary lines each have a slope of unity (1.0); the σ_{II}-intercepts are at S_y and $-S_y$. The sloped lines correspond to the maximum shear stress at a material point being an *in-plane* shear stress; σ_I and σ_{II} have opposite signs. The normalized equations of the boundaries for the Tresca (maximum shear stress) criterion are summarized in *Table 9.2*.

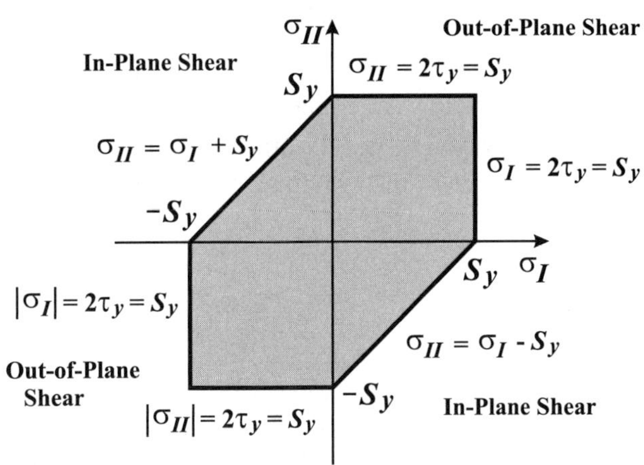

Figure 9.11. Failure Map of *Plane Stress Tresca Yield Criterion*. If the principal stress point (σ_I, σ_{II}) is plotted on the $\sigma_I - \sigma_{II}$ axes, and it falls within the shaded hexagonal region, the material does not yield. The diagonal boundaries are due to *in-plane* shear stresses caused by the principal stresses being of opposite signs.

Table 9.2. Normalized equations for boundaries of Tresca Yield Criterion.

In-Plane Principal Stresses	Shear Failure	Equations that must be satisfied for Principal Stresses to lie within the Tresca boundaries (*Figure 9.11*)	
$\sigma_I > 0,\ \sigma_{II} > 0$ 1st Quadrant	Out-of-plane	$\dfrac{\sigma_I}{S_y} < 1; \quad \dfrac{\sigma_{II}}{S_y} < 1$	[Eq. 9.8]
$\sigma_I < 0,\ \sigma_{II} < 0$ 3rd Quadrant	Out-of-plane	$\dfrac{\left|\sigma_I\right|}{S_y} < 1; \quad \dfrac{\left|\sigma_{II}\right|}{S_y} < 1$	[Eq. 9.9]
$\sigma_I < 0,\ \sigma_{II} > 0$ 2nd Quadrant	In-plane	$\dfrac{\left|\sigma_I\right|}{S_y} + \dfrac{\sigma_{II}}{S_y} < 1$	[Eq. 9.10]
$\sigma_I > 0,\ \sigma_{II} < 0$ 4th Quadrant	In-plane	$\dfrac{\sigma_I}{S_y} + \dfrac{\left|\sigma_{II}\right|}{S_y} < 1$	[Eq. 9.11]

von Mises Failure Criterion (Maximum Distortion Energy)

Another yield criterion, proposed by von Mises, is derived considering the strain energy that develops at a material point. In the ***von Mises Criterion***, the ***Equivalent Stress*** or ***von Mises Stress***, σ_o, is first calculated. In three-dimensions, the von Mises Stress is defined in terms of the principal stresses as follows:

$$2\sigma_o^2 = (\sigma_I - \sigma_{II})^2 + (\sigma_{II} - \sigma_{III})^2 + (\sigma_{III} - \sigma_I)^2$$

$$\sigma_o = \sqrt{\frac{(\sigma_I - \sigma_{II})^2 + (\sigma_{II} - \sigma_{III})^2 + (\sigma_{III} - \sigma_I)^2}{2}}$$

[Eq. 9.12]

When the von Mises Stress (Equivalent Stress) σ_o equals or exceeds the material yield strength S_y, then the material yields. The underlying physics implies that if enough *energy* is put into the system, then the material will yield.

For *plane stress*, one principal stress is zero, so the von Mises stress reduces to:

$$\sigma_o = \left[\sigma_I^2 - \sigma_I \sigma_{II} + \sigma_{II}^2\right]^{1/2}$$

[Eq. 9.13]

Yielding occurs when $\sigma_o = S_y$.

When plotted on the $\sigma_I - \sigma_{II}$ axes, the *plane stress von Mises Yield criterion* is an ellipse that circumscribes the *Tresca criterion* (*Figure 9.12*). Yielding does not occur if the principal stress point (σ_I, σ_{II}) lies within the elliptical boundary.

Experimental results suggest that the von Mises criterion is more accurate than the Tresca criterion for most ductile metals. Computer stress analysis tools generally include the von Mises stress and the maximum shear stress as outputs.

The Tresca and von Mises failure boundaries meet at six points: $(\pm S_y, 0)$, $(0, \pm S_y)$, (S_y, S_y), and $(-S_y, -S_y)$. Where the von Mises ellipse intercepts the $\sigma_I - \sigma_{II}$ axes represents axial-only loading conditions, so yielding occurs when $\sigma = \pm S_y$.

It is of interest to note that in the von Mises criterion, σ_I can exceed S_y when σ_{II} is

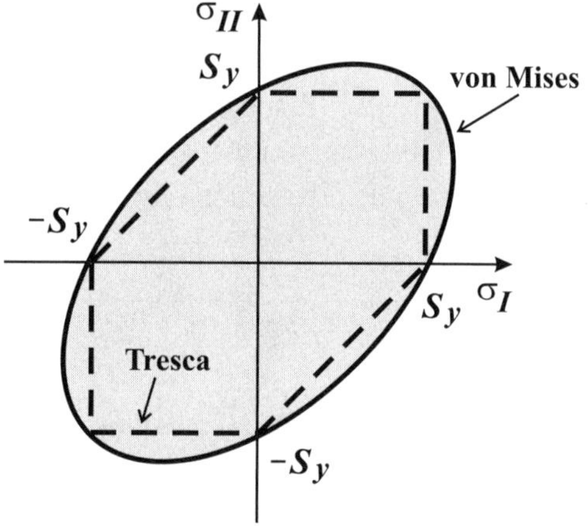

Figure 9.12. Failure Map of *Plane Stress von Mises Yield Criterion*. If the Principal stress point (σ_I, σ_{II}) falls within the elliptical von Mises boundary, the material does not yield. The von Mises ellipse bounds the hexagonal Tresca boundary (dashed), and meets the Tresca boundary at six points. For the von Mises Criterion, a principal stress can exceed the yield strength S_y when the other principal stress has the same sign. This is due to the Poisson effect on the strain energy density.

of the same sign (consider how the Poisson effect influences the strain energy density when the stress element is subjected to biaxial tension).

General von Mises Criterion (Plane Stress)

For a general state of plane stress, σ_x, σ_y, τ_{xy}, the von Mises stress is:

$$\sigma_o = \left[\sigma_x^2 - \sigma_x \sigma_y + \sigma_y^2 + 3\tau_{xy}^2 \right]^{1/2} \qquad \text{[Eq. 9.14]}$$

Yielding occurs when $\sigma_o = S_y$.

In a *uniaxial tension test* along the x-axis, $\sigma_x = \sigma_I$, and $\sigma_{II} = \sigma_{III} = \tau = 0$. The von Mises stress is $\sigma_o = \sigma_x$, so yielding occurs when $\sigma_x = S_y$. *Equation 9.14* for the uniaxial yield case is confirmed.

In the *pure shear test*, $\sigma_x = 0$, $\sigma_y = 0$ and $\tau_{xy} = \tau$. The von Mises stress is:

$$\sigma_o = \left[3\tau_{xy}^2 \right]^{1/2} = \sqrt{3}\,\tau \qquad \text{[Eq. 9.15]}$$

When $\sigma_o = S_y$, the material yields. Thus the relationship between the shear yield strength and axial yield strength, according to the von Mises criterion, is:

$$\tau_y = \frac{S_y}{\sqrt{3}} \qquad \text{[Eq. 9.16]}$$

This relationship was introduced in *Chapter 3* when discussing the torsion test. This is a different relationship than the Tresca assumption for the shear yield strength, where $\tau_y = S_y/2$.

Example 9.5 Yielding in a Cylindrical Pressure Vessel

Given: A thin-walled cylindrical pressure vessel (*Figure 9.13*) contains a gas at pressure p. The vessel radius is R and its wall thickness is t. The axial yield strength from tensile experiments performed on the material is S_y.

Required: Determine the pressure at yield using (a) the Tresca criterion and (b) the von Mises yield criterion.

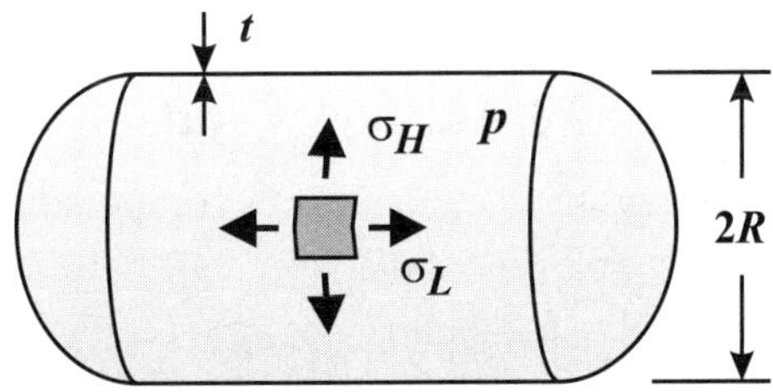

Figure 9.13. Cylindrical pressure vessel.

Solution: *Step 1.* The in-plane principal stresses in a cylindrical pressure vessel are the hoop and longitudinal stresses:

$$\sigma_I = \sigma_H = \frac{pR}{t}; \quad \sigma_{II} = \sigma_L = \frac{pR}{2t}$$

Step 2. Apply the *Tresca criterion.* Since both in-plane stresses are of the same sign, the maximum shear stress is out-of-plane, and is given by:

$$\tau_{max} = \frac{\sigma_I}{2} = \frac{pR}{2t}$$

In the Tresca criterion, yielding occurs when $\tau_{max} = \tau_y = S_y/2$. The pressure to cause yielding is:

$$p_y = \frac{2t\tau_y}{R}$$

$$Answer:\ p_y = \frac{t}{R}S_y$$

Step 3. Apply the *von Mises criterion.* The equivalent stress is:

$$\sigma_o = \left[\sigma_I^2 - \sigma_I\sigma_{II} + \sigma_{II}^2\right]^{1/2} = \left[\left(\frac{pR}{t}\right)^2 - \left(\frac{pR}{t}\right)\left(\frac{pR}{2t}\right) + \left(\frac{pR}{2t}\right)^2\right]^{1/2} = \sqrt{\frac{3}{4}}\left(\frac{pR}{t}\right)$$

Yielding occurs when $\sigma_o = S_y$. The pressure to cause yielding is:

$$p_y = \sqrt{\frac{4}{3}}\left(\frac{t}{R}\right)S_y$$

$$Answer:\ p_y = 1.15\frac{t}{R}S_y$$

For this case, the von Mises Criterion predicts a yield pressure 15% greater than that predicted using Tresca. Although the Tresca criterion is generally easier to apply than the von Mises criterion, the von Mises criterion is generally in better agreement with the actual response of most metals. The Tresca criterion gives conservative results as it usually underestimates the strength of the system.

Example 9.6 Aluminum under a General State of Plane Stress

Given: A stress element in a component made of Aluminum 6061-T6 is subjected to the following plane stress state:

$$\sigma_x = 10.0\ \text{ksi};\quad \sigma_y = 6.0\ \text{ksi};\quad \tau_{xy} = -4.0\ \text{ksi}$$

The axial yield strength of Aluminum 6061-T6 used in design is $S_y = 35$ ksi.

Required: Determine (a) the factor of safety using the Tresca criterion (assuming $\tau_y = S_y/2 = 17.5$ ksi) and (b) the factor of safety using the von Mises criterion.

Step 1. For plane stress states, the maximum shear stress at a point is:

$$\tau_{max} = \max\left[\frac{|\sigma_I - \sigma_{II}|}{2}, \frac{|\sigma_I|}{2}, \frac{|\sigma_{II}|}{2}\right]$$

The in-plane principal stresses are:

$$\sigma_I, \sigma_{II} = \frac{\sigma_x - \sigma_y}{2} \pm \sqrt{\left(\frac{\sigma_x - \sigma_y}{2}\right)^2 + (\tau_{xy})^2}$$

$$= \frac{10.0 + (6.0)}{2} \pm \sqrt{\left(\frac{10.0 - 6.0}{2}\right)^2 + (-4.0)^2} = 12.47, 3.53 \text{ ksi}$$

The in-plane principal stresses arc of the same size sign, thus the maximum shear stress is *out-of-plane*:

$$\tau_{max} = \frac{|\sigma_I|}{2} = \frac{12.47 \text{ ksi}}{2} = 6.23 \text{ ksi}$$

Step 2. For the *Tresca criterion*, with $\tau_y = S_y/2$, the factor of safety is:

$$FS = \frac{\tau_y}{\tau_{max}} = \frac{17.5 \text{ ksi}}{6.23 \text{ ksi}}$$

Answer: FS = 2.81

Step 3. The von Mises stress for a general plane stress element is:

$$\sigma_o = \left[\sigma_x^2 - \sigma_x\sigma_y + \sigma_y^2 + 3\tau_{xy}^2\right]^{1/2}$$

$$= \left[(10)^2 - (10)(6) + (6)^2 + 3(-4.0)^2\right]^{1/2} = 11.13 \text{ ksi}$$

Step 4. For the *von Mises criterion*, the factor of safety is:

$$FS = \frac{S_y}{\sigma_o} = \frac{35 \text{ ksi}}{11.13 \text{ ksi}}$$

Answer: FS = 3.14

The higher *factor of safety* for the von Mises criterion compared to the Tresca criterion implies that the Tresca criterion is more conservative. The von Mises criterion predicts a stronger system.

Comparison of the Yield Criteria

If the axial yield strength S_y is known, then the shear yield strength τ_y as predicted by each method is:

for Tresca: $\tau_y = S_y/2$ for von Mises: $\tau_y = S_y/\sqrt{3}$ [Eq. 9.17]

The ratio of the axial yield strength to the shear yield strength, for each method is then:

for Tresca: $S_y/\tau_y = 2$ for von Mises: $S_y/\tau_y = \sqrt{3} = 1.73$ [Eq. 9.18]

For Aluminum 6061-T6, the tabulated values for design calculations are: $S_y = 35$ ksi and $\tau_y = 20$ ksi. The accepted strength ratio is therefore:

$$S_y/\tau_y = 35/20 = 1.75 \qquad \text{[Eq. 9.19]}$$

The von Mises ratio agrees well with the tabulated ratio, thus, the von Mises criterion is the better model for Aluminum 6061-T6, as it is for most other ductile metals.

The Tresca criterion is generally easier to use in hand calculations, and is a more conservative criterion. Any stress state that satisfies the Tresca criterion automatically satisfies the von Mises criterion (*Figure 9.12*).

3D von Mises Failure Criterion

The von Mises stress is defined by the 3D principal stresses as follows:

$$\sigma_o = \sqrt{\frac{(\sigma_I - \sigma_{II})^2 + (\sigma_{II} - \sigma_{III})^2 + (\sigma_{III} - \sigma_I)^2}{2}} \qquad \text{[Eq. 9.20]}$$

If σ_o equals or exceeds the yield strength S_y, the material yields. An important consequence of *Equation 9.20* is that if the principal stresses are each increased by the same amount, then the value of σ_o remains the same, since the differences in the principal stresses do not change.

The hydrostatic stress is the average of the three principal stresses:

$$p = \frac{\sigma_I + \sigma_{II} + \sigma_{III}}{3} \qquad \text{[Eq. 9.21]}$$

When the three principal stresses at yielding are plotted, the result is the surface of a cylinder whose axis is in the [1,1,1] direction, which makes the same angle with each of the three principal stress axes (*Figure 9.14*). The cylinder intercepts the stress axes at $\pm S_y$. The radius of the cylinder is:

$$R = \sqrt{\frac{2}{3}} S_y \qquad \text{[Eq. 9.22]}$$

As p increases, the location on the cylindrical surface where yielding occurs moves further away from the origin.

In 2D, the intersection of the cylinder with the $\sigma_I - \sigma_{II}$ plane is the plane stress von Mises ellipse.

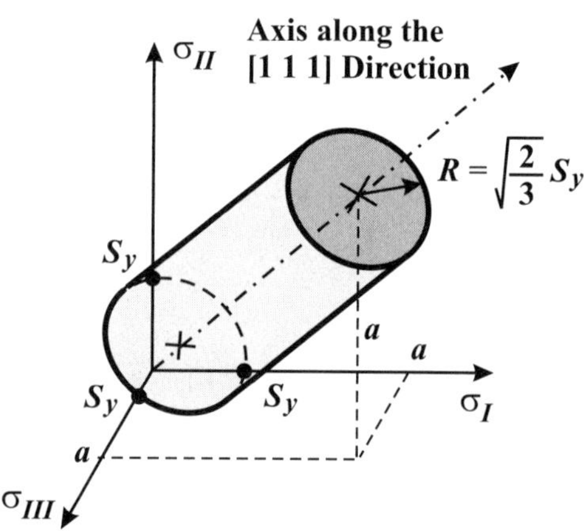

Figure 9.14. In 3D, the *von Mises ellipse* becomes a *cylinder*; the yield boundary is the surface of the cylinder. The cylinder axis is in the [1 1 1] vector direction, and the cylinder intercepts each principal axis at $\sigma = S_y$. The distances a in the figure are included to help illustrate the direction of the cylinder axis.

Example 9.7 Submerged Anchor

Given: At sea-level, the shank of a ship's anchor, with cross-sectional area $A = 5000$ mm^2, yields when a tensile force of 1.25 MN is applied to it. This load corresponds to a yield strength of $S_y = 250$ MPa. At sea, the anchor is submerged and holds a ship in place (*Figure 9.15*).

Required: Determine the applied stress σ_a on the anchor shank to yield the anchor material if it can be submerged to 5.0 km (~3 miles!). Take the density of seawater to be $\rho = 1000$ kg/m^3.

Solution: When submerged at 5000 m, the hydrostatic pressure on the anchor is:

$$p = \rho g h$$
$$= (1000 \text{ kg/m}^3)(9.81 \text{ m/s}^2)(5000 \text{ m})$$
$$= 49.1 \text{ MPa}$$

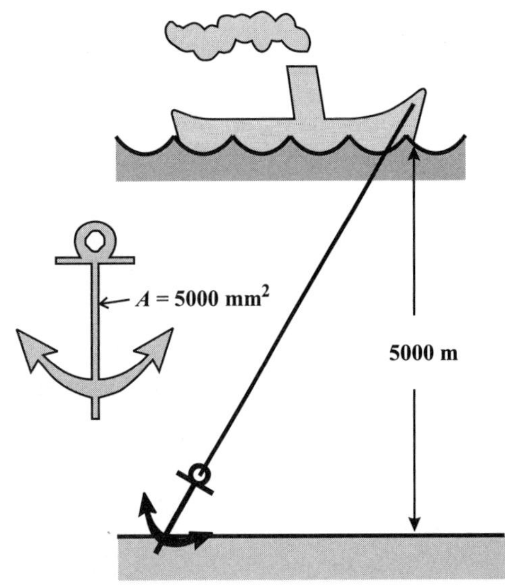

Figure 9.15. Submerged anchor under tension. Copyright ©2008 Dominic J. Dal Bello and licensors. All rights reserved.

The pressure causes a hydrostatic compressive stress, applied equally in all directions.

The three principal stresses in the anchor shank are then:

$$\sigma_I = \sigma_a - p = \sigma_a - 49.1 \text{ MPa}$$
$$\sigma_{II} = -p \quad = -49.1 \text{ MPa}$$
$$\sigma_{III} = -p \quad = -49.1 \text{ MPa}$$

Applying the von Mises condition for yielding, $\sigma_o = S_y$:

$$2S_y^2 = (\sigma_I - \sigma_{II})^2 + (\sigma_{II} - \sigma_{III})^2 + (\sigma_{III} - \sigma_I)^2$$
$$2(250)^2 = [(\sigma_a - p) - (-p)]^2 + [(-p) - (-p)]^2 + [(-p) - (\sigma_a - p)]^2$$
$$= 2\sigma_a^2$$

$$\textit{Answer: } \sigma_a = 250 \text{ MPa}$$

For yielding to occur when the anchor is submerged, the applied axial stress in the shank is the same as when the anchor is at sea-level. The hydrostatic stress has no effect on the stress required to yield the material.

Chapter 10   Buckling

10.0 Introduction

When a slender member is subjected to an axial compressive load, it may fail by a condition called **buckling**. An axial member in compression is a **column** or a *strut*. Buckling is a geometric instability in which the lateral displacement of the axial member can suddenly become very large (*Figure 10.1*). Buckling is easily demonstrated when a steel ruler is compressed. The ruler remains straight as the load is increased; when the load reaches a critical value, the middle of the ruler suddenly deflects sideways. The ruler has *buckled.* Examples of structural members and systems that are subjected to loads that may cause buckling are:

1. Building columns that transfer loads to the ground;
2. Truss members in compression;
3. Micro-machines (MEMS devices);
4. Submarine hulls subjected to water pressure (this type of buckling is beyond the scope of this text).

The following example introduces the concept of *geometric instability.*

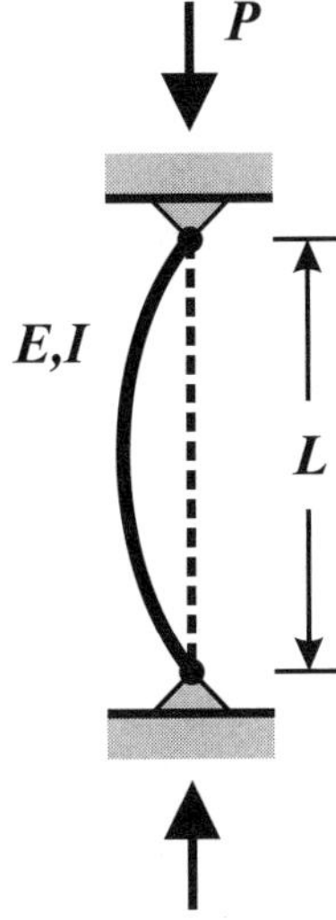

Figure 10.1. Original shape (dashed) and buckled shape (solid) of a pinned–pinned column under compressive load.

Example 10.1 Tower on Spring Supports

Tall buildings in earthquake regions are sometimes supported by special rubber foundations. These large supports isolate the building from the ground's vibrations. A simple schematic and model of such a building is shown in *Figure 10.2.*

The model consists of an inverted rigid T-shape of height H and breadth $2B$; the supports are represented by springs of stiffness k (*Figure 10.2b*). The weight of the building W is applied at its center of gravity distance H above the base. Due to W, the supports displace downward by an amount:

F.A. Leckie, D.J. Dal Bello, *Strength and Stiffness of Engineering Systems*, Mechanical Engineering Series,   **293**
DOI 10.1007/978-0-387-49474-6_10, © Springer Science+Business Media, LLC 2009

$$\delta_C = \frac{W}{2k}$$

as shown in *Figure 10.2c*.

Due to wind, or some other sideways load, the structure may rotate clockwise by angle θ causing point A to move up, and point D to move further down, both by distance $B \sin\theta$ (*Figure 10.2d*). The total downward displacements of points A and D, δ_A and δ_D, are then:

$$\delta_A = \delta_C - B \sin\theta$$

$$\delta_D = \delta_C + B \sin\theta$$

The reaction forces, R_A and R_D, are found from the spring deflections:

$$R_A = k\delta_A = k(\delta_C - B \sin\theta)$$

$$R_D = k\delta_D = k(\delta_C + B \sin\theta)$$

Applying equilibrium in the vertical direction (*Figure 10.2d*):

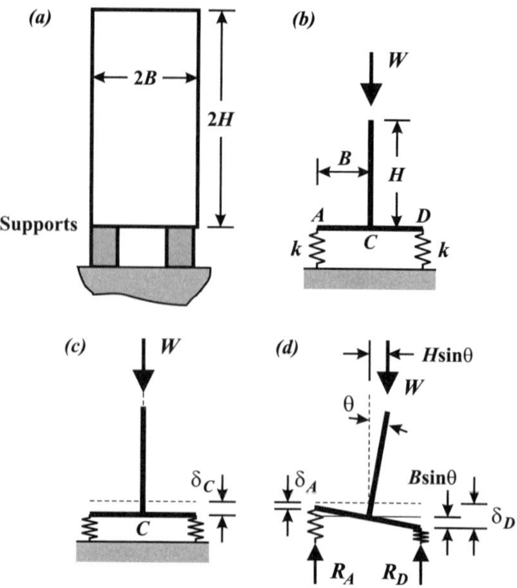

Figure 10.2. (a) Schematic of building. **(b)** Rigid-inverted "T" and two-spring model. **(c)** Downward deflection due to weight *W*. **(d)** Additional deflection due to rotation θ.

$$\sum F_y = 0: \quad R_A + R_D - W = 0$$

Substituting the spring deflection expressions for R_A and R_D into the equilibrium equation, and rearranging, gives the deflection of point C:

$$\delta_C = \frac{W}{2k}$$

which is just the deflection of the system without rotation.

Applying moment equilibrium about point C:

$$\sum M_C = 0: \quad R_A(B \cos\theta) - R_D(B \cos\theta) + W(H \sin\theta) = 0$$

Again substituting for R_A and R_D, and rearranging, gives the equilibrium condition:

$$[WH - 2kB^2 \cos\theta]\sin\theta = 0$$

There are two solutions:

1. $\sin\theta = 0$, from which $\theta = 0°$, i.e., the building stands straight;
2. $[WH - 2kB^2 \cos\theta] = 0$, from which $\cos\theta = \dfrac{WH}{2kB^2}$

Figure 10.3 is a plot of weight *W* against rotation θ, where *W* is normalized by $2kB^2/H$.

For *Solution 1*, when θ = 0°, the building does not tilt but simply deflects downward by δ_C. This result is represented by the thick vertical line on the plot. Provided θ remains zero, *W* may be as large as possible, limited only by the material strength of the building. This is a trivial solution since a non-zero θ is assumed.

Solution 2 is represented by the cosine curve on the plot. For a given angle of tilt θ, if the building weight *W* is greater than $[2kB^2/H]\cos θ$, then the overturning moment is greater than the resisting moment of the spring foundations, and the building tilts catastrophically.

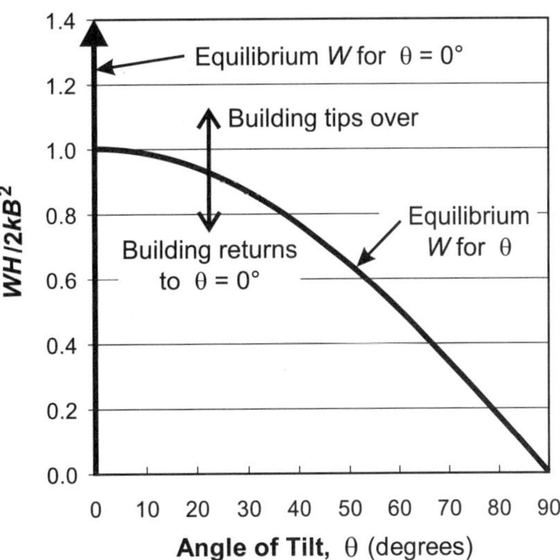

Figure 10.3. Weight versus angle of tilt. The bifurcation point is at $WH/2kB^2 = 1$.

If, on the other hand, *W* is less than $[2kB^2/H]\cos θ$, then the resisting moment is greater than the overturning moment, and the structure returns to θ = 0° (assuming sufficient damping so that any oscillation will die down).

If *W* exactly equals $[2kB^2/H]\cos θ$, and there are no dynamic affects, the building will stay precariously in equilibrium at angle θ, as defined by the cosine curve of *Figure 10.3*.

Solution 1 intersects *Solution 2* when θ = 0°, where:

$$\frac{WH}{2kB^2} = 1$$

The point where the plot diverges is called the **bifurcation point.**

Up to the bifurcation point, $WH/2kB^2 = 1$, any small tilt returns to θ = 0°. If $WH/2kB^2 > 1$, *any* slight tilt will cause the structure to overturn. Failure by such a geometric instability is called *buckling*. In this case, the critical load W_{cr} is:

$$W_{cr} = \frac{2kB^2}{H}$$

If the building load *W* is greater than W_{cr}, the building will tip over due to any slight angle.

10.1 Buckling of a Column

In addition to supporting transverse loads, beams can also carry compressive axial loads. When axial compressive stresses in a long member dominate, the member is referred to as a **column** or *strut*. Columns are subject to **buckling**. When the compressive force reaches a certain critical value, the column buckles sideways (*Figure 10.4b*).

To develop the buckling model of a column, consider a beam of bending stiffness *EI* (*Figure 10.4a*). The beam is simply supported (pinned at both ends) so that no moment is applied at the supports and the ends are free to rotate. *Compressive axial* force *P* is applied through the centroid of the cross-section. Load *P* is increased from zero until the beam – acting as a column – buckles with deflection *v(x)* in the *y*-direction (*Figure 10.4b*).

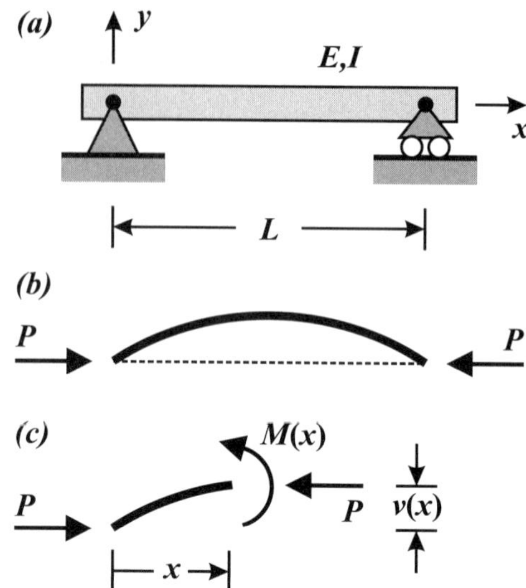

Figure 10.4. (a) Beam pinned at both ends. **(b)** Beam under axial compressive load *P*. **(c)** FBD from 0 to *x*.

Consider the FBD of the beam segment from the left support to a cut at distance *x* (*Figure 10.4c*). Moment equilibrium about the left support (or about the cut) gives:

$$M(x) + Pv(x) = 0 \qquad \text{[Eq. 10.1]}$$

Recall the moment–curvature relation for a beam:

$$M(x) = EI\frac{d^2v(x)}{dx^2} = EIv''(x) \qquad \text{[Eq. 10.2]}$$

Substituting *Equation 10.2* into *Equation 10.1* gives:

$$EIv''(x) + Pv(x) = 0 \qquad \text{[Eq. 10.3]}$$

Dividing by *EI,* and substituting:

$$\alpha^2 = \frac{P}{EI} \qquad \text{[Eq. 10.4]}$$

reduces *Equation 10.3* to:

$$v''(x) + \alpha^2 v(x) = 0 \qquad \text{[Eq. 10.5]}$$

The solution of this *second-order linear differential equation* is:

$$v(x) = A \sin \alpha x + B \cos \alpha x \qquad \text{[Eq. 10.6]}$$

Constants A and B are found by applying the **geometric boundary conditions**. In this case, the *boundary conditions* require that the displacements at each end of the beam (column) be zero. Hence, the boundary conditions are $v = 0$ at $x = 0$ and $v = 0$ at $x = L$, which gives:

- $B = 0$
- $A \sin \alpha L = 0$

The displacement equation is thus reduced to:

$$v(x) = A \sin \alpha x \qquad \text{[Eq. 10.7]}$$

Equation 10.7 has two possible solutions for $v(L) = 0$:

1. $A = 0$
2. $\sin \alpha L = 0$

Solution 1, $A = 0$, means that there is no lateral deflection and the beam (column) remains straight – it does not buckle. The applied load is limited by the material properties (compressive strength) of the column. This is a trivial solution, since the column is assumed to buckle. This solution is analogous to the building in *Example 10.1* remaining vertical.

Solution 2, $\sin \alpha L = 0$, means that the column takes the shape of a sine wave (*Figure 10.4b*). This solution is satisfied whenever $\alpha L = n\pi$, where $n = 0, 1, 2, 3...$ Substituting back into the expression for α (*Equation 10.4*) gives the value of the load P required to cause buckling of the pinned–pinned column:

$$P_n = \frac{(n\pi)^2 EI}{L^2} \qquad \text{[Eq. 10.8]}$$

As P is increased from zero, the first buckling load occurs at $n = 1$. Values for $n > 1$ correspond to higher order buckling loads (sine waves of higher frequency), and give greater values of P_n. Since P corresponding to $n = 1$ is reached first, it is the critical load P_{cr} required to buckle the pinned–pinned column. The **buckling load** is given by the **Euler Buckling formula**:

$$P_{cr} = \frac{\pi^2 EI}{L^2} \qquad \text{[Eq. 10.9]}$$

The buckling load increases with the value of the bending stiffness EI, but decreases with the square of the length L – the distance between pinned ends. Because of the strong effect of L, columns in practice are of limited length.

The **buckling strength** σ_{cr} is the **Euler buckling load** divided by the cross-sectional area A of the column:

$$\sigma_{cr} = \frac{\pi^2 EI}{AL^2} \qquad \text{[Eq. 10.10]}$$

For long columns, the buckling strength is usually less than the compressive strength of the column material (the yield strength S_y for metals), and so buckling is often the limiting factor in design.

Two failure conditions are then possible in columns under compressive stress σ ($\sigma = P/A$):

1. *Buckling*, when the buckling strength is reached: $\sigma = \sigma_{cr} = \dfrac{\pi^2 EI}{AL^2}$ [Eq. 10.11]

2. *Yielding*, when the yield strength is reached: $\sigma = S_y$ [Eq. 10.12]

Both conditions must therefore be checked to ensure a proper design.

If the material of a buckled column remains linear–elastic, then the column will return to its original undeflected shape when the load is removed. However, a large load (or end-displacement) will cause a buckled metal column to yield in bending (resulting in a permanent bend in the column), or a brittle column to fracture.

Example 10.2 Tent Pole

Background: Tent structures support themselves with tension members (fabric and rope) and compression members (struts, poles, etc.) (*Figure 10.5*). The greater the tension in the tensile members, the greater the compression in the compressive members.

Given: A pup-tent has poles made of thin-walled steel tubes:

$$L = 36 \text{ in.}, \ R = 0.25 \text{ in.}, \ t = 0.02 \text{ in.}$$

$$E = 30{,}000 \text{ ksi}$$

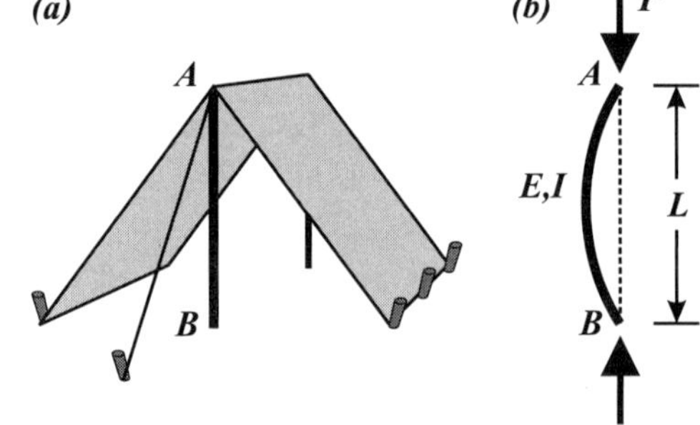

Figure 10.5. (a) Pup-tent. **(b)** Pole *AB* under compressive force *P.*

Required: When campers tighten the fabric and rope, the poles are put into compression. Determine the compressive force required to buckle pole *AB*. The ends of the pole are both pinned (free to rotate).

Solution: The moment of inertia of a hollow tube is:

$$I = \pi R^3 t$$

and is the same about any axis in the plane of the cross-section. The buckling load is:

$$P_{cr} = \frac{\pi^2 EI}{L^2} = \frac{\pi^2 (30 \times 10^6 \text{ psi})[\pi(0.25 \text{ in.})^3 (0.02 \text{ in.})]}{(36 \text{ in.})^2}$$

Answer: $P_{cr} = 224$ lb

10.2 Radius of Gyration and Slenderness Ratio

It is convenient to combine the geometric variables I, A, and L in the buckling strength formula (*Equation 10.10*) into a single parameter that characterizes the column geometry.

The moment of inertia I is first expressed in an equivalent form:

$$I = Ar^2 \qquad \text{[Eq. 10.13]}$$

where r is the **radius of gyration**. The equivalent cross-section has all of its area A concentrated at distance r from the neutral axis (*Figures 10.6b* and *c*).

With $I = Ar^2$, the buckling strength becomes:

$$\sigma_{cr} = \frac{\pi^2 E(Ar^2)}{AL^2} = \pi^2 E\left(\frac{r}{L}\right)^2 \qquad \text{[Eq. 10.14]}$$

The **slenderness ratio** s is the column's length L divided by its radius of gyration r:

$$s = \frac{L}{r} \qquad \text{[Eq. 10.15]}$$

The *slenderness ratio* combines all three geometric terms I, A, and L, into a single variable that describes a column's tendency to buckle. The buckling strength then reduces to:

$$\sigma_{cr} = \frac{\pi^2 E}{s^2} \qquad \text{[Eq. 10.16]}$$

A column is described as **long** or **short** depending on the value of its slenderness ratio s, not its absolute length. In *Figure 10.7*, the two rightmost columns have the same length, but the center one is more slender, so it will buckle under a smaller stress. The center column is *long* and the rightmost column is *short*. The two leftmost columns have the same slenderness ratio; they will buckle under the same stress (but not the same force).

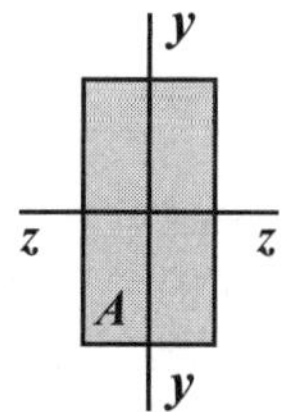

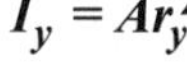

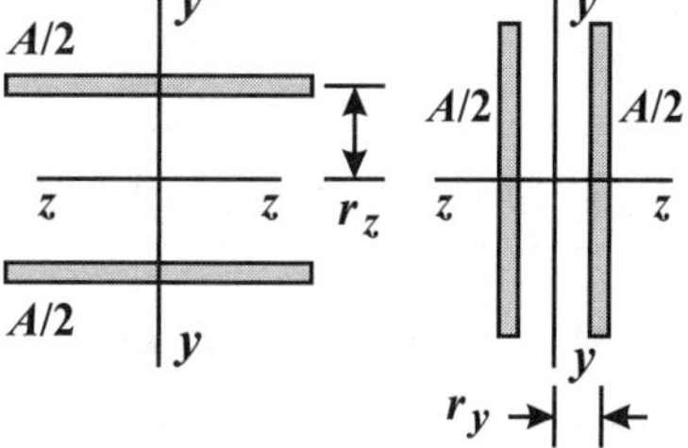

Figure 10.6. (a) Cross-section. (b) Radius of gyration about the z-axis. (c) Radius of gyration about the y-axis.

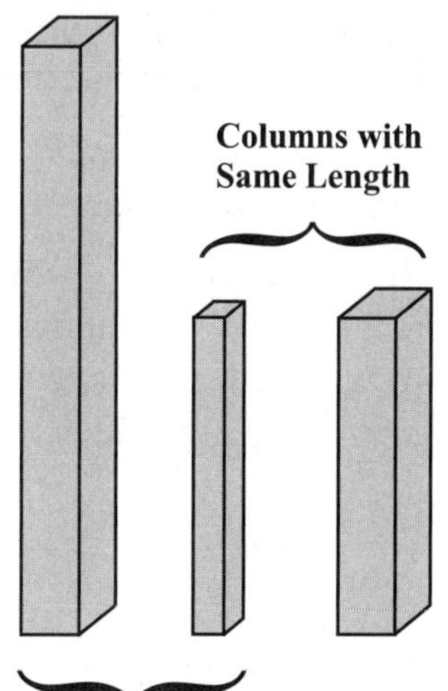

Figure 10.7. Slenderness ratio versus length.

Buckling About Two Axes

For most cross-sections, the moments of inertia I_z and I_y (about the z- and y-axis, respectively), are different (*Figure 10.6*). Therefore, there are generally two radii of gyration:

$$r_z = \sqrt{\frac{I_z}{A}}; \quad r_y = \sqrt{\frac{I_y}{A}}$$

[Eq. 10.17]

The larger radius of gyration (moment of inertia) corresponds to the so-called *strong axis* (it is more resistant to bending and buckling, *Figure 10.6b*); the smaller radius corresponds to the *weak axis* (it is less resistant to buckling, *Figure 10.6c*). It is weak-axis buckling that is generally of concern in design.

Since slenderness ratio depends on radius of gyration (*Equation 10.15*), two slenderness ratios must be considered, s_z corresponding to r_z, and s_y corresponding to r_y.

10.3 Boundary Conditions and Effective Length

The *buckling load* was derived for a *pinned–pinned* column (*Figure 10.8a*). The geometric boundary conditions – the type of supports – of a column affect its buckling load. When the buckling load calculations are repeated for a column with both ends fixed again rotation (*Figure 10.8b*), the buckling load can be shown to be:

$$P_{cr} = \frac{4\pi^2 EI}{L^2} = \frac{\pi^2 EI}{(L/2)^2}$$

[Eq. 10.18]

which is four times the value of the buckling load for the pinned–pinned column.

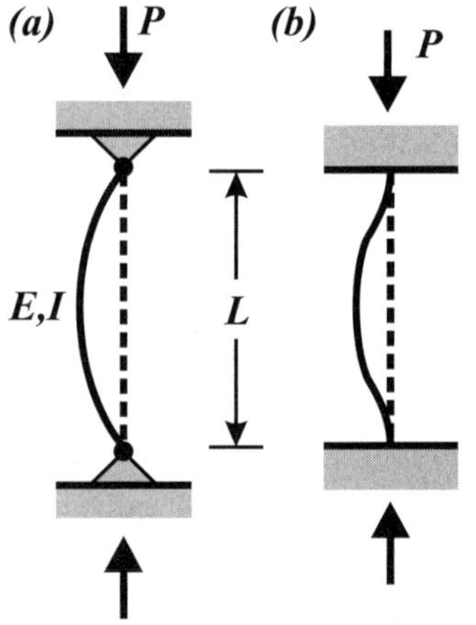

The difference in buckling load caused by different boundary conditions is dealt with by a term known as the *effective length*. The buckling formula (*Equation 10.9*) is rewritten:

$$P_{cr} = \frac{\pi^2 EI}{L_e^2}$$

[Eq. 10.19]

Figure 10.8. Buckled shapes of **(a)** a pinned–pinned column of length L and **(b)** a fixed–fixed column of length L.

where L_e is the *effective length* of the column. The effective length depends on the boundary conditions or geometric constraints. The effective length of a pinned–pinned column is:

$$L_e = L$$

[Eq. 10.20]

From *Equation 10.18*, the effective length of a fixed–fixed column is:

$$L_e = L/2 \qquad \text{[Eq. 10.21]}$$

The effective length is the length the column would be if it buckled as a *pinned–pinned* column.

In general, the effective length is given by:

$$L_e = kL \qquad \text{[Eq. 10.22]}$$

where k is the **effective length factor**. Values of k for various boundary conditions are given in *Table 10.1*.

The *effective slenderness ratio* s_e is:

$$s_e = \frac{L_e}{r} = \frac{kL}{r} \qquad \text{[Eq. 10.23]}$$

and the *buckling strength* is then:

$$\sigma_{cr} = \frac{\pi^2 EI}{A L_e^2} = \frac{\pi^2 E}{s_e^2} \qquad \text{[Eq. 10.24]}$$

Table 10.1. Effective Length for columns with common end conditions.
Dashed line indicates the original column of length L. Solid line indicates buckled shape.

End condition	Pinned–pinned	Fixed–free	Fixed–fixed	Fixed–pinned
The *effective length* is the distance between points on the column where the moment is zero, corresponding to the end conditions of the standard pinned–pinned column. Zero moment also occurs when the curvature of the column is zero (is changing sign). The fixed–free column is mirrored through the fixed end to visualize its effective length.	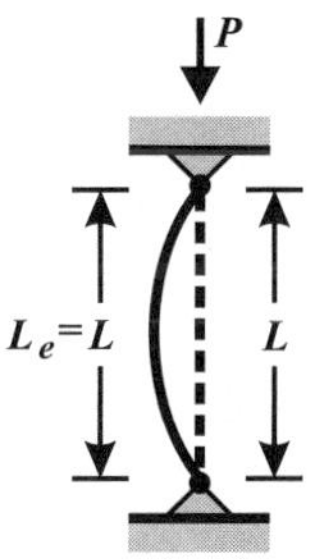	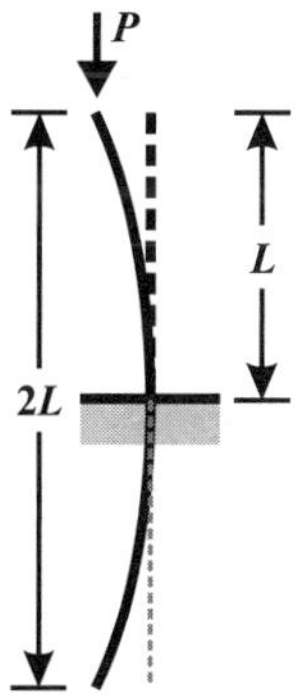	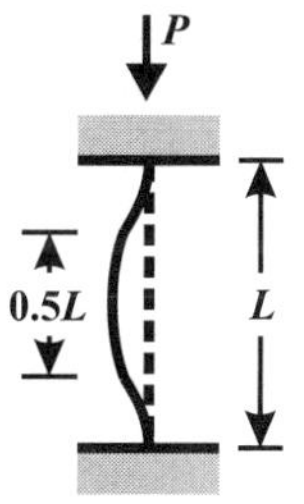	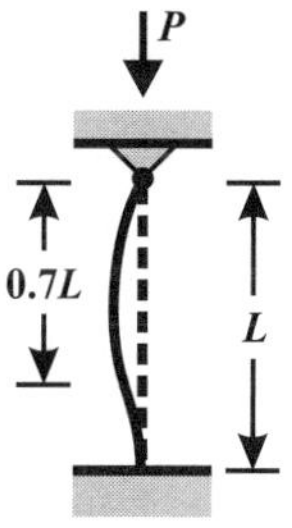
Effective length, L_e	L	$2L$	$0.5L$	$0.7L$
Effective length factor, k	1	2	0.5	0.7
Relative strength for same overall length, L	1	0.25	4	2

Experience is necessary to determine which effective length to apply based on the actual end-conditions of the column. When the column is supported at both ends but not well constrained against rotation, it is generally best to use the pinned–pinned condition for conservative results when determining an allowable compressive stress.

Physically, the *effective length* is the distance between points on the buckled column where the moment goes to zero, i.e., where the column is effectively pinned. Considering the *deflected shape*, the moment is zero where the curvature is zero (from beam theory, $M = EI\kappa$). Zero curvature corresponds to an inflection point in the deflected shape (where the curvature changes sign).

For the *fixed–fixed* column, the curvature of the buckled shape changes sign at the "quarter-points" of the column, $L/4$ from each end; these are the inflection points. Thus, the length of the column between effective pins is $L_e = L/2$. As noted earlier, the *fixed–fixed* column is four times stronger in buckling than the *pinned–pinned* column of the same total length and cross-section.

Example 10.3 Free-Standing Column Under Compressive Force

Given: A steel column ($E = 200$ GPa) built into the ground has length $L = 2.0$ m and supports an axial compressive load P. The dimensions of the cross-section are $b = 50$ mm and $d = 100$ mm (*Figure 10.9*).

Required: Determine (a) the force to buckle the column P_{cr} and (b) the buckling strength σ_{cr}. (c) If the factor of safety against buckling is FS = 2.0, determine the allowable compressive force P_A.

Solution: *Step 1.* The Euler buckling force and the buckling strength are:

$$P_{cr} = \frac{\pi^2 EI}{L_e^2} \quad \text{and} \quad \sigma_{cr} = \frac{\pi^2 EI}{AL_e^2}$$

where L_e is the effective length of the column. Here, the column boundary conditions are *fixed–free* for buckling about both the z- and y-axes. The effective length is $L_e = 2L$ for both z- and y-axes buckling.

The moment of inertia depends on the axis about which buckling takes place. The moment of inertia about each axis (z- and y-) is:

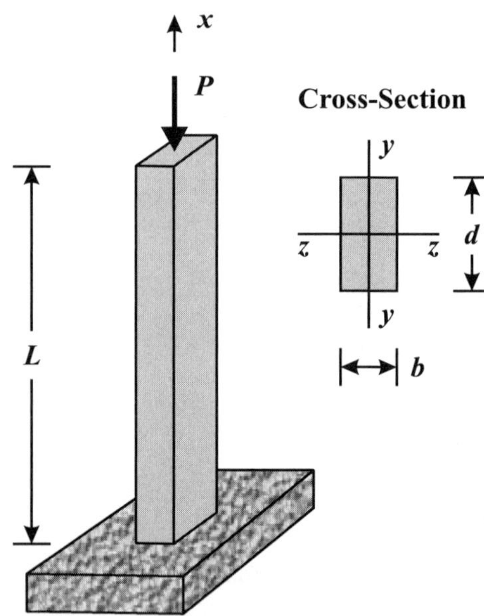

Figure 10.9. Column with rectangular cross-section under compressive load *P*.

$$I_z = \frac{bd^3}{12} = \frac{(0.05\ \text{m})(0.10\ \text{m})^3}{12} = 4.17\times10^{-6}\ \text{m}^4 \leftarrow \text{for strong-axis buckling}$$

$$I_y = \frac{db^3}{12} = \frac{(0.10\ \text{m})(0.05\ \text{m})^3}{12} = 1.042\times10^{-6}\ \text{m}^4 \leftarrow \text{for weak-axis buckling}$$

Step 2. The end condition for buckling about either axis is fixed–free. Thus, buckling about the y-axis governs since I_y gives the smaller Euler buckling load:

$$P_{cr} = \frac{\pi^2 EI}{L_e^2} = \frac{\pi^2 EI_y}{(2L)^2} = \frac{\pi^2(200\times10^9\ \text{Pa})(1.042\times10^{-6}\ \text{m}^4)}{4(2.0\ \text{m})^2} = 128.55\times10^3\ \text{N}$$

Answer: $\underline{P_{cr} = 128.6\ \text{kN}}$

The buckling strength is:

$$\sigma_{cr} = \frac{\pi^2 EI}{AL_e^2} = \frac{P_{cr}}{A} = \frac{128.6\times10^3\ \text{N}}{(0.10\ \text{m})(0.050\ \text{m})}$$

Answer: $\underline{\sigma_{cr} = 25.7\ \text{MPa}}$

This is a very slender member with $s_{e,\,y} = L_e/\sqrt{I_y/A} = 124$. Structural steel has a yield strength of $S_y = 250$ MPa; here, σ_{cr} is only 10% of S_y.

Step 3. The allowable force P_A is the failure load divided by the factor of safety:

$$P_A = \frac{P_{cr}}{\text{FS}} = \frac{128.6\ \text{kN}}{2.0}$$

Answer: $\underline{P_A = 64.3\ \text{kN}}$

Example 10.4 Column with End Conditions Dependent upon Buckling Axis

Given: An aluminum column is made from a standard I-beam cross-section (W8×13) as shown in *Figure 10.10*. The column is built-in at the base. The top of the column sits in a channel; friction forces between the column and the channel structure are negligible. The geometry of the cross-section is: $I_z = 39.6$ in.^{4}, $I_y = 2.73$ in.^{4}, and $A = 3.84$ in.^{2}. For aluminum, $E = 10\times10^6$ psi and $S_y = 35$ ksi. The total height of the column is $a = 12.0$ ft and the height to the channel is $b = 11.0$ ft.

Required: Determine (a) the force to cause buckling about the z-axis, (b) the force to cause buckling about the y-axis, and (c) the force to yield the column. (d) As P increases from zero, determine how the column fails.

Solution: *Step 1. Buckling about the z-axis* (the *strong axis*). The base of the column is *fixed*. The top of the column is *free* to move along the channel (*Figure 10.10b*). Hence, the top of the column is *free* with respect to buckling about the z-axis. Therefore, the column is *fixed–free* for z-axis buckling (strong-axis buckling); the effective length is $L_{e,z} = 2a$.

$$P_{cr,z} = \frac{\pi^2 E I_z}{L_{e,z}^2}$$

$$= \frac{\pi^2(10\times10^6 \text{ psi})(39.4 \text{ in.}^4)}{[2(12 \text{ ft} \times 12 \text{ in./ft})]^2}$$

Answer: $\underline{P_{cr,z} = 46.9 \text{ kips}}$

Step 2. Buckling about the y-axis (the *weak axis*). The top of the column is constrained so it neither move in the *z*-direction, nor can it rotate about the *y*-axis (*Figure 10.10a*). Hence, the top of the column is *fixed* with respect to buckling about the *y*-axis. The column is *fixed–fixed* for *y*-axis buckling; the effective length is $L_{e,y} = 0.5b$.

$$P_{cr,y} = \frac{\pi^2 E I_y}{L_{e,y}^2}$$

$$= \frac{\pi^2(10\times10^6 \text{ psi})(2.73 \text{ in.}^4)}{[0.5(11 \times 12 \text{ in.})]^2}$$

Answer: $\underline{P_{cr,y} = 61.9 \text{ kips}}$

Step 3. The force to *yield* the column is:

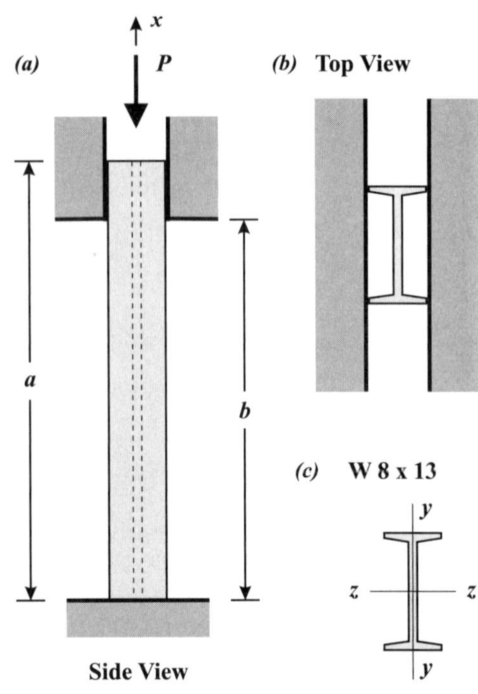

Figure 10.10. (a) Side view of column under axial load. **(b)** Top view. **(c)** Cross-section of W8×13 I-beam. "8" is the nominal depth of the I-beam (in inches), and "13" is the weight per unit length (lb/ft).

$$P_y = S_y A = (35\times10^3 \text{ lb/in.}^2)(3.84 \text{ in.}^2)$$

Answer: $\underline{P_y = 134 \text{ kips}}$

Step 4. The lowest calculated failure load is due to buckling about the *z*-axis. Although the *z*-axis is the *strong axis*, the end conditions are *fixed–free*, which effectively doubles the column length. The *y*-axis is the *weak axis*, but the *fixed–fixed* end conditions halve the length.

Failure is by buckling about the *z*-axis, at a value of:

Answer: $\underline{P_{cr,z} = 46.9 \text{ kips}}$

Bracing

The buckling strength of a column may be increased by providing bracing along the length of the column. Bracing restricts sideways movement, and therefore lowers the effective length of a column. In *Figure 10.11*, bracing at the middle of a pinned–pinned column has reduced its effective length in the plane of the paper by two, essentially creating two pinned–pinned columns, one on top of the other (*Figure 10.11b*). Thus, for

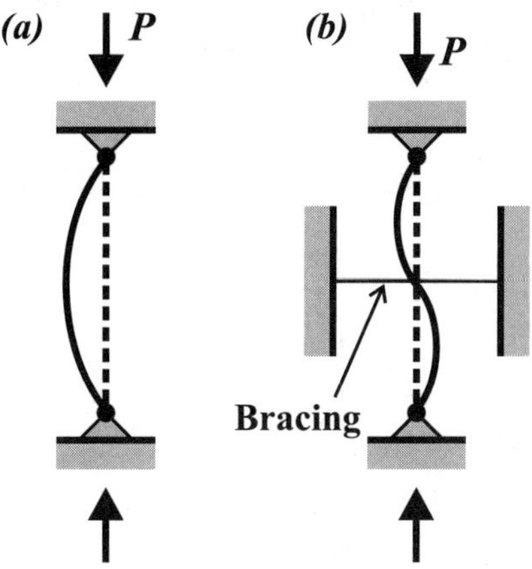

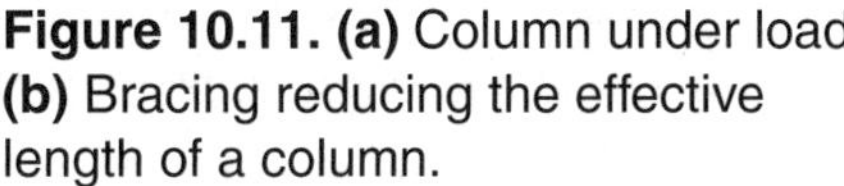

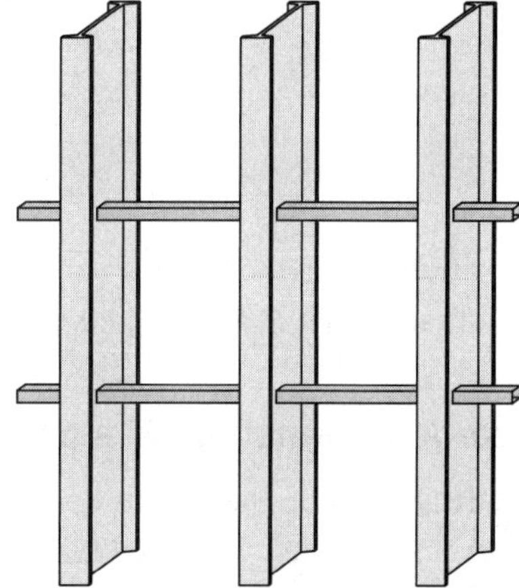

Figure 10.11. (a) Column under load. **(b)** Bracing reducing the effective length of a column.

Figure 10.12. Cross-members constrain each column. Here, bracing reduces the effective length for weak-axis buckling.

buckling in the plane of the paper, $L_e = L/2$, which increases the buckling strength by a factor of 4.

Bracing is usually configured to resist weak-axis buckling, which is often the limiting load in the design of columns (*Figure 10.12*). If the weak-axis buckling strength is increased sufficiently, the column strength is then governed by strong-axis buckling, or by general yielding of the cross-section.

10.4 Transition from Yielding to Buckling

The Euler buckling strength σ_{cr} is plotted against the slenderness ratio, $s = L/r$, in *Figure 10.13*. Because σ_{cr} is proportional to $1/s^2$, the buckling strength decreases rapidly.

For very short columns (small s), the buckling strength is large. However, the column strength cannot exceed the compressive strength of its material S_C. For metals, this is the yield strength S_y, which is drawn as a horizontal line since it is independent of column length (*Figure 10.13*).

Thus, depending on the slenderness ratio, a column *fails* by either:

1. *material failure* (e.g., *yielding* in metals), or
2. *geometric instability* (*buckling*).

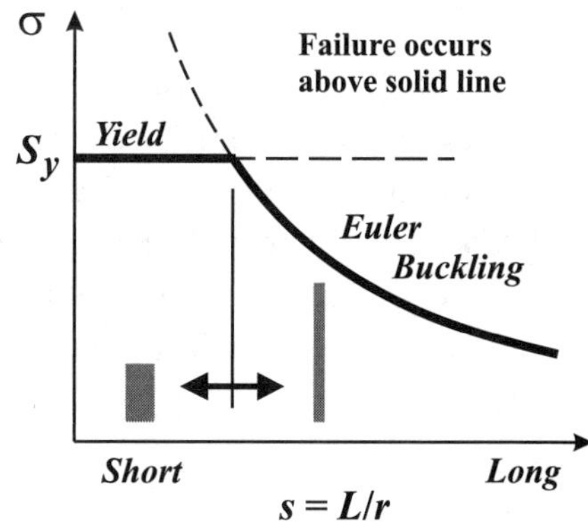

Figure 10.13. Schematic of column strength versus slenderness ratio s.

The transition point between yielding and buckling failure can be determined by setting the *buckling strength* equal to the yield strength: $\sigma_{cr} = S_y$. The *transition slenderness ratio* is:

$$s_{tr} = \pi \sqrt{\frac{E}{S_y}}$$
[Eq. 10.25]

The transition slenderness ratio provides an estimate of what is a *short column* (yield-dominated) and what is a *long column* (buckling-dominated).

Example 10.5 Transition Slenderness Ratio

Required: Determine the transition slenderness ratios for Aluminum 6061-T6 (structural aluminum, $S_y = 35$ ksi, $E = 10,000$ ksi), and Steel A36 (structural steel, $S_y = 36$ ksi, $E = 30,000$ ksi).

Solution: The transition slenderness ratios for the aluminum and steel columns are:

$$s_{tr,al} = \pi \sqrt{\frac{E_{al}}{S_{y,al}}} = \pi \sqrt{\frac{10,000}{35}}$$

Answer: $\underline{s_{tr,al} = 53.1}$

$$s_{tr,st} = \pi \sqrt{\frac{E_{st}}{S_{y,st}}} = \pi \sqrt{\frac{30,000}{36}}$$

Answer: $\underline{s_{tr,st} = 90.7}$

From this calculation, it can be deduced that a steel column can be about 70% longer than an aluminum column of the same cross-section when considering buckling.

Columns of Intermediate Length – A Transition Zone

In reality, there is no sudden change in failure mode as implied by *Figure 10.13* and *Equation 10.25*. Columns of **intermediate length** fail by a combination of material failure (due to axial and bending loads) and Euler buckling. Experiments are used to determine design formulas for the allowable compressive stress in columns. The allowable stress is modeled by different formulas depending on the slenderness ratio and on the column material, e.g., steel, aluminum, wood, etc. Design formulas for specific materials may be found in handbooks for the particular industry of interest.

For aluminum, the allowable stress used in the design of *intermediate columns* is given by a linear function, as shown qualitatively in *Figure 10.14*. The allowable axial compressive stress on the cross-section is governed by the three-part boundary. *Short columns* ($s < s_1$) are limited by material strength (independent of length), while *long columns* ($s > s_2$) are limited by Euler buckling (proportional to $1/s^2$). The values of slenderness ratios s_1 and s_2, and constants C_1 through C_4, depend on the particular aluminum being used, and are given in the current *Aluminum Design Manual* (The Aluminum Association, Inc., 2005).

Allowable Compressive Stress for Aluminum Columns

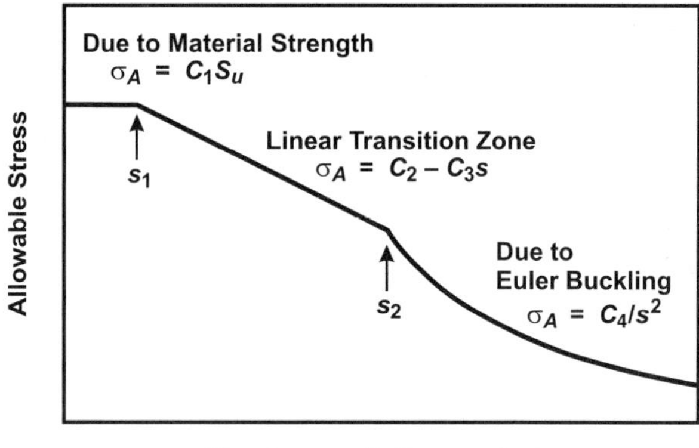

Figure 10.14. Allowable compressive stress versus slenderness ratio. Values of ratios s_1 and s_2 and constants C_1 through C_4 depend on specific material. Per the *Aluminum Design Manual* of The Aluminum Association, Inc. (© 2005); used with permission.

Example 10.6 Aluminum Columns of Various Slenderness Ratios

Given: For the design of Aluminum 6061-T6 columns, the allowable compressive stress on the cross-section is given by two equations, one for short and intermediate columns, and the other for long columns:

$$s < s_2 = 66: \quad \sigma_A = 20.2 - 0.126s \;\; \text{ksi}$$

$$s > s_2 = 66: \quad \sigma_A = \frac{51{,}100}{s^2} \;\; \text{ksi}$$

For many common aluminum alloys, the constant region of allowable stress (for short columns) does not exist, i.e., $s_1 = 0$.

Required: Two Aluminum 6061-T6 columns with the same cross-sectional area have slenderness ratios of (a) $s_a = 50$ and (b) $s_b = 100$. Determine the allowable axial compressive stress σ_A in each column.

Solution: *Step 1.* Slenderness ratio $s_a = 50$ falls between $s_1 = 0$ and $s_2 = 66$; this is an *intermediate column* governed by the linear equation. The allowable compressive stress is:

$$\sigma_{A,a} = 20.2 - 0.126s = 20.2 - 0.126(50)$$

Answer: $\underline{\sigma_{A,a} = 13.9 \text{ ksi}}$

Step 2. Slenderness ratio $s_b = 100 > s_2 = 66$; this is a *long column* governed by Euler buckling. The allowable compressive stress is:

$$\sigma_{A,b} = \frac{51{,}100}{s^2} = \frac{51{,}100}{(100)^2}$$

Answer: $\underline{\sigma_{A,b} - 5.1 \text{ ksi}}$

A column's strength is very sensitive to its slenderness ratio.

10.5 Column Shortening

When a pinned–pinned column buckles, the applied loads move closer together, by distance Δ (*Figure 10.15*). Originally, the loads are distance L apart. When the column buckles, its total length – the length of the neutral axis – remains essentially unchanged, but the points of application of the loads move closer so that the distance between the loads is now $l = L - \Delta$.

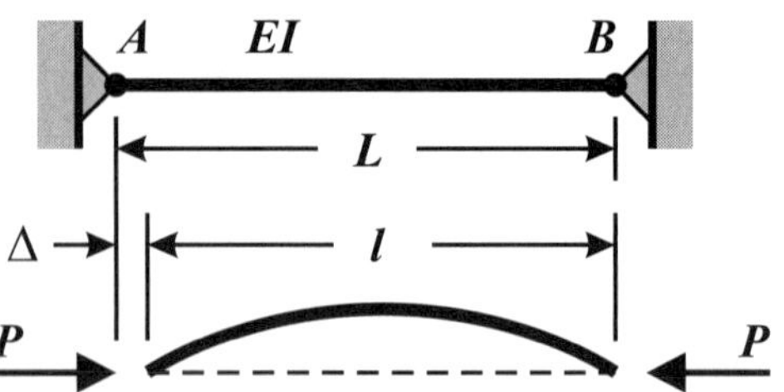

Figure 10.15. The ends of buckled column AB move closer by Δ.

The deflected shape is:

$$y = v(x) = a\,\sin\frac{\pi x}{l} \qquad \text{[Eq. 10.26]}$$

where a is the sideways deflection of the column at its center (*Figure 10.16*).

Using Pythagoras, the increment of distance along the column ds, is given by:

$$ds = \left[1 + \left(\frac{dy}{dx}\right)^2\right]^{0.5} dx$$

$$= \left[1 + \frac{\pi^2 a^2}{l^2}\cos^2\!\left(\frac{\pi x}{l}\right)\right]^{0.5} dx$$

$$\text{[Eq. 10.27]}$$

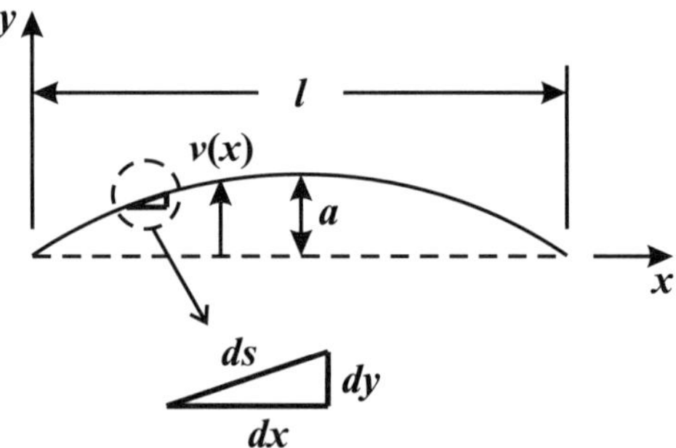

Figure 10.16. Geometry used to determine shortening of a buckled column.

The ratio a/l is small, and using the binomial equation to expand ds:

$$ds = \left[1 + \frac{\pi^2 a^2}{2l^2}\cos^2\!\left(\frac{\pi x}{l}\right)\right] dx \qquad \text{[Eq. 10.28]}$$

The integral of ds is the total length of the column L:

$$L = \int_0^l \left[1 + \frac{\pi^2 a^2}{2l^2}\cos^2\!\left(\frac{\pi x}{l}\right)\right] dx = l + \frac{a^2 \pi^2}{4l} \qquad \text{[Eq. 10.29]}$$

The displacement of the pinned ends in the direction of the applied loads is then:

$$\Delta = L - l = \frac{a^2 \pi^2}{4l} = \frac{a^2 \pi^2}{4(L - \Delta)} \qquad \text{[Eq. 10.30]}$$

Since Δ is much smaller than L when the column begins to buckle, then $L - \Delta \approx L$, and Δ can be simplified to:

$$\Delta = \frac{a^2\pi^2}{4L} \quad \text{or} \quad \frac{\Delta}{a} = \frac{\pi^2 a}{4\,L} \qquad \text{[Eq. 10.31]}$$

The shortening of the column Δ is small compared to the buckled sideways displacement a, since a is much smaller than L. Use is made of this result in micro-mechanical devices to measure small displacements Δ of a load on a micro-column. With L known, the sideways displacement a is measured, and the smaller shortening Δ is then calculated. Note that sideways deflection a must be measured; it cannot be calculated from the applied forces.

10.6 Effect of Imperfections

Real systems are not perfect. The assumed locations and directions of ideally applied column loads – e.g., a purely axial force applied through a column's centroid – cannot be duplicated in practice. Also, the geometry of an as-manufactured column is generally not perfectly straight. The effect of *imperfections* is illustrated in the following two examples.

Example 10.7 Tower on Spring Supports

Given: The load applied to the tower of *Example 10.1* was assumed to be perfectly aligned with the center of the system. In practice, loads are generally applied off-axis, with an eccentricity of e (*Figure 10.17a*).

Required: Determine the effect of eccentricity e on the tendency for the tower to tip.

Solution: *Step 1*. On repeating the equilibrium calculations of *Example 10.1*, vertical equilibrium gives the same displacement of point C as before:

Figure 10.17. **(a)** Model of building on elastic foundation loaded with eccentricity e. **(b)** Deflection of system.

$$\delta_C = \frac{W}{2k}$$

Likewise, the reaction forces R_A and R_D are:

$$R_A = k\delta_A = k(\delta_C - B\sin\theta)$$

$$R_D = k\delta_D = k(\delta_C + B\sin\theta)$$

However, the eccentricity causes the moment equilibrium equation to change:

$$R_A(B\cos\theta) - R_D(B\cos\theta) + W(H\sin\theta + e\cos\theta) = 0$$

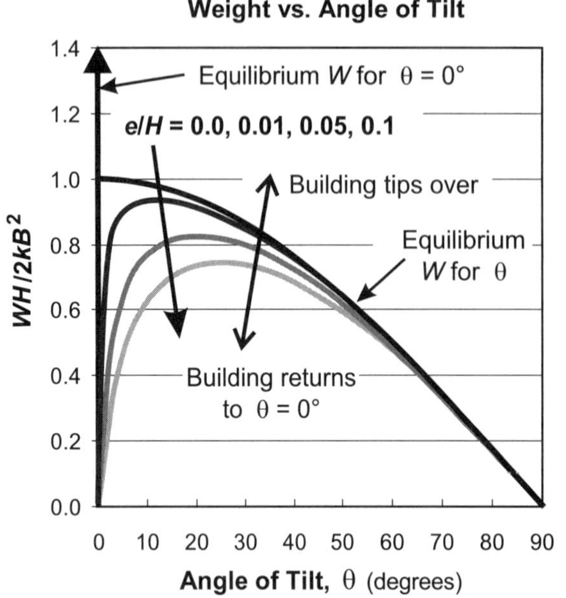

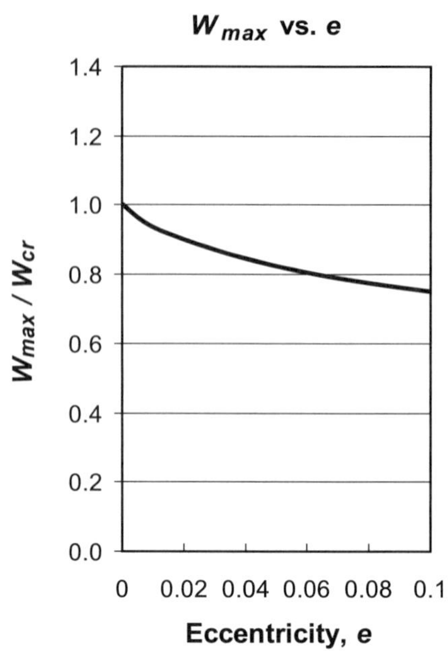

Figure 10.17. (c) Weight versus angle of tilt for various values of *e/H*.

Figure 10.17. (d) Maximum weight versus eccentricity *e/H* that will cause building to return to vertical.

Step 2. Substituting the expressions for R_A and R_D into the equilibrium equation, and rearranging, gives the equilibrium condition:

$$W[H\sin\theta + e\cos\theta] - 2kB^2 \sin\theta \, \cos\theta = 0$$

Solving for W gives:

$$W = W_{cr}\frac{\sin\theta \, \cos\theta}{\sin\theta + \dfrac{e}{H}\cos\theta}$$

where $W_{cr} = \dfrac{2kB^2}{H}$ is the buckling load when $e = 0$.

The relationship between the load W and the angle of tilt θ is plotted for various eccentricities e/H in *Figure 10.17c*. As eccentricity increases, the maximum buckling (tipping) load decreases. This plot can be cross-plotted to give the maximum load in terms of eccentricity e/H as in *Figure 10.17d*. For example, if the eccentricity is $e/H = 1/20$ (0.05), the maximum load is about 82% of the ideal case.

Example 10.8 Column with Initial Bend

Given: The Euler analysis of a buckling column assumes the unloaded column is perfectly straight. However, it is to be expected that a column may be slightly bent on delivery (*Figure 10.18a*).

Required: Determine the effect of the imperfection on the response of the column. Assume the unloaded shape is:

$$v_i(x) = a_i \sin\frac{\pi x}{L}$$

where a_i is the initial offset of the center of the column as supplied.

Solution: *Step 1.* On applying axial load P, the column deflects an *additional* amount $v(x)$ (*Figure 10.18b*). From the FBD of length x (*Figure 10.18c*):

$$M + P(v + v_i) = 0$$

Step 2. The moment in the column is related to the additional deformation $v(x)$ by the bending equation:

$$M = EI\frac{d^2v}{dx^2}$$

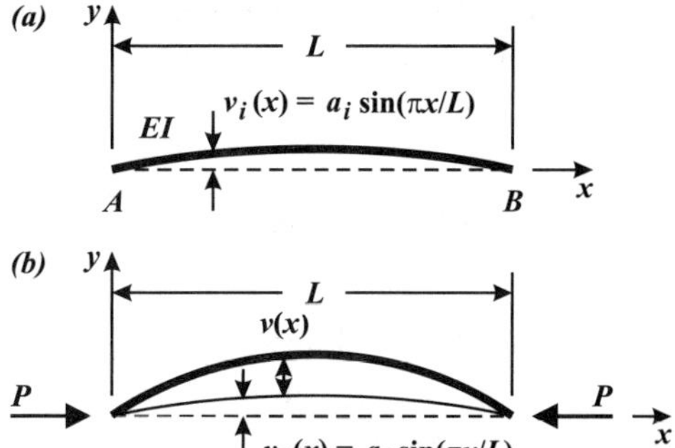

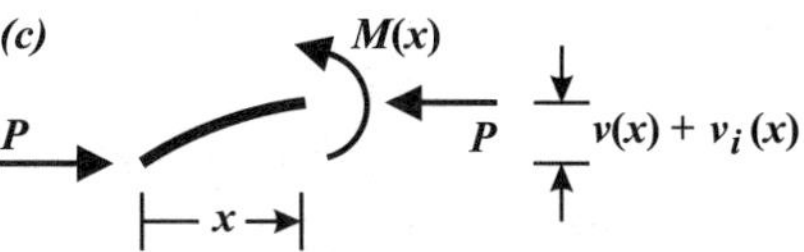

Figure 10.18. **(a)** Column with initial imperfection, $v_i(x)$. **(b)** Additional deflection $v(x)$ due to buckling. **(c)** FBD of column from 0 to x ($x < L/2$).

Eliminating M from the two equations results in a differential equation for the displacement:

$$\frac{d^2v}{dx^2} + \frac{P}{EI}v = -\frac{P}{EI}v_i = -\frac{Pa_i}{EI}\sin\frac{\pi x}{L}$$

With the relation $\alpha^2 = \dfrac{P}{EI}$, the differential equation becomes:

$$\frac{d^2v}{dx^2} + \alpha^2 v = -\alpha^2 a_i \sin\frac{\pi x}{L}$$

Step 3. The solution for the *additional* displacement v due to the initial imperfection v_i is:

$$v(x) = \frac{a_i}{\left[\left(\frac{\pi}{L}\right)^2 - \alpha^2\right]}\sin\frac{\pi x}{L}$$

Or, in terms of the applied load P and the Euler load $P_{cr} = (\pi^2 EI)/L^2$:

$$v(x) = \frac{a_i}{[P_{cr}/P - 1]}\sin\frac{\pi x}{L}$$

Step 4. The additional displacement v_{max} at the center is:

$$v_{max} = \frac{a_i}{[P_{cr}/P - 1]}$$

The value of the additional displacement is a function of load P, with initial displacement a_i being determined from measurement. The response can be rewritten as the amplification of the initial imperfection:

$$\text{Amplification} = \frac{v_{max} + a_i}{a_i} = \frac{P_{cr}/P}{[P_{cr}/P - 1]} = \frac{1}{[1 - P/P_{cr}]} \qquad \text{[Eq. 10.32]}$$

Figure 10.18d shows that as P approaches the Euler buckling load P_{cr}, the amplification of the initial imperfection increases. For example, at $P/P_{cr} = 0.8$, the amplification is five times the initial imperfection. The column has already started to buckle sideways. As P approaches the critical value, the sideways displacement increases rapidly, corresponding to observations of actual systems that are slowly loaded until they buckle excessively (e.g., the steel ruler).

The dashed lines in *Figure 10.18d* represent the response of a *perfectly straight* (ideal) column. The perfectly straight column exhibits no sideways displacement until $P = P_{cr}$, when the amplification instantaneously becomes very large and the system buckles. The imperfect column approaches the straight column for $P = P_{cr}$.

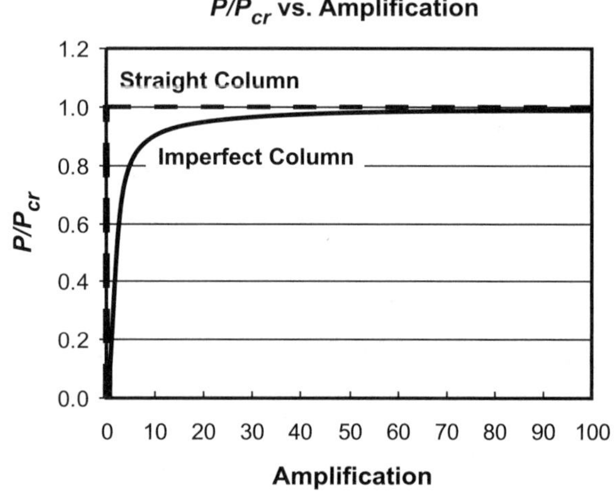

Figure 10.18. **(d)** Amplification of central displacement of a column with initial imperfection (here, a sine curve).

10.7 Effect of Lateral Forces

In addition to axial compressive loads, columns are often subjected to lateral, or transverse, forces. This occurs particularly in braced columns. *Figure 10.19* shows a pinned–pinned column subjected to axial force P and lateral force F applied at its center.

Because force F is applied at the center of the column, the transverse deflection is symmetric about the middle. From the FBD of the left-hand part of the beam ($x < L/2$, *Figure 10.19b*), equilibrium requires that:

$$M + Pv + \frac{F}{2}x = 0 \qquad \text{[Eq. 10.33]}$$

Following the same procedures used to derive the buckling formula results in the following differential equation:

$$\frac{d^2v}{dx^2} + \alpha^2 v = -\beta^2 3x \qquad \text{[Eq. 10.34]}$$

where:

$$\alpha^2 = \frac{P}{EI} \qquad \text{[Eq. 10.35]}$$

and

$$\beta^2 = \frac{F}{2EI} \qquad \text{[Eq. 10.36]}$$

The solution of this equation has the form:

$$v(x) = A\,\sin\alpha x + B\,\cos\alpha x - \frac{\beta^2}{\alpha^2}x \qquad \text{[Eq. 10.37]}$$

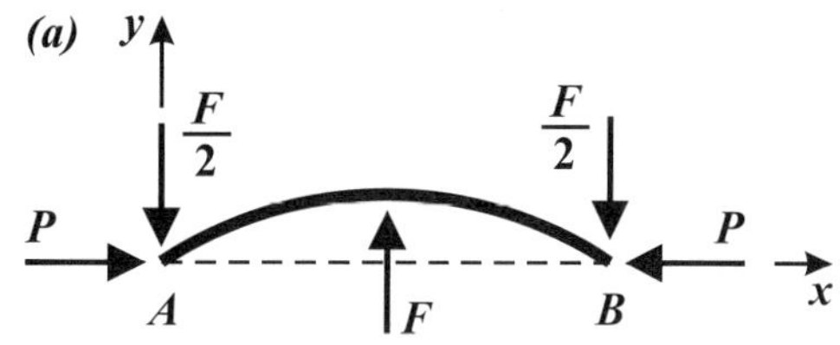

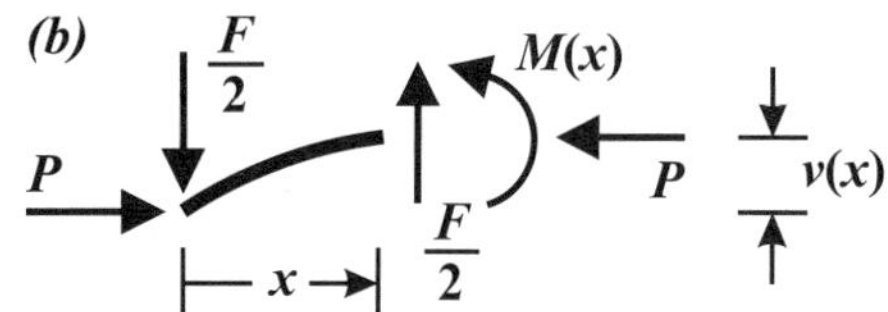

Figure 10.19. **(a)** Column subjected to compressive force P and lateral force F. **(b)** FBD of column for $0 < x < L/2$.

Applying the boundary condition $v = 0$ at $x = 0$ gives $B = 0$. From symmetry, the slope at $x = L/2$ must be zero:

$$v'\!\left(\frac{L}{2}\right) = A\,\alpha\,\cos\!\left(\alpha\frac{L}{2}\right) - \frac{\beta^2}{\alpha^2} = 0 \qquad \text{[Eq. 10.38]}$$

from which:

$$A = \frac{\beta^2}{\alpha^2}\left[\frac{1}{\alpha\,\cos(\alpha L/2)}\right] = \frac{F}{2P}\left[\frac{1}{\alpha\,\cos(\alpha L/2)}\right] \qquad \text{[Eq. 10.39]}$$

Hence, after substitution, the displacement is:

$$v(x) = \frac{F}{2P}\frac{\sin\alpha x}{\alpha\,\cos(\alpha L/2)} - \frac{F}{2P}x \qquad \text{[Eq. 10.40]}$$

At $x = L/2$, the central displacement is:

$$v\!\left(\frac{L}{2}\right) = \frac{F}{2P}\left[\frac{\sin(\alpha L/2)}{\alpha\,\cos(\alpha L/2)} - \frac{L}{2}\right] \qquad \text{[Eq. 10.41]}$$

Substituting the values of α and β gives:

$$v\!\left(\frac{L}{2}\right) = \frac{FL}{4P}\left[\frac{2}{\pi}\sqrt{\frac{P_{cr}}{P}}\tan\!\left(\frac{\pi}{2}\sqrt{\frac{P}{P_{cr}}}\right) - 1\right] \qquad \text{[Eq. 10.42]}$$

The *transverse stiffness* of the column when it is subjected to compressive load P is therefore:

$$k = \frac{F}{v(L/2)} = \frac{4P}{L}\left[\frac{2}{\pi}\sqrt{\frac{P_{cr}}{P}}\tan\!\left(\frac{\pi}{2}\sqrt{\frac{P}{P_{cr}}}\right) - 1\right]^{-1} \qquad \text{[Eq. 10.43]}$$

The transverse stiffness of a beam without compressive load P is:

$$k_o = \frac{F}{\delta_{max}} = \frac{F}{\left(\dfrac{FL^3}{48EI}\right)} = \frac{48EI}{L^3} \qquad \text{[Eq. 10.44]}$$

Therefore, the ratio of the transverse stiffness of the *loaded column* to that of the *unloaded column* is:

$$\frac{k}{k_o} = \frac{\pi^2}{12} \frac{P}{P_{cr}} \left[\frac{2}{\pi} \sqrt{\frac{P_{cr}}{P}} \tan\left(\frac{\pi}{2}\sqrt{\frac{P}{P_{cr}}}\right) - 1 \right]^{-1} \qquad \text{[Eq. 10.45]}$$

Setting $\phi = \dfrac{\pi}{2}\sqrt{\dfrac{P}{P_{cr}}}$, *Equation 10.45* is rewritten:

$$\frac{k}{k_o} = \frac{\phi^2}{3\left[\dfrac{1}{\phi}\tan\phi - 1\right]} \qquad \text{[Eq. 10.46]}$$

The stiffness ratio k/k_o is plotted against the load ratio P/P_{cr} in *Figure 10.20*. The complex *Equation 10.45* reduces to approximately a straight line:

$$\frac{k}{k_o} = 1 - \frac{P}{P_{cr}} \qquad \text{[Eq. 10.47]}$$

A complex problem has been reduced to a straightforward result, a very desirable engineering solution.

When the compressive load on the column is 60% of the buckling load, the transverse stiffness is reduced to 40% of its unloaded value. In general, when the compressive load is $x\%$ of the buckling load, the transverse stiffness is $(100-x)\%$ of the unloaded stiffness.

When columns are highly loaded – near their buckling load – the transverse stiffness is very small. Sideway forces can then have two different effects:

1. A small unbalanced sideways force may induce unstable buckling;
2. Modest constraint forces (from appropriate bracing systems) can resist buckling.

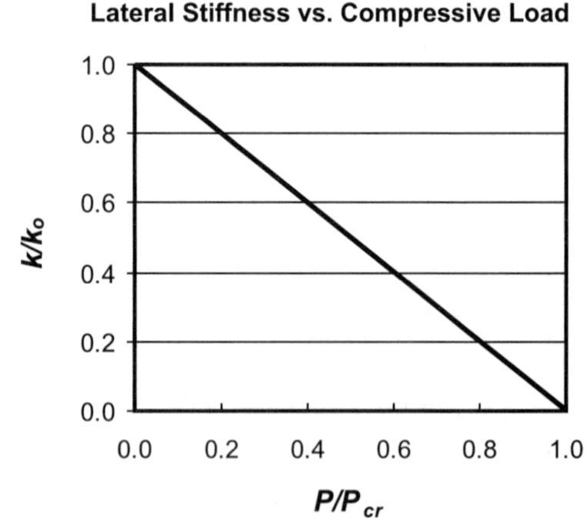

Figure 10.20. Lateral stiffness versus compressive load.

10.8 Oil-Canning Effect

Another phenomenon that resembles buckling is called **oil-canning**, from the *snap-through* (popping) observed/heard when emptying metal oil cans or other thin-walled cans containing liquid. A simple 2D model illustrating oil-canning consists of two identical elastic bars pinned to each other and to rigid supports distance 2L apart, as shown in *Figure 10.21*. Before the load is applied, the bars are at an angle α below the horizontal. The original length of each bar is thus $L_{AB} = L_{BC} = L/\cos\alpha$.

When transverse load F is slowly (statically) applied at the junction of the bars, the system deflects upward by distance δ. The bars shorten, and now are at an angle θ below the horizontal. The length of each bar is now $L_{AB} = L_{BC} = L/\cos\theta$.

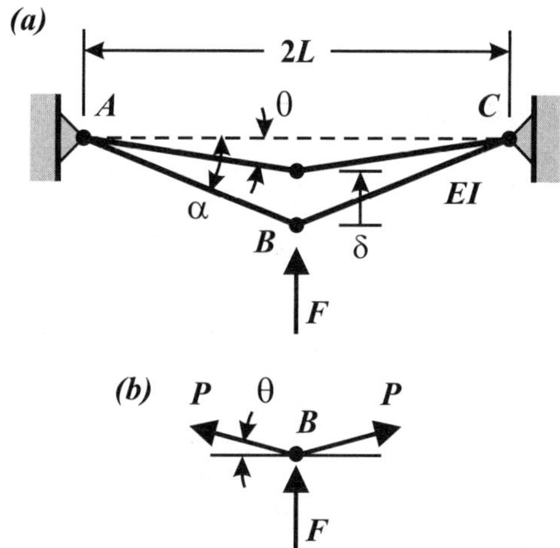

Figure 10.21. **(a)** Two elastic bars *AB* and *BC*, pinned at joint *B*. Transverse force *F* is applied at joint *B*. **(b)** FBD of joint *B*.

The change of length of each bar is:

$$\Delta = L\left(\frac{1}{\cos\theta} - \frac{1}{\cos\alpha}\right)$$

[Eq. 10.48]

The strain in each bar is:

$$\varepsilon = \frac{\Delta}{L/(\cos\alpha)} = \frac{\cos\alpha}{\cos\theta} - 1$$

[Eq. 10.49]

and the stress is:

$$\sigma = E\varepsilon = E\left(\frac{\cos\alpha}{\cos\theta} - 1\right)$$

[Eq. 10.50]

If the original cross-sectional area is A, then the area of each bar in the deflected position is, considering the Poisson effect:

$$A_\theta = (1 - 2\nu\varepsilon)A$$

[Eq. 10.51]

where ν is Poisson's ratio. The force in each bar $P(\theta)$ in the deflected position is then:

$$P(\theta) = A_\theta\sigma = AE\left[1 - 2\nu\left(\frac{\cos\alpha}{\cos\theta} - 1\right)\right]\left(\frac{\cos\alpha}{\cos\theta} - 1\right)$$

[Eq. 10.52]

Resolving the forces in the vertical direction of the FBD of *Figure 10.21b* gives:

$$F + 2P\sin\theta = 0 \qquad \text{[Eq. 10.53]}$$

The applied force F as a function of angle θ, is then:

$$F(\theta) = 2AE\left[1 - 2\nu\left(\frac{\cos\alpha}{\cos\theta} - 1\right)\right]\left(1 - \frac{\cos\alpha}{\cos\theta}\right)\sin\theta \qquad \text{[Eq. 10.54]}$$

The displacement δ of the applied force as a function of θ is:

$$\delta(\theta) = L(\tan\alpha - \tan\theta) \qquad \text{[Eq. 10.55]}$$

Next, normalizing force F by EA, and displacement δ by L, the force–displacement curve is shown in *Figure 10.21c*. For the graph, the initial bar angle is $\alpha = 20°$ (joint B is below the horizontal by $\delta/L = 0.36$) and Poisson's ratio is taken to be $\nu = 0.3$.

The interpretation of *Figure 10.21c* is as follows. The applied force F increases until it reaches a value of $0.017AE$, when *snap-through* occurs. Here, the snap-through angle is $\theta = 11.5°$. At this angle, joint B has moved $\delta/L = 0.17$, approximately halfway from its original position to the horizontal line AC in *Figure 10.21a*.

The slope of the force–displacement curve is the stiffness K of the system to transverse loads F. As δ increases, the stiffness continually decreases and becomes zero when $\delta/L = 0.17$. The displacement immediately jumps to $\delta/L = 0.79$, corresponding to a position about $\theta = 27°$ *above* the horizontal line AC (*Figure 10.21a*). The negative slope of the F–δ curve means the system is no longer resisting the force, but is in fact assisting it. The resulting sudden movement is known as *snap-through* or *oil-canning*.

The process described is the same effect that occurs in metal and plastic oil cans, the snap-through of a can's side being evident by the characteristic popping sound. This effect also occurs when popping out an unwelcome dent in a car panel.

Reaction Force

From equilibrium, the magnitude of the horizontal reaction force R_x in *Figure 10.21d* equals the horizontal component of force P:

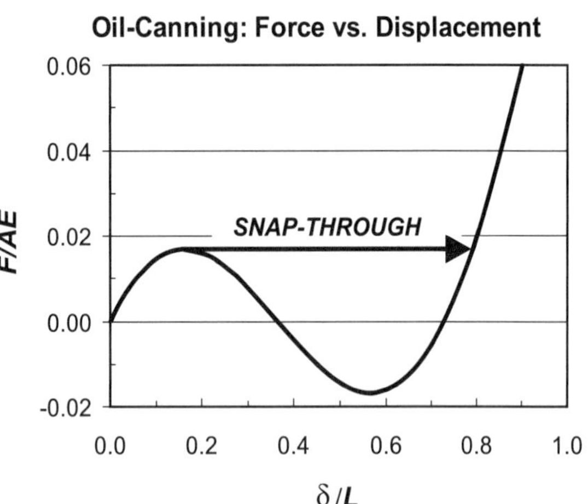

Figure 10.21. **(c)** Force versus displacement for two-bar system, with initial angle $\alpha = 20°$ below the horizontal. The horizontal ($\theta = 0°$) corresponds to $\delta/L \sim 0.36$.

(d)

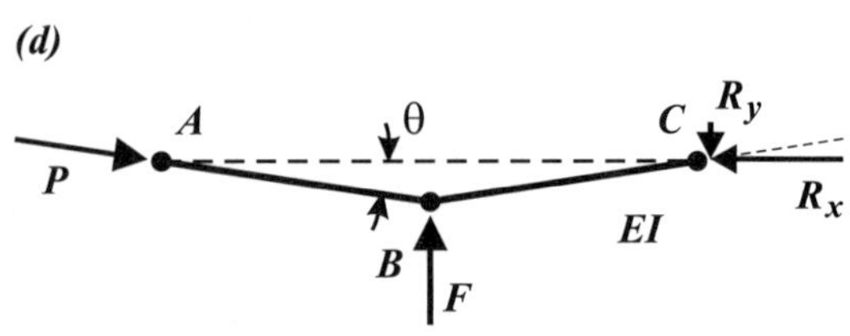

Figure 10.21. **(d)** FBD of system. R_x is maximum when $\theta = 0°$.

$$R_x = P(\theta)\cos\theta \qquad \text{[Eq. 10.56]}$$

The value of the horizontal reaction is maximum when both bars are compressed the maximum amount, which is when $\theta = 0°$. As the system passes through $\theta = 0°$, R_x reaches a maximum value of:

$$R_{max} = P(0°) = AE[1 - 2v(\cos\alpha - 1)](\cos\alpha - 1)$$

$$\text{[Eq. 10.57]}$$

For $\alpha = 20°$ and $v = 0.3$, the maximum value is $R_{max} = 0.0625AE$, which is about 3.7 times greater than the applied load to cause snap-through ($F = 0.017AE$). This illustrates the mechanical advantage that can be achieved with snap-through. An example of the use of this lever-like advantage is a pile-driving machine (*Figure 10.22*). As force F is applied transverse to the horizontal bars, the hammer drives the pile into the ground with a force larger than F.

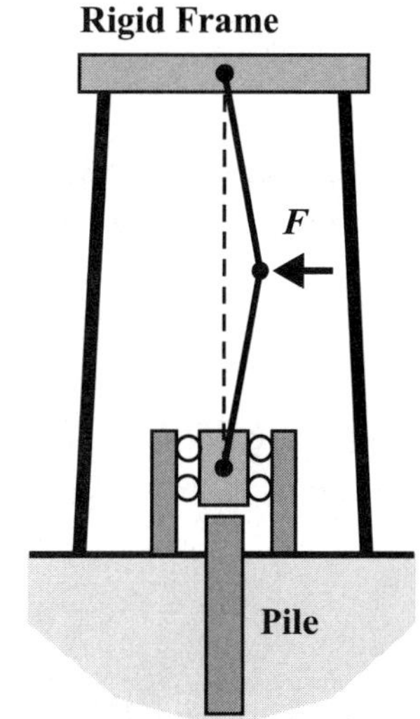

Figure 10.22. A pile-driver.

10.9 Buckling on an Elastic Foundation

Beam Displacement-Distributed Load Relation

Figure 10.23a represents a beam of length L with simply-supported ends loaded by distributed force $w(x)$. The modulus is E and the moment of inertia is I. An element of length dx is shown in *Figure 10.23b*, where dx is so small that $w(x)$ is constant over its length. Element dx is acted on by $w(x)$, shear forces V and $V+dV$, and moments M and $M+dM$.

Applying equilibrium to dx, the rates of change of the shear force and bending moment can be written:

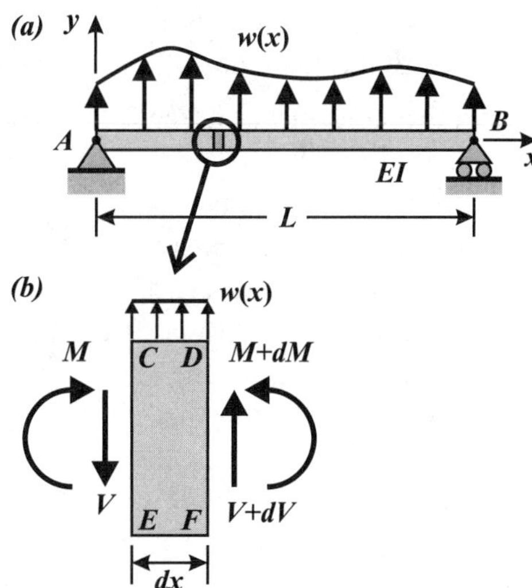

$$\frac{dV}{dx} = -w \qquad \text{[Eq. 10.58]}$$

$$\frac{dM}{dx} = -V \qquad \text{[Eq. 10.59]}$$

Figure 10.23. (a) Beam of length L subjected to distributed load $w(x)$. (b) Beam element dx.

so that:

$$\frac{d^2M}{dx^2} = w \qquad\qquad \text{[Eq. 10.60]}$$

From beam theory (*Chapter 6*), the second derivative of the beam deflection $v(x)$ is proportional to the moment:

$$v''(x) = \frac{d^2v}{dx^2} = \frac{M(x)}{EI} \qquad\qquad \text{[Eq. 10.61]}$$

Combining *Equations 10.60* and *10.61* shows that the fourth derivative of the deflection is proportional to the distributed load:

$$EI\frac{d^4v}{dx^4} = w \qquad\qquad \text{[Eq. 10.62]}$$

Column on an Elastic Foundation

Figure 10.24a represents a column of length L supported on an elastic foundation and having simply-supported ends. The beam material modulus is E and the cross-section moment of inertia is I. The elastic foundation provides a transverse force per unit length on the column when the transverse deflection of the beam is non-zero. At any position x, the load per unit length due to the foundation is:

$$w_e(x) = -kv(x) \qquad \text{[Eq. 10.63]}$$

where k is the foundation stiffness per unit length (force/length per unit length) and $v(x)$ is the transverse deflection of the column.

A column element of length dx, displaced $v(x)$, is shown in *Figure 10.24b*. The total vertical force per unit length transverse to the column w is made up of two parts: (1) w_e due to the displacement of the elastic foundation and (2) the change in the y-component of axial load P over length dx. The slope of $v(x)$, dv/dx ($= \tan\theta$) is small, so the y-component of the axial force P is, using the small angle approximation:

$$P_y = P\sin\theta = P\theta$$

$$= P\tan\theta = P\frac{dv}{dx} \qquad \text{[Eq. 10.64]}$$

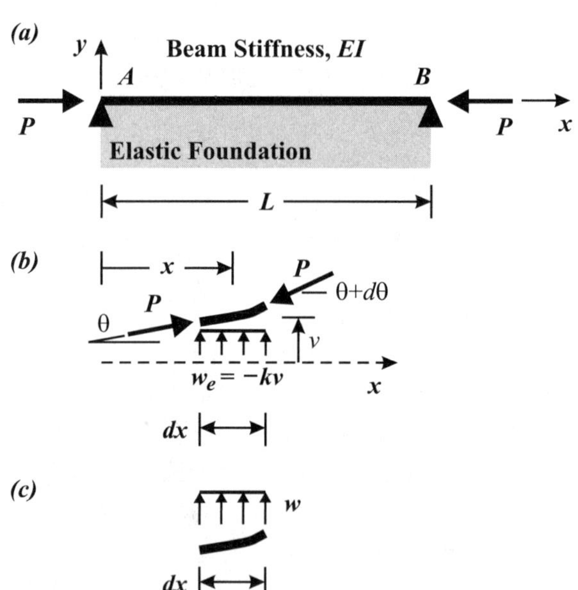

Figure 10.24. **(a)** Beam on elastic foundation, with pinned ends, subjected to compressive force *P*. **(b)** Element *dx* subjected to axial force *P* and transverse distributed force due to elastic foundation w_e. **(c)** Effective force per unit length *w* on element *dx*.

The slope at the right side of dx is:

$$\frac{dv}{dx} + \frac{d^2v}{dx^2}dx \qquad \text{[Eq. 10.65]}$$

The net force on dx is then (*Figure 10.24b*):

$$w\,dx = w_e\,dx + P\frac{dv}{dx} - P\left(\frac{dv}{dx} + \frac{d^2v}{dx^2}dx\right) \qquad \text{[Eq. 10.66]}$$

so that the total force per unit length is (*Figure 10.24c*):

$$w = -kv - P\frac{d^2v}{dx^2} \qquad \text{[Eq. 10.67]}$$

Substituting *Equation 10.67* into *Equation 10.62* results in the deflection equation for a column subjected to axial load P on an elastic foundation of stiffness per unit length k:

$$EI\frac{d^4v}{dx^4} + P\frac{d^2v}{dx^2} + kv = 0 \qquad \text{[Eq. 10.68]}$$

Buckling Load

Again, consider the column of length L supported on an elastic foundation with simply-supported ends (*Figure 10.24*). The column is subjected to compressive force P. When it buckles, a column of length L takes on the shape:

$$v = A\sin\frac{\pi x}{L} \qquad \text{[Eq. 10.69]}$$

where A is unknown. This equation satisfies the boundary conditions: $v = 0$ at $x = 0$ and $x = L$. The second and fourth derivatives of v are:

$$\frac{d^2v}{dx^2} = -A\left(\frac{\pi}{L}\right)^2\sin\frac{\pi x}{L} \qquad \text{[Eq. 10.70]}$$

and

$$\frac{d^4v}{dx^4} = A\left(\frac{\pi}{L}\right)^4\sin\frac{\pi x}{L} \qquad \text{[Eq. 10.71]}$$

Substituting *Equations 10.70* and *10.71* into *Equation 10.68* gives:

$$EI\left(\frac{\pi}{L}\right)^4 - P\left(\frac{\pi}{L}\right)^2 + k = 0 \qquad \text{[Eq. 10.72]}$$

from which the buckling load of the column on an elastic foundation is:

$$P = \frac{\pi^2 EI}{L^2} + k\left(\frac{L}{\pi}\right)^2 \qquad \text{[Eq. 10.73]}$$

The first term is the classical buckling load P_{cr} and the second term is the beneficial effect due to the elastic foundation; the buckling load has been increased due to the lateral support of the foundation. Note that the classical buckling load varies with $1/L^2$, while the effect of the foundation increases with L^2.

Example 10.9 Buckling of a Composite Sandwich Structure

Given: A sandwich structure is subjected to a bending moment M (*Figure 10.25a*). The structure consists of an elastic core of modulus E_c, sandwiched and well bonded between two stiff plates of modulus E, both of cross-sectional area t thick and b wide. The plates are distance d apart, where $t \ll d$, so that d is the depth of the sandwich structure.

The upper and lower plates can each be considered as axial members on an elastic foundation. The ratio of the moduli is $E/E_c = 10$.

Required: (a) Determine the applied moment that will cause buckling of the plate in compression (the upper plate) and (b) describe the buckled shape, i.e., determine the buckling length.

Solution: In this example, the buckling length L is not given. There are also no pins to support the compressive plate. However, from symmetry, the buckled form of the compressed plate will fit the

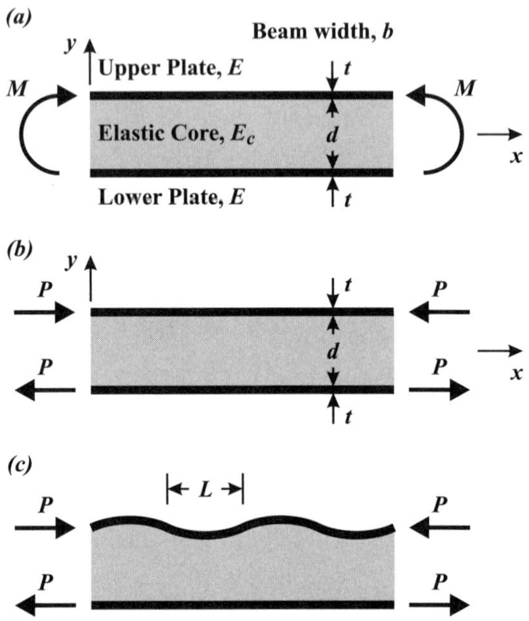

Figure 10.25. (a) Sandwich structure subjected to bending moment M. (b) Axial loads in the stiff plates support the moment; the contribution of the flexible core is neglected. (c) Buckling of compressed (upper) plate.

solution to *Equation 10.73*, where L is the distance between inflection points–points where the bending moment in the column (compressed plate) is zero. The system will seek out a distance L which minimizing the buckling load.

Minimizing *Equation 10.73* by taking the derivative with respect to L gives:

$$\frac{dP}{dL} = -3\frac{\pi^2 EI}{L^3} + \frac{2kL}{\pi^2} = 0$$

where I is the moment of inertia of the plate and k is the stiffness per unit length of the elastic core.

Solving for the buckled length L:

$$L^4 = \frac{3\pi^4 EI}{2k}$$

Substituting L into *Equation 10.73*, and simplifying gives the load to buckle a plate:

$$P_{min} = \sqrt{\frac{11}{3}kEI}$$

Since the core stiffness is much less than that of the plates, the bending moment is assumed to be supported only by the axial loads in the plates (*Figure 10.25b*). The axial load P in each plate is therefore:

$$P = \frac{M}{d}$$

The stiffness per unit length of the sandwich core, with $E/E_c = 10$, is:

$$k = E_c\frac{b}{d} = \frac{Eb}{10d}$$

and the moment of inertia of a single plate is:

$$I = \frac{bt^3}{12}$$

so that the compressive load to buckle a plate is:

$$P_{min} = \sqrt{\frac{11}{360}}(btE)\sqrt{\frac{t}{d}}$$

Neglecting the contribution of the core material, the moment at which buckling occurs is:

$$M_{min} = P_{min}d$$

$$\text{\textit{Answer}: } M_{min} = \sqrt{\frac{11}{360}}(btE)\sqrt{td}$$

The distance between the inflection points of the buckled plate (*Figure 10.25c*) is given by:

$$L^4 = \frac{3\pi^4 EI}{2k} = \frac{3\pi^4 E(bt^3/12)}{2(Eb/10d)}$$

from which:

$$\text{\textit{Answer}: } L = \pi\left[\frac{5t^3d}{4}\right]^{1/4}$$

The compressed plate buckles with a half-wavelength of L. The distance between consecutive crests or consecutive valleys of the buckled plate is $2L$.

Chapter 11   Energy Methods

11.0 Introduction

Energy methods are useful as a general problem-solving approach, and are the basis of computer-based *finite element analysis* (FEA) programs. The energy methods are also used in quick approximate calculations to help making preliminary design decisions. The formulation of the energy methods are based on (1) the *displacement method* and (2) the *force method.*

One advantage of energy methods is that they are generally easier to apply than the traditional displacement and force methods. Energy is a scalar, while displacements and forces are vectors. The energy methods are illustrated by analyzing the three-bar truss shown in *Figure 11.1*.

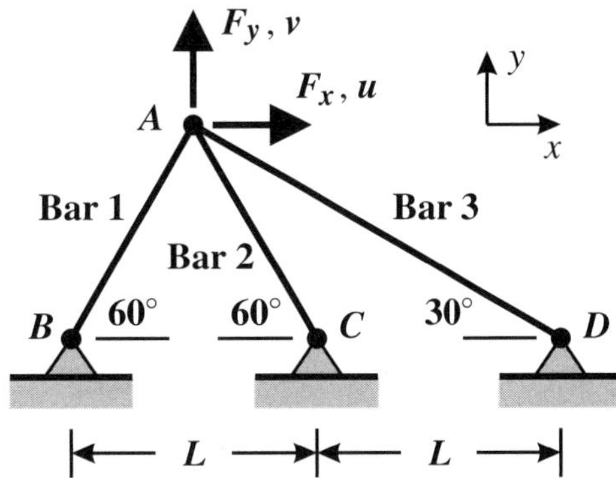

Figure 11.1. Three-bar truss structure.

11.1 Internal and Complementary Energy

Axial Members – Linear–Elastic Response

Axial force P applied to a bar of length L and cross-sectional area A causes the bar to elongate Δ (*Figure 11.2*). For a linear-elastic material with modulus E, the elongation is:

$$\Delta = \frac{PL}{AE} \qquad \text{[Eq. 11.1]}$$

When the load is removed, the bar returns to its original unloaded length ($\Delta = 0$).

The ***internal energy*** U stored in the bar is dependent upon elongation (displacement), and is defined as:

$$U(\Delta) = \int_0^{\Delta} dU = \int_0^{\Delta} P\, d\Delta \qquad \text{[Eq. 11.2]}$$

F.A. Leckie, D.J. Dal Bello, *Strength and Stiffness of Engineering Systems*, Mechanical Engineering Series,   **323**
DOI 10.1007/978-0-387-49474-6_11, © Springer Science+Business Media, LLC 2009

The increment of internal energy, $dU = P\,d\Delta$, is the shaded area in *Figure 11.2c*. Substituting for P from *Equation 11.1*, the internal energy due to elongation Δ is:

$$U(\Delta) = \int_0^{\Delta} \frac{EA\Delta}{L}d\Delta = \frac{EA}{2L}\Delta^2$$

$$[\text{Eq. }11.3]$$

which is the area of the shaded triangle in *Figure 11.2d*.

The ***internal complementary energy*** C is dependent upon force, and is defined as:

$$C(P) = \int_0^P dC = \int_0^P \Delta\,dP$$

$$[\text{Eq. }11.4]$$

The increment of internal complementary energy, $dC = \Delta\,dP$, is the shaded area in *Figure 11.2e*. Substituting for Δ from *Equation 11.1*, the complementary energy due to force P is:

$$C(P) = \int_0^P \frac{PL}{AE}dP = \frac{L}{2AE}P^2$$

$$[\text{Eq. }11.5]$$

which is the area of the shaded triangle in *Figure 11.2f*.

From *Figures 11.2d* and *f*, the sum of $U(\Delta)$ and $C(P)$ is:

$$U(\Delta) + C(P) = P\Delta \qquad [\text{Eq. }11.6]$$

In this treatment, *only linear–elastic materials* are considered. However, *Equations 11.2, 11.4* and *11.6* also hold for a non-linear force–elongation (stress–strain) curve, where $U(\Delta)$ is the area under the curve and $C(P)$ is the area above the curve.

Comparison of Energy States

Figure 11.3 shows the force–elongation curve for a linear material that has been loaded to state A (P_A, Δ_A), and then to state B (P_B, Δ_B). The internal energy corresponding to Δ_A is the lower hatched triangle in *Figure 11.3a*:

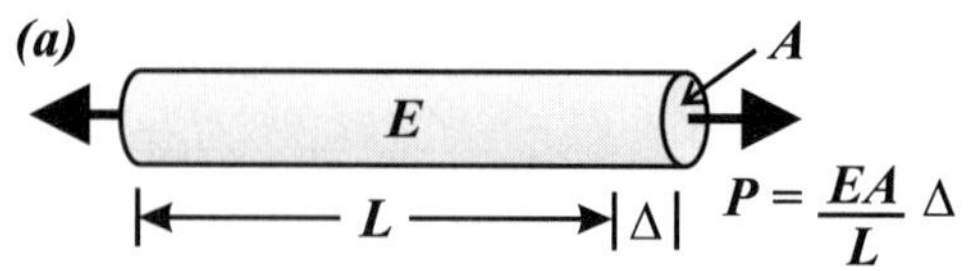

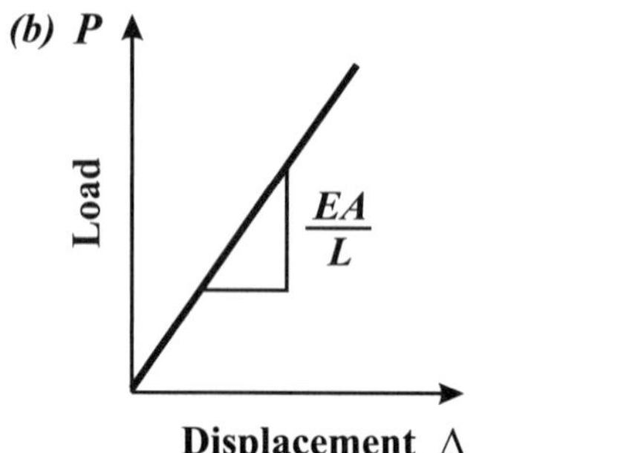

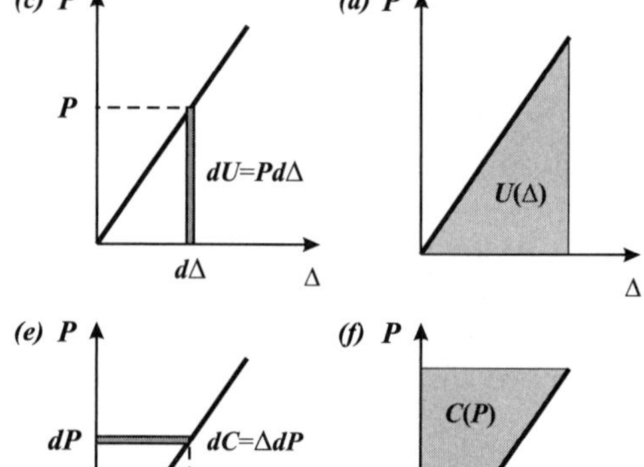

Figure 11.2. (a) Bar subjected to axial force. **(b)** Force–elongation curve for a linear–elastic material. **(c)** Differential change in internal energy dU with differential change in elongation $d\Delta$. **(d)** Total internal energy $U(\Delta)$. **(e)** Differential change in complementary energy dC with differential change in force dP. **(f)** Total internal complementary energy $C(P)$.

$$U(\Delta_A) = \frac{EA}{2L}(\Delta_A)^2 \qquad \text{[Eq. 11.7]}$$

The internal energy corresponding to Δ_B is the entire triangle:

$$U(\Delta_B) = \frac{EA}{2L}(\Delta_B)^2 \qquad \text{[Eq. 11.8]}$$

From the geometry of *Figure 11.3a*:

$$U(\Delta_B) - U(\Delta_A) \geq P_A(\Delta_B - \Delta_A) \qquad \text{[Eq. 11.9]}$$

The term on the right side of the inequality represents the rectangular area of height P_A and width $\Delta_B - \Delta_A$. The inequality of *Equation 11.9* is always valid for a linear material. The relationship is also valid for non-linear materials provided that the slope of the force–elongation (stress–strain) curve is always positive.

A similar inequality applies to the complementary energy (*Figure 11.3b*):

$$C(P_B) - C(P_A) \geq \Delta_A(P_B - P_A) \qquad \text{[Eq. 11.10]}$$

The term on the right of the inequality represents the rectangular area of width Δ_A and height $P_B - P_A$.

These two inequalities are used to establish the *minimum energy theorems*.

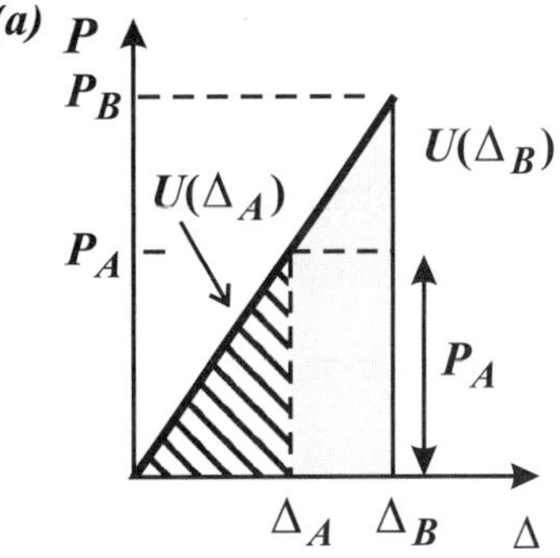

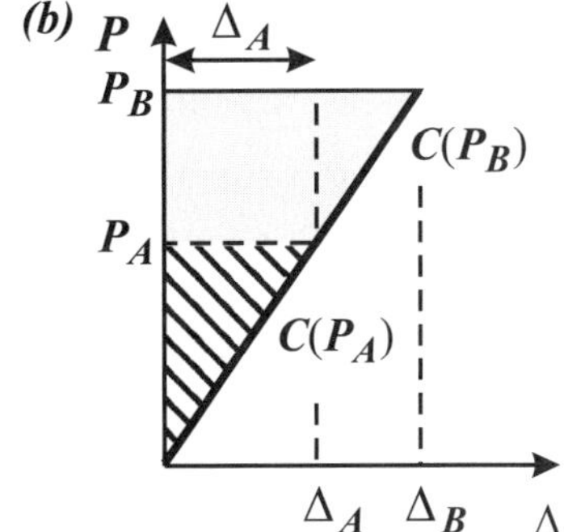

Figure 11.3. (a) Internal energies at states *A* and *B*. **(b)** Complementary energies at states *A* and *B*.

Three-Bar Truss, Displacement and Force Methods

The three-bar truss system shown in *Figure 11.4* will be used to illustrate various applications of the *energy method*. The truss is an assembly of three bars, with geometric and elastic properties given in *Table 11.1*. For this particular structure, the values have been chosen so that the axial stiffness E_iA_i/L_i of each bar is the same; in general, this is not true.

Horizontal load F_x and vertical load F_y are applied at joint *A*. The horizontal and vertical displacements of joint *A* are *u* and *v*. The internal forces in Bars 1, 2, and 3 are P_1,

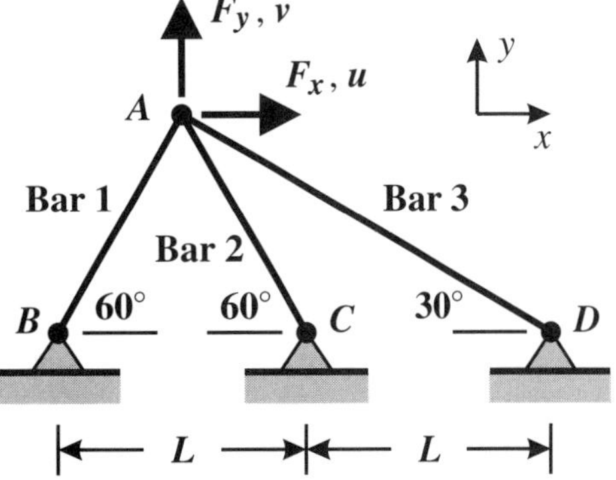

Figure 11.4. Three-bar truss studied in this chapter.

P_2, and P_3, respectively, and their elongations are Δ_1, Δ_2, and Δ_3.

Equilibrium

Equilibrium of joint A (*Figure 11.5*) relates the applied loads to the internal forces:

$$F_x - P_1 \cos 60° + P_2 \cos 60° + P_3 \cos 30° = 0$$

$$F_y - P_1 \sin 60° - P_2 \sin 60° - P_3 \sin 30° = 0$$

[Eq. 11.11]

where $\quad \cos 30° = \sin 60° = \sqrt{3}/2$

$$\cos 60° = \sin 30° = 1/2$$

Thus:

$$F_x = \frac{P_1}{2} - \frac{P_2}{2} - \frac{\sqrt{3}P_3}{2}$$

$$F_y = \frac{\sqrt{3}P_1}{2} + \frac{\sqrt{3}P_2}{2} + \frac{P_3}{2}$$

[Eq. 11.12]

Table 11.1. Geometric and elastic properties of bars of *Figure 11.4*.

Bar	Length	Area	Modulus	Axial Stiffness
1	L	A	E	EA/L
2	L	A	E	EA/L
3	$\sqrt{3}L$	$\sqrt{3}A$	E	EA/L

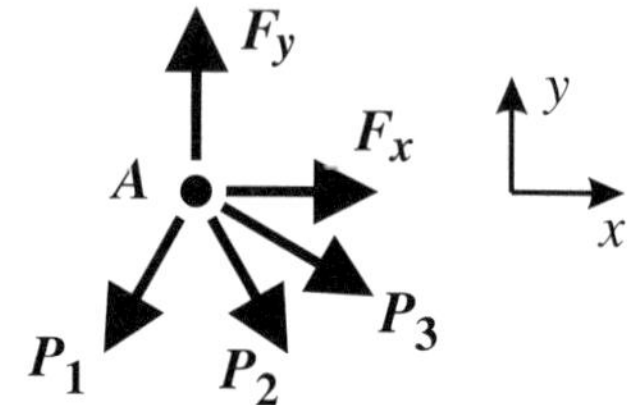

Figure 11.5. FBD of Joint A.

The system is *redundant* or *statically indeterminate* since there are three unknown (internal) forces and only two equations of equilibrium.

Force–Elongation

The force–elongation relationship in each bar is:

$$P_i = \frac{E_i A_i}{L_i} \Delta_i$$

[Eq. 11.13]

Using the values from *Table 11.1*, the force in each bar for this particular case is:

$$P_1 = \frac{EA}{L} \Delta_1$$

$$P_2 = \frac{EA}{L} \Delta_2$$

$$P_3 = \frac{EA}{L} \Delta_3$$

[Eq. 11.14]

Elongation–Displacement (Compatibility)

Consider the elongation Δ_1 of Bar 1 (*AB*) resulting from horizontal displacement u (*Figure 11.6a*). The elongation is found by projecting u onto line AB (i.e., perform the dot-product of vector $u\mathbf{i}$ and a unit vector in the direction of AB):

$$\Delta_1(u) = u \cos 60° = \frac{u}{2}$$

[Eq. 11.15]

The elongation of Bar 1 resulting from displacement v is found by projecting v onto AB:

$$\Delta_1(v) = v\cos 30° = \frac{\sqrt{3}}{2}v \qquad \text{[Eq. 11.16]}$$

The elongation of Bar 1 caused by both displacements simultaneously is then:

$$\Delta_1 = \frac{u}{2} + \frac{\sqrt{3}}{2}v \qquad \text{[Eq. 11.17]}$$

Repeating this process for each bar gives their elongations in terms of the displacement of joint A:

$$\Delta_1 = \frac{u}{2} + \frac{\sqrt{3}}{2}v$$

$$\Delta_2 = -\frac{u}{2} + \frac{\sqrt{3}}{2}v \qquad \text{[Eq. 11.18]}$$

$$\Delta_3 = -\frac{\sqrt{3}}{2}u + \frac{v}{2}$$

The elongation of each bar Δ_i must be *compatible* with the system displacements, u and v. All of the bars must still be joined at joint A under load.

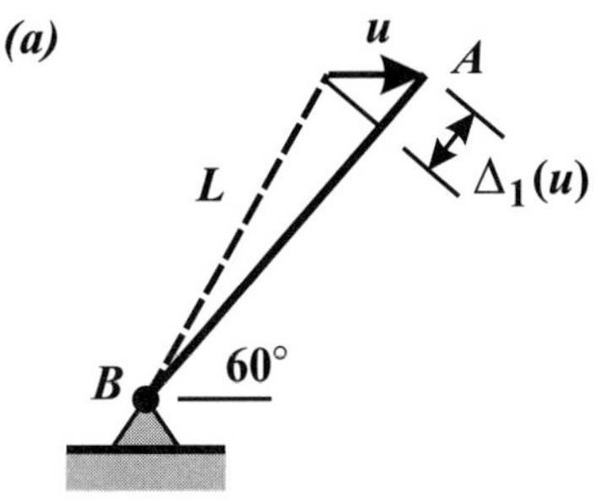

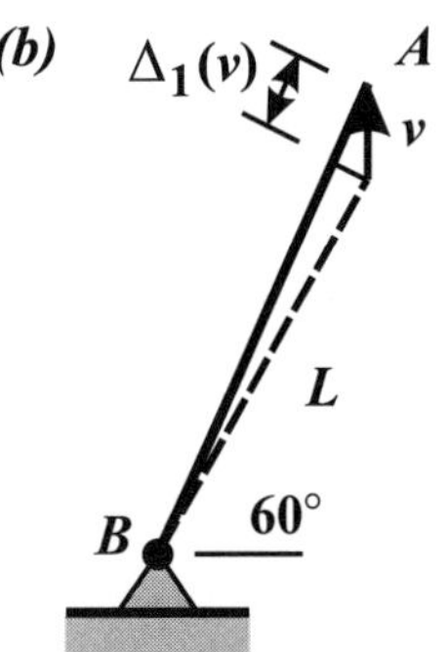

Figure 11.6. (a) Elongation of Bar 1 due to displacement u. **(b)** Elongation of Bar 1 due to displacement v.

The aim now is to determine the relationship between the applied loads (F_x, F_y) and the displacements (u, v) using equilibrium, Hooke's Law, and compatibility (*Equations 11.12, 11.14,* and *11.18,* respectively). To achieve this goal, use is made of the *displacement* and *force methods* described in *Chapter 4.*

Example 11.1 Three-bar Truss: Displacement Method

Required: Using the *displacement method,* determine the relationship between the applied forces and the displacement of joint A of the three-bar truss (*Figure 11.4*).

Solution: In the displacement method, the (unknown) applied forces (F_x and F_y) needed to cause the enforced (known) displacements (u and v) are determined. The three steps are as follows:

1. *Elongation–displacement (compatibility), Equation 11.18.*
2. *Force–elongation, Equation 11.14* (force–stress–strain–elongation relationships combined).
3. *Equilibrium, Equation 11.12.*

Step 1. Starting with *Equation 11.18,* the elongations Δ_1, Δ_2, and Δ_3 are determined in terms of displacements u and v.

Step 2. Substituting the elongations into *Equation 11.14,* the internal forces P_1, P_2, and P_3 are determined in terms of u and v:

$$P_1 = \frac{EA}{L}\Delta_1 = \frac{EA}{L}\left(\frac{u}{2} + \frac{\sqrt{3}}{2}v\right)$$

$$P_2 = \frac{EA}{L}\Delta_2 = \frac{EA}{L}\left(-\frac{u}{2} + \frac{\sqrt{3}}{2}v\right)$$

$$P_3 = \frac{EA}{L}\Delta_3 = \frac{EA}{L}\left(-\frac{\sqrt{3}}{2}u + \frac{v}{2}\right)$$

Step 3. Internal forces P_1, P_2, and P_3, in terms of u and v, are substituted into the equilibrium equations, *Equation 11.12*, to give:

$$\textit{Answer: } F_x = \frac{EA}{4L}[5u - \sqrt{3}v] \quad \text{and} \quad F_y = \frac{EA}{4L}[-\sqrt{3}u + 7v]$$

Writing the results in matrix form:

$$\begin{bmatrix} F_x \\ F_y \end{bmatrix} = \frac{EA}{4L}\begin{bmatrix} 5 & -\sqrt{3} \\ -\sqrt{3} & 7 \end{bmatrix}\begin{bmatrix} u \\ v \end{bmatrix} \quad \text{or} \quad \mathbf{f = Ku} \qquad \text{[Eq. 11.19]}$$

where vectors $\mathbf{f}$ and $\mathbf{u}$ are the force and displacement vectors, respectively, and $\mathbf{K}$ is the stiffness matrix of the system. Notice that the stiffness matrix is symmetric about its diagonal; this is a general property of the stiffness matrix of linear–elastic systems.

Example 11.2 Three-bar Truss: Force Method

Required: Using the *force method*, determine the relationship between the applied forces and the displacement of joint A of the three-bar truss (*Figure 11.4*).

Solution: In the force method, the displacements (u and v) caused by applied forces (F_x and F_y) are determined. The three steps are:

1. *Equilibrium, Equation 11.12.*
2. *Elongation–force, Equation 11.14.*
3. *Displacement–elongation (compatibility), Equation 11.18.*

Step 1. Equilibrium is given by *Equation 11.12*. Since there are only two equilibrium equations for the three unknown internal forces, P_1, P_2, and P_3, the system is *redundant*. To begin, select the force in Bar 3 as the redundant member, so $P_3 = R$. The redundant force is for the time being assumed to be known. The internal forces in terms of applied loads F_x and F_y and redundant force R, are then:

$$P_1 = \frac{\sqrt{3}F_x + F_y + R}{\sqrt{3}}; \quad P_2 = \frac{-\sqrt{3}F_x + F_y - 2R}{\sqrt{3}}; \quad P_3 = R \qquad \text{[Eq. 11.20]}$$

Step 2. The *Elongation–force* relations are given by *Equation 11.14*. The elongation of each bar is:

$$\Delta_1 = \frac{P_1 L}{AE} = \left[\frac{\sqrt{3}F_x + F_y + R}{\sqrt{3}}\right]\frac{L}{AE}$$

$$\Delta_2 = \frac{P_2 L}{AE} = \left[\frac{-\sqrt{3}F_x + F_y - 2R}{\sqrt{3}}\right]\frac{L}{AE} \qquad \text{[Eq. 11.21]}$$

$$\Delta_3 = \frac{P_3\sqrt{3}L}{\sqrt{3}AE} = \frac{RL}{AE}$$

Step 3. The *displacement-elongation* relations of are, from *Equation 11.18*:

$$\Delta_1 = \frac{u}{2} + \frac{\sqrt{3}}{2}v ; \quad \Delta_2 = -\frac{u}{2} + \frac{\sqrt{3}}{2}v ; \quad \Delta_3 = -\frac{\sqrt{3}}{2}u + \frac{v}{2}$$

Eliminating u and v from these elongation equations results in the following *compatibility* condition:

$$\Delta_1 - 2\Delta_2 + \sqrt{3}\Delta_3 = 0$$

Substituting the elongations from *Equation 11.21* into the last equation gives the redundant force:

$$R = \frac{-3\sqrt{3}F_x + F_y}{8}$$

Equating the equations for Δ_i of *Step 3* with those of *Step* 2, and solving for u and v:

$$u = [2F_x + \sqrt{3}R]\frac{L}{AE} \quad \text{and} \quad v = \frac{1}{3}[2F_y - R]\frac{L}{AE}$$

Since R is now known in terms of F_x and F_y, the displacements are:

$$Answer: u = \left[\frac{7}{8}F_x + \frac{\sqrt{3}}{8}F_y\right]\frac{L}{AE} \quad \text{and} \quad v = \left[\frac{\sqrt{3}}{8}F_x + \frac{5}{8}F_y\right]\frac{L}{AE}$$

Writing the results in matrix form:

$$\begin{bmatrix} u \\ v \end{bmatrix} = \frac{L}{8AE}\begin{bmatrix} 7 & \sqrt{3} \\ \sqrt{3} & 5 \end{bmatrix}\begin{bmatrix} F_x \\ F_y \end{bmatrix} \quad \text{or} \quad \mathbf{u} = \mathbf{F}\mathbf{f} \qquad \text{[Eq. 11.22]}$$

where $\mathbf{u}$ and $\mathbf{f}$ are the displacement and force vectors, respectively, and $\mathbf{F}$ is the flexibility matrix of the system. Notice that the flexibility matrix is symmetric about its diagonal; this is a general property of the flexibility matrix for linear–elastic materials.

The flexibility matrix is the inverse of the stiffness matrix found in *Example 11.1*:

$$F^{-1} = \left(\frac{L}{8AE} \begin{bmatrix} 7 & \sqrt{3} \\ \sqrt{3} & 5 \end{bmatrix} \right)^{-1} = \frac{EA}{4L} \begin{bmatrix} 5 & -\sqrt{3} \\ -\sqrt{3} & 7 \end{bmatrix} = K$$

Although several steps of *Examples 11.1* and *11.2* were not shown, the force method is generally more algebraically intense than the displacement method.

11.2 Principle of Virtual Work

The conditions for equilibrium and compatibility can be expressed neatly in compact form by the so-called *principle of virtual work*. In this approach, the only conditions that need to be satisfied are *equilibrium* and *compatibility*. It is not necessary to make use of the elastic force–elongation relationships (i.e., Hooke's Law). The method is illustrated with the three-bar truss in *Figure 11.7*.

The truss is subjected to applied loads F_x and F_y at joint A. The applied loads are related to the internal force of each bar, P_1, P_2, and P_3, by equilibrium (*Equation 11.12*). There are three internal forces and only two equilibrium equations, meaning there are an infinite number of possibilities for the internal forces that satisfy equilibrium.

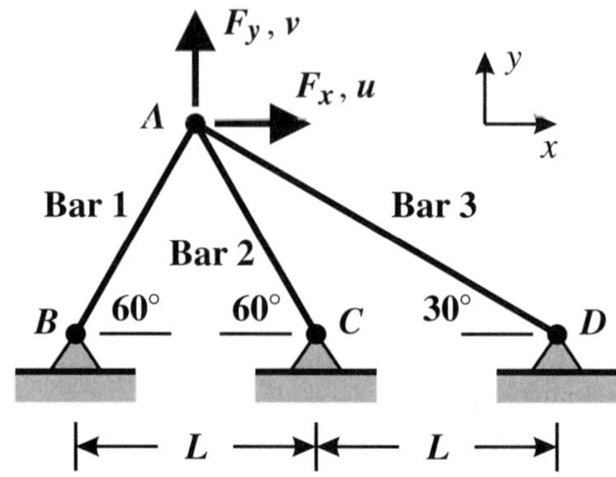

Figure 11.7. The internal forces of the three-bar truss must be in equilibrium with the applied forces. The elongations of each bar must be compatible with the movement of joint *A*.

The displacements, u and v, and the elongations of the bars, Δ_1, Δ_2, and Δ_3, are related by compatibility (*Equation 11.18*).

The force–elongation (Hooke's law) relationships (*Equation 11.14*) are not enforced.

The *virtual work equation* is formulated by multiplying the equilibrium equations for F_x and F_y (*Equation 11.12*) by displacements u and v, respectively, giving:

$$F_x u = \left(\frac{P_1}{2} - \frac{P_2}{2} - \frac{\sqrt{3}P_3}{2} \right) u$$

$$F_y v = \left(\frac{\sqrt{3}P_1}{2} + \frac{\sqrt{3}P_2}{2} + \frac{P_3}{2} \right) v$$

[Eq. 11.23]

Adding the two equations and collecting like force terms gives:

$$F_x u + F_y v = P_1\left(\frac{u}{2} + \frac{\sqrt{3}v}{2}\right) + P_2\left(-\frac{u}{2} + \frac{\sqrt{3}v}{2}\right) + P_3\left(-\frac{\sqrt{3}u}{2} + v\right) \qquad \text{[Eq. 11.24]}$$

The terms in parentheses are the bar elongations Δ_1, Δ_2, and Δ_3 that are compatible with displacements u and v (*Equation 11.18*).

The *virtual work equation* is therefore:

$$F_x u + F_y v = P_1\Delta_1 + P_2\Delta_2 + P_3\Delta_3 \qquad \text{[Eq. 11.25]}$$

The virtual work done by external forces F_x and F_y undergoing displacements u and v equals that done by internal forces P_1, P_2, and P_3 undergoing elongations Δ_1, Δ_2, and Δ_3.

The work is *virtual*. Although the external and internal forces are in equilibrium, and the displacements and elongations are compatible, the force–elongation (stress–strain) relationships have not been satisfied. Since Hooke's Law is not enforced, a solution to *Equation 11.25* is not necessarily the actual solution.

The *method of virtual work* is shown in *Example 11.3*. Virtual work will be a key concept in developing the *minimum energy principles* in *Section 11.3*.

Example 11.3 Three-bar Truss: Virtual Work

Given: The external forces on the three-bar truss in *Figure 11.7* are taken to be $F_x = 5.0$ kN and $F_y = 3.0$ kN and the displacements are taken to be $u = 2.0$ mm and $v = 5.0$ mm.

Required: Show by example that *any* set of internal forces in equilibrium with F_x and F_y satisfies the virtual work equation (*Equation 11.25*).

Solution: *Step 1.* The *external virtual work* is:

$$F_x u + F_y v = (5 \text{ kN})(2 \text{ mm}) + (3 \text{ kN})(5 \text{ mm}) = 25 \text{ N} \cdot \text{m}$$

Step 2. Internal virtual work. From compatibility (*Equation 11.18*), the elongations of the bars due to the given displacements are:

$$\Delta_1 = \frac{u}{2} + \frac{\sqrt{3}}{2}v = 5.33 \text{ mm}$$

$$\Delta_2 = -\frac{u}{2} + \frac{\sqrt{3}}{2}v = 3.33 \text{ mm}$$

$$\Delta_3 = -\frac{\sqrt{3}}{2}u + \frac{v}{2} = 0.768 \text{ mm}$$

A set of internal forces in equilibrium with F_x and F_y were developed in *Example 11.2* as *Equation 11.20*:

$$P_1 = \frac{\sqrt{3}F_x + F_y + R}{\sqrt{3}}; \quad P_2 = \frac{-\sqrt{3}F_x + F_y - 2R}{\sqrt{3}}; \quad P_3 - R$$

Any value of the redundant force R will satisfy equilibrium.

Step 3a. Internal forces Selection 1. Select the redundant force to be zero, $R = 0$:

$$P_1 = \frac{\sqrt{3}F_x + F_y}{\sqrt{3}} = 6.73 \text{ kN}; \quad P_2 = \frac{-\sqrt{3}F_x + F_y}{\sqrt{3}} = -3.287 \text{ kN}; \quad P_3 = 0$$

The *internal virtual work* is:

$$P_1\Delta_1 + P_2\Delta_2 + P_3\Delta_3 = (6.73)(5.33) + (-3.28)(3.33) + (0)(0.768)$$

$$= 25 \text{ N} \cdot \text{m}$$

The internal virtual work equals the external virtual work.

Step 3b. Internal forces Selection 2. Select the redundant force to be $R = 10$ kN:

$$P_1 = 12.5 \text{ kN}; \quad P_2 = -14.8 \text{ kN}; \quad P_3 = 10 \text{ kN}$$

The internal work is:

$$P_1\Delta_1 + P_2\Delta_2 + P_3\Delta_3 = (12.5)(5.33) + (-14.8)(3.33) + (10)(0.768)$$

$$= 25 \text{ N} \cdot \text{m}$$

Again, the internal virtual work equals the external virtual work.

By inference, since R can take on an infinite number of values, any set of internal forces in equilibrium with the applied loads satisfies the virtual work equation. The correct value of R for a given set of displacements (u,v) is the one that minimizes the energy of the system.

11.3 Minimum Energy Principles

The displacement and force methods described in *Examples 11.1* and *11.2*, can be algebraically challenging. Methods using energy concepts are easier to apply. Furthermore, *approximate energy methods* are useful in determining upper and lower bounds on the stiffness and flexibility of load-bearing systems. Approximate methods that are quick and easy to apply are useful to engineers in the early stages of the development of a design.

Potential Energy of a Force

The potential energy V of force **F** with components F_x and F_y is:

$$V = -F_x u - F_y v \qquad \text{[Eq. 11.26]}$$

where u and v are the displacements of the force in the x- and y-directions, respectively (*Figure 11.8*). For example, if mass m is subjected to gravity (in the negative y-direction), the gravitational force on the

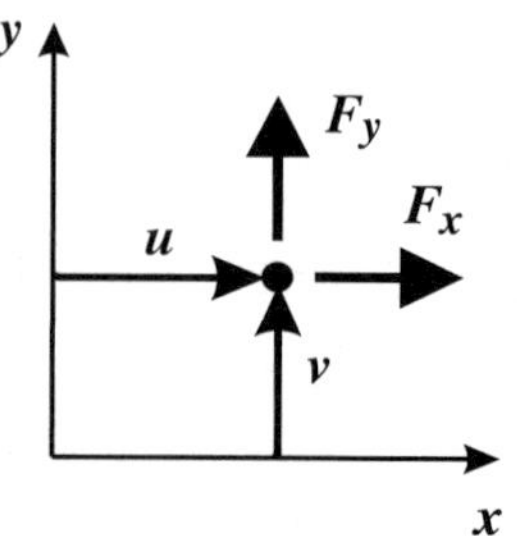

Figure 11.8. Displacement of force.

mass is $F_y = -mg$. When the mass is raised to a height $v = h$, its potential energy is:

$$V = -F_y v = mgh \qquad \text{[Eq. 11.27]}$$

The potential energy depends only on the final position and not on the path to get to that position.

The elastic energy stored in the bars also depends only on the final position (u,v) of joint A (*Figure 11.9*). Each axial member acts as a spring of stiffness $K_i = E_i A_i / L_i$. For the three-bar truss in these examples, the axial stiffness of each bar has been chosen to be the same.

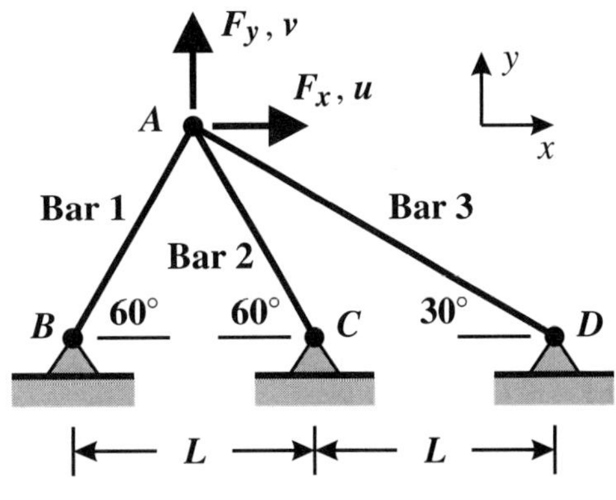

Figure 11.9. Three-bar truss.

Methods of Energy Minimization

Consistent with the approach of earlier chapters, there are two minimum energy methods:

1. The *displacement energy method*: the energy principle is expressed in terms of joint displacements **u**, and compatible bar elongations Δ.

2. The *force energy method*: the energy principle is expressed in terms of applied forces **f**, and equilibrium internal bar forces **P**.

Here, **u**, Δ, **f**, and **P** are the displacement, elongation, applied force, and internal force vectors, respectively. The methods are demonstrated using the three-bar truss previously discussed.

Displacement Method – Minimum Total Energy

The problem posed in *Figure 11.9* is to determine the applied forces (F_x, F_y) in terms of enforced displacements (u, v). The **internal energy** U of the three bars is:

$$U(\Delta_1, \Delta_2, \Delta_3) = \frac{EA}{2L}(\Delta_1^2 + \Delta_2^2 + \Delta_3^2) \qquad \text{[Eq. 11.28]}$$

Substituting for the elongations in terms of the displacements (*Equation 11.18*):

$$U(u, v) = \frac{EA}{2L}\left[\left(\frac{u}{2} + \frac{\sqrt{3}}{2}v\right)^2 + \left(-\frac{u}{2} + \frac{\sqrt{3}}{2}v\right)^2 + \left(-\frac{\sqrt{3}}{2}u + \frac{v}{2}\right)^2\right] \qquad \text{[Eq. 11.29]}$$

The potential energy of the applied loads F_x and F_y is:

$$V(u, v) = -F_x u - F_y v \qquad \text{[Eq. 11.30]}$$

The **total energy** T of the system is the sum of the internal and potential energies:

$$T(u, v) = U(u, v) + V(u, v) \qquad \text{[Eq. 11.31]}$$

so that:

$$T(u, v) = \frac{EA}{2L}\left[\left(\frac{u}{2} + \frac{\sqrt{3}}{2}v\right)^2 + \left(-\frac{u}{2} + \frac{\sqrt{3}}{2}v\right)^2 + \left(-\frac{\sqrt{3}}{2}u + \frac{v}{2}\right)^2\right] - F_x u - F_y v \qquad \text{[Eq. 11.32]}$$

The *total energy* is a function of joint A displacements (u, v) and applied forces (F_x, F_y).

Now consider displacements u^* and v^*, that are close to, but not equal to, the actual (or exact) values of u and v. The difference between the exact and approximate values of u and v are δu and δv, respectively:

$$u^* = u + \delta u; \quad v^* = v + \delta v \qquad \text{[Eq. 11.33]}$$

The approximate elongations Δ_i^* of the bars due to u^* and v^* are:

$$\Delta_1^* = \Delta_1 + \delta\Delta_1; \quad \Delta_2^* = \Delta_2 + \delta\Delta_2; \quad \Delta_3^* = \Delta_3 + \delta\Delta_3 \qquad \text{[Eq. 11.34]}$$

where Δ_i are the exact values and $\delta\Delta_i$ are the differences due to δu and δv.

The total energy corresponding to the nearby displacements is

$$T(u^*, v^*) = U(u^*, v^*) - F_x u^* - F_y v^* \qquad \text{[Eq. 11.35]}$$

By substituting the compatibility equations, it can be shown that the difference between the total energy corresponding to nearby displacements (u^*, v^*) and the total energy corresponding to the exact displacements (u, v) is:

$$T(u^*, v^*) - T(u, v)$$

$$= \frac{EA}{2L}\left[2(\Delta_1\delta\Delta_1 + \Delta_2\delta\Delta_2 + \Delta_3\delta\Delta_3) + ({\delta\Delta_1}^2 + {\delta\Delta_2}^2 + {\delta\Delta_3}^2)\right] - F_x\delta u - F_y\delta v$$

$$\text{[Eq. 11.36]}$$

Since $P_1 = \dfrac{EA}{L}\Delta_1$, etc., then the difference in total energy can be rewritten:

$$T(u^*, v^*) - T(u, v)$$

$$= \left[(P_1\delta\Delta_1 + P_2\delta\Delta_2 + P_3\delta\Delta_3) - F_x\delta u - F_y\delta v\right] + \frac{EA}{2L}(\delta\Delta_1^2 + \delta\Delta_2^2 + \delta\Delta_3^2)$$

$$\text{[Eq. 11.37]}$$

The term in square brackets is the *virtual work equation* with external loads F_x and F_y undergoing displacements δu and δv, and equilibrium internal loads P_1, P_2, and P_3 in members extending by $\delta\Delta_1$, $\delta\Delta_2$, $\delta\Delta_3$ (compatible with δu and δv). Thus, the term in square brackets is zero, as defined in *Equation 11.25*.

Equation 11.37 then reduces to:

$$T(u^*, v^*) - T(u, v) = \frac{EA}{2L}(\delta\Delta_1^2 + \delta\Delta_2^2 + \delta\Delta_3^2) \qquad \text{[Eq. 11.38]}$$

The displacement errors, $\delta\Delta_i$, are all squared and are therefore always positive. As a consequence, the total energy associated with the approximate displacements (u^*, v^*) is always greater than the total energy of the correct solution (u, v):

$$T(u^*, v^*) \geq T(u, v) \qquad \text{[Eq. 11.39]}$$

Hence, the exact solution corresponds to the *minimum of the total energy*.

The total energy as a function of u and v can be plotted in 3D as shown in *Figure 11.10*. The **total energy surface** is concave upward. In mathematical terms, the exact solution is given by the conditions:

$$\frac{\partial T(u, v)}{\partial u} = 0$$

$$\frac{\partial T(u, v)}{\partial v} = 0 \qquad \text{[Eq. 11.40]}$$

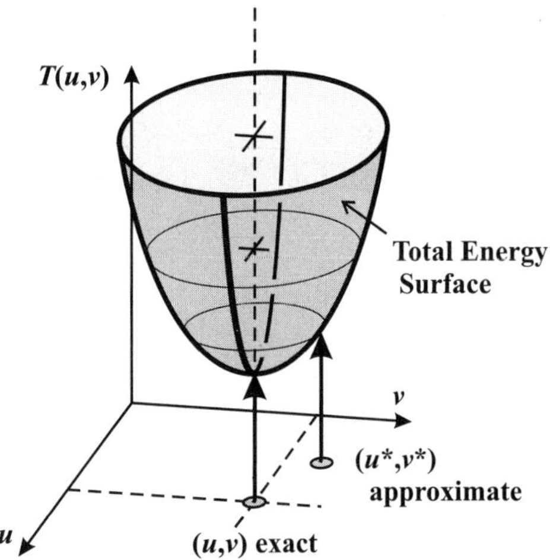

Figure 11.10. The total energy surface $T(u,v)$. The minimum value occurs when the values of u and v correspond to the exact (actual) values.

Applying these conditions will determine the exact solution.

Note that since the displacement error terms are squared, if the nearby solution (u^*, v^*) is close to the exact solution (u, v), the error in estimating the total energy can be smaller than the error in the displacements themselves.

The *minimization of total energy* is the method generally used in the computer-based finite element method, where numerical analysis is used to analyze large engineering systems.

Example 11.4 Three-bar Truss, Displacement Minimum Energy Method

Given: The three-bar truss shown in *Figure 11.11*.

Required: Determine the relationship between applied forces F_x and F_y and displacements u and v using the displacement energy method.

Solution: *Step 1.* In terms of u and v, and F_x and F_y, the total energy is given in *Equation 11.32*, which can be reduced to:

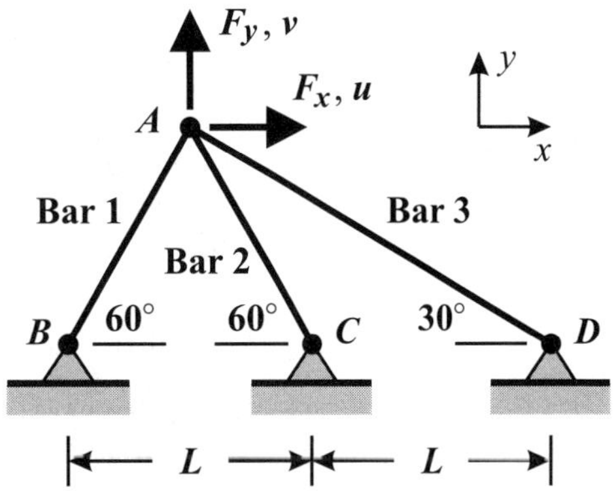

Figure 11.11. Three-bar truss.

$T(u, v)$

$$= \frac{EA}{2L}\left[\frac{5u^2}{4} - \frac{\sqrt{3}\,uv}{2} + \frac{7v^2}{4}\right] - F_x u - F_y v$$

Step 2. Minimizing the total energy with respect to u and v:

$$\frac{\partial T}{\partial u} = \frac{EA}{2L}\left(\frac{5u}{2} - \frac{\sqrt{3}\,v}{2}\right) - F_x = 0$$

$$\frac{\partial T}{\partial v} = \frac{EA}{2L}\left(-\frac{\sqrt{3}\,u}{2} + \frac{7v}{2}\right) - F_y = 0$$

The forces in terms of the displacements are:

$$F_x = \frac{EA}{4L}(5u - \sqrt{3}\,v)$$

Answer:

$$F_y = \frac{EA}{4L}(-\sqrt{3}\,u + 7v)$$

or, in matrix form:

Answer:
$$\begin{bmatrix} F_x \\ F_y \end{bmatrix} = \frac{EA}{4L}\begin{bmatrix} 5 & -\sqrt{3} \\ -\sqrt{3} & 7 \end{bmatrix}\begin{bmatrix} u \\ v \end{bmatrix}$$

This is the same result derived in *Example 11.1* using the traditional displacement method. Note that the energy method is much easier to apply and involves less algebra.

The matrix equation is $\mathbf{f} = \mathbf{Ku}$, where $\mathbf{f}$ and $\mathbf{u}$ are the force and displacement vectors, respectively, and $\mathbf{K}$ is the stiffness matrix of the system (*Equation 11.19*). Note that the stiffness matrix is symmetric about its diagonal; this is a general property of the stiffness matrix of linear–elastic materials.

> ***Symmetry of the Stiffness Matrix***
>
> Mathematically, the stiffness matrix of an elastic system must be symmetric. The internal energy at displacement u and v is U, independent of the path to reach that state. The stiffness matrix is:
>
> $$\mathbf{K} = \begin{bmatrix} k_{11} & k_{12} \\ k_{21} & k_{22} \end{bmatrix}$$
>
> and the forces in the x- and y-directions are:
>
> $$F_x = k_{11}u + k_{12}v = \frac{\partial U(u, v)}{\partial u} \quad \text{(i)}$$
>
> $$F_y = k_{21}u + k_{22}v = \frac{\partial U(u, v)}{\partial v} \quad \text{(ii)}$$
>
> Taking the partial derivative of F_x with respect to v, and of F_y with respect to u, respectively, gives:
>
> $$k_{12} = \frac{\partial^2 U}{\partial u\,\partial v} \quad \text{and} \quad k_{21} = \frac{\partial^2 U}{\partial v\,\partial u}$$
>
> Since U is path independent, $\frac{\partial^2 U}{\partial u\,\partial v} = \frac{\partial^2 U}{\partial v\,\partial u}$. Thus, $k_{12} = k_{21}$; the stiffness matrix is symmetric.

In systems where many members are connected at a few joints, the displacement method can be very powerful. An example follows where the elongation of each member of the system can be described by a single displacement parameter.

Example 11.5 Bicycle Wheel (*Examples 4.15, 4.26*)

Given: A bicycle wheel (*Figure 11.12*) is modeled as a set of elastic spokes, each of length R, and cross-sectional area A, attached to a rigid rim at one end and a rigid hub at the other. The load applied to the hub is F, and the resulting downward deflection of the hub is v. There are N spokes (N is large).

Required: Determine the relationship between the force F and downward displacement v.

Solution: *Step 1.* The hub displaces downward v, so the elongation of any spoke at angle θ to the horizontal is $\Delta = v \sin\theta$ (*Figures 11.12c, d*). The internal energy of a single spoke is:

$$U_i(\theta) = \frac{EA\Delta^2}{2R} = \frac{EA(v\sin\theta)^2}{2R}$$

The number of spokes dN in increment $d\theta$ is:

$$dN = N\frac{d\theta}{2\pi}$$

The internal energy of dN spokes is then:

$$dU = \left(N\frac{d\theta}{2\pi}\right)\left(\frac{EA}{2R}\right)(v\,\sin\theta)^2$$

Integrating from $\theta = 0$ to 2π gives the internal energy of all the spokes:

$$U(v) = \frac{NEAv^2}{4\pi R}\int_0^{2\pi}\sin^2\theta\,d\theta = \frac{NEAv^2}{4R}$$

Step 2. The total energy is:

$$T(v) = U(v) + V(v) = \frac{NEAv^2}{4R} - Fv$$

Step 3. Minimizing the total energy:

$$\frac{\partial T(v)}{\partial v} = \frac{NEAv}{2R} - F = 0$$

gives the force:

$$\text{\textit{Answer:}}\quad F = \frac{NEA}{2R}v$$

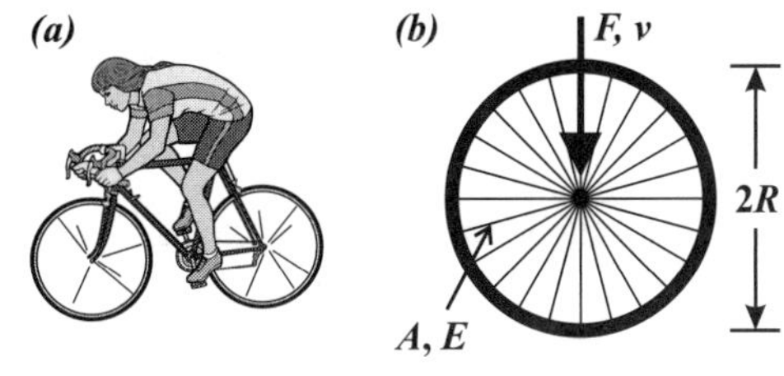

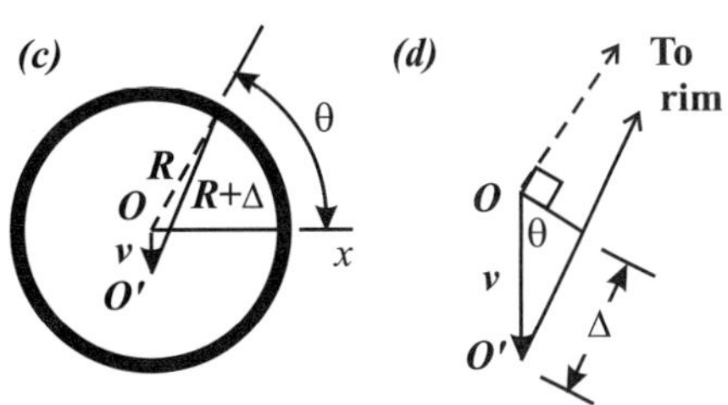

Figure 11.12. Bicycle hub subjected to load F deflects downward distance v. Copyright ©2008 Dominic J. Dal Bello and licensors. All rights reserved.

The stiffness of the wheel for a downward force applied at the hub is: $K = \dfrac{F}{v} = \dfrac{NEA}{2R}$

Force Method – Minimum Total Complementary Energy

The problem of *Figure 11.13* is now to determine the displacements (u, v) in terms of the known applied forces (F_x, F_y). The **internal complementary energy** C of the system is the sum of the complementary energies of each bar:

$$C(P_1, P_2, P_3) = \frac{L}{2EA}(P_1^2 + P_2^2 + P_3^2)$$

$$\text{[Eq. 11.41]}$$

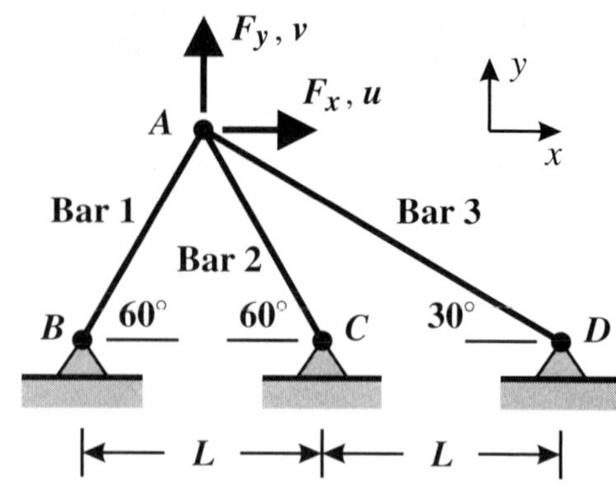

Figure 11.13. Three-bar truss.

From equilibrium (*Equation 11.20*), the internal forces are, with P_3 as the redundant:

$$P_1 = \frac{\sqrt{3}F_x + F_y + R}{\sqrt{3}}$$

$$P_2 = \frac{-\sqrt{3}F_x + F_y - 2R}{\sqrt{3}} \qquad \text{[Eq. 11.42]}$$

$$P_3 = R$$

The potential energy of the applied forces is

$$V(F_x, F_y) = -F_x u - F_y v \qquad \text{[Eq. 11.43]}$$

The ***total complementary energy*** Ω of the system is the sum of the complementary and potential energies:

$$\Omega(F_x, F_y, R) = C(P_1, P_2, P_3) + V(F_x, F_y)$$

$$= \frac{L}{2AE}(P_1^2 + P_2^2 + P_3^2) - F_x u - F_y v \qquad \text{[Eq. 11.44]}$$

Substituting the expressions for the internal forces P_i from *Equation 11.42* gives:

$$\Omega(F_x, F_y, R) = \frac{L}{2AE}\left(\frac{8R^2 + 6F_x^2 + 2F_y^2 + 6\sqrt{3}RF_x - 2RF_y}{3}\right) - F_x u - F_y v$$

$$\text{[Eq. 11.45]}$$

Now consider applied forces $F_x{}^*$ and $F_y{}^*$ that are close to, but not equal to, the exact values of F_x and F_y, but are in equilibrium with the internal forces. The difference between the exact and approximate values of F_x and F_y are δF_x and δF_y, respectively:

$$F_x{}^* = F_x + \delta F_x \quad \text{and} \quad F_y{}^* = F_y + \delta F_y \qquad \text{[Eq. 11.46]}$$

The corresponding internal forces of the bars are then:

$$P_1{}^* = \frac{\sqrt{3}F_x{}^* + F_y{}^* + R^*}{\sqrt{3}} ; \quad P_2{}^* = \frac{-\sqrt{3}F_x{}^* + F_y{}^* - 2R^*}{\sqrt{3}} ; \quad P_3{}^* = R^*$$

$$\text{[Eq. 11.47]}$$

The total complementary energy corresponding to the nearby forces $F_x{}^*$ and $F_y{}^*$, and approximate redundant R^*, is:

$$\Omega(F_x{}^*, F_y{}^*, R^*) = C(P_1{}^*, P_2{}^*, P_3{}^*) + V(F_x{}^*, F_y{}^*) \qquad \text{[Eq. 11.48]}$$

so:

$$\Omega(F_x^*, F_y^*, R^*)$$

$$= \frac{L}{2AE}\left[\frac{8(R^*)^2 + 6(F_x^*)^2 + 2(F_y^*)^2 + 6\sqrt{3}R^*F_x^* - 2R^*F_y^*}{3}\right] - F_x^*u - F_y^*v$$

[Eq. 11.49]

By substitution, it can be shown that the difference in the total complementary energy due to the approximate forces, F_x^*, F_y^*, and R^*, and that due to the exact solution, is:

$$\Omega(F_x^*, F_y^*, R^*) - \Omega(F_x, F_y, R)$$

$$= \frac{L}{2AE}\left[2(P_1\delta P_1 + P_2\delta P_2 + P_3\delta P_3) + (\delta P_1^2 + \delta P_2^2 + \delta P_3^2)\right] - \delta F_x u - \delta F_y$$

[Eq. 11.50]

Since $\Delta_1 = (P_1 L)/(AE)$, etc., then:

$$\Omega(F_x^*, F_y^*, R^*) - \Omega(F_x, F_y, R)$$

$$= \left[(\Delta_1\delta P_1 + \Delta_2\delta P_2 + \Delta_3\delta P_3) - \delta F_x u - \delta F_y v\right] + \frac{L}{2AE}(\delta P_1^2 + \delta P_2^2 + \delta P_3^2)$$

[Eq. 11.51]

The term in square brackets is the *virtual work equation* with external loads δF_x and δF_y undergoing displacements u and v, and equilibrium internal forces, δP_1, δP_2, and δP_3, in members extending by Δ_1, Δ_2, and Δ_3 (compatible with u and v). Thus, the term in the square brackets is zero. *Equation 11.51* reduces to:

$$\Omega(F_x^*, F_y^*, R^*) - \Omega(F_x, F_y, R) = \frac{L}{2AE}(\delta P_1^2 + \delta P_2^2 + \delta P_3^2) \qquad \text{[Eq. 11.52]}$$

The internal force errors, δP_i, are all squared and are therefore always positive. As a consequence, the total complementary energy associated with the approximate loads (F_x^*, F_y^*) is always greater than the total complementary energy of the exact solution (F_x, F_y). Hence:

$$\Omega(F_x^*, F_y^*, R^*) \geq \Omega(F_x, F_y, R) \qquad \text{[Eq. 11.53]}$$

In mathematical terms, the solution to the problem is given by the minimum conditions:

$$\frac{\partial\Omega(F_x, F_y, R)}{\partial F_x} = 0; \quad \frac{\partial\Omega(F_x, F_y, R)}{\partial F_y} = 0; \quad \frac{\partial\Omega(F_x, F_y, R)}{\partial R} = 0 \qquad \text{[Eq. 11.54]}$$

Example 11.6 **Three-bar Truss, Force Minimum Energy Method**

Given: The 3-bar truss shown in *Figure 11.14*.

Required: Determine the force–displacement relationship using the force energy method (complementary energy).

Solution:

Step 1. Taking the redundant force as $P_3 = R$, the total complementary energy is, from *Equation 11.45*:

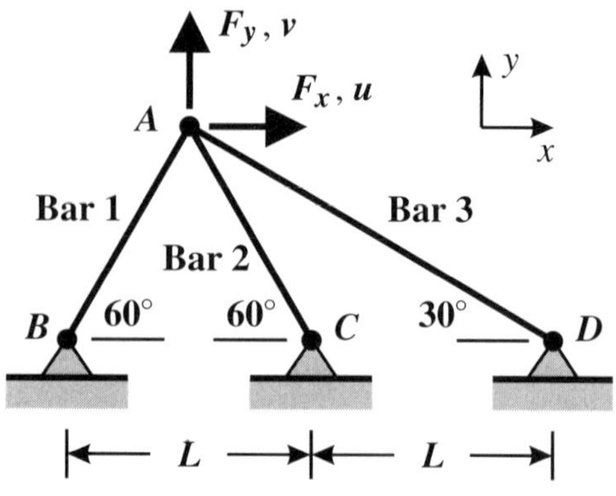

Figure 11.14. Three-bar truss.

$$\Omega(F_x, F_y, R)$$

$$= \frac{L}{2AE}\left(\frac{8R^2 + 6F_x^2 + 2F_y^2 + 6\sqrt{3}RF_x - 2RF_y}{3}\right) - F_x u - F_y v$$

Step 2. Taking partial derivatives of the total complementary energy with respect to applied forces F_x, F_y, and redundant R:

$$\frac{\partial\Omega}{\partial F_x} = \frac{L}{2AE}\left(\frac{12F_x + 6\sqrt{3}R}{3}\right) - u = 0$$

$$\frac{\partial\Omega}{\partial F_y} = \frac{L}{2AE}(4F_y - 2R) - v = 0$$

$$\frac{\partial\Omega}{\partial R} = \frac{L}{2AE}(16R + 6\sqrt{3}F_x - 2F_y) = 0$$

Step 3. From the partial derivative with respect to R, the exact redundant force is:

$$R = \frac{F_y - 3\sqrt{3}F_x}{8}$$

Substituting R into the other two equations, and solving for the exact displacements:

$$u = \frac{L}{8AE}(7F_x + \sqrt{3}F_y)$$

Answer:

$$v = \frac{L}{8AE}(\sqrt{3}F_x + 5F_y)$$

or, in matrix form:

$$Answer: \quad \begin{bmatrix} u \\ v \end{bmatrix} = \frac{L}{8AE} \begin{bmatrix} 7 & \sqrt{3} \\ \sqrt{3} & 5 \end{bmatrix} \begin{bmatrix} F_x \\ F_y \end{bmatrix}$$

which is the same as found in *Example 11.2*.

The matrix equation is $\mathbf{u} = \mathbf{Ff}$, where $\mathbf{u}$ and $\mathbf{f}$ are the displacement and applied force vectors, respectively, and $\mathbf{F}$ is the flexibility matrix of the system (*Equation 11.22*). Note that the flexibility matrix is symmetric about its diagonal; this is a general property of the flexibility matrix for linear–elastic materials. Also, note that the flexibility matrix is the inverse of the stiffness matrix, developed in *Examples 11.1* and *11.4*:

$$\mathbf{F}^{-1} = \left(\frac{L}{8AE} \begin{bmatrix} 7 & \sqrt{3} \\ \sqrt{3} & 5 \end{bmatrix} \right)^{-1} = \frac{EA}{4L} \begin{bmatrix} 5 & -\sqrt{3} \\ -\sqrt{3} & 7 \end{bmatrix} = \mathbf{K}$$

11.4 Bending Energy

Beam problems are often solved using energy methods. To achieve this, the expressions for the *internal energy* and *complementary energy* of a beam in bending must be determined.

The elastic moment–curvature (M–κ, load–displacement) relationship at any position along the beam is (*Figure 11.15*):

$$M = EI\kappa \qquad \text{[Eq. 11.55]}$$

At a given cross-section, the **internal energy per unit length** of a beam, without derivation, is:

$$\frac{dU(\kappa)}{ds} = \int_0^{\kappa} M \, d\kappa = \int_0^{\kappa} (EI\kappa) d\kappa = \frac{EI\kappa^2}{2}$$

$$\text{[Eq. 11.56]}$$

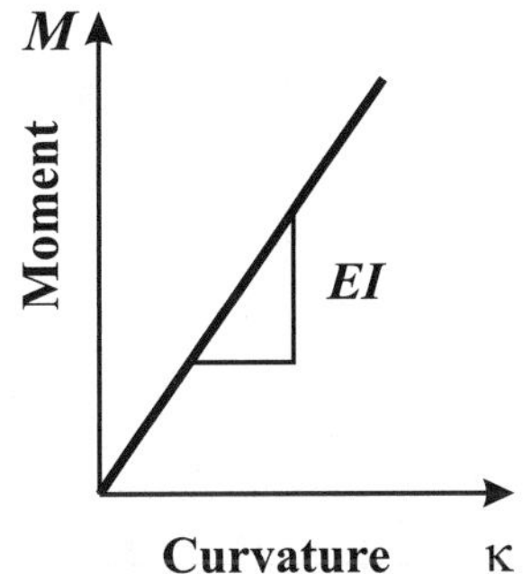

Figure 11.15. Moment–curvature relationship for linear–elastic beams.

The **complementary energy per unit length** of a beam is:

$$\frac{dC(M)}{ds} = \int_0^{M} \kappa \, dM = \int_0^{M} \left(\frac{M}{EI} \right) dM = \frac{M^2}{2EI} \qquad \text{[Eq. 11.57]}$$

The internal energy per unit length is a function of *curvature* κ and the complementary energy per unit length is a function of *bending moment M*. The curvature and moment are the appropriate displacement and load variables for a beam. In general, κ and M vary with distance along the beam. Since the expression for M is easier to find than that for κ, the complementary energy method is most often used.

Example 11.7 Micro-Hinge

Given: Micro-hinges are used in small components of dimensions measured in microns (10^{-6} m). Such a system is shown in *Figure 11.16a*. The material is polysilicone, with a modulus of $E = 170$ GPa. The hinge has thickness t and width B (into the paper), and is bent into a semi-circle of radius R. The arm has thickness T, width B, and length L. Horizontal force F is applied at the end of the arm, causing displacement δ.

Required: Considering bending only, determine the stiffness of the hinge system $K = F/\delta$.

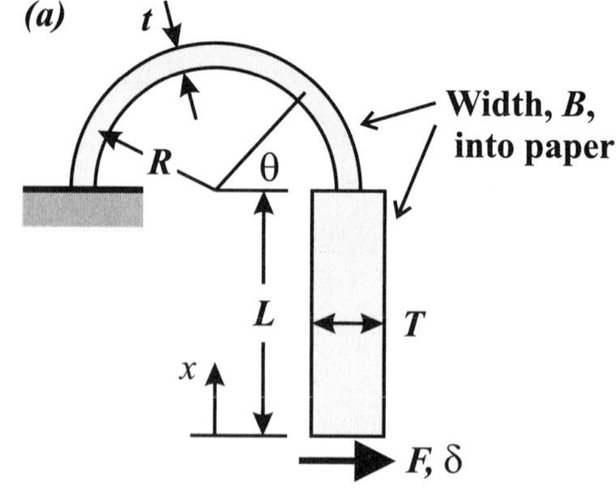

Figure 11.16. (a) A micro-hinge.

Solution: *Step 1.* The complementary energy of differential beam length ds is:

$$dC(M) = \frac{M^2}{2EI}ds$$

The complementary energy of the circular hinge is:

$$C_H(M) = \frac{1}{2EI_H}\int_L M^2\, ds = \frac{1}{2EI_H}\int_0^\pi M^2(R\, d\theta)$$

where $I_H = Bt^3/12$ and $ds = R\, d\theta$.

From the FBD in *Figure 11.16b*, the bending moment in the hinge is a function of θ:

$$M(\theta) = FL\left(1 + \frac{R}{L}\sin\theta\right)$$

Thus, the complementary energy of the hinge is:

$$C_H(M) = \frac{R}{2EI_H}\int_0^\pi [M(\theta)]^2 d\theta = \frac{F^2 L^2 R}{2EI_H}\left[\pi + \frac{4R}{L} + \left(\frac{R}{L}\right)^2\frac{\pi}{2}\right]$$

Step 2. The moment in the arm, as a function of x is, from *Figure 11.16c*:

$$M(x) = Fx$$

so the complementary energy of length dx is:

$$dC_A(M) = \frac{M^2}{2EI_A}dx = \frac{F^2 x^2}{2EI_A}dx$$

where $I_A = BT^3/12$ is the moment of inertia of the arm.

Integrating from $x = 0$ to L gives the complementary energy of the arm:

$$C_A(M) = \frac{F^2 L^3}{6EI_A}$$

Step 3. The potential energy is $V = -F\delta$, so the total complementary energy is:

$$\Omega = C_H + C_A - F\delta$$

Taking the derivative with respect to the load F:

$$\frac{\partial \Omega}{\partial F} = \frac{FL^2 R}{EI_H}\left[\pi + \frac{4R}{L} + \left(\frac{R}{L}\right)^2 \frac{\pi}{2}\right] + \frac{FL^3}{3EI_A} - \delta = 0$$

The displacement of the load is then:

$$\delta = \frac{FL^2 R}{EI_H}\left[\pi + \frac{4R}{L} + \left(\frac{R}{L}\right)^2 \frac{\pi}{2} + \left(\frac{L}{3R}\right)\left(\frac{I_H}{I_A}\right)\right]$$

Step 4. The stiffness of the system is:

Answer:

$$K = \frac{F}{\delta} = \frac{EI_H}{L^2 R}\left[\pi + \frac{4R}{L} + \left(\frac{R}{L}\right)^2 \frac{\pi}{2} + \left(\frac{L}{3R}\right)\left(\frac{t}{T}\right)^3\right]^{-1}$$

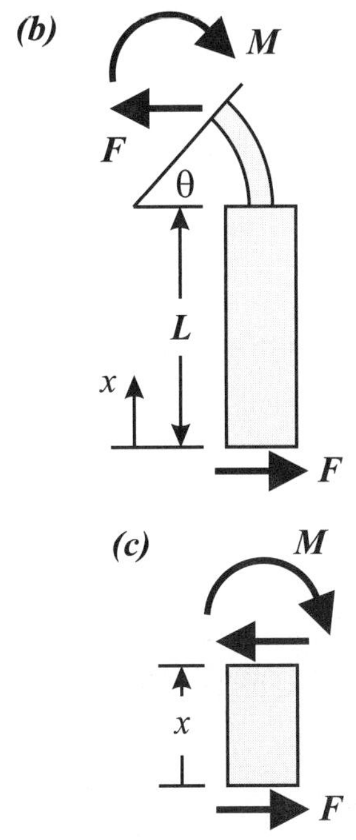

Figure 11.16. (b) The moment in the curve of the hinge varies with angle θ. **(c)** Moment in arm.

Example 11.8 Micro-Hinge, Numerical Example

Given: The micro-hinge of the previous example with numerical values:

$$t = 0.75 \ \mu m; \quad T = 4 \ \mu m; \quad R = 10 \ \mu m; \quad B = 10 \ \mu m; \quad L = 100 \ \mu m$$

The material has a modulus of $E = 170$ GPa.

Required: Determine the stiffness of the hinge system.

Solution: The stiffness is:

$$K = \frac{F}{\delta} = \frac{EI_H}{L^2 R}\left[\pi + \frac{4R}{L} + \left(\frac{R}{L}\right)^2 \frac{\pi}{2} + \left(\frac{L}{3R}\right)\left(\frac{t}{T}\right)^3\right]^{-1}$$

The pertinent geometric properties are:

$$I_H = \frac{Bt^3}{12} = \frac{(10\times10^{-6})(0.75\times10^{-6})^3}{12} = 0.352\times10^{-24} \ m^4$$

where: $\left(\frac{t}{T}\right)^3 = \left(\frac{0.75}{4}\right)^3 = 6.59\times10^{-3} \ m^4$ and $\frac{R}{L} = \frac{1}{10}$

Thus:

$$K = \frac{F}{\delta} = \frac{(170\times10^{9})(0.352\times10^{-24})}{(100\times10^{-6})^{2}(10\times10^{-6})}\left[\pi + \frac{4}{10} + \left(\frac{1}{10}\right)^{2}\frac{\pi}{2} + \left(\frac{10}{3}\right)(6.59\times10^{-3})\right]^{-1}$$

Answer: $K = 0.167$ N/m

The stiffness is very small, as is common in such micro-devices.

Example 11.9 Circular Proof Ring

Background: Tensile testing machines are calibrated on a regular basis to ensure that they are accurate. The calibration can be performed using a *proof ring*, a circular ring whose stiffness is known very accurately (*Figure 11.17a*). These high-strength steel rings come in a range of sizes depending on the load capacity of the test machine being calibrated.

Given: A circular proof ring of average radius R, thickness T, and breadth B, is subjected into two diametrically opposite loads of equal magnitude F (*Figure 11.17a*). The forces move distance Δ away from each other.

Required: Determine the deflection Δ in terms of force F, and dimensions T, R, and B. Use the complementary energy method (force method).

Solution: *Step 1.* Consider the ring cut across its diameter (*Figure 11.17b*). From equilibrium, the axial force at each cut is $F/2$. Due to symmetry, the horizontal force is zero (consider the other half of the ring). The moment M_o, is unknown, and is taken as the redundancy.

The FBD of an arc length of the ring defined by angle θ is shown in *Figure 11.17c*. At the interior surface defined by θ, N is the normal force, V is the shear force, and M is the bending moment.

Since the complementary energy method is to be used, equilibrium must be satisfied:

$$N = \frac{F\cos\theta}{2}$$

$$V = \frac{F\sin\theta}{2}$$

$$M = M_o - \frac{FR}{2}(1 - \cos\theta)$$

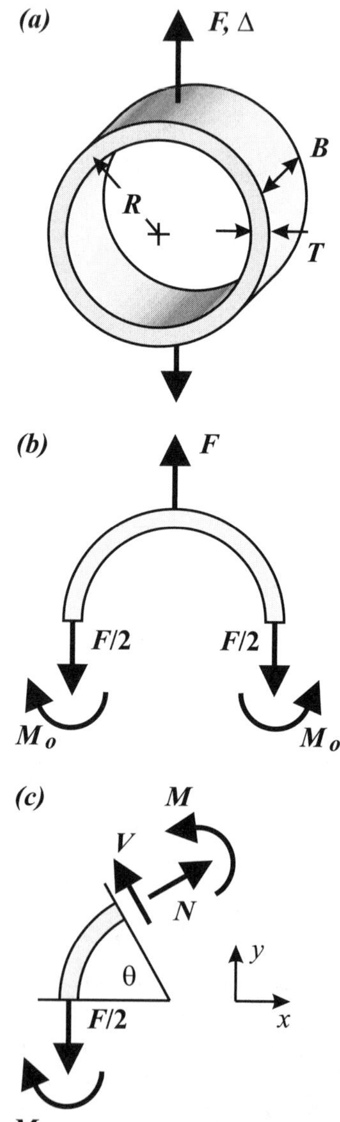

Figure 11.17. (a) Proof ring under tensile load F. **(b)** FBD of half the proof ring. **(c)** FBD of half ring cut at angle θ.

Step 2. Due to internal loads N, V, and M, the internal complementary energy per unit length around the ring circumference is:

$$\frac{dC(N, V, M)}{ds} = \frac{1}{2}\left[\frac{N^2}{EA} + \frac{V^2}{GA} + \frac{M^2}{EI}\right]$$

where $A = TB$ and $I = (BT^3)/12$.

The complementary energy of the ring is, with $ds = R\,d\theta$:

$$C(N, V, M) = 4\int_0^{\pi/2} \frac{1}{2}\left[\frac{N^2}{EA} + \frac{V^2}{GA} + \frac{M^2}{EI}\right] R\,d\theta$$

The ring has four identical quarters and it is necessary only to integrate between 0 and $\pi/2$, and multiply by 4. Substituting the expressions for N, V, and M, and integrating, gives the complementary energy as a function of load F and redundant moment M_o:

$$C(F, M_o)$$

$$= 2R\left[\frac{\pi F^2}{16EA} + \frac{\pi F^2}{16GA} + \frac{1}{EI}\left[\frac{\pi}{2}M_o^2 - \left(\frac{\pi}{2} - 1\right)FM_oR + \left(\frac{3\pi}{4} - 2\right)\frac{F^2R^2}{4}\right]\right]$$

Step 3. The total complementary energy is:

$$\Omega(F, M_o)$$

$$= 2R\left[\frac{\pi F^2}{16EA} + \frac{\pi F^2}{16GA} + \frac{1}{EI}\left[\frac{\pi}{2}M_o^2 - \left(\frac{\pi}{2} - 1\right)FM_oR + \left(\frac{3\pi}{4} - 2\right)\frac{F^2R^2}{4}\right]\right] - F\Delta$$

Minimizing the total complementary energy with respect to M_o:

$$\frac{\partial\Omega}{\partial M_o} = \frac{2R}{EI}\left[\pi M_o - \left(\frac{\pi}{2} - 1\right)FR\right] = 0$$

gives $M_o = \frac{FR}{\pi}\left(\frac{\pi}{2} - 1\right) = 0.182FR$

Substituting the expression for M_o into $\Omega(F, M_o)$, and minimizing with respect to F:

$$\frac{\partial\Omega}{\partial F} = 2R\left[\frac{\pi F}{8EA} + \frac{\pi F}{8GA} + \frac{1}{EI}\left[-\frac{1}{\pi}\left(\frac{\pi}{2} - 1\right)^2 FR^2 + \left(\frac{3\pi}{4} - 2\right)\frac{FR^2}{2}\right]\right] - \Delta = 0$$

Step 4. Since $A = TB$, $I = (BT^3)/12$, and $G \sim 3/8E$, the displacement can be reduced to:

$$\textit{Answer:}\ \Delta = \frac{0.1488FR^3}{EI}\left[0.4399\left(\frac{T}{R}\right)^2 + 1.173\left(\frac{T}{R}\right)^2 + 1.0\right]$$

The first term in the square brackets is due to tension, the second term to shear, and the third term to bending. Taking a typical proof ring value of $T/R = 1/20$, the deflection is:

$$\Delta = \frac{0.1488FR^3}{EI}[0.0011 + 0.0029 + 1.0]$$

The contributions of the tensile and shear loads to deflection can be neglected; together they sum to 0.4% of the deflection due to bending. For beam-like components subjected to axial, shear, and bending loads, it is common to simplify the calculations by neglecting the effect of the axial and shear loads.

11.5 Approximation Methods

A great advantage of energy methods is that useful *approximations* can be made with simple calculations. Simple calculations are very useful at the early stages of a new design, where structural members must be roughly sized. Examples are now given of the ***approximation method*** using the displacement and the force energy methods. To demonstrate the approximation method, the three-bar truss is again studied, for the particular case of $F_x = 0$ (*Figure 11.18*).

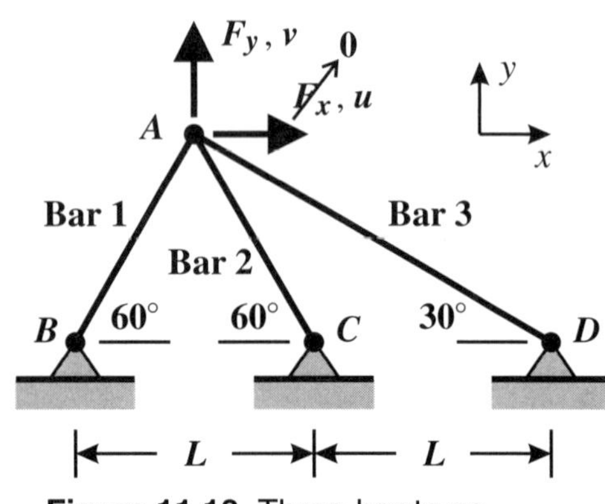

Figure 11.18. Three-bar truss.

Approximate Displacement (Minimum Total Energy) Method

In this method, the displacements are approximated. Since F_x is zero, the horizontal displacement u of joint A is likely to be small. Thus, u is *approximated* as $u^* = 0$ and v as v^*. Although joint A actually moves to the right (u is positive per *Example 11.2* with $F_x = 0$), approximating u as zero simplifies the calculations. The goal now is to approximate the relationship between F_y and v.

From the compatibility conditions of *Equation 11.18*, the elongations in terms of $u^* = 0$ and v^* reduce to:

$$\Delta_1 = \frac{\sqrt{3}}{2}v^*; \quad \Delta_2 = \frac{\sqrt{3}}{2}v^*; \quad \Delta_3 = \frac{v^*}{2} \qquad \text{[Eq. 11.58]}$$

The approximate total energy T is given by *Equation 11.32*, which for $u^* = 0$ and v^* is:

$$T(0, v^*) = \frac{EA}{2L}\left[\left(\frac{\sqrt{3}}{2}v^*\right)^2 + \left(\frac{\sqrt{3}}{2}v^*\right)^2 + \left(\frac{v^*}{2}\right)^2\right] - F_y v^*$$

$$= \frac{EA}{8L}(7v^{*2}) - F_y v^* \qquad \text{[Eq. 11.59]}$$

Minimizing the approximate total energy with respect to v^* gives:

$$\frac{\partial T}{\partial v^*} = \frac{EA}{4L}(7v^*) - F_y = 0 \qquad \text{[Eq. 11.60]}$$

so that:

$$F_y = \frac{7EA}{4L}v^* \rightarrow k^* = \frac{7EA}{4L} \qquad \text{[Eq. 11.61]}$$

where $k^* = F_y/v^*$ is the approximate stiffness of the truss in the y-direction for a vertical load at joint A.

The exact $F_y\text{–}v$ relationship when $F_x = 0$, is (from *Example 11.2*):

$$v = \frac{5L}{8EA}F_y \rightarrow k = \frac{8EA}{5L} \qquad \text{[Eq. 11.62]}$$

The ratio of the approximate stiffness to the exact stiffness is:

$$\frac{k^*}{k} = 1.094 \qquad \text{[Eq. 11.63]}$$

In spite of the crude assumption that $u = 0$, the approximate stiffness in the y-direction is only 9.4% greater than the exact value. The approximate displacement method always predicts a stiffness greater than the exact value.

Approximate Force (Minimum Total Complementary Energy) Method

To use the *minimum total complementary energy method*, an approximation is made about the internal forces. As always, the equilibrium equations must be satisfied. *Equation 11.20* gives the equilibrium forces in each bar when the force in Bar 3 is the redundant: $P_3 = R$. When $F_x = 0$, equilibrium requires that:

$$P_1 = \frac{F_y + R}{\sqrt{3}}; \quad P_2 = \frac{F_y - 2R}{\sqrt{3}}; \quad P_3 = R \qquad \text{[Eq. 11.64]}$$

It is now *approximated* that the force in Bar 3 is zero (Bar 3 is the nearest to perpendicular with applied load F_y, so P_3 is likely the smallest). The internal forces reduce to:

$$P_1^* = \frac{F_y}{\sqrt{3}}; \quad P_2^* = \frac{F_y}{\sqrt{3}}; \quad P_3^* = 0 \qquad \text{[Eq. 11.65]}$$

The total complementary energy is then:

$$\Omega(F_y, P_1^*, P_2^*, P_3^*) = \frac{L}{2EA}[(P_1^*)^2 + (P_2^*)^2 + (P_3^*)^2] - F_y v^*$$

$$= \frac{LF_y^2}{3EA} - F_y v^* \qquad \text{[Eq. 11.66]}$$

where v^* is the displacement that results from the *approximation* that $P_3 = R = 0$.

Minimizing with respect to F_y gives:

$$\frac{\partial \Omega}{\partial F_y} = \frac{2LF_y}{3EA} - v^* = 0 \qquad \text{[Eq. 11.67]}$$

Thus, the approximate stiffness is:

$$k^* = \frac{F_y}{v^*} = \frac{3EA}{2L} \qquad \text{[Eq. 11.68]}$$

The exact expression is (from *Example 11.2*):

$$k = \frac{8EA}{5L} \qquad \text{[Eq. 11.69]}$$

Hence:

$$\frac{k^*}{k} = 0.94 \qquad \text{[Eq. 11.70]}$$

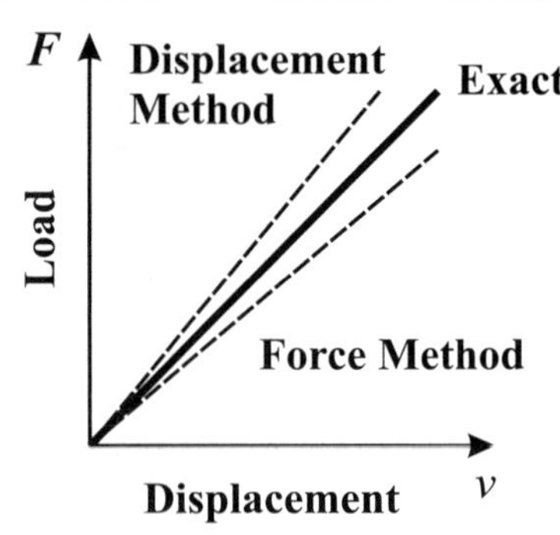

Figure 11.19. The displacement method overestimates the exact stiffness, while the force method underestimates the exact stiffness.

In spite of the crude assumption that $P_3 = 0$, the approximate stiffness is only 6% less than the exact value. The approximate force method always predicts a stiffness less than the exact value.

Qualitative load–displacement graphs for the exact and approximate solutions are shown in *Figure 11.19*. The displacement method overpredicts the stiffness, while the force method underpredicts the stiffness. In spite of the coarseness of the assumptions, the exact stiffness is closely bracketed by the two methods (*Equations 11.63, 11.70*).

Estimate of the Torsional Stiffness of a Solid Shaft

In micro-electromechanical systems (MEMS) devices, it is difficult to make circular shafts. Components in torsion are therefore usually rectangular. Energy methods can be used to approximately determine the torsional stiffness of such systems.

Example 11.10 Square Shaft in Torsion

Given: A shaft of length L and of square cross-section with sides D is subjected to torque T resulting in angle of twist θ (*Figure 11.20a*).

Required: Determine the torsional stiffness of the shaft $k_T = T/\theta$, using the approximate force energy method.

Solution: *Step 1.* The shaft is split up into a series of concentric square thin-walled tubes (*Figure 11.20b*). Each thin-walled tube, distance x from the center of the

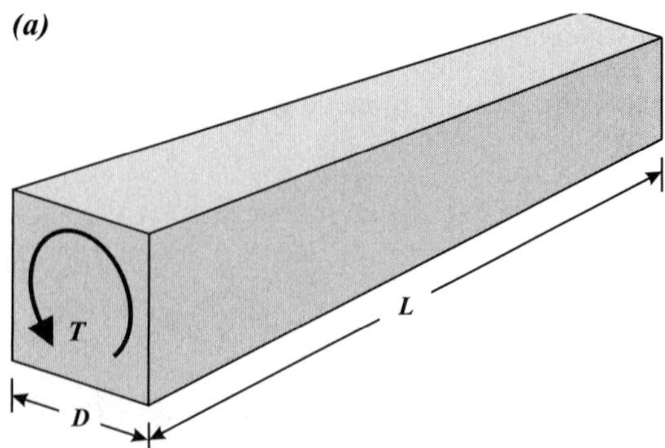

Figure 11.20. (a) A square shaft in torsion.

cross-section, has thickness dx. The shear stress is assumed to be constant on each side of the tube (*Figure 11.20c*), and varies linearly with distance x:

$$\tau = Bx$$

where B is a constant to be determined from equilibrium.

Step 2. The contribution to the torque of one side of the thin-walled square tube is:

$$dT = \tau[dA]x = (Bx)(2x\,dx)x = 2Bx^3\,dx$$

There are four sides to the thin-walled tube, and integrating the contribution of each tube from $x = 0$ to $D/2$ results in:

$$T = \int_0^{D/2} 4[2Bx^3\,dx] = \frac{BD^4}{8}$$

Hence: $B = \dfrac{8T}{D^4}$

from which the shear stress is: $\tau = \dfrac{8Tx}{D^4}$

Step 3. The complementary internal energy of a thin-walled tube subjected to shear stress τ is:

$$dC(\tau) = \frac{\tau^2}{2G}(8xL\,dx)$$

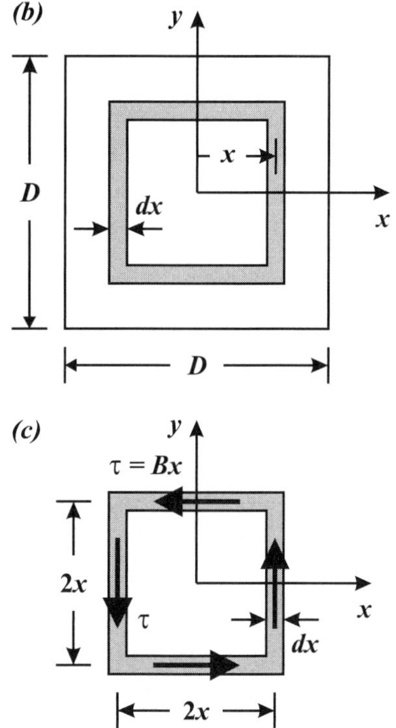

Figure 11.20. (b) Square cross-section. **(c)** Square tube of thickness dx, with constant shear stress $\tau = Bx$ on each side.

where $\tau^2/(2G)$ is the elastic shear strain energy density and $8xL\,dx$ is the volume of a thin-walled tube. The complementary internal energy of all the tubes is:

$$C(T) = \int_0^{D/2} \frac{\tau^2}{2G}(8xL\,dx) = \frac{4L}{G}\int_0^{D/2}\left(\frac{8Tx}{D^4}\right)^2 x\,dx = \frac{4LT^2}{GD^4}$$

The displacement of torque T is θ, so the total complementary energy is:

$$\Omega(T) = \frac{4LT^2}{GD^4} - T\theta$$

Minimizing with respect to torque T gives:

$$\frac{\partial\Omega(T)}{\partial T} = 0 \quad \text{so:} \quad \theta = \frac{8LT}{GD^4}$$

Step 4. The *torsional stiffness* $k_T = T/\theta$ is estimated to be:

$$Answer: k_T = \frac{GD^4}{8L} = 0.125\frac{GD^4}{L}$$

The exact value for the stiffness from a computer analysis is: $k_T = 0.14\frac{GD^4}{L}$

As expected from the approximate force method, the stiffness is underestimated. In this example, the error is 11%.

Example 11.11 **Ultra-Precision Device**

Given: An ultra-precision MEMS device is shown in *Figure 11.21a*. The arms have rectangular cross-section $B = 20\ \mu m$ wide (into the paper) and $t = 3.0\ \mu m$ deep. The length of the arms is $L = 175\ \mu m$, and they are doubled back upon themselves. The connectors are 20 μm long and 18 μm thick. Devices of this thickness are made by a deposition process. The material is silicon for which $E = 160$ GPa. The displacement of the moving platform is Δ, which is caused by force F (*Figure 11.21b*).

Required: Approximate the stiffness, $k = F/\Delta$, using the displacement method. Assume a displaced shape for the arms.

Solution: The connectors are assumed to be rigid with respect to the arms. This assumption is made because the bending stiffness of a beam is proportional to $EI/L^3 \propto EBt^3/L^3$. Since E and B are the same for the connectors and the arms, the ratio of the stiffness of the connectors to the stiffness of the arms is:

$$\left(\frac{t}{L}\right)^3_{connector}\left[\left(\frac{t}{L}\right)^3_{arms}\right]^{-1}$$

$$= \left(\frac{18}{20}\right)^3\left[\left(\frac{3}{175}\right)^3\right]^{-1} = 145\times10^3$$

The stiffness of the connectors is much larger than that of the arms, so the connectors are assumed to be rigid.

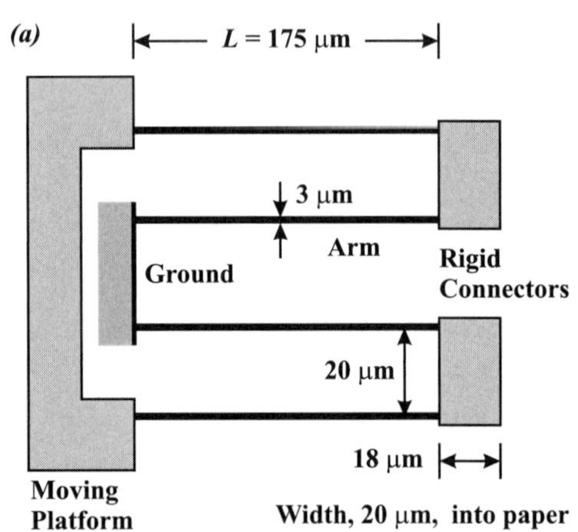

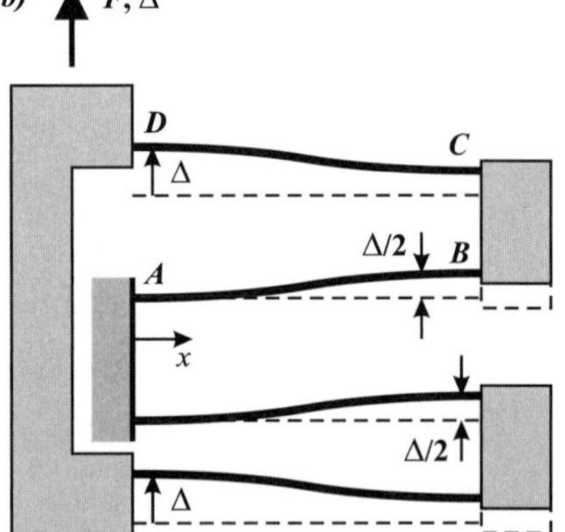

Figure 11.21. **(a)** MEMS device: a moving platform attached to two 175 μm double-cantilever beams. **(b)** Displacement of beams due to movement of platform by distance Δ.

Consider arm AB (*Figure 11.21b*). From geometry, the displacement of point A is zero, and of point B is $\Delta/2$, where Δ is the displacement of the moving platform, point D. The slope of AB at each end is zero. By symmetry, the other arms must have the same shape. The relative displacement of arm CD is $\Delta/2$.

The displaced form (deflected shape) of AB can be assumed to be:

$$v = \frac{\Delta}{4}\left(1 - \cos\frac{\pi x}{L}\right)$$

where $L = 175$ µm is the length of each arm. At $x = 0$ (point A) the displacement equation gives $v = 0$, and at $x = L$ (point B) the equation gives $v = \Delta/2$.

The slope of AB is:

$$\frac{dv}{dx} = \frac{\Delta\pi}{4L}\sin\frac{\pi x}{L}$$

which satisfies the condition that the slope is zero at the endpoints, $x = 0$ and L:

$$\frac{dv(0)}{dx} = \frac{dv(L)}{dx} = 0$$

Although the assumed form of the displacement is not necessarily the actual displacement, it satisfies the geometric boundary conditions; i.e., the approximated deflection is compatible with the required deflections and slopes at the supports.

The curvature of arm AB is:

$$\kappa = \frac{d^2 v}{dx^2} = \frac{\Delta}{4}\left(\frac{\pi}{L}\right)^2\cos\frac{\pi x}{L}$$

The internal energy due to bending of arm AB is then:

$$U(\Delta) = \frac{EI}{2}\int_0^L \kappa^2 dx = \frac{EI}{2}\int_0^L \left[\frac{\Delta}{4}\left(\frac{\pi}{L}\right)^2\cos\frac{\pi x}{L}\right]^2 dx = \frac{EI}{2}\left(\frac{\Delta}{4}\right)^2\left(\frac{\pi}{L}\right)^4\frac{L}{2}$$

The total energy of the system – the four arms and the applied force F – is:

$$T(\Delta) = 4U(\Delta) - F\Delta = \frac{EI\Delta^2}{16}\left(\frac{\pi}{L}\right)^4 L - F\Delta$$

Minimizing with respect to Δ gives:

$$\frac{\partial T(\Delta)}{\partial \Delta} = -F + \frac{EI\Delta}{8}\left(\frac{\pi}{L}\right)^4 L = 0$$

The stiffness of the system is therefore:

$$\textit{Answer:} \quad k = \frac{F}{\Delta} = 12.18\frac{EI}{L^3}$$

The actual stiffness of the system is the combined stiffness of four tip-loaded cantilever beams:

$$k = 12\frac{EI}{L^3}$$

The approximate displacement method overestimates the actual stiffness of the system (here, by about 1.5%).

11.6 Effect of Shear Stress

Classical beam theory is based on the assumption that the shear stresses make no contribution to beam deflection. The assumption is usually valid when the beam is made of a uniform material throughout.

In aerospace applications, where weight is important, the beam flanges are typically made of aluminum and the web is made of an aluminum honeycomb structure or aluminum foam (*Figure 11.22a*). In such situations, the shear modulus of the web material is typically much smaller than that of the flange material. Consider the effect of the more flexible web material on beam deflection.

Figure 11.22b represents the structural components of an aircraft wing of length L subjected to a uniformly distributed load w (force/length) and an end load F. The cross-section is modeled as two metal flanges of width B and thickness T, which are distance D apart (*Figure 11.22a*). The elastic modulus of the flange material is E. The web is a honeycomb material of elastic shear modulus G_W. The aim is to estimate the vertical deflection of the wing tip Δ. The total complementary energy method (based on the force method) is used.

Using the FBDs of *Figures 11.22c and d* gives the shear force V and bending moment M as a function of x. Since $R = wL + F$ and $M_o = wL^2/2 + FL$, then:

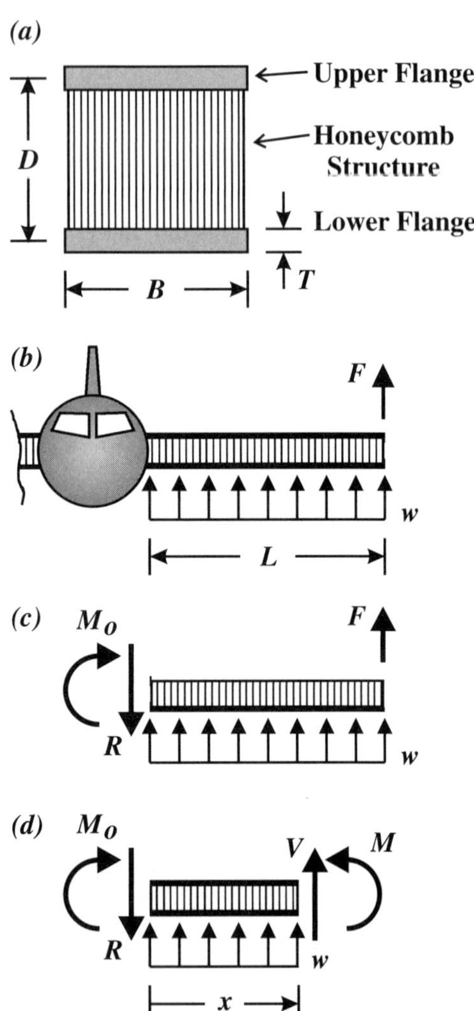

Figure 11.22. (a) Cross-section of beam. **(b)** Air pressure loading on wing w and tip load F. **(c)** FBD of wing. **(d)** FBD of length x of the beam.

$$V(x) = w(L - x) + F$$

$$M(x) = \frac{w(L - x)^2}{2} + F(L - x)$$

[Eq. 11.71]

Assuming that the flanges support the entire bending moment, the bending stress σ_B in the flange is:

$$\sigma_B = \frac{M}{AD} = \frac{1}{(BT)D} = \frac{1}{(BT)D}\left[\frac{w(L - x)^2}{2} + F(L - x)\right]$$

[Eq. 11.72]

The stress in each flange σ_B is assumed to be uniform; in other words, the flanges are considered to be axial members.

Assuming the web supports the entire shear force, the average shear stress τ is:

$$\tau = \frac{V}{A} = \frac{w(L - x) + F}{BD}$$

[Eq. 11.73]

The internal complementary energy due to the bending moment and shear force is:

$$C(w, F) = 2BT \int_0^L \frac{\sigma_B^2}{2E} dx + DB \int_0^L \frac{\tau^2}{2G_W} dx$$

[Eq. 11.74]

where E is the modulus of the flange and G_W is the shear modulus of the web.

The expression for the total complementary energy is therefore:

$$\Omega(w, F) = \frac{BT}{E} \int_0^L \sigma_B^2 \, dx + \frac{DB}{2G_W} \int_0^L \tau^2 \, dx - \int_0^L w(x)v(x)dx - F\Delta$$

[Eq. 11.75]

where:

- $-\displaystyle\int_0^L w(x)v(x)dx$ is the potential energy of load $w(x)$ with deflection $v(x)$, and

- $-F\Delta$ is the potential energy of load F with deflection Δ.

The deflections are determined by minimizing the total complementary energy with respect to w and F. The algebra can be reduced by performing the differentiation before completing the integrals of *Equation 11.75*. For example:

$$\frac{\partial \Omega(w, F)}{\partial F} = \frac{BT}{E} \int_0^L \left[2\sigma_B \frac{\partial \sigma_B}{\partial F}\right]dx + \frac{DB}{2G_W} \int_0^L \left[2\tau \frac{\partial \tau}{\partial F}\right]dx - \Delta = 0$$

[Eq. 11.76]

where, from *Equations 11.72* and *11.73*:

$$\frac{\partial \sigma_B}{\partial F} = \frac{L - x}{BTD} \quad \text{and} \quad \frac{\partial \tau}{\partial F} = \frac{1}{BD}$$

The partial derivative of Ω with respect to w is also determined and set equal to zero.

After performing the mathematics, the tip deflection can be written:

$$\Delta = \frac{2}{EBTD^2}\left(\frac{FL^3}{3} + \frac{wL^4}{8}\right) + \frac{1}{G_W BD}\left(FL + \frac{wL^2}{2}\right) \qquad \text{[Eq. 11.77]}$$

For the case $F = 0$:

$$\Delta = \frac{2}{EBTD^2}\left(\frac{wL^4}{8}\right)\left[1 + 2\frac{E}{G_W}\frac{DT}{L^2}\right] \qquad \text{[Eq. 11.78]}$$

The first term in square brackets is associated with the value for the end deflection using traditional bending theory neglecting shear effects: $\Delta = wL^4/(8EI)$. The second term modifies the displacement to include the shear response of the honeycomb.

A typical shear modulus for the aluminum honeycomb structure is approximately 1/40 that of an aluminum alloy ($E \sim 10{,}000$ ksi and $G \sim 4000$ ksi), so:

$$G_W = \frac{G}{40} = \frac{4000 \text{ ksi}}{40} = 100 \text{ ksi} \qquad \text{[Eq. 11.79]}$$

Let $B = D = L/10$, and $T = D/10 = L/100$. Thus, with $E/G_W = 100$, $D/L = 1/10$, and $T/L = 1/100$, the expression for the tip displacement reduces to:

$$\Delta = \frac{2}{EBTD^2}\left(\frac{wL^4}{8}\right)[1 + 0.2] \qquad \text{[Eq. 11.80]}$$

The correction term due to shear is the second term in the square brackets. Here the calculated displacement is 20% larger than that obtained from classical beam bending.

Note that if the web is of the same material as the flanges, $G_W = G = 4000$ ksi, then:

$$\Delta = \frac{2}{EBTD^2}\left(\frac{wL^4}{8}\right)[1 + 0.005] \qquad \text{[Eq. 11.81]}$$

The deflection due to shear loading is then negligible, as is typically assumed.

However, replacing the entire web area with the flange material makes for a heavy beam. With an I- or box-beam having uniform material properties and of total web width $b = B/10$, then:

$$\Delta = \frac{2}{EBTD^2}\left(\frac{wL^4}{8}\right)[1 + 0.05] \qquad \text{[Eq. 11.82]}$$

In this case, the deflection due to shear loading is 5% of that due to bending.

12.0 Introduction

A typical stress–strain curve for a ductile material (metals) in tension is shown in *Figure 12.1*. In the elastic range, Hooke's law applies, and stress σ is linearly proportional to strain ε with modulus E:

$$\sigma = E\varepsilon \qquad \text{[Eq. 12.1]}$$

The material remains *elastic* – it returns to its initial length upon unloading – provided that the applied stress does not exceed the material's yield strength S_y.

When the stress exceeds the yield strength, the slope of the stress–strain curve begins to decrease significantly. The material is said to have *yielded*. This *non-linear region* is the *inelastic* or *plastic* response of the material. The slope in this region is the *tangent modulus* E_t. During plastic deformation, the material volume remains constant, corresponding to a Poisson's ratio of $\nu = 0.5$.

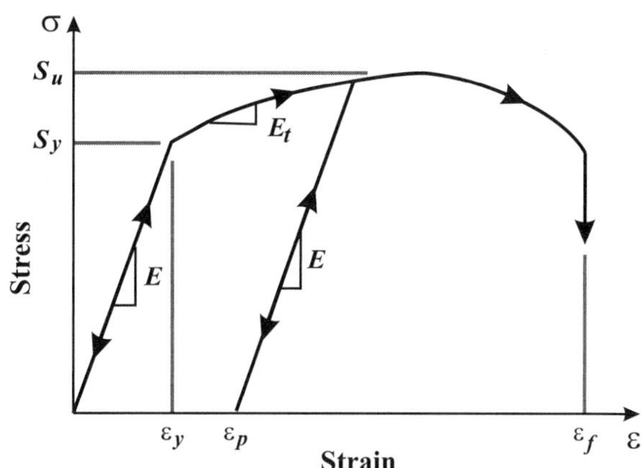

Figure 12.1. Stress–strain curve for a ductile material. Arrows indicate the direction of the stress–strain response.

Table 12.1. Representative properties of two ductile metals.

Material	E, GPa (Msi)	S_y, MPa (ksi)	ε_y, %	$5\varepsilon_y$, %	ε_f, %
Steel A36	207 (30)	250 (36)	0.125	0.625	~24%
Aluminum 6061-T6	70 (10)	240 (35)	0.34	1.70	~15%

If necking has not occurred, the *unloading* stress–strain curve is linear–elastic with slope E. When the stress is completely removed, a permanent plastic strain ε_p remains.

In *elastic design*, a system *fails* when the maximum calculated elastic stress reaches the yield strength S_y. It became evident to engineers that this approach is conservative, and economic gain is possible by taking advantage of a material's *ductility*.

The methods developed to take advantage of plasticity are referred to as **plastic design**. Experience suggests that the plastic strain needed to take advantage of ductility is

F.A. Leckie, D.J. Dal Bello, *Strength and Stiffness of Engineering Systems*, Mechanical Engineering Series, **355**
DOI 10.1007/978-0-387-49474-6_12, © Springer Science+Business Media, LLC 2009

approximately $5S_y/E$. The values of the yield strain, $\varepsilon_y = S_y/E$, for structural steel and aluminum are given in *Table 12.1*. The magnitude of strain, $5S_y/E$, is rarely greater than about 2%, much smaller than the failure strain ε_f (~15%+) of ductile materials, when the material breaks into two pieces. Failure due to *fracture*, or material separation, is not a problem normally considered in plastic design. However certain manufacturing processes, such as sheet forming, require large plastic strains, and then local necking defines the limit of the manufacturing process.

12.1 Elastic–Plastic Idealization

It is possible to model in detailed mathematical terms, the form of a stress–strain curve shown in *Figure 12.1*. However, for most applications, a simplified model of the material response is helpful in understanding system behavior. The simplified stress–strain curve for the idealized ductile material is shown in *Figure 12.2*.

On initial application of stress, the behavior is linear–elastic:

$$\sigma = E\varepsilon \qquad \text{[Eq. 12.2]}$$

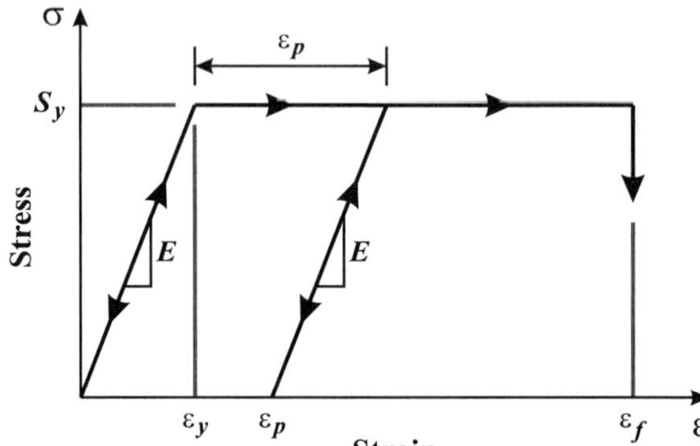

Figure 12.2. Idealized stress–strain curve for a ductile material. Elastic–perfectly plastic model.

until the stress reaches the yield strength S_y. Further strain occurs with no additional load. In this simplified model, the material deforms plastically at constant stress:

$$\sigma = S_y \qquad \text{[Eq. 12.3]}$$

This type of behavior is termed ***elastic–perfectly plastic***. Provided that the material does not break into two, *Figure 12.2* is an adequate model of plastic behavior. The elastic–plastic model is conservative in that the yield strength of ductile materials typically increases during plastic deformation (*Figure 12.1*) due to strain-hardening.

During plastic deformation, the total strain ε is the sum of the ***elastic strain*** ε_e and *plastic strain* ε_p :

$$\varepsilon = \varepsilon_e + \varepsilon_p = \frac{S_y}{E} + \varepsilon_p \qquad \text{[Eq. 12.4]}$$

When the stress is reduced from $\sigma = S_y$, the material response is elastic, with unloading modulus E. The strain at stress σ is:

$$\varepsilon = \frac{\sigma}{E} + \varepsilon_p \qquad \text{[Eq. 12.5]}$$

When the stress is completely removed (to $\sigma = 0$), the elastic strain is removed. The total strain is now composed only of the *plastic strain* ε_p.

If the applied stress is reduced further so that it becomes negative or *compressive* ($\sigma < 0$), the response is shown in *Figure 12.3*. The plastic strain in tension is now represented by $\varepsilon_{p,t}$. As the material unloads from $\sigma = S_y$ to $-S_y$, the material response is elastic with modulus E. The strain at any stress σ is:

$$\varepsilon = \frac{\sigma}{E} + \varepsilon_{p,t} \qquad \text{[Eq. 12.6]}$$

When σ reaches the *compressive yield strength*, $-S_y$, the change in the elastic strain from the tensile yield condition is $\Delta\varepsilon = -2S_y/E$, so the strain when the material just yields in compression is:

$$\varepsilon = \frac{-S_y}{E} + \varepsilon_{p,t} \qquad \text{[Eq. 12.7]}$$

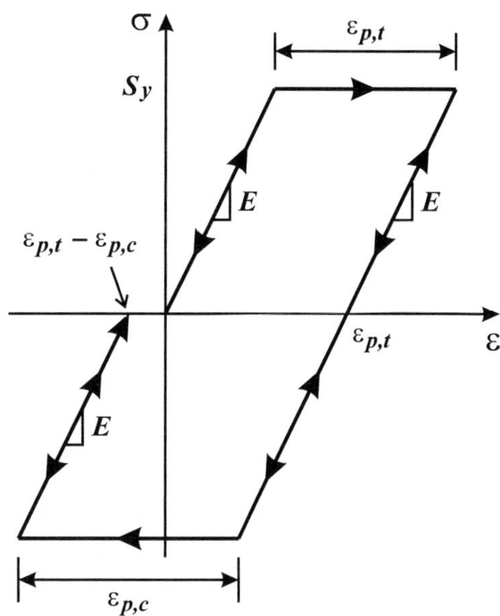

Figure 12.3. Loading cycle with tensile and compressive (reverse) plastic deformation.

In the elastic–perfectly plastic model, the magnitude of the compressive stress cannot be increased further. The material yields in compression at $\sigma = -S_y$, which causes a compressive plastic strain of magnitude $\varepsilon_{p,c}$. The total strain is now:

$$\varepsilon = \frac{-S_y}{E} + \varepsilon_{p,t} - \varepsilon_{p,c} \qquad \text{[Eq. 12.8]}$$

Upon reloading – removing the compressive load – the strain at any stress becomes:

$$\varepsilon = \frac{\sigma}{E} + \varepsilon_{p,t} - \varepsilon_{p,c} \qquad \text{[Eq. 12.9]}$$

Finally, completely removing the applied compressive stress to $\sigma = 0$ results in a total plastic strain of:

$$\varepsilon_p = \varepsilon_{p,t} - \varepsilon_{p,c} \qquad \text{[Eq. 12.10]}$$

To summarize the loading sequence in *Figure 12.3*:

1. elastic loading ($\sigma = E\varepsilon$) to yield strength $+S_y$;
2. yielding (plastic deformation) at constant stress $\sigma = +S_y$, with tensile plastic strain $\varepsilon_{p,t}$;
3. elastic unloading, $\Delta\sigma = E(\Delta\varepsilon)$ (i.e., reducing the tensile stress/increasing the compressive stress) to compressive yield strength $\sigma = -S_y$;
4. compressive yielding at constant stress $\sigma = -S_y$, with compressive plastic strain $\varepsilon_{p,c}$;
5. elastic reloading, $\Delta\sigma = E(\Delta\varepsilon)$ (i.e., reducing the compressive stress) to zero stress, with accumulated plastic strain $\varepsilon_{p,t} - \varepsilon_{p,c}$.

This loading cycle can continue indefinitely. The plastic strains in tension and compression are in opposite directions, and therefore tend to cancel each other. Thus, *reverse plasticity* may be used to attempt to return an axial member to its original length after it has been plastically deformed (i.e., $\varepsilon_{p,t} = \varepsilon_{p,c}$).

12.2 Elastic–Plastic Calculations: Limit Load

For a ductile system subjected to statically (slowly) applied load F, the *elastic limit* F_y is the load at which the system first yields. The *limit load* F_L is the maximum load that an elastic–perfectly plastic system can support. At the *limit load*, every material point on a cross-section has yielded (and supports the maximum possible stress), and unconstrained plastic deformation occurs until the material breaks into two parts.

The ideal *elastic–perfectly plastic behavior* of axial members, torsion members, and beams, is studied below. The *limit load* is calculated for each case. The linear response of each member when the limit load is removed, is also studied. Because unloading is elastic, once a system experiences a permanent plastic strain, *residual stresses* will generally exist in the unloaded system.

Example 12.1 Axial Members in a Two-Bar System

Given: The two-bar system shown in *Figure 12.4a* is subjected to force F and is constrained to move vertically due to the rigid boss. Bar 1 has length L and cross-sectional area A. Bar 2 has length $3L$ and area A. The elastic modulus of both bars is E and the yield strength is S_y. The material is elastic–perfectly plastic.

Required: Determine:

(a) the force–displacement relationship F–δ, and its diagram,

(b) the stress in each bar as a function of displacement δ,

(c) the limit load F_L , and

(d) the residual stresses when the limit load is removed.

Solution: *Equilibrium* requires that the internal forces in the two bars, P_1 and P_2 , balance the applied load F (*Figure 12.4b*):

$$P_1 + P_2 = F$$

Dividing by area A gives:

$$\sigma_1 + \sigma_2 = \frac{F}{A}$$

where σ_1 and σ_2 are the stresses in Bar 1 and Bar 2, respectively. The average stress in the system is:

$$\sigma_{ave} = \frac{F}{2A}$$

Compatibility requires that the elongation of each bar, Δ_1 and Δ_2, be the same as the displacement of the system δ:

$$\Delta_1 = \Delta_2 = \delta$$

$$\varepsilon_1 L = \varepsilon_2(3L) = \delta$$

where ε_1 and ε_2 are the strains in Bar 1 and Bar 2, respectively.

Elastic Behavior

When initially loaded, the behavior of each bar is elastic ($\sigma = E\varepsilon$) and the force–elongation relationships are:

$$P_1 = \frac{EA}{L}\delta \quad \text{and} \quad P_2 = \frac{EA}{3L}\delta$$

Applying equilibrium ($F = P_1 + P_2$) gives the elastic load–displacement relation:

$$\textit{Answer: } F = \frac{4EA}{3L}\delta$$

which is the initial linear region of the force–elongation curve in *Figure 12.4c*.

Solving for the corresponding internal forces, stresses and strains in each bar:

$$P_1 = \frac{3F}{4} = \frac{EA}{L}\delta \quad \text{and} \quad P_2 = \frac{F}{4} = \frac{EA}{3L}\delta$$

$$\textit{Answer: } \sigma_1 = \frac{3F}{4A} = \frac{E}{L}\delta \quad \text{and} \quad \sigma_2 = \frac{F}{4A} = \frac{E}{3L}\delta$$

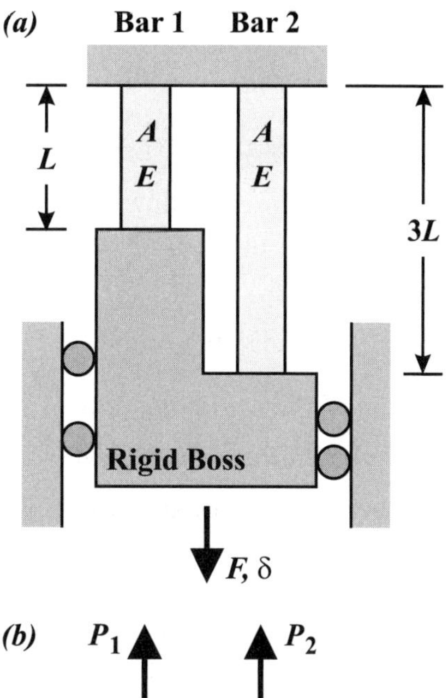

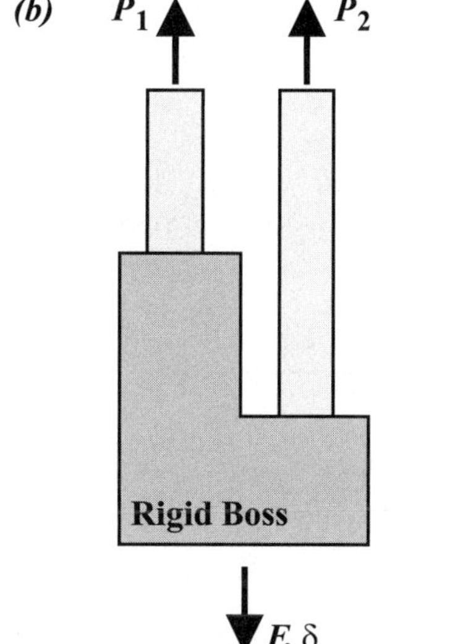

Figure 12.4. (a) Two-bar structure under axial load *F*. Not to scale. **(b)** FBD of system.

$$\varepsilon_1 = \frac{3F}{4AE} = \frac{\delta}{L}$$

and $$\varepsilon_2 = \frac{F}{4AE} = \frac{\delta}{3L}$$

During elastic loading, stress is *concentrated* in the shorter Bar 1 (*Figure 12.4d*):

$$\sigma_1 = 1.5\sigma_{ave} = 3\sigma_2$$

The *stress concentration factor* is 1.5.

Onset of Yielding – First Yield (F_y, δ_y)

As the load is increased, Bar 1 yields first. The stress and strain in Bar 1 are:

$$\textit{Answer: } \sigma_1 = \frac{P_1}{A} = S_y$$

$$\varepsilon_1 = S_y/E$$

At *first yield*, applied load F_y and corresponding system deflection δ_y are (*Figure 12.4c*):

$$F_y = \frac{4AS_y}{3}$$

$$\delta_y = \varepsilon_1 L = \frac{S_y L}{E}$$

The stress in Bar 2 is (*Figure 12.4d*):

$$\textit{Answer: } \sigma_2 = \frac{\sigma_1}{3} = \frac{S_y}{3}$$

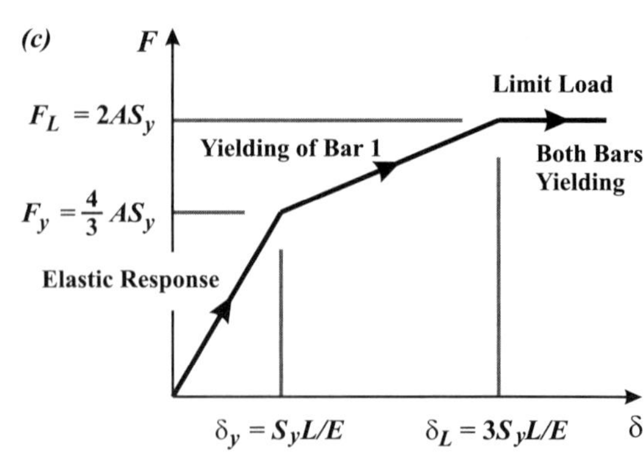

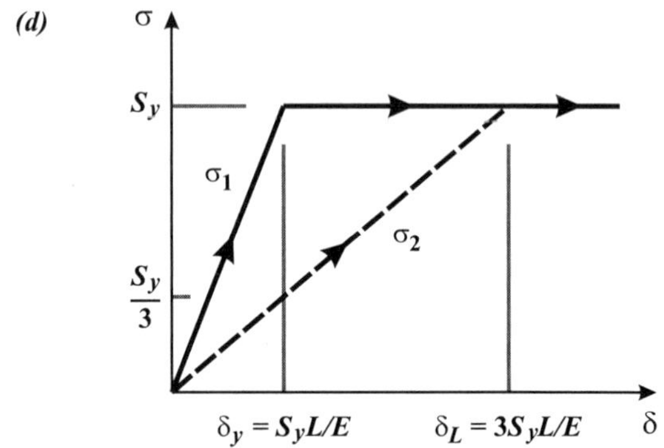

Figure 12.4. (c) Applied load versus displacement. **(d)** Stress in each bar versus displacement.

In *elastic design*, the load at first yield F_y defines the strength of the two-bar system.

Elastic–Plastic Behavior

After Bar 1 has yielded, additional load can be applied to the system. Bar 1 deforms plastically at constant stress $\sigma_1 = S_y$, while Bar 2 continues to deform elastically with stress $\sigma_2 = E\delta/(3L)$ (*Figure 12.4d*). Hence:

$$\textit{Answer: } F = P_1 + P_2 = S_y A + \frac{EA}{3L}\delta$$

which describes the second linear region of *Figure 12.4c*. The stresses and strains in each bar when Bar 1 is yielding and Bar 2 is elastic are:

Answer: $\underline{\sigma_1 = S_y}$ and $\underline{\sigma_2 = \dfrac{E}{3L}\delta}$

$$\varepsilon_1 = \frac{S_y}{E} + \varepsilon_{p1} \quad \text{and} \quad \varepsilon_2 = \frac{\delta}{3L}$$

where ε_{p1} is the plastic strain in Bar 1. As the system elongates, the stress in Bar 1 remains constant, but the overall load increases. The additional load is supported by Bar 2 only (*Figures 12.4c* and *d*). This **stress redistribution** due to plasticity reduces the stress concentration associated with σ_1 (i.e., $\sigma_1 < 1.5\sigma_{ave}$).

The Limit Load (F_L, δ_L)

When Bar 2 yields, both bars are at the yield strength:

Answer: $\underline{\sigma_1 = \sigma_2 = S_y}$

so the applied load is:

$$F = P_1 + P_2 = 2S_y A$$

Load F cannot be increased further. This load is the *limit load* F_L (*Figure 12.4c*):

Answer: $\underline{F_L = 2S_y A}$

The displacement δ at the limit load is when Bar 2 begins to yield ($\varepsilon_2 = S_y /E$):

$$\delta_L = (3L)\varepsilon_2 = \frac{3S_y L}{E}$$

The *limit load* may be calculated directly by simply setting the force in each bar equal to the force required to yield that bar (see *Example 12.4*). The *stress redistribution* is complete; all material points are at yield, and $\sigma_1 = \sigma_{ave} = F/2A$; the stress concentration factor is 1.0.

The limit load is greater than the load to cause first yield F_y. For this two-bar system, $F_L /F_y = 1.5$.

Unconstrained Plastic Deformation

When the applied load reaches the limit load F_L, further plastic elongation can occur without additional load. This deformation is said to be *unconstrained*. While F_L is the maximum load, δ_L is not the maximum displacement (*Figure 12.4c*).

The strains in Bars 1 and 2 are:

$$\varepsilon_1 = \frac{S_y}{E} + \varepsilon_{p1} \quad \text{and} \quad \varepsilon_2 = \frac{S_y}{E} + \varepsilon_{p2}$$

where ε_{p1} and ε_{p2} are the plastic strains in each bar. From compatibility, $\Delta_1 = \Delta_2$, so $\varepsilon_1 = 3\varepsilon_2$, which means the plastic strains can be related:

$$\varepsilon_{p1} = \frac{2S_y}{E} + 3\varepsilon_{p2}$$

Unloading from the Limit Load

When the load is removed, the stresses in both bars decrease *elastically*. The elastic response of the system has already been determined during initial loading. For a change in applied load ΔF, the internal forces in the bars change by:

$$\Delta P_1 = \frac{3(\Delta F)}{4} \quad \text{and} \quad \Delta P_2 = \frac{\Delta F}{4}$$

When the entire load F_L is removed:

$$\Delta F = -F_L = -2S_y A$$

The internal forces following complete removal of the limit load F_L are therefore:

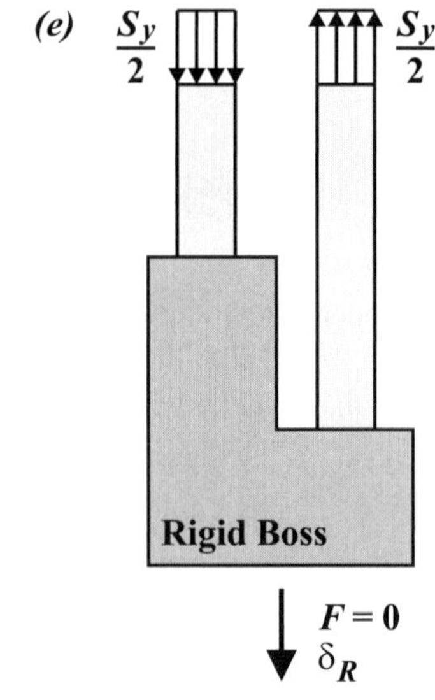

Figure 12.4. (e) *Residual Stresses when unloaded from limit load F_L.*

$$P_1 = S_y A + \frac{3(-F_L)}{4} = \frac{-S_y A}{2} \quad \text{and} \quad P_2 = S_y A + \frac{(-F_L)}{4} = \frac{+S_y A}{2}$$

The internal forces are in equilibrium, i.e., $P_1 + P_2 = 0$, since the applied force $F = 0$.

The stresses in each bar are:

$$\text{Answer: } \sigma_1 = -S_y/2 \quad \text{and} \quad \sigma_2 = +S_y/2$$

Although the applied load has been removed, Bar 1 has a compressive stress $-S_y/2$ and Bar 2 has a tensile stress $+S_y/2$ (*Figure 12.4e*). Such stresses are called **residual stresses**. No external load is applied, but stresses remain in the system due to the loading history. Bar 1 has had more plastic deformation than Bar 2, so in order for the bars to remain the same length when elastically unloaded, Bar 1 must be in compression and Bar 2 in tension.

The **residual strain** in each bar is:

$$\varepsilon_{1,R} = \frac{\sigma_1}{E} + \varepsilon_{p1} = \frac{-S_y}{2E} + \varepsilon_{p1} \quad \text{and} \quad \varepsilon_{2,R} = \frac{\sigma_2}{E} + \varepsilon_{p2} = \frac{S_y}{2E} + \varepsilon_{p2}$$

The *residual elongation* δ_R of the system when unloaded from (F_L, δ_L) is shown in *Figure 12.4f*. For this case, neither bar yields in compression, so unloading is linear, with slope $4EA/3L$. In general, the residual elongation when unloading from (F, δ) is:

$$\delta_R = \delta - F\left(\frac{3}{4}\frac{L}{AE}\right)$$

When unloading from $\delta = \delta_L = 3S_y L/E$ (*Figure 12.4f*):

$$\delta_R = \frac{3S_yL}{E} - (2AS_y)\left(\frac{3}{4}\frac{L}{AE}\right) \qquad (f)$$

$$= \frac{1.5S_yL}{E}$$

Reloading (after unloading from F_L, δ_L)

If load is now reapplied, the stress in each bar starts from its residual value ($\sigma_1 = -0.5S_y$, $\sigma_2 = +0.5S_y$). Both bars now remain elastic and both yield at the same time, Bar 1 experiencing a change in stress ($1.5S_y$) three times that of Bar 2 ($0.5S_y$). Bar 1 elastically strains three times as much as Bar 2.

The system reloading force–displacement (F–δ) curve (*Figure 12.4f*) follows the line from the (δ_R, 0) to (δ_L, F_L). Further elongation is unconstrained. Measured from δ_R, the force–displacement curve is now *elastic-perfectly plastic* (i.e., it is linear up to $F = F_L$, and then remains constant during unconstrained elongation).

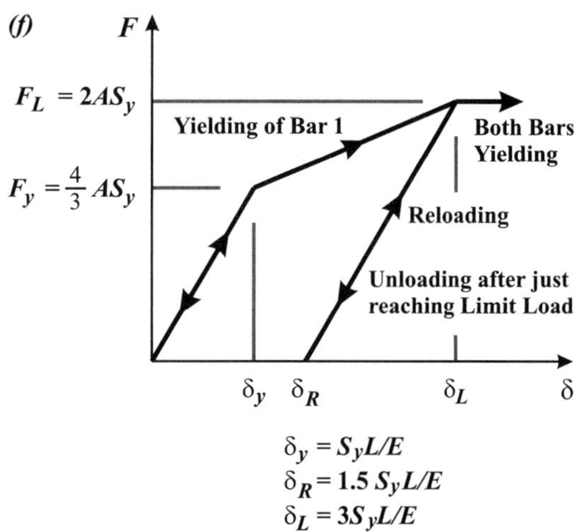

Figure 12.4. (f) Force-displacement curve. If either bar has yielded, there is a residual elongation upon removing the load. Here, the load is removed from $\delta = \delta_L$, when the *Limit Load* has just been reached. Upon reloading, the curve is linear until the limit load F_L is again reached.

Summary

The applied force F and stresses σ_1 and σ_2 as functions of continuously increasing displacement δ (without unloading) are summarized in *Table 12.2*, and shown in *Figures 12.4c* and *d*. The load–displacement response of the two-bar system is, in general, described by a tri-linear model (*Figure 12.4f*).

Table 12.2. Applied force and bar stresses during continuously increasing* displacement δ.

δ	F	σ_1	σ_2
$0 \leq \delta \leq \dfrac{S_yL}{E}$	$\dfrac{4EA}{3L}\delta$	$\dfrac{3F}{4A} = \dfrac{E}{L}\delta$	$\dfrac{F}{4A} = \dfrac{E}{3L}\delta$
$\dfrac{S_yL}{E} \leq \delta \leq \dfrac{3S_yL}{E}$	$S_yA + \dfrac{EA}{3L}\delta$	S_y	$\dfrac{E}{3L}\delta$
$\delta \geq \dfrac{3S_yL}{E}$	$2S_yA$	S_y	S_y

* Without unloading.

Example 12.2 Solid Circular Shaft in Torsion

Given: A solid circular shaft of length L and radius R is subjected to torque T (*Figure 12.5a*). The shear modulus is G and the shear yield strength is τ_y. The material is elastic–perfectly plastic (*Figure 12.5b*).

Required: Determine:

(a) the torque–angle of twist relationship T–θ, and its diagram,

(b) the limit torque T_L, and

(c) the residual stress when the limit load is removed.

Solution: *Elastic Behavior – First Yield*

The initial response of the shaft is elastic. The shear stress of a solid circular shaft is linear with distance r from its axis (*Figure 12.5c*):

$$\tau(r) = \frac{Tr}{J} = \frac{2Tr}{\pi R^4} = \frac{r}{R}\tau_{max}$$

where:

$$\tau_{max} = \frac{2T}{\pi R^3}$$

The angle of twist for a solid shaft is:

$$\theta = \frac{TL}{JG} = \frac{2TL}{\pi R^4 G}$$

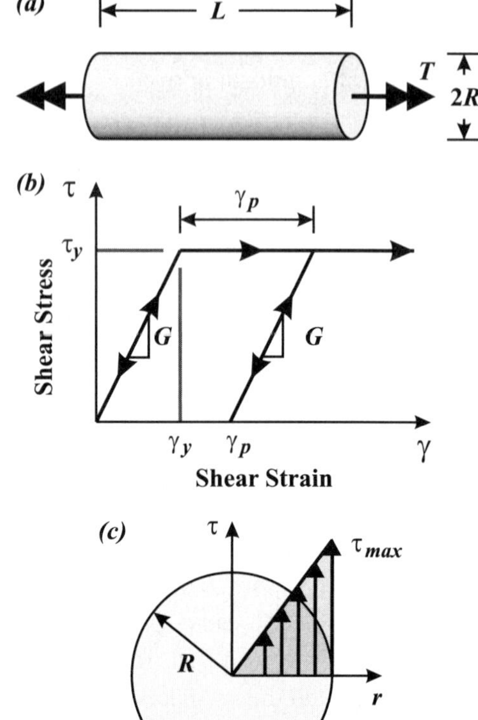

Figure 12.5. (a) Solid shaft under torsion. **(b)** Elastic–perfectly plastic shear stress–strain curve. **(c)** Elastic shear stress distribution.

The torque–angle of twist relationship in the elastic region is:

$$\textit{Answer: } T = \frac{\pi R^4 G}{2L}\theta$$

The maximum shear stress increases with torque until yielding first occurs ($\tau_{max} = \tau_y$). The yield torque is:

$$T_y = \frac{\pi R^3 \tau_y}{2}$$

The angle of twist at first yield is:

$$\theta_y = \frac{\tau_y L}{RG}$$

In *elastic design*, T_y defines the strength of the circular shaft in torsion.

Elastic–Plastic Behavior

When the applied torque exceeds T_y, the shaft begins to plastically deform, starting at the outer radius R and moving inwards to a radius r_e (*Figure 12.5d*). The material within radius r_e ($r < r_e$) remains elastic, and is called the **elastic core**. Between r_e and R, the shear stress is constant and equal to the shear yield strength τ_y. In torsion, any radius on the cross-section remains a straight line due to symmetry. The angle of twist of the elastic core θ_e, and thus the angle of twist of the entire shaft θ, is:

$$\theta = \theta_e = \frac{\tau_y L}{r_e G}$$

The torque supported by the elastic core, T_e, is:

$$T_e = \frac{\pi r_e^3 \tau_y}{2}$$

The torque supported by the plastic portion of the cross-section, T_p, is found by integrating the contribution of the torques of many thin-walled shafts, each of radius r and thickness dr (area $dA = 2\pi r\, dr$), from r_e to R. The shear stress from r_e to R is the shear yield strength τ_y:

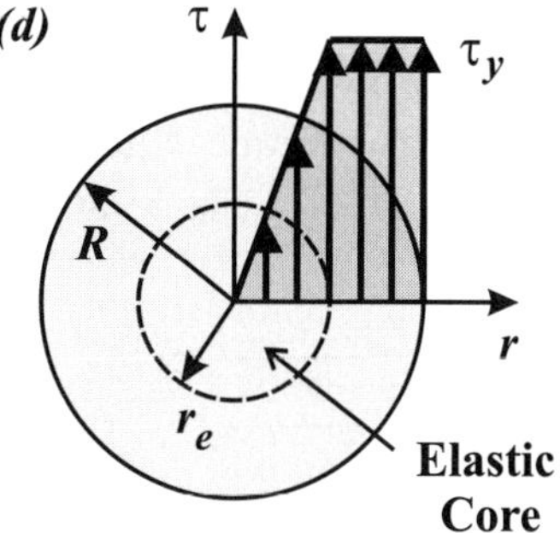

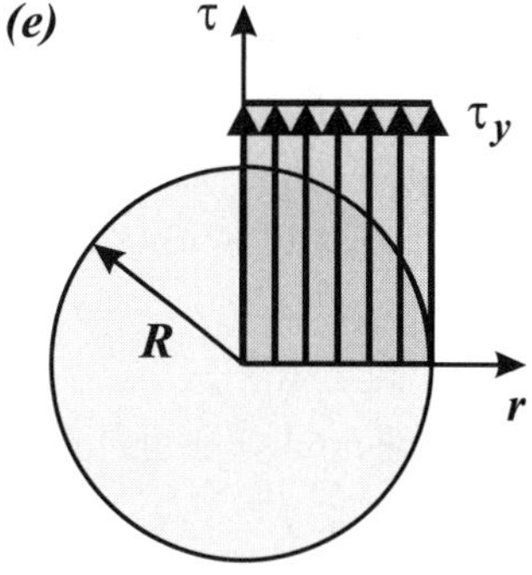

Figure 12.5. **(d)** Shear stress distribution due to elastic-plastic response. **(e)** Shear stress distribution when cross-section is completely plastic.

$$T_p = \int_{r_e}^{R} r[\tau(r)dA] = \int_{r_e}^{R} r[\tau_y(2\pi r\, dr)] = 2\pi\tau_y \int_{r_e}^{R} r^2\, dr = \frac{2\pi\tau_y}{3}[R^3 - r_e^3]$$

The total torque T is the sum of elastic torque T_e and plastic torque T_p:

$$T = T_e + T_p = \frac{2\pi\tau_y}{3}R^3\left[1 - \frac{1}{4}\left(\frac{r_e}{R}\right)^3\right]$$

Limit Torque

When $r_e = 0$, the material is yielding everywhere (*Figure 12.5e*) and the resulting torque is the *limit torque* T_L:

$$Answer:\ T_L = \frac{2\pi\tau_y}{3}R^3 = \frac{4}{3}T_y$$

The *limit torque* is greater than the yield torque T_y by a factor of 4/3. No further torque can be supported by the cross-section.

Substituting the limit torque, T_L, and the angle of twist of the elastic core, $\theta = \theta_e$, into the total torque equation gives the torque–angle of twist relation, T–θ, for $T > T_y$:

$$\text{Answer:}\quad T = T_L\left[1 - \frac{1}{4}\left(\frac{\tau_y L}{GR\theta}\right)^3\right]$$

A plot of the T–θ curve is shown in *Figure 12.5f*. The torque is asymptotic to $T = T_L$. After the cross-section becomes fully plastic, the shaft continues to deform plastically without further load as it is twisted about its axis to large values of θ ($\gg \theta_y$).

First yield occurs when the angle of twist is:

$$\theta_y = \frac{\tau_y L}{RG}$$

So the torque at first yield can also be written:

$$T_y = T_L\left[1 - \frac{1}{4}(1)^3\right] = \frac{3}{4}T_L$$

Unloading from the Limit Torque

Unloading the shaft from the limit torque T_L is *elastic*. The elastic stress as a function of r is:

$$\tau(r) = \frac{2Tr}{\pi R^4}$$

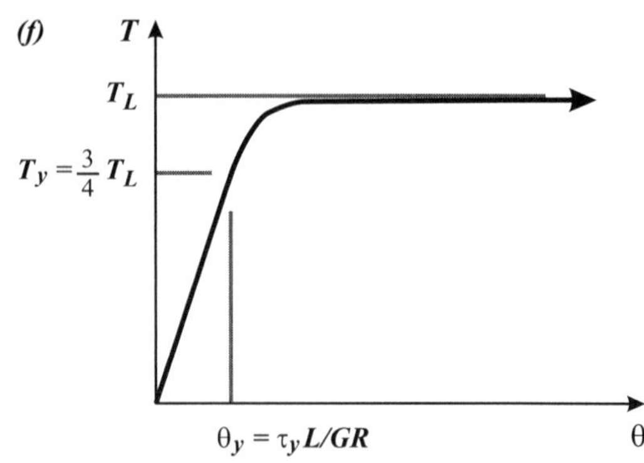

Figure 12.5. (f) Torque versus angle of twist.

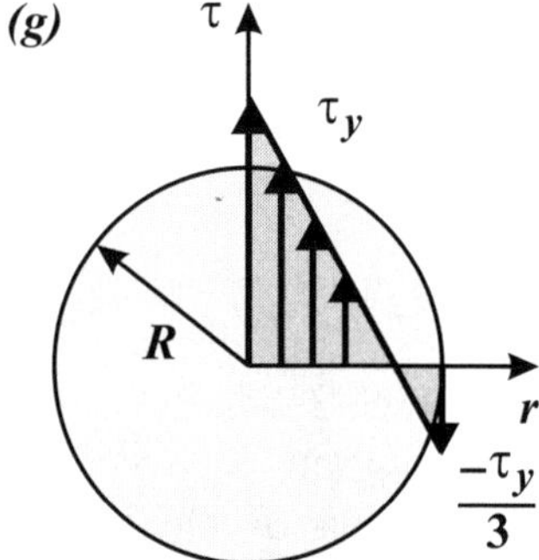

Figure 12.5. (g) *Residual shear stresses* upon elastic unloading from the T_L.

Removing T_L is the equivalent of applying an additional elastic torque that is negative, $T = -T_L$. This gives the *residual stress* distribution $\tau_R(r)$:

$$\tau_R(r) = \tau_y - \frac{2T_L r}{\pi R^4} = \tau_y - \frac{2r}{\pi R^4}\left(\frac{2\pi\tau_y R^3}{3}\right)$$

$$\text{Answer:}\quad \tau_R(r) = \tau_y\left[1 - \frac{4r}{3R}\right]$$

The residual stress is plotted in *Figure 12.5g*. Note that the residual shear stress on the surface of the shaft is opposite to the stress due to the applied (and now unloaded) torque. The axis of the shaft ($r = 0$) has residual stress τ_y since elastic unloading contributes zero to the stress at that location.

Example 12.3 Rectangular Beam in Bending

Given: A rectangular beam of breadth B and depth D is subjected to moment M (*Figure 12.6a*). The material is elastic–perfectly plastic with modulus E and yield strength S_y (in tension and compression).

Required: Determine:

(a) the moment–curvature relationship M–κ, and its diagram,

(b) the limit moment M_L, and

(c) the residual stress when the limit load is removed.

Solution:

Elastic Behavior – First Yield

The initial response of the beam is elastic. The bending stress and curvature are given by:

$$\sigma(y) = -\frac{My}{I} \quad \text{and} \quad \kappa = \frac{1}{R} = \frac{M}{EI}$$

Recall that the curvature κ is the inverse of the radius of curvature R and the moment of inertia for a rectangular cross-section is $I = BD^3/12$.

The moment–curvature relationship in the elastic region is:

$$\textit{Answer: } \underline{M = EI\kappa}$$

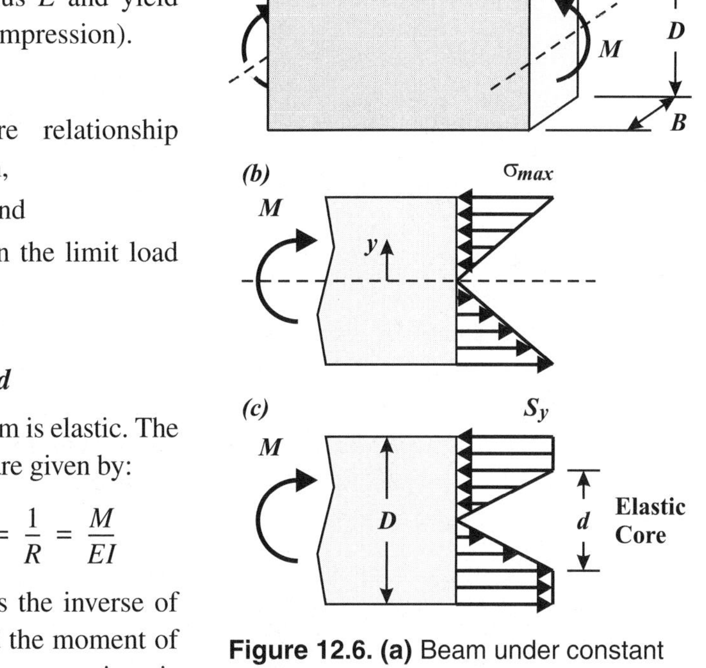

Figure 12.6. (a) Beam under constant moment M. **(b)** Stresses in beam due to elastic response. Yielding occurs when $\sigma_{max} = S_y$. **(c)** Stresses in beam due to elastic–plastic response.

In pure bending, the maximum stress σ_{max} occurs at $y = \pm D/2$ (*Figure 12.6b*), and yielding occurs when σ_{max} reaches the yield strength $\mp S_y$ (compression and tension). For a rectangular beam, the magnitude of the moment at first yield, and the corresponding curvature, are:

$$M_y = \frac{S_y BD^2}{6} \quad \text{and} \quad \kappa_y = \frac{2S_y}{DE}$$

In *elastic design*, M_y defines the strength of the beam in bending.

Elastic–Plastic Behavior

When the applied moment exceeds M_y, the beam begins to yield, starting at $y = \pm D/2$ and moving inwards (*Figure 12.6c*). The *elastic core* is the portion of the beam that remains elastic, having a height of d, with $|y| < d/2$. The material for $|y| > d/2$ is plastic with stress $\pm S_y$.

The moment supported by the elastic core, M_e, is:

$$M_e = \frac{S_y B d^2}{6}$$

Due to symmetry in bending, plane sections continue to remain plane, so the curvature of the beam is defined by the curvature of the elastic core:

$$\kappa_e = \frac{2S_y}{dE}$$

The moment supported by the plastic portion of the cross-section, M_p, is found by integrating the contribution of the moments due to many axial members of area dA having width B and thickness dy, from $y = d/2$ to $D/2$, and from $y = -d/2$ to $-D/2$ (*Figure 12.6c*). The stress in the plastic region is constant and equal to the yield strength S_y. Integrating $dM = y[S_y B \, dy]$ from $y = d/2$ to $D/2$, and multiplying by 2 to account for $y = -d/2$ to $-D/2$, gives:

$$M_p = 2\int_{d/2}^{D/2} y[S_y B \, dy] = 2S_y B \int_{d/2}^{D/2} y \, dy = \frac{S_y BD^2}{4}\left[1 - \left(\frac{d}{D}\right)^2\right]$$

The total moment is then:

$$M = M_e + M_p = \frac{S_y BD^2}{4}\left[1 - \frac{1}{3}\left(\frac{d}{D}\right)^2\right]$$

Limit Moment

When $d = 0$, the material is yielding everywhere (*Figure 12.6d*) and the resulting moment is the *limit moment* M_L. For a rectangular beam:

$$\textit{Answer: } M_L = \frac{S_y BD^2}{4} = 1.5 M_y$$

No further moment can be supported.

The moment–curvature (M–κ) relationship for $M > M_y$ can be written:

$$\textit{Answer: } M = M_L\left[1 - \frac{4}{3}\left(\frac{S_y}{ED\kappa}\right)^2\right]$$

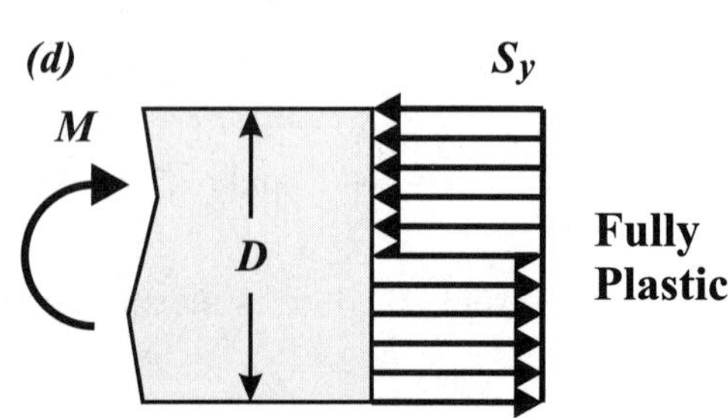

Figure 12.6. (d) Stresses in beam at the limit moment.

A plot of the M–κ curve is shown in *Figure 12.6e*. The moment is asymptotic to M_L. When the cross-section becomes fully plastic the curvature is *unconstrained*. The beam at the limit moment can have a large change in slope (curvature) without further load. This response is called a ***plastic hinge***.

Unloading from the Limit Moment

When the moment is removed, the stress–strain response is elastic. Removing the limit moment is the same as elastically adding $-M_L$. The additional stress caused by elastically unloading moment M_L is:

$$\Delta\sigma(y) = -\frac{(-M_L)y}{I} = \frac{3y}{D}S_y$$

At $y = +D/2$, the change of stress upon unloading is $+1.5S_y$ as shown in *Figure 12.6f*.

Upon removal of the limit moment, *residual stresses* are introduced. These residual stresses are:

Answer:

- for $y > 0$:

$$\sigma_R(y) = -S_y + \frac{3y}{D}S_y = S_y\left(-1 + \frac{3y}{D}\right)$$

- for $y < 0$:

$$\sigma_R(y) = +S_y + \frac{3y}{D}S_y = S_y\left(1 + \frac{3y}{D}\right)$$

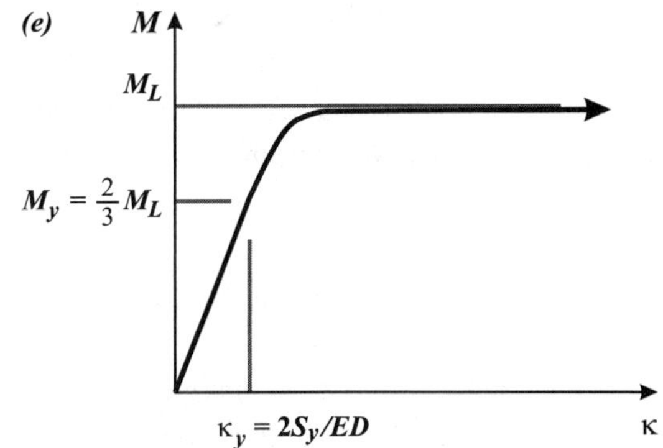

Figure 12.6. (e) Moment–Curvature relationship.

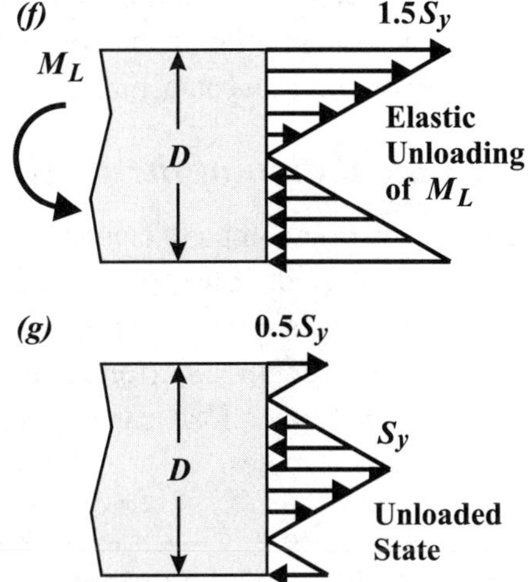

Figure 12.6. (f) Stresses due to elastic unloading of the beam from M_L. **(g)** Residual stresses in beam.

The distribution of the residual stresses is shown in *Figure 12.6g*. The residual stresses are zero at $y = \pm D/3$.

Summary of Examples

In the elastic–perfectly plastic analysis of the previous three examples, each system responds elastically until the most highly stressed point(s) reaches the yield strength. The stress in an elastic–perfectly plastic material cannot exceed the yield strength. As a system continues to deform plastically, *stress redistribution* (non-elastic redistribution of stress) occurs until the limit load (F_L, T_L, M_L) is reached. At the limit load, the stress everywhere on the cross-section is at the yield strength. For the rectangular beam, the yield moment M_y (when yielding first occurs) and the limit moment $M_L = 1.5M_y$ are, respectively:

$$M_y = \frac{S_y BD^2}{6} \quad \text{and} \quad M_L = \frac{S_y BD^2}{4}$$

Designers can take advantage of plastic deformation to create more efficient and lighter structures. For example, in bending applications, the design moment can be 90% of the limit moment. The system is allowed to plastically deform at its maximum expected (and rarely reached) load in service.

In addition, since the material is assumed to be elastic–perfectly plastic, the maximum stress in the analysis is the initial yield strength S_y. Since yield strength actually increases during plastic deformation, then the calculated limit load is smaller than the maximum load that a cross-section can actually carry.

Direct Calculation of the Limit Load

The details of a complete elastic–perfectly plastic analysis, even in simple examples, can be quite involved. However, it is often a straight-forward task to determine the limit load directly. This is done by assuming that the stress at every point on a cross-section equals the yield strength $\pm S_y$ (for normal stress) or τ_y (for shear stress). Once every point has yielded, no further load can be supported. This technique is demonstrated in the following three examples.

Example 12.4 **Two-Bar System**

Given: The two-bar system under load F (*Figure 12.7*). Bar 1 has cross-sectional area A_1 and Bar 2 has area A_2. The yield strength is S_y.

Required: Determine the limit load F_L.

Solution: The limit load is reached when the axial stress in each bar is S_y:

$$\text{Answer: } P_L = A_1 S_y + A_2 S_y$$

For $A_1 = A_2$, $P_L = 2AS_y$, which agrees with the elastic–plastic calculation of *Example 12.1*.

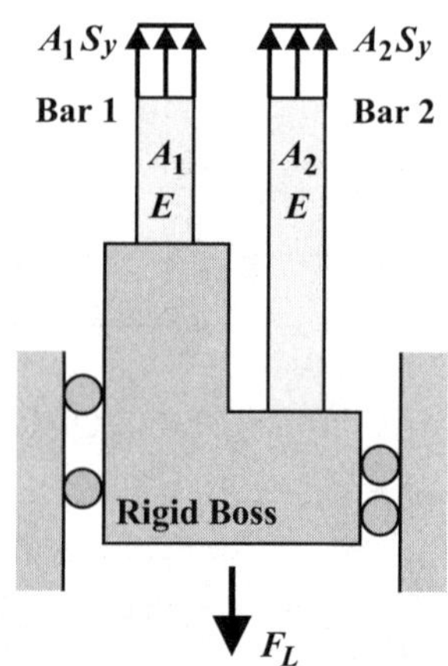

Figure 12.7. Stresses in two-bar system at limit load.

Example 12.5 Solid Circular Shaft in Torsion

Given: The solid circular shaft under torque T (*Figure 12.8*). The radius is R. The shear yield strength is τ_y.

Required: Determine the limit torque T_L.

Solution: The limit torque is reached when the shear stress on the entire cross-section is τ_y. Equilibrium requires:

Answer:

$$T_L = \int_0^R r[\tau_y 2\pi r\, dr] = \frac{2\pi R^3 \tau_y}{3}$$

which agrees with the elastic–plastic calculation of *Example 12.2*.

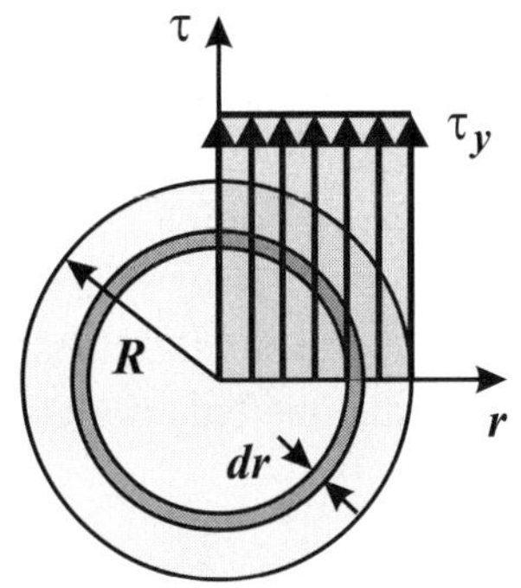

Figure 12.8. Shear stresses on cross-section of a solid shaft at limit torque.

Example 12.6 Rectangular Beam in Bending

Given: The rectangular beam under moment M (*Figure 12.9*). The breadth is B and the depth is D. The tensile and compressive yield strengths are equal, with magnitude S_y.

Required: Determine the limit moment M_L.

Solution: The limit load is reached when the bending stress on the entire cross-section is at the yield strength $\pm S_y$. Equilibrium requires:

Answer:

$$M_L = 2\int_0^{D/2} y[S_y B\, dy] = \frac{BD^2 S_y}{4}$$

which agrees with result of *Example 12.3*.

The limit moment is also the product of the equivalent force acting over each half-area $S_y(BD/2)$, and the distance between the centroids of each S_y load distribution $D/2$ (*Figure 12.9b*):

$$M_L = \left[S_y\!\left(B\frac{D}{2}\right)\right]\!\left(\frac{D}{2}\right) = \frac{BD^2 S_y}{4}$$

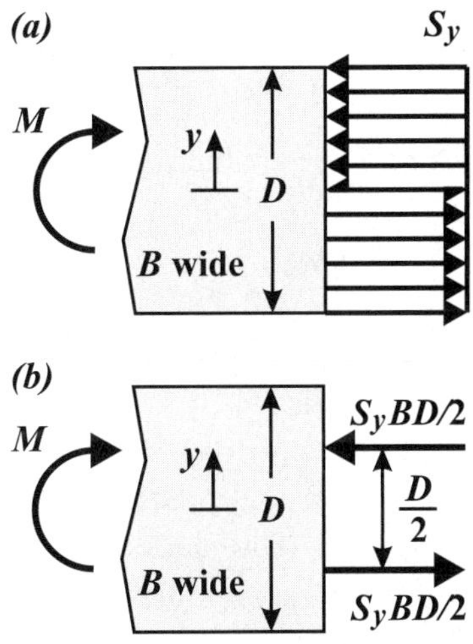

Figure 12.9. (a) Stresses in beam at limit moment (side view). **(b)** Equivalent forces acting at the centroids of the $+S_y$ and $-S_y$ load areas.

12.3 Limit Loads in Beams: Plastic Hinges

When the limit moment of a beam cross-section is reached, the curvature of the beam at that location is unbounded, or unconstrained, as shown in the M–κ relationship of *Figure 12.6e*. Large curvatures cause the beam to act as if it is *hinged* (or pinned) at that location. That position along the beam is therefore referred to as a **plastic hinge**. A nice example of this behavior is in the plastic bending of a paper clip or other thin metal rod. The deflection and slope due to the plastic hinge are large, much greater than those due to elastic deformation. This deflection behavior may be used to calculate the limit loads of beams.

Example 12.7 Simply-Supported Beam under Uniform Load

Given: A simply-supported beam of rectangular cross-section, depth D and breadth B, is subjected to a uniformly distributed load w (force/length), as shown in *Figure 12.10a*. The material yield strength is S_y.

Required: Determine the limit distributed load w_L.

Solution: The maximum moment in a beam under UDL occurs at the center (*Figure 12.10b*):

$$M_{max} = \frac{wL^2}{8}$$

A plastic hinge will form when $M_{max} = M_L$ and the beam will deflect as shown in *Figure 12.10c*. The deflection due to the hinge dominates over the deflection due to elastic deformation.

The limit moment for a rectangular beam is:

$$M_L = \frac{S_y BD^2}{4}$$

Equating the two moment expressions gives the limit load:

$$\textit{Answer: } w_L = \frac{2S_y BD^2}{L^2}$$

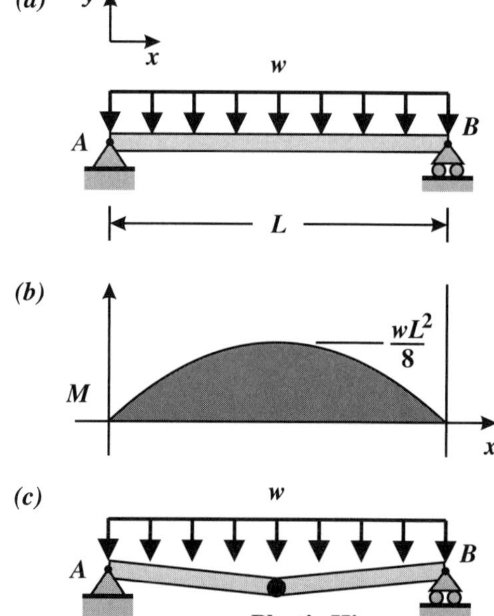

Figure 12.10. (a) Simply supported beam under uniformly distributed load *w*. **(b)** Moment diagram of beam. **(c)** Deformation of beam due to plastic hinge at its center.

In a design limited by elastic behavior, the failure moment is the moment at first yield:

$$M_y = \frac{S_y BD^2}{6}$$

when the distributed load w is:

$$w_y = \frac{4S_y BD^2}{3L^2}$$

The ratio of the limit load to the yield load is:

$$\frac{w_L}{w_y} = \frac{2}{4/3} = 1.5$$

As discovered in *Example 12.3*, the limit load that a rectangular beam can support is 50% greater than that predicted by elastic analysis. Hence the attraction of plastic design methods now widely used in engineering systems.

Example 12.8 Fixed–Fixed Beam under Central Point Load

Given: A fixed–fixed beam of rectangular cross-section, depth D and breadth B, is subjected to a central point load P as shown in *Figure 12.11a*. The yield strength is S_y.

Required: Determine the limit load P_L.

Solution: The *elastic analysis* for this problem is relatively complex. The *limit load analysis* is fairly simple.

When the applied load reaches its limit value, $P = P_L$. At the limit load, the beam deforms in a similar manner as the simply-supported beam of *Example 12.7*, but now three *plastic hinges* must be formed, one under the point load, and one at each wall (*Figures 12.11c* and *d*). At the hinges, the moment must be $M = M_L = S_y BD^2/4$.

By symmetry, half of force P is supported at each wall. A FBD of a segment of the beam (from 0 to x) when the beam is loaded by P_L is shown in *Figure 12.11b*. Equilibrium requires the internal moment to be linear with x (*Figure 12.11c*). At $x = 0$ and $L/2$, the magnitude of the internal moment is

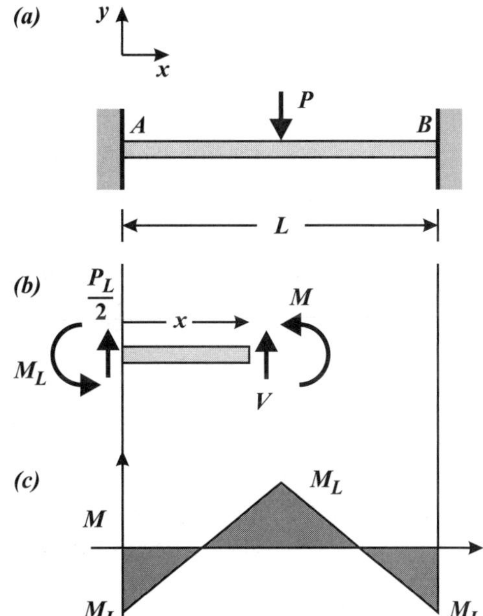

Figure 12.11. (a) Fixed–fixed beam under central load P. **(b)** FBD of beam for $0 < x < L/2$. By symmetry, the moment at the center of the beam equals the moment at each wall. **(c)** The moment diagram of the beam is linear.

$M = M_L$. Thus, moment equilibrium about the right side of the segment from $x = 0$ to $L/2$ (*Figure 12.11b*) gives:

$$M_L - \frac{P_L}{2}\left(\frac{L}{2}\right) + M_L = 0$$

Therefore:

$$\text{Answer: } P_L = \frac{8M_L}{L} = \frac{2S_y BD^2}{L}$$

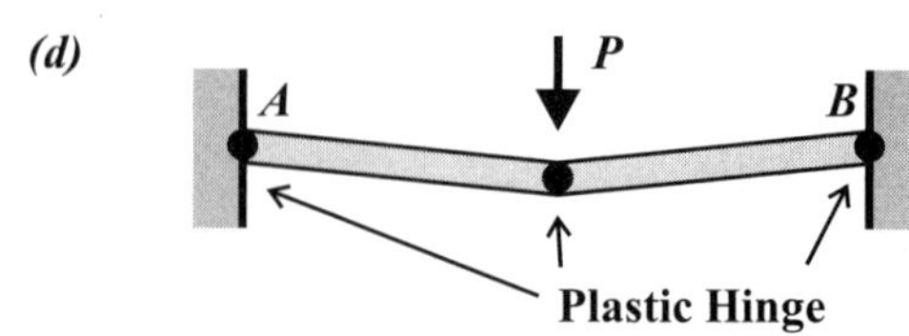

Figure 12.11. (d) Deflection of beam due to plastic hinges at its center and at the walls.

The deflection of the beam is shown in *Figure 12.11d*. The deflection is primarily due to the unbounded curvatures of the plastic hinges. The elastic deflection is negligible, so the beam deflection is linear between hinges. The geometric response at a plastic hinge is similar to that at a pin (*Figure 12.10c*).

Limit Load Estimates: Upper and Lower Bounds

When the calculation of the exact *limit load* is not obvious, it is possible to perform simple calculations that *bound*, or bracket, the limit load. An example of such a situation is when the system is statically indeterminate (*Figure 12.12*). The two methods used to bracket the exact solution are:

1. The *force method*, which *underestimates* the limit load; it provides a *lower bound* to the solution. *Equilibrium* must be satisfied in the force method.

2. The *displacement method*, which *overestimates* the limit load; it provides an *upper bound*. *Compatibility* must be satisfied in the displacement method.

These two methods are illustrated with an example.

Example 12.9 **Fixed-Pinned Beam**

Given: A beam of length L supports a uniformly distributed load w. The beam is built-in (fixed) at one end and simply-supported with a roller at the other (*Figure 12.12a*). The limit moment of the beam cross-section is M_L.

Required: Estimate the limit distributed load w_L.

Solution: The reactions and internal forces cannot be solved by statics alone. The system is *redundant* or *statically indeterminate*.

Force Method – Lower Bound

Guess 1. The value of the reaction force R at the pin is unknown. Let us take a first educated guess that $R = 3w_L/8 = 0.375w_L$ (*Figure 12.12b*), which is actually the reaction for the elastic solution (*Example 6.9*). The bending moment along the beam is then:

$$M(x) = \frac{-w(L-x)^2}{2} + R(L-x) \qquad \text{(a)}$$

$$= \frac{-w(L-x)^2}{2} + \frac{3wL}{8}(L-x)$$

A *local* maximum of the moment occurs where the shear force equals zero:

$$V(x) = -\frac{dM}{dx} = -w(L-x) + \frac{3wL}{8} = 0$$

which occurs at $x = 5L/8$.

So a *possible* value for the maximum moment is:

$$M\left(\frac{5L}{8}\right) = \frac{-w}{2}\left(\frac{3L}{8}\right)^2 + \frac{3wL}{8}\left(\frac{3L}{8}\right)$$

$$= \frac{9wL^2}{128} = 0.0703wL^2$$

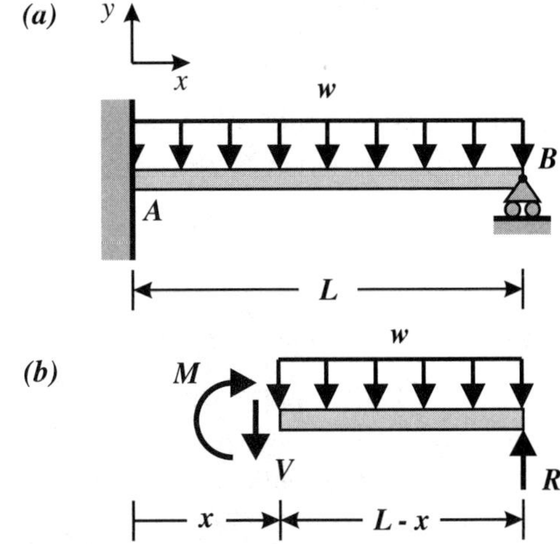

Figure 12.12. (a) Fixed-pinned beam under uniform load *w*. **(b)** Right-hand FBD of beam. The maximum moment in an elastic calculation occurs at the wall.

The moment at the wall ($x = 0$) may also be a maximum:

$$M(0) = \frac{-wL^2}{8} = -0.125wL^2$$

Thus, the moment with the maximum magnitude occurs at the wall. This value provides an estimate of the limit load w_L:

$$w_L = \frac{8M_L}{L^2}$$

Guess 2. The actual reaction R is probably not the elastic reaction since yielding causes stress redistribution within the system. For a second approximation, guess that $R = 2w_L/5 = 0.40w_L$. A repeat of the equilibrium equation, which must always be satisfied in the force method, results in:

$$M(x) = \frac{-w(L-x)^2}{2} + \frac{2wL}{5}(L-x)$$

The local maximum moment occurs where the shear stress is equal to zero ($x = 3L/5$):

$$M\left(\frac{3L}{5}\right) = \frac{wL^2}{12.5}$$

while the moment at the wall is:

$$M(0) = \frac{-wL^2}{10}$$

The maximum moment for this guess occurs at the wall. The second estimate of w_L is:

$$w_L = \frac{10M_L}{L^2}$$

which is larger than the estimate from *Guess 1*. Since the force method provides a lower bound, select the larger value:

$$Answer:\ w_L = \frac{10ML}{L^2}$$

Displacement Method – Upper Bound

Guess 1. In this approximation, the failure mode is taken as two plastic hinges, one at the wall and one at the center of the beam (*Figure 12.12c*). Point *B* is a roller and supports no moment. The equivalent point load over each half of the beam, $F = wL/2$, is shown.

The next step is to equate the *external work* done by applied load w, to the *internal work* required to create the plastic hinges. The elastic energy in bending is assumed to be negligible compared to that of the large plastic moments and angular displacements at the hinges.

The work to form a plastic hinge at the wall (point A) is $M_L\theta$ (*Figure 12.12c*). The work needed to form a plastic hinge at the center of the beam is $2M_L\theta$. The total energy to create both plastic hinges is thus:

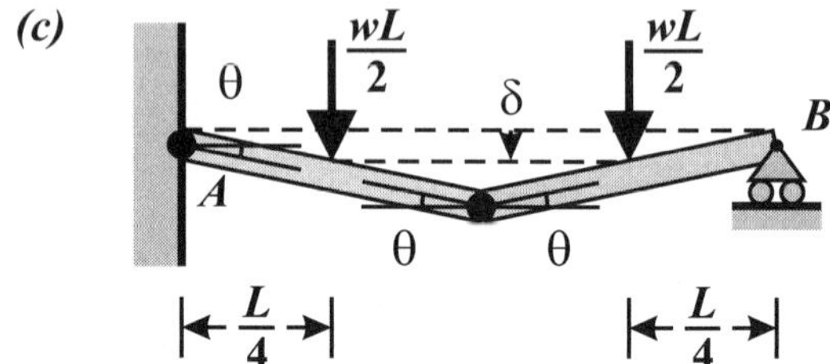

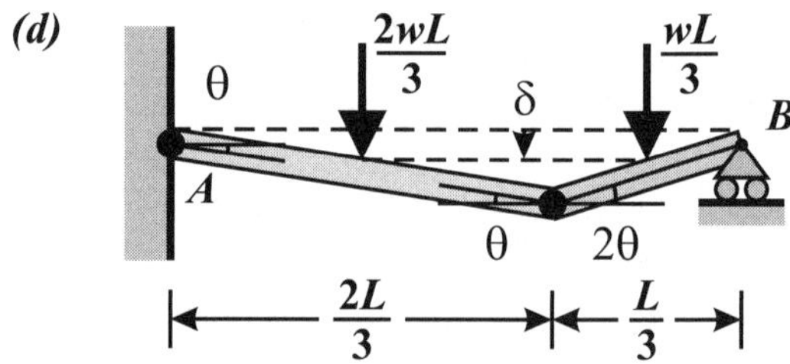

Figure 12.12. **(c)** Deflection of beam assuming plastic hinges at wall and center of beam. The distributed load on each half-beam is converted to equivalent force $wL/2$. These forces move distance $\delta = L\theta/4$. **(d)** Deflection of beam assuming plastic hinges at wall and at $x = 2L/3$. The equivalent loads each move $\delta = L\theta/3$.

$$W_p = 3M_L\theta$$

Due to the hinges, each equivalent load $F = wL/2$ displaces by $\delta = L\theta/4$. Hence, the work done by applied load w during plastic deflection is:

$$W_w = 2F\delta = 2\left(\frac{wL}{2}\right)\left(\frac{L}{4}\theta\right) = \frac{wL^2\theta}{4}$$

Equating the two work expressions gives the estimated upper limit load:

$$w_L = \frac{12M_L}{L^2}$$

which is an *upper bound* to the actual limit load.

Guess 2. Due to the pin at point B, the right side of the beam is less constrained than the left. A second approximation therefore, considers a plastic hinge at the wall, and one $2L/3$ from the wall (*Figure 12.12d*). The total plastic work to create the hinges is:

$$W_p = M_L\theta + 3M_L\theta = 4M_L\theta$$

The equivalent loads in this case are $2wL/3$ and $wL/3$. Both displace by $\delta = L\theta/3$. Hence, the work done by the applied load is:

$$W_w = \left(\frac{2wL}{3}\right)\left[\frac{1}{2}\left(\frac{2L}{3}\theta\right)\right] + \left(\frac{wL}{3}\right)\left[\frac{1}{2}\left[\frac{L}{3}(2\theta)\right]\right] = \frac{wL^2}{3}$$

Equating the two work expressions gives the estimated limit load:

$$\textit{Answer:}\ w_L = \frac{12M_L}{L^2}$$

Guess 1 gives the same result for the upper bound as *Guess 2*.

Lower and Upper Bounds

The true limit load lies somewhere between the calculated estimates of the maximum lower limit and the minimum upper limit. Hence:

$$\textit{Answer:}\ \frac{10M_L}{L^2} \le w_L \le \frac{12M_L}{L^2}$$

Exact Solution

It can be shown that at the true limit load, the reaction R is:

$$R = 0.414w_L$$

and the corresponding limit load is:

$$w_L = 11.67\frac{M_L}{L^2}$$

This value lies within the bounds generated by the approximate solutions. In this case, an estimated value that is midway between the approximated bounds – $11M_L/L^2$ – is within 7% of the exact solution.

General Remarks

The force method always gives a lower bound and the displacement method always gives an upper bound. The force method (equilibrium) calculations are generally more difficult than the displacement method (compatibility) calculations. However, the force method results in an applied load that does not exceed the true limit load. Hence, the force method gives conservative approximations for the actual strength of a system, and thus is useful in design.

12.4 Limit Surface for Rectangular Beam under Combined Loading

The basic element used in the development of the ASME pressure vessel design codes is a rectangular beam subjected to axial (normal) force N and moment M (*Figure 12.13a*). For the case when the moment is zero ($M = 0$), the limit load is (*Figure 12.13b*):

$$N_L = BDS_y \qquad \text{[Eq. 12.11]}$$

For the case when the axial load is zero ($N = 0$), the limit moment is (*Figure 12.13c*):

$$M_L = \frac{BD^2 S_y}{4} \qquad \text{[Eq. 12.12]}$$

When an axial force and a moment are applied simultaneously, the limit state is reached when the stresses are everywhere at the yield strength $\pm S_y$. At some height on the cross-section, $y = y_o$, the stress changes sign (*Figure 12.13d*). For:

- $y < y_o$, the stress is tensile ($+S_y$);

- $y > y_o$, the stress is compressive ($-S_y$).

Applying equilibrium in the x-direction gives:

$$N = S_y B\left(\frac{D}{2} + y_o\right) - S_y B\left(\frac{D}{2} - y_o\right)$$

$$= 2y_o BS_y$$

Normalizing force N by the limit load N_L:

$$\frac{N}{N_L} = 2\left(\frac{y_o}{D}\right) \qquad \text{[Eq. 12.13]}$$

Applying moment equilibrium about the z-axis through the centroid gives:

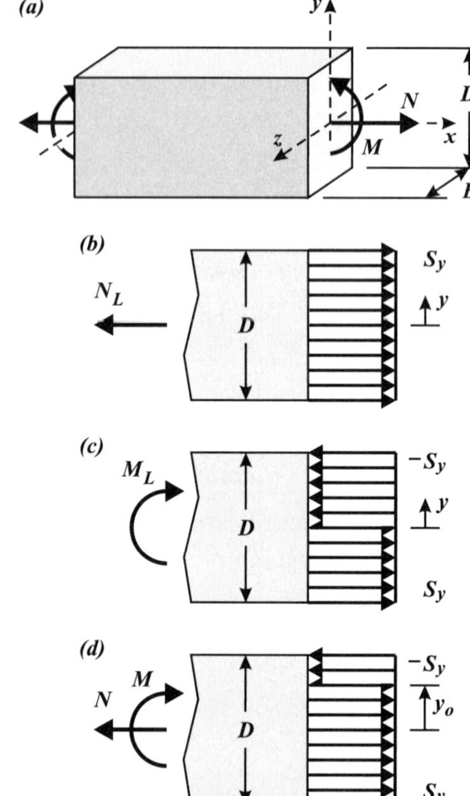

Figure 12.13. (a) Beam under combined axial load N and moment M. (b) Stresses in beam at limit load, with $M = 0$. (c) Stresses in beam at limit moment, with $N = 0$. (d) Stresses in beam at limit load due to combined axial and bending loads N and M.

$$M = \left[\frac{1}{2}\left(\frac{D}{2}+y_o\right)\right]\left[S_yB\left(\frac{D}{2}+y_o\right)\right] + \left[\frac{1}{2}\left(\frac{D}{2}-y_o\right)\right]\left[S_yB\left(\frac{D}{2}-y_o\right)\right]$$

$$= \frac{BD^2S_y}{4}\left[1-4\left(\frac{y_o}{D}\right)^2\right]$$

[Eq. 12.14]

Normalizing moment M by the limit moment M_L:

$$\frac{M}{M_L} = 1-4\left(\frac{y_o}{D}\right)^2$$

[Eq. 12.15]

Eliminating y_o between *Equations 12.13* and *12.15* gives the **limit load condition** in terms of M and N:

$$\frac{M}{M_L} = 1-\left(\frac{N}{N_L}\right)^2$$

[Eq. 12.16]

Alternatively, the limit condition occurs when:

$$\left(\frac{N}{N_L}\right)^2 + \frac{M}{M_L} = 1$$

[Eq. 12.17]

The surface represented by *Equation 12.16* (and *Equation 12.17*) is shown in *Figure 12.14*. The values of *M* and *N* that satisfy *Equation 12.16* are combinations for which the cross-section of the beam becomes completely plastic. The curve in *Figure 12.14* is called the **limit load surface**.

If the combination of the normal force *N* and moment *M* falls outside of the limit surface, the applied loading exceeds the beam's load-carrying capacity.

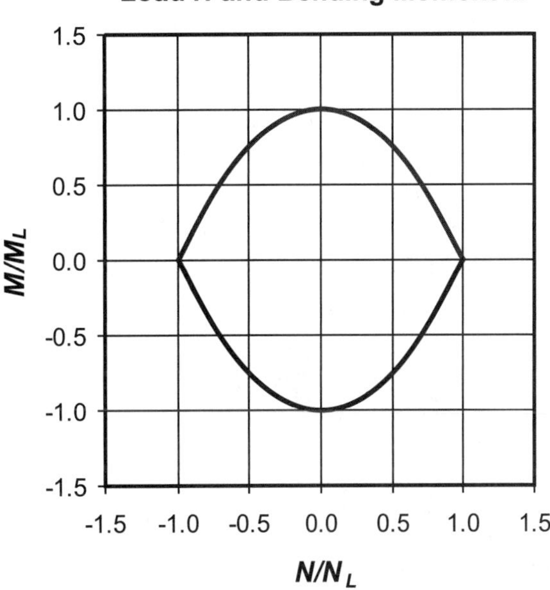

Figure 12.14. Limit Load Surface for combined axial and bending moment.

12.5 Design Applications in Plasticity

Two approaches that demonstrate how plasticity may be used in design are:

1. The area replacement method, and

2. The ASME pressure vessel code method.

These plastic design methods allow design decisions to be made using simple calculations.

Area Replacement Method

Pressure vessels contain fluid under compression. To fill and empty the vessels, holes must be cut into the walls and piping attached. Removing the material to form the hole reduces the strength of the vessel. The question is how to design – reinforce – the junction of the pipe and vessel without loss of strength? The design of a reinforcement often employs a material's ductility. Elastic design is limited by high-stress concentrations at the pipe–vessel interface. By allowing plastic deformation, stresses are redistributed until the stress at every point is at the yield condition.

Design of a Spherical Pressure Vessel with an Attached Pipe

Consider a thin-walled spherical pressure vessel of average radius R and thickness T ($T \ll R$), to which a thin-walled pipe of average radius r and thickness t ($t \ll r$) is attached (*Figure 12.15a*). The pressure of the contained fluid is p. The biaxial state of stress in the vessel is:

$$\sigma_I = \sigma_{II} = \frac{pR}{2T}; \ \sigma_{III} = 0 \qquad \text{[Eq. 12.18]}$$

The stress perpendicular to the vessel surface is $\sigma_{III} \sim 0$.

Using the 3D von Mises Yield condition, yielding occurs when:

$$\sqrt{\frac{[(\sigma_I - \sigma_{II})^2 + (\sigma_{II} - \sigma_{III})^2 + (\sigma_{III} - \sigma_I)^2]}{2}} = S_y$$

$$\text{[Eq. 12.19]}$$

Substituting the stresses of *Equation 12.18* into *Equation 12.19*, the required thickness T of the sphere is:

$$T = \frac{pR}{2S_y} \qquad \text{[Eq. 12.20]}$$

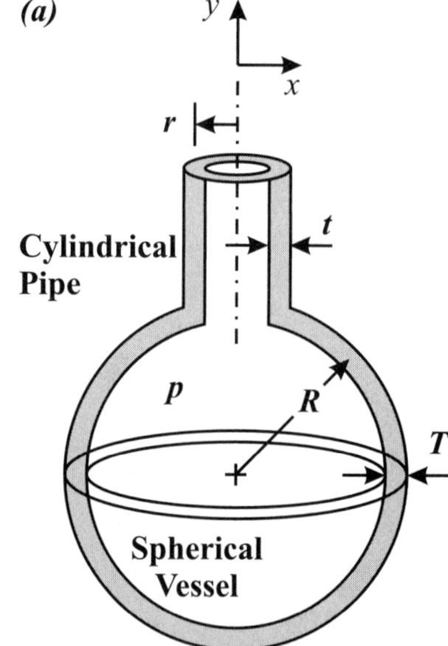

Figure 12.15. (a) Cut-away view of a pipe attached to a spherical pressure vessel. The pipe axis is along the *y*-axis. Not to scale.

Design of the Cylindrical Pipe

The biaxial state of stress in the pipe – a cylindrical pressure vessel – is given by:

$$\sigma_I = \frac{pr}{t}; \quad \sigma_{II} = \frac{pr}{2t}; \quad \sigma_{III} = 0 \qquad \text{[Eq. 12.21]}$$

The required thickness for the pipe using the von Mises condition (*Equation 12.19*) is:

$$t = \frac{\sqrt{3}\,pr}{2\,S_y} \qquad \text{[Eq. 12.22]}$$

The required thicknesses of the spherical vessel and the cylindrical pipe are now known.

Design of the Vessel-Pipe Joint

Consider equilibrium of point B at the intersection of the sphere and the cylinder (*Figure 12.15b*). The angle θ is defined by:

$$\sin \theta = \frac{r}{R} \qquad \text{[Eq. 12.23]}$$

The line load distributions (force per unit length) at the vessel–pipe intersection (*Figure 12.15c*) are:

$$f_v = \sigma_s T = \frac{pR}{2} \qquad \text{[Eq. 12.24]}$$

$$f_p = \sigma_L t = \frac{pr}{2} \qquad \text{[Eq. 12.25]}$$

where f_v is the force per unit length in the spherical vessel and f_p is the force per unit length in the pipe.

Resolving the line loads into the x- and y-directions (*Figure 12.15c*):

$$f_{vx} = \frac{pR}{2}\cos\theta; \quad f_{vy} = \frac{-pR}{2}\sin\theta$$

$$\text{[Eq. 12.26]}$$

$$f_{px} = 0; \quad f_{py} = \frac{pr}{2} \qquad \text{[Eq. 12.27]}$$

Summing the line loads in the y-direction (axial direction of the pipe) gives:

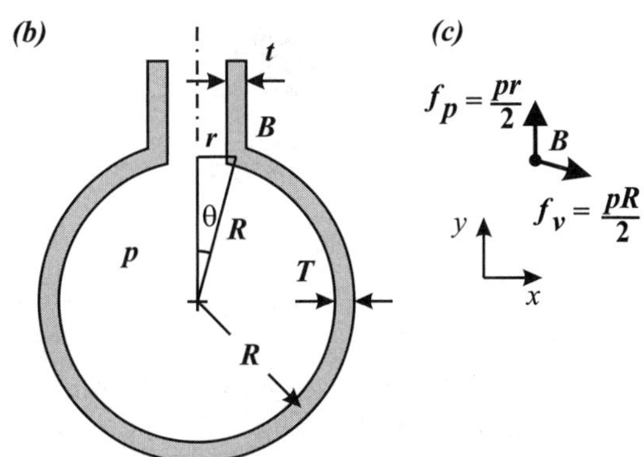

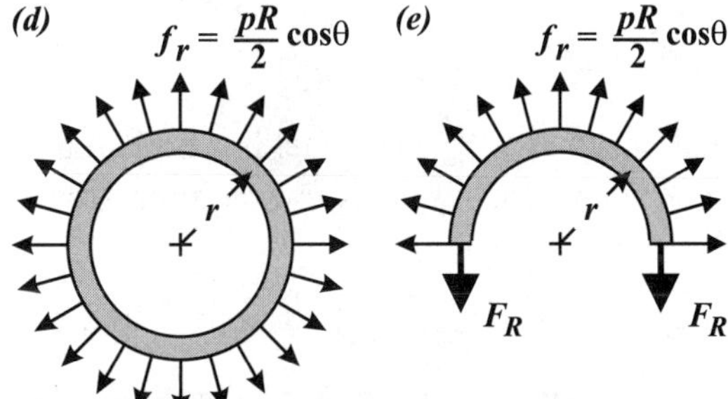

Figure 12.15. (b) Geometry of vessel–pipe intersection. **(c)** Force per unit length at point B, at the vessel–pipe intersection. **(d)** View along axis of pipe showing radial force per unit length at the joint exerted on the pipe due to pressure in the spherical vessel. **(e)** FBD of half-pipe at intersection. The force per unit length exerted on the pipe by the pressure vessel is resisted by tensile forces F_R in the reinforcing ring.

$$\sum f_y = \frac{pr}{2} - \frac{pR}{2}\sin\theta = \frac{pr}{2} - \frac{pR}{2}\left(\frac{r}{R}\right) = 0 \qquad\qquad \text{[Eq. 12.28]}$$

Equilibrium in the y-direction is satisfied.

Summing loads in the x-direction or radial (r-) direction of the pipe (*Figure 12.15c*):

$$\sum f_x = \sum f_r = \frac{pR}{2}\cos\theta \qquad\qquad \text{[Eq. 12.29]}$$

When the cylinder and sphere are brought together, equilibrium in the x-direction is not satisfied – the stresses in the sphere act to stretch the pipe radially at the vessel-pipe interface, expanding the hole (*Figure 12.15d*). To satisfy equilibrium, an equal and opposite horizontal force must be supplied by a reinforcement, a ring of radius r, and cross-sectional area A.

The tensile force F_R supported by the ring reinforcement at the intersection is found by applying equilibrium to *Figure 12.15e*:

$$2F_R = 2r\left(\frac{pR}{2}\cos\theta\right) \qquad\qquad \text{[Eq. 12.30]}$$

$$F_R = \frac{prR}{2}\cos\theta \qquad\qquad \text{[Eq. 12.31]}$$

Assume that the ring is manufactured from the same material as the pipe and the vessel, with yield strength S_y. The cross-sectional area A of the reinforcing ring (*Figure 12.15f*) is then defined by its axial limit load F_L :

$$F_L = AS_y = F_R \qquad \text{[Eq. 12.32]}$$

with $pR/2T = S_y$ at yield. Equating the expressions for F_R gives the required area A of the reinforcing ring:

$$A = rT\cos\theta \qquad \text{[Eq. 12.33]}$$

In practice, the radius of the pipe is much smaller than that of the spherical vessel so that $\cos\theta \sim 1.0$. Or, the maximum possible value of $\cos\theta$ is 1.0. Either way, the required area is:

$$A = rT \qquad \text{[Eq. 12.34]}$$

The total ring area at any cross-section of the interface is $2A = 2rT$. This value is equal to the **projected area** cut through the spherical vessel (thickness T) by the pipe (projected length $2r$, *Figure 12.15f*). This simple design procedure is known as the **area replacement method**.

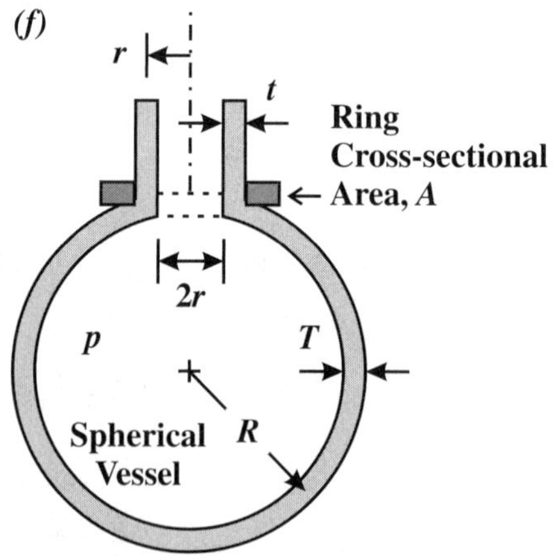

Figure 12.15. **(f)** The reinforcing ring of cross-sectional area A has the same area as the projected area of the vessel wall that was cut away to attach the pipe.

Use of Limit Surface in the ASME Pressure Vessel Code

The ASME design codes are based on the concept of the limit load. The ability of a material to redistribute stresses by yielding allows higher loads to be supported than those determined using elastic-only analysis. But because most stresses are determined using elastic calculations, the question arises on *how to interpret the results of elastic calculations in terms of the limit*

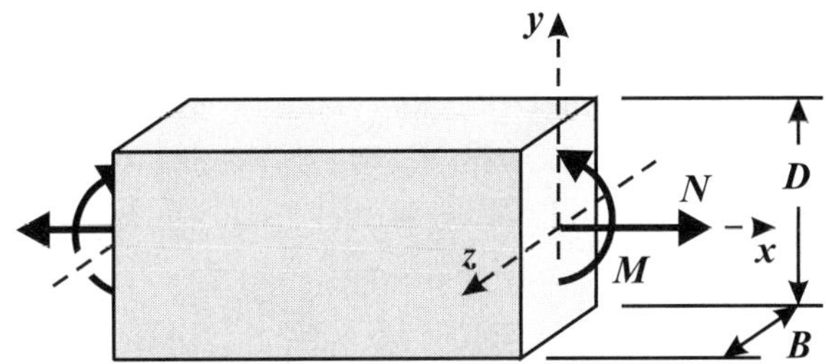

Figure 12.16. Rectangular beam subjected to axial load *N* and bending moment *M*.

surface (*Figure 12.14*). The following approach, pioneered by ASME, illustrates how to integrate plasticity concepts using elastic calculations.

For a beam subjected to axial load *N* and bending moment *M* (*Figure 12.16*), the limit surface is given by *Equation 12.16* and shown in *Figure 12.14*. The equation of the limit surface is repeated here:

$$\frac{M}{M_L} = 1 - \left(\frac{N}{N_L}\right)^2 \qquad \text{[Eq. 12.35]}$$

For a rectangular beam subjected to axial load *N* and bending moment *M* (*Figure 12.16*), an *elastic analysis* gives the axial stress σ_a and bending stress σ_b :

$$\sigma_a = \frac{N}{BD} \quad \text{and} \quad \sigma_b = \frac{6M}{BD^2} \qquad \text{[Eq. 12.36]}$$

The limit loads for axial- and bending-only loading are:

$$N_L = BDS_y \quad \text{and} \quad M_L = \frac{BD^2 S_y}{4} \qquad \text{[Eq. 12.37]}$$

Substituting the expressions for *N*, *M*, N_L, and M_L into *Equation 12.35* gives:

$$\frac{\sigma_b}{S_y} = \frac{3}{2}\left[1 - \left(\frac{\sigma_a}{S_y}\right)^2\right] \qquad \text{[Eq. 12.38]}$$

The plot of the parabolic limit condition in a rectangular beam in terms of the *elastic* axial and bending stresses, σ_a and σ_b, normalized by yield strength S_y, is shown in *Figure 12.17*. This is the *limit surface* in terms of the elastically calculated stress.

When an *elastic calculation* of the stresses is made, the point determined by the elastic stresses (σ_a, σ_b) must fall inside the limit surface if the system is not to reach its limit load. When the calculated elastic stresses lie outside the limit surface, the system has exceeded the limit condition.

Note that the normalized elastic bending stress, σ_b/S_y, can exceed unity (1.0) and still be within the limit surface. For bending-only, the calculated elastic bending stress on a rectangular cross-section at the limit moment is $\sigma_b = 1.5S_y$.

For a design tool, the limit surface is simplified further into two straight lines that include a factor of safety of 1.5 (*Figure 12.17*). The elastically calculated stresses are then limited by two conditions. The first is a line of slope –1, defined by:

$$\frac{\sigma_a}{S_y} + \frac{\sigma_b}{S_y} = 1 \qquad \text{[Eq. 12.39]}$$

The second condition is a vertical line given by:

$$\frac{\sigma_a}{S_y} = \frac{2}{3} \qquad \text{[Eq. 12.40]}$$

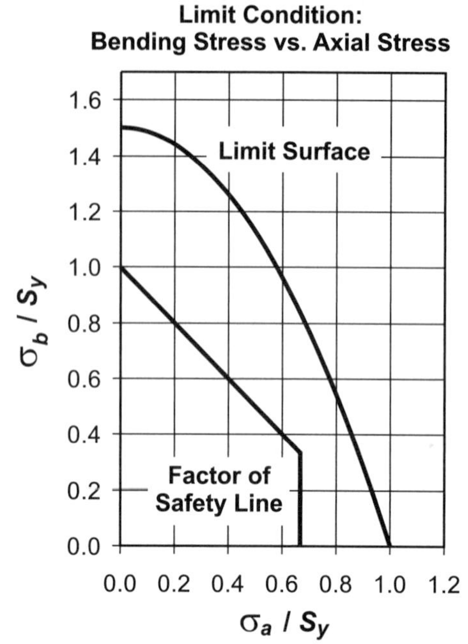

Figure 12.17. Limit condition for rectangular beam using elastic stress calculations for axial stress σ_a and bending stress σ_b, in first quadrant of σ_b–σ_a plot. The straight lines are typical factor of safety lines.

Example 12.10 Pressure Vessel with Tensile and Bending Stresses

Given: An elastic analysis of the stresses at the inside and outside surfaces of a pressure vessel with a thick wall are computed to be $\sigma_1 = 200\ \text{MPa}$ and $\sigma_2 = -100\ \text{MPa}$ (*Figure 12.18a*). The yield strength of the material is $S_y = 200\ \text{MPa}$.

Required: Determine if the ASME Code conditions, including the factor of safety, are satisfied.

Solution: The elastic stress at any point in a system subjected to axial forces and bending moments is the sum of the axial stress (constant over the cross-section) and the bending stress (linear over the cross-section), as shown in *Figure 12.18b*. For elements 1 and 2, at opposite sides of the pressure vessel wall, the stresses are:

$$\sigma_1, \sigma_2 = \sigma_a \pm \sigma_b$$

where σ_a is the *average* or *membrane stress*:

$$\sigma_a = \frac{\sigma_1 + \sigma_2}{2} = \frac{200 + (-100)}{2} = 50\ \text{MPa}$$

and σ_b is the maximum bending stress:

$$\sigma_b = \frac{\sigma_1 - \sigma_2}{2}$$

$$= \frac{200 - (-100)}{2} = 150 \text{ MPa}$$

Checking the conditions from *Equations 12.39* and *12.40*:

Condition 1:

$$\frac{\sigma_a}{S_y} + \frac{\sigma_b}{S_y} = \frac{50}{200} + \frac{150}{200} = 1.0 \le 1.0$$

OK

Condition 2:

$$\frac{\sigma_a}{S_y} = \frac{50}{200} = \frac{1}{4} \le \frac{2}{3} \quad \textbf{OK}$$

Since both conditions are satisfied, the design meets the code requirement.

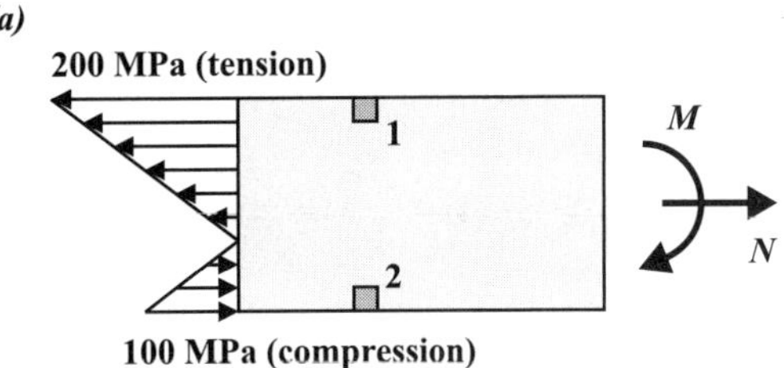

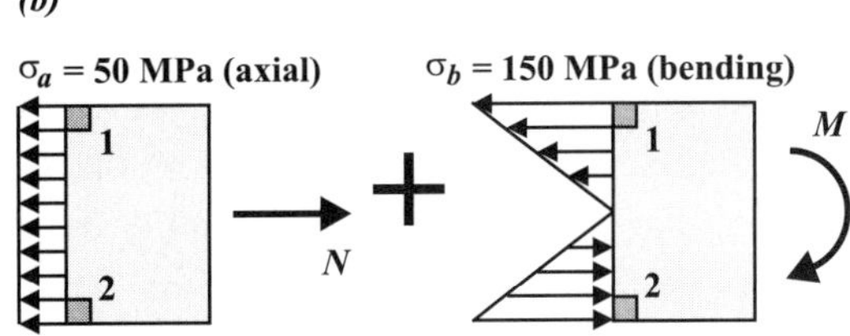

Figure 12.18. (a) Stress state developed in rectangular beam due to axial load *N* and bending moment *M*. (b) Stresses resolved into those due to axial loading only, and those due to bending only.

12.6 Three-Dimensional Plasticity

When an element is subjected to the three principal stresses σ_I, σ_{II}, σ_{III} (*Figure 12.19a*), the *von Mises stress* or *equivalent stress* is:

$$\sigma_o = \left[\frac{(\sigma_I - \sigma_{II})^2 + (\sigma_{II} - \sigma_{III})^2 + (\sigma_{III} - \sigma_I)^2}{2} \right]^{1/2}$$

[Eq. 12.41]

Yielding occurs when the von Mises stress equals the yield strength:

$$\sigma_o = S_y$$

[Eq. 12.42]

For the standard uniaxial tension test, $\sigma_{II} = \sigma_{III} = 0$ (*Figure 12.19b*), and the yield condition reduces to $\sigma_I = S_y$, which is consistent with the original definition of the yield strength.

During plastic deformation, experiments show that the *volumetric strain* is zero; the material volume does not change. The plastic volumetric strain is the sum of the plastic normal strains:

$$\varepsilon_{p1} + \varepsilon_{p2} + \varepsilon_{p3} = 0$$

[Eq. 12.43]

For a perfectly plastic material, the plastic strains in each principal direction due to the principal stresses can be described by:

$$\varepsilon_{p1} = \lambda\left[\sigma_I - \left(\frac{\sigma_{II} + \sigma_{III}}{2}\right)\right]$$

$$\varepsilon_{p2} = \lambda\left[\sigma_{II} - \left(\frac{\sigma_{III} + \sigma_I}{2}\right)\right] \qquad \text{[Eq. 12.44]}$$

$$\varepsilon_{p3} = \lambda\left[\sigma_{III} - \left(\frac{\sigma_I + \sigma_{II}}{2}\right)\right]$$

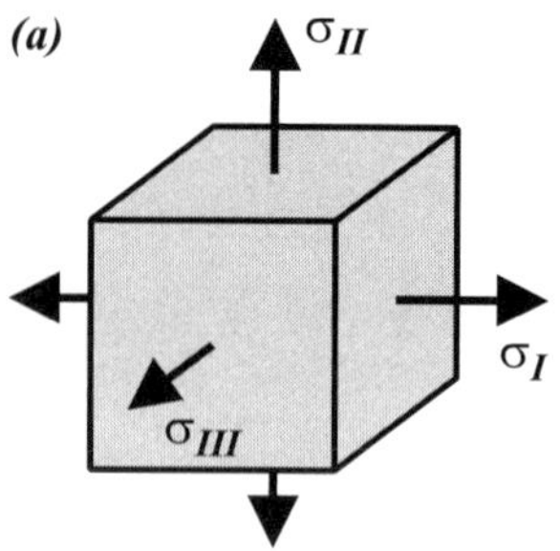

where the value of λ defines the magnitude of the strains applied to the material. The terms in the parentheses correspond to the Poisson effect with $\nu = 0.5$, the value of Poisson's ratio during plastic deformation.

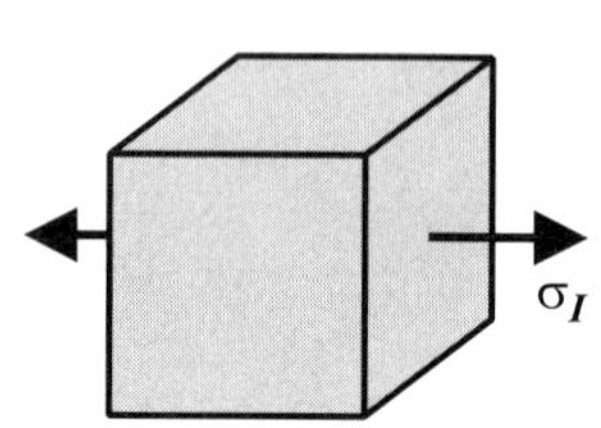

The applied stresses are generally known, so once the plastic strain in any direction is measured, λ is determined. With λ known, the other plastic strains can be calculated.

Taking the difference of the plastic strains gives:

$$\varepsilon_{p1} - \varepsilon_{p2} = \frac{3\lambda}{2}(\sigma_I - \sigma_{II})$$

$$\varepsilon_{p2} - \varepsilon_{p3} = \frac{3\lambda}{2}(\sigma_{II} - \sigma_{III}) \qquad \text{[Eq. 12.45]}$$

$$\varepsilon_{p3} - \varepsilon_{p1} = \frac{3\lambda}{2}(\sigma_{III} - \sigma_I)$$

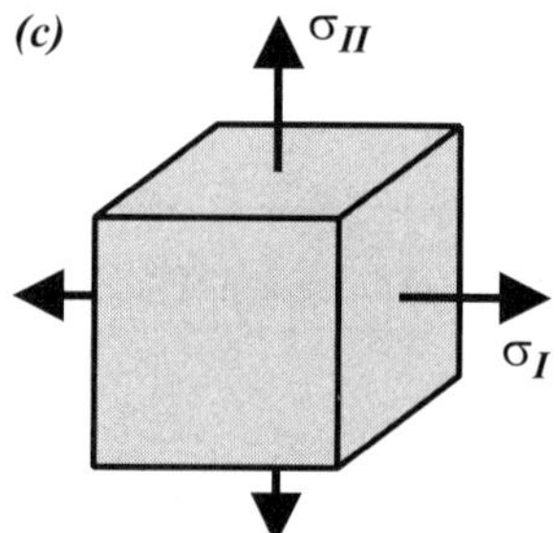

Yield Conditions for Plane Stress

In *plane stress*, the out-of-plane principal stresses is zero, e.g., $\sigma_{III} = 0$ (*Figure 12.19c*). Thus, the plane–stress yield condition reduces to:

$$[\sigma_I^2 - \sigma_I\sigma_{II} + \sigma_{II}^2]^{1/2} = S_y \qquad \text{[Eq. 12.46]}$$

and the plastic strains reduce to:

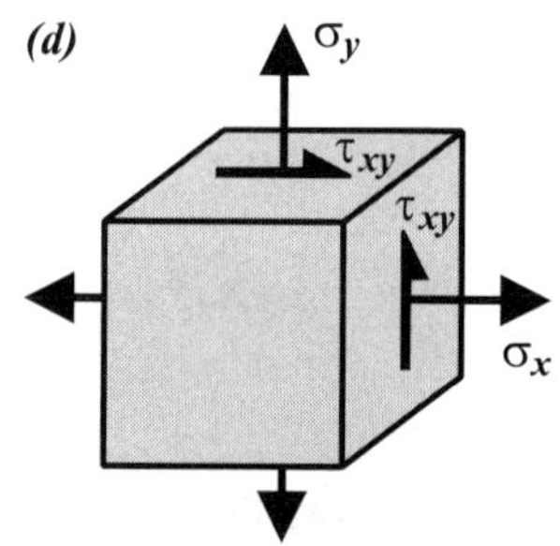

$$\varepsilon_{p1} = \lambda\left[\sigma_I - \frac{\sigma_{II}}{2}\right]$$

$$\varepsilon_{p2} = \lambda\left[\sigma_{II} - \frac{\sigma_I}{2}\right] \qquad \text{[Eq. 12.47]}$$

$$\varepsilon_{p3} = -\lambda\left[\frac{\sigma_I + \sigma_{II}}{2}\right]$$

Figure 12.19. **(a)** A 3D element subjected to principal stresses. **(b)** Element subjected to uniaxial stress. **(c)** Element subjected to plane stress principal stresses. **(d)** Element subjected to a general state of plane stress.

General Plane Stress State

In the x–y plane, the stresses are σ_x, σ_y, and τ_{xy}; the out-of-plane stresses are $\sigma_z = \tau_{yz} = \tau_{zx} = 0$ (*Figure 12.19d*). The in-plane principal stresses are:

$$\sigma_I = \frac{\sigma_x + \sigma_y}{2} + \sqrt{\left(\frac{\sigma_x - \sigma_y}{2}\right)^2 + \tau_{xy}^2}$$

$$\sigma_{II} = \frac{\sigma_x + \sigma_y}{2} - \sqrt{\left(\frac{\sigma_x - \sigma_y}{2}\right)^2 + \tau_{xy}^2}$$

[Eq. 12.48]

Substituting these expressions into the von Mises yield criterion (*Equation 12.46*) gives:

$$\left[\left(\frac{\sigma_x + \sigma_y}{2}\right)^2 + 3\left[\left(\frac{\sigma_x - \sigma_y}{2}\right)^2 + \tau_{xy}^2\right]\right]^{1/2} = S_y$$

[Eq. 12.49]

Applying the strain transformation equations, the corresponding plastic strains are:

$$\varepsilon_{x,p} = \lambda\left[\sigma_x - \frac{\sigma_y}{2}\right]$$

$$\varepsilon_{y,p} = \lambda\left[\sigma_y - \frac{\sigma_x}{2}\right]$$

$$\gamma_{xy,p} = 3\lambda\tau_{xy}$$

[Eq. 12.50]

Uniaxial Stress

In the uniaxial test, the only non-zero stress is σ_I (*Figure 12.19b*). The plastic normal strains are then:

$$\varepsilon_{p1} = \lambda\sigma_I$$

$$\varepsilon_{p2} = \varepsilon_{p3} = \frac{-\lambda}{2}\sigma_I$$

[Eq. 12.51]

which conform to experimental observations that the volumetric change during plastic deformation is zero:

$$\varepsilon_{p1} + \varepsilon_{p2} + \varepsilon_{p3} = 0$$

[Eq. 12.52]

Example 12.11 Over-Pressurized Pipe

Given: A pipe of radius r and thickness t ($t \ll r$, *Figure 12.20*) has been accidentally over-pressurized (it happens!). In an after-event investigation, the plastic strain in the hoop (circumferential) direction is determined to be $\varepsilon_{H,p} = 4.0\%$. The yield strength is S_y.

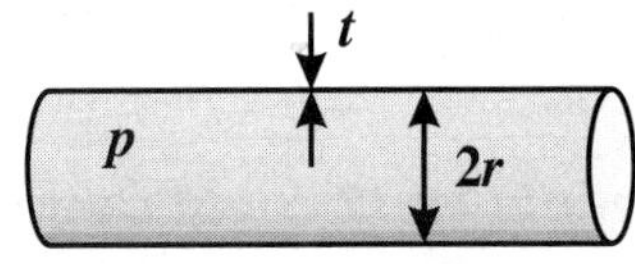

Figure 12.20. Pressurized pipe.

Required: Determine (a) the pressure to yield the pipe and (b) the change in thickness of the pipe due to plastic deformation corresponding to $\varepsilon_{H,p}$.

Solution: The principal stresses in the pipe are:

$$\sigma_H = \frac{pr}{t}; \quad \sigma_L = \frac{pr}{2t} \; ; \; \sigma_{III} = 0$$

Applying the plane–stress yield condition (*Equation 12.46*):

$$\left[\frac{1}{2}\left[\left(\frac{pr}{t} - \frac{pr}{2t}\right)^2 + \left(\frac{pr}{2t} - 0\right)^2 + \left(0 - \frac{pr}{t}\right)^2\right]\right]^{1/2} = \frac{\sqrt{3}}{2}\frac{pr}{t} = S_y$$

so, the pressure at yield is:

$$\text{Answer: } p_y = \frac{2}{\sqrt{3}}\frac{t}{r}S_y$$

The plastic strain in the hoop direction is:

$$\varepsilon_{H,\,p} = \lambda\left[\sigma_H - \frac{1}{2}(\sigma_L + 0)\right] = \lambda\left[\frac{p_y r}{t} - \frac{1}{2}\left(\frac{p_y r}{2t} + 0\right)\right] = \lambda\frac{\sqrt{3}}{2}S_y$$

Since the measured value of the hoop strain is $\varepsilon_{H,p} = 0.04 = 4.0\%$, then:

$$\lambda = \frac{2}{\sqrt{3}}\frac{\varepsilon_{H,\,p}}{S_y} = \frac{2}{\sqrt{3}}\frac{(0.04)}{S_y}$$

Hence, the longitudinal and radial strains are:

$$\varepsilon_{L,\,p} = \lambda\left[\frac{p_y r}{2t} - \frac{1}{2}\left(\frac{p_y r}{t} + 0\right)\right] = 0$$

$$\varepsilon_{r,\,p} = \lambda\left[0 - \frac{1}{2}\left(\frac{p_y r}{t} + \frac{p_y r}{2t}\right)\right] = -\lambda\frac{3pr}{4t} = -\left[\frac{2}{\sqrt{3}}\frac{(0.04)}{S_y}\right]\left(\frac{3}{4}\right)\left(\frac{2}{\sqrt{3}}S_y\right) = -0.04$$

The change in thickness is:

$$\text{Answer: } \Delta t = \varepsilon_{r,\,p}t = -0.04t \text{ or } -4.0\%$$

The pipe does not elongate during plastic deformation ($\varepsilon_{L,p} = 0$), which is beneficial since the end-supports of the pipe will not be disturbed. The increase in radius (positive hoop strain) is taken up by the decrease in thickness of the pipe wall (negative radial-strain). The net change in material volume is zero, i.e., $\Sigma\varepsilon_i = 0$.

Example 12.12 **Yield Condition of a Pressurized Pipe under Torque**

Given: A pipe of radius r and thickness t ($t \ll r$), contains a gas at pressure p (*Figure 12.21*). Due to the movements of the supports, it is also subjected to a torque T; the pipe acts as a thin-walled shaft when subjected to torque. The yield strength is S_y .

Required: Determine the torque to cause yielding T_y as a function of pressure p.

Solution: The stresses due to the pressure are:

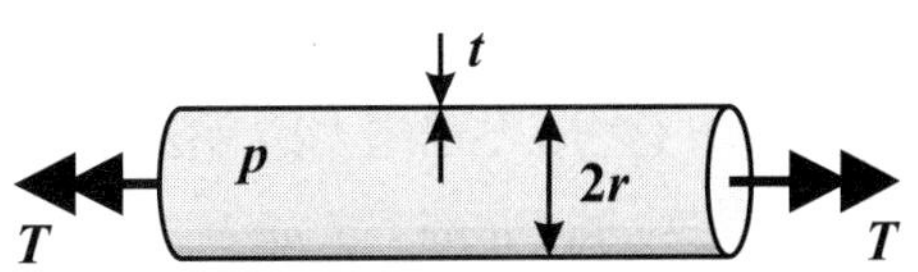

Figure 12.21. Pressurized pipe subjected to torque.

$$\sigma_H = \frac{pr}{t} \quad \text{and} \quad \sigma_L = \frac{pr}{2t}$$

The shear stress due to the torque on a thin-walled shaft is:

$$\tau = \frac{T}{2\pi r^2 t}$$

Due to the shear stress, the hoop and longitudinal stresses are no longer the principal stresses. The general plane stress yield condition (*Equation 12.49*) must be applied:

$$\left[\frac{1}{4}\left(\frac{pr}{t} + \frac{pr}{2} \right)^2 + 3\left[\frac{1}{4}\left(\frac{pr}{t} - \frac{pr}{2t} \right)^2 + \left(\frac{T}{2\pi r^2 t} \right)^2 \right] \right]^{1/2}$$

$$= \left[\left(\frac{3pr}{4t} \right)^2 + 3\left[\left(\frac{pr}{4t} \right)^2 + \left(\frac{T}{2\pi r^2 t} \right)^2 \right] \right]^{1/2} = S_y$$

The torque to cause yielding is then:

$$Answer: \quad \frac{T}{S_y 2\pi r^2 t} = \sqrt{\frac{1}{3} - \frac{1}{4}\left(\frac{pr}{S_y t} \right)^2}$$

The normalized torque T^* is plotted against the normalized pressure p^* in *Figure 12.22*, where:

$$T^* = \frac{T}{S_y 2\pi r^2 t} \quad \text{and} \quad p^* = \frac{pr}{S_y t}$$

The solid line defines the yield surface.

As an example, if a *Factor of Safety* of 1.5 has been applied to the hoop stress, then:

$$\frac{pR}{t} = \frac{2}{3}S_y \quad \text{so} \quad p^* = 0.67$$

For this pressure, the yield torque of the pipe is:

$$T_y^* = \frac{T}{S_y 2\pi r^2 t} = \sqrt{\frac{1}{3} - \frac{1}{9}} = 0.47$$

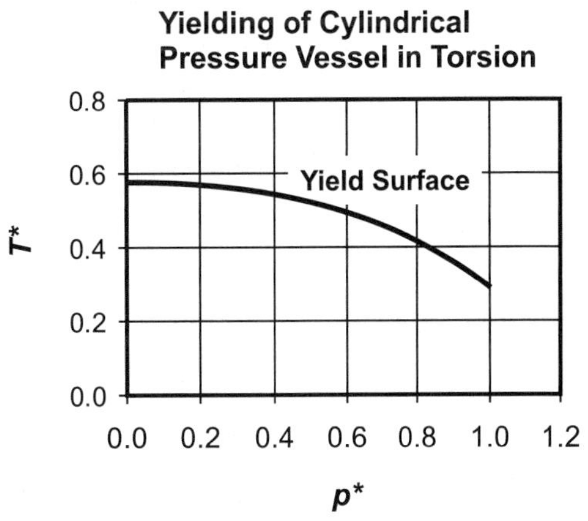

Figure 12.22. Torque versus pressure yield surface.

Example 12.13 Solid Circular Shaft Subjected to Torque and Axial Load

Given: A solid circular shaft of radius R supports torque T and axial force P (*Figure 12.23*). Force P is tensile ($P > 0$) or compressive ($P < 0$). The yield strength in tension or in compression is S_y and the shear yield strength is $\tau_y = S_y / \sqrt{3}$.

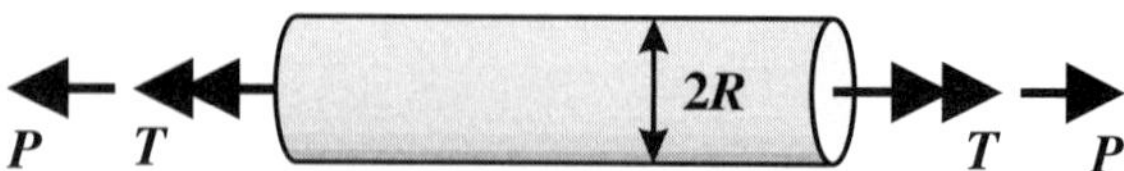

Figure 12.23. Solid circular shaft subjected to torque and axial load.

Required: Determine the load combination, P and T, that defines the limit load condition.

Solution: The axial stress is:

$$\sigma = \frac{P}{\pi R^2}$$

Taking the shear stress as constant over the cross-section at the limit load, the torque is:

$$T = \int_0^R [2\pi r^2 \, dr]\tau = \frac{2\pi R^3}{3}\tau$$

so:
$$\tau = \frac{3T}{2\pi R^3}$$

With $\sigma_x = \sigma$ and $\sigma_y = 0$, and $\tau_{xy} = \tau$, the plane stress von Mises yield criterion (*Equation 12.49*) reduces to:

$$\sigma^2 + 3\tau^2 = S_y^2$$

or

$$\frac{1}{S_y^2}\left(\frac{P}{\pi R^2}\right)^2 + \frac{3}{S_y^2}\left(\frac{3T}{2\pi R^3}\right)^2 = 1$$

The force-only limit load (i.e., $T = 0$) is:

$$P_L = S_y \pi R^2$$

and the torque-only limit torque ($P = 0$) is:

$$T_L = \frac{2\pi R^3}{3}\tau_y = \frac{2\pi R^3}{3}\frac{S_y}{\sqrt{3}}$$

Substituting P_L and T_L into the yield equation gives the limit load condition:

$$Answer: \left(\frac{P}{P_L}\right)^2 + \left(\frac{T}{T_L}\right)^2 = 1$$

Since the force and torque terms are squared, their applied directions do not matter – the force may be tensile or compressive, the torque may be positive or negative.

Yield Conditions for Plane Strain

In *plane strain*, the out-of-plane strain is zero, $\varepsilon_3 = 0$. From *Equation 12.44*:

$$\varepsilon_{p3} = \lambda\left[\sigma_{III} - \left(\frac{\sigma_I + \sigma_{II}}{2}\right)\right] = 0 \quad \text{[Eq. 12.53]}$$

Hence, the out-of-plane stress must be:

$$\sigma_{III} = \frac{\sigma_I + \sigma_{II}}{2} \quad \text{[Eq. 12.54]}$$

To enforce the plane strain condition during yielding, the out-of-plane normal stress is the average of the other two stresses (*Figure 12.24*). Again, this corresponds to a Poisson's ratio during plastic deformation of $v = 0.5$.

Substituting *Equation 12.54* into the general 3D yield condition – *Equation 12.41* and *Equation 12.42* – gives the yield condition:

$$\frac{\sqrt{3}}{2}(\sigma_I - \sigma_{II}) = \pm S_y \quad \text{[Eq. 12.55]}$$

The solutions to this condition are shown in *Figure 12.25*. Within the shaded region, no yielding occurs.

If σ_I and σ_{II} are of the same sign (and hence σ_{III} is of the same sign), the principal stresses can be very much larger than the value of S_y. It is the *difference in the principal stresses*, σ_I and σ_{II}, that causes yielding. An example of extremely large principal stresses occurs in mountains, where earth is subjected to high hydrostatic compressive stresses, but does not yield.

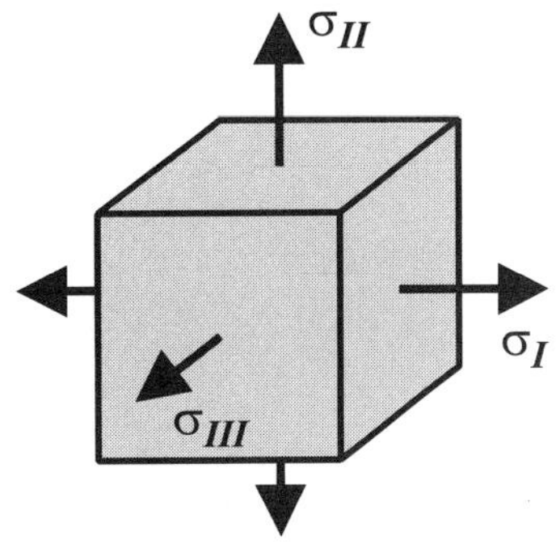

Figure 12.24. Plane strain element.

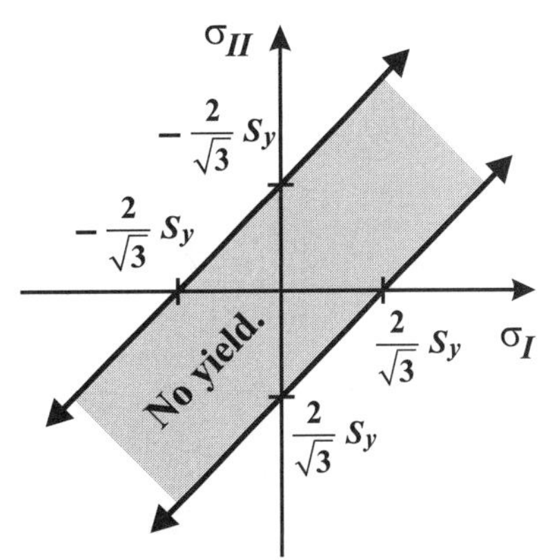

Figure 12.25. Yield condition for plane strain. Yielding does not occur if the principal stresses are within the region bounded by the two lines.

The principal strains for the plane strain case are:

$$\varepsilon_{p1} = \frac{3\lambda}{4}[\sigma_I - \sigma_{II}]$$

$$\varepsilon_{p2} = -\frac{3\lambda}{4}[\sigma_I - \sigma_{II}] \quad \text{[Eq. 12.56]}$$

$$\varepsilon_{p3} = 0$$

General Stress State

If a plane strain element is subjected to stresses σ_x, σ_y, and τ_{xy}, the yield condition gives:

$$[(\sigma_x - \sigma_y)^2 + 4\tau_{xy}^2]^{1/2} = \pm\frac{2}{\sqrt{3}}S_y \qquad \text{[Eq. 12.57]}$$

The corresponding strains are:

$$\varepsilon_x = \frac{3\lambda}{4}[\sigma_x - \sigma_y]$$

$$\varepsilon_y = -\frac{3\lambda}{4}[\sigma_x - \sigma_y] \qquad \text{[Eq. 12.58]}$$

$$\gamma_{xy} = 3\lambda\tau_{xy}$$

Recall that the maximum magnitude of the in-plane shear stress is:

$$\tau_{max} = \frac{\sigma_I - \sigma_{II}}{2} \qquad \text{[Eq. 12.59]}$$

and the yield condition (*Equation 12.55*) becomes two straight lines (*Figure 12.25*):

$$\frac{\sqrt{3}}{2}(\sigma_I - \sigma_{II}) = \pm\sqrt{3}\tau_{max} = \pm S_y \qquad \text{[Eq. 12.60]}$$

Example 12.14 Punch Load on a Half-Space

Given: A hard rectangular punch of width d and infinite depth (into the paper) is pushed into a half-space volume made of an elastic–perfectly plastic material (*Figure 12.26a*). This situation occurs in practice as the Brinell hardness test to measure the hardness of a material. In the model, the punch and material are assumed to be infinitely thick (into the paper), so the system is a 2D *plane strain* problem. The term *half-space* implies the material is infinitely thick into the paper and infinitely deep below the plane surface.

Required: Using approximate methods, determine the value of the punch load per unit depth into the plane of the paper, P_L, that is required to indent (yield) the material.

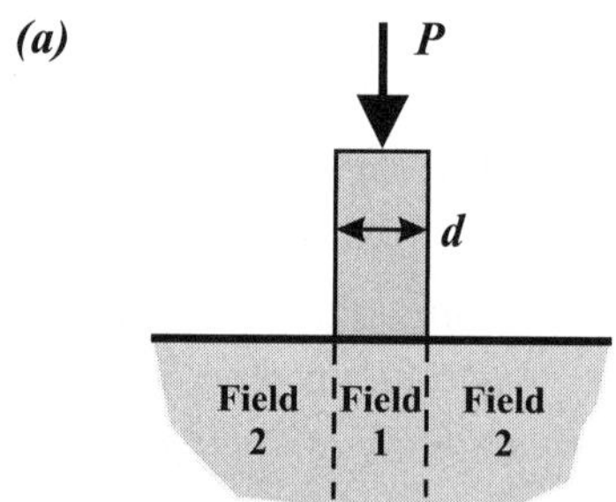

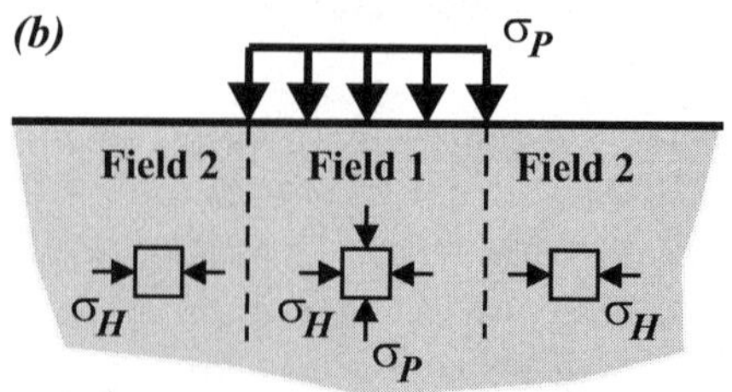

Figure 12.26. **(a)** Rectangular punch bearing on Field 1. **(b)** Approximate stresses in the three fields.

Solution:

Force Method to Determine a Lower Bound

In the force method, an equilibrium stress field is guessed and a lower bound to the load is found.

As a first approximation, the half-space is broken into three fields, Field 1, directly under the punch, and Fields 2, on either side of Field 1 (*Figure 12.26b*). The shear stress in all parts of the half space is assumed to be zero.

Under the punch, in Field 1, a stress element is subjected to the punch stress σ_P. Such an element may also be subjected to a horizontal compressive stress σ_H. The stress on the surface above Field 2 is zero, so a stress element in Field 2 is only subjected to horizontal stress σ_H. Although this simplified stress distribution is unlikely the actual stress response, the stresses are in equilibrium, as required in the force method.

From the plane strain yield condition, an element in Field 2 yields when:

$$\sigma_H = \frac{2}{\sqrt{3}} S_y$$

In Field 1, the horizontal stress for equilibrium across the junction of the fields is σ_H, and the vertical stress from the punch is σ_P. Hence, yielding occurs in Field 1 when:

$$\sigma_I - \sigma_{II} = \sigma_P - \sigma_H = \frac{2}{\sqrt{3}} S_y$$

To yield the material, the applied stress must be:

$$\sigma_P = \sigma_H + \frac{2}{\sqrt{3}} S_y = \frac{4}{\sqrt{3}} S_y$$

or $$\sigma_P = 2.31 S_y$$

Since this is an equilibrium stress field that nowhere exceeds the yield condition ($\sigma_o = S_y$), the estimate for σ_P underestimates the required value of the punch load. However, the result for σ_P indicates that the plane strain condition increases the stress required to indent the material to values above the uniaxial yield strength, i.e., $\sigma_P > S_y$.

Displacement Method to Determine an Upper Bound

In the displacement method, a displacement field is guessed and an upper bound to the load is found. Equilibrium does not necessarily need to be satisfied.

Consider the material near the punch to be made of five rigid equilateral triangles A, B, C, D, and E, each of side d, shown in *Figure 12.26c*. The triangles are restricted to move by either slipping with respect to each other, or slipping along the stationary slip planes represented by the heavy lines in *Figure 12.26d*. The term *slip* is used to describe the physics of plastic deformation (beyond the scope of this text). Slip is caused by the shear stress on the plane reaching a critical value, i.e., $\tau = \tau_y$.

Due to punch stress σ_P, triangle A moves vertically downward distance Δ. To accommodate the movement of triangle A, triangles B and C move horizontally outward distance $\Delta/\sqrt{3}$. Triangles D and E move vertically and horizontally, respectally. Triangle D moves horizontally half as far as triangle B and vertically half as far as triangle A. Thus triangle D moves horizontally $\Delta/(2\sqrt{3})$ and vertically $\Delta/2$ for a net movement of $\Delta/\sqrt{3}$ along the 60° (heavy) slip lines.

The triangles move, or slip, when the shear stress between them equals to the shear yield strength:

$$\tau = \tau_y = \frac{S_y}{\sqrt{3}}$$

The shear force per unit depth (into the paper) on each side of the triangle during slip is then:

$$f = \frac{S_y d}{\sqrt{3}}$$

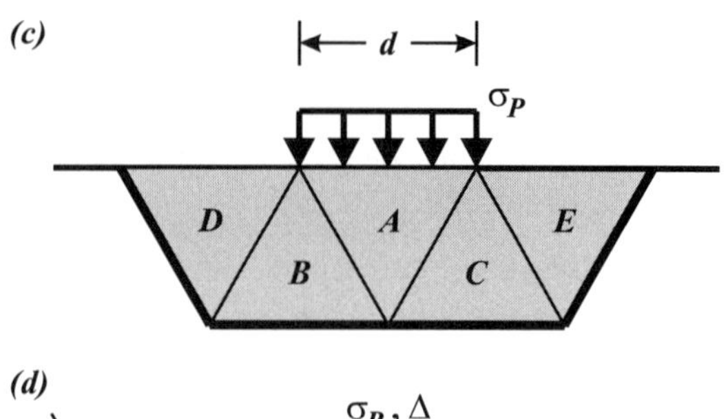

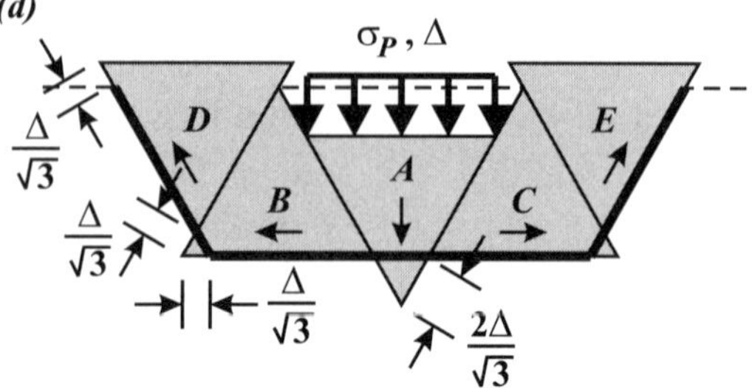

Figure 12.26. **(c)** Material near punch divided into five equilateral triangles. Slip is permitted parallel to the sides of the triangles. **(d)** Motion of triangles due to downward displacement Δ of triangle A.

To determine the energy dissipated as the triangles slip past each other, the relative movement of each pair of triangles must be determined. Also, the movement of the triangles along the stationary slip planes (the heavy lines in *Figure 12.26d*) must be determined. The relative movement between:

- Triangles A and B is $2\Delta/\sqrt{3}$.
- Triangle B and the stationary horizontal slip plane is $\Delta/\sqrt{3}$.
- Triangles B and D is $\Delta/\sqrt{3}$.
- Triangle D and the stationary 60° slip plane is $\Delta/\sqrt{3}$.

The movements of triangles B and D are mirrored by triangles C and E. Thus, the energy dissipated per unit depth by the movement of the triangles is:

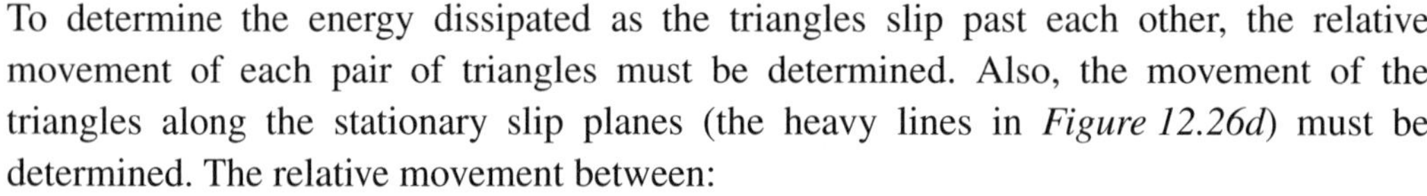

$$U_L = 2\left[\frac{S_y d}{\sqrt{3}}\left(\frac{2\Delta}{\sqrt{3}}\right) + \frac{S_y d}{\sqrt{3}}\left(\frac{\Delta}{\sqrt{3}}\right) + \frac{S_y d}{\sqrt{3}}\left(\frac{\Delta}{\sqrt{3}}\right) + \frac{S_y d}{\sqrt{3}}\left(\frac{\Delta}{\sqrt{3}}\right)\right] = \frac{10}{3}S_y\Delta d$$

The work done per unit depth by the punch in moving downward distance Δ during plastic deformation is:

$$W = (\sigma_p d)\Delta$$

Setting the applied work equal to the dissipated energy gives the stress to indent the material:

$$\sigma_P = \frac{10}{3}S_y = 3.33S_y$$

This stress overestimates the required indentation stress; it is an *upper bound.*

Lower and Upper Bounds on Load

Calculations based on the force (equilibrium) method predict a load less than the actual value (a lower bound). Calculations based on the displacement method predict a load greater than the actual value (an upper bound). In the case of the punch problem, the limit load per unit depth is bound by:

$$\textit{Answer: } 2.31S_y d < P_L < 3.33S_y d$$

More detailed finite element calculations show that P_L for a flat punch is approximately $3.0S_y d$. This simple analysis has bounded the load necessary to indent the plane strain material.

12.7　Cyclic Thermal Loading: Shakedown and Ratcheting

Loads – both mechanical and thermal – are often cyclic. Power plants (on earth and in space), engines, and electronic chips, are all subject to cyclic temperature loads. The purpose of this section is to determine the effect of cyclic thermal loading on the limit load ideas introduced earlier in this chapter.

For an elastic–perfectly plastic ductile system under static mechanical loading, the load at *first yield* F_y is the maximum load it can support before any material point yields. The *limit load* F_L is the maximum load the system can support. At the limit load, all material points have yielded, and unconstrained plastic deformation occurs.

In the case of *static mechanical loading* combined with *cyclic thermal loading*, the corresponding design concepts are *shakedown* and *ratcheting*. **Shakedown** is a condition reached after initial plastic straining produces a beneficial state of residual stress. After initial yielding, subsequent cyclic loading occurs entirely in the elastic range. When the loading exceeds the condition for *shakedown, ratcheting* occurs. **Ratcheting** is the accumulation of plastic strain with each thermal cycle. In microchip technology, this phenomenon is referred to as *thermal crawl.*

Two-Bar System Subjected to Constant Force and Thermal Cycling

Shakedown and *ratcheting* are illustrated by the two-bar system shown in *Figure 12.27a.* Bars 1 and 2 are constrained to elongate vertically by the same amount δ due to the rigid boss. The system is subjected to a constant mechanical load F. Bar 2 is subjected to a cyclic temperature load ΔT (*Figure 12.27b*). For this case, the bars are identical, with length L and cross-sectional area $A/2$. The material properties are modulus E, yield strength S_y, and thermal expansion coefficient α. The material properties are assumed to be constant with temperature, and the material is elastic–perfectly plastic. The goal is to investigate the response of the system to mechanical loading F and thermal loading ΔT.

Elastic Response without Thermal Load

Consider the case when only force F is applied. There is no thermal load (thermal load condition (**a**) in *Figure 12.27b*). The stresses in the bars, σ_1 and σ_2, must satisfy *equilibrium*:

$$F = \frac{A}{2}\sigma_1 + \frac{A}{2}\sigma_2 \qquad \text{[Eq. 12.61]}$$

Compatibility requires that the bars elongate by the same amount. Here, since the bars are the same length, their strains are always equal. For the elastic case:

$$\varepsilon_1 = \varepsilon_2$$

$$\frac{\sigma_1}{E} = \frac{\sigma_2}{E} \qquad \text{[Eq. 12.62]}$$

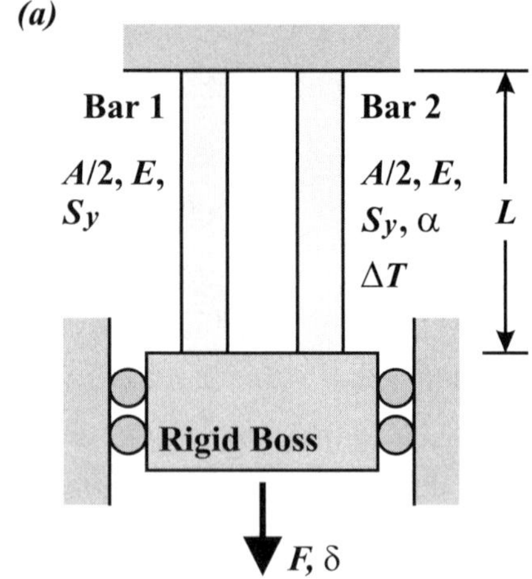

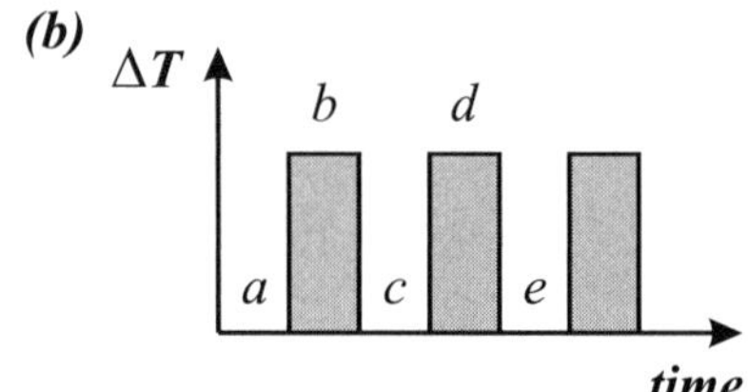

Figure 12.27. (a) Two-bar system subjected to constant load F. (b) Bar 2 is subjected to a cyclic thermal load ΔT.

Equation 12.62 requires that the stresses in the bars also be the same, so:

$$\sigma_1 = \sigma_2 = \frac{F}{A} \qquad \text{[Eq. 12.63]}$$

The system deflection is:

$$\delta = \frac{FL}{AE} \qquad \text{[Eq. 12.64]}$$

Elastic Response with Thermal Load

The application of thermal load ΔT to Bar 2 (thermal load (**b**) in *Figure 12.27b*), causes Bar 2 to elongate due to thermal expansion. The stresses in the bars must still satisfy *equilibrium*:

$$F = \frac{A}{2}\sigma_1 + \frac{A}{2}\sigma_2 \qquad \text{[Eq. 12.65]}$$

Compatibility requires that the strains in the bars always be the same:

$$\varepsilon_1 = \varepsilon_2$$

$$\frac{\sigma_1}{E} = \frac{\sigma_2}{E} + \alpha(\Delta T) \qquad \text{[Eq. 12.66]}$$

Bar 1 has purely elastic (mechanical) strain, while Bar 2 now has both elastic and thermal strains.

Combining *Equations 12.65* and *12.66*, the stresses are:

$$\sigma_1 = \frac{F}{A} + \frac{E\alpha(\Delta T)}{2}$$

$$\sigma_2 = \frac{F}{A} - \frac{E\alpha(\Delta T)}{2} \qquad \text{[Eq. 12.67]}$$

To maintain compatibility during thermal loading, the stress in Bar 1 increases, and that in Bar 2 decreases. To maintain equilibrium, the stresses must change by equal and opposite amounts – the magnitude of the ***thermal stress***, $E\alpha(\Delta T)/2$.

The deflection of the system equals the elastic deflection of Bar 1:

$$\delta = \frac{L}{E}\left[\frac{F}{A} + \frac{E\alpha(\Delta T)}{2}\right] \qquad \text{[Eq. 12.68]}$$

To simplify the algebra, the following dimensionless groups are introduced:

$$s_1 = \frac{\sigma_1}{S_y}; \quad s_2 = \frac{\sigma_2}{S_y}$$

$$p = \frac{F}{S_y A}; \quad t = \frac{E\alpha(\Delta T)}{2S_y} \qquad \text{[Eq. 12.69]}$$

The stresses of *Equation 12.67* can then be rewritten in their normalized form:

$$s_1 = p + t$$

$$s_2 = p - t \qquad \text{[Eq. 12.70]}$$

The stress–strain response in each bar due to thermal loading is shown in *Figure 12.28*; both bars start at point (*a*), and move to shaded points (*b*). Yielding does not occur as long as the normalized stress in Bar 1 is less than unity, $s_1 < 1.0$ (i.e., $\sigma_1 < S_y$).

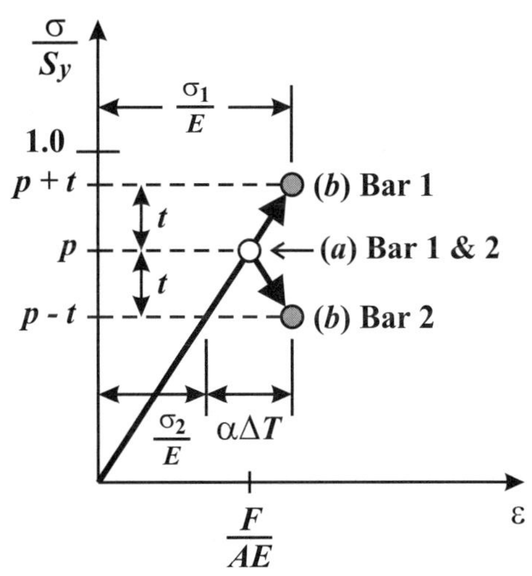

Figure 12.28. Stress–strain responses of bars due to mechanical load only – point (*a*), and due to mechanical and thermal loads – points (*b*). When thermal load is applied, the response of each bar moves from (*a*) to (*b*). At (*b*), Bar 1 has only mechanical strain, while Bar 2 has both mechanical and thermal strains.

Although the strains in the bars are the same, the strain is Bar 1 is purely elastic, while that in Bar 2 has both elastic and thermal components (*Figure 12.28*):

$$\varepsilon_1 = \frac{\sigma_1}{E} = \frac{p+t}{E}S_y$$

$$\varepsilon_2 = \frac{\sigma_2}{E} + \alpha(\Delta T) = \frac{p-t}{E}S_y + \frac{2t}{E}S_y$$

[Eq. 12.71]

Upon removal of the thermal load (to thermal loading (**c**) in *Figure 12.27b*), the stresses and strains of both bars return to their initial values $s = p = F/S_yA$ and $\varepsilon = F/AE$ (point (**a**), *Figure 12.28*). The cyclic thermal load can continue to be applied and the system will continue to remain elastic.

First Yield due to Thermal Load

Bar 1 just yields upon application of thermal load t if:

$$s_1 = p + t = 1.0 \qquad \text{[Eq. 12.72]}$$

as shown in *Figure 12.29*. For equilibrium, the stress is Bar 2 is then:

$$s_2 = p - t = 2p - 1 \qquad \text{[Eq. 12.73]}$$

For a given mechanical load p, the cyclic thermal load t to cause first yielding is, from *Equation 12.72*:

$$t_y = 1 - p \qquad \text{[Eq. 12.74]}$$

This condition is shown in *Figure 12.30* for the case being studied. Provided that the cyclic thermal load t is less than or equal to $1-p$, the system remains elastic upon further cyclic loading.

Elastic–Plastic Response

When the thermal load t equals t_y, Bar 1 reaches the yield condition. Using the ideal elastic–plastic model, the stress in Bar 1 cannot exceed the yield strength.

If the thermal load exceeds t_y $(t > t_y)$, Bar 2 continues to thermally expand, and Bar 1 must deform plastically (at $\sigma_1 = S_y$) to ensure that the elongations of the bars

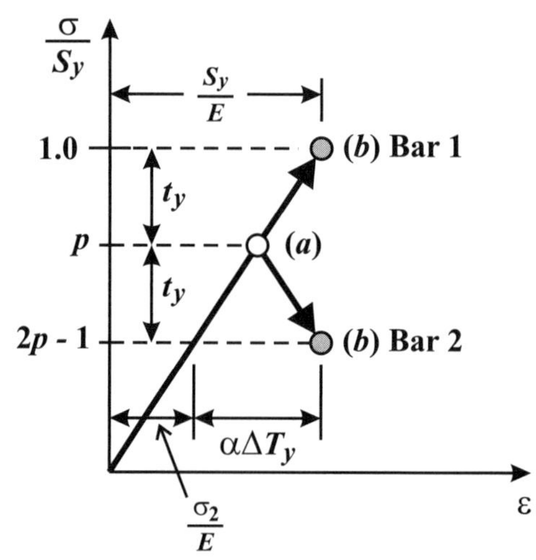

Figure 12.29. Stress–strain responses – point (**a**) to points (**b**) – for value of thermal load to cause Bar 1 just to yield $(t = t_y)$.

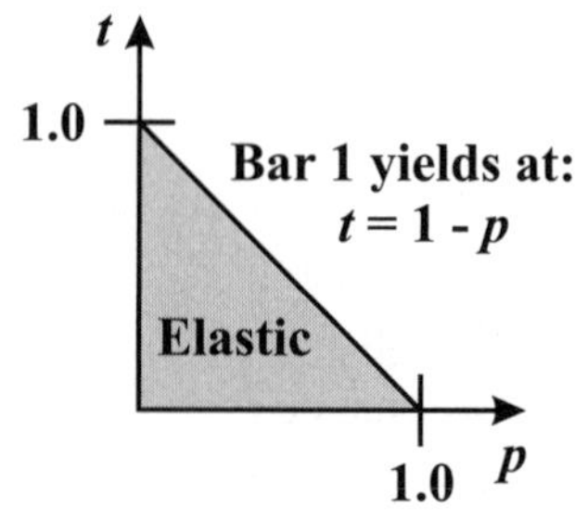

Figure 12.30. Thermal–Mechanical load map to ensure elastic conditions.

remain the same. This response is shown in *Figure 12.31* (point (**a**) to points (**b**)). The stresses are now:

$$s_1 = 1.0$$
$$s_2 = 2p - 1$$

[Eq. 12.75]

Since yielding occurs in Bar 1, its strain has elastic and plastic components:

$$\varepsilon_1 = \frac{S_y}{E} + \varepsilon_{p1}$$

[Eq. 12.76]

where ε_{p1} is the plastic strain in Bar 1. The strain in Bar 2 has elastic and thermal components:

$$\varepsilon_2 = \frac{\sigma_2}{E} + \alpha(\Delta T) = \frac{(2p-1)S_y}{E} + \frac{2tS_y}{E}$$

[Eq. 12.77]

Since strains ε_1 and ε_2 are equal in this system, the plastic strain in Bar 1 must be:

$$\frac{E}{S_y}\varepsilon_{p1} = 2(p + t) - 2$$

[Eq. 12.78]

Cyclic Thermal Loading – Shakedown

With Bar 1 having plastic strain given by *Equation 12.78*, thermal loading ΔT is now removed (to thermal loading (**c**) in *Figure 12.27b*). The stress in Bar 1 decreases linearly and that in Bar 2 increases linearly (from points (**b**) to (**c**) in *Figure 12.32*). The additional behavior is elastic and the changes in the stresses are:

$$\Delta s_1 = \Delta\sigma_1 / S_y = -t$$
$$\Delta s_2 = \Delta\sigma_2 / S_y = +t$$

[Eq. 12.79]

The stresses upon removal of the thermal load are:

$$s_1 = 1 - t$$
$$s_2 = 2p - 1 + t$$

[Eq. 12.80]

as shown in *Figure 12.32*.

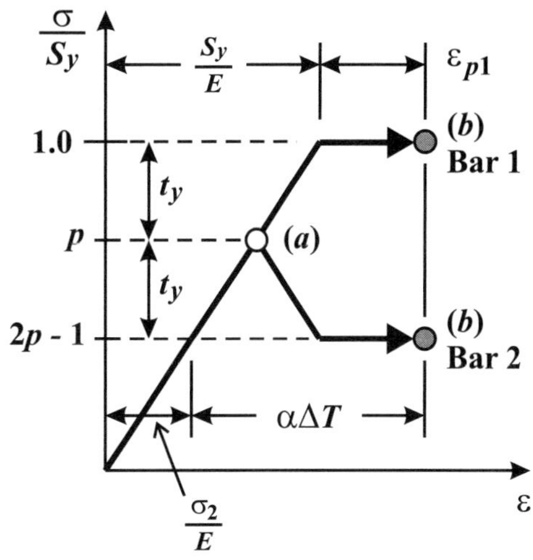

Figure 12.31. Stress–strain responses – point (**a**) to points (**b**) – for values of thermal load large enough to cause Bar 1 to plastically deform ($t > t_y$).

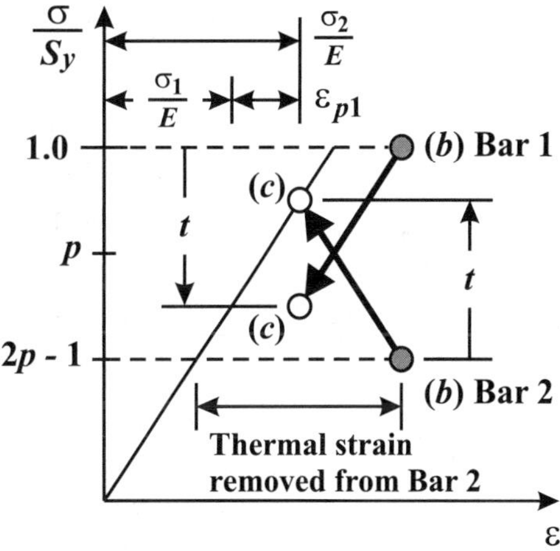

Figure 12.32. Stress–strain response for thermal unloading after Bar 1 has plastically deformed ($t > t_y$). Bar 2 does not yield provided $t < t_s$. The stress–strain responses are from points (**b**) to points (**c**).

Provided that $s_2 \leq 1.0$ after removal of the thermal load, Bar 2 does not yield.

The thermal load t required for Bar 2 to just yield ($s_2 = 1.0$) when the thermal load is removed is:

$$t = t_s = 2(1 - p) \qquad \text{[Eq. 12.81]}$$

The response for Bar 2 to just yield is shown in *Figure 12.33*.

For $t_y \leq t \leq t_s$, except for the first half cycle, the response to cyclic thermal loading is completely elastic. There will be no further plastic strain in either bar with continued thermal cycling. This condition is called **shakedown** and t_s is the **shakedown limit**.

For the particular case being studied, shakedown occurs in the region of the t–p diagram shown in *Figure 12.34*. Here, for a given load p, the shakedown limit t_s is twice the value of the elastic limit t_y.

For shakedown, plastic strain occurs in Bar 1 only during the first thermal loading, with value given in *Equation 12.78*. Further cyclic thermal loading results in only elastic deformation, the bars cycling between points *(b)* and *(c)* in *Figures 12.32* or *12.33*. The shakedown phenomenon is integrated into design codes.

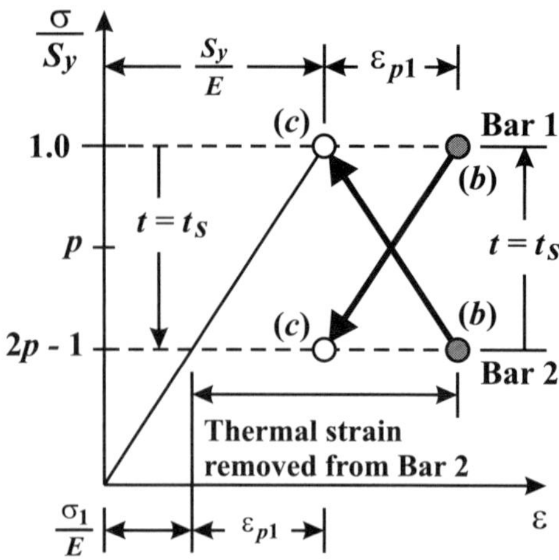

Figure 12.33. Thermal unloading after Bar 1 has plastically deformed; Bar 2 just at yield ($t = t_s$).

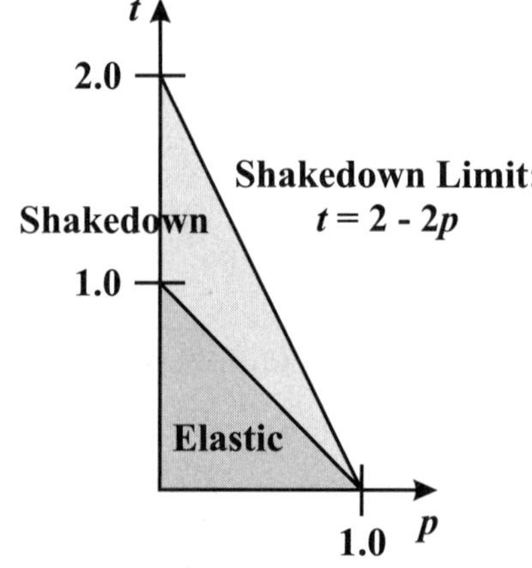

Figure 12.34. Thermal–mechanical load map to ensure shakedown.

Cyclic Thermal Loading – Ratcheting

When the cyclic thermal load exceeds the shakedown condition, $t > t_s$, Bar 1 yields upon first application of the thermal load. In addition, the plastic deformation of Bar 1 is large enough to cause Bar 2 to plastically deform when the thermal load is removed.

The stress–strain response for this condition is shown in *Figure 12.35* as temperature is removed (from thermal load *(b)* to *(c)*). The stresses when the thermal load is removed are:

$$s_1 = 2p - 1$$

$$s_2 = 1$$

[Eq. 12.82]

With the thermal load removed, the strain in Bar 1 has elastic and plastic components:

$$\varepsilon_1 = \frac{\sigma_1}{E} + \varepsilon_{p1} = \frac{s_1 S_y}{E} + \varepsilon_{p1}$$

[Eq. 12.83]

Substituting the values of s_1 from *Equation 12.82* and $E\varepsilon_{p1}/S_y$ from *Equation 12.78*, gives the total strain in Bar 1:

$$\frac{E}{S_y}\varepsilon_1 = [2p - 1] + [2(p + t) - 2]$$

[Eq. 12.84]

The total strain in Bar 2 is:

$$\varepsilon_2 = \frac{S_y}{E} + \varepsilon_{p2}$$ 　　[Eq. 12.85]

so its plastic strain is:

$$\frac{E}{S_y}\varepsilon_{p2} = \frac{E}{S_y}\varepsilon_2 - 1$$ 　　[Eq. 12.86]

Since the total strains in the bars are the same, $\varepsilon_1 = \varepsilon_2$, the plastic strain in Bar 2 is found by subtracting unity (1.0) from *Equation 12.84*, so that:

$$\frac{E}{S_y}\varepsilon_{p2} = 2(2p + t - 2)$$ 　　[Eq. 12.87]

During the first thermal load cycle, both bars experience plastic deformation.

Reapplication of Thermal Load, $t > t_s$

When the thermal load is applied a second time (from thermal load (*c*) to (*d*) in *Figure 12.27b*), Bar 2 unloads elastically, and a second increment of plastic strain is induced in Bar 1 (*Figure 12.36*):

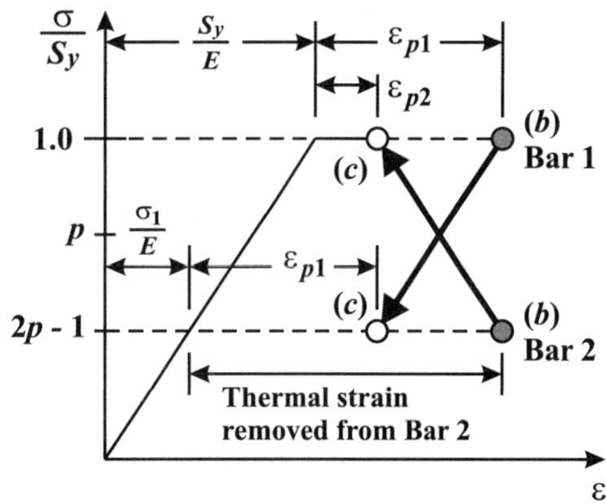

Figure 12.35. Thermal unloading for $t > t_s$. The plastic strain in Bar 1 is large enough to cause Bar 2 to plastically deform when thermal load is removed.

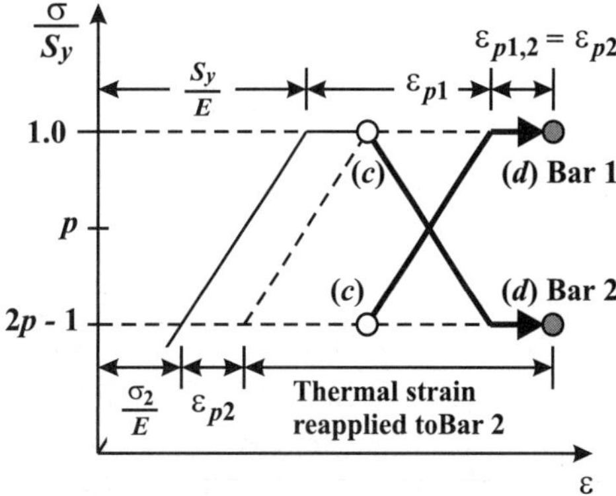

Figure 12.36. Second application of thermal load for $t > t_s$. Bar 1 plastically deforms a second time. The stress–strain responses are from points (*c*) to points (*d*).

$$\varepsilon_{p1,2} = 2(2p + t - 2)\frac{S_y}{E}$$ 　　[Eq. 12.88]

where the second number in the subscript represents Cycle 2. This strain is the same as the plastic strain in Bar 2 during the first thermal unloading half-cycle (*Equation 12.87*). With each subsequent application of the thermal load, Bar 1 deforms plastically by the same amount.

Likewise, during the second thermal unloading, Bar 2 deforms plastically by the same amount as during the first thermal unloading. Hence, during each thermal cycle, each bar, and thus the system itself, experiences an incremental plastic strain of:

$$\Delta\varepsilon_p = 2(2p + t - 2)\frac{S_y}{E} \quad [\text{Eq. 12.89}]$$

This continual elongation of the bars with cyclic thermal loading is called **ratcheting**. There is no limit for the condition; plastic strain will continue to accrue – the system will *ratchet* – with each cycle.

If there are N loading cycles, then the total plastic strain is:

$$\varepsilon(N) = 2(2p + t - 2)\frac{S_y}{E}N \quad [\text{Eq. 12.90}]$$

Thermal–Mechanical Load Map and Strain History

The complete loading t–p map for the two-bar system studied here is shown in *Figure 12.37*. *Figure 12.37* is only valid for two bars of equal length and area, made of the same material, and with thermal loading only on Bar 2 (*Figure 12.27*).

A purely *elastic* response occurs for combinations of p and t in the *elastic region* of *Figure 12.37*. For such load combinations, both bars remain elastic during thermal cycling. The thermal load in Bar 2 is not sufficient to cause Bar 1 to yield.

Elastic shakedown occurs for combinations of p and t in the *shakedown region* of *Figure 12.37*. For such load combinations, Bar 1 yields upon initial application of the temperature in Bar 2, but further thermal cycling produces only elastic responses in both bars, as shown in the strain history diagram of *Figure 12.38b*.

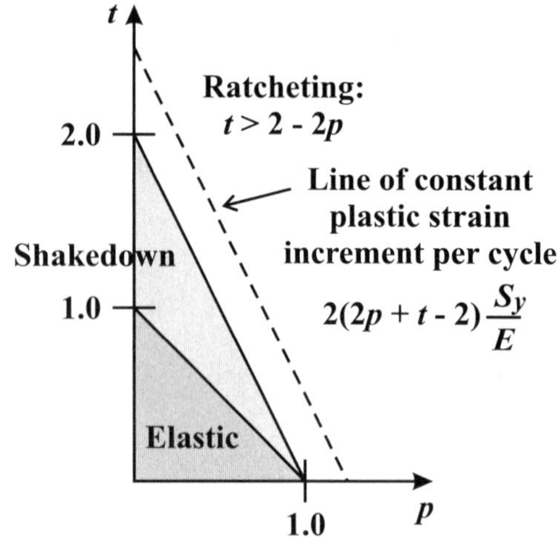

Figure 12.37. Thermal–mechanical load map.

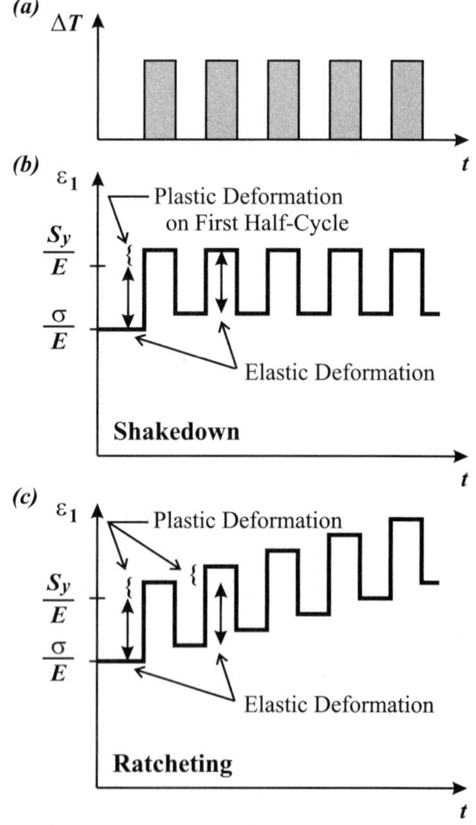

Figure 12.38. (a) Temperature history of Bar 2. **(b)** Strain history of Bar 1 for shakedown condition. **(c)** Strain history of Bar 1 for ratcheting condition.

Ratcheting occurs for combinations of p and t above the shakedown region. Plastic straining occurs during each half of the thermal load cycle. The dashed line in *Figure 12.37* represents points in the *ratcheting* region that have the same value of plastic strain per cycle $\Delta\varepsilon_p$, given by *Equation 12.89*. During ratcheting, the system continuously elongates, as shown in *Figure 12.38c*.

Example 12.15 Thermal Cycling of a Two-Bar System

Given: The two-bar system just studied is made of steel ($E = 30{,}000$ ksi, $S_y = 36$ ksi, $\alpha = 14 \times 10^{-6}/°C$) and is subjected to constant force F and cyclic thermal loading ΔT (*Figure 12.39*). The bars are of equal length $L = 24$ in., and equal cross-sectional area $A = 0.25$ in.^{2}. The mean stress due only to mechanical loading is $\sigma = 10$ ksi. Assume that the material properties remain constant with temperature.

Required: Determine (a) the thermal load ΔT_y to cause first yield, (b) the maximum thermal load ΔT_s that does not exceed the shakedown condition, and (c) the plastic strain per cycle when $\Delta T = 400°C$.

Solution: *Step 1.* The load normalized by the yield strength is:

$$p = \frac{\sigma}{S_y} = \frac{10 \text{ ksi}}{36 \text{ ksi}} = 0.278$$

The elastic strain caused by load F is:

$$\varepsilon = \frac{\sigma}{E} = \frac{10\times10^3}{30\times10^6} = 0.033\%$$

Bar 1 will just yield when the thermal load on Bar 2 is (*Equation 12.74*):

$$t_y = 1 - p = 1 - 0.278 = 0.722$$

or

$$\Delta T_y = \frac{2t_y S_y}{E\alpha} = \frac{2(0.722)(36{,}000 \text{ psi})}{(30\times10^6 \text{ psi})(14\times10^{-6}°C^{-1})}$$

Answer: $\Delta T_y - 124°\ C$

Step 2. The maximum thermal load for *shakedown* for $p = 0.278$ is (*Equation 12.81*):

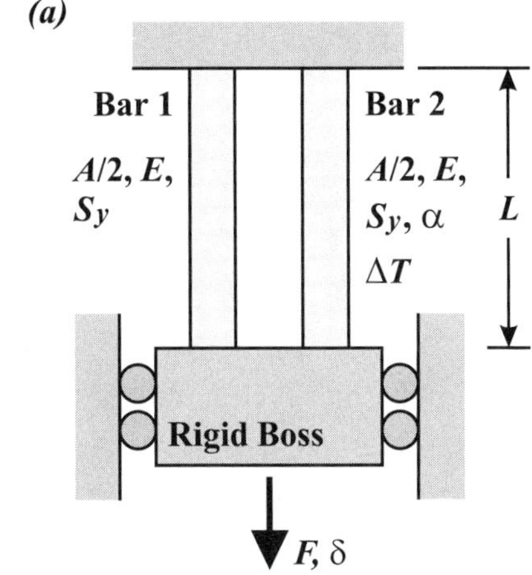

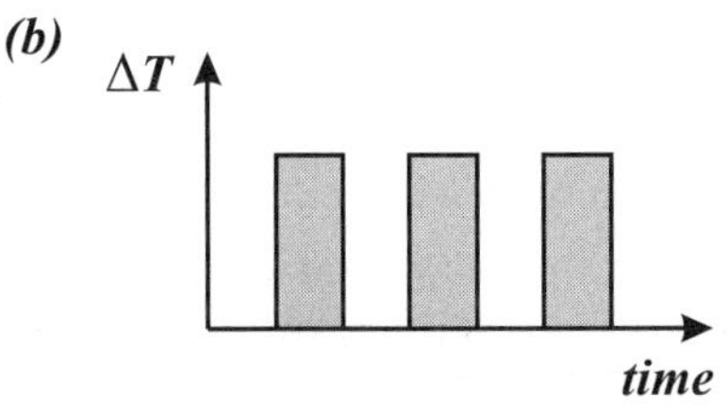

Figure 12.39. (a) Two-bar structure subjected to constant load F and (b) cyclic temperature ΔT.

$$t_s = 2(1-p) = 1.44$$

Answer: $\Delta T_s = 2(\Delta T_y) = 248\,°C$

Step 3. When the thermal load is $\Delta T = 400\,°C$:

$$t = \frac{E\alpha(\Delta T)}{2S_y} = \frac{(30\times10^6\ \text{psi})(14\times10^{-6}\,°\text{C}^{-1})(400\,°\text{C})}{2(36{,}000\ \text{psi})} = 2.33$$

The point $(p,t) = (0.278,\ 2.33)$ corresponds to a point outside of the shakedown region (*Figure 12.37*). The system is *ratcheting*. The plastic strain per cycle is:

$$\Delta\varepsilon_p = 2(2p + t - 2)\frac{S_y}{E} = 2[2(0.278) + (2.33) - 2]\left(\frac{36\times10^3}{30\times10^6}\right) = 0.002$$

Answer: $\Delta\varepsilon_p = 0.2\%$

After 10 thermal cycles, the accumulation of plastic strain is 2%. Accumulated strains due to ratcheting can grow quickly, and are generally much greater than the static elastic strains (here, 0.033%).

12.8 Large Plastic Strains

In manufacturing processes such as metal forming, the plastic strains are much larger than the elastic strains. The presence of large plastic strains affects the load capacity and the in-service failure strain of a component.

When a tensile test is performed on a bar of ductile material, the load–displacement $(P\text{–}\Delta)$ graph has the form shown in *Figure 12.40*. The load reaches a maximum value, when *necking* begins to occur. The load then decreases with increasing displacement until the specimen breaks into two parts. The maximum load defines the point of necking or *plastic instability*.

Strain and stress are generally defined in terms of the *original length L*, and *original cross-sectional area A*, of the bar. The **nominal strain** or **engineering strain** is:

$$\varepsilon_n = \frac{\Delta}{L} \qquad\qquad \text{[Eq. 12.91]}$$

and the **nominal stress** or **engineering stress** is:

$$\sigma_n = \frac{P}{A} \qquad\qquad \text{[Eq. 12.92]}$$

The graph of engineering stress–strain, shown in *Figure 12.40b*, has the same shape as the load–displacement $(P\text{–}\Delta)$ diagram of *Figure 12.40a*. The engineering stress–strain $(\sigma\text{–}\varepsilon)$

curve is simply the load–displacement curve divided by *constant* values – original area A and original length L.

The nominal strain to failure is typically 20% or more in ductile materials, many times greater than the elastic strains which are in the order of 0.2%. Thus the effects of elasticity can be neglected in the current discussion.

During plastic deformation, the volume of the material remains constant (Poisson's ratio is $v = 0.5$ during plastic deformation). If the extended length of the bar is l and the reduced cross-sectional area is a, then neglecting elastic deformation:

$$AL = al \qquad \text{[Eq. 12.93]}$$

In tension, length $l > L$, so that area $a < A$. The actual stress σ_t, or *true stress*, is greater than the nominal value:

$$\sigma_t = \frac{P}{a} > \frac{P}{A} = \sigma_n \qquad \text{[Eq. 12.94]}$$

To account for the effects of geometric changes of the bar, it is convenient to introduce the concepts of *true strain* ε_t and *true stress* σ_t.

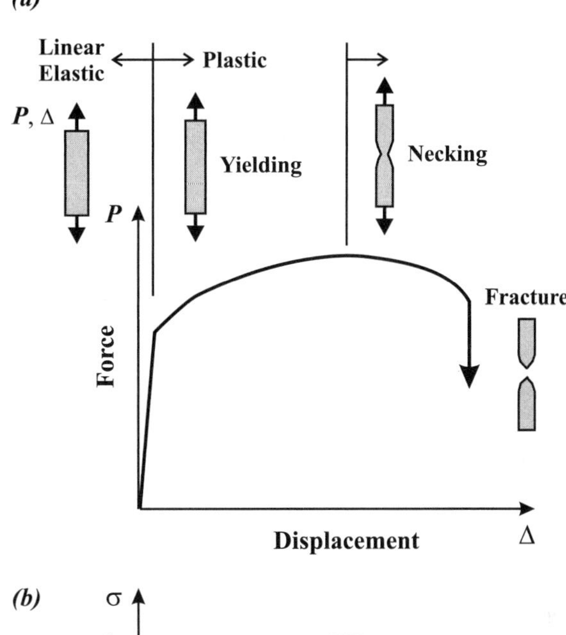

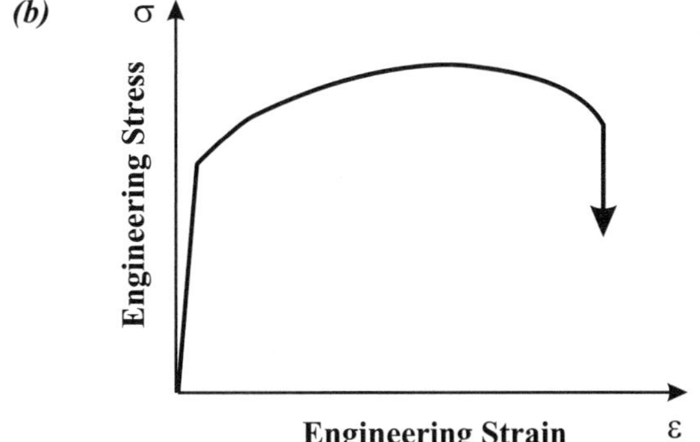

Figure 12.40. (a) The load–displacement diagram for a ductile bar in tension. **(b)** The engineering stress–engineering strain diagram for a ductile bar in tension.

True Strain and True Stress

The length of a tension member in the strained condition is $l = L + \Delta$ and the cross-sectional area is a. When a further increment of elongation δl is applied to the bar, the increment of **true strain** $\delta \varepsilon_t$ is defined as:

$$\delta \varepsilon_t = \frac{\delta l}{l} \qquad \text{[Eq. 12.95]}$$

The increment of *true strain* is the ratio of the incremental change in length δl to the current length l. Integrating $\delta \varepsilon_t$ from the bar's initial length L to any length l, gives the *true strain* at length l:

$$\varepsilon_t = \int_L^l \frac{\delta l}{l} = \ln\left(\frac{l}{L}\right) \qquad \text{[Eq. 12.96]}$$

Since $l = L + \Delta$ and $\varepsilon_n = \Delta/L$, then *true strain* ε_t can be related to *engineering strain* ε_n :

$$\varepsilon_t = \ln(1 + \varepsilon_n) \qquad \text{[Eq. 12.97]}$$

For small (i.e., elastic) values of strain, the error in using engineering strain is negligible since:

$$\lim_{\varepsilon_n \to 0} \varepsilon_t = \lim_{\varepsilon_n \to 0} \ln(1 + \varepsilon_n) \approx \varepsilon_n \qquad \text{[Eq. 12.98]}$$

For example, for strains within the elastic range, e.g., $\varepsilon_n = 0.10\%$, the true strain is 0.09995%, giving an error with respect to the true strain of 0.05%. Hence, it is acceptable to use engineering strain in the elastic range. As the strain increases, so does the error. If the engineering strain is $\varepsilon_n = 10\%$, well beyond the elastic region, the true strain is $\varepsilon_t = 9.5\%$, an error of about 5%.

Alternatively, strain can be expressed as:

$$\frac{l}{L} = e^{\varepsilon_t} \qquad \text{[Eq. 12.99]}$$

True stress σ_t is defined in terms of the current cross-sectional area a:

$$\sigma_t = \frac{P}{a} \qquad \text{[Eq. 12.100]}$$

Combining the constant volume condition *Equation 12.93* ($a = AL/l$) with *Equation 12.99* gives:

$$\sigma_t = \frac{Pl}{AL} = \sigma_n e^{\varepsilon_t} \qquad \text{[Eq. 12.101]}$$

where σ_n is the *nominal* or *engineering stress*. For typical elastic strains, e.g., $\varepsilon_t = 0.10\%$, the error in the nominal stress with respect to true stress is about 0.1%. Hence, it is acceptable to use engineering stress in the elastic range. For large values of strain, e.g., $\varepsilon_t \sim 10\%$, the error in the nominal stress is about 9.5%.

True Stress–True Strain Curve

When the engineering stress–engineering strain (σ_n–ε_n) diagram is converted to the true stress–true strain (σ_t–ε_t) diagram, the resulting curve has the form shown in *Figure 12.41*, plotted up to the point at which the system begins to *neck*. In tension, true stress is greater than its related engineering stress, and true strain is less than its related engineering strain.

For many metals, the true stress–true strain response is described by the relationship:

$$\sigma_t = K\varepsilon_t^n \qquad \text{[Eq. 12.102]}$$

where n and K are material constants. *Table 12.3* gives typical values of n, which is known as the **hardening index**, and constant K. The relationship of *Equation 12.102* is used until necking occurs. Once the test specimen starts to neck locally, stress in the bar is no longer the same throughout, and the true stress–strain relationship is no longer valid.

Condition for Plastic Instability

Plastic instability, or *necking*, occurs when the applied load reaches a maximum. The load–true stress relationship is:

$$P = \sigma_t a \qquad \text{[Eq. 12.103]}$$

The maximum load occurs when $\delta P = 0$, so that:

$$\delta P = \delta(\sigma_t a) = \sigma_t(\delta a) + a(\delta \sigma_t) = 0$$
$$\text{[Eq. 12.104]}$$

from which:

$$\frac{\delta a}{a} = -\frac{\delta \sigma_t}{\sigma_t} \qquad \text{[Eq. 12.105]}$$

Since the volume is constant:

$$V = AL = al \qquad \text{[Eq. 12.106]}$$

then:

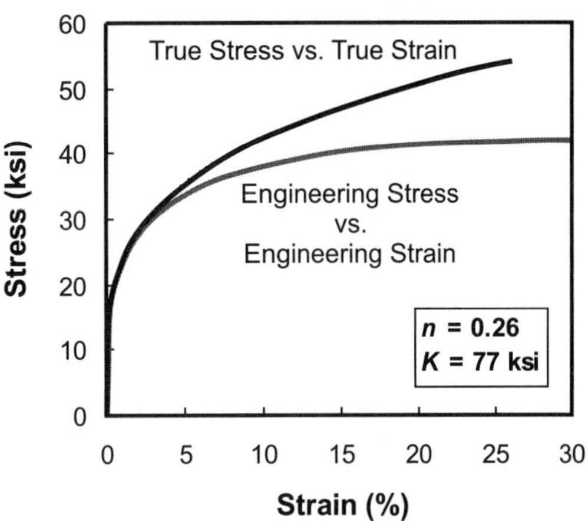

Figure 12.41. Stress–strain curves for engineering and true values of stress and Strain for steel, 0.05% carbon, annealed. Response up to plastic instability (necking).

Table 12.3. Sample values of hardness index n and constant K.

Material	n	K (ksi)
Steel, 0.05% carbon, annealed	0.26	77
Steel, 0.6% C, quenched and tempered, 1000°F	0.10	228
Copper, annealed	0.54	46
70/30 brass, annealed	0.49	130

Source: http://www.key-to-steel.com/articles/ art42.htm   Accessed May 2008.

$$\delta V = \delta(al) = a(\delta l) + l(\delta a) = 0 \qquad \text{[Eq. 12.107]}$$

It follows from *Equations 12.107* and *12.95* ($\delta \varepsilon_t = \delta l/l$) that:

$$\frac{\delta a}{a} = -\frac{\delta l}{l} = -d\varepsilon_t \qquad \text{[Eq. 12.108]}$$

Equating *Equations 12.105* and *12.108* gives the instability condition:

$$\frac{\delta \sigma_t}{\delta \varepsilon_t} = \sigma_t \qquad \text{[Eq. 12.109]}$$

where, from *Equation 12.102*:

$$\sigma_t = K\varepsilon_t^n \qquad \text{[Eq. 12.110]}$$

Differentiating:

$$\frac{\delta\sigma_t}{\delta\varepsilon_t} = nK\varepsilon_t^{(n-1)} \qquad \text{[Eq. 12.111]}$$

Equating *Equation 12.110* and *12.111* gives the true strain at plastic instability (necking):

$$\varepsilon_{t,o} = n \qquad \text{[Eq. 12.112]}$$

The true strain at instability is equal to the *hardening index, n*. Referring to *Table 12.3*, the true instability strain for annealed copper is 54%, and that for a quenched and tempered steel is 10%. Typical experimental data indicate that the ductility (engineering strain to failure) of annealed copper is about five times that of a quenched and tempered steel.

In terms of engineering strain, instability or necking occurs at (from *Equation 12.97*):

$$\varepsilon_{n,o} = e^{\varepsilon_{t,o}} - 1 = e^n - 1 \qquad \text{[Eq. 12.113]}$$

For the stress–strain curves plotted in *Figure 12.41*, necking occurs at true strain $\varepsilon_{t,o} = n = 0.26$, which is equivalent to engineering strain $\varepsilon_{n,o} = 0.30$.

Three-Dimensional True Stress–True Strain Plastic Instability

For 3D problems, the material response is analogous to the true stress–true strain tensile response given in *Equation 12.102*:

$$\overline{\sigma}_t = K\overline{\varepsilon}_t^{\,n} \qquad \text{[Eq. 12.114]}$$

Here, $\overline{\sigma}_t$ is the **effective stress** and $\overline{\varepsilon}_t$ is the **effective strain**. These two terms are defined by the principal values of true stress and true strain:

$$\overline{\sigma}_t = \left[\frac{(\sigma_{I,t} - \sigma_{II,t})^2 + (\sigma_{II,t} - \sigma_{III,t})^2 + (\sigma_{III,t} - \sigma_{I,t})^2}{2}\right]^{1/2} \qquad \text{[Eq. 12.115]}$$

$$\overline{\varepsilon}_t = \left[\frac{2[(\varepsilon_{I,t} - \varepsilon_{II,t})^2 + (\varepsilon_{II,t} - \varepsilon_{III,t})^2 + (\varepsilon_{III,t} - \varepsilon_{I,t})^2]}{9}\right]^{1/2} \qquad \text{[Eq. 12.116]}$$

For a material loaded in the true principal stress state, $\sigma_{I,t}$, $\sigma_{II,t}$, $\sigma_{III,t}$, the true principal strains are:

$$\varepsilon_{I,t} = \frac{\overline{\varepsilon}_t}{\overline{\sigma}_t}\left[\sigma_{I,t} - \frac{1}{2}(\sigma_{II,t} + \sigma_{III,t})\right]$$

$$\varepsilon_{II,t} = \frac{\overline{\varepsilon}_t}{\overline{\sigma}_t}\left[\sigma_{II,t} - \frac{1}{2}(\sigma_{III,t} + \sigma_{I,t})\right] \qquad \text{[Eq. 12.117]}$$

$$\varepsilon_{III,t} = \frac{\overline{\varepsilon}_t}{\overline{\sigma}_t}\left[\sigma_{III,t} - \frac{1}{2}(\sigma_{I,t} + \sigma_{II,t})\right]$$

The factor of 1/2 corresponds to the value of Poisson's ratio during plastic deformation. For the uniaxial case ($\sigma_{II,t} = \sigma_{III,t} = 0$), the effective stress reduces to: $\overline{\sigma}_t = \sigma_{I,t}$.

The 3D plastic instability is shown with a spherical pressure vessel in *Example 12.16*.

Example 12.16 Instability of a Spherical Pressure Vessel

Given: A spherical pressure vessel (*Figure 12.42*) is subjected to increasing pressure p until plastic instability takes place. Instability occurs when the pressure reaches its maximum value. Such a test is often called the *bulge test*. The initial radius is R and thickness is T ($R \ll T$). Under load, the radius is r and the thickness is t ($r \ll t$). The yield strength is S_y.

Required: Using the effective stress and strain equations for 3D systems, determine the true strain in the plane of the pressure vessel wall at plastic instability, $\varepsilon_{1,o}$.

Solution: *Step 1.* For a spherical pressure vessel, the true principal stresses are:

$$\sigma_{I,t} = \frac{pr}{2t}; \quad \sigma_{II,t} = \frac{pr}{2t}; \quad \sigma_{III,t} = 0$$

where $\sigma_{I,t}$ and $\sigma_{II,t}$ are in the plane of the vessel wall and $\sigma_{III,t}$ is perpendicular to the wall. Applying *Equation 12.115*, the effective true stress is:

$$\overline{\sigma}_t = \sigma_{I,t}$$

Step 2. The true strains are related by the observation that the volumetric strain during plastic deformation is zero:

$$\varepsilon_{1,t} + \varepsilon_{2,t} + \varepsilon_{3,t} = 0$$

From symmetry of the spherical vessel:

$$\varepsilon_{1,t} = \varepsilon_{2,t}$$

so the constant volume condition gives the true strain in the *III*-direction:

$$\varepsilon_{3,t} = -2\varepsilon_{1,t}$$

Applying *Equation 12.116*, the effective true strain is:

$$\bar{\varepsilon}_t = 2\varepsilon_{1,t}$$

Step 3. From equilibrium of a spherical half-shell under pressure p:

$$p\pi r^2 = \sigma_{I,t}(2\pi rt)$$

so

$$p = \frac{2t(\sigma_{I,t})}{r} = \frac{2t\bar{\sigma}_t}{r}$$

The condition for maximum pressure p is $dp = 0$:

$$dp = 2\left[\frac{t(d\bar{\sigma}_t)}{r} + \frac{\bar{\sigma}_t(dt)}{r} - \frac{t\bar{\sigma}_t(dr)}{r^2}\right] = 0$$

from which:

$$\frac{d\bar{\sigma}_t}{\bar{\sigma}_t} = \frac{dr}{r} - \frac{dt}{t}$$

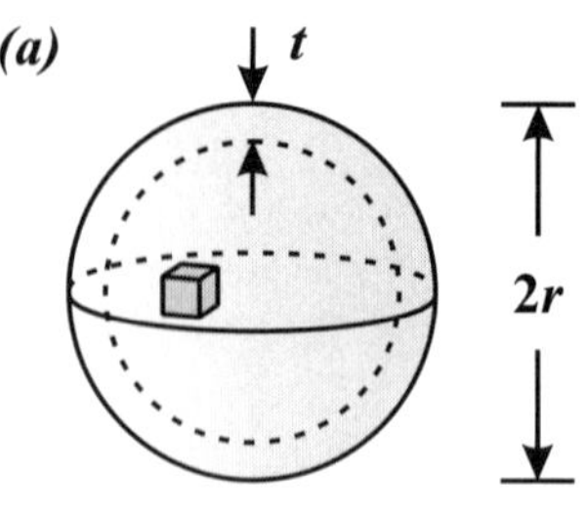

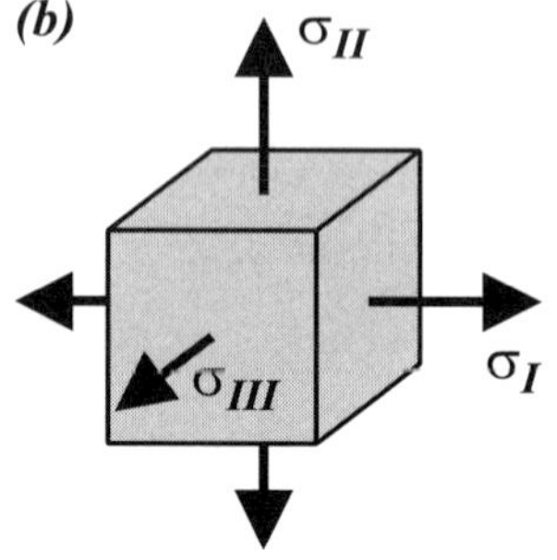

Figure 12.42. **(a)** Spherical pressure vessel. **(b)** 3D element showing the principal stresses.

The increments of the true strains are:

$$d\varepsilon_{1,t} = d\varepsilon_{2,t} = \frac{dr}{r}$$

$$d\varepsilon_{3,t} = \frac{dt}{t}$$

Noting that:

$$d\varepsilon_{3,t} = -2d\varepsilon_{1,t} \quad \text{(since volume does not change during yielding) and}$$

$$d\bar{\varepsilon}_t = 2d\varepsilon_{1,t}$$

then:

$$\frac{d\bar{\sigma}_t}{\bar{\sigma}_t} = \frac{dr}{r} - \frac{dt}{t} = d\varepsilon_{1,t} - d\varepsilon_{3,t} = 3d\varepsilon_{1,t} = \frac{3}{2}d\bar{\varepsilon}_t$$

or

$$\frac{d\bar{\sigma}_t}{d\bar{\varepsilon}_t} = \frac{3}{2}\bar{\sigma}_t$$

which is the condition for maximum load.

Applying the uniaxial true stress–true strain relationships (*Equations 12.110* and *12.111*) to the effective stress and strain gives:

$$\bar{\sigma}_t = K\bar{\varepsilon}_t^{\,n}$$

$$\frac{\delta\bar{\sigma}_t}{\delta\bar{\varepsilon}_t} = nK\bar{\varepsilon}_t^{\,(n-1)}$$

Setting the two expressions for $\delta\bar{\sigma}_t/\delta\bar{\varepsilon}_t$ equal to each other gives:

$$\frac{\delta\bar{\sigma}_t}{\delta\bar{\varepsilon}_t} = nK\bar{\varepsilon}_t^{\,(n-1)} = \frac{3}{2}\bar{\sigma}_t = \frac{3}{2}K\bar{\varepsilon}_t^{\,n}$$

Solving for the effective strain:

$$\bar{\varepsilon}_t = \frac{2}{3}n = 2\varepsilon_{1,t}$$

The corresponding value of the true in-plane strain at failure is therefore:

$$\textit{Answer:} \ \ \varepsilon_{1,o} = \frac{n}{3}$$

This is an important result. When manufacturing spherical vessels, load instability, or necking, occurs at a strain one-third that measured in a uniaxial tension test. The ductility for this loading and geometry is effectively one-third that of the uniaxial case.

Chapter 13   Effect of Flaws: Fracture

13.0 Introduction

The normal stress in a straight bar of cross-sectional area A and length L subjected to axial force P (*Figure 13.1*) is:

$$\sigma = \frac{P}{A} \qquad \text{[Eq. 13.1]}$$

If the working stress is half the yield strength, $S_y/2$, then the required cross-sectional area is:

$$A = \frac{2P}{S_y} \qquad \text{[Eq. 13.2]}$$

The weight of the bar is then:

$$W = \gamma A L = 2\frac{\gamma}{S_y}PL \qquad \text{[Eq. 13.3]}$$

where γ is the weight density of the material.

The yield strength of steel S_y can be modified by suitable heat and/or mechanical treatment, or by the addition of alloying elements. The weight density γ of various types of steels is essentially constant, irrespective of the production process. Thus, the weight of a bar can best be decreased by selecting a high-strength steel in the design of an engineering structure.

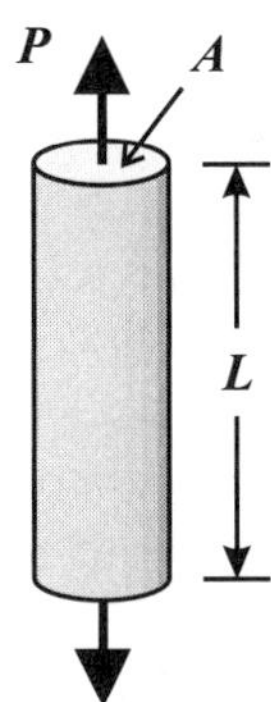

Figure 13.1. Bar subjected to axial force.

Figure 13.2. Some World War II Liberty Ships suffered catastrophic fracture. The USS *Esso Manhattan* failed in New York harbor in March, 1943, and was subsequently repaired (US government photograph. Lane, F.C., *Ships for Victory.* Baltimore: The John Hopkins University Press, 1951. Reprinted 2001).

However, it was learned in practice that components made of high-strength steel had a tendency to fail by **fracture** – breaking into two pieces – with little or no plastic deformation. This phenomenon is also known as *fast fracture*. Investigations of the failure surfaces indicated that failures initiated at material locations having small pre-existing *cracks*.

F.A. Leckie, D.J. Dal Bello, *Strength and Stiffness of Engineering Systems*, Mechanical Engineering Series,   **413**
DOI 10.1007/978-0-387-49474-6_13, © Springer Science+Business Media, LLC 2009

In response to the high demand for ships during World War II, welds replaced riveted connections in many cases, particularly in the rapidly produced supply vessels called Liberty Ships. During the welding process, cracks could be unintentionally introduced, providing initiation points for sudden and catastrophic fracture, as shown in *Figure 13.2*. Cracks could begin at welds, and other areas where the geometry was discontinuous (e.g., at sharp geometries and notches). Since the steel plates that made up the ship's hull were now welded together, a crack could grow through one steel plate (which becomes more brittle at lower temperatures), through the weld, and into the next plate. In riveted construction, a crack could only grow through a single plate.

> A Board of Investigation was convened by the Secretary of the Navy in 1943 to inquire into the design and methods of construction of welded steel merchant vessels, after several failures of inspected vessels.
>
> ...A specific case of structural failure cited in the [interim] Report [of June 3, 1944] was the case of the *Esso Manhattan*... which, on March 29, 1943 [in New York Harbor] broke in two. The fracture started in a butt weld between plates A-9 and A-10 at the crown of the deck. With a sound described variously as a thump, thud, bang, crash or explosion, the fracture ran across the deck... and down both sides, progressing to the bilge port and starboard. The vessel jack-knifed and the bow dug under an oncoming wave. The crew abandoned in lifeboats and were picked up... The vessel was repaired and returned to service.
>
> ...Contrary to popular impression, hull fractures were not confined to Liberty ships but were shared by other types of vessels. Practically all fractures originated in discontinuities occasioned by design details and notch effects incidental to imperfect welding.
>
> from: *The Coast Guard at War: December 7,1941-July 18, 1944, Marine Inspection*, Vol. XIII.
> Source: http://www.uscg.mil/history/REGULATIONS/CGWar_13_Marine_Inspection.html
> Accessed May 2008.

Although manufacturing quality improves with experience and better technology, it is difficult to ensure an as-manufactured crack-free component. Because the presence of cracks is inevitable, analysis methods are needed to assess the effect of these flaws on component strength, and to develop appropriate design tools.

Fracture Mechanics is the branch of engineering mechanics used to predict the strength of a component with a crack. The study of how components failed by fracture was accelerated due to the Liberty Ship problems.

Cracks in most *metallic (ductile) systems* can generally be detected before they are large enough to lead to catastrophic failure. Aircraft, for example, are routinely examined for the presence of fatigue cracks that may cause fast fracture.

In *ceramic (brittle) materials*, the crack sizes that cause failure are very small, so the detection of flaws of a critical size is difficult. In such cases, the design strategy is based on statistical methods. Both approaches for dealing with cracks are discussed in this chapter.

13.1 An Introduction to Fracture Mechanics

Early attempts to understand the onset of fracture were based on stress criteria. Consider a plate made of a *linear elastic* material subjected to applied tensile stress σ (*Figure 13.3*). The stress at the edge of an elliptical hole of length $2a$ and height $2b$ in a large plate ($a << W$, $b << W$) is:

$$\sigma_{max} = \sigma\left(1 + 2\frac{a}{b}\right) \qquad \text{[Eq. 13.4]}$$

Large values of a/b cause high **stress concentrations**, σ_{max}/σ.

As a/b increases, the elliptical hole approaches the geometry of a crack. A criterion for fracture assumes that when σ_{max} is equal to the *ultimate tensile strength* S_u of the material, fracture occurs. Experiments indicate that this simple method of modeling cracks does not accurately predict failure. *Equation 13.4* depends only on the ratio a/b, but the applied stress at fracture was found to depend on the absolute length of the crack $2a$.

It was subsequently discovered that an energy criterion for fracture could explain the experimental observations.

Critical Energy Release Rate

To understand the energy approach, consider a long strip of adhesive tape of width b, attached to a flat surface (*Figure 13.4a*). The tape is removed by applying a constant force P normal to the surface. The required force depends on the strength of the tape's adhesive with the surface.

The work to remove tape length a from the surface is:

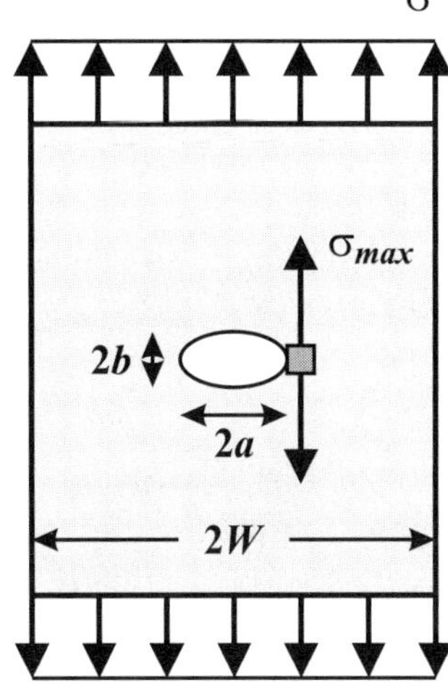

Figure 13.3. Large plate of width $2W$ with elliptical hole, $2a$ by $2b$. The plate is considered infinitely large if $a << W$, $b << W$. The plate is in a state of plane stress if its thickness is also small compared to its width.

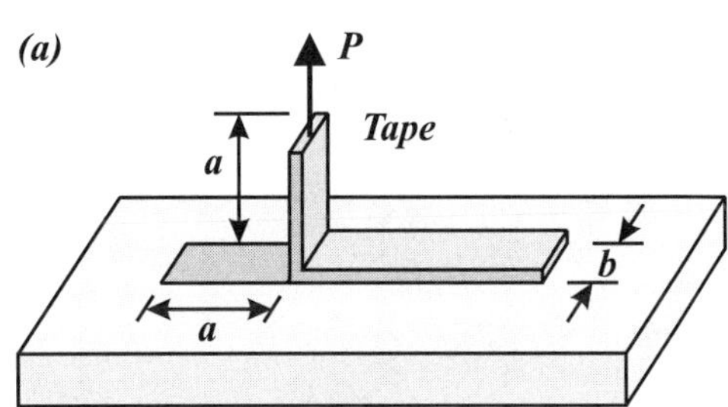

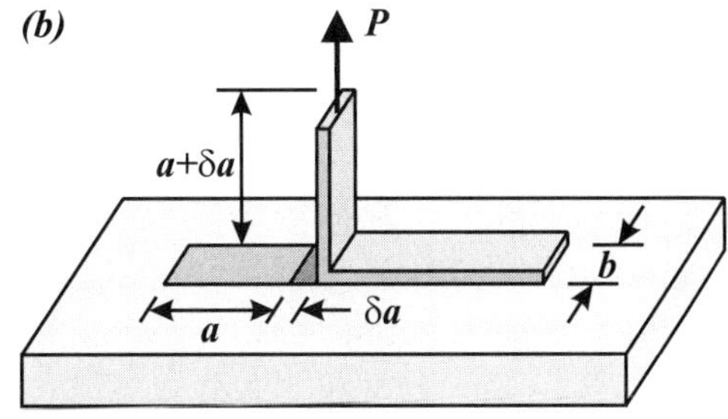

Figure 13.4. (a) Tape pulled from a surface by constant force P, exposing area ab. **(b)** The work done to grow exposed surface by δa is $P\,\delta a = G_c b\,\delta a$.

$$W = Pa \qquad \text{[Eq. 13.5]}$$

When a increases to $a+\delta a$, the additional work done by force P is:

$$\delta W = P\,\delta a \qquad \text{[Eq. 13.6]}$$

If the energy needed to separate a unit surface of the tape is G_c, then the energy dE required to detach an additional length of tape δa is

$$dE = G_c b\,\delta a \qquad \text{[Eq. 13.7]}$$

Equating the work done by force P to the energy required to separate the tape from the surface (*Figure 13.4b*) gives:

$$P\,\delta a = G_c b\,\delta a \qquad \text{[Eq. 13.8]}$$

from which:

$$G_c = \frac{P}{b} \qquad \text{[Eq. 13.9]}$$

In a simple experiment, a constant force of 20 N was needed to detach a strip of cellophane tape $b = 20$ mm wide, adhered to the surface of a ceramic tile. The energy per unit area to separate the tape is thus:

$$G_c = \frac{20\ \text{N}}{0.02\ \text{m}} = 1.0\ \text{kJ/m}^2$$

The quantity G_c is the ***critical energy release rate***, or ***toughness***, of a material. It is the energy required to create a new pair of free surfaces of unit area. The dimensions of toughness are energy per area. A value of 1.0 kJ/m^2 is typical of tape-adhesives.

The toughness of a material is determined using standardized test methods. Some representative toughness values are given in *Table 13.1*. The energy required to create a new surface of steel is substantially greater than that of an aluminum alloy. Brittle materials such as cast iron and ceramics, and wood parallel to the grain, have relatively

Table 13.1. Representative values of Toughness G_c and Fracture Toughness K_c.

Material	G_c (kJ/m^2)	K_c (MPa/m$^{1/2}$)
Mild steel	100	140
Rotor steels	200–240	200–215
Pressure vessel steels	150	170
Aluminum alloys, high to low strength	8–30	23–45
Cast iron	0.2–3.0	6–20
Cement	0.03	0.2
Engineering ceramics	0.02–0.1	3–5
Wood, perpendicular to grain	8.0–20	11–13
Wood, parallel to grain	0.5–2.0	0.5–1.0

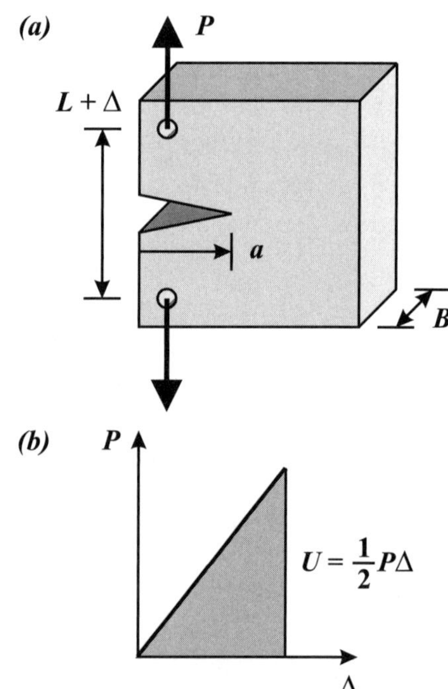

Figure 13.5. (a) A component subjected to force P with pre-existing edge crack of length a.
(b) Load–displacement diagram of component.

small toughness values. It is often necessary to perform tests on a particular material to find its toughness.

Elastic Energy Release Rate

Consider a component (compact test specimen) of thickness B having a crack of length a (*Figure 13.5a*). A pair of equal and opposite loads, P, are applied at points initially L apart. The corresponding relative displacement of the loads is Δ. The elastic law for this condition is:

$$P = K\Delta \qquad \text{[Eq. 13.10]}$$

where K is the component stiffness.

The displacement in terms of the force is:

$$\Delta = \frac{P}{K} = CP \qquad \text{[Eq. 13.11]}$$

where C is the **compliance**, or *flexibility*, of the system.

When the response is linear–elastic, the *work* done on the system equals the *internal elastic strain energy* stored in the system:

$$W = U = \frac{1}{2}P\Delta \qquad \text{[Eq. 13.12]}$$

which is the triangular area under the P–Δ curve in *Figure 13.5b*.

Energy Release Rate at Constant Load

Consider that the load on the component is such that the material begins to tear, extending the crack by δa, where δa is small relative to a (*Figure 13.6a*). During crack extension, load P is kept constant. With the extended crack, the component is less stiff, so for the same force P, the displacement of the component is larger by $\delta\Delta$. The internal strain energy is then:

$$U + \delta U = \frac{1}{2}P(\Delta + \delta\Delta) \qquad \text{[Eq. 13.13]}$$

as shown in *Figure 13.6b*.

The increment of work done by constant load P as the crack extends is:

$$\delta W = P\,\delta\Delta \qquad \text{[Eq. 13.14]}$$

The change in internal strain energy of the component is:

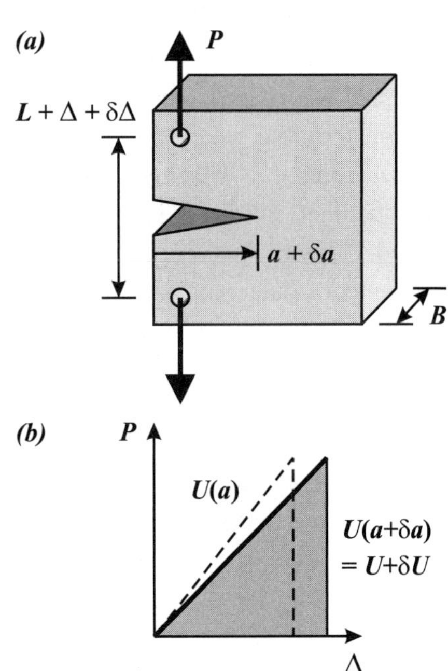

Figure 13.6. (a) A component with crack growth δa subjected to load P. **(b)** Load–displacement curves for components with crack lengths a and $a+\delta a$, subjected to the same load P.

$$\delta U = \frac{1}{2}P\,\delta\Delta \qquad \text{[Eq. 13.15]}$$

Hence, half of the work done by load P is used to increase the internal strain energy of the component, which leaves an excess energy δE of:

$$\delta E = \delta W - \delta U = \frac{1}{2}P\,\delta\Delta = \delta U \qquad \text{[Eq. 13.16]}$$

This energy is the energy made available to extend the crack (i.e., to break atomic bonds to create new free surfaces).

The *energy released per unit area of new crack* surface is:

$$G = \frac{\delta E}{B\,\delta a} = \frac{\delta U}{B\,\delta a} = \frac{1}{B}\frac{\partial U}{\partial a}\bigg|_{P\,=\,const} \qquad \text{[Eq. 13.17]}$$

G is the **elastic energy release rate**. This is the energy available per unit area (kJ/m^2) to grow a crack of width B by distance δa. If the calculated value of G exceeds the material *toughness* G_c, the crack will grow. If $G < G_c$, there is not enough energy available to grow the crack; the crack length remains constant, i.e., $\delta a = 0$.

Energy Release Rate at Constant Displacement

Consider that the load on the component is such that the material begins to tear, extending the crack by δa, where δa is small relative to a (*Figure 13.7a*). During crack extension, displacement Δ is held constant. With the extended crack, the component is less stiff, so for the same displacement Δ, the required force is changed by δP. The internal energy is then:

$$U + \delta U = \frac{1}{2}(P + \delta P)\Delta \qquad \text{[Eq. 13.18]}$$

as shown in *Figure 13.7b*. The system with the larger crack is less stiff, so δP is negative.

Since Δ is constant, the increment of work done by the applied load is $\delta W = 0$. The change in internal energy of the component is:

$$\delta U = \frac{1}{2}(\delta P)\Delta \qquad \text{[Eq. 13.19]}$$

Hence, the energy available to be released from the component to extend the crack is:

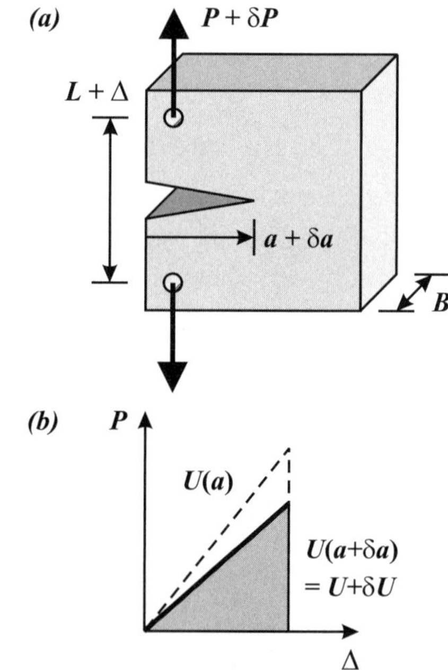

Figure 13.7. (a) A component with crack growth δa subjected to constant displacement. **(b)** Load–displacement curves for components with crack lengths a and $a+\delta a$, subjected to the same displacement Δ.

$$\delta E = \delta W - \delta U = 0 - \frac{1}{2}(\delta P)\Delta = -\delta U \qquad \text{[Eq. 13.20]}$$

Note that δP, and thus δU, is negative, so δE is positive.

The energy released per unit area of new crack surface is:

$$G = \frac{\delta E}{B\,\delta P} = \frac{-\delta U}{B\,\delta P} = -\frac{1}{B}\frac{\partial U}{\partial P}\bigg|_{\Delta\,=\,const} \qquad \text{[Eq. 13.21]}$$

Again, G is the *energy release rate*.

Energy Release Rate in terms of Compliance

The internal strain energy is:

$$U = \frac{1}{2}P\Delta = \frac{1}{2}CP^2 = \frac{1}{2}\frac{\Delta^2}{C} \qquad \text{[Eq. 13.22]}$$

where C is the compliance or flexibility of the component, the inverse of its stiffness K.

For constant load P:

$$G = \frac{1}{B}\frac{\partial U}{\partial a}\bigg|_{P\,=\,const} = \frac{P^2\,\delta C}{2B\,\delta a} \qquad \text{[Eq. 13.23]}$$

For constant displacement Δ:

$$G = -\frac{1}{B}\frac{\partial U}{\partial a}\bigg|_{\Delta\,=\,const} = -\left(\frac{\Delta^2}{2B}\right)\frac{\delta(C^{-1})}{\delta a} = \frac{P^2\,\delta C}{2B\,\delta a} \qquad \text{[Eq. 13.24]}$$

Thus, using either the constant–load condition, or constant–displacement condition, the energy release rate calculation gives the same result.

Conditions for Crack Initiation and Growth

The calculated *energy release rate* (the load) of a system is G, and the *critical energy release rate* or *toughness* (the material property) is G_c. The criterion for *crack growth*, or *fracture*, is $G > G_c$. The three possible relationships for G and G_c, and the corresponding system responses are:

- $G < G_c$: the crack does not initiate (it does not start to grow);
- $G = G_c$: the crack initiates, but does not continue to grow; and
- $G > G_c$: the crack initiates and grows (fast fracture).

$$\text{[Eq. 13.25]}$$

Crack **initiation** is when the crack begins to grow by a small amount, but quickly stops; fast fracture does not occur. The condition $G = G_c$ is analogous to the uniaxial stress in a ductile material when it reaches the yield strength $\sigma = S_y$; the material just yields but does not plastically deform.

Example 13.1 Calculating *G*: Center Crack in Infinite Plate

Given: A common fracture problem is a center crack of length $2a$ in an *infinite plate* of thickness B (*Figure 13.8*). The plate is $2W$ wide and L long, where both $2W$ and L are much greater than crack length $2a$ ($a \ll W$, $a \ll L$) and B. Thus, the crack and nearby material responses are unaffected by the overall geometry of the plate. The plate is subjected to stress σ.

The internal strain energy for a plate with center crack $2a$ is given by:

$$U = 2WLB\left(\frac{\sigma^2}{2E}\right) + \frac{\pi a^2 B \sigma^2}{2E} \qquad \text{[Eq. 13.26]}$$

The first term is the energy of the plate without a crack. The second term is the increase of strain energy due to the presence of the center crack. The mathematical expression for the second term is obtained using the *Theory of Elasticity*, not shown here.

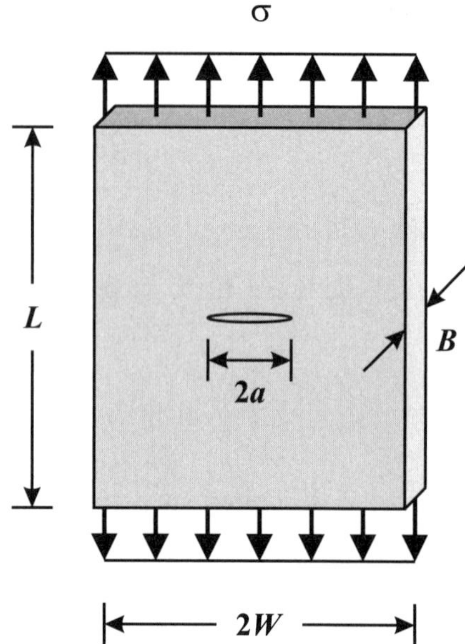

Figure 13.8. Plate with center crack of length 2a. The plate is infinitely large if *a* << *L* and *a* << *W*.

Required: Determine the energy release rate G for the center crack.

Solution: The energy release rate G for a center crack in an infinite plate is:

$$G = \frac{1}{B}\frac{\partial U}{\partial a}\bigg|_{P = const}$$

$$\textit{Answer:}\ \ G = \frac{\pi a \sigma^2}{E} \qquad \text{[Eq. 13.27]}$$

Note that G increases linearly with crack length a. The longer the crack length, the more energy is available to grow the crack further. Larger cracks are thus more likely to be the cause of failure than smaller cracks.

Example 13.2 Double Cantilever Beams under Constant Load *P*

Given: Double cantilever beams are often used in experiments to determine the toughness of a material. A double cantilever is shown in *Figure 13.9*. The solid beam is $2h$ deep and B wide. A crack of length a on the center-line of the double cantilever is the cantilever length. Equal and opposite loads P are applied to open the crack.

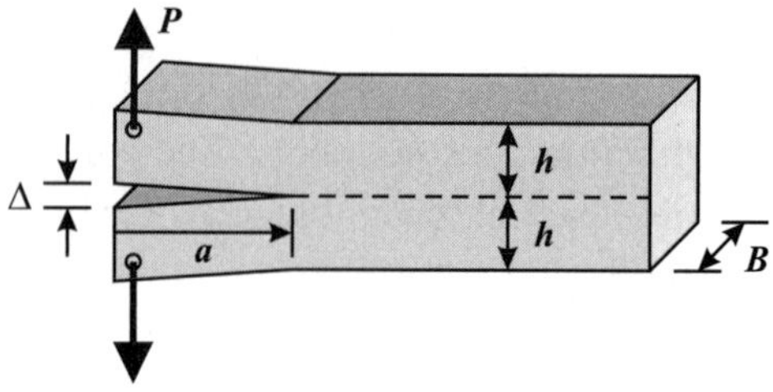

Figure 13.9. Double cantilever.

Required: Under constant load P, determine the energy release rate G.

Solution: *Step 1.* The crack opening is Δ, so each cantilever deflects $\Delta/2$. From beam theory, the tip displacement is:

$$\frac{\Delta}{2} = \frac{Pa^3}{3EI} \quad \text{where} \quad I = \frac{bh^3}{12}$$

The internal energy due to force P is then:

$$U = 2\left[\frac{1}{2}P\left(\frac{\Delta}{2}\right)\right] = 2\left[\frac{1}{2}P\left(\frac{Pa^3}{3EI}\right)\right] = \frac{P^2a^3}{3EI}$$

Step 2. Hence, the energy release rate G is:

$$G = \frac{1}{B}\frac{\partial U}{\partial a}\bigg|_{P\,=\,const}$$

$$\text{Answer: } G = \frac{P^2a^2}{BEI}$$

Example 13.3 Double Cantilever Beams under Constant Displacement

Given: The double cantilever is shown in *Figure 13.10*.

Required: Under constant displacement Δ, determine the energy release rate G.

Solution: *Step 1.* The crack opening is Δ, so each cantilever deflects by $\Delta/2$:

$$\frac{\Delta}{2} = \frac{Pa^3}{3EI} \quad \text{where} \quad I = \frac{bh^3}{12}$$

The internal energy due to constant displacement Δ is:

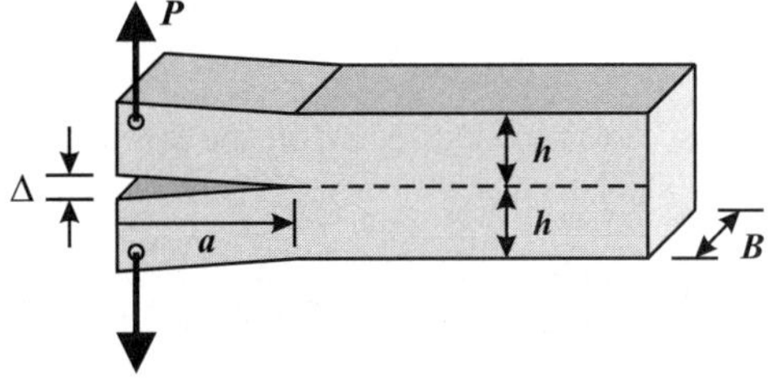

Figure 13.10. Double cantilever.

$$U = 2\left[\frac{1}{2}P\left(\frac{\Delta}{2}\right)\right] = 2\left[\frac{1}{2}\left(\frac{3EI}{2a^3}\right)\left(\frac{\Delta}{2}\right)\right] = \frac{3\Delta^2 EI}{4a^3}$$

Step 2. Hence, the energy release rate G is:

$$G = -\frac{1}{B}\frac{\partial U}{\partial a}\bigg|_{\Delta\,=\,const} = -\frac{3\Delta^2 EI}{4B}\left[\frac{\delta(a^{-3})}{\delta a}\right]$$

$$\text{Answer: } G = \frac{9\Delta^2 EI}{4Ba^4}$$

Substituting the load–displacement expression, $\dfrac{\Delta}{2} = \dfrac{Pa^3}{3EI}$, into the solution gives the same expression for G found in *Example 13.2*:

$$G = \frac{9\Delta^2 EI}{4Ba^4} = \frac{9EI}{Ba^4}\left(\frac{Pa^3}{3EI}\right)^2 = \frac{P^2a^2}{BEI}$$

Example 13.4   **Stress to Cause Crack Growth**

Given: An infinitely wide plate ($a \ll W$) made of an aluminum alloy is subjected to stress σ (*Figure 13.11*). A central crack of length $2a = 30$ mm exists in the plate. The toughness of the aluminum is $G_c = 25$ kJ/m^2 and the modulus is $E = 70$ GPa.

Required: Determine the minimum stress required to cause failure by fracture (crack initiation and growth).

Solution: *Step 1*. The energy release rate G for an infinite plate with a center crack $2a$ is (*Equation 13.27*):

$$G = \frac{\pi a \sigma^2}{E}$$

Step 2. The crack begins to grow when $G \geq G_c = 25$ kJ/m^2.

Thus, fracture occurs when the stress exceeds:

$$\sigma = \sqrt{\frac{G_c E}{\pi a}} = \sqrt{\frac{(25 \times 10^3 \text{ J/m}^2)(70 \times 10^9 \text{ Pa})}{\pi(0.015 \text{ m})}}$$

Answer: $\underline{\sigma = 193 \text{ MPa}}$

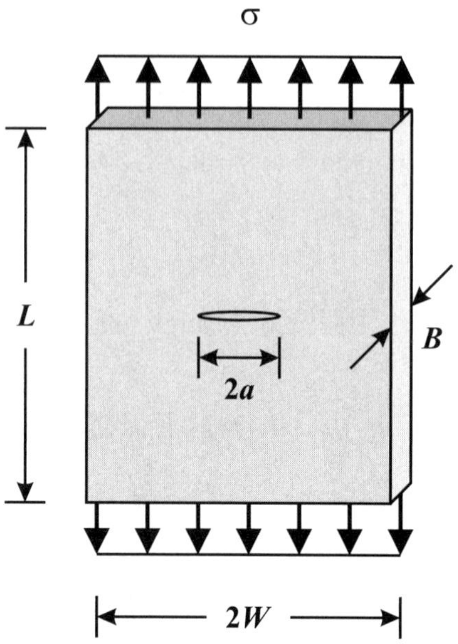

Figure 13.11. Plate with center crack.

If the material is Aluminum 6061-T6, then $S_y = 240$ MPa. The plate will fail by fracture before it starts to yield. This situation is undesirable since fracture occurs before plastic deformation becomes evident.

Example 13.5 **Steel Plate**

Given: An infinite plate made of pressure vessel steel is to be subjected to an applied stress $\sigma = 2/3 S_y$. The steel has properties:

$$E = 200 \text{ GPa}, \quad S_y = 1600 \text{ MPa}, \quad G_c = 150 \text{ kJ/m}^2$$

Required: Determine the maximum permissible center crack length $2a$ so that fracture does not occur.

Solution: *Step 1*. The energy release rate G for an infinite plate with a center crack $2a$ is:

$$G = \frac{\pi a \sigma^2}{E}$$

The applied stress is: $\sigma = \frac{2}{3}S_y = \frac{2}{3}(1600\times10^6 \text{ Pa}) = 1067 \text{ MPa}$

Step 2. Under this stress, for crack initiation, the initial crack half-length is:

$$a = \frac{G_c E}{\pi\sigma^2} = \frac{(150\times10^3 \text{ J/m}^2)(200\times10^9 \text{ Pa})}{\pi(1067\times10^6 \text{ Pa})^2} = 8.4 \text{ mm}$$

If the crack length $2a$ is less than 16.8 mm, the crack will not initiate under the applied load.

$$\textit{Answer: } (2a)_{max} = 16.8 \text{ mm}$$

Note that if the applied stress is half that used above, then the initial crack length may be up to four times longer without failure by fracture.

Stress Intensity Factor

An equivalent method of representing the *energy release rate G* is the **stress intensity factor K**, where K is related to G by the following equations from the Theory of Elasticity:

$$K^2 = GE \quad \text{for plane stress (e.g., for a thin specimen)} \qquad \text{[Eq. 13.28]}$$

$$K^2 = \frac{GE}{1 - v^2} \quad \text{for plane strain (e.g., for a thick specimen)} \qquad \text{[Eq. 13.29]}$$

Stress intensity factor equations for various loadings and geometries are generally given in tabular form and are available in printed references. Some stress intensity factors are given in *Table 13.2*. Stress intensity factor calculators are also available online (e.g., http://www.fatiguecalculator.com).

The *energy release rate* for a center crack in an infinite plate is (*Example 13.1*):

$$G = \frac{\pi a \sigma^2}{E} \qquad \text{[Eq. 13.30]}$$

The corresponding expression for the plane stress *stress intensity factor* is:

$$K = \sqrt{EG} = \sigma\sqrt{\pi a} \qquad \text{[Eq. 13.31]}$$

The units of stress intensity factor are MPa$\sqrt{\text{m}}$ (ksi$\sqrt{\text{in.}}$).

The advantage of the stress intensity factor approach is that the expression for K is dependent only upon load σ and crack length a; unlike G, K is *independent of material*. Consequently, expressions for K determined by mathematical techniques can be given in tabular form (e.g., *Table 13.2*), which is convenient in engineering practice. In research, however, it is the energy release rate G that is generally used to study crack propagation.

Table 13.2. Stress Intensity Factors for common geometries.

Geometry	Elastic stress intensity factor, K
Central Crack (plane stress)	Infinite plate ($a << W$): $\quad K = \sigma\sqrt{\pi a}$ Finite plate (width $2W$): $$K = \sigma\sqrt{\pi a}\left[\sec\frac{\pi a}{2W}\right]^{1/2} \qquad \text{[Source 1a]}$$
Single Edge Crack (plane stress)	Infinite plate ($a << W$): $\quad K = 1.12\sigma\sqrt{\pi a}$ Finite plate (width W): $$K = \sigma\sqrt{\pi a}\left[f_E\!\left(\frac{a}{W}\right)\right] \qquad \text{[Source 1b]}$$ $$f_E = 1.122 - 0.231\left(\frac{a}{W}\right) + 10.55\left(\frac{a}{W}\right)^2 - 21.71\left(\frac{a}{W}\right)^3 + 30.38\left(\frac{a}{W}\right)^4$$
Semi-elliptical Surface Crack in *semi-infinite body* (body $>> a$)	$$K = \sigma\sqrt{\pi a}\left(\frac{1.12}{E^*}\right) \qquad \text{[Source 2]}$$ Cubic fit for E^* by present authors: $\quad < 0.6\%$ error up to $a/2c = 0.5$ $$E^* \approx 1.0 + 0.30\left(\frac{a}{2c}\right) + 2.67\left(\frac{a}{2c}\right)^2 - 1.99\left(\frac{a}{2c}\right)^3$$
Three-Point Bending (plane stress)	$$K = \frac{3FL}{2BW^2}\sqrt{a}\left[f_3\!\left(\frac{a}{W}\right)\right] \qquad \text{[Source 3]}$$ $$f_3 = 1.93 - 3.07\left(\frac{a}{W}\right) + 14.53\left(\frac{a}{W}\right)^2 - 25.11\left(\frac{a}{W}\right)^3 + 25.8\left(\frac{a}{W}\right)^4$$
Double Cantilever Beam (plane stress)	$$K = \frac{Pa}{\sqrt{BI}} \qquad \text{where:} \quad I = \frac{Bh^3}{12}$$ See: *Examples 13.2 and 13.3* for derivation; $K = \sqrt{EG}$

Sources and *Notes*:

1. Tada, H., et al., *The Stress Analysis of Cracks Handbook*, 2nd ed., Paris Productions, Inc., 1985.

   a. Accuracy: 0.3% up to $a/W = 0.7$; 1% at $a/W = 0.8$. Based on fit by C.E. Feddersen (1966).

   b. Accuracy: 0.5% up to $a/W = 0.6$. Based on least square fit, B. Gross and J.E. Srawley (1964); W.F. Brown and J.E. Srawley (1966).

2. Kåre, H., *Introduction to Fracture Mechanics*, McGraw-Hill, 1984.

   E^* is the complete elliptic integral with argument $\sqrt{1 - (a/c)^2}$.

3. Agarwal, B.D., and Broutman, L.J., *Analysis and Performance of Fiber Composites*, J.Wiley and Sons (1980).

   Note: $\sigma = (3FL)/(2BW^2)$ is the maximum bending stress in a beam in three-point bending.

Note that the term ***infinite plate*** (e.g., in *Table 13.2*) refers to cases when the crack size, a, is small compared to the overall specimen geometry (i.e., $a << W$). The specimen is so large that its dimensions do not affect the response of the crack and the nearby material. If the crack length is not small, results for a *finite* geometry must be used. In general, the stress intensity factor can be written:

$$K = f \cdot \sigma \sqrt{\pi a}$$
[Eq. 13.32]

where f is a function of component geometry and σ is an applied stress. For a center crack in an infinite plate, $f = 1.0$, and for an edge crack, $f = 1.12$.

Critical Stress Intensity Factor

Fracture occurs when the *stress intensity factor* K (the load) reaches a critical value, called the **critical stress intensity factor**, or **fracture toughness** K_c. *Fracture toughness* is a material property. A few representative values of fracture toughness are given in *Table 13.1*.

For plane stress conditions:

$$K = \sqrt{EG}$$
[Eq. 13.33]

At fracture, $K = K_c$ and $G = G_c$, so:

$$K_c = \sqrt{EG_c}$$
[Eq. 13.34]

The *fracture toughness* K_c for various materials can be calculated from material modulus E and *toughness* G_c. Note that K_c and G_c are often both generally called *toughness*. To distinguish between the two, K_c is called the *fracture toughness*.

Conditions for Crack Initiation and Growth

The condition for fracture initiation and growth using the stress intensity approach is exactly the same as the energy release condition given in *Equation 13.25* except that the stress intensity factor is used in place of the energy release rate, and the material critical stress intensity factor (fracture toughness) is used in place of the material critical energy release rate (toughness).

The calculated *stress intensity factor* (the load) is K and the *critical stress intensity factor* or *fracture toughness* (the material property) is K_c. The criterion for *crack growth*, or *fracture*, is $K > K_c$. The three possible relationships for K and K_c, and their corresponding system responses are:

- $K < K_c$: the crack does not initiate;
- $K = K_c$: the crack initiates but does not grow; and
- $K > K_c$: the crack initiates and grows (fast fracture).

[Eq. 13.35]

Example 13.6 Inspection of Cracks

Given: During a routine maintenance inspection at an industrial plant, an edge crack of length $a = 50$ mm is discovered in a large (infinite) plate of mild steel. The plate is subjected to a stress of $\sigma = 100$ MPa perpendicular to the crack. The value of K_c for the steel is 140 MPa$\sqrt{\text{m}}$.

Required: Determine if the presence of the crack will likely cause fracture.

Solution: The stress intensity factor for an edge crack in an infinitely large plate is, from *Table 13.2*:

$$K = 1.12\sigma\sqrt{\pi a} = 1.12(100 \text{ MPa})\sqrt{\pi(0.05 \text{ m})} = 44.4 \text{ MPa}\sqrt{\text{m}}$$

The critical stress intensity or fracture toughness of the steel is $K_c = 140$ MPa$\sqrt{\text{m}}$. Since $K < K_c$, the crack is stable; it will not grow. The plate may continue to be used, but should continue to be inspected.

Example 13.7 Inspection of Fatigue Cracks

Given: When metals are subjected to cyclic loading, it is known that cracks slowly grow with each cycle. Consider a steel plate subjected to cyclic loading with a maximum tensile stress of $\sigma = 100$ MPa. A center crack of initial length $2a$ exists in the plate. The fracture toughness of the steel is $K_c = 140$ MPa$\sqrt{\text{m}}$.

Required: Determine the crack length $2a_f$ that will cause the plate to fail by fracture.

Solution: At fracture initiation, $K = K_c$, so:

$$a_f = \frac{1}{\pi}\left(\frac{K_c}{\sigma}\right)^2 = \frac{1}{\pi}\left(\frac{140 \text{ MPa}\sqrt{\text{m}}}{100 \text{ MPa}}\right)^2$$

$$\Rightarrow a_f = 0.624$$

Answer: $2a_f = 1.25$ m

The center crack can reach a length of $2a_f = 1.25$ m (!) before fracture occurs. Such a crack, as might occur in a bridge or in a ship, can be observed during routine inspection.

13.2 Design Considerations

Critical Crack Length

Example 13.6 illustrates how to determine if a crack is large enough to cause fracture. The **critical crack length** $2a_c$ is defined as the length of the crack for fracture when a ductile material is subjected to a stress equal to the yield strength $\sigma = S_y$.

For a center crack of length $2a$ in a large thin plate (plate width $2W \gg 2a$), the applied stress at facture is:

$$\sigma = \frac{K_c}{\sqrt{\pi a}} \qquad \text{[Eq. 13.36]}$$

For failure by yielding:

$$\sigma = S_y \qquad \text{[Eq. 13.37]}$$

These failure conditions are plotted in *Figure 13.12*. The intersection of the two conditions defines the critical crack length:

$$\sigma = \frac{K_c}{\sqrt{\pi a_c}} = S_y \qquad \text{[Eq. 13.38]}$$

Solving for the critical half crack length a_c:

$$a_c = \frac{1}{\pi}\left(\frac{K_c}{S_y}\right)^2 \qquad \text{[Eq. 13.39]}$$

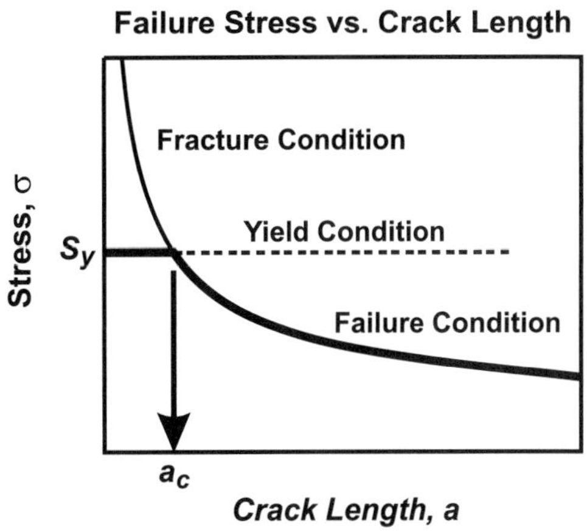

Failure Stress vs. Crack Length

Figure 13.12. Stress versus crack length. For $a < a_c$, the system yields; for $a > a_c$, the system fractures.

The value of the critical crack length depends only upon the material properties of the material. If the crack is shorter than a_c, the material will fail by yielding. If the crack is longer than a_c, the material fails by fracture.

Example 13.8 **Critical Crack Length**

Required: Determine the critical crack length for rotor steel with $K_c = 210 \text{ MPa}\sqrt{m}$ and $S_y = 240 \text{ MPa}$.

Solution: The critical crack length is:

$$a_c = \frac{1}{\pi}\left(\frac{K_c}{S_y}\right)^2 = \frac{1}{\pi}\left(\frac{210 \text{ MPa}\sqrt{m}}{240 \text{ MPa}}\right)^2$$

Answer: $a_c = 0.244 \text{ m}$

For a center crack in a large plate of rotor steel, the critical crack length is $2a_c = 488 \text{ mm}$. Such a crack can generally be detected visually.

Representative Critical Crack Lengths

Estimates of the critical crack length of various materials are given in *Table 13.3*. The table has been created using representative material properties. In practice, scatter of the

Table 13.3. Representative Critical Center Crack Lengths, $2a_c$, below which fracture does not occur.

Material	G_c (kJ/m^2)	K_c (MPa/m$^{1/2}$)	S_y (MPa)	Critical Crack Length, $2a_c$ (mm)
Mild steel	100	140	220	258
Rotor steels	220	210	240	487
Pressure vessel steels	150	170	1600	7.2
Aluminum alloys, high strength	10	26	500	1.7
Cast iron	1	13	500	0.43
Cement	0.03	0.2	3	2.8

material properties, and thus of the critical crack length, is expected. For a given application, calculations should be done using the properties of the actual material.

The following observations may be made:

- The critical crack length for rotor- and mild-steels are often large enough to be noticed by visual observation.
- Pressure vessels steels, although having a high yield strength, are more susceptible to the presence of cracks, so careful inspection is required.
- High strength aluminum alloys are also sensitive to small cracks. Inspection for cracks is a standard activity in aircraft maintenance.
- Cast iron, cement, and ceramics, are brittle; it is difficult to detect cracks of the critical length. The design of brittle materials is based on statistical methods, which are described in *Section 13.6*.

Proof Test to Guarantee Design

When components are difficult to manufacture and analyze, when design experience is limited, or when crack detection is difficult, *proof tests* can be used to demonstrate that a design does not fail by fracture.

Consider a large plate that is to be subjected to a maximum working stress $\sigma = \sigma_w$, with a center crack of length $2a$ (*Figure 13.13*). A proof test is performed at stress $\sigma_p = k\sigma_w$, where k is a constant greater than 1.0 (typically $k \sim 4/3$). If the component does not fail during the proof test, then the applied stress intensity factor of the proof load is less than the fracture toughness:

$$K = \sigma_p\sqrt{\pi a} < K_c \qquad \text{[Eq. 13.40]}$$

Hence the maximum length of any pre-existing crack must be smaller than:

$$a < \frac{1}{\pi}\left(\frac{K_c}{\sigma_p}\right)^2 \qquad \text{[Eq. 13.41]}$$

For the plate to fracture at the working stress σ_w, the crack length is:

$$a_w = \frac{1}{\pi}\left(\frac{K_c}{\sigma_w}\right)^2 = \frac{k^2}{\pi}\left(\frac{K_c}{\sigma_p}\right)^2 > k^2 a \qquad \text{[Eq. 13.42]}$$

For $k = 4/3$ the crack length $2a_w$ to cause fracture at the working load σ_w is at least $k^2 = 1.78$ times longer than any pre-existing crack in the panel $2a$. Hence, it can be deduced that under working conditions, the existing crack length is less than that to cause fracture, with a typical factor of safety. This procedure has been used in rocket design.

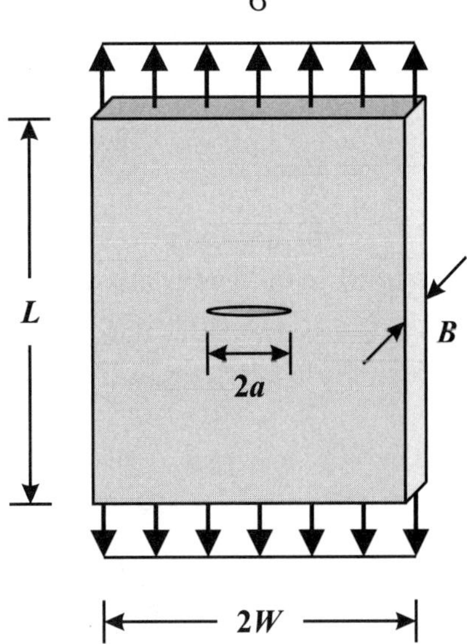

Figure 13.13. Large plate with center crack 2a.

Leak-Before-Break Design

Thin-walled pressure vessels ($t << R$) may have thumb nail surface cracks (*Figure 13.14*). Under pressure, these cracks may grow through the vessel wall. Fast fracture of these vessels, especially those that contain gasses, can be violent. However, if the crack penetrates the wall *before* fast fracture can occur, then failure will be by leakage of the contained gasses. Hence the name *Leak-Before-Break Design*.

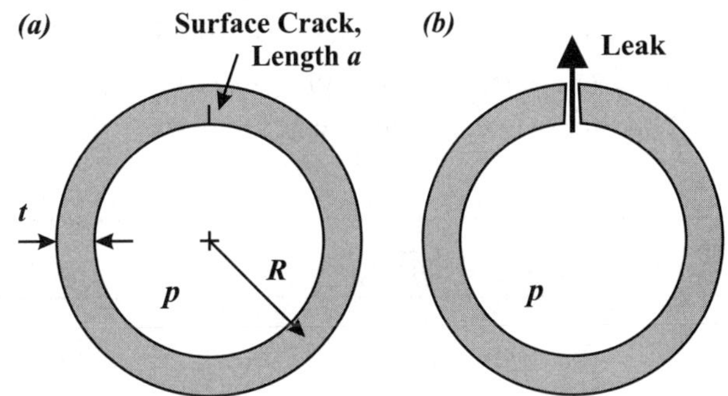

Figure 13.14. (a) Cross-section of pressure vessel. **(b)** Leak-before-break.

Due to the pressure, the stress intensity factor of a semi-elliptical surface crack, for $a << 2c$, is estimated to be (from *Table 13.2*):

$$K = 1.12\sigma\sqrt{\pi a} \qquad \text{[Eq. 13.43]}$$

where σ is the maximum stress in a pressure vessel (the hoop stress in a cylindrical vessel and the spherical stress in a spherical vessel). Fast fracture occurs when the crack length is:

$$a_f = \frac{1}{1.25\pi}\left(\frac{K_c}{\sigma}\right)^2 \qquad \text{[Eq. 13.44]}$$

Leakage occurs when the crack has grown through the thickness of the wall without fast facture occurring. The vessel will leak before it fractures provided that:

$$t < a_f = \frac{1}{1.25\pi}\left(\frac{K_c}{\sigma}\right)^2 \qquad \text{[Eq. 13.45]}$$

In other words, the thickness of the pressure vessel t is less than the crack size required to cause fracture a_f. The crack will never reach length a_f.

The stress at yield is the material yield strength S_y. For leak-before-break to apply for all possible elastic stresses, the wall thickness t must satisfy the condition:

$$t < \frac{1}{1.25\pi}\left(\frac{K_c}{S_y}\right)^2 = \frac{1}{1.25}a_c = 0.8a_c \qquad \text{[Eq. 13.46]}$$

where a_c is the critical crack length of the material, defined by *Equation 13.39*.

If $t < 0.8a_c$, then the crack cannot grow large enough to reach the fracture condition; there is not enough material.

If $t > 0.8a_c$, fracture can occur. Failure of pressure vessels by fast fracture can be violent and dangerous.

From *Table 13.3*, the critical crack lengths, a_c, for mild and rotor steels are typically more than 125 mm. However, for pressure vessel steel and aluminum alloys that have small a_c values (in the order of 1.0 mm), leak-before-break design may not be used, as the wall thickness is generally greater than a_c.

13.3 Crack Stability

Two conditions are required to extend an existing crack:
1. crack growth must initiate (the crack must start to grow), and
2. there must be sufficient driving force for the crack to continue to grow.

When a **stable crack** initiates, it stops growing after a short extension. An **unstable crack** initiates and grows instantaneously (fast-fracture) and the system fails catastrophically.

This section discusses the question: *Does a crack grow in a stable or unstable manner?* To investigate this question, the *resistance curve* and *tearing modulus* are introduced.

The Resistance Curve or R-Curve

In a fracture test, crack growth initiates when the energy release rate G reaches the material critical energy release rate, or toughness, G_c.

For brittle materials, the toughness of a material with increasing crack extension δa is constant (*Figure 13.15a*). A brittle material remains elastic.

For ductile and semi-ductile materials, however, it is observed in experiments that toughness *increases* with crack growth δa. The G_c versus δa relationship is a rising curve (*Figure 13.15b*), known as the **resistance curve** or **R-curve**. A ductile material yields near the crack tip; energy must now be put into plastic deformation as well as crack growth.

For this discussion, the resistance curve is simply taken as a straight line with positive slope T. The toughness $G_c(\delta a)$ as a function of crack extension δa in a ductile material is:

$$G_c(\delta a) = G_{ci} + T\,\delta a \qquad \text{[Eq. 13.47]}$$

where G_{ci} is the material toughness for crack initiation and T is the **tearing modulus**, the slope of the resistance curve. The tearing modulus is a material property.

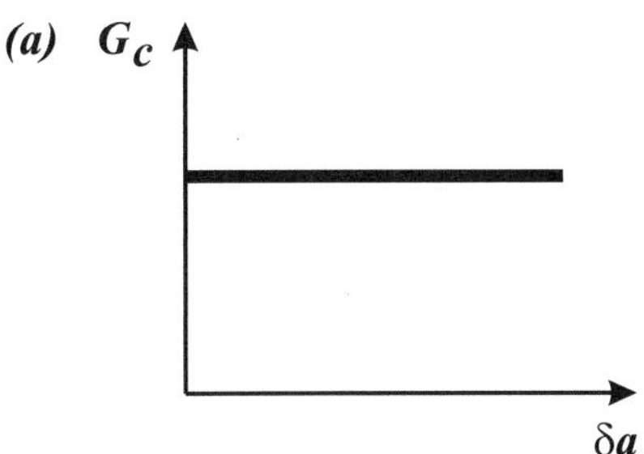
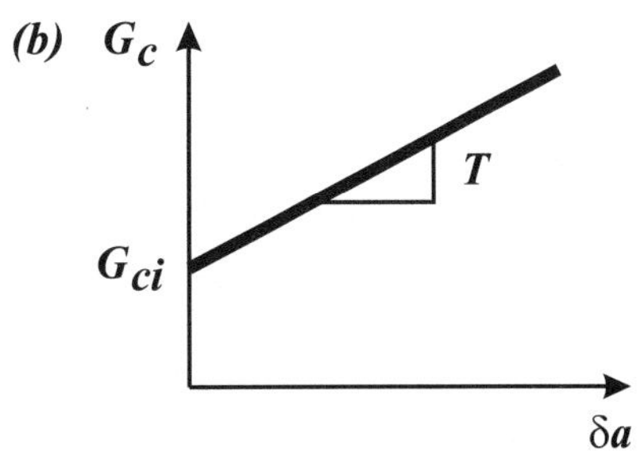

Figure 13.15. (a) G_c versus δa for a brittle material. (b) G_c versus δa (*R*-curve) for a ductile material. As the crack grows, the toughness increases.

Stability of a Crack

Consider a center crack of length $2a$ in an infinite plate of a ductile material subjected to stress σ. The energy release rate is:

$$G = \frac{K^2}{E} = \frac{\sigma^2 \pi a}{E} \qquad \text{[Eq. 13.48]}$$

Crack growth initiates when the energy release rate equals the initial material toughness $G = G_{ci}$ (at $\delta a = 0$). For a given stress σ, the crack length at initiation a_i is:

$$a_i = \frac{E G_{ci}}{\sigma^2 \pi} \qquad \text{[Eq. 13.49]}$$

If the crack grows by δa, the new energy release rate of the system is:

$$G + \delta G = \frac{\sigma^2 \pi}{E}(a_i + \delta a) = G_{ci} + \frac{\sigma^2 \pi}{E}\delta a \qquad \text{[Eq. 13.50]}$$

If this new energy release rate is less than the material toughness at the new crack length $G_c(\delta a)$ (*Equation 13.47*), the crack is *stable* and will not grow:

$$G + \delta G < G_c(\delta a)$$

$$G_{ci} + \frac{\sigma^2 \pi}{E}\delta a < G_{ci} + T\delta a \qquad \text{[Eq. 13.51]}$$

$$\frac{\sigma^2 \pi}{E}\delta a < T\delta a$$

From the condition for crack initiation (*Equation 13.49*):

$$\frac{\sigma^2 \pi}{E} = \frac{G_{ci}}{a_i} \qquad \text{[Eq. 13.52]}$$

Substituting this expression into *Equation 13.51* gives the condition for a *stable* crack. The crack does not grow provided that:

$$\frac{G_{ci}}{a_i} < T \qquad \text{[Eq. 13.53]}$$

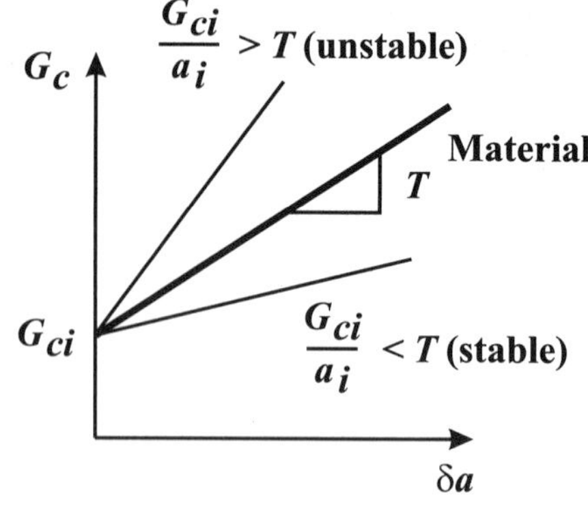

Figure 13.16. Although crack growth may initiate in ductile materials, if $G_{ci}/a_i < T$, crack growth is arrested; the crack is *stable*.

as illustrated qualitatively in *Figure 13.16*.

The crack is *unstable* – it initiates and propagates (grows) – when:

$$\frac{G_{ci}}{a_i} > T \qquad\qquad \text{[Eq. 13.54]}$$

Example 13.9 Cracks Stability in Steel and Aluminum

Given: Consider a proof test performed on two similar components, both with center cracks, one made of steel and the other of aluminum. The proof test is performed at a stress $\sigma_p = 0.9S_y$. The pertinent material properties are given in *Table 13.4*.

Table 13.4. Yield Strength, Toughness, Fracture Toughness, and Tearing Modulus.

Material	S_y (MPa)	G_c (MJ/m²)	K_c (MPa/m$^{1/2}$)	T (MJ/m³)
Steel, ASTM A533	483	0.2	0.2	200
Aluminum, 6061-T6	280	0.02	37	2.7

Required: For each material, (a) determine the size of an existing center crack so that crack growth just initiates due the proof load $\sigma_p = 0.9S_y$, and (b) determine if the crack is *stable* or *unstable*.

Solution: *Step 1.* For a center crack to initiate due to $\sigma_p = 0.9S_y$, it must have a length of:

$$a_i = \frac{G_c E}{\pi \sigma_p^2} = \frac{G_c E}{0.81 \pi S_y^2}$$

Step 2. For the steel:

$$a_{i,st} = \frac{(0.20 \times 10^6 \text{ J/m}^2)(200 \times 10^9 \text{ Pa})}{0.81 \pi (483 \times 10^6 \text{ Pa})^2}$$

Answer: $\underline{a_{i,st} = 67.4 \text{ mm}}$

The slope of the energy release rate for the steel at crack initiation is:

$$\frac{\delta G}{\delta a} = \frac{G_{ci}}{a_i} = \frac{0.20 \times 10^6 \text{ J/m}^2}{0.0674 \text{ m}} = 3.0 \text{ MJ/m}^3$$

The tearing modulus of the steel is: $T_{st} = 200 \text{ MJ/m}^3$.

Since $T > \delta G / \delta a$ at initiation, there is no crack growth:

Answer: the crack in the steel is stable.

Step 3. For the aluminum:

$$a_{i,al} = \frac{(0.02 \times 10^6 \text{ J/m}^2)(70 \times 10^9 \text{ Pa})}{0.81 \pi (280 \times 10^6 \text{ Pa})^2}$$

Answer: $\underline{a_{i,al} = 7.0 \text{ mm}}$

The slope of the energy release rate for the aluminum at crack initiation is:

$$\frac{\delta G}{\delta a} = \frac{G_{ci}}{a_i} = \frac{0.02 \times 10^6 \text{ J/m}^2}{0.007 \text{ m}} = 2.9 \text{ MJ/m}^3$$

The tearing modulus of the aluminum is: $T_{al} = 2.7 \text{ MJ/m}^3$.

Since $T < \delta G / \delta a$ at initiation, the crack grows by fast fracture:

Answer: the crack in the aluminum is unstable.

This example illustrates one reason why aluminum is rarely used as a pressure vessel material.

13.4 Modes of Fracture

The three ***modes*** of fracture are shown in *Figure 13.17* in terms of a double cantilever with a crack length a. The double cantilever can be loaded by three sets of forces P, F, and S, with each load set applied along a primary axis. Loads applied to the crack in these ways are referred to as ***Mode I, II*** and ***III*** loading, respectively. In each case, the crack extends along the center plane of the double-cantilever.

The work of previous sections has concentrated on Mode I, with the load applied normal to the crack surface. Mode I is generally the easiest to analyze and usually the most critical.

Since structures can be loaded in many ways, the loading on a crack is generally a combination of the three modes. This situation is described as **mixed mode** fracture.

The energy release rate G for each of the three basic cases is developed here. Recalling *Equation 13.17*, for constant load P:

$$G = \frac{1}{B}\frac{\partial U}{\partial a}\bigg|_P \qquad \text{[Eq. 13.55]}$$

Recalling *Equation 13.21*, for constant displacement Δ:

$$G = -\frac{1}{B}\frac{\partial U}{\partial a}\bigg|_\Delta \qquad \text{[Eq. 13.56]}$$

The internal strain energies U are calculated using beam-bending theory, or the energy stored in an axial bar.

Mode I

In *Mode I* loading, a pair of loads P is applied normal to the plane of the crack and normal to the direction of crack growth (*Figure 13.17a*). The relative displacement Δ of the loads is given by beam deflection:

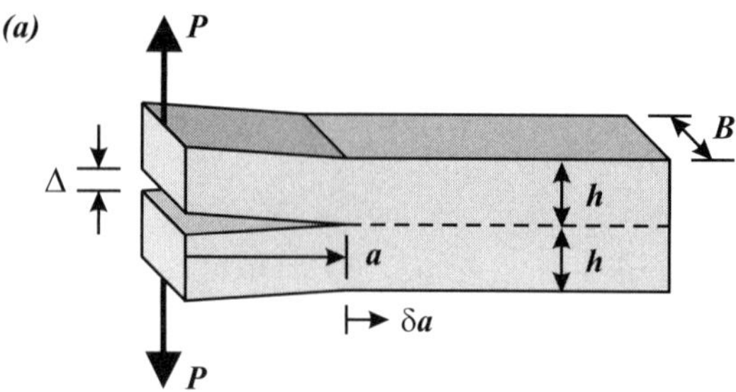
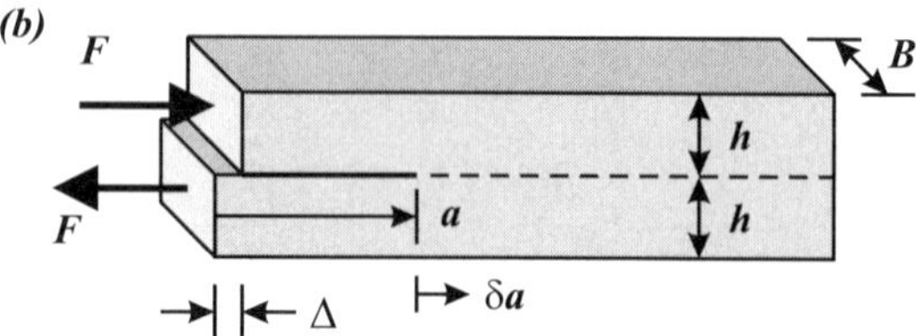
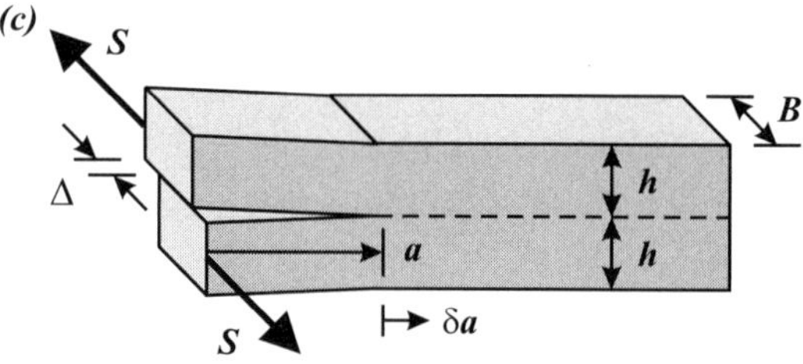

Figure 13.17. The three modes of fracture: **(a)** Mode I, **(b)** Mode II, and **(c)** Mode III.

$$\frac{\Delta}{2} = \frac{Pa^3}{3EI_1} \qquad \text{[Eq. 13.57]}$$

where $I_1 = Bh^3/12$. The internal strain energy of the double cantilever is:

$$U = 2\left[\frac{1}{2}P\left(\frac{\Delta}{2}\right)\right] = \frac{P^2a^3}{3EI_1} = \frac{3\Delta^2EI_1}{4a^3}$$

$$\text{[Eq. 13.58]}$$

For constant load, the *Mode I energy release rate* is:

$$G_I = \frac{1}{B}\frac{\partial U}{\partial a}\bigg|_P = \frac{P^2a^2}{BEI_1} \qquad \text{[Eq. 13.59]}$$

For constant displacement:

$$G_I = -\frac{1}{B}\frac{\partial U}{\partial a}\bigg|_{\Delta} = \frac{9\Delta^2 E I_1}{4Ba^4}$$ [Eq. 13.60]

By substituting the expression for Δ in terms of P, it can be shown that the expressions for G_I derived by the two methods are the same.

To distinguish between the failure modes, the corresponding material toughness – for applied loading perpendicular to the crack surface – is referred to as G_{Ic}, and the initiation condition is: $G_I = G_{Ic}$. Likewise, the Mode I fracture condition may be expressed in terms of stress intensity factor and fracture toughness, with $K_I = K_{Ic}$ at crack initiation.

Mode II

In *Mode II* loading, a pair of loads F is applied parallel to the crack plane, and along the direction of crack growth (*Figure 13.17b*). The relative displacement Δ of the loads is given by the deflection of an axial member:

$$\frac{\Delta}{2} = \frac{Fa}{AE}$$ [Eq. 13.61]

where $A = BH$. The internal strain energy of the two bars is:

$$U = 2\left[\frac{1}{2}F\left(\frac{\Delta}{2}\right)\right] = \frac{F^2 a}{AE} = \frac{\Delta^2 EA}{4a}$$ [Eq. 13.62]

For constant load:

$$G_{II} = \frac{1}{B}\frac{\partial U}{\partial a}\bigg|_{P} = \frac{F^2}{BAE}$$ [Eq. 13.63]

For constant displacement:

$$G_{II} = -\frac{1}{B}\frac{\partial U}{\partial a}\bigg|_{\Delta} = \frac{\Delta^2 EA}{4Ba^2}$$ [Eq. 13.64]

By substituting the expression for Δ in terms of F, it can be shown that the expressions for G_{II} derived by the two methods are the same.

Note that G_{II} is *independent* of the crack length a (*Equation 13.63*); the amount of energy released with crack growth δa is proportional to δa ($\delta U = GB\,\delta a$).

The criterion for crack initiation is: $G_{II} = G_{IIc}$, where G_{IIc} is the material toughness corresponding to Mode II fracture. Toughness G_{IIc} is generally different from Mode I toughness G_{Ic}. The Mode II conditions can also be expressed in terms of stress intensity, with crack initiation occurring when $K_{II} = K_{IIc}$.

Mode III

In *Mode III* loading, a pair of loads S is applied parallel to the plane of the crack, normal to the direction of crack growth (*Figure 13.17c*). The relative displacement Δ of the loads is given by beam deflection:

$$\frac{\Delta}{2} = \frac{Sa^3}{3EI_3} \qquad \text{[Eq. 13.65]}$$

where $I_3 = hB^3/12$. The internal energy of the two cantilevers is:

$$U = 2\left[\frac{1}{2}S\left(\frac{\Delta}{2}\right)\right] = \frac{S^2a^3}{3EI_3} = \frac{3\Delta^2 EI_3}{4a^3} \qquad \text{[Eq. 13.66]}$$

For constant load:

$$G_{III} = \frac{1}{B}\frac{\partial U}{\partial a}\bigg|_P = \frac{S^2a^2}{BEI_3} \qquad \text{[Eq. 13.67]}$$

For constant displacement:

$$G_{III} = -\frac{1}{B}\frac{\partial U}{\partial a}\bigg|_\Delta = \frac{9\Delta^2 EI_3}{4Ba^2} \qquad \text{[Eq. 13.68]}$$

By substituting the expression for Δ in terms of S, it can be shown that the expressions for G_{III} derived by the two methods are the same.

The criterion for crack initiation is: $G_{III} = G_{IIIc}$, where G_{IIIc} is the material toughness corresponding to Mode III fracture. The Mode III conditions can also be expressed in terms of stress intensity, with crack initiation occurring when $K_{III} = K_{IIIc}$.

Note that the Mode III solution for G is the same as the Mode I solution, except for the definition of the moment of inertia I.

Mixed Mode Fracture

When the crack is subjected to all three loads simultaneously, the energy release rate is the sum of the individual energy release rates:

$$G_{total} = G_I + G_{II} + G_{III} \qquad \text{[Eq. 13.69]}$$

The simplest failure criterion is that the material toughness is independent of the mode mixity, so that:

$$G_c = G_{Ic} + G_{IIc} + G_{IIIc} \qquad \text{[Eq. 13.70]}$$

Thus, the condition for crack initiation is:

$$G_{total} = G_c \qquad \text{[Eq. 13.71]}$$

More complex criteria are discussed in fracture mechanics texts and in the research literature.

The most critical loading is generally Mode I, when the crack is being opened by loads applied normal to the crack.

13.5 Thin Film on a Substrate: Spalling

Electronic components and protective coatings on metals are sometimes made by depositing a thin film on a thick substrate at elevated temperature, followed by cooling to room temperature by temperature ΔT. Because the *coefficients of thermal expansion* of the coating and of the metal substrate are different, a **residual stress** σ_R is induced upon cooling. A 2D representation is shown in *Figure 13.18a*.

The coating thickness h is much smaller than the thickness of the substrate d ($h \ll d$); the width of both components is B (into the paper, *Figure 13.18*).

If the coating and substrate were each free to expand or contract due to the temperature change ($\Delta T < 0$), the strain in the coating would be:

$$\varepsilon_{c,t} = \alpha_c(\Delta T) \qquad \text{[Eq. 13.72]}$$

and that in the substrate would be:

$$\varepsilon_{s,t} = \alpha_s(\Delta T) \qquad \text{[Eq. 13.73]}$$

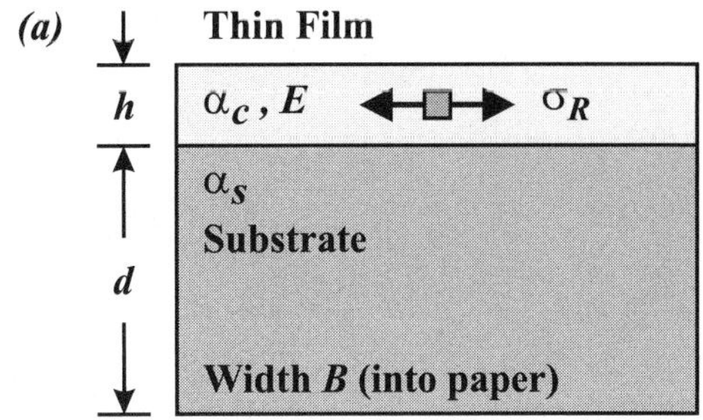

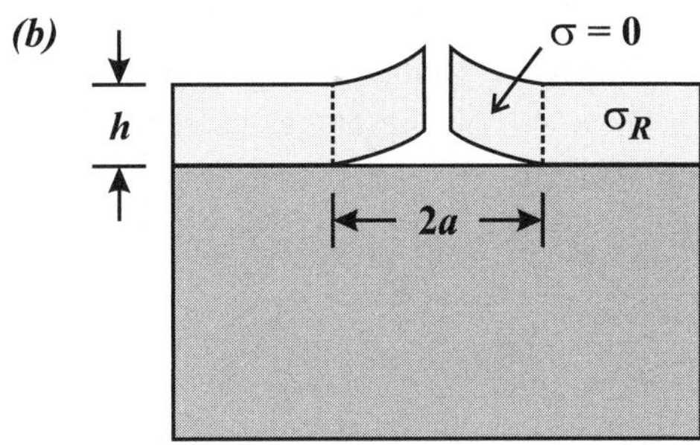

Figure 13.18. (a) Thin film on a thick substrate, $h \ll d$. **(b)** *Spalling* of thin film. Energy released from the thin film goes into creating a crack at the film–substrate interface.

where α_c and α_s are the coefficients of thermal expansion of the coating and the substrate, respectively.

However, the coating must continue to be attached to the substrate at the lower temperature; i.e., the total strain in each material must be the same. Since the coefficients of thermal expansion of the coating and the substrate are generally different, mechanical (elastic) stresses must exist in both materials after cooling.

If the stress in the coating upon cooling is the residual stress, σ_R, then the mechanical strain in the coating is.

$$\varepsilon_{c,m} = \frac{\sigma_R}{E_c}$$
[Eq. 13.74]

where E_c is the modulus of the coating. To satisfy equilibrium, there is also a stress in the substrate, but since the substrate is much thicker than the coating, the average substrate stress is small and can be neglected, i.e., $\sigma_s = 0$.

Equating the total strains after cooling, $\varepsilon_c = \varepsilon_s$, gives:

$$\frac{\sigma_R}{E_c} + \alpha_c(\Delta T) = \alpha_s(\Delta T)$$
[Eq. 13.75]

from which the residual stress in the thin coating is:

$$\sigma_R = E_c(\alpha_s - \alpha_c)(\Delta T)$$
[Eq. 13.76]

Usually $\alpha_s > \alpha_c$, so that the residual stress in the coating is compressive, $\sigma_R < 0$ (recall that $\Delta T < 0$). A large compressive stress causes the coating to **spall** (debond and peel) away from the substrate (*Figure 13.18b*). To determine the condition for spalling, the energy release rate must be calculated.

Referring to the 2D representation of the system in *Figure 13.18b*, the spall has a half length a. When the crack advances by δa, the stress in the coating material of length δa reduces from σ_R to zero. Considering only one-half of the 2D spall, the loss of internal energy in the coating is taken to be:

$$\delta U = \left(\frac{\sigma_R^2}{2E_c}\right)(Bh\,\delta a)$$
[Eq. 13.77]

where B is the breadth of the coating. This energy is available to drive the crack, and the energy release rate is:

$$G = \frac{1}{B}\frac{\partial U}{\partial a}\bigg|_\sigma = \frac{\sigma_R^2 h}{2E_c}$$
[Eq. 13.78]

If this value exceeds the toughness G_c of the *coating–substrate interface*, the crack grows. Note that the energy release rate is independent of crack length a, but is proportional to coating thickness h. Hence, it is advantageous to keep the layer thickness as small as possible. Such a phenomenon is observed with paint; a thin coat of paint is less likely to spall than a thick coat of paint.

Example 13.10 Coating Thickness to Prevent Spalling

Given: A ceramic coating is applied to a thick metal substrate at 900°C, and subsequently cooled to 20°C. The thermal coefficient of expansion of the coating is $\alpha_c = 6\times10^{-6}\ °C^{-1}$ and of the substrate is $\alpha_s = 12\times10^{-6}\ °C^{-1}$. The coating modulus is $E_c = 400$ GPa and the toughness of the coating–substrate interface is $G_c = 0.05$ kJ/m^2.

Required: (a) Approximate the residual stress. (b) Determine the maximum thickness of the coating to avoid spalling.

Solution: *Step 1.* The residual stress is taken from *Equation 13.76*:

$$\sigma_R = E_c(\alpha_s - \alpha_c)(\Delta T) = (400 \times 10^9 \text{ Pa})[(12 - 6) \times 10^{-6} {}^\circ\text{C}^{-1}](-880\,^\circ\text{C})$$

Answer: $\sigma_R = -2.11$ GPa

Step 2. The energy release rate is: $G = \dfrac{\sigma_R^2 h}{2E_c}$

The interface starts to crack when $G = G_c$. The maximum coating thickness is therefore:

$$h_{max} = \frac{2E_c G_c}{\sigma_R^2} = \frac{2(400 \times 10^9 \text{ Pa})(0.05 \times 10^3 \text{ J/m}^2)}{(2.11 \times 10^9 \text{ Pa})^2} = 19.0 \times 10^{-6} \text{ m}$$

Answer: $h_{max} = 19.0$ μm

It is, therefore, not surprising that coatings on electronic substrates are very thin. The use of thin films is the basis of the thermo-mechanical reliability of these systems.

13.6 Statistical Design with Brittle Materials

The superchargers used in high performance cars are often made from ceramics. During the manufacturing process, internal flaws are inherently introduced. Because ceramics have low toughnesses, the strength of ceramic components is controlled by flaw size. If the toughness of a ceramic is $K_{Ic} = 1.0$ MPa$\sqrt{\text{m}}$, and the applied tensile stress is 200 MPa, the maximum allowable size of a center crack of length $2a$ is approximated by the fracture condition:

$$a_{max} = \frac{1}{\pi}\left(\frac{K_{Ic}}{\sigma}\right)^2 = \frac{1}{\pi}\left(\frac{1}{200}\right)^2 = 8.0 \text{ μm} \qquad \text{[Eq. 13.79]}$$

The maximum center crack length is $2a_{max} = 16.0$ μm, which is small, and cannot be seen with the naked eye. Poor preparation of ceramic materials leads to a subpar product with relatively large cracks; such components are weak. Careful preparation is expensive, but produces strong products with only small cracks.

Despite rigorous manufacturing procedures, internal cracks are introduced randomly into the material. These inherent cracks are of various sizes and positions as indicated by the pictorial of a ceramic plate in *Figure 13.19a*. The strength of the plate σ_f is dictated by the longest crack a_{max}. The failure stress σ_f is:

$$\sigma_f = \frac{K_{Ic}}{\sqrt{\pi a_{max}}} \qquad \text{[Eq. 13.80]}$$

The larger the crack size, the smaller the stress at fracture.

Now suppose the plate is divided into several smaller specimens (*Figure 13.19b*). Only one of the smaller plates includes the large crack a_{max}, so that plate fails at σ_f. In each of the other smaller plates, the largest crack is shorter than a_{max}, so each can support a stress greater than σ_f. Thus, the *average strength* of the smaller specimens is greater than the strength of the original large specimen.

Said another way, the larger the specimen, the more likely a large flaw will be found, and thus the smaller the specimen's strength.

The volume of a ceramic component is therefore important in determining its strength. In practice, highly stressed ceramic components tend to be small, typically measured in the order of 10 mm or less.

The Weibull Distribution of Strength

Consider a tensile test performed on a large number of ceramic specimens (about 1000). Because strength is dependent on volume, the tests are all carried out on specimens of the same **reference volume** V_o. The fraction of the specimens $P_s(V_o, \sigma)$ that survive a given applied stress σ is plotted against σ. The shape of the *survival probability* curve P_s–σ has the form shown in *Figure 13.20*.

At low stress levels, the survival rate P_s is 1.0 (all specimens survive). As the stress level is increased, the survival fraction decreases to zero.

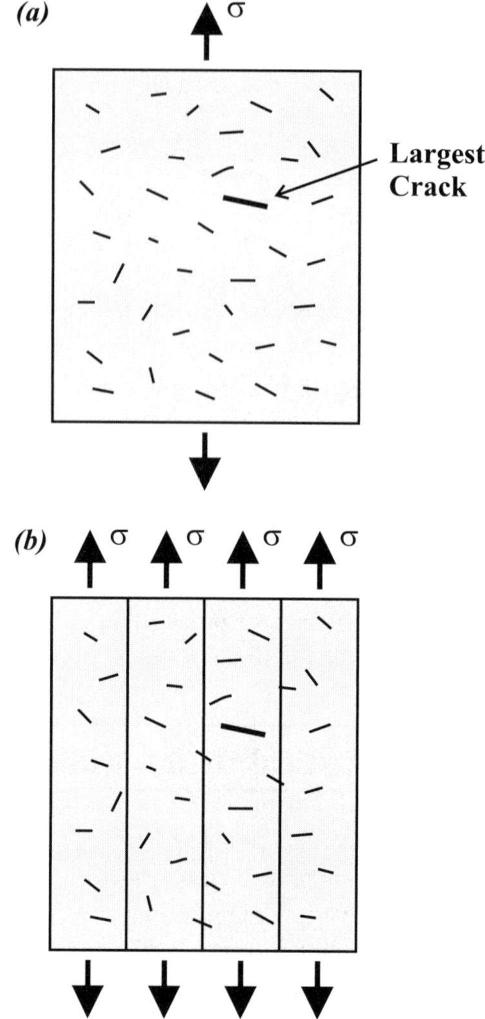

Figure 13.19. **(a)** The ceramic plate fails when the applied stress σ reaches the fracture criterion for the largest crack. **(b)** Same plate divided into four plates. Only one plate will fail at the stress governed by the largest crack. The average stress to fail all four plates will be higher than the stress to fail the single large plate.

The fraction of specimens that have *failed* at any stress level σ is:

$$P_f(V_o, \sigma) = 1 - P_s(V_o, \sigma)$$

[Eq. 13.81]

The *failure probability* curve P_f–σ has the form shown in *Figure 13.20*.

An expression for $P_s(V_o, \sigma)$ that fits experimental data well for ceramic materials is the **Weibull distribution**, which is given by:

$$P_s(V_o, \sigma) = \exp\left[-\left(\frac{\sigma}{S_o}\right)^m\right]$$

[Eq. 13.82]

and for $P_f(V_o, \sigma)$:

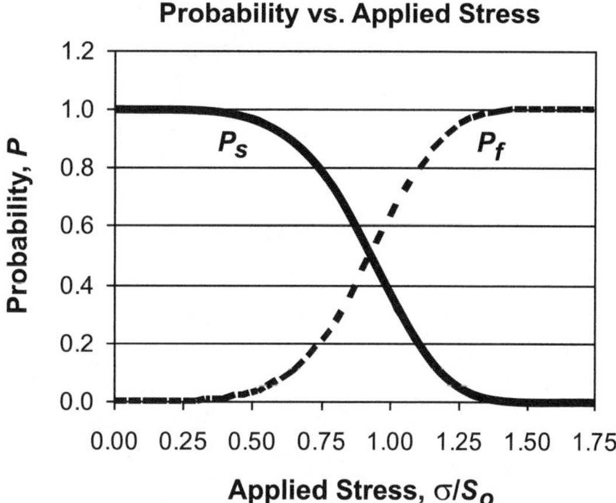

Figure 13.20. Probability of survival P_s and probability of failure P_f as a function of applied stress σ/S_o.

$$P_f(V_o, \sigma) = 1 - \exp\left[-\left(\frac{\sigma}{S_o}\right)^m\right]$$

[Eq. 13.83]

where S_o and m are material parameters that are fitted to experimental data. The term S_o is the **reference stress** and m is the **Weibull modulus**. Variables V_o, S_o, and m are the **Weibull parameters**.

When the applied stress σ is S_o, then the survival fraction is independent of m, and is given by:

$$P_s(V_o, S_o) = \exp(-1) = e^{-1} = 0.368$$

[Eq. 13.84]

The value of S_o is the stress σ corresponding to a survival fraction of 0.368; it is determined from experimental data (*Figure 13.20*).

To determine the constant m requires more work. Taking the natural logarithm of both sides of *Equation 13.82* gives:

$$\ln(P_s) = -\left(\frac{\sigma}{S_o}\right)^m$$

[Eq. 13.85]

Inverting P_s to remove the negative sign, and taking logs again, gives:

$$\ln\left[\ln\left(\frac{1}{P_s}\right)\right] = m \ln\left(\frac{\sigma}{S_o}\right)$$

[Eq. 13.86]

A plot of $\ln[\ln(1/P_s)]$ against $\ln(\sigma/S_o)$ is shown in *Figure 13.21*. The experimental data are fitted by a straight line. The slope of the line is the *Weibull modulus m*.

A more detailed description of determining the Weibull parameters is found in *Section 13.8*.

The variations of $P_S(V_o, \sigma)$ against the stress for $m = 5$ (a typical value for bulk ceramics, ~ 10 mm), $m = 9$, and $m = 15$ (a typical value for ceramic fibers, ~100 μm) are shown in *Figure 13.22*.

When m is high, the P_S–σ curve has a sharp decrease around stress S_o. A high value of m means the dispersion (variation, or spread) of specimen strengths is small; the majority of specimens have about the same strength. Conversely, when m is low, the P_S–σ curve decreases gradually. A low value of m means the dispersion of strengths is wide.

Mean Strength

The ***mean strength*** σ_m of the distribution is given by:

$$\frac{\sigma_m}{S_o} = \int_0^\infty \frac{\sigma}{S_o} dP_s$$

$$= \int_0^\infty m\left(\frac{\sigma}{S_o}\right)^m \exp\left[-\left(\frac{\sigma}{S_o}\right)^m\right] d\left(\frac{\sigma}{S_o}\right)$$

$$= \Gamma\left(1 + \frac{1}{m}\right)$$

[Eq. 13.87]

where $\Gamma(x)$ is the ***gamma function***, the values of which are given in mathematical tables or software. The mean strengths for various values of m are given in *Table 13.5*.

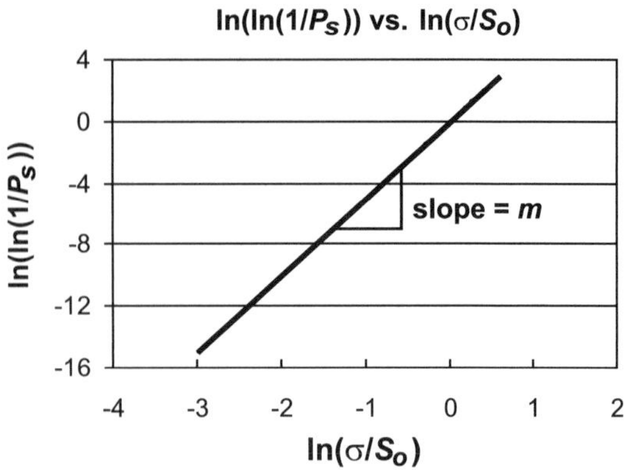

Figure 13.21. Experimental data are plotted on the ln[ln(1/P_S)] versus ln(σ/S_o) graph. The slope is the Weibull modulus m.

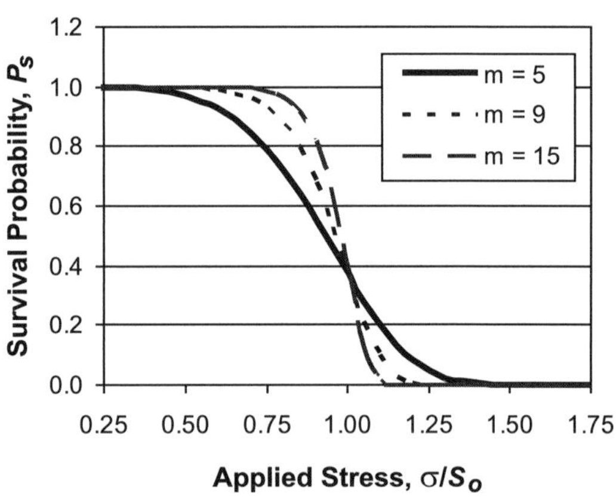

Figure 13.22. Probability of survival P_s as a function of applied stress σ/S_o for various Weibull moduli m.

Table 13.5. Mean strength as a function of m.

	$m = 5$	7	9	15
σ_m/S_o	0.918	0.935	0.947	0.966

Effect of Volume

Consider a brittle tensile specimen divided into n elements, each of volume V_o as shown in *Figure 13.23a*. The total volume is $V = nV_o$. The probability of survival of the system of n elements is the product of the survival probabilities $P_s(V_o, \sigma)$ of all n elements, each of which is subjected to tensile stress σ. Thus:

$$
\begin{aligned}
P_s(V, \sigma) &= [P_s(V_o, \sigma)]^n \\
&= [P_s(V_o, \sigma)]^{V/V_o} \\
&= \left[\exp\left[-\left(\frac{\sigma}{S_o}\right)^m \right] \right]^{V/V_o} \\
&= \exp\left[-\frac{V}{V_o}\left(\frac{\sigma}{S_o}\right)^m \right]
\end{aligned}
$$

[Eq. 13.88]

This formulation is also valid when the elements are aligned, as in a chain (*Figures 13.23b* and *c*), since all elements experience the same stress. Hence, this approach is often referred to as ***weakest link statistics***. As soon as the weakest link fails, the entire systems fails.

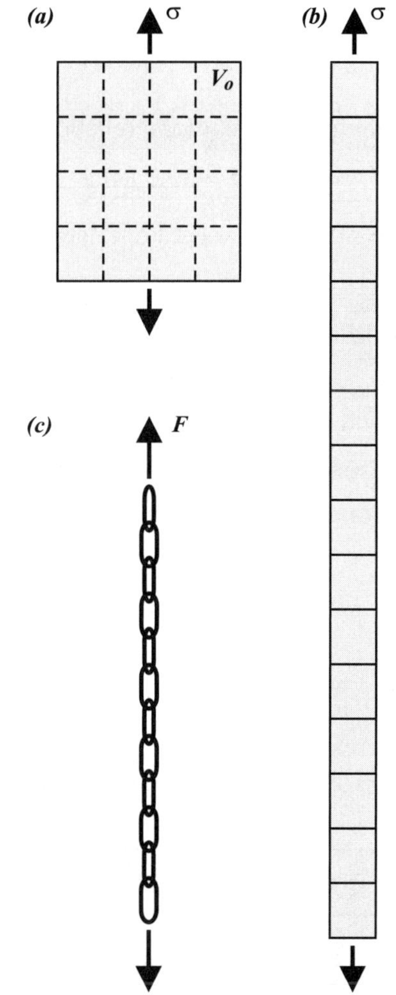

Figure 13.23. **(a)** A plate divided into n elements, each of volume V_o. **(b)** A long specimen with same volume as that in *Part (a)*. **(c)** A chain is only as strong as its weakest link.

Example 13.11 **Probability of Survival: Volume V_o**

Given: A bar of volume $V = V_o$ is subjected to applied stress σ. The *Weibull modulus m* is 5 and the *reference stress S_o* is 200 MPa.

Required: Determine the stress σ that has a reliability of 999/1000, i.e., only 1 in 1000 specimens breaks at stress σ.

Solution: The survival rate is:

$$P_s(V_o, \sigma) = \frac{999}{1000} = \exp\left[-\left(\frac{\sigma}{S_o}\right)^m\right] = \exp\left[-\left(\frac{\sigma}{200 \text{ MPa}}\right)^5\right]$$

Solving for the stress for a reliability of 0.999 gives:

Answer: $\sigma = 50.2$ MPa

At $\sigma = 50.2$ MPa, at least one specimen in 1000 will probably fail.

***Example 13.12* Probability of Survival: Volume 100 V_o**

Given: A bar of volume $V = 100 V_o$ is subjected to applied stress σ. The value of m is 5 and S_o is 200 MPa.

Required: Determine the stress σ that has a reliability of 999/1000.

Solution: The survival rate is:

$$P_s(V, \sigma) = \frac{999}{1000} = \exp\left[-\frac{V}{V_o}\left(\frac{\sigma}{S_o}\right)^m\right] = \exp\left[-100\left(\frac{\sigma}{200 \text{ MPa}}\right)^5\right]$$

Answer: $\sigma = 20.0$ MPa

By increasing the volume by a factor of 100 from *Example 13.11*, the allowable stress for $P_s = 99.9\%$ drops from 50 to 20 MPa. The larger the volume, the lower the strength.

13.7 Effect of Non-Uniform Stress in Statistical Design

In general, the stress in a component varies from point-to-point. Typical components are beams in bending, shafts in torsion, and disks that are rotating. The aim now is to find the survival probability $P_s(V, F)$, where V is the total volume of the component and F is the applied load.

The influence of the stress distribution on strength is calculated by dividing the component of volume V into a large number of constant–stress elements, each of volume dV, and then calculating the survival probability of each element. The survival probability of element dV subjected to tensile stress σ is:

$$P_s(dV, \sigma) = \exp\left[-\frac{dV}{V_o}\left(\frac{\sigma}{S_o}\right)^m\right] \qquad \text{[Eq. 13.89]}$$

Stress σ varies throughout the component. The survival probability of the entire component is found by multiplying together the survival probability of all elements dV. Recalling the rule for exponential multiplication (e.g., $e^a e^b = e^{a+b}$), the expression for the survival probability of the entire component is:

$$P_s(V, F) = \exp\left[-\int_V \frac{1}{V_o}\left(\frac{\sigma}{S_o}\right)^m dV\right]$$ [Eq. 13.90]

where V is the total volume and F is the load that causes the varying stress σ on differential elements dV. In addition, since compressive stresses ($\sigma < 0$) do not cause cracks to grow, elements dV that are in compression do not contribute to the integral of *Equation 13.90*.

Example 13.13 Beam Under Constant Moment

Given: A beam with a square cross-section of sides b, and having length L, is subjected to a constant negative moment M over its entire length (*Figure 13.24*).

Required: Determine the expression for the probability of survival, $P_s(V, M)$.

Solution: *Step 1.* Referring to the survival fraction, *Equation 13.90*, the stress as a function of location must be found, along with an appropriate element dV, and the limits of integration.

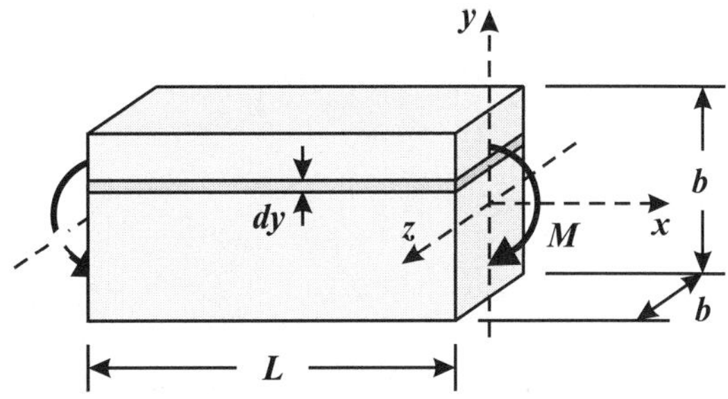

Figure 13.24. Beam subjected to constant moment M.

Bending stress varies linearly with distance y from the neutral axis:

$$\sigma(y) = \sigma_{max}\frac{2y}{b}$$

where the maximum bending stress for a square cross-section is:　$\sigma_{max} = \dfrac{6M}{b^3}$

Since $M(x)$ is constant, the expression for stress as a function of position (x,y) is known.

A rectangular beam in bending can be considered as a stack of axial members, each b wide and dy deep, each supporting normal stress $\sigma(y)$. The differential volume dV is:

$$dV = Lb(dy)$$

Also for $y < 0$, the stresses are compressive. The compressive side of the beam can be neglected in the probability calculation since compressive stresses do not cause fracture.

Step 2. Substituting σ and dV into *Equation 13.90*, and setting up the integral from $y = 0$ to $b/2$, gives:

$$P_s(V, M) = \exp\left[-\int_0^{b/2} \frac{1}{V_o}\left[\frac{\sigma_{max}}{S_o}\left(\frac{2y}{b}\right)\right]^m [Lb(dy)]\right]$$

Step 3. Integrating:

$$\text{Answer: } P_s(V, M) = \exp\left[\frac{-V}{2(m+1)V_o}\left(\frac{\sigma_{max}}{S_o}\right)^m\right]$$

where $V = b^2 L$ is the total volume of the beam and $\sigma_{max} = \dfrac{6M}{b^3}$.

The term $\dfrac{1}{2(m+1)}$ is known as the **load factor** λ and depends on the volume geometry and how it is loaded. The survival probability can then be rewritten to include λ:

$$P_s(V, M) = \exp\left[-\lambda\frac{V}{V_o}\left(\frac{\sigma_{max}}{S_o}\right)^m\right] \qquad \text{[Eq. 13.91]}$$

For a beam in pure bending:

$$\lambda = \frac{1}{2(m+1)}$$

The following examples determine the load factors for different loads and geometries.

Example 13.14 Three-Point Bend Test

Given: A simply-supported rectangular beam, breadth b, depth d, and length L is subjected to a central load P (*Figure 13.25*). The beam is made of a ceramic with Weibull modulus m, reference stress S_o, and reference volume V_o.

Required: Determine the survival probability $P_s(V, P)$ for the beam in three-point bending.

Solution: *Step 1.* The moment of inertia is $I = bd^3/12$. Bending stress varies linearly with distance x along the beam and with distance y from the neutral axis. The stress distribution from $x = 0$ to $L/2$ is:

$$\sigma(x, y) = -\frac{M(x)y}{I}$$

$$= -\left(\frac{Px}{2}\right)\frac{y}{I} = -\sigma_{max}\frac{4xy}{Ld}$$

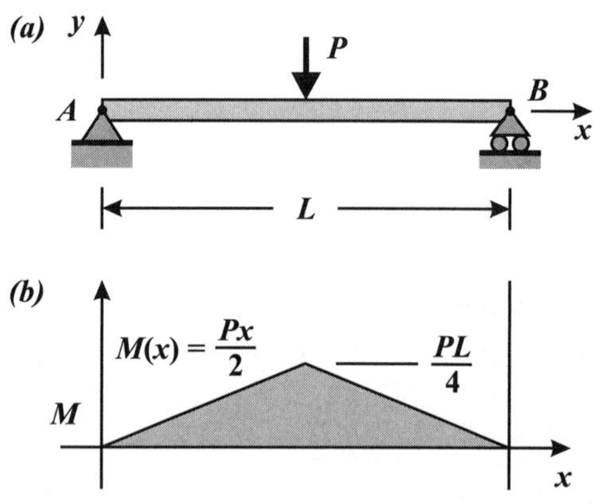

Figure 13.25. (a) Simply supported beam subjected to central point load *P.* **(b)** Moment diagram.

The maximum bending stress σ_{max} for a beam of rectangular cross-section in three-point bending is:

$$\sigma_{max} = \frac{M_{max}c}{I} = \left(\frac{PL}{4}\right)\frac{d/2}{I} = \frac{3PL}{2bd^2}$$

Note that σ_{max} only occurs at two points ($x = L/2$, $y = \pm d/2$), and only one of those points is in tension. For the present case, this point is at the bottom center of the three-point beam, and is the location where fracture will most likely (but not necessarily) occur.

Step 2. Substituting the expression for stress $\sigma(x,y)$ into the probability equation:

$$P_s(V, P) = \exp\left[-2\int_0^{L/2}\int_{-d/2}^0 \frac{1}{V_o}\left(\frac{\sigma_{max}}{S_o}\frac{4xy}{Ld}\right)^m (b\,dydx)\right]$$

The differential volume element is $dV = b\,dydx$. The x-integral is taken from $x = 0$ to $L/2$ (half the beam), so the expression within the exponential function is multiplied by 2 to included the entire volume in tension. The stresses in the system are tensile only for $y < 0$; the compressive stresses in the beam ($y > 0$) do not cause fracture. Thus, the y-integral is taken from $-d/2$ to 0.

Step 3. Upon integration, the probability of survival is:

$$Answer:\ P_s(V, P) = \exp\left[-\frac{V}{2(m+1)^2 V_o}\left(\frac{\sigma_{max}}{S_o}\right)^m\right]$$

where $V = (bd)L$ is the total volume and $\sigma_{max} = \dfrac{3PL}{2bd^2}$.

The load factor λ for the rectangular beam in three-point bending is:

$$\lambda = \frac{1}{2(m+1)^2}$$

Example 13.15 Spinning Rod

Given: A ceramic rod of length L and radius R, spins about a vertical axis perpendicular to the rod's axis, with angular velocity ω in radians per second (*Figure 13.26*). The ceramic has mass density ρ and Weibull parameters m, S_o, and V_o.

Required: Determine the survival probability $P_s(V, \omega)$ for the spinning rod.

Solution: *Step 1.* The stress induced in the rod by the centripetal force is a function of x. Consider a FDB of a differential element $dV = \pi R^2\,dx$ (*Figure 13.26b*). The net force on dV is the product of its mass $dm = \rho\,dV$ and its centripetal acceleration $\omega^2 x$, which is in the negative x-direction:

$$dF = [\rho(\pi R^2)dx](-\omega^2 x)$$

Dividing both sides by the cross-sectional area gives:

$$d\sigma = -\rho\omega^2 x\, dx$$

Integrating both sides gives the stress along the rod:

$$\sigma(x) = \frac{-\rho\omega^2 x^2}{2} + C$$

Applying the boundary condition that the ends of the rod are stress free, $\sigma(x = \pm L/2) = 0$, gives the constant C:

$$C = \frac{\rho\omega^2 L^2}{8}$$

The maximum stress occurs at the axis of rotation $(x = 0)$:

$$\sigma_{max} = \frac{\rho\omega^2 L^2}{8}$$

The stress distribution throughout the rod can then be written:

$$\sigma(x) = \sigma_{max}\left[1 - \left(\frac{2x}{L}\right)^2\right]$$

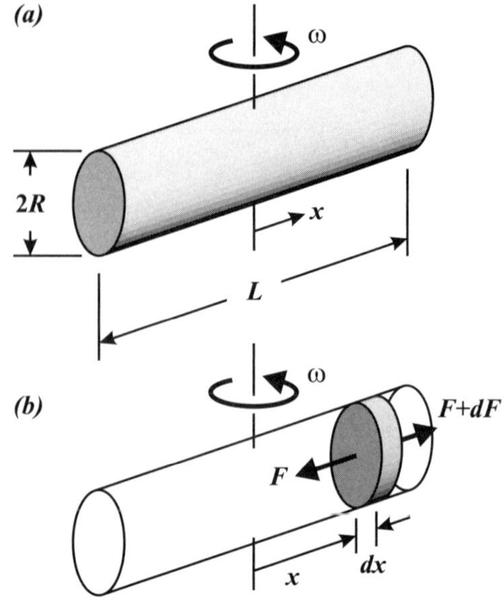

Figure 13.26. (a) Rod spinning with angular velocity ω. **(b)** FBD of length of rod dV.

Step 2. Substituting σ and dV into *Equation 13.90* gives:

$$P_s(V, \omega) = \exp\left[-2\int_0^{L/2} \frac{1}{V_o}\left[\frac{\sigma_{max}}{S_o}\left[1 - \left(\frac{2x}{L}\right)^2\right]\right]^m [\pi R^2\, dx]\right]$$

The differential volume is $dV = \pi R^2\, dx$. The integral is taken from $x = 0$ to $L/2$, and the expression within the exponential is multiplied by 2 since the entire volume is in tension.

Step 3. Integrating, and substituting $\bar{x} = 2x/L$, gives:

$$P_s(V, \omega) = \exp\left[-\frac{V}{V_o}\left(\frac{\sigma_{max}}{S_o}\right)^m \int_0^1 [1 - \bar{x}^2]^m\, d\bar{x}\right]$$

where $V = \pi R^2 L$ and $\sigma_{max} = \dfrac{\rho\omega^2 L^2}{8}$.

The integral is the *load factor*:

$$\lambda = \int_0^1 [1 - \bar{x}^2]^m\, d\bar{x}$$

Performing the standard integral:

$$\lambda = \frac{2^{2m}(m!)^2}{(2m+1)!}$$

Thus:

$$\textit{Answer: } P_s(V, \omega) = \exp\left[-\left(\frac{2^{2m}(m!)^2}{(2m+1)!}\right)\left(\frac{V}{V_o}\right)\left(\frac{\sigma_{max}}{S_o}\right)^m\right]$$

Based on: G.G. Trantina and C.A. Johnson "Spin Testing of Ceramic Materials," *Fracture Mechanics of Ceramics*, vol.3, Plenum Publishing Corporation, 1978.

Load Factor

The probability of survival P_s of a ceramic component has the general form:

$$P_s(V, F) = \exp\left[-\lambda\frac{V}{V_o}\left(\frac{\sigma_{max}}{S_o}\right)^m\right] \qquad \text{[Eq. 13.92]}$$

where λ is the *load factor*. The **load factor** is a measure the *effective volume* of the component. *Equation 13.92* is the survival probability of a component of volume λV subjected to uniform stress σ_{max} (*Equation 13.88*). Factor λ is a function of Weibull modulus m.

Table 13.6 summarizes the load factors for the four components and loadings discussed in this section. The load factor depends on the distribution of stress in the material, and thus depends on the component geometry and how the component is loaded.

For the uniaxial bar, the stress is uniform – all points in the component are subjected to the same stress, $\sigma(x, y) = \sigma_{max}$. Hence, the bar's load factor is $\lambda = 1.0$.

For a beam subjected to a constant moment, only the top (or bottom) of the cross-section is subjected to the maximum tensile stress. All other material points are subjected to less tensile stress, and only half of the beam is actually in tension. The effective volume of a beam in bending is much smaller than that of the axial bar; e.g., for $m = 5$, $\lambda = 0.083$.

Table 13.6. Load Factor and Maximum Stress for various components and loadings.

Member, Loading	Uniaxial Bar, Tension	Rectangular Beam, Constant Moment	Rectangular Beam, Three-Point Bending	Bar, Spinning
Load factor, λ	1.0	$\dfrac{1}{2(m+1)}$	$\dfrac{1}{2(m+1)^2}$	$\dfrac{2^{2m}(m!)^2}{(2m+1)!}$
Maximum stress, σ_{max}	$\dfrac{P}{A}$	$\dfrac{6M}{bd^2}$	$\dfrac{3PL}{2bd^2}$	$\dfrac{\rho\omega^2 L^2}{8}$

A graph of load factor λ against Weibull modulus m for various geometries and loads is shown in *Figure 13.27*. The figure shows that increasing the Weibull modulus decreases the load factor and thus decreases the effect of component volume on survival probability. For the spinning bar, if m increases from $m = 5$ to 15, the load factor decreases from $\lambda \sim 0.4$ to 0.2, decreasing the effective volume by nearly a factor of 2.

A high modulus means that the dispersion of strength in the ceramic is small, so that volume becomes statistically less important to the failure strength. For large values of m, λ approaches a constant value for each configuration.

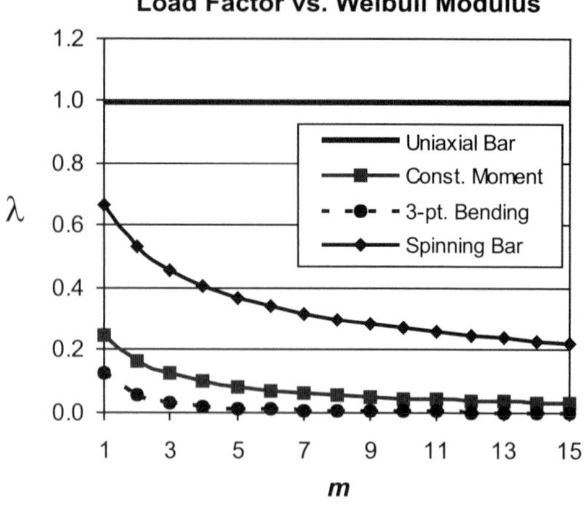

Figure 13.27. Load factor as a function of Weibull modulus for various geometries and loads: bar in tension, rectangular beam with constant moment, rectangular beam in three-point bending, and spinning bar.

Example 13.16 Spinning Rod 2

Given: A rod of length $L = 100$ mm and cross-section $b = 5$ mm square spins about an axis through its center of mass (*Figure 13.28*). The density of the material is 3200 kg/m^3. The Weibull parameters are:

$$m = 5, \quad S_o = 300 \text{ MPa}, \quad V_o = 0.11 \times 10^{-6} \text{ m}^3$$

Required: Determine (a) the maximum stress in the rod for a probability of survival of 999 in 1000 and (b) the rotational speed ω corresponding to that stress.

Solution: *Step 1.* The volume of the spinning rod is $V = 2.5 \times 10^{-6}$ m^3. For a spinning rod, with $m = 5$, the load factor is:

$$\lambda = \frac{2^{2m}(m!)^2}{(2m+1)!} = \frac{2^{10}(5!)^2}{11!} = 0.369$$

Step 2. The probability function is:

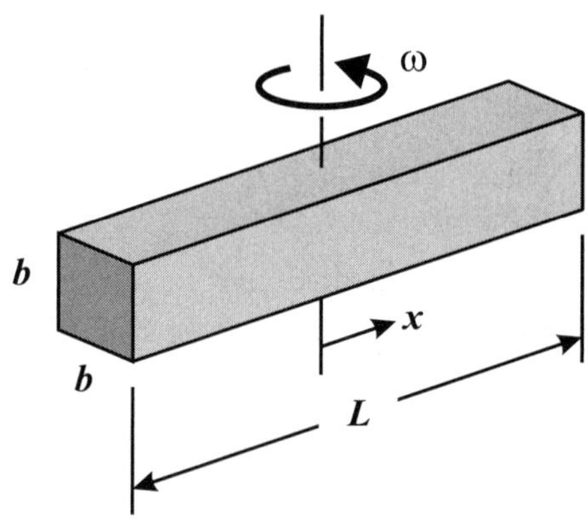

Figure 13.28. Spinning rod with square cross-section.

$$P_s(V, \omega) = \exp\left[-\lambda \frac{V}{V_o}\left(\frac{\sigma_{max}}{S_o}\right)^m\right]$$

$$0.999 = \exp\left[-0.369\left(\frac{2.5\times10^{-6}\ \text{m}^3}{0.11\times10^{-6}\ \text{m}^3}\right)\left(\frac{\sigma_{max}}{300\ \text{MPa}}\right)^5\right]$$

Step 3. Solving for the maximum stress:

$$\Rightarrow \sigma_{max} = 49\ \text{MPa}$$

Step 4. For a spinning rod: $\sigma_{max} = \dfrac{\rho\omega^2 L^2}{8}$

Thus the angular velocity is:

$$\omega = \left[\frac{8\sigma_{max}}{\rho L^2}\right]^{1/2} = \left[\frac{8(49\times10^6\ \text{Pa})}{(3200\ \text{kg/m}^3)(0.1\ \text{m})^2}\right]^{1/2}$$

Answer: $\omega = 3500\ \text{rad/s} = 33{,}420\ \text{rev/min}$

The rotational speed of small ceramic components can be very high.

13.8 Determining Weibull Parameters

Tensile tests on brittle materials are difficult to perform, so it is usual to measure their material properties from standard three-point bend tests. A series of *three-point bend* tests is performed on standard beams of length 80 mm and square cross-section of area 100 mm² (*Figure 13.29*). Representative results of such a test performed on 1000 samples are given in *Table 13.7* and the survival probability is plotted in *Figure 13.30*.

The *Weibull parameters* that must be determined are: the *reference volume* V_o, the *reference stress* S_o, and the *Weibull modulus* m.

The probability of survival for the three-point bend test is:

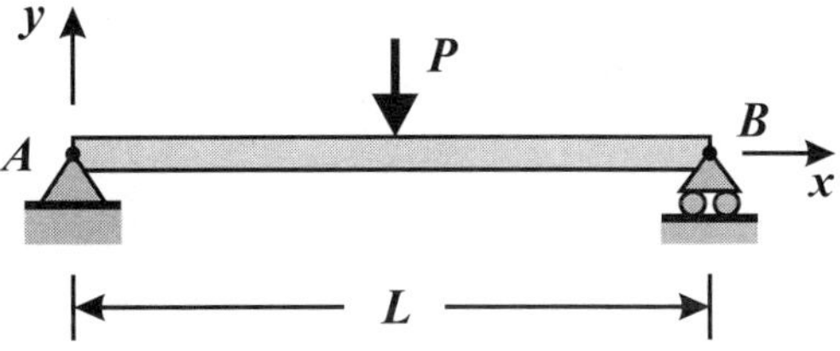

Figure 13.29. Three-point bend test.

Table 13.7. Maximum bending stress σ_{max}, and corresponding number of surviving beams n.

σ_{max} (MPa)	n	σ_{max} (MPa)	n	σ_{max} (MPa)	n
75	999	200	882	350	115
100	994	225	788	375	47
125	987	250	665	400	15
150	969	300	368	425	3
175	931	325	240	450	0

$$P_s(V, \sigma_{max}) = \exp\left[-\frac{1}{2(m+1)^2}\left(\frac{V}{V_o}\right)\left(\frac{\sigma_{max}}{S_o}\right)^m\right]$$

[Eq. 13.93]

where V is the volume of one sample:

$$V = d^2L = 8.0\times10^{-6}\ \text{m}^2$$

[Eq. 13.94]

If the reference volume V_o is selected to be:

$$V_o = \frac{V}{2(m+1)^2}$$

[Eq. 13.95]

then the probability distribution of the three-point bend test reduces to:

$$P_s(V, \sigma_{max}) = \exp\left[-\left(\frac{\sigma_{max}}{S_o}\right)^m\right]$$

[Eq. 13.96]

The reference stress S_o is defined by the stress σ_{max} for a survival probability of:

$$P_s = e^{-1} = 0.368$$ [Eq. 13.97]

Reading directly from the experimental graph (*Figure 13.30*) gives $S_o = 300$ MPa.

To find the Weibull modulus, repeat *Equation 13.86*:

$$\ln\left[\ln\left(\frac{1}{P_s}\right)\right] = m\ \ln\left(\frac{\sigma}{S_o}\right)$$

[Eq. 13.98]

Using the experimental data, the expression $\ln[\ln(1/P_s)]$ is plotted against $\ln(\sigma/S_o)$ and a straight line fitted through the data points as shown in *Figure 13.31*. The Weibull modulus is the slope of the straight line. Here, $m = 5$.

The material properties that fit the experiment are: $m = 5$ and $S_o = 300$ MPa. The reference volume V_o is $V_o = 0.11\times10^{-6}\ \text{m}^3$.

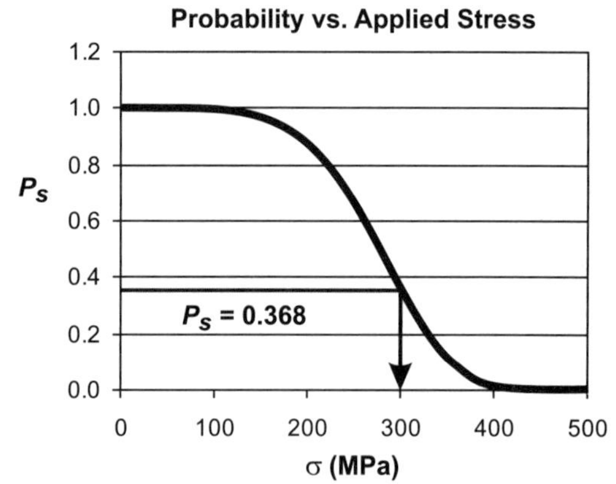

Figure 13.30. Probability of survival versus stress.

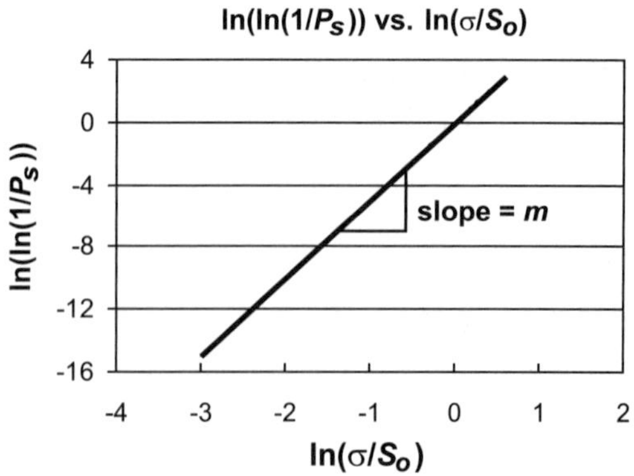

Figure 13.31. Determination of Weibull modulus.

13.9 Strength of Fiber Bundles: Global Load Sharing

Ceramic fibers are used in composite materials because of their strength and stiffness. For example, 100 µm diameter silicon carbide fibers can have a strength of 2000 MPa. To attain such strength, the maximum flaw or crack length must be less than 0.7 µm (assuming $K_c = 3$ MPa$\sqrt{m}$). Variation in fiber strength is a consequence of geometric imperfections introduced during the process of drawing the fiber. Fiber processing, therefore, is an important area of research and development.

Following the probability ideas used to determine the strength of ceramics, tests are performed on many fibers to find the parameters of the Weibull distribution. Because the geometry of the fibers is long and slender, length is used in place of volume. The fraction of the fibers of length L that survive an applied stress σ is therefore described by the Weibull relation:

$$P_s(L, \sigma) = \exp\left[-\frac{L}{L_o}\left(\frac{\sigma}{S_o} \right)^m \right]$$

[Eq. 13.99]

where L_o and S_o are the *reference length* and *reference stress*, respectively. The values L_o and S_o correspond to a fiber length and stress level for a survival fraction of 0.368. In high tech applications, the value of L_o is often chosen to be 25.4 mm (1.0 in.).

Bundle Strength

Fibers can be formed into *fiber bundles* that consist of many fibers acting in parallel, as represented in *Figure 13.32*. The fibers are not connected, but are constrained so that they each have the same strain as the bundle. The effect of this constraint is key to the overall response of the fiber bundle system.

For convenience, let each fiber have a length of L. The probability of survival for one fiber subjected to stress σ is:

$$P_s(L, \sigma) = \exp\left[-\frac{L}{L_o}\left(\frac{\sigma}{S_o} \right)^m \right]$$

[Eq. 13.100]

If there are n individual fibers in the bundle, weakest link statistics gives the probability of survival of the system:

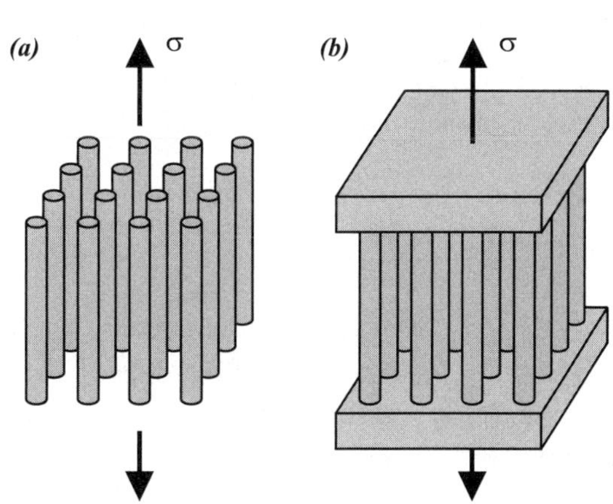

Figure 13.32. (a) Unconstrained fiber bundle under stress. **(b)** Constrained bundle where all fibers have the same strain.

$$P_s(nL, \sigma) = \exp\left[-\frac{nL}{L_o}\left(\frac{\sigma}{S_o}\right)^m\right]$$
[Eq. 13.101]

Note that nL is the total length of the fibers. Using weakest link statistics, the probability of survival of the system would decrease rapidly with increasing numbers of fibers, since if one link fails, the entire system fails. For fiber bundles, this does not happen in practice.

In bundles where the fibers are constrained to have the same strain (*Figure 13.32b*), when one fiber breaks, the system does not fail. The crack does not immediately grow into the neighboring fiber. Instead, the load that was supported by the broken fiber is transferred to the remaining unbroken fibers. This phenomenon is known as **global load sharing**.

If the nominally applied stress on a fiber bundle of length L is σ_n, then when some of the fibers break, the stress σ supported by each of the unbroken fibers is:

$$\sigma = \frac{\sigma_n}{P_s}$$
[Eq. 13.102]

where P_s is the survival probability of a fiber of length L at stress σ. For example, if the applied stress is σ_n, and 20% of the fibers have failed ($P_s = 0.8$), the surviving fibers must now support 125% of the nominal stress ($\sigma = \sigma_n/P_s = 1.25\sigma_n$).

Rearranging *Equation 13.102*:

$$\sigma_n = \sigma P_s = \sigma \exp\left[-\frac{L}{L_o}\left(\frac{\sigma}{S_o}\right)^m\right]$$
[Eq. 13.103]

Normalizing by S_o gives the relationship between the nominal stress σ_n and the actual strain of the bundle, $\varepsilon = \sigma/E$:

$$\frac{\sigma_n}{S_o} = \frac{\varepsilon E}{S_o}\exp\left[-\frac{L}{L_o}\left(\frac{\varepsilon E}{S_o}\right)^m\right]$$
[Eq. 13.104]

The resulting stress–strain curve, for the case $m = 5$ and $L/L_o = 1.0$, is shown in *Figure 13.33*. Initially, the linear slope of the curve is defined by the elastic modulus of the ceramic fibers. As stress increases, fibers begin to break and the *effective modulus* of the fiber bundle decreases.

The maximum value of the applied stress is determined by taking the derivative of *Equation 13.104* and setting it to zero. The maximum occurs when:

$$\frac{\varepsilon E}{S_o} = m^{-1/m}$$
[Eq. 13.105]

and the corresponding maximum applied stress is the **fiber bundle strength** σ_b:

$$\frac{\sigma_b}{S_o} = \left(\frac{L_o}{mL}\right)^{1/m} \exp\left(-\frac{1}{m}\right)$$

$$= \left(\frac{L_o}{L}\right)^{1/m} f_b$$

[Eq. 13.106]

where f_b is:

$$f_b = \left(\frac{1}{me}\right)^{1/m}$$

[Eq. 13.107]

Values of f_b for various moduli m are given in *Table 13.8*. The strength σ_b scales with $L^{-1/m}$ so that the *fiber bundle strength* decreases as the length increases. This result is expected since there is more material for a crack of critical length to exist in.

Figure 13.33 gives the stress–strain curves of a fiber bundle for $m = 5$, 9, and 15. As m increases, the spread of the individual fiber strengths decreases. Thus, more fibers survive to higher values of σ/S_o, which means the fiber bundle strength σ_b increases. The survival probability P_s of the fibers at $\sigma = \sigma_b$ is:

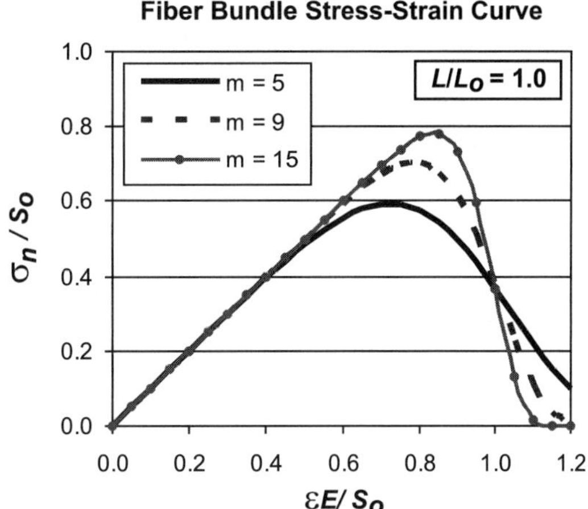

Figure 13.33. Fiber bundle stress–strain curves for bundles with fibers of various Weibull moduli.

Table 13.8. Dependence of f_b on m.

m	5	9	15
f_b	0.593	0.701	0.781

$$P_s = 1 - e^{-1/m}$$

[Eq. 13.108]

This explains the increase in height of the stress–strain curves with increasing values of m.

The distribution of fiber strengths also explains the shape of the stress–strain curves (*Figure 13.33*). For large values of m, the fibers tend to fail at approximately the same stress. This is evidenced by the stress–strain curve for $m = 15$, which is linear to higher values of stress and strain than the curve for $m = 5$; relatively few fibers in the $m = 15$ material have failed before the bundle strength σ_b is reached. Since the distribution of fiber strength is small for $m = 15$, as soon as fibers begin to fail, they nearly all fail over a small strain range, causing a steep drop in the stress–strain curve. Conversely, for $m = 5$, which has a larger spread of fiber strengths, there is still a significant number of fibers intact beyond the strain corresponding to the bundle strength. Failure is a more gradual process for the $m = 5$ material, as evidenced by the less-steep drop in its stress–strain curve.

Example 13.17 Fiber Bundle Strength

Given: A bundle of fibers has a total fiber length of 2.0 m. For reference length $L_o = 25.4$ mm, the mean strength of the fibers is $\sigma_m = 1800$ MPa. The Weibull modulus is $m = 9$.

Required: Determine the strength of the fiber bundle.

Solution: *Step 1.* The fiber bundle strength is:

$$\sigma_b = S_o \left(\frac{L_o}{L} \right)^{1/m} f_b$$

where $f_b(m = 9) = 0.701$.

Step 2. The relationship between the mean fiber strength σ_m and S_o for $m = 9$ is from *Table 13.5*:

$$\sigma_m = 0.947 S_o$$

Hence: $S_o = \dfrac{\sigma_m}{0.947} = \dfrac{1800 \text{ MPa}}{0.947} = 1901 \text{ MPa}$

Step 3. The bundle strength is:

$$\sigma_b = (1901 \text{ MPa}) \left(\frac{25.4 \times 10^{-3} \text{ m}}{2.0 \text{ m}} \right)^{1/9} (0.701)$$

Answer: $\sigma_b = 820$ MPa

Chapter 14 — Joints

14.0 Introduction

Fasteners (connectors such as bolts, rivets, pins, nails, and screws) and adhesives hold a structure together by transferring load from one component to another. *Figure 14.1* is a *lap joint*, where two members overlap each other at the connection. Connections are often the weakest link in a system, and thus are of major concern when accessing system integrity. Connections can also be difficult to analyze; there are many parts and stresses to consider, and the advanced theories of joining can be very complex.

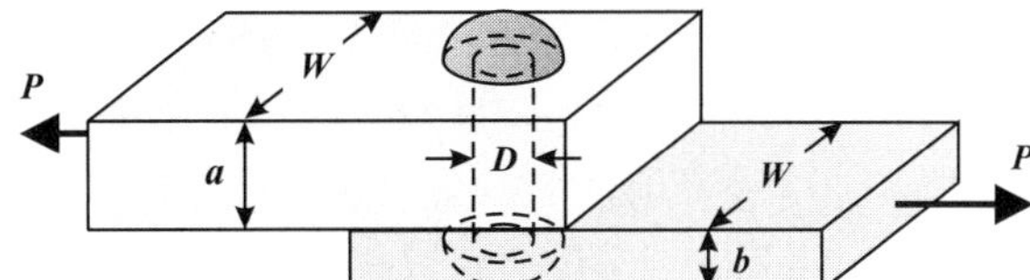

Figure 14.1. A simple connection – one rivet joining two plates.

Bolted joints are generally associated with traditional construction methods. Bolts are also used to join components made of composites, providing new design challenges.

The discussion in this chapter begins with a basic analysis of simple bolt-like connections, such as bolts, rivets, and pins. This is followed by a more complex analysis of a bolted connection, which includes stress concentrations due to the holes at the joint. Finally, an analysis of adhesive joints is developed.

Bearing Stress

When two objects are in contact with each other – or *bear* against each other – a stress exists between them. This contact stress is the **bearing stress** σ_B. In *Figure 14.2*, a footing of a system transfers force F to the floor. The contact area A is called the *bearing area*. The *bearing stress* is the bearing force over the bearing area:

$$\sigma_B = \frac{F}{A} \qquad \text{[Eq. 14.1]}$$

Bearing stress is compressive – two objects can only press against each other. Failure in bearing is due to yielding of one of the materials, resulting in a surface indentation.

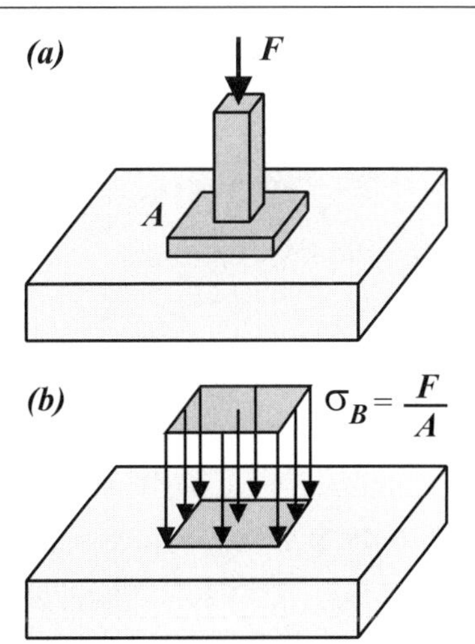

Figure 14.2. The footing bears on the floor with force F.

F.A. Leckie, D.J. Dal Bello, *Strength and Stiffness of Engineering Systems*, Mechanical Engineering Series, **457**
DOI 10.1007/978-0-387-49474-6_14, © Springer Science+Business Media, LLC 2009

14.1 Simple Fasteners

Consider the simple connection of one rivet (or bolt, etc.) holding two plates together (*Figures 14.1* and *14.3*). The upper and lower plates have thickness a and b, respectively, and both have width W into the paper. Force P on the upper plate acts to the left, and that on the lower plate acts to the right. The force must be transferred through the rivet of diameter D.

The stresses that act on the rivet are found by taking FBDs of the rivet and applying equilibrium.

Average Bearing Stress

The upper plate *bears* on the top part of the rivet, pulling it towards the left. The lower plate bears on the lower part of the rivet, pulling it towards the right. By Newton's Third Law, the rivet also bears on each plate.

The ***average bearing stress*** is the bearing force divided by the *projected area* of the rivet. This area is the rivet diameter multiplied by its imbedded length. The bearing stress (*Figure 14.3b*) on the upper part of the rivet is:

$$\sigma_{B,a} = \frac{P}{aD} \qquad \text{[Eq. 14.2]}$$

And on the lower part:

$$\sigma_{B,b} = \frac{P}{bD} \qquad \text{[Eq. 14.3]}$$

It is the bearing stress that transfers load P from the plate to the rivet (and from the rivet to the plate).

The next step is to determine how the load is transferred from the upper part of the rivet to the lower part.

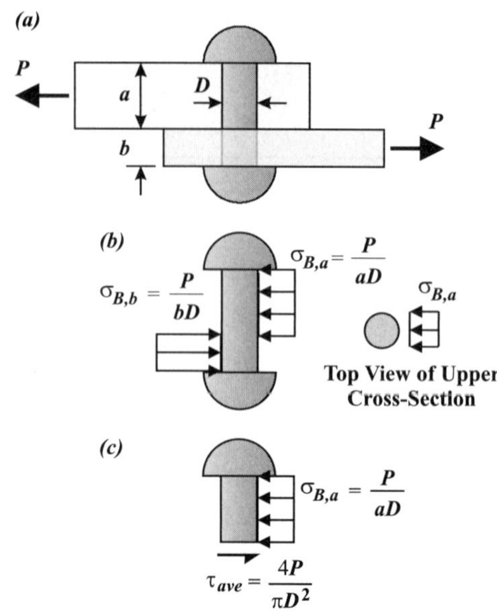

Figure 14.3. **(a)** A 2D FBD of connection. **(b)** Sketch of rivet showing average bearing stresses; top view of the bearing stress acting on the upper part of the rivet. **(c)** Stresses acting on the part of the rivet in the upper plate. The rivet transfers the load via shear stress.

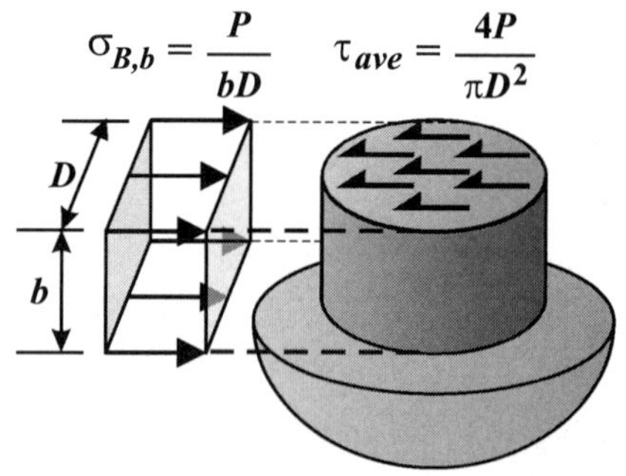

Figure 14.4. Stresses acting on the lower part of the rivet.

Average Shear Stress

Consider the top part of the rivet only (*Figure 14.3c*). The bearing force acts to the left. To maintain equilibrium, a shear force must act to the right on the cross-section of the rivet having *shear area A_s*. It is the shear stress in the rivet that transfers the force from the upper to the lower part of the rivet.

The ***average shear stress*** on a cross-section of the rivet varies from zero at the top of the rivet to the maximum at the interface of the two plates. The ***maximum average shear stress*** in the rivet is:

$$\tau_{ave} = \frac{P}{A_s} = \frac{4P}{\pi D^2} \qquad \text{[Eq. 14.4]}$$

Figure 14.4 shows a 3D view of the lower part of the rivet, with the average bearing and shear stresses acting on it. Detailed calculations (or physical argument) show that the bearing stress actually varies around the rivet–plate interface, and that the shear stress varies over the circular cross-section. However, in typical design calculations, the *average* values of the bearing and shear stresses are determined. The average values are taken into account in listings of the shear strength for various rivets, bolts, etc. In fact, it is the allowable shear force (or allowable average shear stress) for a bolt that is often tabulated, and so calculating average stress values are acceptable design practice.

Single and Double Shear

The bolt joining the members in *Figure 14.5a* is in a state of ***single shear***. The entire load P is transferred in shear over one cross-section of the bolt (*Figure 14.5b*).

The bolt in *Figure 14.5c* is in ***double shear***. Load P is transferred over two cross-sections. Thus, while the bolt transfers the same force P, the maximum shear stress is half that of the bolt in *single shear* (*Figure 14.5d*).

The system in *double shear*, often seen in pin-and-clevis connections, has the additional advantage of being balanced due to its symmetry. By contrast, the single shear system introduces a small moment at the joint since the axial forces in the top and bottom plates are not coaxial. The axial loads will tend to align, possibly causing the connected members to deform (*Figure 14.5e*).

14.2 Failure in Bolt-type Connections: A Basic Analysis

Bolted joints have several possible ***failure modes***. In joint design, each failure mode must be evaluated to ensure that the joint has sufficient strength.

In *Figure 14.6*, two members of the same width W and thickness a are joined by a single connector (e.g., a rivet or a bolt) of diameter D. The center of the connector hole is distance c from the free-end of the members; this is the ***edge distance***. The joint transfers force P.

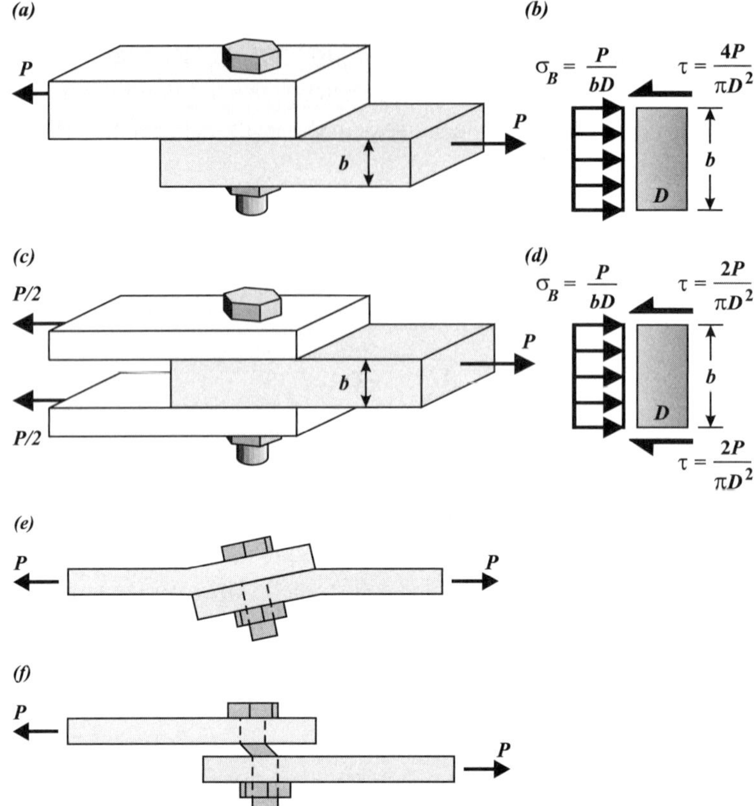

Figure 14.5. **(a)** Bolt in *single shear*. **(b)** Part of bolt embedded in lower plate. The shear force is transferred between plates over one cross-section. **(c)** Bolt in *double shear*. **(d)** Part of bolt embedded in middle plate. The shear force is transferred over two cross-sections, resulting in a shear stress half that of the *single shear* case. **(e)** Failure of single shear joint due to axial loads aligning. **(f)** Failure of joint by bolt yielding in shear.

In this section, the joined members are assumed to be *ductile* and ***elastic–perfectly plastic***. As the material at a cross-section yields, stress is redistributed until the entire cross-section is at the yield strength S_y at failure.

The cross-sectional area of the member away from the joints is known as the ***gross area*** A, which is used to determine the ***gross stress*** in the member. The stress in each member between joints is generally checked before the joints are analyzed.

The joints themselves may fail by one of the modes discussed below.

Tensile Failure at Net-Section

Because holes are drilled in each member, its cross-sectional area is reduced at the bolt locations (*Figure 14.6b*). The force must pass through the remaining area before being transferred in bearing to the bolt. The average axial stress through the reduced cross-section is the ***net-section stress***:

$$\sigma_{net} = \frac{P}{A_{net}} = \frac{P}{a(W-D)} \quad \text{[Eq. 14.5]}$$

The ***net area*** is the *gross area* minus the area removed by the holes. For the single-bolt connection, $A_{net} = a(W-D)$. The net-section stress must not exceed the allowable axial stress of the plate material. Due to plastic deformation and stress-redistribution at the net-section, the stress at failure everywhere on the net-section equals the yield strength S_y.

To make up for the area lost due to the holes, the gross area of a member is generally increased at the joint, as illustrated in *Example 14.1*. Sometimes the increased area (without the holes) is known as the *gross area*.

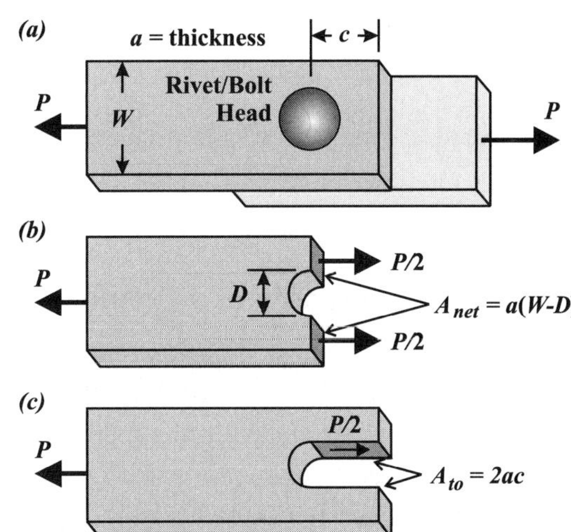

Figure 14.6. **(a)** Rivet/bolt connection joining two members. **(b)** FBD of upper member cut at net section, $A_{net} = a(W-D)$. **(c)** FBD of upper member cut where connector may tear out. The tear-out area is $A_{to} = 2ac$.

Bearing Failure at Connector-Plate Interface

The average bearing stress between the plate and the bolt is:

$$\sigma_B = \frac{P}{aD} \quad \text{[Eq. 14.6]}$$

The bearing stress at failure is the ***bearing yield strength*** S_B of the plate material or the connector material, whichever is the smaller.

Tear-out Failure

A connector may tear out of the plate as shown in *Figure 14.6c*, especially if the *edge distance c* is small. The force required to cause ***tear-out*** (or *shear-out*) of the single bolt is the product of the plate shear strength $\tau_f = \tau_y$ and the area over which the shear acts, $2ac$:

$$P_{to} = \tau_f[2(ac)] = 2ac\tau_f \quad \text{[Eq. 14.7]}$$

The factor 2 appears in the equation since the shear stress acts over two surfaces, each of area ac.

Shear Failure in Connector

The maximum shear stress in the connector is:

$$\tau_{ave} = \frac{P}{A_s} = \frac{4P}{\pi D^2} \qquad \text{[Eq. 14.8]}$$

This shear stress must be smaller than the allowable shear stress of the connector material. Failure of the bolt due to yielding in shear is shown in *Figure 14.5f*.

Example 14.1 Integrity of a Connection

Given: Consider the single-bolt lap joint in *Figure 14.7*. The thickness of the left (upper) and right (lower) axial members are a and b, respectively. The width of the main part of the left member is W_1, increasing gradually to W_2 at the joint; the width of the right member is W_4 increasing to W_3. The bolt diameter is D; the edge distance is c. Assume that all the members are made of different materials.

The *allowable stresses* (the *factor of safety* has already been applied to the strengths) are:

- <u>in the left member</u> (material a):
  normal stress: S_a;
  shear: τ_a; bearing: $S_{B,a}$;
- <u>in the right member</u> (material b):
  normal stress: S_b;
  shear: τ_b; bearing: $S_{B,b}$;
- <u>in the bolt</u>: shear: τ_{bolt};
  bearing: $S_{B,bolt}$.

Required: Derive expressions for the stresses acting at critical sections of the member, and relate each stress to the appropriate allowable stress.

Solution:

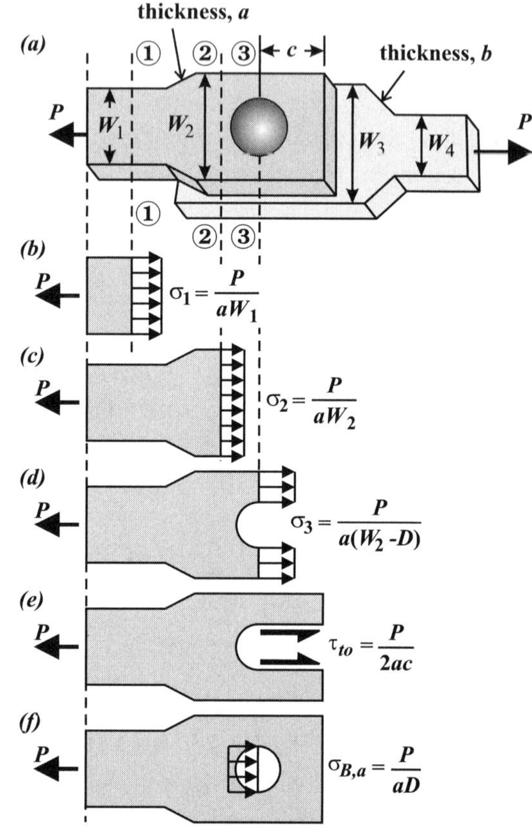

Figure 14.7. **(a)** Bolt joining two members with increased areas at the connection. **(b)** Stress at *Section 1-1*. **(c)** Stress at *2-2*. **(d)** Stress at *3-3* (the net-section). **(e)** Shear stress over tear-out area. **(f)** Bearing stress of bolt acting on the upper member.

Left (Upper) Member

Consider FBDs of the left (upper) member cut at various cross-sections, and determine the stresses at each section.

The *gross area* of the main length of the left member is $A = aW_1$, thus:

Section 1-1 (Figure 14.7b): $\sigma_1 = \dfrac{P}{aW_1} < S_a$ allowable normal stress

The axial stress σ_1 must be less than the allowable normal stress S_a. The dimensions may be varied to insure that the stress is less than the allowable, or the applied force must be reduced. At the widened section:

Section 2-2 (Figure 14.7c): $\sigma_2 = \dfrac{P}{aW_2} < S_a$ allowable normal stress

Stress σ_2 need not be calculated; since $W_2 > W_1$ then $\sigma_2 < \sigma_1$.

Section 3-3 is the *net-section* through the hole center:

Section 3-3 (Figure 14.7d): $\sigma_3 = \dfrac{P}{a(W_2 - D)} < S_a$ allowable normal stress

The bolt may *tear-out* or cause *bearing failure* in the plate:

Tear-out area (Figure 14.7e): $\tau_{to} = \dfrac{P}{2ac} < \tau_a$ allowable shear stress

Bearing area (Figure 14.7f): $\sigma_{B,a} = \dfrac{P}{aD} < S_{B,a}$ allowable bearing stress

The Bolt

The upper part of the bolt must take the bearing stress from the left (upper) member, transfer it via shear to its lower part, and then via bearing stress to the right (lower) member. Thus:

Bearing stress, upper: $\sigma_{B,a} = \dfrac{P}{aD} < S_{B,\,bolt}$ allowable bearing (bolt)

Shear stress: $\tau_{bolt} = \dfrac{4P}{\pi D^2} < \tau_{bolt}$ allowable shear (bolt)

Bearing stress, lower: $\sigma_{B,b} = \dfrac{P}{bD} < S_{B,\,bolt}$ allowable bearing (bolt)

Right (Lower) Member

The right member is subjected to the same types of stress as the upper member, but the values are generally different if W_1 is not equal to W_4, W_2 is not equal to W_3, or a is not equal to b. The allowable stresses in the right member are different from those in the left member if they are made of different materials.

Summary

At the joint, each member must satisfy at least four strength requirements (gross tension, net-section tension, bearing, and tear-out). The connector must satisfy three requirements (shear, two bearing). Even in such a simple joint, the possibility of failure in each of the 11 conditions must be checked.

Example 14.2 Multiple Bolts in a Single Row

Given: Two Aluminum 6061-T6 plates are joined by five 1/2-in. diameter bolts (*Figure 14.8*). The width and thickness of each plate are $W = 8.5$ in. and $t = 3/16$ in., respectively. The *edge distance* – from the line passing through the bolt centers to the edge of each plate – is $c = 2.0$ in. The bolt-to-bolt distance is $e = 1.0$ in. (1.5 in. center-to-center). The material properties for the plate and bolt material are the same. The allowable stresses (the factor of safety has already been applied) are:

> Normal: $S_a = 19$ ksi;
> Shear: $\tau_a = 11$ ksi; and
> Bearing: $S_{B,a} = 34$ ksi.

Required: Verify that the connection can support a load of $P = 10{,}000$ lb without exceeding the allowable stresses.

Solution:

Bolts

Each bolt has a cross-sectional area of:

$$A_s = \frac{\pi D^2}{4} = \frac{\pi(0.5)^2}{4} = 0.196 \text{ in.}^2$$

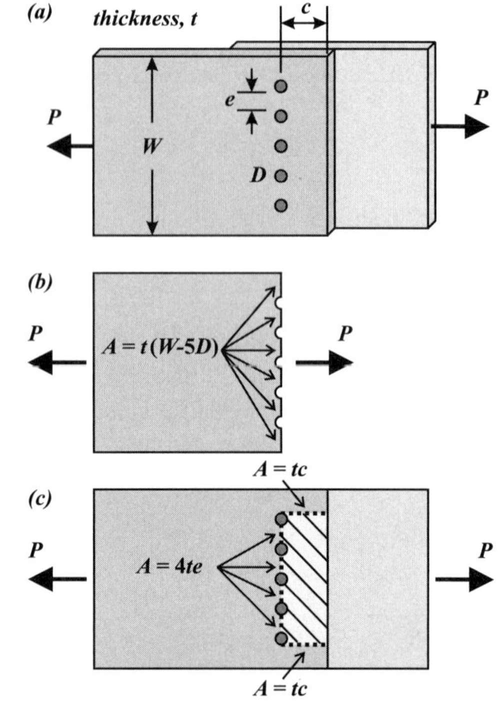

Figure 14.8. **(a)** Five bolts joining two plates. **(b)** FBD of the left/upper plate cut through the bolt holes (net section). **(c)** Cross-hatched area is part of upper plate that may tear-out.

and a bearing area of:

$$A_B = Dt = (0.5 \text{ in.})(3/16 \text{ in.}) = 0.0938 \text{ in.}^2$$

There are five bolts, so each bolt must transfer a force of:

$$P_b = \frac{P}{5} = \frac{10{,}000 \text{ lb}}{5} = 2000 \text{ lb}$$

Shear stress, bolt. The shear stress in each bolt is:

$$\tau_{bolt} = \frac{P_b}{A_s} = \frac{2000 \text{ lb}}{0.196 \text{ in.}^2} = 10.2 \text{ ksi}$$

$$\Rightarrow \tau_{bolt} < \tau_a \quad \textbf{OK}$$

The shear stress in the bolt is within the allowables.

Bearing stress, bolt.

$$\sigma_{B,\,bolt} = \frac{P_b}{A_B} = \frac{2000\text{ lb}}{0.0938\text{ in.}^2} = 21.3\text{ ksi}$$

$$\Rightarrow \underline{\sigma_{B,\,bolt} < S_{B,\,a}} \quad \textbf{OK}$$

Plates

Axial stress, plate. The highest axial stress occurs at the cross-section that has the least area; this is where the holes are drilled (*Figure 14.8b*). The entire force P must pass through the net-section, $t(W-5D)$. Thus:

$$\sigma_{net} = \frac{P}{t(W-5D)} = \frac{10,000\text{ lb}}{(3/16\text{ in.})[8.5\text{ in.} - 5(0.5\text{ in.})]} = 8.9\text{ ksi}$$

$$\Rightarrow \underline{\sigma_{net} < S_a} \quad \textbf{OK}$$

Bearing stress, plate. The bearing stress is the same as that on the bolts, and they are of the same material:

$$\sigma_B = \frac{P_b}{A_B} = \frac{2000\text{ lb}}{0.09375\text{ in.}^2} = 21.3\text{ ksi}$$

$$\Rightarrow \underline{\sigma_B < S_{B,\,a}} \quad \textbf{OK}$$

Tear-out, plate. For the five bolts to tear-out of the upper plate, the cross-hatched area in *Figure 14.8c* may be the part of the left/upper plate removed. The areas that connect this cross-hatched area to the rest of the upper plate are $2tc$ and $4te$, where c is the edge distance and e is the distance between bolt surfaces. Shear stress acts over area $2tc$ and normal stress acts over $4te$. To ensure the *tear-out* requirement is satisfied, the tear-out force is calculated by letting all the stresses reach their allowables:

$$P_{to} = S_a(4te) + \tau_a(2tc)$$

$$= (19\text{ ksi})[4(3/16\text{ in.})(1.0\text{ in.})] + (11\text{ ksi})[2(3/16\text{ in.})(2.0\text{ in.})] = 22.5\text{ kips}$$

$$\Rightarrow \underline{P < P_{to}} \quad \textbf{OK}$$

Each bolt may also plow-through the upper plate individually, the plate failing in shear:

$$P_{to,\,2} = 5[(2tc)(\tau_a)] = 5[2(3/16\text{ in.})(2\text{ in.})(11\text{ ksi})] = 41.25\text{ kips}$$

$$\Rightarrow \underline{P < P_{to,\,2}} \quad \textbf{OK}$$

Since the lower plate is the same as the upper plate, its analysis has already been completed.

Summary

All the stresses are within the allowables, and the allowable tear-out force is greater than the applied load. Therefore, the connection can support the load $P = 10$ kips.

Example 14.3 Bolts in Regular Grid-Pattern

Given: Four bolts in a square-pattern join two members as shown in (*Figure 14.9*). The connection supports load *P*.

Required: Determine (a) the force acting on each bolt and (b) the forces acting through *Sections 1-1* and *2-2* (*Figure 14.9b*).

Solution: *Step 1.* Since the bolt is in a regular square pattern, the load is taken to be carried equally by all four bolts. Each bolt must support a force of:

$$Answer:\quad P_b = \frac{P}{4}$$

Step 2. The force acting across *Section 1-1* is the entire force transferred by the connection:

$$Answer:\quad F_1 = P$$

Step 3. At *Section 2-2*, half of the load in the upper plate has already been removed by the bolts at *Section 1-1*. The force that is not removed by the first set of bolts is the *by-pass* (or *pass-through*) *force* P_t (*Figure 14.9c*). This force must be transferred over net *Section 2-2*:

$$Answer:\quad F_2 = P_t = \frac{P}{2}$$

The stresses in the plates and bolts are calculated as described previously.

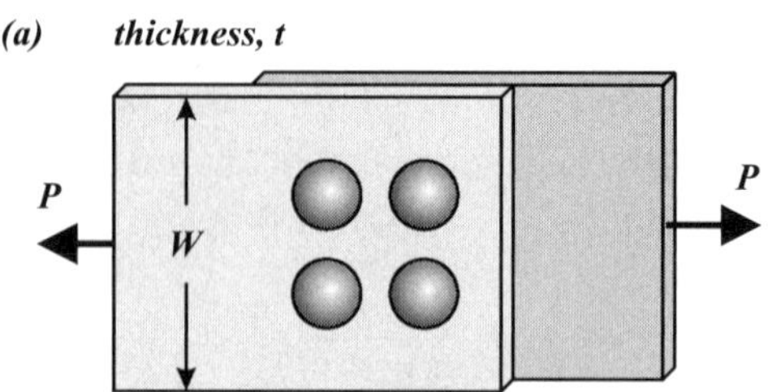

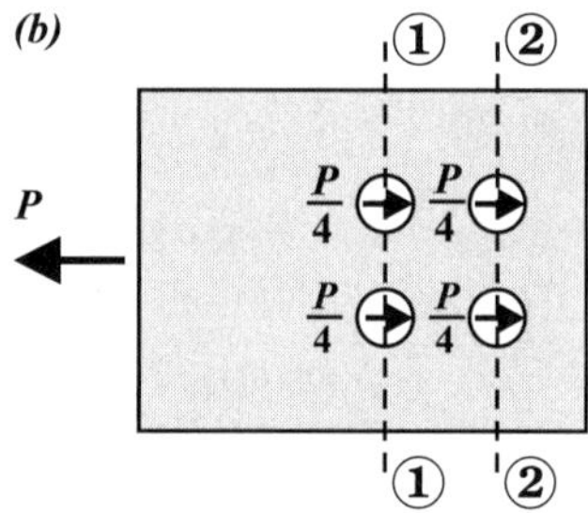

Figure 14.9. (a) Four bolts in a square pattern. **(b)** Force *P* is assumed to be equally distributed among the bolts.

Bolts in a regular grid-pattern of any size can be analyzed as illustrated in the above examples. Bolts in an offset grid-pattern must be analyzed using a more complicated formulation, not presented here, but available in design manuals.

14.3 Failure of Bolt-type Connections: An Advanced Analysis

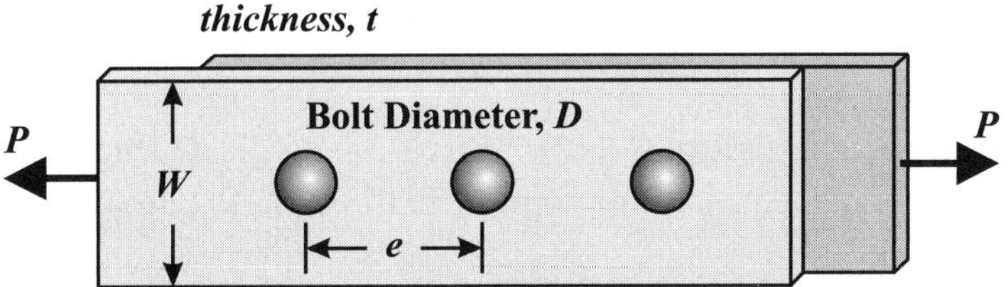

Figure 14.10. Two members joined by multiple bolts.

Structural members made of polymer matrix composites are often joined with metal bolts (*Figure 14.10*). Since polymeric components are generally brittle, the beneficial effect associated with yielding in ductile materials cannot be assumed.

In the previous section, it was assumed that only average stresses needed be calculated to determine joint strength. This approach is satisfactory when the plate material is ductile, and high local stresses are relieved by plastic deformation. Assuming that the material is *elastic–perfectly plastic,* the entire net-section is at the yield strength when the joint fails.

For materials that remain essentially elastic until failure – brittle materials – the maximum elastic stresses caused by the presence of the holes must be calculated. For basic geometries, the stresses can be determined using *stress concentration factors*, which are known for many geometries and can be found on the web (e.g., www.fatiguecalculator.com) and in printed references.

A detailed presentation of the following approach is found in the paper by L.J. Hart-Smith, "Mechanically-Fastened Joints for Advanced Composites – Phenomenological Considerations and Simple Analyses" (*Fibrous Composites in Structural Design*, Plenum Press, 1980, pp. 543–574).

Stress Concentration Factor

Consider a thin elastic plate of width W subjected to axial stress σ (*Figure 14.11a*). If a hole of diameter D is drilled into the center of the plate, then the stress at the cross-section through the hole center is not uniform (*Figure 14.11b*). The *average stress* on the *net-section* is:

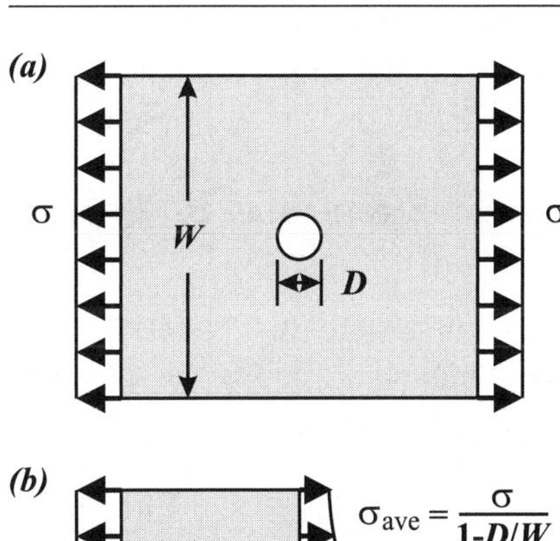

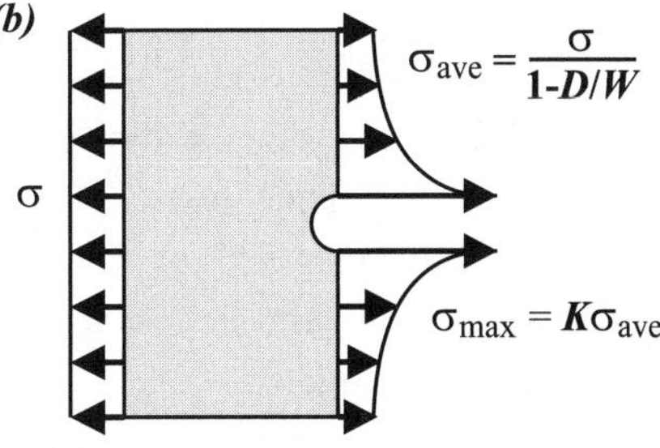

Figure 14.11. **(a)** Plate with a hole. **(b)** The stress concentration factor of a plate with a hole for $D << W$ is $K = 3$.

$$\sigma_{ave} = \frac{\sigma}{1 - D/W} \qquad \text{[Eq. 14.9]}$$

The maximum stress at the hole surface is:

$$\sigma_{max} = K\sigma_{ave} \qquad \text{[Eq. 14.10]}$$

where K is the *stress concentration factor* (SCF). The SCF depends on the shape and dimensions of the component, and the type of load. As noted above, the equations for many SCFs are known.

When the hole is small compared to the plate ($D/W \ll 1$) and the material remains elastic, the maximum stress reduces to the well-known result:

$$\sigma_{max} = 3\sigma \qquad \text{[Eq. 14.11]}$$

Lap Joint with n In-line Bolts

Consider a lap joint with n bolts, arranged in a single line in the direction of the load (*Figure 14.12a*). The joined plates each have width W and thickness t; the bolt diameter is D and the bolts are placed e apart, center-to-center. The joint transmits load P by shear in the bolts, and the bolts are assumed to share the load equally. In the upper (left) plate, the load reduces from P at the loaded left end of the plate to zero at the right end.

Consider a length of the upper plate formed by two cuts, one on either side of the first bolt (*Figure 14.12b*). Load P is supported by the **bolt load** P_b and by the **bypass load** P_t which passes through to the rest of the upper plate. This loading condition can be broken up into a plate with a hole loaded by P_t, and a plate loaded at the hole by bearing force P_b (*Figure 14.12c*). The maximum bypass load is at the first bolt.

Elastic Stress due to Bypass Load

If the material is elastic, the maximum stress due to the average *bypass stress* σ_t is:

$$\sigma_{max,t} = K_{te}\sigma_t \qquad \text{[Eq. 14.12]}$$

and occurs at the side of the hole (*Figure 14.13a*). The elastic stress concentration factor K_{te} for a plate of width W with a center hole of diameter D, is modeled by:

$$K_{te} = 2 + \left(1 - \frac{D}{W}\right)^3 \qquad \text{[Eq. 14.13]}$$

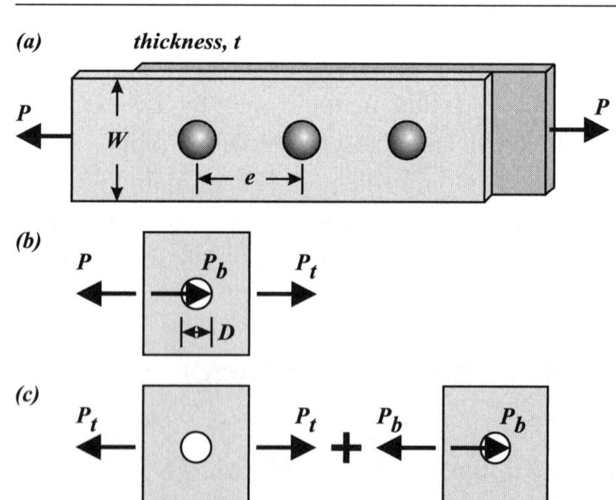

Figure 14.12. **(a)** Two plates joined by multiple bolts. **(b)** FBD of length of upper plate around first bolt. **(c)** Loads on FBD considered as the superposition of a (1) bypass force P_t and (2) a bearing force P_b.

which reduces to $K_{te} = 3$ for large plates. The average bypass stress is:

$$\sigma_t = \frac{P_t}{(W-D)t} \qquad \text{[Eq. 14.14]}$$

Elastic Stress due to Bearing Load

The maximum stress in the plate due to the bearing stress σ_b is:

$$\sigma_{max,b} = K_{be}\sigma_b \qquad \text{[Eq. 14.15]}$$

and also occurs at the side of the hole (*Figure 14.13b*). The elastic stress concentration factor K_{be} for a plate of width W with a series of center holes of diameter D each subjected to bearing stress is modeled:

$$K_{be} = 1 + \frac{2}{(W/D)-1} - \frac{1.5\theta}{(W/D)+1}$$

$$\text{[Eq. 14.16]}$$

Variable θ is a correction factor for the bolt spacing e. In this development, the effect of spacing will be neglected, which means $\theta = 1.0$. The average bearing stress is:

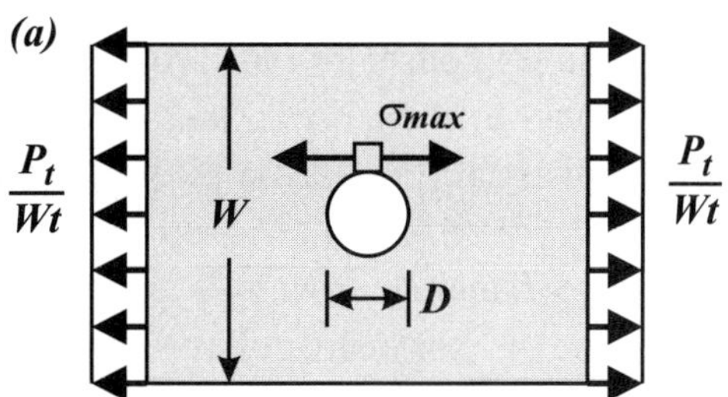

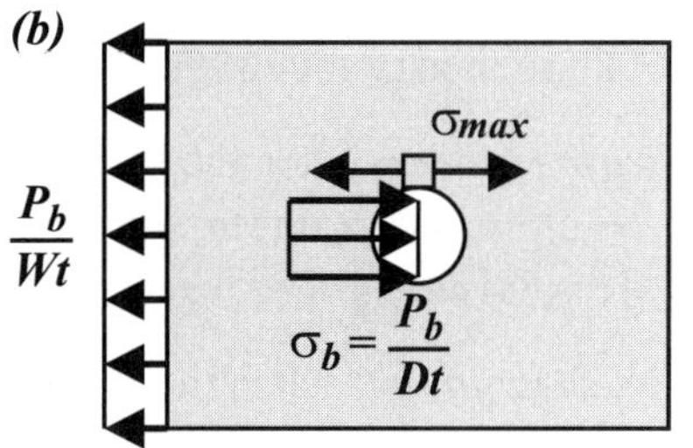

Figure 14.13. **(a)** Plate with a hole subjected to applied stress $\sigma_t = P_t/Wt$. **(b)** Plate with a hole subjected to bearing stress $\sigma_b = P_b/Dt$.

$$\sigma_b = \frac{P_b}{Dt} \qquad \text{[Eq. 14.17]}$$

Total Maximum Stress

The maximum tensile stress in the plate is the sum of the stresses due to the bypass and bearing loads:

$$\sigma_{max} = K_{te}\sigma_t + K_{be}\sigma_b \qquad \text{[Eq. 14.18]}$$

Plasticity and Delamination

Equation 14.18 is only valid if the stresses remain elastic. High stresses cause plastic deformation in metals and delamination in polymer composites. The resulting **stress redistribution** due to these non-linear responses can be expressed by modifying the stress concentration factors. The modifications are determined from the results of experiments. The modified stress concentration factors are:

$$K_{tp} = 1 + c(K_{te} - 1)$$
$$K_{bp} = 1 + c(K_{be} - 1)$$

$$\text{[Eq. 14.19]}$$

K_{tp} is the *plastic stress concentration factor* for the pass-through stress and K_{bp} is the plastic stress concentration factor for the bearing stress; these SCFs replace the *elastic stress concentration factors* in *Equation 14.18*. Variable c is a correlation factor depending on the material. For elastic (brittle) materials, $c = 1$, and the plastic SCF reduces to the elastic SCF. For fully plastic (ductile) materials, yielding occurs over the entire net section so that $c = 0$ and $K_{tp} = K_{bp} = 1$ (i.e., at failure, the stress everywhere is the yield strength). The value of c for many polymeric composites is approximately $c = 0.25$.

Failure Modes

Two failure modes are currently considered:

1. *tensile failure*: when the maximum calculated stress in the plate reaches the ultimate tensile strength of a ceramic or polymeric material, $\sigma_{max} = S_u$, or the yield strength S_y for ductile materials;

2. *bearing failure*: when the bearing stress σ_b on the plate reaches the bearing strength, $\sigma_{b,max} = S_B$, causing the hole to elongate.

For metals, $S_B \sim 1.73 S_y$; for polymer composites, $S_B \sim S_u$, and for ceramics $S_B \gg S_u$. These values are assumed in further discussion. Because the bearing strength of ceramics is much larger than their tensile strength, bearing failure for ceramics is not considered.

As a final assumption, the bolts are considered to be strong enough to support the applied load, both in bearing and in shear; the bolts do not limit the design.

Joint Efficiency

In the absence of the joint, the plate can support an ultimate load:

$$P_u = S_u W t \qquad \text{[Eq. 14.20]}$$

Joint failure occurs either by tensile failure at the hole, or in bearing at the bolt–plate interface. The **efficiency** of the joint, η, is:

$$\eta = \frac{P_f}{P_u} \qquad \text{[Eq. 14.21]}$$

where P_f is the lowest calculated load for the various failure modes. The efficiency is the strength of the joint compared to the host system. Efficiency calculations are illustrated in the following examples.

Example 14.4 Single-Bolt Connection

Given: A single-bolt connection joins two plates, both of width W and thickness t (*Figure 14.14*). The bolt is of diameter D.

Required: Determine the efficiency η of the connection.

Solution: *Step 1.* For a single-bolt connection, the bypass load is zero, $P_t = 0$, and the bolt load is $P_b = P$. Hence, the maximum tensile stress in the plate is:

$$\sigma_{max} = K_b \sigma_b$$

where $\sigma_b = P/(Dt)$ and K_b is the plastic (general) stress concentration factor.

Step 2. The plate can fail either in tension or in bearing. For the plate to fail in tension:

$$K_b \sigma_b = S_u \quad (S_y \text{ for metals})$$

Applying *Equation 14.21*, the efficiency for *tensile failure* of a joint with a single bolt is found to be:

$$\text{Answer: } \eta_t = \frac{P_f}{P_u} = \frac{1 - \dfrac{D}{W}}{1 + c\left[\dfrac{W}{D} - 1.5\left[\dfrac{1 - (D/W)}{1 + (D/W)}\right]\right]} \qquad \text{[Eq. 14.22]}$$

where the value of c depends on the plate material.

Step 3. For the plate to fail in bearing:

$$\sigma_b = S_B$$

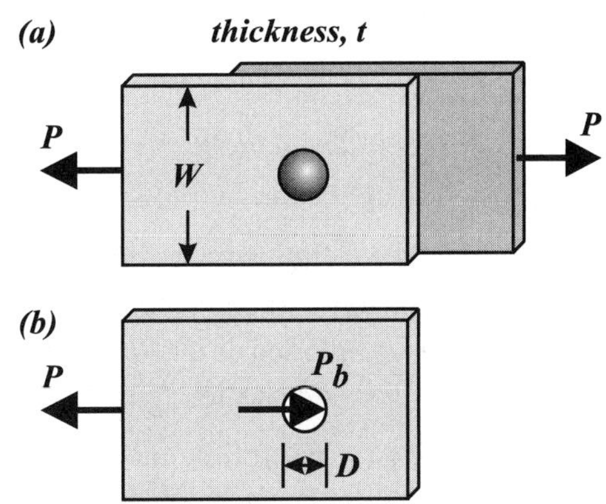

Figure 14.14. (a) Single bolt joint. **(b)** FBD of upper (*left*) member.

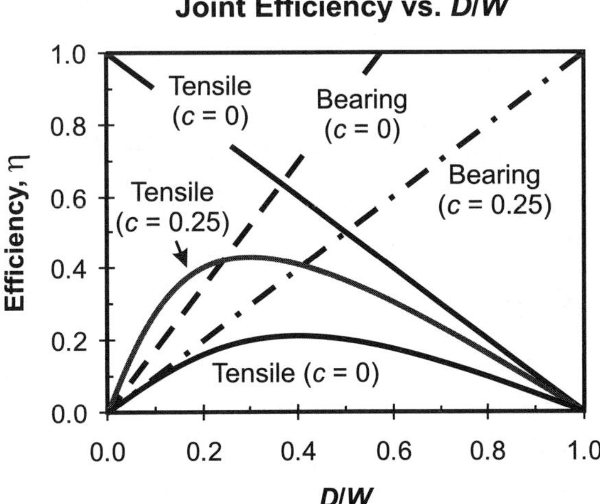

Figure 14.15. Efficiency of a single-bolt connection for joining various materials: ductile materials ($c = 0$), brittle materials ($c = 1.0$), and polymer matrix composites ($c = 0.25$).

Applying *Equation 14.21*, the joint efficiency for *bearing failure* is:

$$Answer:\ \eta_b = \left(\frac{D}{W}\right)\left(\frac{S_B}{S_u}\right)$$

[Eq. 14.23]

Discussion: The efficiency of the single-bolt connection for any given *D/W* is the minimum of the tensile and bearing efficiencies. Joint efficiency versus *D/W* for the various failure modes are plotted in *Figure 14.15* for $c = 0$ (perfectly plastic metal, $S_B = 1.73S_y$), $c = 0.25$ (polymer composite, $S_B = S_u$), and $c = 1.0$ (elastic ceramic, $S_B \gg S_u$).

From *Figure 14.15*, the following observations may be made for single-bolt connections:

1. In metals that are perfectly plastic ($c = 0$), both efficiencies are represented by straight lines. The maximum possible efficiency is $\eta = 0.64$ when D/W 0.37; tensile and bearing failure occur simultaneously.
2. In ceramics that remain elastic ($c = 1.0$) and do not fail in bearing, the tensile efficiency plot has negative curvature. The maximum possible efficiency is $\eta = 0.21$ when $D/W = 0.4$; tensile failure occurs at the sides of the hole.
3. In polymer composites ($c = 0.25$), bearing efficiency is linear, while tensile efficiency has a negative curvature. For the assumptions made, the maximum possible efficiency is $\eta = 0.4$ when $D/W = 0.4$; tensile and bearing failure occur simultaneously.

Example 14.5 Multi-Bolt Connection

Given: A multi-bolt (*n*-bolt) connection joins two members, both of width *W* and thickness *t* (*Figure 14.16*). The bolts each have diameter *D*.

Required: Determine the efficiency η of the connection.

Solution: *Step 1.* For the multi-bolt connection, the bypass load P_t is maximum at the first bolt:

$$P_t = P - P_b \quad \text{[Eq. 14.24]}$$

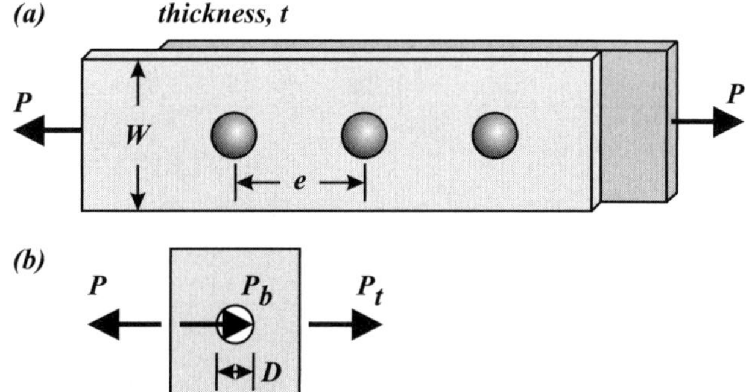

Figure 14.16. (a) Multi-bolt connection. **(b)** The maximum pass-through load is at the first bolt.

To determine the actual force distribution among the bolts requires a lengthy analysis. However, yielding in metals and delamination in polymer composites allow stress redistribution to occur, and it is therefore assumed in this analysis that load *P* is distributed equally among the bolts. While this is a major assumption, it leads to observations which are consistent with experiment.

If *n* is the number of bolts, then the bearing load on each bolt is:

$$P_b = \frac{P}{n} \qquad \text{[Eq. 14.25]}$$

The bypass load P_t is largest at the first bolt:

$$P_t = P - P_b = \frac{n-1}{n}P \qquad \text{[Eq. 14.26]}$$

Step 2. The maximum tensile stress in the plate is therefore at the first bolt, and is:

$$\sigma_{max} = K_t\sigma_t + K_b\sigma_b = K_t\left(\frac{n-1}{n}\right)\left[\frac{P}{(W-D)t}\right] + K_b\left[\frac{P}{Dt}\right] \qquad \text{[Eq. 14.27]}$$

Step 3. Applying the tensile failure condition:

$$\sigma_{max} = K_t\sigma_t + K_b\sigma_b = S_u \qquad \text{[Eq. 14.28]}$$

and the bearing failure condition:

$$\sigma_b = S_B \qquad \text{[Eq. 14.29]}$$

The efficiency for tensile failure is:

$$Answer: \; \eta_t = \left[K_t\left(\frac{n-1}{n}\right)\left[\frac{1}{1-(D/W)}\right] + K_b\frac{W}{nD}\right]^{-1} \qquad \text{[Eq. 14.30]}$$

Step 4. The efficiency for bearing failure is:

$$Answer: \; \eta_b = n\left(\frac{D}{W}\right)\left(\frac{S_B}{S_u}\right) \qquad \text{[Eq. 14.31]}$$

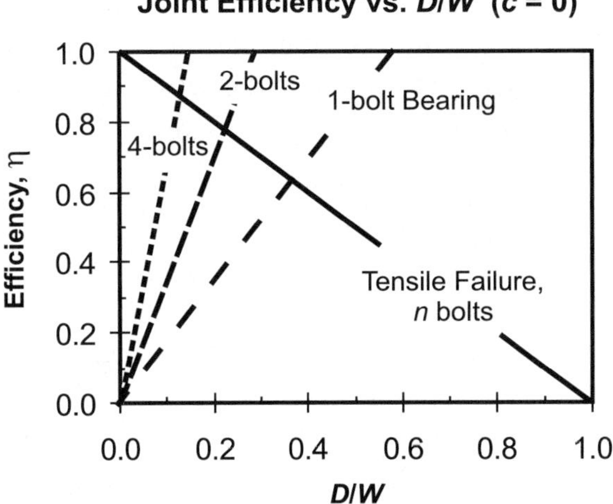

Figure 14.17. Efficiency of multi-bolt connection for elastic–perfectly plastic materials, $c = 0$.

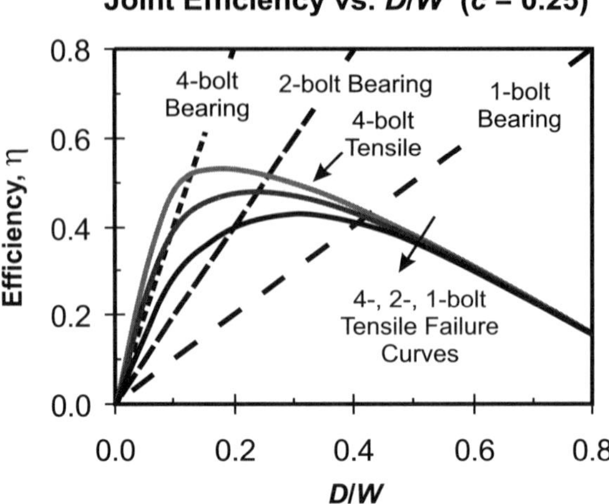

Figure 14.18. Efficiency of multi-bolt connection for polymer–matrix composites, $c = 0.25$.

Discussion: The efficiency of the connection for any given D/W is the minimum of the two efficiencies. Graphs of joint efficiency versus D/W are given for perfectly plastic materials in *Figure 14.17* and for polymer matrix composites in *Figure 14.18*.

For *ductile materials*, $c = 0$, $S_B = 1.73 S_y$ (*Figure 14.17*), the following observations may be made:

1. The maximum efficiency for one bolt is $\eta = 0.64$ when $D/W \sim 0.37$. Tensile and bearing failure occur simultaneously.
2. The maximum efficiency increases with the number of bolts. For four bolts, the maximum possible efficiency is 0.87 when $D/W \sim 0.12$. Tensile and bearing failure occur simultaneously.
3. The increase in efficiency is not linear with the number of bolts (cost). For each additional bolt, less additional efficiency is gained. The increase in efficiency is eventually offset by the increased joint cost; an economic limit is reached.

For *polymeric materials*, $c = 0.25$, $S_B = S_u$ (*Figure 14.18*):

1. The maximum efficiency for one bolt is $\eta = 0.4$ when $D/W \sim 0.4$. Tensile and bearing failure occur simultaneously.
2. The maximum efficiency increases with the number of bolts. For four bolts, the maximum efficiency is 0.52 when $D/W \sim 0.2$. The failure mode is tensile.
3. Increasing the number of bolts from 1 to 4 only increases the efficiency from 0.4 to 0.52. The maximum efficiency of two bolts ($\eta = 0.5$ at $D/W \sim 0.26$) is approximately the same as the maximum efficiency for four bolts. In practice, the number of bolts in a composite joint is therefore limited.

Final bolt designs are confirmed by performing mechanical tests on the actual joint.

14.4 Stress Distribution in Adhesive Lap Joints in Shear

Adhesives are used to join metal, polymer and composite plates and panels (*Figure 14.19*). The adhesive transmits load P from the upper plate to the lower plate by means of shear stresses in the adhesive. Adhesives are strong when loaded in shear, but have little strength when loaded by normal stresses.

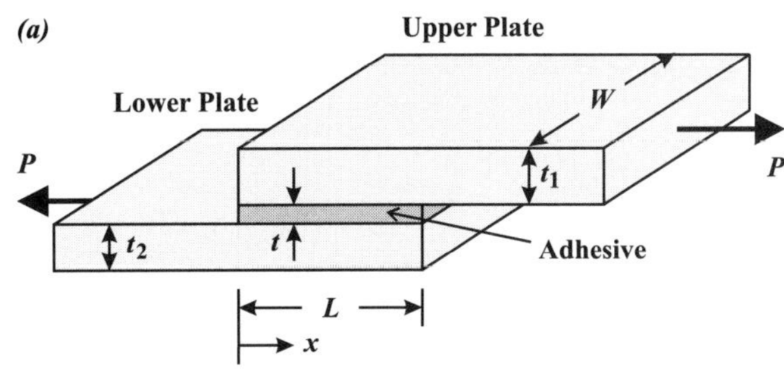

Figure 14.19. (a) Adhesive lap joint.

For the lap joint in *Figure 14.19a*, the upper (right) plate has width W, thickness t_1, and elastic modulus E_1. The lower (left) plate has width W, thickness t_2, and elastic modulus E_2. The adhesive has thickness t and shear modulus G. The applied load transmitted through the joint is P. The length of the bonded joint (overlap) is L.

The axial stress in each plate varies with distance x, measured from the left end of the joint. The normal (axial) stress and displacement at position x of the upper plate are $\sigma_1(x)$ and $u_1(x)$, respectively. The corresponding values of the lower plate are $\sigma_2(x)$ and $u_2(x)$. At their free ends, the plates are stress-free: $\sigma_1(0) = \sigma_2(L) = 0$. The adhesive is assumed to be subjected to only shear stress $\tau(x)$.

From the FBD formed by taking a cut at position x in the joint (*Figure 14.19b*):

$$P = \sigma_1(Wt_1) + \sigma_2(Wt_2) \qquad \text{[Eq. 14.32]}$$

Consider an element dx in the upper plate (*Figure 14.19c*). Equilibrium requires that the change in σ_1 over distance dx be:

$$d\sigma_1(Wt_1) = \tau(x)(W\,dx) \qquad \text{[Eq. 14.33]}$$

or:

$$\frac{d\sigma_1}{dx} = \frac{\tau(x)}{t_1} \qquad \text{[Eq. 14.34]}$$

Now consider an element dx in the lower plate (*Figure 14.19d*). Equilibrium requires that:

$$\frac{d\sigma_2}{dx} = -\frac{\tau(x)}{t_2} \qquad \text{[Eq. 14.35]}$$

From the last two equations, $(Wt_1)d\sigma_1$ and $(Wt_2)d\sigma_2$ are equal but opposite in sign. At any distance x, the force in the upper plate $(Wt_1\sigma_1)$ increases at the same rate that the force in the lower plate $(Wt_2\sigma_2)$ decreases; the load is being transferred from one plate to the other.

Using the definition for strain, $\varepsilon = du/dx$, and Hooke's Law, the strains in the upper and lower plates are:

$$\varepsilon_1(x) = \frac{du_1}{dx} = \frac{\sigma_1(x)}{E_1} \qquad \text{[Eq. 14.36]}$$

$$\varepsilon_2(x) = \frac{du_2}{dx} = \frac{\sigma_2(x)}{E_2} \qquad \text{[Eq. 14.37]}$$

The shear strain in the adhesive is (*Figure 14.19e*):

$$\gamma(x) = \frac{u_1(x) - u_2(x)}{t} = \frac{\tau(x)}{G} \qquad \text{[Eq. 14.38]}$$

Differentiating *Equation 14.38* with respect to x:

$$\frac{1}{G}\frac{d\tau}{dx} = \frac{1}{t}\left(\frac{du_1}{dx} - \frac{du_2}{dx}\right) \qquad \text{[Eq. 14.39]}$$

and then substituting *Equations 14.36* and *14.37* gives:

$$\frac{1}{G}\frac{d\tau}{dx} = \frac{1}{t}\left(\frac{\sigma_1}{E_1} - \frac{\sigma_2}{E_2}\right) \qquad \text{[Eq. 14.40]}$$

Substituting for σ_2 (*Equation 14.32*) and for τ (*Equation 14.34*) gives a second-order differential equation for stress σ_1:

$$\frac{d^2\sigma_1}{dx^2} - \lambda^2\sigma_1 + C = 0 \qquad \text{[Eq. 14.41]}$$

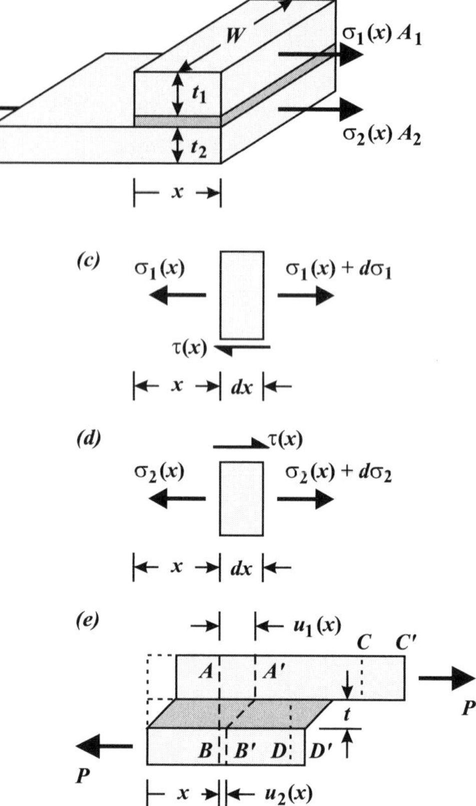

Figure 14.19. **(b)** FBD of joint cut at distance x. **(c)** Element of upper plate, dx long, t_1 tall, W wide (into paper). Stresses drawn in positive directions. **(d)** Element of lower plate, dx long, t_2 tall, W wide (into paper). **(e)** Deformation of system, with left edge of adhesive ($x = 0$) taken to be fixed. The original system profile is indicated by the dotted lines. Any vertical line AB (*dashed*) displaces to $A'B'$.

where $\lambda^2 = \dfrac{G}{t}\left(\dfrac{1}{E_1 t_1} + \dfrac{1}{E_2 t_2}\right)$ and $C = \dfrac{G}{t}\dfrac{P}{E_2 W t_2 t_1}$

The solution of *Equation 14.41* has the form:

$$\sigma_1(x) = A\,\cosh\lambda x + B\,\sinh\lambda x + \frac{C}{\lambda^2} \qquad \text{[Eq. 14.42]}$$

Constants A and B are found by applying the boundary conditions. In the upper plate, at $x = 0$ (the free-surface), $\sigma_1 = 0$; at $x = L$, $\sigma_1 = P/Wt_1$.

Completing the algebra, the expression for the normal stress σ_1 is:

$$\sigma_1(x) = \frac{C}{\lambda^2}(1 - \cosh \lambda x) + \frac{\sinh \lambda x}{\sinh \lambda L}\left[\frac{P}{Wt_1} - \frac{C}{\lambda^2}(1 - \cosh \lambda L)\right] \qquad \text{[Eq. 14.43]}$$

The shear stress $\tau(x)$ is calculated by taking the derivative of $\sigma_1(x)$ (*Equation 14.43*), and substituting the result into *Equation 14.34*. Stress $\sigma_2(x)$ is determined from equilibrium in *Equation 14.32*. After some mathematical manipulation, the stresses are:

$$\sigma_1(x) = \frac{P}{Wt_1 k \,\sinh \lambda L}[(k-1)[\sinh \lambda L - \sinh \lambda(L-x)] + \sinh \lambda x] \qquad \text{[Eq. 14.44]}$$

$$\sigma_2(x) = \frac{P}{Wt_2 k \,\sinh \lambda L}[(k-1)[\sinh \lambda(L-x)] + \sinh \lambda L - \sinh \lambda x] \qquad \text{[Eq. 14.45]}$$

$$\tau(x) = \frac{P\lambda L}{WLk \,\sinh \lambda L}[(k-1)[\cosh \lambda(L-x)] + \cosh \lambda x \,] \qquad \text{[Eq. 14.46]}$$

where:

$$k = 1 + \frac{E_1 t_1}{E_2 t_2} \qquad \text{[Eq. 14.47]}$$

and

$$\lambda^2 = \frac{G}{t}\left(\frac{1}{E_1 t_1} + \frac{1}{E_2 t_2}\right) \qquad \text{[Eq. 14.48]}$$

Constant k is a measure of the relative axial stiffnesses, $EA = E(tW)$, of the plates. Constant λ^2 is a measure of the ratio of the adhesive shear stiffness to the plate stiffnesses.

A plot of the variation of the shear stress with position for various values of λL is given in *Figure 14.20* for $k = 3$, i.e., $E_1 t_1 = 2E_2 t_2$. The shear stress is normalized by the average shear stress $\bar{\tau} = P/(WL)$, and the position by the joint length L. The shear stress is maximum at the ends of the joint, $x = 0$ and L. In general, the shear stresses at each end are not equal.

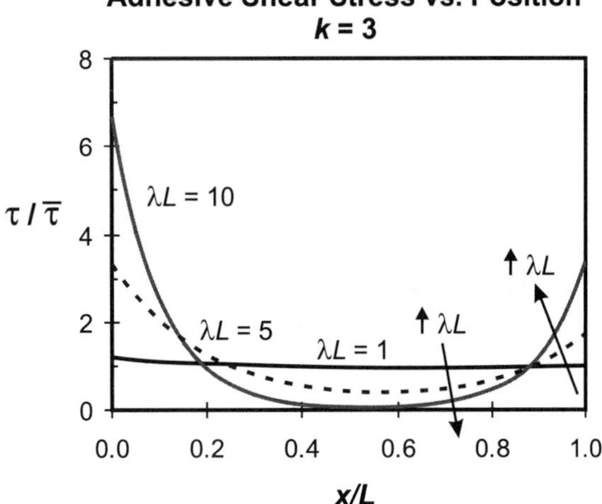

Figure 14.20. Unbalanced Joint, $k = 3$. Shear stress in the adhesive $\tau/\bar{\tau}$ versus position along the joint x/L.

As λL increases, the less effective the center of the adhesive area is in supporting stress. For $\lambda L = 10$, approximately 25% of the center region supports essentially zero stress.

Balanced Joints

A ***balanced joint*** is achieved when the shear stresses at the ends are equal, $\tau(0) = \tau(L)$. From *Equation 14.46*, the shear stresses at the ends are:

$$\tau(0) = \frac{P\lambda L}{WLk \ \sinh \ \lambda L}[(k-1)\cosh \ \lambda L + 1]$$

$$\tau(L) = \frac{P\lambda L}{WLk \ \sinh \ \lambda L}[(k-1) + \cosh \ \lambda L]$$

[Eq. 14.49]

These stresses are equal provided $k - 1 = 1$ or $k = 2$. Substituting $k = 2$ into *Equation 14.47*, gives:

$$\frac{E_1 t_1}{E_2 t_2} = 1$$

[Eq. 14.50]

Since W is the same for both plates, then the axial stiffnesses, $EA = EWt$, in a balanced joint are equal:

$$E_1 W t_1 = E_2 W t_2$$

[Eq. 14.51]

Hence the stresses for a *balanced joint* become:

$$\frac{\sigma_1}{\overline{\sigma}_1} = \frac{1}{2 \ \sinh \ \lambda L}[\sinh \ \lambda L - \sinh \ \lambda(L-x) + \sinh \ \lambda x]$$

[Eq. 14.52]

$$\frac{\sigma_2}{\overline{\sigma}_2} = \frac{1}{2 \ \sinh \ \lambda L}[\sinh \ \lambda L + \sinh \ \lambda(L-x) - \sinh \ \lambda x]$$

[Eq. 14.53]

$$\frac{\tau}{\overline{\tau}} = \frac{\lambda L}{\sinh \ \lambda L}[\cosh \ \lambda(L-x) + \cosh \ \lambda x]$$

[Eq. 14.54]

The axial stresses are normalized by the average stresses in the plates away from the joint and the shear stress is normalized by the average shear stress in the adhesive:

$$\overline{\sigma}_1 = \frac{P}{Wt_1} \ ; \quad \overline{\sigma}_2 = \frac{P}{Wt_2} \ ; \quad \overline{\tau} = \frac{P}{WL}$$

[Eq. 14.55]

Plots of the normalized stress distributions in a balanced joint are given in *Figures 14.21* and *14.22*.

Effect of Parameter λ

For the case $E_1 = E_2$ and $t_1 = t_2$, the parameter λ reduces to:

$$\lambda = \sqrt{\frac{2G}{tE_1t_1}} = \sqrt{\frac{2G}{tE_2t_2}}$$

[Eq. 14.56]

and the dimensionless parameter λL is:

$$\lambda L = \sqrt{\frac{2G}{tE_1t_1}} L = \sqrt{\frac{2G}{tE_2t_2}} L$$

[Eq. 14.57]

In practice, λL lies between 1 and 15.

Normalized stress distribution plots for the shear stress in the adhesive (*Figure 14.21*) and for the axial stress in the lower plate σ_2 (*Figure 14.22*) are given for various values of λL.

When $\lambda L = 1$, the shear stress in the adhesive is essentially constant and equal to the average value (*Figure 14.21*). In addition, the axial stress in the lower plate is linear, evidence that the shear stress is constant over the entire length of the joint (*Figure 14.22, Equation 14.35*).

As λL increases, the less effective is the central region of the joint. For $\lambda L = 10$, the shear stress in the central 30% of the joint is almost zero (between $x/L = 0.35$ and 0.65, $\tau/\bar{\tau} < 0.160$, *Figure 14.21*), and so it does not contribute significantly to load transfer. The central region of the adhesive is essentially stress-free. As a result, for $\lambda = 10$, the axial stress in the lower plate is essentially constant over the central 30% of the joint (between at $x/L = 0.35$ and 0.65, $\sigma/\bar{\sigma_2}$ varies by less than 6%, from 0.514 to 0.486, *Figure 14.22*).

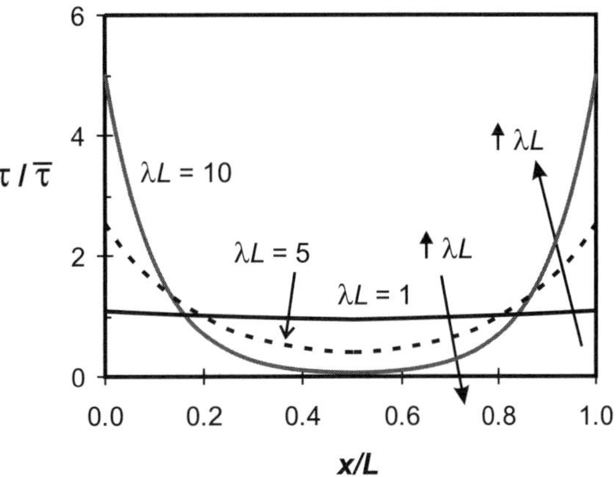

Figure 14.21. Balanced joint, $k = 2$. Shear stress in the adhesive $\tau/\bar{\tau}$ versus position along joint x/L.

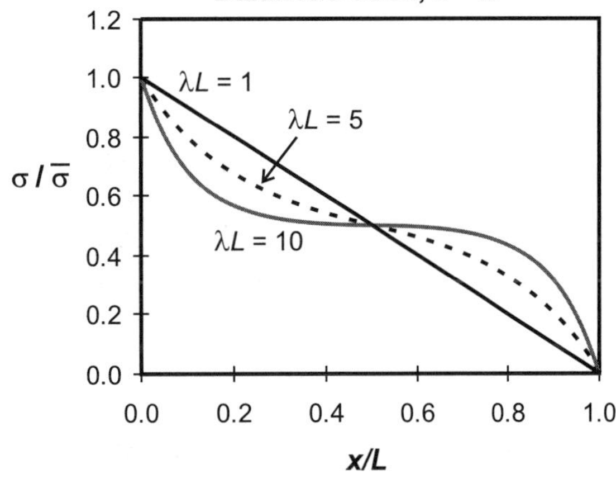

Figure 14.22. Balanced joint, $k = 2$. Axial stress in the lower plate $\sigma_2/\bar{\sigma_2}$ versus position along joint x/L.

Shear Stress Concentration Factor

The *shear stress concentration factor* (SCF) is:

$$K_\tau = \frac{\tau_{max}}{\bar{\tau}}$$ [Eq. 14.58]

The maximum shear stresses occur at the ends of the joint ($x = 0$ and L), and are equal when the joint is balanced.

The shear SCF for a *balanced joint* is calculated using *Equation 14.52* at $x = 0$ or L. The balanced joint SCF is plotted in *Figure 14.23* against λL. The SCF can be approximated with a bilinear model, indicated by the dotted line.

Provided that $\lambda L < 2$, the SCF is approximately constant and equal to $K_\tau = 1$. For this case, the shear stress is essentially uniform in the adhesive (*Figure 14.21*).

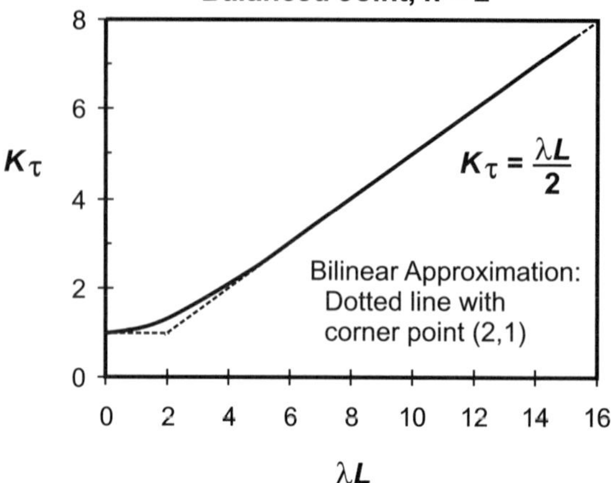

Figure 14.23. The shear stress concentration factor for a *balanced joint* is approximated by a bilinear model (*dotted line*): for $\lambda L < 2$, $K_\tau = 1$; for $\lambda L > 2$, $K_\tau = 0.5(\lambda L)$.

For values of λL greater than 2, the SCF for a balanced joint is approximately:

$$K_\tau = \frac{\lambda L}{2}$$ [Eq. 14.59]

For $\lambda L = 10$, $K_\tau = 5$, so $\tau_{max} = 5\bar{\tau}$, as shown in *Figure 14.21*. Likewise, for $\lambda L = 5$, $K_\tau = 2.5$.

Joint Strength – Adhesive Failure

The adhesive fails when the maximum shear stress reaches the adhesive's shear strength $\tau_{max} = \tau_f$. The failure load P_f is related to τ_f through the SCF:

$$P_f = \bar{\tau}_f (WL) = \frac{\tau_f}{K_\tau} WL$$ [Eq. 14.60]

Provided $\lambda L > 2$, $K_\tau = 0.5\lambda L$. The maximum load P_f that can be transmitted through the joint is then:

$$P_f = \frac{2\tau_f W}{\lambda}$$ [Eq. 14.61]

An interesting feature of this equation is that the failure load is independent of adhesive length L (!).

When the design stress is exceeded, a natural reaction is to increase the length of the joint in order to decrease the average stress and thus increase the failure load. With bonded joints, this strategy does not work. Increasing the length increases the SCF by the same factor, and no advantage is gained. For example, if L is doubled, the average shear stress $\bar{\tau}$ is halved, but the SCF, $K_\tau = 0.5\lambda L$, is also doubled; the maximum shear stress remains the same.

Substituting the expression for λ from *Equation 14.56* into *Equation 14.61* gives the failure load per unit width of a balanced joint:

$$\frac{P_f}{W} = \tau_f\sqrt{\frac{tE_1t_1}{G}} = \tau_f\sqrt{\frac{tE_2t_2}{G}} \qquad \text{[Eq. 14.62]}$$

Besides modifying τ_f, strength can be increased by selecting an adhesive with a lower shear modulus G, or by increasing the adhesive thickness t.

The shear modulus for available epoxy adhesives is approximately $G \sim 1$ GPa. Changing the adhesive thickness t may be the only strategy available to increase the strength of the joint. However, fracture is more likely in a thicker adhesive; thickness is often limited by the fracture condition.

Example 14.6 Double Lap Joint

Given: A double lap joint consists of three aluminum plates ($E = 10\times10^6$ psi) joined with an adhesive (*Figure 14.24*). The thickness of each of the outer plates is $t_o = 0.05$ in. and of the inner plate is $t_i = 0.2$ in.; the width of each plate is $W = 2.0$ in. The total load to be transferred by the joint is $P = 3000$ lb. The thickness of the adhesive is $t = 0.005$ in. and the shear modulus is $G = 250,000$ psi. The length of the adhesive joint is $L = 1.5$ in.

Required: Determine the shear stress at each end of the adhesive joint.

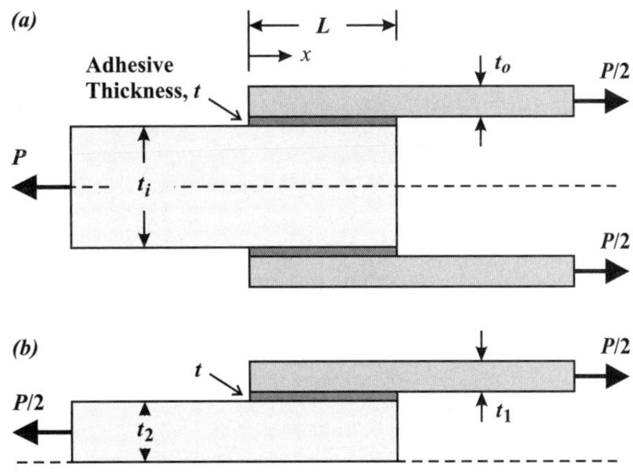

Figure 14.24. (a) Double lap joint. **(b)** By symmetry, a double lap joint is analyzed as a single lap joint.

Solution: Since the joint is symmetric, only half needs be analyzed (*Figure 14.24b*): $t_1 = t_o$ and $t_2 = t_i/2$. This configuration does not satisfy the balanced condition and consequently the full set of relationships of *Equations 14.44* through *14.48* is required.

The value of λ is given by *Equation 14.48*:

$$\lambda^2 = \frac{G}{t}\left(\frac{1}{E_1 t_1} + \frac{1}{E_2 t_2}\right)$$

$$= \frac{250\times10^3 \text{ psi}}{(5.0\times10^{-3} \text{ in.})}\left[\frac{1}{(10\times10^6 \text{ psi})(0.05 \text{ in.})} + \frac{1}{(10\times10^6 \text{ psi})(0.1 \text{ in.})}\right] = 150 \frac{1}{\text{in.}^2}$$

so:

$$\lambda = 12.25 \text{ in.}^{-1}$$

and

$$\lambda L = (12.25 \text{ in.}^{-1})(1.5 \text{ in.}) = 18.37$$

Equation 14.47 gives the value of k:

$$k = 1 + \frac{E_1 t_1}{E_2 t_2} = 1.5$$

The average shear stress in the adhesive is:

$$\bar{\tau} = \frac{P/2}{WL} = \frac{1500 \text{ lb}}{(1.5 \text{ in.})(2.0 \text{ in.})} = 500 \text{ psi}$$

Equation 14.46 gives the shear stress in the adhesive, subjected to force $P/2$:

$$\tau(x) = \frac{(P/2)\lambda L}{WLk \, \sinh \lambda L}[(k-1)[\cosh \lambda(L-x)] + \cosh \lambda x]$$

At the ends of the adhesive, the shear stresses are maxima:

$$\tau(0) = (500 \text{ psi})\frac{18.4}{1.5\sinh 18.4}[(1.5-1)(\cosh 18.4) + \cosh 0] = 3062 \text{ psi}$$

$$\tau(1.5) = (500 \text{ psi})\frac{18.4}{1.5\sinh 18.4}[(1.5-1)(\cosh 0) + \cosh(18.4)] = 6123 \text{ psi}$$

Answer: $\underline{\tau(0) = 3.1 \text{ ksi}}$

Answer: $\underline{\tau(1.5) = 6.1 \text{ ksi}}$

The *stress concentration factors* at $x=0$ and 1.5 in. are $K_\tau = \tau/\bar{\tau} = 6.13$ and 12.3, respectively. Since the joint is unbalanced, the shear stresses at the ends are unequal, as shown by the solid line in *Figure 14.24c*. The transfer of the load between the plates and the adhesive occurs primarily at the ends of the joint. The contribution of the central 60% of the adhesive is small.

To achieve a joint with a more uniform shear stress, the value of λL should be decreased. One method is to reduce the length of the adhesive to $L = 0.5$ in.

Then $\lambda L = (12.2)(0.5) = 6.1$ and the average shear stress increases to:

$$\bar{\tau} = \frac{1500 \text{ lb}}{(0.5 \text{ in.})(2.0 \text{ in.})} = 1500 \text{ psi}$$

Again using *Equation 14.46* with the new values of λL and $\bar{\tau}$, the end shear stresses are:

$$\tau(0) = (1500)\frac{6.1}{1.5 \sinh 6.1}[(1.5-1)(\cosh 6.1) + \cosh 0] = 3077 \text{ psi}$$

$$\tau(0.5) = (1500)\frac{6.1}{1.5 \sinh 6.1}[(1.5-1)(\cosh 0) + \cosh(6.1)] = 6114 \text{ psi}$$

Answer: $\tau(0) = 3.1$ ksi

Answer: $\tau(1.5) = 6.1$ ksi

By decreasing L, the average shear stress increases. At the same time, the SCFs have decreased to $K_\tau(x=0) = 2.05$ and $K_\tau(x=0.5) = 4.1$. The new normalized shear stress distribution is the dashed line in *Figure 14.24c*.

The increase in $\bar{\tau}$ and the decrease in SCF offset each other. While the shear stress distribution has become more uniform, the maximum shear stresses remain the same.

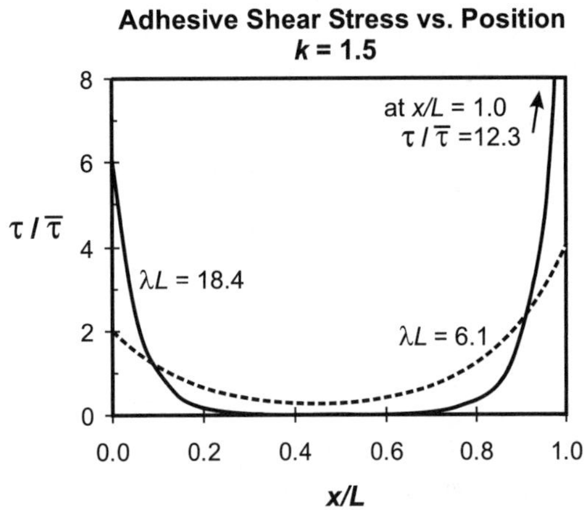

Figure 14.24. (c) Shear stress distribution in joint.

14.5 Design Problem

In a space vehicle, a thin panel is attached to a primary member by means of an aluminum T-joint (*Figure 14.25a*). The panel is aluminum, with a thickness of 0.080 in., and is to support a maximum tensile force per unit width (into the paper) of $P/W = 600$ lb/in. An adhesive of thickness $t = 0.005$ in. is to be used.

The goal is to design the joint by specifying the thickness of the angle members t_o, and the length of the joint L. The shear strength of the adhesive is unknown, but can be determined with tests.

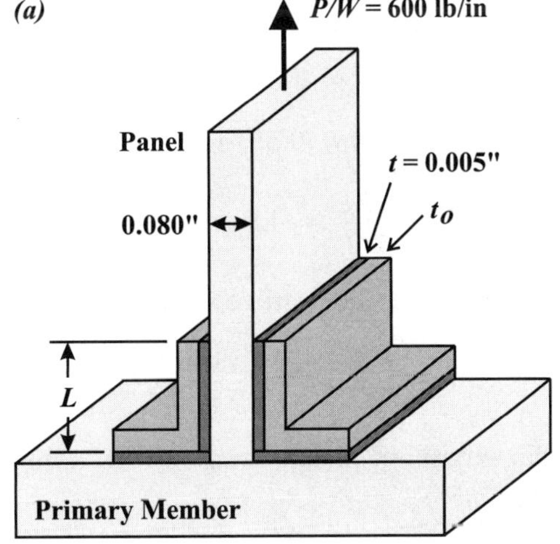

Figure 14.25. (a) Secondary member in space vehicle.

Experiment to Determine Shear Strength of Adhesive

To estimate the shear strength of the adhesive, the test coupon (specimen) shown in *Figure 14.25b* is tested to failure. The coupon is a 1.0 in. wide aluminum ($E = 10.0$ Msi) joint with joint length 1.5 in. The thickness of each outer plate is 0.09375 in. (3/32 in.), the thickness of the center plate is 0.1875 in. (3/16 in.), and the thickness of the adhesive is $t = 0.005$ in. The shear modulus of the adhesive is $G = 100$ ksi.

During the test, failure occurs by shear in the adhesive when the applied load is $P = 1150$ lb. The shear strength of the adhesive must now be determined.

The test specimen is *symmetric* about the center so it may be analyzed as a single lap joint. In addition, since the inner plate is twice the width of each outer plate, and all are made of the same material, the joint is *balanced*.

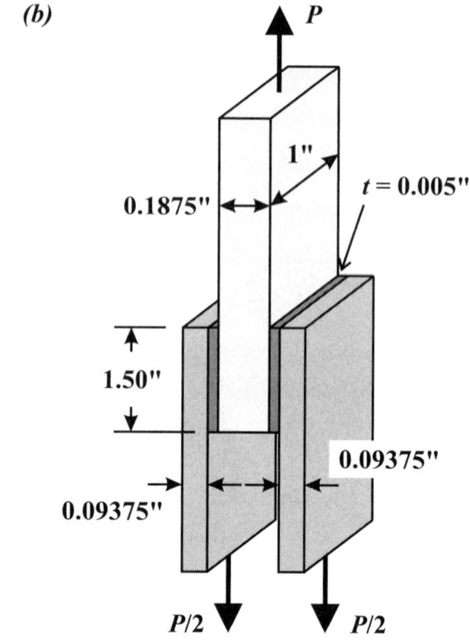

Figure 14.25. (b) Test coupon.

From *Equation 14.48*, with $t_1 = t_2 = 0.09375$ in.:

$$\lambda^2 = \frac{G}{t}\left(\frac{1}{E_1 t_1} + \frac{1}{E_2 t_2}\right) = \frac{0.1\times10^6 \text{ psi}}{0.005 \text{ in.}}\left[\frac{2}{(10\times10^6 \text{ psi})(0.09375 \text{ in.})}\right]$$

so $\lambda = 6.532$ in.$^{-1}$

and $\lambda L = 9.80$

From *Equation 14.59* or *Figure 14.23* the stress concentration factor is:

$$K_\tau = \frac{\lambda L}{2} = 4.9$$

At failure, the average shear stress is:

$$\bar{\tau} = \frac{P/2}{WL} = \frac{(1150 \text{ lb})/2}{(1.0 \text{ in.})(1.5 \text{ in.})} = 383 \text{ psi}$$

The maximum shear stress in the adhesive at failure is the shear strength τ_f of the adhesive:

$$\tau_f = K_t \bar{\tau} = (4.9)(383 \text{ psi}) = 1877 \text{ psi}$$

Design of Joint

The joint is shown in *Figure 14.25a*. The thickness of the angles t_o and the length of the joint L must be determined. A *balanced design* is a practical choice, so the thickness of the angle connections is half the inner plate member thickness:

$$t_o = \frac{t_i}{2} = \frac{0.080 \text{ in.}}{2} = 0.04 \text{ in.}$$

the value of λ for the T-joint is then given by:

$$\lambda^2 = \frac{G}{t}\left[\frac{1}{E_o t_o} + \frac{1}{E_i(t_i/2)}\right] = \frac{0.1 \times 10^6 \text{ psi}}{0.005 \text{ in.}}\left[\frac{2}{(10 \times 10^6 \text{ psi})(0.040 \text{ in.})}\right] = 100 \text{ in.}^{-2}$$

from which: $\lambda = 10.0 \text{ in.}^{-1}$

The failure load per unit width for a *single lap joint* is (*Equation 14.61*):

$$\frac{P_f}{W} = \frac{2\tau_f}{\lambda}$$

For a *double lap joint*, twice the area transfers the load, so the failure load is:

$$\frac{P_f}{W} = \frac{4\tau_f}{\lambda} = \frac{4(1877 \text{ psi})}{10 \text{ in.}^{-1}} = 750 \text{ lb/in.} > 600 \text{ lb/in.} \textbf{ OK}$$

The maximum load per unit width on the member is to be $P/W = 600$ lb/in. The T-joint that joins the panel to the structural member is sufficient.

The failure strength is independent of joint length for $L > 0.2$ in. ($\lambda L = 2$). For ease of manufacture, select a joint length of about 0.5 in. ($\lambda L = 5$).

Thus, for the design of the panel T-joint: $t_o = 0.04$ in. and $L = 0.5$ in.

15.0 Introduction

Weight-saving is a significant driver in the development of new products. Reducing the weight of an airplane allows its payload to be increased. Minimizing weight reduces the amount of energy required to launch spacecraft. Weight reduction also improves performance in automobiles, bicycles, and sports equipment such as racquets and golf clubs.

Consider a bar of cross-sectional area A and length L subjected to tensile load P (*Figure 15.1*). The elongation of the bar is taken as the primary design consideration. The mass of the bar is:

$$m = \rho A L \qquad \text{[Eq. 15.1]}$$

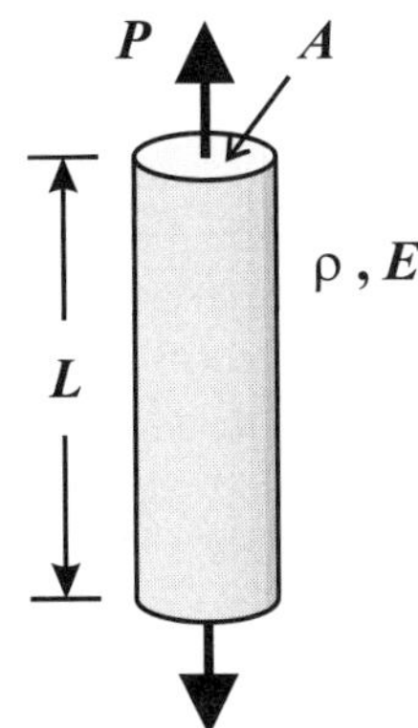

Figure 15.1. Tension bar subjected to load P.

where ρ is the mass density of the material. The elongation of the bar is:

$$\Delta = \frac{PL}{AE} \qquad \text{[Eq. 15.2]}$$

Eliminating A from *Equations 15.1* and *15.2* gives the mass:

$$m = \frac{P}{\Delta}\left(\frac{\rho}{E}\right)L^2 \qquad \text{[Eq. 15.3]}$$

For a given system stiffness P/Δ (system performance) and length L, the mass of the bar is minimized by selecting a material with a low value of ρ/E, or a high value of E/ρ. The ratio E/ρ is the **specific modulus** (stiffness per unit density).

Table 15.1 gives typical values for E/ρ for five representative **composite materials**, as well as for structural steel and structural aluminum. The table illustrates that a bar can be made significantly lighter if made of a composite material such as graphite/epoxy (graphite fibers in an epoxy matrix) instead of steel or aluminum.

If strength is the primary design consideration, the load at failure is:

$$P_u = A S_u \qquad \text{[Eq. 15.4]}$$

F.A. Leckie, D.J. Dal Bello, *Strength and Stiffness of Engineering Systems*, Mechanical Engineering Series, **487**
DOI 10.1007/978-0-387-49474-6_15, © Springer Science+Business Media, LLC 2009

Table 15.1. Representative values of Specific Modulus and Specific Strength for various composites, steel, and aluminum.

Material [source]	Fiber Volume Fraction, f	Density, ρ (kg/m³)	Modulus in Fiber Direction, E (GPa)	Tensile Strength, S_u (MPa)	E/ρ (rank)	S_u/ρ (rank)
Graphite/Epoxy (PMC) [1]	0.57	1.59	294	589	185 (1)	370 (4)
Carbon/Epoxy (PMC) [2]	0.60	1.58	142	1830	90 (2)	1160 (1)
E-glass/Epoxy (PMC) [2]	0.60	2.10	45	1020	21 (7)	486 (3)
Boron/Aluminum (MMC) [1]	0.50	2.65	235	1370	89 (3)	517 (2)
SiC/CAS (CMC) [1]	0.39	2.72	121	393	44 (4)	144 (5)
Steel A36	n/a	7.85	210	450	25 (6)	57 (7)
Aluminum 6061-T6	n/a	2.70	70	310	26 (5)	115 (6)

Notes: PMC: Polymer Matrix Composite; MMC: Metal Matrix Composite; CMC: Ceramic Matrix Composite.

Sources for *Tables 15.1, 15.2, 15.3, 15.4,* and *15.5*:

1. Daniel, I.M., and Ori, I., *Engineering Mechanics of Composite Materials*, Oxford University Press, 1994, pp. 34–35. Reproduced with permission of Oxford University Press.
2. Barbero, E.J., *Introduction to Composite Materials Design*, Taylor and Francis, 1999, p. 8. Reproduced with permission of Routledge, Inc., a division of Informa plc.
3. Agarwal, B.D., and Broutman, L.J., *Analysis and Performance of Fiber Composites*, J.Wiley and Sons, 1980.
4. Mallick, P.K., *Fiber-Reinforced Composites: Materials Manufacturing and Design*, 2nd ed., Marcel Dekker, 1993.
5. Jones, R.M., *Mechanics of Composites Materials*, 2nd ed., Taylor and Francis, 1999.
6. Chawla, K.K., *Composite Materials*, Springer-Verlag, 1987.
7. Chawla, K.K., *Fibrous Materials*, Cambridge University Press, 1998.

where S_u is the ultimate strength of the material. Eliminating A from *Equations 15.1* and *15.4* gives the mass:

$$m = P_u \left(\frac{\rho}{E}\right) L \qquad \text{[Eq. 15.5]}$$

For a specified failure load P_u and length L, the mass of the bar is minimized by selecting a material with a low value of ρ/S_u, or a high value of S_u/ρ. The ratio S_u/ρ is the **specific strength**.

Table 15.1 lists typical values of S_u/ρ for a number of representative materials. Although the specific strength rankings are different from those for specific modulus, composites again rank high.

The specific properties of composites tend to be higher than those of conventional materials. What is a *composite*? And what is it about *composites* that gives them superior specific properties?

15.1 Composite Materials

A ***composite*** is a material system in which two physically distinct materials are combined to create a new material. An everyday example of a composite is steel–reinforced concrete; the steel rebar gives tensile strength to the concrete, which is otherwise weak in tension.

There are different types of composites, but the most efficient in terms of stiffness and strength are ***continuous fiber composites***, which are the focus of this chapter. Long ***fibers*** of small diameter (~100 μm) are embedded in a ***matrix*** or host material (*Figure 15.2*).

Composites are identified by their constituents. The convention is to first indicate the fiber material and then the matrix material. For example, from *Table 15.1*, the polymer matrix composite *graphite/epoxy* consists of graphite fibers in an epoxy matrix, while the metal matrix composite *boron/aluminum* (B/Al) consists of boron fibers in an aluminum matrix. Composites are normally produced in the form of a long tape or as a sheet called a ***lamina*** or a ***ply*** (*Figure 15.2*). Several *laminae* (*plies*) are stacked together and subjected to heat and pressure to bond them into a single unit called a ***laminate*** (*Figure 15.3*).

Fiber

Continuous fiber composites consist of long fibers of small diameter typically between 10 and 150 μm. Glass, carbon, and graphite fibers are of smaller diameter, while ceramic fibers such as silicon carbide and boron are generally of larger diameter.

Ceramic fibers are strong and stiff, and tend to be used in high performance materials. Typical properties for boron and silicon carbide fibers are given in *Table 15.2*; S_u and E are both high. If the fiber diameter is made less than the

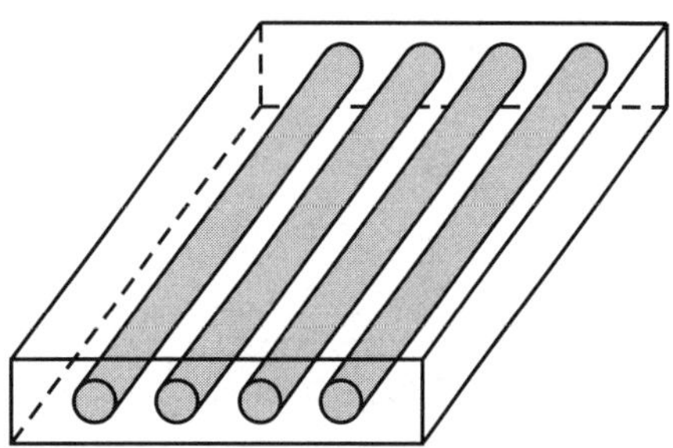

Figure 15.2. Continuous fibers are embedded in a matrix to form a *lamina* or *ply.*

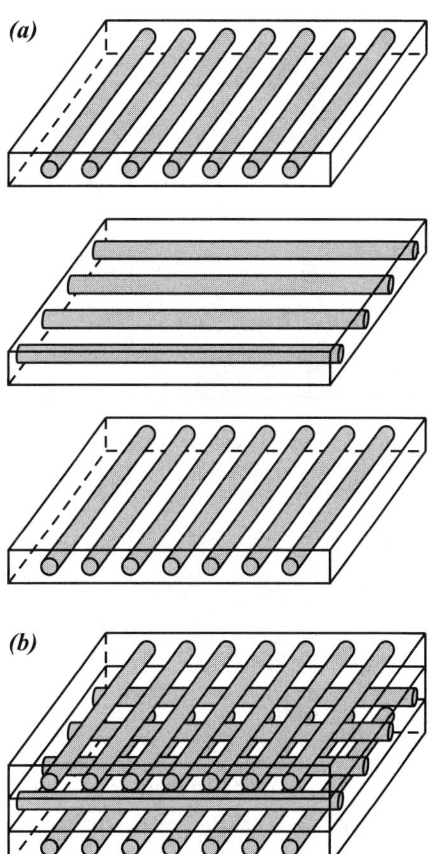

Figure 15.3. (a) Three *laminae* or *plies* formed into **(b)** a *laminate.*

Table 15.2. Representative values for Fiber materials.

Fiber [source]	Density, ρ (kg/m³)	Modulus, E (GPa)	Strength, S_u (MPa)	Poisson's ratio, v	Diameter, d (µm)
Graphite [3,4,6]	1.9	390	2100	0.2	8
Carbon (AS-4) [2,4]	1.8	240	4000	0.2	7
E-glass [1-6]	2.55	72	3400	0.2	10
Boron [1,2,4-6]	2.6	400	3400	0.2	140
Silicon Carbide [1,2,4,6]	3.1	400	3500	0.2	140
Steel wire [5,7]	7.9	210	4000	0.3	100

Table 15.3. Representative values for Matrix materials.

Matrix [source]	Density, ρ (kg/m³)	Modulus, E (GPa)	Strength, S_u (MPa)	Poisson's ratio, v
Epoxy [4]	1.3	3	100	0.4
Aluminum [4,6,7]	2.7	70	300	0.33
Titanium 6Al-4V [4,6]	4.5	110	900	0.3

Sources for *Tables 15.2* and *15.3*: see *Sources* for *Table 15.1*.

---

critical crack length for the material, then fracture of the fibers due to internal flaws is avoided.

Glass fibers are commonly used in low- to medium-performance components because of their high strength and low cost; however, their moduli are typically small ($E_{glass} \sim E_{aluminum}$). Graphite fibers are stiff but relatively weak; conversely, carbon fibers are strong, but not stiff.

Matrix

The most common matrix material is some form of polymer or epoxy, resulting in a *polymer matrix composite* or PMC. Polymers typically have low stiffness and low strength, and must be used at moderate temperatures. Young's modulus for a polymer is typically 5 GPa or less and the ultimate strength is 120 MPa or less. Typical epoxy matrix properties are given in *Table 15.3*.

Metal matrix composites, MMCs, are used in applications involving high temperatures upwards of 700°C. *Ceramic matrix composites*, CMCs, are used in higher temperature applications, such as in gas turbines where temperatures exceed 1000°C. Although ceramics are inherently brittle, careful materials engineering can produce CMCs that exhibit ductile behavior.

Volume Fraction

The role of the fiber is to carry the stress and that of the matrix is to locate the fibers. The fibers are sometimes referred to as the *reinforcement* since they make the matrix or host material stronger. The amount of reinforcement is defined by the fiber *volume fraction*, which is the ratio of the volume of the fibers V_f to the total volume of the composite V:

$$f = \frac{V_f}{V} \qquad \text{[Eq. 15.6]}$$

For continuous fiber composites, the fiber is the same length as the composite, so the volume fraction reduces to the *area fraction*:

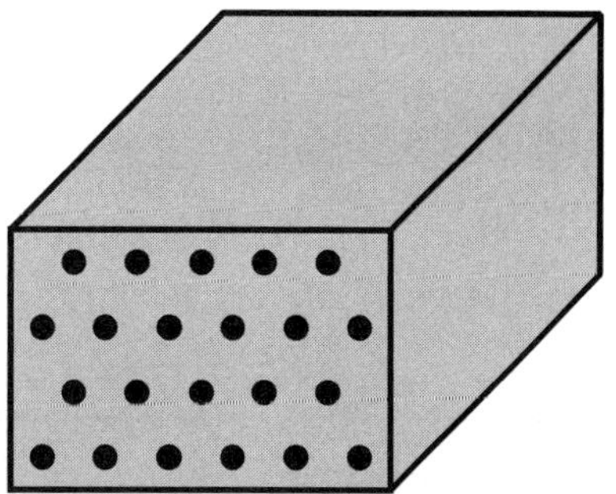

Figure 15.4. For continuous fiber composites, the fiber area fraction equals the fiber volume fraction.

$$f = \frac{A_f}{A} \qquad \text{[Eq. 15.7]}$$

where A_f is the cross-sectional area of the fibers and A is the total cross-sectional area of the composite (*Figure 15.4*).

The volume (area) fraction of the matrix is then:

$$f_m = 1 - f \qquad \text{[Eq. 15.8]}$$

The mass of the composite is:

$$m = V_f \rho_f + (V - V_f)\rho_m \qquad \text{[Eq. 15.9]}$$

where ρ_f is the density of the fiber and ρ_m is the density of the matrix. Dividing by the total volume gives the density of the composite, a weighted average of the fiber and matrix densities:

$$\rho = f\rho_f + (1 - f)\rho_m \qquad \text{[Eq. 15.10]}$$

This form of equation – a weighted average based on the volume fractions – is called the *rule of mixtures*.

Example 15.1 Density and Mass of a Composite Engine Rod

Given: An engine rod is made of the composite SiC/Ti. The fibers are silicon carbide $(\rho_f = 3100 \text{ kg/m}^3)$ and the matrix is titanium $(\rho_m = 4500 \text{ kg/m}^3)$. The fiber volume fraction is $f = 0.32$. The rod is of solid circular cross-section with diameter $D = 12$ mm and length, $L = 0.75$ m.

Required: Determine (a) the density of the composite and (b) the mass of the composite rod.

Solution: The density of the composite and the mass of the rod are:

$$\rho = f\rho_f + (1-f)\rho_m = (0.32)(3100 \text{ kg/m}^3) + (1-0.32)(4500 \text{ kg/m}^3)$$

Answer: $\rho = 4050 \text{ kg/m}^3$

$$m = \rho A L = (4050 \text{ kg/m}^3)\left[\frac{\pi(0.012 \text{ m})^2}{4}\right](0.75 \text{ m}) = 0.344 \text{ kg}$$

Answer: $m = 0.344 \text{ kg}$

15.2 Properties of a Lamina (Ply)

A plan view of a lamina subjected to *in-plane* stresses σ_x, σ_y, and τ_{xy}, is shown in *Figure 15.5*. The *local axes x–y* are selected so that the *x*-axis coincides with the **fiber direction**. Because the fibers have higher modulus and strength than the matrix, it is expected that the stiffness and strength of the lamina measured in the *x*-direction, or **longitudinal direction**, are greater than those in the *y*-direction, or **transverse direction**.

The elastic properties of composite lamina are provided by manufacturers. Representative values are given in *Table 15.4*. E_x is the modulus in the longitudinal (fiber) direction; E_y is the modulus in the transverse direction, perpendicular to the fibers. For comparison purposes, structural steel and aluminum are also included.

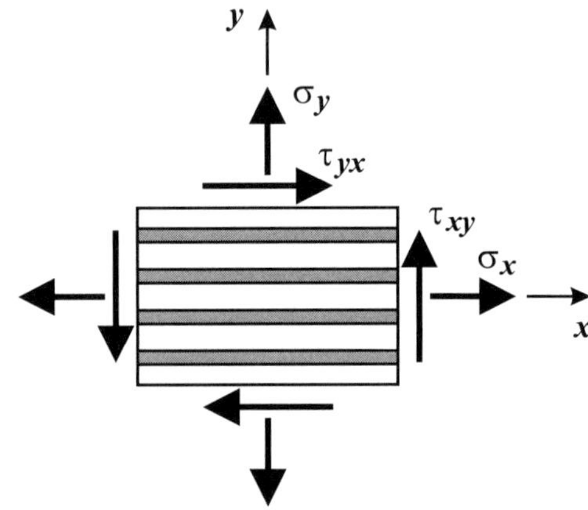

Figure 15.5. Plan view of a lamina showing its coordinate system, having applied longitudinal stress σ_x, transverse stress σ_y, and shear stress τ_{xy}.

Strength properties are given in *Table 15.5*.

When the material properties in the *x*- and *y*-directions differ, the material is said to be **anisotropic**. The modulus or stiffness ratio, E_x/E_y, is a measure of the *degree of anisotropy* of a composite material. Metals are **isotropic** since their properties are the same irrespective of the direction of applied stress.

For the PMC graphite/epoxy lamina, the elastic moduli are:

$$E_x = 294 \text{ GPa}, \quad E_y = 6.4 \text{ GPa}, \quad G_{xy} = 4.9 \text{ GPa}$$

The **longitudinal modulus** E_x is 46 times greater than the **transverse modulus** E_y. The in-plane shear modulus G_{xy} is also small compared to the longitudinal modulus. Design using materials of high anisotropy is difficult because of the need to match the stiffnesses of joined components.

Table 15.4. Representative Elastic Properties of Composite Lamina.

Material [source]	Fiber Volume Fraction, f	Density, ρ (kg/m³)	Longitudinal Modulus, E_x (GPa)	Transverse Modulus, E_y (GPa)	Shear Modulus, G_{xy} (GPa)	Major Poisson's Ratio, ν_{xy}	Minor Poisson's Ratio, ν_{yx}
Graphite/Epoxy [1]	0.57	1.59	294	6.4	4.9	0.23	0.005
Carbon/Epoxy [2]	0.60	1.58	142	10.3	7.2	0.27	0.02
E-glass/Epoxy [2]	0.60	2.10	45	12	5.5	0.19	0.05
Boron/Aluminum [1]	0.50	2.65	235	137	47	0.3	0.17
SiC/CAS [1]	0.39	2.72	121	112	44	0.2	0.18
Steel A36	n/a	7.85	200	200	77	0.3	0.3
Aluminum 6061-T6	n/a	2.70	70	70	26	0.33	0.33

Table 15.5. Representative Ultimate Strengths of Composite Lamina.

Material [source]	Fiber Volume Fraction, f	Longitudinal Tensile $S_{u,x}$ (MPa)	Longitudinal Compressive, $S_{c,x}$ (MPa)	Transverse Tensile, $S_{u,y}$ (MPa)	Transverse Compressive, $S_{c,y}$ (MPa)	In-plane Shear, τ_u (MPa)
Graphite/Epoxy [1]	0.57	589	491	29	98	49
Carbon/Epoxy [2]	0.60	1830	1096	57	228	71
E-glass/Epoxy [2]	0.60	1020	620	40	140	60
Boron/Aluminum [1]	0.50	1373	1573	118	157	128
SiC/CAS [1]	0.39	393	–	22	–	–
Steel A36	n/a	250 (yield)	250 (yield)	250 (yield)	250 (yield)	144 (yield)
Aluminum 6061-T6	n/a	240 (yield)	240 (yield)	240 (yield)	240 (yield)	139 (yield)

Sources for composite values in *Tables 15.4* and *15.5*:

1. Daniel, I.M., and Ori, I., *Engineering Mechanics of Composite Materials*, Oxford University Press, 1994, pp. 34–35. Reproduced with permission.

2. Barbero, E.J., *Introduction to Composite Materials Design*, Taylor and Francis, 1999, p. 8. Reproduced with permission.

For the MMC boron/aluminum, the stiffness ratio E_x/E_y is 1.72, while for the CMC SiC/CAS, the stiffness ratio is 1.08. The relatively stronger and stiffer matrices of MMCs and CMCs generally make them less anisotropic than PMCs.

Elastic Relationships

The strains in the x- and y-directions of a lamina caused by applied longitudinal and transverse stresses, σ_x and σ_y, are:

$$\varepsilon_x = \frac{\sigma_x}{E_x} - v_{yx}\frac{\sigma_y}{E_y} \quad \text{and} \quad \varepsilon_y = \frac{\sigma_y}{E_y} - v_{xy}\frac{\sigma_x}{E_x} \qquad \text{[Eq. 15.11]}$$

In the Poisson's ratio terms, the first subscript specifies the stress direction that causes the Poisson effect, and the second specifies the direction of the appropriate strain. Thus, v_{yx} is the Poisson's ratio relating the stress in the y-direction to the resulting transverse strain in the x-direction. Because the composite ply is anisotropic, the two terms v_{yx} and v_{xy} are different.

The shear strain–shear stress relationship is:

$$\gamma_{xy} = \frac{\tau_{xy}}{G_{xy}} \qquad \text{[Eq. 15.12]}$$

Mathematical descriptions in matrix form are very convenient for composite calculations. The **flexibility matrix** (or *compliance matrix*) **F** of a lamina is defined in terms of the strain and stress vectors, ε and σ:

$$\varepsilon = \mathbf{F}\sigma \qquad \text{[Eq. 15.13]}$$

or

$$\begin{bmatrix} \varepsilon_x \\ \varepsilon_y \\ \gamma_{xy}/2 \end{bmatrix} = \begin{bmatrix} 1/E_x & -v_{yx}/E_y & 0 \\ -v_{xy}/E_x & 1/E_y & 0 \\ 0 & 0 & 1/2G_{xy} \end{bmatrix} \begin{bmatrix} \sigma_x \\ \sigma_y \\ \tau_{xy} \end{bmatrix} \qquad \text{[Eq. 15.14]}$$

The term $\gamma_{xy}/2$ is used for the shear strain to simplify calculations, as was done in *Chapter 8 (Transformation of Stress and Strain)*. The flexibility matrix is symmetric, so that:

$$\frac{v_{xy}}{E_x} = \frac{v_{yx}}{E_y} \qquad \text{[Eq. 15.15]}$$

The **stiffness matrix K** is defined by:

$$\sigma = \mathbf{K}\varepsilon \qquad \text{[Eq. 15.16]}$$

where:

$$\mathbf{K} = \mathbf{F}^{-1} = \begin{bmatrix} \dfrac{E_x}{(1 - v_{xy}v_{yx})} & \dfrac{v_{xy}E_y}{(1 - v_{xy}v_{yx})} & 0 \\[3mm] \dfrac{v_{yx}E_x}{(1 - v_{xy}v_{yx})} & \dfrac{E_y}{(1 - v_{xy}v_{yx})} & 0 \\[3mm] 0 & 0 & 2G_{xy} \end{bmatrix} \qquad \text{[Eq. 15.17]}$$

The *stiffness matrix* is always symmetric, so that:

$$v_{yx}E_x = v_{xy}E_y \qquad \text{[Eq. 15.18]}$$

which is the same relationship given in *Equation 15.15*.

Since the moduli are different in each direction, there are two values for Poisson's ratio. The larger ratio is the **major Poisson's ratio** and the smaller ratio is the **minor Poisson's ratio**. Since $E_x > E_y$, then the major Poisson's ratio is v_{xy} and the minor Poisson's ratio is v_{yx}.

Example 15.2 Flexibility and Stiffness of a Graphite/Epoxy Lamina

Given: A graphite/epoxy composite lamina having the properties given in *Tables 15.4* and *15.5*.

Required: Determine (a) the flexibility matrix, (b) the strains when the applied stresses are: $\sigma_x = 300$ MPa, $\sigma_y = 15$ MPa, and $\tau_{xy} = 25$ MPa, and (c) if the applied stresses cause failure.

Solution: *Step 1.* From *Table 15.4*, the elastic properties are:

$$E_x = 294 \text{ GPa}; \quad E_y = 6.4 \text{ GPa}; \quad G = 4.9 \text{ GPa}; \quad v_{xy} = 0.23; \quad v_{yx} = 0.005$$

and from *Table 15.5* the tensile and shear strengths are:

$$S_{u,x} = 589 \text{ MPa}; \quad S_{u,y} = 29 \text{ MPa}; \quad \tau_u = 49 \text{ MPa}$$

Note that the condition $v_{yx}E_x = v_{xy}E_y$ is satisfied:

$$(0.005)(294 \text{ GPa}) = 1.47 \text{ GPa} \quad \text{and} \quad (0.23)(6.4 \text{ GPa}) = 1.47 \text{ GPa}$$

Step 2. The flexibility matrix $\mathbf{F}$ is:

$$\mathbf{F} = \begin{bmatrix} \dfrac{1}{E_x} & \dfrac{-v_{yx}}{E_y} & 0 \\[3mm] \dfrac{-v_{xy}}{E_x} & \dfrac{1}{E_y} & 0 \\[3mm] 0 & 0 & \dfrac{1}{2G_{xy}} \end{bmatrix} = \dfrac{1}{E_x} \begin{bmatrix} 1 & \dfrac{-v_{yx}E_x}{E_y} & 0 \\[3mm] -v_{xy} & \dfrac{E_x}{E_y} & 0 \\[3mm] 0 & 0 & \dfrac{E_x}{2G_{xy}} \end{bmatrix}$$

$$\textit{Answer: } \mathbf{F} = \frac{1}{294 \text{ GPa}} \begin{bmatrix} 1 & -0.23 & 0 \\ -0.23 & 45.9 & 0 \\ 0 & 0 & 30 \end{bmatrix}$$

Step 3. The strains are determined from the stress–strain law, $\varepsilon = \mathbf{F}\sigma$:

$$\begin{bmatrix} \varepsilon_x \\ \varepsilon_y \\ \gamma_{xy}/2 \end{bmatrix} = \mathbf{F} \begin{bmatrix} \sigma_x \\ \sigma_y \\ \tau_{xy} \end{bmatrix} = \frac{1}{294 \text{ GPa}} \begin{bmatrix} 1 & -0.23 & 0 \\ -0.23 & 45.9 & 0 \\ 0 & 0 & 30 \end{bmatrix} \begin{bmatrix} 300 \\ 15 \\ 25 \end{bmatrix} \text{ MPa}$$

$$\textit{Answer: } \begin{bmatrix} \varepsilon_x \\ \varepsilon_y \\ \gamma_{xy}/2 \end{bmatrix} = \begin{bmatrix} 1.01 \times 10^{-3} \\ 2.11 \times 10^{-3} \\ 2.55 \times 10^{-3} \end{bmatrix} = \begin{bmatrix} 0.10 \\ 0.21 \\ 0.26 \end{bmatrix} \%$$

Although stress σ_x is over 10 times greater than either σ_y or τ_{xy}, the strain in the *x*-direction has the smallest value since E_x is very large ($E_x/E_y = 46$).

Step 4. The applied stresses σ and relevant strengths $\mathbf{S}$ of the laminate are, respectively:

$$\begin{bmatrix} \sigma_x \\ \sigma_y \\ \tau_{xy} \end{bmatrix} = \begin{bmatrix} 300 \\ 15 \\ 25 \end{bmatrix} \text{ MPa} \quad \text{and} \quad \begin{bmatrix} S_{u,\,x} \\ S_{u,\,y} \\ \tau_u \end{bmatrix} = \begin{bmatrix} 589 \\ 29 \\ 49 \end{bmatrix} \text{ MPa}$$

The relevant strength components are all greater than the applied stresses, so:

Answer: The ply does not fail.

Comparison of Ply Properties

There are numerous composite materials. Suppliers provide tables of properties for their particular products. The material properties given in *Tables 15.4* and *15.5* are representative of various types of composites.

Because glass fiber composites are inexpensive, they are used in many applications. They have high strength, but low stiffness. In terms of specific properties (*Table 15.1*), they rate well in specific strength, but are at the bottom of the list of specific stiffness. Hence they are used when strength and cost are important design considerations. The ratio of moduli, or degree of anisotropy, is $E_x/E_y = 3.8$. A common composite is fiberglass, which consists of *non-continuous fibers* or *whiskers* (the analysis of such composites is not presented here).

A carbon fiber reinforced polymer (CFRP) composite – a polymer matrix composite (PMC) – consists of carbon fibers in an epoxy matrix. CFRPs are of moderate cost and have the highest specific strength and good specific stiffness (*Table 15.1*). The degree of anisotropy is $E_x/E_y = 13.8$. Their application is in aircraft and automobile components.

Graphite fibers reinforced polymers (GFRP) are generally expensive. They are very stiff with the highest specific stiffness in *Table 15.1*, and are often used in space applications. The degree of anisotropy is high with $E_x/E_y = 46$. Their typical strength, however, is not as great as that of carbon/epoxy.

Boron/aluminum (B/Al) is a metal matrix composite (MMC). The composite has good specific stiffness and specific strength. The degree of anisotropy of B/Al is small with $E_x/E_y = 1.7$; MMCs are more isotropic than PMCs. Boron/aluminum has been used in space applications. Another common class of MMCs are silicon carbide/titanium (SiC/Ti) composites, which are generally used in high-temperature applications (>400°C) due to titanium's high-melting temperature and resistance to oxidation.

SiC/CAS is a ceramic matrix composite (CMC). Although not as strong or stiff as the other composites, because they are made of ceramics, they are used in high-temperature applications (>800°C). The stiffness ratio is $E_x/E_y = 1.08$, so that they are almost isotropic. However, they have poor transverse strength; their strength ratio is high $S_{u,x}/S_{u,y} = 18$.

15.3 Approximating the Elastic Properties of a Lamina

The properties of a lamina or ply are generally obtained from mechanical tests on the actual material. In the absence of hard data, the anisotropic properties of a lamina can be estimated by simple calculations.

A lamina with continuous longitudinal fibers is shown in *Figure 15.6*. The fiber has modulus E_f, shear modulus G_f, Poisson's ratio v_f, and ultimate tensile strength $S_{u,f}$. The corresponding values of the matrix are E_m, G_m, v_m, and $S_{u,m}$. The fiber volume fraction is f, and the matrix volume fraction is $(1-f)$.

The elastic properties of a lamina (ply) are estimated below.

Longitudinal Elastic Properties

Consider a lamina, with total cross-sectional area normal to the fibers A, subjected to applied stress in the fiber-direction σ_x (*Figure 15.6*). The stress in the fibers is σ_f and in the matrix is σ_m; these stresses are not equal. The lamina strain in the longitudinal direction is ε_x and in the transverse direction is ε_y. Because they are bonded together, the fibers and matrix have the same longitudinal strain as the lamina:

$$\varepsilon_x = \varepsilon_f = \varepsilon_m \qquad \text{[Eq. 15.19]}$$

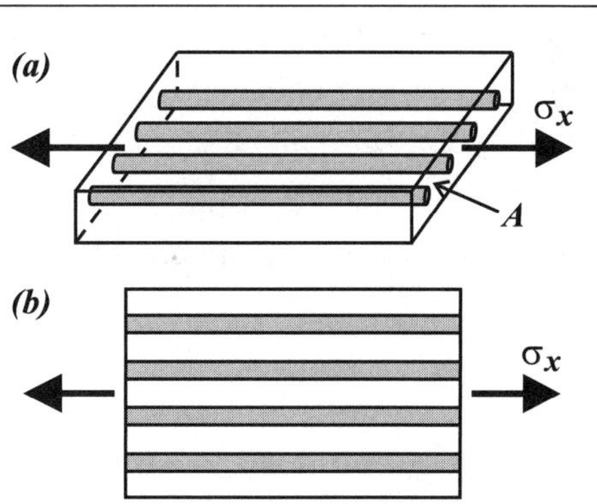

Figure 15.6. (a) A lamina subjected to tensile stress in the fiber direction. **(b)** Top view of lamina.

From Hooke's Law:

$$\varepsilon_x = \frac{\sigma_x}{E_x} = \varepsilon_f = \frac{\sigma_f}{E_f} = \varepsilon_m = \frac{\sigma_m}{E_m} \qquad \text{[Eq. 15.20]}$$

where E_x is the longitudinal modulus of the lamina. Considering equilibrium at any cross-section, the applied force equals the sum of the forces in the fibers and matrix:

$$A\sigma_x = A_f\sigma_f + A_m\sigma_m \qquad \text{[Eq. 15.21]}$$

where A_f and A_m are the total cross-sectional areas of the fibers and the matrix, respectively. Dividing both sides of *Equation 15.21* by A gives:

$$\sigma_x = f\sigma_f + (1-f)\sigma_m \qquad \text{[Eq. 15.22]}$$

Rewriting σ_f and σ_m in terms of strain and moduli (*Equation 15.20*) results in:

$$\begin{aligned} \sigma_x &= f\varepsilon_f E_f + (1-f)\varepsilon_m E_m \\ &= \varepsilon_x[\,fE_f + (1-f)E_m\,] \end{aligned} \qquad \text{[Eq. 15.23]}$$

The lamina modulus in the x-direction, $E_x = \sigma_x/\varepsilon_x$, is thus a weighted average – rule of mixtures – of the moduli of the composite constituents:

$$E_x = fE_f + (1-f)E_m \qquad \text{[Eq. 15.24]}$$

The relation between longitudinal modulus E_x and volume fraction f is linear, as shown in *Figure 15.7*.

The stresses in the fiber and matrix in terms of the longitudinal strain ε_x are:

$$\sigma_f = E_f\varepsilon_x$$

$$\sigma_m = E_m\varepsilon_x \qquad \text{[Eq. 15.25]}$$

The constituent with the larger modulus – usually the fiber – is subjected to the larger stress since both constituents must strain by the same amount.

Due to the Poisson effect, the transverse strains in the fiber and matrix are:

$$\varepsilon_{y,f} = -\nu_f\varepsilon_x$$

$$\varepsilon_{x,m} = -\nu_m\varepsilon_x \qquad \text{[Eq. 15.26]}$$

Since the matrix and fiber are aligned, their individual transverse strains both

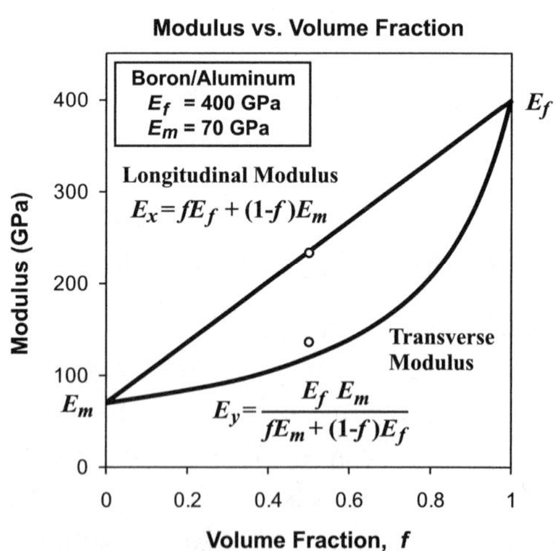

Figure 15.7. Graph of modulus versus volume fraction f, for the MMC B/Al ($E_f/E_m \sim 5.7$). The hollow circles represent tabulated moduli E_x and E_y (*Table 15.4*).

contribute to the total transverse strain according to their volume fractions. The total transverse strain is:

$$\varepsilon_y = -f v_f \varepsilon_f - (1-f) v_m \varepsilon_m \qquad \text{[Eq. 15.27]}$$

The Poisson's ratio for loading in the fiber (x-) direction, causing strain in the transverse (y-) direction is:

$$v_{xy} = -\frac{\varepsilon_y}{\varepsilon_x} = f v_f + (1-f) v_m \qquad \text{[Eq. 15.28]}$$

Transverse Elastic Properties

Consider a lamina with applied transverse stress σ_y in the y-direction, perpendicular to the fibers (*Figure 15.8a*). To simplify the calculations, but still approximate the physics, the composite is modeled as shown in *Figure 15.8b*. In the model, the load is no longer shared in a complex way between the cylindrical fiber and the surrounding matrix; both are now simply subjected to σ_y.

The fiber and matrix strains in the load direction are:

$$\varepsilon_f = \frac{\sigma_y}{E_f} \quad \text{and} \quad \varepsilon_m = \frac{\sigma_y}{E_m} \qquad \text{[Eq. 15.29]}$$

The total strain ε_y is the weighted sum of the strains of the constituents:

$$\varepsilon_y = f\varepsilon_f + (1-f)\varepsilon_m = \sigma_y\left[\frac{f}{E_f} + \frac{(1-f)}{E_m}\right] \qquad \text{[Eq. 15.30]}$$

Hence, the composite modulus in the transverse direction, E_y, is:

$$\frac{1}{E_y} = \frac{\varepsilon_y}{\sigma_y} = \left[\frac{f}{E_f} + \frac{(1-f)}{E_m}\right] \qquad \text{[Eq. 15.31]}$$

or

$$E_y = \frac{E_f E_m}{f E_m + (1-f) E_f} \qquad \text{[Eq. 15.32]}$$

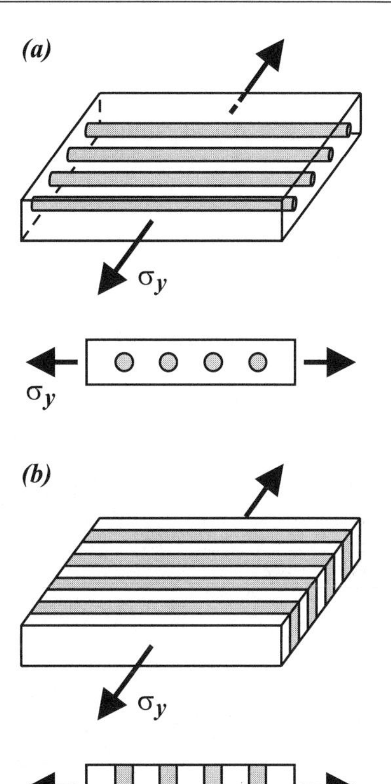

Figure 15.8. (a) Lamina under transverse stress; isometric view and view along fiber direction. **(b)** Simplified model of lamina under transverse stress; isometric view and view along fiber direction.

The variation of transverse modulus E_y is shown in *Figure 15.7*. For small values of f, the transverse modulus E_y is generally weakly dependent on E_f, i.e., the change in E_y from E_m at $f = 0$ to its value at $f = 0.5$ is relatively small. The transverse properties are generally dominated by the matrix properties.

Since the stiffness and flexibility matrices are symmetric (*Equations 15.14, 15.17*), the Poisson's ratio for stresses in the transverse (y-) direction causing strains in the fiber (x-) direction is:

$$v_{yx} = \frac{E_y}{E_x} v_{xy}$$

[Eq. 15.33]

In-Plane Shear Elastic Properties

When a lamina, as modeled in *Figure 15.8b*, is subjected to in-plane shear stress τ_{xy}, both fiber and matrix are subjected to τ_{xy} (*Figure 15.9a*). Hence, the shear stresses in the fiber and matrix are equal:

$$\tau_f = \tau_m = \tau_{xy}$$

[Eq. 15.34]

The corresponding shear strains are different since the shear moduli of the fibers and matrix are different (*Figure 15.9b*). The shear strains are:

$$\gamma_f = \frac{\tau_{xy}}{G_f} \quad \text{and} \quad \gamma_m = \frac{\tau_{xy}}{G_m}$$

[Eq. 15.35]

Adding the contribution of each constituent according to their respective volume fractions gives:

$$\gamma_{xy} = f\gamma_f + (1-f)\gamma_m$$

$$= \tau_{xy}\left[\frac{f}{G_f} + \frac{(1-f)}{G_m}\right] = \frac{\tau_{xy}}{G_{xy}}$$

[Eq. 15.36]

Hence, the composite shear modulus is:

$$\frac{1}{G_{xy}} = \left[\frac{f}{G_f} + \frac{(1-f)}{G_m}\right]$$

[Eq. 15.37]

or

$$G_{xy} = \frac{G_f G_m}{fG_m + (1-f)G_f}$$

[Eq. 15.38]

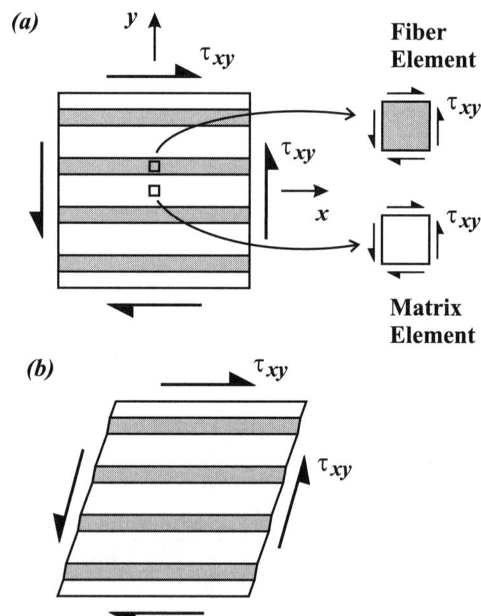

Figure 15.9. (a) Lamina under shear stress. **(b)** The fiber is generally stiffer than the matrix, and so has smaller shear strains.

Summary of Elastic Constants

The components of a laminate's flexibility and stiffness matrices (*Equations 15.14, 15.17*) can be determined from the approximate equations:

$$E_x = fE_f + (1-f)E_m \qquad\qquad \text{[Eq. 15.39]}$$

$$E_y = \frac{E_f E_m}{fE_m + (1-f)E_f} \qquad\qquad \text{[Eq. 15.40]}$$

$$v_{xy} = f v_f + (1-f) v_m \qquad\qquad \text{[Eq. 15.41]}$$

$$v_{yx} = \frac{E_y}{E_x} v_{xy} \qquad\qquad \text{[Eq. 15.42]}$$

$$G_{xy} = \frac{G_f G_m}{fG_m + (1-f)G_f} \qquad\qquad \text{[Eq. 15.43]}$$

The equations for E_x and E_y are plotted in *Figure 15.7* ($E_f/E_m{\sim}5.7$) and in *Figure 15.10* ($E_f/E_m{\sim}80$). Tabulated moduli are also included in the graphs.

The expression for E_x approximates actual composite values well; both fiber and matrix must strain by the same amount since the constituents act in parallel. The expression for E_x is generally an *upper bound* to actual data.

As E_f/E_m increases, the transverse response E_y becomes less dependent on volume fraction; i.e., E_y is approximately constant over a larger range of values (compare *Figures 15.7* and *15.10*). Although the expressions for E_y and G_{xy} are based on a simplified model (the constituents are assumed to act in series), they provide a *lower bound* to actual values, and generally have the same trend as experimental data.

Poisson's ratio v_{xy} is the *major Poisson's ratio*; it is associated with the transverse (*y*-) strain caused by longitudinal stress σ_x. Ratio v_{yx} is the *minor Poisson's ratio* (it is less than v_{xy}); it is associated with the longitudinal (*x*-) strain caused by transverse stress σ_y.

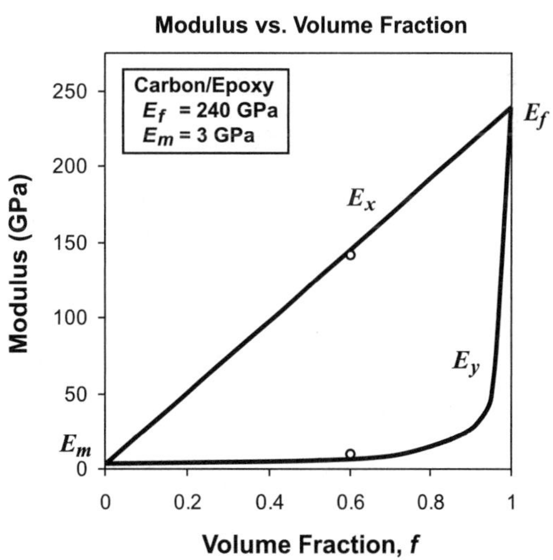

Figure 15.10. Modulus versus volume fraction a carbon/epoxy PMC ($E_f/E_m{\sim}80$). The matrix properties dominate the transverse properties. The hollow circles represent tabulated moduli E_x and E_y (*Table 15.4*).

The theoretical volume fraction limit for cylindrical fibers all having the same radius is 0.907, when the fibers are touching (*Figure 15.11a*). The circular cross-sections form a hexagonal array, with the side of each hexagon equal to the diameter of the fiber. Such extreme packing can damage the fibers and does not provide for good bonding between the fiber and matrix. Practical values of fiber reinforcement are typically between $f = 0.30$ and 0.60 (*Figure 15.11b*).

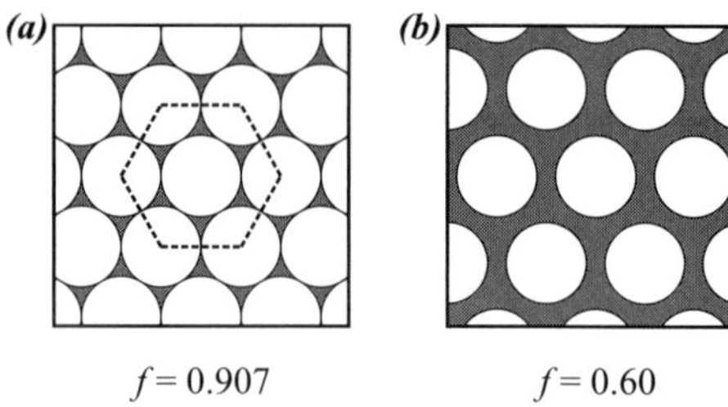

Figure 15.11. (a) Maximum theoretical volume fraction. **(b)** Typical volume fraction for PMC composites.

Example 15.3 Elastic Properties of a Carbon/Epoxy Lamina (Ply)

Given: A composite is made with carbon fibers in an epoxy matrix and has a volume fraction of $f = 0.60$. The elastic properties of the fiber and matrix are, from *Tables 15.2, 15.3*:

Fibers: $E_f = 240$ GPa, $v_f = 0.2$, $G_f = 100$ GPa, approximated from: $G = \dfrac{E}{2(1+v)}$

Matrix: $E_m = 3$ GPa, $v_m = 0.3$, $G_m = 1.2$ GPa, approximated

Required: Estimate the elastic properties of a unidirectional carbon/epoxy lamina with $f = 0.60$, and compare with those found in *Table 15.4*.

Solution: *Step 1.* The longitudinal modulus is calculated to be:

$$E_x = fE_f + (1-f)E_m = (0.6)(240 \text{ GPa}) + (0.4)(3 \text{ GPa})$$

Answer: $\underline{E_x = 145 \text{ GPa}}$

The other elastic constants are:

$$E_y = \frac{E_f E_m}{fE_m + (1-f)E_f} = \frac{(240)(3)}{(0.6)(3) + (0.4)(240)}$$

Answer: $\underline{E_y = 7.4 \text{ GPa}}$

$$v_{xy} = fv_f + (1-f)v_m = (0.6)(0.2) + (0.4)(0.3)$$

Answer: $\underline{v_{xy} = 0.24}$

$$v_{yx} = \frac{E_y}{E_x}v_{xy} = \frac{7.4}{145}(0.24)$$

Answer: $\underline{v_{yx} = 0.012}$

$$G_{xy} = \frac{G_f G_m}{fG_m + (1-f)G_f} = \frac{(100)(1.2)}{(0.6)(1.2) + (0.4)(100)}$$

Answer: $\underline{G_{xy} = 3 \text{ GPa}}$

Table 15.6. Comparison of tabulated and approximated lamina properties.

Material Carbon/Epoxy	f	E_x (GPa)	E_y (GPa)	G_{xy} (GPa)	ν_{xy}	ν_{yx}
Tabulated	0.60	142	10.3	7.2	0.27	0.02
Approximate	0.60	145	7.4	3	0.24	0.012

Step 2. Table 15.6 compares the tabulated (*Table 15.4*) and approximated elastic properties of the composite. The formula for the longitudinal modulus E_x approximates the actual modulus well, and in general overestimates the modulus. The formulae for the transverse and shear properties are not as accurate, and in general underestimate the actual properties. The main point to note, however, is that the transverse and shear properties for the epoxy–matrix composite are low, which is reflected in the results of the approximate calculations. The primary goal of the estimates is to quickly calculate ply properties during the design process, before the composite is actually manufactured. Mechanical tests on the as-manufactured composite are used to determine their actual properties.

Polymer Matrix Lamina Approximations

In polymer matrix composites, the matrix modulus E_m is much smaller than the fiber modulus E_f. The matrix stiffness can therefore be neglected in the longitudinal stiffness E_x expression of *Equation 15.40*, resulting in:

$$E_x = fE_f \qquad \text{[Eq. 15.44]}$$

The E_x versus f response is shown in *Figure 15.10* for $E_f/E_m = 80$ (carbon/epoxy). The modulus starts at essentially zero (3 MPa) and increases linearly with a slope of E_f ($\Delta E_x = 237$ GPa and $E_f = 240$ GPa differ by only 1.3%). The stress in the longitudinal direction is supported primarily by the much stiffer fibers; the contribution of the matrix is negligible. Thus the longitudinal ply strength can be approximated by:

$$S_{u,x} = fS_{u,f} \qquad \text{[Eq. 15.45]}$$

where $S_{u,f}$ is the strength of the fibers.

For the transverse modulus E_y, since $E_f \gg E_m$, *Equation 15.41* can be reduced to the approximate expression:

$$E_y = \frac{E_m}{1-f} \qquad \text{[Eq. 15.46]}$$

The E_y versus f response is also shown in *Figure 15.10* for $E_f/E_m = 80$. At $f = 0.75$, the approximate equation gives $E_y = 4E_m$ ($3.9E_m$ if the full expression is used, *Equation 15.41*), which is only $0.05E_f$. In the transverse direction, the strength is also governed by the weaker matrix, and is approximated by:

$$S_{u,y} = S_{u,m} \qquad \text{[Eq. 15.47]}$$

where $S_{u,m}$ is the strength of the matrix.

The shear modulus and shear strength are also governed by the more flexible and weaker matrix:

$$G_{xy} = \frac{G_m}{1-f}$$

[Eq. 15.48]

$$\tau_u = \tau_{u,m}$$

[Eq. 15.49]

15.4 Laminates

A single composite lamina (ply) has good properties in the fiber or longitudinal direction, but not so good properties in the transverse and shear directions. This is especially true of PMCs for which the transverse stiffness and strength are approximately 5–10% of the longitudinal values. The properties in the longitudinal direction are *fiber-dominated*, while those in the transverse and shear directions are *matrix-dominated*.

When composites are used in cylindrical pressure vessels (*Figure 15.12*), the material must have sufficient strength in both the hoop and axial (*H-* and *L-*) directions. To support the biaxial stresses, it is necessary to add fibers in a second direction. Two or more *laminae*, or *plies*, are bonded together, and the resulting material system is known as a *laminate*. A cylindrical pressure vessel is examined in *Example 15.4*.

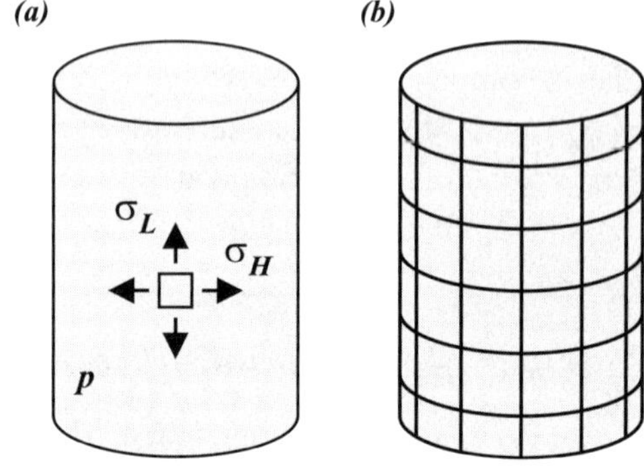

Figure 15.12. **(a)** Pressure vessels are subjected to biaxial stresses. **(b)** Several laminae (plies) may be formed into a *laminate* so that fibers are aligned in more than one direction. Possible (but not best) reinforcement directions for a cylindrical pressure vessel are shown.

Two-Ply Laminate

A *laminate* consists of two or more laminae or plies, with various fiber orientations, stacked together and bonded to form one unit. In practice, laminates can be quite complex, having several layers of plies at various orientations and of various thicknesses.

The *global stresses* applied to a laminate system are generally known from design requirements. The resulting *global strains* need to be determined, as well as the *local stresses and strains* within a single ply (laminate). The global (overall) stress–strain response is found by determining the *global stiffness of the laminate*, which depends on the individual ply stiffnesses, their orientations with respect to the global laminate coordinate system, and the number and thicknesses of the plies in each orientation. The stresses and strains developed in an individual ply depend on the ply's *local stiffness* and its orientation with respect to the global coordinate system.

The essential ideas can be illustrated with a two-ply laminate, made of two identical plies oriented symmetrically about the global *X*-axis, as shown in *Figure 15.13*. A stress element of the laminate in the *global X–Y coordinate system* is subjected to *global stresses* σ_X, σ_Y, and τ_{XY} (*Figure 15.13a*). The corresponding global strains in the *X–Y* system are ε_X, ε_Y, and γ_{XY}.

The laminate is made of two plies, A and B, of equal thickness *t*, and oriented symmetrically about the *X*-axis at angles $\pm\theta$ (*Figures 15.13b,c*). The aim is to determine the **stiffness of the laminate** as a whole, as well as the **in-ply (local) stresses and strains** of the individual plies due to the applied global stresses.

Local Stiffness of Each Ply

Consider ply B, whose fibers are oriented at angle θ clockwise from the global *X*-axis (*Figure 15.13c*). The *local coordinate system* of ply B is x_B–y_B, where the x_B-axis is parallel to the fibers, and the y_B-axis is perpendicular to the fibers. The **local stresses** and **strains** in ply B are:

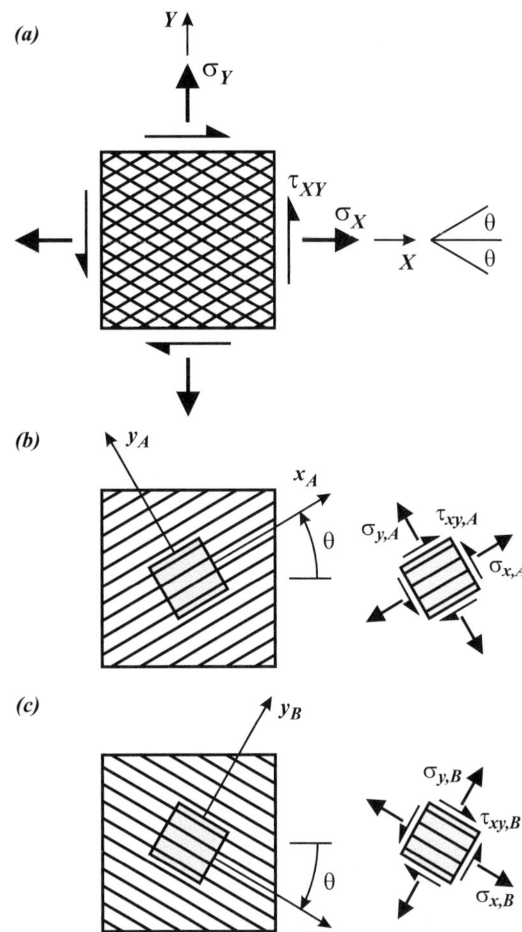

$$\sigma_B = \begin{bmatrix} \sigma_x \\ \sigma_y \\ \tau_{xy} \end{bmatrix}_B \qquad \text{[Eq. 15.50]}$$

$$\varepsilon_B = \begin{bmatrix} \varepsilon_x \\ \varepsilon_y \\ \gamma_{xy}/2 \end{bmatrix}_B \qquad \text{[Eq. 15.51]}$$

Figure 15.13. **(a)** A laminate of two plies (laminae), subjected to global stresses σ_X, σ_Y, and τ_{XY}. The plies are oriented at $\pm\theta$. **(b)** Ply A is at $+\theta$, and is subjected to local ply stresses $(\sigma_x, \sigma_y, \tau_{xy})_A$. **(c)** Ply B is at $-\theta$, and is subjected to local ply stresses $(\sigma_x, \sigma_y, \tau_{xy})_B$.

The **local stiffness** of ply B in its local x–y coordinate system is found from *Equation 15.17*:

$$\mathbf{K}_B = \begin{bmatrix} \dfrac{E_x}{(1 - v_{xy}v_{yx})} & \dfrac{v_{xy}E_y}{(1 - v_{xy}v_{yx})} & 0 \\[2ex] \dfrac{v_{yx}E_x}{(1 - v_{xy}v_{yx})} & \dfrac{E_y}{(1 - v_{xy}v_{yx})} & 0 \\[2ex] 0 & 0 & 2G_{xy} \end{bmatrix}_B \qquad \text{[Eq. 15.52]}$$

The *local* stress–strain relationship in matrix form is then:

$$\sigma_B = \mathbf{K}_B \varepsilon_B \qquad \text{[Eq. 15.53]}$$

Similarly for ply A, the local stresses σ_A are related to the local strains ε_A by the *local stiffness* of ply A, $\mathbf{K}_A$:

$$\sigma_A = \mathbf{K}_A \varepsilon_A \qquad \text{[Eq. 15.54]}$$

For this analysis, the plies are taken to be identical so that their local stiffnesses are the same:

$$\mathbf{K} = \mathbf{K}_A = \mathbf{K}_B \qquad \text{[Eq. 15.55]}$$

Global Stiffness of Laminate

Stress and Strain Transformation

The local stresses and strains in ply B (σ_B and ε_B) are transformed to the global directions, X–Y, using the transformation relationships:

$$\begin{aligned} \sigma_{B,\,G} &= \mathbf{T}_B \sigma_B \\ \varepsilon_{B,\,G} &= \mathbf{T}_B \varepsilon_B \end{aligned} \qquad \text{[Eq. 15.56]}$$

where $\sigma_{B,G}$ and $\varepsilon_{B,G}$ are the global stresses and strains in ply B in the global X–Y system. The **transformation matrix** $\mathbf{T}_B$ corresponds to a *counterclockwise rotation* of θ:

$$\mathbf{T}_B = \begin{bmatrix} c^2 & s^2 & 2sc \\ s^2 & c^2 & -2sc \\ -sc & sc & c^2 - s^2 \end{bmatrix} \qquad \text{[Eq. 15.57]}$$

where $c = \cos\theta$ and $s = \sin\theta$ (see *Chapter 8: Transformation of Stress and Strain, Equation 8.22*). Angle θ is between 0 and 90°.

Performing the same operations on ply A, but noting that the rotation θ must be *clockwise*, then the local stresses and strains in ply A (σ_A and ε_A) are transformed to the global directions, X–Y:

$$\sigma_{A,G} = \mathbf{T}_A \sigma_A$$
$$\varepsilon_{A,G} = \mathbf{T}_A \varepsilon_A$$

[Eq. 15.58]

where $\sigma_{A,G}$ and $\varepsilon_{A,G}$ are the global stresses and strains in ply A in the global directions, X and Y. The transformation matrix $\mathbf{T}_A$ corresponds to a *clockwise rotation* of θ:

$$\mathbf{T}_A = \begin{bmatrix} c^2 & s^2 & -2sc \\ s^2 & c^2 & 2sc \\ sc & -sc & c^2 - s^2 \end{bmatrix}$$

[Eq. 15.59]

where $c = \cos\theta$ and $s = \sin\theta$. Angle θ is between 0 and 90°.

Since matrices $\mathbf{T}_B$ and $\mathbf{T}_A$ correspond to equal but opposite rotations of θ (the sign of θ is taken as positive in both matrices), then:

$$\mathbf{T}_A = \mathbf{T}_B^{-1}$$

[Eq. 15.60]

Compatibility

Compatibility requires that the global strains of the individual plies, $\varepsilon_{B,G}$ and $\varepsilon_{A,G}$, be the same as the global strains of the laminate ε_G, so:

$$\varepsilon_{B,G} = \varepsilon_{A,G} = \varepsilon_G$$

[Eq. 15.61]

The global stresses in ply B, $\sigma_{B,G}$ are related to the local stresses in ply B, σ_B, by *Equation 15.56*:

$$\sigma_{B,G} = \mathbf{T}_B \sigma_B$$

[Eq. 15.62]

Substituting the local stress–strain relationship from *Equation 15.53* ($\sigma_B = \mathbf{K}_B \varepsilon_B$), and the strain transformation from *Equation 15.56* ($\varepsilon_B = \mathbf{T}_B^{-1} \varepsilon_{B,G}$), relates the global stresses in ply B to the global strains ε_G:

$$\sigma_{B,G} = \mathbf{T}_B \mathbf{K}_B \mathbf{T}_B^{-1} \varepsilon_{B,G}$$
$$= \mathbf{K}_{B,G} \varepsilon_G$$

[Eq. 15.63]

Matrix $\mathbf{K}_{B,G}$ is the *global stiffness* of ply B, i.e., its stiffness in the X–Y coordinate system:

$$\mathbf{K}_{B,G} = \mathbf{T}_B \mathbf{K}_B \mathbf{T}_B^{-1}$$

[Eq. 15.64]

Following the same procedures for ply A gives its global stiffness $\mathbf{K}_{A,G}$:

$$\mathbf{K}_{A,G} = \mathbf{T}_A \mathbf{K}_A \mathbf{T}_A^{-1}$$

[Eq. 15.65]

Since the plies are of the same thickness, and the global stiffness of each ply is known, the *global stiffness of the laminate* $\mathbf{K}_G$ is the average of the two:

$$\mathbf{K}_G = \frac{1}{2}(\mathbf{K}_{A,G} + \mathbf{K}_{B,G}) = \frac{1}{2}(\mathbf{T}_A \mathbf{K}_A \mathbf{T}_A^{-1} + \mathbf{T}_B \mathbf{K}_B \mathbf{T}_B^{-1}) \qquad \text{[Eq. 15.66]}$$

Note that the actual number of plies and their thicknesses can vary. *Equation 15.66* is valid as long as there are only two symmetric orientations, and the total ply thickness in each orientation is the same.

Without derivation, *Equation 15.66* can be extended for the general case of many plies with various orientations. The laminate stiffness is:

$$\mathbf{K}_G = \frac{1}{t}\sum_i^n t_i (\mathbf{T}_i \mathbf{K}_i \mathbf{T}_i^{-1}) \qquad \text{[Eq. 15.67]}$$

where t is the total thickness of n plies and t_i, $\mathbf{T}_i$ and $\mathbf{K}_i$ are the thickness, transformation matrix, and local stiffness of the *i*th ply.

Given the applied global stresses, σ_G, the global strains of the laminate are calculated:

$$\varepsilon_G = \mathbf{K}_G^{-1}\sigma_G \qquad \text{[Eq. 15.68]}$$

Stresses in Individual Plies

The local strains in the x_B–y_B plane of ply B are found using the transformations of *Equation 15.56* and the stress–strain relationship of *Equation 15.68*:

$$\varepsilon_B = \mathbf{T}_B^{-1}\varepsilon_G = \mathbf{T}_B^{-1}\mathbf{K}_G^{-1}\sigma_G \qquad \text{[Eq. 15.69]}$$

Applying *Equation 15.53*, the **local stresses** in ply B, σ_B, in terms of the *applied global stresses* σ_G, are thus:

$$\sigma_B = \mathbf{K}_B \varepsilon_B = \mathbf{K}_B \mathbf{T}_B^{-1} \mathbf{K}_G^{-1}\sigma_G \qquad \text{[Eq. 15.70]}$$

Likewise, the *local stresses* in ply A, σ_A, are:

$$\sigma_A = \mathbf{K}_A \varepsilon_A = \mathbf{K}_A \mathbf{T}_A^{-1} \mathbf{K}_G^{-1}\sigma_G \qquad \text{[Eq. 15.71]}$$

These matrix calculations are easy to do using computer software such as MATLAB or Mathcad. The examples below demonstrate how the calculations are performed.

Failure of a Ply

Once the local stresses in each ply are determined, ply integrity can be assessed. *Equations 15.70* and *15.71* each represent three in-plane ply stresses: σ_x, σ_y, and τ_{xy}. Failure occurs in a ply when any of the stresses exceed its corresponding ply strength (*Table 15.5*). The possible ply failure modes are: longitudinal tension (e.g., fiber failure), longitudinal compression (e.g., buckling of fibers or compressive failure of system), transverse tension (e.g., debonding of fiber and matrix, or matrix failure), transverse compression (e.g., matrix failure), and in-plane shear failure (e.g., matrix failure).

When the first ply fails, the laminate response changes. The system becomes less stiff but continues to support load, analogous to yielding in metals. In composites with many plies, the plies progressively fail until the system breaks in to two. For elastic design, the global load to cause the first ply to fail is taken as the strength of the laminate.

Failure due to *out-of-plane* stresses (perpendicular to the plies) is not considered in this treatment. Also not considered is the effect of non-symmetric ply lay-ups about the center plane of the laminate (a plane parallel to the plies).

Example 15.4 Cylindrical Pressure Vessel

Given: A cylindrical pressure vessel is modeled as a two-ply laminate of carbon/epoxy. The plies are wound so that they are at angles $\theta = \pm 35°$ to the hoop direction (*Figure 15.14*). The thicknesses of the plies in each orientation are equal. The actual composite is made of many plies, but as long as there are only two ply orientations, each having the same total thickness, the two-ply laminate model can be used.

The stresses in the pressure vessel wall are $\sigma_H = 360$ MPa in the hoop direction and $\sigma_L = 180$ MPa in the longitudinal direction. The diameter and length of the vessel are $D = 0.30$ m and $L = 1.0$ m, respectively.

The local elastic properties of each ply are, from *Table 15.4*:

$$E_x = 142 \text{ GPa}, \quad E_y = 10.3 \text{ GPa}$$

$$G = 7.2 \text{ GPa}$$

$$\nu_{xy} = 0.27, \quad \nu_{yx} = 0.02$$

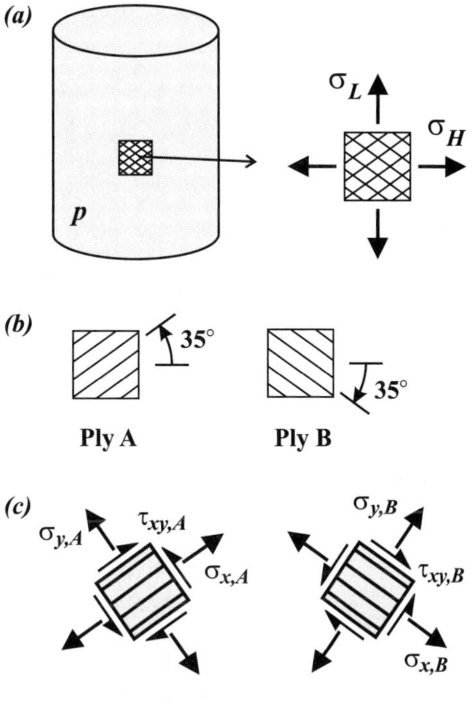

The local ply strengths in each direction are, from *Table 15.5*:

Longitudinal (tension, compression): $S_{u,x} = 1830$ MPa, $S_{c,x} = 1096$ MPa

Transverse (tension, compression): $S_{u,y} = 57$ MPa, $S_{c,y} = 228$ MPa

Shear: $\tau_u = 71$ MPa

Figure 15.14. (a) Pressure vessel with (b) plies oriented at ±35°. (c) Ply stresses.

Required: Determine (a) the stresses in plies A and B, (b) the factor of safety of the system for failure of the first ply, and (c) the changes in diameter and length of the vessel.

Solution: *Step 1.* Since the plies are of the same material, their local stiffnesses are equal:

$$\mathbf{K}_A = \mathbf{K}_B = \begin{bmatrix} \dfrac{E_x}{(1 - v_{xy} v_{yx})} & \dfrac{v_{xy} E_y}{(1 - v_{xy} v_{yx})} & 0 \\[2mm] \dfrac{v_{yx} E_x}{(1 - v_{xy} v_{yx})} & \dfrac{E_y}{(1 - v_{xy} v_{yx})} & 0 \\[2mm] 0 & 0 & 2G_{xy} \end{bmatrix} = \begin{bmatrix} 143 & 2.80 & 0 \\ 2.80 & 10.4 & 0 \\ 0 & 0 & 14.4 \end{bmatrix} \text{ GPa}$$

Step 2. Since the plies have equal thickness, the stiffness of the laminate is:

$$\mathbf{K}_G = \frac{1}{2}(\mathbf{T}_A \mathbf{K}_A \mathbf{T}_A^{-1} + \mathbf{T}_B \mathbf{K}_B \mathbf{T}_B^{-1})$$

where $\mathbf{T}_B$ and $\mathbf{T}_A$ are the transformation (rotation) matrices for $\theta = \pm 35°$, respectively:

$$\mathbf{T}_B = \begin{bmatrix} c^2 & s^2 & 2sc \\ s^2 & c^2 & -2sc \\ -sc & sc & c^2 - s^2 \end{bmatrix} = \begin{bmatrix} 0.6710 & 0.3290 & 0.9397 \\ 0.3290 & 0.6710 & -0.9397 \\ -0.4698 & 0.4698 & 0.3420 \end{bmatrix}$$

where $s = \sin 35°$ and $c = \cos 35°$;

$$\mathbf{T}_A = \mathbf{T}_B^{-1} = \begin{bmatrix} 0.6710 & 0.3290 & -0.9397 \\ 0.3290 & 0.6710 & 0.9397 \\ 0.4698 & -0.4698 & 0.3420 \end{bmatrix}$$

With these matrices determined, the global stiffness and flexibility matrices, $\mathbf{K}_G$ and $\mathbf{F}_G$, are calculated:

$$\mathbf{K}_G = \begin{bmatrix} \dfrac{E_H}{(1 - v_{HL} v_{LH})} & \dfrac{v_{HL} E_L}{(1 - v_{HL} v_{LH})} & 0 \\[2mm] \dfrac{v_{LH} E_H}{(1 - v_{HL} v_{LH})} & \dfrac{E_L}{(1 - v_{HL} v_{LH})} & 0 \\[2mm] 0 & 0 & 2G_{HL} \end{bmatrix} = \begin{bmatrix} 73.0 & 29.0 & 0 \\ 29.0 & 27.7 & 0 \\ 0 & 0 & 66.8 \end{bmatrix} \text{ GPa}$$

$$\mathbf{F}_G = \begin{bmatrix} 1/E_H & -v_{LH}/E_L & 0 \\ -v_{HL}/E_H & 1/E_L & 0 \\ 0 & 0 & 1/(2G_{HL}) \end{bmatrix} = \begin{bmatrix} 0.0235 & -0.0246 & 0 \\ -0.0246 & 0.0618 & 0 \\ 0 & 0 & 0.0150 \end{bmatrix} \frac{1}{\text{GPa}}$$

From $\mathbf{K}_G$ or $\mathbf{F}_G$, the elastic *global* properties of the *laminate* are found to be:

$$E_H = 42.6 \text{ GPa} \qquad E_L = 16.2 \text{ GPa} \qquad G_{HL} = 33.4 \text{ GPa}$$

$$v_{HL} = 1.05 \qquad v_{LH} = 0.40$$

Step 3. The stresses in each ply are:

$$\sigma_A = \mathbf{K}_A \mathbf{T}_A^{-1} \mathbf{K}_G^{-1} \sigma_G \quad \text{and} \quad \sigma_B = \mathbf{K}_B \mathbf{T}_B^{-1} \mathbf{K}_G^{-1} \sigma_G$$

where $\quad \sigma_G = \begin{bmatrix} \sigma_H \\ \sigma_L \\ \tau \end{bmatrix} = \begin{bmatrix} 360 \\ 180 \\ 0 \end{bmatrix} \text{MPa}$

Solving for the local stresses in each ply:

$$Answer\!: \ \sigma_A = \begin{bmatrix} \sigma_x \\ \sigma_y \\ \tau_{xy} \end{bmatrix}_A = \begin{bmatrix} 501 \\ 39.2 \\ 11.8 \end{bmatrix} \text{MPa} \quad \text{and} \quad \sigma_B = \begin{bmatrix} \sigma_x \\ \sigma_y \\ \tau_{xy} \end{bmatrix}_B = \begin{bmatrix} 501 \\ 39.2 \\ -11.8 \end{bmatrix} \text{MPa}$$

The longitudinal stress in each ply is 501 MPa in tension, the transverse stress is 39.2 MPa in tension, and the shear stress is ±11.8 MPa. The ply normal stresses in its local coordinate system are all tensile, and are the same in each ply.

Since the plies have equal and opposite rotations, the local shear stresses are of equal magnitude but of opposite sign, which is necessary to maintain compatibility between the two individual plies.

Step 4. The corresponding tensile and shear strengths for each ply are:

$$S_u = \begin{bmatrix} 1830 \\ 57 \\ 71 \end{bmatrix} \text{MPa}$$

The factor of safety for each mode of failure is then:

$$\text{FS} = \begin{bmatrix} 1830/501 \\ 57/39.2 \\ 71/11.8 \end{bmatrix} = \begin{bmatrix} 3.65 \\ 1.45 \\ 6.01 \end{bmatrix}$$

The smallest value determines the system factor of safety:

$$Answer\!: \ \underline{\text{FS} = 1.45}$$

corresponding to failure in tension in the transverse direction.

Step 5. The global strains are:

$$\varepsilon_G = \mathbf{K}_G^{-1} \sigma_G = \begin{bmatrix} \varepsilon_H \\ \varepsilon_L \\ \gamma_{HL}/2 \end{bmatrix} = \begin{bmatrix} 403 \\ 228 \\ 0 \end{bmatrix} \times 10^{-6}$$

The changes in diameter and length of the vessel are:

$$\Delta D = \varepsilon_H D = (403 \times 10^{-6})(0.3 \text{ m})$$

Answer: $\underline{\Delta D = 0.129 \text{ mm}}$

$$\Delta L = \varepsilon_L L = (228 \times 10^{-6})(1.0 \text{ m})$$

Answer: $\underline{\Delta L = 0.228 \text{ mm}}$

Effect of Ply Angle on Laminate Stiffness

It is necessary when designing a composite lay-up to understand the effect of ply angle θ on the global stiffness. *Equation 15.66* defines the global stiffness of a laminate, having an equal number of identical plies at $+\theta$ and $-\theta$, in terms of the ply stiffnesses and angles. The equation is repeated here:

$$\mathbf{K}_G = \frac{1}{2}(\mathbf{K}_{A,G} + \mathbf{K}_{B,G}) = \frac{1}{2}(\mathbf{T}_A \mathbf{K}_A \mathbf{T}_A^{-1} + \mathbf{T}_B \mathbf{K}_B \mathbf{T}_B^{-1}) \qquad \text{[Eq. 15.72]}$$

Using a spreadsheet program such as Excel, or a mathematical analysis tool such as Mathcad or MATLAB, the laminate stiffness and flexibility matrices can be calculated, and the effect of ply orientation studied.

The global stress–strain relationship is:

$$\varepsilon_G = \mathbf{F}_G \sigma_G = \mathbf{K}_G^{-1} \sigma_G$$

$$\begin{bmatrix} \varepsilon_X \\ \varepsilon_Y \\ \gamma_{XY}/2 \end{bmatrix} = \begin{bmatrix} 1/E_X & -v_{YX}/E_Y & 0 \\ -v_{XY}/E_X & 1/E_Y & 0 \\ 0 & 0 & 1/(2G_{XY}) \end{bmatrix} \begin{bmatrix} \sigma_X \\ \sigma_Y \\ \tau_{XY} \end{bmatrix} \qquad \text{[Eq. 15.73]}$$

where E_X, E_Y, G_{XY}, v_{XY}, and v_{YX} are the global elastic properties of the composite. These properties all depend upon the orientation of the plies θ.

Effect of Ply Angle on Graphite/Epoxy

Figure 15.15 shows the variation of elastic properties in a two-ply graphite/epoxy composite with ply angle $\pm\theta$. The elastic properties for a single ply are found in *Table 15.4*: $E_x = 294$ GPa, $E_y = 6.4$ GPa, and $G_{xy} = 4.9$ GPa.

The vertical axis is normalized (divided) by 294 MPa, the longitudinal modulus of a single ply. The global flexibility matrix of the laminate for different orientations $\pm\theta$ (*Equation 15.73*), can be calculated from the local ply stiffnesses and the transformation equations (*Equation 15.72*). Once the flexibility matrix is known for a given orientation $\pm\theta$, determining the global elastic constants of the laminate, E_X, E_Y, G_{XY}, v_{XY}, and v_{YX}, is straightforward.

When $\theta = 0°$, the laminate stiffness is the same as for a single ply since plies A and B are both oriented in the global X-direction, i.e.: $E_X = 294$ GPa, $E_Y = 6.4$ GPa, and $G_{XY} = 4.9$ GPa.

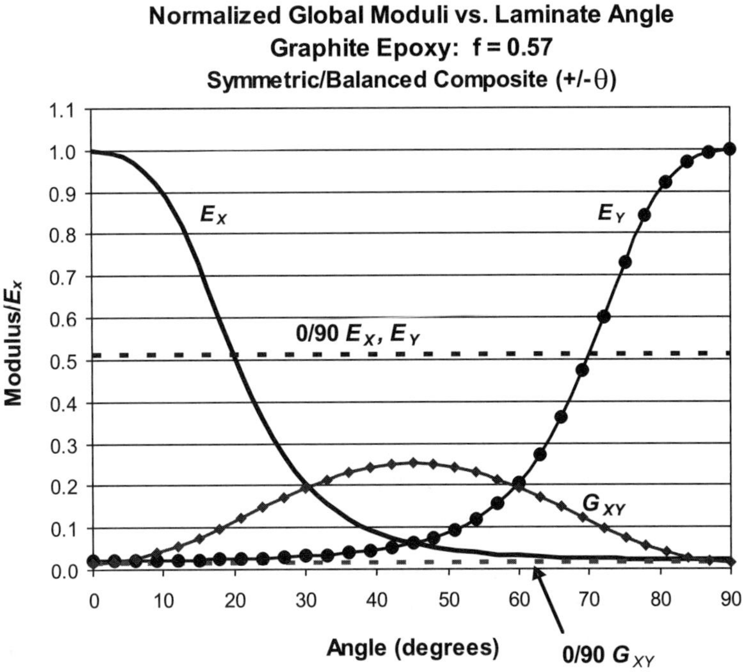

Figure 15.15. Variation of elastic moduli in a symmetric laminate, fibers oriented at ±θ. Dashed line "0/90 E_x, E_y" represents the modulus of a 0/90 composite in one of the primary fiber directions.

When $\theta = \pm35°$ – the ideal angle for a cylindrical pressure vessel – the stiffness in the global X-direction is reduced by a factor of more than 6 compared to the $\theta = 0°$ orientation, while the global shear stiffness is increased by a factor of over 12. There is a trade-off – a gain in the shear modulus is achieved at the expense of the longitudinal modulus.

The largest global shear stiffness is realized at $\theta = \pm45°$, where $G_{XY} \sim 0.25E_X$. At this angle, the global shear loads are carried parallel and perpendicular to the fiber direction of each ply. Since the local ply longitudinal modulus E_x is dominated by the fibers, then the global shear modulus G_{XY} at $\theta = \pm45°$ is also dominated by the fibers.

Example 15.5 Carbon/Epoxy Torsion Tube

Given: A thin-walled torsion tube is modeled as a two-ply laminate of carbon/epoxy (*Figure 15.16*). The thicknesses of the plies are equal. The plies are wound at angles $\theta = \pm45°$. The diameter and length of the tube are $D = 40$ mm and $L = 1.0$ m, respectively.

The elastic properties of each lamina, or ply, are, from *Table 15.4*:

$$E_x = 142 \text{ GPa}, \quad E_y = 10.3 \text{ GPa}, \quad G = 7.2 \text{ GPa}$$

$$v_{xy} = 0.27, \quad v_{yx} = 0.02$$

The strengths are, from *Table 15.5*:

Longitudinal (tension, compression):

$$S_{u,x} = 1830 \text{ MPa}, \quad S_{c,x} = 1096 \text{ MPa}$$

Transverse (tension, compression):

$$S_{u,y} = 57 \text{ MPa}, \quad S_{c,y} = 228 \text{ MPa}$$

Shear:

$$\tau_u = 71 \text{ MPa}$$

The global directions, X- and Y- are the circumferential (hoop) and axial directions of the tube, respectively. Note that since $\theta = \pm45°$, the longitudinal direction of one ply is along the transverse direction of the other; i.e., the fibers in the ply A are perpendicular to those in ply B.

Required: Determine (a) the elastic properties of the laminate, (b) the maximum shear stress that can be applied to the tube, and the corresponding failure mode, and (c) the angle of twist of the tube at failure.

Solution: *Step 1.* The stiffness of each ply in its local coordinate system is:

$$\mathbf{K}_A = \mathbf{K}_B = \begin{bmatrix} 143 & 2.8 & 0 \\ 2.8 & 10.3 & 0 \\ 0 & 0 & 14.4 \end{bmatrix} \text{GPa}$$

Step 2. The stiffness of the laminate is (*Equation 15.72*):

$$\mathbf{K}_G = \frac{1}{2}(\mathbf{T}_A \mathbf{K}_A \mathbf{T}_A^{-1} + \mathbf{T}_B \mathbf{K}_B \mathbf{T}_B^{-1})$$

where $\mathbf{T}_A$ and $\mathbf{T}_B$ are the transformation matrices:

$$\mathbf{T}_B = \begin{bmatrix} c^2 & s^2 & 2sc \\ s^2 & c^2 & -2sc \\ -sc & sc & c^2 - s^2 \end{bmatrix} \qquad \text{where } s = \sin45° \text{ and } c = \cos45°$$

Hence

$$\mathbf{T}_B = \begin{bmatrix} 0.5 & 0.5 & 1 \\ 0.5 & 0.5 & -1 \\ -0.5 & 0.5 & 0 \end{bmatrix}$$

Following similar calculations, or, since $\mathbf{T}_A = \mathbf{T}_B^{-1}$:

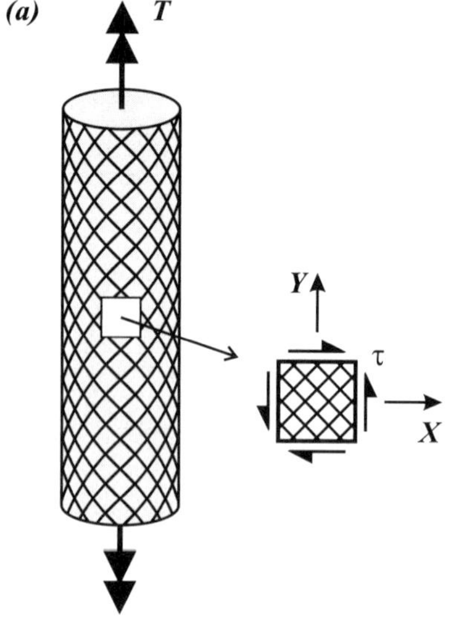

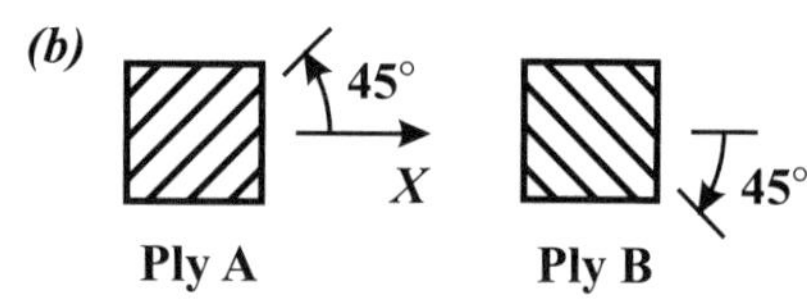

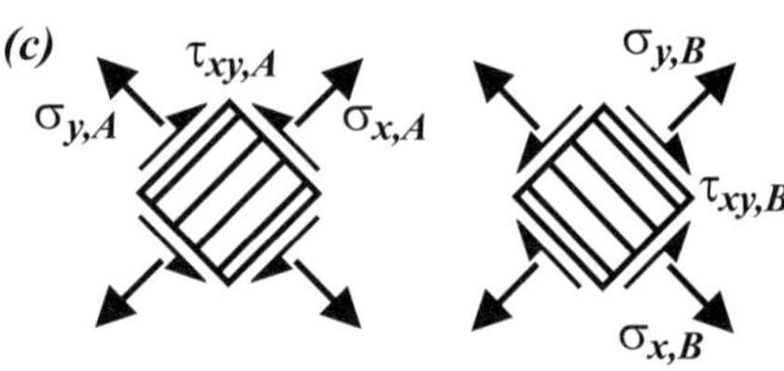

Figure 15.16. (a) Composite shaft in torsion bar with **(b)** plies oriented at ±45°. **(c)** Local ply stresses.

$$\mathbf{T}_A = \begin{bmatrix} 0.5 & 0.5 & -1 \\ 0.5 & 0.5 & 1 \\ 0.5 & -0.5 & 0 \end{bmatrix}$$

Substituting these values into *Equations 15.72* and *15.73* gives the global stiffness and flexibility matrices, $\mathbf{K}_G$ and $\mathbf{F}_G$.

From the global flexibility matrix, the elastic properties of the laminate are:

Answer: $\underline{E_X = E_Y = 24.4 \text{ GPa}}$

Answer: $\underline{G_{XY} = 36.8 \text{ GPa}}$

Answer: $\underline{v_{XY} = v_{YX} = 0.69}$

Step 3. Substituting to find the stresses in each ply:

$$\sigma_A = \mathbf{K}_A \mathbf{T}_A^{-1} \mathbf{K}_G^{-1} \sigma_G \quad \text{and} \quad \sigma_B = \mathbf{K}_B \mathbf{T}_B^{-1} \mathbf{K}_G^{-1} \sigma_G$$

To determine the strength of the system, the stresses in each ply must be determined from the applied load. The stresses in each ply are then compared to the corresponding strength of the ply (longitudinal tension, longitudinal compression, transverse tension, transverse compression, and shear) to determine if the laminate fails.

Step 4. One approach to find the strength is to apply a global test load, calculate the in-ply stresses, and then determine the factor of safety for that load based on the strengths of each ply. Multiplying the test load by the factor of safety gives the actual strength.

Assume that the tube supports a global shear stress of $\tau_{XY} = 400$ MPa (*Figure 15.16d*), so the global stress vector is:

$$\sigma_G = \begin{bmatrix} 0 \\ 0 \\ 400 \end{bmatrix} \text{MPa}$$

The stresses in each ply are then calculated to be:

$$\sigma_A = \begin{bmatrix} 759 \\ -41 \\ 0 \end{bmatrix} \text{MPa} \quad \text{and}$$

$$\sigma_B = \begin{bmatrix} -759 \\ 41 \\ 0 \end{bmatrix} \text{MPa}$$

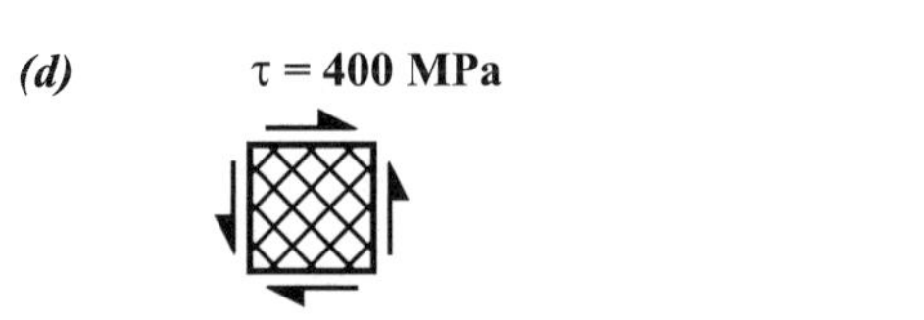

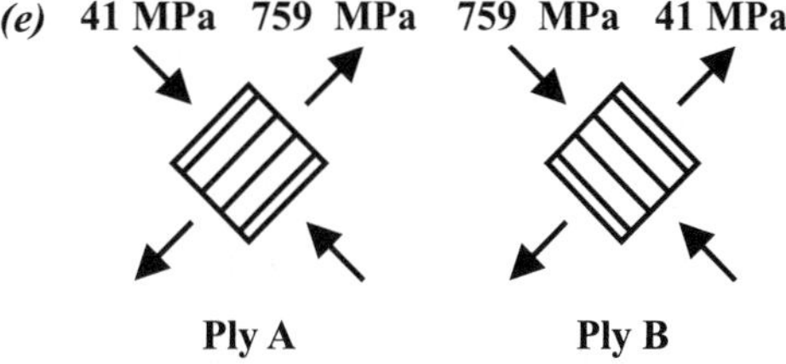

Figure 15.16. (d) Applied shear stress. **(e)** Stresses in ply A and B due to applied shear stress.

In ply A, the longitudinal stress is 759 MPa in tension, the transverse stress is 41 MPa in compression, and the shear stress is zero (*Figure 15.16e*). The corresponding strengths are $S_{u,x} = 1830$ MPa (longitudinal tension), $S_{c,y} = 228$ MPa (transverse compression), and $\tau_u = 71$ MPa (shear). Comparing the appropriate strengths of each ply to the corresponding stresses, the factors of safety for ply A are:

$$FS_A = \begin{bmatrix} 1830/759 \\ 228/41 \\ 71/0 \end{bmatrix} = \begin{bmatrix} 2.4 \\ 5.5 \\ \infty \end{bmatrix}$$

In ply B, the longitudinal stress is 759 MPa in compression, the transverse stress is 41 MPa in tension and the shear stress is zero (*Figure 15.16e*). The corresponding strengths are $S_{c,x} = 1096$ MPa (longitudinal compression), $S_{u,y} = 57$ MPa (transverse tension), and $\tau_u = 71$ MPa (shear). Comparing the corresponding strengths and stresses, the factors of safety in ply B are:

$$FS_B = \begin{bmatrix} 1096/759 \\ 57/41 \\ 71/0 \end{bmatrix} = \begin{bmatrix} 1.45 \\ 1.39 \\ \infty \end{bmatrix}$$

The lowest factor of safety, 1.39, corresponds to the *failure mode*, or *failure mechanism*, of the composite. As the applied shear stress is increased from zero, the first factor of safety to reach unity (1.0) is that for *transverse tension* of ply B. Composite failure begins in ply B, transverse to its fibers.

Step 5. The tube fails when the transverse stress in ply B equals 57 MPa. The original value of the applied test stress $\tau_{XY} = 400$ MPa can be scaled up to give the applied global shear stress at failure:

$$\tau_{XY,f} = (1.39)(400 \text{ MPa})$$

Answer: $\underline{\tau_f = 550 \text{ MPa}}$

Step 6. The angle of twist of the shaft when $\tau = \tau_f$ is:

$$\theta = \frac{\tau L}{G_{XY} R} = \frac{(550 \times 10^6 \text{ Pa})(1.0 \text{ m})}{(36.8 \times 10^9 \text{ Pa})(0.02 \text{ m})}$$

Answer: $\underline{\theta = 0.75 \text{ rad.} = 43°}$

The angle of twist is very large for this composite. Carbon epoxies are strong, but not stiff.

Failure of Composite Tubes in Torsion

The failure mode of a laminate depends on the properties and orientations of the plies that make up the laminate. The ply properties depend on the fiber and matrix properties and the fiber volume fraction.

For the thin-walled tube in torsion, with fibers oriented at $\theta = \pm 45°$, then the principal stresses in each ply are parallel and perpendicular to the fibers, as shown in

Table 15.7. Shear Modulus, In-Plane Shear Strength, and Failure Mode of a Two-Ply composite, $\theta = \pm 45°$, when loaded in In-Plane Shear.

Composite	G (GPa)	τ_u (MPa)	Failure Mode in Plies
Graphite/Epoxy	74.5	250	Axial (fiber-direction) compression
Carbon/Epoxy	36.8	550	Transverse (cross-ply) tension
E-glass/Epoxy	13.2	108	Transverse tension
Boron/Al	76.5	178	Transverse tension

Figure 15.16e. The magnitudes of the in-ply stresses depend on the elastic properties of the materials. Failure may occur by axial tension or compression of the fibers, or by transverse tension or compression of a ply.

The analysis of *Example 15.5* is repeated for several composite systems, and the results are shown in *Table 15.7*. A weak matrix usually causes the plies loaded in transverse (cross-ply) tension to fail first. For the graphite/epoxy system, failure is by compression of the fibers. The very stiff graphite fibers in one ply shield the transversely loaded matrix in the other ply. Since graphite fibers are weaker in compression than in tension, the plies with fibers loaded in compression fail first.

Orientation of Plies in Design

The above discussions demonstrate that to take full advantage of the stiffness of the fibers, they must be properly oriented. For example, ski-boards made from PMC laminates have fibers that are in the ski-direction at the center where there is maximum flexural bending moment, and fibers that flare out at the ends to resist twisting.

Simple rules can be suggested for composite design:

- Uniform uniaxial loading is supported efficiently by fibers oriented with the stress. Examples include axial members and the material furthest from the neutral axis of beams in bending (e.g., the flanges of I-beams). In order to support compressive loads and resist buckling, a few plies oriented at 90° from the load direction are also included.
- Shear loading is supported efficiently using plies at ±45° to the shear direction; examples include shafts in torsion and webs of beams in shear.

An advantage of laminate composites is that the plies can be oriented to most efficiently support the anticipated loads on the structure and the number of plies can be varied as necessary. In other words, the material can be engineered.

Cylindrical Pressure Vessels

In the pressure vessel example (*Example 15.4*), it was assumed that the ply angle was $\theta = \pm 35°$ from the hoop direction (*Figure 15.14*). Applying the laminate analysis for various ply angles shows that for polymer matrix composites (PMCs), the greatest strength is achieved when the ply angle is approximately $\theta = \pm 35°$.

Torsion Tubes

In the torsion tube example (*Example 15.5*), the ply angle is ±45° (*Figure 15.16*). Applying the laminate analysis for various ply angles shows that for PMCs, the greatest strength is achieved when the ply angle is $\theta = \pm 45°$. While the globally applied laminate load is a shear stress, each ply primarily supports the load in the local fiber direction in tension or compression.

Beams in Bending and in Shear

A beam supports both bending moments and shear forces (*Figure 15.17*). The bending moment is carried primarily by tensile and compressive stresses at the top and bottom flanges of the beam, while the shear force is carried primarily by the web. Consequently, most of the plies in the flanges are aligned with the axis of the beam, i.e., $\theta = \pm 0°$, with a few at 90° to resist buckling. Most of the plies in the web are aligned at $\theta = \pm 45°$ to the beam axis because of the superior shear properties of this layup.

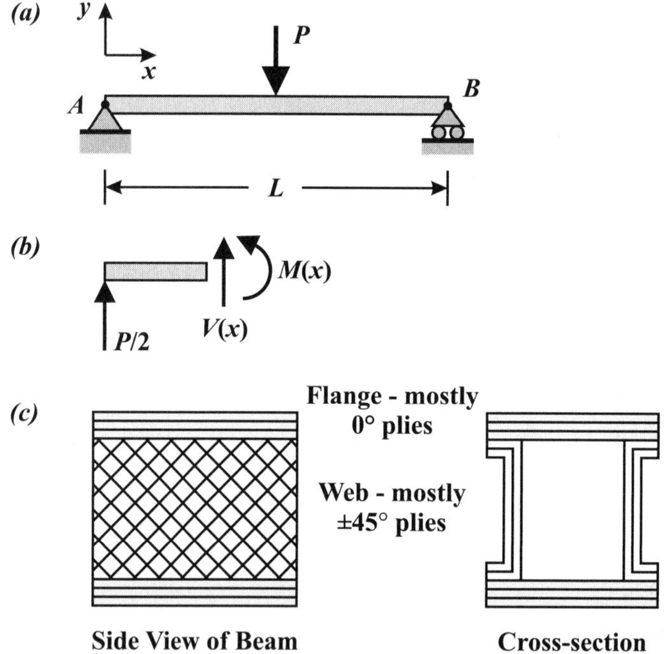

Figure 15.17. **(a)** Simply supported beam under center load. **(b)** FBD of beam segment from 0 to *x*. **(c)** Side and cross-sectional views of beam. In the flanges, the plies are primarily oriented at 0° to support the tensile and compressive bending stresses. In the web, the plies are primarily oriented at ±45°, the best way to support shear loads. Joining the flange plies to the web plies is an important part of the manufacturing process.

16.0 Introduction

Many engineering systems are designed to incorporate the coupled mechanical and electrical (or thermal) characteristics of a material. Such systems are known as **smart systems**.

A smart system may exhibit a mechanical response to an electrical or a thermal input, or an electrical response to a mechanical input. Example applications include air bag accelerometers, engine valve actuators, morphing structures that change shape due to input voltages (*Figure 16.1*), and instruments that measure the small forces in biological systems.

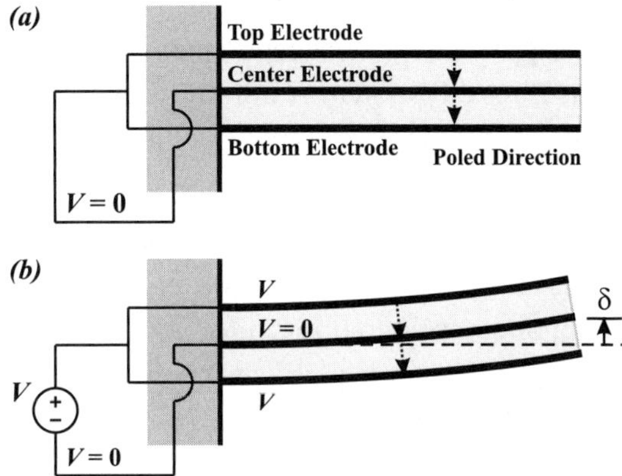

Figure 16.1. A voltage applied to a bimorph beam causes it to deflect. This piezoelectric system is acting as an *actuator.*

Devices that change shape and apply forces due to input voltages are known as *actuators*. Devices that produce electrical signals due to forces or other physical phenomena are known as *sensors*.

Three types of smart systems are introduced in this chapter:

1. *micro-electromechanical systems* (MEMS),
2. *piezoelectric materials*, and
3. *shape memory alloys*.

MEMS and piezoelectric systems are used as small-scale and often high-precision electro-mechanical actuators and sensors. Shape memory alloys have stress–strain characteristics that depend on temperature; they can also be used in actuators and sensors. Understanding the coupled effects of mechanical forces, electrical potentials, and thermal loads are necessary to determine the total response of a smart system.

When two disciplines are combined, the symbols that describe various physical quantities may conflict with each other. Different vocabularies and conventions cause basic communication challenges in all interdisciplinary studies. It is important to explicitly define the meaning of the symbols used. In this chapter, Y represents Young's

F.A. Leckie, D.J. Dal Bello, *Strength and Stiffness of Engineering Systems*, Mechanical Engineering Series, **519**
DOI 10.1007/978-0-387-49474-6_16, © Springer Science+Business Media, LLC 2009

Modulus, while E represents the electric field. The symbol S_j represents strain, where j represents the direction (1-, 2-, 3-) or the type of strain (mechanical or electrical, m or e). The symbols ε and ε_o represent the permittivity, or dielectric constant, of a material. By using electrical variables in this chapter in favor of mechanical ones, the reader will be more familiar with the electrical properties of materials given in manufacturer's catalogs.

16.1 MEMS

MEMS is short for ***micro-electromechanical systems***. MEMS used to apply forces or displacements are *actuators*, those used to measure motion are *sensors*. MEMS are used in such applications as accelerometers for deploying air bags, and in testing machines for measuring the properties of thin films. The entire device is typically on the scale of a few millimeters, and the components in the device are measured in micrometers (μm).

The behavior of MEMS devices depends on the electrostatic properties of parallel plates subjected to a *voltage* difference. Two common configurations are as follows: (1) the *parallel plate capacitor (Figure 16.2)* and (2) the *comb drive (Figure 16.3)*.

Parallel Plate Capacitor

In this device, two parallel plates are separated by a small distance and move normal to the plate areas (*Figure 16.2a*). The non-conducting material between the plates is known as the **dielectric**. In most mechanical devices, the dielectric is air.

In an electric circuit, *current* (moving charge) can flow to and from a capacitor, but not through it since the plates are not connected by a conducting material. When current does not have a closed circuit or path to follow, the system is an *open circuit*. If an RC (resistor–capacitor) circuit is subjected to a constant voltage source, the capacitor will eventually charge to the value of the voltage source (*Figure 16.2b*). One plate will be positively charged, while the other will be negatively charged. The separation of charge creates an *electric field* between the plates, and forces within the system. The charge on

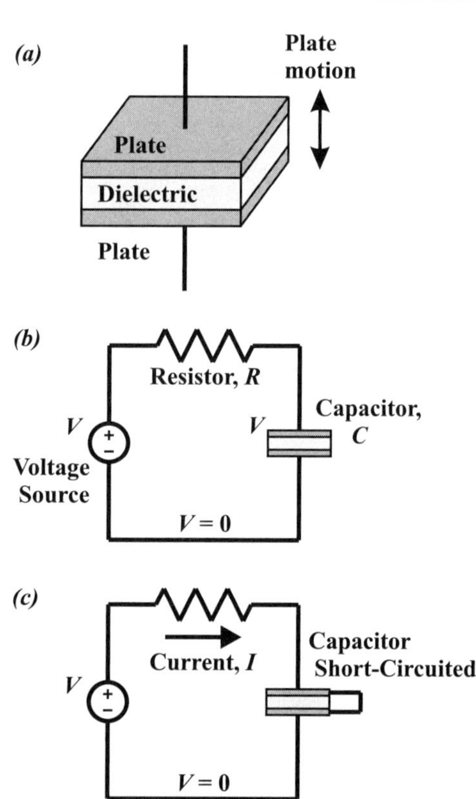

Figure 16.2. (a) Parallel plate capacitor. **(b)** Due to constant voltage *V*, the capacitor will charge to voltage *V* and current will drop to 0. **(c)** A short-circuited capacitor stores no voltage and does not impede current flow.

the plates is key to the workings of the microdevice.

When current has a continuous path, the system is a *closed circuit* or *loop*. For a constant current to freely flow through a capacitor, the plates must be *short-circuited*, or connected by a conductor (*Figure 16.2c*).

Comb Drive

In this configuration, many parallel plates make up the teeth of a comb, and the teeth of two combs overlap (*Figure 16.3*). The distance between the parallel plates is constant and the combs move parallel to the plate areas. Comb drives are used for both actuation and sensing.

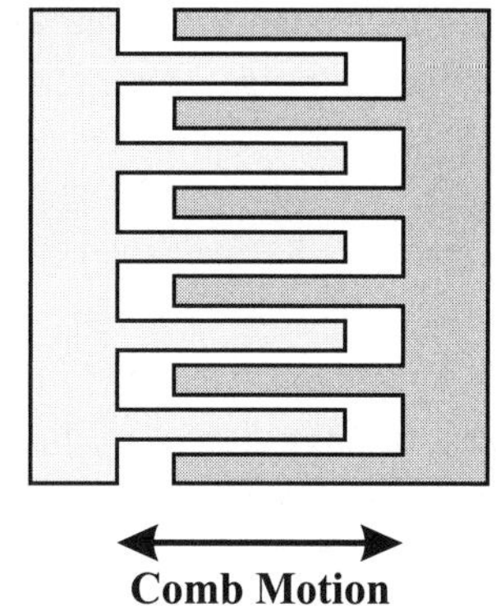

Figure 16.3. Comb drive.

16.2 Parallel Plate Capacitors

Capacitance and Internal Energy

Consider two parallel plates, each of area A, separated by distance d (*Figure 16.4a*). Distance d is much smaller than the plate size $\sqrt{A}$. The space between the plates may or may not contain a dielectric material. *Voltage V* (measured in volts, V) is applied across the plates. The application of V causes an *electric charge Q* (in coulombs, C) on the plates, one being positive, the other being negative.

Experiments that measure V and Q in capacitors show a linear relationship:

$$Q = CV \qquad \text{[Eq. 16.1]}$$

where C is the *capacitance* (*Figure 16.4b*). The unit of capacitance is coulombs per volt, known as a *farad* (F). The capacitance for practical capacitors is small and in the picofarad (1 pF = 10^{-12} F) to microfarad (1 μF = 10^{-6} F) range.

The region between the plates stores energy in an *electric field*. The energy stored in a parallel plate capacitor with voltage V and charge Q is the area under the V–Q curve.

(a)

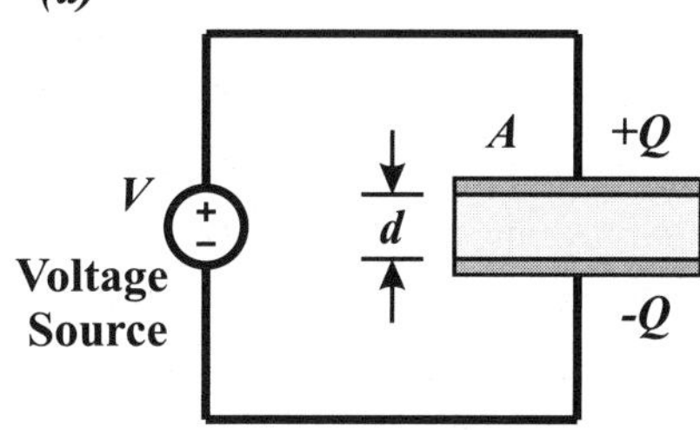

(b)

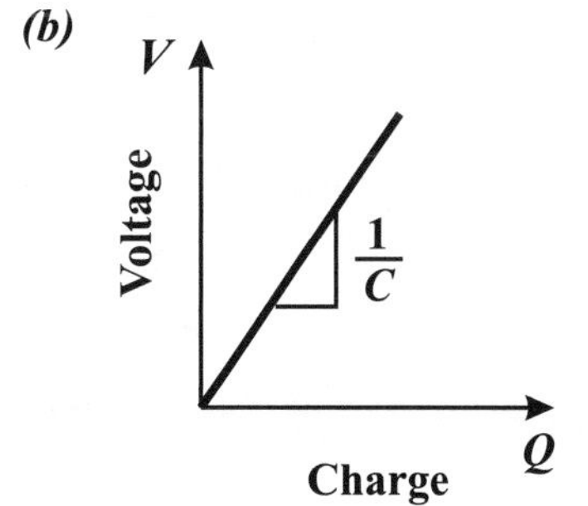

Figure 16.4. (a) Capacitor with applied voltage *V.* **(b)** Capacitor voltage–charge relationship.

The energy needed to increase the charge of a capacitor by an incremental amount δQ at voltage V (*Figure 16.5*) is:

$$\delta U = V \, \delta Q = \frac{Q}{C} \delta Q \qquad \text{[Eq. 16.2]}$$

Integrating gives the *internal energy* stored in the capacitor in terms of charge Q. No energy is stored when the charge is zero, so:

$$U(Q) = \int_0^Q \frac{Q}{C} \delta Q = \frac{1}{2} \frac{Q^2}{C} \qquad \text{[Eq. 16.3]}$$

which is the area under the V–Q curve. The internal energy in terms of voltage V is:

$$U(V) = \frac{1}{2} C V^2 \qquad \text{[Eq. 16.4]}$$

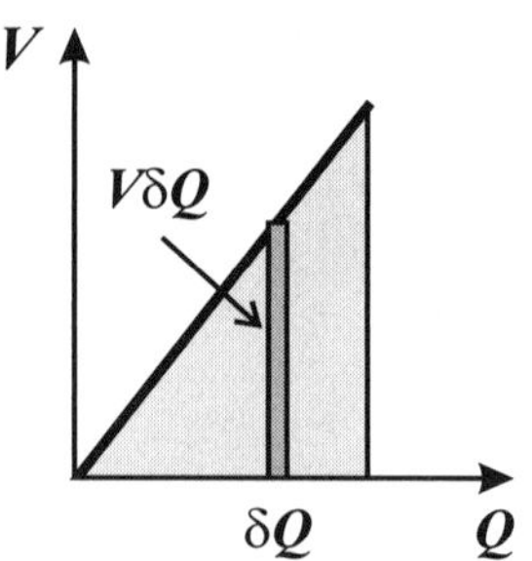

Figure 16.5. The energy stored in the capacitor is the area under the V–Q curve.

Capacitance in Terms of Material Properties

It is convenient to calculate capacitance in terms of the electrical properties of the *dielectric*, the material between the plates. The properties of the dielectric are found from experiments using the set-up shown in *Figure 16.4*, in which V and Q are measured. To determine the properties independent of the area and distance of the plates, two definitions, analogous to stress and strain, are introduced.

The ***electric field*** E (analogous to stress) is the ratio of voltage to distance d, i.e., the *voltage gradient*:

$$E = \frac{V}{d} \qquad \text{[Eq. 16.5]}$$

The ***electric (charge) flux density***, or ***electric displacement*** D (analogous to strain) is:

$$D = \frac{Q}{A} \qquad \text{[Eq. 16.6]}$$

A graph of the relationship between electric field E and flux density D (*Figure 16.6*) follows the linear relation:

$$D = \varepsilon E \qquad \text{[Eq. 16.7]}$$

where ε is the ***permittivity***, or ***dielectric constant***, of the material between the plates. Permittivity is the electric flux density per unit electric field stored in the dielectric material between the plates. Permittivity has dimensions of farads per meter (F/m), as shown by a check of the units:

$$\varepsilon = \frac{D}{E} = \frac{Q \, d}{A \, V} = \frac{Q \, d}{V A} = C \frac{d}{A} \rightarrow \text{farad} \frac{1}{\text{m}} \qquad \text{[Eq. 16.8]}$$

The permittivity of a material is usually given in terms of the permittivity of a vacuum, ε_0:

$$\varepsilon = \varepsilon_r \varepsilon_0 \qquad \text{[Eq. 16.9]}$$

where $\varepsilon_0 = 8.85 \times 10^{-12}$ F/m and ε_r is the *relative permittivity* or *relative dielectric constant*. The relative permittivity is the dimensionless ratio of the material permittivity to that of a vacuum. For mica, $\varepsilon_r = 5.4$, so its permittivity is $\varepsilon = 5.4\varepsilon_0$. In mechanical applications, the gap between the plates is air for which the relative permittivity is taken to be unity ($\varepsilon_r = 1.0$).

Equating *Equations 16.6* and *16.7*, and substituting V/d for E, gives the charge:

$$Q = \frac{\varepsilon A}{d} V \qquad \text{[Eq. 16.10]}$$

Comparing this result with *Equation 16.1*, the capacitance is:

$$C = \frac{\varepsilon A}{d} \qquad \text{[Eq. 16.11]}$$

The linear Q–V relation holds for a typical air gap capacitor ($\varepsilon = \varepsilon_0$) provided that the electric field $E = V/d$ is less than about 3×10^6 V/m. Above this value, electric breakdown can occur by sparking, or *arcing*, across the plates, transferring charge across the air gap.

In MEMS devices, the plates are coated, and an electric field with a strength in order of magnitude higher (~30 to 50×10^6 V/m) may generally be sustained before breakdown.

The variation of the capacitance per plate area C/A with air gap distance d is shown in *Figure 16.7*. Capacitance decreases with increasing gap distance.

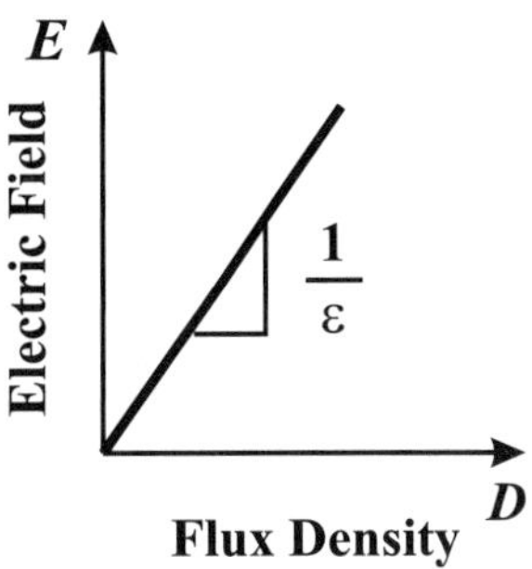

Figure 16.6. Electric field versus flux density (electric displacement).

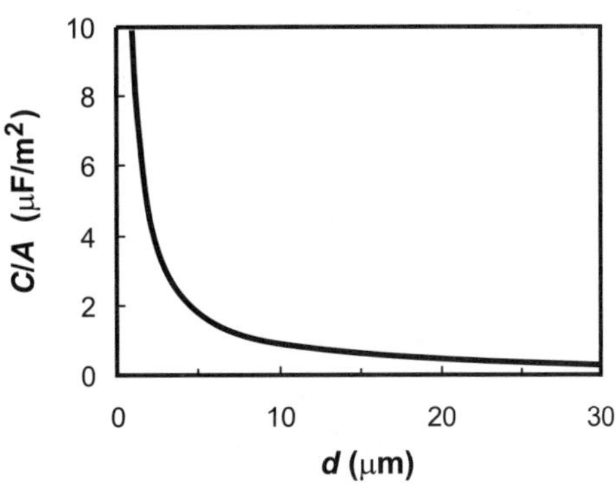

Figure 16.7. Air-gap capacitance per unit area versus plate distance: $C/A = \varepsilon_0/d$.

Example 16.1 Capacitance

Required: Determine the capacitance of a parallel plate capacitor of area $10\ \mu m \times 20\ \mu m$ with an air gap $d = 3\ \mu m$.

Solution: For air, $\varepsilon = \varepsilon_o$, so:

$$C = \frac{\varepsilon_o A}{d} = \frac{(8.85 \times 10^{-12}\ \text{F/m})(10 \times 10^{-6}\ \text{m})(20 \times 10^{-6}\ \text{m})}{3 \times 10^{-6}\ \text{m}}$$

Answer: $\underline{C = 590 \times 10^{-18}\ \text{F}}$

Force on a Parallel Plate Capacitor

Due to the attraction between the two oppositely charged capacitor plates, a resisting force F is required to keep them in place. In the absence of a sufficient force, the plates will snap together. The required force can be calculated using energy methods, in which the gap d is increased to $d+\delta d$, where δd is a small increment of displacement. There are two standard ways of performing the energy calculations:

1. the charge Q remains constant during displacement. This can be achieved by charging the plates to Q, and then electrically isolating each plate so that charge cannot be added or removed;
2. the voltage V remains constant. This is achieved by connecting the plates across a constant voltage source; current ($i = dQ/dt$) is permitted to flow to and from the plates so that charge may be added or removed.

The work done on a capacitor by force F moving incremental distance δd is:

$$\delta W = F\ \delta d \qquad\qquad \text{[Eq. 16.12]}$$

and the electrical energy supplied due to voltage V experiencing incremental charge δQ is (*Equation 16.2*):

$$\delta U_E = V\ \delta Q \qquad\qquad \text{[Eq. 16.13]}$$

Hence, the increase in total stored energy is:

$$\delta U = \delta W + \delta U_E = F\ \delta d + V\ \delta Q \qquad\qquad \text{[Eq. 16.14]}$$

Method 1. Energy Calculation for Constant Charge

Charge Q is held constant ($\delta Q = 0$) by electrically isolating the capacitor so that it cannot discharge. The capacitor gap is then increased from d to $d+\delta d$. The total internal energy $U(d)$ for constant Q is:

$$U(d) = \frac{Q^2}{2C} = \frac{Q^2}{2}\frac{d}{\varepsilon_o A} \qquad\qquad \text{[Eq. 16.15]}$$

The internal energy $U(d+\delta d)$ corresponding to gap $d+\delta d$ is:

$$U(d + \delta d) = \frac{Q^2(d + \delta d)}{2 \quad \varepsilon_o A} \qquad \text{[Eq. 16.16]}$$

so the change in total internal energy during the displacement is:

$$\delta U = U(d + \delta d) - U(d) = \frac{Q^2 \, \delta d}{2 \, \varepsilon_o A} \qquad \text{[Eq. 16.17]}$$

The work done by F to displace the plates by distance δd is:

$$\delta W = F \, \delta d \qquad \text{[Eq. 16.18]}$$

Since the plates are electrically isolated, there is no additional electrical energy supplied by an external voltage source ($\delta U_E = 0$). The increase in internal energy is due only to the mechanical work $F \, \delta d$. Thus, $\delta W = \delta U$, which gives:

$$F = \frac{Q^2}{2\varepsilon_o A} \qquad \text{[Eq. 16.19]}$$

Using *Equation 16.10*, the force is also given in terms of the voltage V by:

$$F = \frac{\varepsilon_o A}{2}\left(\frac{V}{d}\right)^2 \qquad \text{[Eq. 16.20]}$$

Method 2. **Energy Calculation for Constant Voltage**

The capacitor voltage V is held constant ($\delta V = 0$) by connecting it to a constant voltage source. The capacitor gap is then increased from d to $d+\delta d$. The total internal energy $U(d)$ for constant V is:

$$U(d) = \frac{CV^2}{2} = \frac{\varepsilon_o A V^2}{2d} \qquad \text{[Eq. 16.21]}$$

The change in total internal energy δU due to an increase in gap distance of δd is:

$$\delta U = U(d + \delta d) - U(d) = \frac{\varepsilon_o A V^2}{2(d + \delta d)} - \frac{\varepsilon_o A V^2}{2d} = \frac{\varepsilon_o A V^2}{2}\left[\frac{d - (d + \delta d)}{(d + \delta d)d}\right] \qquad \text{[Eq. 16.22]}$$

Since, $\delta d/d \ll 1$, the increment of internal energy can be reduced to:

$$\delta U = \frac{-\varepsilon_o A V^2}{2d^2}\delta d \qquad \text{[Eq. 16.23]}$$

The change in charge δQ due to increment δd is:

$$\delta Q = C(d + \delta d)V - C(d)V = V\left[\frac{\varepsilon_o A}{d + \delta d} - \frac{\varepsilon_o A}{d}\right] = \varepsilon_o A V\left[\frac{d - (d + \delta d)}{(d + \delta d)d}\right] \qquad \text{[Eq. 16.24]}$$

Since $\delta d/d \ll 1$, δQ reduces to:

$$\delta Q = \frac{-\varepsilon_o A V}{d^2}\delta d \qquad \text{[Eq. 16.25]}$$

Since capacitance decreases with increasing gap distance, and V is constant, the capacitor loses charge ($Q = CV$).

Repeating *Equation 16.14*, the incremental energy balance is:

$$F\,\delta d + V\,\delta Q = \delta U \qquad \text{[Eq. 16.26]}$$

Substituting for δQ and δU:

$$F\,\delta d + V\left(\frac{-\varepsilon_o A V}{d^2}\delta d\right) = \frac{-\varepsilon_o A V^2}{2d^2}\delta d \qquad \text{[Eq. 16.27]}$$

results in the force required to keep the plates apart:

$$F = \frac{\varepsilon_o A}{2}\left(\frac{V}{d}\right)^2 \qquad \text{[Eq. 16.28]}$$

which is the same result found using *Method 1* (*Equation 16.20*). In practice, voltage V is controlled, so the expression for force F in terms of V is generally used, as opposed to the expression for F in terms of charge Q (*Equation 16.19*).

Electrostatic breakdown (short-circuiting of the plates by arcing across the air-gap) occurs when the electric field ($E = V/d$) in the MEMS device is approximately 40×10^6 V/m. The maximum value of the force per area on the plates with an air gap is therefore:

$$\left(\frac{F}{A}\right)_{max} = \frac{\varepsilon_o}{2}\left(\frac{V}{d}\right)^2_{max}$$

$$= \frac{(8.85\times10^{-12})}{2}(40\times10^6)^2$$

$$= 7080 \text{ N/m}^2$$

The variation of the force per plate area F/A with air gap distance d is shown in *Figure 16.8*; force decreases with increasing gap distance. The maximum force is limited by electrostatic breakdown.

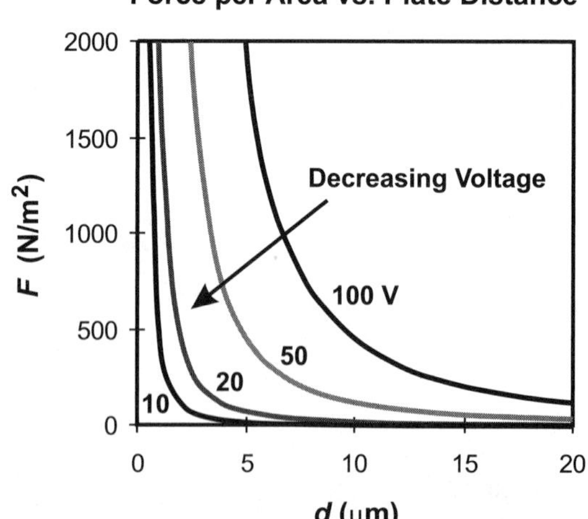

Figure 16.8. Force required to keep parallel plates apart for an air-gap capacitor.

Example 16.2 **Parallel Plate**

Given: A parallel plate capacitor 10 μm × 20 μm with an air gap $d = 3$ μm.

Required: Determine (a) the attractive force between the plates if the voltage is 5 V, (b) the maximum voltage that the capacitor can tolerate without electrostatic breakdown at $E_{max} = 40 \times 10^6$ V/m, and (c) the corresponding force at breakdown.

Solution: *Step 1.* The attractive force is:

$$F = \frac{\varepsilon_o A}{2}\left(\frac{V}{d}\right)^2 = \frac{(8.85 \times 10^{-12}\text{ F/m})(10 \times 10^{-6}\text{ m})(20 \times 10^{-6}\text{ m})}{2}\left(\frac{5\text{ V}}{3 \times 10^{-6}\text{m}}\right)^2$$

Answer: $\underline{F = 0.00246\ \mu N = 2.46\text{ nN}}$

Step 2. Electrostatic breakdown occurs when $E = V/d = 40 \times 10^6$ V/m, so:

$$V_{max} = d\left(\frac{V}{d}\right)_{max} = (3 \times 10^{-6}\text{ m})(40 \times 10^6\text{ V/m})$$

Answer: $\underline{V_{max} = 120\text{ V}}$

Step 3. The corresponding force at breakdown is:

$$F_{max} = \frac{\varepsilon_o A}{2}\left(\frac{V}{d}\right)^2_{max} = \frac{(8.85 \times 10^{-12})(10 \times 10^{-6})(20 \times 10^{-6})}{2}(40 \times 10^6)^2$$

Answer: $\underline{F_{max} = 1.42\ \mu N}$

16.3 Capacitive Accelerometer

One application of a capacitor in MEMS technology is as an accelerometer. In this design, two plates are fixed to a MEMS device and a movable or suspended plate of mass m is mounted on elastic supports of stiffness k (*Figure 16.9a*, elastic supports not shown). The suspended plate can have motion with respect to the device. The MEMS device itself is attached to the system (e.g., a car), and so it has the same acceleration as the system.

Each plate has area A and the plates are initially distance d apart. The three plates form two capacitors that make up part of an energized AC (alternating current) electrical circuit. For now, the details of the electric circuit are omitted.

With no acceleration, the upper fixed plate and suspended plate have a capacitance of:

$$C_o = \frac{\varepsilon_o A}{d} \qquad\qquad \text{[Eq. 16.29]}$$

Likewise, the suspended plate and lower fixed plate have a capacitance of:

$$C_o = \frac{\varepsilon_o A}{d} \qquad \text{[Eq. 16.30]}$$

The capacitors have the same values. Hence, when subjected to the same AC current, the AC voltage across each capacitor is the same. The difference in peak voltages across the capacitors is zero.

The system is now subjected to acceleration a, as shown in *Figure 16.9b*. Due to its inertia (mass), the suspended plate moves closer to the lower plate by distance x $(x \ll d)$, which equals the inertial force ma divided by the spring stiffness of the elastic supports k:

$$x = \frac{ma}{k} \qquad \text{[Eq. 16.31]}$$

The distance x is proportional to acceleration, and is small compared to the overall plate distance d.

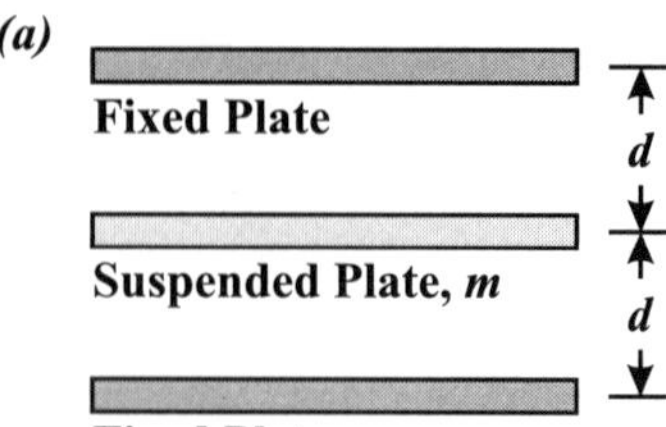

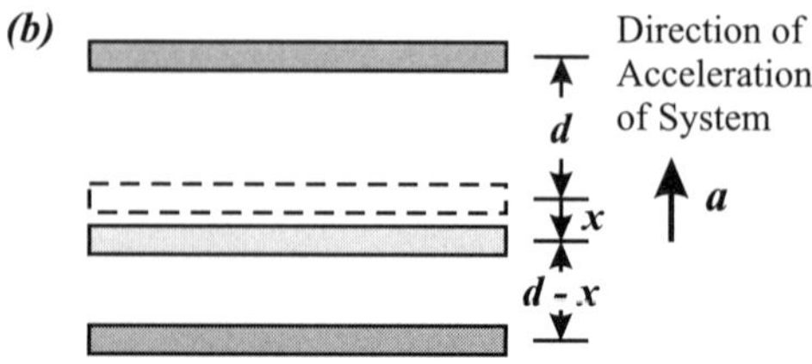

Figure 16.9. (a) Suspended plate of mass m between two fixed plates (side view). **(b)** System subjected to acceleration a.

Due to the acceleration, the upper fixed plate and the suspended plate have a new capacitance:

$$C_1 = \frac{\varepsilon_o A}{d + x} = \frac{d}{d + x} C_o \qquad \text{[Eq. 16.32]}$$

Likewise, the suspended plate and lower fixed plate have a new capacitance:

$$C_2 = \frac{\varepsilon_o A}{d - x} = \frac{d}{d - x} C_o \qquad \text{[Eq. 16.33]}$$

The difference in the new capacitor values is:

$$\Delta C = C_2 - C_1 = \frac{d}{d - x} C_o - \frac{d}{d + x} C_o = \frac{2xd}{d^2 - x^2} C_o \qquad \text{[Eq. 16.34]}$$

Since x is small compared to d, the x^2 term in the denominator is negligible. Thus the difference in capacitance between the two capacitors is proportional to x, which in turn is proportional to a:

$$\Delta C = \frac{2x}{d} C_o = \frac{2ma}{dk} C_o \qquad \text{[Eq. 16.35]}$$

Since there is now a difference in capacitance, there is a difference in the output voltage of the AC circuit.

Another way of understanding the basic system response is as follows. When subjected to an AC current, the peak voltage across a capacitor is inversely proportional to its capacitance. Thus, when subjected to the same AC current, the peak voltage across the upper and suspended plates (V_{p1}) and across the suspended and lower plates (V_{p2}) during acceleration are:

$$V_{p1} = \frac{B}{C_1} = \frac{B(d+x)}{C_o d} \quad \text{and} \quad V_{p2} = \frac{B}{C_2} = \frac{B(d-x)}{C_o d} \qquad \text{[Eq. 16.36]}$$

Constant B depends on the voltage and frequency of the AC source and the characteristics of other circuit elements. The difference in the peak voltages is proportional to movement x, and thus to acceleration a:

$$\Delta V_p = V_{p1} - V_{p2} = \frac{2B}{C_o d}x = \frac{2B}{C_o d}\frac{ma}{k} \qquad \text{[Eq. 16.37]}$$

With appropriate circuitry, the peak voltage difference is amplified. This amplified signal is the output of the circuit. If the resulting output voltage due to the movement of the plate is large enough, indicating a critical acceleration, the output will trigger a system response, e.g., the deployment of an air bag, or the lock-down of a laptop's hard drive.

Example 16.3 Accelerometer

Given: The accelerometer in *Figure 16.10a* consists of two fixed plates. Between the fixed plates is a suspended plate with dimensions $700 \times 70 \times 10\ \mu m$. The system is made of silicon oxide with Young's modulus $Y = 130\ GPa$ and density $\rho = 2328\ kg/m^3$. The suspended plate is $d = 3\ \mu m$ from each fixed plate, and is supported by four silicon oxide beams that are doubled back upon themselves and fixed to the moving system. The beams are $150\ \mu m$ long, with square cross-sections that are $25\ \mu m^2$.

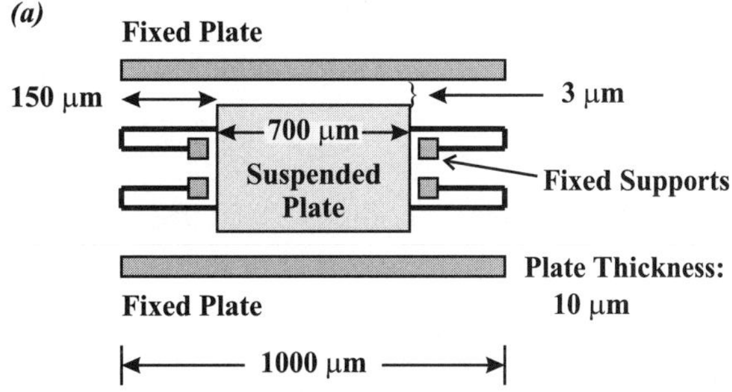

Figure 16.10. (a) Side view of two fixed plates and a suspended plate. The suspended plate is connected to the device base by four double cantilevers. Not to scale.

Required: Determine (a) the stiffness of the system k, (b) the natural frequency of the system $\omega_o = \sqrt{k/m}$, (c) the maximum deflection of the mass when the system is subjected to a negative acceleration (e.g., during a collision) of magnitude $5g$ ($\sim 49\ m/s^2$), and (d) the difference in capacitor values ΔC due to the $5g$ acceleration.

Solution: *Step 1.* The mass of the suspended plate is:

$$m = \rho L H t = \left(2328\times10^3\,\frac{\text{g}}{\text{m}^3}\right)(700\times10^{-6})(70\times10^{-6})(10\times10^{-6})\ \text{m}^3 = 1.141\times10^{-6}\ \text{g}$$

Step 2. Spring stiffness. The moment of inertia of each cantilever's square cross-section is:

$$I = \frac{b^4}{12} = \frac{\left(5\times10^{-6}\ \text{m}\right)^4}{12}\ \text{m}^4 = 52.1\times10^{-24}\ \text{m}^4$$

An enlargement of the upper left double-cantilever *ABCD* is shown in *Figure 16.10b*. The stiffness of one double-cantilever is calculated using *Figures 16.10b, c* and *d*. *AB* and *CD* are elastic cantilevers, while member *BC* is rigid. The upper cantilever *AB* is attached to the suspended plate at point *A*, and point *D* is fixed to the system. When the suspended plate displaces distance δ with respect to the system, point *A* displaces δ and points *B* and *C* displace $\delta/2$. The midpoint of *AB* must therefore deflect by $3\delta/4$, and the midpoint of *CD* by $\delta/4$. Due to the rigid connector *BC*, the slope of the beams at *B* and *C* are zero. The dynamic load on *ABCD* to cause displacement δ is $P = ma/4$ (there are four cantilevers).

By symmetry, the left half of cantilever *AB*, *BF,* is a cantilever beam of length *L*/2 under tip load *P*, having relative deflection $\delta/4$ (*Figure 16.10d*). By observation, point *F* is an inflection point of beam *AB* (the curvature at *F* is zero) so there is no internal moment there. Thus, the load–displacement equation is:

$$\frac{\delta}{4} = \frac{P(L/2)^3}{3YI}$$

where *Y* is Young's modulus. The stiffness of one double-cantilever is therefore:

$$k_1 = \frac{P}{\delta} = \frac{6YI}{L^3}$$

The total spring stiffness of the four cantilevers is:

$$k = 4\left(\frac{6YI}{L^3}\right)$$

$$= 24\frac{(130\times10^9\ \text{Pa})(52.1\times10^{-24}\ \text{m}^4)}{(150\times10^{-6}\ \text{m})^3}$$

Answer: k = 48.1 N/m

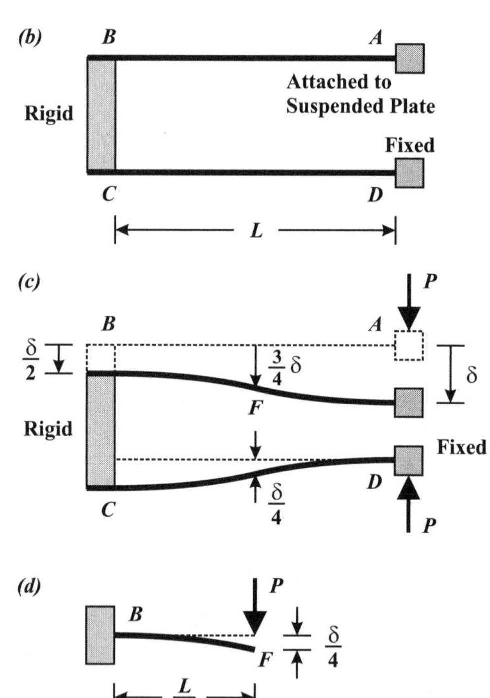

Figure 16.10. (b) A double-cantilever. **(c)** Displacement of double-cantilever δ due to force *P* at point *A*. **(d)** The relationship between *P* and δ is determined by observing that double-cantilever *ABCD* with all four ends fixed against rotation is four tip-loaded cantilevers back-to-back.

Step 3. The *natural frequency* of the system is:

$$\omega_o = \sqrt{\frac{k}{m}} = \sqrt{\frac{48.1 \text{ N/m}}{1141 \times 10^{-12} \text{ kg}}}$$

Answer: $\omega_o = 205 \times 10^3$ rad/s = 32.6 kHz

The natural frequency of practical microsystems is high (20+ kHz) since they must be insensitive to background noises, which are of low frequencies.

Step 4. Response to acceleration. If the deceleration is 5g, the inertial force on the movable mass is:

$$4P = ma = m(5g) = (1140 \times 10^{-12} \text{ kg})[5(9.81 \text{ m/s}^2)] = 55.9 \times 10^{-9} \text{ N}$$

Due to this force, the static deflection of the movable mass is:

$$\delta = \frac{ma}{k} = \frac{55.9 \times 10^{-9} \text{ N}}{48.1 \text{ N/m}} = 1.16 \times 10^{-9} \text{ m}$$

Answer: $\delta = 1.16$ nm

The difference in capacitor values ΔC due to the 5g-magnitude acceleration is, from *Equation 16.35*:

$$\Delta C = \frac{2\delta}{d} C_o = \frac{2ma}{dk} C_o$$

Here:

$$C_o = \frac{\varepsilon_o A}{d} = \frac{(8.85 \times 10^{-12} \text{ F/m})(700 \times 10^{-6} \text{ m})(10 \times 10^{-6} \text{ m})}{3.00 \times 10^{-6} \text{ m}}$$

$$= 0.02065 \times 10^{-12} \text{ F} = 0.02065 \text{ pF}$$

so:

$$\Delta C = \frac{2\delta}{d} C_o = \frac{2(1.16 \times 10^{-9} \text{ m})}{3.00 \times 10^{-6} \text{ m}}(0.02065 \times 10^{-12} \text{ F})$$

Answer: $\Delta C = 16.0 \times 10^{-18}$ F

Due to the acceleration, the values of the capacitances change. When subjected to the same AC current, the voltage across each capacitor will be different. This voltage difference is amplified, and if the output voltage is large enough, it will trigger action within the larger system; e.g., an airbag will deploy.

16.4 Electrostatic Snap-Through in MEMS Devices

A problem with parallel plate capacitors that occurs in practice is when the support structure is unable to withstand the attractive electrostatic forces. If the electrostatic force between plates is sufficiently large, the system will **snap-through**, with the capacitor plates binding together.

The design problem can be modeled as a suspended capacitor plate supported by a spring of stiffness k; the other end of the spring is fixed (*Figure 16.11*). The lower capacitor plate is fixed in position. The plates have area A and are initially distance d apart; the spring is initially unstretched. Both plates have an initial voltage of $V = 0$ so that there is no attraction between them.

Voltage V is slowly applied across the plates, which causes the plates to attract and the spring to deflect by distance x. The upward force applied by the spring due to the displacement of the suspended plate is:

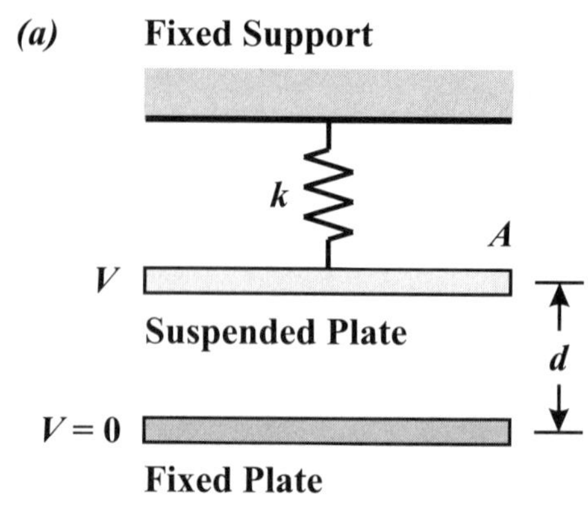

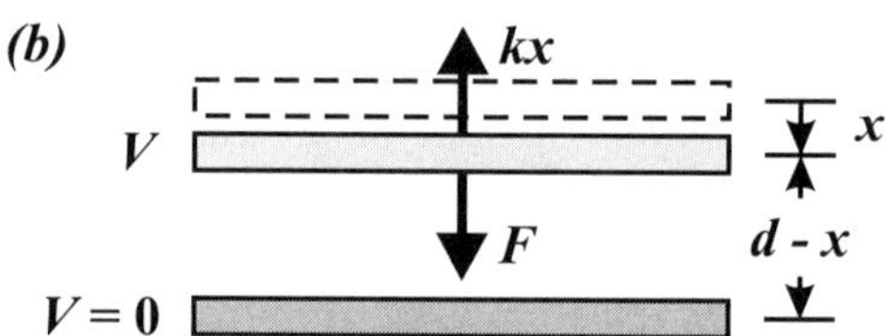

Figure 16.11. **(a)** The suspended plate of a capacitor is supported by spring k. The initial voltage of the suspended plate is $V = 0$. **(b)** For equilibrium the electrostatic force F must be balanced by the spring force kx.

$$F_s = kx \qquad \text{[Eq. 16.38]}$$

The downward electrical force is:

$$F = \frac{\varepsilon_o A}{2}\left(\frac{V}{d-x}\right)^2 \qquad \text{[Eq. 16.39]}$$

From equilibrium, the displacement–voltage relationship is:

$$kx = \frac{\varepsilon_o A}{2}\left(\frac{V}{d-x}\right)^2 \qquad \text{[Eq. 16.40]}$$

Rearranging to solve for V in terms of displacement x gives:

$$V = \sqrt{\frac{2kd^3}{\varepsilon_o A}\left[\frac{x}{d}\left(1-\frac{x}{d}\right)^2\right]} \qquad \text{[Eq. 16.41]}$$

A normalized plot of voltage V against resulting displacement x for equilibrium conditions is shown in *Figure 16.12*. The voltage is normalized by dividing by:

$$V' = \sqrt{(2kd^3)/(\varepsilon_o A)} \quad \text{[Eq. 16.42]}$$

As the applied voltage is slowly applied across the capacitor plates, the displacement increases until $x/d \sim 0.35$ when $V/V' \sim 0.385$. Beyond this point, the voltage actually required to maintain the equilibrium displacement decreases. However, since the applied voltage is increasing, the attractive force of the capacitor overcomes the restoring force of the spring, and the two plates snap together. This problem is of practical significance in MEMS devices since adjacent surfaces can become stuck together by electrostatic forces; the system will not perform as designed.

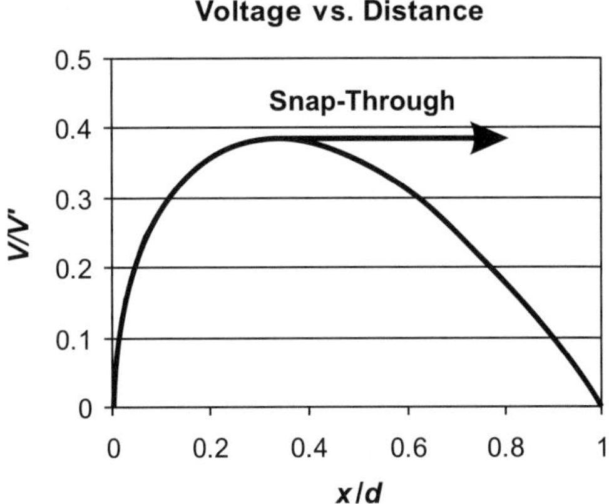

Figure 16.12. Normalized voltage as a function of distance required to keep the suspended capacitor plate in equilibrium. If the applied voltage exceeds the maximum, *snap-through* occurs and the plates will stick together.

Example 16.4 Snap-Through of Spring-Capacitor

Given: The suspended plate of a capacitor is supported by four double-cantilevers (*Figure 16.13*), each 150 μm long. The thickness of the system is 10 μm.

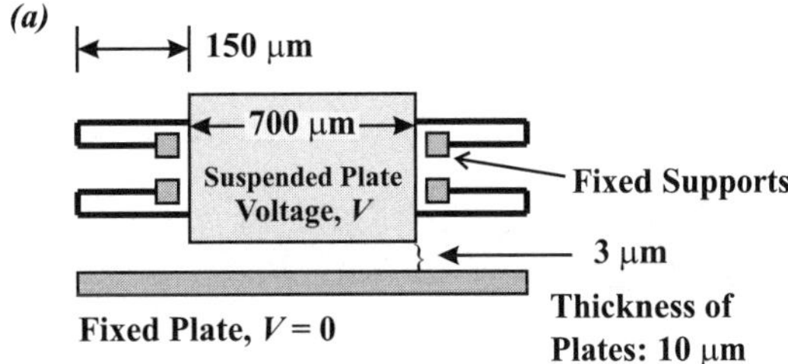

Required: Determine the voltage V_{max} to cause snap-through, when the two plates will snap together. In other words, at what voltage does the electric force overcome the spring force?

Solution: *Step 1.* When voltage V is applied, the suspended plate moves toward the fixed plate by distance x. The upward force applied by the spring to the suspended plate is:

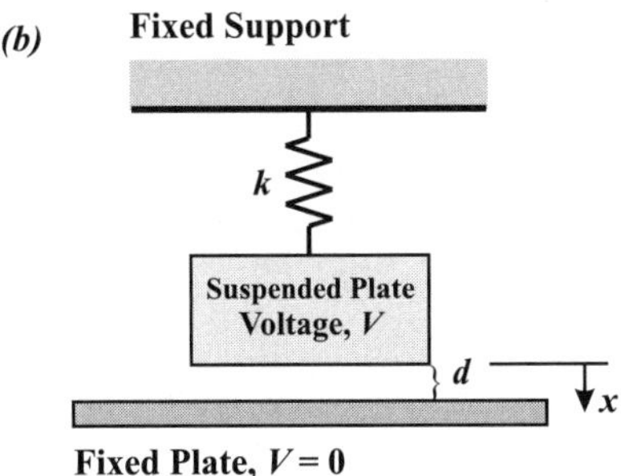

Figure 16.13. **(a)** The mass suspended on four double cantilevers is the upper plate of the capacitor. **(b)** Spring model of system.

$$F_s = kx$$

where from *Example 16.3*:

$$k = \frac{24EI}{L^3} = 48.1 \text{ N/m}$$

The downward attractive force applied by the lower plate to the free plate is:

$$F = \frac{\varepsilon_o A}{2}\left(\frac{V}{d-x}\right)^2$$

The voltage is then:

$$V = \sqrt{\frac{2kd^3}{\varepsilon_o A}\left[\frac{x}{d}\left(1-\frac{x}{d}\right)^2\right]}$$

Step 2. The maximum voltage in *Figure 16.12* is the voltage at snap-through:

$$V_{max} = 0.385\sqrt{\frac{2kd^3}{\varepsilon_o A}}$$

Using the same values for *k*, *d* and *A* as in *Example 16.3*:

$$V_{max} = 0.385\sqrt{\frac{2(48.1 \text{ N/m})(3\times10^{-6} \text{ m})^3}{(8.85\times10^{-12} \text{ F/m})[(700\times10^{-6} \text{ m})(10\times10^{-6} \text{ m})]}}$$

$$\textit{Answer: } \underline{V_{max} = 78.8 \text{ V}}$$

Note that in this case, the voltage gradient at snap-through ($V/d = 26.3\times10^6$ V/m) is smaller than the gradient approximated for electrostatic breakdown ($\sim$40$\times10^6$ V/m).

16.5 Comb Drive

Comb drive capacitors move parallel to their plate areas (*Figure 16.14a*), so that plate distance *d* remains constant. The width of the comb teeth into the paper is *h*, and they overlap by a distance *x* (*Figure 16.14b*).

Consider one tooth of the moving comb (*Figure 16.14b*). The applied voltage between the moving and fixed combs is *V*, and the charge *Q* is:

$$Q = CV \qquad \text{[Eq. 16.43]}$$

where, as earlier, $C = \varepsilon_o A/d$. The area is now $A = 2hx$. The factor of 2 enters the equation since the moving tooth acts as a lower plate to the upper fixed tooth, and an upper plate to the lower fixed tooth. Thus:

$$Q = \frac{\varepsilon_o(2hx)}{d}V \qquad \text{[Eq. 16.44]}$$

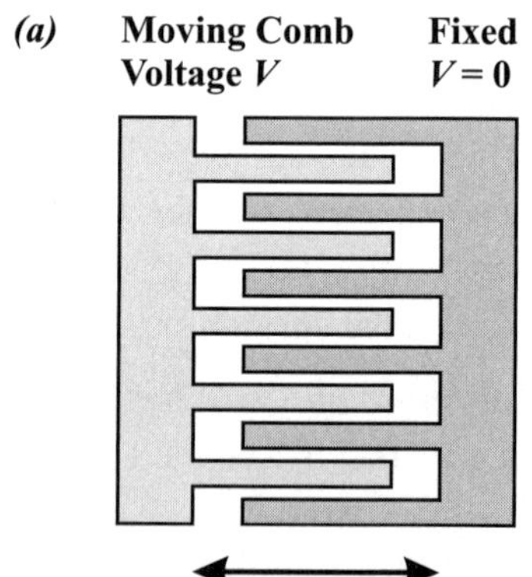

Figure 16.14. (a) Interlocking combs, one stationary with voltage *V* = 0, and the other moving with voltage *V*.

The electrical internal energy associated with a single tooth is:

$$U = \frac{1}{2}CV^2 = \frac{\varepsilon_o h x}{d}V^2 \qquad \text{[Eq. 16.45]}$$

The moving comb is now displaced towards the fixed comb by δx, during which the voltage difference remains constant (*Figure 16.14c*). The electrical energy supplied by the voltage source is $V\,\delta Q$. The mechanical work done by the force is $F\,\delta x$.

From energy balance, the increment of applied energy equals that stored:

$$F\,\delta x + V\,\delta Q = \delta U$$

$$F\,\delta x + V\left(\frac{2\varepsilon_o h}{d}V\,\delta x\right) = \frac{\varepsilon_o h}{d}V^2\,\delta x \qquad \text{[Eq. 16.46]}$$

Solving for the force on a single moving tooth:

$$F = -\frac{\varepsilon_o h V^2}{d} \qquad \text{[Eq. 16.47]}$$

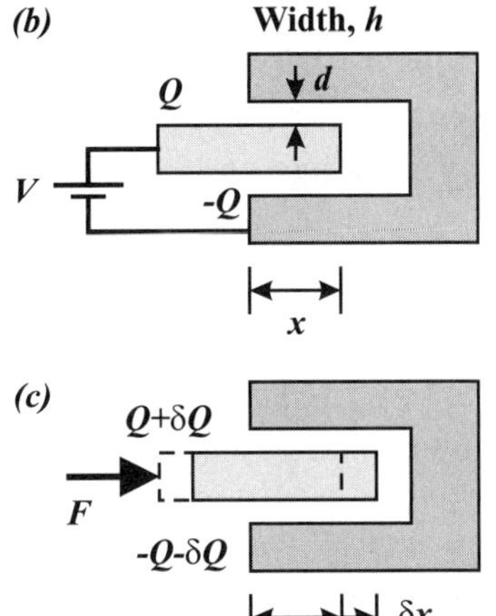

Figure 16.14. **(b)** A moving tooth acts both as a lower and an upper plate to stationary teeth. **(c)** Force *F* and change in charge δQ with movement δx.

The sign of the force is negative, which means that the force is in the opposite direction of that drawn in *Figure 16.14c*. The electrical attraction attempts to close the combs, and a force is required to keep them apart.

The total force between the combs is the force on a single tooth (*Equation 16.47*) multiplied by the total number of teeth on a single comb.

Example 16.5 Multiple-Tooth Comb

Given: The tooth of a moveable comb has dimensions: $h = 12\ \mu m$, $d = 1.5\ \mu m$ (*Figure 16.14*).

Required: Determine the force that can be applied by a 9000-tooth comb due to a voltage difference of 50 V.

Solution: *Step 1.* For one tooth, the magnitude of the force is:

$$F = \frac{\varepsilon_o h V^2}{d} = \frac{(8.85\times10^{-12}\ \text{F/m})(12\times10^{-6}\ \text{m})(50\ \text{V})^2}{1.5\times10^{-6}\ \text{m}} = 177\times10^{-9}\ \text{N}$$

Step 2. For 9000 teeth, the force is 9000 times that value, or:

Answer: $F_{9000} = 1.59\times10^{-3}\ \text{N} = 1.59\ \text{mN}$

Machines have been made to apply such small forces on micro-structures.

Example 16.6 Micro-test Frame

Given: Micro-test frames are used to apply small loads such as those required for testing biological systems.

A set of two 60-tooth combs, 360 μm long, are arranged as shown in *Figure 16.15*. The upper comb is on a cantilever beam, and the lower comb is fixed. The teeth of the comb set have width $h = 12$ μm (into the paper), and the distance between teeth is $d = 1.5$ μm (as in *Example 16.5*). When a voltage is applied between the combs, the electrostatic forces cause the cantilever beam to deflect. The beam has bending stiffness $YI = 0.164 \times 10^{-9}$ N·m^2. The applied voltage is 50 V. Assume there is no contact (short-circuiting) between the combs.

Required: Approximate (a) the maximum shear force, (b) the maximum bending moment, and (c) the maximum deflection of the beam backbone.

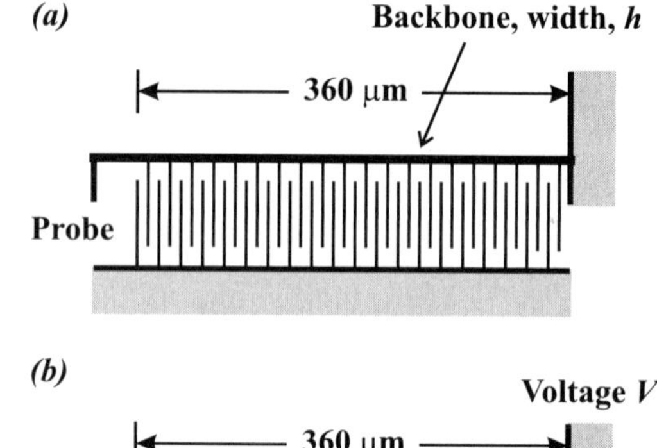

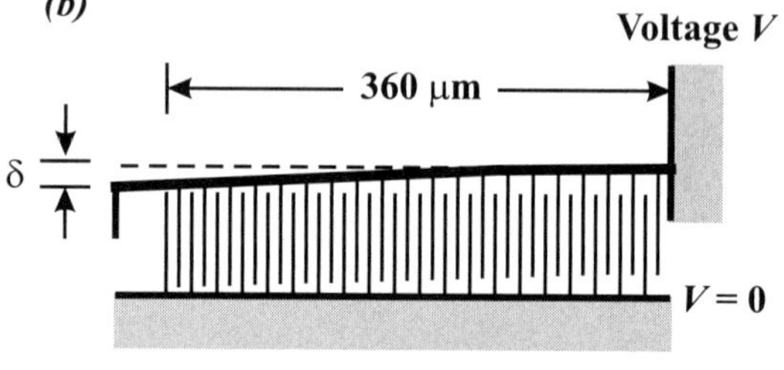

Figure 16.15. (a) Two combs, one on a cantilever backbone, the other fixed. **(b)** When voltage *V* is applied to the cantilever, the beam deflects.

Solution: *Step 1.* In the previous example, the force on each comb tooth was calculated to be:

$$F = \frac{\varepsilon_o h V^2}{d} = 177 \times 10^{-9} \text{ N}$$

The total force on the 60-tooth comb is then:

$$F_{total} = 60(177 \times 10^{-9} \text{ N}) = 10.6 \times 10^{-6} \text{ N}$$

This is the maximum shear force supported by the beam, and it occurs at the fixed end, so:

$$\text{Answer: } \underline{F_{max} = 10.6 \times 10^{-6} \text{ N}}$$

Step 2. The total load is assumed to be uniformly distributed along the beam:

$$w = \frac{F_{total}}{L} = \frac{10.6 \times 10^{-6} \text{ N}}{360 \times 10^{-6} \text{ m}} = 29.4 \times 10^{-3} \text{ N/m}$$

Step 3. The maximum moment in a cantilever beam under a uniformly distributed load occurs at the fixed end:

$$M_{max} = \frac{wL^2}{2} = \frac{(29.4 \times 10^{-3}\ \text{N/m})(360 \times 10^{-6}\ \text{m})^2}{2} = 1.91 \times 10^{-9}\ \text{N·m}$$

Answer: $\underline{M_{max} = 1.91 \times 10^{-9}\ \text{N·m}}$

Step 4. The tip deflection of a cantilever beam under distributed load is:

$$\delta = \frac{wL^4}{8YI} = \frac{(29.4 \times 10^{-3}\ \text{N/m})(360 \times 10^{-6}\ \text{m})^4}{8(0.164 \times 10^{-9}\ \text{m}^4)} = 376 \times 10^{-9}\ \text{m}$$

Answer: $\underline{\delta = 376\ \text{nm}}$

or

$$\frac{\delta}{L} = \frac{376 \times 10^{-9}\ \text{m}}{360 \times 10^{-6}\ \text{m}} = 0.00104 = \frac{1}{975}$$

Here, the deflection-to-span ratio is small, and satisfies the beam deflection requirements of most traditional engineering systems (~1/240).

16.6 Piezoelectric Behavior

Piezoelectric materials change dimension when subjected to a voltage. Mechanical output due to electrical input is called the ***converse*** (or ***reverse***) ***piezoelectric effect***. This phenomenon is used to accurately control displacements of valves in automobile engines and nozzles in inkjet printers. Such piezoelectric systems are called actuators or motors.

When a piezoelectric material is subjected to stress, a voltage is created. Electrical output due to mechanical input is the ***direct piezoelectric effect***. This response is used to provide the voltage output of a force sensor, the spark in a gas ignitor, and the energy to the flickering lights in childrens' shoes. Such piezoelectric systems are called sensors or generators.

Piezoelectric materials occur in nature (e.g., a quartz single crystal), but those commonly used in engineering systems are different forms of lead zirconium titanate (PZT) and other polarized engineering ceramics that have a perovskite crystalline structure. The materials properties depend on the processes used in manufacture. Material properties may be found in manufacturers' catalogs.

Elastic Law

The mechanical strain S_m in a linear elastic material with Young's modulus Y subjected to stress σ is:

$$S_m = \frac{\sigma}{Y} \qquad\qquad \text{[Eq. 16.48]}$$

Note that in this chapter, Young's modulus is *Y*, while the electric field is *E*; strain is *S*, while the dielectric constant is ε.

Electrostatic Law

Consider a ceramic material with electrodes on its upper and lower surfaces (*Figure 16.16*). The material *permittivity*, or *dielectric constant*, is ε; the electrode plates have area *A* and are separated by distance *t*. The capacitor is subjected to voltage source V_s, represented by the circle with "+" and "−" indicating its polarity. The *electric field E* goes from the positive charge +*Q* on the upper plate to the negative charge −*Q* on the lower plate, which is also the direction of the *voltage drop V.*

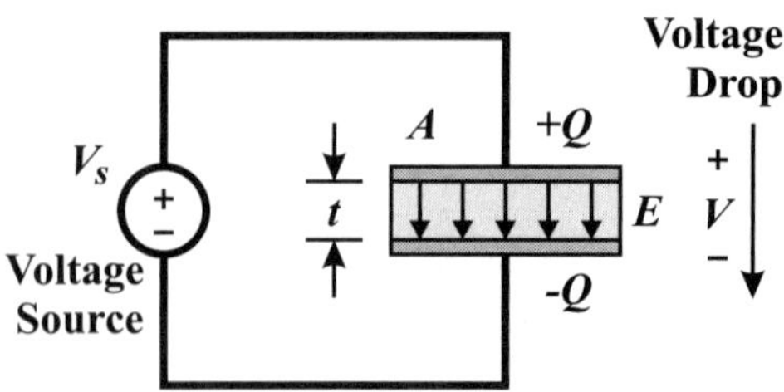

Figure 16.16. Voltage V_s applied across the electrodes of a capacitor. The capacitor's electric field *E* is in the direction of the voltage drop *V.* Applying a positive voltage source V_s results in a downward (positive) electric field *E* in − and a positive voltage drop V across − the capacitor.

The *electric flux density*, or *electric displacement*, is the charge per unit plate area:

$$D = \frac{Q}{A} \qquad \text{[Eq. 16.49]}$$

Recalling the capacitor charge–voltage relationship (*Equation 16.1*), and the definition of capacitance *C* (*Equation 16.11*), the electric displacement can be rewritten:

$$D = \frac{CV}{A} = \frac{\varepsilon A}{t}\left(\frac{V}{A}\right) = \varepsilon\frac{V}{t} \qquad \text{[Eq. 16.50]}$$

Recall that the material permittivity is (*Equation 16.9*):

$$\varepsilon = \varepsilon_r \varepsilon_o \qquad \text{[Eq. 16.51]}$$

where ε_o is the permittivity of air (assumed to be equal to that of a vacuum: $\varepsilon_o = 8.85\times10^{-12}$ F/m), and ε_r is the dimensionless *relative permittivity* of the material.

The magnitude of *electric field E* is the gradient of the *voltage drop V*:

$$E = \frac{V}{t} \qquad \text{[Eq. 16.52]}$$

Thus, the electric displacement is:

$$D = \varepsilon E \qquad \text{[Eq. 16.53]}$$

The direction of the electric field and the direction of the voltage drop across the capacitor are the same. In *Figure 16.16*, if the voltage source V_s is positive (i.e., the voltage above the source is higher than below it), then the applied electric field *E* in the capacitor is downward.

In this treatment, the *positive sense of an electric field E* in a piezoelectric material (from $+Q$ to $-Q$) is in the same direction as the *poling direction* (or *poling electric field*), defined below. A *positive voltage drop V* is also in the *poling direction*.

Poling

Piezoelectric materials have a crystal structure that exhibits an electrical polarity or **dipole moment** below its *Curie temperature* (named for Jacques and Pierre Curie who discovered the piezoelectric effect in the late 1800s). In a single crystal, the entire material has a dipole moment (*Figure 16.17a*). In practice, a piezoelectric material is composed of many *domains*, or regions, each with a dipole moment. The domains are randomly oriented so that overall the material has no net polarity (*Figure 16.17b*).

Before being used in an engineering system, a piezoelectric material is subjected to a high voltage V_p that tends to align the domain dipoles (and slightly lengthens the material) in the direction of the applied electric field E_p. This process is called **poling** (*Figure 16.17c*). Upon removal of the poling voltage V_p, the material retains an overall dipole moment. The direction of the poling electric field sets the **poling direction**, which here is represented by a dotted arrow (*Figure 16.17d*).

It is important to identify the poling direction because piezoelectric properties differ in different directions. In this treatment, the poled direction is aligned with the local 3-direction.

The response of the internal dipole to applied stresses and electric fields is what gives piezoelectric materials their special properties. The direction of the applied stress or electric field with respect to the poling direction governs the material response.

Piezoelectric materials are limited by applied stress (they are ceramics), by voltage (a large voltage can reverse or reduce the poling, causing a non-linear response), and by temperature (above the Curie temperature, the material's structure does

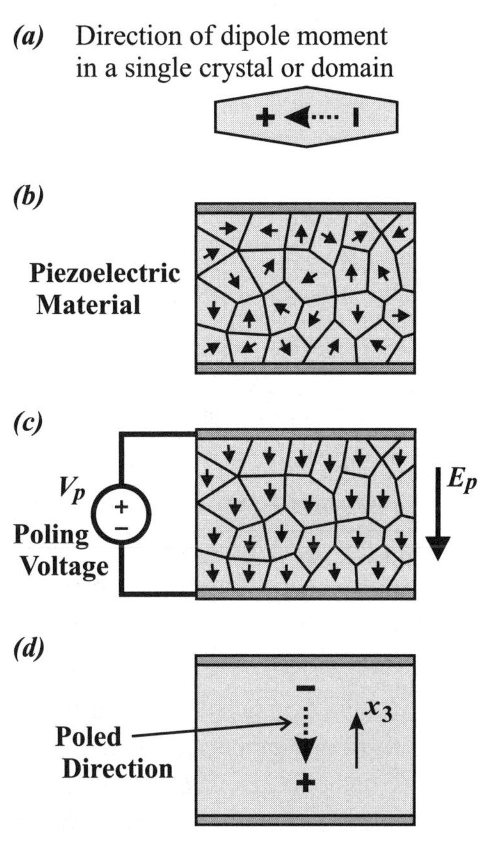

Figure 16.17. **(a)** Direction of a dipole moment; the arrow points from negative to positive charge. **(b)** Piezoelectric material with randomly oriented domains. **(c)** Material subjected to poling voltage V_p (poling electric field E_p). **(d)** Upon removal of V_p, the piezoelectric material has a permanent dipole moment (domain boundaries not shown).

not have a dipole moment). It is assumed here that the system remains within the linear operating range, and below the Curie temperature.

Piezoelectric Law

The mechanical and electrical responses of a piezoelectric material are *coupled.* Piezoelectric materials strain when subjected to either a stress or an electric field. The total strain S due to the *stress* σ (causing mechanical strain $S_m = \sigma/Y$) and to the *total electric field E* (causing electrical strain $S_e = d{\times}E$) is:

$$S = S_m + S_e = \frac{\sigma}{Y} + dE \qquad \text{[Eq. 16.54]}$$

where d is the **piezoelectric charge constant**, discussed below. Stress σ is positive in tension, and electric field E is positive in the poling direction.

When subjected to both stress and an electric field, the electrostatic response is given by the *electric displacement D*:

$$D = d\sigma + \varepsilon E \qquad \text{[Eq. 16.55]}$$

In matrix form, the **piezoelectric law** equations are then:

$$\begin{bmatrix} S \\ D \end{bmatrix} = \begin{bmatrix} 1/Y & d \\ d & \varepsilon \end{bmatrix}\begin{bmatrix} \sigma \\ E \end{bmatrix} \qquad \text{[Eq. 16.56]}$$

It should be noted that the electric field E has two components:

   (1) the *applied* electric field E_a , and

   (2) an *induced* electric field E_i , caused by the applied stress (discussed below).

Equation 16.56 describes the mechanical and the electrical response of the *poled* material. The coupled electromechanical material properties (i.e., d) are determined from the results of two tests, described below. In the first test, a voltage-only load is applied to the system; in the second test, a stress-only load is applied.

Equation 16.56 is a reduced form of a larger matrix equation. In 3D, the stress vector σ has six stress components (three normal and three shear), and the electric field E vector has three components. Likewise, the strain vector S has six components, and the electric displacement vector D has three. Symbol $1/Y$ represents the material compliance or flexibility matrix (6×6), d is the dielectric constant matrix and its transpose (6×3 and 3×6), and ε is the permittivity matrix (3×3). For simplicity, only electric fields in the poling direction, and the three normal stresses, are considered in this treatment.

Application of Voltage Only

Consider a piezoelectric material (e.g., PZT) of length, width and thickness L, W, and t, between two electrodes (*Figure 16.18a*). A 3D coordinate system is defined, with directions 1-2-3. The poled direction is the *negative* 3-direction (downward), and the electrode plates are on the positive and negative 3-surfaces of the material.

With no applied stress, a positive (downward) electric field E_{3a} (E is positive in the poling direction) is applied across the electrodes by voltage V_{3a}, as shown in *Figure 16.18a*. No current flows through the ceramic PZT material so the circuit is open.

The applied electric field causes changes in lengths ΔL, ΔW, Δt, in the 1-, 2- and 3-directions, respectively. The strain S_i in each direction due to the electric field is:

$$S_{1,e} = \frac{\Delta L}{L} = d_{31}E_{3a} = d_{31}\frac{V_{3a}}{t}$$

$$S_{2,e} = \frac{\Delta W}{W} = d_{32}E_{3a} = d_{32}\frac{V_{3a}}{t}$$

$$S_{3,e} = \frac{\Delta t}{t} = d_{33}E_{3a} = d_{33}\frac{V_{3a}}{t}$$

[Eq. 16.57]

The strain $S_{j,e}$ in the j-direction is the product of the *piezoelectric charge constant* d_{3j} and the electric field $E_{3a} = V_{3a}/t$. In the notation d_{3j}, the first subscript (3) is the direction of the electric field and the second subscript (1, 2, or 3) is the direction of the strain caused by the applied field. In PZT materials, constants d_{31} and d_{32} are equal and negative, while d_{33} is unique and positive.

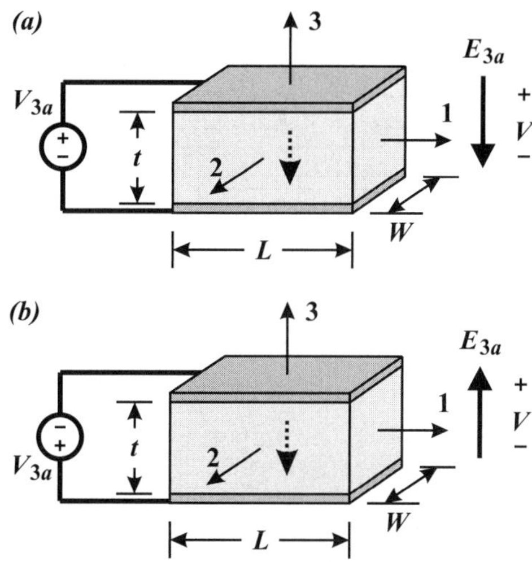

Figure 16.18. **(a)** A piezoelectric material subjected to a *positive* electric field E_{3a} (in the poling direction). Here, $E_{3a} > 0$, voltage drop $V > 0$ (in the poling direction), and $V_{3a} > 0$. The changes in dimensions are: $\Delta t > 0$, $\Delta L < 0$, $\Delta W < 0$. **(b)** A piezoelectric material subjected to a *negative* electric field E_{3a} (opposite the poling direction). $E_{3a} < 0$ and voltage drop $V < 0$. Thus, $\Delta t < 0$, $\Delta L > 0$, $\Delta W > 0$.

In *Figure 16.18a*, the electric field E_{3a} and the voltage drop across the capacitor, $V = V_{3a}$, are both in the poling direction and are thus both *positive*. In *Figure 16.18b*, E_{3a} is opposite the poling direction and is thus *negative*. General observations can be made about the direction of the electric field and the resulting strain.

When the applied electric field is in the poled direction (here the negative 3-direction, or downward), thickness t increases; i.e., $E_{3a} > 0$ so $S_{3,e} > 0$ (*Figure 16.18a, Equation 16.57, $d_{33} > 0$*). The voltage source V_{3a} gives the upper electrode a positive charge and the lower electrode a negative charge. The negative end of the dipole moment – the tail of the dotted arrow – is attracted to the upper electrode, and the positive end is attracted to the lower electrode. At the same time, both length and width decrease (analogous to the Poisson effect). An electric field in the poling direction has the same direction as the original poling electric field E_p.

When the applied electric field E_{3a} is *opposite* the poled direction, the thickness decreases; i.e., $E_{3a} < 0$ (upwards) so $S_{3,e} < 0$ (*Figure 16.18b, Equation 16.57*). In

Table 16.1. Typical Piezoelectric Constants for piezoelectric material PZT4.

Piezoelectric Charge Constants, d_{3i}	Piezoelectric Voltage Constants, g_{3i}
$d_{33} = 285\times10^{-12}$ m/V	$g_{33} = 24.9\times10^{-3}$ V m/N
$d_{31} = d_{32} = -122\times10^{-12}$ m/V	$g_{31} = g_{32} = -10.3\times10^{-3}$ V m/N

Figure 16.18b, the upper electrode is negatively charged, and the lower electrode is positively charged. The ends of the dipole moment are repulsed by the electrodes, decreasing the material thickness t. At the same time, the length and width increase. An electric field opposite the poling direction is opposite the original poling electric field E_p .

Piezoelectric Constants, d, g and ε

The piezoelectric constants, d_{ij} and g_{ij} , relate the mechanical stresses and strains to the electrical fields and displacements. They are material properties, and for a typical PZT4 material are given in *Table 16.1*.

Constants d_{ij} are the **piezoelectric charge constants**. From *Equation 16.57*, they are the strain in the j-direction per unit electric field in the i-direction. From the general piezoelectric law (*Equation 16.56*), it can be deduced that d_{ij} is also the induced electric displacement in the i-direction per unit stress in the j-direction.

Constants g_{ij} are the **piezoelectric voltage constants**, introduced in the next subsection. They are the electric field (voltage gradient) induced in the i-direction per unit stress in the j-direction. Constants g_{ij} are also the strain in the j-direction per unit electric displacement in the i-direction.

The **dielectric constant** – the **permittivity** – is related to the piezoelectric constants:

$$\varepsilon = \frac{d_{ij}}{g_{ij}}$$

[Eq. 16.58]

Example 16.7 **Strain due to Applied Voltage**

Given: A piezoelectric actuator is made of a PZT4 material, $t = 50$ μm thick, between two electrodes. The applied voltage is $V_{3a} = 100$ V with polarity as shown in *Figure 16.19*. The poling direction is in the negative 3-direction. The piezoelectric properties of the dielectric material are given in *Table 16.1*.

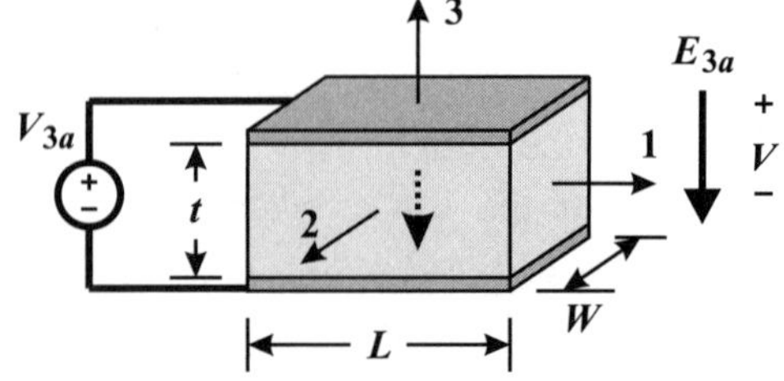

Figure 16.19. Piezoelectric material subjected to applied voltage V_{3a}.

Required: Determine (a) the strain in the poling direction S_3 and (b) the strain in the transverse direction S_1 (c) Repeat Part (a) for $V_{3a} = -60$ V.

Solution: *Step 1.* The strain is due to the applied voltage. Due to the polarity of the source, the electric field in the material is in the negative-3 direction, which is the poling direction. Thus, E_{3a} is positive. The voltage drop V is also in the poling direction, so it is also positive: $V = V_{3a} = 100$ V.

From *Equation 16.57* and *Table 16.1*:

$$S_3 = d_{33}\frac{V}{t} = (285 \times 10^{-12} \text{ m/V})\frac{100 \text{ V}}{50 \times 10^{-6} \text{ m}}$$

Answer: $\underline{S_3 = 570 \times 10^{-6}}$

The positive charge at the upper electrode and the negative charge at the lower electrode cause the material thickness to increase.

Step 2. The transverse strain in the 1-direction is:

$$S_1 = d_{31}\frac{V}{t} = (-122 \times 10^{-12} \text{ m/V})\frac{100 \text{ V}}{(50 \times 10^{-6} \text{ m})}$$

Answer: $\underline{S_1 = -244 \times 10^{-6}}$

The transverse strain S_1 is negative, analogous to the Poisson effect in mechanical loading.

Step 3. For $V_{3a} < 0$, E_{3a} and V are both *negative*, i.e., opposite the poling direction. Thus, $V = V_{3a} = -60$ V, and strain in the 3-direction is negative:

Answer: $\underline{S_3 = -342 \times 10^{-6}}$

The thickness decreases.

Application of Stress Only (Isolated Electrodes)

Consider the piezoelectric material sandwiched between two electrodes, with the poling direction in the negative 3-direction (*Figure 16.20a*). Normal stresses are applied in each direction σ_1, σ_2, and σ_3 . The electrodes at top and bottom are electrically isolated from each other so that no charge (and thus no current) may flow to or from them (i.e., $D = 0$).

The application of the stress elongates or shortens the dipole moment, and thus induces a voltage in the 3-direction (the direct piezoelectric effect). The ***induced voltage*** due to each stress acting individually is measured. The relationships between the induced voltage and the applied stresses are:

$$\frac{V_{31i}}{t} = -g_{31}\sigma_1$$

$$\frac{V_{32i}}{t} = -g_{32}\sigma_2 \qquad \text{[Eq. 16.59]}$$

$$\frac{V_{33i}}{t} = -g_{33}\sigma_3$$

Notation V_{31i} is the induced (i) voltage in the 3-direction caused by a stress in the 1-direction. Constants g_{ij} are the *piezoelectric voltage constants* that depend on the material, and for PZT4 are given in *Table 16.1*. Values $g_{31} = g_{32}$ are negative, and g_{33} is positive.

Physically, the induced voltage has a polarity that tries to return the thickness to its original poled dimension; it reduces the effect of the applied stress. As illustration, consider σ_3 only.

A positive (tensile) stress σ_3 increases the thickness, while the induced voltage V_{33i} reduces the thickness. From *Equation 16.59*, for $\sigma_3 > 0$, since $g_{33} > 0$, then $V_{33i} < 0$. A negative induced voltage V_{33i} gives the upper electrode a negative charge and the lower electrode a positive charge. The induced voltage tends to shorten the dipole. The induced electrical field E_{3i} in the material is negative since it is opposite the poled direction. The induced voltage V_{3i} is also negative for a negative (compressive) transverse stress (σ_1 or σ_2), as given by *Equation 16.59*, since $g_{31} = g_{32} < 0$.

A negative (compressive) stress σ_3 shortens the thickness, while the induced (positive) voltage V_{33i} increases the thickness. From *Equation 16.59*, for $\sigma_3 < 0$, since $g_{33} > 0$, $V_{33i} > 0$. The induced voltage causes a positive charge on the upper electrode, and a negative charge on the lower electrode. The induced voltage lengthens the dipole. The induced electrical field E_{3i} in the material is positive since it is in the poling direction. A similar voltage response occurs for a positive (tensile) transverse stress (σ_1 or σ_2).

Since the induced voltage opposes the change in dimension, the PZT material has a higher effective modulus than the material without the piezoelectric effect. For the same applied stress σ, the resulting strain S and strain energy density ($U_D = \sigma S/2$) are smaller. Part of the work done by the applied stress is now stored as electrical energy.

The total ***induced electric field*** E_{3i} (negative voltage gradient) caused by the three stresses is:

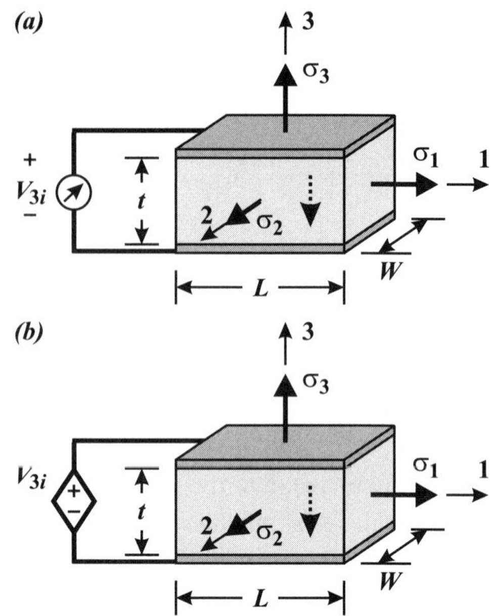

Figure 16.20. **(a)** Piezoelectric material subjected to stresses in three directions. Induced voltage V_{3i} is measured across the electrodes. **(b)** Simple model of electrical response. The diamond is a standard symbol in circuits for a dependent-voltage source; V_{3i} depends on the applied stress.

$$E_{3i} = \frac{V_{3i}}{t} = \frac{V_{31i}}{t} + \frac{V_{32i}}{t} + \frac{V_{33i}}{t} = -(g_{31}\sigma_1 + g_{32}\sigma_2 + g_{33}\sigma_3) \qquad \text{[Eq. 16.60]}$$

A simple model for the induced voltage is given in *Figure 16.20b*; the diamond is a common symbol in circuits for a dependent (controlled) source. Here, V_{3i} is dependent on the applied stress.

The induced electric field V_{3i}/t causes electrical strains of the form of *Equation 16.57*:

$$S_{1,e} = \frac{\Delta L}{L} = d_{31}\frac{V_{3i}}{t}$$

$$S_{2,e} = \frac{\Delta W}{W} = d_{32}\frac{V_{3i}}{t} \qquad \text{[Eq. 16.61]}$$

$$S_{3,e} = \frac{\Delta t}{t} = d_{33}\frac{V_{3i}}{t}$$

The normal stresses cause mechanical strains:

$$S_{1,m} = \frac{\sigma_1}{Y} - v\frac{\sigma_2}{Y} - v\frac{\sigma_3}{Y}$$

$$S_{2,m} = -v\frac{\sigma_1}{Y} + \frac{\sigma_2}{Y} - v\frac{\sigma_3}{Y} \qquad \text{[Eq. 16.62]}$$

$$S_{3,m} = -v\frac{\sigma_1}{Y} - v\frac{\sigma_2}{Y} + \frac{\sigma_3}{Y}$$

Superimposing the mechanical strains and the electrical strains caused by the induced voltage gives:

$$S_1 = \frac{\sigma_1}{Y} - v\frac{\sigma_2}{Y} - v\frac{\sigma_3}{Y} + d_{31}\frac{V_{3i}}{t}$$

$$S_2 = -v\frac{\sigma_1}{Y} + \frac{\sigma_2}{Y} - v\frac{\sigma_3}{Y} + d_{32}\frac{V_{3i}}{t} \qquad \text{[Eq. 16.63]}$$

$$S_3 = -v\frac{\sigma_1}{Y} - v\frac{\sigma_2}{Y} + \frac{\sigma_3}{Y} + d_{33}\frac{V_{3i}}{t}$$

If the two electrodes are short-circuited so that they must have the same voltage (e.g., *Figure 16.21a*), then V_{3i} is zero, and the strains are due only to the mechanical loads. *Equation 16.63* then reduces to the standard form of Hooke's Law (*Equation 16.62*).

Application of Uniaxial Stress Only

To further understand the physical interaction between the mechanical and the electrical properties without evaluating a large number of terms, consider a piezoelectric material subjected to stress σ_3 in the poling direction only (*Figure 16.21*).

Short-Circuited Electrodes

If the electrodes are short-circuited so that they must have the same voltage (*Figure 16.21a*), then no induced voltage can exist ($E_3 = V_{3i} = 0$). The strain is simply:

$$S_3 = \frac{\sigma_3}{Y}$$

[Eq. 16.64]

Isolated Electrodes

When the plates are electrically isolated so that no charge can flow ($D = 0$, *Figure 16.21b*), the induced voltage across the plates due to stress σ_3 is:

$$\frac{V_{3i}}{t} = -g_{33}\sigma_3$$

[Eq. 16.65]

The total strain is then:

$$S_3 = \frac{\sigma_3}{Y} + d_{33}\frac{V_{3i}}{t} = \frac{\sigma_3}{Y} - d_{33}(g_{33}\sigma_3)$$

$$= \left(\frac{1}{Y} - d_{33}g_{33}\right)\sigma_3$$

[Eq. 16.66]

Rearranging, the *effective Young's Modulus* of the material in the 3-direction when the plates are each isolated is:

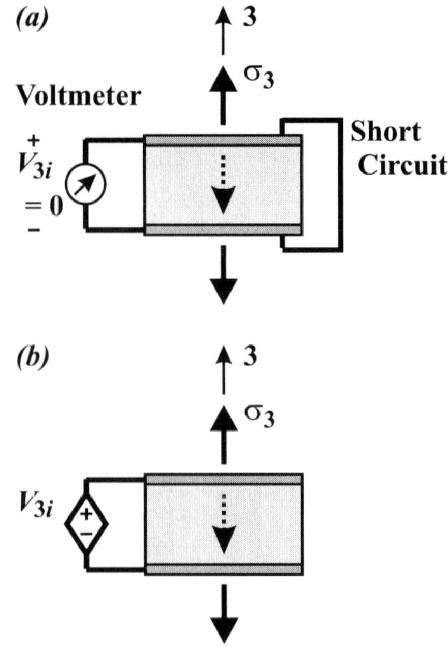

Figure 16.21. **(a)** PZT under stress with plates short-circuited; no voltage is induced across the plates. **(b)** PZT with plates electrically isolated. Voltage V_{3i} is induced across the electrodes.

$$Y^* = \frac{\sigma_3}{S_3} = \frac{Y}{1 - d_{33}g_{33}Y}$$

[Eq. 16.67]

For PZT4 materials, the pertinent properties are: $Y = 66$ GPa, $d_{33} = 285 \times 10^{-12}$ m/V, and $g_{33} = 24.9 \times 10^{-3}$ V m/N. Substituting these values into the effective modulus equation gives:

$$Y^* = 124 \text{ GPa}$$

[Eq. 16.68]

The induced voltage causes the PZT material to have a higher effective modulus than the material without the piezoelectric effect. In this case, the modulus is nearly doubled.

Simultaneous Application of Voltage and Axial Stress

We will now derive the general *piezoelectric law*, first stated in *Equation 16.56*.

Consider the case when both voltage V_{3a} and stress σ_3 are applied to the open-circuit system at the same time (*Figure 16.22*). The stress induces voltage V_{3i}. The **total voltage** is the sum of the *applied voltage* and the *induced voltage*:

$$V_3 = V_{3a} + V_{3i} \qquad \text{[Eq. 16.69]}$$

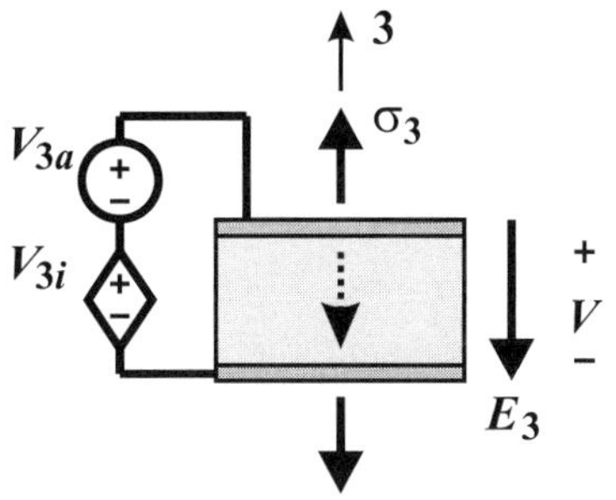

The electric field in the piezoelectric material is in the poling direction (downward):

$$E_3 = \frac{V_3}{t} = \frac{V_{3a} + V_{3i}}{t} \qquad \text{[Eq. 16.70]}$$

Figure 16.22. Simultaneous application of V_{3a} and σ_3, produces total voltage: $V = V_{3a} + V_{3i}$.

The electric displacement due to the *applied* voltage source V_{3a} only, or *applied* electric field E_{3a} (*Equation 16.53*) is:

$$D_3 = \varepsilon \frac{V_{3a}}{t} = \varepsilon E_{3a} \qquad \text{[Eq. 16.71]}$$

The strain caused by the applied voltage V_{3a} (*Equation 16.57*) is:

$$S_{3,e} = d_{33} \frac{V_{3a}}{t} = d_{33} E_{3a} \qquad \text{[Eq. 16.72]}$$

The stress acting alone causes no electric displacement D. The strain due to the applied stress σ_3 (*Equations 16.66* and *16.67*) is:

$$S_{3,m} = \frac{\sigma_3}{Y^*} = \left(\frac{1}{Y} - d_{33} g_{33} \right) \sigma_3 \qquad \text{[Eq. 16.73]}$$

where Y^* is the effective Young's modulus caused by the induced voltage.

The total strain is thus:

$$S_3 = \frac{\sigma_3}{Y^*} + d_{33} E_{3a} \qquad \text{[Eq. 16.74]}$$

while the total electric displacement is given in *Equation 16.71*.

The *piezoelectric law* can be written in matrix form in terms of the *applied stress* σ_3 and the *applied electric field* E_{3a}:

$$\begin{bmatrix} S_3 \\ D_3 \end{bmatrix} = \begin{bmatrix} 1/Y^* & d_{33} \\ 0 & \varepsilon \end{bmatrix} \begin{bmatrix} \sigma_3 \\ E_{3a} \end{bmatrix}$$

[Eq. 16.75]

Equation 16.75 can be reformulated to give the standard piezoelectric matrix equation (*Equation 16.56*). The total strain, in terms of stress σ_3 and *applied* voltage V_{3a} is:

$$S_3 = \left(\frac{1}{Y} - d_{33}g_{33}\right)\sigma_3 + d_{33}\frac{V_{3a}}{t}$$

[Eq. 16.76]

From *Equation 16.65*:

$$g_{33} = -\frac{V_{3i}}{t}\frac{1}{\sigma_3}$$

[Eq. 16.77]

Substituting *Equation 16.77* into *Equation 16.76* gives the strain in terms of the stress and the *total* electric field $E_3 = E_{3a} + E_{3i}$:

$$S_3 = \left(\frac{\sigma_3}{Y} + d_{33}\frac{V_{3i}}{t}\right) + d_{33}\frac{V_{3a}}{t} = \frac{\sigma_3}{Y} + d_{33}\frac{V_{3a} + V_{3i}}{t}$$

$$= \frac{\sigma_3}{Y} + d_{33}E_3$$

[Eq. 16.78]

From *Equation 16.75*, the electric displacement is:

$$D_3 = \varepsilon\frac{V_{3a}}{t} = \varepsilon\frac{V_3 - V_{3i}}{t} = \varepsilon\frac{V_3}{t} - \varepsilon\frac{V_{3i}}{t}$$

[Eq. 16.79]

Again, applying *Equation 16.65*:

$$D_3 = \varepsilon\frac{V_3}{t} - \varepsilon(-g_{33}\sigma_3) = \varepsilon\frac{V_3}{t} + \varepsilon g_{33}\sigma_3$$

[Eq. 16.80]

The dielectric constant is the ratio of the piezoelectric constants $\varepsilon = d_{33}/g_{33}$ (*Equation 16.58*), so *Equation 16.80* can be rewritten:

$$D_3 = d_{33}\sigma_3 + \varepsilon\frac{V_3}{t} = d_{33}\sigma_3 + \varepsilon E_3$$

[Eq. 16.81]

Finally, the relationship between the displacements (the strain S_3 and the electric displacement D_3) and the total loads (the applied stress σ_3 and the total electric field E_3) can be summarized by the *piezoelectric law*:

$$\begin{bmatrix} S_3 \\ D_3 \end{bmatrix} = \begin{bmatrix} 1/Y & d_{33} \\ d_{33} & \varepsilon \end{bmatrix} \begin{bmatrix} \sigma_3 \\ E_3 \end{bmatrix}$$

[Eq. 16.82]

which was introduced in *Equation 16.56*.

Example 16.8 **Piezoelectric Stack: Valve Lifters**

Given: The electrodes in a stack actuator are in the form of a pair of interlocking capacitor combs (*Figure 16.23a*). The material between the combs is poled by applying poling voltage V_p across the combs. As a consequence, the poling direction in the material alternates, pointing towards the negative electrodes (*Figures 16.23b, c*).

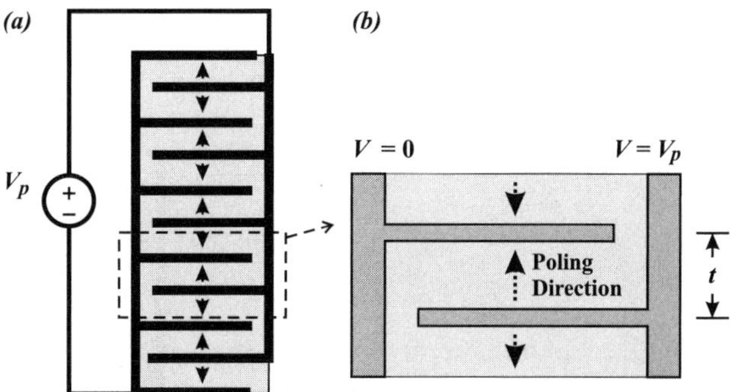

Figure 16.23. **(a)** Piezoelectric stack subjected to poling voltage V_p (not to scale). **(b)** Close-up of two comb teeth and piezoelectric material sandwiched between them.

After poling, compressive load W is applied to the stack causing compressive stress σ_3. The stress causes the stack to shorten, and induces a voltage V_i between the combs (*Figure 16.23d*). The loads on stacks can be significant because the material is in compression, not in tension where it is weak. The compressive strength of piezoelectric materials is approximately 250 MPa, but to ensure reliability during operation, maximum working stresses of 150 MPa are normal.

In this design, the applied stress due to load W is $\sigma_3 = -80$ MPa. The effective modulus for PZT4 is $Y^* = 124$ GPa, and the piezoelectric constants are given in *Table 16.1*. There are $n = 240$ teeth in each comb and the thickness of each PZT layer is $t = 125$ μm.

Required: For the given load, determine (a) the deflection of the stack Δ, (b) the induced voltage V_i, (c) the additional voltage V_a that needs to be *applied* to return the stack to its original height, and (d) the total voltage across the combs due to both the induced and the applied voltages.

Solution: *Step 1.* Apply stress $\sigma_3 = -80$ MPa. The strain due to the mechanical load $\varepsilon_{3,m}$ for the configuration where the electrodes are isolated is:

$$S_{3,\,m} = \frac{\sigma_3}{Y^*} = \frac{-80 \times 10^6 \text{ Pa}}{124 \times 10^9 \text{ Pa}} = -645 \times 10^{-6}$$

The deflection of the stack due to the stress is (*Figure 16.23d*):

$$\Delta = S_{3,\,m} L = S_{3,\,m}(nt) = (-645 \times 10^{-6})[240(125 \times 10^{-6} \text{ m})]$$

Answer: $\underline{\Delta \,-\, -19.4 \text{ μm}}$

Step 2. The induced voltage is (with g_{33} from *Table 16.1*):

$$V_i = (-g_{33}\sigma_3)t = -(24.9\times10^{-3}\ \text{V m/N})(-80\times10^6\ \text{Pa})(125\times10^{-6}\ \text{m})$$

Answer: $\underline{V_i = 249\ \text{V}}$

The induced voltage gradient (electric field) is:

$$\frac{V_i}{t} = -g_{33}\sigma_3$$

$$= 1.99\times10^6\ \text{V/m}$$

Step 3. To return the actuator to its original height, an applied voltage V_a must be added to the induced voltage. The strain due to the applied voltage V_a is:

$$S_{3,e} = d_{33}\frac{V_a}{t}$$

This strain must be equal and opposite the strain caused by the mechanical load $S_{3,m}$, so:

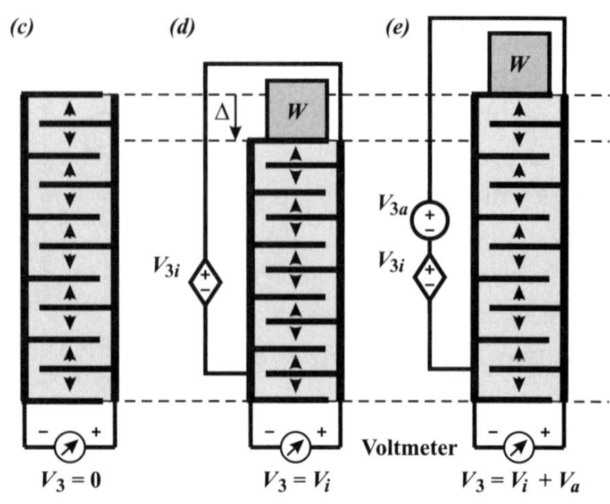

Figure 16.23. (c) Unloaded stack. **(d)** Applied load causes stack to deflect by Δ and induces voltage V_i. **(e)** Additional voltage V_a is added to raise stack back to its original height.

$$V_a = \frac{t}{d_{33}}(S_{3,e}) = \frac{t}{d_{33}}(-S_{3,m}) = \frac{125\times10^{-6}\ \text{m}}{285\times10^{-12}\ \text{m/V}}(645\times10^{-6})$$

Answer: $\underline{V_a = 283\ \text{V}}$

The applied voltage could also have been solved by directly applying *Equation 16.75*:

$$S_3 = \frac{\sigma_3}{Y^*} + d_{33}\frac{V_{3a}}{t} = 0$$

$$V_{3a} = -\frac{\sigma_3}{Y^*}\left(\frac{t}{d_{33}}\right) = -\left(\frac{-80\times10^6\ \text{Pa}}{124\times10^9\ \text{Pa}}\right)\left(\frac{125\times10^{-6}\ \text{m}}{285\times10^{-12}\ \text{m/V}}\right)$$

$$\Rightarrow \underline{V_a = 283\ \text{V}}$$

Step 4. The total voltage V_3 across the electrodes for there to be no net change in stack height with $\sigma = -80$ MPa is (*Figure 16.23e*):

$$V_3 = V_i + V_a = 249 + 283\ \text{V}$$

Answer: $\underline{V_3 = 532\ \text{V}}$

The total voltage could also have been solved using *Equation 16.82*, with $Y = 66$ GPa, the short-circuited Young's modulus of the PZT4 material:

$$S_3 = \frac{\sigma_3}{Y} + d_{33}\frac{V_3}{t} = 0$$

so:

$$V_3 = -\frac{\sigma_3}{Y}\left(\frac{t}{d_{33}}\right) = -\left(\frac{-80\times10^6 \text{ Pa}}{66\times10^9 \text{ Pa}}\right)\left(\frac{125\times10^{-6} \text{ m}}{285\times10^{-12} \text{ m/V}}\right)$$

$$\Rightarrow \underline{V_3 = 532 \text{ V}}$$

Example 16.9 Actuator to Lift Atomic Force Microscope

Given: A thin-walled piezoelectric tube of average diameter $D = 10$ mm, thickness $t = 125$ μm, and length $L = 50$ mm is used to control displacement in the 1-direction (*Figure 16.24a*). Electrodes are attached to the inner and outer surfaces of the tube, and the poling direction is radially outward, in the positive 3-direction (*Figure 16.24b*). The material is PZT4, so that $d_{31} = -122\times10^{-12}$ m/V. Such a device might be used to control the vertical position of an atomic force microscope (*Figure 16.24c*).

Required: Determine (a) the change in length of the cylinder if the applied voltage is 250 V, with the polarity of the voltage source applied as shown in *Figure 16.24d* and (b) the applied voltage V required to lengthen the tube by 10 μm.

Solution: *Step 1.* The polarity of the applied voltage shown in *Figure 16.24d* causes an electric field in the same direction as the poling direction. Thus, $E_{3a} > 0$, and $V_{3a} = 250$ V. The cylinder thickness increases, which implies that its length decreases.

The strain in the axial direction is:

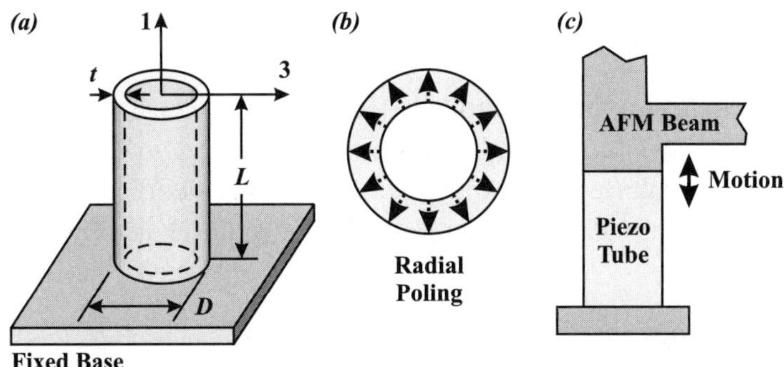

Figure 16.24. **(a)** Thin-walled piezoelectric tube with electrodes on the inner and outer surfaces. **(b)** The tube is poled in the positive 3-direction, from the inner to outer electrode. **(c)** Tube used as an actuator to adjust height of AFM beam.

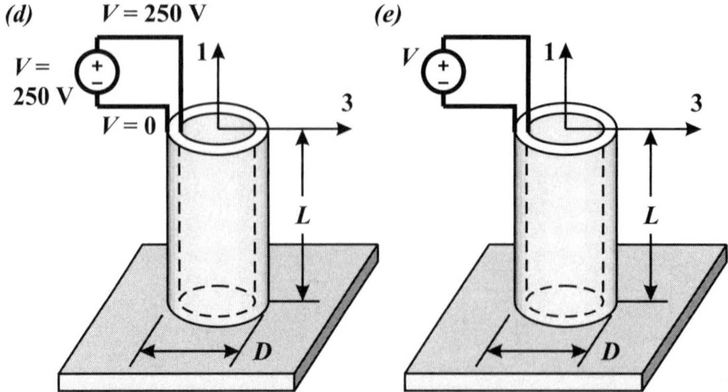

Figure 16.24. (d) With a positive applied voltage $V = +250$ V (the inner electrode subjected to positive voltage, outer electrode at ground), the cylinder decreases in length by 12.2 µm. **(e)** General case for positive applied voltage. The voltage source causes an electric field radially outward.

$$S_1 = \frac{\Delta L}{L} = d_{31}\frac{V_{3a}}{t}$$

Thus:

$$\Delta L = d_{31}\left(\frac{V_{3a}}{t}\right)L = (-122\times10^{-12}\ \text{m/V})\left(\frac{250\ \text{V}}{125\times10^{-6}\ \text{m}}\right)(50\times10^{-3}\ \text{m})$$

Answer: $\underline{\Delta L = -12.2\ \mu\text{m}}$

The cylinder shortens due to a positive voltage V_{3a} being applied to the inner electrode with voltage $V = 0$ at the outer electrode.

Step 2. The voltage necessary to lengthen the tube by 10 µm is:

$$V = \left(\frac{\Delta L}{L}\right)\left(\frac{t}{d_{13}}\right) = \left(\frac{10\times10^{-6}\ \text{m}}{50\times10^{-3}\ \text{m}}\right)\left(\frac{125\times10^{-6}\ \text{m}}{-122\times10^{-12}\ \text{m/V}}\right)$$

Answer: $\underline{V = -205\ \text{V}}$

The negative sign indicates that the applied electric field (voltage drop) must be opposite the poled direction; here the applied electric field must act towards the center of the cylinder. The applied voltage must be applied in the opposite direction of *Part (a)*, i.e., opposite the source polarity drawn in *Figure 16.24e*, meaning V is negative.

Example 16.10 Force Sensor

Given: The piezoelectric thin-walled tube of *Example 16.9* is used to measure small forces. By measuring voltage, the applied stress and thus force, can be determined (*Figure 16.25*).

Required: Determine the applied load in the 1-direction (the axial direction) if the measured induced voltage is (a) $V_{3i} = 50$ mV and (b) $V_{3i} = -200$ mV.

Solution: *Step 1.* The induced electric field in the 3-direction due to applied stress in the 1-direction is:

$$\frac{V_{3i}}{t} = -g_{31}\sigma_1$$

The force that causes the induced voltage V_{3i} is therefore:

$$P = \sigma_1 A = -\frac{V_{3i}}{g_{31}t}(2\pi Rt) = -\frac{V_{3i}}{g_{31}}(2\pi R)$$

Step 2. For $V_{3i} = 0.050$ V, $\sigma_1 = 38.8$ MPa, so the applied load is:

$$P = -\frac{(0.050 \text{ V})}{(-10.3\times10^{-3} \text{ V m/N})}[2\pi(5\times10^{-3} \text{ m})]$$

Answer: $\underline{P = 0.153 \text{ N}}$

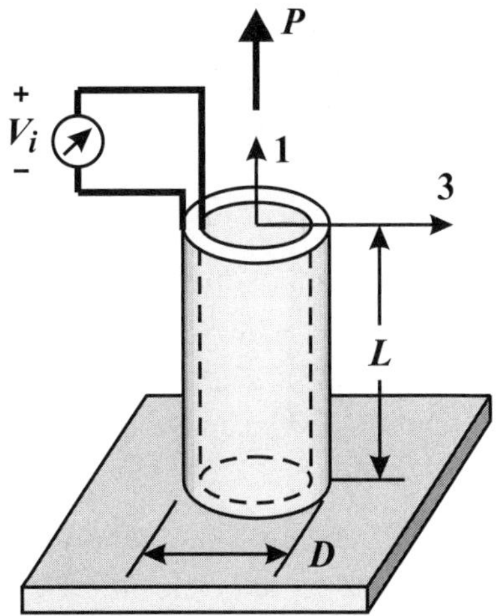

Figure 16.25. Piezo tube used as a sensor to measure axial loads.

Since the poling direction is radially outward, if $V_i > 0$ ($E_i > 0$), then the induced voltage thickens the tube. Since the induced voltage acts to counter the strain due to the stress, $V_i > 0$ implies that the mechanical load reduces the tube thickness. Relating this to the Poisson effect, the applied load must be tensile ($P > 0$).

Step 3. For $V_{3i} = -0.200$ V, $\sigma_1 = -155$ MPa so:

Answer: $\underline{P = -0.610 \text{ N}}$

16.7 Piezoelectric Bending

Piezoelectric effects can be used to induce bending in beams. Such a beam is called a **bimorph**. A bimorph consists of two layers of piezoelectric material, with electrodes on the top and bottom surfaces and one at the interface of the two layers (*Figure 16.26*). In this example, the piezoelectric layers have been poled in the same direction. Other configurations are possible, provided that the application of voltage causes one layer to expand, and the other to contract.

When voltage is applied to the bimorph as shown in *Figure 16.26b*, the top layer thickness increases and its length decreases (the applied electric field E is in the poling direction). The bottom layer thickness decreases and its length increases (E is opposite the poling direction). The bimorph bends upward and acts as an *actuator*.

If a mechanical load is used to deflect the beam, a voltage is induced across the electrodes. The bimorph acts as a *sensor*.

The design loads on bimorphs are typically small due to the limiting stress of ceramics in tension, however, they are capable of much larger displacements than piezoelectric stacks.

For bending, the appropriate stresses and strains are in the x_1-direction, and the electric field is in the x_3-direction. The appropriate laws for strain in the x_1-direction (1-direction), and electric displacement in the x_3-direction (3-direction) are, at any point, from *Equation 16.56*:

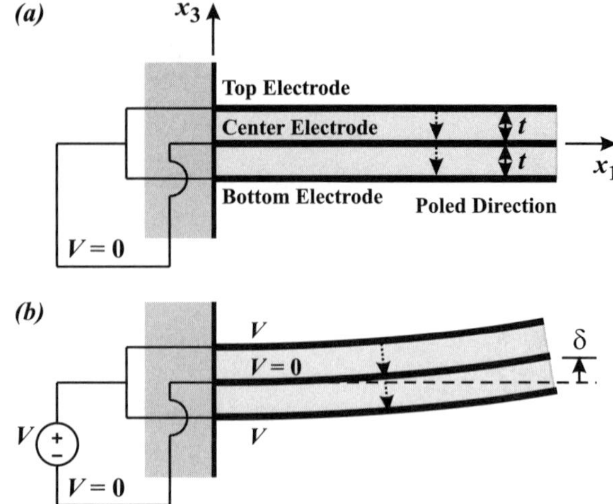

Figure 16.26. **(a)** Bimorph beam system. The layers are poled in the same direction, but **(b)** the applied voltages (electric fields) are in opposite directions.

$$\begin{bmatrix} S_1 \\ D_3 \end{bmatrix} = \begin{bmatrix} 1/Y & d_{31} \\ d_{31} & \varepsilon \end{bmatrix} \begin{bmatrix} \sigma_1 \\ E_3 \end{bmatrix}$$

[Eq. 16.83]

where σ_1 is the bending stress and E_3 is the total electric field due to both the stress-induced electric field E_{3i} and the applied electric field E_{3a}.

The strain and electric displacement can also be written in terms the *applied electric field E_{3a}* analogous to *Equation 16.75*:

$$\begin{bmatrix} S_1 \\ D_3 \end{bmatrix} = \begin{bmatrix} 1/Y_1^* & d_{31} \\ 0 & \varepsilon \end{bmatrix} \begin{bmatrix} \sigma_1 \\ E_{3a} \end{bmatrix}$$

[Eq. 16.84]

where Y_1^* is the effective modulus in the x_1-direction:

$$Y_1^* = \frac{Y}{1 - d_{31} g_{31} Y}$$

[Eq. 16.85]

The piezoelectric constants, g_{31} and d_{31}, are related by the dielectric constant (*Equation 16.60*):

$$\varepsilon = \frac{d_{31}}{g_{31}}$$

[Eq. 16.86]

Response to Moment and Applied Voltage

The loads on a bimorph beam are the applied moment M, and the total voltage, which is the sum of the applied and induced voltages: $V = V_a + V_i$. Voltage V is measured at the outer electrodes, with the internal electrode maintained at $V = 0$ (*Figure 16.27*). When voltage V_a is applied to the system, the electric fields in the layers are equal but opposite. Each layer is t thick, and has constant width (breadth) of B into the paper.

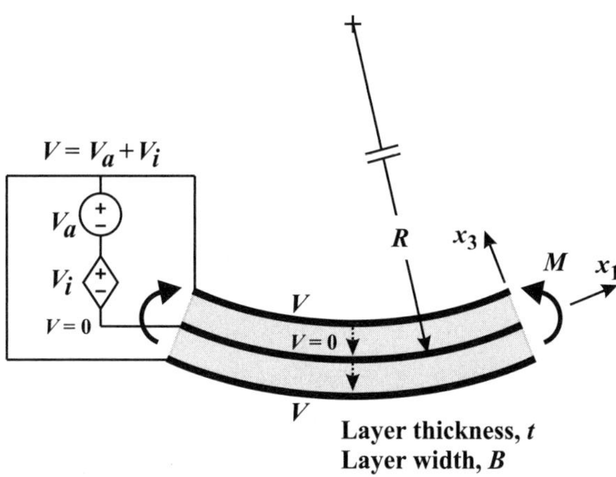

Figure 16.27. Bimorph subjected to moment M and total voltage V. Moment M induces voltage V_i across the electrodes. V_a is the applied voltage.

Considering the applied moment M only, the top layer contracts and the bottom layer expands in the x_1-direction. The bending stresses induce an electric field $E_{3i} = V_{3i}/t$ in the PZT material.

Considering the applied voltage V_{3a} only (due to applied electric field E_{3a}), the top layer contracts and the bottom layer expands in the x_1-direction, so that the beam develops a radius of curvature. Compatibility of the two layers at their interface requires that there be stresses in the x_1-direction at the center of the beam cross-section.

The strain in the top layer ($x_3 > 0$) due to applied bending stress σ_1 and applied electric field E_{3a} is, from *Equation 16.84*:

$$S_{1,t} = \frac{\sigma_1}{Y_1{}^*} + d_{31}E_{3a} \qquad \text{[Eq. 16.87]}$$

In the top layer, E_{3a} is in the poling direction, so it is positive. Since the electric displacement is $D_3 = \varepsilon E_{3a}$, the strain becomes:

$$S_{1,t} = \frac{\sigma_1}{Y_1{}^*} + \frac{d_{31}}{\varepsilon}D_3 \qquad \text{[Eq. 16.88]}$$

Solving for the Moment M

Applying the same displacement condition as in traditional beam theory, that plane-sections remain plane, then the strain in the top layer is linear with x_3 :

$$S_{1,t} = \frac{-x_3}{R} \qquad \text{[Eq. 16.89]}$$

where R is the radius of curvature of the bimorph beam. At $x_3 = 0$ (the centroid of the beam) the strain is zero. Setting the previous two equations equal to each other:

$$\frac{-x_3}{R} = \frac{\sigma_1}{Y_1^*} + \frac{d_{31}}{\varepsilon} D_3 \qquad \text{[Eq. 16.90]}$$

Solving for the stress in the top layer, $\sigma_{1,t}$:

$$\sigma_{1,t} = Y_1^*\left(\frac{-x_3}{R} - \frac{d_{31}}{\varepsilon} D_3\right) \qquad \text{[Eq. 16.91]}$$

The stress has two components, one which varies linearly with x_3, and the other is a constant depending upon the applied electric field.

Multiplying the stress by $x_3\, dA = x_3 B\, dx_3$, where B is the breath of the beam and then integrating from $x_3 = 0$ to t, gives the moment supported in the top layer:

$$M_t = -\int_0^t \sigma_1 B x_3\, dx_3 = -Y_1^* B \int_0^t \left(\frac{-x_3}{R} - \frac{d_{31}}{\varepsilon} D_3\right) x_3\, dx_3 \qquad \text{[Eq. 16.92]}$$

Performing the integration, the moment supported by the top layer is:

$$M_t = Y_1^* B\left(\frac{t^3}{3R} + \frac{d_{31} t^2}{\varepsilon\ 2} D_3\right) \qquad \text{[Eq. 16.93]}$$

The stress in the lower or bottom layer ($x_3 < 0$) is given by:

$$\sigma_{1,b} = Y_1^*\left(\frac{-x_3}{R} + \frac{d_{31}}{\varepsilon} D_3\right) \qquad \text{[Eq. 16.94]}$$

This stress is the same as the stress in the top layer (*Equation 16.91*) except the term due to the applied electric field is opposite in sign (E_{3a} is negative since it is opposite the poling direction in the bottom layer; thus D_3 is also negative).

Performing similar calculations, the moment supported by the bottom layer is:

$$M_b = Y_1^* B\left(\frac{t^3}{3R} + \frac{d_{31} t^2}{\varepsilon\ 2} D_3\right) \qquad \text{[Eq. 16.95]}$$

which is the same moment supported by the top layer.

The total moment is the sum of the moments supported by both layers:

$$M = 2Y_1^* B\left(\frac{t^3}{3R} + \frac{d_{31} t^2}{\varepsilon\ 2} D_3\right) \qquad \text{[Eq. 16.96]}$$

This equation gives the moment in terms of radius of curvature R and electric displacement D_3.

As a check, when the applied electric field is zero (i.e., $D_3 = 0$), then:

$$M = 2Y_1^* B \frac{t^3}{3R} = \frac{Y_1^*}{R}\left[\frac{B(2t)^3}{12}\right] = \frac{Y_1^* I}{R} \qquad \text{[Eq. 16.97]}$$

This is the moment–radius of curvature relationship for a rectangular beam of modulus Y_1^*, depth $2t$, and breadth B.

Solving for the Total Voltage V

From *Equation 16.83*, the total electric field E_3 in the top layer is:

$$\varepsilon E_3 = D_3 - d_{31}\sigma_1 \qquad \text{[Eq. 16.98]}$$

Substituting the expression for the stress from *Equation 16.91* results in:

$$\varepsilon E_3 = D_3 - d_{31} Y_1^*\left(\frac{-x_3}{R} - \frac{d_{31}}{\varepsilon} D_3\right) \qquad \text{[Eq. 16.99]}$$

or

$$E_3 = \frac{d_{31}}{\varepsilon}\frac{Y_1^*}{R} x_3 + \left(1 + \frac{Y_1^* d_{31}^2}{\varepsilon}\right)\frac{D_3}{\varepsilon} \qquad \text{[Eq. 16.100]}$$

The total electric field, like the stress, has two components, one that varies linearly with x_3, and the other is constant and depends upon the charge (electric displacement D_3).

The electric field is the voltage gradient:

$$E_3 = -\frac{dV_3}{dx_3} \qquad \text{[Eq. 16.101]}$$

E_3 is positive in the direction of decreasing voltage, in the direction of the voltage drop.

Combining *Equations 16.100* and *16.101*, and multiplying both sides by $(-dx_3)$:

$$dV_3 = -\left[\frac{d_{31}}{\varepsilon}\frac{Y_1^*}{R} x_3 + \left(1 + \frac{Y_1^* d_{31}^2}{\varepsilon}\right)\frac{D_3}{\varepsilon}\right]dx_3 \qquad \text{[Eq. 16.102]}$$

At $x_3 = 0$, $V = 0$. Integrating *Equation 16.102* from $x_3 = 0$ to t gives the change in voltage from the center electrode to the top electrode, which is simply the total voltage V. Thus:

$$V = -\int_0^t\left[\frac{d_{31}}{\varepsilon}\frac{Y_1^*}{R} x_3 + \left(1 + \frac{Y_1^* d_{31}^2}{\varepsilon}\right)\frac{D_3}{\varepsilon}\right]dx_3 \qquad \text{[Eq. 16.103]}$$

Performing the integral and dividing by t, gives the average voltage gradient:

$$\frac{V_3}{t} = \frac{d_{31}}{\varepsilon}\frac{Y_1^*}{R}\frac{t}{2} + \left(1 + \frac{Y_1^* d_{31}^2}{\varepsilon}\right)\frac{D_3}{\varepsilon} \qquad \text{[Eq. 16.104]}$$

This equation gives the total voltage in terms of radius of curvature R and charge density D. By algebraically manipulating the expression for Y_1^* (*Equation 16.85*), it can be shown that the quantity in parenthesis is:

$$\left(1 + \frac{Y_1^* d_{31}^2}{\varepsilon}\right) = \frac{Y_1^*}{Y}$$

[Eq. 16.105]

Combined Moment and Voltage Solution

Equations 16.96 and *16.104* can be written in matrix form for the applied moment M and total voltage V_3:

$$\begin{bmatrix} \dfrac{M}{2B} \\[2ex] V_3 \end{bmatrix} = Y_1^* \begin{bmatrix} \dfrac{t^3}{3} & \dfrac{d_{31} t^2}{\varepsilon}\dfrac{1}{2} \\[2ex] \dfrac{d_{31} t^2}{\varepsilon}\dfrac{1}{2} & \dfrac{1}{Y\varepsilon} \end{bmatrix} \begin{bmatrix} \dfrac{1}{R} \\[2ex] D_3 \end{bmatrix}$$

[Eq. 16.106]

Example 16.11 Voltage to Cause Displacement (Actuator)

Given: An unloaded bimorph is used to cause an upward displacement δ by applying voltage $V_{3a} = V$ across its electrodes (*Figure 16.28*). The dimensions are length $L = 25$ mm, layer thickness $t = 0.25$ mm, and layer breadth $B = 1.0$ mm. The bimorph is made of a PZT4 material. The short-circuit elastic modulus is $Y = 66$ GPa, and the piezoelectric charge constant is:

$$d_{31} = -122 \times 10^{-12} \text{ m/V}$$

Required: Determine the voltage needed to displace the tip of the cantilever by $\delta = 10$ μm.

Solution: *Step 1.* Since there is no applied moment, $M = 0$, the curvature-load relationship of *Equation 16.106* is:

$$0 = Y_1^*\left(\frac{t^3}{3R} + \frac{d_{31} t^2}{\varepsilon}\frac{1}{2} D_3\right)$$

so:

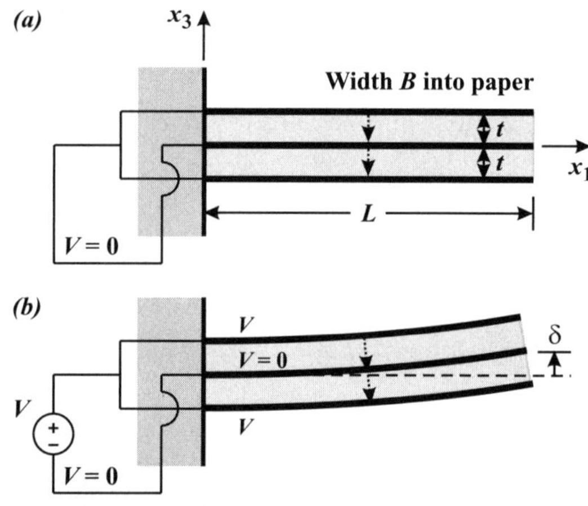

Figure 16.28. (a) A bimorph beam. (b) The beam deflects when subjected to applied voltage V.

$$\frac{1}{R} = -\frac{3}{t^3}\left(\frac{d_{31}}{\varepsilon}\frac{t^2}{2}D_3\right) = -\frac{3}{2}\frac{d_{31}}{t}\frac{V_{3a}}{t}$$

Step 2. From beam theory, curvature is the second derivative of the beam displacement:

$$\frac{1}{R} = \frac{d^2v}{dx^2}$$

Integrating once to find the slope:

$$\frac{dv}{dx_1} = -\left(\frac{3}{2}\frac{d_{31}}{t^2}V_{3a}\right)x_1 + C_1$$

The slope at $x_1 = 0$ is $dv/dx_1 = 0$, so $C_1 = 0$. Integrating to find the beam deflection:

$$v(x_1) = -\left(\frac{3}{4}\frac{d_{31}V_{3a}}{t^2}\right)x_1^2 + C_2$$

The deflection at $x_1 = 0$ is $v = 0$, so $C_2 = 0$. The expression for the displaced shape of the bimorph is:

$$v(x_1) = -\left(\frac{3}{4}\frac{d_{31}V_{3a}}{t^2}\right)x_1^2$$

Step 3. Rearranging, to solve for the applied voltage:

$$V_{3a} = \frac{-4t^2v}{3d_{31}x_1^2}$$

To cause a tip displacement of $\delta = v(x_1 = L) = 10$ μm, the applied voltage is:

$$V_{3a} = \frac{-4t^2\delta}{3d_{31}L^2} = \frac{-4(0.25\times10^{-3}\text{ m})^2(10\times10^{-6}\text{ m})}{3(-122\times10^{-12}\text{ m/V})(25\times10^{-3}\text{ m})^2}$$

Answer: $V_{3a} = 10.9$ V

The average voltage gradient is:

$$\frac{V_{3a}}{t} = \frac{10.9\text{ V}}{0.25\times10^{-3}\text{ m}} = 43.6\times10^3\text{ V/m}$$

Example 16.12 **Application of Moment to Induce Voltage (Sensor)**

Given: A bimorph beam is subjected to moment M only (*Figure 16.29*). The electrodes are isolated so that no charge may flow to or away from them. The thickness of each layer is t and the width is B.

Required: Determine the voltage induced by the applied load.

Solution: From *Equation 16.106*:

$$\begin{bmatrix} \dfrac{M}{2B} \\[2ex] V_3 \end{bmatrix} = Y_1^* \begin{bmatrix} \dfrac{t^3}{3} & \dfrac{d_{31}t^2}{\varepsilon}\dfrac{}{2} \\[2ex] \dfrac{d_{31}t^2}{\varepsilon}\dfrac{}{2} & \dfrac{1}{Y\varepsilon} \end{bmatrix} \begin{bmatrix} \dfrac{1}{R} \\[2ex] D_3 \end{bmatrix}$$

With no applied voltage, the charge density is $D_3 = 0$. Thus, the moment is:

$$M = \frac{2Bt^3}{3}(Y_1^*)\frac{1}{R} + 0$$

The induced voltage as a function of radius of curvature is:

$$V_3 = V_i = \frac{d_{31}t^2}{\varepsilon}\frac{}{2}(Y_1^*)\frac{1}{R} + 0$$

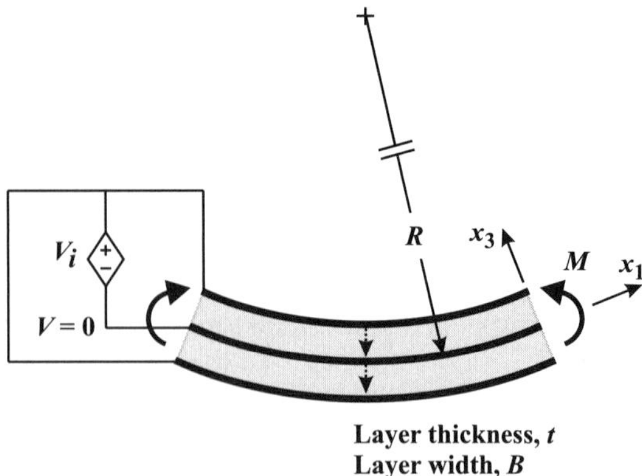

Figure 16.29. Bimorph beam subjected to constant moment M induces voltage V_i.

Eliminating R from the previous two equations gives V_i in terms of moment M:

$$V_i = \frac{d_{31}t^2}{\varepsilon}\frac{}{2}(Y_1^*)\left[\frac{3}{2}\frac{M}{Bt^3}\frac{1}{Y_1^*}\right] = \frac{3}{4}\frac{d_{31}}{\varepsilon}\frac{1}{Bt}M$$

Recalling that $g_{31} = d_{31}/\varepsilon$, then:

$$\textit{Answer: } V_i = \frac{3}{4}\frac{d_{31}}{\varepsilon}\frac{1}{Bt}M = \frac{3}{4}g_{31}\frac{1}{Bt}M$$

Example 16.13 Force Sensor

Given: A cantilever beam is used to measure force P at its tip. The cantilever is made of a PZT4 bimorph (*Figure 16.30*). The dimensions are: $L = 25$ mm, $t = 0.25$ mm, $B = 1.0$ mm.

The appropriate dielectric constant is:

$$g_{31} = -10.6 \times 10^{-3} \text{ V·m/N}$$

Required: Determine the voltage output due to applied force $P = 0.10$ N.

Solution:

The maximum bending moment is:

$$PL = (0.10 \text{ N})(0.025 \text{ m})$$
$$= 0.0025 \text{ N·m}$$

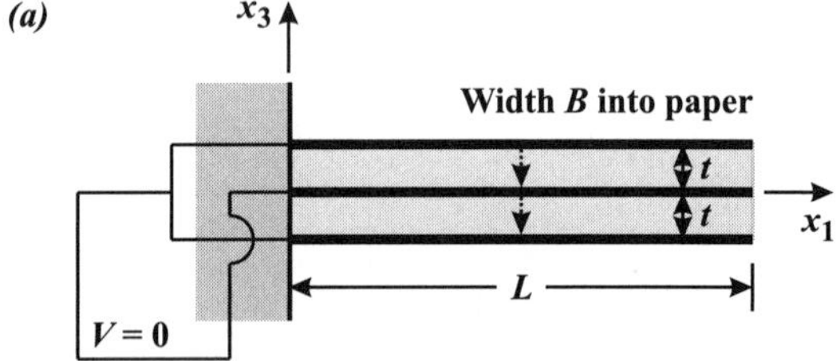

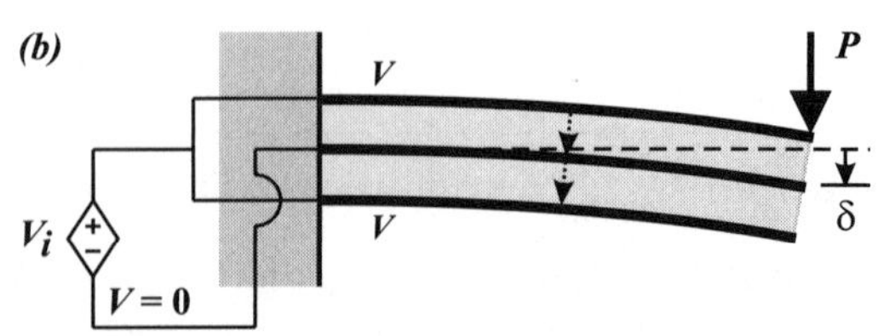

Figure 16.30. Force P induces a voltage across the electrodes which is maximum at the base of the cantilever.

The induced voltage is:

$$V_i = \frac{3}{4}g_{31}\frac{M}{Bt} = \frac{3}{4}(-10.6\times10^{-3}\ \text{V·m/N})\frac{(0.0025\ \text{N·m})}{(0.001\ \text{m})(0.25\times10^{-3}\ \text{m})}$$

Answer: $V_i = -79.5$ V

16.8 Shape Memory Alloys

The stress–strain response of ***shape memory alloys*** (SMA) depends on the applied stress and temperature. The strength of these materials can be significant and thus it has been suggested that they be used as actuators to lift relatively large loads.

The most common SMA is composed of nearly equal amounts of nickel and titanium (NiTi). The response of NiTi was discovered and studied at the Naval Ordnance Laboratory (NOL), so the alloy is generally called ***nitinol*** (NiTi–NOL). Slight modifications to the atomic proportions, and additions of small amounts of other alloys, cause variations in the stress–strain–temperature response of the material.

The key to nitinol's stress–strain response is that it can take on three different crystalline (solid) phases depending upon stress and temperature – ***twinned martensite, austenite***, and ***stress-induced martensite***.

Under no load, the phase is *twinned martensite* at low temperatures and *austenite* at high temperatures. When low-temperature *martensite* is heated, *austenite* begins to form at temperature A_s (austenite start), and the nitinol is completely transformed at $T = A_f$ (austenite finish) (*Figure 16.31*). Upon cooling, *martensite* begins to form at $T = M_s$ and the transformation is complete at $T = M_f$. Transformation temperatures can range from approximately –50 to 100°C depending on the specific SMA. The *hysteresis H* is the difference between the heating and the cooling transformation temperatures; typically $H \sim 30$°C. The overall dimensions of a nitinol sample are unchanged during this phase transformation.

When nitinol is subjected to stress, the temperatures required for transformation increase.

Shape Memory

Idealized stress–strain curves for nitinol are shown in *Figure 16.32*. At low temperatures ($T < A_s$, *Figure 16.32a*), the response of the *twinned martensite* is linear until it reaches a critical stress, when the material transforms into *deformed* or *stress-induced martensite*. This new crystalline

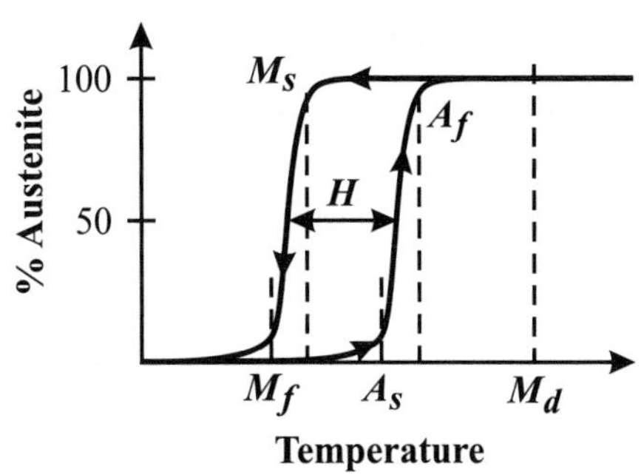

Figure 16.31. Phase-temperature curves for the martensite–austenite transformation.

phase is larger in volume than the twinned martensite, resulting in a transformation strain ε_T of approximately 5%. After the transformation is complete, the stress–strain response is again linear (the right side of the stress–strain curve). When the stress is removed, the transformed strain remains.

If the deformed nitinol is then heated to $T > A_f$, austenite forms and the original geometry is recovered (the dotted line in *Figure 16.32a*). This response to heating is the **shape memory effect**. The austenite can then be cooled into the twinned martensitic phase without further change in overall geometry. The cycle just described is illustrated in *Figure 16.33a*.

Superelasticity

At temperatures $T > A_f$ (*Figure 16.32b*) the unloaded NiTi is austenite. As the stiffer austenite is loaded, at a critical stress level, stress-induced martensite begins to form. Again, a transformation strain ε_T of approximately 5% occurs with stress remaining essentially constant. After the transformation, the stress–strain response is again linear.

Upon removal of the stress, the stress-strain curve plateaus at a lower stress level as the stress-induced martensite transforms back into austenite (the lower branch of *Figure 16.32b*). The overall system response is *elastic* since the material returns to $\varepsilon = 0$ when the load it completely removed. This *non-linear elastic* response due to phase changes is known as *superelasticity* or *pseudoelasticity*.

Nitinol that has been transformed from austenite into stress-induced martensite at $T > A_f$ can exhibit the *shape memory effect* while under load. This is done by heating the material to a sufficient temperature. Under

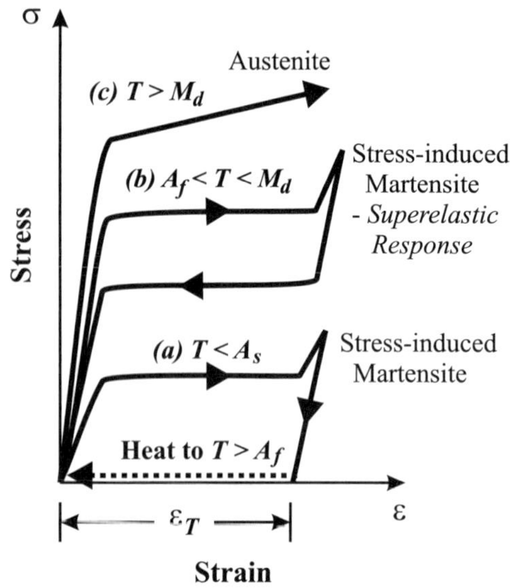

Figure 16.32. Idealized stress–strain curves for NiTi at three temperature levels.

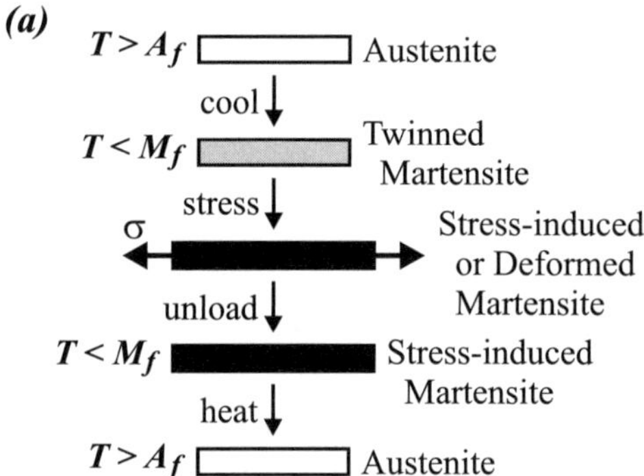

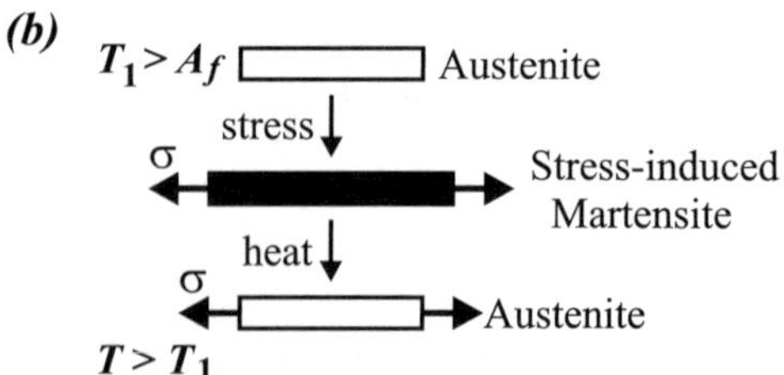

Figure 16.33. **(a)** Temperature-load steps showing shape memory effect. **(b)** Shape memory effect under loaded condition.

stress, the temperature for the austenite-to-stress-induced martensite transformation increases, so further heating the stress-induced martensite can recover the austenitic phase (*Figures 16.33, 16.34*).

High Temperature Response

At temperatures above a critical value, $T = M_d$, austenite does not transform into stress-induced martensite. The stress–strain curve is linear until the austenite yields; the material fails similar to other ductile metals (*Figure 16.32c*).

Applications

Shape memory alloys are generally used in the stress-induced martensitic phase. Upon heating to sufficient temperature, the alloy returns to its austenitic phase, decreasing in length at a constant stress (as at 37°C in *Figure 16.34*). Nitinol can, therefore, be used in actuators, applying motion at a constant force (stress) during its transformation. Or, a nitinol wire can be used as a protective sensor in an electrical circuit. If sufficient current flows through the wire, the heated nitinol will contract, opening (breaking) the circuit.

Nitinol is biocompatible and can be fabricated on the small scale, so it has extensive uses in current and future medical applications. Nitinol is often engineered such that $A_f < 37°C$ (98.6°F), the temperature of the human body. Orthodontic nitinol arch wires are deformed at low temperatures into stress-induced martensite. In the mouth, the heated wires contract into the austenite phase, applying a force on the tooth. Because of the large transformation strain recovered at constant stress, the force on the tooth remains constant over relatively large movements (*Figure 16.34*). Nitinol braces do not need to be adjusted (retightened) as often as those made of traditional linear–elastic metals.

A more significant and complex biomedical application of nitinol is for stents used to open arteries. Nitinol stents, initially smaller in diameter than arteries, are designed to expand (unfold) when inserted into the human body.

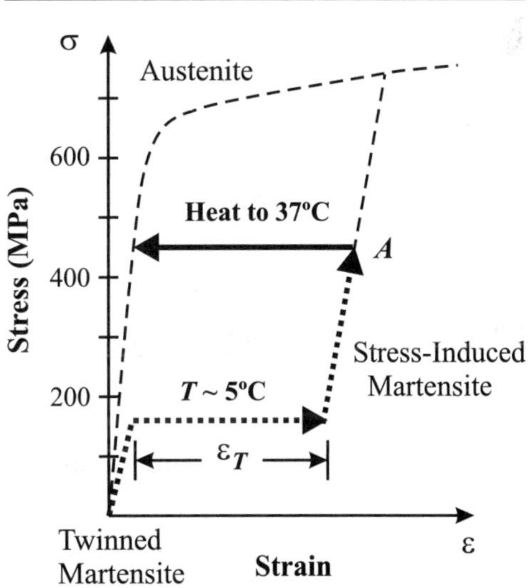

Figure 16.34. Representative stress–strain response for biomedical nitinol, $A_f \sim 35°C$. Low-temperature twinned martensite is deformed into stress-induced martensite. The system, e.g., an arch wire for braces, is inserted and subjected to stress *A*. Due to body temperature $T \sim 37°C$, the wire begins to transform into austenite at constant stress. The contracting nitinol gives the tooth its required motion.

A common superelastic application is in eyeglass frames, which can be accidentally subjected to large deformations. Where typical metals would permanently yield, nitinol experiences a non-linear but elastic response (*Figure 16.32b*).

The transformation strain of nitinol can be large (~5%), and can occur at relatively large stresses (e.g., $\sigma > 400$ MPa). It has, therefore, been suggested that this material can be used to lift relatively large loads over large displacements. Consider the following example.

Example 16.14 Morphing Truss System with Shape Memory Member

Given: The pinned-jointed truss system shown in *Figure 16.35a* is subjected to a tip load *W*. The angles of the webs are all 60° from the horizontal. Each member has cross-sectional area $A = 5.0$ mm^2 and length $L = 25$ mm. The load is $W = 750$ N.

Member *AB* is 12.5 mm long and is made of the shape memory alloy NiTi. Before load *W* is applied, it is assumed that *AB* is horizontal, and that it has already transformed from austenite to stress-induced martensite, and so has already expanded by $\varepsilon_T = 5\%$

After application of load *W*, member *AB* is heated to a sufficient temperature so that it experiences the shape memory phase transformation (back to austenite). The accompanying strain is $-\varepsilon_T = -5\%$.

Required: Approximate the distance δ_T that load *W* is lifted due to the transformation of member *AB*.

Solution: Since member *AB* is horizontal, the reaction at joint *A* equals the force in *AB*, P_{AB} (*Figure 16.35b*):

$$A_x = P_{AB}$$

Taking moments about the joint *C* gives:

$$P_{AB} = 2\sqrt{3}\,W$$

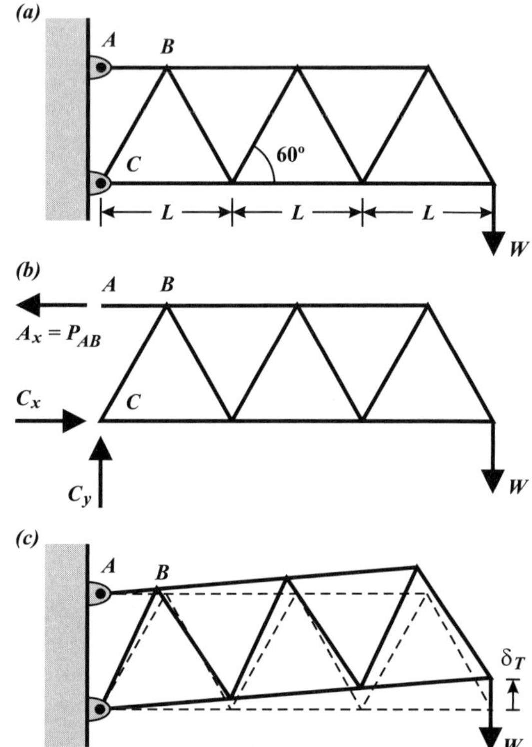

Figure 16.35. (a) Truss supporting weight *W*. Member *AB* is a shape memory alloy. After the load is applied, *AB* is heated so that it undergoes a phase transformation, making it shorter. **(b)** FBD of truss. **(c)** When *AB* shortens, load *W* is lifted by δ_T (here, *AB* is shortened by about 13% to magnify the response).

The temperature is now increased to cause AB to transform back into austenite. The temperature T must be greater than A_f for the particular stress:

$$\sigma_{AB} = \frac{P_{AB}}{A} = \frac{2\sqrt{3}\,W}{A} = \frac{2\sqrt{3}(750\text{ N})}{5\times10^6\text{ m}^2} = 520\text{ MPa}$$

Due to the strain transformation, the change in length of AB is:

$$\Delta_T = -\varepsilon_T L_{AB} = -\varepsilon_T\left(\frac{L}{2}\right)$$

The tensile force in P_{AB} does negative work as the bar shortens. Since the nitinol transforms at constant stress, and thus at constant force, the work done during shortening the bars is:

$$U = P_{AB}\Delta_T = -(2\sqrt{3}\,W)\varepsilon_T\left(\frac{L}{2}\right) = -\sqrt{3}\,W\varepsilon_T L$$

The work done by downward load W moving upward by δ_T is:

$$\text{Work} = -W\delta_T$$

Equating the work terms, the displacement of the tip due to the transformation is (*Figure 16.35c*):

$$\delta_T = \sqrt{3}\,\varepsilon_T L$$

For $L = 25$ mm, and $\varepsilon_T = 0.05$, the upward movement of load W is:

$$\delta_T = \sqrt{3}(0.05)(25\text{ mm})$$

Answer: $\underline{\delta_T = 2.2\text{ mm}}$

The deflection index is the ratio of the deflection to the overall length of the system:

$$f = \frac{\delta}{3L} = \frac{2.2\text{ mm}}{3(25\text{ mm})} = \frac{1}{34} = 0.029 = 2.9\%$$

Typical deflection indices in traditional beam systems are $f\sim1/240$–0.4%. Thus, shape memory alloys actuators may prove useful in lifting loads significant distances.

Problems: Chapter 1   Opening Remarks

1.1 Rank the loads in order of magnitude, from highest to lowest: *proof load, ultimate load, working load.*

1.2 A drum-winch slowly lifts a large bucket full of ore from a mine. The ore has a density of 1600 kg/m^3. The bucket volume is 8.00 m^3. Neglect the mass of the bucket.

Determine (a) the mass of the ore supported by the cable, (b) the force in the cable in newtons, and (c) the force in pounds. (d) if the cable can support 500 kN, determine the factor of safety for this load.

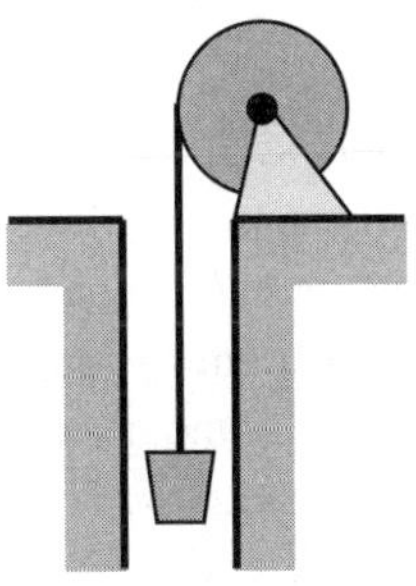

1.3 A residential structure (house) is built in New England. The size of the house is 2000 square feet, half of which is located on the second floor. The effective area of the roof is 1300 square feet. Since it is a residence, each floor must be designed to support a load of 40 psf. Due to snow, a load of 30 psf must be supported on the roof.

Considering the second-floor load and the snow load, determine the total load that the first floor walls must be designed to support.

1.4 The concrete counterweight of a large crane weighs $W_C = 350,000$ lb. Concrete has a weight density of $\gamma = 150$ lb/ft^3.

If the counterweight is approximately a cube, determine the length of each side (a) in feet and (b) in meters.

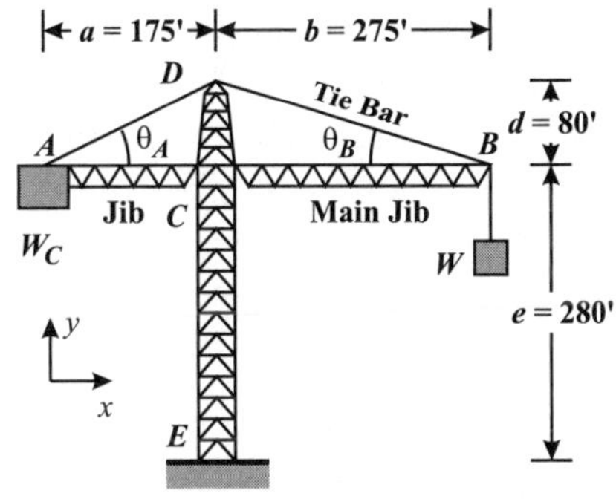

1.5 At the end of Stearns Wharf in Santa Barbara, California, large wooden logs at the edge of the wharf are used as seats and as physical barriers between pedestrians and the ocean, 20 ft below. A typical log has a diameter of $2R = 2.5$ ft and a length of 24 ft. The density of the wood is estimated to be 40 lb/ft^3.

Determine the minimum rating of the crane required to move the cylinders. Assume a factor of safety of 2.0, and that cranes come in 2000-lb increments (i.e., 2000, 4000, 6000 lb, etc.)

1.6 A pressure vessel is designed to contain a working pressure of 45 MPa.

Per the ASME Pressure Vessel Code, determine the minimum pressure required for a proof test.

1.7 The nominal strength of a production run of steel bars is 50 kN.

(a) If the working load is 20 kN, determine the factor of safety. (b) Your boss asks you to perform proof tests on a number of the bars at 60% above the working load. What fraction of the nominal strength is the proof load?

1.8 A crane slowly lifts a 30 ft length of 12 in. diameter standard pipe. Lift points D and E are 8.0 ft apart. A 12 in. standard steel pipe has an inside diameter of 12.0 in., a wall thickness of 0.375 in., and weighs 49.56 lb/ft (see *Appendix D*).

(a) If the main strap AB that supports the pipe can support 10,000 lb, determine the factor of safety for this case. (b) Determine the weight density (lb/ft^3) of the pipe material.

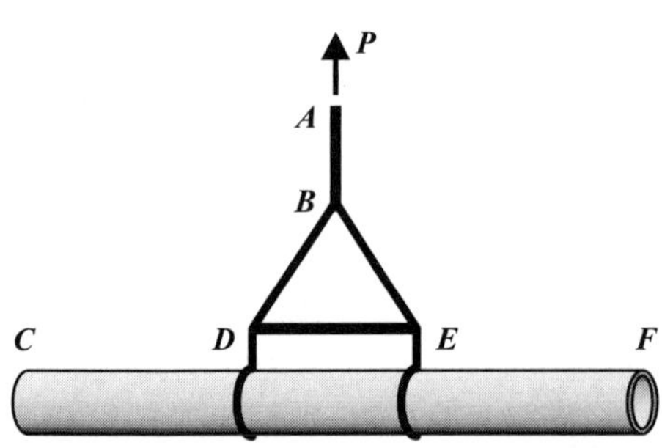

1.9 The plan (top) view of the second story of an *office building* is shown. Each support column (lettered) is represented by a shaded square and the columns are arrayed in a rectangular grid pattern at varying distances. The lines between columns represent primary floor beams.

The load on each column can be approximated using a basic load–distribution model. The *tributary area* of each column – the area contributing to its load – is the area the column supports. Each column supports one-fourth of the rectangular area bound by it and the three other columns that form the rectangle. The total load on a single column is the product of the area load (psf) and the column's total tributary area (ft^2).

Floor loads for some building-types are given in *Table 1.1* of the text.

Determine (a) the load (force) on columns C, G, and I, respectively, (b) the column(s) with the least load, and the value of the load, and (c) the column(s) with the greatest load, and the value of the load. (d) If this is a plan view of the second story of a *library stack room*, determine the load on column G.

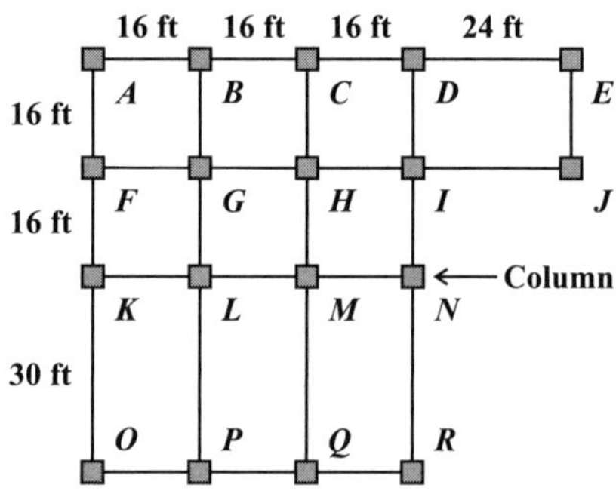

1.10 *Traditional wind-load analysis.* A steel frame building is constructed with four parallel *arches* or *frames*, A–D, separated by purlins (*Figure (a)*). The frames support the building *envelope* (the walls and roof). The span is $L = 80$ ft, the wall height is $H = 12$ ft, the arch spacing is $B = 15$ ft, and the half-roof length is $R = 50$ ft. The building is subjected to wind normal to the ridge line, with a velocity of $V = 70$ mph.

The equivalent *pressure* on any surface (wall or roof) due to the wind is given by:

$$p_i = C_q(0.00256V^2)$$

where p_i is in pounds per square foot (psf) and V is in mph. The variable C_q depends on the building surface, and is defined for a particular case by *Figure (b)*. A positive C_q means the pressure acts against the surface and a negative value means the pressure acts away from the surface (suction).

Determine (a) the total force on the entire left (windward) wall of the structure and (b) the total load on the right-half (leeward) roof. (c) Determine the distributed line load w (lb/ft) that acts on *each segment* of *arch C* – the windward wall, the windward roof, the leeward roof and the leeward wall. Draw the distributed load pattern with values and units on a 2D sketch of *arch C* (e.g., *Figure (c)*). Draw the distributed loads proportional to their values.

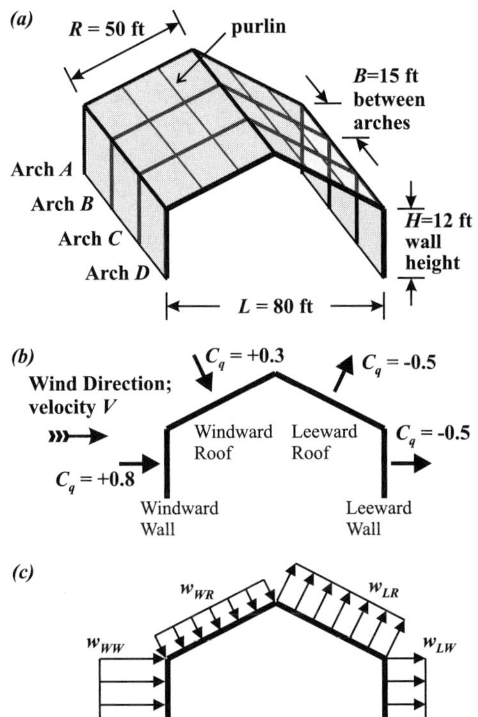

Hint: The load on each arch segment (wall or roof) can be estimated using a basic load–distribution model. The *tributary area* on each arch segment – the area contributing to its load – is half the wall or roof area that the arch segment shares with the neighboring arch(es).

The product of the appropriate pressure and the segment's tributary area is the total load (force) on each segment (e.g., the load on the windward roof of arch C). This load acts uniformly over the length of the segment.

1.11 A 16.0 kN truck (including its load) is stopped on a bridge. The bridge is $L = 14.0$ m long and the truck's wheelbase is $s = 4.0$ m. Assume that the weight of the truck is evenly distributed between the front and the rear axles. The rear of the truck is $a = 4.0$ m from the left end of the bridge.

Model the bridge: Create a free body diagram including the reactions at the supports Do not solve for the reactions.

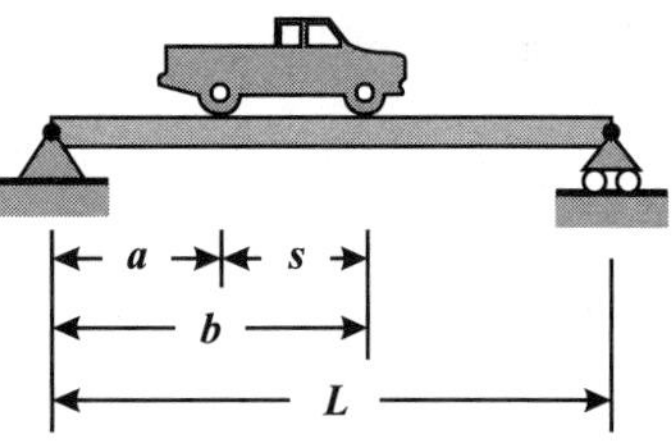

1.12 A 2000 lb weight hangs from a 14 ft long beam. The beam is supported by a roller at the left end, and is built-in to the wall at the right end.

Model the beam: Create a free body diagram including the reactions at the supports Do not to solve for the reactions.

Note: This is an indeterminate system; it cannot be solved using statics alone.

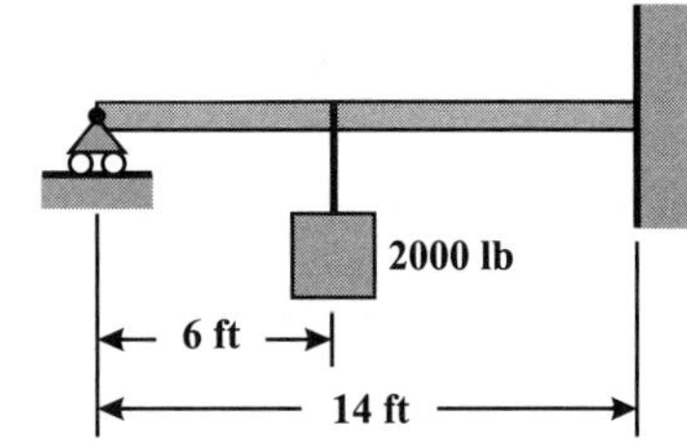

Problems: **Chapter 2 Statics**

2.1 Axial Members

2.1 A cantilever truss supports a point load at joint *D*.

Calculate the forces in members *CD*, *BC*, *BE*, and *FE*. Assume all member forces are in tension so that a positive value indicates *tension* (T) and a negative value indicates *compression* (C).

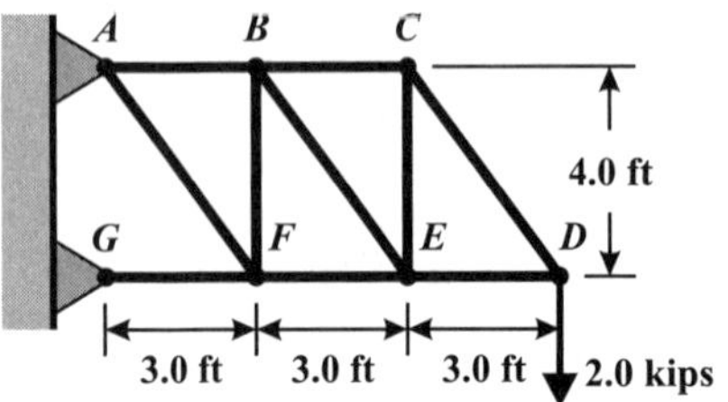

2.2 All of the members of simply supported truss *AE* are *L* = 3.0 m long. Forces *H* = 10.0 kN and *V* = 20.0 kN are applied as shown.

Calculate the force in members *AB*, *BC*, *IC*, and *IG*. Assume all member forces are in tension so that a positive value indicates *tension* (T) and a negative value indicates *compression* (C).

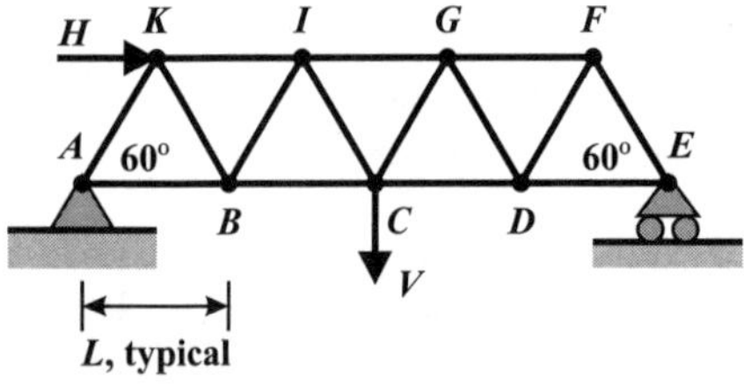

2.2 Torsion Members

2.3 A set of gears transfers torque from one shaft to another.

If the radius of the large gear is twice the radius of the small gear, determine the ratio of the torques, T_2 to T_1.

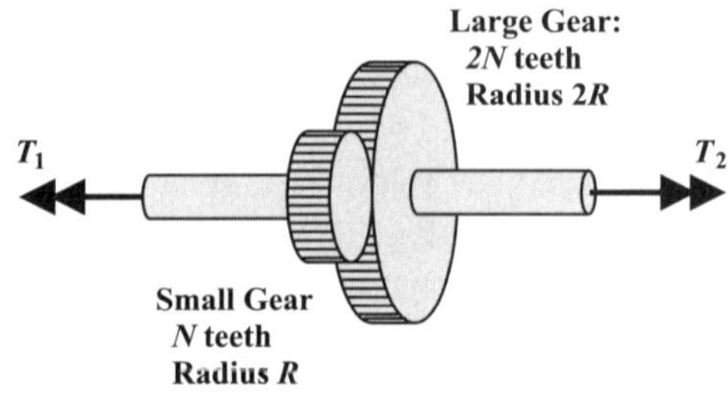

2.4 A shaft has four gears, *A*, *B*, *C*, and *D* (shown without gear teeth). Gears *A* through *C* are subjected to torques T_A, T_B, and T_C, acting as shown.

(a) If the system is in equilibrium, determine the torque on gear *D*, T_D. Write the answer as positive or negative with respect to the curved arrow shown. (b) Determine the torque supported inside the shaft between gears *A* and *B*, T_{AB}, between *B* and *C*, T_{BC}, and between *C* and *D*, T_{CD}.

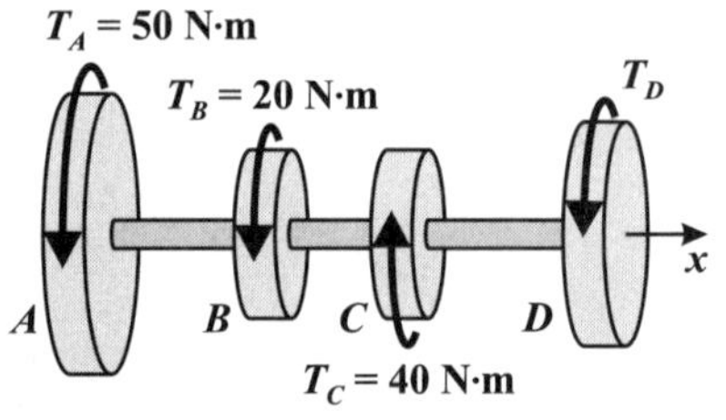

2.3 Beams

2.5 The face of a highway sign, 10 ft wide by 6.0 ft tall, is acted on by a uniform wind pressure of 10 psf (perpendicular to the sign). The center of the sign is *H* = 8.0 ft above the ground. Two symmetrically placed posts support the sign. The posts are positioned in such a way that the wind load does not cause them to twist.

Calculate the magnitude of the shear force and moment reactions on each post at the ground (due to symmetry of the geometry and of the load, the reaction of the ground on both posts is the same).

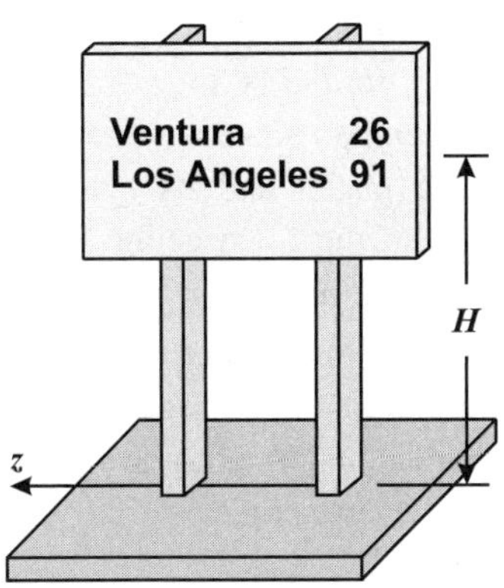

2.6 A 16.0 kN truck (including its load) slowly crosses a bridge that is modeled as a simply supported beam. The bridge is $L = 14.0$ m long and the truck's wheelbase is $s = 4.0$ m. Assume that the weight of the truck is evenly distributed between the front and the rear axles.

Draw the shear and moment diagrams for the beam, and determine the maximum moment (a) when the rear axle just reaches the bridge (left support) and (b) when the rear axle is $a = 5.0$ m from the left support (i.e., the truck is at the center).

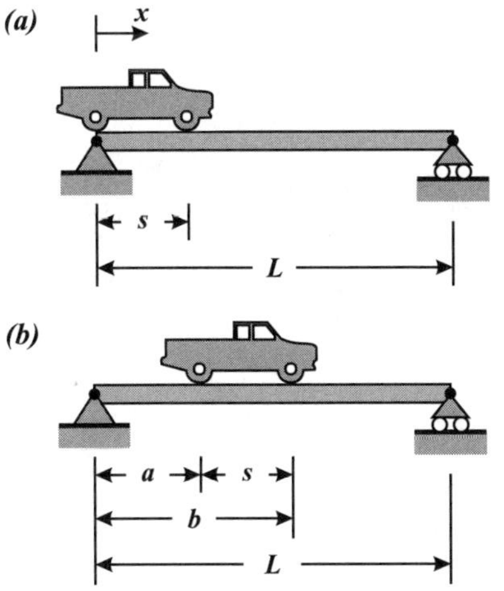

2.7 Consider the truck and bridge of *Prob. 2.6*.

Determine (a) the value of a that gives the largest moment as the truck crosses the bridge and (b) magnitude of that moment.

Hint: It is not when the truck is at the center of the beam.

2.8 A crane slowly lifts a 30 ft length of 12 in. diameter standard pipe that weighs 49.56 lb/ft. Lift points D and E are 8.0 ft apart.

(a) Draw the shear and moment diagram of the pipe. (b) Determine the maximum bending moment in the pipe.

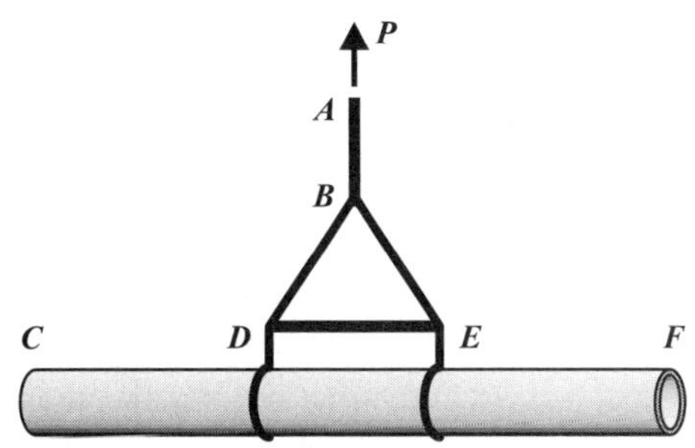

2.9 You are designing a deck for a house outside of Chicago, Illinois. The deck is $c = 12$ ft wide by $a = 6.0$ ft deep, and is supported by three beams as shown. The ground snow load around Chicago is 30 psf.

Determine (a) the total snow load (force) on the deck assuming the deck load is the same as the ground load, (b) the uniformly distributed load, w, on each beam, Beams 1, 2, and 3 (*hint*: what fraction of the deck does each beam support, assuming the deck remains level?), and (c) the magnitude of the maximum bending moment in the center beam (Beam 2).

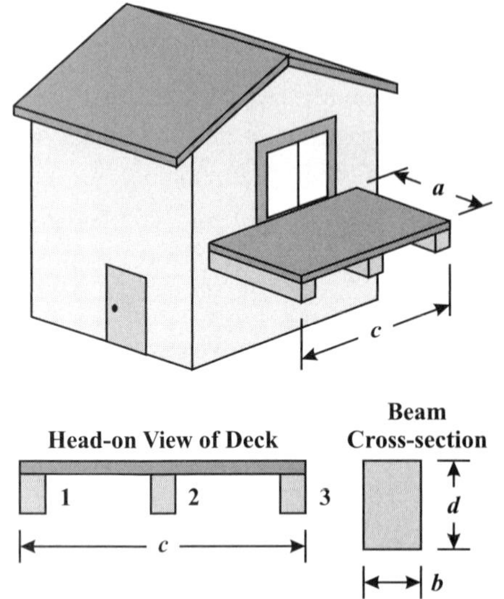

Head-on View of Deck

Beam Cross-section

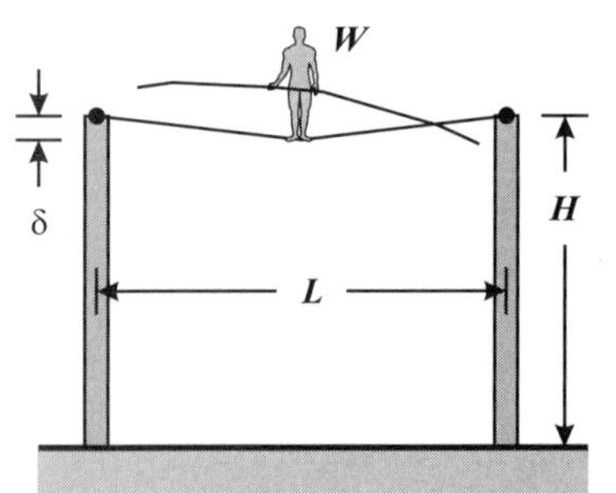

2.4 Combined Loading

2.10 A tight-rope walker weighs $W = 160$ lb. He walks on a rope $H = 20.0$ ft above the ground. The rope spans two poles that are $L = 30.0$ ft apart. When he reaches the center, the rope is taut and has deflected $\delta = 1.0$ ft.

Determine the horizontal and vertical reaction forces and the reaction moment at the base of the left pole.

2.11 The *piston, connecting rod* (*AB*) and *crank* (*BC*) of an engine system are pinned together as shown. As the piston moves left-and-right, the crank rotates and provides torque T about an axis (out of the paper) at shaft C (the resisting torque is shown). Rod AB is a two-force member. Assume there is no friction between the piston and its cylinder. At the instant shown, the force acting on the piston from the gas pressure is $F = 3.0$ kips.

Determine (a) the force in rod AB, and (b) the magnitude of the torque about the C-axis. (c) Determine the normal force on the cylinder. Is it upward or downward?

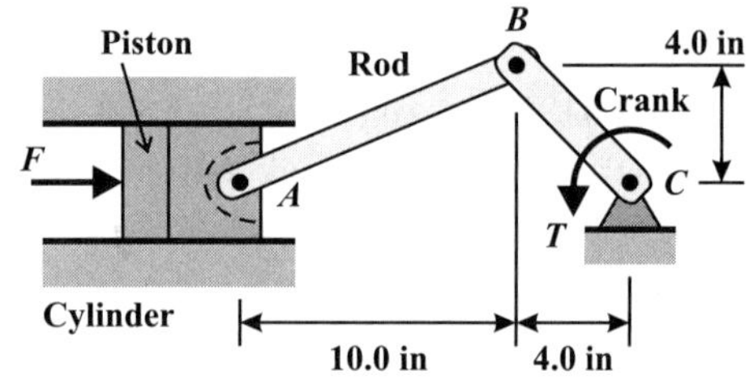

2.12 A boom is subjected to downward force F. The load and geometry are: $F = 400$ N, $a = 8.0$ m, $b = 2.0$ m, $D = 100$ mm.

Determine the reactions at cross-section ABC.

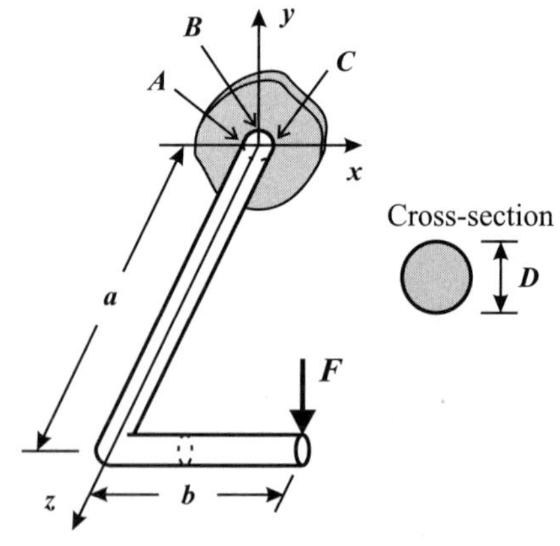

Problems: Chapter 3 Strain and Stress

3.1 Axial Properties

3.1 A steel bar is 8.00 ft long. When a 4000 lb tensile load is applied to the bar, it elongates 0.100 in. Steel has a Young's modulus of E = 30,000 ksi (kilopounds per square inch).

Determine (a) the axial strain in the bar, (b) the axial stress in the bar, and (c) the cross-sectional area of the bar. (d) If the yield strength is S_y = 60.0 ksi, determine the factor of safety.

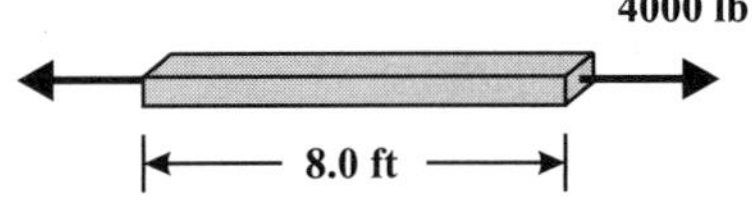

3.2 Truss *ABC* is subjected to a downward force of 20.0 kN at joint *B*. The truss is made of 25.0 mm diameter solid aluminum rods, *AB* and *BC*. *BC* is 1.00 m long. For aluminum, Young's modulus is 70 GPa.

Determine (a) the stress in each bar σ_{AB} and σ_{BC} and (b) the elongation of each bar Δ_{AB} and Δ_{BC}.

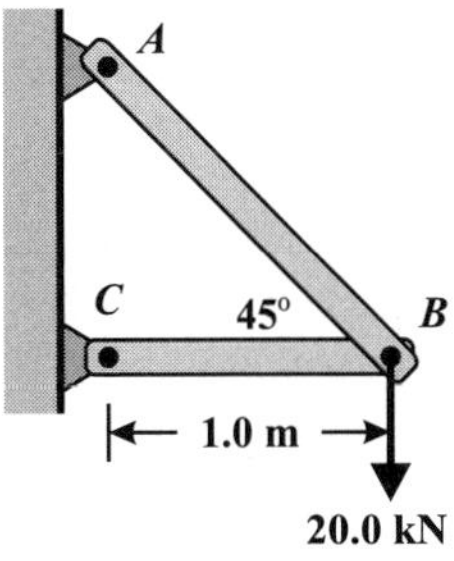

3.3 The *piston, connecting rod (AB),* and *crank (BC)* of an engine system are pinned together as shown. As the piston moves left-and-right, the crank rotates and provides torque T about an axis at *C* (the resisting torque is shown). Rod *AB* is a two-force member and has a cross-sectional area of A = 0.8 in.^{2}. Assume there is no friction between the piston and its cylinder. At the instant shown, the force acting on the piston from the gas pressure is: F = 3.0 kips.

For the instant shown, determine the axial stress in rod *AB*.

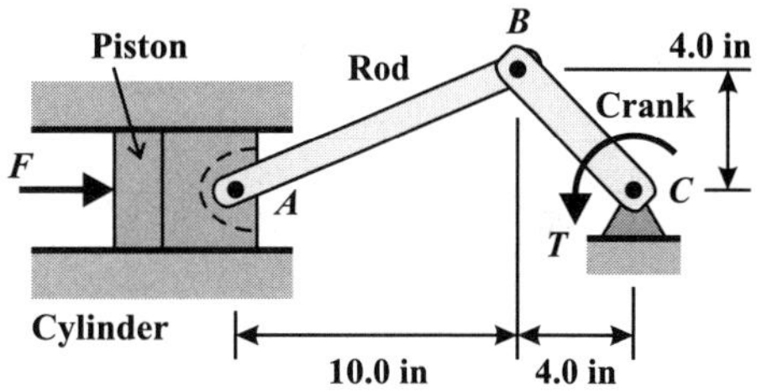

3.4 The stress–strain curve for a metal bar subjected to a tension test is shown below: the entire curve, and a detailed curve for $0 < \varepsilon < 0.010$.

Determine (a) Young's modulus E, (b) the yield strength S_y using the 0.2%-offset approximation, (c) the ultimate strength of the material S_u and the strain at which it occurs, and (d) the failure strain, ε_f. (e) Calculate the resilience (the maximum elastic strain energy density). (f) Estimate the total area under the curve, which is the *toughness* of the material (the total area is the energy density required to break the material into two).

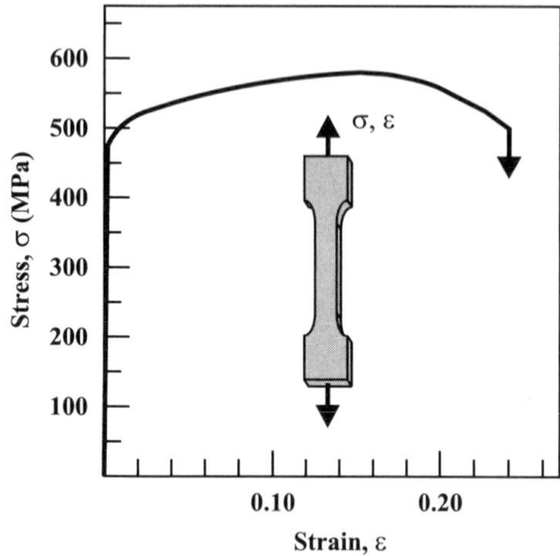

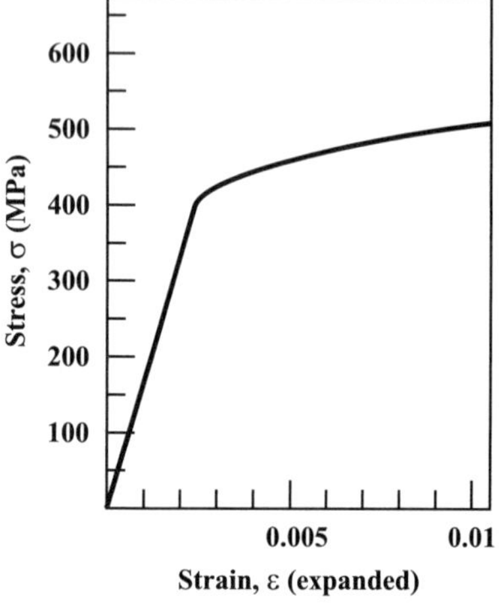

Strain, ε (expanded)

Excel Hint: use a "X–Y (Scatter) Plot". Do not use a "Line Plot." Line Plots evenly distribute the data along the *x*-axis as if the data were categories (e.g., months of the year). The given stress–strain data are not evenly spaced.

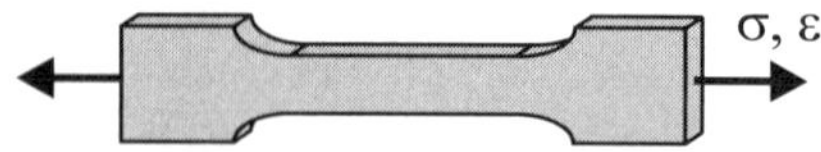

Stress (MPa)	Strain	Stress (MPa)	Strain
0	0.0	250	0.009
30	0.0001	300	0.025
70	0.0003	340	0.05
90	0.0004	380	0.09
125	0.0006	435	0.15
165	0.0008	450	0.25
200	0.0010	440	0.30
245	0.0014	400	0.36
255	0.0025	325	0.40
250	0.0050	0	0.40

3.5 A steel bar is tested in tension, resulting in the stress–strain data given in the table.

(a) Using Excel (or another graphing program), or using *engineering/graph paper* (draw to scale), plot the stress–strain curve. Fit a smooth curve through the points. Two plots are necessary: one to focus on the initial part of the curve (e.g., $0 < \varepsilon < 0.01$) and one for the entire range of strains (see *Prob. 3.4*).

From the curves, determine (b) the proportional limit, (c) the modulus of elasticity, (d) the yield strength using the 0.2% offset approach, and (e) the ultimate tensile strength.

Note: Many steels exhibit a "shelf" in the stress–strain curve due to a mechanism beyond the scope of this course.

3.6 From a tension test on a metal, it is reported that the modulus is $E = 15 \times 10^3$ ksi, the proportional limit is $S_p = 45$ ksi, the yield strength is $S_y = 55$ ksi, the tensile strength is $S_u = 70$ ksi, and the failure strain is $\varepsilon_f = 34\%$. However, no stress–strain curve was included in the technical memorandum of the experiment.

(a) On engineering/graph paper, sketch, to scale, a possible stress–strain curve. Label pertinent points/values (E, S_y, S_u, etc.).
(b) Estimate the resilience of the material.

3.7 An aluminum bar is subjected to an axial force of $P = 10.0$ kN. The bar has a solid circular cross-section with diameter $D = 10$ mm and length $L = 0.80$ m. The modulus is $E = 70$ GPa, the yield strength is $S_y = 240$ MPa and Poisson's ratio is $\nu = 0.33$.

Determine (a) the stress in the bar, (b) the factor of safety against yielding, (c) the change in length of the bar, and (d) the new diameter of the bar.

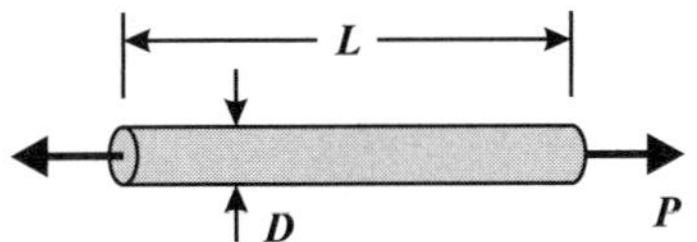

3.8 The aluminum bar of *Prob. 3.7* is subjected to an axial force of $P = -14.0$ kN. The compressive yield strength is $S_y = -240$ MPa.

Determine (a) the stress in the bar, (b) the factor of safety against yielding, (c) the change in length of the bar, and (d) the new diameter of the bar.

3.9 An aluminum bar of length $L = 36.0$ in. and diameter $D = 1.400$ in. is subjected to a tensile force of $P = 24.0$ kips. Aluminum has a modulus of $E = 10,000$ ksi and a Poisson's ratio of $v = 0.33$.

Determine (a) the change in diameter of the bar and (b) the change in volume of the bar (answer to two significant digits).

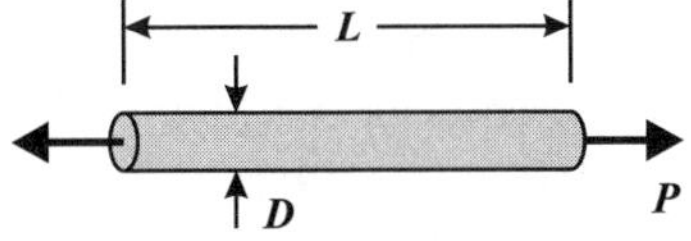

3.10 The modulus of resilience U_R of a material is the strain energy density that it can store without undergoing plastic deformation (i.e., the maximum elastic strain energy).

(a) Determine the resilience of the steel, aluminum, nickel, and titanium alloys listed below. (b) For the same size structure, which material can absorb the most energy without plastically deforming?

Alloy	E (GPa)	S_y (MPa)
Steel	200	250
Aluminum	70	240
Nickel	210	600
Titanium	115	800

3.11 Structural steel has yield strength $S_y = 250$ MPa and modulus $E = 200$ GPa. A steel bar with a volume of 0.002 m^3 is loaded slowly in tension.

Determine (a) the modulus of resilience of the steel and (b) the total energy (N·m) required to yield the component.

3.12 When an aluminum bar of constant cross-sectional area is subjected to tensile force $P = 20.0$ kN, it elongates $\Delta = 3.0$ mm. The volume of the bar is 0.00032 m^3. The material properties are $E = 70$ GPa and $S_y = 240$ MPa.

(a) Assuming the bar responds linearly, use energy concepts to determine the stress in the bar. (b) What is the maximum tensile force P_{max} such that the system remains elastic.

3.13 A steel bar of diameter of $D = 20$ mm is loaded in fatigue, with maximum and minimum loads $P_{max} = 60$ kN and $P_{min} = -60$ kN; the load is sinusoidal with zero mean stress. The steel's S–N curve is shown below and its yield strength is 480 MPa.

(a) Estimate the *fatigue life, N_f*, of the bar for the given loading. (b) What fraction of the yield strength is the cyclic stress amplitude? (c) Estimate the maximum amplitude of the cyclic force so that the specimen has *infinite fatigue life* (i.e., so that it does not fail by fatigue).

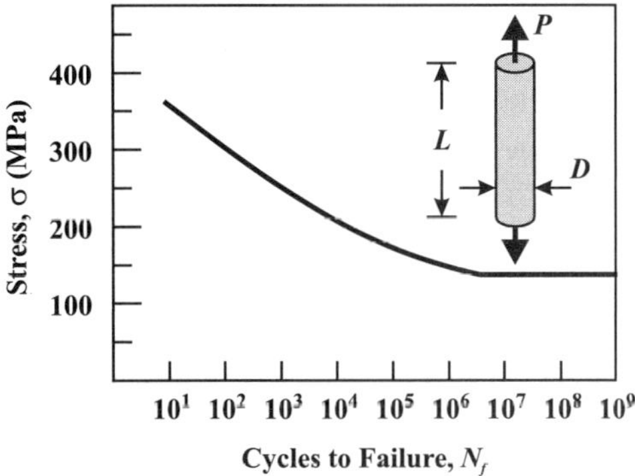

3.14 Use the internet to connect to: http://fatiguecalculator.com/cgi-bin/StressShowMatProp.pl

(a) Using data for steel 1040, determine the fatigue strength for $N_f = 10^6$ cycles with zero mean stress. (b) If the mean stress is $\sigma_m = 100$ MPa, determine the amplitude of the new cyclic stress that gives the same fatigue life.

3.15 For aluminum 7075-T651, $S_u = 580$ MPa, $S_f' = 1131$ MPa, and $b = -0.122$.

(a) Determine the fatigue strength for $N_f = 10^7$ cycles with zero mean stress. (b) If the mean stress is $\sigma_m = 100$ MPa, determine the amplitude of the new cyclic stress that gives the same fatigue life.

3.2   Shear Properties

3.16 A *thin-walled* steel shaft of average diameter $D = 50$ mm and thickness $t = 3.0$ mm is subjected to torque $T = 350$ N·m. From a tension test, the axial yield strength is $S_y = 320$ MPa.

(a) Determine the average shear stress in the shaft τ. (b) If the shear modulus is $G = 75$ GPa, determine the shear strain. (c) Determine the maximum torque so that the system remains linear–elastic.

3.17 A rubber pad is sandwiched between two steel plates subjected to shear force $V = 100{,}000$ lb. The dimensions of the plate are $a = 9.0$ in. and $b = 12.0$ in. The thickness of the rubber is $c = 4.50$ in. After the force is applied, the top plate is found to have displaced laterally by $\delta = 0.055$ in.

Determine the shear modulus G of the rubber.

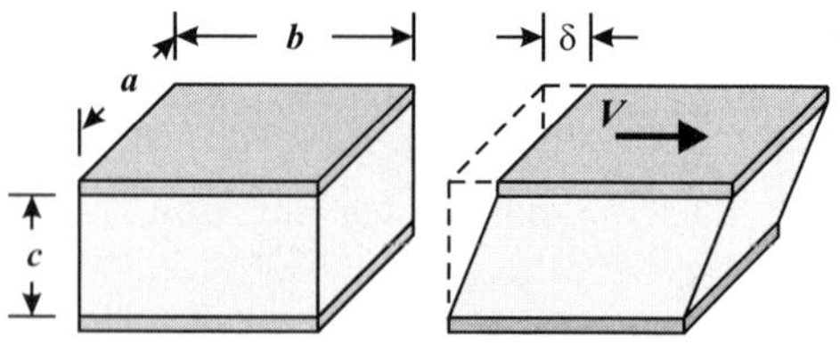

3.18 A thin-walled steel shaft is 20 in. long, has an average radius of $R = 1.20$ in. and a thickness of 0.05 in. The shear modulus is $G = 12{,}000$ ksi and the shear yield strength is $\tau_y = 30$ ksi.

(a) Determine the torque that will cause the thin-walled shaft to yield. (b) If the shaft is limited to an angle of twist of $1.0°$, determine the maximum value of the torque. (c) Which condition limits the design? Explain.

3.19 A thin-walled tube is to be used for two different applications: (1) tension-only and (2) torsion-only. The yield strengths of the material are $S_y = 250$ MPa and $\tau_y = 160$ MPa. The average radius of the tube is $R = 30$ mm and the thickness is $t = 4.0$ mm.

(a) When loaded by an axial force-only, determine the tensile force required to yield the tube, P_y. (b) For the axial load-only case, if the factor of safety against yielding is FS = 2.0, determine the maximum allowable force that may be applied to the tube, P_{allow}. (c) When the tube is loaded by a torque-only, determine the torque required to yield the tube, T_y. (d) For the torsion load-only case, if the factor of safety is FS = 2.0, determine the maximum allowable torque that may be applied to the tube, T_{allow}.

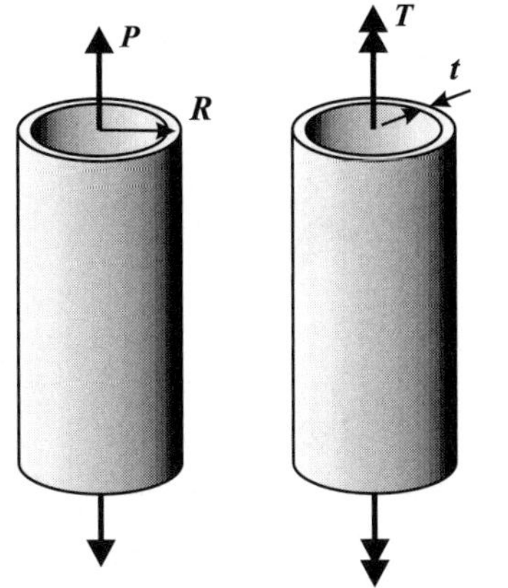

3.3   *General Stress and Strain*

3.20 A 2D state of stress, σ_x, σ_y, and τ_{xy}, exists at a point in an aluminum plate. The state of stress is *plane stress* – the *out-of-plane* stresses are zero. For aluminum, $E = 70$ GPa, $G = 28$ GPa, and $v = 0.33$.

Determine the normal strains in the *x*-, *y*-, and *z*-directions, ε_x, ε_y and ε_z, and the shear strain γ_{xy} for the following stress states (σ_x, σ_y, τ_{xy}):

(a) (200 MPa, 0 MPa, 75 MPa)

(b) (200 MPa, 100 MPa, 0 MPa)

(c) (200 MPa, 100 MPa, 75 MPa)

(d) (200 MPa, 200 MPa, 75 MPa)

(e) (100 MPa, –150 MPa, 100 MPa)

(f) (–100 MPa, 200 MPa, –85 MPa)

(g) (–200 MPa, –100 MPa, 100 MPa)

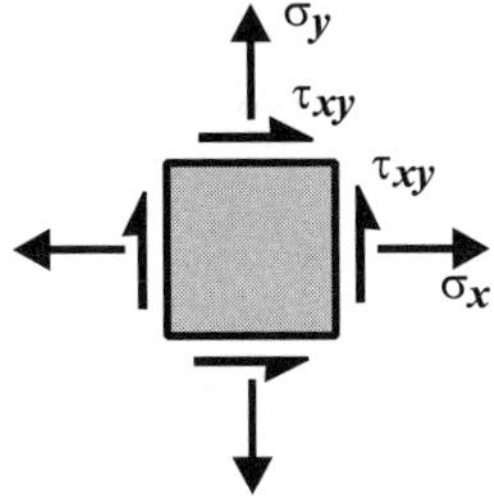

3.21 A tri-axial state of stress, σ_x, σ_y, and σ_z, exists in a steel machine part. For steel, $E = 30{,}000$ ksi and $v = 0.3$.

Determine the normal strains in the *x*-, *y*-, and *z*-directions, ε_x, ε_y, and ε_z for the following stress states:

(a) $\sigma_x = 15.0$ ksi, $\sigma_y = 5.00$ ksi, $\sigma_z = 10.0$ ksi

(b) $\sigma_x = -15.0$ ksi, $\sigma_y = 10.0$ ksi, $\sigma_z = 5.00$ ksi

(c) $\sigma_x = 15.0$ ksi, $\sigma_y = 15.0$ ksi, $\sigma_z = 15.0$ ksi

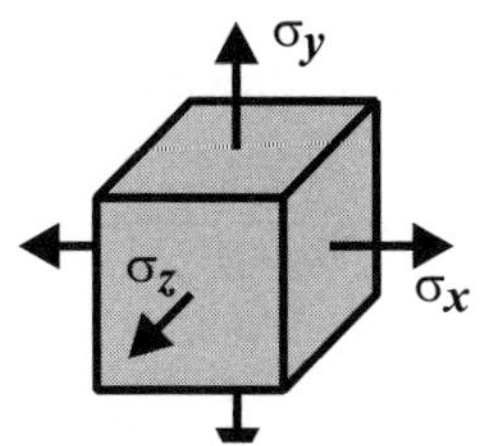

3.22 A rubber has modulus $E = 200$ MPa and Poisson's ratio $v = 0.5$. A uniaxial state of stress, $\sigma_x = 4.0$ MPa, exists in a part made of this rubber.

(a) Determine the normal strains in the *x*-, *y*-, and *z*-directions, ε_x, ε_y, and ε_z. (b) The new volume of a cube of material with strains ε_x, ε_y, and ε_z can be expressed as:

$$V_{new} = V(1 + \varepsilon_x)(1 + \varepsilon_y)(1 + \varepsilon_z)$$

where V is the original volume. Using the strains calculated in *Part (a)*, what can you conclude about the change in volume of rubber (or any material with $v = 0.5$) when subjected to uniaxial stress? Is your statement also true for a biaxial state of stress? for a triaxial state of stress? Explain.

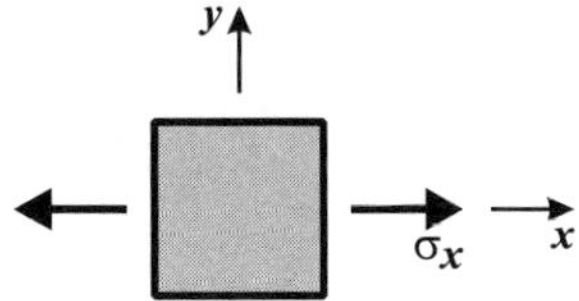

Problems: **Chapter 4 Axial Members and Pressure Vessels**

4.1   Axial Members – Force Method

4.1 An aluminum bar is subjected to a compressive force of 10.0 kN. Its rectangular cross-section is 10.0 mm by 25.0 mm, and it is 1.20 m long. Aluminum has a modulus of $E = 70$ GPa.

Determine (a) the stress in the bar, (b) the axial strain, and (c) the elongation of the bar.

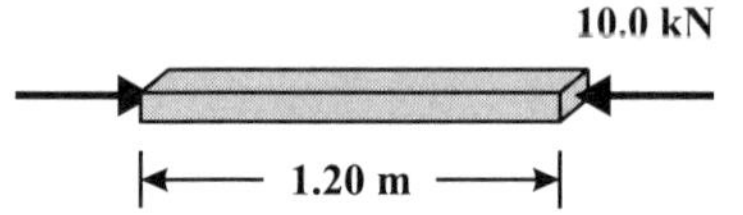

4.2 A steel bar is 8.00 ft long. When a 4000 lb tensile load is applied to the bar, it stretches 0.100 in.   Steel   has   a   modulus   of $E = 30,000$ ksi.

Determine (a) the axial strain in the bar and (b) the cross-sectional area of the bar. (c) If the yield strength of the steel is $S_y = 60.0$ ksi, what is the factor of safety for this load?

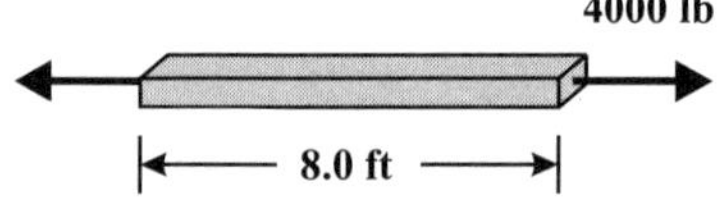

4.3 Truss *ABC* is subjected to a downward force of 20.0 kN. The truss is made of 25.0 mm diameter solid aluminum bars, *AB* and *BC*. Rod *BC* is 1.00 m long. Aluminum has a modulus of $E = 70$ GPa.

Determine (a) the stress in each bar σ_{AB} and σ_{BC} and (b) the elongation of each bar Δ_{AB} and Δ_{BC}.

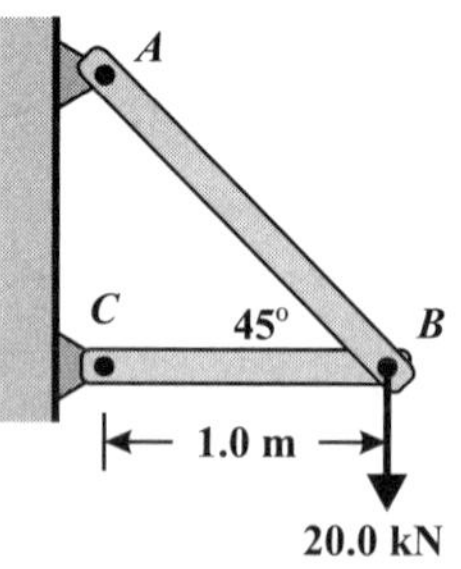

4.4 Truss *ABC* is subjected to a horizontal force of 20.0 kN. The truss is made of 25.0 mm diameter solid aluminum rods, *AB* and *BC*. Rod *BC* is 1.00 m long. Aluminum has a modulus of $E = 70$ GPa.

Determine (a) the stress in each bar σ_{AB} and σ_{BC} and (b) the elongation of each bar Δ_{AB} and Δ_{BC}.

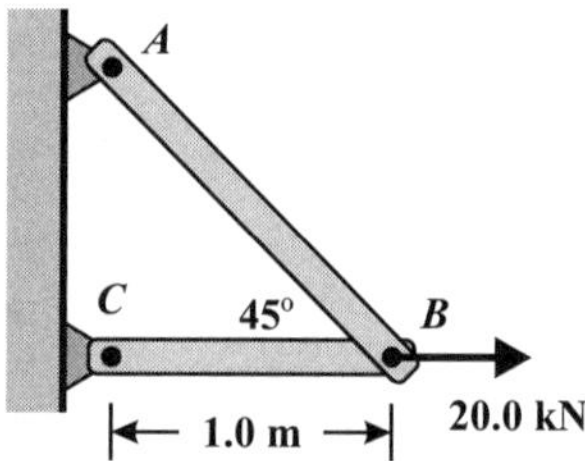

4.5 A set of steel rods, each of cross-sectional area *A*, is used to support two levels of a hanging walkway. A schematic of one of these rods, *ABC,* and its loading, is shown. The total load supported at the lower level is *2F* (point *C*) and at the upper level is *4F* (point *B*).

For $E = 30,000$ ksi, $L = 10.0$ ft, $A = 0.800$ in.^{2}, and $F = 1.20$ kips, determine (a) the stress in each segment of the rod, σ_{AB} and σ_{BC}, (b) the elongations of each segment, Δ_{AB} and Δ_{BC}, and (c) the displacements (movements) of points *B* and *C*, δ_B and δ_C.

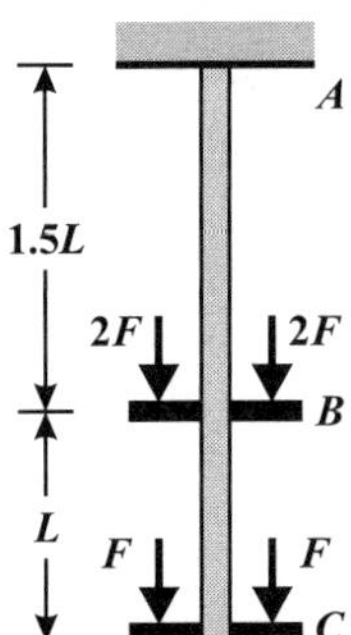

Determine (a) the stress in the lowest pipes due to the design load and (b) the total shortening of the tower due to the design load.

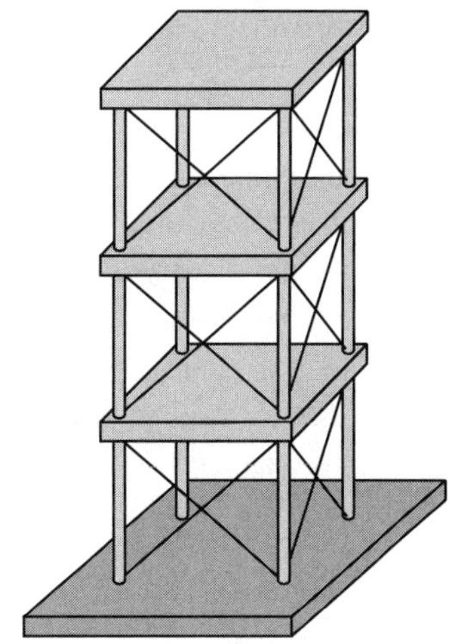

4.6 Solid stepped concrete column, *ABC*, supports downward load $F_1 = 2.0$ MN at its top. A second downward load F_2 acts uniformly on the step. The cross-sectional areas of the upper and lower portions are $A_{AB} = 0.08$ m^2 and $A_{BC} = 0.12$ m^2. The lengths are $L_{AB} = 1.5$ m and $L_{BC} = 2.0$ m. The modulus of concrete is $E = 20$ GPa. Assume there are no limitations due to strength.

Determine (a) the normal stress in the upper segment of the column, σ_{AB}, (b) the force F_2 so that the stresses are the same in each segment, i.e. $\sigma_{AB} = \sigma_{BC}$, and (c) the force F_2 if the strain in the lower segment is to be $\varepsilon_{BC} = -0.10\%$ (i.e., *BC* is in compression).

4.8 An aluminum bar with cross-sectional area 0.5 in.^{2} is subjected to loads $F_1 = 2.5$ kips, $F_2 = 5.5$ kips, and $F_3 = 2.0$ kips. The loads are applied in the directions shown. The dimensions of the bar are: $L_{AB} = 20$ in., $L_{BC} = 10$ in., and $L_{CD} = 15$ in. The modulus is $E = 10,000$ ksi.

Determine the total change in length of the bar.

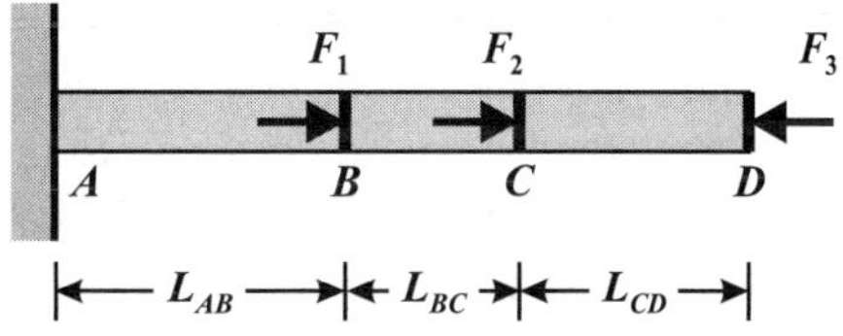

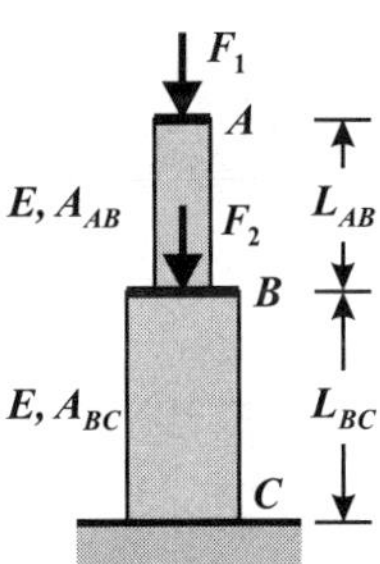

4.7 An observation tower has three rigid elevated floors. Each floor is 6.00 ft by 6.00 ft and the distance between floors is 8.00 ft. The floors are supported by four steel pipes ($E = 30,000$ ksi), each with outer diameter 3.00 in. and inner diameter 2.75 in. Each floor (including the topmost) is designed to carry a uniformly distributed load of 40.0 psf. Assume the pipes share the load of each floor equally, and that the tower does not bend or twist.

4.9 Stepped titanium bar *ABC* having two different cross-sectional areas, $A_{AB} = 1.0$ in.^{2} and $A_{BC} = 2.0$ in.^{2}, is held between two rigid supports. The segments have lengths $L_{AB} = 8.0$ in. and $L_{BC} = 12.0$ in. Load $F = 15$ kips is applied at joint *B*. The modulus of titanium is $E = 15,000$ ksi.

Determine (a) the reaction forces at point *A* and point *C* and (b) the movement δ of point *B* due to the applied force.

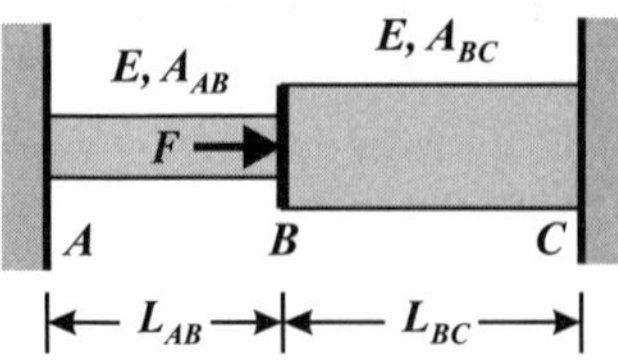

4.10 A concrete column has a square cross-sectional area 16 by 16 in. and is 10.0 ft tall. Concrete has a weight density of $\gamma = 150$ lb/ft^3 and a modulus of $E = 4000$ ksi.

(a) Determine the change in length of the column due to its own weight. (b) Repeat the problem for a column that is 24 by 24 in. in cross-section.

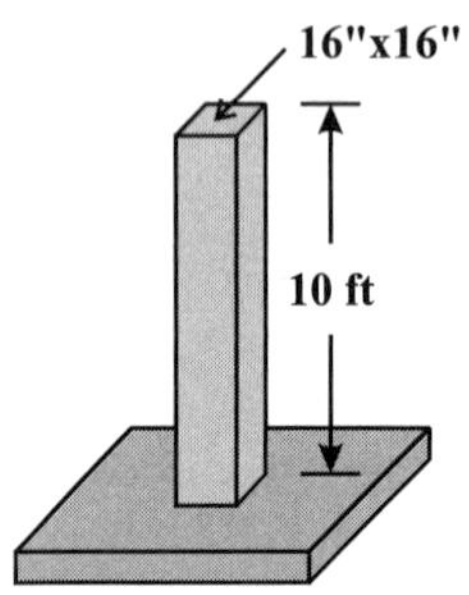

4.11 A wooden pile, driven into the ground, supports a load F. The load is transferred to the surrounding earth entirely by friction along the side of the pile. The interfacial shear stress, τ, between the pile and earth, is assumed to be uniform (constant) over the surface of the pile. The pile has length L, diameter D, cross-sectional area A, and modulus E.

(a) Derive a formula for the change in length of the pile, Δ, in terms of F, L, A, and E. Do not include τ in the solution (it can be substituted out).

Hint: Measure x from the bottom, where the internal force is $P(x=0) = 0$; review the nail pull-out problem in the text. (b) Plot the compressive stress, σ_c versus distance along the length of the pile. (c) Repeat *Part (a)* for a linear interfacial friction distribution: $\tau(x) = kx$ (where k is constant and x is measured from the bottom). Do not include τ or k in your final solution.

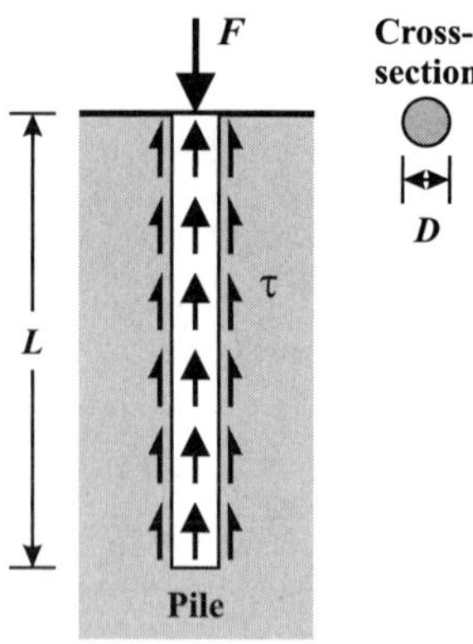

4.12 A tapered steel ($E = 30,000$ ksi) bar hangs from the ceiling and supports a $W = 1200$ lb weight. The bar has constant thickness $t = 0.25$ in. and tapers from $a = 4.00$ in. at the ceiling to $b = 2.00$ in. The bar is $L = 3.00$ ft long.

Determine (a) the stress in the bar as a function of distance from the load point (take the bottom of the bar to be $x = 0$) and (b) the total change in length of the bar.

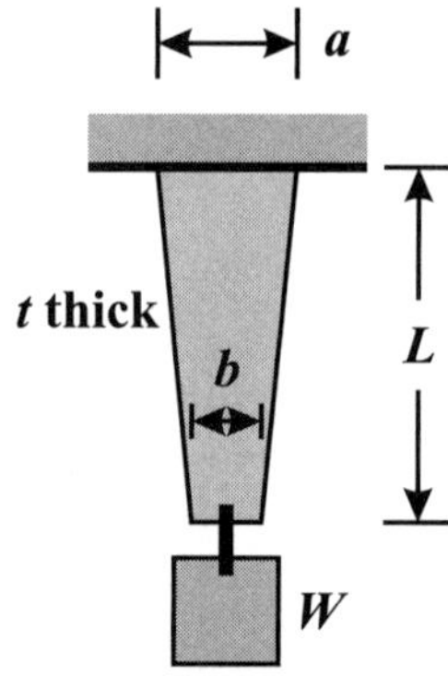

4.13 A concrete pile is poured into the shape of a truncated cone and supports compressive force of magnitude $F = 500$ kN. The diameter at the top is 0.40 m and at the bottom is 1.00 m. The pile is 0.60 m tall. Young's modulus is 28 GPa.

Determine (a) the stress $\sigma(x)$ in the pile as a function of distance from the top of the pile and (b) the total change in length of the pile Δ.

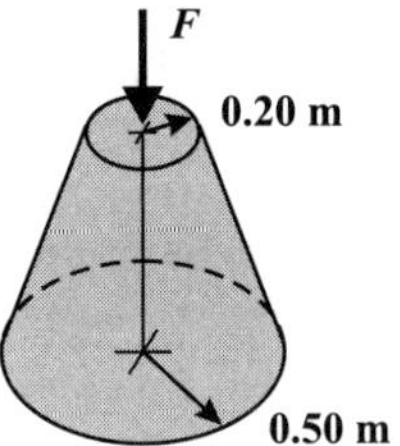

4.14 Truss *ABC* is subjected to a downward force of 20.0 kN at joint *B*. The truss is made of 25.0 mm diameter solid aluminum bars, *AB* and *BC*. *BC* is 1.00 m long. Aluminum has a modulus of $E = 70$ GPa.

Apply equilibrium and use compatibility to determine the deflection of joint *B* – its horizontal and vertical displacements, *u* and *v*.

Hint: Construct the extension of each bar individually, then draw perpendiculars to the end of each extension (this essentially rotates the bars).

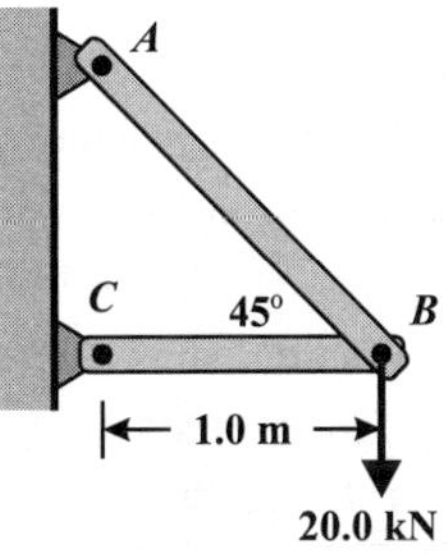

4.15 Truss *ABC* is subjected to a horizontal force of 20.0 kN at joint *B*. The truss is made of 25.0 mm diameter solid aluminum rods, *AB* and *BC*. *BC* is 1.00 m long. Aluminum has a modulus of $E = 70$ GPa.

Apply equilibrium and use compatibility to determine the deflection of joint *B* – its horizontal and vertical displacements, *u* and *v*.

Hint: Construct the extension of each bar individually, then draw perpendiculars to the end of each extension (this essentially rotates the bars).

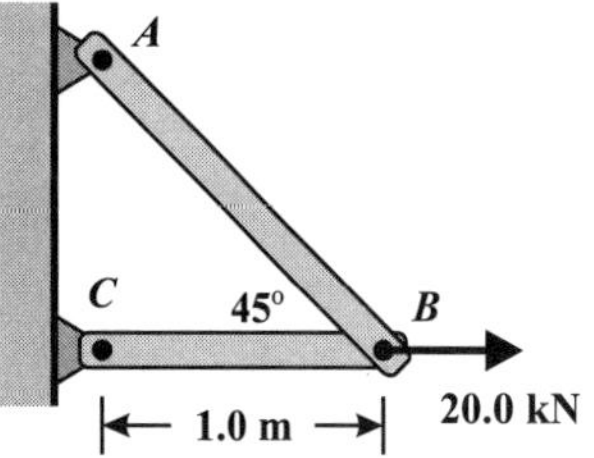

4.16 Truss *ABC* is subjected to downward force $F = 10.0$ kN at joint *B*. The truss is made of aluminum with $E = 70.0$ GPa and $S_y = 240$ MPa (in either tension or compression). The cross-sectional area of each bar is $A = 125 \times 10^{-6}$ m^2. Length $L = 1.00$ m.

Determine (a) the stress in each bar, σ_{AB} and σ_{BC}, (b) the factor of safety against yielding for bar *AB*, (c) the change in length of each bar, and (d) the downward displacement of joint *B*, *v*.

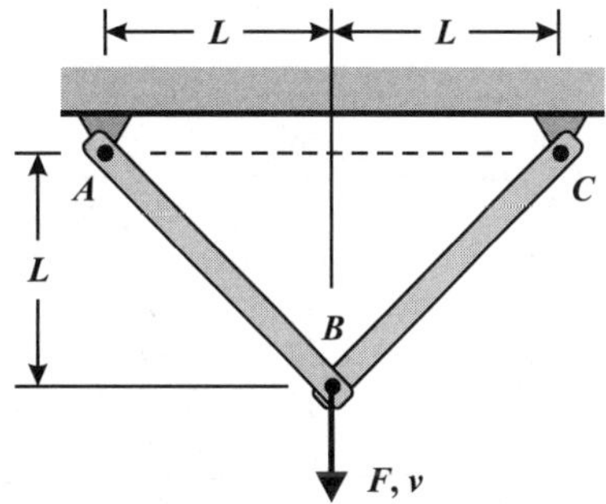

4.17 Truss *ABC* is subjected to horizontal force $F = 10.0$ kN at joint *B*. The truss is made of aluminum with $E = 70.0$ GPa and $S_y = 240$ MPa (in either tension or compression). The cross-sectional area of each bar is $A = 125 \times 10^{-6}$ m^2. Length $L = 1.00$ m.

Determine (a) the stress in each bar, σ_{AB} and σ_{BC}, (b) the factor of safety against yielding for bar *AB*, (c) the change in length of each bar, and (d) the horizontal and vertical displacements, *u* and *v*, of joint *B*.

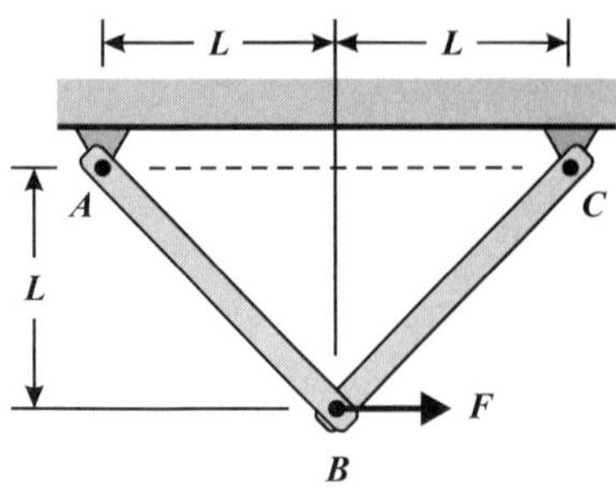

4.18 Truss *ABC* is subjected to downward force $F = 10.0$ kN at joint *B*. The truss is made of aluminum with $E = 70.0$ GPa and $S_y = 240$ MPa (in either tension or compression). The cross-sectional area of each bar is $A = 125 \times 10^{-6}$ m^2. Length $L = 1.00$ m.

Determine (a) the stress in each bar, σ_{AB} and σ_{BC}, (b) the Factor of Safety against yielding for bar *AB*, (c) the change in length of each bar, and (d) the downward displacement of joint *B*, *v*.

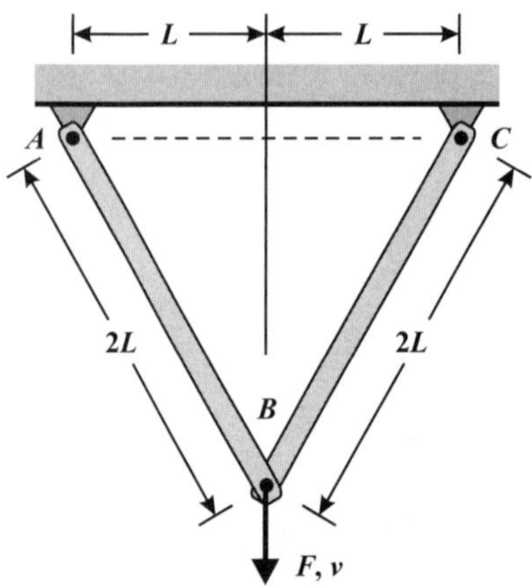

4.19 Truss *ABC* is subjected to horizontal force $F = 10.0$ kN at joint *B*. The truss is made of aluminum with $E = 70.0$ GPa and $S_y = 240$ MPa (in either tension or compression). The cross-sectional area of each bar is $A = 125 \times 10^{-6}$ m^2. Length $L = 1.00$ m.

Determine (a) the stress in each bar, σ_{AB} and σ_{BC}, (b) the factor of safety against

yielding for bar *AB*, (c) the change in length of each bar, and (d) the horizontal and vertical displacement of joint *B*, *u* and *v*.

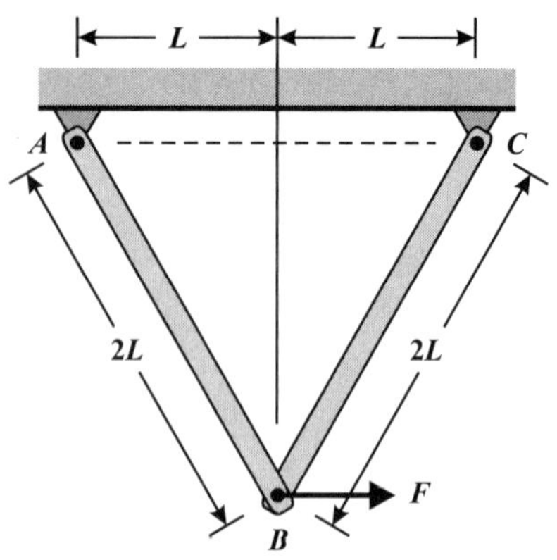

4.2   Axial Members – Displacement Method

4.20 Truss *ABCD* is subjected to downward force $F = 10.0$ kN at joint *B*. The truss is made of aluminum with $E = 70.0$ GPa and $S_y = 240$ MPa (in either tension or compression). The cross-sectional area of each bar is $A = 125 \times 10^{-6}$ m^2. Length $L = 1.00$ m.

Determine (a) the stress in each bar, σ_{AB}, σ_{BC} and σ_{BD}, (b) the factor of safety against yielding for bar *AB*, (c) the change in length of each bar, and (d) the downward displacement of joint *B*, *v*.

Hint: Due to symmetry, the horizontal movement *u* is zero.

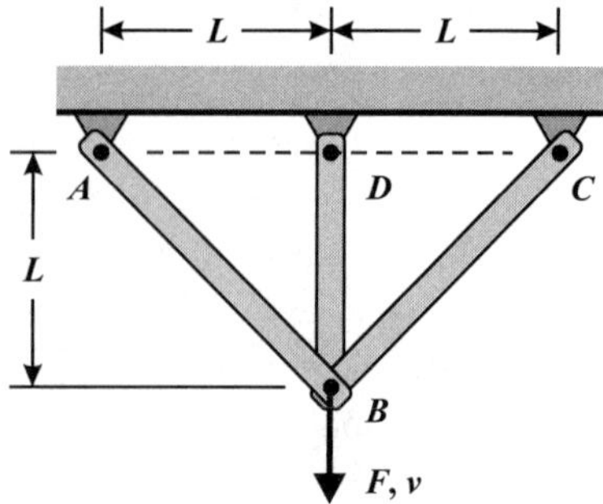

4.21 Truss *ABCD* is subjected to horizontal force $F = 10.0$ kN at joint *B*. The truss is made of aluminum with $E = 70.0$ GPa and $S_y = 240$ MPa (in either tension or compression). The cross-sectional area of each bar is $A = 125 \times 10^{-6}$ m^2. Length $L = 1.00$ m.

Determine (a) the stress in each bar, σ_{AB}, σ_{BC} and σ_{BD}, (b) the change in length of each bar, and (c) the horizontal and vertical displacement of joint *B*, *u* and *v*.

Hint: Apply displacement *u* only, and determine the elongations of, and the force in, each member as a function of *u*. Repeat for *v*. Add (superimpose) the individual solutions; note that the net force in the *y*-direction should be zero.

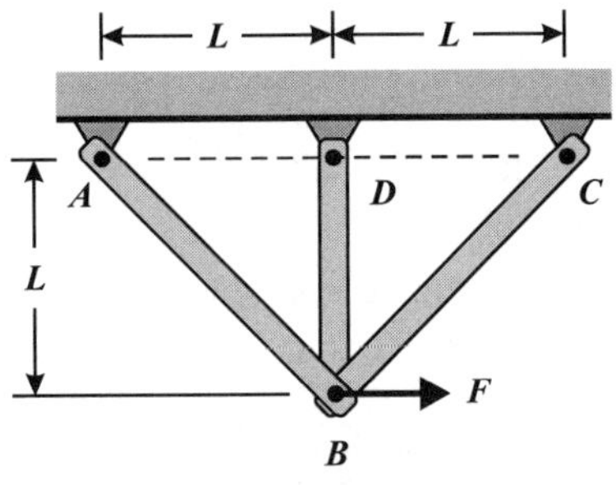

4.22 Truss *ABCD* is subjected to downward force $F = 10.0$ kN at joint *B*. The truss is made of aluminum with $E = 70.0$ GPa and $S_y = 240$ MPa (in either tension or compression). The cross-sectional area of each bar is $A = 125 \times 10^{-6}$ m^2. Length $L = 1.00$ m.

Determine (a) the stress in each bar, σ_{AB}, σ_{BC} and σ_{BD}, (b) the Factor of Safety against yielding for bar *AB*, (c) the change in length of each bar, and (d) the downward displacement of joint *B*, *v*.

Hint: Due to symmetry, the horizontal movement *u* is zero.

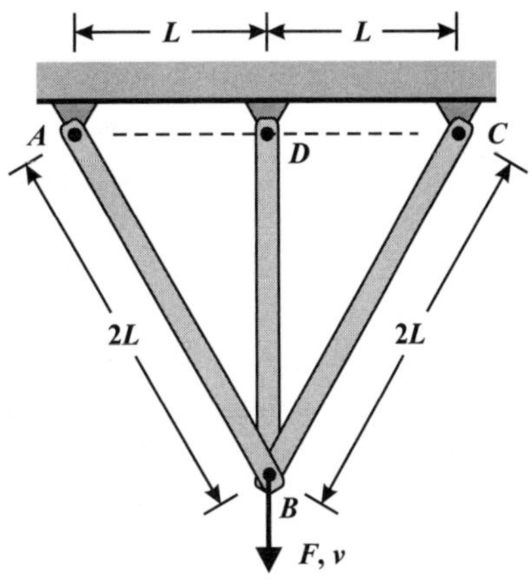

4.23 A steel reinforced concrete column is subjected to a load of $F = 48,000$ lb which is applied through a rigid plate at the top of the column. The steel reinforcements are six steel members that have a total area of $A_s = 6.0$ in.^{2}. The total cross-sectional area of the column is $A = 36$ in.^{2}. The material properties are:

Steel: $E_s = 30 \times 10^6$ psi

Concrete: $E_c = 4 \times 10^6$ psi.

Neglecting the Poisson effect, determine (a) the stress in the steel σ_s, (b) the stress in the concrete σ_c, (c) the change in length of the column, and (d) the effective Young's modulus of the system E_{eff}, i.e., the applied stress on the column (*F/A*) divided by its strain.

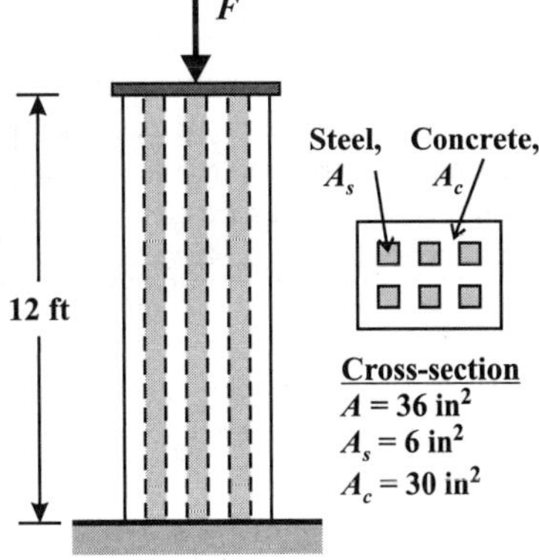

4.24 Composite materials are designed to take advantage of the properties of two different materials. A titanium matrix composite is reinforced with unidirectional silicon carbide (ceramic) fibers in the direction of the applied load (*Figure (a)*). These fibers are on the order of 100 μm in diameter. The total cross-sectional area of a composite strut is A, while the area of the fibers is A_f. The fiber area fraction is $f = A_f/A$ which is also the fiber volume fraction since a unidirectional composite is just an extrusion of the cross-section.

The modulus of titanium is 115 GPa and of silicon carbide is 360 GPa. The volume fraction of the fibers is $f = 0.35$. The applied stress on the composite is $\sigma = 400$ MPa in the direction of the fibers and the total cross-sectional area of the composite is 5.00 mm by 100 mm.

Neglecting the Poisson effect, determine (a) the stress in the titanium matrix, σ_m, (b) the stress in the ceramic fiber, σ_f, (c) the strain in the fiber and the matrix ε (they are the same), and (d) the Effective Young's modulus of the system E_{eff}, i.e., the stress applied to the composite divided by its strain.

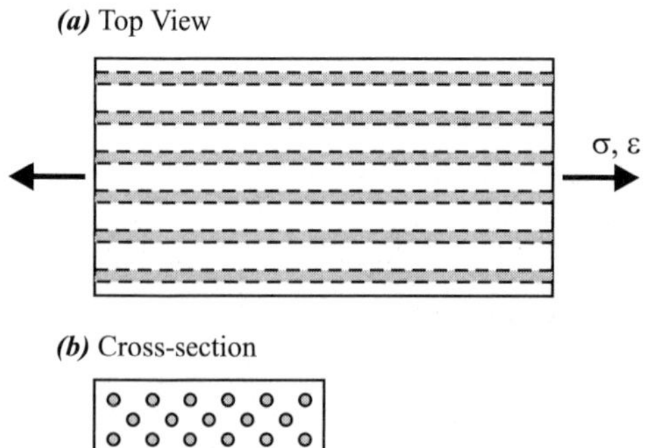

4.25 A 20 ft long rigid beam hangs level from two steel rods ($E = 30,000$ ksi). A $P = 10$ kip load is applied at point C; $a = 14.0$ ft. The cross-sectional area of rod B is $A_B = 0.5$ in.^{2}.

The unstretched length of each rod is $L = 12.0$ ft.

(a) If the rigid beam is to remain perfectly level as the load is applied, determine the cross-sectional area of rod A. (b) What is the elongation of each rod due to the load?

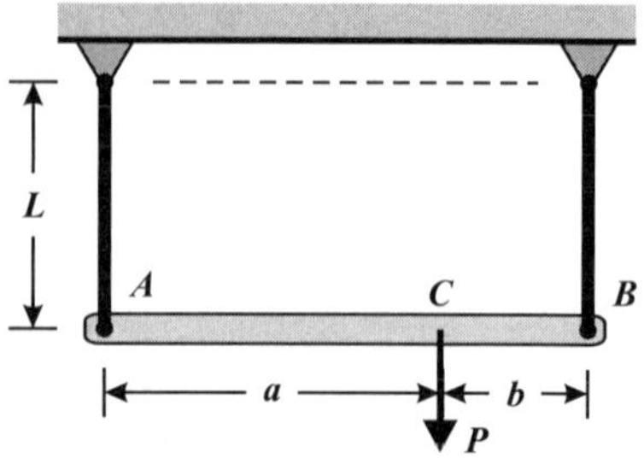

4.3 Thermal Loading

4.26 A steel bar is 1.000 m long at room temperature (25°C) and is then heated to 100°C. The coefficient of thermal expansion is $\alpha = 14\times10^{-6}/°C$.

Determine the change in length of the bar if it is free to expand (i.e., it is unconstrained).

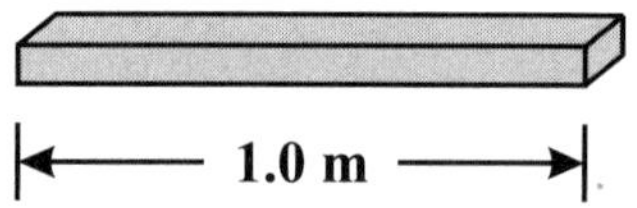

4.27 An aluminum bar is $L = 1.200$ m long and is placed tightly between two rigid supports when the temperature is 10°C. The modulus is $E = 70$ GPa, and the coefficient of thermal expansion of aluminum is $\alpha = 23\times10^{-6}/°C$.

If the temperature on a warm day is 35°C, and the supports are not affected by the temperature, determine the stress developed in the aluminum bar.

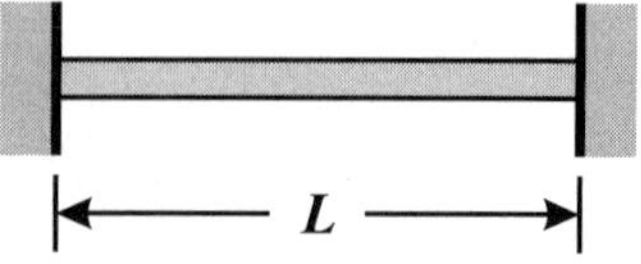

4.28 Two bars are attached to a rigid boss. Both bars are $L = 0.4$ m long with cross-sectional area $A = 0.002$ m^2. Bar 1 is aluminum and Bar 2 is steel; the bars have the following properties:

Bar	E (GPa)	S_y (MPa)	α ($\times 10^{-6}/°$C)
1	70	240	23
2	200	320	14

Assume that the properties do not change with temperature.

(a) If only Bar 2 is heated by 50°C, determine the stress in each bar. (b) Determine the thermal load ΔT (in °C) on Bar 2 required to yield Bar 1. (c) If only Bar 1 is heated by 50°C, determine the stress in each bar.

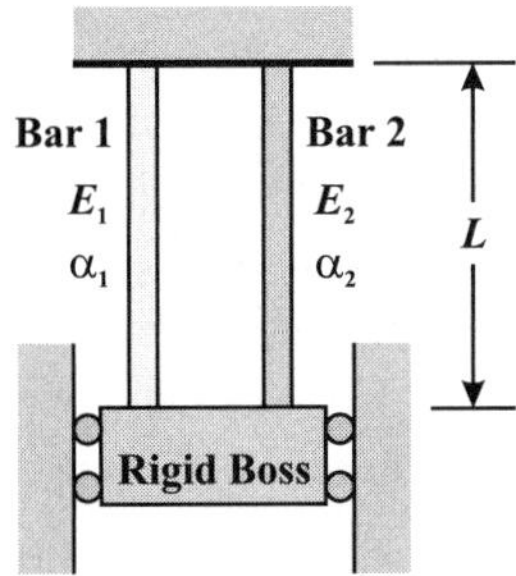

4.29 Two aluminum bars are attached to a rigid boss. Bar 1 is $L_1 = 36$ in. long and has a cross-sectional area of 0.5 in.^{2}. Bar 2 is $L_2 = 48$ in. long and has area of 0.5 in.^{2}. A downward force of $F = 10.0$ kips is applied to the rigid boss. Aluminum has a yield strength of $S_y = 35.0$ ksi and a coefficient of thermal expansion of $\alpha = 13 \times 10^{-6}/°$F.

(a) Determine the stress in each bar due only to the mechanical load. (b) If thermal load ΔT is applied to only Bar 2, determine the maximum thermal load (in °F) that can be applied so that Bar 1 does not yield.

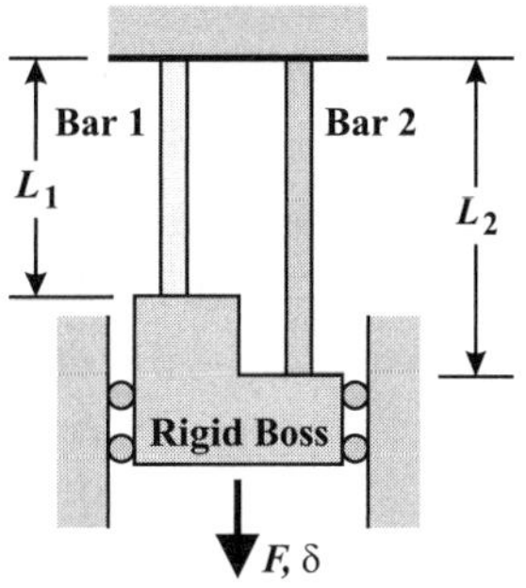

4.30 A steel pipe of outer diameter $D = 150$ mm and thickness $t = 10$ mm, carries steam at temperature ΔT above the outside temperature. Steel has properties $E = 200$ GPa, $S_y = 400$ MPa, and $\alpha = 14 \times 10^{-6}/°$C. Assume for the moment that the internal pressure is zero.

Use a two bar-system to *model* the pipe as follows:

two parallel bars of length equal to the average of the inner and outer circumferences, both having a width of half the pipe thickness. Both bars have the same depth (into the paper); take the depth as unity (1.0). Only the inner bar is loaded by temperature ΔT. Since pressure is neglected, there is no stress due to mechanical load.

Approximate the temperature ΔT_y to cause yielding in the cool bar, corresponding to yielding of the outside of the pipe.

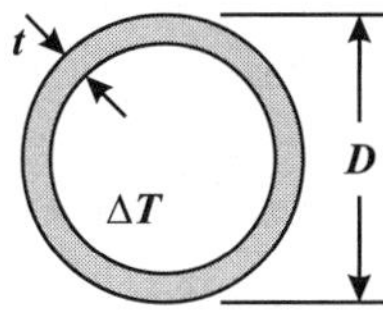

4.31 A unidirectional SiC/Ti (silicon carbide fiber/titanium matrix) composite is processed at 900°C. The strain of both components at that temperature is taken as zero. The composite is cooled to room temperature (25°C). Because of the difference in thermal expansion coefficients, residual stresses exist in the composite. The material properties are:

   SiC: $E_f = 360$ GPa $\alpha_f = 4.8 \times 10^{-6}/°C$

   Ti: $E_m = 115$ GPa $\alpha_m = 11 \times 10^{-6}/°C$

The volume fraction of the SiC fibers is $f = A_f/A = 0.35$, where A_f is the cross-sectional area of the fibers and A is the total cross-sectional area of the composite. Here, the cross-sectional area is 5.00 mm by 100 mm and the fibers have diameter $D = 100$ μm.

Using a two-bar model, approximate the residual axial stresses in the fiber and in the matrix when the composite is cooled to room temperature.

Hint: Due to compatibility, the axial strain in each component is always the same. Neglect the Poisson effect.

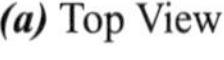

(a) Top View

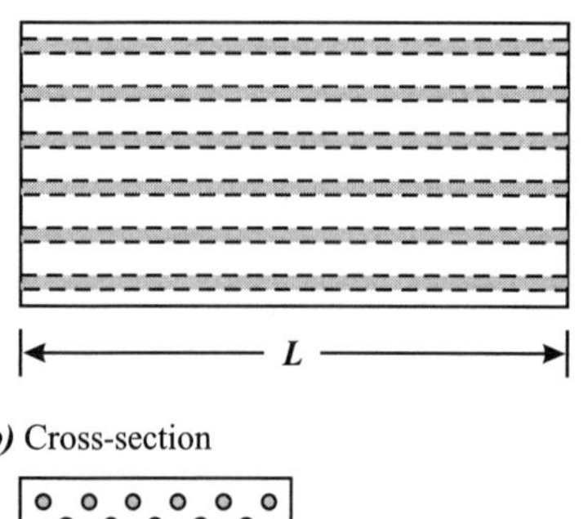

$$\longleftarrow \qquad L \qquad \longrightarrow$$

(b) Cross-section

4.4   *Pressure Vessels*

4.32 A steel cylindrical pressure vessel at an industrial plant has a diameter of $D = 1.60$ m and wall thickness of $t = 20.0$ mm. The cylinder body has length $L = 6.00$ m. For steel, $E = 200$ GPa, $S_y = 480$ MPa, and Poisson's ratio $v = 0.30$.

(a) Using a factor of safety against yielding of 3.0, determine the allowable (maximum working) pressure of the contained gas. (b) If the contained pressure is 600 kPa, determine the change in length of the cylinder from its unpressurized (unloaded) state. (c) If the contained pressure is 600 kPa, determine the change in radius of the cylinder from its unpressurized (unloaded) state.

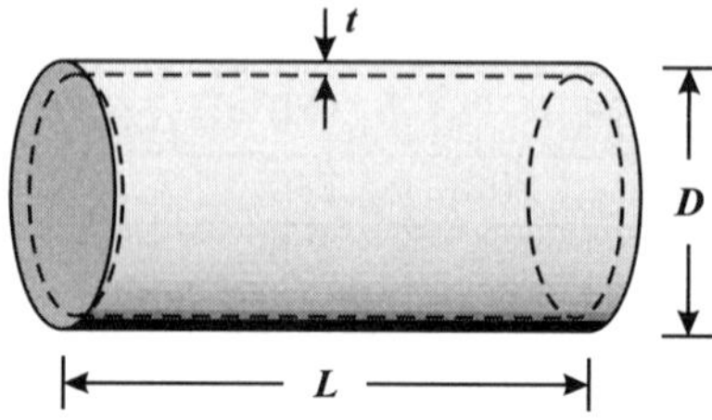

4.33 A spherical propane tank has radius $R = 400$ mm and a thickness of $t = 5$ mm. The material has yield strength $S_y = 400$ MPa, modulus $E = 200$ GPa, and Poisson's ratio $v = 0.3$.

(a) Determine the stresses in the walls of the vessel if $p = 1.0$ MPa. Draw the stresses on a plane–stress stress element. (b) If the factor of safety against yielding is 2.0, determine the allowable pressure that can be contained in the vessel. (c) If $p = 1.0$ MPa, determine the strain around the circumference of the vessel.

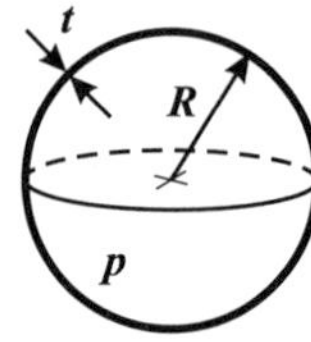

4.34 Grain silos are large thin-walled cylinders. As with water, the pressure increases linearly with depth below the surface of the "fluid" (the grain). The mass density of the grain is ρ.

Derive formulas for the hoop stress and axial stress anywhere in the wall of the silo, distance x below the surface of the grain. Note that silos are not pressurized like gas cylinders.

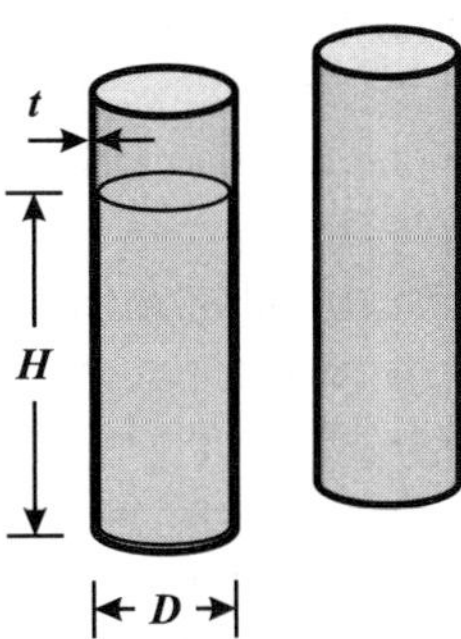

4.35 A factory uses a cylindrical pressure vessel with an inner diameter of 700 mm. The vessel operates at an internal pressure of 6.0 MPa. The steel used in the construction of the vessel has an allowable tensile stress of 280 MPa (the factor of safety has already been applied). The lid of the pressure vessel is attached to the body using 36 equally spaced bolts on a circle of diameter $D_{circle} = 800$ mm around the circumference of the pressure vessel. The allowable stress in tension for the bolts is 245 MPa.

(a) Calculate the minimum allowable wall thickness to support the load. (b) Determine the minimum bolt diameter at the root of the threads of the bolts (*hint*: what force does each bolt need to carry)? Assume the bolt-heads and nuts are strong enough.

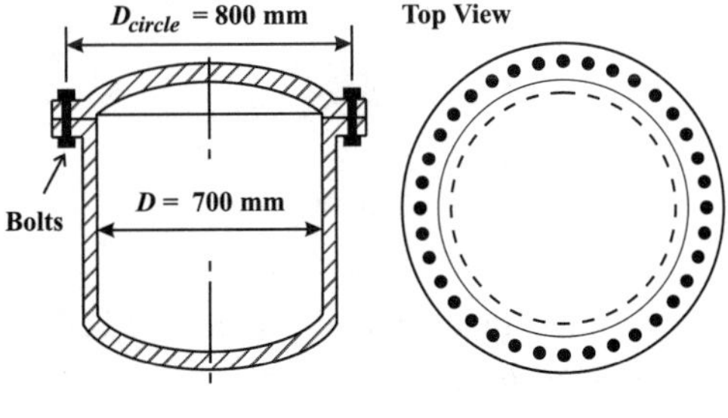

4.36 A pressure vessel has a cylindrical body and hemispherical end caps. The body and end caps are made of the same material and have radius R. The thickness of the body is T and the thickness of the end caps is t. Young's modulus is E and Poisson's ratio is v.

(a) Determine the ratio of the thicknesses t/T if the hoop strain at the cylinder/end cap joint is to be the same in each part (i.e., so the hoop strain is the same as the spherical strain at the joint). Matching these strains prevents excessive stresses at the joint. (b) If the yield strength is S_y, determine the pressure required to cause yielding p_y in the vessel. Based on the results of *Part (a)*, does failure occur in the cylindrical body or in the spherical end-cap?

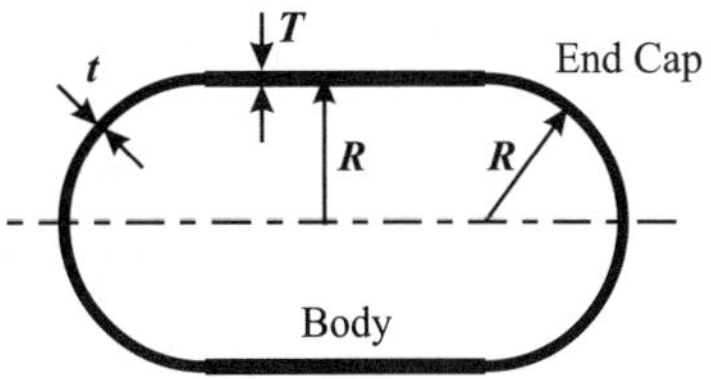

Cut-away of pressure vessel

4.37 A six-pack is shaken so that the pressure in each can is $p = 20$ psi. A board is placed on the six pack and a student(s) weighing W stands on the board so his weight is evenly distributed to the six cans. The aluminum properties are $E = 10,000$ ksi, $S_y = 35$ ksi, and $v = 0.33$. Each can has an average radius of $R = 1.25$ in., thickness $t = 0.01$ in., and length $L = 4.50$ in.

(a) Determine the critical weight of the student, W_{cr}, so that a stress element on the surface of the can is in a state of uniaxial stress. Assume that no support is provided by the contained liquid and the cans do not fail by buckling (being crushed). (b) Determine the strain in the longitudinal direction of the can due to pressure $p = 20$ psi and force W_{cr}. (c) Determine the strain in the hoop direction of the can due to pressure $p = 20$ psi and force W_{cr}.

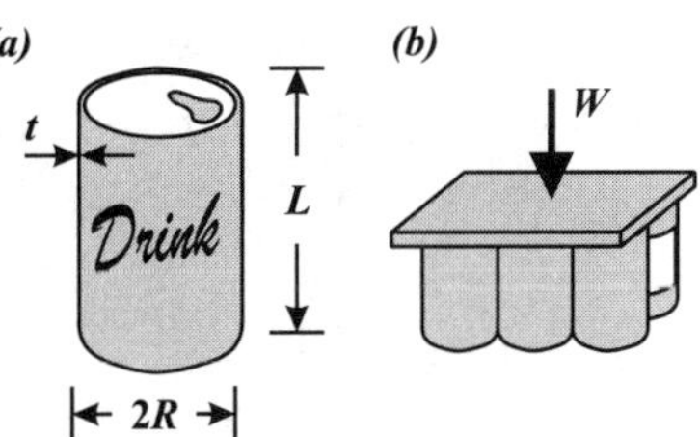

4.38 Pressurized air in inflatable structures allows them to be roughly handled. Without air pressure, a basketball can be crumpled into a relatively small space. Another application of inflatable structures are bounce-houses that give children untold hours of joy. Air pressure may also be used to deploy space structures such as unfolding antennae.

Air beams are large pressurized cylinders made of industrial plastic fabric, and are used as support frames in temporary structures. After inflation, the unfolded structure becomes a thin wall pressure vessel. Failure occurs under two circumstances:

(1) when the stress in the fabric membrane becomes compressive, causing the membrane to wrinkle or kink;

(2) when the tensile stress in the fabric reaches the fabric strength, S_u .

A pressurized fabric tube is subjected to a compressive axial force F. Determine the load that causes compressive failure of the tube. Assume the load does not cause bending.

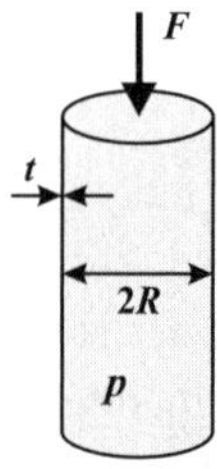

4.39 A thin-walled cylindrical pressure vessel is made by rolling a steel plate, and welding along a longitudinal line. End caps are also welded onto each end of the cylinder body.

Material Properties:

    Steel: $S_{y,s} = 550$ MPa, $E = 200$ GPa

    Weld: $S_{y,w} = 450$ MPa, $E = 200$ GPa

Geometry:

    Diameter: $2R = 1.6$ m

    Thickness: $t = 12.0$ mm.

(a) For a factor of safety is 2.0, determine the allowable pressure, p_{allow}. (b) A Pressure release valve (safety valve) is attached to the pressure vessel. The valve is a hollow cylinder, encasing a piston (or plug) attached to a spring. Assume there is no leakage around the piston, and friction between the piston and the cylinder is negligible. With no pressure, the bottom of the piston is aligned with the inner wall of the vessel. The valve cylinder has inner length $L = 120$ mm and inner diameter $b = 20$ mm. The pressure release opening is $\Delta = 20$ mm above the inside wall of the vessel. If the valve is to release the gas when the internal pressure is $p_{rel} = 4.0$ MPa, determine the required stiffness k of the spring.

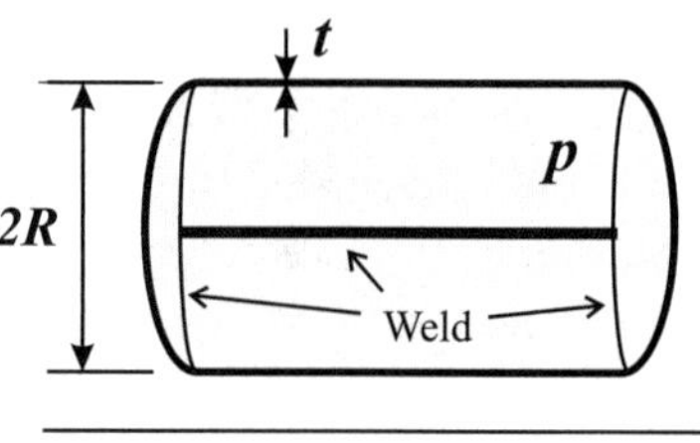

Safety Valve, Part (b)

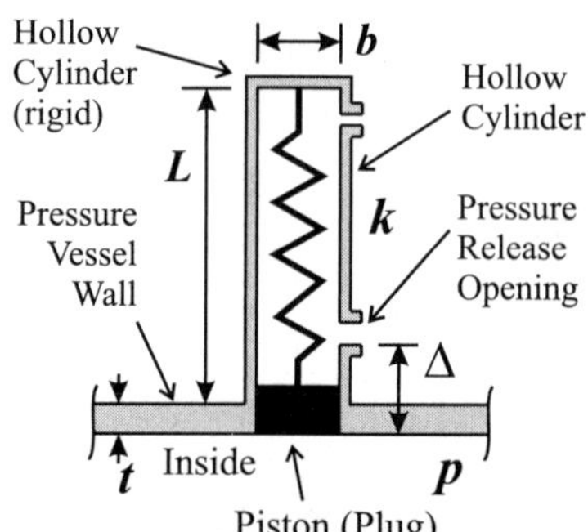

4.40 A steel pipe, outer diameter $D = 150$ mm and thickness $t = 10$ mm, carries steam at temperature ΔT above the outside temperature. The pressure in the pipe is 20 MPa. Steel has properties $E = 200$ GPa, $S_y = 400$ MPa, and $\alpha = 14 \times 10^{-6}/°C$. Use a two bar-system to model the pipe as follows:

two parallel bars of length equal to the average of the inner and outer circumferences, both having a width of half the pipe thickness. Both bars have the same depth (into the paper); take the depth as unity (1.0). Only the inner bar is loaded by temperature ΔT. The mechanical stress in each bar is approximated by the hoop stress.

Approximate the temperature ΔT_y to cause yielding in the cool bar, corresponding to yielding of the outside of the pipe.

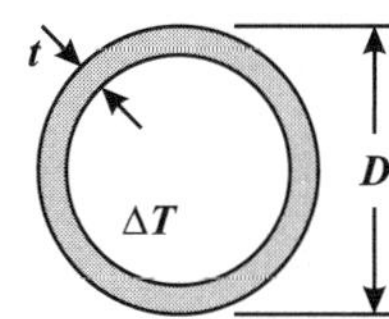

4.4   Stress Concentration Factors

4.41 A steel plate supports force F. The plate is $W = 20$ in. wide, $L = 50$ in. long, and $t = 0.50$ in. thick, and has a hole $D = 1.0$ in. in diameter drilled though it. The yield strength is 36 ksi.

Determine the force F_y to cause yielding.

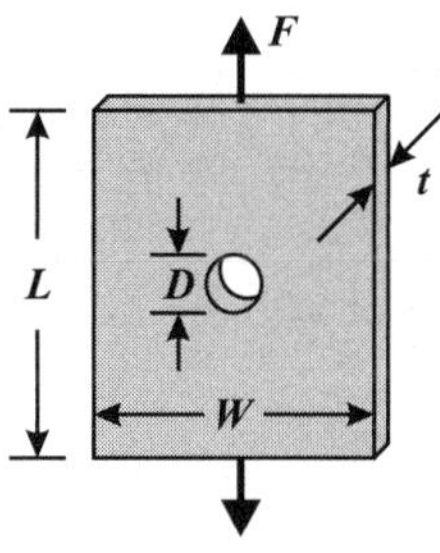

4.42 An aluminum plate has an elliptical hole. The plate is $W = 300$ mm wide, $L = 500$ mm long, and $t = 5.00$ mm thick. The hole is $2a = 4.0$ mm wide and $2b = 2.0$ mm high. The applied stress is $\sigma = 40$ MPa and aluminum has a yield strength of $S_y = 240$ MPa.

(a) Determine the stress concentration factor SCF at the tip of the elliptical hole. (b) Keeping b constant, determine the maximum length of the hole $2a$, so that the aluminum will not yield.

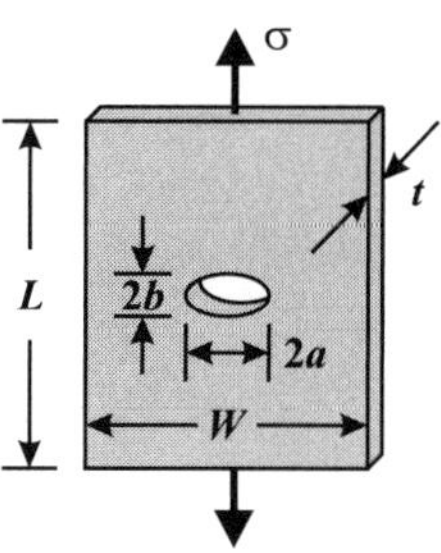

4.43 A hole of diameter $2a$ is drilled in an infinite plate (the plate width and height are much larger than the hole diameter). The plate is loaded by stress σ_{ave}.

(a) Determine the stress at the hole surface, $r = a$, one radius from the center of the hole. (b) How many radii n from the center of the hole is the stress in the plate only 10% higher than the average stress? i.e., when the stress function equals $\sigma(na) = 1.1\,\sigma_{ave}$.

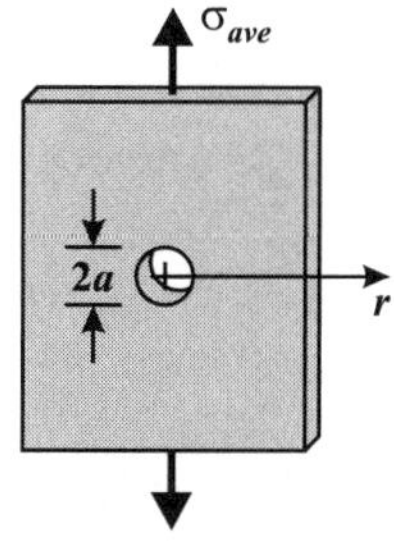

4.5   Energy Methods

4.44 A steel bar is 8.00 ft long and has a cross-sectional area of 0.5 in.^{2}. The bar supports a 4000 lb tensile load. Young's modulus is $E = 30,000$ ksi and the yield strength is $S_y = 36$ ksi.

Determine (a) total energy stored in the bar and (b) the elastic strain energy density.

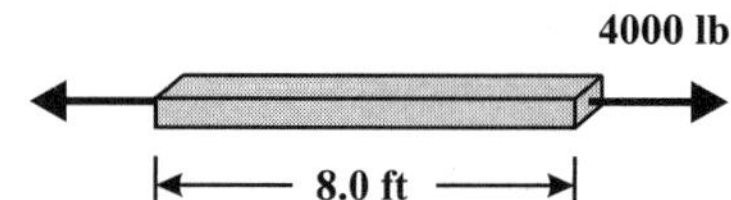

4.45 Truss ABC is subjected to downward force $F = 20.0$ kN at joint B. The truss is made of aluminum with $E = 70.0$ GPa and $S_y = 240$ MPa (in either tension or compression). The cross-sectional area of each bar is $A = 125 \times 10^{-6}$ m^2. Length $L = 1.00$ m.

Determine (a) the energy stored in each bar (N·m), (b) the downward displacement (movement) v of joint B, and (c) the stiffness of the assembly when subjected to a downward force at joint B, $k = F/v$.

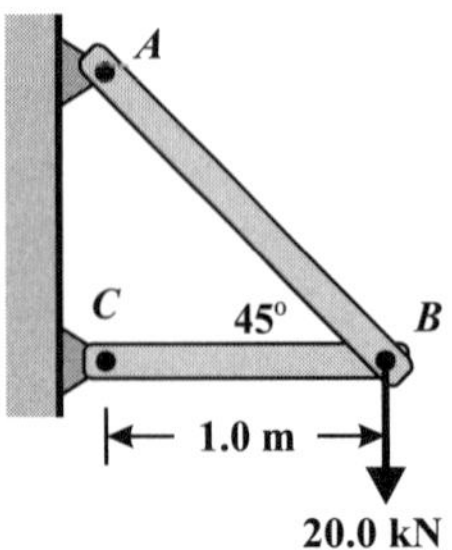

4.46 Truss ABC is subjected to horizontal force $F = 20.0$ kN at joint B. The truss is made of aluminum with $E = 70.0$ GPa and $S_y = 240$ MPa (in either tension or compression). The cross-sectional area of each bar is $A = 125 \times 10^{-6}$ m^2. Length $L = 1.00$ m.

Determine (a) the energy stored in each bar, (b) the horizontal displacement u of Joint B, and (c) the stiffness of the assembly when subjected to a horizontal force at joint B, $k = F/u$.

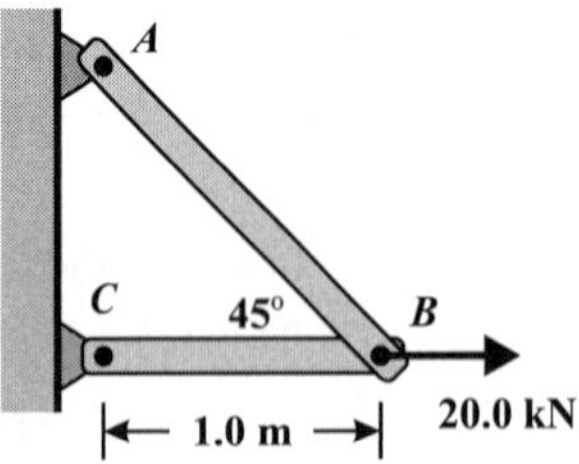

4.47 Truss ABC is subjected to horizontal force $F = 20.0$ kN at joint B. The truss is made of steel with $E = 200$ GPa and $S_y = 250$ MPa (in either tension or compression). The cross-

sectional area of each bar is $A = 125 \times 10^{-6}$ m^2. Length $L = 1.00$ m.

Determine (a) the energy stored in each bar, (b) the horizontal displacement u of joint B, and (c) the stiffness of the assembly when subjected to a horizontal force at joint B, $k = F/u$.

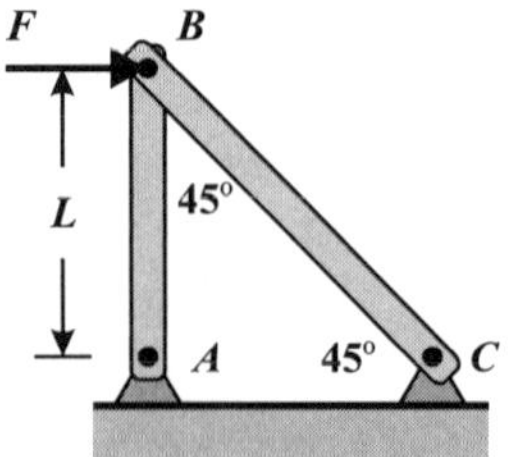

4.48 Steel truss ABC is subjected to a horizontal force of $F = 2000$ lb at joint B. The cross-sectional area of each bar, AB and AC, is $A = 0.60$ in.^{2}. Length $L = 36$ in., $E = 30,000$ ksi.

Determine (a) the energy in each bar, (b) the horizontal displacement u of Joint B, and (c) the stiffness of the assembly when subjected to a horizontal force at joint B, $k = F/u$.

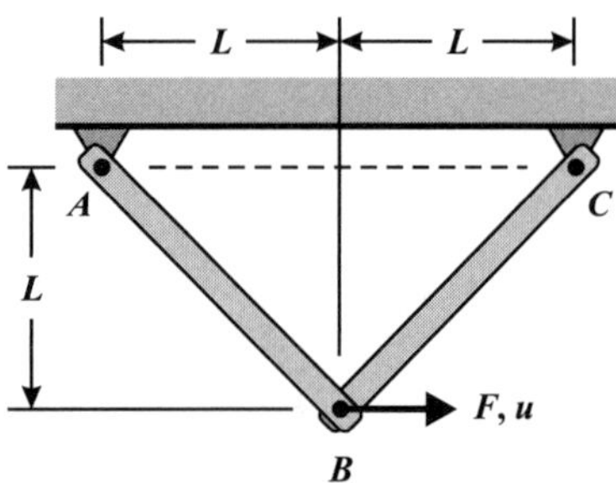

4.49 Truss ABC supports downward force F at joint B. Bar AB is of length $2L$ and bar BC is of length L. Each bar has cross-sectional area A, modulus E, and mass density ρ. The weight of the truss is much less than the applied load, so the weight can be neglected in the stress analysis.

Determine (a) the vertical displacement v in terms of F, L, A, and E, (b) the stiffness $k = F/v$ of the frame, and (c) the mass M of the bars in terms of E, ρ, L, and specified stiffness k. (d) Determine the function of material property(ies) that you would use to select a material to minimize the mass of the system for a specified value of k.

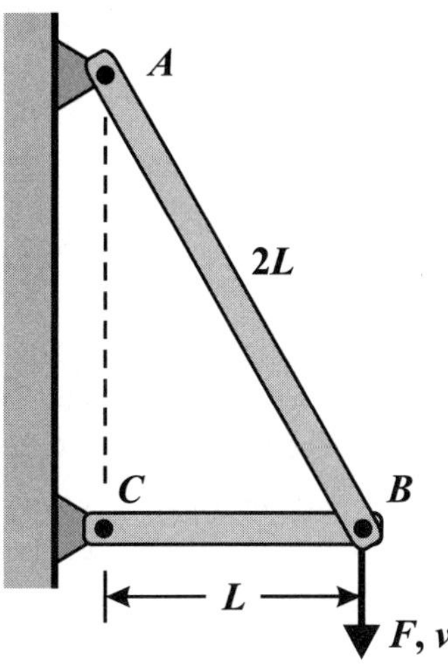

4.50 Steel truss ABC is subjected to a vertical force of $F = 5000$ lb at joint B. The cross-sectional area of each bar, AB and AC, is $A = 0.60$ in.^{2}, $L = 20$ in., and the modulus of steel is $E = 30,000$ ksi.

Determine (a) the algebraic expression for the energy in each bar in terms of F, L, A and E. (b) Determine the magnitude of the downward displacement v of joint B, and (c) the value of the assembly stiffness $k = F/v$ when subjected to a vertical force at joint B.

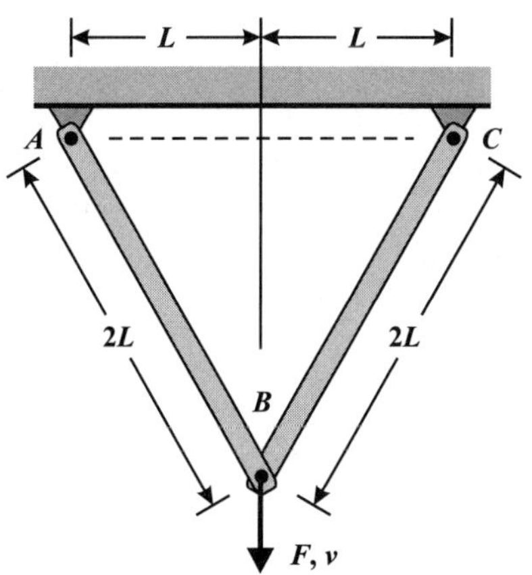

4.51 Steel truss $ABCD$ is subjected to a vertical force of $F = 5000$ lb at joint B vertical displacement v. The cross-sectional area of

each bar is $A = 0.60$ in.^{2}, $L = 20$ in., and the modulus of steel is $E = 30,000$ ksi.

(a) Determine the algebraic expression for the energy in each bar in terms of displacement v, A, L, and E. (b) Determine the magnitude of the displacement v, and (c) the value of the assembly stiffness $k = F/v$ when subjected to a vertical force at joint B

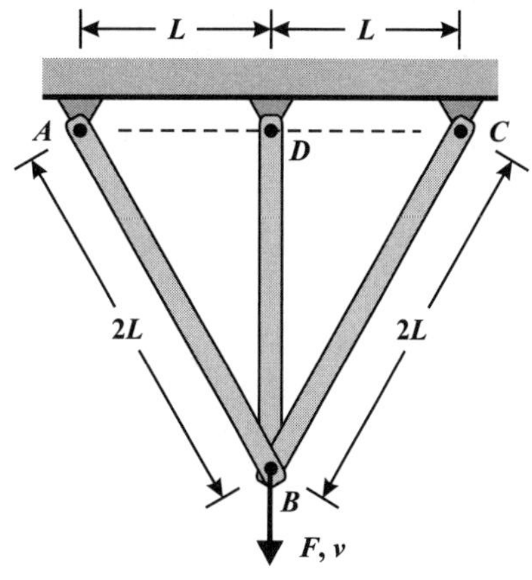

4.52 A computer finite element analysis (FEA) is performed on the truss shown. Each member is steel ($E = 30,000$ ksi), with a cross-section of 2.0 in.^{2}. The downward displacement of joint B when loaded by a 5.0 kip force is found to be $v_B = 0.0147$ in. The forces (in kips) in each member are calculated to be (a negative sign indicates a compressive force):

AB: +2.500 kips	AH: −4.507 kips
BC: +1.667 kips	HB: +3.750 kips
CD: +0.833 kips	BG: +1.502 kips
DE: +0.833 kips	CG: +1.250 kips
HG: −2.500 kips	CF: +1.502 kips
GF: −1.667 kips	FD: 0 kips
	EF: −1.502 kips

Using the energy method, verify that the energy stored in the members is equal to the work done by the 5.0 kip force (a spreadsheet may simplify your work).

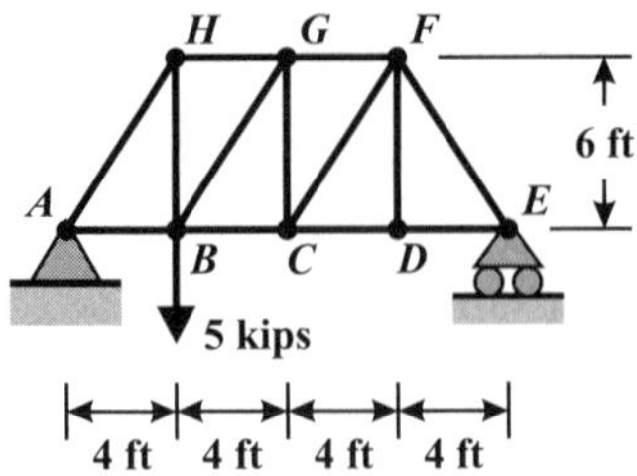

4.53 All the members of the steel truss are $L = 3.0$ m long, with rectangular cross-section 100 by 50 mm. Vertical force $V = 40$ kN is applied as shown. $E = 200$ GPa.

Using the energy method, (a) determine the total energy stored in the bars, and (b) the vertical displacement of joint B.

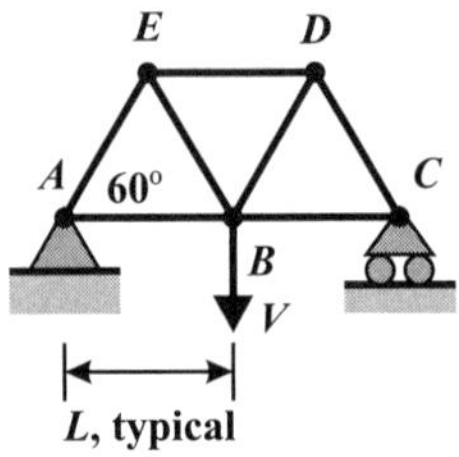

4.54 All the members of the truss are $L = 3.0$ m long, with rectangular cross-section 100 by 50 mm. Horizontal force $H = 40$ kN is applied as shown. $E = 200$ GPa.

Using the energy method, (a) determine the total energy stored in the bars, and (b) the horizontal displacement of joint E.

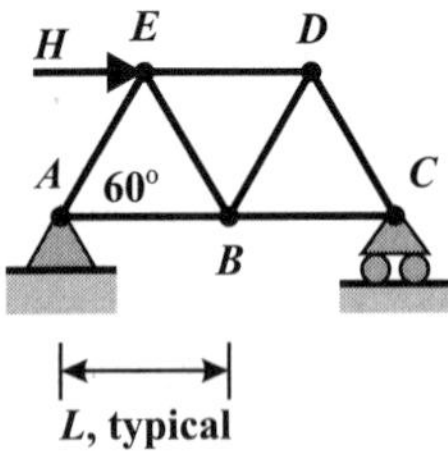

4.55 A steel truss is loaded at joint D by a downward force of 2.0 kips. The area of each member is 2.0 in.^{2}, and $E = 30,000$ ksi.

Using the energy method, (a) determine the total energy stored in the bars, and (b) the downward displacement of joint D.

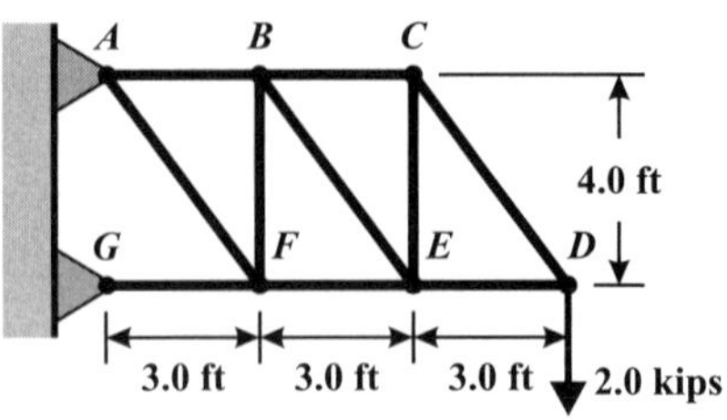

4.56 An aluminum truss is shown in the figure. Joint B is forced to move downward by $v = 2.0$ mm; there is no horizontal displacement, $u = 0$. Length $L = 1.00$ m and the cross-sectional area of each bar is $A = 0.0008$ m^2. The modulus is 70 GPa.

Determine (a) the force in each member (b) the components, F_x and F_y, of the applied force that causes the displacement, and (c) the magnitude of the applied force.

Hint: Determine the change in length of each bar as a function of u and v. The energy method is not necessarily required.

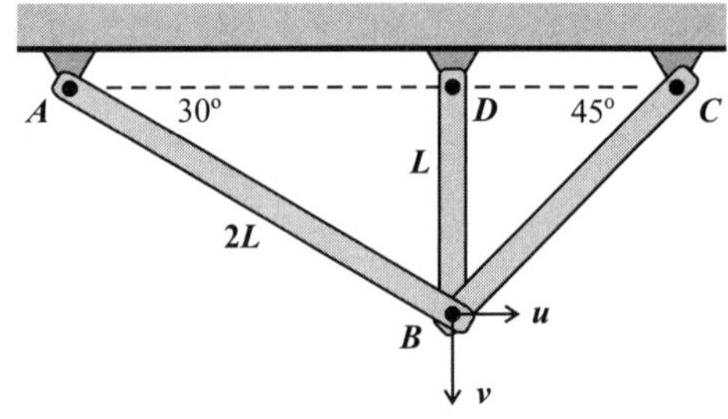

4.57 Consider the aluminum truss of *Prob. 4.56*. Joint B is forced to move to the right by $u = 3.0$ mm and downward by $v = 2.0$ mm. Length $L = 1.00$ m and the cross-sectional area of each bar is $A = 0.0008$ m^2. The modulus is 70 GPa.

Determine (a) the force in each member, (b) the components F_x and F_y of the applied force that causes the displacements, and (c) the magnitude of the applied force in the direction of the displacement of joint B.

Hint: Determine the change in length of each bar as a function of u and v. The energy method is not necessarily required.

Problems: **Chapter 5 Torsion Members**

5.1 Shafts of Circular Cross-Section

5.1 An $L = 5.0$ ft long solid steel shaft has a diameter of $D = 4.0$ in. and is subjected to a torque of $T = 50$ kip-in. The modulus is $G = 11,000$ ksi and the shear yield strength is $\tau_y = 20$ ksi.

Determine (a) the polar moment of inertia J, (b) the maximum shear stress in the shaft τ_{max}, (c) the Factor of Safety against yielding, and (d) the angle of twist between the ends of the shaft θ.

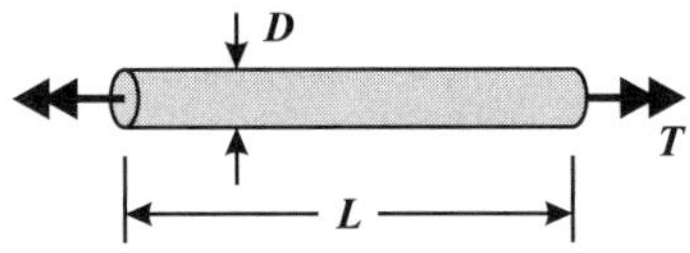

5.2 A thick-walled steel shaft is subjected to a torque of $T = 2.0$ kN·m. The inner radius is $R_i = 30$ mm and the outer radius is $R = 40$ mm. The shear yield strength is $\tau_y = 200$ MPa.

Due to the applied torque, determine (a) the stress at the inner surface of the shaft and (b) the stress at the outer surface of the shaft. (c) Determine the torque required to cause yielding.

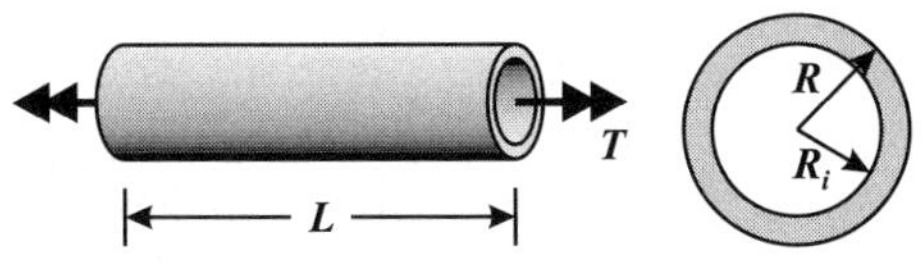

5.3 A thick-walled shaft has inner radius $R_i = 40$ mm and outer radius $R = 50$ mm.

Calculate the polar moment of inertia J using (a) the thick-walled shaft formula and (b) the thin-walled formula (use the average radius for *Part (b)*). (c) For this case ($t/R_{ave} = 22\%$), determine the percent error in the thin-walled formula with respect to the exact value (the thick-walled formula).

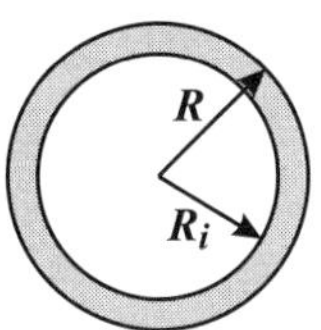

5.4 A solid shaft of radius R is to be replaced by a thick-walled shaft of outer radius R and inner radius $R/2$. Both shafts are made of the same material, have the same allowable shear stress τ_{allow} and modulus G.

When replacing the solid shaft with the hollowly shaft, determine (a) the percent change in weight, (b) the percent change in torque carrying capacity (for $\tau_{max} = \tau_{allow}$), and (c) the percent change in torsional stiffness, $K_T = T/\theta$.

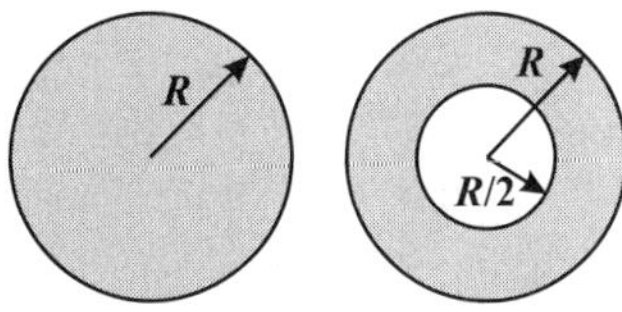

5.5 A steel shaft is 1.5 m long and is to support a 15 kN·m torque without exceeding a shear stress of 100 MPa.

Design, i.e., select dimensions for: (a) a solid shaft of radius R_s, (b) a thick-walled shaft of inner and outer radii, R_i and R_o, where $R_o = 4R_i$, and (c) a thin-walled shaft of average radius R_t and thickness t; let $t = 0.1R_t$. (d) Determine the weight savings (in percent) when replacing the solid shaft with the thick-walled shaft. (e) Determine the weight savings (in percent) when replacing the solid shaft with the thin-walled shaft.

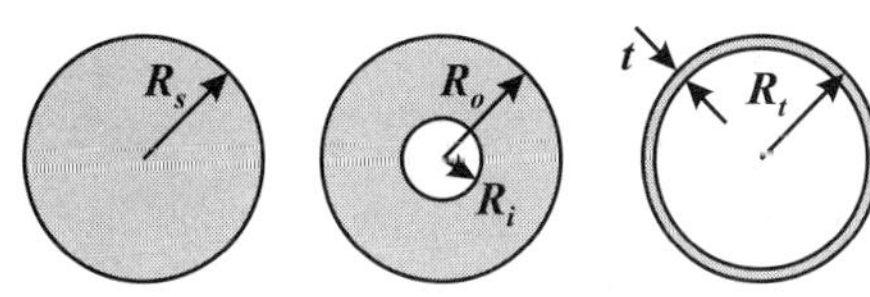

5.2 Torsion Members – Force Method

5.6 A 1.0 m by 2.0 m sign is subjected to a wind pressure of 1.2 kPa, perpendicular to its face. The sign is supported on one side by a hollow circular steel pipe of outer diameter 160 mm and inner diameter 140 mm. The shear modulus is $G = 77$ GPa. For simplicity, assume the load is transferred from the sign to the pipe at a single point, 4.0 m above the ground.

Determine (a) the torque carried in the pipe due to the wind and (b) the shear stress due to torsion only. (c) If the allowable angle of twist from the ground to the middle of the sign is $1.0°$, determine the allowable torque that can be applied. Does the current pipe need to be replaced?

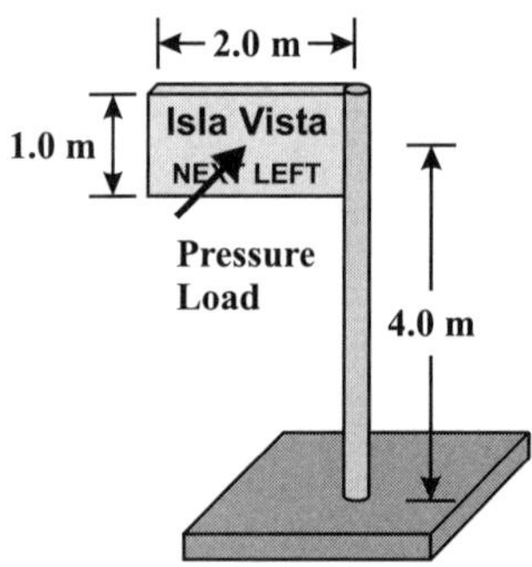

5.7 Stepped shaft *ABC* consists of two solid circular segments, and is fixed at section *A*. The shaft is subjected to torques T_B and T_C, acting in opposite directions. Segment *AB* has diameter $D_{AB} = 2.0$ in. and length $L_{AB} = 20$ in.; segment *BC* has diameter $D_{BC} = 1.0$ in. and length $L_{BC} = 12$ in. The material is an aluminum alloy with shear modulus $G = 4.0 \times 10^6$ psi. The magnitudes of the applied torques are:

$$T_B = 5000 \text{ lb-in.} \quad \text{and} \quad T_C = 2000 \text{ lb-in.}$$

Determine (a) the maximum shear stress τ_{max} in the shaft and the segment in which it occurs (*AB* or *BC*) and (b) the angle of rotation of cross-section *C*, θ_{CA} (in degrees) with respect to the fixed section *A*.

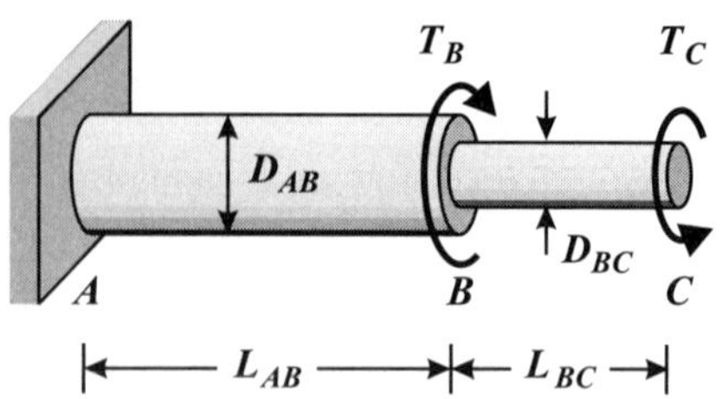

5.8 A solid shaft with four gears, *A, B, C* and *D* (shown without gear teeth), turns at a constant velocity. Gears *A* through *C* are subjected to applied torques T_A, T_B and T_C.

If the allowable shear stress is 60 MPa, determine the required diameter of the shaft.

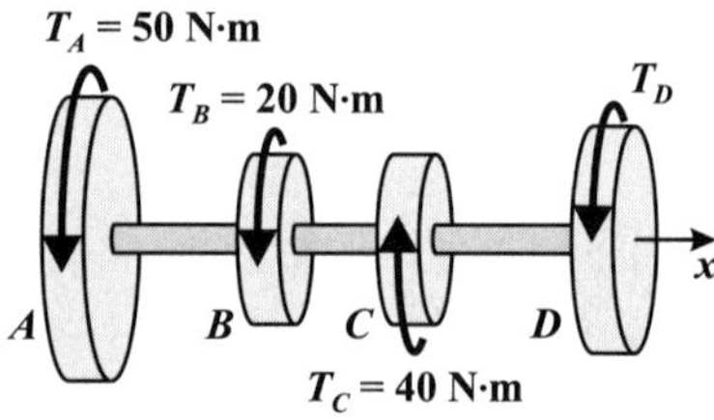

5.9 A sign is $a = 6.0$ ft wide by $b = 3.0$ ft tall, and is supported by a stepped steel shaft *ABC*. The sign is subjected to a wind load of 15 psf perpendicular to its face. Segment *AB* is $L_{AB} = 6.0$ ft long and made of Standard 3½-in. pipe (outer diameter $D_{AB} = 4.000$ in. inner diameter 3.548 in.). Segment *BC* is $L_{BC} = 8.0$ ft long and made of Standard 4 in. pipe (outer diameter $D_{BC} = 4.500$ in., inner diameter 4.026 in.). Segment *AB* and *BC* are welded together. For simplicity, assume the load is transferred from the sign to *AB* at a single point, point *A*, and that joint *B* is rigid, of sufficient strength, and of negligible dimension compared to the lengths.

Consider the torque supported by the shaft. Determine (a) the maximum shear stress in each segment *AB* and *BC*, and (b) the total angle of twist from the ground to point *A* if $G = 11{,}000$ ksi.

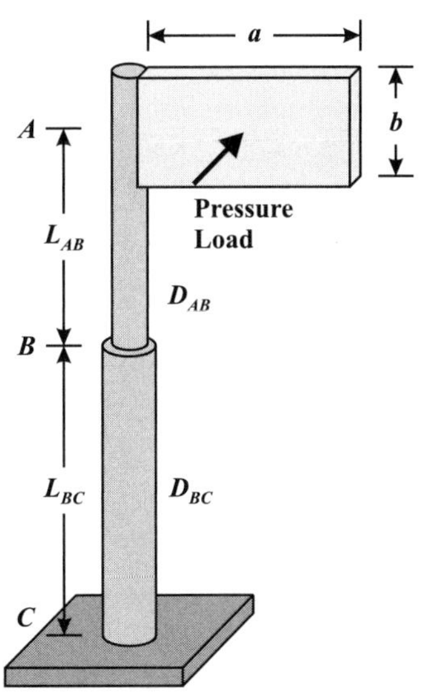

5.10 The cylindrical length of a screwdriver blade is $L = 120$ mm long, with a diameter of $D = 7.00$ mm. The screwdriver is made of steel: $G = 75$ GPa, $\tau_y = 250$ MPa. As a person tightens the screw, he applies a torque of $T = 10.0$ N·m to the screwdriver.

Determine (a) the maximum shear stress in the cylindrical length L of the screwdriver and (b) the angle of twist of the cylindrical length.

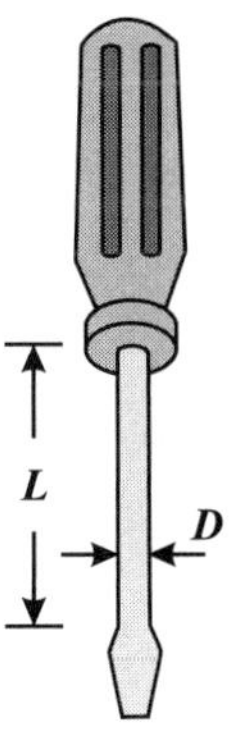

5.11 There is clearance in an industrial machine for a torsional shaft up to 3.0 in. in diameter. The current shaft is solid with a diameter of 2.0 in. The maximum allowable shear stress is 10.0 ksi. To save weight, you are asked to design a hollow shaft that will provide the greatest weight savings while still supporting the same torque at the same allowable shear stress.

(a) Determine the allowable torque that can be carried by the current shaft. (b) Determine the inner and outer diameters of the proposed shaft, and (c) the percent weight savings.

5.3 Torsion Members – Displacement Method

5.12 The aluminum $(G = 4000$ ksi) stepped shaft ABC is fixed at both ends. Segment AB is $L_{AB} = 18$ in. long with a diameter of $D_{AB} = 1.2$ in. Segment BC is $L_{BC} = 12$ in. long with a radius of $D_{BC} = 1.4$ in. Torque $T = 500$ lb-in. is applied at joint B.

Determine (a) the reactions at A and C, and (b) the angle of twist of section B.

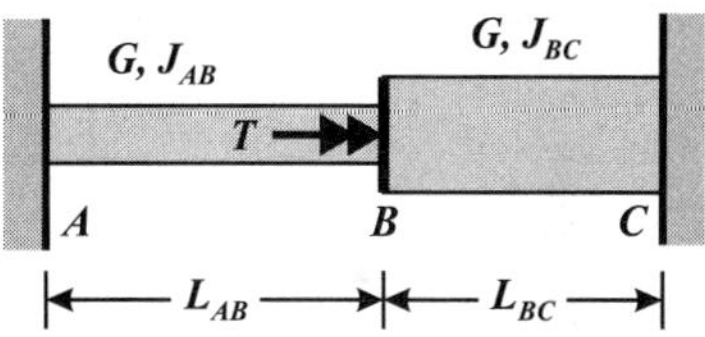

5.13 The stepped solid shaft has diameters $D_{AB} = 50$ mm and $D_{BC} = 30$ mm. The length of each segment is $L_{AB} = 600$ mm and $L_{BC} = 400$ mm. For the present case, $T_B = 0$. Torque T_C causes section C to rotate by a total of $0.8°$. The material is steel, with $G = 75$ GPa.

Determine (a) torque T_C and (b) the angle that section B rotates θ_B (i.e., the angle of twist of AB).

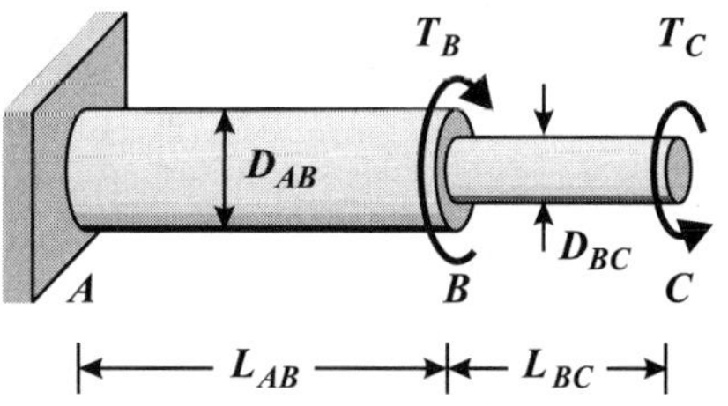

5.4   *Closed Thin-Walled Members in Torsion*

5.14 Box beams that support airplane wings in bending also support torsion loads. Thus, the beam is also called a *torsion box*. The torsion box in an aircraft has outer dimensions $b = 1.0$ m wide and $d = 0.6$ m deep, and has constant thickness of $t = 10$ mm. The box is made of aluminum with an allowable shear stress of $\tau_{allow} = 100$ MPa and modulus $G = 26$ GPa. The length of the beam is $L = 15$ m.

When the torsion box is operating at its allowable stress, determine (a) the allowable torque that the box can support, (b) the shear flow anywhere in the torsion box, and (c) the angle of twist of the beam.

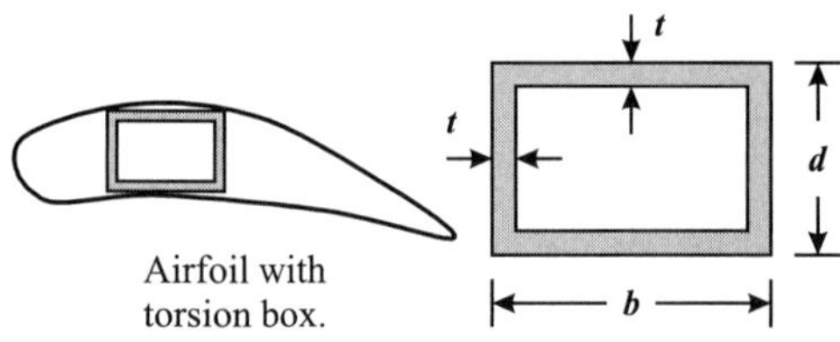

Airfoil with torsion box.

5.15 A thin-walled tube of elliptical cross-section is made by extruding aluminum ($\tau_y = 20$ ksi) through a hardened steel die. The outer major and minor diameters of the ellipse at the centerline of the wall are $2a = 3.0$ in. and $2b = 2.0$ in., respectively, and the wall thickness is $t = 0.10$ in. thick. The length of the tube is $L = 72$ in.

The area of an ellipse is: $A = \pi ab$

and the perimeter is:

$$P = \pi(a+b)\left[1 + \frac{1}{4}\left(\frac{a-b}{a+b}\right)^2 + \frac{1}{64}\left(\frac{a-b}{a+b}\right)^4 + \frac{1}{256}\left(\frac{a-b}{a+b}\right)^6 + \dots \right]$$

or, approximately:

$$P \approx 2\pi\sqrt{\frac{1}{2}(a^2 + b^2)}$$

The latter equation is within 5% of the actual value for $0.36 < b/a < 2.95$

If the applied torque is 1200 lb-in., estimate (a) the shear stress anywhere on the cross-section and (b) the angle of twist.

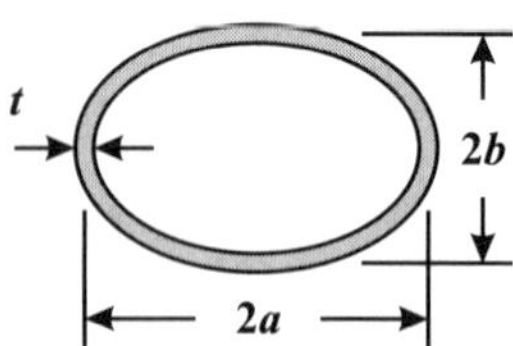

5.16 A hollow steel box-beam is 6.0 in. wide and 8.0 in. deep (centerline-to-centerline of wall). The thickness of the sides is 0.3 in. and of the top and bottom is 0.5 in. The beam is subjected to an applied torque of 200 kip-in. $G = 11,000$ GPa.

Determine (a) the stress along each side of the cross-section and (b) the angle of twist per unit length along the beam.

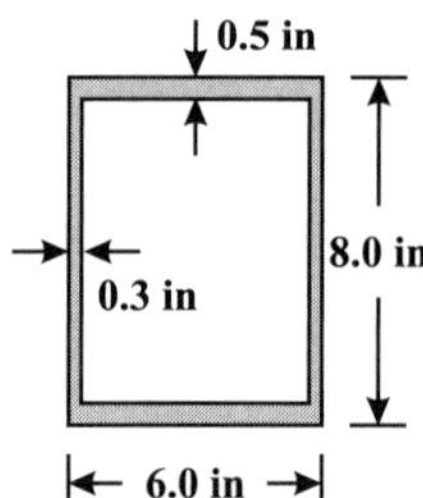

5.5   *Power Transmission*

5.17 The maximum torque in a vehicle's solid steel drive shaft occurs when it transmits 240 hp at 3000 rpm.

Determine the minimum required diameter for the solid shaft if the allowable shear stress is $\tau_{allow} = 8.0$ ksi.

5.18 Solid steel shaft ABC is driven by a 240 kW motor at 1800 rpm (revolutions per minute). The gears at B and C power machines that require 80 and 160 kW, respectively. The lengths of the two segments are $L_{AB} = 1.0$ m and $L_{BC} = 0.50$ m. The modulus is $G = 75$ GPa. The diameter of the shaft D is constant. The bearing is frictionless.

Determine the required diameter D if both of the following conditions must be satisfied:

1. the allowable maximum shear stress is 60 MPa and

2. the allowable angle of twist between sections A and C is 3.0°.

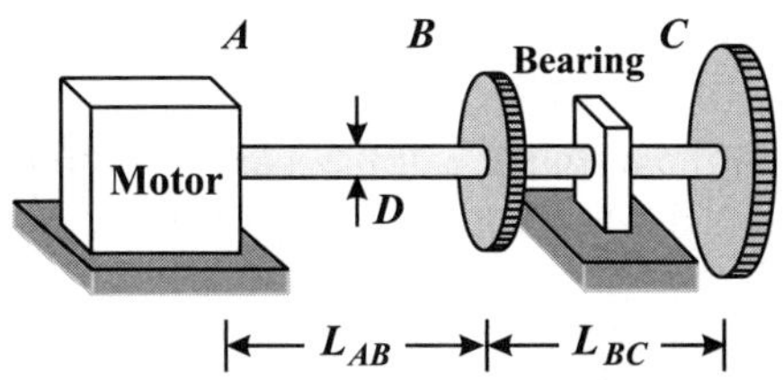

5.19 A propeller-driven boat moves forward at constant velocity. A solid drive shaft transmits 2400 kW to the propeller at 1600 rpm. The diameter of the shaft is 100 mm and the length from the engine to the propeller is 3.0 m. The shear modulus of the steel is 75 GPa.

Determine (a) the torque carried in the shaft, (b) the maximum shear stress in the shaft, and (c) the angle of twist of the shaft from the engine to the propeller. (d) If the allowable shear stress is limited to 35 MPa, the shaft must be replaced with a larger shaft, also of solid cross-section. Determine the required diameter of the new shaft.

5.20 A solid circular shaft is to be designed to transmit 135 hp at 300 rpm. The shaft is 6.0 ft long. It is made of steel with a shear modulus of 12,000 ksi. The allowable angle of twist between the ends of the shaft is 1.2°.

Determine (a) the torque transmitted, (b) the required shaft radius to satisfy the design requirement (assume the material is strong enough), and (c) the maximum shear stress due to the torque.

5.21 A solid circular shaft is to be designed to transmit 40 kW at 2.0 rev/s. The shaft is 2.0 m long. It is to be made of steel with a shear modulus of 75 GPa. The design requirements are:

1. the maximum shear stress is to be no more than 60 MPa;

2. the twist of the shaft is to be no more than 1.0°.

Determine (a) the torque transmitted, (b) the shaft radius required to satisfy Requirement 1, (c) the shaft radius required to satisfy Requirement 2 and (d) the shaft radius that satisfies both criteria.

Problems: **Chapter 6: Bending Members: Beams**

6.1 A simply supported beam supports a force of $F = 10.0$ kN.

(a) Determine equations for the shear force and bending moment over the entire length of the beam. Measure x from the left support. (b) Draw the shear force and bending moment diagrams. (c) Determine the magnitudes of the maximum shear force and maximum moment. Where are they located?

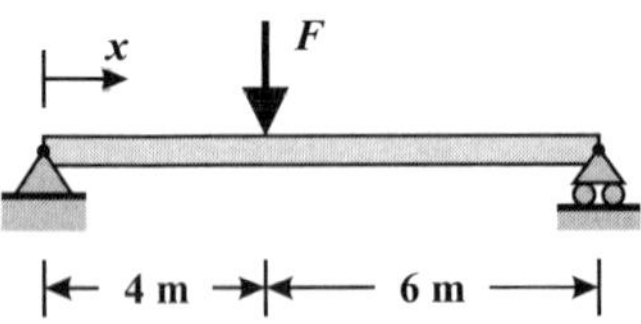

6.2 A 16.0 kN truck (including its load) crosses a bridge. The bridge is $L = 14.0$ m long and the truck's wheelbase is $s = 4.0$ m. Assume that the weight of the truck is evenly distributed between the front and the rear axles.

Determine the location of the rear axle a such that the moment is maximum. What is the moment?

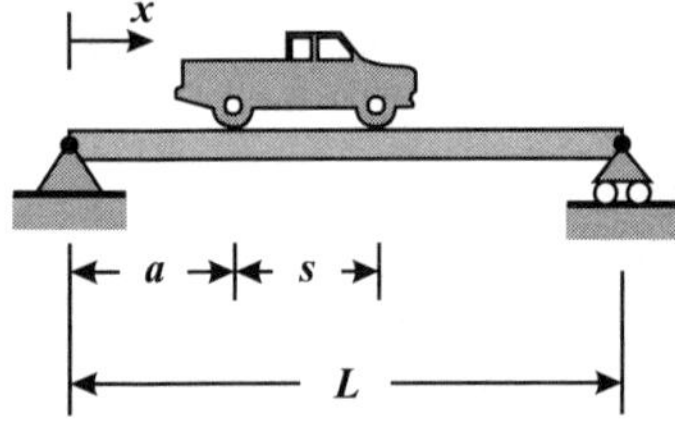

6.1   Bending Strain and Stress

6.3 A plastic pipe is bent into a 90° turn. The length of the curve is $L = 3.0$ ft and the outside diameter of the pipe is $D = 1.0$ in.

Determine (a) the radius of curvature of the pipe and (b) the magnitude of the maximum compressive strain.

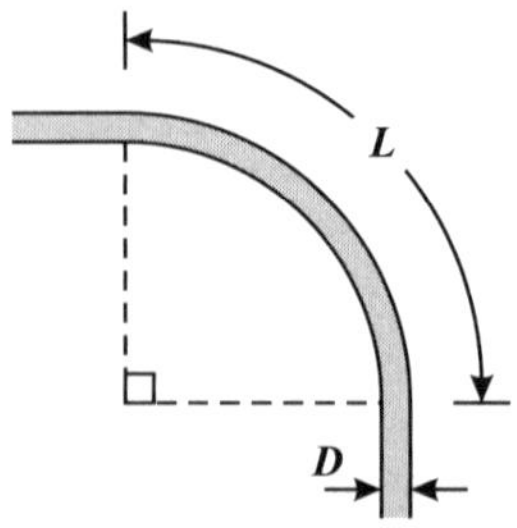

6.4 A flexible cantilever beam is $L = 0.500$ m long and is turned back upon itself to form a semi-circle.

If the magnitude of the maximum strain is $\varepsilon = 0.008$, determine the thickness t of the beam.

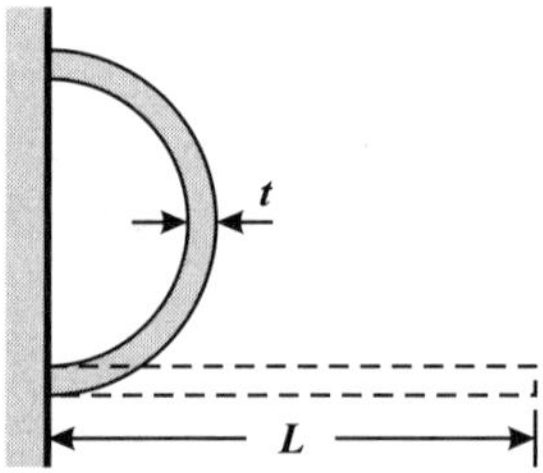

6.5 Structural steel has a yield strength of $S_y = 36$ ksi. Four I-beams are available for use in a particular project:

Beam	A (in.^{2})	I (in.^{4})	y_{max} (in.)
S12×31	9.3	218	6.00
W10×30	8.8	170	5.75
W10×33	9.7	170	4.87
W14×26	7.7	245	6.95

(a) Which I-beam(s) can support a bending moment of 1260 kip-in. without yielding? (be strict). (b) Cost drives choice of structural components. The cost of steel I-beams are generally proportional to their weight. Which beam (that can support the load) is the best to select from an economic standpoint?

6.6 The beam from *Prob. 6.1* is made of a rectangular cross-section, $b = 100$ mm wide by $d = 200$ mm deep. Take $E = 200$ GPa (steel).

Determine (a) the maximum bending stress in the beam and (b) the radius of curvature of the beam at the location of the maximum moment.

6.7 A cantilever beam supports a uniformly distributed load $w = 120$ lb/ft $= 10$ lb/in. The beam is 10 ft long and has a rectangular cross-section of width $b = 2.0$ in. and depth $d = 6.0$ in.

Determine (a) the maximum moment M_{max} and (b) the maximum bending stress σ_{max}. (c) If the allowable stress is $\sigma_{allow} = 2.0$ ksi and the width of the beam b is to remain 2.0 in., determine the minimum required depth of the beam d_{new} so that the allowable stress is not exceeded.

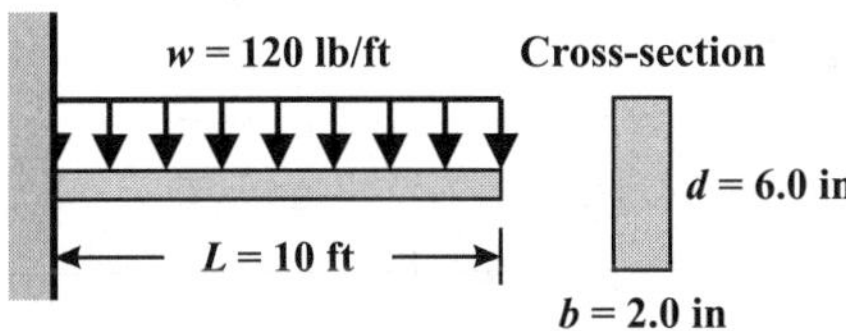

6.8 Three proposed cross-section configurations for joining three boards are shown. Each board is $b{\times}3b$ ($b = 50$ mm, $3b = 150$ mm).

(a) Determine the moment of inertia I about the z-axis for each configuration. Use the Parallel-Axis Theorem as necessary. (b) If the bending moment about the z-axis is 5000 N·m, what is the maximum bending stress in each cross-section? (c) Which cross-section is the most effective use of the material in bending?

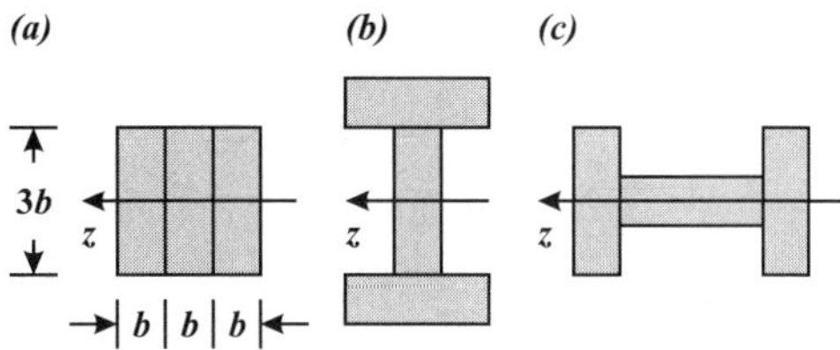

6.9 A new material is tested in four-point bending. The geometry is $L = 200$ mm, $a = 40$ mm, $b = 10$ mm, and $d = 5.0$ mm.

(a) If $P = 500$ N, determine the maximum bending stress in the beam. (b) What is the advantage of the four-point bend test?

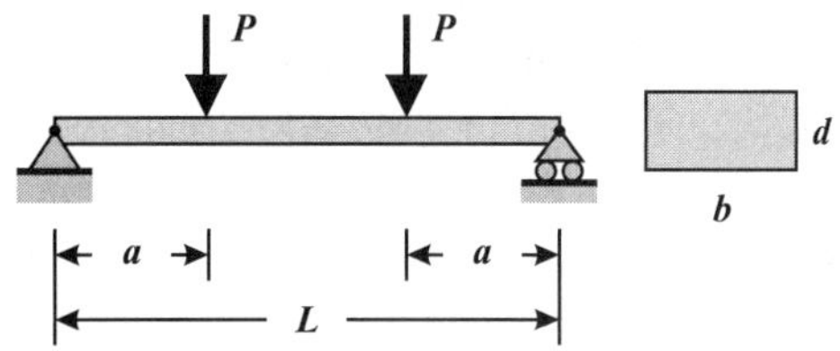

6.10 You are designing a deck for a house outside Chicago, Illinois. The deck is $c = 12$ ft wide by $a = 6.0$ ft deep, and is supported by three beams as shown. Each beam has a cross-section $b = 4.0$ in. wide by $d = 8.0$ in. deep. The ground snow load around Chicago is 30 psf (pounds per square foot). Assume the deck supports the same snow load.

Determine the maximum bending stress in the center beam, Beam 2.

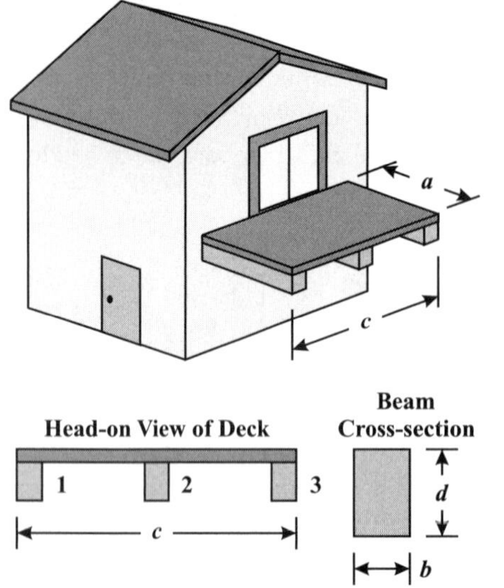

Head-on View of Deck

Beam Cross-section

6.11 A 40,000 lb truck crosses a 28 ft long single lane bridge. The truck's wheelbase is $L = 14$ ft, and at the instant shown, the truck is in the middle of the bridge. One-fourth of the total weight is distributed to the front axle. The bridge is supported by four steel I-beams, S20×75 (see *Appendix C*); assume each beam shares equally in supporting the load.

Determine the maximum bending stress for the condition shown. Where is the maximum bending stress located?

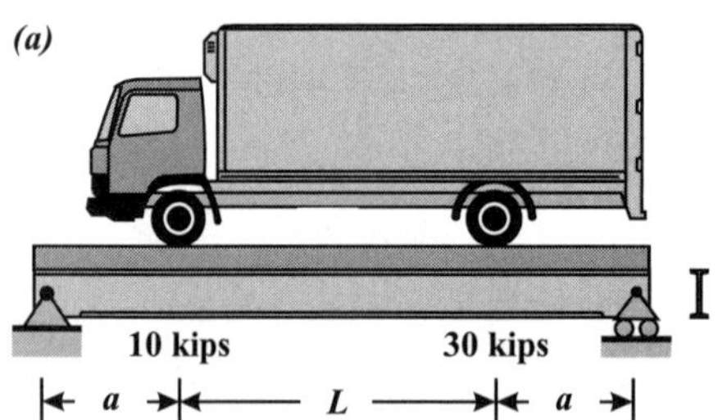

(a)

(b) Head-on view of bridge.

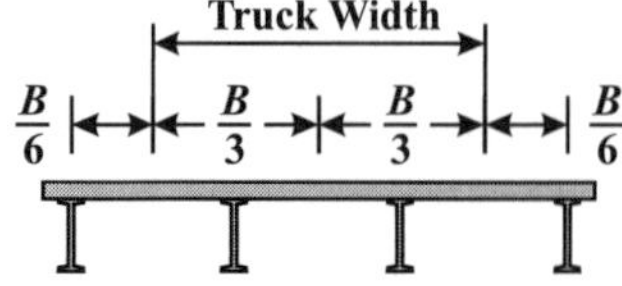

6.12 The front face of a highway sign, 10 ft wide by 6.0 ft tall, is acted on by a uniform wind pressure of 10 psf, perpendicular to the sign. The center of the sign is $H = 8.0$ ft above the ground. Two posts (symmetrically placed) support the sign. The posts are positioned in such a way that the wind load does not cause them to twist.

If each post has a cross-section of $b = 4.0$ in. and $d = 6.0$ in., determine the maximum bending stress in either post.

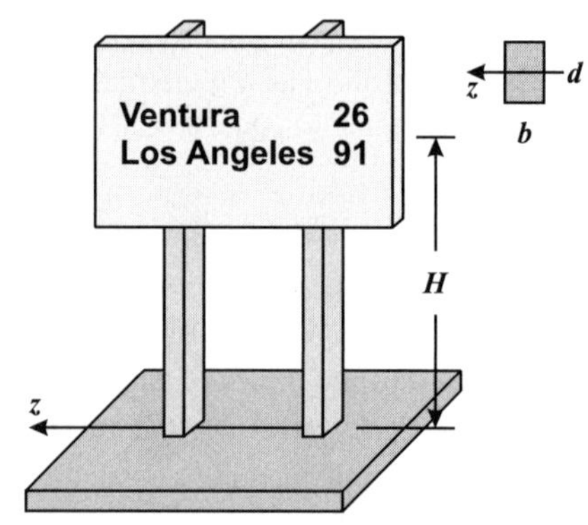

6.13 A crane slowly lifts a 30 ft length of 12 in. diameter standard pipe. Lift points D and E are 8.0 ft apart. A 12 in. standard steel pipe weighs 49.56 lb/ft, has outer diameter $D = 12.75$ in. and moment of inertia $I = 289$ in.[4] (see *Appendix D*).

Determine the maximum bending stress in the pipe.

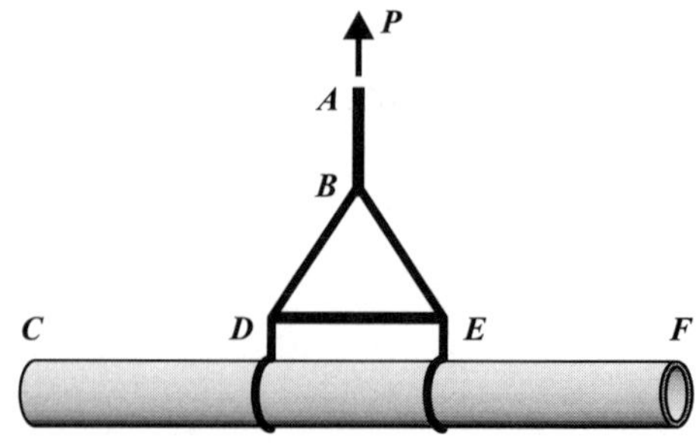

6.14 Inflatable tubes made of industrial plastic fabric are used as beams in temporary structures. Such a tube of radius R, thickness t, and length L and is inflated to internal pressure p. The airbeam supports a uniformly distributed load w (force per length). Failure occurs when the maximum stress reaches the fabric's tensile strength S_u, or when any point is subjected to compression, when the tube will kink or buckle.

Determine the maximum load w_{max} that can be supported if failure is by compression. Give the load in terms of p, R, t, and L.

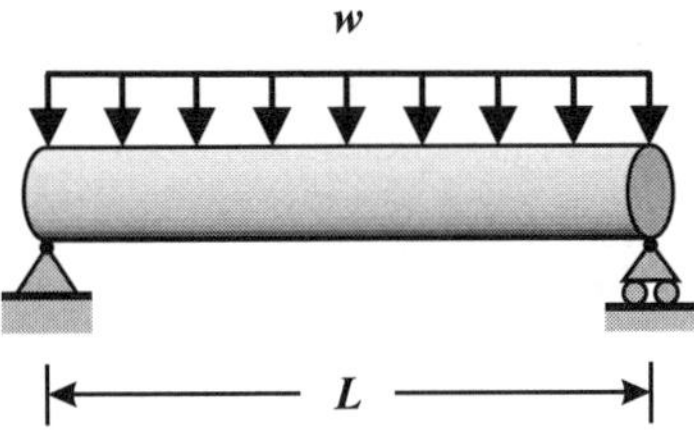

6.15 A simply supported aluminum beam, $L = 4.0$ m long, is subjected to a uniformly distributed load over half of its length, $w = 5.0$ kN/m. The beam is of rectangular cross-section $b = 50$ mm wide by $d = 80$ mm deep. The material properties are $E = 70$ GPa and $S_y = 240$ MPa.

Determine (a) magnitude of the maximum bending stress and (b) the factor of safety against yielding.

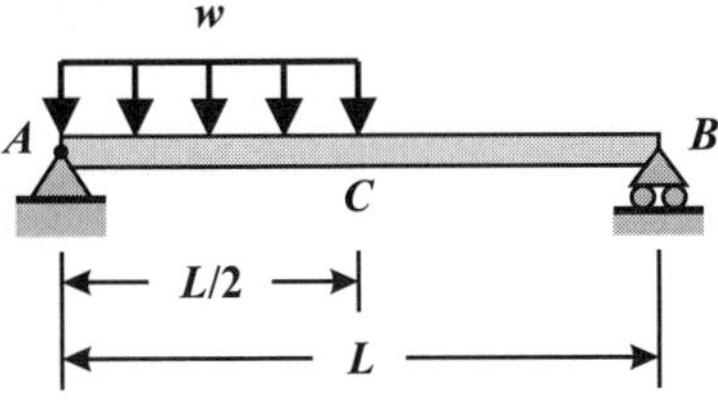

6.16 A simply supported beam, $L = 12.0$ feet long, is subjected to a linearly increasing distributed load, from $w(0) = 0$ to $w(L) = w_o = 3000$ lb/ft. The beam is an S10×35 I-beam (see *Appendix C*), made of steel with $E = 30,000$ ksi and $S_y = 50.0$ ksi.

Determine (a) the moment as a function of x, $M(x)$ and (b) the maximum bending stress.

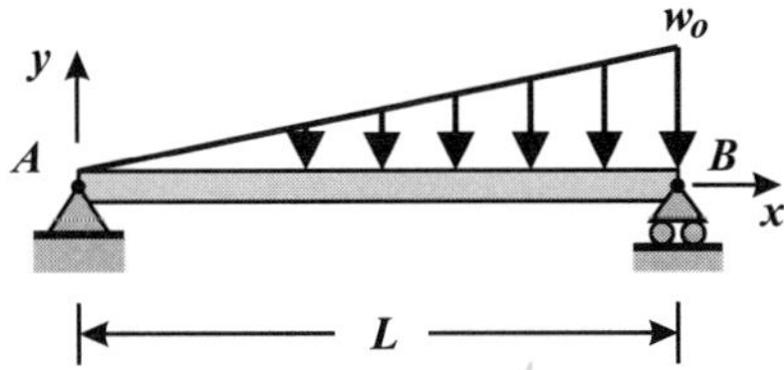

6.17 A simply supported beam, $L = 12.0$ ft long, is subjected to a linearly decreasing load from $w(0) = w_o = 3000$ lb/ft to $w(L) = 0$. The beam is an S10×35 I-beam, made of steel with $E = 30,000$ ksi and $S_y = 50.0$ ksi.

Determine (a) the moment as a function of x, $M(x)$, and (b) the maximum bending stress.

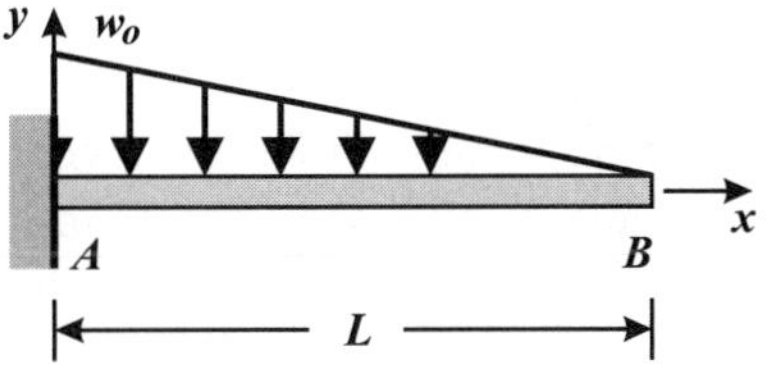

6.18 The beam supported and loaded as shown in *Figure (a)* has the moment diagram shown in *Figure (b)*. The cross-section of the beam is the T-section of *Figure (c)*. The centroid of the cross-section is 55 mm above the base and the moment of inertia is $I = 29.3 \times 10^6$ mm. The beam has an *allowable tensile stress* of 400 MPa and an *allowable compressive stress* of 250 MPa.

Determine the allowable load P. Check all possibilities.

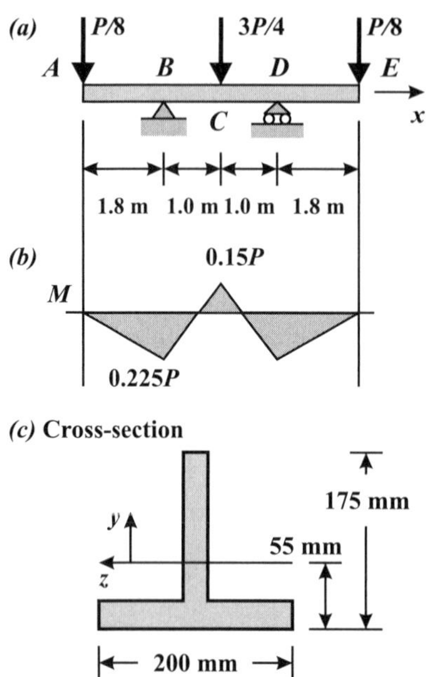

6.19 A "T"-beam supports a bending moment that places its cap (cross-bar) in tension and its foot in compression.

If the allowable stresses are 42 MPa in *tension* and 84 MPa in *compression*, determine the width of the web w so that both allowable stresses occur simultaneously. In other words, so that the maximum tensile stress equals the allowable tensile stress and the maximum compressive stress equals the allowable compressive stress.

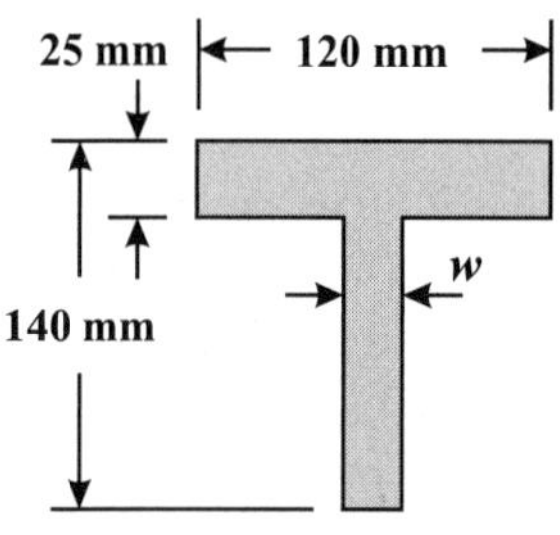

6.20 A ceramic beam used in a hydrogen converter is to made from a brittle ceramic for which the compressive strength $S_c = 1.2$ GPa is 10 times that of the tensile strength S_u. The

beam is subjected to a positive moment (compression at the top).

(a) If the beam has a square cross-section of side 20 mm (*Figure (a)*), determine the bending moment at failure. (b) If the beam has an inverted T-beam cross-section (*Figure (b)*), having the same area as that of *Figure (a)*, determine the bending moment at failure. (c) Determine the percent change in strength when replacing cross-section *(a)* with *(b)*.

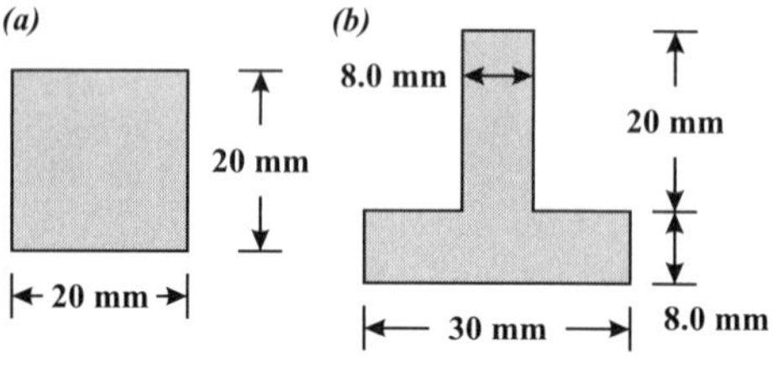

6.2   Beam Deflection

6.21 A steel ($E = 200$ MPa) water pipe is supported across a length $L = 8.0$ m. The pipe has an outside diameter of 200 mm and a thickness of 10.0 mm. The density of steel is $\rho_s = 7900$ kg/m^3 and that of water is $\rho_w = 1000$ kg/m^3. The water does not support any of the load. Assume that the pipe is free to rotate at its supports (i.e., it is simply supported).

Determine (a) the maximum bending stress in the pipe due to the weight of the water and the self-weight of the pipe and (b) the *difference* between the maximum deflection of the pipe when it is full and when it is empty.

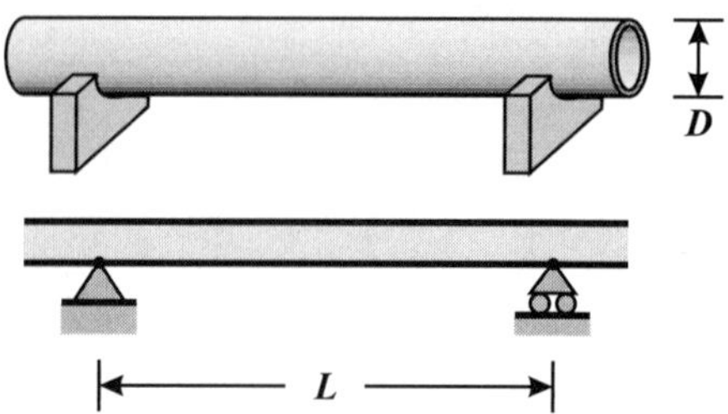

6.22 Cantilever beam AB is L long and is subjected to uniformly distributed load w. The beam has constant bending stiffness EI.

Derive (develop) the equations for (a) the moment $M(x)$ and (b) for the deflection $v(x)$.

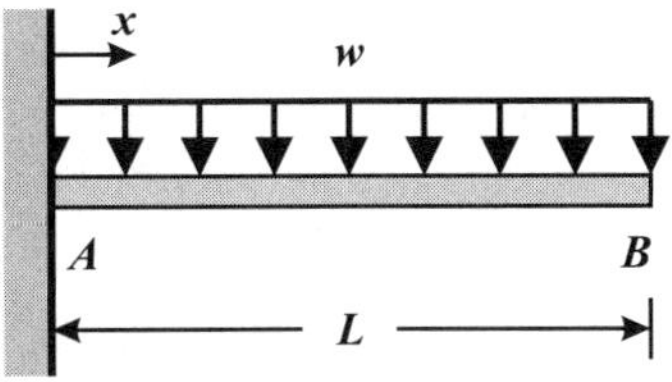

6.23 An aluminum ($E = 10,000$ ksi) cantilever beam has moment of inertia $I = 120$ in.^{4} and length $L = 96$ in. The beam is tip-loaded by a force of $P = 4.0$ kips. A structural component (represented by the triangle) is placed under the center of the beam at point B, distance h below the bottom surface of the beam.

(a) Derive the algebraic expression for the displacement of the beam $v(x)$ in terms of P, L, E, and I. Assume for now that the beam does not contact the structural component. (b) Using the given values, if $h = 0.200$ in., will the cantilever beam hit the structural component? Justify your answer with appropriate calculations.

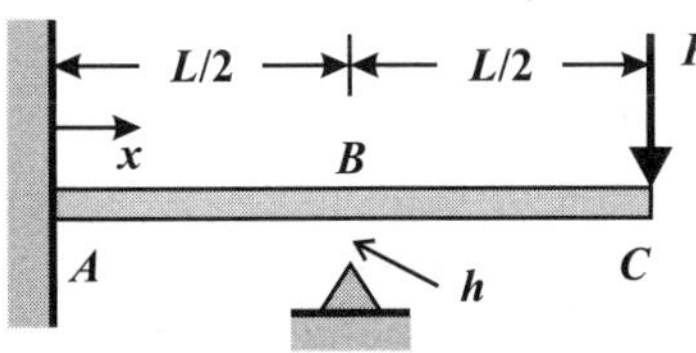

6.24 Beam AB is L long and is subjected to a linearly increasing distributed load $w(x)$. The beam has constant bending stiffness EI.

Determine (a) the expression for the moment $M(x)$, (b) the location and value of the maximum moment, (c) the expression for the deflection $v(x)$, and (d) the location of the maximum deflection.

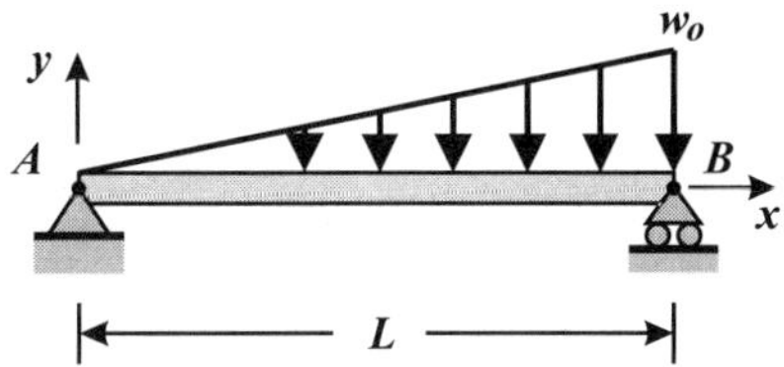

6.25 Point load P is applied at distance a from the left support of simply supported beam AB of total length $L = a + b$. The beam has constant bending stiffness EI.

(a) Derive (develop) equations for the deflection and slope, $v(x)$ and $v'(x)$, over the entire length of the beam. (b) Determine the the maximum displacement δ_{max}.

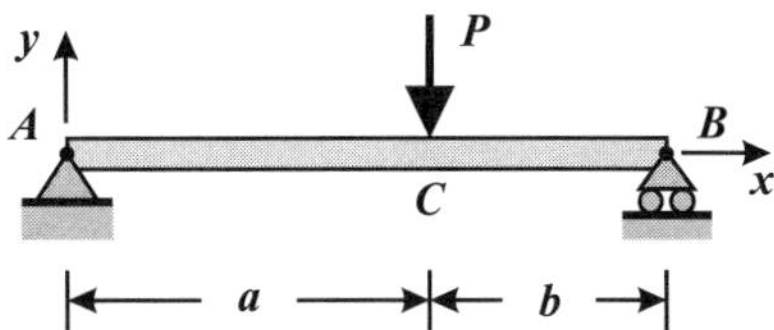

6.26 Point load P is applied at tip point B on an overhanging beam of total length $L = a + b$. The beam has constant bending stiffness EI.

(a) Derive (develop) equations for the deflection and slope, $v(x)$ and $v'(x)$, over each segment of the beam, AC and CB. (b) Determine the maximum value of the deflection between points A and C, and (c) the deflection at point B.

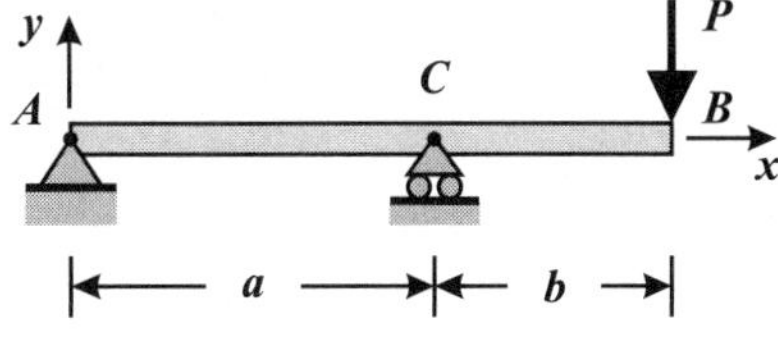

6.27 A concentrated load P is placed at distance L from the built-in end of a cantilever beam of total length $3L/2$. The constant bending stiffness is EI.

Derive (develop) equations for the deflection and slope, $v(x)$ and $v'(x)$, over each length of the beam, AC and CB.

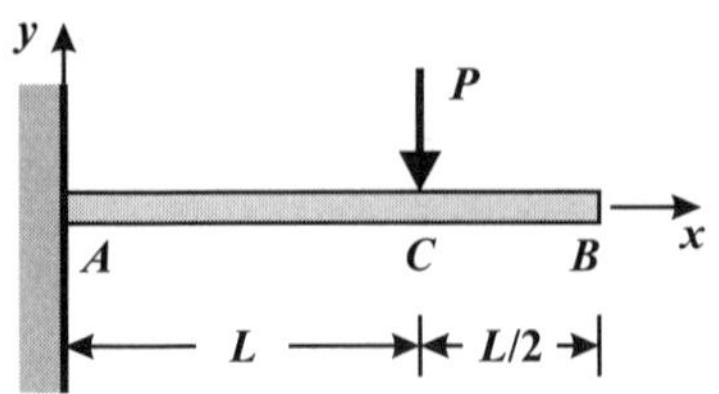

6.28 A uniformly distributed load w is applied over the built-in half of a cantilever beam of length L and constant bending stiffness EI.

(a) Derive (develop) equations for the deflection and slope, $v(x)$ and $v'(x)$, over each segment of the beam, AC and CB. (b) Determine the deflection and slope at C, v_C and θ_B, and (c) the deflection at B, v_B.

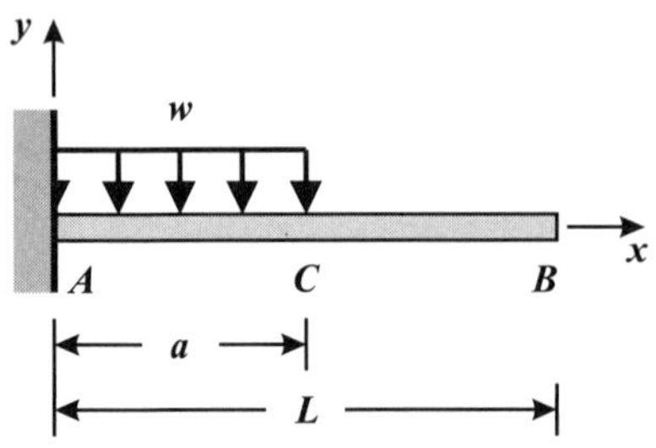

6.29 A truck (including its load) weighs $2W$. The simply supported bridge is L long, and truck's wheelbase is s. Assume that the weight of the truck is evenly distributed between the front and rear axles. The bridge has bending stiffness EI.

When the truck is at the center of the bridge, $a = (L-s)/2$, determine expressions for the displacement of the bridge from $x = 0$ to $x = L/2$.

Hint: the deflection is symmetric.

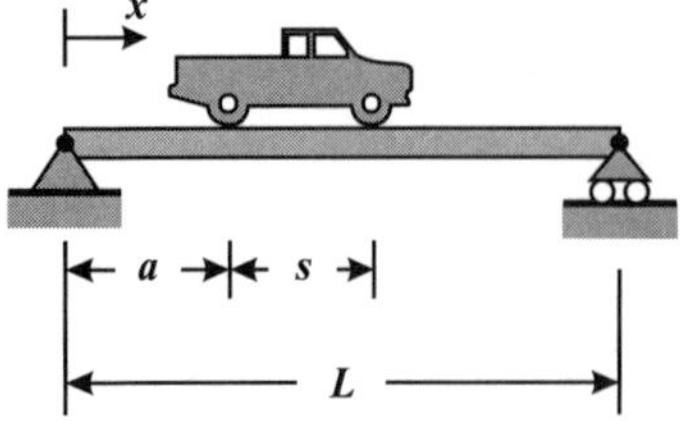

6.30 A steel I-beam AB of length $L_1 = 20$ ft supports a uniform load $w = 2.0$ kips/ft. The

beam has a moment of inertia $I = 300$ in.^{4}. At point B, the beam is supported by a steel rod of cross-sectional area $A_{BC} = 1.0$ in.^{2} and length $L_2 = 5.0$ ft. The modulus is $E = 30,000$ ksi.

(a) Determine the elongation of the rod. (b) Determine algebraic expressions for the deflection $v(x)$ and the slope $v'(x)$ of the beam. (*Hint*: the geometric boundary conditions are the displacements of each end of the beam.) (c) Determine the maximum deflection of the beam (an Excel table/plot or other computer tool might be useful).

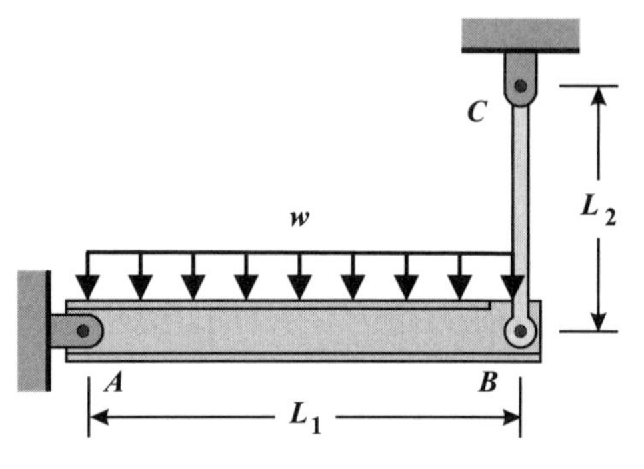

6.3   Statically Indeterminate (Redundant) Beams

6.31 Beam AB of length L is simply supported at the left end ($x = 0$) and fixed at the right end ($x = L$), and supports uniformly distributed load w. The constant bending stiffness is EI.

Determine (a) the reaction force at support A, (b) the reaction force and moment at the wall, (c) the equation of the beam's deflection $v(x)$, and (d) the deflection at the center of the beam.

Hint: The system is redundant. Take the reaction force at point A as the redundant force R. Solve for the moment $M(x)$ in terms of R, w, and L and integrate. Apply the three geometric boundary conditions (one slope and two displacement conditions) to solve for R and the two constants of integration.

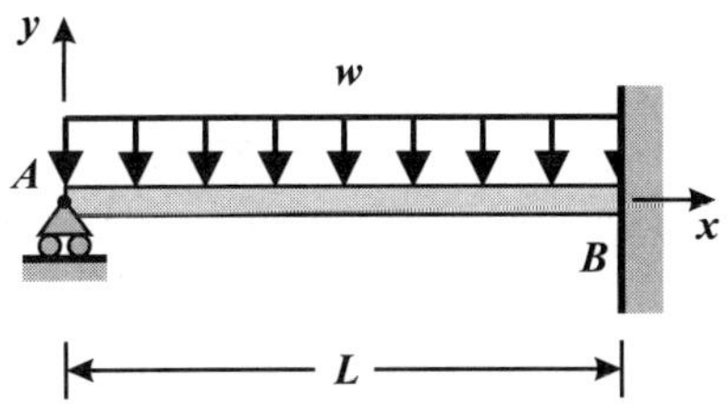

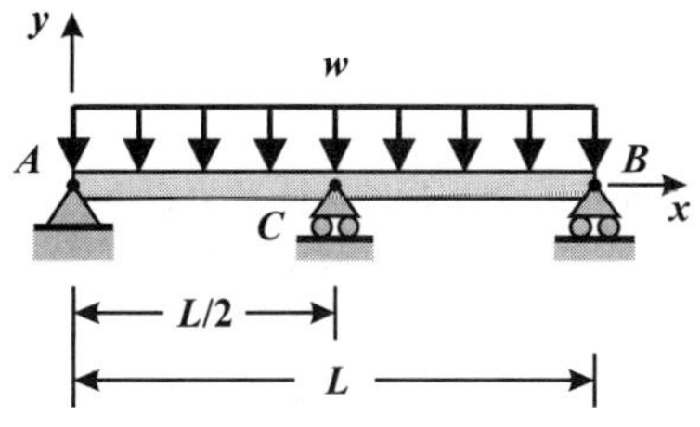

6.32 Repeat *Prob. 6.31* using the *Method of Superposition*.

Break the beam into two simpler (and statically determinate) problems: (1) a cantilever beam free at $x = 0$ under UDL w and (2) a cantilever beam free at $x = 0$ subjected to upward tip load R at point A (the redundant force). Compatibility requires that the sum of the tip displacements from the two sub-problems equal the actual displacement at point A, which is zero. This solves for R.

6.33 A beam of length L is fixed at both ends and subjected to a uniformly distributed load w. The constant bending stiffness is EI.

Determine (a) the reactions at point A and B, and (b) the maximum beam deflection.

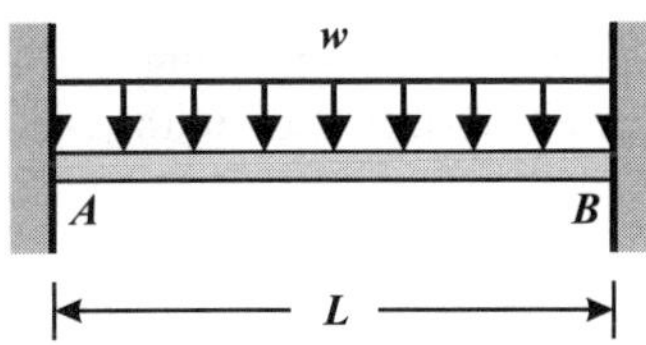

6.34 A beam of length L is simply supported at both ends, with a roller support at the center. The beam is subjected to a uniformly distributed load w. The constant bending stiffness is EI.

Determine (a) the reactions at each support and (b) the maximum deflection.

Hint: Do not assume that roller C supports half of the load.

6.4 Shear Stress

6.35 A simply supported cantilever beam $L = 4.0$ m long supports UDL $w = 2000$ N/m. The beam has rectangular cross-section $b = 0.200$ m wide and $d = 0.100$ m deep.

Determine (a) the bending stress having the greatest magnitude and where it exists (give the x- and y-coordinates), (b) the maximum shear stress and where it exists (x, y). *Note*: there may be more than one location for each type of stress.

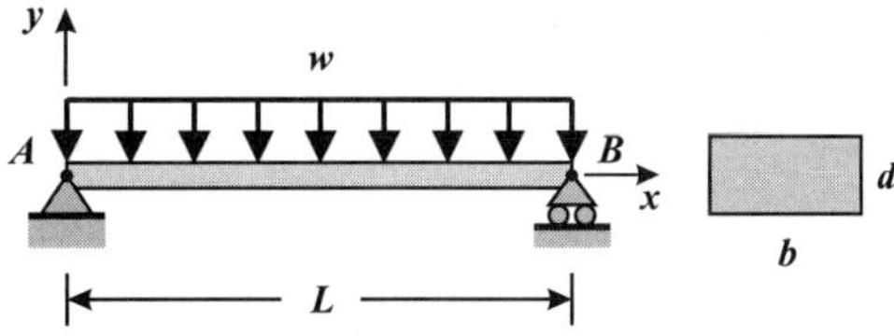

6.36 A cantilever beam supports UDL $w = 120$ lb/ft $= 10$ lb/in. The beam is 10 ft long and has a rectangular cross-section $b = 2.0$ in. wide and $d = 6.0$ in. deep.

Determine the maximum shear stress and its location.

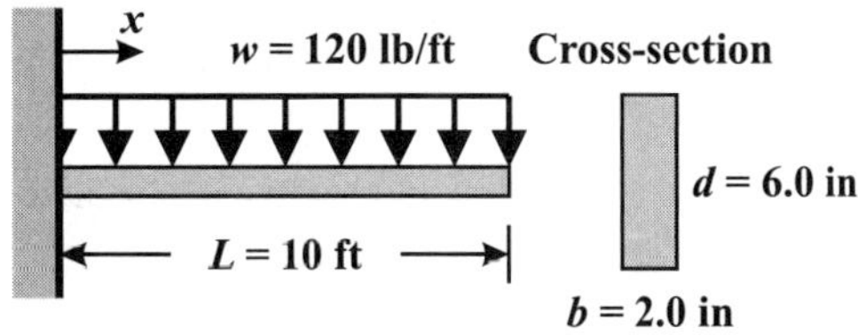

6.37 Three $L = 3.0$ m long wood metric "2 by 4" beams – actual dimensions $a \times b = 38 \times 89$ mm – are glued together to form a solid beam. The beam is simply supported. Assume the glue limits the design.

(a) If the glue joints have an allowable shear stress of $\tau_g = 1.0$ MPa, determine the allowable load P_{allow} that may be applied at the midpoint of the span. (b) Determine the maximum bending stress for the load found in *Part (a)*.

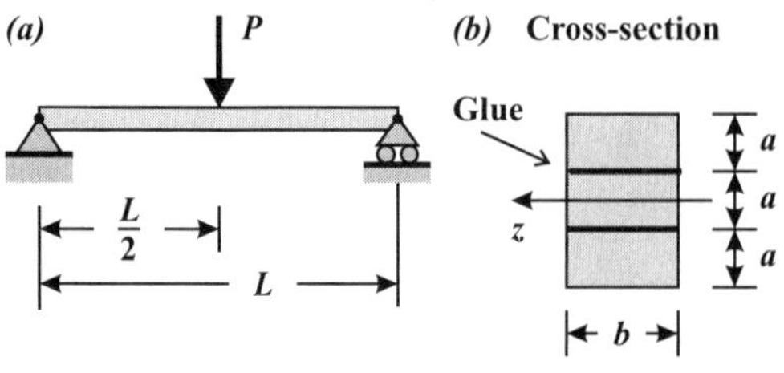

6.38 Two 8.0 ft long wood "4 by 4"s – actual dimensions $b{\times}b = 3.5$ by 3.5 in. – are glued together to form a solid beam. The weight density of wood is $\gamma = 0.02$ lb/in.^{3}. The beam is simply supported.

If the glued joint has an allowable shear stress of $\tau_g = 200$ psi, determine the allowable load P_{allow} that may be applied to the midpoint of the span (take into account the weight of the beams).

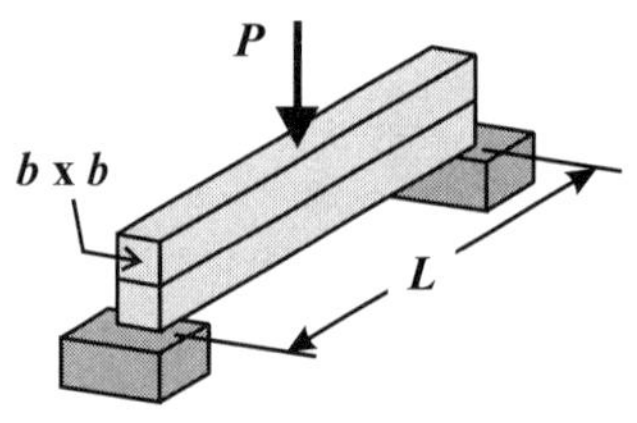

6.39 Three proposed cross-section configurations for joining three boards are shown. Each board is $b{\times}3b$ ($b = 50$ mm, $3b = 150$ mm). Each cross-section is to support a shear force of 5.0 kN.

Determine the maximum shear stress for each cross-section.

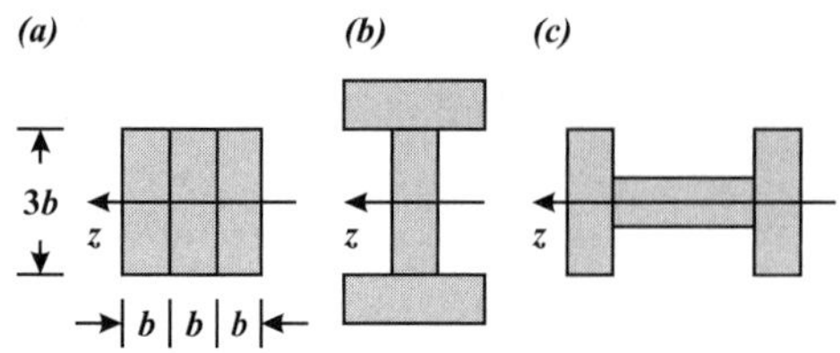

6.40 The built-up I beam supports a maximum shear force of $V = 200$ kN. The cross-section has breadth $b = 120$ mm, web height $d = 100$ mm, flange and web thicknesses $f = w = 20.0$ mm,

Determine the shear stress (a) at cut A–A, the centroid of the beam, (b) at cut B–B, just below the flange, and (c) at cut C–C, through the flange at $z = 30.0$ mm.

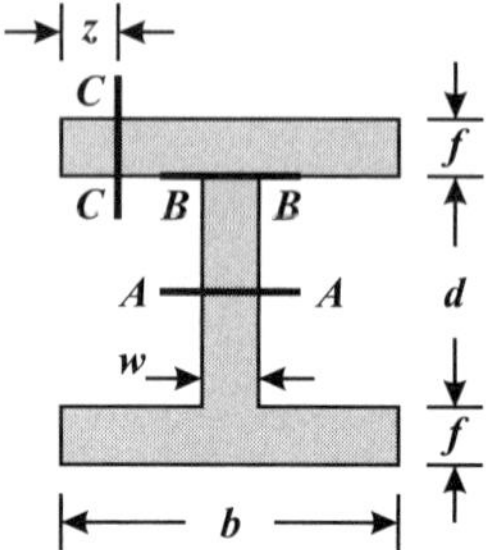

6.41 A hex wrench is subjected to downward force $F = 40$ N. The wrench is not being twisted. The side of each hexagon is $a = 3.00$ mm and the wrench has lengths $b = 40$ mm and $L = 100$ mm.

Determine (a) the maximum bending stress and (b) the maximum shear stress in the wrench. Bending occurs about the z-axis.

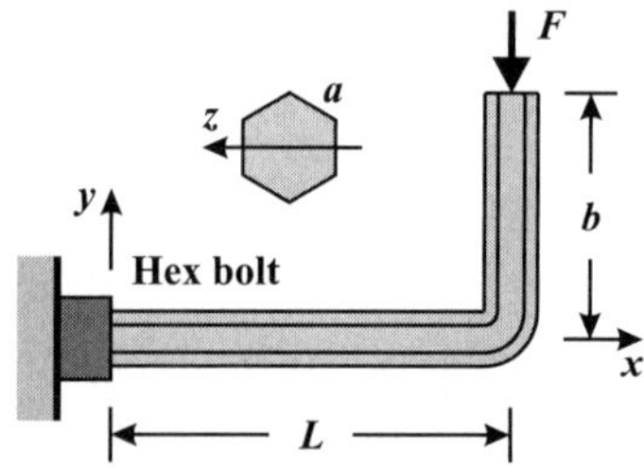

6.42 A hex wrench is subjected to downward force $F = 40$ N. The wrench is not being twisted. The side of each hexagon is $a = 3.00$ mm and the wrench has lengths $b = 40$ mm and $L = 100$ mm.

Determine (a) the maximum bending stress and (b) the maximum shear stress in the wrench. Bending occurs about the z-axis.

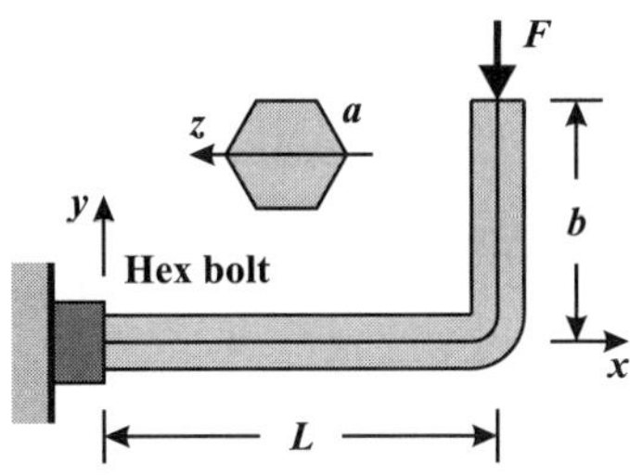

6.43 A box beam is constructed of four wooden boards. Each board has a cross-section of $b \times d = 1 \times 6$ in. (actual dimensions) as shown in the figure. The boards are joined together by nails. The allowable shear force in each nail is 700 lb.

Determine the maximum allowable spacing s of the nails if the maximum vertical shear force V on the cross-section is 2000 lb.

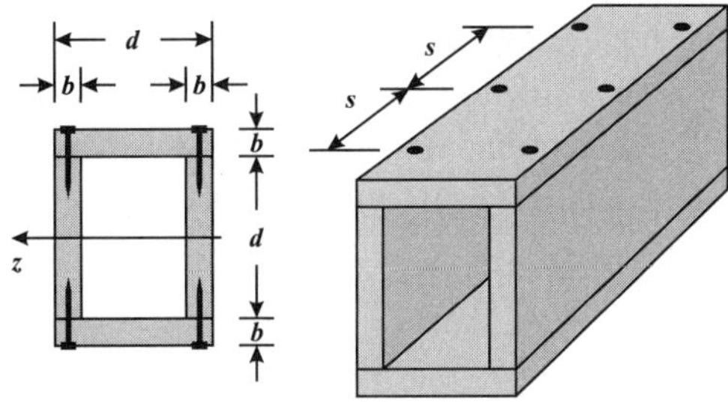

6.5 *Shape: Section Modulus and Shape Factor*

6.44 Structural steel has a yield strength of 36 ksi. A beam is to support a bending moment of 1260 kip-in. without yielding.

(a) Determine the section modulus of each I-beam cross-section listed below. (b) Which of beam(s) can support the applied moment without yielding?

Beam	A (in.^{2})	I (in.^{4})	y_{max} (in.)
S12×31	9.3	218	6.00
W10×30	8.8	170	5.75
W10×33	9.7	170	4.87
W14×26	7.7	245	6.95

6.45 Determine the section modulus about the z- and y-axes for (a) a rectangle of width $b = 4.0$ in. and depth $d = 8.0$ in. and (b) a solid circle with diameter $D = 4.0$ in.

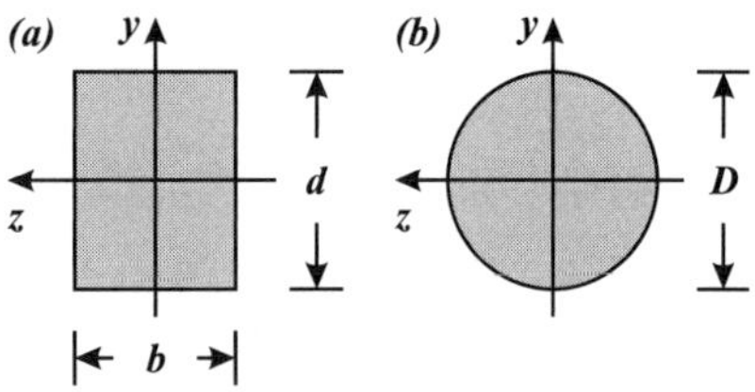

6.46 Determine the section modulus about the z-axis for (a) the built-up I-beam: $w = f = 0.5$ in., $b = 6.0$ in. and $d = 7.0$ in. and (b) the built-up T-beam: $w = f = 0.5$ in., $b = 6.0$ in., and $d = 8.0$ in.

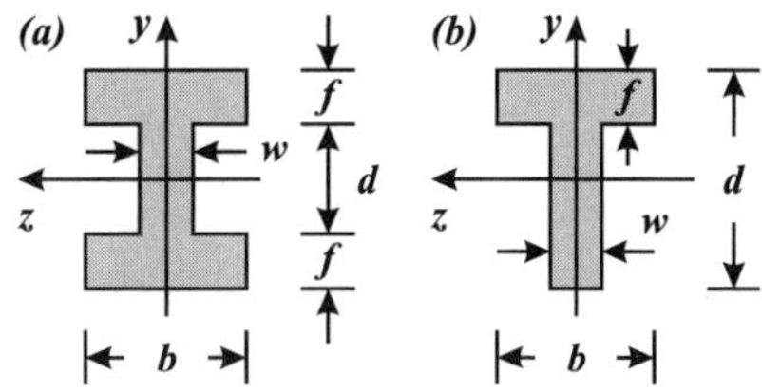

6.47 Determine the section modulus about the z-axis for a truss, with half of its area assumed to be concentrated at each chord.

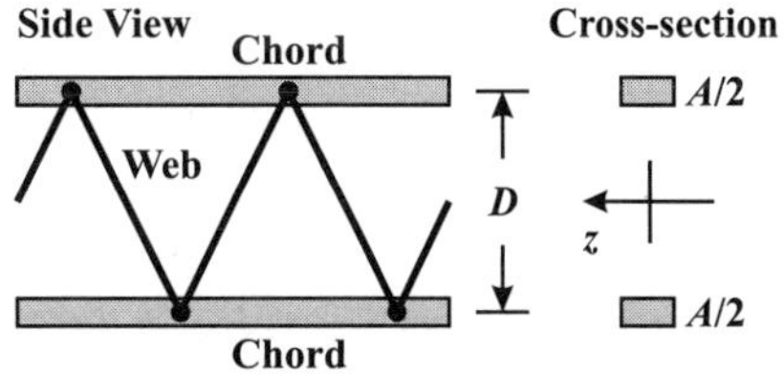

6.48 Determine the numerical elastic shape factor about the z-axis for (a) the rectangle and (b) the circle in *Prob. 6.45*.

6.49 Determine the numerical elastic shape factor about the z-axis for the built-up I-beam in *Prob. 6.46*.

6.50 **(a)** Derive the algebraic elastic shape factor about the z-axis for the truss in *Prob. 6.47.* If $A/2 = 4.0$ in.^{2} and $D = 24$ in., determine the numerical shape factor for the truss.

6.7 Design of Beams

6.51 A steel cantilever beam is subjected to a downward tip load $P = 10.0$ kN. The beam is 4.0 m long. The allowable bending stress is 100 MPa and the allowable deflection is 1/250 of the span. The modulus of steel is 200 GPa.

(a) If the beam is to be three times as deep as it is wide ($d = 3b$), determine the size of the beam. (b) Select the *smallest* S-shape I-beam from *Appendix C* (in terms of cross-sectional area) that will also satisfy the design requirements. What is the weight savings (as a percent) compared to the rectangular beam?

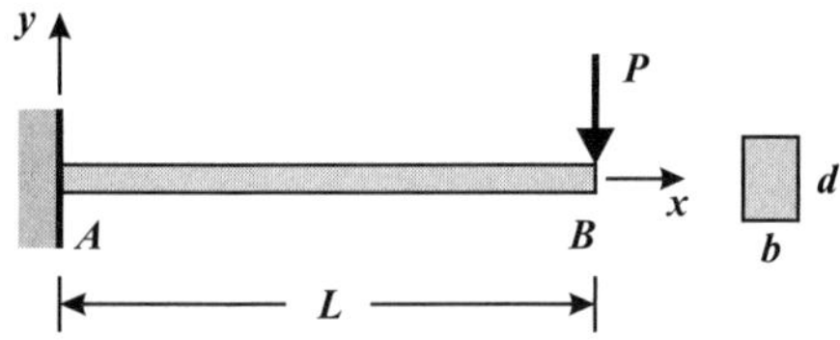

6.52 A steel ($E = 30,000$ ksi) cantilever beam of length $L = 8.0$ ft is subjected to a uniformly distributed load $w = 1000$ lb/ft.

(a) Select the *smallest* W-shape I-beam from *Appendix C* (in terms of cross-sectional area) so that both of the following conditions are satisfied:

1. the allowable bending stress is 16 ksi and
2. the allowable deflection is 0.50 in.

(b) Reanalyze the beam to include the weight of your beam in the distributed load. Do you need to change the beam cross-section? If so, to what W-shape? (Assume space considerations require using the same nominal depth.)

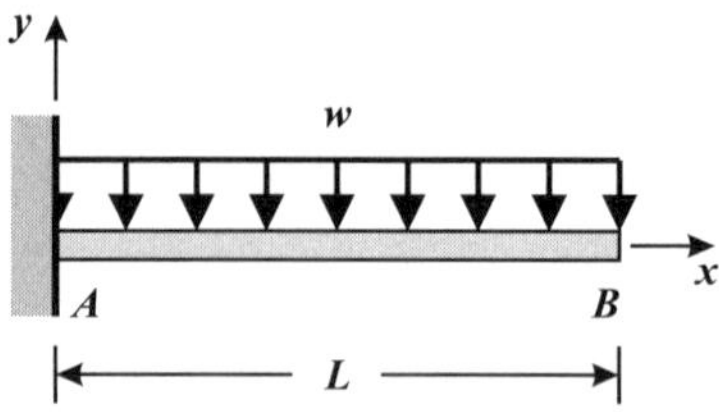

6.53 A student rides a skateboard at constant velocity across the campus. The student's effective weight (due to static and dynamic loading) is assumed to be applied to the board as shown. The skateboard is constructed of a wooden laminate (i.e., plywood). Neglect the weight of the board, and assume the skateboard is simply supported at the wheels.

In order to avoid failure, the board must:

1. not touch the ground,
2. not break in bending, and
3. not delaminate (fail in shear).

Determine the maximum effective weight W_{max} so that the board does not fail by any of the three failure criteria listed above.

Skateboard dimensions
Length:	$L = 42.0$ in.
Width:	$b = 10.0$ in.
Thickness:	$t = 0.50$ in.
Height above ground:	$c = 3.0$ in.

Plywood properties:
Elastic modulus:	$E = 1400$ ksi
Bending strength:	$S_f = 6.4$ ksi
Shear strength:	$\tau_f = 1.0$ ksi

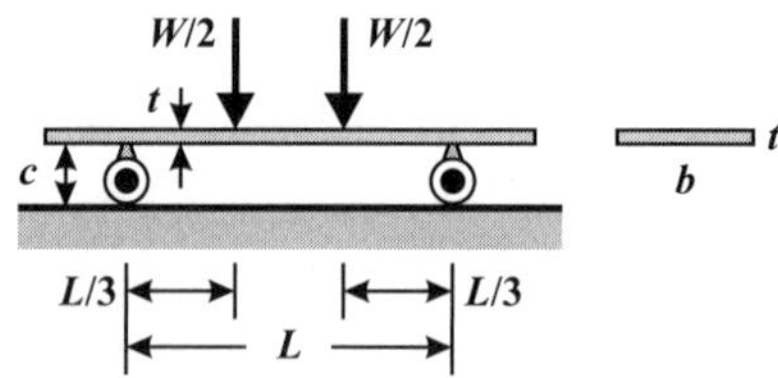

6.54 An aluminum ($E = 10,000$ ksi) cantilever is $L = 12$ ft long, and is subjected to tip load $P = 4000$ lb. The beam has a rectangular cross-section, b wide by d deep.

The allowable normal stress is $\sigma_{allow} = 10.0$ ksi and the allowable shear stress is $\tau_{allow} = 6.0$ ksi. If the depth of the beam is to be *twice* the width ($d = 2b$), determine the minimum required depth, d_{min}, to satisfy both requirements.

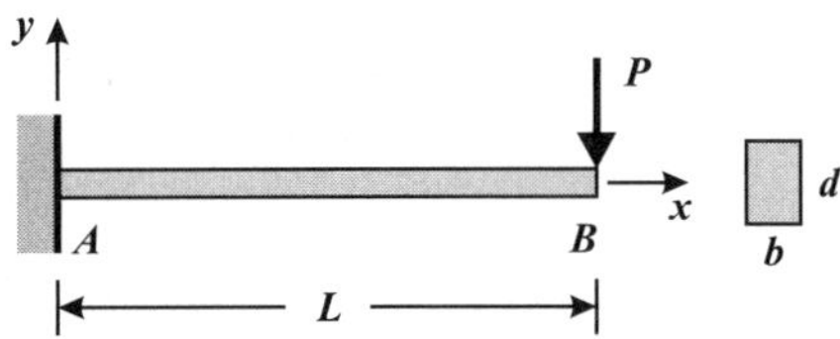

6.55 Select a material for a tip-loaded cantilever beam that will minimize the weight (mass) for a specified stiffness $k = P/\delta$. Assume that the cross-sectional shape of each beam is the same. Length L is fixed.

Material	$\rho\,(\mathrm{kg/m^3})$	E (GPa)
Steel	7800	200
Aluminum	2700	70
Silicon carbide	2700	450
Wood	600	12

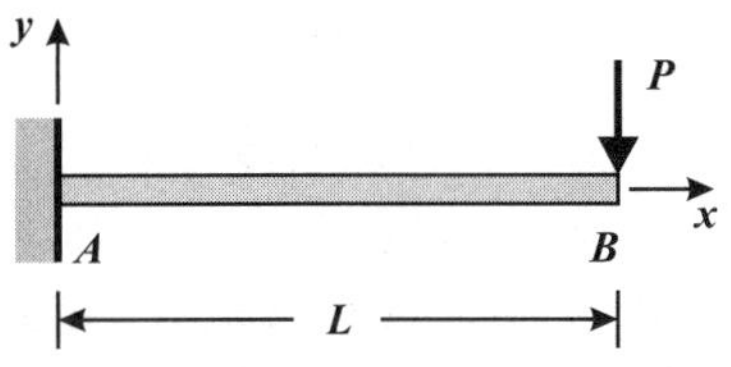

6.56 Consider a tip-loaded cantilever beam of length L. For a given stiffness, $k = P/\delta$, to minimize the mass, the following combination of material properties and elastic shape factor must be maximized:

$$\left(\frac{E}{\rho^2}\right)(\phi)$$

(a) From the below table, which combination will result in a beam of least mass? (b) If the cantilever beam is 4.0 m long, and is to be made of the wooden section ($\phi = 6$), determine the cross-sectional dimensions if the stiffness is to be 2×10^6 N/m

(e.g., the beam deflects 1.0 mm for a load of 2000 N).

Material	ρ $(\mathrm{kg/m^3})$	E (GPa)	ϕ (shape)
Wood	600	12	6 (e.g, rectangle)
Aluminum	2700	70	24 (e.g., box beam)
Steel	7800	200	30 (e.g, I-beam)

6.57 A simply supported beam is subjected to a uniformly distributed load w. The beam has length L, cross-sectional area A, and moment of inertia I. The material properties are E and ρ.

(a) The downward deflection at the center of the beam is $\delta = 5wL^4/(384EI)$. For a given stiffness $k = wL/\delta$, determine the equation for the mass of the beam in terms of w, L, ρ, E, and elastic shape factor ϕ. (b) Three different materials are proposed (see table) for a beam having specified length L and stiffness wL/δ. If the shape factor for the titanium cross-section is $\phi = 24$, determine the shape factors for the aluminum and steel cross-sections so that the mass of each beam is the same.

Materials	$\rho\,(\mathrm{kg/m^3})$	E (GPa)
Steel	7800	200
Aluminum	2700	70
Titanium	3600	115

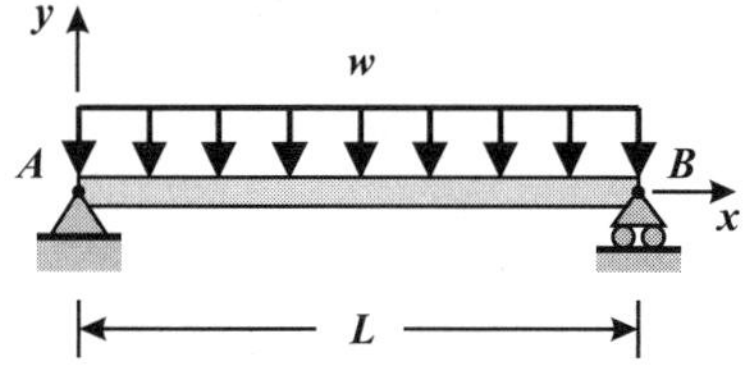

Problems: **Chapter 7 Combined Loading**

7.1 The cylindrical length of a screwdriver blade is $L = 120$ mm long, with a diameter of $D = 7.00$ mm. The screwdriver is made of steel:

$$E = 200 \text{ GPa}, \quad G = 75 \text{ GPa},$$

$$S_y = 400 \text{ MPa}, \tau_y = 250 \text{ MPa}.$$

As a person drives the screw, she applies a torque of $T = 6.0$ N·m, and an axial compressive force of $F = 100$ N.

(a) Determine the maximum shear stress in the cylindrical length of the screwdriver due only to torsion. (b) Determine the normal stress in the screwdriver due to force F. (c) Draw a stress element that exists on the front surface of the shaft (in the x–y plane). Draw and label (with values and units) the stresses that exist on the element.

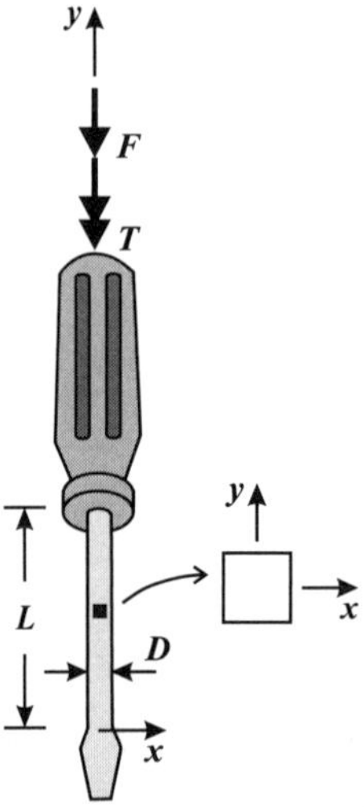

7.2 An L-bracket is made of a hollow rectangular cross-section of width $b = 40$ mm, depth $d = 80$ mm, and thickness $t = 5.0$ mm. Lengths $L_1 = 400$ mm and $L_2 = 300$ mm.

If $P = 2000$ N, determine the state of stress (the nature and values and senses of the stresses) at points A and B. Point B is opposite point A. Draw the stress elements at points A

and B as viewed from the outside of the bracket (e.g., material point A is shown below in its coordinate system). Force P is in the z-direction and passes through the centroid of the cross-section.

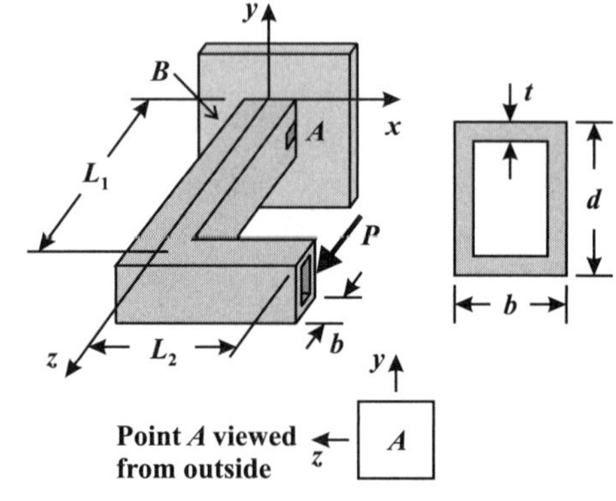

7.3 A sign weighing 40 lb is $L = 6.0$ ft long and is held up by two cables that are pulled nearly horizontal in front of the entrance of a parking lot.

Determine the stress states at the top and bottom of the sign (points A and B) due only to the horizontal load in the cables. Neglect any vertical loading.

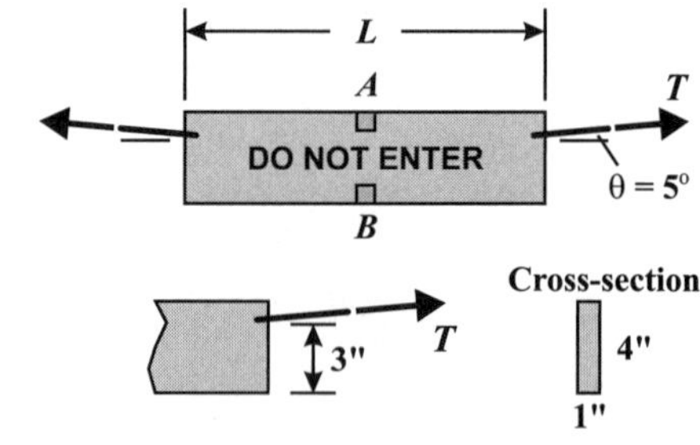

7.4 Force F is applied at the end of a cantilever beam as shown.

For any angle α, determine the stress states at the top, middle, and bottom of the beam, points S, O, and Q, respectively. Draw and label the stress elements at points S, O, and Q, in terms of F, b, d, L, and α.

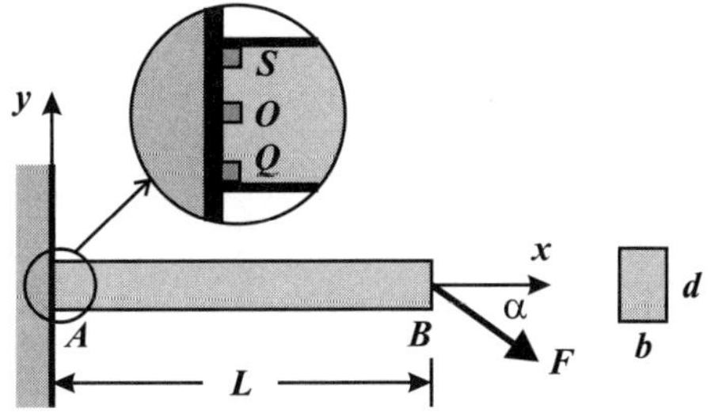

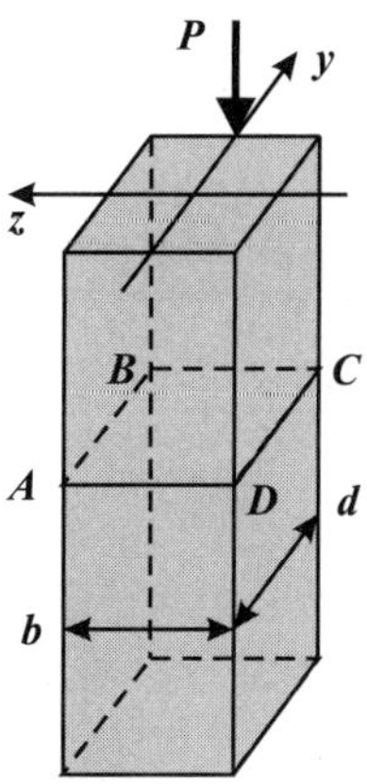

7.5 A tight-rope walker weighs $W = 160$ lb. He walks on a rope $H = 20.0$ ft above the ground. The rope spans two poles that are $L = 30.0$ ft apart. When he reaches the center, the rope is taut and has deflected $\delta = 1.0$ ft. The diameter of each solid pole is $2R = 8.0$ in.

Determine the stress states at the base of the left pole – on its left side (A), right side (B) and front (C). Draw the three stress elements as viewed from outside of the pole (the z-axis points out of the paper).

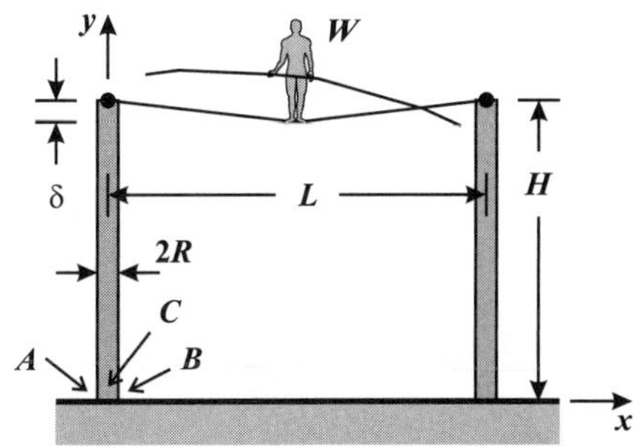

7.6 Axial compressive load $P = 200$ kN is applied on the y-axis of a column's cross-section at $y = +d/2$. The dimensions of the cross-section are $b = 300$ mm and $d = 500$ mm.

Determine the stresses at points A and C, at any cross-section within the column.

7.7 Axial compressive load $P = 200$ kN is applied on the corner of a column's rectangular cross-section at $y = +d/2$ and $z = -b/2$. The dimensions of the cross-section are $b = 300$ mm and $d = 500$ mm.

Determine the stresses at points A, B, C, and D, at a cross-section within the column

Note: Bending occurs about two axes.

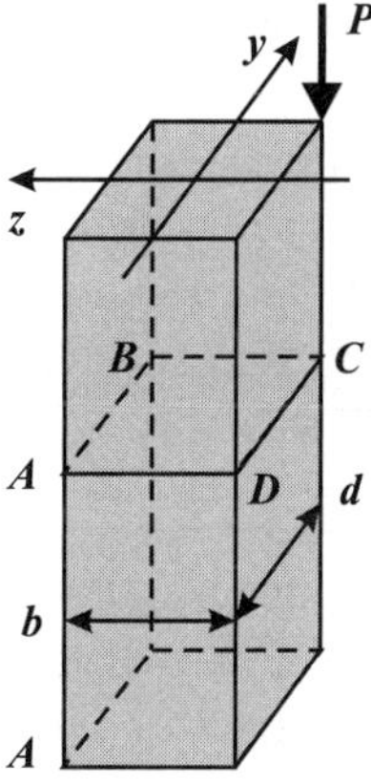

7.8 A highway sign is subjected to wind pressure $p = 1.2$ kPa. The sign is supported by a pipe 4.0 m long, having outside diameter $D = 150$ mm and thickness $t = 10$ mm.

Determine (a) the torque, shear force and moment acting at the base due to the wind load. Neglect the weight of the sign and pipe. (b) Determine the stresses at points A, B, C, and D – the front, left, right, and rear of the pipe at the base. (c) Draw the stress elements as viewed from the outside of the pipe. Include the numerical values and appropriate units.

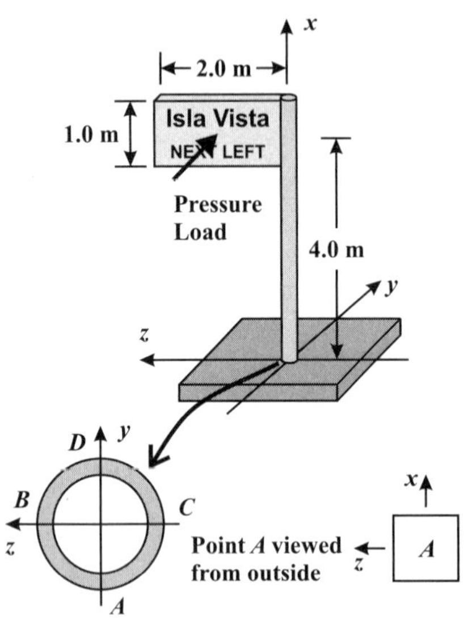

7.9 A propeller-driven ship moves forward at constant velocity against a drag force of 300 kN. The solid drive shaft transmits 2500 kW to the propeller at 1000 rpm. The radius of the shaft is 200 mm.

Determine (a) the normal stress along the axis of the shaft and (b) the shear stress due to torsion only. (c) Draw a stress-element aligned with the axial and hoop (circumferential) directions of the shaft.

7.10 A steel column has a W12×96 cross-section. The column has length $L = 12.0$ ft and supports a compressive force $P = 100$ kips. Load P is applied eccentrically on the y-axis at distance e from the centroid. Structural steel has modulus $E = 30,000$ ksi and yield strength $S_y = 36$ ksi (in either tension or compression).

Determine the maximum value of e such that there are no tensile stresses in the column.

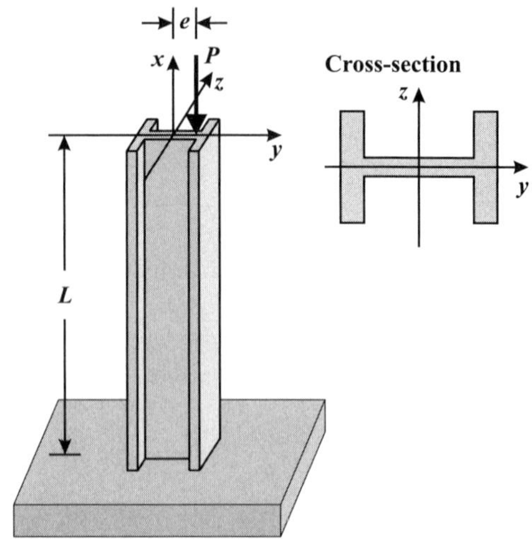

7.11 Show that the kern of a solid circular column is a circle of radius $r = R/4$, where R is the radius of the column. In other words, show that a compressive load P must have an eccentricity e less than $R/4$ so that no tension is developed on the cross-section.

7.12 At Disneyland, guests were once able to get back and forth from Tomorrowland to Fantasyland by riding the *Skyway*, a gondola reminiscent of those in the Alps. There was even a dedication plaque from the American Society of Mechanical Engineers (ASME).

A similar set-up is shown. The gondola is supported by two bent arms (one being shown in the figure). The arms are round with a diameter D. The main vertical section of each arm is offset by $L = 7.0$ in. from the line of action of the gondola's weight force. The force on each arm is W. The *allowable normal stress* in the arm material is ±15.0 ksi and the *allowable shear stress* is 8.0 ksi.

If the gondola is required to support a total weight of $2W = 1300$ lb, determine the diameter D of the arms.

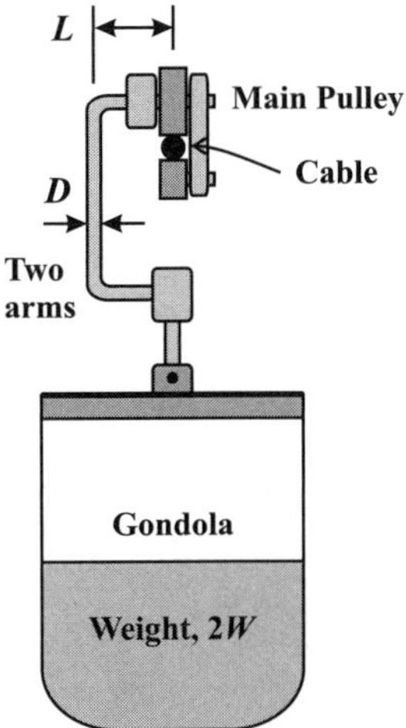

7.13 In design, it is often assumed that the tensile strength of a brittle material is zero. Components made of brittle materials can be strengthened by *pre-stressing* so that the brittle material is in compression before application of the load.

A rectangular concrete beam of length L, depth d, and breadth b is pre-stressed by a steel rod that is pulled in tension through the center of the beam with load P, and clamped while in tension to the ends of the beam. The rod under tension force P places the concrete in compression with force $-P$; stresses exist in the beam before the externally applied load is introduced. The beam is then loaded with a uniformly distributed load w.

Estimate the load w_{max} (in terms of P) when cracking first appears in the beam, i.e., when the stress anywhere in the beam becomes tensile. Neglect the area of the steel reinforcement, as well as its effect in supporting the bending load (the rod is at the centroid of the cross-section).

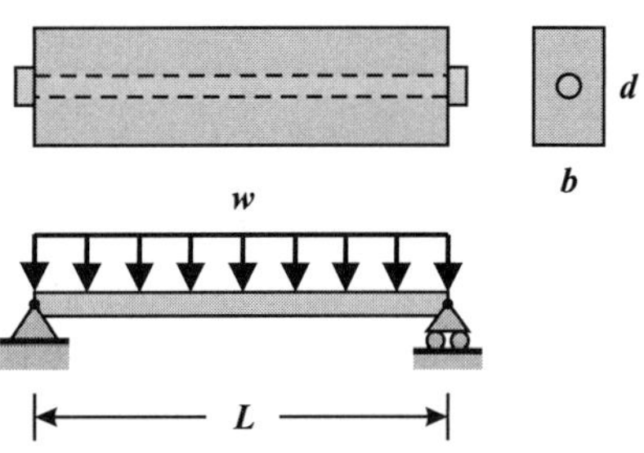

7.14 A six-pack is shaken so that the pressure in each can is $p = 20$ psi. A board is placed on the six pack and a student(s) weighing W stands on the board so his weight is evenly distributed to the six cans. The aluminum properties are $E = 10,000$ ksi, $S_y = 35$ ksi, and $v = 0.33$. Each can has an average radius of $R = 1.25$ in., thickness $t = 0.01$ in., and length $L = 4.50$ in.

Determine the critical weight of the student, W_{cr}, so that a stress element on the surface of the can is in a state of uniaxial stress. Assume that no support is provided by the contained liquid, and that the cans do not fail by buckling (being crushed).

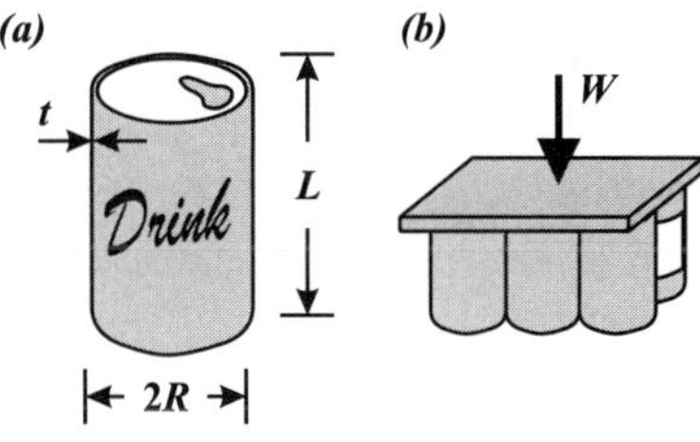

7.15 A boom extending out of the Space Shuttle is modeled below. The boom is required to support force F applied vertically at the tip. Assume the boom has a hollow circular cross-section of inner and outer diameters D_i and D. $F = 400$ N, $a = 12$ m, $b = 2.0$ m, $D_i = 80$ mm, $D = 100$ mm.

(a) Determine the stress states at points A, B, and C – at the left, top and right of the boom at the built-in end. (b) Draw the stress elements at each point as viewed from the outside of the boom.

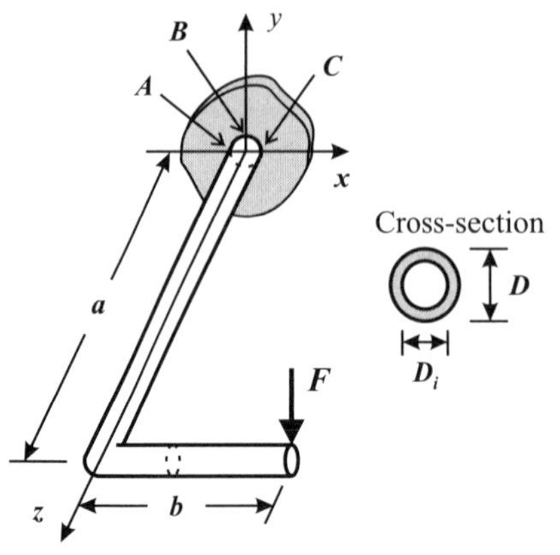

7.16 In attempting to open a locked door, a person applies a downward force F to the door handle. The handle, with solid circular cross-section of diameter D, does not perceptively move (the door is locked). Thus, cross-section ABC is fixed at the doorplate. The force and geometry are $F = 20$ lb, $e = 2.5$ in., $h = 4.0$ in., and $D = 0.75$ in.

(a) Determine the algebraic stress states at surface points A, B, C – at the left, top, and right of the handle at the built-in end. Give the stresses in terms of F, e, h, D, A, I, and/or J. Draw and label the stress element at each point as viewed from the outside of the handle. (b) Determine the numerical stresses at points A, B, and C. Draw and label the stress element at each point.

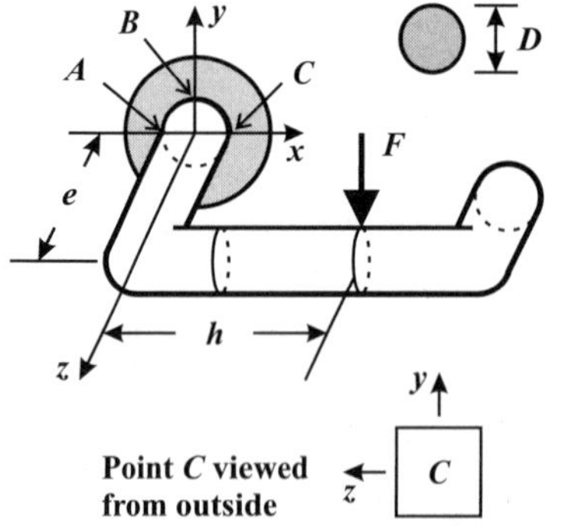

7.17 In attempting to open a locked door, a person applies a downward force F and an inward force P to the door handle. The handle, with solid circular cross-section of diameter D, does not perceptively move (the door is locked). Thus, cross-section ABC is fixed at the doorplate. The forces and geometry are $F = 20$ lb, $P = 30$ lb, $e = 2.5$ in., $h = 4.0$ in., and $D = 0.75$ in.

(a) Determine the algebraic stress states at surface points A, B, C – at the left, top, and right of the handle at the built-in end. Give the stresses in terms of F, P, e, h, D, A, I, and/or J. Draw and label the stress element at each point as viewed from the outside of the handle. (b) Determine the numerical stresses at points A, B, C. Draw and label the stress element at each point.

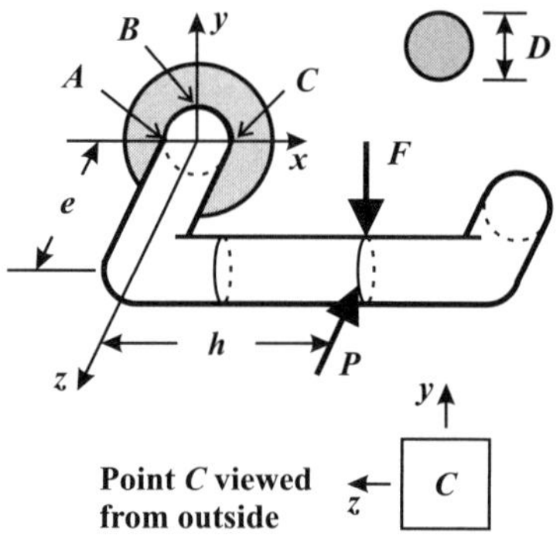

7.18 Structure $ABCD$ has a hollow circular cross-section, outer diameter $2R = 6.0$ in. and inner diameter $2R_i = 5.5$ in. The bracket consists of vertical arm AB, and horizontal arms BC (along the z-axis) and horizontal arm CD (along the x-axis). The lengths of segments AB, BC, and CD are: $a = 8.0$ ft, $b = 4.0$ ft, and $c = 2.0$ ft, respectively. A downward force of $P = 2.0$ kips acts at point D.

Consider only 3D stress elements where the three faces are aligned with the x-, y-, and z-axes. Determine the maximum tensile stress, the maximum (least tensile) compressive stress, and the maximum shear stress in segment (a) CD, (b) BC, and (c) AB.

Hint: Consider how each segment acts as an axial, a torsional, and/or a bending member – first CD, then BC, and finally AB.

(d) Determine the maximum tensile and compressive stresses in AB (the faces of the 3D element are not necessarily aligned with the x-, y-, and z-axes).

Note: The shear force acting in a beam of hollow circular cross-section causes a maximum shear stress of:

$$\tau_{max} = \frac{4}{3}\frac{V}{A}\left(\frac{R^2 + RR_i + R_i^2}{R^2 + R_i^2}\right)$$

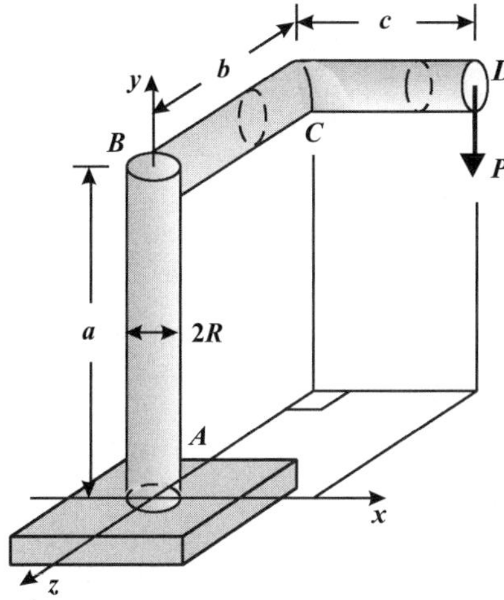

7.19 A steel I-beam of length $L_1 = 20$ ft supports UDL $w = 2.0$ kips/ft. The beam has a moment of inertia $I = 300$ in.^{4}. At point B, the beam is supported by a steel rod of cross-sectional area $A_{BC} = 1.0$ in.^{2} and length $L_2 = 5.0$ ft. Young's modulus is $E = 30,000$ ksi.

(a) Determine algebraic expressions for the deflection $v(x)$ and the slope $v'(x)$ of the beam. Do this using the method of superposition:

- find $v_1(x)$ and $v_1'(x)$ of a simply supported beam under distributed load w;
- find $v_2(x)$ and $v_2'(x)$ that a *rigid* beam AB would experience due to stretching of rod BC;

– sum the results to find the total, e.g., $v(x) = v_1(x) + v_2(x)$.

(b) Determine the location and value of the maximum deflection of the beam.

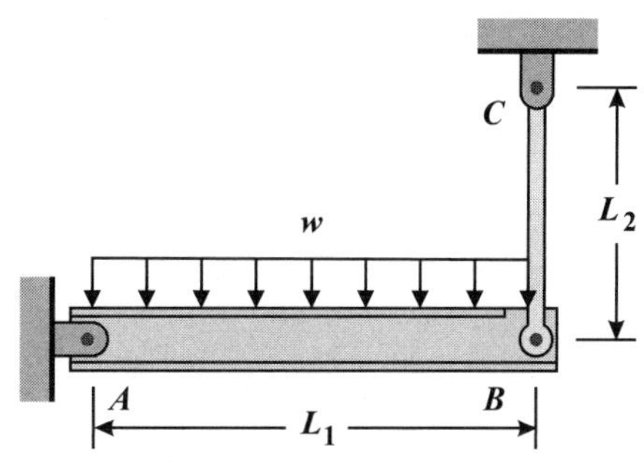

7.20 A beam with bending stiffness EI supports UDL w. The beam is simply supported at point A by a roller and built-in at point B.

Derive expressions for the deflection $v(x)$ and the slope $v'(x)$ of the beam by using the method of superposition. Superimpose the solutions of two cases: (1) a cantilever beam fixed at point B under UDL w and (2) a cantilever beam fixed at point B loaded by redundant force R at point A. Enforce the required deflection at roller A.

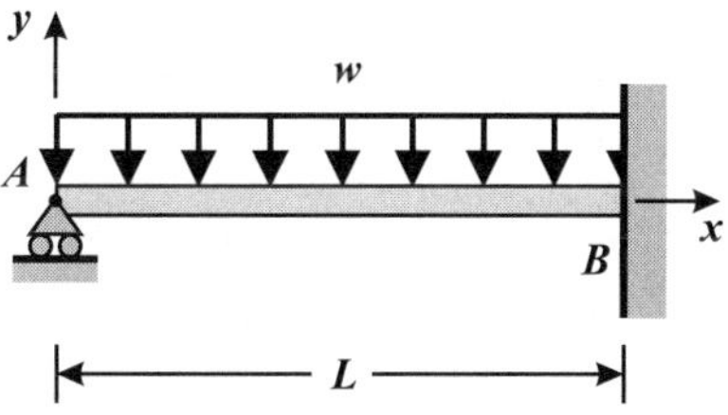

7.21 Repeat *Prob. 7.20* by superimposing the solutions of (1) a simply supported beam under UDL w and (2) a simply supported beam under redundant couple M_B applied at point B. Enforce the required slope at point B.

Problems: **Chapter 8 Transformation of Stress and Strain**

8.1 Stress Transformation (Plane Stress)

8.2 Principal Stresses

8.3 Maximum (In-Plane) Shear Stress

8.1 The stress states (σ_x, σ_y, τ_{xy}) at several points in a material system are listed below.

Determine the transformed stresses ($\sigma_{x'}$, $\sigma_{y'}$, $\tau_{x'y'}$) on each element when it is rotated by the given angle θ (positive counterclockwise). Drawn and label the stress element in its new orientation.

(a) (200 MPa, 100 MPa, 0 MPa), $\theta = 20°$
(b) (–200 MPa, 0 MPa, 80 MPa), $\theta = 20°$
(c) (200 MPa, 100 MPa, 80 MPa), $\theta = 30°$
(d) (200 MPa, 100 MPa, 80 MPa), $\theta = -30°$
(e) (200 MPa, –100 MPa, 80 MPa), $\theta = 35°$
(f) (200 MPa, 100 MPa, –80 MPa), $\theta = 35°$
(g) (200 MPa, 100 MPa, –80 MPa), $\theta = -35°$
(h) (–200 MPa, 100 MPa, 80 MPa), $\theta = 75°$
(i) (–200 MPa, –100 MPa, –80 MPa), $\theta = 65°$

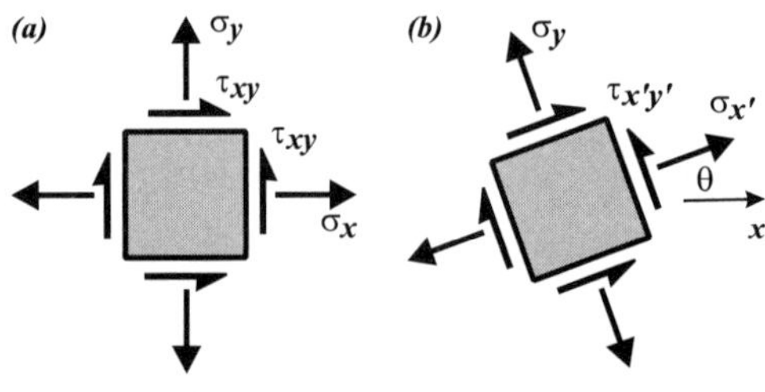

8.2 Consider a state of stress (σ_x, σ_y, τ_{xy}). As the stress element is rotated, (a) what is the shear stress when the normal stresses equal the principal stresses? (b) What are the normal stresses when the in-plane shear stress is the maximum shear stress?

8.3 The stress states (σ_x, σ_y, τ_{xy}) at several elements in a material system are listed below.

Determine the principal stresses and the directions in which they act. Draw the stress element in its new orientation.

(a) (15 ksi, 0 ksi, –6 ksi)
(b) (15 ksi, 10 ksi, 6 ksi)
(c) (15 ksi, –10 ksi, 6 ksi)
(d) (–15 ksi, –10 ksi, 6 ksi)
(e) (15 ksi, 10 ksi, –6 ksi)
(f) (–10 ksi, –10 ksi, 6 ksi)
(g) (10 ksi, –10 ksi, –6 ksi)

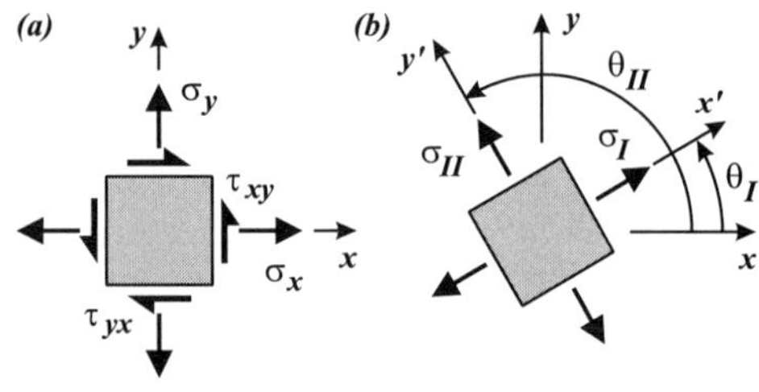

8.4 The stress states (σ_x, σ_y, τ_{xy}) at several elements in a material system are listed below.

Determine the maximum in-plane shear stresses and the associated normal stresses. Draw the stress element in its new orientation.

(a) (15 ksi, 0 ksi, –6 ksi)
(b) (15 ksi, 10 ksi, 6 ksi)
(c) (15 ksi, –10 ksi, 6 ksi)
(d) (–15 ksi, –10 ksi, 6 ksi)
(e) (15 ksi, 10 ksi, –6 ksi)
(f) (–10 ksi, –10 ksi, 6 ksi)
(g) (10 ksi, –10 ksi, –6 ksi)

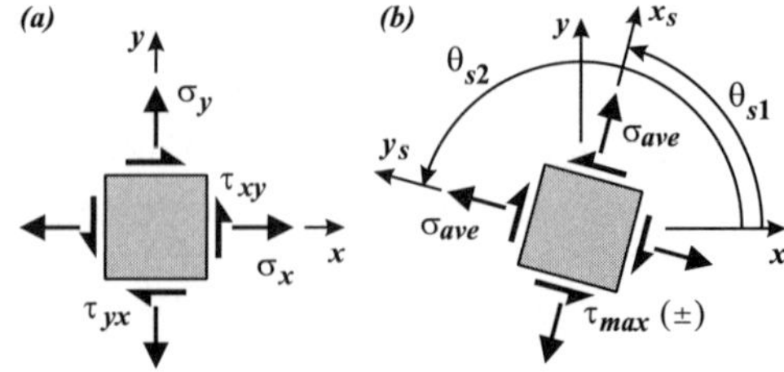

8.5 To determine if a weld is strong enough, the stress perpendicular to the weld-line, σ_w, and the shear stress parallel to the weld-line, τ_w, must be determined.

A 400 by 240 mm test plate is formed by welding two trapezoidal plates together. A uniaxial load of 60 MPa is applied to the plate.

(a) Determine the normal stress σ_w acting perpendicular to the weld, and the shear stress τ_w acting parallel to the weld (*note*: σ_w is positive if it acts in tension and τ_w is positive if it acts on the new x-face in the new y-direction). (b) Determine the maximum shear stress in the main volume of the plate.

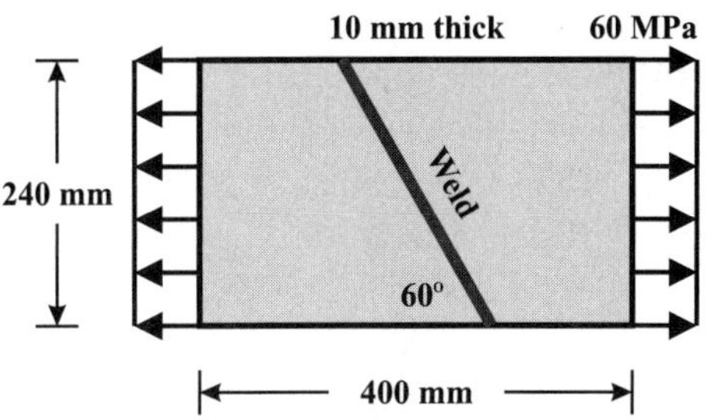

8.6 A 9.0 by 15 in. plate formed by welding two triangular plates together. The loads applied to the plate result in a tensile stress of 5.0 ksi in the x-direction and a compressive stress of 1.0 ksi in the y-direction.

(a) Determine the normal stress σ_w acting perpendicular to the weld and the shear stress τ_w acting parallel to the weld (*note*: σ_w is positive if it acts in tension and τ_w is positive if it acts on the new x-face in the new y-direction). (b) Determine the maximum In-plane shear stress in the main volume of the plate.

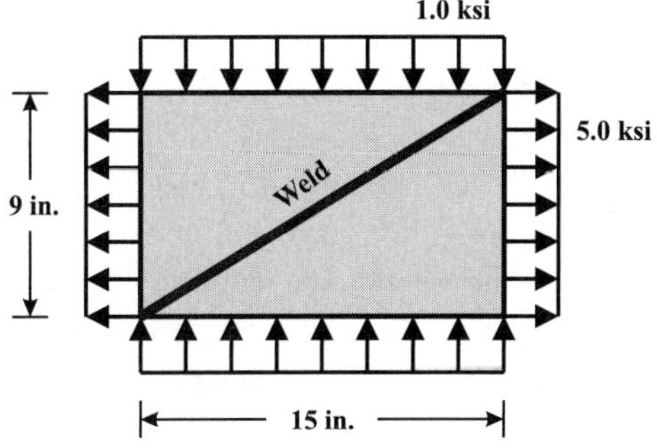

8.7 The cylindrical length of a screwdriver blade is $L = 120$ mm long with a diameter of $D = 7.00$ mm. The screwdriver is made of steel:

$$E = 200 \text{ GPa}, G = 75 \text{ GPa},$$
$$S_y = 400 \text{ MPa}, \tau_y = 250 \text{ MPa}.$$

As a person drives the screw, she applies a torque of $T = 6.0$ N·m and an axial compressive force of $F = 100$ N.

(a) Draw and label a stress element on the front surface of the shaft (in the x–y plane). (b) Determine the principal stresses for this stress state, and (c) the maximum in-plane shear stress.

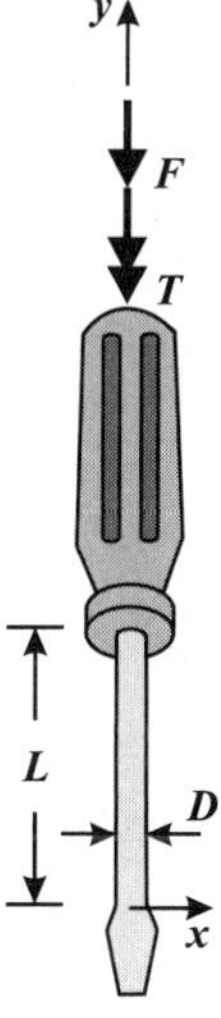

8.8 A propeller-driven ship moves forward at constant velocity against a drag force of 300 kN. The solid drive shaft transmits 2500 kW to the propeller at 1000 rpm. The radius of the shaft is 200 mm.

Determine (a) the maximum principal stress on the surface of the shaft and the angle it makes with the shaft axis, (b) the minimum principal stress on the surface of the shaft, and (c) the maximum in-plane shear stress on the surface of the shaft.

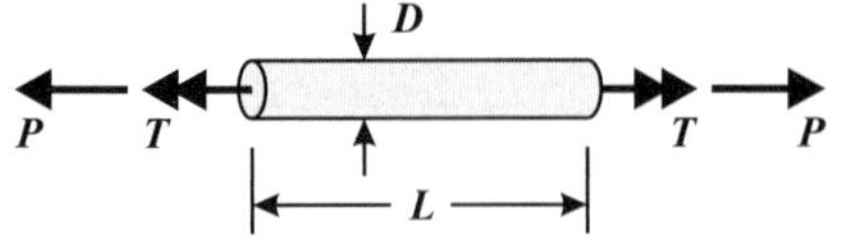

8.9 An L-bracket is made of a hollow rectangular cross-section of width $b = 40$ mm, depth $d = 80$ mm, and thickness $t = 5.0$ mm. Lengths $L_1 = 400$ mm and $L_2 = 300$ mm.

If $P = 2000$ N, determine (a) the principal stresses at points A and B and (b) the maximum in-plane shear stresses at points A and B. Point B is opposite point A. Force P is in the z-direction and passes through the centroid of the cross-section.

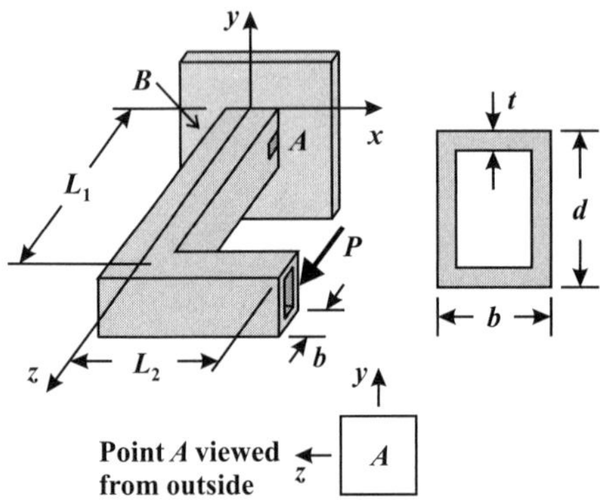

Point A viewed from outside

8.10 A boom extending out of the Space Shuttle is modeled below. The boom is required to support force F applied vertically at the tip. Assume the boom has a hollow circular cross-section of inner and outer diameters D_i and D. $F = 400$ N, $a = 12$ m, $b = 2.0$ m, $D_i = 80$ mm, $D = 100$ mm.

Determine (a) the numerical values of the principal stresses at point A and (b) the numerical value of the maximum shear stress at point B.

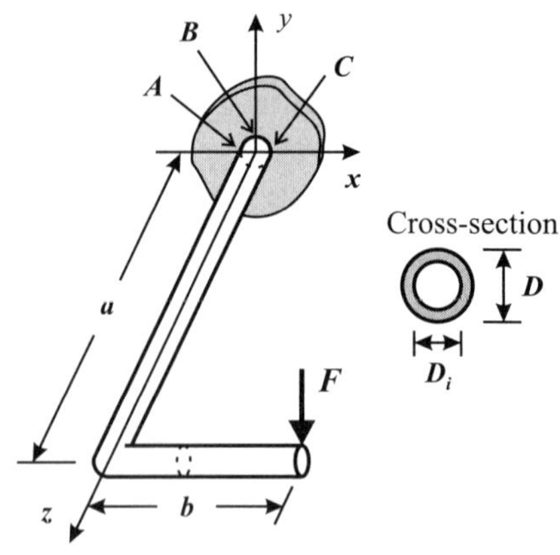

Cross-section

8.11 In attempting to open a locked door, a student applies a downward force F to the door handle. The handle, with solid circular cross-section of diameter D, does not perceptibly move (the door is locked). Cross-section ABC is fixed at the doorplate. The force and geometry are $F = 20$ lb, $e = 2.5$ in., $h = 4.0$ in., and $D = 0.75$ in.

Considering only points A, B, and C, determine (a) the numerical value of the maximum principal stress at cross-section ABC and (b) the numerical value of the maximum shear stress at cross-section ABC.

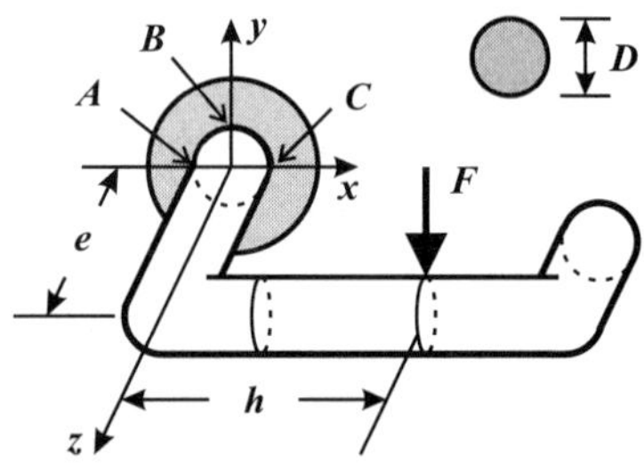

8.12 Structure $ABCD$ has a hollow circular cross-section, outer diameter $2R = 6.0$ in., and inner diameter $2R_i = 5.5$ in. The bracket consists of vertical arm AB, and horizontal arms BC (along the z-axis) and horizontal arm CD (along the x-axis). The lengths of segments AB, BC and CD are: $a = 8.0$ ft, $b = 4.0$ ft, and $c = 2.0$ ft, respectively. A downward force of $P = 2.0$ kips acts at point D.

Combined stresses. Consider only 3D stress elements where the three faces are aligned with the x-, y-, and z-axes. Determine the maximum tensile stress, the maximum (least tensile) compressive stress, and the maximum shear stress in segment (a) *CD*, (b) *BC*, and (c) *AB*.

Hint: Consider how each segment acts as an axial, a torsional, and/or a bending member – first *CD*, then *BC*, and finally *AB*.

(d) Determine the maximum tensile and compressive stresses in *AB* (the faces of the 3D element are not necessarily aligned with the x-, y-, and z-axes).

Note: The shear force acting in a beam of hollow circular cross-section causes a maximum shear stress of:

$$\tau_{max} = \frac{4}{3}\frac{V}{A}\left(\frac{R^2 + RR_i + R_i^2}{R^2 + R_i^2}\right)$$

where R is outer radius and R_i is inner radius.

(e) *Stress transformation.* Determine the maximum in-plane shear stress from the stress states found in *Parts (a)–(d)*.

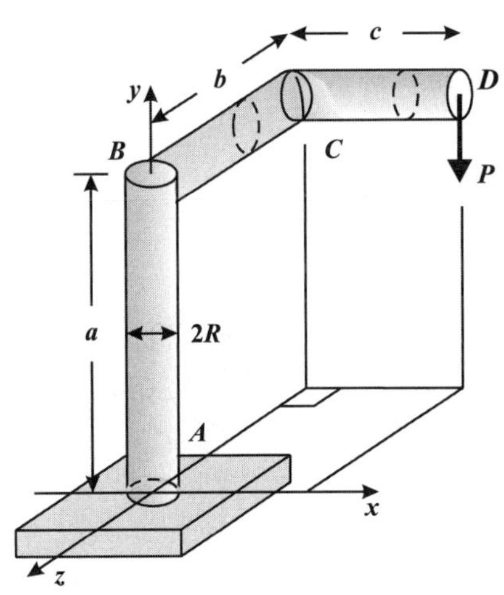

8.4 Mohr's Circle

8.13 The stress states $(\sigma_x, \sigma_y, \tau_{xy})$ at several points in a material system are listed below.

Use Mohr's circle to determine the new stress state $(\sigma_{x'}, \sigma_{y'}, \tau_{x'y'})$ for each element when it is rotated by the given angle θ (positive counterclockwise). Draw the stress element in its new orientation.

(a) (15 ksi, 10 ksi, 0 ksi), $\theta = 20°$
(b) (–15 ksi, 0 ksi, 6 ksi), $\theta = 20°$
(c) (15 ksi, –10 ksi, 6 ksi), $\theta = -30°$
(d) (10 ksi, 10 ksi, 0 ksi), $\theta = 30°$
(e) (120 MPa, 80 MPa, 40 MPa), $\theta = 20°$
(f) (–120 MPa, –80 MPa, 40 MPa), $\theta = 20°$
(g) (120 MPa, –80 MPa, –40 MPa), $\theta = 20°$

8.14 The stress states $(\sigma_x, \sigma_y, \tau_{xy})$ at several points in a material system are listed below.

For each stress state, (i) determine the principal stresses and the directions in which they act. Draw the stress element in its new orientation. (ii) Determine the maximum in-plane shear stress and the associated normal stresses and angles. Draw the stress element in its new orientation.

(a) (15 ksi, 10 ksi, 0 ksi)
(b) (–15 ksi, 0 ksi, 6 ksi)
(c) (15 ksi, –10 ksi, 6 ksi)
(d) (10 ksi, 10 ksi, 0 ksi)
(e) (120 MPa, 80 MPa, 40 MPa)
(f) (–120 MPa, –80 MPa, 40 MPa)
(g) (120 MPa, –80 MPa, –40 MPa)

8.5 Strain Transformation

8.6 Strain Gages

8.15 The strain states $(\varepsilon_x, \varepsilon_y, \gamma_{xy})$ at several points in a material system are listed below.

Determine the transformed strains $(\varepsilon_{x'}, \varepsilon_{y'}, \gamma_{x'y'})$ for each element when it is rotated by the given angle θ (positive counterclockwise).

(a) $(150 \times 10^{-6}, 100 \times 10^{-6}, 80 \times 10^{-6})$, $\theta = 20°$
(b) $(200 \times 10^{-6}, -150 \times 10^{-6}, 60 \times 10^{-6})$, $\theta = 30°$
(c) $(150 \times 10^{-6}, 100 \times 10^{-6}, -100 \times 10^{-6})$, $\theta = 50°$

8.16 The strain states (ε_x, ε_y, γ_{xy}) at several points in a material system are listed below.

(i) Determine the principal strains and the directions in which they act. (ii) Determine the magnitude of the maximum shear strain. (iii) If $G = 75$ GPa, determine the maximum shear stress.

(a) $(150{\times}10^{-6}, 100{\times}10^{-6}, 80{\times}10^{-6})$

(b) $(200{\times}10^{-6}, -150{\times}10^{-6}, 60{\times}10^{-6})$

(c) $(150{\times}10^{-6}, 100{\times}10^{-6}, 100{\times}10^{-6})$

8.17 In a uniaxial test, two strain gages measure the strain in the $0°$-direction (the load-direction) and the $90°$-direction (transverse to the load). When the applied stress is 80 MPa:

$$\varepsilon_0 = +660{\times}10^{-6}, \quad \varepsilon_{90} = -185{\times}10^{-6}$$

Determine the Young's modulus and the Poisson's ratio of the material.

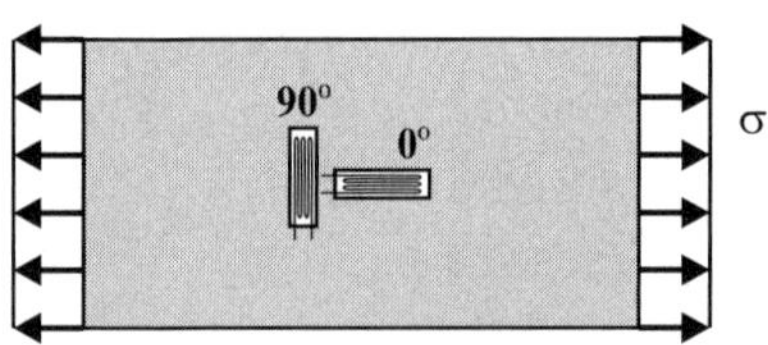

8.18 A 0–45–90° strain gage rosette records the strains states (ε_0, ε_{45}, ε_{90}) = (ε_A, ε_B, ε_C) listed below.

Determine the principal strains and the maximum in-plane shear strain for each system.

(a) $(+200{\times}10^{-6}, +120{\times}10^{-6}, +100{\times}10^{-6})$

(b) $(+150{\times}10^{-6}, +50{\times}10^{-6}, -40{\times}10^{-6})$

(c) $(-100{\times}10^{-6}, +20{\times}10^{-6}, -30{\times}10^{-6})$

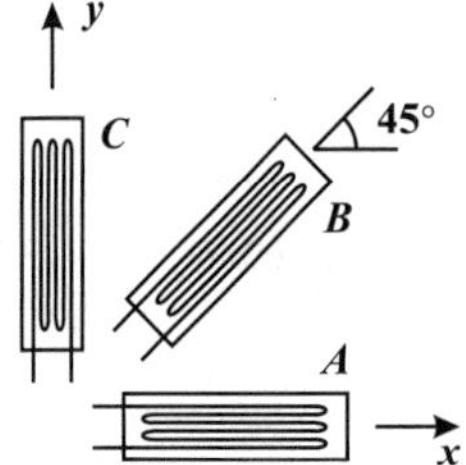

8.19 Derive the expressions for the principal strains for a 0–60–120° strain gage rosette, with measured strains $\varepsilon_0 = \varepsilon_A$, $\varepsilon_{60} = \varepsilon_B$, and $\varepsilon_{120} = \varepsilon_C$.

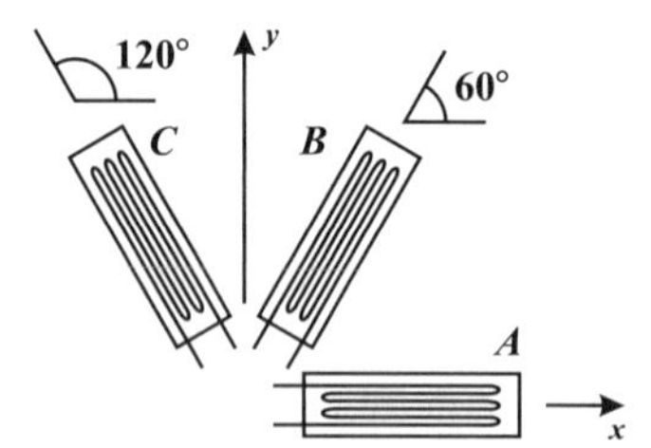

8.7   Three-Dimensional Stress

8.20 The 3D principal stresses (σ_I, σ_{II}, σ_{III}) at several points are listed below.

For each element, determine (i) the maximum *in-plane* (*I–II* plane) shear stress, and (ii) the two maximum *out-of plane* shear stresses, and (iii) the maximum shear stresses in the system. (iv) Determine the plane (*I–II*, *II–III* or *I–III*) in which the maximum shear stress in the system acts.

(a) (15 ksi, 5.0 ksi, 0 ksi)

(b) (15 ksi, −10 ksi, 0 ksi)

(c) (15 ksi, 10 ksi, 8.0 ksi)

(d) (10 ksi, −10 ksi, −8.0 ksi)

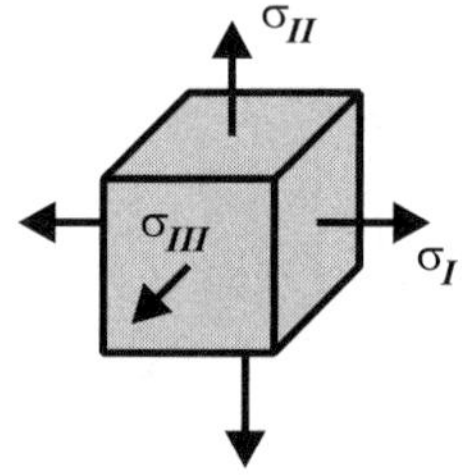

Problems: Chapter 9 Failure Criteria

9.1 What failure theory would you use to determine if a ductile material (metal) fails? What failure theory would you use to determine if a brittle material (ceramic) fails?

9.1 Failure Condition for Brittle Materials

9.2 A solid circular shaft of diameter $D = 10$ mm is subjected to both axial force F and torque T. The shaft is made of a ceramic with an ultimate tensile strength of 100 MPa and a compressive strength of 900 MPa.

(a) If the axial force is 500 N in tension, what is the maximum torque that can be applied without failure? (b) If the axial force in 500 N in compression, what is the maximum torque that can be applied without failure?

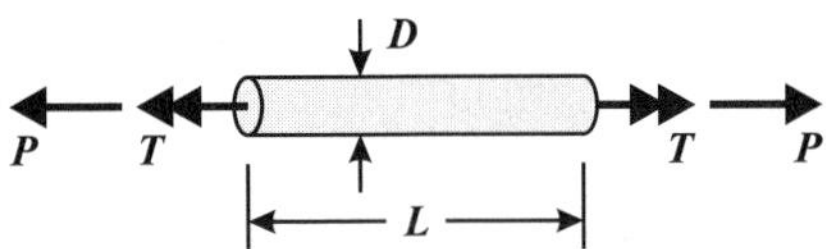

9.3 A femur bone has an outer diameter of 1.2 in. and an inner diameter of 0.70 in. The femur is subjected to a torque of 20 lb-ft and a moment of 30 lb-ft. The tensile strength of the bone is 16 ksi.

Determine if the bone fails.

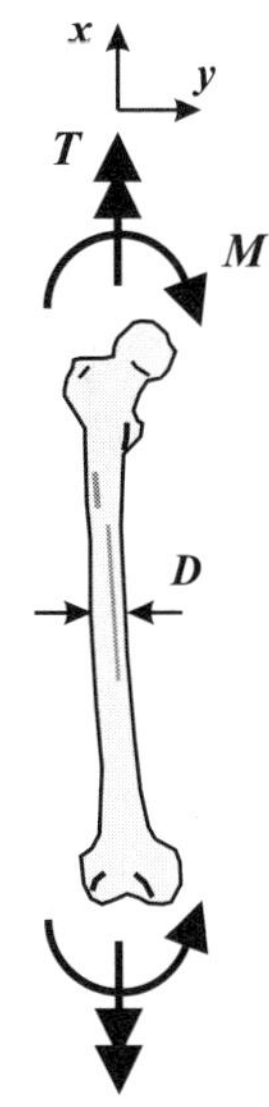

9.2 Failure Condition for Onset of Yielding of Ductile Materials

9.4 The cylindrical length of a screwdriver blade is $L = 120$ mm long with a diameter of $D = 7.00$ mm. The screwdriver is made of steel:

$$E = 200 \text{ GPa}, G = 75 \text{ GPa},$$
$$S_y = 400 \text{ MPa}, \tau_y = 250 \text{ MPa}.$$

As a person drives the screw, she applies a torque of $T = 6.0$ N·m, and an axial compressive force of $F = 100$ N.

(a) Using the Tresca condition, determine if the screwdriver yields. (b) Using the von Mises condition, determine if the screwdriver yields.

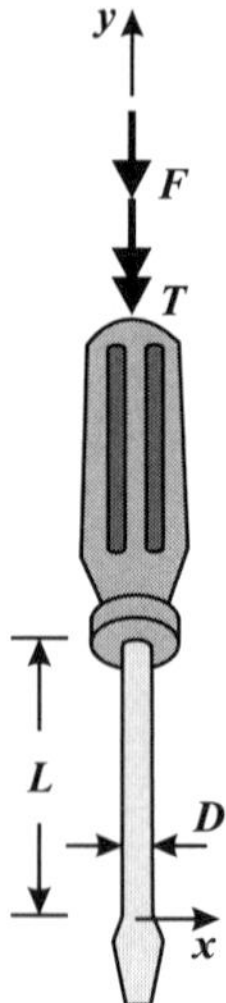

9.5 The strain energy density stored in a stress element under a triaxial state of stress $(\sigma_I, \sigma_{II}, \sigma_{III})$ is:

$$U_D = \sum \frac{1}{2}\sigma_i\varepsilon_i = \frac{1}{2}[\sigma_I\varepsilon_I + \sigma_{II}\varepsilon_{II} + \sigma_{III}\varepsilon_{III}]$$

When the element is subjected to a uniaxial load, the stresses are σ_I and $\sigma_{II} = \sigma_{III} = 0$. At yield, $\sigma_I = S_y$ and the strain energy density equals the resilience:

$$U_D = \frac{1}{2}\frac{\sigma_I^2}{E} = \frac{1}{2}\frac{S_y^2}{E} = U_R$$

Per the von Mises criterion, this is the value of the strain energy density required to yield a material no matter how the material is loaded. In general, a 3D state of stress is converted into the *von Mises* or *equivalent stress*, σ_o. Yielding occurs when $\sigma_o = S_y$, or when:

$$\frac{1}{2}\frac{\sigma_o^2}{E} = U_R$$

(a) For triaxial loading $(\sigma_I, \sigma_{II}, \sigma_{III})$, determine the strain energy density in terms of stresses σ_I, σ_{II}, and σ_{III}, and elastic constants E and v. (b) When a metal yields, its Poisson's ratio becomes $v = 0.5$. Simplify your expression from *Part (a)* to include this value of v. (c) From *Part (b)* and the definition of resilience, show that at yielding, the *von Mises stress* is given by:

$$\sigma_o = \sqrt{\frac{(\sigma_I - \sigma_{II})^2 + (\sigma_{II} - \sigma_{III})^2 + (\sigma_{III} - \sigma_I)^2}{2}}$$

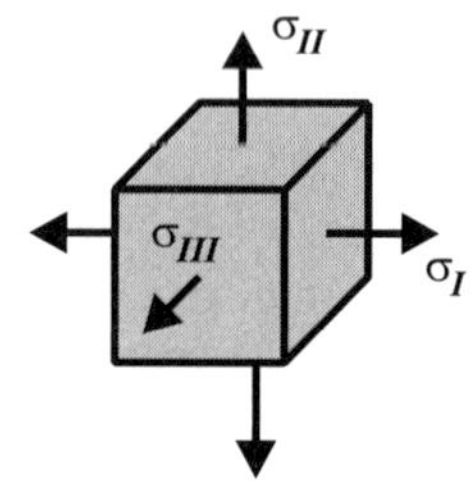

9.6 Determine the maximum power that can be transmitted by the following shafts without failure.

(a) A steel shaft of diameter 100 mm rotating at 550 rev/s. The yield strength in axial tension is $S_y = 250$ MPa. (b) A ceramic silicon carbide (SiC) shaft of diameter 3.0 mm rotating at a speed of 55,000 rev/s. The tensile strength of SiC is taken to be $S_u = 100$ MPa.

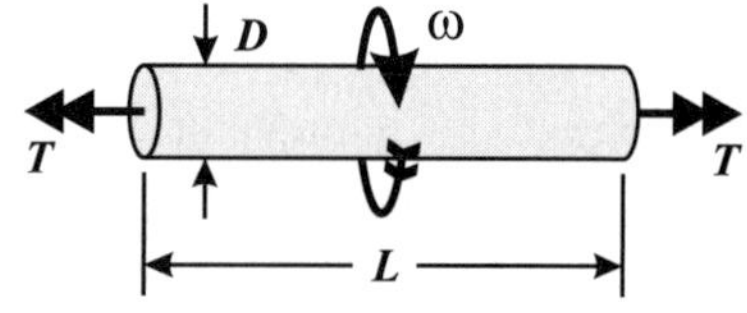

Problems: **Chapter 10 Buckling**

10.1 Buckling of a Column

10.2 Radius of Gyration and Slenderness Ratio

10.3 Boundary Conditions and Effective Length

10.4 Transition from Yielding to Buckling

10.1 A steel column has a W12×96 cross-section. The column has length L and supports a compressive force P through the centroid. The column is unconstrained (free) at the top, and built-in at the ground. Structural steel has modulus $E = 30{,}000$ ksi and yield strength $S_y = 36$ ksi.

Determine the length of the column, L (in feet), so that Euler buckling and yielding will occur simultaneously as load P is increased from zero.

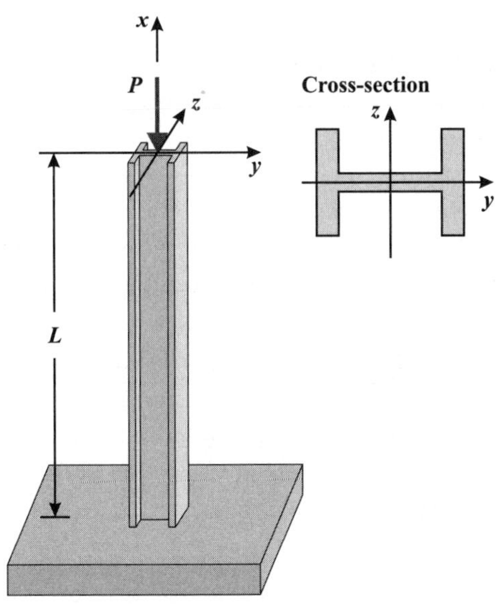

10.2 Truss *ABC* supports downward force F at joint B. Each bar has a solid round cross-

section of diameter $D = 15.0$ mm and is made of steel with $E = 200$ GPa. Length $L = 1.0$ m.

What is the minimum value of F to cause *BC* to buckle.

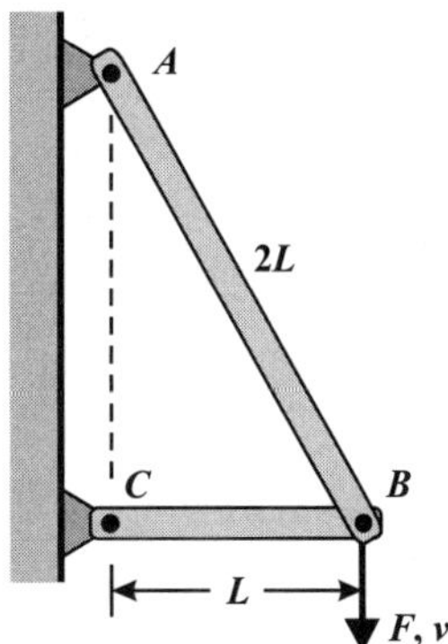

10.3 All the members of a steel truss are $L = 3.0$ m long, with rectangular cross-section 100 by 50 mm. Force $V = 40$ kN is applied at joint B. $E = 200$ GPa.

Determine the applied load V_{cr} that will cause any member to buckle. Indicate which member(s) buckles. Assume the ends of each member are pinned and buckling occurs about the weak axis.

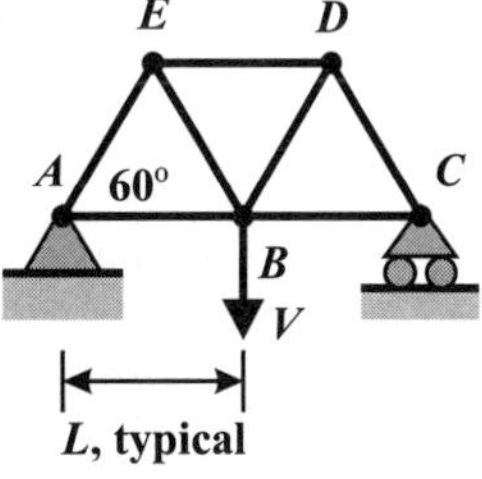

10.4 An aluminum frame ($E = 70$ GPa, $S_y = 240$ MPa) consists of two members, *AB* and *CD*, joined at point *C* as shown (not to scale). Point *D* rests on the rough ground, and is not constrained except for sliding horizontally (assume it is pinned). At joint *C*, *CD* is pinned for buckling in the plane of the paper, and fixed for buckling out of the plane of the paper. The frame is loaded by force F at point A. The frame dimensions are: $a = 100$ mm, $L = 200$ mm, $d = 40$ mm, $t = 30$ mm,   and $b = 10$ mm (b is the thickness of the members into the paper).

Calculate (a) the load F_{yA} to yield beam AB, (b) the load F_{yD} to yield CD, and (c) the load F_B to buckle CD. (d) Indicate how the system fails as F is increased from zero.

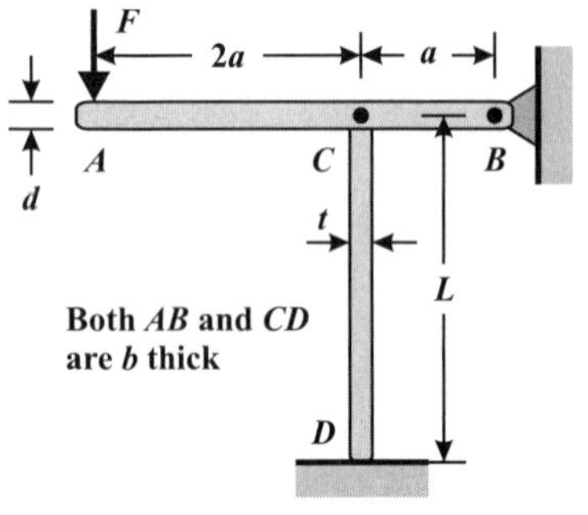

10.5 A steel pole is made of standard 3 in. pipe (see *Appendix D*) and supports a load of $P = 20.0$ kips at its top.

Determine the length L that will cause the column to buckle.

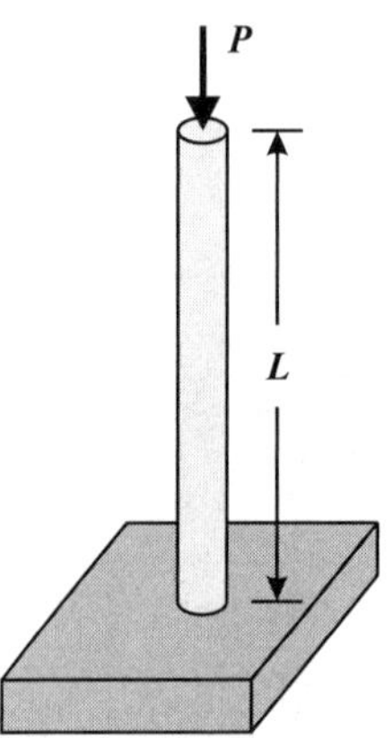

10.6 A steel column, built into a building foundation, helps to support the second floor. The column has a rectangular cross-section of width $b = 70$ mm and depth $d = 100$ mm. The material properties are: $E = 200$ GPa and $S_y = 250$ MPa.

Determine (a) the load P_y to yield the column and (b) the load P_{cr} to buckle the column. (c) Which criterion limits the design?

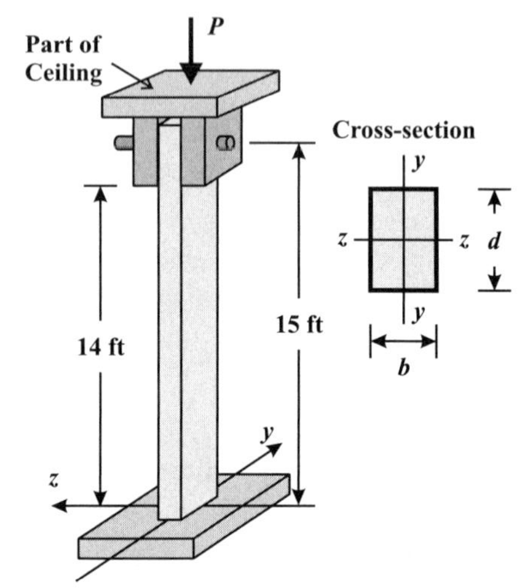

10.7 The columns of a building are built of the steel W-shape W18×76, with $E = 30,000$ ksi and $S_y = 36$ ksi. In a preliminary design, the columns are to be an uninterrupted 30 ft long, i.e., purlin CD is not in the original design. Assume the columns at AB are pinned for buckling in the plane of the paper (about the weak axis), but free for buckling out of the plane (about the strong axis).

(a) Determine the load (on each column) to cause buckling in the original design (without member CD). (b) Purlin CD is added in the second design iteration. Determine the new load to cause buckling.

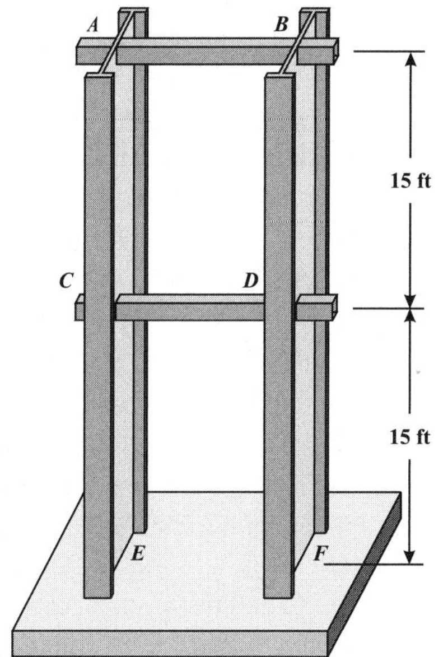

10.6 Effect of Imperfections

10.7 Effect of Lateral Forces

10.8 A vertical rigid strut is supported at its base by a spring of torsional stiffness $K_T = M/\theta$.

If a vertical force P is applied at the top of the strut, determine the buckling load, i.e., the force at which the column will tip over when there is even the slightest tilt θ of the strut.

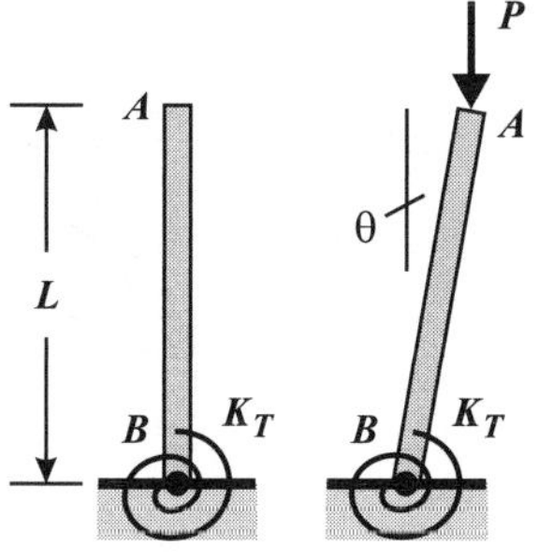

10.9 (a) If vertical force P is applied to the strut of *Prob. 10.8* with an eccentricity e, determine the relation between P and the angle of tilt θ so that the strut remains in equilibrium

at θ. (b) If $e/L = 0.05$, plot the P/K_T versus θ relationship, and determine the maximum load P for stability (so that the strut does not tip over).

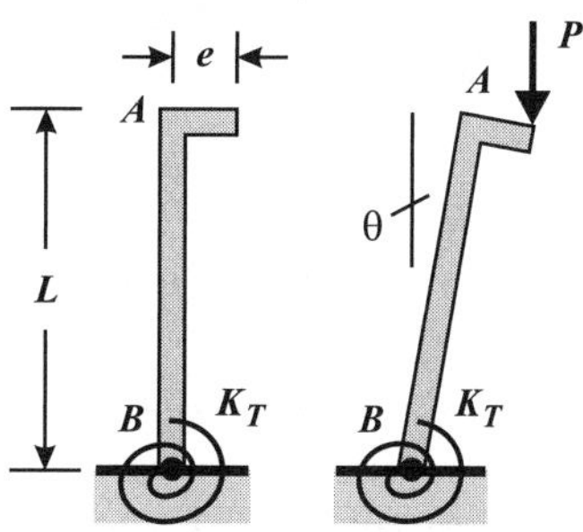

10.10 (a) If vertical force P and horizontal force H are applied at the top of the rigid strut of *Prob. 10.8*, determine P in term of θ and H so that the system remains in equilibrium. (b) If $H/P = 0.1$, plot the P/K_T versus θ relationship, and determine the maximum load P for stability (so that the strut does not tip over).

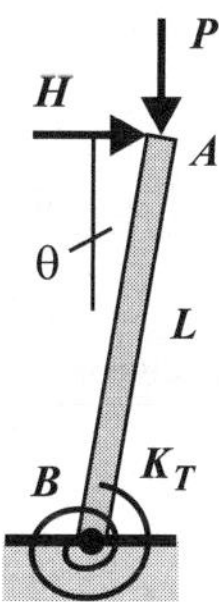

10.11 A 1000 μm long column of silicon has a rectangular cross-section 15 μm wide (into the paper) and 1.0 μm deep. The modulus is 160 GPa. The column has an initial deformed shape as shown, with unloaded central displacement a. In an experiment, the column is found to buckle due to an axial compressive load of $P = 7.6$ μN, with an additional transverse central displacement of $\Delta = 9.0$ μm.

Estimate the initial center imperfection a of the beam.

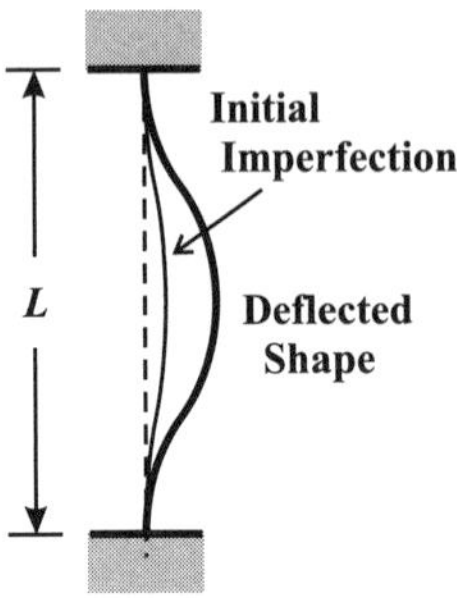

10.12 An aluminum ($E = 10{,}000$ ksi) column pinned at both ends has length $L = 10$ ft. The minor moment of inertia is $I = 2.73$ in.^{4}. During construction, it is found that the center of the column has been displaced $a = 1.0$ in. from the design axis. The compressive force on the column is $P = 0.7 P_{cr}$.

Determine (a) the buckling load of the column and (b) the transverse load F that must be applied at the center to bring the column back towards true, i.e., to return the center to the (dotted) line between A and B.

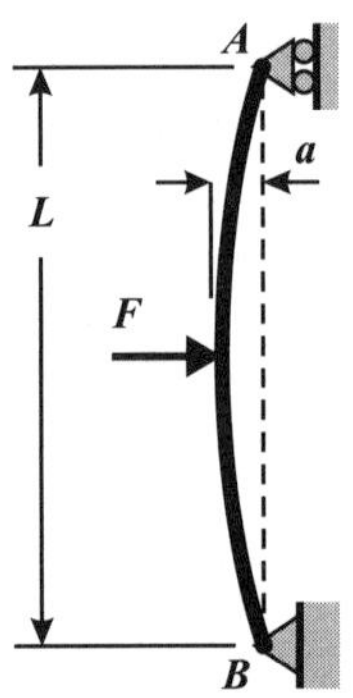

10.13 The direction of thrust of a jet engine is adjusted by a series of actuators in compression. The actuators are modeled as thin-walled circular columns in compression. The columns are of length L, radius R, and thickness t, and are pinned at both ends.

(a) Determine an expression for the mass of the column in terms of the critical buckling load P_{cr}, material properties ρ and E, and column length L and radius R (see *Section 6.6 – Design of Beams*). (b) Using the material properties in *Table 6.7*, select the material that will give the lightest design.

10.14 Continuous rails without expansion joints arc subjcctcd to buckling in hot weather. For steel, $E = 200$ GPa, $S_y = 250$ MPa, and $\alpha = 14 \times 10^{-6}/°$C. The cross-section of a light rail is $A = 6500$ mm^2, $I_z = 19.7 \times 10^6$ mm^4, and $I_y = 3.00 \times 10^6$ mm^4, where I_z corresponds to buckling out of the plane of the tracks (vertically) and I_y corresponds to buckling in the plane of the tracks (horizontally). The rails are connected to the ties and the ties are pinned to the ground every $L = 4.0$ m.

(a) Determine the stress in the rail when it is heated by 35°C from normal temperatures. (b) Will the thermal stress yield the rail? (c) Will the thermal stress buckle the rail?

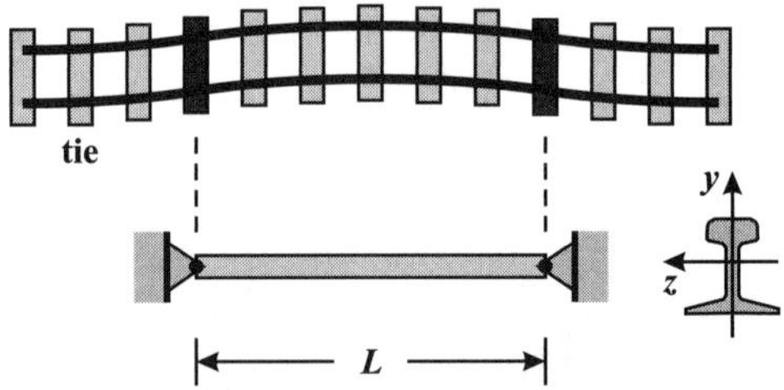

Problems: Chapter 11 Energy Methods

11.1 Internal and Complementary Energy

11.2 Principle of Virtual Work

11.3 Minimum Energy Principles

11.1 Truss ABC is loaded at joint B by forces F_x and F_y. Joint B displaces by u and v. The modulus and cross-sectional area of both members are E and A. Assume the system remains elastic.

Determine (a) the internal energy stored in each bar in terms of displacements u and v (b) the internal complementary energy in each bar in terms of forces F_x and F_y, and (c) the relationship between applied loads F_x and F_y and displacements u and v. Give the results in matrix form.

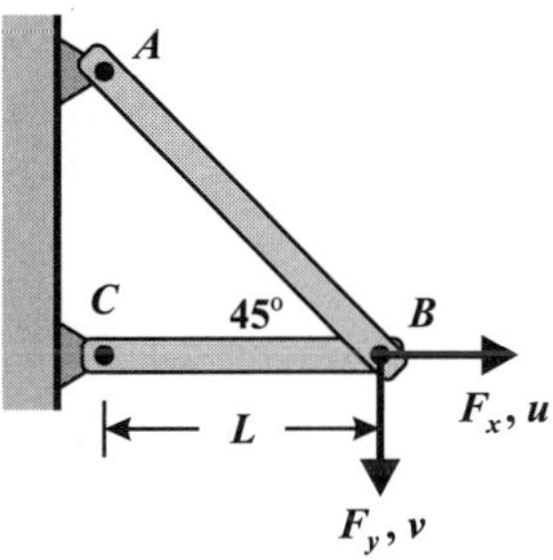

11.2 Stepped bar ABC is subjected to loads F_1 and F_2 applied at points B and C. By minimizing the total complementary energy, determine the downward displacements of point B and point C, u_B and u_C, respectively, in terms of F_1 and F_2.

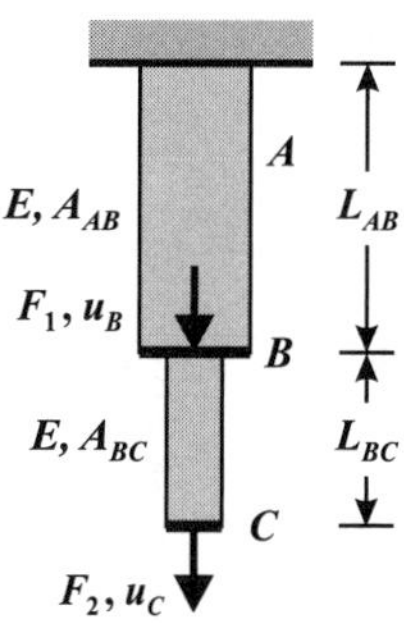

11.3 Joint B of truss $ABCD$ is caused to displace u and v due to forces F_x and F_y. The modulus and cross-sectional area of all members are E and A. The truss is redundant.

(a) Determine the internal energy in each member in terms of u and v. (b) By minimizing the total energy of the system, determine the applied forces F_x and F_y in terms of the displacements. (c) Determine the stiffness matrix of the system. (d) For the case: $u = 2.0$ mm (right), $v = 3.0$ mm (down), $A = 0.001$ m^2, $L = 1.0$ m, and $E = 70$ GPa, determine F_x and F_y.

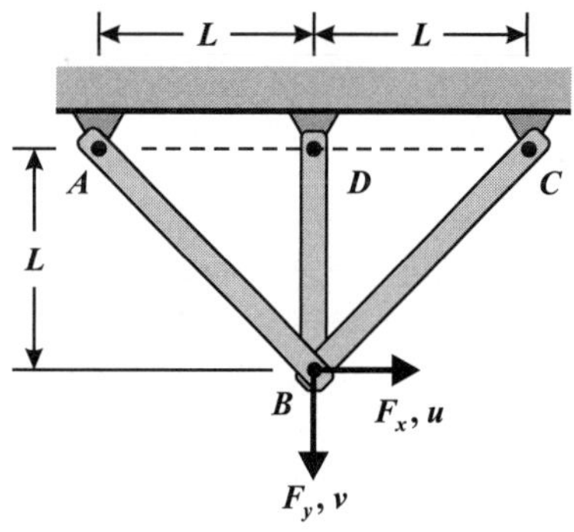

11.4 Truss $ABCD$ is loaded at joint B by forces F_x and F_y. Joint B deflects by u and v. The modulus and cross-sectional area of each member are E and A, respectively.

(a) Determine the force in each member in terms of F_x and F_y and redundant force R; assume the redundant force is in bar BD.

(b) By minimizing the total complementary Energy of the system, determine the displacements u and v in terms of the applied forces. (c) Determine the flexibility matrix of the system.

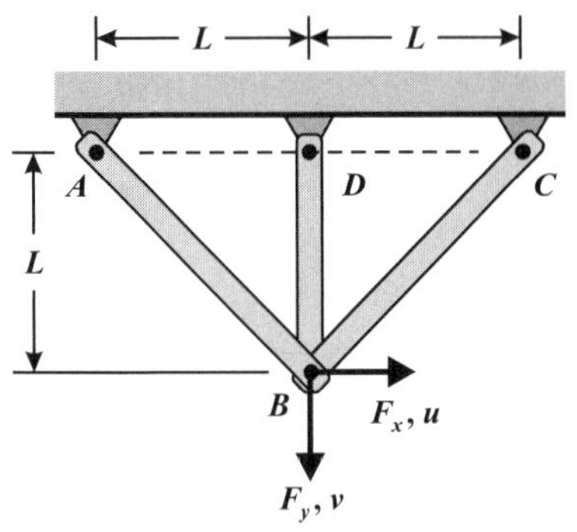

11.5 Repeat *Prob. 11.3* for the following truss.

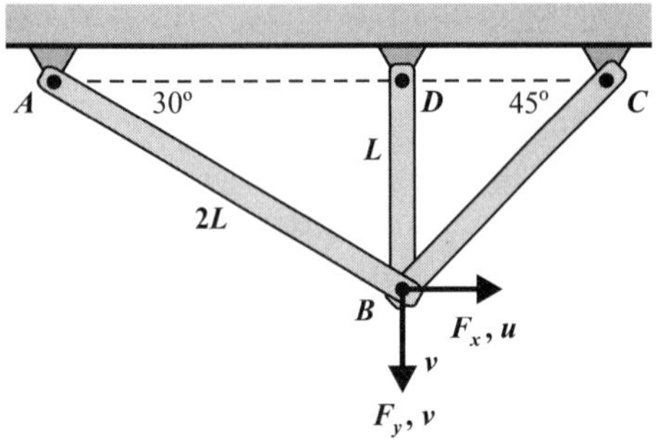

11.6 Repeat *Prob. 11.4* for the truss shown in *Prob. 11.5*.

11.7 Steel truss *ABCD* is loaded at joint B by force F with a magnitude of 10,000 lb. The force F has a slope of 4/3 downward and to the right. The modulus is $E = 30,000$ ksi. The cross-sectional area of each bar is $A = 0.5$ in.^{2}.

Minimize the total complementary energy to determine the horizontal and vertical displacements, u and v, of point B.

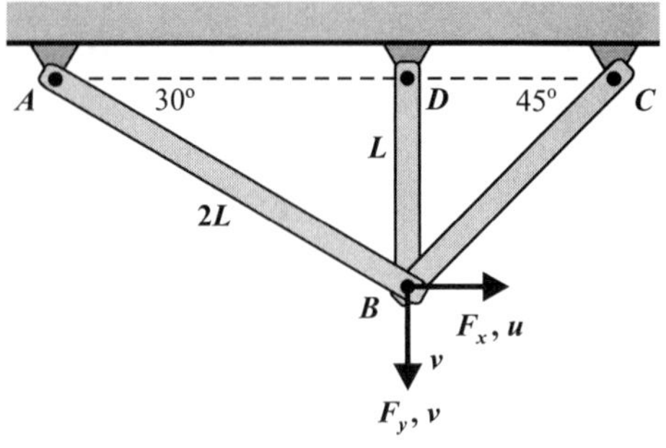

11.8 Truss *ABCD* is subjected to downward force F at joint C. The truss is in the form of a square, with sides of length L. Members AB, BC and CD each have cross-sectional area A and diagonals AC and BD each have cross-sectional area $\sqrt{2}A$. The diagonals are not joined at their centers. The modulus of each member is E.

Using the force in AB as the redundant ($P_{AB} = R$), determine (a) the total complementary energy in the system. (b) Minimize the total complementary energy to determine the value of the redundancy R, (c) the downward deflection v of the load F, and (d) the stiffness of the system in the vertical direction, $k = F/v$, for loads at point B.

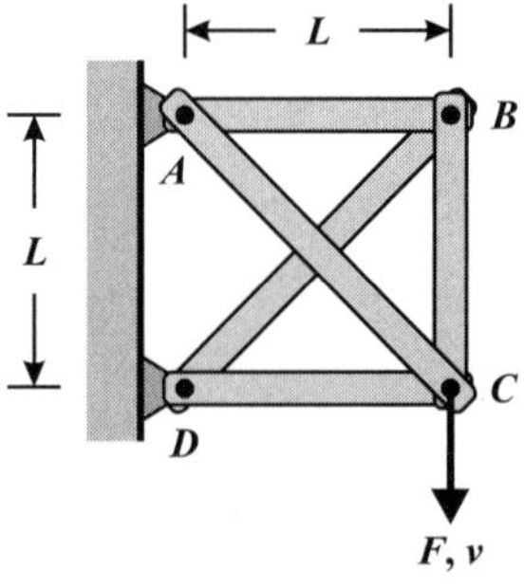

11.9 Truss *ABCD* is subjected to downward force F at joint C. The truss is in the form of a square, with sides of length L. Members AB, BC, and CD each have cross-sectional area A, diagonals AC and BD each have cross-sectional area αA (α is a constant). The diagonals are not joined at their centers. The modulus of each member is E.

Using the force in *AB* as the redundant $(P_{AB} = R)$, determine (a) the Total complementary energy in the system. (b) Minimize the total complementary energy to determine the value of the redundancy *R*, (c) the downward deflection *v* of the load *F*, and (d) the stiffness of the system in the vertical direction $k = F/v$. (e) Plot the stiffness as function of α. Is there an optimal value for α? Or is there a practical value for α beyond which there is less significant gain in system stiffness as α is increased.

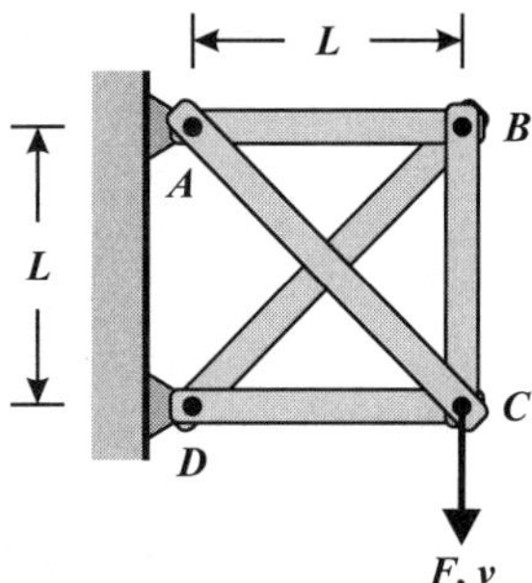

11.10 Space structures are made of 3D trusses. The 3D truss *ABCD* is made of three elements, each of length $L = 1.0$ m and modulus $E = 70$ GPa. All elements are supported by ball-and-socket joints (3D pins). joints *B*, *C* and *D* are supported at the ground, and loaded at joint *A*.

Develop a matrix expression for the displacements of Point *A*, *u*, *v*, and *w*, in terms of applied loads F_x, F_y and F_z.
Hint: use total complementary energy.

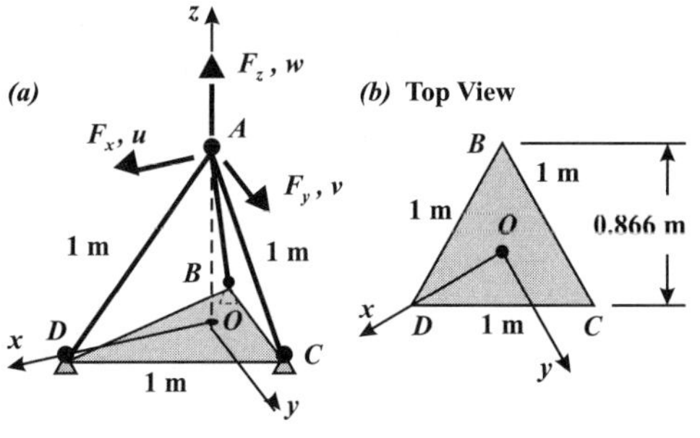

11.4 Bending Energy

11.11 A cantilever beam of length $2L$ is subjected to point load *P* at its tip. The modulus is *E*, and the moment of inertia is *I*.

Use total complementary energy to determine the tip deflection (at $x = 0$). Does the result agree with the deflection expression for a tip-loaded cantilever beam of length $2L$ (see *Appendix F*)? *Note*: *x* is measured from the tip of the beam.
Hint: The complementary energy in a beam of length *l* is:

$$C = \int_0^l \frac{M^2}{2EI} dx$$

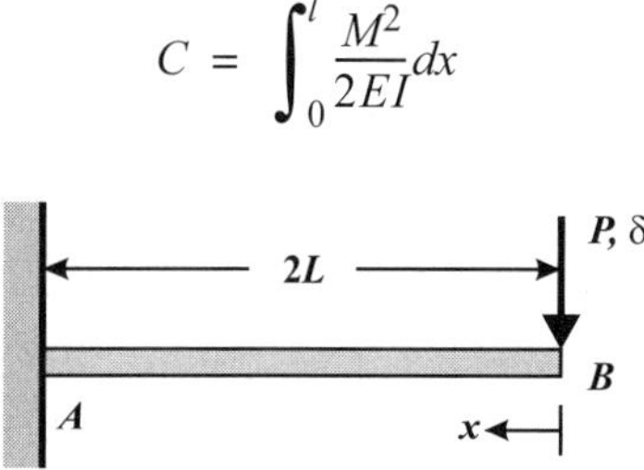

11.12 A cantilever beam, length $2L$, is subjected to two point loads: P_1 at its center and P_2 at its tip. The modulus is *E* and the moment of inertia is *I*.

Use total complementary energy to determine the tip deflection (at $x = 0$).
Hint: The complementary energy in a beam of length *l* is:

$$C = \int_0^l \frac{M^2}{2EI} dx$$

11.13 A MEMS element is made of a rectangular beam and a solid circular shaft. Torque *T* is applied to the shaft at point C, and rotates by angle θ. The beam is of length L_1 and has moment of inertia *I*. The shaft is of length L_2 and has polar moment of inertia J. The elastic moduli are E and G.

Determine the total angle of twist due to the twisting of the shaft and the bending of the beam using total complementary energy.

Hint: The complementary energies in a beam in bending and in a shaft in torsion, both of length l, are:

$$C_M = \int_0^l \frac{M^2}{2EI}dx \quad \text{and} \quad C_T = \int_0^l \frac{T^2}{2GJ}dx$$

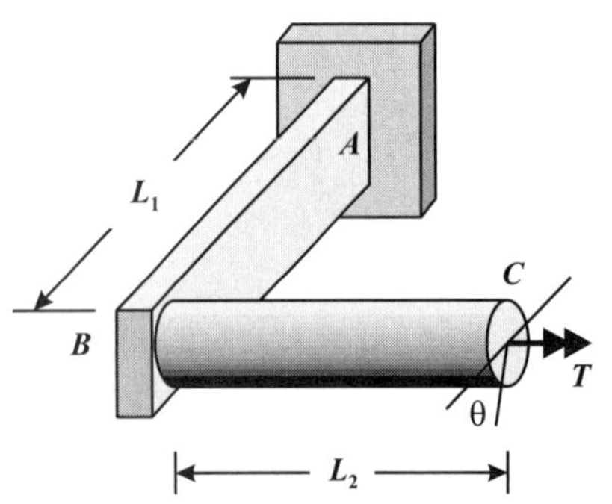

11.14 An arrangement often used to strengthen and stiffen a beam system is to reinforce the beam with a frame of rods. An example of this technique is the roof beams at the Louvre Museum in Paris.

Force F is applied at the center of the beam, point C. Tensile rods AD and BD are held away from the beam by compressive strut CD. The bending stiffness of the beam is EI and the axial stiffness of the axial members is EA.

By applying the minimum complementary energy principle, with the force in the strut as the redundant $P_{CD} = R$ and the forces in the rods $P_{AD} = P_{BD} = T$, determine (a) the magnitude of R, (b) the maximum bending moment in the beam, and (c) the downward deflection of point C. Compare this deflection to that of a simply supported beam (with stiffness EI) without the truss-support.

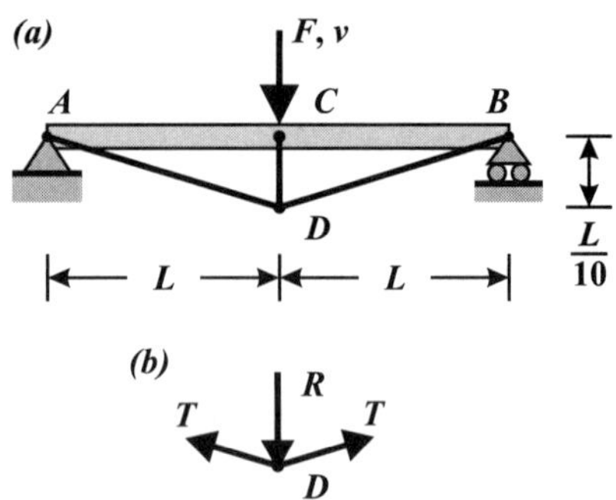

11.15 The mechanical properties of biological cells can be investigated by indenting a probe into the cell surface. The displacement of the probe can be readily measured with a standard microscope, but the resistive forces of the cell are very small and difficult to measure. A special force sensor designed for this purpose is schematically shown. Sensor ABC is manufactured from the ceramic silicon carbide (SiC). Element BC is much thicker than that of AB, and so BC may be assumed to be rigid. When the probe is indented against the cell, the reaction force has components in both the x- and y-directions. The material modulus is E, and the moment of inertia of AB is I.

Considering only bending of AB, determine the stiffness matrix relating the cell forces F_x and F_y to the probe displacements u and v (all are drawn in the directions of the positive axes). Use the minimum complementary energy method.

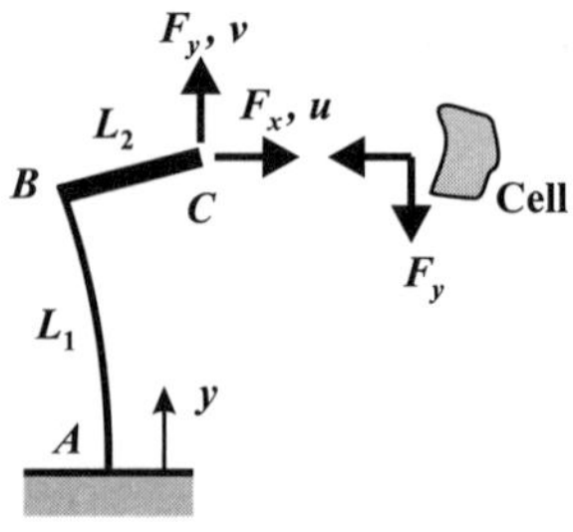

11.5 Approximation Methods

11.16 High-frequency torsional oscillators are made using small MEMS devices.

Approximate the torsional stiffness of a SiC shaft of square cross-section of side $D = 15\ \mu m$ and of length $L = 600\ \mu m$. The material has a modulus of 360 GPa.

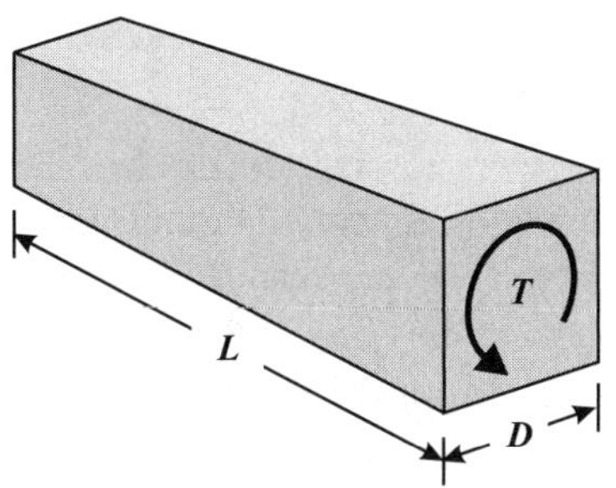

11.17 The tip of a cantilever beam is supported by a pin. The beam is subjected to a uniformly distributed load w. Assume the beam deflection has the following form:

$$v(x) = A + Bx + Cx^2 + Dx^3$$

(a) Minimize the total energy T to determine constants A, B, C, and D. The Complementary Energy is a function of curvature ($\kappa = v''$):

$$U = \frac{EI}{2}\int_0^L (v'')^2 dx$$

The potential of the load is:

$$V = -w \int_0^L v(x)\, dx$$

(b) Plot your solution, and the actual solution (*Example 6.9*) on the same graph to compare the estimate.

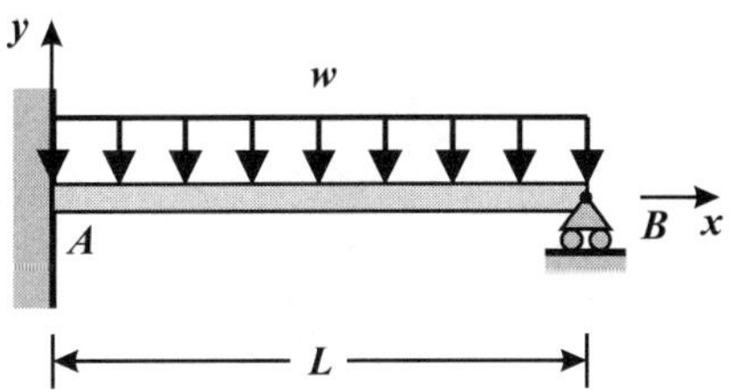

11.18 Consider a cantilever beam on an elastic foundation. Assume a displaced form, and minimize the total energy in the system.

Assume the elastic curve has the form:

$$v(x) = \left[1 - \cos\left(\frac{\pi x}{2L}\right)\right]\delta$$

This form satisfies the displacement boundary conditions at each end, and the slope boundary condition at $x = 0$: $v(0) = v(L) = 0$, and $v'(0) = 0$. The elastic foundation has a stiffness per unit length k (force/length^2); the elastic energy stored in length dx of the elastic foundation is:

$$dU_f = \frac{1}{2}(k\, dx)[v(x)]^2$$

(a) Minimize the total energy to solve for the force required to displace the tip by distance δ. (b) Consider a micro-cantilever beam used to measure the stiffness of biological tissues. The beam is $L = 1000\ \mu m$ long, $t = 2\ \mu m$ thick, and $b = 10\ \mu m$ wide. The beam rests on an elastic foundation (biological tissue). The force required to displace the cantilever tip by $\delta = 1\ \mu m$ is $P = 10\ \mu N$. Determine the stiffness per unit length of the tissue.

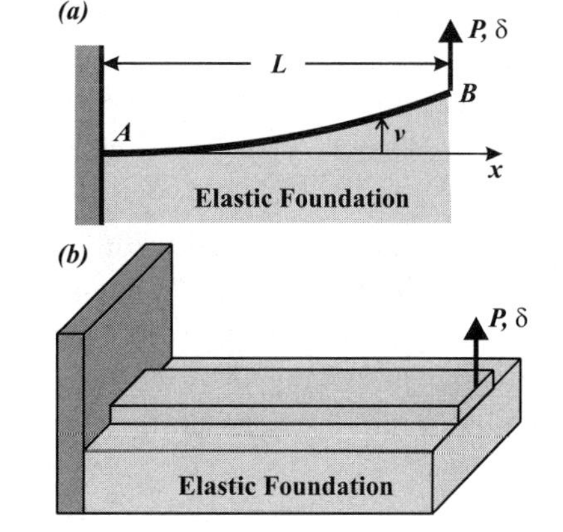

Problems: **Chapter 12 Ductile Materials and Design**

12.2 Elastic–Plastic Calculations: Limit Load

12.1 Steel truss *ABCD* is subjected to a downward force *F* at joint *B*. The cross-sectional area of each bar is $A = 0.002$ m^2, and $L = 1.0$ m. The modulus of steel is $E = 200$ GPa and its yield strength is $S_y = 250$ MPa.

Determine (a) the load to cause yielding F_y and (b) the limit load F_L.

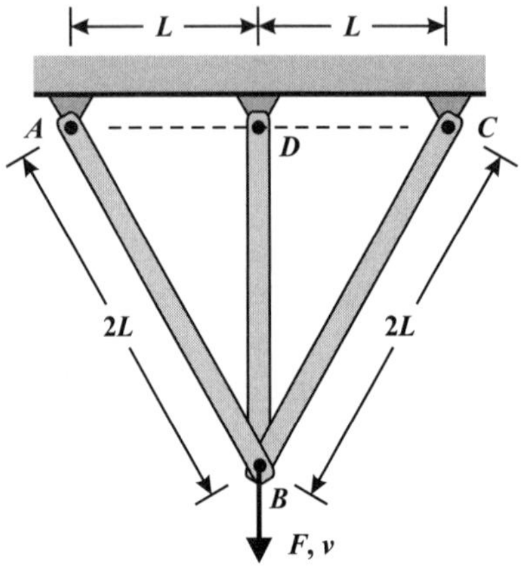

12.2 A simply supported steel I-beam, 10 ft long, has a cross-section 5 in. wide by 10 in. deep (total depth), with flange thickness 0.5 in. The yield strength of the material is $S_y = \pm 36$ ksi.

(a) Determine the limit moment assuming that only the flanges support bending stress. (b) After installation, is discovered that the maximum uniformly distributed load that is to be applied is $w = 8.0$ kips/ft, a greater value than originally specified. Determine if the beam can support the load. (c) A simple modification to increase the bending strength of a beam is to weld plates to its flanges. If plates 5.0 in. wide and 0.5-in. thick are welded to the flanges, determine the new limit moment.

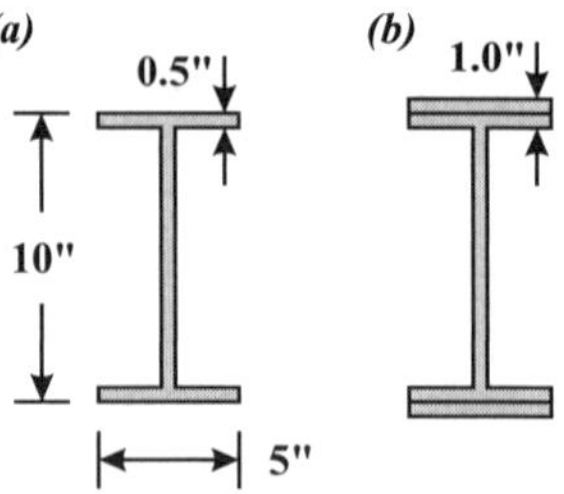

12.3 A simply supported beam is subjected to point load *P* at its center. The beam is of rectangular cross-section *B* wide by *D* deep, and has yield strength S_y.

Determine the limit load P_L.

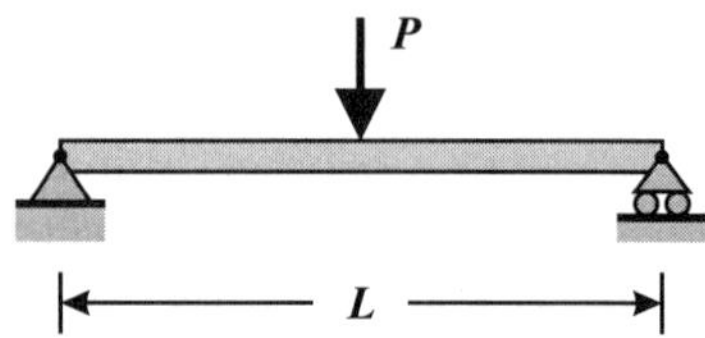

12.4 An aluminum beam has rectangular cross-section 1.0 in. wide by 4.0 in. deep. The beam supports maximum moment *M*. The modulus is $E = 10,000$ ksi and the yield strength is $S_y = 35$ ksi.

Determine (a) the magnitude of the moment when yielding first occurs M_y, and (b) the limit moment M_L. (c) Determine the size of the elastic core *d*, when $M_{max} = 10,000$ lb-ft and (d) the curvature κ at that location.

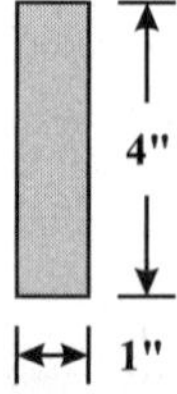

12.5 An aluminum beam with rectangular cross-section 30 mm wide by 150 mm deep is loaded by a pure moment, causing compression at the top of the beam. The maximum magnitude of the moment is $M = 35.0$ kN·m, which causes plastic deformation. The modulus is $E = 70$ GPa.

(a) Determine the size of the elastic core. (b) If the 35.0 kN·m moment is removed, determine the residual stress at the top of the cross-section.

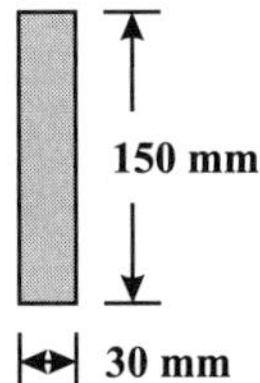

12.6 A simply supported built-up steel I-beam supports distributed load w. Steel has modulus $E = 200$ GPa and yield strength $S_y = 250$ MPa.

Determine the limit load w_L (a) neglecting the contribution of the web and (b) including the web (i.e., the entire cross-section is yielding).

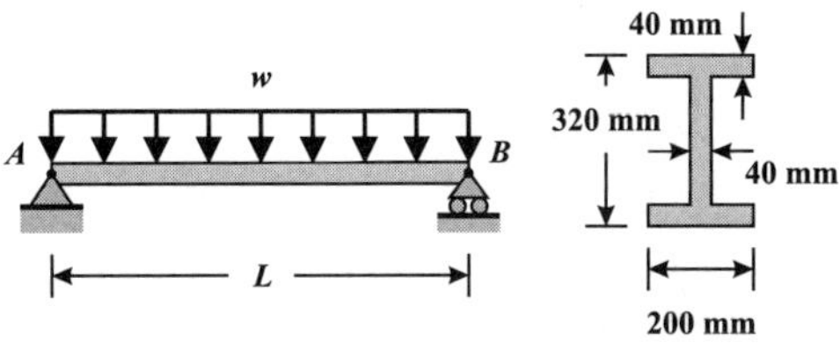

12.7 A steel shaft, 100 mm in diameter and 1.0 m long, is loaded in torsion with torque T. For steel, $G = 77$ MPa and $\tau_y = 150$ MPa.

Determine the (a) torque at which plastic deformation first occurs T_y and (b) the limit torque T_L. (c) If $T = 30$ kN·m, determine the size of the elastic core, r_e and (d) the angle of twist of the shaft.

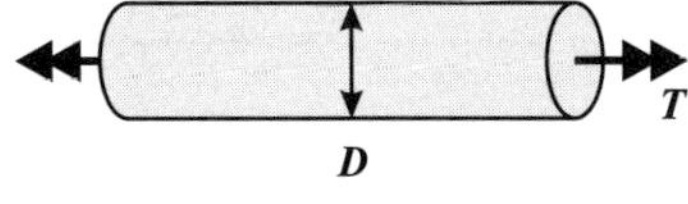

12.8 Determine the limit torque of a hollow shaft, outer diameter $D = 75$ mm and inner diameter $D_i = 50$ mm. The shear yield strength is $\tau_y = 150$ MPa.

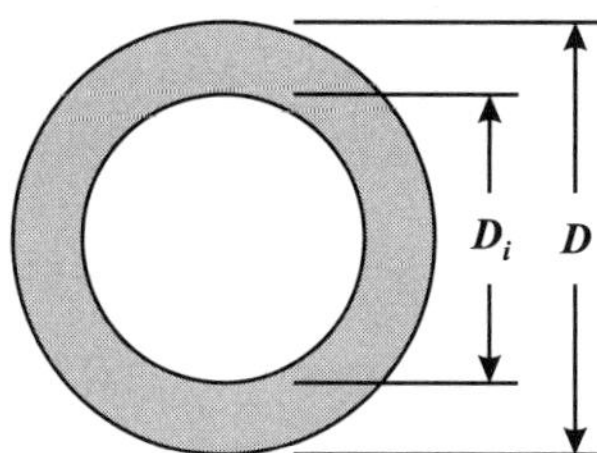

12.9 A solid copper shaft ($D_c = 10$ mm) is in a protective steel sleeve ($D_s = 18$ mm). The system is subjected to a torque T. The material properties are:

Steel: $G_s = 75$ GPa, $\tau_{ys} = 150$ MPa

Copper: $G_c = 45$ GPa, $\tau_{yc} = 100$ MPa

The steel–copper interface is well bonded. Recall that in torsion, due to symmetry, radii must remain straight; thus, strain γ is linear with distance r from the shaft axis.

(a) Determine the total torque applied to the shaft so that the steel at the interface just reaches its yield strength. (b) If the shaft is now unloaded from the condition given in *Part (a)*, determine the residual stress as a function of radius r.

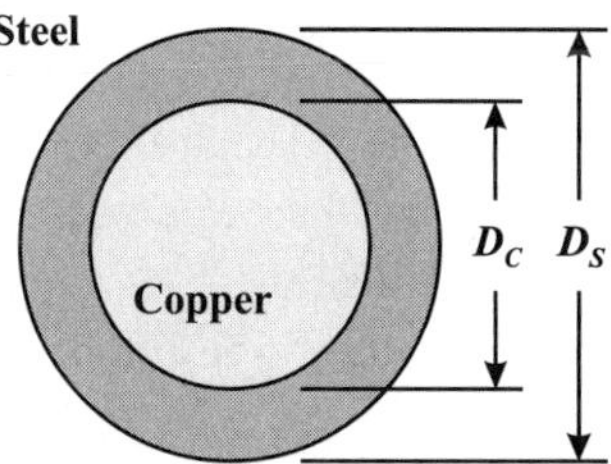

12.10 Truss $ABCD$ is subjected to downward force F at joint C. The truss is in the form of a square, with sides of length L. Members AB, BC, CD, and DA each have cross-sectional area A, and diagonals AC and BD each have cross-sectional area $\sqrt{2}A$. The diagonals are not joined at their centers.

Determine the load–displacement diagram, F–v, as the load is increased from zero to the limit load F_L.

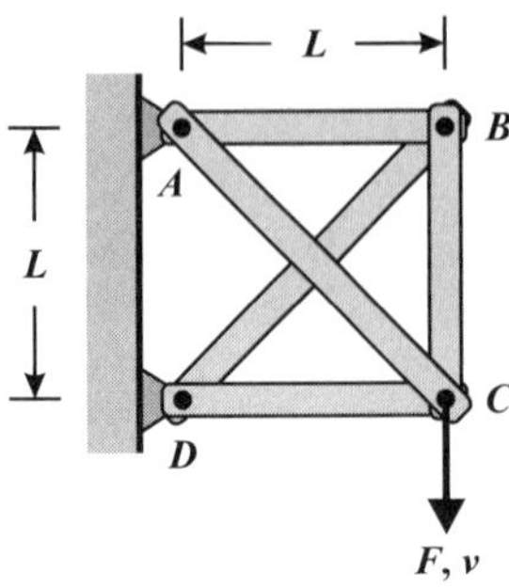

12.3 Limit Loads in Beams: Plastic Hinges

12.11 The free end of a cantilever is supported by a roller. The beam is subjected to central point load P. The limit moment for the cross-section is M_L.

Using plastic hinges, estimate (a) a lower bound and (b) an upper bound to the limit load P_L.

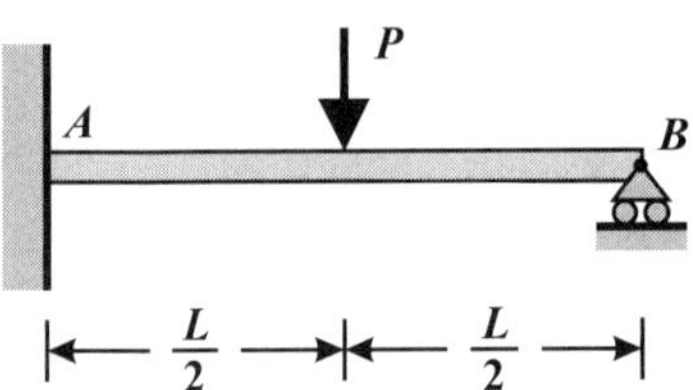

12.12 Two cantilever beams, both of length 3.0 m, are joined at their free ends by a roller bearing. Such a situation arises when building a bridge of beams having insufficient length to cross the entire span. Load P is applied 1 m to the right of the connection.

If the limit bending moment of the section is $M_L = 40$ kN·m, estimate the limit load P_L using plastic hinges. Bracket P_L between a lower and an upper bound.

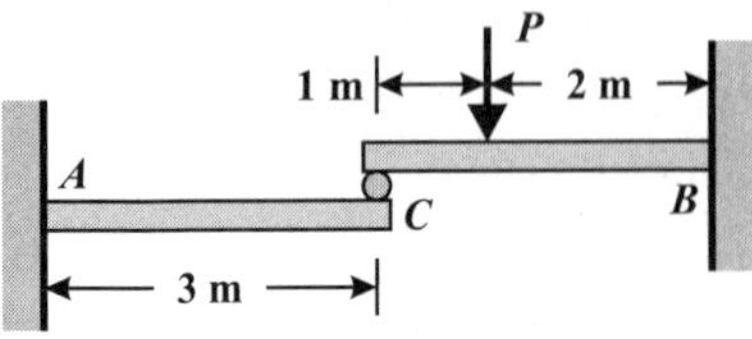

12.4 Limit Surface for Rectangular Beam – Combined Loading

12.13 Axial compressive load N is applied with eccentricity e on the y-axis of a square cross-section having sides D. The yield strength is S_y.

If $e = D/2$, determine the values of the force N/N_L and moment M/M_L at the limit condition.

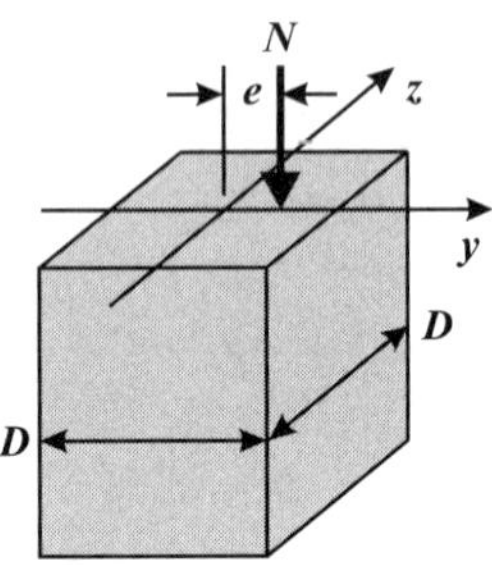

12.14 Using the ASME limit surface conditions, determine if the following rectangular aluminum beams:

    (1) fall within the factor of safety line,

    (2) fall within the limit surface, or

    (3) fall outside of the limit surface.

$S_y = 35$ ksi. The bending moment acts about the z-axis. Dimensions b and d are defined by the figure below.

	Breadth b (in.)	**Depth** d (in.)	N (lb)	M_z (lb-in.)
(a)	2	4	200	200
(b)	2	4	100	240
(c)	4	2	100	100
(d)	4	2	200	150
(e)	2	6	250	420

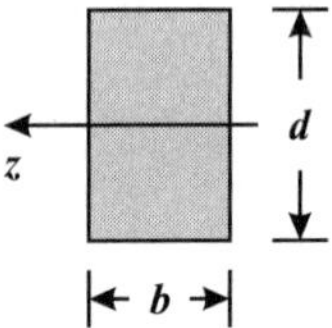

12.5 Design Applications in Plasticity

12.6 Three-Dimensional Plasticity

12.15 Force F = 16.0 kN is applied at the end of a cantilever beam, angle α below the horizontal. The system geometry is b = 50 mm, d =100 mm, and L = 2.0 m. The yield strength is S_y = 250 MPa.

Determine the maximum angle α such that yielding does not occur.

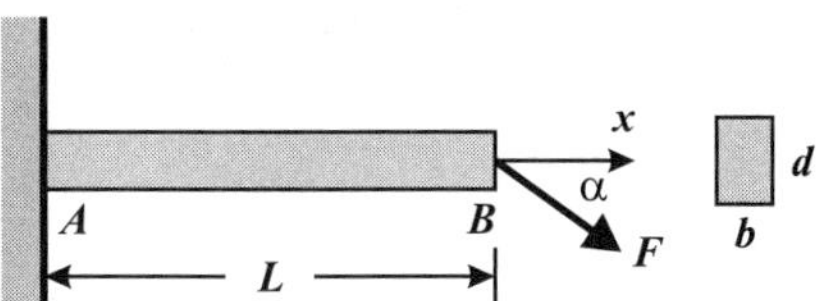

12.16 A triangular soil (earth) element at angle θ (from the horizontal) will begin to slide off a squared-off embankment of height h when the weight of the triangle overcomes the shear strength τ_y of the soil. The soil has mass density ρ. Assume the soil has a thickness of unity (1.0) into the paper.

Determine the maximum height of a squared-off embankment, h_{max} , in terms of τ_y, ρ and θ.

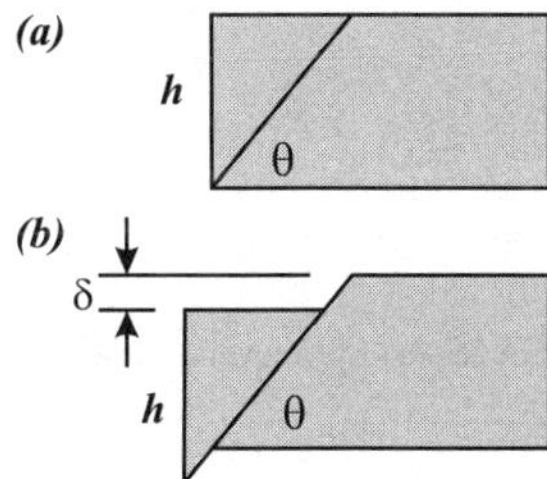

12.17 Pressure is applied at the edge of a square-off set of soil.

Determine the angle of repose θ of the soil in terms of pressure p and shear strength τ_y. Neglect the self-weight of the soil.

Hint: Equate the work done by the pressure to the work required to cause slipping.

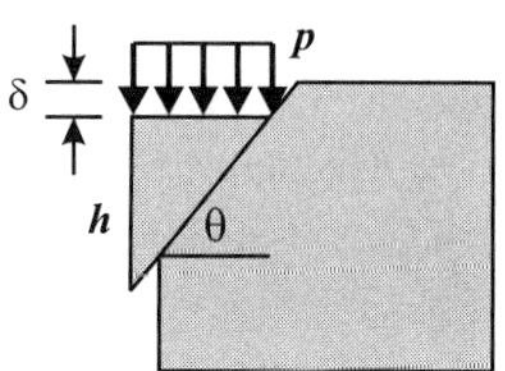

12.18 Consider the punch problem of *Example 12.14* and the half-cylinder element bound by the dotted line shown below. Assume that under pressure p, applied over punch width d, that a half-cylinder rotates out. Let $\tau_y = S_y/\sqrt{3}$.

Determine the upper bound for pressure p. Compare this to the upper bound determined in *Example 12.14*.

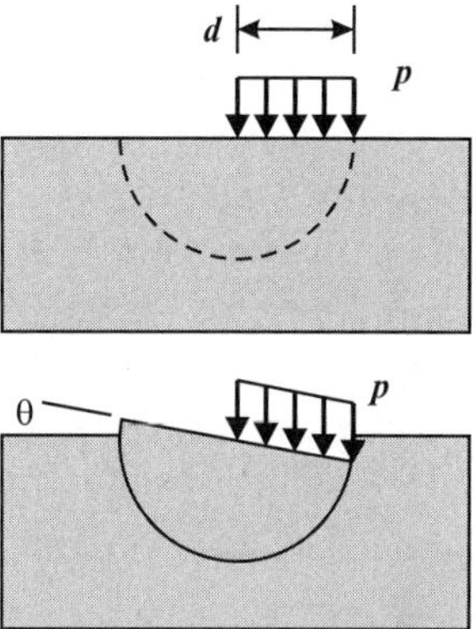

12.7 Thermal Cyclic Loading: Shakedown and Ratcheting

12.19 A steel pipe, outer diameter D = 150 mm and thickness t = 10 mm, carries steam at temperature ΔT above the outside temperature. The pressure in the pipe is 20 MPa. Steel has properties E = 200 GPa, S_y = 400 MPa and α = 14×10^{-6}/°C. Use a two bar-system to model the pipe as follows:

two parallel bars of length equal to the average of the inner and outer circumferences, both having a width of half the pipe thickness. Both bars have the same depth (into the paper); take the depth as unity (1.0). Only the inner bar is loaded by temperature ΔT. The mechanical stress in each bar is approximated by the hoop stress. Assume that the mechanical stress (due to pressure) is constant with time, and that the thermal load cycles.

(a) Approximate the temperature ΔT_y to cause yielding in the cool bar, corresponding to yielding of the outside of the pipe. (b) Estimate the shakedown limit ΔT_s of the system, assuming the pressure remains constant and the temperature cycles by ΔT_s.

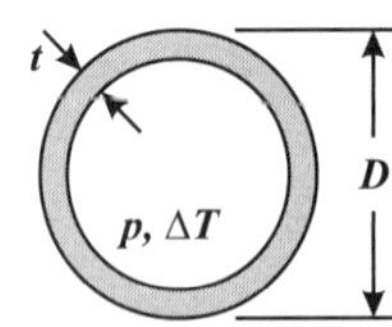

12.20 Bar 2 is twice as long as Bar 1; both have the same cross-sectional area A. Both bars are made of the same material with properties E, S_y, and α. A downward force F is applied to the rigid boss; the boss ensures both bars have the same elongation. Bar 2 is subjected to thermal cyclic loading that varies from 0 to ΔT.

Hint: Define normalized quantities:

$$s_1 = \sigma_1/S_y \qquad s_2 = \sigma_2/S_y$$

$$p = F/A \qquad t = E\alpha\,\Delta T$$

(a) Develop expressions for the mechanical stress in each bar σ_{m1} and σ_{m2}, due only to the applied force. (b) Develop an expression for the temperature load such that Bar 1 will just yield ΔT_y. (c) Develop an expression for the shakedown limit ΔT_s. (d) Plot the elastic, shakedown, and ratcheting regions on a t–p diagram (map). (e) Consider the case for $L_1 = 0.4$ m, $L_2 = 0.8$ m, with each bar having area $A = 0.0001$ m^2. Take the material to be aluminum, $E = 70$ GPa,

$S_y = 240$ MPa, and $\alpha = 23.6\times10^{-6}/°$C. If a downward force of $F = 20.0$ kN is applied to the rigid boss, determine the maximum temperature load ΔT to avoid shakedown, and the maximum temperature load to avoid ratcheting.

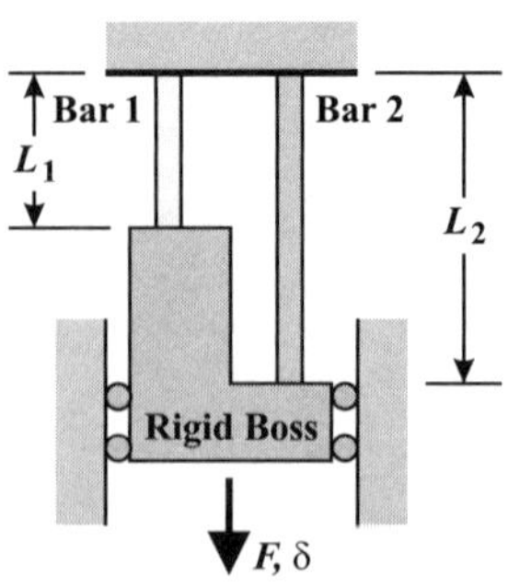

12.8 Large Plastic Strains

12.21 A cylindrical pressure vessel is subjected to increasing pressure p until plastic instability takes place. The unpressurized radius and thickness are R and T, respectively. The pressurized radius and thickness are r and t. The yield strength is S_y.

Considering 3D plasticity, determine the strain in the hoop direction at plastic instability.

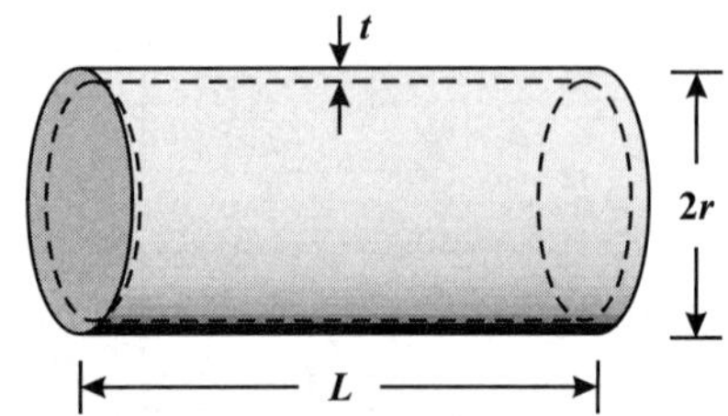

Problems: **Chapter 13 Effect of Flaws: Fracture**

13.1 An Introduction to Fracture Mechanics

13.1 An adhesive tape $w = 40$ mm wide is pulled off of a substrate by distance $a = 100$ mm. The energy of adhesion of the tape with the substrate is $G_c = 1.2$ kJ/m^2.

Determine the load to peel off the tape for the following two conditions: (a) constant force P_a applied normal to the substrate and (b) constant load P_b applied parallel to the substrate.

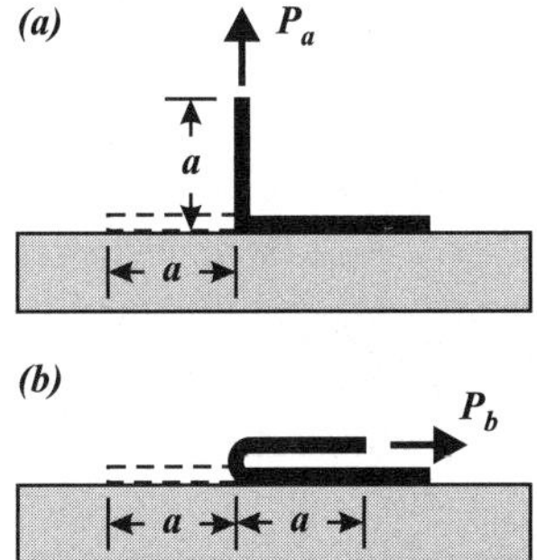

13.2 A cylindrical metal bar of radius R is coated by a thin protective ceramic layer of thickness t. The system is subjected to average axial stress σ applied over the entire cross-section. The modulus of the bar is E_b, and of the coating is E_c ($E_c > E_b$). Neglect any Poisson effect.

If the interphase (the intermetallic compound formed at the boundary of the two materials) has toughness G_c, determine the applied stress σ at which debond occurs between the coating and the metal bar. When debonding occurs, take the stress in the coating to be zero over debond length a, and that in the bar to be σ'.

Hint: Before debonding (crack growth), compatibility requires the axial strain in each component to be the same, $\varepsilon_b = \varepsilon_c$. The axial stresses are different and must be in equilibrium with the applied load: $A\sigma = A_b\sigma_b + A_c\sigma_c$.

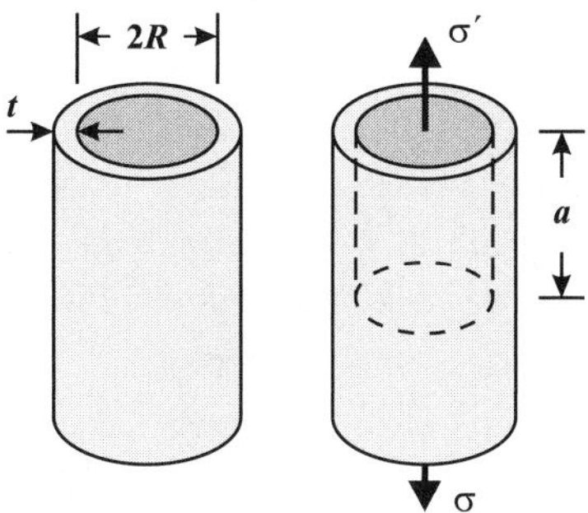

13.3 A thin steel plate ($E = 200$ GPa, $S_y = 250$ MPa) of width $2W$ has a center crack of width $2a = 3.0$ mm.

If the applied stress is $\sigma = 200$ MPa, determine the stress intensity factor K for the following geometries (a) $2W \gg 2a$, (b) $2W = 8(2a)$, and (c) $2W = 4(2a)$.

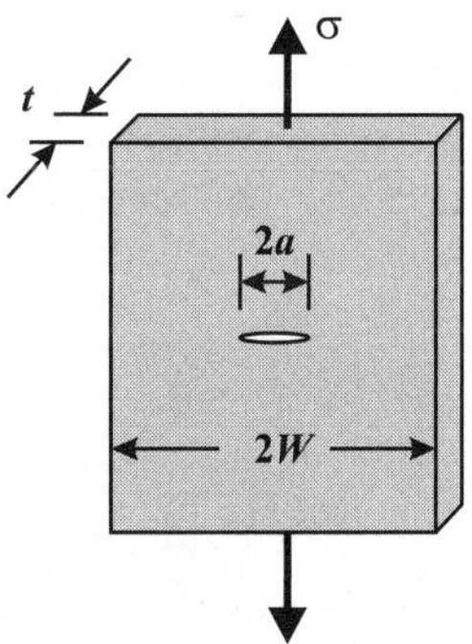

13.4 A thin plate with an edge crack $a = 2.0$ mm long is subjected to an applied stress $\sigma = 170$ MPa.

Determine the stress intensity factor if (a) $2W \gg 2a$ and (b) $2W = 4(2a)$.

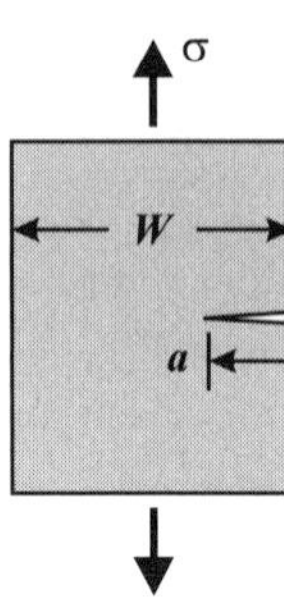

13.5 A steel pipe, diameter $2R$ and thickness t, transports gas at pressure p. A longitudinal crack, length $2a$, exists through the thickness of the wall (gas is leaking, but there is hoop stress in the wall). Young's modulus is E.

Determine (a) the energy release rate G and (b) the stress intensity factor K.

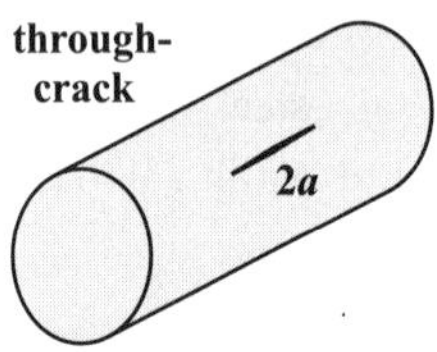

13.2 Design Considerations

13.6 Consider a thin-walled pipe made of pressure vessel steel, with $S_y = 500$ MPa and $K_{Ic} = 150$ MPa$\sqrt{\text{m}}$. A semi-circular (semi-elliptical) surface crack may grow from the outside (or inside) of the pipe, through the pipe wall. The crack has length $2a$ and the crack depth a is less than the wall thickness t.

Estimate the maximum value of $2a$ so that the pipe does not fracture (i.e., it leaks before breaks).

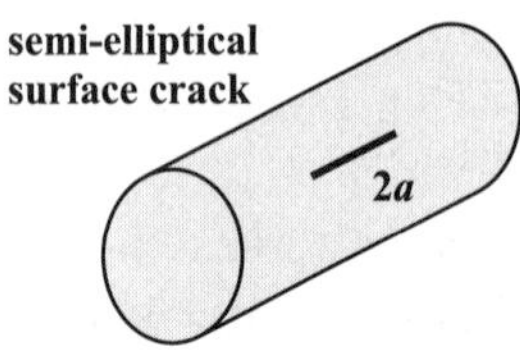

13.7 Consider a steel pressure vessel of diameter $D = 1.0$ m and wall thickness $t = 15$ mm. Inspection of the vessel finds a semi-elliptical surface crack 10 mm long. The yield strength is 1400 MPa, and the fracture toughness is $K_{Ic} = 75$ MPa$\sqrt{\text{m}}$.

Determine if fracture will occur due to an internal pressure of $p = 20$ MPa.

13.8 Before a rocket is launched, a pressure test is performed on a (thin-wall) cylindrical rocket casing. The yield strength is $S_y = 300$ MPa, and the fracture toughness is $K_{Ic} = 150$ MPa$\sqrt{\text{m}}$. To establish that no large cracks exist, a proof test is performed at 7/8 of the pressure at yielding (at failure).

If the casing does not fail during the test, determine the largest possible crack that can exist in the system.

13.9 An aluminum plate is $2W = 500$ mm wide and $t = 6.0$ mm thick. The yield strength is $S_y = 300$ MPa and the fracture toughness is $K_{Ic} = 24$ MPa$\sqrt{\text{m}}$. A center crack of total length $2a$ is found in the plate. The plate is subjected to an applied stress σ, which is increased from zero until failure.

(a) If the applied stress required to fracture the plate is $\sigma_a = 150$ MPa, determine the crack length $2a$. (b) Determine the maximum value of $2a$ such that the plate will fail by yielding (plastic deformation) and not by fracture.

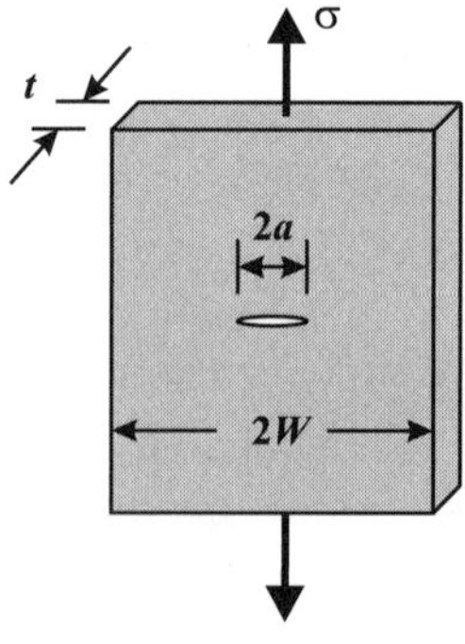

13.10 To make an electronic device, a thin layer of copper, thickness t, is deposited on a substrate of silicon, thickness D. The layer is very thin, $t \ll D$. The modulus of the substrate is E and of the copper is E_c. The interphase (the intermetallic compound between the copper and the substrate) has a toughness of G_c. The system is subjected to bending moment M.

If the width of the system into the paper is B, determine the moment required to initiate debonding of the copper from the substrate.

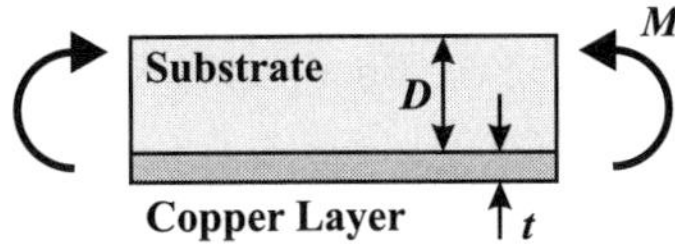

13.3 Crack Stability

13.11 A steel cantilever beam has a central longitudinal crack of length $a = 13$ mm. The breath of the beam is $b = 10$ mm and its total depth is $d = 20$ mm. Steel has a modulus of $E = 200$ MPa and a critical energy release rate of $G_c = 0.2$ MJ/m^2.

(a) Determine the load P_c to cause crack initiation. (b) If the tearing modulus is $T = 200$ MJ/m^3, will the crack continue to grow if $P = P_c$ remains constant? (c) Repeat *Parts (a)* and *(b)* for aluminum ($E = 70$ GPa, $G_c = 0.02$ MJ/m^2, and $T = 2.7$ MJ/m^3).

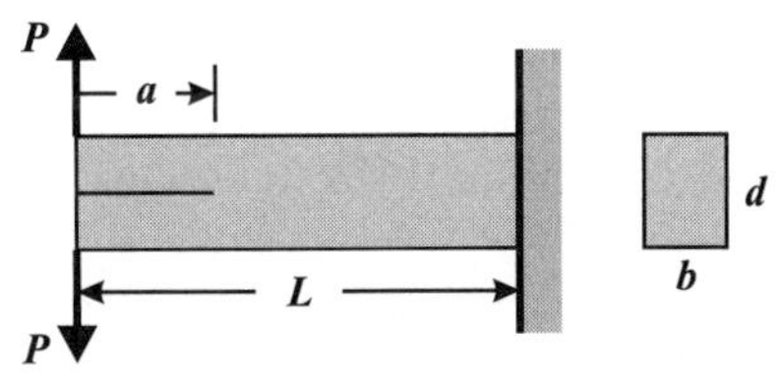

13.4 Modes of Fracture

13.5 Thin Film on a Substrate: Spalling

13.6 Statistical Design with Brittle Materials

13.12 A plate to be tested in tension is made of a ceramic with the following Weibull parameters: $S_o = 300$ MPa, $V_o = 0.11 \times 10^{-6}$ m^3, and $m = 5$. The test specimen has dimensions 40 mm wide by 80 mm long, with a thickness of 2 mm.

(a) Determine the survival probability for an applied tensile stress of 100 MPa. (b) If the volume is doubled, determine the new survival probability at 100 MPa.

13.7 The Effect of Non-Uniform Stress in Statistical Design

13.13 A MEMS cantilever beam has length $L = 1000$ μm, breadth $b = 10$ μm, and depth $d = 4$ μm. The system is made of silicon carbide, with Weibull parameters $S_o = 300$ MPa, $V_o = 0.11 \times 10^{-6}$ m^3, and $m = 5$.

Determine the tip load P that can be applied to assure a reliability of 0.999.

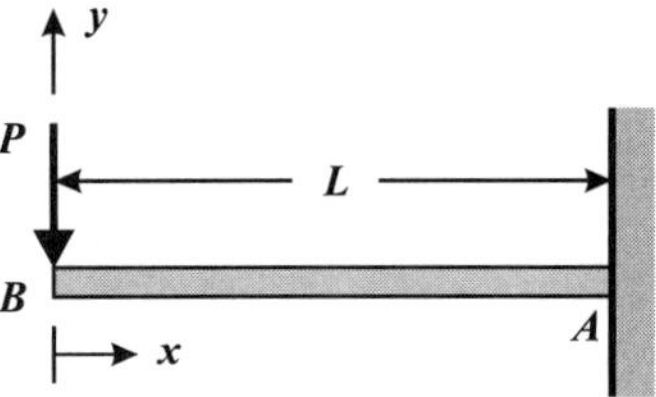

13.14 Consider a ceramic shaft of radius R and length L which transmits a torque T. The shear stress at any distance for the axis of the shaft is $\tau(r) = (r/R)\tau_{max}$. Failure occurs due to tensile stress. At 45° from the torsion axis, the tensile stress is $\sigma(r) = \tau(r)$; this is the maximum principal stress at that stress point.

Determine the load factor λ for the shaft in torsion. The reference stress, reference volume, and Weibull modulus are S_o, V_o, and m.

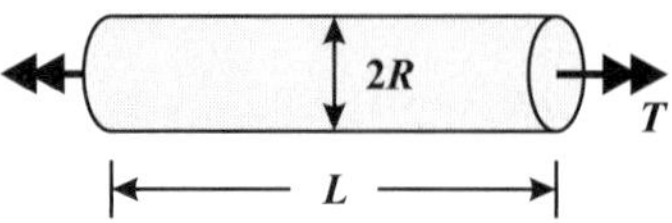

13.8 Determining Weibull Parameters

13.15 One-hundred three-point bend tests are performed on a new ceramic material. The maximum bending stress in the beam is σ_{max}, and the corresponding number of surviving samples n, are given in the table. The volume tested is the standard beam: 80 mm long and 10×10 mm square. Take the reference volume V_o to be the volume of the beam.

Estimate (a) the reference stress S_o and (b) the Weibull modulus m.

σ_{max} (MPa)	n	σ_{max} (MPa)	n
200	100	380	60
220	99	400	48
240	98	420	36
260	96	440	24
280	94	460	12
300	90	480	5
320	85	500	2
340	79	520	1
360	70	540	0

13.9 Strength of Fiber Bundles: Global Load Sharing

13.16 A bundle of fibers has a total length of 3.0 m. For a reference length of $L_o = 25.4$ mm, the reference stress is $S_o = 1600$ MPa and $m = 11$.

Determine the fiber bundle strength.

13.17 The strength of a bundle of fibers is to be at least 900 MPa. For a reference length of $L_o = 25.4$ mm, the mean stress is $\sigma_m = 1500$ MPa and $m = 9$.

Determine the maximum total length of the fibers in the bundle.

Problems: **Chapter 14 Joints**

14.1 Simple Fasteners

14.2 Failure in Bolt-type Connections: A Basic Analysis

14.1 Two axial members are joined by a single bolt of diameter $D = 0.375$ in. The members are $W = 2.0$ in. wide, and $a = 0.375$ in. and $b = 0.50$ in. thick, respectively. The members support a force of $P = 8000$ lb.

Determine (a) the net section stress in the left member, (b) the bearing stress between the bolt and the left member, (c) the bearing stress between the bolt and the right member, and (d) the maximum average shear stress in the bolt.

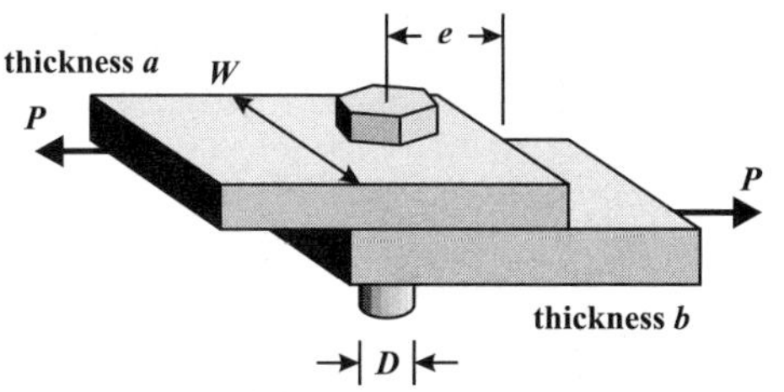

14.2 Two aluminum members are joined by a single steel bolt. The geometry of the connection is: bolt diameter $D = 0.375$ in., plate width $W = 2.0$ in., plate thicknesses $a = 0.375$ in. and $b = 0.50$ in., and edge distance $e = 0.5$ in. The yield, shear yield, and bearing yield strengths are:

Aluminum:

$S_y = 35$ ksi, $\tau_y = 20$ ksi, $S_{By} = 65$ ksi

Steel bolt:

$\tau_y = 35$ ksi, $S_{By} = 90$ ksi

Determine the strength of the joint.

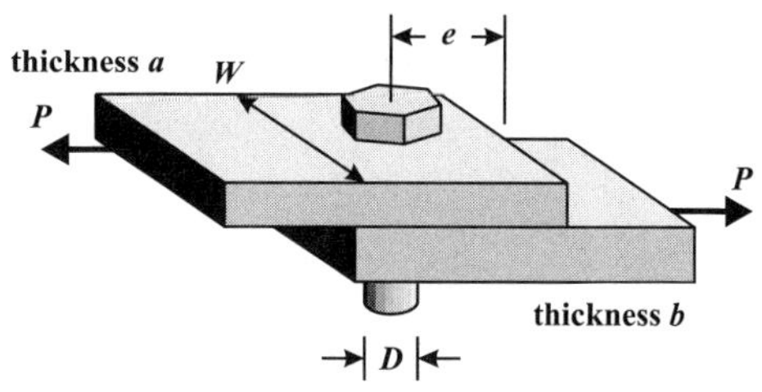

14.3 Aluminum truss ABC is subjected to a downward force of 20.0 kN. Members AB and BC are $W = 36$ mm wide and $t = 8.0$ mm thick. BC is 1.00 m long. The modulus of aluminum is 70 GPa. The diameter of each pin is $D = 10.0$ mm.

Determine (a) the shear stress in the pin at joint A (assume single shear) and (b) the bearing stress between pin A and member AB.

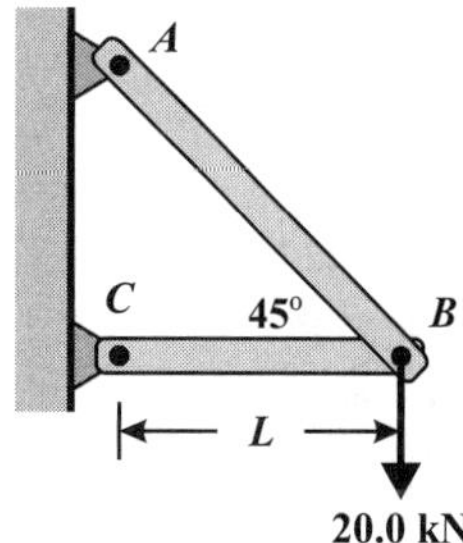

14.4 A bridge is $L = 20$ ft long and is supported at four points by large pins, 4.0 in. in diameter (two pins and two rollers). Each pin has an allowable average shear stress of 12 ksi, and is in double shear.

Considering only shear in the pins, determine the allowable distributed load on the bridge.

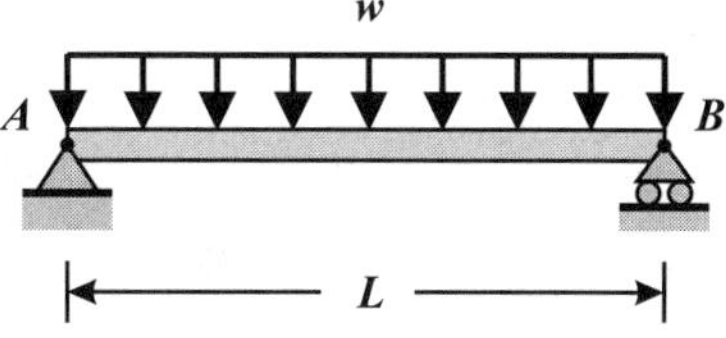

14.5 Two steel bolts join two aluminum plates. The geometry is: $a = 0.500$ in., $b = 0.750$ in., $D = 0.375$ in., and $W = 5.00$ in. The allowable stresses in the plates are:

axial: 20 ksi, shear: 12 ksi, bearing: 36 ksi,

and for the bolts:

shear: 12 ksi, bearing: 36 ksi.

(a) Considering only the plates, determine the allowable load P_{Ap}. (b) Considering only the bolts, determine the allowable load P_{Ab}. (c) It is desired to modify the joint geometry so that the allowable load is at least $P = 10,000$ lb. If necessary, determine the new required bolt diameter and plate thicknesses. If stock bolts are available in 1/16th increments, select the appropriate bolt. If thicker plates are required, and are available in 1/32th increments, select the appropriate plate thickness. Plate width W cannot be changed.

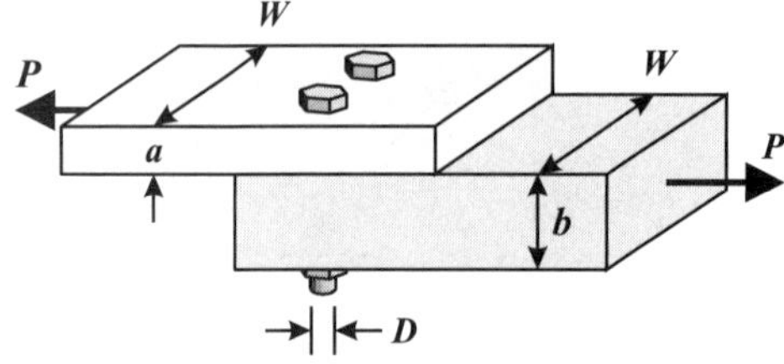

14.6 Two 100 mm wide plates are joined using two bolts. The plate thicknesses are $a = 12$ mm and $b = 10$ mm. The diameter of each fastener is $D = 8.0$ mm.

(a) Determine the shear and bearing stresses on the bolts. (b) Replace the left member by two plates, each 6.0 mm thick, and redo the calculations. (c) Explain the advantage(s) of the second configuration.

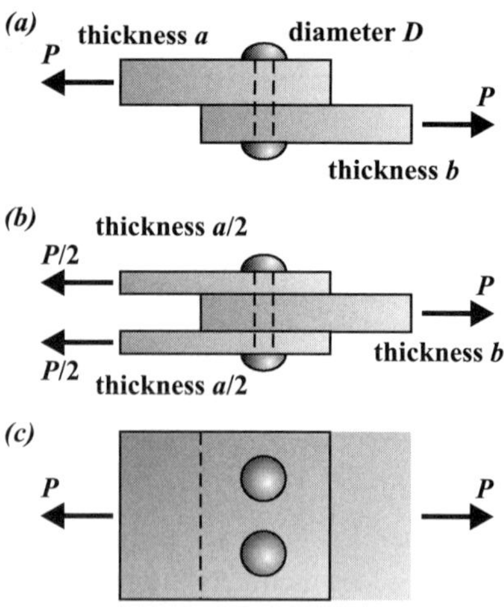

14.7 Three steel plates, each 0.90 in. thick, are joined by three bolts, each of diameter 0.75 in.

(a) If $P = 15$ kips, what is the maximum bearing stress acting on the bolts. (b) What force, P_{ult}, is required to fail the bolts in shear if the ultimate shear strength of the bolt material is 32 ksi.

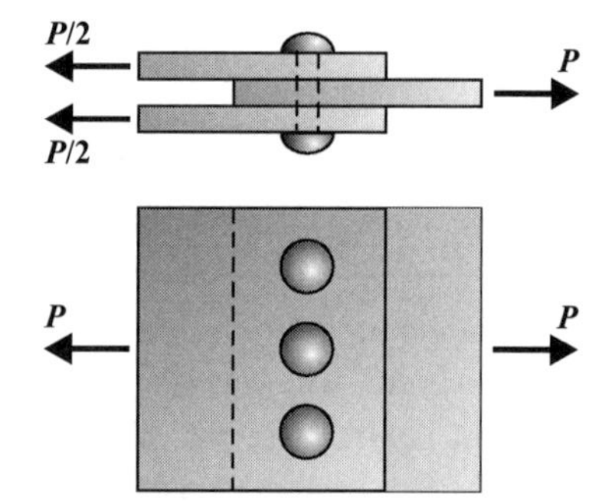

14.8 Two axial members are joined together by two rivets. At the joint, the members have an increased area. The geometry is: $W_1 = 1.0$ in., $W_2 = 2.0$ in., $D = 0.25$ in., and $t = 0.30$ in.

Assume the rivets are strong enough. If the joined members have yield strength 35 ksi and bearing yield strength 65 ksi, determine the strength of the joint.

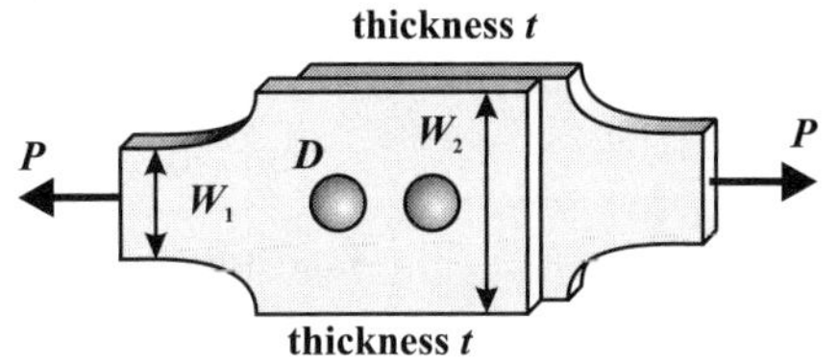

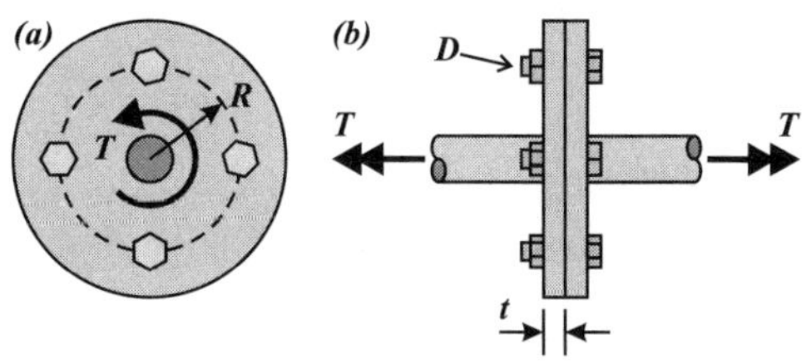

14.9 Vintage pressure vessels and boilers were made by rolling metal sheets into a cylinder, and riveting the seams together. Such a vessel has a diameter of $2R = 3.0$ ft, a thickness of 0.5 in., and contains steam at 200 psi. The vessel and rivets are both steel, with allowable normal stress $S_y = 10$ ksi, allowable shear stress 6.0 ksi, and allowable bearing stress 18 ksi. The rivets have a diameter of $D = 0.75$ in.

Determine (a) the stresses in the vessel wall and (b) the maximum and minimum allowable center-to-center distance between rivet centers, e_{max} and e_{min}.

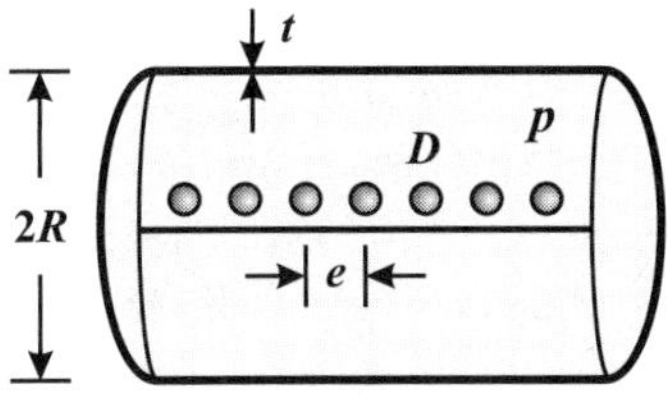

14.10 Two shafts are joined together by four bolts at their mating circular flanges. The shaft transfers 25.0 kW of power at 120 rad/s. The bolt-circle (circle of bolts) has a radius of $R = 100$ mm and the diameter of each bolt is $D = 15$ mm. The thickness of each flange is $t = 10$ mm.

Determine (a) the force on each bolt, (b) the bearing stress on each bolt, and (c) the shear stress in each bolt.

Hint: The force supported by each bolt is proportional to its distance from the centroid of the bolt distribution (here the center of the shaft). Here, all bolts are equidistant from centroid.

14.11 The knuckle joint is made of a steel yoke (clevis) and rod: $E = 30{,}000$ ksi, $S_y = 36$ ksi, $\tau_y = 20$ ksi, and bearing yield strength $S_{By} = 65$ ksi. The pin is made of steel: $E = 30{,}000$ ksi, $S_y = 50$ ksi, $\tau_y = 27$ ksi, and bearing yield strength $S_{By} = 90$ ksi.

Design the joint (select all dimensions, d, w, a, b, e, and D) so that the joint can support an allowable force of 20.0 kips. Use a factor of safety of 2 against all strengths.

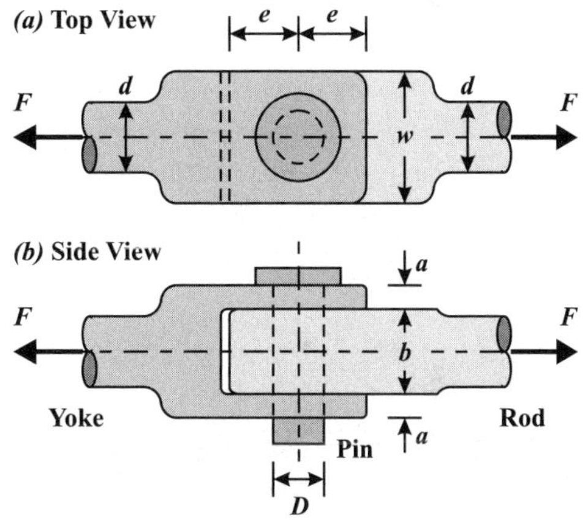

14.3 Failure of Bolt-Type Connections: An Advanced Analysis

14.12 A lap joint with five bolts in line with the applied load joins two metal members. The geometry is $D = 10$ mm, $W = 80$ mm, and $t = 10$ mm. Distance e is large enough so one hole does not affect the stress concentration at its neighboring hole(s). The applied load is $P = 20.0$ kN.

Determine (a) the average bearing stress on each bolt and (b) the by-pass stress in the first bolt. (c) Determine the maximum tensile stress at the hole surface at the first bolt due to the bolt load and (d) due to the bypass stress. (e) Determine the efficiency of the connection considering only the tensile stress at the hole surface if $S_y = 250$ MPa.

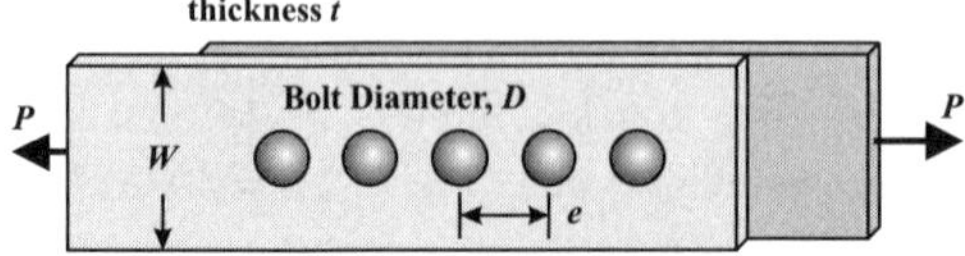

14.4 Stress Distribution in Adhesive Lap Joints in Shear

14.13 An adhesive ($G = 2.0$ GPa) is used to join two aluminum plates ($E = 70$ GPa). Each plate is $t_1 = t_2 = 5.0$ mm thick and $W = 300$ mm wide. The plates overlap by $L = 50$ mm. The thickness of the adhesive is $t = 0.100$ mm. The applied load is $P = 15.0$ kN.

Determine (a) the maximum stress in the adhesive and (b) the shear stress concentration factor.

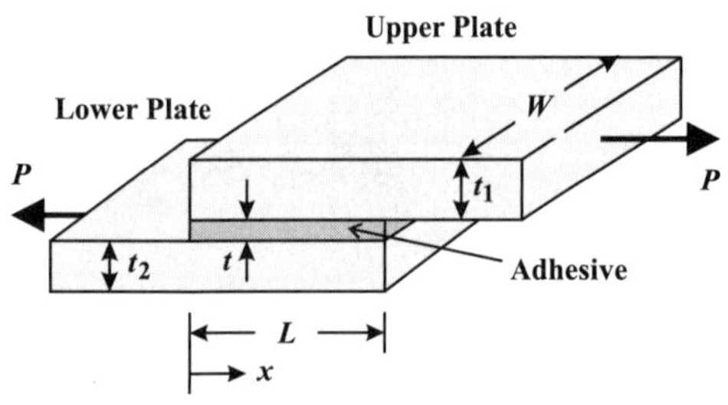

14.14 Repeat *Prob. 14.13* for $t_1 = 7.5$ mm and $t_2 = 5.0$ mm. Determine the stress and stress concentration factor at each end of the joint.

14.5 Design Problem

14.15 Two cylinders that make up part of the fuselage of a small airplane are joined together. The cylinders are made of sheet aluminum ($E = 10,000$ ksi), and are 5.0 ft in diameter and 0.20 in. thick. A lap joint is to be used to join the cylinders as detailed in *Figure (b)*. The cylinders are $t_1 = t_2 = 0.10$ in. thick at the joint. The shear modulus and allowable shear stress of the adhesive are $G = 250$ ksi and $\tau_{allow} = 1.0$ ksi, respectively. The joint must support an allowable tensile force of 5000 lb.

Determine the minimum length of the overlap L.

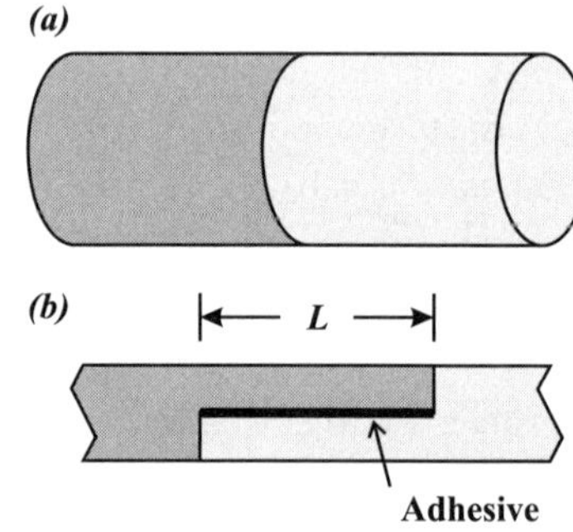

14.16 Two cylinders that make up part of the fuselage of a small airplane are joined together. The cylinders are made of a carbon/epoxy composite, with modulus in the fuselage direction of $E = 6200$ ksi. The cylinders are 5.0 ft in diameter and 0.40 in. thick. The cylinders are $t_1 = t_2 = 0.20$ in. thick at the lap joint. The shear modulus and allowable shear stress of the adhesive are $G = 250$ ksi and $\tau_{allow} = 1.0$ ksi, respectively. The joint must support an allowable tensile force of 5000 lb.

Determine the minimum length of the overlap L.

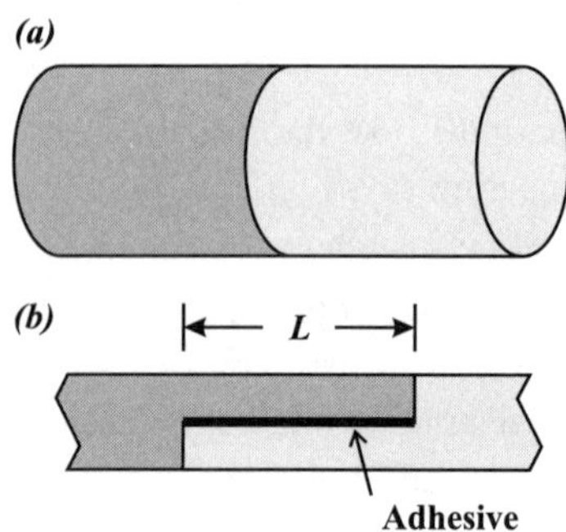

14.17 Two cylinders that make up part of the fuselage of a small airplane are joined together. The cylinders are made of sheet aluminum ($E = 10,000$ ksi) and are 5.0 ft in diameter and 0.20 in. thick. A lap joint is to be used to join the cylinders as detailed in *Figure (b)*. The cylinders are riveted together with rivets having an allowable shear stress of $\tau_{allow} = 10$ ksi. The diameter of each rivet is $D = 0.20$ in. The joint must support an allowable tensile force of 5000 lb.

Determine the maximum center-to-center spacing of the rivets.

(a)

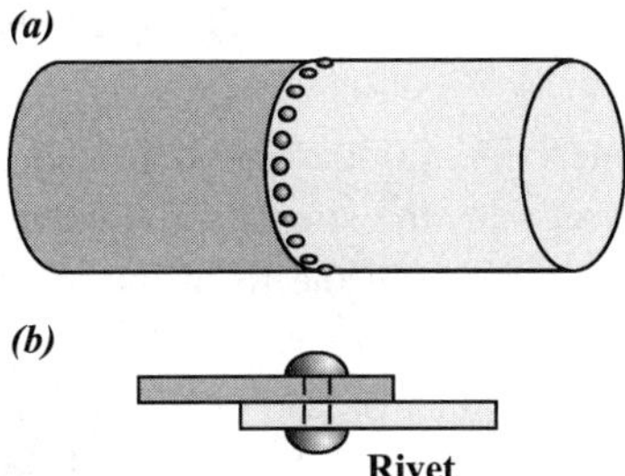

(b)

Problems: **Chapter 15 Composites**

15.1 Composite Materials

15.2 Properties of a Lamina (Ply)

15.3 Approximating the Elastic Properties of a Lamina

15.1 A steel reinforced concrete column is subjected to a compressive load of $F = 48.0$ kips, applied through a rigid plate at the top of the column. The steel reinforcements are six steel members that have a total area of $A_s = 6.0$ in.^{2}. The total cross-sectional area of the column is $A = 36$ in.^{2}. The material properties are:

> Steel: $E_s = 30 \times 10^6$ psi
>
> Concrete: $E_c = 4 \times 10^6$ psi.

Neglecting the Poisson effect, determine (a) the stress in the steel σ_s, (b) the stress in the concrete σ_c, (c) the change in length of the column, and (d) the effective Young's modulus of the system E_{eff}, i.e., the average applied stress (F/A) divided by the axial strain.

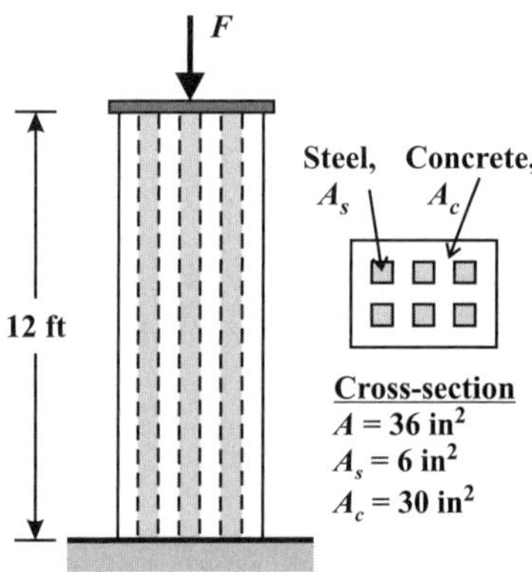

15.2 Composite materials are designed to take advantage of the properties of two different materials. A titanium matrix composite is reinforced with unidirectional silicon carbide (ceramic) fibers in the direction of the applied load (*Figure (a)*). These fibers are on the order of 100 μm in diameter. The total cross-sectional area of a composite system is A, while the area of the fibers is A_f. The fiber area fraction is $f = A_f/A$ which is also the fiber volume fraction since a unidirectional composite can be thought of as an extrusion of the cross-section.

The modulus of titanium is 115 GPa and of silicon carbide is 360 GPa. The volume fraction of the fibers is $f = 0.35$. The applied stress is $\sigma = 400$ MPa and the total cross-sectional area of the composite is 5.00 mm by 100 mm.

Determine (a) the stress in the titanium matrix σ_m, (b) the stress in the ceramic fiber σ_f, (c) the strain in the fiber and the matrix ε (they are the same), and (d) the effective Young's modulus of the system E_{eff}, i.e., the applied stress σ divided by the axial strain ε.

(a) Top View

(b) Cross-section

15.3 Approximate the elastic properties – E_x, E_y, G, v_{xy}, and v_{yx} – for a graphite/epoxy lamina (ply) with fiber volume fraction $f = 0.57$.

> Graphite: $E_f = 400$ GPa, $v_f = 0.2$,
> $G_f = 170$ GPa
>
> Epoxy: $E_m = 3$ GPa, $v_m = 0.4$,
> $G_m = 1.2$ GPa

Compare the results with those found in *Table 15.4.*

15.4 Approximate the elastic properties – E_x, E_y, G, v_{xy} and v_{yx} – for an E-glass/epoxy lamina (ply) with fiber volume fraction $f = 0.60$.

     Carbon: $E_f = 72$ GPa, $v_f = 0.2$,
                       $G_f = 30$ GPa

     Epoxy: $E_m = 3$ GPa, $v_m = 0.4$,
                       $G_m = 1.2$ GPa

Compare the results with those found in *Table 15.4*.

15.5 Approximate the elastic properties – E_x, E_y, G, v_{xy}, and v_{yx} – for a boron/aluminum lamina (ply) with fiber volume fraction $f = 0.50$.

     Boron: $E_f = 400$ GPa, $v_f = 0.2$,
                      $G_f = 170$ GPa

     Aluminum: $E_m = 70$ GPa, $v_m = 0.33$,
                      $G_m = 26$ GPa

Compare the results with those found in *Table 15.4*.

15.6 The metal matrix composite (MMC) consists of alumina (Al_2O_3) fibers in a matrix of an aluminum alloy (Al–4.5 Mg) with fiber volume fraction $f = 0.40$. The modulus of each constituent is:

     Alumina: $E_f = 285$ GPa

     Aluminum: $E_m = 70$ GPa

The yield strength of the matrix is $S_y = 250$ MPa. Neglect the Poisson effect.

(a) Determine the elastic modulus E_x of a unidirectional lamina in the fiber-direction. (b) An average stress of $\sigma = 200$ MPa is applied to the composite in the fiber direction. Determine the stress in the fiber σ_f and in the matrix σ_m.

(a) Top View

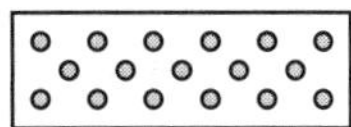

(b) Cross-section

15.7 Repeat *Prob. 15.6* but include the residual stress due to processing the composite as described below. Neglect the Poisson effect.

A metal matrix composite (MMC) consists of alumina (Al_2O_3) fibers in a matrix of an aluminum alloy (Al–4.5 Mg). The fiber fraction is $f = 0.40$. The modulus and coefficient of thermal expansion α of each constituent are:

     Alumina: $E_f = 285$ GPa
                    $\alpha_f = 6.0 \times 10^{-6}/°C$

     Aluminum: $E_m = 70$ GPa
                    $\alpha_m = 23 \times 10^{-6}/°C$

The yield strength of the matrix is $S_y = 250$ MPa. Neglect the Poisson effect.

(a) The composite is formed at 600°C, and then cooled to room temperature (20°C). Assuming there is no stress at the processing temperature, estimate the axial residual stresses at room temperature in each constituent $\sigma_{f,R}$ and $\sigma_{m,R}$. (b) An average stress of $\sigma = 200$ MPa is now applied to the composite in the fiber direction. Determine the stress in the fiber σ_f and in the matrix σ_m.

15.4 Laminates

15.8 You are to design fiber lay-ups for different composite systems, i.e., specify the ply angles: "0°" (a ply in the load direction), "90°" (a ply transverse to the load), or ±θ° (two plies at ±θ to the load direction.

Note: If you have $+\theta$, you also need $-\theta$ to avoid shear incompatibility).

Assume that each structure is to be four plies thick. Also, for combined loading, assume that the cross-section can be designed such that each type of load is supported independently by different parts of the section. E.g., in flanged beams, the flange supports bending and the web supports shear.

Qualitatively design the lay-up of:

(a) an I-beam (four-ply lay-up in both the web and the flange), (b) a torsion box that supports both torsion, and bending about the z-axis and (c) a tube that is to support both torsion and tension.

15.9 The two-ply lay-up model of a carbon/epoxy composite ($f = 0.60$) panel is $\theta = \pm 20°$ (the plies are $\pm 20°$ from the load direction). The carbon/epoxy ply properties are given in *Tables 15.4* and *15.5*.

(a) Determine the global stiffness of the composite. (b) If the applied load is $\sigma = 250$ MPa, determine the stresses in each ply. (c) Determine the factor of safety of the system.

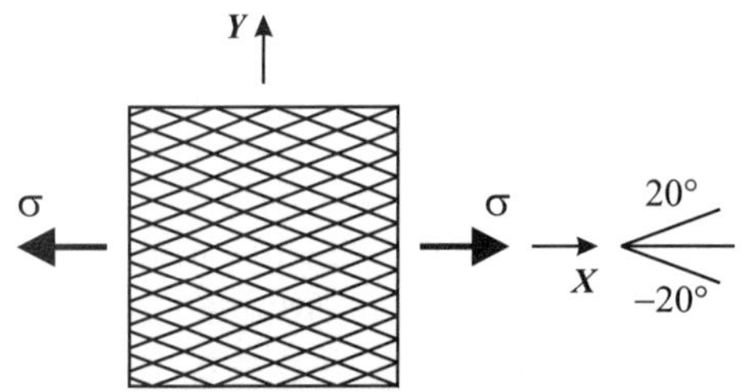

15.10 The two-ply lay-up model of a graphite/epoxy composite ($f = 0.57$) panel is $\theta = \pm 20°$ (the plies are $\pm 20°$ from the load direction, see *Prob. 15.9*). The graphite/epoxy ply properties are given in *Tables 15.4* and *15.5*.

(a) Determine the global stiffness of the composite. (b) If the applied load is $\sigma = 250$ MPa, determine the stresses in each

ply. (c) Determine the factor of safety of the system.

15.11 The two-ply lay-up model for a SiC/Ti (silicon carbide fiber/titanium matrix) composite is $0°/90°$. The modulus of titanium is 115 GPa, and of silicon carbide is 360 GPa. The volume fraction of the fibers is $f = 0.35$.

(a) Determine the elastic properties of the laminate. (b) Determine the stress in each ply if the applied stress is $\sigma = 400$ MPa.

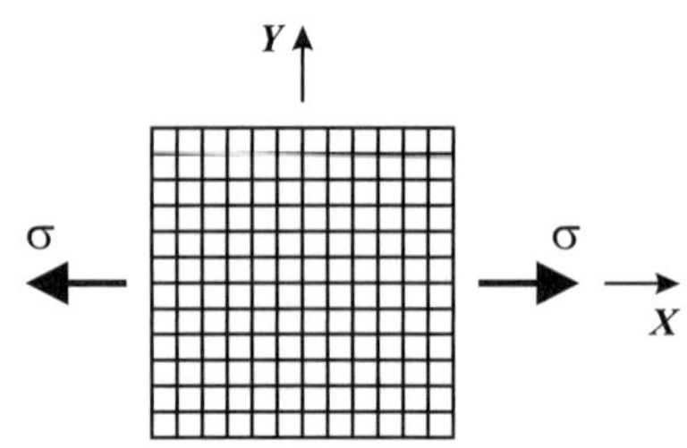

15.12 A thin-walled cylindrical pressure vessel is manufactured from a glass/epoxy composite (*Tables 15.4* and *15.5*). Half of the plies are oriented at $+\theta$ from the hoop direction, and half at $-\theta$. The vessel has average radius R and thickness t. The tensile strength of the fibers is $S_{u,t}$.

If the fibers are to be oriented so that each ply is subjected to tensile stresses only (no compressive or shear stresses), determine the optimum value of the fiber angle θ.

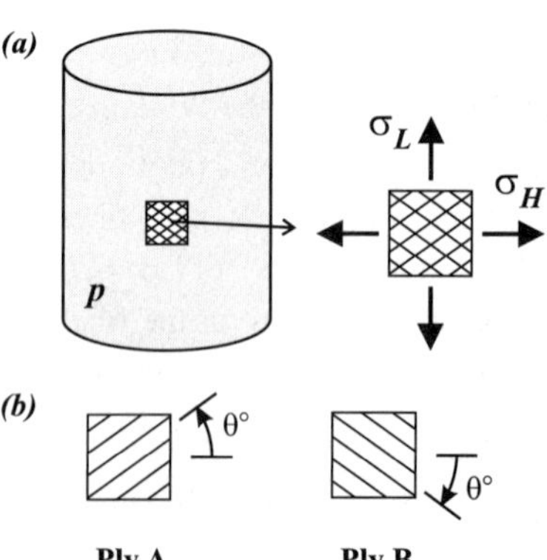

Problems: Chapter 16 Smart Systems

16.1 MEMS

16.1 Because the dimensions of MEMS beams are measured in microns, a variation in size of just a micron means that the calculated and actual stiffnesses of as-manufactured beams can differ significantly. The actual stiffness can be calibrated against a beam whose stiffness is known accurately (*Prob. 16.2*).

(a) A manufactured SiC (silicon carbide) MEMS cantilever beam has length 967 μm, breadth 9.7 μm, and depth 1.8 μm. The modulus is $E = 360$ GPa. Determine the transverse stiffness P/δ of the beam with respect to a point load at its tip. (b) Compare the as-manufactured stiffness to the nominal stiffness of the beam having ideal length 950 μm, breadth 10.0 μm, and depth 2.00 μm.

16.2 The stiffness of a manufactured MEMS beam is calibrated by comparing it with a beam whose stiffness to a tip load P is known to be $k_o = P/\delta = 327$ nN/μm. Known beam AC is built-in at fixed point A, and unknown beam BC is built-in at movable support B. The beams contact each other at their mutual tip, point C. The support of BC is moved normal to the beam by distance δ_B, causing tip displacement δ_C. Displacements δ_B and δ_C are measured.

(a) In terms of known stiffness k_o, and displacements δ_B and δ_C, determine the stiffness of the manufactured beam, BC. (b) If $\delta_B = 9.2$ μm and $\delta_C = 3.4$ μm, determine the stiffness of BC.

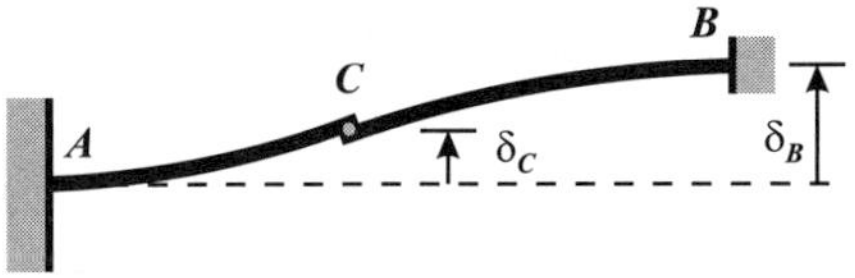

16.3 A MEMS cantilevered system is point-loaded at point C. Beams AB and BC both have bending stiffness EI. The connector at point B is rigid so that the slopes of AB and BC at that point B are equal. Beam AB is L long. The design is such that the slope of the structure is to be zero at the load point (point C).

Determine the length of the overhang, s, to achieve this state.

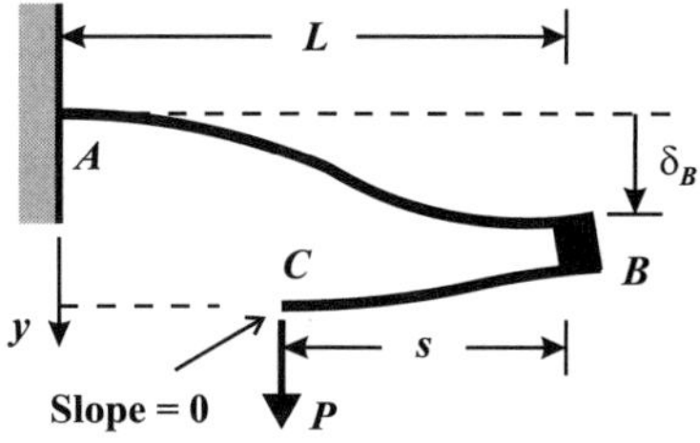

16.4 A device to measure lateral (horizontal) accelerations has the form shown. The suspended mass is attached to the fixed base (the vehicle) through four beams, each of modulus E, area A, moment of inertia I, and thermal expansion coefficient α. The beams are much longer than the heights of the beam connector and moving mass, so the latter two can be considered rigid. The moving mass is vertical distance h above the fixed base.

Determine the change in spacing, Δh, when the device is subjected to a temperature ΔT. What is the advantage of this set-up?

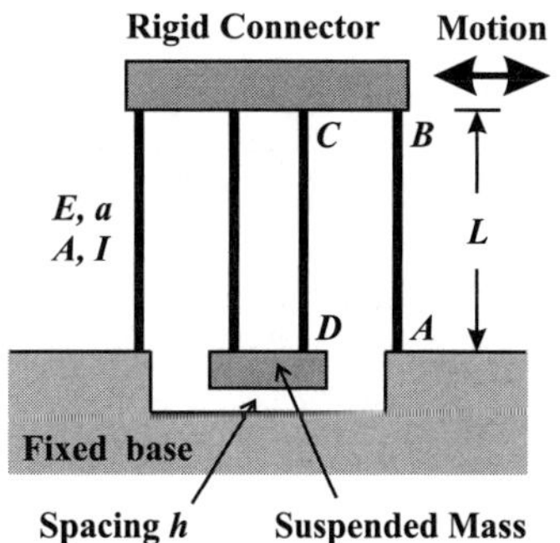

16.5 A device to measure lateral accelerations has the form shown (see also *Prob. 16.4*). The beams are typically of length $L = 1000$ μm, and have square cross-section $b = d = 10$ μm. An inertial force ma is exerted on the suspended mass when the fixed base (i.e., the vehicle) accelerates by a.

Manufacture is such that thicknesses b and d can vary by ±1 μm. Determine the percentage variation of the stiffness $k = ma/\delta$ of such manufactured devices due to this variation.

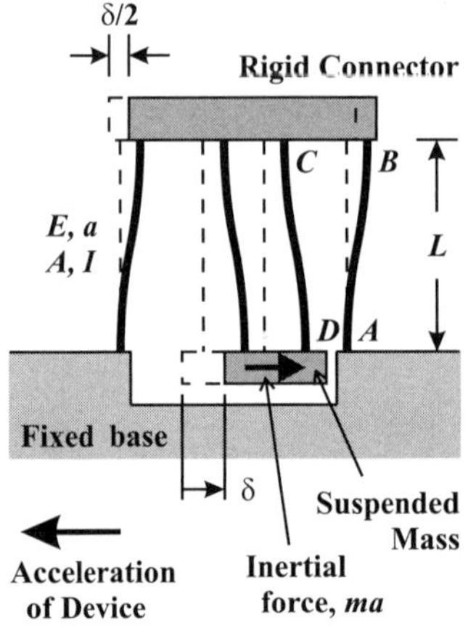

16.2 Parallel Plate Capacitors

16.6 The capacitance C of parallel plates can be used to measure the distance d between them.

Plot capacitance versus distance for the case of $a = 10$ mm, $b = 10$ mm, with d varying from 5 to 50 μm. Only air exists between the plates.

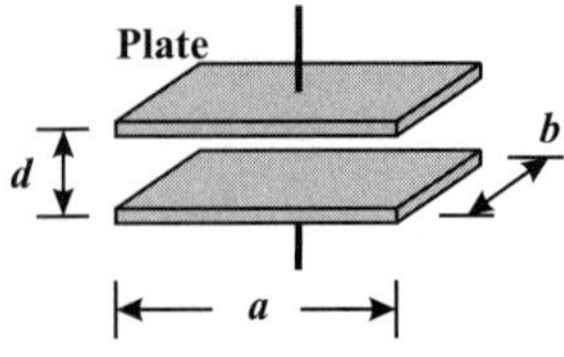

16.7 Two parallel plates each have an area of $A = 4.0$ mm^2 and an air gap between them of x.

(a) Determine the extreme values of the capacitance if x can have extreme values of 1.0 and 25 μm. (b) If the plates are initially 5.0 μm apart, determine the change in capacitance if the plates are moved apart an additional 1.0 μm.

16.8   Two   parallel   plates,   both 1000 by 200 μm, are separated by distance x.

(a) A voltage of $V = 10$ V is applied across the plates. Determine the force on the plates $F(x)$ required to keep them from moving closer together. (b) For this case, determine the effective stiffness of the plate system (i.e., $k = dF/dx$).

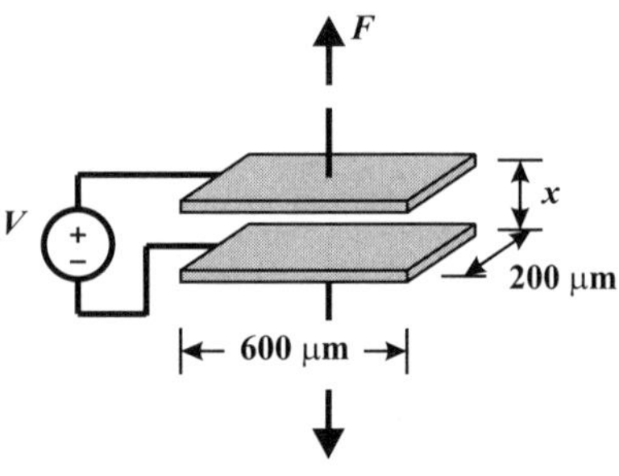

16.9 Two parallel plates are distance x apart. The plates have an area of $A = 4.0$ mm^2 and are subjected to a voltage of $V = 15$ V.

(a) Plot the force F to keep the plates from moving versus plate distance x, for values of x from 1 to 25 μm. (b) Estimate the value of the force when electrostatic breakdown (arcing) is expected to occur, assuming a breakdown voltage gradient of 40×10^6 V/m.

16.3 Capacitive Accelerometer

16.10 A suspended plate is supported by four cantilever beams. The suspended plate has mass 3×10^{-9} kg, and area $A = 120 \times 10^{-9}$ m^2.

The beams each have length $L = 200$ μm and square cross-section of sides $B = 6$ μm. With no acceleration, the suspended plate is $d = 5$ μm from each of the fixed plates (the fixed plates are attached to a moving vehicle). When the vehicle accelerates, the suspended plate moves closer to one of the fixed plates (*Figure (b)*).

(a) Determine the stiffness of the cantilever system. (b) Determine the difference in capacitances $\Delta C = C_1 - C_2$ due to an acceleration of magnitude $|a| = 5g$. C_1 is the capacitance between the upper and lower plates and C_2 is the capacitance between the suspended and lower plates.

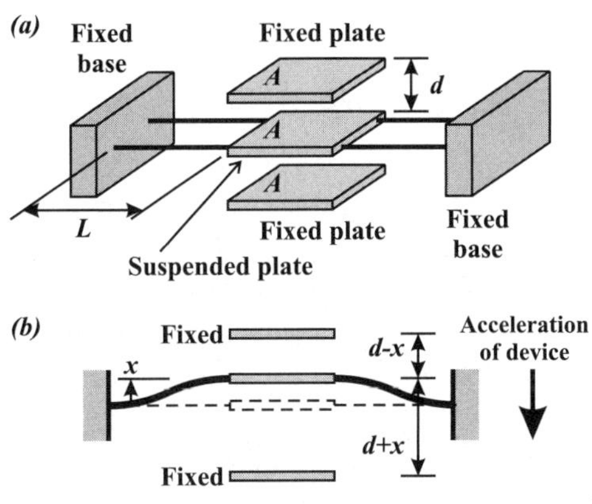

16.4 Electrostatic Snap-Through in MEMs Devices

16.11 A suspended plate is supported by four cantilever beams. Each plate has an area of $A = 120 \times 10^{-9}$ m^2. The beams each have length $L = 200$ μm and square cross-section of sides $B = 6$ μm. In the equilibrium condition, the plates are $d = 5$ μm from each other. Voltage V is applied across the plates.

(a) Determine the stiffness of the cantilever system. (b) Determine the voltage to cause snap-through.

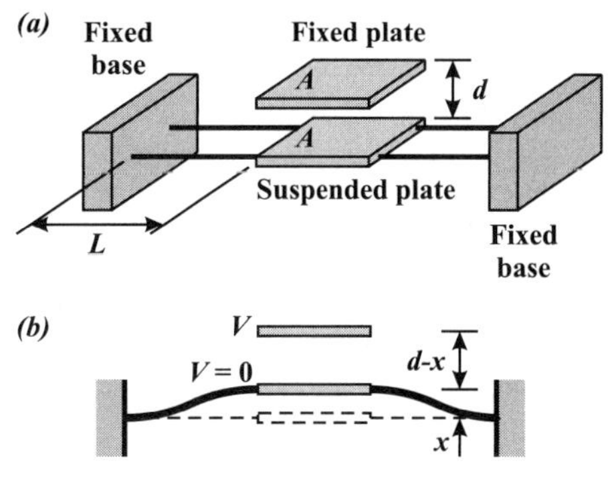

16.5 Comb Drive

16.12 Two plates are b wide and are separated by an air gap d. They overlap by distance a.

When voltage V is applied across the plates, determine the ratio of the vertical attractive force to the horizontal attractive force.

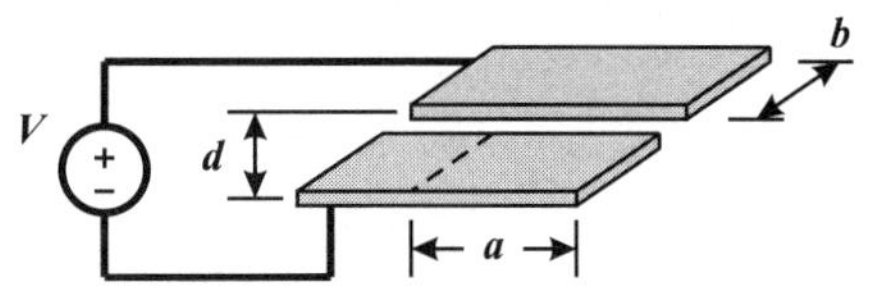

16.13 A suspended plate of an accelerometer is supported by four cantilever beams. Each plate has area $a = 600$ μm by $b = 200$ μm. The suspended plate has a mass of 3×10^{-9} kg. The beams each have length $L = 200$ μm, and square cross-section of sides $B = 6$ μm. In the equilibrium condition, the plates are $d = 5$ μm from each other.

(a) Determine the stiffness of the cantilever system. (b) Due to acceleration of the device, the suspended plate moves parallel to the fixed plate by distance x (*Figure (b)*). Determine the change in capacitance of the parallel plates if $x = 40$ μm. (c) If the acceleration of the device has magnitude $|a| = 5g$, determine the capacitance of the parallel plates.

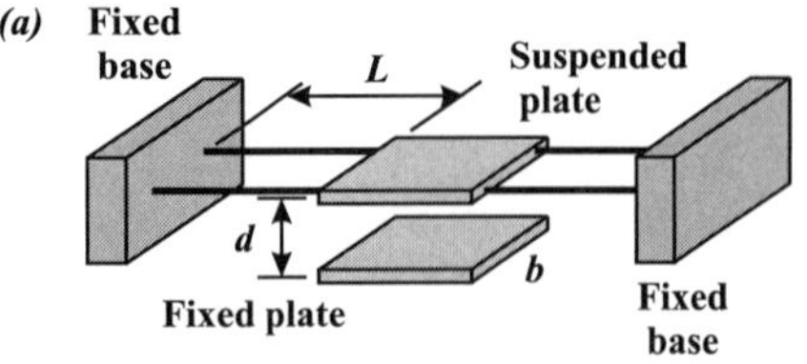

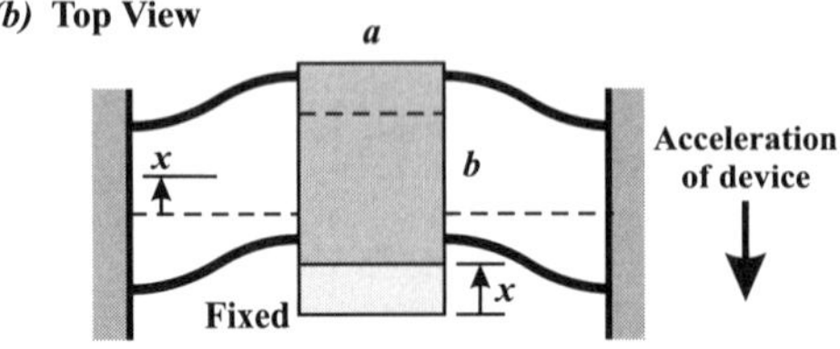

16.14 A schematic of a device used to apply small compressive loads to small specimens is shown. The device consists of several sets of comb pairs. Only two pairs are shown.

Each comb pair consists of cantilevered arm AB fixed at point A, and wing CD which is cantilevered from point D on rigid and moveable trunk ED. The arms and wings are 355 and 400 µm long, respectively, and are 23 µm apart. The comb teeth are 15 µm long, at 7 µm intervals. The arm and wing combs overlap by 7 µm, and are 1.5 µm apart. The breadth of the combs is 12 µm (into the paper). The bending stiffness of the arms and wings is $EI = 8.12 \times 10^{-10}$ N·m^2.

Voltage is applied across the combs, and they attract each other. Wings CD approach arms AB, moving trunk ED upward, compressing the micro-specimen.

(a) Determine the force generated by one pair of combs on specimen EF, when the applied voltage difference between the arms and the wings is 50 V. (b) Approximate the deflection of tip point B of arm AB.

(based on M.T.A. Saif and N.C. MacDonald, "A Millinewton Microloading Device," *Sensors and Actuators*, A52, Elsevier Science B.V., 1996).

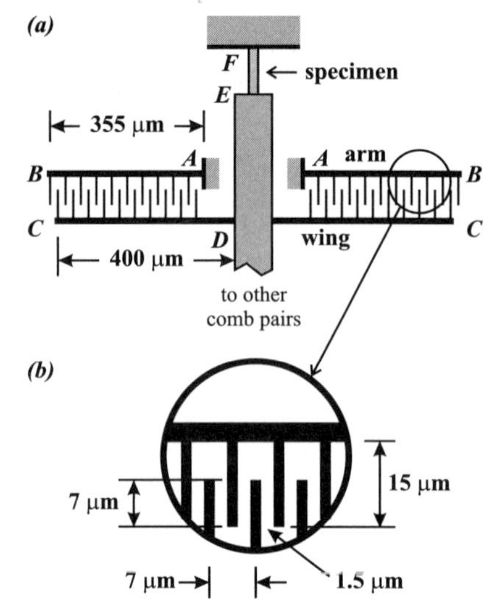

16.6 Piezoelectric Behavior

16.15 A piezoelectric material having the piezoelectric constants given in *Table 16.1* is placed between two electrodes. The system is loaded by a compressive stress of $\sigma_3 = -200$ MPa along the poling direction, as shown.

Determine the magnitude of the voltage induced across the electrodes.

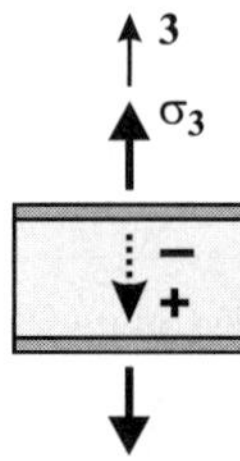

16.16 A piezoelectric material having the piezoelectric constants given in *Table 16.1* is placed between two electrodes. The short-circuited modulus is $Y = 66$ GPa. An electric field of $E_{3a} = -50$ kV/m is applied across the electrodes, causing the distance between electrodes to decrease. In other words, E is opposite the poled direction, the voltage drop is negative, or the source voltage V_{3a} is negative.

Determine the strain (a) in the poling (3-) direction and (b) in the transverse directions.

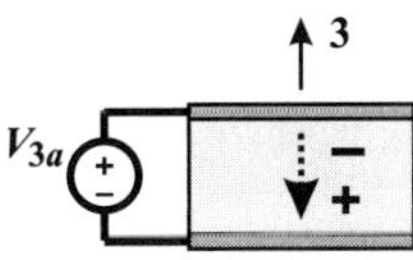

16.17 Combine the applied stress and applied electric field from *Prob. 16.15* and *Prob. 16.16*. Take Poisson's ratio to be $v = 0.2$.

Determine the strain (a) in the poling (3-) direction and (b) in the transverse (1-, 2-) directions.

16.18 Piezoelectric materials can be used in micro-pump systems. A stack actuator is subjected to an AC electrical signal, $v(t) = 12 \sin(360t)$ V. The stack is made of PZT4, with $d_{33} = 285 \times 10^{-12}$ m/V. The stack has a total height of 8 mm and each PZT layer is $t = 100$ μm thick.

Determine the double amplitude $2\delta_{max}$ (the stroke length) of the piezoelectric stack.

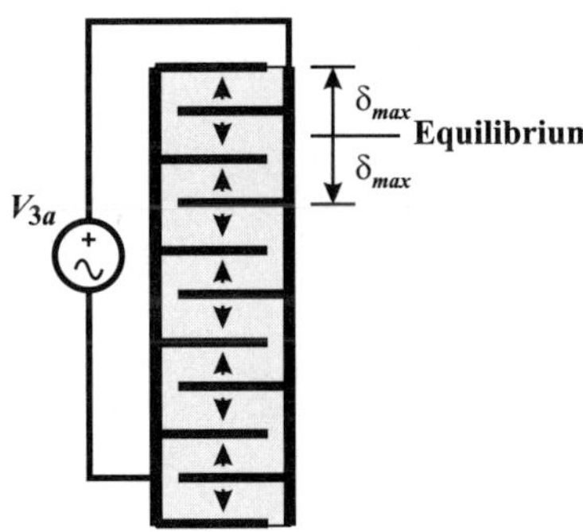

16.7 Piezoelectric Bending

16.19 A cantilever bimorph made of PZT4 is used to cause an upward displacement of δ by applying voltage V. The bimorph has length $L = 20$ mm, layer thickness $t = 0.150$ mm, and width $B = 1.2$ mm (into the board). The PZT materials has short-circuit modulus $Y = 66$ GPa

and piezoelectric charge constant $d_{31} = -122 \times 10^{-12}$ m/V.

Determine the displacement δ for an applied voltage of $V = 15$ V.

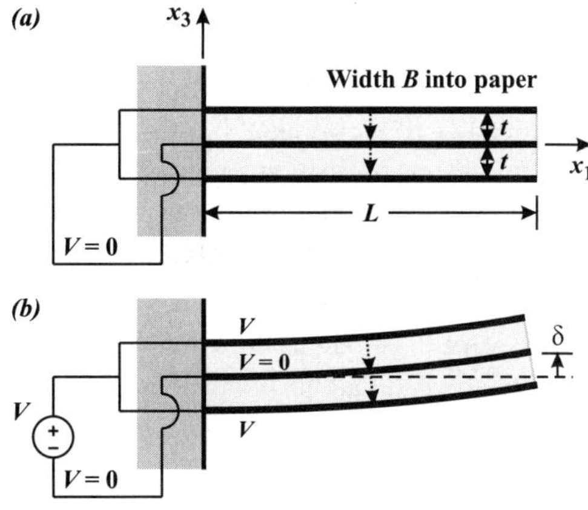

16.8 Shape Memory Alloys

16.20 Two wires are tested in tension in a strain-controlled test at room temperature. The first wire is steel, with $E_s = 200$ GPa and $S_{y,s} = 500$ MPa. The second wire is nitinol, $E_n = 70$ GPa and $S_{y,n} = 450$ MPa (assumed to be the same for all phases). The low-temperature *twinned-martensite* transforms into *stress-induced martensite* at room-temperature when $\sigma = S_{tr} = 350$ MPa, with a transformation strain of $\varepsilon_T = 4.9\%$.

(a) If both systems are loaded to a stress of $\sigma = 400$ MPa, determine the difference in strain between the steel and transformed nitinol. (b) Determine the strain in the nitinol when the stress is removed.

16.21 Truss *ABC* is subjected to downward force $F = 3.0$ kN at joint B at room temperature. The truss is made of nitinol (NiTi). This nitinol has modulus $E = 70$ GPa (assume E is the same for all phases), and transforms to stress-induced martensite at $S_{tr} = 350$ MPa. The transformation into stress-induced martensite causes the nitinol to undergo a strain transformation of $\varepsilon_T = +5\%$.

The cross-sectional area of each bar is $A = 4.0 \times 10^{-6} \, \text{m}^2$ and distance $L = 15$ mm.

Determine (a) the stress in each bar, (b) the elongation of each bar including the 5% transformation strain (do the bars actually transform?), and (c) the downward displacement v of joint B from its unloaded position. (d) If the system is now heated to 100°C, and transformation to austenite occurs (the 5% transformation strain is recovered), determine the new downward movement of joint B from its unloaded position.

Assume the modulus does not change with temperature, and that the elongations are small enough that the changes in slope of the truss members are negligible.

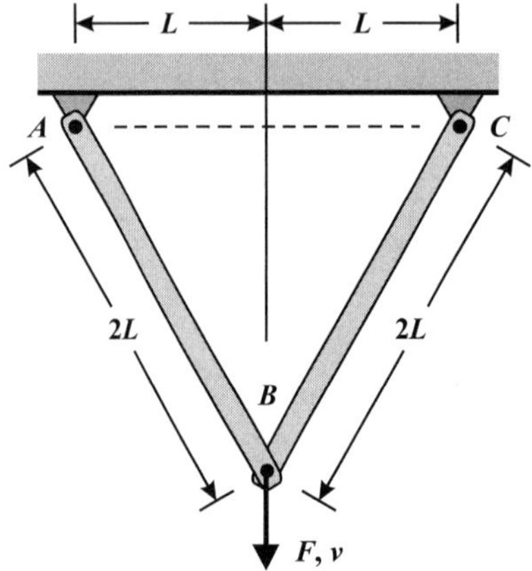

APPENDICES

Appendix Contents

A. Trigonometric Formulas and Geometric Properties

Tables A.1 and *A.2* give trigonometric formulas and geometric properties for common shapes. The terms in *Table A.2* are defined as follows:

A:	Cross-sectional area	c_z :	Distance from z-axis through centroid to furthest material point in y-direction
$\bar{z}$:	Location of centroid on z-axis in figure	$c_{z,max}$, $c_{z,min}$:	c-Values for sections not symmetric about z-axis
$\bar{y}$:	Location of centroid on y-axis in figure	c_y :	Distance from y-axis through centroid to furthest material point in z-direction
J:	Polar moment of inertia for round sections (for torsion)	Z_z , Z_y :	Section modulus about z- and y-axis, respectively: $Z_z = I_z/c_z;\;\; Z_y = I_y/c_y$
I_z , I_y :	Moment of inertia about z- and y-axis, respectively (for bending)	r_z , r_y :	Radius of gyration about z- and y-axis, respectively: $r_z = \sqrt{I_z/A};\;\; r_y = \sqrt{I_y/A}$

Not all terms are given for all shapes, especially those without an axis of symmetry.

Table A.1. Trigonometric formulas.

<table>
<tr>
<td>

Right Triangle

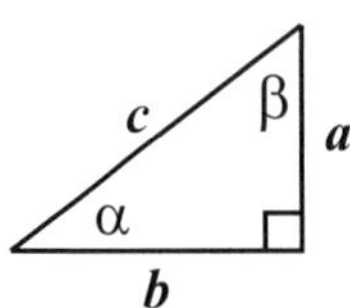

</td>
<td>

$$\text{Area} = \frac{ab}{2} \qquad\qquad \alpha + \beta = 90°$$

Pythagorean relationships

$$c^2 = a^2 + b^2$$
$$b^2 = c^2 - a^2$$
$$a^2 = c^2 - b^2$$

Definitions

$$\sin\alpha = \cos\beta = \frac{a}{c} \qquad\qquad \csc\alpha = \sec\beta = \frac{c}{a}$$

$$\cos\alpha = \sin\beta = \frac{b}{c} \qquad\qquad \sec\alpha = \csc\beta = \frac{c}{b}$$

$$\tan\alpha = \frac{1}{\tan\beta} = \frac{a}{b} \qquad\qquad \cot\alpha = \frac{1}{\cot\beta} = \frac{b}{a}$$

</td>
</tr>
<tr>
<td>

General Triangle

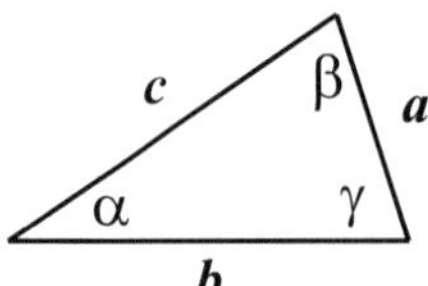

</td>
<td>

$$\text{Area} = \frac{ab}{2}\sin\gamma = \frac{ac}{2}\sin\beta = \frac{bc}{2}\sin\alpha$$

$$\text{Area} = \sqrt{s(s-a)(s-b)(s-c)} \quad \text{where } s = \frac{a+b+c}{2}$$

$$\alpha + \beta + \gamma = 180°$$

Law of Sines

$$\frac{\sin\alpha}{a} = \frac{\sin\beta}{b} = \frac{\sin\gamma}{c}$$

Law of Cosines

$$c^2 = a^2 + b^2 - 2ab\,\cos\gamma$$
$$b^2 = a^2 + c^2 - 2ac\,\cos\beta$$
$$a^2 = b^2 + c^2 - 2bc\,\cos\alpha$$

</td>
</tr>
<tr>
<td>

Positive quadrants for sine, cosine, and tangent

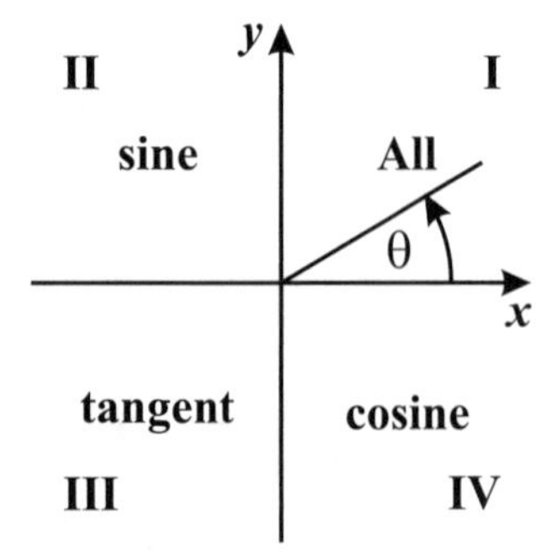

</td>
<td>

Identities

$$\sin^2\theta + \cos^2\theta = 1$$
$$1 + \tan^2\theta = \sec^2\theta$$
$$1 + \cot^2\theta = \csc^2\theta$$

Half and Double-Angle Formulas

$$\sin^2\theta = \frac{1 - \cos 2\theta}{2} \qquad \sin 2\theta = 2\sin\theta\cos\theta$$

$$\cos^2\theta = \frac{1 + \cos 2\theta}{2} \qquad \cos 2\theta = \cos^2\theta - \sin^2\theta$$

Sums and Differences of Angles

$$\sin(\alpha \pm \beta) = \sin\alpha\cos\beta \pm \cos\alpha\sin\beta$$
$$\cos(\alpha \pm \beta) = \cos\alpha\cos\beta \mp \sin\alpha\sin\beta$$

</td>
</tr>
</table>

Table A.2. Geometric properties of cross-sections.

Square	$A = d^2$
	$I_z = I_y = \dfrac{d^4}{12}$
	$c_z = c_y = \dfrac{d}{2}$
	$Z_z = Z_y = \dfrac{d^3}{6}$
	$r_z = r_y = \dfrac{d}{\sqrt{12}} = 0.28868d$

Diamond	$A = d^2$
	$I_z = I_y = \dfrac{d^4}{12}$
	$c_z = c_y = \dfrac{d}{\sqrt{2}} = 0.70711d$
	$Z_z = Z_y = \dfrac{d^3}{6\sqrt{2}} = 0.11785d^3$
	$r_z = r_y = \dfrac{d}{\sqrt{12}} = 0.28868d$

Rectangle		
	$A = bd$	
	$I_z = \dfrac{bd^3}{12}$	$I_y = \dfrac{db^3}{12}$
	$c_z = \dfrac{d}{2}$	$c_y = \dfrac{b}{2}$
	$Z_z = \dfrac{bd^2}{6}$	$Z_y = \dfrac{db^2}{6}$
	$r_z = \dfrac{d}{\sqrt{12}} = 0.28868d$	$r_y = \dfrac{b}{\sqrt{12}} = 0.28868b$

Hollow Rectangle		
	$A = bd - b_i d_i$	
	$I_z = \dfrac{bd^3 - b_i d_i^3}{12}$	$I_y = \dfrac{db^3 - d_i b_i^3}{12}$
	$c_z = \dfrac{d}{2}$	$c_y = \dfrac{b}{2}$
	$Z_z = \dfrac{bd^3 - b_i d_i^3}{6d}$	$Z_y = \dfrac{db^3 - d_i b_i^3}{6b}$
	$r_z = \sqrt{\dfrac{bd^3 - b_i d_i^3}{12A}}$	$r_y = \sqrt{\dfrac{db^3 - d_i b_i^3}{12A}}$

<table>
<tr>
<td>Solid Circle
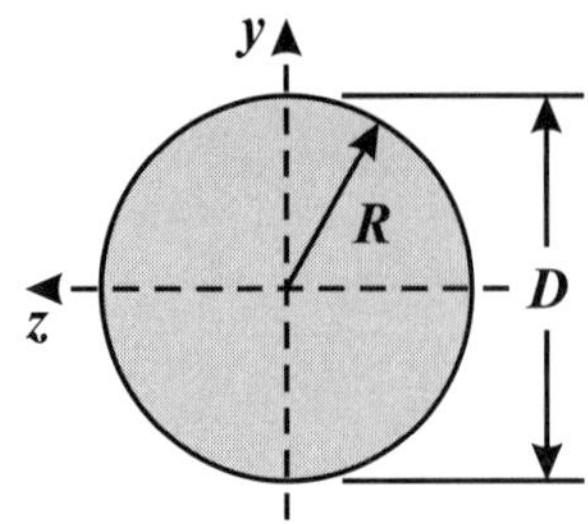</td>
<td>

$$A = \pi R^2 = \frac{\pi D^2}{4}$$

$$I_z = I_y = \frac{\pi R^4}{4} = \frac{\pi D^4}{64} \qquad J = 2I = \frac{\pi R^4}{2} = \frac{\pi D^4}{32}$$

$$c_z = c_y = R = \frac{D}{2}$$

$$Z_z = Z_y = \frac{\pi R^3}{4} = \frac{\pi D^3}{32}$$

$$r_z = r_y = \frac{R}{2} = \frac{D}{4}$$

</td>
</tr>
<tr>
<td>Hollow Circle
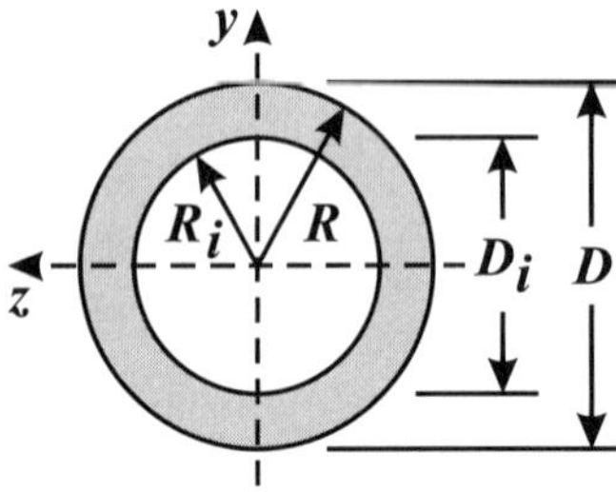</td>
<td>

$$A = \pi(R^2 - R_i^2) = \frac{\pi}{4}(D^2 - D_i^2)$$

$$I_z = I_y = \frac{\pi(R^4 - R_i^4)}{4} = \frac{\pi(D^4 - D_i^4)}{64} \qquad J = 2I$$

$$c_z = c_y = R = \frac{D}{2}$$

$$Z_z = Z_y = \frac{\pi(R^4 - R_i^4)}{4R} = \frac{\pi(D^4 - D_i^4)}{32D}$$

$$r_z = r_y = \frac{\sqrt{R^2 + R_i^2}}{2} = \frac{\sqrt{D^2 + D_i^2}}{4}$$

</td>
</tr>
<tr>
<td>Thin-walled Circle
R = outer radius
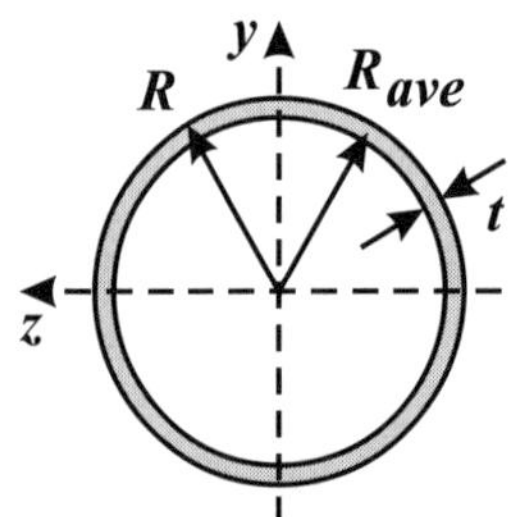</td>
<td>

$$A = 2\pi R_{ave}t = \pi D_{ave}t \qquad R_{ave} = R - \frac{t}{2} \qquad D_{ave} = D - t$$

$$I_z = I_y = \pi R_{ave}^3 t = \frac{\pi D_{ave}^3 t}{8} \qquad J = 2I = 2\pi R_{ave}^3 t = \frac{\pi D_{ave}^3 t}{4}$$

$$c_z = c_y = R = \frac{D}{2}$$

$$Z_z = Z_y = \frac{\pi R_{ave}^3 t}{R} = \frac{\pi D_{ave}^3 t}{4D}$$

$$r_z = r_y = \frac{R_{ave}}{\sqrt{2}} = \frac{D_{ave}}{2\sqrt{2}}$$

</td>
</tr>
<tr>
<td>Semi-Circle
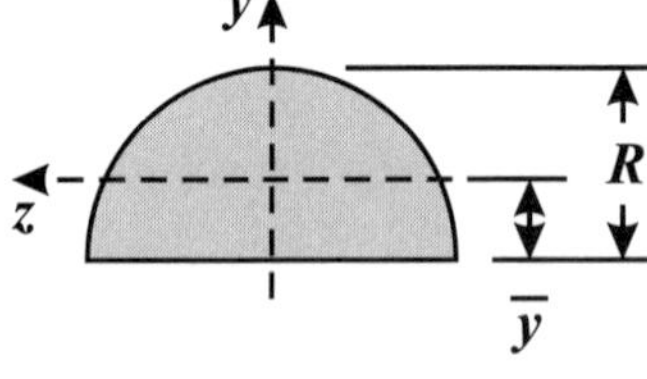</td>
<td>

$$A = \frac{\pi R^2}{2} = \frac{\pi D^2}{8} \qquad\qquad \overline{y} = \frac{4R}{3\pi}$$

$$I_z = R^4\left(\frac{\pi}{8} - \frac{8}{9\pi}\right) = 0.10976R^4$$

$$c_{z,\,max} = \left(1 - \frac{4}{3\pi}\right)R = 0.57559R \qquad c_{z,\,min} = \frac{4R}{3\pi}$$

$$Z_{z,\,min} = \frac{R^3(9\pi^2 - 64)}{24\,(3\pi - 4)} = 0.19069R^3$$

$$r_z = R\frac{\sqrt{9\pi^2 - 64}}{6\pi} = 0.26434R$$

</td>
</tr>
</table>

Built-up I-beam d = total depth 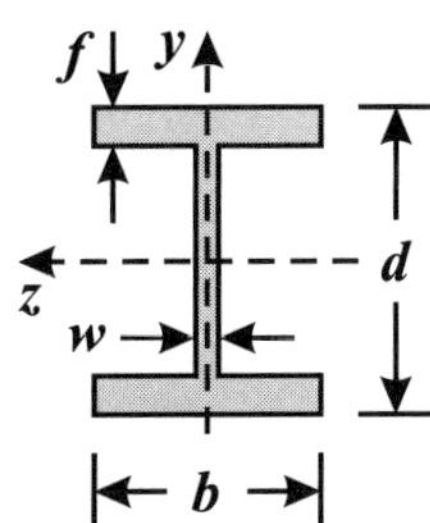	$A = w(d - 2f) + 2(bf) = bd - (b - w)(d - 2f)$ $I_z = \dfrac{bd^3 - (b - w)(d - 2f)^3}{12}$, outer rectangle minus inner, or $I_z = \dfrac{w(d - 2f)^3}{12} + 2\left[\dfrac{bf^3}{12} + bf\left(\dfrac{d}{2} - \dfrac{f}{2}\right)^2\right]$, parallel-axis theorem $I_y = \dfrac{(d - 2f)w^3}{12} + 2\left(\dfrac{fb^3}{12}\right)$ $c_z = \dfrac{d}{2}$ $c_y = \dfrac{b}{2}$ $Z_z = \dfrac{2I_z}{d}$ $Z_y = \dfrac{2I_y}{b}$ $r_z = \sqrt{I_z/A}$ $r_z = \sqrt{I_y/A}$
Channel, constant thickness 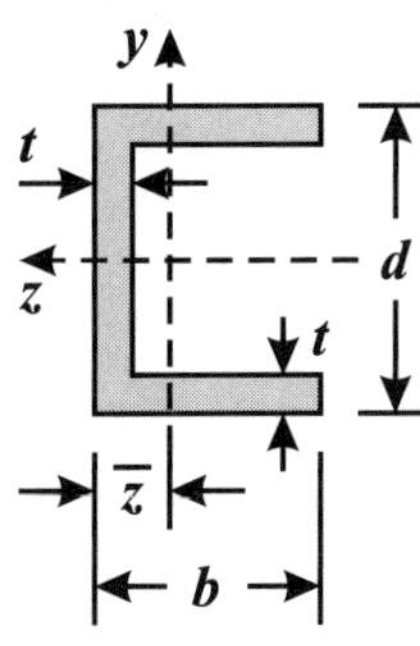	$A = t[(d - 2t) + 2b]$ $\overline{z} = \dfrac{t(d - 2t) + 2b^2}{2[(d - 2t) + 2b]}$ $I_z = \dfrac{bd^3 - (b - t)(d - 2t)^3}{12}$, outer rectangle minus inner, or $I_z = \dfrac{t(d - 2t)^3}{12} + 2\left[\dfrac{bt^3}{12} + bt\left(\dfrac{d}{2} - \dfrac{t}{2}\right)^2\right]$, parallel-axis theorem $I_y = \left[\dfrac{(d - 2t)t^3}{12} + t(d - 2t)\left(\overline{z} - \dfrac{t}{2}\right)^2\right] + 2\left[\dfrac{tb^3}{12} + (bt)\left(\dfrac{b}{2} - \overline{z}\right)^2\right]$ $c_z = \dfrac{d}{2}$ $c_{y,\,max} = b - \overline{z}$ $Z_z = \dfrac{2I_z}{d}$ $Z_{y,\,min} = \dfrac{I_y}{b - \overline{z}}$ $r_z = \sqrt{I_z/A}$ $r_z = \sqrt{I_y/A}$
Angle, equal leg thicknesses 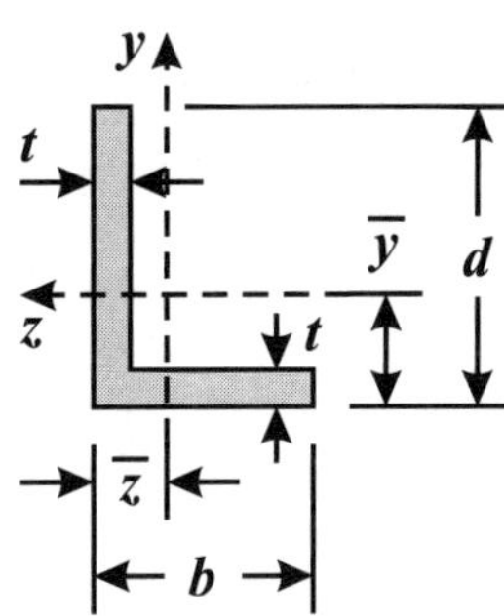	$A = t(d + b - t)$ $\overline{z} = \dfrac{t(d - 2t) + b^2}{2(d + b - t)}$ $\overline{y} = \dfrac{t(b - 2t) + d^2}{2(d + b - t)}$ $I_z = \dfrac{1}{3}[t(d - \overline{y})^3 + b\,\overline{y}^3 - (b - t)(\overline{y} - t)^3]$ $I_y = \dfrac{1}{3}[t(b - \overline{z})^3 + d\,\overline{z}^3 - (d - t)(\overline{z} - t)^3]$

<table>
<tr>
<td>

Truss/Space Frame

Two areas $A/2$ concentrated d apart

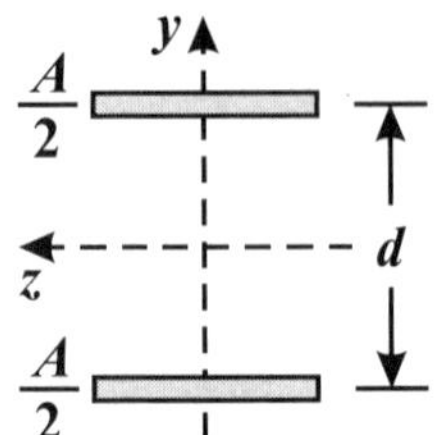

</td>
<td>

$A = A$

$I_z = \dfrac{Ad^2}{4}$

$c_z = \dfrac{d}{2}$

$Z_z = \dfrac{Ad}{2}$

$r_z = \dfrac{d}{2}$

</td>
</tr>
<tr>
<td>

Triangle

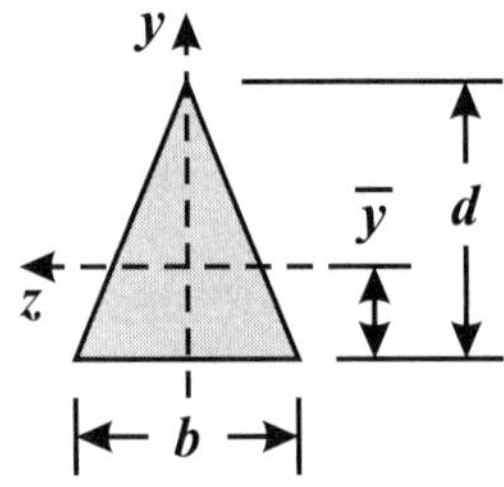

</td>
<td>

$A = \dfrac{bd}{2}$ $\qquad\qquad \overline{y} = \dfrac{d}{3}$

$I_z = \dfrac{bd^3}{36}$

$c_{z,\,max} = \dfrac{2d}{3}$ $\qquad\qquad c_{z,\,min} = \dfrac{d}{3}$

$Z_{z,\,min} = \dfrac{bd^2}{24}$

$r_z = \dfrac{d}{\sqrt{18}} = 0.23570d$

</td>
</tr>
<tr>
<td>

Hexagon

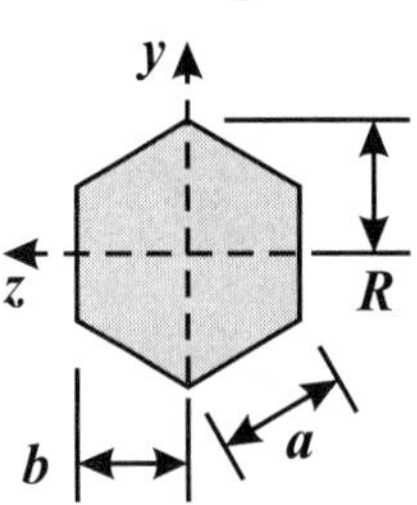

</td>
<td>

$A = \dfrac{3\sqrt{3}}{2}a^2 = 2.5981a^2$ $\qquad R = a \qquad b = \dfrac{\sqrt{3}}{2}a$

$I_z = I_y = \dfrac{5\sqrt{3}}{16}a^4 = 0.54127a^4$

$c_z = R = a$ $\qquad\qquad c_y = b$

$Z_z = \dfrac{5\sqrt{3}}{16}a^3 = 0.54127a^3$ $\qquad Z_y = \dfrac{5}{8}a^3 = 0.625a^3$

$r_z = r_y = \sqrt{\dfrac{5}{24}}\,a = 0.45643a$

</td>
</tr>
<tr>
<td>

Half Ellipse

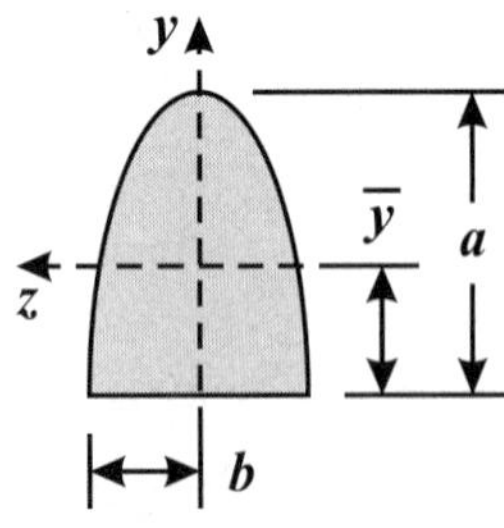

</td>
<td>

$A = \dfrac{\pi ab}{2}$ $\qquad\qquad \overline{y} = \dfrac{4a}{3\pi}$

$I_z = ba^3\left(\dfrac{\pi}{8} - \dfrac{8}{9\pi}\right) = 0.10976ba^3$ $\qquad I_y = \dfrac{\pi}{8}ab^3$

$c_{z,\,max} = \left(1 - \dfrac{4}{3\pi}\right)a = 0.57559a$ $\qquad c_{z,\,min} = \dfrac{4a}{3\pi}$

$Z_{z,\,min} = \dfrac{ba^2}{24}\dfrac{(9\pi^2 - 64)}{(3\pi - 4)} = 0.19069ba^2$

$r_z = a\dfrac{\sqrt{9\pi^2 - 64}}{6\pi} = 0.26434a$

</td>
</tr>
</table>

Quarter Ellipse 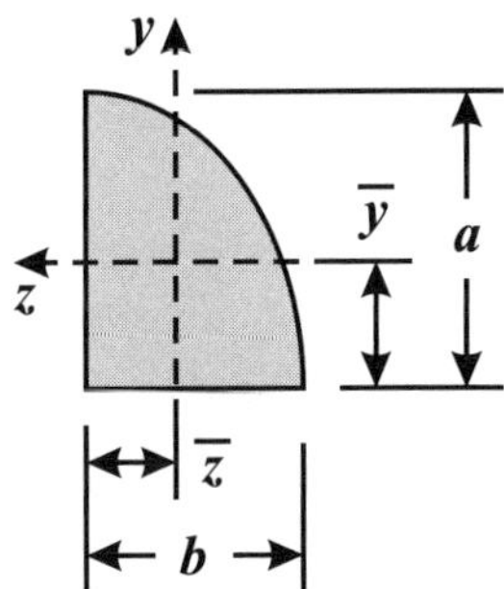	$A = \dfrac{\pi ab}{4}$ $\qquad \bar{z} = \dfrac{4b}{3\pi}$ $\qquad \bar{y} = \dfrac{4a}{3\pi}$ $I_z = ba^3\left(\dfrac{\pi}{16} - \dfrac{4}{9\pi}\right) = 0.054878\,ba^3$ $I_y = ab^3\left(\dfrac{\pi}{16} - \dfrac{4}{9\pi}\right) = 0.054878\,ab^3$
Parabola 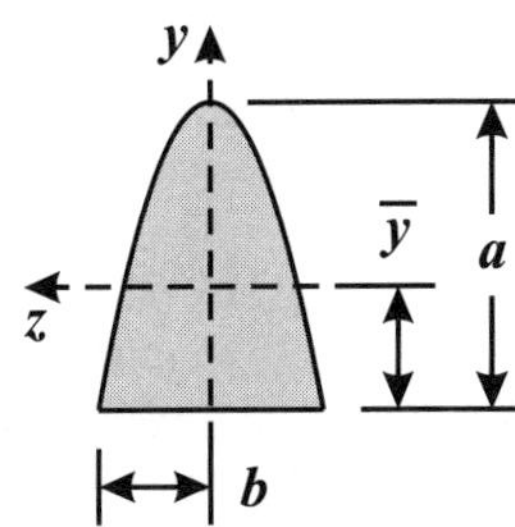	$A = \dfrac{4}{3}ab$ $\qquad\qquad \bar{y} = \dfrac{2}{5}a$ $I_z = \dfrac{16}{175}ba^3$ $\qquad I_y = \dfrac{4}{15}ab^3$ $c_{z,\,max} = \dfrac{3}{5}a$ $\qquad c_{z,\,min} = \dfrac{2}{5}a$ $Z_{z,\,min} = \dfrac{16}{105}ba^2 = 0.15238\,ba^2$ $r_z = \sqrt{\dfrac{12}{175}}\,a = 0.26186\,a$
Half Parabola 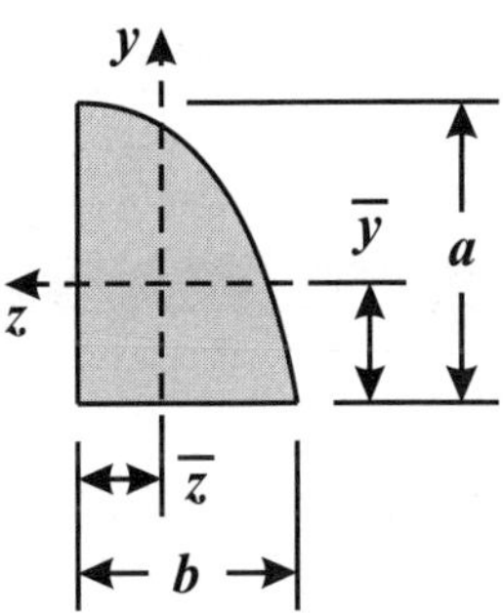	$A = \dfrac{2}{3}ab$ $\bar{z} = \dfrac{3}{8}a$ $\qquad \bar{y} = \dfrac{2}{5}a$ $I_z = \dfrac{8}{175}ba^3$ $\qquad I_y = \dfrac{19}{480}ab^3$
Complement of Half Parabola 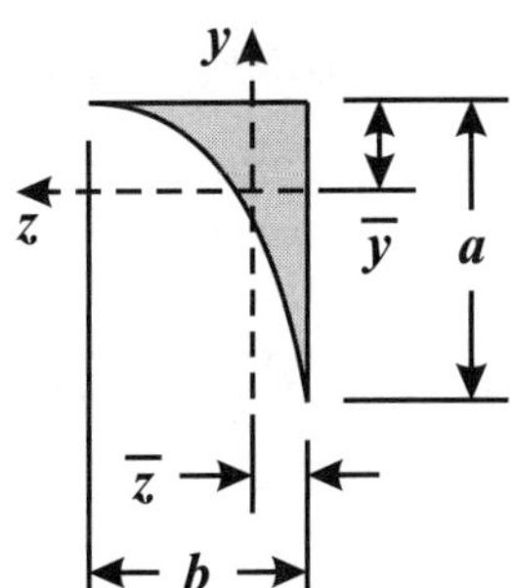	$A = \dfrac{1}{3}ab$ $\bar{z} = \dfrac{b}{4}$ $\qquad \bar{y} = \dfrac{3}{10}a$ $I_z = \dfrac{37}{2100}ba^3$ $\qquad I_y = \dfrac{1}{80}ab^3$

B. Representative Properties: Physical, Elastic, and Thermal

Representative densities, elastic moduli, Poisson's ratios, and thermal coefficients are given in *Table B.1*, in both SI and US units. There are literally thousands of materials. The quantities presented here are for comparison of various material systems only.

The values listed here should not be used in the actual design of engineering systems; the designer should use values provided by the material supplier and/or manufacturer.

Key:
- ρ : density (kg/m^3)
- γ : weight density (lb/in.^{3})
- E : Young's modulus (GPa, Msi)
- G : shear modulus (GPa, Msi)
- ν : Poisson's ratio
- α : thermal expansion coefficient ($10^{-6}/^\circ$C, $10^{-6}/^\circ$F)

Table B.1. Representative properties of materials, S.I. units (U.S. units).

Material	$\rho(\gamma)$ kg/m^3 (lb/in.^{3})	E GPa (Msi)	G GPa (Msi)	ν	α $10^{-6}/^\circ$C ($10^{-6}/^\circ$F)
Steels and Iron					
ASTM A36 (structural)	7850 (0.284)	200 (29)	79 (11.5)	0.3	12 (6.7)
ASTM A710, Grade A	7850 (0.284)	205 (30)	80 (11.6)	0.29	12 (6.7)
Stainless 316	8000 (0.289)	193 (28)	77 (11.2)	0.3	16 (8.9)
Gray cast iron	7150 (0.258)	100 (15)	40 (5.8)	0.26	11.4 (6.3)
Ductile cast iron, annealed	7150 (0.258)	169 (25)	65 (9.4)	0.29	11.2 (6.2)
Aluminum alloys					
Alloy 2014-T6	2800 (0.101)	72 (10.4)	28 (4.1)	0.33	22.9 (12.7)
6061-T6 (structural)	2700 (0.098)	69 (10)	26 (3.8)	0.33	23.6 (13.1)
7075-T6 (aircraft)	2800 (0.101)	72 (10.4)	28 (4.1)	0.33	23.4 (13.0)
Other non-ferrous alloys and metals					
Gold, pure, annealed	19,300 (0.697)	77 (11)	27 (3.9)	0.42	14 (7.8)
Nickel Superalloy	8220 (0.297)	207 (30)	80 (11.6)	0.3	13.3 (7.4)
Titanium alloy, 6Al-4V	4430 (0.160)	114 (17)	43 (6.2)	0.34	8.6 (4.8)
Polymers					
Epoxy	1250 (0.045)	2.4 (0.35)	–	–	80–115 (44–64)
Nylon 66	1100 (0.040)	3 (0.44)	–	0.39	140 (78)
PVC	1400 (0.051)	3 (0.44)	–	0.38	120 (69)
Rubber	1200 (0.043)	0.01–0.1 (0.0015–0.015)	0.003–0.03 (0.0005–0.005)	0.45–0.5	160 (89)
Glass and Engineering Ceramics					
Glass	2600 (0.094)	69 (10)	28 (4)	0.23	9 (5)
Alumina, Al$_2$O$_3$	3900 (0.141)	360 (52)	–	0.22	7.4 (4.1)
Silicon carbide, SiC	3200 (0.116)	400 (58)	–	0.17	4.6 (2.6)
Silicon nitride, Si$_3$N$_4$	3200 (0.116)	300 (44)	–	0.3	4 (2.2)
Wood and Concrete					
Douglas Fir (in bending, parallel ‖, perpendicular ⊥ to grain)	480 (0.017)	12 ‖, 0.6⊥ (1.7 ‖, 0.1⊥)	–	–	4.0 ‖, 28⊥ (2.2 ‖, 16⊥)
Concrete, medium strength	2400 (0.087)	25 (3.6)	–	0.20	9.9 (5.5)
Concrete, high strength	2400 (0.087)	30 (4.4)	–	0.20	9.9 (5.5)

B. Representative Properties: Strength and Failure Strain

Representative yield strength, ultimate strength, and failure strains are given in *Table B.2*. There are literally thousands of materials. The quantities presented here are for comparison of various material systems only.

The values listed here should not be used in the actual design of engineering systems; the designer should use values provided by the material supplier and/or manufacturer.

Key:

S_y : yield strength (MPa, ksi)

τ_y : shear yield strength (MPa, ksi); estimated from S_y for metals

S_u : ultimate strength (MPa, ksi) [T] tension, [C] compression

ε_f : failure strain, ductility (%), in tension

Table B.2. Representative properties of materials, SI units (US units).

Material	S_y MPa (ksi)	τ_y MPa (ksi)	S_u MPa (ksi)	ε_f (%)
Steels and Iron				
ASTM A36 (structural)	250 (36)	140 (20)	450 (65)	23
ASTM A710, Grade A	450 (65)	260 (38)	495 (72)	20
Stainless 316	310 (45)	180 (26)	620 (90)	30
Gray cast iron	–	–	200 (29) [T] 650 (94) [C]	–
Ductile cast iron, annealed	330 (48)	170 (25)	460 (67)	15
Aluminum alloys				
Alloy 2014-T6	410 (59)	240 (35)	480 (70)	12
6061-T6 (structural)	276 (40) [a]	160 (23)	310 (45)	17
7075-T6 (aircraft)	500 (73)	190 (28)	570 (83)	11
Other non-ferrous alloys and metals				
Gold, pure, annealed	–	–	130 (19)	45
Nickel superalloy	900 (131)	520 (75)	1275 (185)	25
Titanium alloy, 6Al-4V	850 (121)	490 (71)	950 (138)	14
Polymers				
Epoxy	30–100 (4–14)	–	30–120 (4–17)	3–6
Nylon 66	60 (9)	–	75 (11)	150–300
PVC	45 (6.5)	–	50 (7)	40–80
Rubber	–	–	12 (2)	100–800
Glass and Engineering Ceramics				
Glass	–	–	69 (10)	–
Alumina, Al_2O_3	–	–	2500 (360) [C]	–
Silicon carbide, SiC	–	–	4000 (580) [C]	–
Silicon nitride, Si_3N_4	–	–	3500 (510) [C]	–
Wood and Concrete				
Douglas fir (in bending for tension; parallel ‖ and perpendicular ⊥ to grain)	–	–	100(15) ‖ ; 2.7(0.4)⊥ [T] 45(7) ‖ , 7(1))⊥ [C]	–
Concrete, medium strength	–	–	28 (4) [C]	–
Concrete, high strength	–	–	40 (6) [C]	–

Notes: [a]: in design with Al 6061-T6, the value used for S_y is the minimum expected value, $S_y = 35$ ksi $= 240$ MPa (*Aluminum Design Manual*, The Aluminium Association, Inc., 2005).

C. Rolled Steel Cross-Sections

Steel beams with cross-sectional shapes such as *I-beams*, *Channels*, and *Angles* (with equal and unequal legs), are manufactured by hot-rolling. These beams are usually seen during construction of a commercial building or bridge. They may be purchased "off-the-shelf" from various steel companies. The American Institute of Steel Construction's *Steel Construction Manual* provides tables of the many standardized sections available and their geometric properties.

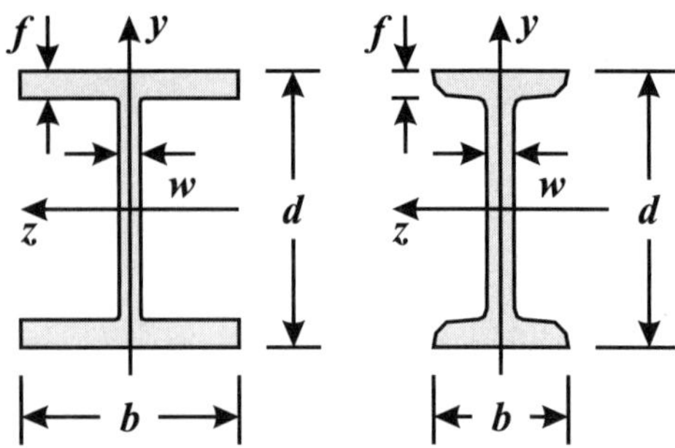

Left: W-shape (wide flange) and
Right: S-shape (American Standard).

The figure at right shows a *W-Shape* or *wide-flange shape* and the *S-shape* or *American Standard Shape*. The vertical section of the cross-section is the *web*; the web is generally designed to support shear forces applied in the *y*-direction. The horizontal sections at the top and bottom are the *flanges* (from the French word for flank). The flanges are generally designed to support bending moments about the horizontal (*z*-) axis.

I-beam shapes are designated as follows: $W12 \times 40$ or $S20 \times 75$

- The *letter* signifies the type of cross-section, *W* for wide flange and *S* for standard.
- The *first number* is the *nominal depth*, in inches (the exact depth for many S-beams).
- The *second number* is the *weight in pounds per linear foot* of the beam, which is proportional to its area. The second number is the self-weight of the beam as a distributed load. The total weight of a beam is calculated by multiplying the second number by the beam's length in feet.

For metric I-beams, the depth is given in millimeters and the weight/length (mass/length) is given in kgf/m.

Tables C.1 and *C.2* are a sampling of the many standardized W and S shapes available (US units); there are many more sizes than presented here. The terms in the tables are defined as follows:

A: gross cross-sectional area	w: thickness of web
d: total depth of beam	I_z, I_y: moment of inertia about z- and y-axis, respectively
b: width (or breadth) of flange	Z_z, Z_y: section modulus about z- and y-axis, respectively
f: thickness of flange	r_z, r_y: radius of gyration about z- and y-axis, respectively

Note: The values listed in *Tables C.1* and *C.2* should not be used for design; the appropriate manufacturer's data should be used.

Table C.1. Selected *W-shapes* (Wide Flange Shapes), US units.

Cross-section			Flange		Web	About z-axis			About y-axis		
Designation	A (in.^{2})	d (in.)	b (in.)	f (in.)	w (in.)	I_z (in.^{4})	Z_z (in.^{3})	r_z (in.)	I_y (in.^{4})	Z_y (in.^{3})	r_y (in.)
W24 × 104	30.6	24.06	12.750	0.750	0.500	3100	258	10.1	259	40.7	2.91
W24 × 68	20.1	23.73	8.965	0.585	0.415	1830	154	9.55	70.4	15.7	1.87
W18 × 106	31.1	18.73	11.200	0.940	0.590	1910	204	7.84	220	39.4	2.66
W18 × 76	22.3	18.21	11.035	0.680	0.425	1330	146	7.73	152	27.6	2.61
W18 × 50	14.7	17.99	7.495	0.570	0.355	800	88.9	7.38	40.1	10.7	1.65
W16 × 57	16.8	16.43	7.120	0.715	0.430	758	92.2	6.72	43.1	12.1	1.60
W16 × 40	11.8	16.01	6.995	0.505	0.305	518	64.7	6.63	28.9	8.25	1.57
W16 × 31	9.12	15.88	5.525	0.440	0.275	375	47.2	6.41	12.4	4.49	1.17
W12 × 96	28.2	12.71	12.160	0.900	0.550	833	131	5.44	270	44.4	3.09
W12 × 72	21.1	12.25	12.040	0.670	0.430	597	97.4	5.31	195	32.4	3.04
W12 × 50	14.7	12.19	8.080	0.640	0.370	394	64.7	5.18	56.3	13.9	1.96
W12 × 40	11.8	11.94	8.005	0.515	0.295	310	51.9	5.13	44.1	11.0	1.93
W12 × 30	8.79	12.34	6.520	0.440	0.260	238	38.6	5.21	20.3	6.24	1.52
W12 × 16	4.71	11.99	3.990	0.265	0.220	103	17.1	4.67	2.82	1.41	0.773
W10 × 112	32.9	11.36	10.415	1.250	0.755	716	126	4.66	236	45.3	2.68
W10 × 68	20.0	10.40	10.130	0.770	0.470	394	75.7	4.44	134	26.4	2.59
W10 × 45	13.3	10.10	8.020	0.620	0.350	248	49.1	4.32	53.4	13.3	2.01
W10 × 33	9.71	9.73	7.960	0.435	0.290	170	35.0	4.19	36.6	9.20	1.94
W10 × 30	8.84	10.47	5.810	0.510	0.300	170	32.4	4.38	16.7	5.75	1.37
W10 × 15	4.41	9.99	4.000	0.270	0.230	68.9	13.8	3.95	2.89	1.45	0.810
W8 × 58	17.1	8.75	8.220	0.810	0.510	228	52.0	3.65	75.1	18.3	2.10
W8 × 48	14.1	8.50	8.110	0.685	0.400	184	43.3	3.61	60.9	15.0	2.08
W8 × 31	9.13	8.00	7.995	0.435	0.285	110	27.5	3.47	37.1	9.27	2.02
W8 × 28	8.25	8.06	6.535	0.465	0.285	98.0	24.3	3.45	21.7	6.63	1.62
W8 × 24	7.08	7.93	6.495	0.400	0.245	82.8	20.9	3.42	18.3	5.63	1.61
W8 × 21	6.16	8.28	5.270	0.400	0.250	75.3	18.2	3.49	9.77	3.71	1.26
W8 × 18	5.26	8.14	5.250	0.330	0.230	61.9	15.2	3.43	7.97	3.04	1.23
W8 × 13	3.84	7.99	4.000	0.255	0.230	39.6	9.91	3.21	2.73	1.37	0.843

Table C.2. Selected Standard *S-shapes* (American Standard Shapes), US units.

Cross-section			Flange		Web	About z-axis			About y-axis		
Designation	A (in.^{2})	d (in.)	b (in.)	f (in.)	w (in.)	I_z (in.^{4})	Z_z (in.^{3})	r_z (in.)	I_y (in.^{4})	Z_y (in.^{3})	r_y (in.)
S20 × 86	25.3	20.30	7.060	0.920	0.660	1580	155	7.89	46.8	13.3	1.36
S20 × 75	22.0	20.00	6.385	0.795	0.635	1280	128	7.62	29.8	9.32	1.16
S20 × 66	19.4	20.00	6.255	0.795	0.505	1190	119	7.83	27.7	8.85	1.19
S18 × 70	20.6	18.00	6.251	0.691	0.711	926	103	6.71	24.1	7.72	1.08
S18 × 54.7	16.1	18.00	6.001	0.691	0.461	804	89.4	7.07	20.8	6.94	1.14
S15 × 50	14.7	15.00	5.640	0.622	0.550	486	64.8	5.75	15.7	5.57	1.03
S15 × 42.9	12.6	15.00	5.501	0.622	0.411	447	59.6	5.95	14.4	5.23	1.07
S12 × 50	14.7	12.00	5.477	0.659	0.687	305	50.8	4.55	15.7	5.74	1.03
S12 × 40.8	12.0	12.00	5.252	0.659	0.462	272	45.4	4.77	13.6	5.16	1.06
S12 × 35	10.3	12.00	5.078	0.544	0.428	229	38.2	4.72	9.87	3.89	0.980
S12 × 31.8	9.35	12.00	5.000	0.544	0.350	218	36.4	4.83	9.36	3.74	1.00
S10 × 35	10.3	10.00	4.944	0.491	0.594	147	29.4	3.78	8.36	3.38	0.901
S10 × 25.4	7.46	10.00	4.661	0.491	0.311	124	24.7	4.07	6.79	2.91	0.954
S8 × 23	6.77	8.00	4.171	0.426	0.441	64.9	16.2	3.10	4.31	2.07	0.798
S8 × 18.4	5.41	8.00	4.001	0.426	0.271	57.6	14.4	3.26	3.73	1.86	0.831
S6 × 17.25	5.07	6.00	3.565	0.359	0.465	26.3	8.77	2.28	2.31	1.30	0.675
S6 × 12.5	3.67	6.00	3.332	0.359	0.232	22.1	7.37	2.45	1.82	1.09	0.705

D. Standard Steel Pipe

Steel pipe comes in several grades: *standard, extra strong* (XS), and *double-extra strong* (XXS). The stronger pipes have thicker walls, e.g., a "2-in. pipe" of each grade has an outer diameter of 2.375 in., but the wall thicknesses are: 0.154 (Standard), 0.218 (XS), and 0.436 in. (XXS).

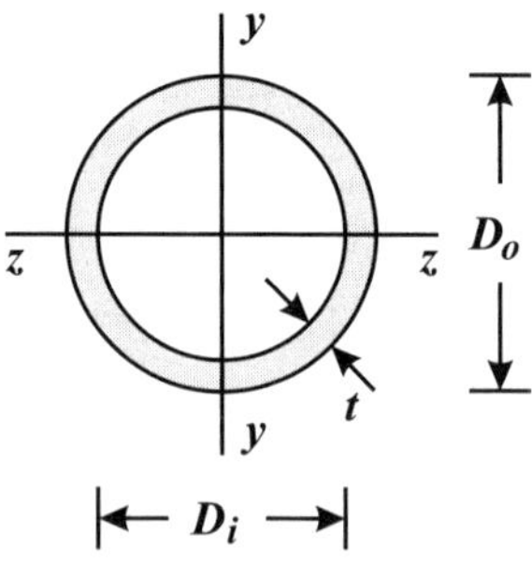

Pipe designation is also given by *schedule numbers.* Schedule 40 is equivalent to standard and Schedule 80 to extra strong. Schedule 160 falls between extra strong and double-extra strong (e.g., for a "2-in. pipe," the wall thickness of Schedule 160 is 0.343 in.).

Table D.1 gives geometric values of *Standard Steel Pipe* (Schedule 40), "1-in." and above. The actual outside diameter is larger than the nominal size. Geometric properties for other schedules may be found in other references (do not interpolate thicknesses from the "2-in. pipe" examples above).

Table D.1. Selected Standard Steel Pipe (Schedule 40).

Nominal Diameter (in.)	Outer D_o (in.)	Inner D_i (in.)	Thickness t (in.)	Area A (in.^{2})	Weight per Foot (lb)	I (in.^{4})	Z (in.^{3})	r (in.)
1	1.315	1.049	0.133	0.494	1.68	0.087	0.494	0.42
1 $^1/_4$	1.660	1.380	0.140	0.669	2.27	0.195	0.669	0.54
1 $^1/_2$	1.900	1.610	0.145	0.799	2.72	0.310	0.799	0.62
2	2.375	2.067	0.154	1.075	3.65	0.666	1.07	0.79
2 $^1/_2$	2.875	2.469	0.203	1.704	5.79	1.53	1.70	0.95
3	3.500	3.068	0.216	2.23	7.58	3.02	2.23	1.16
3 $^1/_2$	4.000	3.548	0.226	2.68	9.11	4.79	2.68	1.34
4	4.500	4.026	0.237	3.17	10.79	7.23	3.17	1.51
5	5.563	5.047	0.258	4.30	14.62	15.2	4.30	1.88
6	6.625	6.065	0.280	5.58	18.97	28.1	5.58	2.25
8	8.625	7.981	0.322	8.40	28.55	72.5	8.40	2.94
10	10.750	10.020	0.365	11.91	40.48	161	11.9	3.67
12	12.750	12.000	0.375	14.58	49.56	279	14.6	4.38

E. Structural Lumber

Lumber is referred to by its cross-section's nominal dimensions. Wooden studs in residential buildings are made with "2 × 4" lumber ("2 by 4s"). The actual dimensions are 1.5 × 3.5 in. because the board has been milled down.

After being cut to nominal size, lumber is *dressed,* generally *surfaced four sides* (S4S). Lumber is typically dressed 0.5 in. below nominal; i.e., an "6 × 6" is actually 5.5 × 5.5 in.

Table E.1 gives a selection of typical lumber cross-sections. The weight density of lumber is approximated to be 40 pounds per cubic foot (lb/ft^3).

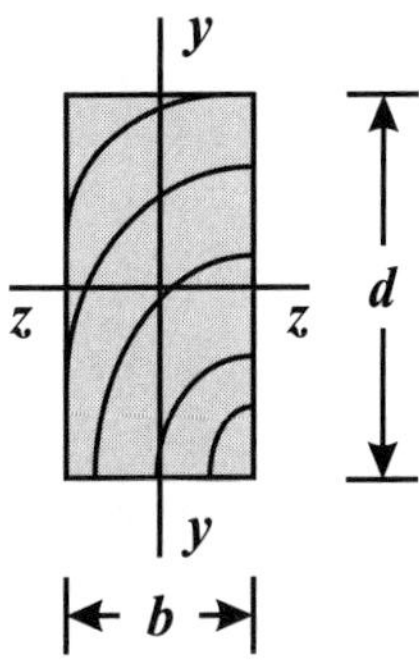

Table E.1. Selected Section Properties for Structural Lumber. American Standard Dressed Sizes (S4S).

Nominal Size $b \times d$ (in.)	Dressed Size $b \times d$ (in.)	Area A (in.^{2})	About Strong Axis		Weight per Foot (lb)*
			I (in.^{4})	Z (in.^{3})	
2 × 4	1.5 × 3.5	5.25	5.36	3.06	1.46
2 × 6	1.5 × 5.5	8.25	20.8	7.56	2.29
2 × 8	1.5 × 7.25	10.9	47.6	13.1	3.02
2 × 10	1.5 × 9.25	13.9	98.9	21.4	3.85
2 × 12	1.5 × 11.25	16.9	178	31.6	4.69
3 × 4	2.5 × 3.5	8.75	8.93	5.10	2.43
3 × 6	2.5 × 5.5	13.8	34.7	12.6	3.82
3 × 8	2.5 × 7.25	18.1	79.4	21.9	5.04
4 × 4	3.5 × 3.5	12.3	12.5	7.15	3.40
4 × 6	3.5 × 5.5	19.3	48.5	17.6	5.35
4 × 8	3.5 × 7.25	25.4	111	30.7	7.05
4 × 10	3.5 × 9.25	32.4	231	49.9	8.94
4 × 12	3.5 × 11.25	39.4	415	73.8	10.9
6 × 6	5.5 × 5.5	30.3	76.3	27.7	8.40
6 × 10	5.5 × 9.5	52.3	393	82.7	14.5
6 × 12	5.5 × 11.5	63.3	697	121	17.5
6 × 16	5.5 × 15.5	85.3	1707	220	23.6
8 × 8	7.5 × 7.5	56.3	264	70.3	15.6
8 × 12	7.5 × 11.5	86.3	951	165	23.9
8 × 16	7.5 × 15.5	116.3	2327	300	32.0

* Based on 40 lb/ft^3.

F. Beams

Reaction Forces and Moments; Deflection and Slope Equations

Tables F.1, F.2, and *F.3* give formulas for basic beams under various loads.

In the figures, point A is the left end of the beam and point B is the right end. The overall length of the beam is L. The applied loads are positive *downward* (with gravity). If a load acts upward, a negative value for the load should be used in the equations.

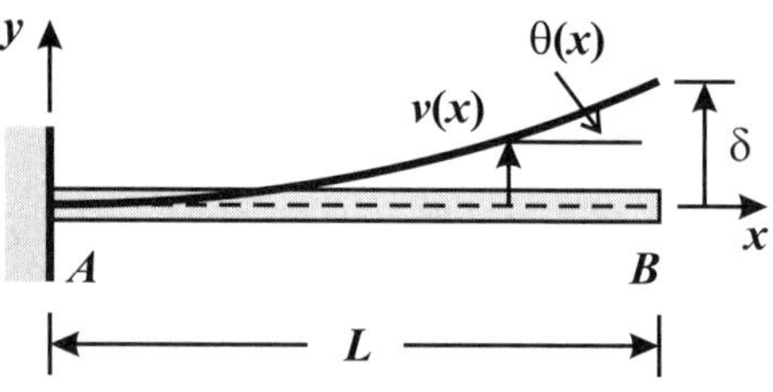

Positive sense of beam deflections and slopes.

The positive sense for deflections is shown in the above figure. Deflections are positive upward; slopes (angles) are positive counterclockwise.

The loads are:

P:	point force (positive in direction drawn).
w:	uniformly distributed load (positive in direction drawn).
w_o:	maximum value of distributed load.

The formulas give:

R_A:	reaction force at point A (positive upwards).
R_B:	reaction force at point B (positive upwards).
M_A:	reaction moment at point A (positive for compression at top of beam and negative for compression at bottom).
M_B:	reaction moment at point B (positive for compression at top).
M_{max}:	sense and magnitude of maximum bending moment (positive for compression at top), and location.
$M(x)$:	moment at x.
$M(x=a)$:	moment at $x = a$.
$v(x)$:	deflection of beam (positive upwards and negative downwards).
$\theta(x)$:	slope of beam (positive counterclockwise and negative clockwise).
δ_{max}:	maximum deflection (positive upwards and negative downwards), and location.
θ_{max}:	maximum slope (positive counterclockwise and negative clockwise) and location.

Notes:

- Not all quantities are provided for each beam/load configuration.
- For statically indeterminate beams, the largest positive and negative moments are listed.
- With the reactions and loads known, shear and moment diagrams may be constructed.

Table F.1. Simply-Supported Beams.

Central Point Load (three-point bending) 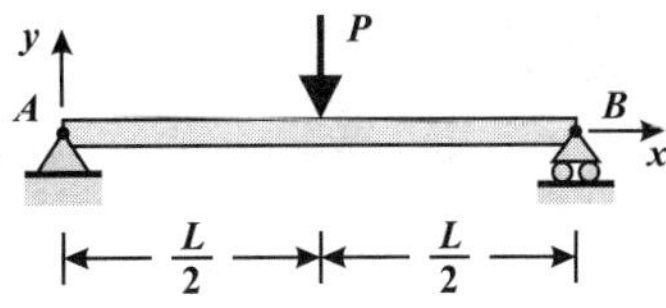	$R_A = R_B = \dfrac{P}{2}$ $M_{max} = \dfrac{PL}{4}$ at center, $x = L/2$ $v(x) = \dfrac{-Px}{48EI}(3L^2 - 4x^2)$ for $x \le L/2$ $\theta(x) = \dfrac{-P}{16EI}(L^2 - 4x^2)$ for $x \le L/2$ $\delta_{max} = \dfrac{-PL^3}{48EI}$ at $x = L/2$
Offset Point Load 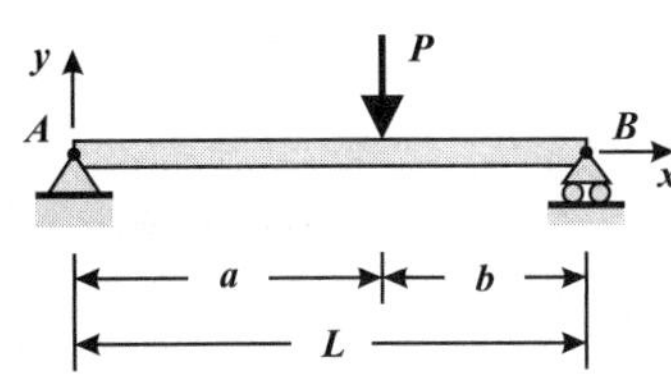	$R_A = \dfrac{Pb}{L};\quad R_B = \dfrac{Pa}{L}$ $M_{max} = \dfrac{Pab}{L}$ at point of load, $x = a$ $v(x) = \dfrac{-Pbx}{6EIL}(L^2 - b^2 - x^2)$ for $x \le a$ $\theta(x) = \dfrac{-Pb}{6EIL}(L^2 - b^2 - 3x^2)$ for $x \le a$ If $a \ge b$: $\delta_{max} = \dfrac{-Pb(L^2 - b^2)^{3/2}}{9\sqrt{3}EIL}$ at $x = \sqrt{\dfrac{L^2 - b^2}{3}}$
Two Equal Point Loads, Symmetrically Placed (four-point bending) 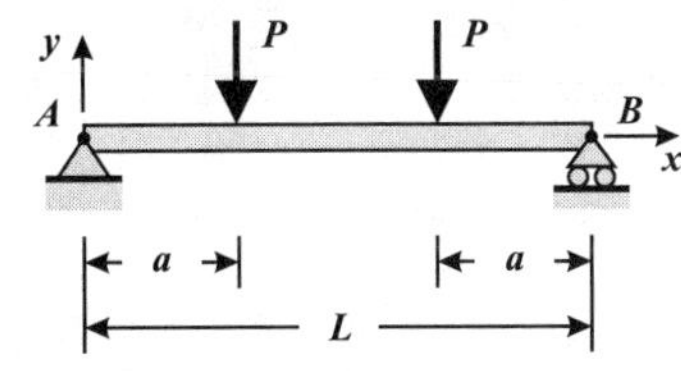	$R_A = R_B = P$ $M_{max} = Pa$ for $a \le x \le L - a$ $v(x) = \dfrac{-Px}{6EI}(3La - 3a^2 - x^2)$ for $x \le a$ $v(x) = \dfrac{-Pa}{6EI}(3Lx - 3x^2 - a^2)$ for $a \le x \le L - a$ $\delta_{max} = \dfrac{-Pa}{24EI}(3L^2 - 4a^2)$ at center, $x = L/2$
Two Equal Point Loads, Non-Symmetrically Placed 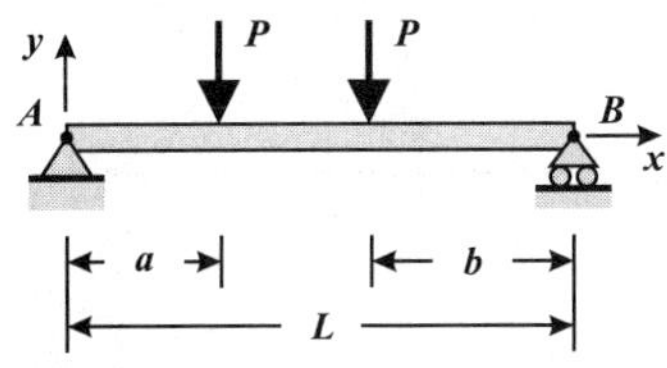	$R_A = \dfrac{P(L - a + b)}{L};\quad R_B = \dfrac{P(L - b + a)}{L}$ $M(x = a) = R_A a$ $M(x = L - b) = R_B b$ If $a > b$, $M_{max} = R_A a$ at $x = a$ If $a < b$, $M_{max} = R_B b$ at $x = L - b$

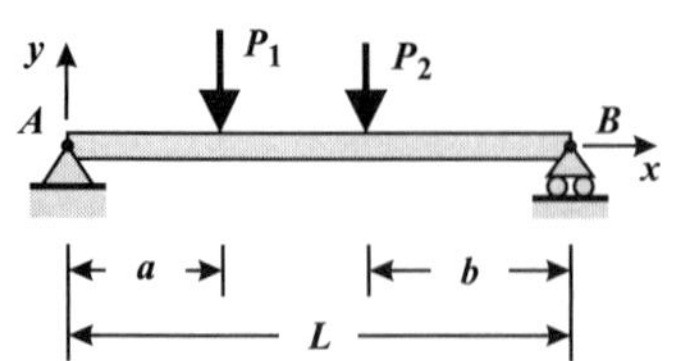

Two Unequal Loads, Non-Symmetrically Placed

$$R_A = \frac{P_1(L-a) + P_2 b}{L} \quad ; \quad R_B = \frac{P_1 a + P_2(L-b)}{L}$$

$$M(x = a) = R_A a$$

$$M(x = L - b) = R_B b$$

If $R_A < P_1$: $M_{max} = R_A a$ at $x = a$

If $R_B < P_2$: $M_{max} = R_B b$ at $x = L - b$

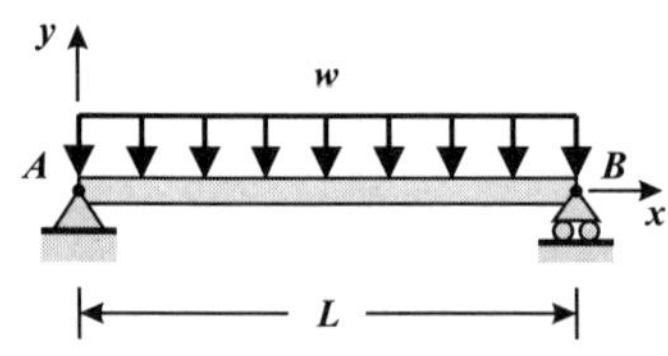

Uniformly Distributed Load (UDL)

$$R_A = R_B = \frac{wL}{2}$$

$$M_{max} = \frac{wL^2}{8} \quad \text{at center, } x = L/2$$

$$v(x) = \frac{-wx}{24EI}(L^3 - 2Lx^2 + x^3)$$

$$\theta(x) = \frac{-w}{24EI}(L^3 - 6Lx^2 + 4x^3)$$

$$\delta_{max} = \frac{-5wL^4}{384EI} \quad \text{at } x = L/2$$

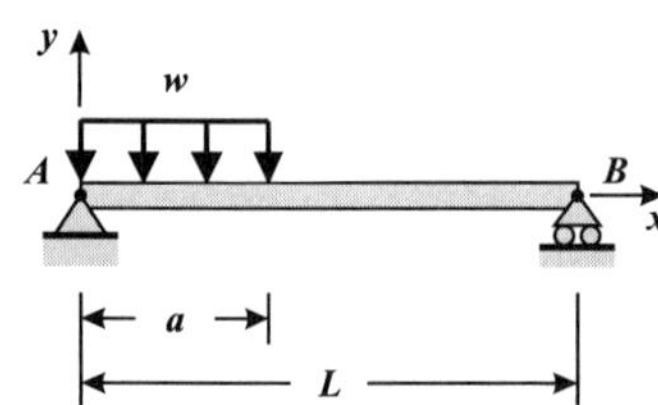

UDL over part of Beam at one end

$$R_A = \frac{wa}{2L}(2L - a) \quad ; \quad R_B = \frac{wa^2}{2L}$$

$$M_{max} = \frac{R_A^2}{2w} \quad \text{at } x = \frac{R_A}{w}$$

$$v(x) = \frac{-wx}{24EIL}\left[a^2(2L-a)^2 - 2ax^2(2L-a) + Lx^3\right] \quad \text{for } x \le a$$

$$v(x) = \frac{-wa^2(L-x)}{24EIL}(4xL - 2x^2 - a^2) \quad \text{for } x \ge a$$

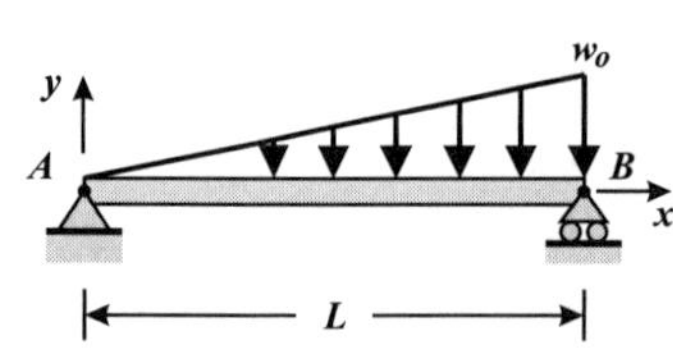

Linearly Increasing Distributed Load

$$R_A = \frac{w_o L}{6} \quad ; \quad R_B = \frac{w_o L}{3}$$

$$M_{max} = \frac{w_o L^2}{9\sqrt{3}} \quad \text{at } x = \frac{L}{\sqrt{3}} = 0.5774L$$

$$v(x) = \frac{-w_o x}{360EIL}(3x^4 - 10L^2 x^2 + 7L^4)$$

$$\delta_{max} = -0.00652\frac{w_o L^4}{EI} \quad \text{at } x = 0.5193L$$

Table F.2. Cantilever Beams.

Point Load at free end 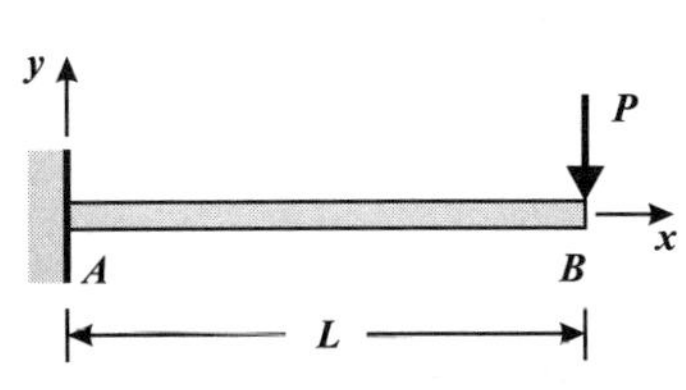	$R_A = P$; $M_A = -PL$ $M_{max} = -PL$ at wall, $x = 0$ $v(x) = \dfrac{-Px^2}{6EI}(3L - x)$ $\theta(x) = \dfrac{-Px}{2EI}(2L - x)$ $\delta_{max} = \dfrac{-PL^3}{3EI}$ at $x = L$; $\theta_{max} = \dfrac{-PL^2}{2EI}$ at $x = L$
Point Load at any point 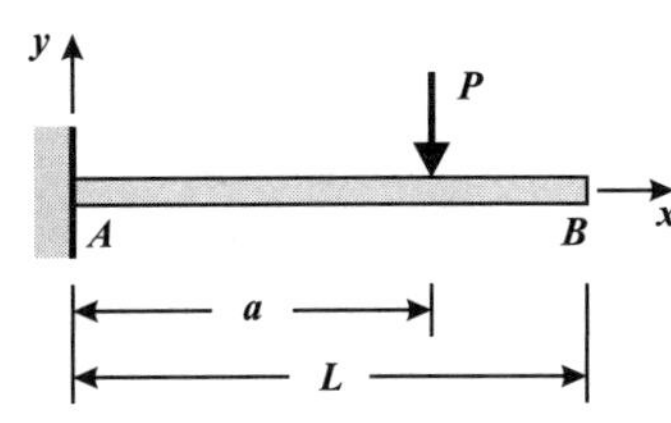	$R_A = P$; $M_A = M_{max} = -Pa$ $v(x) = \dfrac{-Px^2}{6EI}(3a - x)$; $\theta(x) = \dfrac{-Px}{2EI}(2a - x)$ for $x \le a$ $v(x) = \dfrac{-Pa^2}{6EI}(3x - a)$; $\theta(x) = \dfrac{-Pa^2}{2EI}$ for $a \le x \le L$ $v(x = a) = \dfrac{-Pa^3}{3EI}$; $\theta(x = a) = \dfrac{-Pa^2}{2EI}$ $\delta_{max} = \dfrac{-Pa^2}{6EI}(3L - a)$ at $x = L$ $\theta_{max} = \dfrac{-Pa^2}{2EI}$ for $a \le x \le L$
Uniformly Distributed Load (UDL) 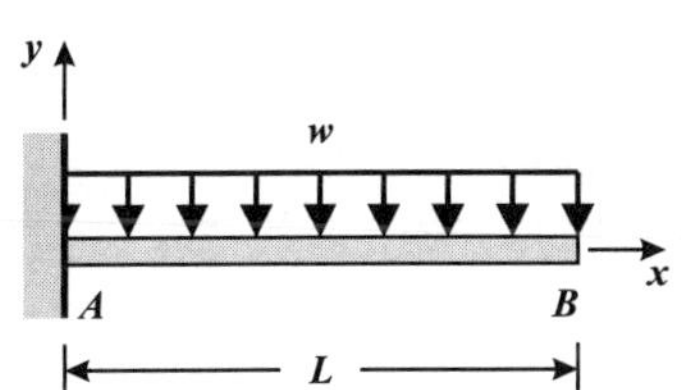	$R_A = wL$; $M_A = M_{max} = \dfrac{-wL^2}{2}$ $v(x) = \dfrac{-wx^2}{24EI}(6L^2 - 4Lx + x^2)$ $\theta(x) = \dfrac{-wx}{6EI}(3L^2 - 3Lx + x^2)$ $\delta_{max} = \dfrac{-wL^4}{8EI}$ at $x = L$; $\theta_{max} = \dfrac{-wL^3}{6EI}$ at $x = L$
Linearly Increasing Distributed Load 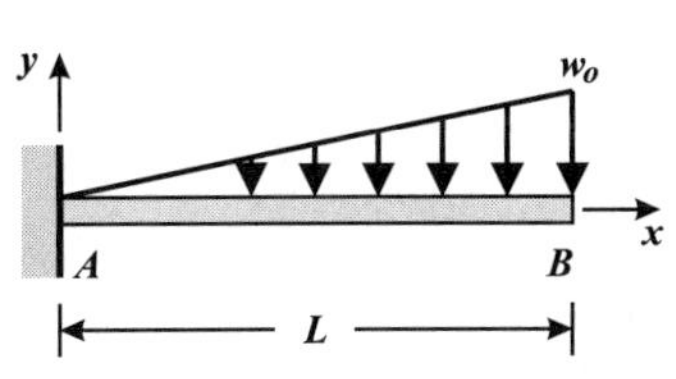	$R_A = \dfrac{w_o L}{2}$; $M_A = M_{max} = \dfrac{-w_o L^2}{3}$ $v(x) = \dfrac{-w_o x^2}{120EIL}(20L^3 - 10L^2 x + x^3)$ $\theta(x) = \dfrac{-w_o x}{24EIL}(8L^3 - 6L^2 x + x^3)$ $\delta_{max} = \dfrac{-11 w_o L^4}{120EI}$ at $x = L$; $\theta_{max} = \dfrac{-w_o L^3}{8EI}$ at $x = L$

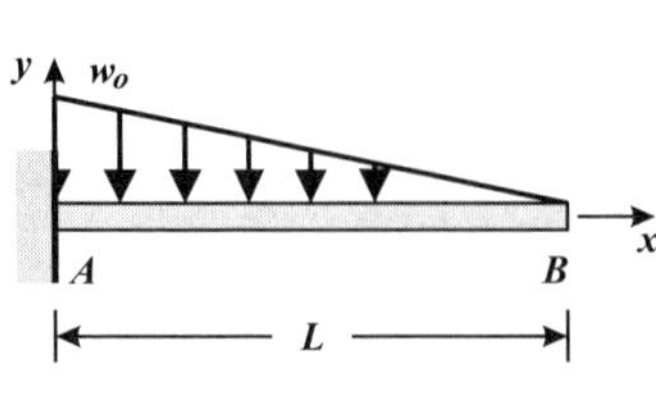

Linearly Decreasing Distributed Load

$$R_A = \frac{w_o L}{2} \quad ; \quad M_A = M_{max} = \frac{-w_o L^2}{6}$$

$$v(x) = \frac{-w_o x^2}{120 EIL}(10L^3 - 10L^2 x + 5Lx^2 - x^3)$$

$$\theta(x) = \frac{-w_o x}{24 EIL}(4L^3 - 6L^2 x + 4Lx^2 - x^3)$$

$$\delta_{max} = \frac{-w_o L^4}{30 EI} \quad \text{at } x = L \quad ; \quad \theta_{max} = \frac{-w_o L^3}{24 EI} \quad \text{at } x = L$$

Table F.3. Statically Indeterminate Beams: Pinned–Fixed and Fixed–Fixed.

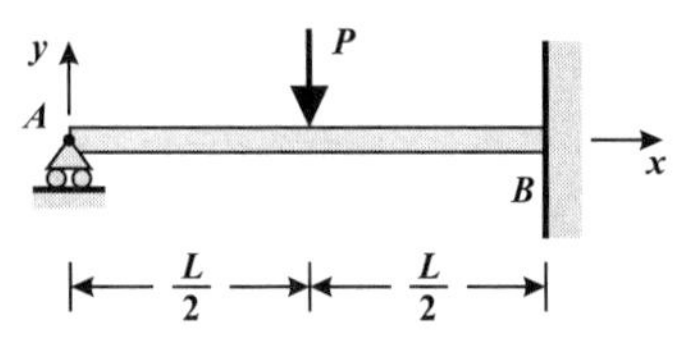

Pinned–Fixed: Central Point Load

$$R_A = \frac{5P}{16}; \quad R_B = \frac{11P}{16}; \quad M_B = M_{max} = \frac{-3PL}{16}$$

$$M(x = L/2) = \frac{5PL}{32} \quad \text{(max. positive moment)}$$

$$v(x) = \frac{-Px}{96EI}(3L^2 - 5x^2) \quad \text{for } x \leq L/2$$

$$v(x) = \frac{-P}{96EI}(x - L)^2(11x - 2L) \quad \text{for } x \geq L/2$$

$$\delta_{max} = -0.009317\frac{PL^3}{EI} \quad \text{at } x = \frac{L}{\sqrt{5}} = 0.4472L$$

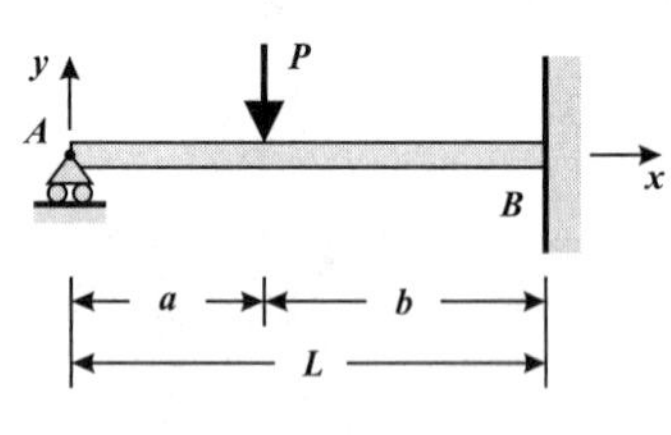

Pinned–Fixed: Offset Point Load

$$R_A = \frac{Pb^2}{2L^3}(a + 2L); \quad R_B = \frac{Pa}{2L^3}(3L^2 - a^2)$$

$$M_B = \frac{-Pab}{2L^2}(a + L); \quad M(x = a) = R_A a$$

$$M_{max} = M_B \quad \text{or} \quad M(x = a)$$

$$v(x) = \frac{-Pb^2 x}{12EIL^3}(3aL^2 - 2Lx^2 - ax^2) \quad \text{for } x \leq a$$

$$v(x) = \frac{-Pa}{12EIL^3}(L - x)^2(3L^2 x - a^2 x - 2a^2 L) \quad \text{for } x \geq a$$

$$\text{If } a < 0.414L: \quad \delta_{max} = \frac{-Pa}{3EI}\frac{(L^2 - a^2)^3}{(3L^2 - a^2)^2} \quad \text{at } x = L\left(\frac{L^2 + a^2}{3L^2 - a^2}\right)$$

$$\text{If } a > 0.414L: \quad \delta_{max} = \frac{-Pab^2}{6EI}\sqrt{\frac{a}{2L + a}} \quad \text{at } x = L\sqrt{\frac{a}{2L + a}}$$

<table>
<tr>
<td>

**Pinned–Fixed:
Uniformly Distributed Load**

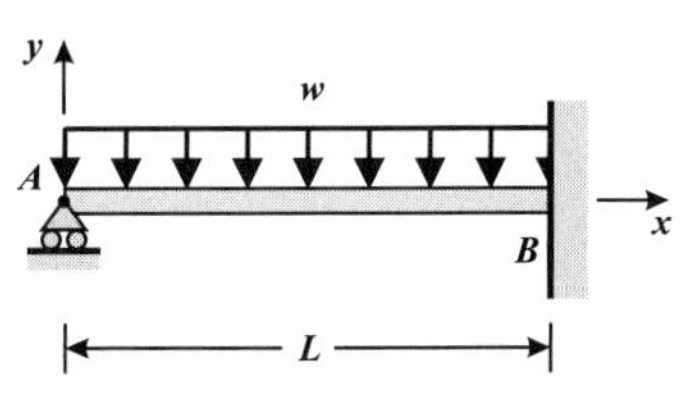

</td>
<td>

$$R_A = \frac{3wL}{8}; \quad R_B = \frac{5wL}{8}$$

$$M_B = M_{max} = \frac{-wL^2}{8}$$

$$M(x = 3L/8) = \frac{9wL^2}{128} \quad \text{(max. positive moment)}$$

$$v(x) = \frac{-wx}{48EI}(L^3 - 3Lx^2 + 2x^3)$$

$$\delta_{max} = \frac{-wL^4}{185EI} \quad \text{at } x = \frac{L}{16}(1 + \sqrt{33}) = 0.4215L$$

</td>
</tr>
<tr>
<td>

**Fixed–Fixed Beam:
Central Point Load**

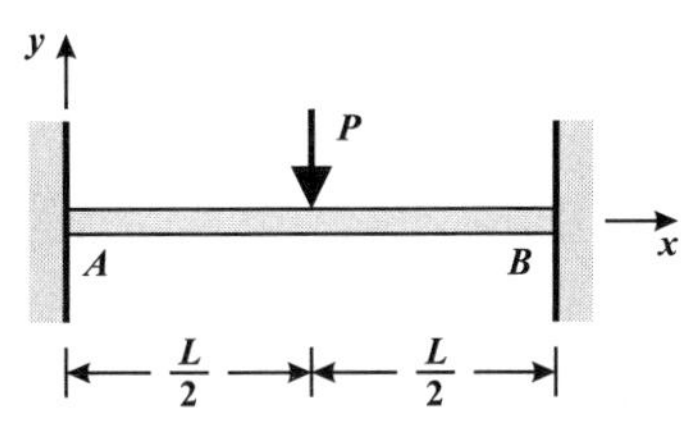

</td>
<td>

$$R_A = R_B = \frac{P}{2}$$

$$M_A = M_B = \frac{-PL}{8}$$

$$M_{max} = \frac{PL}{8} \quad \text{at center and at ends (negative at ends)}$$

$$v(x) = \frac{-Px^2}{48EI}(3L - 4x) \qquad \text{for } x \le L/2$$

$$\delta_{max} = \frac{-PL^3}{192EI} \quad \text{at center, } x = L/2$$

</td>
</tr>
<tr>
<td>

**Fixed–Fixed Beam:
Uniformly Distributed Load**

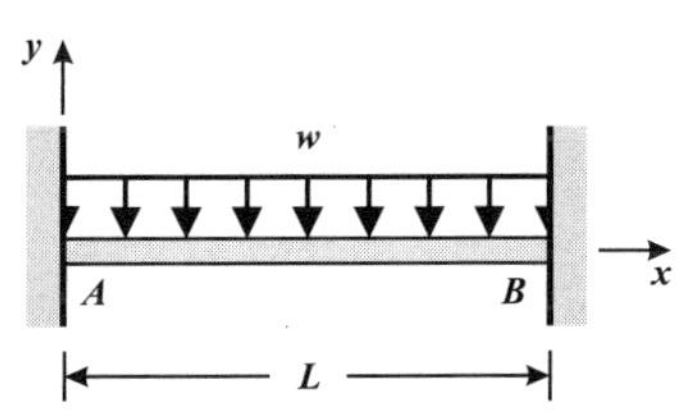

</td>
<td>

$$R_A = R_B = \frac{wL}{2}$$

$$M_A = M_B = M_{max} = \frac{-wL^2}{12}$$

$$M(x = L/2) = \frac{wL^2}{24} \quad \text{(max. positive moment)}$$

$$M(x) = \frac{w}{12}(6Lx - L^2 - 6x^2)$$

$$v(x) = \frac{-wx^2}{24EI}(L - x)^2$$

$$\delta_{max} = \frac{-wL^4}{384EI} \quad \text{at center}$$

</td>
</tr>
</table>

REFERENCES

Books and Standards

2000 International Building Code, International Code Council, 2000.

Agarwal, B.D., and Broutman, L.J., *Analysis and Performance of Fiber Composites,* J.Wiley & Sons, 1980.

Aluminum Design Manual, The Aluminium Association, Inc., 2005.

A.S.M.E. Boiler and Pressure Vessel Code, American Society of Mechanical Engineers, 2007.

Ashby, M.F., and Jones, D.R.H., *Engineering Materials: An Introduction to Their Properties and Applications*, Pergamon Press, 1987.

Ashby, M.F., *Materials Selection in Mechanical Design*, 2nd ed., Butterworth Heinemann, 1999.

Barbero, E.J., *Introduction to Composite Materials Design*, Taylor and Francis, 1999.

Callister, W.D., *Materials Science and Engineering: An Introduction*, 7th ed., J.Wiley & Sons, 2007.

Chawla, K.K., *Composite Materials*, Springer-Verlag, 1987.

Chawla, K.K., *Fibrous Materials*, Cambridge University Press, 1998.

Daniel, I.M., and Ori, I., *Engineering Mechanics of Composite Materials*, Oxford University Press, 1994.

Galilei, Galileo, *Dialogues Concerning Two New Sciences*, 1638. Translated by H. Crew and A. de Salvio, 1914. Reissue Northwestern University Press, 1968.

Jones, R.M., *Mechanics of Composites Materials*, 2nd ed., Taylor and Francis, 1999.

Kåre, H., *Introduction to Fracture Mechanics*, McGraw-Hill, 1984.

Lane, F.C., *Ships for Victory: A History of Shipbuilding under the U.S. Maritime Commission in World War II*, The John Hopkins University Press, 1951, reprint 2001.

Mallick, P.K., *Fiber-Reinforced Composites: Materials Manufacturing and Design*, 2nd ed., Marcel Dekker, 1993.

Moheimimani, S.O.R., and Fleming, A.J., *Piezoelectric Transducers for Vibration Control and Damping*, Spinger-Verlag, London, 2006.

Roark, R.J., and Young, W.C., *Formulas for Stress and Strain*, 5th ed., McGraw-Hill, 1975.

Steel Construction Manual, 13th ed., American Institute of Steel Construction, Inc., 2006.

Tada, H., Paris, P.C., and Irwin, G.R., *The Stress Analysis of Cracks Handbook*, 2nd ed., Paris Productions, Inc., 1985.

Timoshenko, S., and Goodier, J.N., *Theory of Elasticity*, McGraw-Hill, 1970.

Internet Sites – Accessed May 2008

"The Coast Guard at War, December 7, 1941–July 18, 1944: Marine Inspection, Vol. XIII." United States Coast Guard, http://www.uscg.mil/history/REGULATIONS/CGWar_13_Marine_Inspection.html

Hodgson, D.E., Wu, M.H., and Biermann, R.J., "Shape Memory Alloys." Shape Memory Applications, Inc., http://web.archive.org/web/20030605085042http://www.sma-inc.com/SMAPaper.html

"K-10000 tower crane by Kroll Giant Towercranes." Tower Cranes of America, Inc., http://www.towercrane.com/

"Piezoelectricity." APC International, Ltd., http://www.americanpiezo.com/piezo_theory/

Piezo Systems, Inc., http://www.piezo.com/

"Pipeline Facts." Alyeska Pipeline Service Company, http://www.alyeska-pipe.com/Pipelinefacts/Pipe.html

Radmacher, M., "Single Molecules Feel the Force." IOP Publishing, physicsweb.org/articles/world/12/9/9

Ryhänen, J., "Biocompatibility Evaluation of Nickel-Titanium Shape Memory Metal Alloy." University of Oulu, http://herkules.oulu.fi/isbn9514252217/html/x317.html

Socie, D.F., "Fatigue Calculator." D.F. Socie, http://www.fatiguecalculator.com/

Spinrad, H., "Saturn." NASA, http://www.nasa.gov/worldbook/saturn_worldbook.html; World Book, Inc., http://www.worldbookonline.com/wb/Article?id=ar492440

"True Stress – True Strain Curve." Key to Metals, http://www.key-to-steel.com/articles/art42.htm

Papers

Hart-Smith, L.J., "Mechanically-Fastened Joints for Advanced Composites – Phenomenological Considerations and Simple Analyses," *Fibrous Composites in Structural Design*, Plenum Press, 1980.

Saif, M.T.A., and MacDonald, N.C., "A Millinewton Microloading Device," *Sensors and Actuators*, A52, Elsevier Science B.V., 1996.

Trantina, G.G., and Johnson, C.A., "Spin Testing of Ceramic Materials," *Fracture Mechanics of Ceramics*, v.3, Plenum Publishing Corporation, 1978.

T

Mechanical Engineering Series

(continued from page ii)